Aspergillus fumigatus and Aspergillosis

Aspergillus fumigatus and Aspergillosis

EDITED BY

Jean-Paul Latgé

Unité des Aspergillus, Institut Pasteur
Paris, France

William J. Steinbach

Division of Pediatric Infectious Diseases
and Department of Molecular Genetics and Microbiology
Duke University Medical Center
Durham, North Carolina USA

ASM
PRESS

WASHINGTON, D.C.

American Society for Microbiology
1752 N Street, N.W.
Washington, DC 20036-2904

Library of Congress Cataloging-in-Publication Data

Aspergillus fumigatus and aspergillosis / edited by Jean-Paul Latgé, William J. Steinbach.
p. ; cm.
Includes bibliographical references and index.
ISBN 978-1-55581-438-0
1. Aspergillus fumigatus. 2. Aspergillosis. I. Latgé, Jean-Paul, 1948- II. Steinbach, William J.
[DNLM: 1. Aspergillosis. 2. Aspergillus fumigatus—pathogenicity. WC 450 A8395 2009]
QR201.A85A87 2009
579.5′657—dc22

2008031177

Printed in the United States of America

10 9 8 7 6 5 4 3 2 1

Address editorial correspondence to: ASM Press, 1752 N St., N.W., Washington, DC 20036-2904, U.S.A.

Send orders to: ASM Press, P.O. Box 605, Herndon, VA 20172, U.S.A.
Phone: 800-546-2416; 703-661-1593
Fax: 703-661-1501
Email: Books@asmusa.org
Online: estore.asm.org

To the patients with aspergillosis, and their families, who have been told there is little hope

To our friends, colleagues, lab members, students, and families with whom we share this simultaneous love and hatred of *Aspergillus fumigatus*

And to Michelle, who has taught us both about surviving illness with dignity

Contents

Contributors

Barbara D. Alexander
Dept. of Medicine/Infectious Diseases and Dept. of Pathology, Duke University Medical Center, DUMC Box 3035, Durham, NC 27710

Tsunihiro Ando
Dept. of Surgical Pathology, Toho University School of Medicine, 6-11-1 Omori-Nishi, Ota-Ku, Tokyo 143-8541, Japan

David S. Askew
Dept. of Pathology and Laboratory Medicine, University of Cincinnati College of Medicine, Cincinnati, OH 45267-0529

S. Arunmozhi Balajee
Mycotic Diseases Branch, Centers for Disease Control and Prevention, Mail Stop G11, 1600 Clifton Rd., Atlanta, GA 30333

Rosemary A. Barnes
Dept. of Medical Microbiology, Cardiff University, University Hospital of Wales, Cardiff CF14 4XN, United Kingdom

Anne Beauvais
Unité des Aspergillus, Institut Pasteur, 25, Rue du Docteur Roux, 75015 Paris, France

Elaine Bignell
Dept. of Microbiology, Centre for Molecular Microbiology and Infection, Imperial College London, Armstrong Road, London SW7 2AZ, United Kingdom

Emilio Bouza
Clinical Microbiology and Infectious Diseases Dept., Hospital General Universitario Gregorio Marañón, Spanish Study Group of Infection in Transplant Patients (GESITRA), and CIBER de Enfermedades Respiratorias (CIBERES), 28007 Madrid, Spain

Gordon D. Brown
Institute of Infectious Disease and Molecular Medicine, Division of Immunology, University of Cape Town, Observatory, 7925, South Africa

José Antonio Calera
Instituto de Microbiología-Bioquímica, Centro mixto CSIC/USAL, Departamento de Microbiología y Genética, Universidad de Salamanca, Plaza Doctores de la Reina s/n, 37007 Salamanca, Spain

Elio Castagnola
Infectious Diseases Unit, Dept. of Hematology and Oncology, "G. Gaslini" Children Hospital, 16147 Genoa, Italy

Michel Chignard
Institut Pasteur, Unité de Défense innée et Inflammation, and Inserm, U874, 25, Rue du Dr Roux, 75015 Paris, France

Robert A. Cramer, Jr.
Dept. of Veterinary Molecular Biology, Montana State University–Bozeman, Bozeman, MT 59718

David W. Denning
Medicine and Medical Mycology, University of Manchester, and Education and Research Centre, Wythenshawe Hospital, Manchester M23 9LT, United Kingdom

Paul S. Dyer
School of Biology, University of Nottingham, Nottingham NG7 2RD, United Kingdom

Scott G. Filler
Division of Infectious Diseases, Dept. of Medicine, Los Angeles Biomedical Research Institute at Harbor-UCLA Medical Center, Torrance, CA, and The David Geffen School of Medicine at UCLA, Los Angeles, CA 90502

Thierry Fontaine
Unité des Aspergillus, Institut Pasteur, 25, Rue du Dr Roux, 75015 Paris, France

Amandine Gastebois
Unité des Aspergillus, Institut Pasteur, 25, Rue du Dr Roux, 75015 Paris, France

Lisa M. Graham
Institute of Infectious Disease and Molecular Medicine, Division of Immunology, University of Cape Town, Observatory, 7925, South Africa

Reginald Greene
Dept. of Radiology, Massachusetts General Hospital, Harvard Medical School, Boston, MA 02114

Andreas H. Groll
Infectious Disease Research Program, Center for Bone Marrow Transplantation and Dept. of Pediatric Hematology/Oncology, University Children's Hospital, 48149 Muenster, Germany

Hubertus Haas
Biocenter, Division of Molecular Biology, Innsbruck Medical University, Fritz-Pregl-Str. 3, A-6020 Innsbruck, Austria

Raoul Herbrecht
Dept. of Oncology and Hematology, Hôpital de Hautepierre, 67098 Strasbourg, France

Yainitza Hernández-Rodríguez
Dept. of Plant Biology, University of Georgia, Athens, GA 30602

William W. Hope
School of Translational Medicine, The University of Manchester, Manchester, United Kingdom

Wenqi Hu
Merck Frosst Center for Fungal Genetics, Merck & Co., Inc., Montreal, Quebec, H2X 3Y8, Canada

Bo Jiang
GlycoFi, Inc., Lebanon, NH 03766

Olivier Jousson
Centre for Integrative Biology, University of Trento, 38100 Trento, Italy

Nancy P. Keller
Dept. of Plant Pathology and Dept. of Medical Microbiology and Immunology, University of Wisconsin–Madison, Madison, WI 53706

Corné H. W. Klaassen
Dept. of Medical Microbiology and Infectious Diseases, Canisius Wilhelmina Hospital, Weg door Jonkerbos 100, 6532 SZ Nijmegen, The Netherlands

Dimitrios P. Kontoyiannis
Dept. of Infectious Diseases, Infection Control and Employee Health and Dept. of Bone Marrow Transplantation, The University of Texas M. D. Anderson Cancer Center, Houston, TX 77030

Sven Krappmann
Research Center for Infectious Diseases, Julius-Maximilians-University Würzburg, Würzburg, Germany

Karine Lambou
Unité des Aspergillus, Institut Pasteur, 25, Rue du Dr Roux, 75015 Paris, France

Jean-Paul Latgé
Unité des Aspergillus, Institut Pasteur, 25, Rue du Dr Roux, 75015 Paris, France

Russell E. Lewis
Dept. of Clinical Sciences and Administration, The University of Houston College of Pharmacy, and The University of Texas M.D. Anderson Cancer Center, 1441 Moursund St., #424, Houston, TX 77030

Johan Maertens
Dept. of Haematology, Acute Leukaemia and Hematopoietic Stem Cell Transplantation Unit, University Hospitals Leuven, Campus Gasthuisberg, Herestraat 49, B-3000 Leuven, Belgium

Vincent Maertens
Dept. of Haematology, Acute Leukaemia and Hematopoietic Stem Cell Transplantation Unit, University Hospitals Leuven, Campus Gasthuisberg, Herestraat 49, B-3000 Leuven, Belgium

Kieren A. Marr
Comprehensive Transplant and Oncology Infectious Diseases Program, Dept. of Medicine, Johns Hopkins University, Baltimore, MD 21205

Gregory S. May
Division of Pathology and Laboratory Medicine, Dept. of Laboratory Medicine, Research, The University of Texas, M. D. Anderson Cancer Center, Houston, TX 77030

Emilia Mellado
Servicio de Micologia, Centro Nacional de Microbiologia, Instituto de Salud Carlos III, Carretera Majadahonda-Pozuleo Km2, (28.220) Majadahonda, Madrid, Spain

Michelle Momany
Dept. of Plant Biology, University of Georgia, Athens, GA 30602

Michel Monod
Service de Dermatologie, Laboratoire de Mycologie, Centre Hospitalier Universitaire Vaudois, 1011 Lausanne, Switzerland

Richard B. Moss
Center for Excellence in Pulmonary Biology, Dept. of Pediatrics, Stanford University, Palo Alto, CA 94304

Isabelle Mouyna
Unité des Aspergillus, Institut Pasteur, 25, Rue du Dr Roux, 75015 Paris, France

Frank-Michael Müller
Päd. Pneumologie & Speziale Infektiologie, Zentrum für Kinder- u. Jugendmedizin, INF 153, D-69120 Heidelberg, Germany

Patricia Muñoz
Clinical Microbiology and Infectious Diseases Dept., Hospital General Universitario Gregorio Marañón, Spanish Study Group of Infection in Transplant Patients (GESITRA), and CIBER de Enfermedades Respiratorias (CIBERES), 28007 Madrid, Spain

Yasmine Nivoix
Dept. of Pharmacy and Pharmacology, Hôpital de Hautepierre, 67098 Strasbourg, France

Nir Osherov
Dept. of Human Microbiology, Sackler School of Medicine, Tel-Aviv University, Ramat Aviv, Tel-Aviv 69978, Israel

Eric G. Pamer
Infectious Disease Service, Immunology Program, Memorial Sloan Kettering Cancer Center, New York, NY 10021

Paraskevi Panagopoulou
Dept. of Pediatric Oncology, Hippokration Hospital, Thessaloniki 54642, Greece

David S. Perlin
Public Health Research Institute, New Jersey Medical School-UMDNJ at the International Center for Public Health, 225 Warren St., Newark, NJ 07103

Utz Reichard
Dept. of Medical Microbiology and National Reference Center for Systemic Mycoses, University Hospital of Goettingen, 37075 Goettingen, Germany

Judith C. Rhodes
Dept. of Pathology and Laboratory Medicine, University of Cincinnati College of Medicine, Cincinnati, OH 45267-0529

Amariliz Rivera
Immunology Program, Memorial Sloan Kettering Cancer Center, New York, NY 10021

Geoffrey D. Robson
Faculty of Life Sciences, 1.800 Stopford Building, University of Manchester, Manchester M13 9PT, United Kingdom

Terry Roemer
Dept. of Infectious Disease, Merck & Co. Inc., 126 East Lincoln Ave., Rahway, NJ 07065

Emmanuel Roilides
3rd Dept. of Pediatrics, Aristotle University Medical School, Hippokration Hospital, Thessaloniki 54642, Greece

Luigina Romani
Microbiology Section, Dept. of Experimental Medicine and Biochemical Sciences, and Fondazione "Istituto di Ricovero e Cura per le Biotecnologie Trapiantologiche" I.B.i.T., Perugia, Italy

Markus Ruhnke
Medizinische Klinik und Poliklinik II, Charité – Universitätsmedizin Berlin, Campus Mitte, 10117 Berlin, Germany

Robert A. Samson
CBS Fungal Biodiversity Centre, Uppsalalaan 8, 3584 CT, Utrecht, The Netherlands

Taylor Schoberle
Division of Pathology and Laboratory Medicine, Dept. of Laboratory Medicine, Research, The University of Texas, M. D. Anderson Cancer Center, Houston, TX 77030

E. Keats Schwab
Dept. of Plant Pathology and Dept. of Medical Microbiology and Immunology, University of Wisconsin–Madison, Madison, WI 53706

Stefan Schwartz
Medizinische Klinik III, Charité – Universitätsmedizin Berlin, Campus Benjamin Franklin, 12200 Berlin, Germany

Brahm H. Segal
Dept. of Medicine and Dept. of Immunology, Roswell Park Cancer Institute, Buffalo, NY 14263

Kazutoshi Shibuya
Dept. of Surgical Pathology, Toho University School of Medicine, 6-11-1 Omori-Nishi, Ota-Ku, Tokyo 143-8541, Japan

Nina Singh
Dept. of Medicine, Division of Infectious Diseases, University of Pittsburgh, Pittsburgh, PA

William J. Steinbach
Division of Pediatric Infectious Diseases and Department of Molecular Genetics and Microbiology, Duke University Medical Center, Durham, NC 27710

David A. Stevens
Dept. of Medicine, Division of Infectious Diseases, Santa Clara Valley Medical Center, and Stanford University Medical School, San Jose, CA 95128-2699

Hsin-Yun Sun
Dept. of Internal Medicine, National Taiwan University Hospital and National Taiwan University College of Medicine, Taipei, Taiwan

Fredj Tekaia
Unité de Génétique Moléculaire des Levures (URA 2171 CNRS and UFR927 Univ. P. et M. Curie), Institut Pasteur, 25, Rue du Dr Roux, 75724 Paris Cedex 15, France

János Varga
CBS Fungal Biodiversity Centre, Uppsalalaan 8, 3584 CT, Utrecht, The Netherlands

Paul E. Verweij
Dept. of Medical Microbiology, University Medical Center St. Radboud, and Nijmegen University Center for Infectious Diseases, Inflammation and Immunity, Nijmegen, The Netherlands

Claudio Viscoli
Division of Infectious Disease, University of Genoa, San Martino University Hospital, 16132 Genoa, Italy

Thomas J. Walsh
Immunocompromised Host Section, Pediatric Oncology Branch, National Cancer Institute, National Institutes of Health, Bethesda, MD 20892

P. Lewis White
NPHS Microbiology Cardiff, University Hospital of Wales, Cardiff CF14 4XN, United Kingdom

Joanne Wong Sak Hoi
Unité des Aspergillus, Institut Pasteur, 25, Rue du Dr Roux, 75015 Paris, France

Jo-Anne H. Young
Division of Infectious Disease and International Medicine, Dept. of Medicine, University of Minnesota, MMC 250, 420 Delaware St. S.E., Minneapolis, MN 55455

Aimee K. Zaas
Dept. of Medicine/Infectious Diseases, Duke University Medical Center, DUMC Box 3355, Durham, NC 27710

Preface

Medicine has advanced astonishingly fast in the past several decades. As newer medical technologies are repairing the human body in ways never imagined only a few years ago, infectious diseases are emerging as a leading cause of morbidity and mortality among these newly susceptible patients. Invasive and chronic fungal infections have become a formidable clinical opponent, and foremost among them is *Aspergillus fumigatus*.

This volume has been carefully engineered to offer the latest insights into the fundamental biology and pathogenesis of this organism and how it establishes disease, as well as the newest strategies for characterizing, diagnosing, and treating its spectrum of clinical infection. Other great textbooks have been published which address general fungal physiology, overall medical mycology, the biology of the aspergilli specifically, or antifungal therapy, yet this is the first volume which specifically merges the newest scientific knowledge with the latest clinical experience and data to yield a combined synopsis of this organism and its diseases.

Here we have assembled chapters from a large and international contingent of experts in the field to explore every major aspect of *A. fumigatus* and how it kills so many patients. This includes chapters on the species itself, including morphology and unique essential genes. We also discuss the importance of growth and germination as well as the organism's response to environmental stress by moving from a saprophyte to a pathogen. The interface with the host immune system is paramount to disease phenotype, and this is outlined, as are the many faces of disease created by *A. fumigatus*. Newer diagnostic strategies are covered, including advances from the molecular age, the optimal timing of antifungal therapy, and the strategic choice of which agent to use.

It is our intent that *Aspergillus fumigatus and Aspergillosis* will encompass the current state of knowledge to serve as a resource guide for the next decade of study on this organism and the many diseases it causes. We hope that this volume will also serve as a catalyst for future young investigators to begin their own explorations in this field, to challenge the unproven dogmas and define the mechanisms of disease. This textbook is purposefully designed to unify the world's *Aspergillus* experts and collaborate toward our common goal. Here we outline today's state of the art and propose tomorrow's difficult and unanswered questions.

It has been our pleasure to serve as the coeditors, and also as authors, for this exciting new book merging science and medicine. We thank the many contributors to this new voyage; their expertise and efforts have forged a complete volume summarizing the many facets of *A. fumigatus*. We are also grateful to Gregory Payne, our editor at ASM Press, for his tireless energy and dedication to producing such an excellent book. We hope that our effort not only outlines the best thinking, but inspires creative solutions to the growing problem of *A. fumigatus* and aspergillosis.

Jean-Paul Latgé
Institut Pasteur

William J. Steinbach
Duke University

Aspergillus fumigatus and Aspergillosis
Edited by J.-P. Latgé and W. J. Steinbach
©2009 ASM Press, Washington, DC

Chapter 1

Introduction

DAVID A. STEVENS

In 1939, a scientist wrote in a prominent medical journal, "In man...*Aspergillus* infections are so rare as to be of little practical importance" (Henrici, 1939). After World War II, a number of developments caused a rise in the frequency of invasive mycoses. A significant change has been in the use of more intensive chemotherapy for cancer, resulting in transient episodes of neutropenia and enhanced survival of patients, but also in an immunosuppressed state. Another significant medical advance is the increased use of transplantation as a modality to remedy organ dysfunction, used for an increasing variety of solid organs and for bone marrow beset by hypoplasia or invaded by malignant lineages. Aggressive immunosuppression is now utilized for a variety of diseases, particularly collagen-vascular and rheumatic diseases, as well as inflammatory and interstitial lung diseases. Lastly, since World War II there has been a substantial increased survival of premature and low birth weight infants in specialized intensive care units, a sudden epidemic of an immunosuppressive infectious disease called AIDS, improvements in the care and survival of burn patients, and a rise in intravenous drug abuse.

For a genus such as *Aspergillus*, which is ubiquitous in the environment, normally found in air (including, particularly, hospital air), water, soil, decaying plant matter, food, dust, and human habitats, this was "the perfect storm." Experiments of nature (selected congenital immunodeficiencies), and now inadvertent "experiments" brought about by the advances in medical care, eventually taught us that the reason humans were able to survive to 1939 without dying off from aspergillosis had to do in large measure with intact neutrophil and macrophage functions, their arming by immune lymphocytes, and the ability of these defensive cells to deal with the vast numbers of conidia our sinuses and lungs encounter every day. We soon learned the price of the ablation of these arms of innate and acquired immunity. Aside from the rare congenital immunodeficiencies, aspergillosis had until then largely been confined in humans to a variety of allergic disorders, triggered by inhaled conidia, in susceptible hosts, including those with disorders such as asthma, bronchopulmonary aspergillosis, and allergic sinusitis, and to toxin contamination of the food chain. Perhaps we should have anticipated the rise in disease if we had noted the ability of the genus to cause focal disease in lung cavities and invasive disease in sinuses and ears, even in immunocompetent individuals, and if we had noted the extent of disease potential in our immunocompetent animal friends, including invasive sinus disease in dogs, abortion in cattle, invasive airway diseases in birds and in dolphins, infection of the guttural pouch in equine species, and mortality among bee populations and marine organisms.

During these decades of profound medical and technological progress, epidemiologists noted a steadily rising frequency of aspergillosis in autopsy series in cancer hospitals and then later among living patients with leukemia and transplant recipients diagnosed while alive by new emboldened invasive procedures. In the United States in 1996, there were an estimated 10,190 aspergillosis-related hospitalizations; these resulted in 1,970 deaths, 176,300 hospital days, and $633 million in costs. The average hospitalization lasted 17.3 days and cost ~$62,000 (Dasbach et al., 2000). Fast forwarding to the present, we now recognize aspergillosis as afflicting 10 to 15% of patients with solid organ or allogeneic hematopoietic stem cell transplants and of leukemic patients, and it is also not an uncommon infection in advanced AIDS. The overall mortality for these groups of patients when developing aspergillosis ranges from 45 to 90% (Lin et al., 2001). While infections due to the other major opportunistic group of fungi, *Candida* species, have reached a plateau, the infection rate due to *Aspergillus* species keeps rising. Moreover, we are now recognizing *Aspergillus* pathogenicity in intensive care patients who are not immunocompromised in the classic senses of neutropenia or

David A. Stevens • Dept. of Medicine, Division of Infectious Diseases, Santa Clara Valley Medical Center, and Stanford University Medical School, San Jose, CA 95128-2699.

neutrophil dysfunction or having received cytotoxic or immunosuppressive therapy.

Biomedical scientists and physicians have now had to make large investments in the effort to respond to this rising wave of cases, tragic waste of lives, and burden of health care costs. Some recent advances have laid an infrastructure upon which a broad scientific effort, as depicted in books such as the present effort, can be built. I would list, in the forefront of these, three particular developments: (i) advances in our diagnostic capabilities, (ii) the sequencing of the genomes of the major species, and (iii) availability of new therapeutic modalities.

Because even the newer treatments are unable to help a large group of immunocompromised patients with advancing invasive disease, it is essential for the clinician to detect clues to the possibility that a deterioration in a patient's condition may be due to aspergillosis, so that appropriate therapy can begin before this fungus, which not only grows rapidly in vivo but is vasculotropic, thus easily spreading and causing infarction and necrosis in its path, can make the situation irreversible. Appreciation of patterns in imaging with newer radiographic techniques and/or detection of galactomannan, glucan, or DNA by PCR in samples of blood or other body fluids has brought us to the position of being able to use our therapeutic armory at a more opportune, early time in a patient's infection.

Second, the delineation of the genes of these fungi and elucidation of their expression and functions make possible the development of a comprehensive approach to understanding their morphogenesis, growth, metabolism, and biochemistry, as well as virulence factors, and allow discovery of targets for new drugs, new diagnostic tests, and targets for immunotherapy, including the development of vaccines. Third, in the late 1950s we had the first effective weapon for treatment of this growing problem, amphotericin B, an agent that is still effective and in use today, although delivery systems have been advanced that improve its pharmacology and lessen the severe and unwanted side effects. In the 1980s a major advance was made with the development of the first orally effective agent, the azole itraconazole, which had few side effects (Denning et al., 1989). Further development of the azoles has brought us voriconazole and posaconazole, the former studied in a landmark multicenter international trial that proved advantages for azoles over conventional deoxycholate amphotericin B (Herbrecht et al., 2002). The favorable side effect profile of the azoles not only made prolonged therapy possible but also the safety plus the oral availability made outpatient treatment more feasible and opened the possibility of administering drugs to high-risk groups to prevent the establishment of aspergillosis. The additional introduction of the echinocandin class of drugs, which have a unique target, offers us now an array of therapeutic choices, with polyenes and azoles for initial therapy and for the patient who has failed to respond to the initial choice made, and the opportunity to explore using drugs with different mechanisms of action in combinations. These therapeutic advances were made possible by advances in laboratory sciences through standardization of susceptibility testing in vitro, testing in applicable robust animal models that mimic human disease (Clemons and Stevens, 2005), and advances in clinical sciences through efforts to develop a standard nomenclature to assign certainty to clinical diagnoses (Stevens and Lee, 1997; Ascioglu et al., 2002), as well as the efforts of cooperative clinical treatment groups to standardize treatment protocols and report their results. The models have also been critical in the advancement of our understanding of the immune response, and they have opened the possibility of immune modulation to reverse the predisposing immune defects.

It is the steps along the way to these three advances that have stimulated scientists and clinicians to gather together, exchange information, and synthesize information in order to lay the groundwork for the next steps. Although there are numerous review articles, chapters in medical textbooks, and chapters in infectious disease texts, by many authors, that could be cited as important syntheses that introduced new generations to the breadth of the field and to advances within it, I would cite what I view as the important antecedents to the present volume. First was a meeting held in Antwerp in 1987, whose proceedings were published as a book that spanned contributions in both the basic and clinical sciences relevant to *Aspergillus* (Vanden Bossche et al., 1988). Second was a grouping that began with a supplement in the journal *Clinical Infectious Diseases* (Steinbach et al., 2003), which was a precursor to another favorable development, now a series of biennial meetings, the Advances Against Aspergillosis conferences, devoted solely to *Aspergillus* and aspergillosis, that bring basic scientists and clinicians together for the express purpose of engendering and enhancing translational research. Three of the meetings held to date were each followed by publication of the proceedings (Steinbach et al., 2005, 2006). A third meeting took place in 2008, and its proceedings are in press. These bring us to the present effort, which is an integrated text that can fill any gaps between the topics of prior outstanding papers and which is thus yet more ambitious than the four preceding volumes cited (Vanden Bossche et al., 1988; Steinbach et al., 2003, 2005, 2006).

What follows in this volume are expositions, by current leaders, of the present knowledge. They tell us of our present understanding of the structure of the taxon, the methods of reproduction and growth, and its genes and how they are expressed in metabolism and development. We learn about host defenses and pathogenic virulence factors as well. The patient groups at risk are delineated. The varieties of clinical presentations of disease are to be laid out, as are the diagnostic tools we can bring to bear for the individual patient and the strengths and weaknesses of the therapeutic and prophylactic alternatives now available. The two coeditors of this volume have each given us what to my mind are, among the many prior helpful review articles mentioned above, previous comprehensive efforts worthy of citation (Latgé, 1999; Steinbach and Stevens, 2003). Thus, they are well-positioned to orchestrate the current effort, which takes its place as a new tool in the ongoing effort to understand this organism and to battle the diseases it causes.

We can all hope that these efforts will enable readers to strengthen our future understanding of how this organism survives in its ecological niches and replicates. We would like to come even closer to that ideal drug that is fungicidal in vivo, has good tissue penetration (especially of the central nervous system), no toxicity, no drug interactions, engenders no development of resistance, is available for both intravenous and oral use, and has a low cost. We may also hope for diagnostic tests of greater sensitivity, specificity, predictive value, and accessibility and standardization. We would all like to return to the day when we can say "*Aspergillus* infections are so rare as to be of little practical importance."

REFERENCES

Ascioglu, S., J. H. Rex, B. de Pauw, J. E. Bennett, J. Bille, F. Crokaert, D. W. Denning, J. P. Donnelly, J. E. Edwards, Z. Erjavec, E. Fiere, O. Lortholary, J. Maertens, J. F. Meis, T. Patterson, J. Ritter, D. Sellesag, P. M. Shah, D. A. Stevens, and T. J. Walsh. 2002. Defining opportunistic invasive fungal infections in immunocompromised patients with cancer and hematopoietic stem cell transplantation: an international consensus. *Clin. Infect. Dis.* **34:**7–14.

Clemons, K. V., and D. A. Stevens. 2005. Contributions of animal models of aspergillosis to understanding pathogenesis, therapy and virulence. *Med. Mycol.* **43**(Suppl. 1):S101–S110.

Dasbach, E. J., G. M. Davies, and S. M. Teutsch. 2000. Burden of aspergillosis-related hospitalizations in the United States. *Clin. Infect. Dis.* **31:**1524–1528.

Denning, D. W., R. M. Tucker, L. H. Hanson, and D. A. Stevens. 1989. Treatment of invasive aspergillosis with itraconazole. *Am. J. Med.* **89:**791–800.

Henrici, A. T. 1939. An endotoxin from *Aspergillus fumigatus*. *J. Immunol.* **36:**319–338.

Herbrecht, R., D. W. Denning, T. F. Patterson, J. E. Bennett, R. E. Greene, J. W. Oestmann, W. V. Kern, K. A. Marr, P. Ribaud, O. Lortholary, R. Sylvester, R. H. Rubin, J. R. Wingard, P. Stark, C. Durand, D. Caillot, E. Thiel, P. H. Chandrasekar, M. R. Hodges, H. T. Schlamm, P. F. Troke, and B. de Pauw. 2002.Voriconazole versus amphotericin B for primary therapy of invasive aspergillosis. *N. Engl. J. Med.* **347:**408–415.

Latgé, J.-P. 1999. *Aspergillus fumigatus* and aspergillosis. *Clin. Microbiol. Rev.* **12:**310–350.

Lin, S. J., J. Schranz, and S. M. Teusch. 2001. Aspergillosis case-fatality rate: systematic review of the literature. *Clin. Infect. Dis.* **32:**358–366.

Steinbach, W. J., J.-P. Latgé, and D. A. Stevens (ed.). 2005. Advances against aspergillosis. Proceedings of the Advances Against Aspergillosis Conference 9–11 September, 2004, San Francisco, USA. *Med. Mycol.* **43**(Suppl. 1):S1–S319.

Steinbach, W. J., and D. A. Stevens. 2003. Review of newer antifungal and immunomodulatory strategies for invasive aspergillosis. *Clin. Infect. Dis.* **37**(Suppl. 3):S157–S187.

Steinbach, W. J., D. A. Stevens, K. V. Clemons, J.-P. Latgé, and R. B. Moss (ed.). 2006. Proceedings of the 2nd Advances Against Aspergillosis Conference, 22–25 February, 2006, Athens, Greece. *Med. Mycol.* **44**(Suppl. 1):S1–S392.

Steinbach, W. J., D. A. Stevens, D. W. Denning, and R. B. Moss (ed.). 2003. Advances against aspergillosis. *Clin. Infect. Dis.* **37**(Suppl. 3): S155–S292.

Stevens, D. A., and J. Y. Lee. 1997. Analysis of compassionate use itraconazole therapy of invasive aspergillosis by the NIAID Mycoses Study Group criteria. *Arch. Intern. Med.* **157:**1857–1862.

Vanden Bossche, H., D. W. R. Mackenzie, and G. Cauwenbergh (ed.). 1988. Aspergillus *and Aspergillosis*. Plenum Press, New York, NY.

I. THE SPECIES

Aspergillus fumigatus and Aspergillosis
Edited by J.-P. Latgé and W. J. Steinbach

Chapter 2

Morphology and Reproductive Mode of *Aspergillus fumigatus*

Robert A. Samson, János Varga, and Paul S. Dyer

Aspergillus fumigatus, which was described by Johann Baptist Georg Wolfgang Fresenius in 1863 (Fresenius, 1863), is a cosmopolitan filamentous fungus (Pringle et al., 2005; Rydholm et al., 2006). This fungus plays an important role under natural conditions in the aerobic decomposition of organic materials and in recycling environmental carbon and nitrogen. The natural niche of *A. fumigatus* is soil, but it has been reported in various substrata from all over the world. This species is also an important opportunistic human pathogen which may cause several diseases, such as allergic bronchopulmonary aspergillosis, aspergilloma, and invasive aspergillosis, a usually fatal disease particularly in immunocompromised patients. In this review, we will give a general overview of the morphology of *A. fumigatus* and related species and discuss the reproductive mode of this important fungus.

MORPHOLOGY OF *A. FUMIGATUS* AND RELATED SPECIES

A. fumigatus belongs to the *Aspergillus* subgenus *Fumigati* section *Fumigati* (Gams et al., 1985; *A. fumigatus* group according to Raper and Fennell, 1965), which includes 10 other anamorphic species and 23 species which are able to reproduce sexually (Table 1) (Samson et al., 2007; Peterson et al., 2008). The teleomorphs of these species have been assigned to the *Neosartorya* genus (Malloch and Cain, 1972). Within this section, three species exhibit heterothallic (self-incompatible) breeding systems, and all other species are homothallic (self-fertile).

A. fumigatus is generally regarded as a single homogeneous species, but strains may vary in their cultural characteristics and to a lesser degree in their micromorphology. Typical *A. fumigatus* isolates produce dark blue-green colonies on Czapek agar (CZA) and malt extract agar (MEA) plates, consisting of a dense felt of conidiophores intermingled with aerial hyphae. Microscopically, the conidial heads are columnar, compact, and often densely crowded, and conidiophores are short, smooth, up to 300 μm in length, usually more or less green colored, and arise directly from submerged hyphae or as very short branches from aerial hyphae. Vesicles are subclavate, up to 20 to 30 μm in diameter, often colored similar to the conidiophores, and usually fertile on the upper half only. Phialides are directly borne on the vesicle and also pigmented and are usually about 6 to 8 μm by 2 to 3 μm. Conidia mostly are green in mass, smooth to finely echinulate, globose to subglobose, and mostly 2.5 to 3.0 μm in diameter with extremes ranging from 2.0 to 3.5 μm. Strains isolated from human or animal tissues tend to sporulate restrictedly and may show differences in micromorphology, including branched conidiophores, elongate or septate phialides, and conidia extremely variable in size and shape (Raper and Fennell, 1965; Leslie et al., 1988; Rinyu et al., 1995). This variability was frequently misinterpreted and led to the description of several "new" species from clinical sources, including *Aspergillus aviarius* Peck, *Aspergillus bronchialis* Blumentritt, *Aspergillus septatus* Sartory & Sartory, *A. fumigatus* var. *ellipticus* Raper & Fennell (for references, see Raper and Fennell, 1965), *Aspergillus phialiseptatus* Kwon-Chung (also known as *A. phialoseptus*) (Kwon-Chung, 1975), *Aspergillus neoellipticus* Kozakiewicz (1989), *Aspergillus acolumnaris* Kozakiewicz (Kozakiewicz, 1989; Varshney and Sarbhoy, 1981), and *Aspergillus arvii* Aho et al. (Aho et al., 1994). Later molecular and phylogenetic work clarified that all these species are synonyms of *A. fumigatus* (Rinyu et al., 1995; Geiser et al., 1998; Varga et al., 2000; Hong et al., 2005, 2006; Yaguchi et al., 2007).

The taxonomy of *A. fumigatus* and related species has recently been revised (Hong et al., 2006; Samson et

Robert A. Samson and János Varga • CBS Fungal Biodiversity Centre, Uppsalalaan 8, 3584 CT, Utrecht, The Netherlands. **Paul S. Dyer** • School of Biology, University of Nottingham, Nottingham NG7 2RD, United Kingdom.

Table 1. List of accepted species belonging to *Aspergillus* section *Fumigati* (Samson et al., 2007)

Strict anamorphic species

- *Aspergillus brevipes* Smith
- *Aspergillus duricaulis* Raper & Fennell
- *Aspergillus fumigatiaffinis* Hong, Frisvad & Samson
- *Aspergillus fumigatus* Fresenius
  - = *A. anomalus* Pidoplichko & Kirilenko
  - = *A. fumigatus* var. *acolumnaris* Rai *et al*
  - = *A. fumigatus* var. *ellipticus* Raper & Fennell
  - = *A. fumigatus* mut. *helvola* Rai *et al*
  - = *A. phialiseptus* Kwon-Chung
  - = *A. neoellipticus* Kozakiewicz
  - = *Aspergillus arvii* Aho, Horie, Nishimura & Miyaji
- *Aspergillus fumisynnematus* Horie, Miyaji, Nishimura, Taguchi & Udagawa
- *Aspergillus lentulus* Balajee & Marr
- *Aspergillus novofumigatus* Hong, Frisvad & Samson
- *Aspergillus turcosus* Hong, Frisvad & Samson
- *Aspergillus unilateralis* Thrower
  - ≡ *A. brevipes* var. *unilateralis* (Thrower) Kozakiewicz
- *Aspergillus viridinutans* Ducker & Thrower
  - = *A. fumigatus* var. *sclerotiorum* Rai, Agarwal & Tewari

Homothallic *Neosartorya* species

- *Neosartorya assulata* Hong, Frisvad & Samson [anamorph: *A. assulatus* Hong, Frisvad & Samson]
- *Neosartorya aurata* (Warcup) Malloch & Cain [anamorph: *A. igneus* Kozakiewicz]
- *Neosartorya aureola* (Fennell & Raper) Malloch & Cain [anamorph: *A. aureoluteus* Samson & Gams]
- *Neosartorya australensis* Samson, Hong & Varga
- *Neosartorya coreana* Hong, Frisvad & Samson [anamorph: *A. coreanus* Hong, Frisvad & Samson]
- *Neosartorya denticulata* Samson, Hong & Frisvad [anamorph: *A. denticulatus* Samson, Hong & Frisvad]
- *Neosartorya ferenczii* Varga & Samson
- *Neosartorya fischeri* (Wehmer) Malloch & Cain [anamorph: *A. fischeranus* Kozakiewicz]
- *Neosartorya galapagensis* Frisvad, Hong & Samson [anamorph: *A. galapagensis* Frisvad, Hong & Samson]
- *Neosartorya glabra* (Fennell & Raper) Kozakiewicz [anamorph: *A. neoglaber* Kozakiewicz]
- *Neosartorya hiratsukae* Udagawa, Tsubouchi & Horie [anamorph: *A. hiratsukae* Udagawa, Tsubouchi & Horie]
- *Neosartorya laciniosa* Hong, Frisvad & Samson [anamorph: *A. laciniosus* Hong, Frisvad & Samson]
- *Neosartorya multiplicata* Yaguchi, Someya & Udagawa [anamorph: *A. muliplicatus* Yaguchi, Someya & Udagawa]
- *Neosartorya papuensis* Samson, Hong & Varga
- *Neosartorya pseudofischeri* Peterson [anamorph: *A. thermomutatus* (Paden) Peterson]
- *Neosartorya quadricincta* (Yuill) Malloch & Cain [anamorph: *A. quadricingens* Kozakiewicz]
  - = *Neosartorya primulina* Udagawa, Toyazaki & Tsubouchi [anamorph: *A. primulinus* Udagawa, Toyazaki & Tsubouchi]
- *Neosartorya spinosa* (Raper & Fennell) Kozakiewicz [anamorph: *A. spinosus* Kozakiewicz]
  - ≡ *Aspergillus fischeri* var. *spinosus* Raper & Fennell 1965 (basionym)
  - = *Sartorya fumigata* var. *verrucosa* Udagawa & Kawasaki
  - = *Neosartorya botucatensis* Horie, Miyaji & Nishimura [anamorph: *A. botucatensis* Horie, Miyaji & Nishimura]
  - = *Neosartorya paulistensis* Horie, Miyaji & Nishimura [anamorph: *A. paulistensis* Horie, Miyaji & Nishimura]
  - = *Neosartorya takakii* Horie, Abliz & Fukushima [anamorph: *A. takakii* Horie, Abliz & Fukushima]
- *Neosartorya stramenia* (Novak & Raper) Malloch & Cain [anamorph: *A. paleaceus* Samson & Gams]
- *Neosartorya tatenoi* Horie, Miyaji, Yokoyama, Udagawa & Campos-Takagi [anamorph: *A. tatenoi* Horie, Miyaji, Yokoyama, Udagawa & Campos-Takagi]
  - = *Neosartorya delicata* Kong [anamorph: *A. delicatus* Kong]
- *Neosartorya warcupii* Peterson, Varga & Samson

Heterothallic *Neosartorya* species

- *Neosartorya fennelliae* Kwon-Chung & Kim [anamorph: *A. fennelliae* Kwon-Chung & Kim]
  - = *Neosartorya otanii* Takada, Horie & Abliz [anamorph: *A. otanii* Takada, Horie & Abliz]
- *Neosartorya spathulata* Takada & Udagawa [anamorph: *A. spathulatus* Takada & Udagawa]
- *Neosartorya udagawae* Horie, Miyaji & Nishimura [anamorph: *A. udagawae* Horie, Miyaji & Nishimura]

Doubtful species

- *Neosartorya sublevispora* Someya, Yaguchi & Udagawa [anamorph: *A. sublevisporus* Someya, Yaguchi & Udagawa]
- *Neosartorya indohii* Horie [anamorph: *A. indohii* Horie]
- *Neosartorya tsurutae* Horie [anamorph: *A. tsurutae* Horie]
- *Neosartorya nishimurae* Takada, Horie & Abliz [anamorph: *A. nishimurae* Takada, Horie & Abliz]

al., 2007). Based on molecular, morphological, and physiological parameters, *A. fumigatus* sensu lato is divided into five taxa: *A. fumigatus* sensu stricto, *Aspergillus lentulus, Aspergillus fumigatiaffinis, Aspergillus novofumigatus*, and *Aspergillus viridinutans*. Another species, *Aspergillus fumisynnematus* (Horie et al., 1993) has also been validated by Yaguchi et al. (2007) and is closely related to *A. lentulus*. Strains of *A. fumigatus* sensu stricto, *A. lentulus, A. fumisynnematus, A. fumigatiaffinis*, and *A. novofumigatus* are macroscopically similar (Color Plates 1 and 2). However, strains of *A. lentulus, A. fumisynnematus, A. fumigatiaffinis*, and *A. novofumigatus* usually show less sporulation compared to typical *A. fumigatus* isolates. With respect to morphology, the width of conidiophore stipes in *A. fumigatus* sensu stricto ranges from 3.5 to 10 μm, while those of *A. lentulus, A. fumigatiaffinis, A. fumisynnematus*, and *A. novofumigatus* are 2 to 7 μm, 5 to 8 μm, 6 to 8.5 μm, and 4 to 7 μm, respectively. Most strains of *A. fumigatus* sensu stricto have subclavate vesicles, while the vesicles in most strains of *A. lentulus, A. fumisynnematus*, and *A. fumigatiaffinis* are (sub)globose and in *A. novofumigatus* they are subglobose to subclavate. The diameters of the vesicles in *A. fumigatus* sensu stricto, *A. lentulus, A. fumisynnematus, A. fumigatiaffinis*, and *A. novofumigatus* are 10 to 26, 6 to 25, 16 to 20, 16 to 24, and 15 to 30 μm, respectively. In comparison, most vesicles of *A. fumigatus* sensu stricto strains are wider than 22 μm, while in the other taxa the vesicles are narrower. All strains of *A. fumigatus* sensu stricto grow at 50°C but do not grow at 10°C on MEA or CZA medium. Strains of *A. lentulus, A. fumisynnematus, A. fumigatiaffinis*, and *A. novofumigatus* grow or germinate at 10°C on MEA and CZA but do not grow at 50°C. The growth ratio at 25°C versus 45°C on MEA for *A. fumigatus* sensu stricto strains falls in the range of 0.5 to 1.1, whereas for strains of *A. lentulus, A. fumigatiaffinis*, and *A. novofumigatus* they are 1.3 to 4.0, 1.2 to 1.9, and 1.1 to 1.6, respectively (Hong et al., 2005). A further cryptic species, *A. fumigatus* "*occultum*," has also been postulated within the *A. fumigatus* grouping, but no consistent morphological differences were evident between isolates of this proposed species and the majority of isolates described as *A. fumigatus* (Pringle et al., 2005).

Regarding the clinical significance of these and some other related species, molecular studies have revealed that several clinical and soil isolates previously identified as *A. fumigatus* belong to one of these species which exhibit altered antifungal susceptibility profiles against several drugs compared to *A. fumigatus* (Balajee et al., 2004, 2005a, 2005b, 2006; Mellado et al., 2006). These clinical isolates were found to belong to either *A. lentulus* (Balajee et al., 2004, 2005b; Hong et al., 2006; Alhambra et al., 2006; Yaguchi et al., 2007; Lau et al., 2007), *Neosartorya udagawae* (Balajee et al., 2006; Moragues et al., 2006), *Neosartorya pseudofischeri* (Balajee et al., 2005a; Lau et al., 2007; Barrs et al., 2007), *A. viridinutans* (Katz et al., 1998, 2005; Yaguchi et al., 2007), or *A. fumisynnematus* (Yaguchi et al., 2007).

The identification of *A. fumigatus*-like strains, which are anamorphs of heterothallic *Neosartorya* species, using morphological criteria is often difficult. The anamorphs of *N. udagawae* and *N. fennelliae* (Color Plate 3) show a strong resemblance to *A. fumigatus*, and crossing experiments are necessary to obtain the cleistothecia. These fruiting bodies may only be developed when the cultures are older than 14 days. Therefore, for a correct and faster identification, a sequence analysis should be carried out. Our experience with sequencing the calmodin and β-tubulin genes revealed good species delimitation and recognition (Samson et al., 2007).

Recently, strains of *A. fumigatus* sensu lato which were previously identified as *A. fumigatus* in the CBS fungal culture collection based on morphology have been reexamined based on β-tubulin sequence data and randomly amplified polymorphic DNA PCR (S. B. Hong et al., unpublished data). Of the 147 strains examined, 141 strains (95.8%) were reidentified as *A. fumigatus* sensu stricto and 2 strains (1.4%) were identified as *A. lentulus*, while 4 were found to belong to the *A. viridinutans* complex, *N. udagawae, Neosartorya nishimurae*, and related to *A. fumisynnematus*, respectively. These results indicate that *A. fumigatus* is the predominantly occurring species.

GENETIC CONTROL OF CONIDIAL DEVELOPMENT

Screening of the genome of *A. fumigatus* by BLAST analysis has revealed the presence of a series of genes which are known to be involved in conidial development of *Aspergillus nidulans* (based on Table 5 of the supplementary data of Pel et al., 2007). This includes the presence of key transcriptional regulators controlling conidiophore development, such as *brlA, abaA, stuA, wetA*, and *medA*, and the hydrophobin spore coat gene *rodA* (although *dewA* was not detected). In addition, 16 genes involved with signal transduction relating to asexual reproduction were identified, including the *fphA* gene encoding a red light phytochrome (Blumenstein et al., 2005). Furthermore, the developmental regulator VeA, first identified from *A. nidulans*, has been shown to be required for sporulation on certain growth media for *A. fumigatus* (Krappman et al., 2005). Therefore, it is argued that the genetic mechanisms controlling asexual sporulation in *A. fumigatus* are likely to be essentially the same as those observed in *A. nidulans* (reviewed by Fisher and Kües, 2003).

REPRODUCTIVE MODE AND SEXUALITY

A. fumigatus, as with the majority of species of *Aspergillus*, is only known to reproduce by asexual means. However, there is accumulating evidence that gene flow through recombination has occurred, or is occurring, in natural populations of *A. fumigatus* (Dyer and Paoletti, 2005; Paoletti et al., 2005). This has raised the tantalizing possibility that *A. fumigatus* might have the ability to undergo sexual reproduction, involving the development of a teleomorph state predicted to be morphologically similar to the cleistothecia of previously described *Neosartorya* species, which contain crested ascospores (Raper and Fennell, 1965; Malloch and Cain, 1972).

There are various lines of evidence to support the possible occurrence of sexual reproduction in *A. fumigatus*. These were reviewed in detail by Dyer and Paoletti (2005), with subsequent experimental work adding further support to these arguments, as will now be described. Briefly, the evidence comes from four main areas as follows.

(i) Population genetic analyses. Many studies have detected a high degree of genetic variation within natural populations of *A. fumigatus* consistent with sexual reproduction and recombination. For example, Debeaupuis et al. (1997) found over 400 genetically unique isolates (based on retrotransposon fingerprinting) in samples from European clinical and environmental samplings, while a similar study of over 700 samples from hospitals in France found 85% of isolates to be genetically unique (Chazalet et al., 1998). These data, together with other reported observations, were pooled and analyzed by Varga and Tóth (2003), who concluded that there was evidence for recombination in some populations. More recently, Pringle et al. (2005) sequenced five polymorphic loci from 53 worldwide samples of *A. fumigatus* and found sufficient variation to conclude that recombination was occurring, although clonality dominated and recombination was rejected by certain statistical tests. Finally, Paoletti et al. (2005) sequenced three intergenic loci from 106 isolates from five subpopulations from Europe and North America and again found sufficient variation to conclude that recombination had occurred within the test samples. In contrast, Rydholm et al. (2006) reported markedly lower sequence variation in 103 global samples of *A. fumigatus* than in the known sexual *Neosartorya* species *N. fischeri* and *N. spinosa*.

(ii) Genome analysis. The genome of *A. fumigatus* clinical isolate AF293 was one of the first aspergilli to be sequenced. Results of the genome annotation and analysis were reported by Galagan et al. (2005) and Nierman et al. (2005). A special area of interest concerned possible reasons for asexuality in *A. fumigatus*. It was predicted that the genome analysis might reveal possible mutations within coding regions (e.g., point mutations leading to stop codons or frameshifts) for genes required for sexual development, which might explain the absence of sexual reproduction in this species due to lack of an essential gene. However, BLAST analysis revealed that the genome contained a complement of apparently functional genes known to be required for sexual development in ascomycete fungi (Galagan et al., 2005; Nierman et al., 2005). This included genes involved with sexual identity and mating processes, signaling pathways for mating response, and fruit body development and meiosis (such genes linked to sexual reproduction in the aspergilli were detailed in a review by Dyer [2007]). Similar results were obtained when the genome sequence of a second isolate (A1163) of *A. fumigatus* was screened for genes involved in sexual reproduction (N. Fedorova et al., submitted for publication). Thus, from a genomic perspective, *A. fumigatus* appears to contain the sexual machinery necessary for sexual reproduction (Dyer et al., 2003), although it remains possible that there is a defect in an as-yet-unidentified key gene required for sexual development. This contrasts with the genome analysis of *Aspergillus niger*, in which mutations in certain genes relating to sexual development have been reported (Pel et al., 2007). Meanwhile, the presence of defective copies of the *AFUT1* retrotransposon repeat sequence in the genome of *A. fumigatus* suggests that, at least in recent evolutionary history, the species possessed a repeat-induced point mutation process linked to sex (Neuvéglise et al., 1996; Dyer and Paoletti, 2005).

(iii) Mating-type gene presence and distribution. Mating processes in ascomycete fungi are governed by so-called mating-type (*MAT*) genes, which act as key transcriptional activators to control sexual development (Casselton, 2002). *MAT* genes are required for sexual development in both heterothallic and homothallic species and have been shown to be functional in *A. nidulans* (Debuchy and Turgeon, 2006; Paoletti et al., 2007). In the case of heterothallic species, in which complementary MAT1-1 and MAT1-2 isolates are required for sexual reproduction, *MAT* genes are located at a single *MAT* locus within the genome. By convention, isolates of the *MAT1-1* genotype contain a *MAT* locus with a gene encoding a characteristic α-domain protein, whereas isolates of the *MAT1-2* genotype contain a *MAT* locus with a gene encoding a characteristic high-mobility group (HMG) domain protein (Debuchy and Turgeon, 2006). Due to high sequence divergence between the *MAT1-1* and *MAT1-2* loci, these regions have been termed idiomorphs, to distinguish them from homologous alleles found at the same locus. Genome analysis of *A. fumi-*

gatus isolate AF293 revealed the presence of a *MAT1-2*-type idiomorph containing an HMG domain-encoding gene (Galagan et al., 2005). Simultaneous experimental work revealed the presence of isolates of *A. fumigatus* of the *MAT1-1* genotype, containing an idiomorph with an α-domain-encoding gene (Paoletti et al., 2005), that is, the organization of genes at the *MAT* locus of *A. fumigatus* resembled a typical heterothallic sexually reproducing species. This allowed a multiplex PCR diagnostic test for mating type to be devised, which was applied to 290 worldwide samples of *A. fumigatus*. This test revealed the presence of a near 1:1 ratio of *MAT1-1* and *MAT1-2* genotypes, consistent with a sexually reproducing species (Paoletti et al., 2005). More recently, a detailed investigation of population structure from spore trap aerial samplings from Dublin, Ireland, was undertaken. Again, a 1:1 ratio of *MAT1-1* and *MAT1-2* genotypes was detected (O'Gorman et al., 2007). Significantly, these isolates of complementary mating types were found in the same geographic location, meaning that physical separation of mating types does not provide an explanation for lack of sexual reproduction, as observed elsewhere (Snyder et al., 1975). Finally, the recent genome analysis of *A. fumigatus* A1163 has revealed this isolate to be of the *MAT1-1* genotype, encoding an α-domain MAT protein (Fedorova et al., 2008).

(iv) Expression of sex-related genes. Paoletti et al. (2005) were able to show that certain key genes (detected by BLAST genome screening, as described above) relating to sexual reproduction are expressed at the mRNA level in *A. fumigatus*, again consistent with sexual reproduction in the species. Reverse transcription-PCR evidence was presented for expression of *MAT* genes (*MAT1-1* and *MAT1-2*) and genes encoding a pheromone precursor (*ppgA*) and pheromone receptor (*preA* and *preB*). This was the first such report of expression of pheromone-signaling genes in a supposedly asexual fungus, although there were previous reports of *MAT* gene expression in other asexual fungi (Arie et al., 2000; Yun et al., 2000).

Taken as a whole, these results provide compelling evidence that recombination has occurred within the recent evolutionary history of *A. fumigatus* and that recombination might still be occurring in the wild. Such recombination may not necessarily be linked to sexual reproduction. BLAST screening of the *A. fumigatus* genome has revealed the presence of a series of genes linked to heterokaryon (in)compatibility processes, which indicates the possible occurrence of a parasexual cycle (Pál et al., 2007; Fedorova et al., 2008). However, the extent of recombination is arguably more consistent with sexual reproduction.

SIGNIFICANCE OF REPRODUCTIVE MODE AND FUTURE PROSPECTS

In terms of its global distribution and prevalence, *A. fumigatus* is without doubt an evolutionarily successful organism. One reason for the success of the species is the ability to produce prolific numbers of conidia by asexual reproduction. The relatively small diameter (2 to 3 μm) of the conidia would be expected to aid dispersal and is also thought to allow penetration deep into the lung alveoli, a key factor linked to pathogenesis (Anderson et al., 2003). The possible presence of a sexual state leading to ascospore formation is also potentially of great significance to the species. Sexual reproduction through outcrossing would be expected to generate and maintain genetic variation within populations, which is essential to allow long-term survival and evolution of the species (Dyer and Paoletti, 2005). Indeed, asexuality has been viewed as an evolutionary dead end (Normark et al., 2003). Sexual recombination may be particularly important in the clinical setting, should this lead to the production of strains with increased virulence or resistance to antifungal agents (Dyer and Paoletti, 2005). In addition, sexually derived ascospores often have greater resistance to adverse environmental conditions than asexual conidia, allowing dormancy during conditions unfavorable for growth (Paoletti et al., 2007).

There are various explanations for why a teleomorph state might exist but has not yet been identified. For example, sexual reproduction may be infrequent in nature and therefore has not yet been observed by field investigators. This might be due to the need for a particular set of environmental conditions that are rarely encountered in the wild, or the fact that many isolates in nature show a slow decline in sexual fertility, meaning that there are few sufficiently fertile isolates remaining (Dyer and Paoletti, 2005). The latter phenomenon has been observed in *Magnaporthe grisea* (Notteghem and Silué, 1992). Alternatively, sex may be occurring in an environmental niche(s) which is yet to be monitored (e.g., perhaps within compost heaps), so the teleomorph has been overlooked. Work is now under way to see if it is possible to induce sexual reproduction in vivo under controlled conditions in the laboratory, crossing isolates of known *MAT1-1* and *MAT1-2* genotypes. Hyphal aggregates resembling immature cleistothecia have been formed under certain conditions, whereas similar aggregates were not formed in control experiments involving growth of single mating type isolates (P. S. Dyer, J. Carmichael, and S. T. Hibbert, unpublished data). This raises the exciting future prospect of being able to perform classical genetic crosses with *A. fumigatus* in order to study the inheritance patterns, and therefore the genetic basis, of traits of interest, such as those pertaining to pathogenicity and secondary metabolite production.

Acknowledgments. We thank S.-B. Hong for the illustrations from his mating experiments of *N. udagawae* and *N. fennelliae*.

REFERENCES

Aho, R., Y. Horie, K. Nishimura, and M. Miyaji. 1994. *Aspergillus arvii* spec. nov., a new animal pathogen? *Mycoses* **37:**389–392.

Alhambra, A., J. M. Moreno, M. Dolores Moragues, S. Brena, G. Quindós, J. Pontón, and A. del Palacio. 2006. Aislamiento de *Aspergillus lentulus* en un enfermo crítico con EPOC y aspergilosis invasora, abstr. D14. *VIII Congreso Nacional Micologia*, Barcelona, Spain.

Anderson, M. J., J. L. Brookman, and D. W. Denning. 2003. *Aspergillus*, p. 1–39. *In* R. A. Prade and H. J. Bohnert (ed.), *Genomics of Plants and Fungi*. Marcel Dekker, New York, NY.

Arie, T., I. Kaneko, T. Yoshida, M. Noguchi, Y. Nomura, and I. Yamaguchi. 2000. Mating-type genes from asexual phytopathogenic ascomycetes *Fusarium oxysporum* and *Alternaria alternata*. *Mol. Plant Microbe Interact.* **13:**1330–1339.

Balajee, S. A., D. Nickle, J. Varga, and K. A. Marr. 2006. Molecular studies reveal frequent misidentification of *Aspergillus fumigatus* by morphotyping. *Eukaryot. Cell* **5:**1705–1712.

Balajee, S. A., J. Gribskov, M. Brandt, J. Ito, A. Fothergill, and K. A. Marr. 2005a. Mistaken identity: *Neosartorya pseudofischeri* and its anamorph masquerading as *Aspergillus fumigatus*. *J. Clin. Microbiol.* **43:**5996–5999.

Balajee, S. A., J. L. Gribskov, E. Hanley, D. Nickle, and K. A. Marr. 2005b. *Aspergillus lentulus* sp. nov., a new sibling species of *A. fumigatus*. *Eukaryot. Cell* **4:**625–632.

Balajee, S. A., M. Weaver, A. Imhof, J. Gribskov, and K. A. Marr. 2004. *Aspergillus fumigatus* variant with decreased susceptibility to multiple antifungals. *Antimicrob. Agents Chemother.* **48:**1197–1203.

Barrs, V. R., J. A. Beatty, A. E. Lingard, R. Malik, M. B. Krockenberger, P. Martin, C. O'Brien, J. M. Angles, M. Dowden, and C. Halliday. 2007. Feline sino-orbital aspergillosis: an emerging clinical syndrome? *Austr. Vet. J.* **85:**N23.

Blumenstein, A., K. Vienken, R. Tasler, J. Purschwitz, D. Veith, N. Frankenberg-Dinkel, and R. Fischer. 2005. The *Aspergillus nidulans* phytochrome FphA represses sexual development in red light. *Curr. Biol.* **15:**1833–1838.

Casselton, L. A. 2002. Mate recognition in fungi. *Heredity* **88:**142–147.

Chazalet, V., J. P. Debeaupuis, J. Sarfati, J. Lortholary, P. Ribaud, P. Shah, M. Cornet, H. V. Thien, E. Gluckman, G. Brücker, and J. P. Latgé. 1998. Molecular typing of environmental and patient isolates of *Aspergillus fumigatus* from various hospital settings. *J. Clin. Microbiol.* **36:**1494–1500.

Debeaupuis, J. P., J. Sarfati, V. Chazalet, and J. P. Latgé. 1997. Genetic diversity among clinical and environmental isolates of *Aspergillus fumigatus*. *Infect. Immun.* **65:**3080–3085.

Debuchy, R., and B. G. Turgeon. 2006. Mating-type structure, evolution, and function in euascomycetes, p. 293–323. *In* U. Kües and R. Fischer (ed.), *The* Mycota. *I: Growth, Differentiation and Sexuality*. Springer-Verlag, Berlin, Germany.

Dyer, P. S. 2007. Sexual reproduction and significance of *MAT* in the aspergilli, p. 123–142. *In* J. Heitman, J. W. Kronstad, J. W. Taylor, and L. A. Casselton (ed.), *Sex in Fungi: Molecular Determination and Evolutionary Principles*. ASM Press, Washington, DC.

Dyer, P. S., and M. Paoletti. 2005. Reproduction in *Aspergillus fumigatus*: sexuality in a supposedly asexual species? *Med. Mycol.* **43**(Suppl.1)**:**S7–S14.

Dyer, P. S., M. Paoletti, and D. B. Archer. 2003. Genomics reveals sexual secrets of *Aspergillus*. *Microbiology* **149:**2301–2303.

Fedorova, N. D., N. Khaldi, V. S. Joardar, R. Maiti, P. Amedeo, M. J. Anderson, J. Crabtree, J. C. Silva, J. H. Badger, A. Albarraq, S. Anguioli, H. Bussey, P. Bowyer, P. J. Cotty, P. S. Dyer, A. Egan, K. Galens, C. M. Fraser-Liggett, B. J. Haas, J. M. Inman, R. Kent, S. Lemieux, I. Malavazi, J. Orvis, T. Roemer, C. M. Ronning, J. P. Sundaram, G. Sutton, G. Turner, J. C. Venter, O. R. White, B. R. Whitty, P. Youngman, K. H. Wolfe, G. H. Goldman, J. R. Wortman, B. Jiang, D. W. Denning, and W. C. Nierman. 2008. Genomic islands in the pathogenic filamentous fungus *Aspergillus fumigatus*. *PloS Genet.* **11:**e1000046.

Fisher, R., and U. Kües. 2003. Developmental processes in filamentous fungi, p. 41–118. *In* R. A. Prade and H. J. Bohnert (ed.), *Genomics of Plants and Fungi*. Marcel Dekker, New York, NY.

Fresenius, G. 1863. *Beiträge zur Mykologie*. Brönner, Frankfurt am Main, Germany.

Galagan, J., S. E. Calvo, C. Cuomo, L.- J. Ma, J. Wortman, S. Batzoglou, S.- I. Lee, M. Brudno, B. Meray, C. C. Spevak, J. Clutterbuck, V. Kapitonov, J. Jurka, C. Scazzocchio, M. Farman, J. Butler, S. Purcell, S. Harris, G. H. Braus, O. Draht, S. Busch, C. D'Enfert, C. Bouchier, G. H. Goldman, D. Bell-Pedersen, S. Griffiths-Jones, J. H. Doonan, J. Yu, K. Vienken, A. Pain, M. Freitag, E. U. Selker, D. B. Archer, M. A. Penalva, B. R. Oakley, M. Momany, T. Tanaka, T. Kumagai, K. Asai, M. Machida, W. C. Nierman, D. W. Denning, M. Caddick, M. Hynes, M. Paoletti, R. Fischer, B. Miller, P. Dyer, M. S. Sachs, S. A. Osmani, and B. W. Birren. 2005. Sequencing and comparative analysis of *Aspergillus nidulans*. *Nature* **438:**1105–1115.

Gams, W., M. Christensen, A. H. Onions, J. I. Pitt, and R. A. Samson. 1985. Infrageneric taxa of *Aspergillus*, p. 55–62. *In* R. A. Samson and J. I. Pitt (ed.), *Advances in* Penicillium *and* Aspergillus *Systematics*. Plenum Press, New York, NY.

Geiser, D. M., J. C. Frisvad, and J. W. Taylor. 1998. Evolutionary relationships in *Aspergillus* section *Fumigati* inferred from partial β-tubulin and hydrophobin DNA sequences. *Mycologia* **90:**831–845.

Hong, S. B., H. D. Shin, J. B. Hong, J. C. Frisvad, P. V. Nielsen, J. Varga, and R. A. Samson. 2007. New taxa of *Neosartorya* and *Aspergillus* in *Aspergillus* section *Fumigati*. *Antonie Leeuwenhoek* **93:**87–98.

Hong, S. B., H. S. Cho, H. D. Shin, J. C. Frisvad, and R. A. Samson. 2006. Novel *Neosartorya* species isolated from soil in Korea. *Int. J. Syst. Evol. Microbiol.* **56:**477–486.

Hong, S. B., S. J. Go, H. D. Shin, J. C. Frisvad, and R. A. Samson. 2005. Polyphasic taxonomy of *Aspergillus fumigatus* and related species. *Mycologia* **97:**1316–1329.

Horie, Y., M. Miyaji, K. Nishimura, H. Taguchi, and S. Udagawa. 1993. *Aspergillus fumisynnematus*, a new species from Venezuelan soil. *Trans. Mycol. Soc. Japan* **34:**3–7.

Katz, M. E., A. M. Dougall, K. Weeks, and B. F. Cheetham. 2005. Multiple genetically distinct groups revealed among clinical isolates identified as atypical *Aspergillus fumigatus*. *J. Clin. Microbiol.* **43:**551–555.

Katz, M. E., M. Mcloon, S. Burrows, and B. F. Cheetham. 1998. Extreme DNA sequence variation in isolates of *Aspergillus fumigatus*. *FEMS Immunol. Med. Microbiol.* **20:**283–288.

Kozakiewicz, Z. 1989. *Aspergillus* species on stored products. *Mycol. Papers* **161:**1–188.

Krappmann, S., O. Bayram, and G. H. Braus. 2005. Deletion and allelic exchange of the *Aspergillus fumigatus veA* locus via a novel recyclable marker module. *Eukaryot. Cell* **4:**1298–1307.

Kwon-Chung, K. J. 1975. A new pathogenic species of *Aspergillus* in the *Aspergillus fumigatus* series. *Mycologia* **67:**770–779.

Lau, A., S. Chen, T. Sorrell, D. Carter, R. Malik, P. Martin, and C. Halliday. 2007. Development and clinical application of a panfungal PCR assay to detect and identify fungal DNA in tissue specimens. *J. Clin. Microbiol.* **45:**380–385.

Leslie, C. E., B. Flannigan, and L. J. R. Milne. 1988. Morphological studies on clinical isolates of *Aspergillus fumigatus*. *J. Med. Vet. Mycol.* 26:335–341.

Malloch, D., and R. F. Cain. 1972. The *Trichocomataceae*: ascomycetes with *Aspergillus, Paecilomyces*, and *Penicillium* imperfect states. *Can. J. Bot.* 50:2613–2628.

Mellado, E., L. Alcazar-Fuoli, G. Garcia-Effro, N. A. Alastruey-Izquierdo, M. Cuenca-Estrella, and J. L. Rodríguez-Tudela. 2006. New resistance mechanisms to azole drugs in *Aspergillus fumigatus* and emergence of antifungal drugs-resistant *A. fumigatus* atypical strains. *Med. Mycol.* 44:S367–S371.

Moragues, M. D., S. Brena, I. Miranda, R. Laporta, M. Muñoz, J. Pontón, and A. del Palacio. 2006. Colonización pulmonar por *Neosartorya udagawae* en un paciente trasplantado de pulmón, abstr. D15. *VIII Congreso Nacional Micologia*, Barcelona, Spain.

Neuvéglise, C., J. Sarfati, J. P. Latgé, and S. Paris. 1996. *Afut1*, a retrotransposon-like element from *Aspergillus fumigatus*. *Nucleic Acids Res.* 24:1428–1434.

Nierman, W., A. Pain, M. J. Anderson, J. Wortman, S. Kim, J. Arroya, M. Berriman, K. Abe, D. Archer, C. Bermejo, J. Bennett, P. Bowyer, D. Chen, M. Collins, R. Coulsen, R. Davies, P. S. Dyer, M. Farman, N. Fedorova, N. Fedorova, T. V. Feldblyum, R. Fischer, N. Fosker, A. Fraser, J. L. García, M. J. García, A. Goble, G. H. Goldman, K. Gomi, S. Griffith-Jones, R. Gwilliam, B. Haas, H. Haas, D. Harris, H. Horiuchi, J. Huang, S. Humphray, J. Jiménez, N. Keller, H. Khouri, K. Kitamoto, T. Kobayashi, S. Konzack, R. Kulkarni, T. Kumagai, A. Lafon, J. P. Latgé, W. Li, A. Lord, C. Lu, W. H. Majoros, G. S. May, B. L. Miller, Y. Mohamoud, M. Molina, M. Monod, I. Mouyna, S. Mulligan, L. Murphy, S. O'Neil, I. Paulsen, M. A. Peñalva, M. Pertea, C. Price, B. L. Pritchard, M. A. Quail, E. Rabbinowitsch, N. Rawlins, M. A. Rajandream, U. Reichard, H. Renauld, G. D. Robson, S. Rodriguez de Córdoba, J. M. Rodríguez-Peña, C. M. Ronning, S. Rutter, S. L. Salzberg, M. Sanchez, J. C. Sánchez-Ferrero, D. Saunders, K. Seeger, R. Squares, S. Squares, M. Takeuchi, F. Tekaia, G. Turner, C. R. Vazquez de Aldana, J. Weidman, O. White, J. Woodward, J. H. Yu, C. Fraser, J. E. Galagan, K. Asai, M. Machida, N. Hall, B. Barrell, and D. W. Denning. 2005. Genomic sequence of the pathogenic and allergenic *Aspergillus fumigatus*. *Nature* 438:1151–1156.

Normark, B. B., O. P. Judson, and N. A. Morgan. 2003. Genomic signatures of ancient asexual lineages. *Biol. J. Linn. Soc.* 79:69–84.

Notteghem, J. L., and D. Silué. 1992. Distribution of the mating type alleles in *Magnaporthe grisea* populations pathogenic on rice. *Phytopathology* 82:421–424.

O'Gorman, C. M., H. T. Fuller, and P. S. Dyer. 2007. *Aspergillus fumigatus*: mating-type distribution hints at sex in the city?, p. 44. *In Ecology of Fungal Communities.* Annual Scientific Meeting of the British Mycological Society. British Mycological Society, Manchester, United Kingdom.

Pál, K., A. D. van Diepeningen, J. Varga, R. F. Hoekstra, P. S. Dyer, and A. J. M. Debets. 2007. Sexual and vegetative compatibility genes in the aspergilli. *Stud. Mycol.* 59:19–31.

Paoletti, M., C. Rydholm, E. U. Schwier, M. J. Anderson, G. Szakacs, F. Lutzoni, J. P. Debeaupuis, J. P. Latgé, D. W. Denning, and P. S. Dyer. 2005. Evidence for sexuality in the opportunistic fungal pathogen *Aspergillus fumigatus*. *Curr. Biol.* 15:1242–1248.

Paoletti, M., F. A. Seymour, M. J. C. Alcocer, N. Kaur, A. M. Calvo, D. B. Archer, and P. S. Dyer. 2007. Mating type and the genetic basis of self-fertility in the model fungus *Aspergillus nidulans*. *Curr. Biol.* 17:1384–1389.

Pel, H. J., J. H. de Winde, D. B. Archer, P. S. Dyer, G. Hofmann, P. J. Schaap, G. Turner, R. P. de Vries, R. Albang, K. Albermann, M. R. Andersen, J. D. Bendtsen, J. A. Benen, M. van den Berg, S. Breestraat, M. X. Caddick, R. Contreras, M. Cornell, P. M. Coutinho, E. G. Danchin, A. J. Debets, P. Dekker, P. W. van Dijck, A. van Dijk, L. Dijkhuizen, A. J. Driessen, C. d'Enfert, S. Geysens, C. Goosen G. S. Groot, P. W. de Groot, T. Guillemette, B. Henrissat, M. Herweijer, J. P. van den Hombergh, C. A. van den Hondel, R. T. van der Heijden, R. M. van der Kaaij, F. M. Klis, H. J. Kools, C. P. Kubicek, P. A. van Kuyk, J. Lauber, X. Lu, M. J. van der Maarel, R. Meulenberg, H. Menke, M. A. Mortimer, J. Nielsen, S. G. Oliver, M. Olsthoorn, K. Pal, N. N. van Peij, A. F. Ram, U. Rinas, J. A. Roubos, C. M. Sagt, M. Schmoll, J. Sun, D. Ussery, J. Varga, W. Vervecken, P. J. van de Vondervoort, H. Wedler, H. A. Wosten, A. P. Zeng, A. J. van Ooyen, J. Visser, and H. Stam. 2007. Genome sequencing and analysis of the versatile cell factory *Aspergillus niger* CBS 513.88. *Nat. Biotechnol.* 25:221–231.

Peterson, S. W., J. Varga, J. C. Frisvad, and R. A. Samson. 2008. Phylogeny and subgeneric taxonomy of *Aspergillus*, p. 33–56. *In* J. Varga and R. A. Samson (ed.), Aspergillus *in the Genomic Era.* Wageningen Academic Publishers, Wageningen, The Netherlands.

Pringle, A., D. M. Baker, J. L. Platt, J. P. Wares, J. P. Latgé, and J. W. Taylor. 2005. Cryptic speciation in the cosmopolitan and clonal human pathogenic fungus *Aspergillus fumigatus*. *Evolution* 59: 1886–1899.

Raper, K. B., and D. I. Fennell. 1965. *The Genus* Aspergillus. Williams & Wilkins, Baltimore, MD.

Rinyu, E., J. Varga, and L. Ferenczy. 1995. Phenotypic and genotypic analysis of variability in *Aspergillus fumigatus*. *J. Clin. Microbiol.* 33: 2567–2575.

Rydholm, C., G. Szakacs, and F. Lutzoni. 2006. Low genetic variation and no detectable population structure in *Aspergillus fumigatus* compared to closely related *Neosartorya* species. *Eukaryot. Cell* 5: 650–657.

Samson, R. A., S. B. Hong, S. W. Peterson, J. C. Frisvad, and J. Varga. 2007. Polyphasic taxonomy of *Aspergillus* section *Fumigati* and its teleomorph, *Neosartorya*. *Stud. Mycol.* 59:147–203.

Snyder, W. C., S. G. Georgopoulos, R. K. Webster, and S. N. Smith. 1975. Sexuality and genetic behaviour in the fungus *Hypomyces* (*Fusarium*) *solani* f. sp. *cucurbitae*. *Hilgardia* 43:161–185.

Varga, J., and B. Tóth. 2003. Genetic variability and reproductive mode of *Aspergillus fumigatus*. *Infect. Genet. Evol.* 3:3–17.

Varga, J., Z. Vida, B. Tóth, F. Debets, and Y. Horie. 2000. Phylogenetic analysis of newly described *Neosartorya* species. *Antonie Leeuwenhoek* 77:235–239.

Varshney, J. L., and A. K. Sarbhoy. 1981. A new species of *Aspergillus fumigatus* group and comments upon its taxonomy. *Mycopathologia* 73:89–92.

Yaguchi, T., Y. Horie, R. Tanaka, T. Matsuzawa, J. Ito, and K. Nishimura. 2007. Molecular phylogenetics of multiple genes on *Aspergillus* section *Fumigati* isolated from clinical specimens in Japan. *Jpn. J. Med. Mycol.* 48:37–46.

Yun, S. H., T. Arie, I. Kaneko, O. C. Yoder, and B. G. Turgeon. 2000. Molecular organisation of mating type loci in heterothallic, homothallic and asexual *Gibberella/Fusarium* species. *Fungal Genet. Biol.* 31:7–20.

Aspergillus fumigatus and Aspergillosis
Edited by J.-P. Latgé and W. J. Steinbach

Chapter 3

Molecular Methods for Species Identification and Strain Typing of *Aspergillus fumigatus*

S. Arunmozhi Balajee and Corné H. W. Klaassen

The order of fungi, a scandal to art is still chaos with botanists not knowing what is a species, what a variety?
Carl Linnaeus, 1751, *Philosophia Botanica*

Centuries after Carl Linnaeus wrote this description, the species concept and the criteria used to delineate fungal species continue to be the focus of research, discussion, and debate. Fungal species are commonly designated according to a binomial nomenclature as laid out by Carl Linnaeus, whereby an organism is denoted by a genus name and a species name. Regardless of the system of classification, a mycologist should be able to give a specific name to a sample fairly rapidly and with precision without the necessity of complex experimental research, except in cases where new species are being described. Historically, morphology, both gross and microscopic, has been the main criterion for fungal species classification. In this identification strategy, the assumption is that organisms at the species level have a unique shape and structure that will distinguish the species from another. Although the morphology of a fungus is the ultimate expression of its growth processes and is the final display of its complex relationships with its normal habitat, morphological criteria alone are of limited use in *Aspergillus* species discrimination, largely due to the great morphological plasticity of these organisms. Molecular approaches, such as comparative sequence-based methods, offer a more rapid, accurate approach that is applicable to isolates with ambiguous phenotype. Thus, molecular biology has irreversibly shaped the field of mycology, where in the last decade a new phylogenetic species, *Coccidioides posadasii*, has been recognized, the artificial phylum Deuteromycota has been eliminated, and cryptic recombination in phenotypically asexual fungi has been demonstrated largely based on molecular methods. Over the last decade, molecular technology has also played a significant role in resolving the relationships between species within the section *Fumigati* in the genus *Aspergillus*, has enabled species identification within this section, and has facilitated rapid *Aspergillus fumigatus* strain discrimination.

Since *A. fumigatus* remains the predominant etiological agent of invasive aspergillosis in the immunocompromised patient population and delineation of this opportunistic pathogen from its relatives within the section *Fumigati* continues to be challenging, the chapter is arranged as follows: an overview of species within the section *Fumigati*, including *A. fumigatus*, is presented along with an outline of the recently advocated genealogical concordance phylogenetic species recognition concept to recognize members of the section *Fumigati*. In addition, *A. fumigatus* species identification strategies useful for applications in clinical microbiology laboratories are elaborated. Finally, molecular methods currently available for subtyping of *A. fumigatus* strains are described.

THE SECTION *FUMIGATI*

Chapter 2 details the morphology and new species within the section *Fumigati*, genus *Aspergillus*. Here, we give a brief overview of recently described species within the section *Fumigati* based on studies that have used molecular approaches to support species description.

Aspergillus lentulus was described as a new species in 2005 as an etiological agent of invasive infection in immunocompromised patient populations (Balajee et al., 2005b). Interestingly, this newly discovered species was at first described as a slow-sporulating variant of *A. fumigatus*, since both these species share similar mac-

S. Arunmozhi Balajee • Mycotic Diseases Branch, Centers for Disease Control and Prevention, Mail Stop G11, 1600 Clifton Rd., Atlanta, GA 30333. **Corné H. W. Klaassen** • Dept. of Medical Microbiology and Infectious Diseases, Canisius Wilhelmina Hospital, Weg door Jonkerbos 100, 6532 SZ Nijmegen, The Netherlands.

roscopic morphology with minor differences in microscopic morphology. For instance, microscopically, *A. lentulus* isolates have thinner stipes with smaller and predominantly (sub)globose vesicles, while *A. fumigatus* isolates are characterized by wider stipes bearing subclavate vesicles. Detailed phylogenetic analyses of several protein-coding loci and combined locus analysis (Fig. 1) revealed that isolates represent a new species within the section *Fumigati*. Post hoc analyses of the *A. lentulus* and *A. fumigatus* sensu stricto isolates revealed two significant phenotypical differences between these clinically relevant species. All *A. lentulus* isolates tested so far are unable to grow at temperatures ≥48°C, and this may be an important phenotype that can be used in clinical microbiology laboratories to distinguish the two species, since *A. fumigatus* thrives at and above 48°C (Balajee et al., 2005b). Both these species also differ significantly in their secondary metabolite profiles (Larsen et al., 2007). Notably, *A. lentulus* isolates do not produce the purported "virulence" factor gliotoxin, but they produce other metabolites, such as cyclopiazonic acid and neosartorin, in vitro (Larsen et al., 2007). Thus, secondary metabolite profiling revealed that both species are distinct from each other; *A. lentulus* produces six extrolite families never detected in *A. fumigatus*, while *A. fumigatus* produces more than nine extrolite families as yet undiscovered in *A. lentulus* (Larsen et al., 2007).

Although *A. lentulus* was first described in one medical center in the United States, subsequent studies have demonstrated recovery of *A. lentulus* from crop-cultivated soil in Korea, from a dolphin nostril in The Netherlands, from clinical isolates in Australia and Japan (Balajee et al., 2006; Hong et al., 2005; Katz et al., 2005; Yaguchi et al., 2007), and from several other medical centers in the United States (Balajee et al., 2006). Thus, *A. lentulus* appears to have a wide geographic distribution and could be ubiquitous in the environment. Regardless of the place or source of isolation, all *A. lentulus* isolates studied so far have lower susceptibilities in vitro to multiple antifungals that are available for therapy compared to *A. fumigatus*, thereby underlining the clinical relevance of species identification in this section (Balajee et al., 2006; Hong et al., 2005; Katz et al., 2005; Yaguchi et al., 2007).

Apart from *A. lentulus*, eight new species have been described within the section *Fumigati* by Hong et al., who employed polyphasic analyses that included morphology, secondary metabolite profiles, and comparative sequence analyses of at least three loci (Hong et al., 2005, 2006). The eight novel species include three anamorphic species, *A. turcosus, A. fumigatiaffinis*, and *A. novofumigatus*, and five *Neosartorya* teleomorphs, *N. denticulata, N. galapagensis, N. assulata, N. laciniosa*, and *N. coreana*. All of these species have been isolated from soil and have not yet been implicated in clinical infections. Another species, *A. fumisynnematus*, was previously described as a unique species within the section *Fumigati* because of its short conidiophores borne on funiculose aerial hyphae. Earlier comparative analyses with cytochrome *b* sequences clustered these isolates into a separate clade (Wang et al., 2000), and more recently, detailed phylogenetic analyses have demonstrated clearly that these isolates represent new taxa within the section *Fumigati* (Hong et al., 2008).

A. fumigatus var. *occultum* is another recent "cryptic" species described by phylogenetic analyses of five loci encompassing a microsatellite repeat and surrounding nucleotides (Pringle et al., 2005). This species currently consists of five members from Illinois, Colombia, France, and South Africa and is morphologically indistinguishable from *A. fumigatus*, hence the name "occultum," which is Latin for hidden or secret. *A. fumigatus* var. *occultum* was revealed as an independent evolutionary lineage that was well-supported in the majority of single-locus genealogies generated from five different loci (this concept of species recognition is explained in the next section). However, *A. fumigatus* var. *occultum* has not yet been formally and validly described as a species, and the taxonomic position of this cryptic species within the section *Fumigati* is unclear and remains to be elucidated.

Molecular studies have also revealed the risk of relying on morphological features alone for species identification within the section *Fumigati*. Several members in the section *Fumigati* are sexual, and the teleomorphic (sexual) states of these isolates within section *Fumigati* are assigned to the genus *Neosartorya*. Thus, on artificial media in the laboratory, the *Neosartorya* species within the section *Fumigati* produce sexual fruiting bodies called cleistothecia, which serve as diagnostic markers for these taxa. Unfortunately, clinical isolates often do not produce these sexual structures on laboratory media and are thus refractory to traditional methods of identification. Morphological methods are thus not always useful in species delimitation of such isolates. For instance, a recent molecular investigation of several clinical *A. fumigatus* isolates revealed that these isolates were not *A. fumigatus* but *N. pseudofischeri* (Balajee et al., 2005a). The anamorphic state of *N. pseudofischeri, A. thermomutatus*, is indistinguishable from *A. fumigatus*, and since initially none of the isolates produced diagnostic fruiting bodies, the fungi were misidentified as *A. fumigatus*. Sequence comparison of the *benA* and *rodA* regions revealed that these isolates were 100% homologous to *N. pseudofischeri* and they were eventually identified as *A. thermomutatus* (anamorphic state). Likewise, a previously recognized *Aspergillus* species with a known sexual state, *Neosartorya udagawae*, was also misidentified as *A. fumigatus*, as revealed by molecular studies (Balajee et al., 2006). Again, none of these isolates produced the diagnostic cleistothecia and they had overlapping morphological characteristics with *A. fumigatus*. Since the clinical isolates of *N. pseudofischeri* and *N. udagawae* did not produce fruiting bodies, these isolates were identified by their anamorphic states, *A. thermomutatus* and *A. udagawae*, respectively, in keeping with the mycological convention.

At the same time that molecular strategies have helped describe new species within the section *Fumigati*, comparative sequence analysis methodologies have also clarified the status of several isolates that were previously described as "variants" of *A. fumigatus* based on differences in morphological characteristics. Molecular studies have shown that *A. arvii*, *A. fumigatus* var. *helvola*, *A. fumigatus* var. *ellipticus*, and *A. fumigatus* var. *acolumnaris* should be recognized as *A. fumigatus* since they cluster together with *A. fumigatus* in several loci.

PHYLOGENETIC SPECIES RECOGNITION CONCEPT

Innovation in sequencing technology has made the generation of sequence data from multiple genes relatively easy and fairly economical. This has resulted in the increasing popularity of the genealogical concordance phylogenetic species recognition (GCPSR) concept as a method of choice for fungal species determination. GCPSR is a recently recognized concept by which species are recognized based on genetic isolation demonstrated by congruence of gene genealogies (Taylor et al., 2000). Thus, according to GCPSR, if gene trees constructed from sequences are concordant, these branches connect species, but conflict among the gene trees and the transition from concordance to conflict determines the limits of species (Taylor et al., 2000). Species recognition by GCPSR is thus based on phylogenetic analysis of variable molecular characteristics (like polymorphic nucleotide positions) and offers consistency in the delineation of species. GCPSR is especially useful in species recognition in asexual organisms, where the biological species concept fails, in organisms with aberrant phenotypes, or organisms with overlapping phenotypic features, where the morphological species concept has limited utility. It must be emphasized that species delineation by GCPSR does not rely on information from a single locus; rather, GCPSR recognizes species boundaries by using genealogical concordance of multiple independent loci. Thus, a clade in a phylogeny can be recognized as a species if this clade is present in the majority of the single-locus genealogies studied (genealogical concordance) and is well-supported in at least one single-locus genealogy as revealed by strong bootstrap support (genealogical nondiscordance). To illustrate the GCPSR concept, *A. lentulus* was recognized as a new species within section *Fumigati* as follows: five gene regions were selected and primers were designed for PCR amplification and sequencing of these select gene regions. The resultant sequences were aligned, and a genealogy for each gene was constructed using the maximum likelihood (ML) algorithm. Support for nodes was evaluated through bootstrapping with the neighbor-joining algorithm under the same ML model, and bootstrap values were generated from 1,000 pseudoreplicates. Combined phylogeny was also performed from all of the five gene regions with two different algorithms, ML and maximum parsimony. Results demonstrated that *A. lentulus* isolates clustered together in all five single-locus genealogies, was present as a monophyletic clade in all five loci, and had high bootstrap support in three of five single-locus genealogies (Balajee et al., 2005b). Further, as shown in Fig. 1, combined evidence phylogeny trees generated by two different phylogenetic methods had a topology similar to that of trees generated from single loci, with strong bootstrap support. Thus, the *A. lentulus* clade fulfilled the criteria set forth by GCPSR and was described as a new species within the section *Fumigati*. Thus far, for all the new species described within the section *Fumigati*, GCPSR has been used at least in part to support the findings of the new species. For a detailed understanding of the principles of GCPSR and its use in species recognition for fungi, the reader is directed to previous excellent reviews on this subject (Dettman et al., 2003; Taylor et al., 1999, 2000).

IDENTIFICATION OF *A. FUMIGATUS* IN A CLINICAL MICROBIOLOGY LABORATORY

Fungal species identification remains an important activity in a clinical microbiology laboratory. Multilocus phylogenetic methods are excellent tools for understanding species relationships and the population structure of a given organism. However, these methods require time, expertise, and resources that most routine clinical microbiology laboratories may not possess. Here, it would be desirable to have an identification system that could enable a medical mycologist to readily apply a specific epithet to an *Aspergillus* isolate grown from a clinical specimen. Although many of the features used in taxonomical "classification" schemes (as described above) can also be used in an *Aspergillus* "identification" system, it must be remembered that while it is meaningful to spend a great deal of time and effort in establishing a sound *Aspergillus* species classification, "identification" in a clinical microbiology laboratory needs to be swift, especially where pathogens recovered

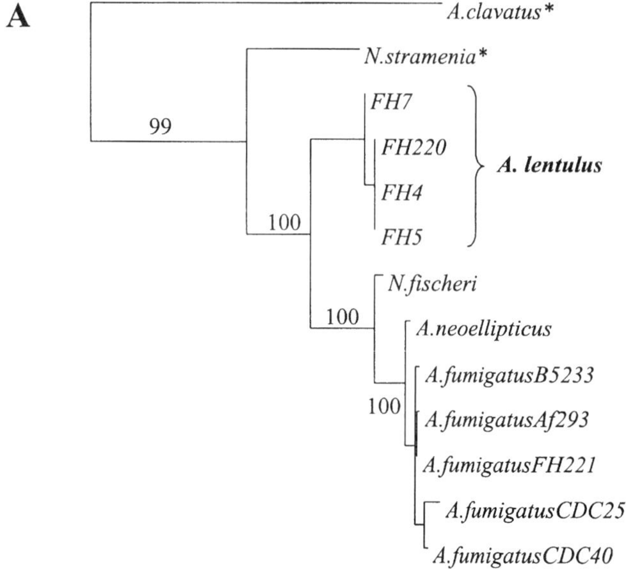

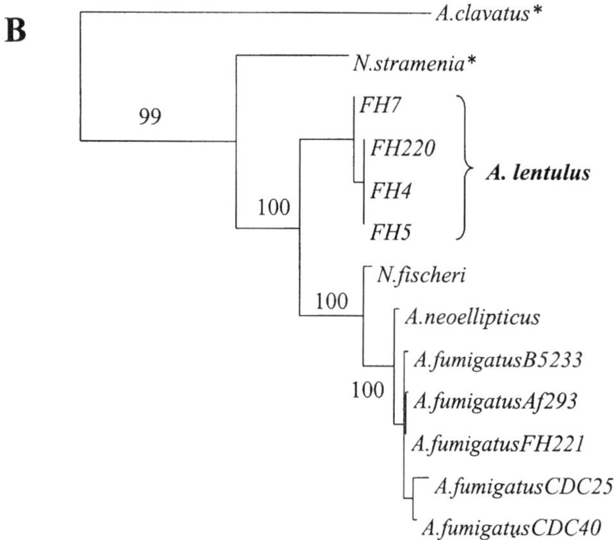

Figure 1. Combined ML trees (a) and parsimony tree (one of three) (b) generated from sequences from five protein-coding loci, showing *A. lentulus* as a separate species with high bootstrap values. Reprinted from *Eukaryotic Cell* with permission of the publisher (Balajee et al., 2005b).

from clinical specimens have to be recognized and reported and action taken.

In a clinical microbiology laboratory, identification of *A. fumigatus* isolates based singly on morphology has the following disadvantages:

(i) Overlapping morphological features of members of this section

(ii) Lack of or poor sporulation of some clinical isolates

(iii) Aberrant phenotypes in clinical isolates that are not consistent with those of "type" isolates

(iv) Requirement of several days to weeks to accomplish identification

(v) Need for experienced personnel

Correct identification of species within the section *Fumigati* is clinically relevant and may inform therapeutic decision making, since some of these species have variable susceptibilities to several currently available antifungal drugs (Balajee et al., 2005b; Mellado et al., 2006; Yaguchi et al., 2007). Specifically, *A. lentulus* appears to have higher MICs in vitro to several antifungal drugs compared to *A. fumigatus*, and thus species information could be important for the clinician to decide which drug to choose for successful treatment. Several members of the section *Fumigati* cause invasive fungal infections in immunocompromised hosts, and thus far nine species within the section *Fumigati* have been implicated in disease, including *A. fumigatus, A. lentulus, A. udagawae, A. thermomutatus, N. fischeri, A. viridinutans, N. spinosa*, and *N. hiratsukae* (Balajee et al., 2007c).

The recently concluded International Workshop on *Aspergillus* Systematics in the Genomic Era Working Group meeting recommended the use of internal transcribed spacer (ITS) regions for inter-section-level identification and the β-tubulin locus for identification of individual species within the various *Aspergillus* sections (Balajee et al., 2007c). The section on molecular tools, below, focuses on such an identification scheme that will be useful for species identification within the section *Fumigati*.

Comparative Sequence Analysis of ITS Regions

The ITS1 and ITS2 with 5.8S rRNA regions of the rRNA operon can be useful markers for *Aspergillus* species discrimination at the subgenus or section level. The ITS region primers make use of conserved regions of the 18S (ITS1) and the 28S (ITS4) rRNA genes to amplify the intervening 5.8S gene and the ITS1 and ITS2 noncoding regions (forward primer, 5′-GGA AGT AAA AGT CGT AAC AAG G; reverse primer, 5′-TCC TCC GCT TAT TGA TAT GC) (Henry et al., 2000). PCR amplification using these primer sets yields a 550- to 600-bp amplicon; the resultant sequence can be used to query the GenBank database for species identification at the section level. Advantages of employing the ITS region for species identification include the universality of the primer sets, amenability of the regions to PCR amplification and sequencing, and the availability of a large and ever-expanding ITS sequence database in the publicly available GenBank database. Although at present there are no consensus cutoff values for *Aspergillus* species, delineation using the ITS regions, and a percent identity match of 100% with a type isolate will place the isolate in the respective section. However, comparative sequence analyses of the ITS regions are not useful to clearly discriminate the several species within the section *Fumigati* (Balajee et al., 2005b). Thus, the ITS regions have limited utility for finer levels of discrimination, such as delineation of *A. lentulus* from *A. udagawae* unambiguously within the section *Fumigati*, and comparative sequence analyses of the protein-coding regions can be more useful.

Comparative Sequence Analysis of Protein-Coding Regions

Several loci have been evaluated, and these include coding gene regions for proteins such as β-tubulin (*benA*), calmodulin (*cal*), rodlet A (*rodA*), and other variable loci that encompass microsatellite repeat motifs (Pringle et al., 2005) and intergenic regions between genes, including *inter1* (found between *sec61* and *ecm40* on chromosome 5), *inter2* (found between *yll034C* and *erb1* on chromosome 1), and *inter3* (found between *glc3* and *atp2* on chromosome 5) (Rydholm et al., 2006). Additionally, a multilocus sequence typing scheme (MLST; explained in detail below in the section on molecular tools) specific for *A. fumigatus* has also been evaluated with the following gene regions: annexin (*ANX4*), β-1,3-glucanosyl transferase (*BGT1*), catalase (*CAT*), lipase (*LIP*), mating-type protein (*MAT1-2*), superoxide dismutase (*SOD*), and zinc transporter (*ZRF2*) (Bain et al., 2007). Most of the evaluated genes have been polymorphic and suitable for species delimitation, recombination studies, and population structure analyses.

In our experience, the *benA* gene is a useful marker for species delimitation within the section *Fumigati*, and the primer set *benA* forward (5′-AAT TGG TGC CGC TTT CTG G) and *benA* reverse (5′-AGT TGT CGG GAC GGA ATA G) can be used for this purpose. A study in our laboratory that analyzed several protein coding loci revealed that the *benA* region can provide enough information to identify a previously recognized species within the section *Fumigati* (unpublished data). Thus, sequencing of the *benA* region and subsequent GenBank analysis will identify an isolate to a species within the section *Fumigati*. However, it must be emphasized that for the purposes of defining a new species and for better understanding of the relationship of the newly described species to other species within the section *Fumigati*, detailed phylogenetic analyses using multiple coding regions as described above should be performed.

Figure 2 shows an example of a work flow for identification of *A. fumigatus* using comparative sequence-based methods that can be used in clinical microbiology laboratories.

POPULATION STRUCTURE OF *A. FUMIGATUS*

Knowledge of the population structure of *A. fumigatus* will greatly enable our understanding of the epidemiology of aspergillosis caused by this organism. To that end, studies have shown a lack of genetic differentiation in *A. fumigatus* isolates when using multiple different markers and from sampling of both clinical and environmental isolates from North America and Europe (Bain et al., 2007; Pringle et al., 2005; Rydholm et al., 2006). Thus, sequence data from the β-tubulin (*benA*) ITS regions of fungal ribosomal DNA and three intergenic regions of a number of *A. fumigatus* isolates revealed only 9 to 18 polymorphic sites (Rydholm et al., 2006), while another study using multilocus sequence data from seven gene fragments found that among the 100 *A. fumigatus* isolates studied, 93% of the isolates differed from the other isolates by only one allele sequence, thus forming a single clonal cluster (Bain et al., 2007b). These studies imply that *A. fumigatus* is a widespread species exhibiting little variation either within geographic regions or on a global scale, and this ubiquity could be attributed in part to the continual gene flow across continents facilitated by the movement of the buoyant *A. fumigatus* conidia across the continents. Thus, *A. fumigatus* appears to be a truly cosmopolitan fungus, and this may be due to the fact that *A. fumigatus* is an opportunistic pathogen of humans, with its natural ecological niche being decaying plant debris, soil, and compost.

MOLECULAR TOOLS FOR SUBTYPING *A. FUMIGATUS* ISOLATES

Advances in molecular biology have led to the use of techniques that are based on the principle that there are genotypic differences from strain to strain but not within a given strain. The use of such molecular markers represents a quantum leap in the field of epidemiology, since biological markers can delineate the continuum of events between an exposure and a resultant infection due to exposure. Over the years, a plethora of molecular typing methods that have provided invaluable information about the genetic diversity in *Aspergillus* isolates have been reported. Because of its clinical relevance, most of these methods were applied to *A. fumigatus*, and as such some of these methods are highly specific for *A. fumigatus*. An aspect that needs to be taken into

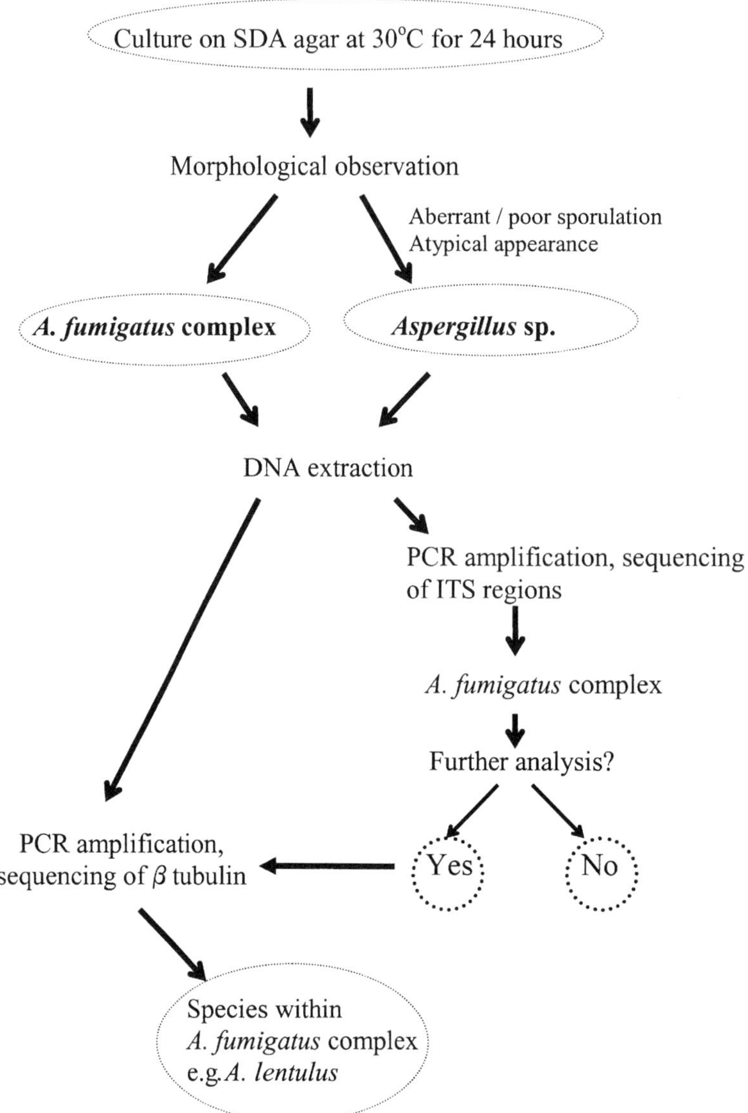

Figure 2. Proposed work flow for identification of *A. fumigatus* using comparative sequence-based methods amenable for use in clinical microbiology laboratories.

consideration is the genome coverage of the different methods: some methods may cover large parts of the genome, whereas targeted typing methods are restricted to a relatively low number of sites in the fungal genome. Accordingly, this explains why different typing methods need not necessarily give similar results with a given set of isolates. Furthermore, all these methods differ in terms of discriminatory power, reproducibility, complexity, throughput, and ease of use, but all of them may prove useful to a certain extent under specific circumstances. Since most of the earlier molecular methods have been extensively reviewed elsewhere (Varga, 2006), only the most recent techniques are outlined below.

Afut1 Hybridization

Multicopy elements were among the first and most popular targets for developing strain typing methods. When dispersed throughout a genome (rather than being clustered in a particular region, as is usually the case with a ribosomal gene cluster), they serve as excellent markers for genome variability. After going through extensive selection procedures, Girardin et al. (1994) identified a multicopy element that consisted of three open reading frames flanked by two long terminal repeats (LTRs), a structure that highly resembled the Gypsy family of retrotransposon elements. *Afut1* carries the characteristics of a retrotransposon-like element but with no apparent functionality left (Neuveglise et al., 1996). The *Afut1* element is present in ~20 copies in the *A. fumigatus* genome and can be targeted by random fragment length polymorphism (RFLP)-Southern blotting procedures. This method is quite straightforward: first, genomic DNA is digested with a restriction enzyme, generating several thousand DNA fragments that are separated according to size using agarose gel electrophoresis. Subsequently, the DNA fragments are transferred to a membrane by capillary transfer and hybridized to a radioactively labeled *Afut1* probe. Only DNA fragments containing (part of) the *Afut1* probe will hybridize, and thus visualization of the fragments by autoradiography can identify specific banding patterns that enable discrimination of unrelated isolates. Differences in sizes between reactive DNA fragments in unrelated isolates are the result of different copy numbers and/or sequence variations in the regions flanking the locations of the *Afut1* element in the chromosome. So far, the *Afut1* element appears to be specific for *A. fumigatus*. Retrotransposon (-like) elements that have been identified in other fungal species also qualify for similar approaches in strain typing (Nielsen et al., 2001; Okubara et al., 2003). Our understanding of the genomic diversity within the *A. fumigatus* population has greatly improved because of strain typing studies that used the *Afut1* RFLP typing strategy (Bart-Delabesse et al., 2001; Chazalet et al., 1998; Neuveglise et al., 1996). Accordingly, *Afut1* RFLP analysis has been considered the "gold standard" in *A. fumigatus* strain typing for many years. Despite the fact that *Afut1* RFLP is an excellent typing method, its use has not disseminated widely, most likely due to the time- and labor-intensive procedures necessary to complete the analyses and the need for highly standardized analyses in order to be able to compare data from different experiments.

In an attempt to alleviate some of the drawbacks of the Southern blotting approach, retrotransposon insertion site context (RISC) typing was introduced as a PCR alternative method for the RFLP procedure (de Ruiter et al., 2007). RISC typing eliminates most of the disadvantages of the RFLP approach, since RISC typing only requires small amounts of genomic DNA, does not require the use of radioactive probes, and allows an entire analysis to be completed within one working day. By studying genomic sequence data from Af293, it became clear that part of the current genetic variability of *A. fumigatus* may be the result of inter- and/or intrachromosomal recombination events involving these retrotransposon element sequences. Specifically, orphan LTR elements reveal recombination events between 5′- and 3′-LTR elements. This may very well explain why at least part of the genetic variability within this species may go unnoticed when using MLST or other such targeted approaches. Both RFLP and RISC typing compare favorably to other high-resolution fingerprinting methods, but one of the main disadvantages of any pattern-based typing approach is that it is extremely difficult (if not impossible) for data exchange between labs. From a unified point of view it makes much more sense to develop fingerprinting methods that yield unambiguous typing data that are readily exchangeable between labs. Microsatellite typing methods offer such a format and are discussed in the following paragraphs in detail.

Microsatellite Typing

Microsatellites are short (up to 10 bp) directly repeated stretches of DNA that are abundantly present in the genomes of higher organisms, including fungi. Microsatellites are often also referred to as short tandem repeats. An interesting aspect about microsatellites is that they are inherently unstable by nature, as during DNA replication, the phenomenon of slipped-strand mispairing may alter the number of repetitions in any given marker. As a result, many microsatellite loci (especially the ones with high repeat numbers) are highly polymorphic between isolates, making them ideal targets for strain discrimination. Among the many advantages of microsatellites over other fingerprinting methods are specificity, the modular concept, multiplex options, reproducibility, discriminatory power, and the numerical output of results. By using primers based on the flanking sequences of a microsatellite, the method is easily amenable to yield reproducible and robust PCR amplification under high-stringency conditions (de Valk et al., 2007a). Subsequently, the size of the amplicon can be easily and very precisely determined by capillary electrophoresis and converted into the corresponding repeat number. It is this repeat number that represents the typing result (in contrast to fingerprints consisting of complex banding patterns). By using combinations of different fluorescent labels for different markers, multiple microsatellites can be combined in a single amplification reaction mixture and capillary run, allowing high-throughput analyses (Fig. 3). The modular concept allows easy incorporation of more markers, if needed. Thus, all these aspects render microsatellites an ideal typing system. Not surprisingly, microsatellites represent the gold standard in human (and animal) forensics, where virtually each and every individual can be identified by its own unique genotype (monozygotic twins and multiples excluded).

Multiple panels of microsatellites have been proposed for *A. fumigatus* (Bart-Delabesse et al., 1998; de Valk et al., 2005) and have performed well in comparative studies (de Valk et al., 2007b; Lasker, 2002). Microsatellite analysis of *A. fumigatus* has confirmed much of the data that were previously generated by the *Afut1* RFLP method, albeit in a format that is much faster and

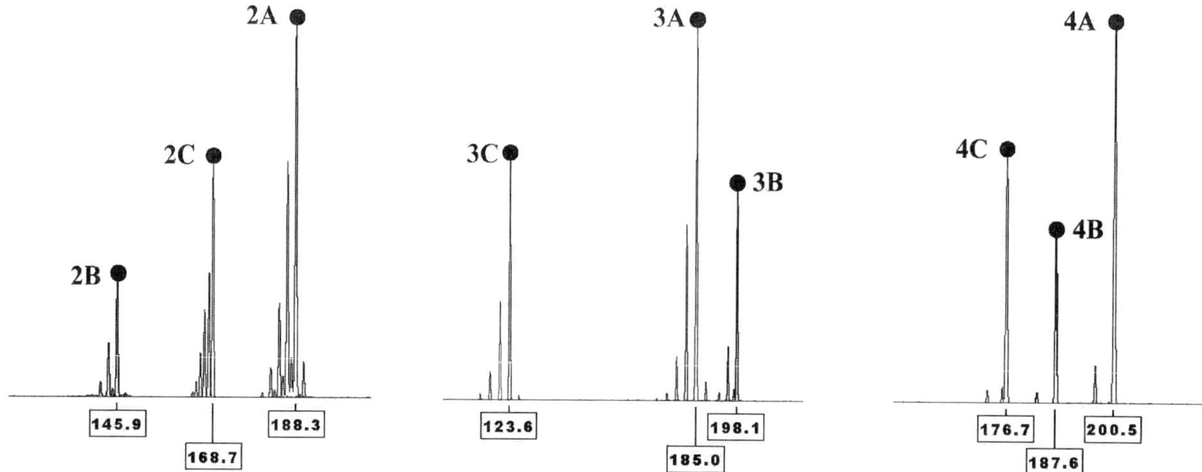

Figure 3. Examples of short tandem repeat amplification results. The principal peak used for analysis is indicated with a dot. Numbers with each dot represent the repeat unit size of the marker, and the letter represents the fluorescent label: A, 6-carboxyfluorescein (FAM); B, hexachlorofluorescein (HEX); C, tetrachlorofluorescein (TET). The boxed numbers below the graph indicate the size of the amplified products (in nucleotides) as determined by high-resolution capillary electrophoresis.

much easier to generate and to interpret (Bart-Delabesse et al., 1999). In addition, it was shown that microsatellites perform very well in epidemiological studies (de Valk et al., 2007b). In Fig. 4, a graphical interpretation of genotypes from about 250 randomly selected *A. fumigatus* isolates reveals isolates that yielded identical or similar microsatellite patterns. Such analyses would be immensely helpful in outbreak situations, where genetically related and unrelated *A. fumigatus* isolates could be rapidly and efficiently identified among epidemiologically linked isolates (branches that are highlighted represent genetically related genotypes).

One shortcoming of a microsatellite-based typing method may be that most microsatellite loci are species specific and, thus, once designed can only be used for strain typing in a single species. Apart from this drawback, the overall benefits of microsatellites over many other genotyping methods are the ability to recognize mixtures of strains and the highly standardized concept that allows exchange of results between labs and deposition of typing data in worldwide databases.

AFLP

Amplified fragment length polymorphism (AFLP) analysis is a molecular fingerprinting method based on selective amplification of restriction fragments in a random but reproducible manner (Vos et al., 1995). Amplification of these fragments is enabled by ligation of synthetic double-stranded DNA adapters to the cohesive ends generated by restriction enzyme digestion. These known adapter sequences serve as primer-binding sites for the corresponding amplification primers. Usually two restriction enzymes are used: one with a 6-bp recognition sequence (cutting once every ~4,000 bp) and one with a 4-bp recognition sequence (cutting once every ~256 bp). When applied to a fungal genome, this approach would lead to the formation of about 100,000 restriction fragments. However, only the portion of the restriction fragments with two different termini is amplified: fragments with identical ends may form a so-called panhandle structure that will prevent them from being amplified efficiently. Still, some 5,000 restriction fragments will remain that can be amplified. By extending the AFLP primers with one or more selective residues into the unknown part of the restriction fragments, the large number of fragments that are generated this way can easily be reduced to a more accessible number: each subsequent selective residue reduces the number of amplified fragments by a factor of about 4. Theoretically, an almost infinite number of combinations of restriction enzymes and selective residues can be made, making AFLP analysis an extremely flexible typing method.

An infection by *A. fumigatus* is thought to be acquired after exposure and subsequent inhalation of conidia from the air in the environment. However, AFLP analysis has demonstrated that water should also be considered as a potential source of *A. fumigatus* (Warris et al., 2003). Furthermore, based on AFLP analysis and confirmation by microsatellite typing and other molecular methods, it was demonstrated that individuals can be colonized by multiple different *A. fumigatus* genotypes in the respiratory tract (Bart-Delabesse et al., 1999; de Valk et al., 2007; Neuveglise et al., 1997). Unlike isolates from deep sites within a body that all

Figure 4. Minimum spanning tree of 250 random *A. fumigatus* isolates based on a categorical analysis of nine microsatellite repeat markers. Each circle represents a unique genotype. The size of each circle corresponds to the number of isolates with the same genotype. Genotypes connected by a shaded background differ by a maximum of two of the nine markers and could be considered a "clonal complex." Thick connecting line, one marker difference; regular connecting line, two marker differences; interrupted line, three or more marker differences.

proved to be indistinguishable within a patient, a variety of genotypes was observed in isolates from sputum and bronchoalveolar lavage fluid. This could have important implications in the way clinical samples are processed in the laboratory.

AFLP is also a particularly interesting typing method because it can be used for typing as well as for identification of microorganisms. In a way, AFLP can also be considered a true PCR alternative for the classical DNA-DNA reassociation method that is still considered to be the gold standard in the definition and recognition process of microbial species. Any two fungi with a high percentage of sequence similarity should also reveal a high percentage of corresponding fragments in an AFLP analysis. This is exactly what is usually observed when multiple isolates from the same species are being analyzed. In Fig. 5, AFLP analysis results of a number of medically relevant species from the section *Fumigati* are shown. It is clear that the calculated similarities of fingerprints from isolates belonging to the same species are much higher than that of the fingerprints of isolates of a different species. In addition, within each species, a number of constant bands are visible that could serve as species-specific markers. These bands could be regarded as markers belonging to the "core" genes of a certain species. Likewise, if there is a clear correlation between certain variable bands and a certain phenotype, these bands could be used to identify the specific DNA elements involved in that phenotype.

Despite the fact that AFLP also has a very high discriminatory power, its use as a typing method is generally restricted to a relatively low number of specialized laboratories. As with RFLP and RISC typing data, AFLP fingerprints are not very suitable for interlaboratory comparisons. Moreover, its long-term reproducibility has not been demonstrated yet. Indeed, this could prove to be problematic because of the multitude of enzymatic steps that are involved in the generation of the final fingerprint (restriction enzyme digestion, adapter ligation, and PCR amplification, respectively).

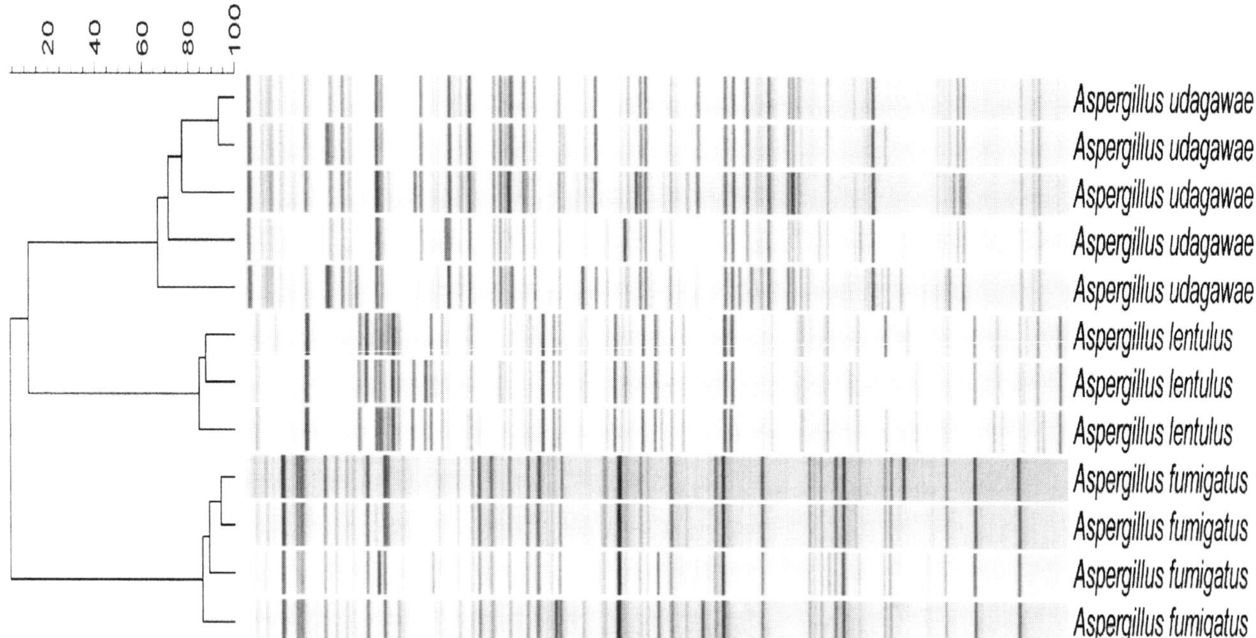

Figure 5. AFLP analysis of five isolates of *A. udagawae*, three isolates of *A. lentulus*, and four isolates each of *A. fumigatus*, revealing species-specific banding patterns.

MLST

MLST was originally developed as a molecular alternative for multilocus enzyme electrophoresis. In the early days of strain discrimination, it was demonstrated that certain housekeeping enzymes display a different electrophoretic mobility in nondenaturing applications as a result of certain amino acid substitutions. These substitutions are in turn the result of nucleic acid substitutions. Analysis of these nucleic acid substitutions are the underlying principle of MLST analysis. Thus, instead of performing phenotypical analyses, DNA sequence analysis of part of a housekeeping gene can now be performed to discriminate between isolates. DNA sequence analysis may even identify nucleotide substitutions that do not affect the mobility of the corresponding proteins. Thus, the number of alleles as determined by DNA sequence analysis (MLST) should be higher than that determined by conventional multilocus enzyme electrophoresis procedures. MLST data are processed as follows: each different allele from a given gene is assigned a unique number. By combining sequence data from multiple genes, a combination of allele numbers is converted into a sequence type (ST). The ST can be traced back to the underlying sequences and vice versa. One of the main advantages of MLST is that this method is quite straightforward. In addition, DNA sequence analysis is a universal concept that allows easy comparisons between laboratories and deposition of sequence data in public repositories. Furthermore, the method is fully amenable to automation, thus taking the burden off laboratory personnel.

The MLST strategy has been very useful in strain typing of numerous bacteria and recently for yeasts, including *Candida albicans, Candida glabrata, Candida tropicalis*, and *Cryptococcus neoformans* var. *grubii*, among others (www.mlst.net). However, MLST has not been useful for strain typing of *A. fumigatus* due to low variability in the genes studied. This could be due to a direct consequence of the mechanisms responsible for the observed differences between isolates, for example, nucleotide transitions and transversions that render the given DNA sequence extremely stable if not subject to selective pressure. A recent study screened 27 candidate genes for *A. fumigatus* and selected a panel of the 7 most discriminatory genes to use for MLST (Bain et al., 2007). The selected gene set only identified 30 STs in a collection of 100 isolates from diverse origins, reiterating the limitations of such a strategy for strain typing of *A. fumigatus* isolates. Since MLST seems to be far more discriminatory for several other fungal species, this result is probably also indicative of *A. fumigatus* being a relatively young species, having accumulated little sequence variation (Bain et al., 2007). In spite of the fact that MLST was not useful for *A. fumigatus* strain typing, this strategy appears to be quite useful for the classification of isolates at the species level and has proven invaluable for the recognition and identification of new species within the section *Fumigati*, as detailed above in the section discussing GCPSR.

```
65_OB1      ACT TCT GTC CCG ACT TCT GTC CCA --- --- --- --- --- --- --- ---
607_OB1     ... ... ... ... ... ... ... ... --- --- --- --- --- --- --- ---
B5855_OB2   ... ... ... ... ... ... ... ..G ACT TCT GTC CCG --- --- --- ---
B5856_OB2   ... ... ... ... ... ... ... ..G ACT TCT GTC CCG --- --- --- ---
B5358_OB3   ... ... ... ... ... ... ... ..G ACT TCT GTC CCG ACT TCT GTC CCG
B5355_OB3   ... ... ... ... ... ... ... ..G --- --- --- --- --- --- --- ---
33_OB5      ... ... ... ... ... ... ... ... --- --- --- --- --- --- --- ---
97_OB6      ... ... ... ... ... ... ... ..G ACT TCT GTC CCG ACT TCT GTC CCG
103_OB6     ... ... ... ... ... ... ... ..G --- --- --- --- --- --- --- ---
N.fischeri  ... ... ... ... ... ... ... ..G --- --- --- --- --- --- --- ---

65_OB1      --- --- --- --- ACT CAA AAC GCG ACT TCA ATC CCG --- --- --- ---
607_OB1     --- --- --- --- ... ... ... ... ... ... ... ... ACT TTT GTC CCG
B5855_OB2   --- --- --- --- --- --- --- --- --- --- --- --- --- --- --- ---
B5856_OB2   ACT TCT GTC CCA ... ... ... ... ... ... ... ... ACT TTT GTC CCG
B5358_OB3   --- --- --- --- --- --- --- --- --- --- --- --- ACT TTT GTC CCG
B5355_OB3   ACT TCT GTC CCA ... ... ... ... ... ... ... ... ACT TTT GTC CCG
33_OB5      --- --- --- --- ... ... ... ... ... ... ... ... --- --- --- ---
97_OB6      ACT TCT GTC CCA ... ... ... ... ... ... ... ... ACT TTT GTC CCG
103_OB6     --- --- --- --- --- --- --- --- --- --- --- --- ACT TTT GTC CCG
N. fischeri --- --- --- --- ... ..G ... ... --- --- --- --- --- --- --- ---

65_OB1      --- --- --- --- --- --- --- --- ACT TCA GTC CCG ACT CAA AAC GCG
607_OB1     ACT CAA AAC GCG ACT TCT GTC CCG ... ... ... ... ... ... ... ...
B5855_OB2   ACT CAA AAC GCG ACT TCT GTC CCG ... ... ... ... ... ... ... ...
B5856_OB2   ACT CAA AAC GCG ACT TCT GTC CCG ... ... ... ... ... ... ... ...
B5358_OB3   ACT CAA AAC GCG ACT TCT GTC CCG ... ... ... ... ... ... ... ...
B5355_OB3   ACT CAA AAC GCG ACT TCT GTC CCG ... ... ... ... ... ... ... ...
33_OB5      --- --- --- --- --- --- --- --- ... ... ... ... ... ... ... ...
97_OB6      ACT CAA AAC GCG ACT TCT GTC CCG ... ... ... ... ... ... ... ...
103_OB6     ACT CAA AAC GCG ACT TCT GTC CCG ... ... ... ... ... ... ... ...
N. fischeri --- --- --- --- --- --- --- --- --- --- --- --- --- --- --- ---

65_OB1      ACT ACT ATT GTG CCA
607_OB1     ... ... ... ... ...
B5855_OB2   ... ... ... ... ...
B5856_OB2   ... ... ... ... ...
B5358_OB3   ... ... ... ... ...
B5355_OB3   ... ... ... ... ...
33_OB5      ... ... ... ... ...
97_OB6      ... ... ... ... ...
103_OB6     ... ... ... ... ...
N. fischeri ... ... ... ... ..G
```

Figure 6. Alignment of the repeat region of the Afu3g08990 locus (CSP locus). Representative sequence types and sequences of *N. fischeri* are shown. A dash indicates an insertion or deletion, and a dot indicates a nucleotide that is identical to the nucleotide in the top sequence. Reprinted from *Eukaryotic Cell* with permission of the publisher (Balajee et al., 2007b).

Single-Locus Sequence Typing Strategy

By analyzing the genomic sequence of Af293, surface protein genes were described that also contain tandemly repeated sequences with some open reading frames containing repeat units as long as 285 nucleotides (Levdansky et al., 2007). These targets may display sequence variation by means of two mechanisms: repeat copy number variation (as the result of insertion or deletion of repeats), as well as DNA sequence variation through nucleotide substitutions. One of the main advantages of these targets is the fact that they are analyzed by DNA sequence analysis, rather than through sizing using a high-resolution capillary platform. Hence, this technique is much more accessible to other laboratories. Recently, Balajee and coworkers investigated the usefulness of one such target, a gene encoding a putative cell surface protein, Afu3g08990 (denoted as CSP by the authors) for *A. fumigatus* strain typing (Balajee et al., 2007b). PCR amplification and sequencing of a partial region of this gene revealed the presence of a highly variable region consisting of tandem repeats (Fig. 6). This repeat region was found to be highly informative for subtyping, as shown by the presence of multiple, well-supported clusters inferred from detailed phylogenetic analysis (Fig. 7) of sequences generated from a large number of *A. fumigatus* isolates recovered from several invasive aspergillosis outbreaks. Thus, as proposed, comparative sequence analyses of the CSP gene can be employed for *A. fumigatus* outbreak source tracking. The main limitation of this target is the relatively low discriminatory power. However, by expanding the

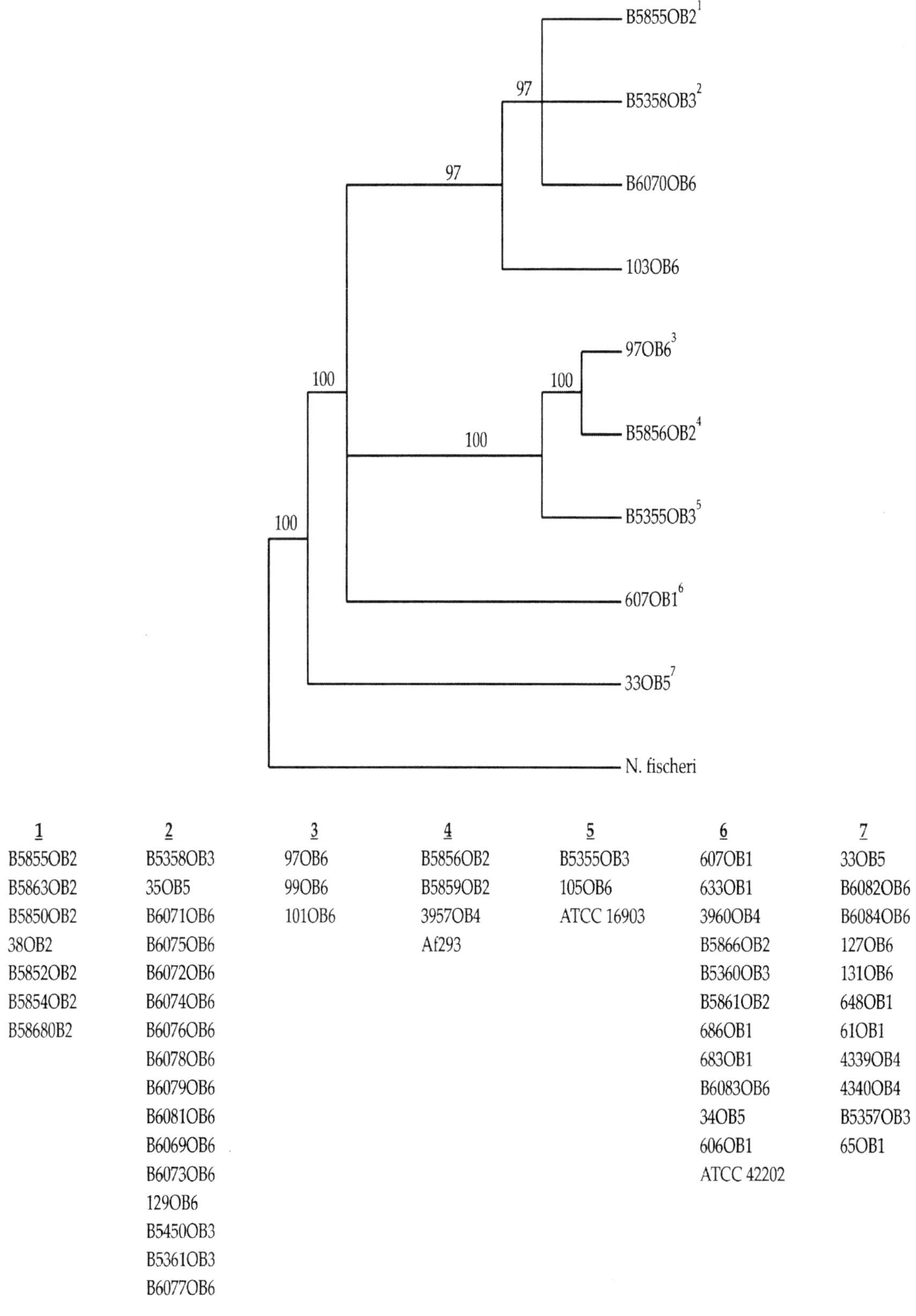

Figure 7. Phylogeny of unique *A. fumigatus* sequence types based on the 564-bp CSP gene fragment. *N. fischeri* is the outgroup taxon used to root the tree. Only unique STs were used to construct the tree. Strains listed below the tree are grouped with other strains that possess identical CSP STs; the numbers above each grouping correspond to the numerical superscripts for each representative ST included in the phylogeny. Isolates from six invasive aspergillosis outbreaks were included in the analysis and are represented as OB1 to -6. Reprinted from the journal *Eukaryotic Cell* with permission of the publisher (Balajee et al., 2007b).

number of similar targets, the discriminatory power of such a typing system could be increased, but this has not yet been attempted. Interestingly, the same targets may be used for different species from the *Fumigati* section, since a very similar sequence also turns out to be present in the genome of *N. fischeri*, the closest relative of *A. fumigatus*.

SUMMARY

Aspergillus species identification by morphological methods continues to have an important place in the clinical microbiology laboratory. However, as detailed in this chapter, these methods have a number of practical limitations, rendering them largely unsuitable for intrasection species identification, comprehensive studies of population structure, and epidemiological and surveillance studies. Recently, polyphasic taxonomy, a method that utilizes a combination of different phenotypic or genotypic data sets to define genera, species, and even taxonomically relevant subspecies, has been proposed and may be a more relevant method of defining species and subspecies within the genus *Aspergillus*.

DNA sequence-based methodologies yield information on different taxonomic levels, including genus and species and levels below species/subspecies. As outlined above, by choosing to sequence different targets, different levels of identification can be secured. Thus, if the goal is to generate a section-level identification, sequencing of the ITS regions would suffice. However, if the goal is to identify the unknown organism to an individual species within the *A. fumigatus* complex, then a protein-coding locus such as *benA* would meet the requirement (Fig. 3). Here, it must be emphasized that for taxonomical classification schemes and phylogenetic relationships, multilocus genealogies need to be elucidated and single-locus genealogy is not appropriate.

Discrimination of individual strains within the *A. fumigatus* population can be performed by banding pattern-based methods, microsatellite typing, or sequencing methodologies, as explained in detail above. As we move forward, sequence-based or microsatellite typing schemes may replace pattern-based molecular methods because of ease in sharing data and objectivity. Although valuable information may be lost by choosing purely sequence-based approaches, enlarging the number of sequencing targets and/or investigating other hypervariable targets may generate sufficient data useful for strain typing.

With the availability of the two fully sequenced *A. fumigatus* genomes, species identification and strain typing strategies using molecular mycology are poised to enter a new era in which next-generation methods, such as multianalyte profiling systems and single-nucleotide polymorphism microarrays, will be designed, validated, and implemented.

REFERENCES

Bain, J. M., A. Tavanti, A. D. Davidson, M. D. Jacobsen, D. Shaw, N. A. Gow, and F. C. Odds. 2007. Multilocus sequence typing of the pathogenic fungus *Aspergillus fumigatus*. *J. Clin. Microbiol.* **45:** 1469–1477.

Balajee, S. A., J. Gribskov, M. Brandt, J. Ito, A. Fothergill, and K. A. Marr. 2005a. Mistaken identity: *Neosartorya pseudofischeri* and its anamorph masquerading as *Aspergillus fumigatus*. *J. Clin. Microbiol.* **43:**5996–5999.

Balajee, S. A., J. L. Gribskov, E. Hanley, D. Nickle, and K. A. Marr. 2005b. *Aspergillus lentulus* sp. nov., a new sibling species of *A. fumigatus*. *Eukaryot. Cell* **4:**625–632.

Balajee, S. A., D. Nickle, J. Varga, and K. A. Marr. 2006. Molecular studies reveal frequent misidentification of *Aspergillus fumigatus* by morphotyping. *Eukaryot. Cell* **5:**1705–1712.

Balajee, S. A., L. Sigler, and M. E. Brandt. 2007a. DNA and the classical way: identification of medically important molds in the 21st century. *Med. Mycol.* **45:**475–490.

Balajee, S. A., S. T. Tay, B. A. Lasker, S. F. Hurst, and A. P. Rooney. 2007b. Characterization of a novel gene for strain typing reveals substructuring of *Aspergillus fumigatus* across North America. *Eukaryot. Cell* **6:**1392–1399.

Balajee, S. A., J. Houbraken, P. E. Verweij, S. B. Hong, T. Yaghuchi, J. Varga, and R. A. Samson. 2007c. *Aspergillus* species identification in the clinical setting. *Stud. Mycol.* **59:**39–46.

Bart-Delabesse, E., C. Cordonnier, and S. Bretagne. 1999. Usefulness of genotyping with microsatellite markers to investigate hospital-acquired invasive aspergillosis. *J. Hosp. Infect.* **42:**321–327.

Bart-Delabesse, E., J. F. Humbert, E. Delabesse, and S. Bretagne. 1998. Microsatellite markers for typing *Aspergillus fumigatus* isolates. *J. Clin. Microbiol.* **36:**2413–2418.

Bart-Delabesse, E., J. Sarfati, J. P. Debeaupuis, W. van Leeuwen, A. van Belkum, S. Bretagne, and J. P. Latgé. 2001. Comparison of restriction fragment length polymorphism, microsatellite length polymorphism, and random amplification of polymorphic DNA analyses for fingerprinting *Aspergillus fumigatus* isolates. *J. Clin. Microbiol.* **39:**2683–2686.

Chazalet, V., J. P. Debeaupuis, J. Sarfati, J. Lortholary, P. Ribaud, P. Shah, M. Cornet, H. Vu Thien, E. Gluckman, G. Brucker, and J. P. Latgé. 1998. Molecular typing of environmental and patient isolates of *Aspergillus fumigatus* from various hospital settings. *J. Clin. Microbiol.* **36:**1494–1500.

de Ruiter, M. T., H. A. de Valk, J. F. Meis, and C. H. Klaassen. 2007. Retrotransposon insertion-site context (RISC) typing: a novel typing method for *Aspergillus fumigatus* and a convenient PCR alternative to restriction fragment length polymorphism analysis. *J. Microbiol. Methods* **70:**528–534.

Dettman, J. R., D. J. Jacobson, E. Turner, A. Pringle, and J. W. Taylor. 2003. Reproductive isolation and phylogenetic divergence in *Neurospora*: comparing methods of species recognition in a model eukaryote. *Evolution* **57:**2721–2741.

de Valk, H. A., J. F. Meis, I. M. Curfs, K. Muehlethaler, J. W. Mouton, and C. H. Klaassen. 2005. Use of a novel panel of nine short tandem repeats for exact and high-resolution fingerprinting of *Aspergillus fumigatus* isolates. *J. Clin. Microbiol.* **43:**4112–4120.

de Valk, H. A., J. F. Meis and C. H. Klaassen. 2007a. Microsatellite based typing of *Aspergillus fumigatus*: strengths, pitfalls and solutions. *J. Microbiol. Methods* **69:**268–272.

de Valk, H. A., J. F. Meis, B. E. de Pauw, P. J. Donnelly, and C. H. Klaassen. 2007b. Comparison of two highly discriminatory molecular fingerprinting assays for analysis of multiple *Aspergillus fumigatus* isolates from patients with invasive aspergillosis. *J. Clin. Microbiol.* 45:1415–1419.

Girardin, H., J. Sarfati, H. Kobayashi, J. P. Bouchara, and J. P. Latgé. 1994. Use of DNA moderately repetitive sequence to type *Aspergillus fumigatus* isolates from aspergilloma patients. *J. Infect. Dis.* **169:** 683–685.

Henry, T., P. C. Iwen, and S. H. Hinrichs. 2000. Identification of *Aspergillus* species using internal transcribed spacer regions 1 and 2. *J. Clin. Microbiol.* **38:**1510–1515.

Hong, S. B., H. S. Cho, H. D. Shin, J. C. Frisvad, and R. A. Samson. 2006. Novel *Neosartorya* species isolated from soil in Korea. *Int. J. Syst. Evol. Microbiol.* **56:**477–486.

Hong, S. B., S. J. Go, H. D. Shin, J. C. Frisvad, and R. A. Samson. 2005. Polyphasic taxonomy of *Aspergillus fumigatus* and related species. *Mycologia* **97:**1316–1329.

Hong, S. B., H. D. Shin, J. Hong, J. C. Frisvad, P. V. Nielsen, J. Varga, and R. A. Samson. 2008. New taxa of *Neosartorya* and *Aspergillus* in *Aspergillus* section *Fumigati*. *Antonie Leeuwenhoek* **93:**87–98.

Katz, M. E., A. M. Dougall, K. Weeks, and B. F. Cheetham. 2005. Multiple genetically distinct groups revealed among clinical isolates identified as atypical *Aspergillus fumigatus*. *J. Clin. Microbiol.* **43:** 551–555.

Larsen, T. O., J. Smedsgaard, K. F. Nielsen, M. A. Hansen, R. A. Samson, and J. C. Frisvad. 2007. Production of mycotoxins by *Aspergillus lentulus* and other medically important and closely related species in section *Fumigati*. *Med. Mycol.* **45:**225–232.

Lasker, B. A. 2002. Evaluation of performance of four genotypic methods for studying the genetic epidemiology of *Aspergillus fumigatus* isolates. *J. Clin. Microbiol.* **40:**2886–2892.

Levdansky, E., J. Romano, Y. Shadkchan, H. Sharon, K. J. Verstrepen, G. R. Fink, and N. Osherov. 2007. Coding tandem repeats generate diversity in *Aspergillus fumigatus* genes. *Eukaryot. Cell* **6:**1380–1391.

Mellado, E., L. Alcazar-Fuoli, G. Garcia-Effron, A. Alastruey-Izquierdo, M. Cuenca-Estrella, and J. L. Rodriguez-Tudela. 2006. New resistance mechanisms to azole drugs in *Aspergillus fumigatus* and emergence of antifungal drugs-resistant *A. fumigatus* atypical strains. *Med. Mycol.* **44**(Suppl.):367–371.

Neuveglise, C., J. Sarfati, J. P. Latgé, and S. Paris. 1996. Afut1, a retrotransposon-like element from *Aspergillus fumigatus*. *Nucleic Acids Res.* **24:**1428–1434.

Neuveglise, C., J. Sarfati, J. P. Debeaupuis, H. Vu Thien, J. Just, G. Tournier, and J. P. Latgé. 1997. Longitudinal study of *Aspergillus fumigatus* strains isolated from cystic fibrosis patients. *Eur. J. Clin. Microbiol. Infect. Dis.* **16:**747–750.

Nielsen, M. L., T. D. Hermansen, and A. Aleksenko. 2001. A family of DNA repeats in *Aspergillus nidulans* has assimilated degenerated retrotransposons. *Mol. Genet. Genomics* **265:**883–887.

Okubara, P. A., B. K. Tibbot, A. S. Tarun, C. E. McAlpin, and S. S. Hua. 2003. Partial retrotransposon-like DNA sequence in the genomic clone of *Aspergillus flavus*, pAF28. *Mycol. Res.* **107:**841–846.

Pringle, A., D. M. Baker, J. L. Platt, J. P. Wares, J. P. Latgé, and J. W. Taylor. 2005. Cryptic speciation in the cosmopolitan and clonal human pathogenic fungus *Aspergillus fumigatus*. *Evolution* **59:** 1886–1899.

Rydholm, C., G. Szakacs, and F. Lutzoni. 2006. Low genetic variation and no detectable population structure in *Aspergillus fumigatus* compared to closely related *Neosartorya* species. *Eukaryot. Cell* **5:** 650–657.

Taylor, J., D. Jacobson, and M. Fisher. 1999. The evolution of asexual fungi: reproduction, speciation and classification. *Annu. Rev. Phytopathol.* **37:**197–246.

Taylor, J. W., D. J. Jacobson, S. Kroken, T. Kasuga, D. M. Geiser, D. S. Hibbett, and M. C. Fisher. 2000. Phylogenetic species recognition and species concepts in fungi. *Fungal Genet. Biol.* **31:**21–32.

Varga, J. 2006. Molecular typing of aspergilli: recent developments and outcomes. *Med. Mycol.* **44:**149–161.

Vos, P., R. Hogers, M. Bleeker, M. Reijans, T. van de Lee, M. Hornes, A. Frijters, J. Pot, J. Peleman, M. Kuiper, et al. 1995. AFLP: a new technique for DNA fingerprinting. *Nucleic Acids Res.* **23:**4407–4414.

Wang, L., K. Yokoyama, M. Miyaji, and K. Nishimura. 2000. Mitochondrial cytochrome *b* gene analysis of *Aspergillus fumigatus* and related species. *J. Clin. Microbiol.* **38:**1352–1358.

Warris, A., C. H. Klaassen, J. F. Meis, M. T. De Ruiter, H. A. De Valk, T. G. Abrahamsen, P. Gaustad, and P. E. Verweij. 2003. Molecular epidemiology of *Aspergillus fumigatus* isolates recovered from water, air, and patients shows two clusters of genetically distinct strains. *J. Clin. Microbiol.* **41:**4101–4106.

Yaguchi, T., Y. Horie, R. Tanaka, T. Matsuzawa, J. Ito, and K. Nishimura. 2007. Molecular phylogenetics of multiple genes on *Aspergillus* section *Fumigati* isolated from clinical specimens in Japan. *Nippon Ishinkin Gakkai Zasshi* **48:**37–46.

Aspergillus fumigatus and Aspergillosis
Edited by J.-P. Latgé and W. J. Steinbach

Chapter 4

Aspergillus fumigatus Specificities as Deduced from Comparative Genomics

AMANDINE GASTEBOIS, KARINE LAMBOU, JOANNE WONG SAK HOI, AND FREDJ TEKAIA

The vast amount of data generated by genome projects makes comparative sequence analysis a powerful approach with which to detect similar or distinct functions in different species. Since the completion of the *Aspergillus fumigatus* genome sequence (Nierman et al., 2005), many completely sequenced genomes have been published, including sequences for various fungal species. With the great diversity, comparative genomic analyses over the entire fungal kingdom at both short and long evolutionary distances can now be robustly performed. The availability of genomes from closely related *Aspergillus* species allows identification of orthologous gene classes between these species. Identification of orthologs is crucial for the reconstruction of the evolutionary histories of genes and species. It is also crucial for addressing specific questions, such as, how many genes are shared by the considered species? How many genes are specific to each or to combinations of these species? What is the extent of the core of genes that share a common history? Which genes underwent duplication or were lost? Answers to such questions may help our understanding of the evolution and the pathogenicity of the *Aspergillus* species.

This review is mainly focused on specific genes of *A. fumigatus* as well as genes missing in *A. fumigatus* but present in other *Aspergillus* species, with the aim to identify genes that encode particular functions specifically associated with the life-style of *A. fumigatus*.

GLOBAL COMPARISON OF THE *A. FUMIGATUS* GENOME WITH OTHER PROKARYOTE AND EUKARYOTE SPECIES

In light of the newly available genomic information, we have reanalyzed *A. fumigatus* to assess its genome specificities. The proteome of *A. fumigatus* was compared to that of each of 182 surveyed species, including 49 eukaryotes (including other fungi and vertebrates), 45 archaeal species, and 88 bacterial species, following a methodology used for large-scale proteome comparisons that has previously been described in detail (Tekaia and Dujon, 1999).

Comparisons of *A. fumigatus* to seven of its closest relatives allowed us to confidently determine *A. fumigatus*-specific and nonspecific genes and families. The species sequenced are *A. flavus* (12,587 genes), *A. oryzae* (12,063 genes), *A. terreus* (10,406 genes) *A. niger* (8,592 genes), *Neosartorya fischeri* (10,407 genes), *A. fumigatus* (9,630 genes), *A. clavatus* (9,124 genes), and *A. nidulans* (10,701 genes). Their genomes have been published (Payne et al., 2006; Machida et al., 2005; Nierman et al., 2005; Galagan et al., 2005; Baker, 2006), and sequences are available at The Institute for Genomic Research and the Broad Institute servers (http://www.broad.mit.edu/annotation/genome/aspergillus_group/MultiHome.html and http://www.tigr.org/tdb/fungal/index.shtml). A total of 83,510 *Aspergillus* proteins were considered. The list and characteristics of all considered genomes can be found at http://www.pasteur.fr/~tekaia/asfu_genomeslist.html. Data presented here are based on the latest version available on these servers as of February 2008.

Many methods for comparative genomics are used. These comparative methods have allowed the detection and classification of paralogs and of orthologs. Species-specific comparisons shown in this review were based on the BLASTP program (Altschul et al., 1997), with the pam250 substitution matrix to favor recognition of large similar segment pairs and the seg filter (Wootton and Federhen, 1993) to mask compositionally biased regions in the query sequences. We considered BLASTP

Amandine Gastebois, Karine Lambou, and Joanne Wong Sak Hoi • Unité des Aspergillus, Institut Pasteur, 25, Rue du Dr Roux, 75724 Paris Cedex 15, France. **Fredj Tekaia** • Unité de Génétique Moléculaire des Levures, URA 2171 CNRS and UFR927, Univ. P. et M. Curie, Institut Pasteur, 25, Rue du Dr Roux, 75724 Paris Cedex 15, France.

e-values of $\leq 1.e^{-9}$ as the significance limit for eukaryotic species and e-values of $\leq 1.e^{-5}$ for all archaeal and bacterial species (Tekaia and Dujon, 1999; Tekaia et al., 1999, 2000; Tekaia and Yeramian, 2006).

Table 1 shows sets of exclusively conserved proteins in each species, in the whole set of the considered *Aspergillus* species, and in the three phylogenetic domains and their combinations. It appears that about 2% of each of the eight genomes are *Aspergillus* specific, whereas about 4 to 11% (from *A. oryzae* and *A. niger*, respectively) are species specific. At least 50% of the *Aspergillus* genes have an ortholog in eukaryotes as distinct as a human, a fish, or a worm, indicating that these genes must be responsible for the basal metabolism of any eukaryotic cell. Very few genes are found in common between bacteria and *Aspergillus*, suggesting that horizontal transfer between prokaryotes and *Aspergillus* does not occur in spite of their permanent contact in the same ecological niche, the soil and decaying vegetation. A low number of horizontal gene transfer events from bacteria to fungi is indeed documented in the literature (Wenzl et al., 2005; Hall et al., 2005). The three *A. fumigatus* genes detected as exclusively conserved in bacteria correspond to an ankyrin repeat protein (AFUA_7G08400), a putative serine protease (AFUA_7G06220), and a DOC family protein (AFUA_8G06180) that is a prophage (a prokaryote virus) maintenance system killer protein.

Table 1 shows also that 258 genes were found exclusively conserved in all eight surveyed *Aspergillus* species, including 225 singletons mostly annotated as hypothetical proteins. Only three classes with three, two, and two genes were shown to be specific to aspergilli. The first cluster codes for thaumatin-like proteins that have a role in conidial germination and encodes antigens like the two other clusters.

SINGLETONS AND PARALOGS IN THE *A. FUMIGATUS* GENOME

The genome of *A. fumigatus* contains 9,630 genes annotated in the latest version available, as of February 2008. Self-comparison of *A. fumigatus* showed that more than 39% of the proteins are not unique (Table 2). Classification of nonunique proteins is essential in genome analysis. In this analysis, classification was performed using three analytical steps as described previously (Tekaia and Latgé, 2005):

- In a first step, reciprocal significant hits were determined in each of the considered species and were partitioned into subsets of proteins. Partitions are disjoint subsets, and each includes proteins significantly similar to at least another protein in the same subset and has no similarity with proteins not included in the same subset. The subset itself is minimal, that is, it cannot be subdivided into subsets. Each nonunique protein was assigned to a partition denoted *Pn.m*, where *n* is the number of proteins in the partition and *m* is an arbitrary order.
- In a second step, the same set was clustered into classes using the mcl program (Enright et al., 2002, 2003; http://micans.org/mcl/), using the –log of the BLASTP e-value and an inflation index (*I*) of 3.0. Each nonunique protein was assigned to a cluster denoted *Cp.q*, where *p* is the number of proteins it includes and *q* is an arbitrary order.
- In a third step, each protein is assigned to both its partition and mcl cluster: *Pn.m.Cp.q*. This is what corresponds to a class (or family) of paralogs.

Table 1. Specific conservation in each considered *Aspergillus* species[a]

Species[b]	Size	Spec (%)	E_Asp	8Asp (%)	Univ	E	A	B	EA	EB	AB	EAB
ASFL	12,587	934 (7.4)	1,459	214 (1.7)	61	6,823	0	5	231	1,701	3	2,890
ASOR	12,063	453 (3.8)	1,701	291 (2.4)	63	6,846	0	1	240	1,657	0	2,866
ASTE	10,406	712 (6.8)	908	189 (1.8)	59	5,550	0	8	222	1,366	4	2,544
NEFI	10,407	503 (4.8)	1,438	244 (2.3)	59	5,596	1	1	237	1,250	3	2,416
ASFU	9,630	497 (5.2)	1,271	258 (2.7)	58	5,532	0	3	230	1,143	1	2,224
ASCL	9,124	568 (6.2)	1,104	235 (2.6)	54	5,249	0	1	229	1,041	0	2,036
ASNG	8,592	719 (8.4)	722	176 (2.0)	31	4,469	1	11	150	1,180	4	2,058
ASNI	10,701	1,217 (11.4)	723	156 (1.5)	51	5,526	0	7	229	1,343	3	2,376

[a] Genome species-specific comparisons were performed for each pair of species (i.e., four whole-proteome runs per species pair, two runs both ways and two self-self runs). Column heads: size, the number of proteins per genome; Spec, the number of specific genes (i.e., genes that have no homolog outside their own species), with the percent specific genes indicated in parentheses; E_Asp, the number of genes specific to *Aspergillus* species (conserved exclusively in one or several *Aspergillus* species); 8Asp, the number of genes exclusively and jointly conserved in the eight considered *Aspergillus* species (with the percentage of such genes in parentheses); Univ, the number of genes conserved in all considered species; E, the number of genes exclusively conserved in eukaryotes; A, the number of genes exclusively conserved in archaeal species; B, the number of genes exclusively conserved in bacterial species; EA, EB, AB, and EAB are the numbers of genes exclusively conserved in one of the combinations of domains. Note also that the difference (Size – Spec) corresponds to nonspecific proteins in each species.

[b] Species codes: ASFL, *A. flavus*; ASOR, *A. oryzae*; ASTE, *A. terreus*; NEFI, *N. fischeri*; ASFU, *A. fumigatus*; ASCL, *A. clavatus*; ASNG, *A. niger*; ASNI, *A. nidulans*.

Table 2. Gene duplication and conservation in *Aspergillus species*[a]

Species[b]	Proportion of duplication in each species and conservation between pairs of distinct species							
	ASFL	ASOR	ASTE	ASNG	NEFI	ASFU	ASCL	ASNI
ASFL	<u>49.9</u>	93.6	86.5	84.3	84.5	84.8	85.1	81.8
ASOR	87.1	<u>49.7</u>	82.8	81.4	80.8	81.3	81.6	78.8
ASTE	80.2	81.5	<u>47.5</u>	81.4	82.6	83.2	83.4	80.0
ASNG	69.3	71.3	71.5	<u>48.2</u>	68.6	69.1	69.0	67.5
NEFI	79.2	80.5	83.9	80.8	<u>42.7</u>	92.9	90.1	80.2
ASFU	77.4	78.9	82.0	79.0	89.8	<u>39.3</u>	88.5	78.6
ASCL	76.4	77.7	81.1	78.2	85.7	87.0	<u>36.6</u>	77.3
ASNI	78.7	80.3	83.3	80.4	82.1	83.1	83.1	<u>43.9</u>

[a] Underscored values on the diagonal (e.g., proportion of ASFL genes conserved in ASFL) correspond to duplication proportions (number of nonunique proteins divided by the total number of proteins in the species) in the indicated species. Nondiagonal values correspond to the proportion of proteins of the species indicated at the top of the column that are conserved in the species indicated for each row. For example, there is 39.3% duplication in *A. fumigatus* (i.e., the number of nonunique genes divided by the total number of genes in *A. fumigatus*); 92.9% is the conservation rate of *A. fumigatus* in *N. fischeri* (i.e., the number of genes in *A. fumigatus* that are conserved in *N. fischeri*, divided by the total number of genes in *A. fumigatus*). Note that the table is not symmetrical because of the normalization with the total number of genes in species shown on the top of the columns.

[b] Species codes: ASFL, *A. flavus*; ASOR, *A. oryzae*; ASTE, *A. terreus*; ASNG, *A. niger*; NEFI, *N. fischeri*; ASFU, *A. fumigatus*; ASCL, *A. clavatus*; ASNI, *A. nidulans*.

The partition method subdivides the whole set into subsets of proteins that are either tightly or loosely related, whereas the mcl method clusters only tightly related proteins. The main advantage of the *Pn.m.Cp.q* notation is that it highlights the presence of different clusters of paralogs in the same partition and hence shows their possible weak relationship to other proteins in the partition; otherwise, the clusters are deemed independent. In each considered species, each nonunique protein is assigned to a class of paralogs determined by the above procedure. Otherwise, the label "singleton" is assigned to a protein without any significant match.

The distribution of the obtained mcl classes according to their sizes is shown in Fig. 1. The distribution is similar for all *Aspergillus* species analyzed. The graph highlights the predominance of small-sized mcl classes. Most mcl classes include less than 10 proteins. As expected, mcl classes are included in large partitions that encode protein kinases (partition of 171 members) and transcription factors (partition of 92 members) or cytochrome P450 (partition of 67 members) and oxidoreductases (159 genes). Interestingly, and as noticed previously (Tekaia and Latgé, 2005), transporter families of the major facilitator superfamilies (four partitions of 105, 80, 52, and 22 genes) and ABC multidrug transporters (partition of 40 proteins) include many members per family. Since these transporters are involved in multidrug efflux or nutrient transport, their presence in high numbers is in agreement with the saprotrophic life of this fungus and allows this fungus to withstand adverse conditions encountered in its soil environment or to exchange rapidly a nutrient with its external milieu. Another family of proteins of interest is the PTH-11 family, with an eight-cysteine domain and which contains 43 members in *A. fumigatus*. Described initially in a study of appressorium differentiation (DeZwaan et al., 1999), this family has been identified in many filamentous fungi. The fact that *A. fumigatus* does not produce any appressorium suggests that such a family may be simply involved in sensing the external milieu and transducing the signal to central regulators (Kulkarni et al., 2003; Lafon et al., 2006).

Duplication and conservation in and between the eight *Aspergillus* species considered are shown in Table 2. The mean duplication rate is about 45%, with the lowest rates in *A. clavatus* (36.6%) and *A. fumigatus* (39.3%), whereas the highest duplication rates are in *A. flavus* (49.9%) and *A. oryzae* (49.7%).

The mean conservation rate is very high (81.5%), varying from 67.5% (conservation of *A. nidulans* in *A. niger*) to 93.6% (conservation of *A. oryzae* in *A. flavus*). Interestingly and in agreement with the taxonomy of the section *Fumigati* (see chapter 2), a high level of conservation is seen between *A. fumigatus*, *N. fischeri*, and *A. clavatus*. This conservation indicates a saprotrophic behavior of these species rather than explains the opportunistic pathogenic behavior of *A. fumigatus*.

ORTHOLOGS OF *A. FUMIGATUS*

Reciprocal best hit (RBH) proteins from pairwise distinct species are potential orthologs (Moreno-Hagelsieb and Latimer, 2008; Poptsova and Gogarten, 2007). For each pair of the eight considered *Aspergillus* species, RBH values were determined among protein pairs from all species comparisons. The set of RBH pairs obtained from the species pair-wise comparisons included 72,605 distinct proteins that are distributed as follows: *A. flavus* (ASFL), 11,066; *A. oryzae* (ASOR),

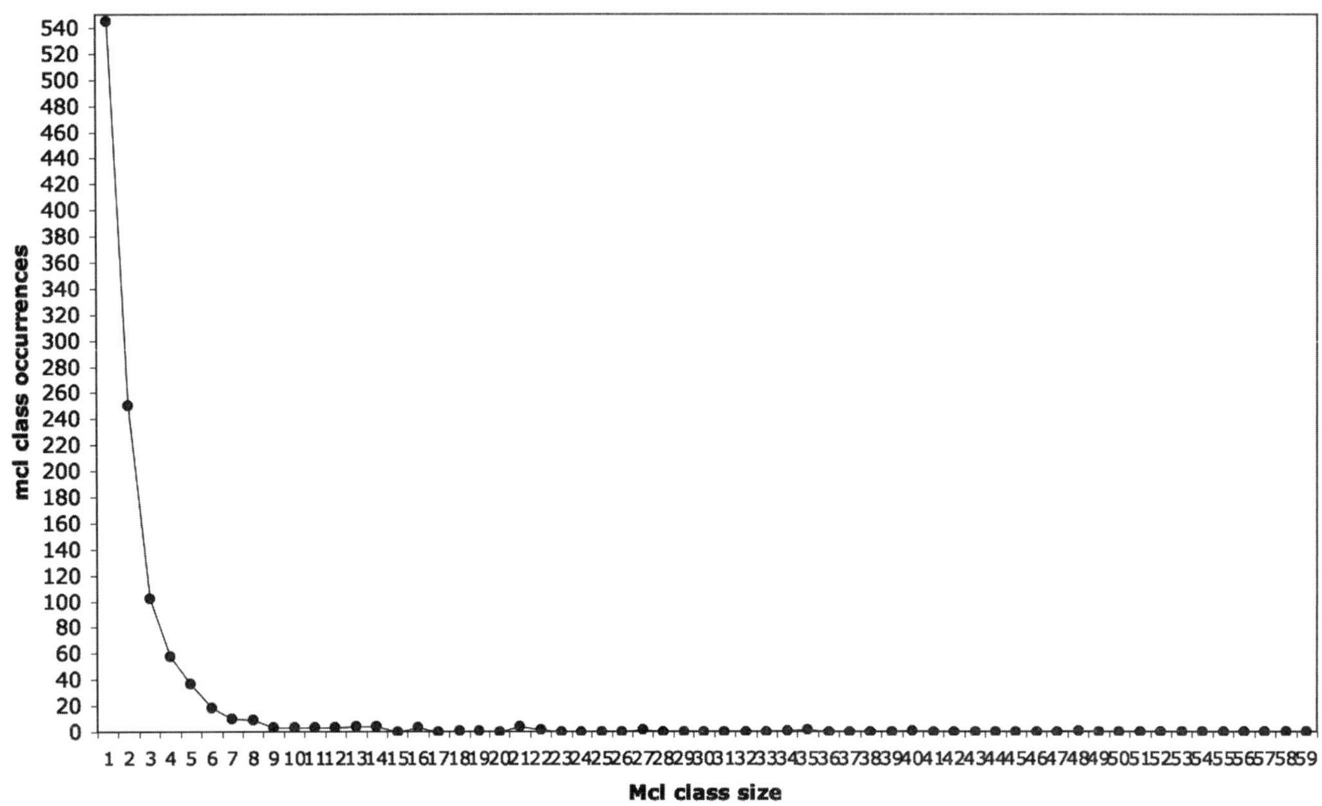

Figure 1. Distribution of the sizes of mcl classes of nonunique proteins in *A. fumigatus* (ASFU) species. The horizontal axis indicates the sizes of the mcl families, i.e., the number of proteins per mcl family. On the vertical axis are the occurrences of families per size.

10,771; *A. terreus* (ASTE), 8,738; *A. niger* (ASNG), 6,929; *N. fischeri* (NEFI), 9,421; *A. fumigatus* (ASFU), 8,894; *A. clavatus* (ASCL), 8,245; *A. nidulans* (ASNI), 8,541. Table 3 shows the distribution of shared RBH proteins between each pair of the considered *Aspergillus* species.

MOTIF SEARCHES

Motif searches using the MEME and MAST programs (Bailey and Elkan, 1995) are very efficient for following gene evolution. A few examples are given below to illustrate the use of this method, which can have also an impact on functional genomic analysis.

The first example is the *FKS* gene, which is the catalytic subunit of the β-1,3-glucan synthase that codes for an enzymatic activity essential for cell wall construction and fungal life itself (Fig. 2A). One gene is present in *A. fumigatus* and other filamentous fungi, whereas 3 are detected in *Saccharomyces cerevisiae* and 12 in the plant *Arabidopsis thaliana*. The structure of all orthologous and paralogous *FKS* genes in fungi is very similar. In all aspergilli, the FKS protein is extremely well-conserved, with duplications of motifs seen at the C and N termini (motifs 12, 14, and 9). Two of the repeated motifs (12 and 14) are absent in the *S. cerevisiae* proteins. The C terminus is almost perfectly conserved in all considered fungal FKS proteins. In the left part (N terminus), although shuffling of motifs is observed, most motif anchors (with the exception of motifs 14, 13, and 8) are also conserved in all considered proteins. In plant (*A. thaliana*) proteins, a major rearrangement occurs in the central part of the protein from domains 13 to 15, suggesting that this region of the gene is not involved in the enzymatic activity of FKSp because both plant and fungal genes use UDP-glucose as substrate to produce β-1,3-glucans. Motif 13 is absent from *A. thaliana* proteins, and motif 15 is present only in two *A. thaliana* proteins. This result suggests that two large general conserved domains at the C and N termini are necessary for the activity of this enzyme.

Another example that uses the method by working in the opposite direction is related to the chitinases: 2 genes are found in *S. cerevisiae*, 3 in humans, and 18 in *A. fumigatus*. Two main clusters can be visualized. Cluster 1 (Fig. 2B, upper portion) contains 12 genes and one ortholog of *S. cerevisiae* and the three human orthologs. It could be deduced from this clustering that these genes may be more involved in degradation than the other

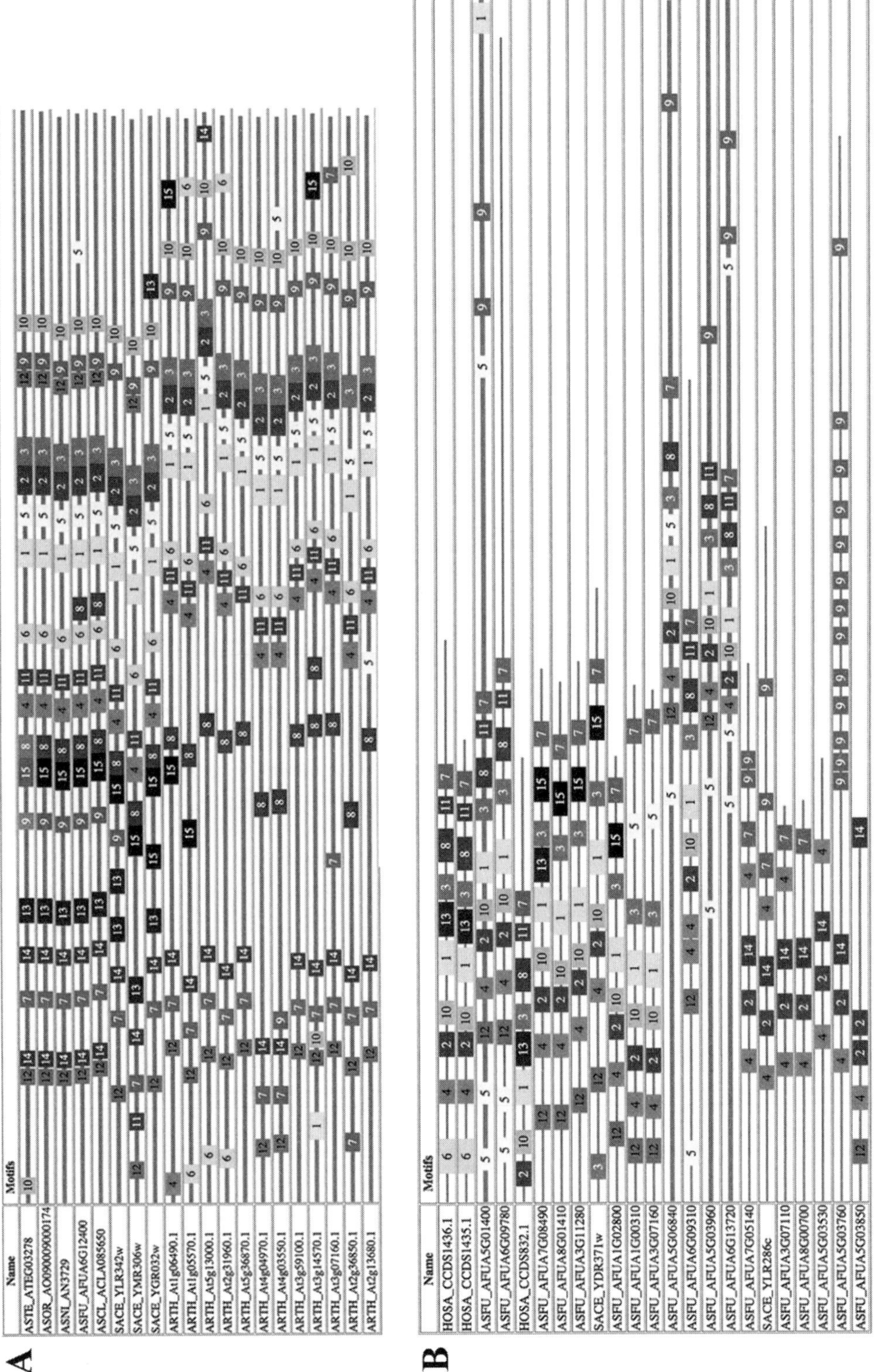

Figure 2. Motif distributions in two sets of homologous protein sequences as obtained with the MEME and MAST programs for FKS (A) and chitinases (B). Sequence identifications are preceded by their species code, as indicated in footnotes for Tables 2 and 3, with additional species corresponding to *Homo sapiens* (HOSA), *Saccharomyces cerevisiae* (SACE), and *Arabidopsis thaliana* (ARTH).

clusters, since a putative function of the chitinases in the human host and in the phagocytes where these enzymes have been localized is to degrade the invading microorganisms that contain chitin. The motif 4-2-10-1-3-7 is conserved in most sequences and appears to anchor their structure. Most rearrangements occur in the N terminus. Cluster 2 contains five *Aspergillus* genes and one *S. cerevisiae* ortholog, with only three conserved motifs, arranged as 4-2-14. This organization is very different from cluster 1 and suggests that the function of this partition is more involved in cell wall modification, as was shown for the yeast ortholog. This cluster is extremely heterogeneous, especially in terms of repetition of motif 9 and the size of the protein. This suggests that this gene cluster is presently evolving, with the singleton AFUA_5G03850 being the most recently evolved gene of this cluster, in agreement with the most profound rearrangements seen in this gene.

SPECIFIC GENES

As shown in Table 1, our comparison procedure identified 497 specific genes for *A. fumigatus*, that is, genes with no significant hit outside their own genome. We have further compared this subset to the seven *Aspergillus* species by using the same comparison procedure but considering the blosum62 substitution matrix instead of pam250. These comparisons and manual curation allowed the detection of new significant hits, particularly for small proteins, reducing the number of specific proteins in *A. fumigatus* to 377 (3.9% of the whole genome). The large majority of them (97%) correspond to hypothetical proteins, and only 13 (3% of the specific set of genes) display known domains (AFUA_1G00170, AFUA_1G00250, AFUA_2G09420, AFUA_3G13800, AFUA_4G13113, AFUA_4G13810, AFUA_5G00170, AFUA_5G14910, AFUA_6G09360, AFUA_6G14070, AFUA_8G02230, AFUA_8G06250, and AFUA_8G06780).

This subset corresponds to proteins enriched in one or several particular amino acids without any identification of their putative function. Two proteins with collagen triple-helix-type repeats (GXY), which are generally found in extracellular structural proteins involved in formation of connective tissue structure (Mayne and Brewton, 1993), were identified in the genome. Furthermore, we identified the presence of two F-box domain proteins (out of 44 in the total *A. fumigatus* predicted proteins). This domain generally has a role in mediating protein-protein interactions. We also found a protein with a DUF614 domain (unknown function) and two putative glycosylphosphatidyl inositol (GPI)-anchored proteins, among the 85 predicted GPI proteins in *A. fumigatus* identified by big-PI (http://mendel.imp.ac.at/sat/gpi/gpi_server.html). Since these proteins were all annotated as hypothetical proteins and were singletons, no functional specificity could be assigned to these specific genes that could be associated with the *A. fumigatus* life-style. However, the analysis of the genome showed two interesting features.

A NEW GENE CLUSTER, NOT INVOLVED IN PRODUCTION OF SECONDARY METABOLITES, IN THE *A. FUMIGATUS* GENOME

In filamentous fungi, genes participating in common catabolic pathways (such as quinate, ethanol, proline, or nitrate utilization) or in secondary metabolite biosynthesis pathways (such as mycotoxins or antibiotic synthesis) are often clustered (Keller and Hohn, 1997). Secondary metabolites are low-molecular-weight natural products which have a restricted taxonomic distribution. These types of chemical compounds are often bioactive and are produced during restricted periods of the life cycle. They can influence both the infection process and the environmental niche outside of the host (Keller et al., 2005). Clustering of genes responsible for secondary metabolite synthesis have been already identified in *A. fumigatus* (Perrin et al., 2007).

The survey of the genome led to the identification of a cluster of genes that seems specific to *A. fumigatus*. It is located within a 100-kb region of chromosome I, in the subtelomeric region, and contains 11 genes coding for hypothetical proteins among 17 genes (Fig. 3). This putative cluster identified only in *A. fumigatus* does not include catabolic or secondary metabolite biosynthesis enzymes, such as polyketide synthase or nonribosomal peptide synthetase, suggesting that they are not involved in such biosynthesis pathways. Recently, in the fungal plant pathogen *Ustilago maydis*, 12 gene clusters were identified and shown to play a crucial role in the infection process (Kämper et al., 2006), but they did not code for secondary metabolites. The proteins identified in the putative cluster specific to *A. fumigatus* show no predicted signal peptides based on SignalP 3.0 analysis (Emanuelsson et al., 2007). Most proteins in this cluster are enriched in serine up to 19% (the serine composition of six proteins is greater than the whole *A. fumigatus* protein serine composition [8.4%]).

These observations suggest that undetected clusters with a structure different from the secondary metabolite clusters described so far in filamentous fungi can be found in the *A. fumigatus* genome. The functional analysis of such clusters in *A. fumigatus* is currently being investigated to determine if their genes are coexpressed and if this subtelomeric chromosomal region plays an essential role in the life cycle of *A. fumigatus*.

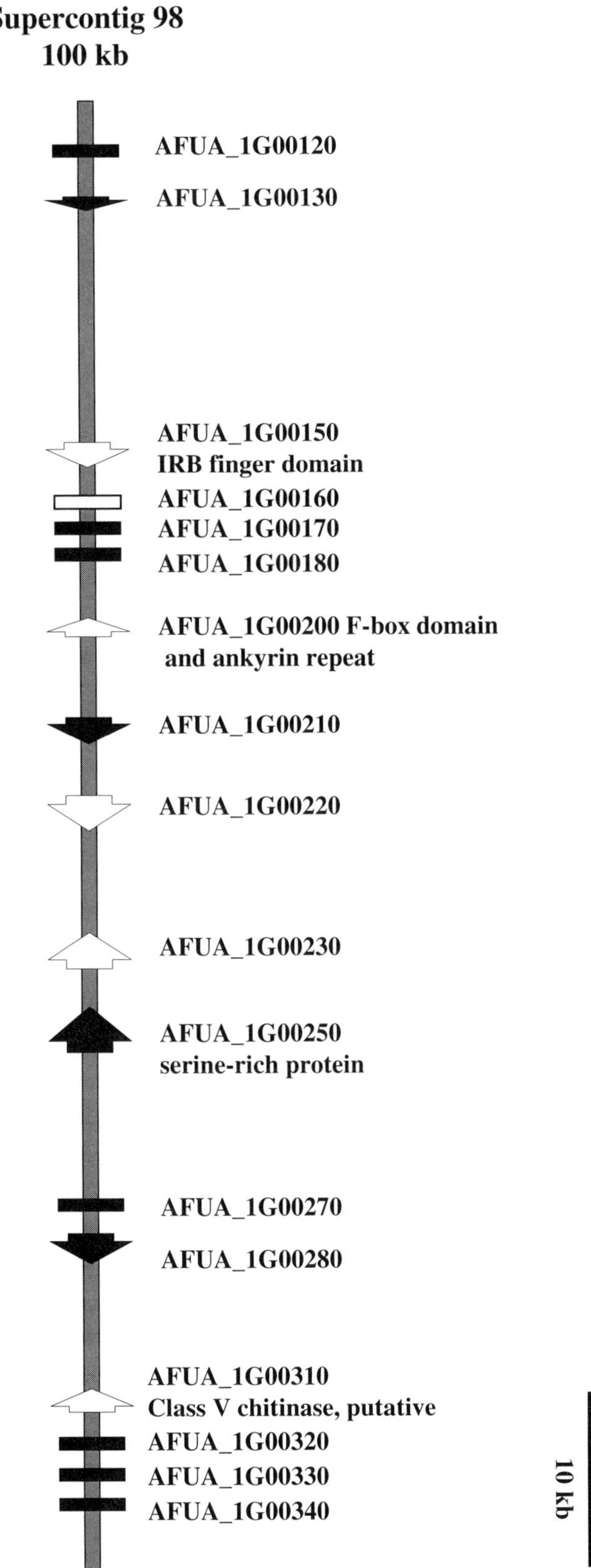

Figure 3. Locus organization of several genes specific to the *A. fumigatus* genome. The genes indicated by black arrows or bars are specific to *A. fumigatus*, and those indicated by white arrows or bars are nonspecific. Eleven genes specific to *A. fumigatus* are located within a 100-kb subtelomeric region of chromosome I (supercontig 98).

SPECIFICITY OF THE SPECIES *A. FUMIGATUS* AND THE SUBSET OF SMALL PROTEINS OF THE GENOME

Among the 377 genes specific to *A. fumigatus*, 74% encode small proteins (shorter than 150 amino acids), whereas only 8% of the genes in the whole *A. fumigatus* genome encode such small proteins. Small proteins are also overrepresented within genes specific to *A. clavatus*, to *N. fisheri*, and to *S. cerevisiae* (71% [284 of 399], 58% [200 of 343], and 65% [295 of 455] are shorter than 150 amino acids long, respectively). The function of these small proteins in *A. fumigatus* remains unknown, but the species specificity for *A. fumigatus*, like other fungal species, could be due to this population of proteins that has never been analyzed exhaustively in any fungal species. In plant pathogenic fungi, many small proteins have been shown to be involved in disease symptom development as well as in disease resistance (Rep, 2005). For example, in the ascomycete *Magnaporthe grisea*, the causal agent of rice blast, a metallothionein-like small protein of 22 amino acids (Mmt1) has been shown to be essential for pathogenicity (Tucker et al., 2004). This cysteine-rich protein is distributed on the inner side of the *M. grisea* fungal cell wall and could be involved in its modification (Tucker et al., 2004). In *A. fumigatus*, only 3% of these small proteins are predicted to be secreted based on SignalP 3.0 analysis (Emanuelsson et al., 2007). Secretion signals for small proteins may be different. Analysis of small proteins could open the way to a new field of investigation for fungal genetics.

MISSING GENES

Characteristics of a species do not only lie in the specific genes but can also result from the absence of certain genes (Dujon, 2006). Accordingly, we have looked at the gene families that are missing in *A. fumigatus*.

The set of shared RBH between the closely related *Aspergillus* species (Table 3) allowed the determination of superfamilies of orthologs (including also probably paralogs) and their conservation profiles. A conservation profile is a vector with eight dimensions, corresponding to the eight considered species, where each position represents the presence or absence of an ortholog in the corresponding species. Eighteen gene superfamilies of orthologs are found to be absent from the genome of *A. fumigatus* but present in all other surveyed *Aspergillus* species. While 9 of 18 of these superfamilies encode hypothetical proteins without any assigned function or known conserved domain, the remaining ones contain a sequence consensus that suggests putative roles in cell

Table 3. Shared reciprocal best hits (RBH) between pairs of species

Species[a]	No. of proteins shared by the two indicated species[b]							
	ASFL	ASOR	ASTE	ASNG	NEFI	ASFU	ASCL	ASNI
ASFL	11,066							
ASOR	10,017	10,771						
ASTE	7,408	7,075	8,738					
ASNG	5,771	5,559	5,255	6,929				
NEFI	7,464	7,174	7,127	5,320	9,421			
ASFU	7,190	6,927	6,888	5,154	8,540	8,894		
ASCL	6,923	6,711	6,687	4,978	7,640	7,450	8,245	
ASNI	7,281	6,973	6,984	5,216	7,130	6,936	6,716	8,541

[a]Species codes: ASFL, *A. flavus*; ASOR, *A. oryzae*; ASTE, *A. terreus*; ASNG, *A. niger*; NEFI, *N. fischeri*; ASFU, *A. fumigatus*; ASCL, *A. clavatus*; ASNI, *A. nidulans*.
[b]Numbers on the diagonal correspond to proteins in a given species that are shared with at least one protein from another species, as obtained by the RBH method. RBH proteins from distinct species are potential orthologs, since the RBH method requires strong conservative relationships among the orthologs so that if a gene 1 from species 1 selects a gene 2 from species 2 as a best hit when performing a BLASTP search with genome 1 against genome 2, then gene 2 must in turn select gene 1 as the best hit when genome 2 is searched against genome 1.

signaling (protein kinase domain, stress-responsive A/B barrel), cell cycle regulation (chromosome segregation ATPase, mitosis regulatory protein), or secondary metabolite pathways (trichothecene transferase family).

In heterothallic filamentous fungi, the mating-type locus conferring mating behavior consists of dissimilar DNA sequences in the mating partners. The locus is either termed *MAT*a or *MAT*α, according to the single mating-type DNA sequence. In *A. fumigatus*, it was observed that one mating-type gene, *MAT*a, was present in the genome of the sequenced strain AF293 (Pöggeler, 2002; Paoletti et al., 2005). Consistent with this, the *MAT*α mating-type sequence was found among genes absent from *A. fumigatus*. This observation validates our method of searching for missing genes.

The largest family absent from *A. fumigatus* corresponds to a transferase family involved in trichothecene biosynthesis. Secreted mainly in *Fusarium* spp., trichothecenes are major mycotoxins that belong to a large family of sesquiterpenoid secondary metabolites. Interestingly, there is as yet no evidence that *A. fumigatus* can produce trichothecenes (J. C. Frisvad, personal communication).

The trichothecene biosynthesis pathway has been unraveled in *Fusarium* spp. The proteins involved in this biosynthesis pathway are encoded by the *Tri* gene family. Table 4 lists the number of orthologs detected when comparing the *Fusarium* sequences against the *Aspergillus* sequences. This table highlights the reduced number of trichothecene biosynthesis gene orthologs in the *A. fumigatus* genome. In *Fusarium* spp., most *Tri* biosynthesis genes are described as part of a cluster. In contrast, orthologs of these genes in *A. fumigatus* are found scattered throughout the genome. Table 4 (footnote b) compiles the gene accession numbers corresponding to each *Tri* homolog. Similar to *Fusarium* spp., in *A. fumigatus Tri6* is an intronless gene that encodes a zinc finger transcription factor. In *Fusarium*, *Tri6* acts as a positive regulator of *Tri* genes (Proctor et al., 1995). Which genes are regulated by the *A. fumigatus* Tri6 homolog remains to be determined. Interestingly, some of the *Tri* orthologs are located within secondary metabolite clusters regulated by the transcription factor LaeA (Perrin et al., 2007). This indicates a possible involvement of *A. fumigatus Tri* homologs in the biosynthesis of secondary metabolites other than trichothecenes.

CONCLUSIONS

In silico comparative genome analyses are powerful in describing similarities and differences between protein sequences in and between completely sequenced organisms. We have identified proteins that are conserved specifically among eukaryotes in *A. fumigatus* and in *Aspergillus* species and those that are common to different species. In this review we were specifically interested in *A. fumigatus* genes.

Genome comparison studies in *Aspergillus* have shown that species- or genus-specific genes constitute a tiny part of the genome compared to other fungal species. Moreover, the analyses of *A. fumigatus*-specific genes have not been very informative in terms of gene functions and do not highlight any particular pathway which could be unique to this fungus. This study confirms that the eight analyzed *Aspergillus* species are closely related and shows that no major particular event of function loss or gain took place in *A. fumigatus* during its evolution.

If the survey of genes conserved exclusively in *Aspergillus* spp. has not been informative in terms of function, further analysis of the genomic specificity of *A. fumigatus* could have other interesting applications. For

Table 4. Trichothecenes: genes in *Aspergillus* spp. determined to be homologous to *Tri* family genes[a]

Protein[b]	No. of orthologs to the *Fusarium Tri* genes found by comparison in the indicated *Aspergillus* species:							
	A. flavus	*A. oryzae*	*A. terreus*	*A. niger*	*N. fischeri*	*A. fumigatus*	*A. clavatus*	*A. nidulans*
Tri5	1	1	3	2	1	0	0	1
Tri4	16	20	11	15	7	3	10	18
Tri6	2	3	1	1	1	1	2	1
Tri3	0	0	0	0	0	0	0	0
Tri11	13	21	9	21	8	5	11	13
Tri12	14	18	8	17	11	9	14	9
Tri7	3	2	6	3	5	3	2	4
Tri10	1	2	0	1	1	2	1	0
Tri13	1	2	2	4	5	3	5	1
Tri8	6	7	5	1	3	2	1	2
Tri101	2	4	3	4	1	0	5	3
Tri1	12	13	11	10	12	8	15	14
Tri16	2	2	1	3	0	0	1	0
Total	73	95	60	82	55	36	67	66

[a]The family of *Tri* genes as defined for *Fusarium* (Kimura et al., 2007). In parentheses are the *Tri* orthologs found in *A. fumigatus*. Among these, underlined are the *Tri* genes located within secondary metabolite clusters regulated by the transcription factor LaeA (Perrin et al., 2007).

[b]*Tri5*, trichodiene synthase; *Tri4*, multifunctional oxygenase responsible for conversion of trichodiene to isotrichotriol (AFUA_5g01360, AFUA_4g09980, and AFUA_1g01270); *Tri6*, trichothecene biosynthesis regulatory protein (AFUA_1g02860); *Tri3*, 3-acetyltrichothecene 15-O-acetyltransferase; *Tri11*, isotrichodermin C-15 hydroxylase (AFUA_7G00290, AFUA_4g09980, AFUA_1g01270, AFUA_4g09470, and AFUA_8g00190); *Tri12*, thichothecene efflux pump (AFUA_6g02400, AFUA_6g03320, AFUA_3g14720, AFUA_6g09710, AFUA_5g14490, AFUA_3g08530, AFUA_2g08230, AFUA_4g03920, and AFUA_1g16910); *Tri7*, 3-acetyltrichothecene 4-O-acetyltransferase (AFUA_6g14000, AFUA_8g02360, and AFUA_8g05970); *Tri10*, trichothecene synthesis regulator (AFUA_3g15290 and AFUA_3g02480); *Tri13*, 3-acetyltrichothecene C-4 hydroxylase (AFUA_1g17725, AFUA_3g14760, and AFUA_2g04290); *Tri8*, trichothecene C-3 deacetylase; *Tri101*, trichothecene 3-O-acetyltransferase; *Tri1*, 3-acetyltrichothecene C-8 hydroxylase (AFUA_2g18010, AFUA_6g08140, AFUA_8g00240, AFUA_6g13940, AFUA_5g03100, AFUA_6g09730, and AFUA_6g01810); *Tri16*, C-8 acyltransferase.

example, these genomic data could provide novel candidate targets for the detection of aspergilli in a diagnostic context, since genes specific and common to *A. fumigatus* or all aspergilli have been now identified. Moreover, a transcriptome analysis of the genes specific to *A. fumigatus* and to the *Aspergillus* genus during growth in vivo may be an interesting avenue of research to pinpoint the metabolic pathways that are unique to the *Aspergillus* mode of life.

To date, comparative genome analyses have not identified genes that could code for virulence factors favoring the establishment of *A. fumigatus* inside its mammalian host. This is in agreement with the opportunistic pathogenicity of this fungus. The ecological niche of this fungus is the soil, where it grows as a saprotrophic organism. It becomes pathogenic only when the host is immunocompromised and, in contrast to plant pathogens, there is no genetic selective pressure for this fungus to invade humans.

Acknowledgments. We thank Jean-Paul Latgé for his support and thoughtful discussions during the analysis and writing of this chapter. F.T. acknowledges Bernard Dujon for his constant support. A.G., K.L., and J.W.S.H. contributed equally to this work.

REFERENCES

Altschul, S. F., T. L. Madden, A. A. Schaffer, J. Zhang, Z. Zhang, W. Miller, and D. J. Lipman. 1997. Gapped BLAST and PSI-BLAST: a new generation of protein database search programs. *Nucleic Acids Res.* **25:**3389–3402.

Bailey, T. L., and C. Elkan. 1995. Unsupervised learning of multiple motifs in biopolymers using EM. *Mach. Learn.* **21:**51–80.

Baker, S. E. 2006. *Aspergillus niger* genomics: past, present and into the future. *Med. Mycol.* **44**(Suppl. 1)**:**S17–S21.

DeZwaan, T. M., A. M. Carroll, B. Valent, and J. A. Sweigard. 1999. *Magnaporthe grisea* pth11p is a novel plasma membrane protein that mediates appressorium differentiation in response to inductive substrate cues. *Plant Cell* **11:**2013–2030.

Dujon, B. 2006. Yeasts illustrate the molecular mechanisms of eukaryotic genome evolution. *Trends Genet.* **22:**375–387.

Emanuelsson, O., S. Brunak, G. von Heijne, and H. Nielsen. 2007. Locating proteins in the cell using TargetP, SignalP, and related tools. *Nat. Protoc.* **2:**953–971.

Enright, A. J., V. Kunin, and C. A. Ouzounis. 2003. Protein families and TRIBES in genome sequence space. *Nucleic Acids Res.* **31:**4632–4638.

Enright, A. J., S. Van Dongen, and C. A. Ouzounis. 2002. An efficient algorithm for large-scale detection of protein families. *Nucleic Acids Res.* **30:**1575–1584.

Galagan, J. E., S. E. Calvo, C. Cuomo, L. J. Ma, J. R. Wortman, S. Batzoglou, S. I. Lee, M. Baştürkmen, C. C. Spevak, J. Clutterbuck, V. Kapitonov, J. Jurka, C. Scazzocchio, M. Farman, J. Butler, S. Purcell, S. Harris, G. H. Braus, O. Draht, S. Busch, C. D'Enfert, C. Bouchier, G. H. Goldman, D. Bell-Pedersen, S. Griffiths-Jones, J. H. Doonan, J. Yu, K. Vienken, A. Pain, M. Freitag, E. U. Selker, D. B. Archer, M. A. Peñalva, B. R. Oakley, M. Momany, T. Tanaka, T. Kumagai, K. Asai, M. Machida, W. C. Nierman, D. W. Denning, M. Caddick, M. Hynes, M. Paoletti, R. Fischer, B. Miller, P. Dyer, M. S. Sachs, S. A. Osmani, and B. W. Birren. 2005. Sequencing of *Aspergillus nidulans* and comparative analysis with *A. fumigatus* and *A. oryzae*. *Nature* **438:**1105–1115.

Hall, C., S. Brachat, and F. S. Dietrich. 2005. Contribution of horizontal gene transfer to the evolution of *Saccharomyces cerevisiae*. *Eukaryot. Cell* 4:1102–1115.

Kämper, J., R. Kahmann, M. Bölker, L. J. Ma, T. Brefort, B. J. Saville, F. Banuett, J. W. Kronstad, S. E. Gold, O. Müller, M. H. Perlin, H. A. Wösten, R. de Vries, J. Ruiz-Herrera, C. G. Reynaga-Peña, K. Snetselaar, M. McCann, J. Pérez-Martín, M. Feldbrügge, C. W. Basse, G. Steinber, J. I. Ibeas, W. Holloman, P. Guzman, M. Farman, J. E. Stajich, R. Sentandreu, J. M. González-Prieto, J. C. Kennell, L. Molina, J. Schirawski, A. Mendoza-Mendoza, D. Greilinger, K. Münch, N. Rössel, M. Scherer, M. Vranes, O. Ladendorf, V. Vincon, U. Fuchs, B. Sandrock, S. Meng, E. C. Ho, M. J. Cahill, K. J. Boyce, J. Klose, S. J. Klosterman, H. J. Deelstra, L. Ortiz-Castellanos, W. Li, P. Sanchez-Alonso, P. H. Schreier, I. Häuser-Hahn, M. Vaupel, E. Koopmann, G. Friedrich, H. Voss, T. Schlüter, J. Margolis, D. Platt, C. Swimmer, A. Gnirke, F. Chen, V. Vysotskaia, G. Mannhaupt, U. Güldener, M. Münsterkötter, D. Haase, M. Oesterheld, H. W. Mewes, E. W. Mauceli, D. DeCaprio, C. M. Wade, J. Butler, S. Young, D. B. Jaffe, S. Calvo, C. Nusbaum, J. Galagan, and B. W. Birren. 2006. Insights from the genome of the biotrophic fungal plant pathogen *Ustilago maydis*. *Nature* 444: 97–101.

Keller, N. P., and T. M. Hohn. 1997. Metabolic pathway gene clusters in filamentous fungi. *Fungal Genet. Biol.* 21:17–29.

Keller, N. P., G. Turner, and J. W. Bennett. 2005. Fungal secondary metabolism: from biochemistry to genomics. *Nat. Rev. Microbiol.* 3:937–947.

Kimura, M., T. Tokai, N. Takahashi-Ando, S. Ohsato, and M. Fujimura. 2007. Molecular and genetic studies of *Fusarium* trichothecene biosynthesis: pathways, genes, and evolution. *Biosci. Biotechnol. Biochem.* 71:2105–2123.

Kulkarni, R. D., H. S. Kelkar, and R. A. Dean. 2003. An eight-cysteine-containing CFEM domain unique to a group of fungal membrane proteins. *Trends Biochem. Sci.* 23:118–121.

Lafon, A., K. H. Han, J. A. Seo, J. H. Yu, and C. D'Enfert. 2006. G-protein and cAMP-mediated signaling in aspergilli: a genomic perspective. *Fungal Genet. Biol.* 43:490–502.

Machida, M., K. Asai, M. Sano, T. Tanaka, T. Kumagai, G. Terai, K. Kusumoto, T. Arima, O. Akita, Y. Kashiwagi, K. Abe, K. Gomi, H. Horiuchi, K. Kitamoto, T. Kobayashi, M. Takeuchi, D. W. Denning, J. E. Galagan, W. C. Nierman, J. Yu, D. B. Archer, J. W. Bennett, D. Bhatnagar, T. E. Cleveland, N. D. Fedorova, O. Gotoh, H. Horikawa, A. Hosoyama, M. Ichinomiya, R. Igarashi, K. Iwashita, P. R. Juvvadi, M. Kato, Y. Kato, T. Kin, A. Kokubun, H. Maeda, N. Maeyama, J. Maruyama, H. Nagasaki, T. Nakajima, K. Oda, K. Okada, I. Paulsen, K. Sakamoto, T. Sawano, M. Takahashi, K. Takase, Y. Terabayashi, J.R. Wortman, O. Yamada, Y. Yamagata, H. Anazawa, Y. Hata, Y. Koide, T. Komori, Y. Koyama, T. Minetoki, S. Suharnan, A. Tanaka, K. Isono, S. Kuhara, N. Ogasawara, and H. Kikuchi. 2005. Genome sequencing and analysis of *Aspergillus oryzae*. *Nature* 438:1157–1161.

Mayne, R., and R. G. Brewton. 1993. New members of the collagen superfamily. *Curr. Opin. Cell Biol.* 5:883–890.

Moreno-Hagelsieb, G., and K. Latimer. 2008. Choosing BLAST options for better detection of orthologues as reciprocal best hits. *Bioinformatics* 3:319–324.

Nierman, W. C., A. Pain, M. J. Anderson, J. R. Wortman, H. S. Kim, J. Arroyo, M. Berriman, K. Abe, D. B. Archer, C. Bermejo, J. Bennett, P. Bowyer, D. Chen, M. Collins, R. Coulsen, R. Davies, P. S. Dyer, M. Farman, N. Fedorova, N. Fedorova, T. V. Feldblyum, R. Fischer, N. Fosker, A. Fraser, J. L. García, M. J. García, A. Goble, G. H. Goldman, K. Gomi, S. Griffith-Jones, R. Gwilliam, B. Haas, H. Haas, D. Harris, H. Horiuchi, J. Huang, S. Humphray, J. Jiménez, N. Keller, H. Khouri, K. Kitamoto, T. Kobayashi, S. Konzack, R. Kulkarni, T. Kumagai, A. Lafon, J. P. Latgé, W. Li, A. Lord, C. Lu, W. H. Majoros, G.S. May, B. L. Miller, Y. Mohamoud, M. Molina, M. Monod, I. Mouyna, S. Mulligan, L. Murphy, S. O'Neil, I. Paulsen, M. A. Peñalva, M. Pertea, C. Price, B. L. Pritchard, M. A. Quail, E. Rabbinowitsch, N. Rawlins, M. A. Rajandream, U. Reichard, H. Renauld, G. D. Robson, S. Rodriguez de Córdoba, J. M. Rodríguez-Peña, C. M. Ronning, S. Rutter, S. L. Salzberg, M. Sanchez, J. C. Sánchez-Ferrero, D. Saunders, K. Seeger, R. Squares, S. Squares, M. Takeuchi, F. Tekaia, G. Turner, C. R. Vazquez de Aldana, J. Weidman, O. White, J. Woodward, J. H. Yu, C. Fraser, J. E. Galagan, K. Asai, M. Machida, N. Hall, B. Barrell, and D. W. Denning. 2005. Genomic sequence of the pathogenic and allergenic filamentous fungus *Aspergillus fumigatus*. *Nature* 438:1151–1156.

Paoletti, M., C. Rydholm, E. U. Schwier, M. J. Anderson, G. Szakacs, F. Lutzoni, J. P. Debeaupuis, J. P. Latgé, D.W. Denning, and P. S. Dyer. 2005. Evidence for sexuality in the opportunistic fungal pathogen *Aspergillus fumigatus*. *Curr. Biol.* 15:1242–1248.

Payne, G. A., W. C. Nierman, J. R. Wortman, B. L. Pritchard, D. Brown, R. A. Dean, D. Bhatnagar, T. E. Cleveland, M. Machida, and J. Yu. 2006. Whole genome comparison of *Aspergillus flavus* and *A. oryzae*. *Med. Mycol.* 44:9–11.

Perrin, R. M., N. D. Fedorova, J. W. Bok, R. A. Cramer, Jr., J. R. Wortman, H. S. Kim, W. C. Nierman, and N. P. Keller. 2007. Transcriptional regulation of chemical diversity in *Aspergillus fumigatus* by LaeA. *PLoS Pathog.* 3:e50.

Pöggeler, S. 2002. Genomic evidence for mating abilities in the asexual pathogen *Aspergillus fumigatus*. *Curr. Genet.* 42:153–160.

Poptsova, M. S., and J. P. Gogarten. 2007. BranchClust: a phylogenetic algorithm for selecting gene families. *BMC Bioinformatics* 8: 120.

Proctor, R. H., T. M. Hohn, S. P. McCormick, and A. E. Desjardins. 1995. *Tri6* encodes an unusual zinc finger protein involved in regulation of trichothecene biosynthesis in *Fusarium sporotrichioides*. *Appl. Environ. Microbiol.* 61:1923–1930.

Rep, M. 2005. Small proteins of plant-pathogenic fungi secreted during host colonization. *FEMS Microbiol. Lett.* 253:19–27.

Tekaia, F., G. Blandin, A. Malpertuy, B. Llorente, P. Durrens, C. Toffano-Nioche, O. Ozier-Kalogeropoulos, E. Bon, C. Gaillardin, M. Aigle, M. Bolotin-Fukuhara, S. Casarégola, J. de Montigny, A. Lépingle, C. Neuvéglise, S. Potier, J. Souciet, M. Wésolowski-Louvel, and B. Dujon. 2000. Genomic exploration of the hemiascomycetous yeasts. 3. Methods and strategies used for sequence analysis and annotation. *FEBS Lett.* 487:17–30.

Tekaia, F., and B. Dujon. 1999. Pervasiveness of gene conservation and persistence of duplicates in cellular genomes. *J. Mol. Evol.* 49: 591–600.

Tekaia, F., and J. P. Latgé. 2005. *Aspergillus fumigatus*: saprophyte or pathogen? *Curr. Opin. Microbiol.* 8:385–392.

Tekaia, F., A. Lazcano, and B. Dujon. 1999. The genomic tree as revealed from whole proteome comparisons. *Genome Res.* 9:550–557.

Tekaia, F., and E. Yeramian. 2006. Evolution of proteomes: fundamental signatures and global trends in amino acid compositions. *BMC Genomics* 7:307.

Tucker, S. L., C. R. Thornton, K. Tasker, C. Jacob, G. Giles, M. Egan, and N. J. Talbot. 2004. A fungal metallothionein is required for pathogenicity of *Magnaporthe grisea*. *Plant Cell* 16:1575–1588.

Wenzl, P., L. Wong, K. Kwang-won, and R. A. Jefferson. 2005. A functional screen identifies lateral transfer of β-glucuronidase (*gus*) from bacteria to fungi. *Mol. Biol. Evol.* 22:308–316.

Wootton, J. C., and S. Federhen. 1993. Statistics of local complexity in amino acid sequences and sequence databases. *Comput. Chem.* 17:149–163.

Aspergillus fumigatus and Aspergillosis
Edited by J.-P. Latgé and W. J. Steinbach
©2009 ASM Press, Washington, DC

Chapter 5

Essential Genes in *Aspergillus fumigatus*

WENQI HU, BO JIANG, AND TERRY ROEMER

This chapter will discuss recent developments in the identification of essential genes and validation of potential antifungal drug targets in *Aspergillus fumigatus*. As an opportunistic fungal pathogen, *A. fumigatus* causes life-threatening invasive aspergillosis (IA) in immunocompromised hosts, especially in patients with acute leukemia, bone marrow or solid organ transplantation, or AIDS (Brakhage, 2005; Latgé, 1999; Marr et al., 2002). Over the past decades, the incidence of IA has increased dramatically to overtake candidiasis as the most prevalent fungal infection in immunodeficient patients in developed countries (Brookman and Denning, 2000; Denning, 1998; McNeil et al., 2001). Current treatment of IA is mainly based on the use of polyenes (e.g., amphotericin B), azoles (e.g., itraconazole and voriconazole), and caspofungin. The polyene amphotericin B targets fungal membranes by intercalating with ergosterol (Charbonneau et al., 2001; Fournier et al., 1998) and has for 30 years remained the front-line drug for treatment of IA (Latgé, 1999), despite its severe toxic side effects (Costa and Nucci, 2001; Gerbaud et al., 2003; Pathak et al., 1998). Azoles such as voriconazole and itraconazole block ergosterol biosynthesis and display less toxicity and side effects but exhibit relatively poor activity against *A. fumigatus* (Kauffman, 2006; Spanakis et al., 2006). The newly developed echinocandin drugs, such as caspofungin and micafungin, belong to a new class of antifungal agents that inhibit β-1,3-glucan biosynthesis, a fungus-specific and essential component of the cell wall (Groll and Walsh, 2001; Hope et al., 2007). As antifungal drugs are limited and their mode of action is restricted to a few targets, their continued use has inevitably produced resistance issues now emerging in the clinical setting (Akins, 2005; Mann et al., 2003; Perlin, 2007). Thus, a more efficacious arsenal of antifungal drugs with novel modes of action is needed.

A rational approach to developing new antifungal agents relies on the identification of novel molecular targets which are involved in different aspects of fungal biology essential for growth and/or pathogenicity and whose function can be chemically inhibited (Firon and d'Enfert, 2002; Jiang et al., 2002). Although virulence factors have long been proposed as potential antifungal targets (Perfect, 1996), no genuine virulence factors have been identified that have been directly shown to be associated with the pathogenicity of *A. fumigatus* (Latgé, 2001; Rementeria et al., 2005; Ronning et al., 2005). Furthermore, recent analyses of *A. fumigatus, Candida albicans*, and *Cryptococcus neoformans* genome sequences failed to identify candidate virulence genes shared by these pathogenic fungi (Nierman et al., 2005; Tekaia and Latgé, 2005). This is in striking contrast to a large number of bacterial pathogens which have evolved to contain pathogenicity islands that encode various virulence factors and are normally absent from nonpathogenic strains of the same or closely related species (Gal-Mor and Finlay, 2006). Conceptually, virulence factors are also more likely to be targets for prophylactic rather than therapeutic agents, as their inactivation should prevent the establishment of an infection but may fail to treat a preexisting infection (Odds, 2005). Moreover, gene products whose function indirectly impairs *A. fumigatus* virulence are generally not required for cell viability (Odds, 2005). Thus, an inhibitor targeting a virulence factor would likely have minor efficacy and a narrow therapeutic spectrum as an antifungal drug.

Essential genes that are critical for fungal survival and growth serve as good drug target candidates, since antifungal agents that inhibit these essential biological processes should phenocopy these effects, namely, cell death and/or growth arrest (Firon and d'Enfert, 2002; Haselbeck et al., 2002; Jiang et al., 2002; Odds, 2005; Tekaia and Latgé, 2005). As such, clearance of the pathogen or a reduction in its abundance could be subsequently neutralized by the weakened host immune system (Haselbeck et al., 2002). Systematic identification of all essential genes has been achieved in the nonpath-

Wenqi Hu, Bo Jiang, and Terry Roemer • Merck Frosst Center for Fungal Genetics, Merck & Co., Inc., Montreal, Quebec H2X 3Y8, Canada.

ogenic yeast *Saccharomyces cerevisiae* (Giaever et al., 2002). Notwithstanding the minor clinical significance of *S. cerevisiae*, defining its complete catalog of the essential gene set has profoundly impacted our basic knowledge of yeast biology and provided an important framework for predicting antifungal drug targets in fungal pathogens. Indeed, a functional genomics approach based on genetic inactivation of homologous genes conserved in *C. albicans* has identified over 500 genes required for growth in this important fungal pathogen (Roemer et al., 2003). In addition, genetic screens to empirically identify essential genes in *C. albicans* include antisense-based gene inactivation (De Backer et al., 2001), transposon-based heterozygote screening (Uhl et al., 2003), and other means (Bruno et al., 2006).

Historically, identification of essential genes in *A. fumigatus* has been limited largely for technical reasons, including the following: (i) *A. fumigatus* lacks a natural sexual cycle, thus precluding the use of classic genetics as performed in *S. cerevisiae*; (ii) it has a much larger genome which contains ~9,900 genes, versus the ~6,000 genes in both *S. cerevisiae* and *C. albicans*; (iii) DNA transformation efficiencies and homologous recombination frequencies are typically low; (iv) molecular reagents for facilitating targeted gene mutations, such as selectable markers, plasmids, transposons, and conditional promoters, are limited; and until recently (v) a complete genome sequence and annotation of *A. fumigatus* was unavailable. Notwithstanding such technical difficulties, determination of gene essentiality in *A. fumigatus* has been investigated using strategies based on RNA interference (RNAi) (Bromley et al., 2006; Henry et al., 2007; Mouyna et al., 2004), parasexual genetics (Firon and d'Enfert, 2002; Firon et al., 2003), and more recently, conditional promoter replacement methods (Hu et al., 2007; Romero et al., 2003). Recently, the *A. fumigatus* genome (strain Af293) has been sequenced using a whole-genome random sequencing method (Nierman et al., 2005). This seminal achievement provided the first important step to systematically determine all essential genes in this principal filamentous fungal pathogen.

ESSENTIAL GENES OF MODEL FUNGAL ORGANISMS AND *C. ALBICANS*

Due to its genetic tractability, *S. cerevisiae* has long been used as a model organism to study gene function. The *S. cerevisiae* genome is 12 Mb in size and contains ~6,600 predicted open reading frames (ORFs) (as of 24 May 2007 [lsqb]http://www.yeastgenome.org[rsqb]) (Goffeau et al., 1996). Importantly, this genome annotation made it possible, for the first time, to study gene function on a genome scale and thus determine all essential genes in yeast. Early efforts to achieve this goal were led by Smith et al. (1996), who used a "genetic footprinting" strategy to assess the phenotypic effects of Ty1 transposon insertions in 268 predicted genes of yeast chromosome V. Phenotypic analysis of this collection identified 51 genes (20%) to be essential for yeast viability. An alternative transposon-based approach principally designed to construct green fluorescent protein (GFP) fusion proteins globally across the yeast genome yielded 11,000 transposon insertional mutants, each carrying a transposon inserted within a region of the genome expressed during vegetative growth and/or sporulation (Ross-Macdonald et al., 1999). These insertional mutants, representing nearly one-third of the genome (approximately 2,000 genes), were screened for growth phenotype under 20 different growth conditions. Essential genes were identified based on (i) the inability to construct haploid insertional mutants or (ii) identification of temperature-sensitive conditional mutants.

A complete identification of all essential genes in *S. cerevisiae* was ultimately achieved by systematically deleting nearly all the ORFs in the genome (Giaever et al., 2002; Winzeler et al., 1999). Deletion mutants, representing 96% of the genome, were constructed by precisely replacing the target ORF from ATG with a stop codon with a PCR-generated deletion cassette which contained a KanMX selectable marker flanked with 43 bp of DNA homologous to the target gene. Transformation of a wild-type diploid strain (BY4743) with the deletion cassette allowed construction of a heterozygous deletion mutant in which one allele of the target gene was precisely deleted and the other allele was untouched. Following sporulation and tetrad dissection of the meiotic progenies, gene essentiality was directly determined by evaluating the growth of haploid deletion mutants on rich medium. Using this systematic method, 1,105 genes, representing 18.7% of the genes examined, were shown to be essential for growth on rich medium. Interestingly, only about half of these (57%) were previously known to be essential. Although *S. cerevisiae* is not a human pathogen, a catalog of its essential gene set provides important insight into predicting potential antifungal drug targets in fungal pathogens.

Identification of essential genes has also been investigated in *Neurospora crassa* and *Aspergillus nidulans*, two other model organisms belonging to the filamentous fungi. Nargang and coworkers (1995) demonstrated the utility of a sheltered disruption strategy in identifying essential genes in *N. crassa*. In this method, a heterokaryotic strain is created which harbors two nuclei, one with a null allele of the target gene and the other with a wild-type allele. Using this strategy, they demonstrated that *MOM22*, a gene encoding a component of the protein import complex of the mitochondrial outer membrane, was essential because homokaryons bearing the *MOM22* disruption could not be isolated. A similar strategy termed the heterokaryon rescue technique was used to identify essential genes in *A. nidulans* (Osmani et al., 2004, 2006). In this technique, gene disruptions are created using the *pyrG* selectable marker. In the case of an essential gene being deleted, the null allele is maintained in spontaneously generated heterokaryons containing two distinct nuclei. One nucleus contains the null allele as well as a functional *pyrG* allele (*pyrG*$^+$). The other one contains the wild-type allele of the essential gene but lacks a functional *pyrG* allele (*pyrG*$^-$). Thus, a simple growth test applied to the uninucleate asexual spores formed from primary transformants can identify deletions of genes that are nonessential from those that are essential (Osmani et al., 1988). Using this approach, Osmani et al. (2004) systematically deleted 30 genes involved in the biogenesis of the nuclear pore complex and demonstrated 18 to be essential in *A. nidulans*. More recently, a high-throughput gene knockout strategy was developed in *N. crassa* which takes advantage of the *S. cerevisiae* recombinational cloning, a *Neurospora* mutant depleted of *Ku70*/*Ku80* (Ninomiya et al., 2004), custom-written software tools, and robotics (Colot et al., 2006). By applying such a high-throughput approach to 103 *Neurospora* genes encoding transcription factors, they demonstrated the following: (i) four genes were essential, as no viable ascospore progeny were observed; (ii) of 99 viable mutants, 40 were shown to be required for normal vegetative or sexual growth and development; and (iii) overall, this methodology achieved a high degree of high-throughput gene disruption at an efficiency >90% in *N. crassa*.

The naturally diploid *C. albicans* is the most commonly encountered human fungal pathogen, causing skin and mucosal infections in generally healthy individuals and life-threatening systemic infections in immunocompromised patients. The completion of the *C. albicans* genome sequence (Jones et al., 2004), combined with the knowledge of the yeast essential gene set, made it possible to systematically identify essential genes in this important pathogenic fungus. To achieve this goal, Roemer and coworkers developed a large-scale target validation strategy, termed GRACE (*g*ene *r*eplacement *a*nd *c*onditional *e*xpression) (Roemer et al., 2003), which involved two steps to construct conditional mutants (Fig. 1). In step one, the first allele of the target gene was precisely deleted with a PCR-generated deletion cassette which contained a *HIS3* selection marker flanked with ~50 bp of DNA sequence. In step two, the native promoter of the second allele was precisely replaced by a PCR-amplified tetracycline promoter (pTet) cassette containing a nourseothricin selectable marker. The pTet promoter is tightly repressed by the addition

STEP 1: Gene Replacement for one allele

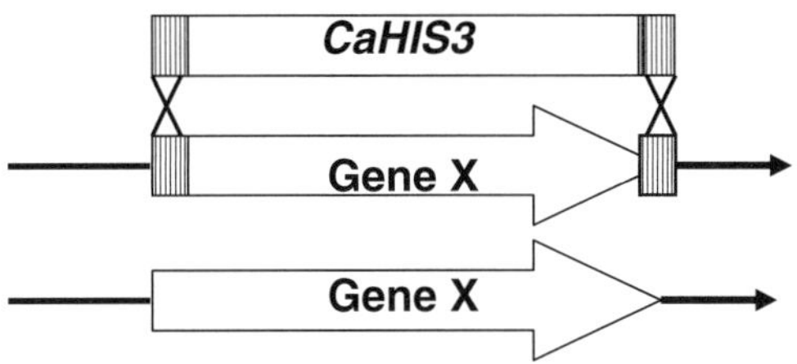

STEP 2: Promoter Replacement for the 2nd allele

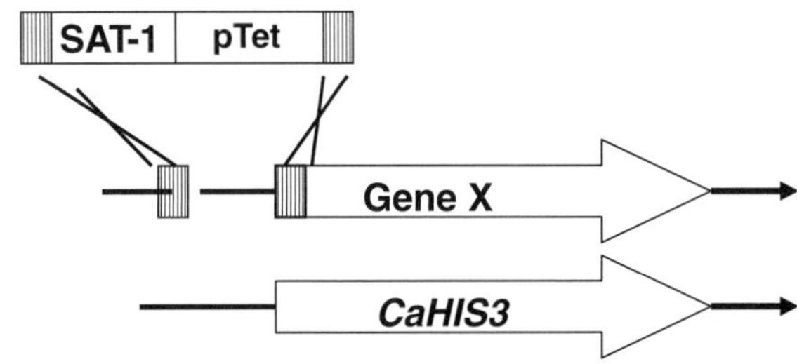

Figure 1. Outline of the GRACE method of target validation in *C. albicans*. Step 1: A heterozygote strain of the target gene was constructed by transforming a wild-type *C. albicans* strain with a PCR-generated *HIS3* disruption cassette flanked with homologous sequences to precisely delete one copy of the target gene. Step 2: The heterozygote strain obtained in step 1 was further transformed with a PCR-generated conditional promoter replacement cassette. Each cassette contains a SAT-1 dominant selectable marker and a conditional promoter (pTet) flanked with homologous DNA sequences to precisely replace the endogenous promoter of the remaining wild-type allele with pTet. Gene essentiality of the target gene was directly assessed by comparing the growth phenotype under inducing and repressing conditions (Roemer et al., 2003).

of tetracycline, thus permitting direct evaluation of gene essentiality under repressing conditions. Applying this approach to 1,152 *C. albicans* genes, Roemer and coworkers demonstrated 567 genes to be essential for growth (Roemer et al., 2003). This PCR-based GRACE approach facilitates precise, efficient, and systematic construction of conditional mutants, thus offering a functional genomics strategy to conduct large-scale essential gene identifications in *C. albicans*. In addition, this GRACE strategy also allows in vivo validation of candidate targets, whereby Tet-regulatable conditional mutants may be examined in a murine candidiasis model, and the addition of tetracycline to the drinking water of mice is used to effectively repress the target gene expression and examine its consequences in establishing or maintaining a *C. albicans* infection (Roemer et al., 2003).

The compendium of recently defined conserved essential genes in *S. cerevisiae* (Giaever et al., 2002), *C. albicans* (Roemer et al., 2003), and other fungi has provided important insights for predicting essential genes in *A. fumigatus*. Notably, over half (61%) of the *S. cerevisiae* essential genes which have an identifiable homolog in *C. albicans* and have been evaluated as GRACE conditional mutants were verified as essential for growth of this pathogen (Roemer et al., 2003). Similarly, Hu et al. (2007) recently showed that, among 54 genes displaying an essential phenotype in both *S. cerevisiae* and *C. albicans* and predicted to have only a single ortholog in *A. fumigatus*, 73% were demonstrated to be essential by conditional promoter replacement. These results demonstrate the clear impact of even a limited compendium of conserved essential genes in predicting gene essentiality in other fungi. Thus, large-scale functional genomics efforts across yeasts and filamentous fungi will continue to refine essential gene annotations in *A. fumigatus*. However, intrinsic differences in gene function across fungi due to diverging biological and physiological functions, genetic redundancies, or pathway duplications or rewirings undoubtedly can confound such predictions, thus emphasizing the importance of directly performing essential gene studies in *A. fumigatus*. Below, we will discuss the application of different genetic strategies to identify essential genes in *A. fumigatus* and some recent studies in which they have been successfully employed.

TARGETED GENE DISRUPTION AND KNOCKOUT STRATEGY

One of the most conventional strategies for studying gene function is to construct a disruption or knockout mutant in which the target gene is disrupted or deleted. In this manner one can study the gene function based on the phenotype observed with the null mutant. Standard gene knockout methods are ideal for identifying nonessential genes and studying their functions, particularly with regard to virulence and pathogenicity. However, the value of this approach in identifying essential genes is more limited. Since *A. fumigatus* is a natural haploid fungus in which each gene has only one allele, no viable null mutant will be recovered if the target gene encodes an essential function. Thus, a conclusion that the target gene is essential can only be inferred statistically based on the fact that no viable mutants can be recovered from a large population of transformants.

A gene disruption mutant is constructed by transforming *A. fumigatus* with a circular (uncut) disruption plasmid which contains a selectable marker and an internal fragment of the target gene, often lying toward the 5′-region (Fig. 2A). Homologous recombination with the target gene via a single-crossover event leads to the construction of a tandem mutant in which the two truncated copies of the target gene are separated by a selection marker. In principal, neither copy of the gene is functional, since one lacks the 3′-region of the gene while the other lacks its promoter and 5′-region. Various selectable markers have been used in *A. fumigatus*, including the auxotrophic marker pyrG (Bok et al., 2006; Cramer et al., 2006; d'Enfert, 1996; Schrettl et al., 2004; Steinbach et al., 2006), as well as dominant selectable markers, such as hygromycin (Beauvais et al., 2005; Fortwendel et al., 2005; Lamarre et al., 2007; Sheppard et al., 2005) and phleomycin (Bhabhra et al., 2004; Kupfahl et al., 2006). Although this method is relatively simple and requires minimal subcloning, it does have several shortcomings (Brakhage and Langfelder, 2002). First, since the gene disruption mutant contains tandem repeats on either side of the selection marker (Fig. 2B), this mutant is genetically unstable, especially in the absence of any selective pressure. Once the selection marker is removed (pop-out), the target gene will switch back to the wild-type allele. Second, this gene disruption method might be problematic in some cases where a truncated peptide is still functional. Third, the gene disruption method might not be applicable to small genes, since efficient homologous recombination in *A. fumigatus* usually requires a large homologous fragment (>0.5 to 1.0 kb). Despite these disadvantages, gene disruption has been used as a simple and powerful method in studying gene functions and evaluating gene essentiality in *A. fumigatus*. Using this method, Schrettl and coworkers (2004) demonstrated that L-ornithine-N^5-monooxygenase (*SidA*), which catalyzes the first committed step of hydroxamate-type siderophore biosynthesis, is essential for virulence. In contrast, disruption of *FtrA*, a gene encoding the high-affinity iron permease, had no effect on virulence in a murine model of invasive aspergillosis. Similarly, Bhabhra et al. (2004) constructed a disruption mutant of *CgrA*, an ortholog of yeast nucleolar protein that functions in ribosome biosynthesis, and which caused a temperature-sensitive growth phenotype at 37°C. More recently, Cramer et al. (2006) demonstrated that disruption of a nonribosomal peptide synthetase (encoded by *gliP*) in *A. fumigatus* impaired the production of gliotoxin, a secondary metabolite which had long been hypothesized to be important in the development of IA. However, this Δ*gliP* mutant displayed no growth defects in vitro and had no impact on survival or tissue burden in a murine model of IA. Thus, these data suggest that gliotoxin is not required for virulence in an immunosuppressed host with an invasive pulmonary infection (Cramer et al., 2006).

Similar to gene disruption, gene knockout strategies are commonly used to study gene functions in *A. fumigatus*. In this method, a linearized knockout cassette containing a selectable marker flanked with approximately 1.0 to 1.5 kb of DNA sequence homologous to

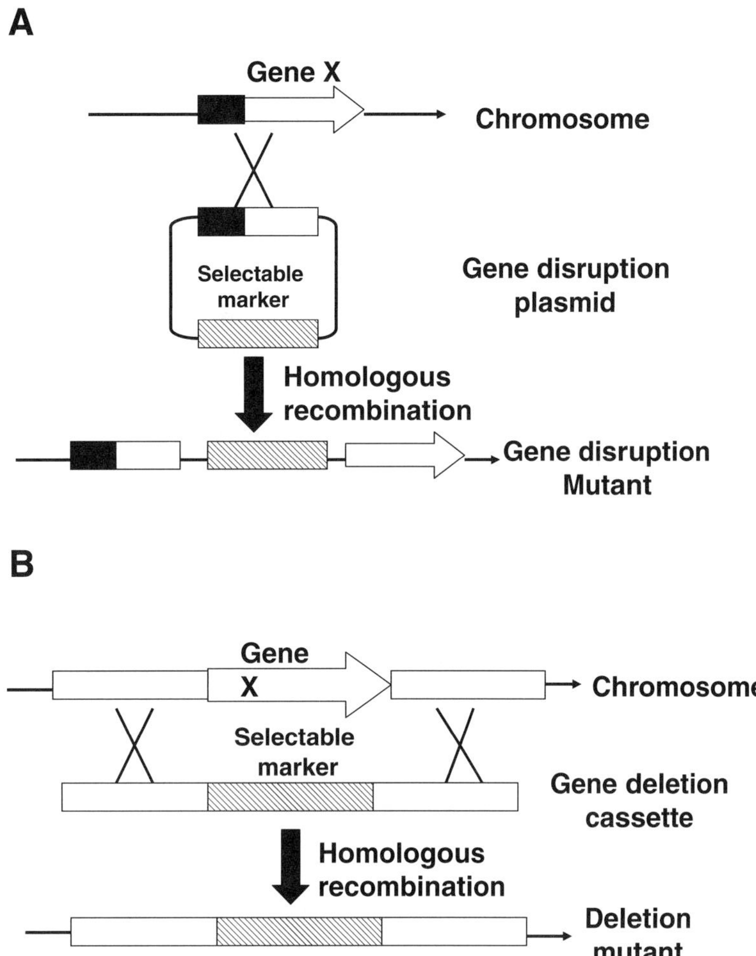

Figure 2. Schematic overview of gene disruption and deletion methods. (A) Schematic representation of a gene disruption event. The open arrow represents the ORF of a gene of interest. A gene disruption plasmid was constructed by cloning a truncated fragment of the target gene into a plasmid containing a selectable marker. Following a single-crossover homologous recombination, the plasmid integrates into the target gene, leading to a disruption of the gene. (B) Schematic representation of the gene deletion method. The open arrow represents the ORF of a gene of interest. A gene deletion cassette containing a selectable marker flanked with appropriate homologous DNA sequence was used to transform *A. fumigatus*. Following a double-crossover homologous recombination, the ORF of the target gene was precisely replaced by the selectable marker.

the target gene is used to transform *A. fumigatus* (Fig. 2B). Homologous recombination between the homologous flanking sequences via a double-crossover event results in the creation of a knockout mutant in which the entire ORF of the target gene is precisely replaced by the selectable marker. Unlike the disruption mutants discussed above, a knockout mutant is genetically stable without any risk of reversion.

Gene knockout methods have been widely used to evaluate gene function of nonessential genes in *A. fumigatus*. Fortwendel and coworkers (2005) constructed an *A. fumigatus* deletion mutant of *rasB*, a gene encoding a fungus-specific *Ras* subfamily homolog. Deletion of *rasB* caused decreased germination and growth rates on solid medium but caused no significant growth impairment in liquid culture, and its virulence was slightly reduced in an animal model. Steinbach and coworkers studied the calcineurin pathway in *A. fumigatus* by constructing a mutant depleted of *cnaA*, a gene encoding the catalytic subunit of calcineurin (Steinbach et al., 2006). Interestingly, they showed that, despite not absolutely required for growth, deletion of *cnaA* did result in severe defects in the growth and development of the fungus. These defects substantially affected the ability of *A. fumigatus* to cause IA, as multiple animal models of IA show the *cnaA* deletion mutant to be avirulent. Recently, Bok and coworkers (2006) investigated the potential virulence effects of *GliZ*, a global transcriptional regulator of gliotoxin biosynthesis. Although a *GliZ* deletion mutant failed to produce any detectable gliotoxin and reduced expression levels in other *gli* cluster genes were detected, no significant consequences to its virulence were observed in a murine pulmonary model of infection.

Although a very powerful tool for nonessential gene characterizations, the gene knockout strategy suffers the same disadvantage as the gene disruption method, namely, its inherent difficulty with essential genes. Nevertheless, Liebmann and coworkers (2004) demonstrated the utility of a gene knockout strategy to identify conditional essential genes in *A. fumigatus.* A deletion mutant of *A. fumigatus lysF*, a gene encoding a homolog of the *A. nidulans* homoaconitase *LysF*, was constructed which led to lysine auxotrophy. Interestingly, they showed that this *lysF* deletion mutant of *A. fumigatus* was avirulent in a low-dose intranasal mouse infection model of IA, thus suggesting that a functional homoaconitase encoded by *lysF* is important for survival of *A. fumigatus* in vivo and a potential target for antifungal therapy.

The application of such a gene disruption or knockout strategy to study gene function and to identify essential genes in *A. fumigatus* has long been restricted by its low efficiency in obtaining positive transformants. In eukaryotes, there are two main recombination pathways which control the integration of exogenous DNA into chromosomes (Ninomiya et al., 2004). The first one utilizes double-strand break (DSB) repair, which is dependent on DNA sequence homology. The second uses the nonhomologous end-joining (NHEJ) pathway, which involves direct ligation of the strand ends independent of DNA homology. Unlike *S. cerevisiae*, which mainly uses an homologous recombination system of DSB repair, many filamentous fungi preferably use the NHEJ pathway. As a consequence, exogenous DNA could be integrated randomly into the genome during the transformation of filamentous fungi (even in the case of long flanking homologous sequences), thus leading to a low frequency of targeted integrations. To overcome this difficulty, Ninomiya and coworkers attempted to disrupt *N. crassa KU70* and *KU80* genes, which encode proteins that function in NHEJ of DSBs (Ninomiya et al., 2004). Strikingly, they demonstrated that the *KU70* and *KU80* disruption mutants, named *mus-51* and *mus-52*, yielded 100% transformants exhibiting integration at the homologous site, compared to only 10 to 30% for the wild type. This seminal achievement clearly demonstrated the powerfulness of *KU70* and *KU80* disruption mutants in significantly improving the efficiency of gene targeting in filamentous fungi. Expanding this into *A. fumigatus*, Krappmann et al. (2006) and da Silva Ferreira et al. (2006) independently demonstrated that *KU70* and *KU80* deletion mutants yielded about 80% homologous integration efficiency compared to less than 5% efficiency in the wild type. In addition, these two studies showed that, when *KU70* and *KU80* deletion mutants were used, the flanking sequences needed could be reduced to about 0.5 kb, compared to >1 kb for the wild type. The reduction of flanking sequences may facilitate the generation of cassettes by PCR, thus avoiding the need for laborious subcloning. Combining the use of the *KU70* and *KU80* deletion mutants with the PCR-based generation of knockout cassettes may facilitate the adaptation of a high-throughput strategy for the large-scale identification of essential genes in *A. fumigatus.*

PARASEXUAL GENETICS APPROACH

A. fumigatus is a haploid filamentous fungus that reproduces asexually through condiospores. However, it has long been known that a parasexual cycle exists in *Aspergillus* species, including *A. nidulans* and *A. fumigatus* (Pontecorvo et al., 1953; Stromnaes and Garber, 1963). Stable diploid strains can be obtained using different selectable markers and appropriate selection procedures (Firon and d'Enfert, 2002; Stromnaes and Garber, 1963). Unlike that from sexual reproduction, diploid strains produced by parasexual reproduction do not undergo meiosis, but haploidization does occur via mitotic chromosomal nondisjunction induced by microtubule-destabilizing agents, such as benomyl and *p*-fluorophenyl-alanine (Clutterbuck, 1992; Timberlake and Marshall, 1988). Som and Kolaparthi (1994) demonstrated the utility of parasexual genetics in identifying essential genes in *A. nidulans.* A heterozygote diploid strain was constructed by deleting one allele of the target gene (*Aras*, a *Ras* homolog) in a diploid strain obtained through the parasexual cycle. This heterozygote diploid strain was then treated with benomyl to form a haploid strain. Analysis of the haploid segregants showed absence of any segregants with the inactivated allele, indicating that *Aras* is essential for growth in *A. nidulans.*

Recently, Firon and coworkers applied a similar parasexual genetics approach to directly identify essential genes in *A. fumigatus* (Firon and d'Enfert, 2002; Firon et al., 2003). They first created a diploid *A. fumigatus* strain via parasexual cycle in which each gene has two alleles. One allele of the target gene was then disrupted with a disruption cassette or transposon, both of which contained a selectable marker (*pyrG*). The heterozygous diploid strains were subjected to haploidization with or without the selective pressure corresponding to the introduced mutation. The absence of haploid segregants under selective conditions is indicative of the inactivation of an essential gene (Firon and d'Enfert, 2002; Firon et al., 2003). Using such a parasexual genetics approach, they showed that *FKS1*gene, encoding the β-1,3-D-glucan synthase, and the *smcA* gene, encoding a member of the SMC (structural maintenance of chromosome) protein family, are essential in *A. fumigatus* (Firon and d'Enfert, 2002).

Combining parasexual genetics with random transposon mutagenesis, Firon and coworkers further developed a useful strategy for identifying essential genes in *A. fumigatus* (Firon et al., 2003). In this method, an artificial *A. fumigatus* diploid strain with one copy of an engineered *impala160* transposon from *Fusarium oxysporum* integrated into its genome was used to generate a collection of 2,386 diploid strains by random in vivo transposon mutagenesis. These 2,386 heterozygous diploid strains were used to perform haploidization analysis using benomyl as an inducer on selective and nonselective media in two independent tests (Fig. 3). It was shown that, among 2,386 diploid strains screened by parasexual genetics, 1.2% had a copy of the transposon integrated into a locus essential for *A. fumigatus* growth. Sequencing analysis revealed 20 previously uncharacterized *A. fumigatus* genes to be essential in this species (Table 1).

Despite its demonstrated usefulness in identifying essential genes in *A. fumigatus* (Firon and d'Enfert, 2002; Firon et al., 2003), the parasexual genetics approach described above might not be suitable for high-throughput screening of essential genes in this species. First, irregularities commonly associated with the haploidization process of the *A. fumigatus* diploid strain might significantly affect the high-throughput screening results (Romero et al., 2003; Stromnaes and Garber, 1963). Second, Firon and d'Enfert (2002) reported that

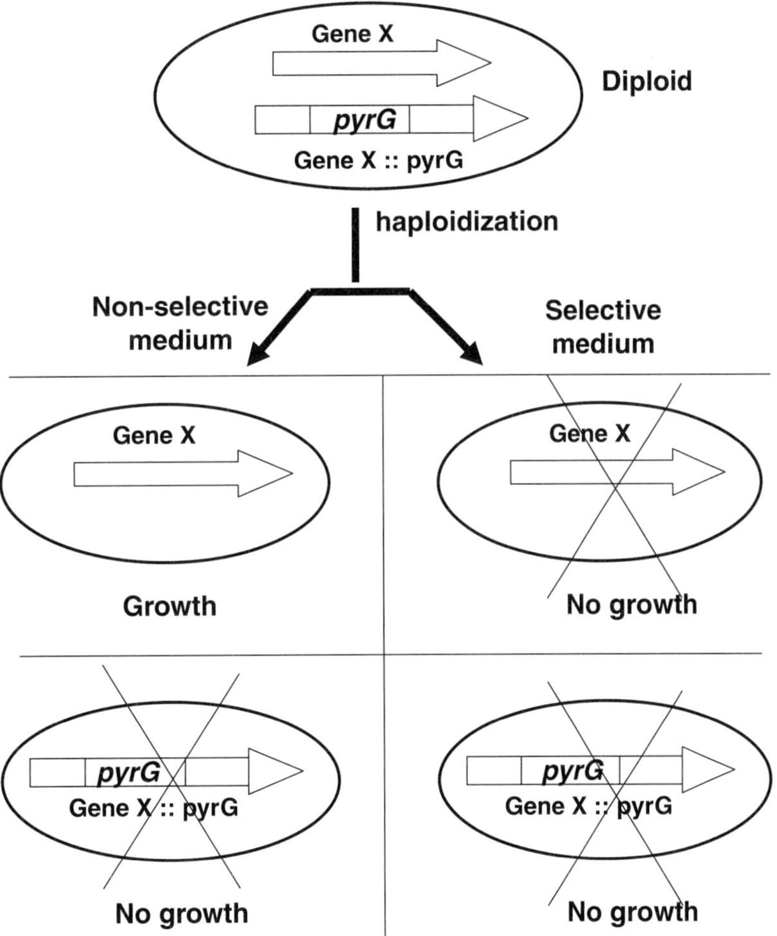

Figure 3. Schematic overview of the parasexual strategy. A diploid (heterozygous) *A. fumigatus* strain was first created by gene disruption or transposon mutagenesis and contains one inactivated allele of the target gene (gene X) as well as a wild-type allele. The heterozygous strain is used to perform haploidization analysis using benomyl as an inducer on selective and nonselective medium in two independent tests. The haploidization process will result in two subpopulations of haploid cells: one bearing the inactivated allele of the target gene and one bearing a wild-type allele. If gene X is essential for *A. fumigatus* growth, haploid progenies cannot be obtained from selective medium, as haploids with the inactivated allele will not be viable and haploids with the wild-type allele lack the selectable marker. Replicated from Firon and d'Enfert (2002) with permission from the publisher and the authors.

Table 1. Experimentally validated *A. fumigatus* essential genes

Biological process	Gene name	Gene description in *S. cerevisiae*	GenBank accession no.[a]	*S. cerevisiae* null mutant phenotype[b]	*C. albicans* GRACE phenotype[c]	*A. fumigatus* pNiiA-CPR mutant phenotype[d]	Reference(s)
Amino acid biosynthesis	*HIS3*[e]	Imidazole-glycerol-phosphate dehydratase	XP_747608	Viable[f]	NA[g]	4+	Hu et al., 2007
	LYS4[e]	Homoaconitate hydratase	XP_753748	Viable[f]	4+	4+	Liebmann et al. 2004; Hu et al., 2007
	LYS9[e]	Saccharopine dehydrogenase	XP_751695	Viable[f]	4+	4+	Hu et al., 2007
	MET16[e]	3′-Phospho-adenylylsulfate reductase	XP_754952	Viable[f]	3+	4+	Hu et al., 2007
	MET2[e]	Homoserine *O*-acetyltransferase	XP_753913	Viable[f]	4+	4+	Hu et al., 2007
	TRP5[e]	Tryptophan synthase	XP_755657	Viable[f]	4+	4+	Hu et al., 2007
Cell cycle control	*CDC27*	Component of anaphase-promoting complex	XP_748098	Lethal	ND	ND	Firon et al., 2003
	RIM11	Serine/threonine protein kinase	XP_747565	Viable	ND	ND	Firon et al., 2003
Cell redox homeostasis	*TRR1*	Thioredoxin reductase	XP_751532	Lethal	4+	3.5+	Hu et al., 2007
Cell wall organization and biogenesis	*FKS1*[h]	β-1,3-glucan synthase	XP_751118	Viable[f]	4+	3.5+	Firon and d'Enfert, 2002; Hu et al., 2007
	GFA1	Glutamine-fructose-6-phosphate aminotransferase	XP_750525	Lethal	4+	4+	Hu et al., 2007
Cellular metabolism	*AUR1*	Phosphatidyl inositol:ceramide phosphoinositol transferase	XP_754623	Lethal	4+	4+	Hu et al., 2007
	CDS1	CDP-diacylglycerol synthase	XP_750449	Lethal	4+	4+	Hu et al., 2007
	COX10	Protoheme IX farnesyltransferase	XP_751991	Viable[f]	ND	ND	Firon et al., 2003
	GUA1	Glutamine amidotransferase	Q4WFT3	Viable[f]	4+	4+	Rodriguez-Suarez et al., 2007
	HEM15	Ferrochelatase	XP_753861	Viable[k]	—[l]	4+	Firon et al., 2003; Hu et al., 2007
	IPP1	Inorganic pyro-phosphatase	XP_754776	Lethal	4+	3+	Hu et al., 2007
	OLE1	Stearoyl-CoA desaturase	XP_748918	Lethal	4+	4+	Hu et al., 2007
	PyrG	Orotidine-5′-monophosphate decarboxylase	CAA72161	Viable[f]	ND	ND	d'Enfert et al., 1996
	SPE2	*S*-Adenosyl-methionine decarboxylase	XP_747977	Viable[k]	0+	3.5+	Firon et al., 2003; Hu et al., 2007
Chromatin structure	*RSC9*	Remodels the structure of chromatin	EAL88634	Lethal	4+	ND	Firon et al., 2003

Cytoskeleton organization and biogenesis	*TUB1*	Tubulin α-1 chain	XP_750005	Lethal	4+	4+	Hu et al., 2007
DNA replication	*PRI1*	DNA primase small subunit	XP_754716	Lethal	4+	4+	Hu et al., 2007
Ergosterol biosynthesis	*ERG10*	Acetyl-CoA acetyltransferase	XP_747207	Lethal	4+	4+	Hu et al., 2007
	ERG11[b]	Lanosterol 14α-demethylase	XP_752137	Lethal	3.5+	0+	Hu et al., 2007
	ERG12	Mevalonate kinase	XP_752047	Lethal	4+	3+	Hu et al., 2007
Nuclear architecture	*NAR1*	Nuclear architecture-related protein	EAL89597	Lethal	3+	ND	Firon et al., 2003
Nuclear distribution	*NudC*	Nuclear distribution protein C homolog	AJ430231	ND	ND	ND	Romero et al., 2003
Nucleotide metabolism	*GUK1*	Guanylate kinase	AAS02092	Lethal	3+	ND	Firon et al., 2003
Protein biogenesis	*MSW1*	Tryptophanyl-tRNA synthetase	XP_747543	Viable	2+	ND	Firon et al., 2003
	RPL11	Ribosomal protein L11	XP_752052	Viable	ND	ND	Firon et al., 2003
	RPL14	Ribosomal protein	XP_747694	Viable	ND	ND	Firon et al., 2003
	RPL17	Ribosomal protein L17	Q6MY48	Viable	ND	ND	Firon et al., 2003
Protein modification	*ALG7*	UDP-*N*-acetyl-glucosamine-1-P transferase	XP_755457	Lethal	4+	3+	Hu et al., 2007
	WBP1	Oligosaccharyltransferase β-subunit	XP_753740	Lethal	4+	NA	Firon et al., 2003
Protein translation	*GCD6*	eIF2B	XP_751131	Lethal	3.5+	4+	Hu et al., 2007
	GUS1	Glutamyl-tRNA synthetase	XP_747988	Lethal	4+	4+	Firon et al., 2003; Hu et al., 2007
	PAB1	Poly(A)-binding protein in cytoplasm and nucleus	XP_750167	Lethal	4+	4+	Hu et al., 2007
	TIF35	eIF3	XP_755320	Lethal	4+	4+	Hu et al., 2007
Protein transport	*GOS1*	SNARE protein with a C-terminal membrane anchor	EAL90378	Viable	2+	NA	Firon et al., 2003
	SEC31	Component of COPII coat of secretory pathway vesicles	XP_755629	Lethal	4+	4+	Hu et al., 2007
	SLY1	Protein involved in vesicle trafficking	XP_754194	Lethal	4+	4+	Hu et al., 2007
	SRP101	Signal recognition particle receptor, α-subunit	XP_747692	Lethal	2+	ND	Firon et al., 2003
	TOM40	Mitochondrial import receptor subunit	XP_747566	Lethal	4+	4+	Hu et al., 2007

Continued on following page

Table 1. *Continued*

Biological process	Gene name	Gene description in *S. cerevisiae*	GenBank accession no.[a]	*S. cerevisiae* null mutant phenotype[b]	*C. albicans* GRACE phenotype[c]	*A. fumigatus* pNiiA-CPR mutant phenotype[d]	Reference(s)
Ribosome biogenesis	*BRX1*	Protein required for rRNA maturation	XP_749971	Lethal	4+	4+	Hu et al., 2007
	DBP10	DEAD box protein 10	EAL88958	Lethal	2+	ND	Firon et al., 2003
	ESF1	Protein required for 18S rRNA biogenesis	XP_749672	Lethal	3.5	4+	Hu et al., 2007
	KRR1	Component of 90S preribosomal particles	XP_755471	Lethal	4+	4+	Hu et al., 2007
	MAK5	Probable RNA helicase of the DEAD box family	XP_750779	Lethal	3.5+	4+	Hu et al., 2007
	NOB1	Essential protein that functions in 20S proteasome maturation and 26S proteasome assembly, component of pre-40S ribosomal particle	XP_747944	Lethal	3+	3.5+	Hu et al., 2007
	NOC3	Protein involved in biogenesis of 60S ribosomal subunit	XP_756039	Lethal	4+	4+	Hu et al., 2007
	NOP4	Nucleolar protein required for ribosome biogenesis	XP_752199	Lethal	3+	4+	Hu et al., 2007
RNA splicing	*LUC7*	U1 snRNA-associated protein	XP_750789	Lethal	4+	4+	Hu et al., 2007
Unknown	*YFL034w*	Probable membrane protein	XP_750126	Viable	0+	ND	Firon et al., 2003

[a] Amino acid sequences can be retrieved from GenBank (http://www.ncbi.nlm.nih.gov/), using the specified accession number(s).
[b] *S. cerevisiae* null mutant phenotypes as annotated by SGD (http://www.yeastgenome.org/).
[c] *C. albicans* phenotypes were determined using a tetracycline-regulatable promoter system (Roemer et al., 2003) in minimal medium (YNB, yeast nitrogen base medium).
[d] In all but three cases, two or more independent p*NiiA*-CPR mutants were recovered.
[e] Phenotype scored as conditional essential on ACM (Aspergillus rich medium) containing ammonium.
[f] These mutants display conditional essential phenotypes.
[g] NA, not applicable; all GRACE strains contain an extraneous *HIS3* gene as an integral part of the tetracycline-regulatable promoter system (Roemer et al., 2003), rendering them unsuitable for evaluating the endogenous *HIS3* gene phenotype.
[h] Underlined genes indicate *A. fumigatus* gene families.
[i] ND, not determined.
[j] The *fks1*Δ *fks2*Δ double mutation is lethal.
[k] Null mutants are lethal in certain genetic backgrounds.
[l] —, no apparently homologous genes exist in this species.

the random insertional mutagenesis protocols currently used for *A. fumigatus*, which rely on the integration of a heterologous DNA into the fungal chromosomes, led to frequent genomic rearrangements. Third, Firon et al. showed that the *impala* transposable element integrates preferentially into noncoding regions of the genome (Firon et al., 2003), thus significantly restricting its utility.

RNAi APPROACHES

RNAi is a process in which double-stranded RNA induces the specific degradation of mRNA to which it is homologous (Agrawal et al., 2003; Fire, 2007). The double-stranded RNA "trigger" is thought to be cleaved into shorter fragments of 21 to 25 nucleotides which then guide specific degradation of the corresponding mRNA (Baulcombe, 2001; Zamore and Aronin, 2003). RNAi has been extensively used as a genetic means to downregulate gene functions in *Caenorhabditis elegans* (Fire, 2007), plants (Watson et al., 2005), and vertebrate systems (Caplen et al., 2001). Liu and coworkers (2002) showed for the first time the utility of an RNAi strategy to study fungal gene function in *C. neoformans*, demonstrating that expression of double-stranded RNA corresponding to portions of the cryptococcal *CAP59* and *ADE2* genes resulted in reduced mRNA levels and phenotypic consequences similar to those of their respective gene disruption. Extending this study, simultaneous interference of *CAP59* and *ADE2* using a chimeric double-stranded RNA produced phenotypes consistent with RNAi blocking expression of both genes, thus demonstrating that more complex genetic analyses between genes, including synthetic lethality, may be achieved by this strategy (Liu et al., 2002).

The application of RNAi to identify essential genes in *A. fumigatus* was first investigated by Latgé and coworkers (Mouyna et al., 2004). An RNAi cassette was constructed by subcloning into a circular plasmid with inverted repeats of 500 bp of coding sequence of the target gene separated by a spacer segment of the GFP sequence (Fig. 4A). The RNAi cassette was introduced into an *A. fumigatus* wild-type strain by transforming protoplasts with this circular plasmid which contains a hygromycin resistance selection marker gene (*hph*). Expression of double-stranded RNA was controlled by the *Aspergillus niger* glucoamylase promoter (p*GlaA*), which is induced by maltose and repressed by xylose; thus, growth phenotypes could be assessed under both conditions. Expression of double-stranded RNA corresponding to portions (approximately 500 bp) of the *Alb1* and *FKS1* genes resulted in reduced mRNA levels for both genes. *Alb1* is a gene involved in melanin and conidia pigmentation production, and its deletion mutant produced white colonies (Tsai et al., 1999). *FKS1* is a known essential gene which encodes the β-1,3-glucan synthase. RNAi of *Alb1* and *FKS1* produced phenotypes similar to those of their respective gene disruption. Bromley et al. (2006) also investigated the utility of the RNAi method to downregulate gene functions using the *A. fumigatus* cellobiohydrolase *cbhB* promoter, which is induced by carboxymethylcellulose and repressed by glucose. An RNAi construct driven by the *cbhB* promoter was used to downregulate the *Alb1* gene; transformants showed low *Alb1* message levels and a loss-of-function phenotype (white colony) in the presence of carboxymethylcellulose (Bromley et al., 2006). More recently, Henry et al. (2007) investigated RNAi-based silencing of *KRE6* and *CRH1*, two genes putatively involved in fungal cell wall biosynthesis, by using a constitutive glyceraldehyde-3-phosphate dehydrogenase promoter (p*gpdA*). Reduced expression of *KRE6* and *CRH1* was obtained; however, complete silencing was not achieved, and no growth phenotype was observed among any RNAi mutants (Henry et al., 2007).

Although RNAi provides an important molecular strategy to downregulate *A. fumigatus* gene expression, further refinements appear necessary. Considering the historically low efficiency of gene targeting in *A. fumigatus*, the RNAi strategy provides a rapid means to evaluate gene functions following an ectopic integration of the silencing plasmid (Henry et al., 2007). However, it was observed that different RNAi mutants displayed a wide range of phenotypes on induction, from partial to full knockout. It was suggested that different sites of integration in the chromosome of the RNAi construct may lead to variations in the efficiency of gene silencing (Mouyna et al., 2004; Bromley et al., 2006). It is also possible that this was caused by DNA rearrangements of the construct, which resulted in the loss of the transcriptional unit of RNAi during integration. This could be overcome by using a nonintegrative plasmid, such as the *AMA1*-based plasmid (Fig. 4B), or by targeted integration using a *KU70*/*KU80* mutant as a starting strain. It was also observed that the insertion of the RNAi plasmid in the genome was often unstable regardless of the promoter used and the target gene investigated (Henry et al., 2007). During the successive transfers of mutants, the RNAi plasmid is frequently lost or modified, leading to the loss of RNAi-based gene silencing. To overcome these issues, a recent report exploited an *AMA1*-based episomal RNAi construct to silence the target gene (Khalaj et al., 2007). In this approach, an RNAi cassette for silencing the *Alb1* gene was constructed using a self-replicating *AMA1*-based plasmid (Aleksenko and Clutterbuck, 1995; Aleksenko et al., 1996). It was shown that the episomal inducible expression of an RNAi construct using the *AMA1* system coupled with the *cbhB* promoter is able to efficiently downregulate *Alb1* gene expression while avoiding the variable levels of gene si-

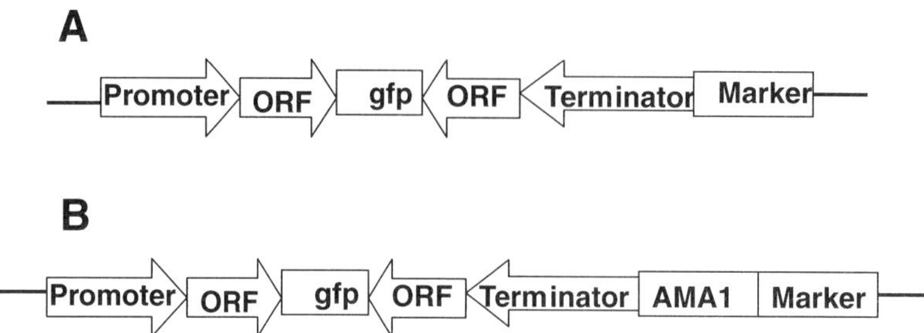

Figure 4. Schematic overview of the RNAi method. (A) Representative example of the RNAi cassette used to silence the target gene in *A. fumigatus*. The RNAi cassette was constructed with inverted repeats of ~500 bp of the coding region of the target gene separated by a spacer segment of GFP sequence. (B) Representative example of the *AMA1*-based RNAi cassette.

lencing encountered with the integrative plasmids (Khalaj et al., 2007).

CONDITIONAL PROMOTER REPLACEMENT APPROACHES

Conditional promoter replacement strategies offer a direct approach to identifying essential genes. With this method, the native promoter of the target gene is precisely replaced with a tightly regulatable promoter, thus enabling a direct evaluation of its phenotype under inducing or repressing conditions (Fig. 5). This conditional promoter replacement strategy has been successfully used to identify essential genes in *S. cerevisiae* (Davydenko et al., 2004; Mnaimneh et al., 2004), *C. albicans* (Roemer et al., 2003), and the filamentous fungus *A. nidulans* (Felenbok et al., 2001). Extending this approach to *A. fumigatus* requires well-characterized endogenous and/or heterologous promoters that are tightly regulated according to differential growth conditions and can effectively shut off gene expression with minimal basal transcription, or "leakiness." For example, Romero and coworkers (2003) employed a heterologous *A. nidulans alcA* promoter to demonstrate that *nudC*, a gene that had previously been shown to be essential in *A. nidulans*, is also essential for growth in *A. fumigatus*. The tetracycline-regulatable promoter system is also likely suitable for *A. fumigatus*, although its direct application to evaluating essential genes remains to be demonstrated (Vogt et al., 2005).

More recently, we reported a conditional promoter replacement system based on the endogenous *NiiA*/*NiaD* promoter (p*NiiA*) of *A. fumigatus* (Hu et al., 2007). This promoter is located between *niiA* and *niaD*, which are divergently expressed and encode the nitrate and nitrite reductases, respectively (Amaar and Moore, 1998). It has been shown that the p*NiiA* promoter in both *A. fumigatus* and *A. nidulans* is tightly regulated by nitrogen sources (Amaar and Moore, 1998; Muro-Pastor et al., 1999; Punt et al., 1995). Two operationally independent signals regulate p*NiiA* expression: an inducer (e.g., nitrate and another secondary nitrogen source) and a repressor (e.g., ammonium and another primary nitrogen source). Expression is achieved solely in the absence of ammonium and in the presence of nitrate, while repression is achieved by the presence of a repressor regardless of the presence of any inducers (Amaar et al., 1998; Punt et al., 1995; Muro-Pastor et al., 1999). Using a p*NiiA* conditional promoter replacement (p*NiiA*-CPR) cassette, marked with *pyrG*, the p*NiiA*-CPR mutants were constructed by homologous recombination-mediated deletion of the native promoter (approximately 250 bp of the 5′-region immediately preceding the start codon of the target gene) and replacement with the p*NiiA* promoter (Fig. 5). Therefore, gene expression is controlled by the nitrogen source provided, and direct determination of the target gene's phenotype is achieved under inducing (aspergillus minimal medium plus nitrate) compared to repressing conditions (aspergillus minimal medium plus ammonium). Applying this p*NiiA*-CPR strategy to 55 genes which had previously been shown to be essential in *S. cerevisiae* and *C. albicans*, 35 genes were directly demonstrated to be essential for *A. fumigatus* growth (Table 1). This essential gene set includes genes involved in diverse biological functions, including intermediary metabolism, ergosterol and protein biosynthesis, glycosylation, secretion, and RNA processing, as well as novel genes of unknown function.

Unlike gene disruption, this p*NiiA*-CPR method employs a directed strategy to systematically evaluate *A. fumigatus* essential genes on a large scale and permits distinction between fungicidal or static terminal phenotypes. Since the terminal phenotype can be directly analyzed under repressing conditions, this approach circumvents the need to complement the mutants with their corresponding wild-type allele, an often time-consuming and difficult step in *A. fumigatus*. Importantly, the p*NiiA*-CPR system also permits phenotypic

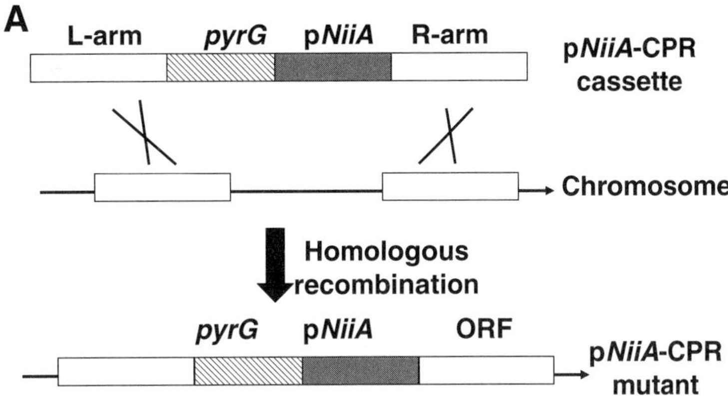

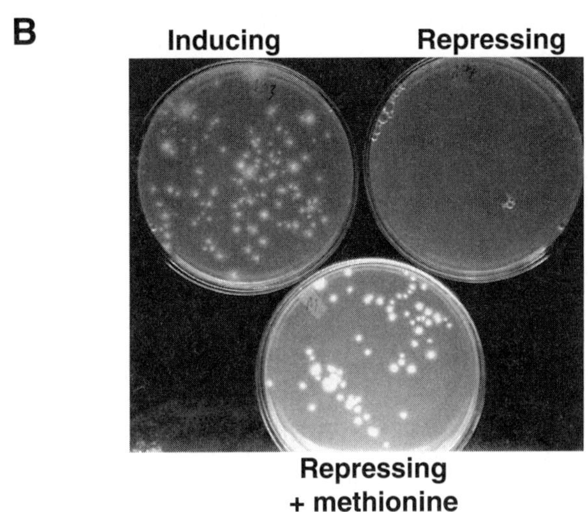

Figure 5. Conditional promoter replacement strategy. (A) Schematic overview of the p*NiiA*-CPR strategy. A conditional promoter replacement cassette containing a *PyrG* selectable marker and a p*NiiA* conditional promoter flanked with 1.5 kb of homologous DNA sequence (L-arm and R-arm) was used to transform the *A. fumigatus* CEA17 strain (*PyrG*⁻). Following homologous recombination, the endogenous promoter of the target gene was precisely replaced by the p*NiiA* condition promoter (Hu et al., 2007). (B) Representative example of gene essentiality validation with a p*NiiA-MET2* mutant. The p*NiiA-MET2* mutant displayed a no-growth phenotype under repressing conditions, suggesting its essential role for growth. Images are reprinted from Hu et al. (2007) with permission from the publisher and the authors.

analyses to be performed in a murine model of IA, as the ammonium level within the murine host is sufficient to repress target expression of p*NiiA*-CPR mutants (Fig. 6). In this way, p*NiiA*-CPR mutants of multiple genes, including *GFA1*, *GCD6*, *SEC31*, and *TUB1*, as well as an *ERG11A/B* double mutant, were shown to be avirulent in a host infection model. Despite these advantages, however, the p*NiiA*-CPR approach also has limitations. Construction of the p*NiiA*-CPR cassette presently requires multiple subcloning steps, which limits its suitability to high-throughput genetic analysis. As with any other conditional promoters, in some instances the p*NiiA* promoter may not be sufficiently repressed below the endogenous gene expression level to achieve a phenotype resembling that of a gene deletion, thus obscuring the gene's true essentiality. Finally, some *A. fumigatus* genes can contain extremely short 5′-exons (fewer than 10 bp), which makes the precise prediction of the start ATG codon, a prerequisite for any successful promoter replacement, technically very challenging.

ESSENTIAL GENES IDENTIFIED IN *A. FUMIGATUS*

Essential genes that are required for fungal survival and growth provide potential antifungal drug targets. In prioritizing antifungal drug targets, a few considerations should be carefully taken into account. First, an ideal antifungal target should be highly conserved among all

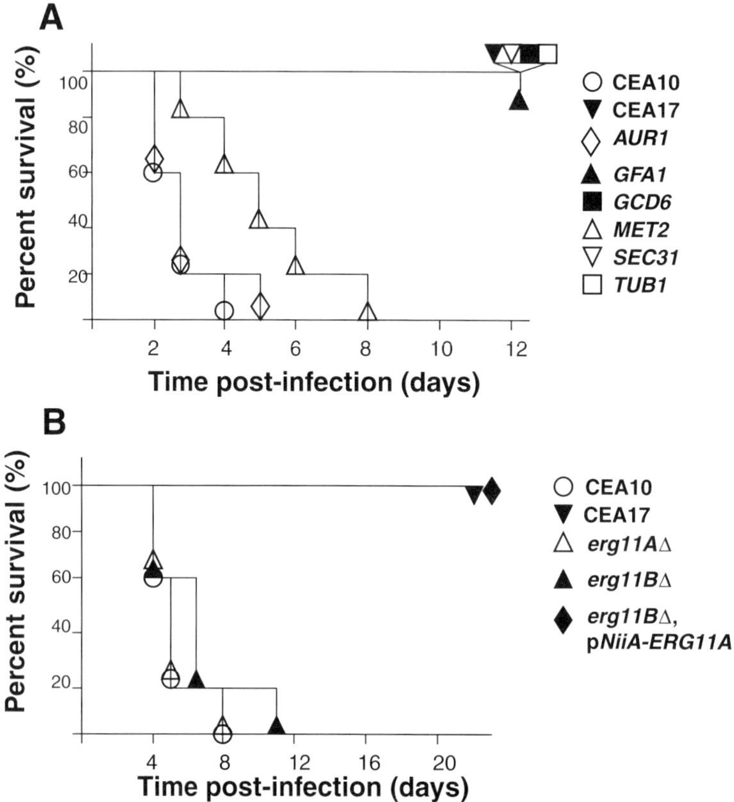

Figure 6. In vivo validation of gene essentiality using p*NiiA*-CPR mutants. (A) ICR male mice were immunocompromised by administrating cyclophosphamide at 150 mg/kg of body weight twice prior to infection and then 100 mg/kg twice a week after infection. Approximately 10^5 viable conidia from individual p*NiiA*-CPR mutants were injected into the tail vein of immunocompromised mice (five mice per group). CEA10 (wild-type) and CEA17 (a $PyrG^-$ auxotroph of CEA10) were included as controls for virulence and avirulence, respectively. (B) Genetic inactivation of the *ERG11* gene family promotes avirulence in an immunocompromised murine model of systemic infection. Pathogenesis of *erg11A*Δ, *erg11B*Δ, and an *ERG11* double mutant (*erg11B*Δ p*NiiA-ERG11A*) was similarly analyzed but over a longer postinfection period (22 days), and animal survival was compared to CEA10 and CEA17 control strains. Figures are reprinted from Hu et al. (2007) with permission from the publisher and the authors.

major human pathogenic fungi, including *A. fumigatus*, *C. albicans*, and non-albicans *Candida* species (e.g., *C. glabrata* and *C. krusei*), as well as *C. neoformans* and other rare molds (e.g., *A. flavus, Fusarium* spp., *Rhizomucor* spp.). Therefore, an inhibitor targeting such a mechanism may provide as wide an antifungal spectrum as can be rationally designed. This criterion is important not only from the viewpoint of pharmaceutical companies but also from that of clinical practice. Development of an antifungal drug that meets the clinical needs will usually take 10 to 15 years, so a niche product that is only efficacious against one species is unlikely to recoup its development cost. From the point of view of clinical practice, treatment of invasive infections is usually empirical, rather than specifically targeted, since many invasive fungal infections are caused by an opportunistic fungus which is often difficult to accurately diagnose in a timely manner. Second, an antifungal target ideally should be absent in humans or should be one that has significantly diverged from its human counterpart so as to minimize target-based toxicity issues. Notwithstanding this view, no "hard rules" apply: existing antifungal therapeutic classes, namely, echinocandins and azoles, are indeed clinically efficacious, yet azoles target an activity conserved in humans while echinocandins do not. Third, a molecular target displaying a fungicidal terminal phenotype when depleted is advantageous as it causes outright cell death, while inhibition of a static target will merely stop cellular growth and keep the cells in a dormant state. However, it is of interest that antifungal azoles such as fluconazole have largely static effects yet display satisfactory clinical effects and continue to serve as a frontline antifungal agent against candidiasis. In addition, a molecular target with known biochemical functions will be viewed favorably since such a target will facilitate the development of in vitro enzymatic assays for high-throughput screening. Moreover, formal demonstration that an essential gene identified in vitro indeed confers an avirulent phenotype in an animal model provides useful insights in prioritizing antifungal targets. Animal models of IAs have been established to evaluate *A. fumigatus* gene essentiality using the conditional promoter replacement mutants constructed with the p*NiiA*-CPR strategy (Hu et al., 2007). From the perspective of drug discovery, those genes that have been validated as essential for growth both in vitro and in vivo are of special interest, since an inhibitor hitting such a molecular target will more likely translate into in vivo efficacy.

A. fumigatus genes which have been experimentally demonstrated to be essential for growth are summarized in Table 1. This essential gene set includes genes involved in various biological and biochemical functions, such as amino acid, cell wall, ergosterol, heme, and lipid biosynthesis, as well as cell cycle control, cellular metabolism, protein transport, ribosome biogenesis, and RNA splicing. Discussed below are a number of *A. fumigatus* essential genes that merit further investigation, particularly from the perspective of their use as potential antifungal drug targets.

Ergosterol Biosynthesis Pathway

The ergosterol pathway is required for biosynthesis of a major constituent of the fungal plasma membrane (ergosterol) and is the target for antifungal azoles (e.g., fluconazole and itraconazole) and allylamine drugs, such as terbinafine. Ergosterol contributes to a variety of cellular functions, including membrane fluidity and integrity, as well as the proper function of membrane-bound enzymes, such as proteins associated with nutrient transport and chitin synthesis (Lupetti et al., 2002). In *S. cerevisiae* and *C. albicans*, essentiality for all genes involved in this pathway has been systematically investigated and has shown that some genes are dispensable while others are essential for viability (Giaever et al., 2002; Roemer et al., 2003). However, until recently, gene essentiality for most genes involved in the ergosterol pathway in *A. fumigatus* was largely unexplored.

ERG11, encoding the lanosterol 14α-demethylase which catalyzes the C-14 demethylation of lanosterol to form 4,4″-dimethyl cholesta-8,14,24-triene-3-beta-ol in the ergosterol biosynthesis pathway, is the known drug target for antifungal azoles. *ERG11* was experimentally shown to be essential in *S. cerevisiae* and *C. albicans* (Kalb et al., 1987; Roemer et al., 2003). Unlike these yeasts, in which *ERG11* is a single gene, *A. fumigatus* contains two copies of *ERG11*, namely, *ERG11A* and *ERG11B* (Mellado et al., 2001). Functionally, *ERG11A* and *ERG11B* are redundant and nonessential, since single deletion mutants are viable while an *ERG11B*Δ *ERG11A*-p*NiiA* double mutant is inviable under repressing conditions, thus demonstrating that the *ERG11* gene family is essential for growth (Hu et al., 2007; Mellado et al., 2001). Interestingly, an *ERG11A* deletion mutant is hypersensitive to fluconazole and itraconazole, suggesting that either *ERG11A* encodes the main 14α-demethylase enzyme functions or the *ERG11A* protein is intrinsically more resistant to azoles (Hu et al., 2007). Additional *A. fumigatus* essential genes involved in ergosterol biosynthesis include *ERG10, ERG12, ERG7, ERG8*, and *ERG20* (Hu et al., 2007, and unpublished data) and as such, provide new targets for therapeutic intervention.

TOM40

TOM40 encodes a translocase of the outer mitochondrial membrane (Tom40p) which is the major component of the TOM complex and forms the pore through which preproteins traverse the outer membrane (Ahting et al., 2001; Rapaport, 2005). The TOM complex is responsible for the recognition of mitochondrial preproteins and for the first stages of their import into mitochondria (Sherman et al., 2006). The TOM complex is composed of at least seven different subunits, of which Tom20 and -70 are the primary receptors, whereas the subunits Tom40, -22, -7, -6, and -5 form the stable TOM core complex (Rapaport, 2005). Tom40p is essential for viability in *S. cerevisiae* (Baker et al., 1990), *N. crassa* (Taylor et al., 2003), *C. albicans* (Roemer et al., 2003), and *A. fumigatus* (Hu et al., 2007). Sequence homology analyses (BLAST) showed that *A. fumigatus* Tom40p is 44, 42, 41, and 29% identical to its homologs in *C. albicans* (3e-74), *C. glabrata* (4e-73), *Schizosaccharomyces pombe* (4e-72), *S. cerevisiae* (2e-70), and *Homo sapiens* (4e-11), suggesting that fungal Tom40p is significantly diverged from its human homolog. Although no specific inhibitors targeting TOM40p are known, the three-dimensional structure for the yeast TOM complex is available (Model et al., 2002) and would assist structure-based molecular modeling of inhibitors to this target complex.

GFA1

GFA1 encodes glutamine-fructose-6-phosphate aminotransferase, which catalyzes the first and rate-limiting step in the hexosamine biosynthesis pathway. The hexosamine biosynthesis pathway is responsible for generating the precursors for N- and O-linked glycosylation. Further, *GFA1* and the hexosamine biosynthesis pathway are involved in the biosynthesis of chitin, a cell wall component in yeast (LaGorce et al., 2002). *GFA1* has been experimentally shown to be essential for growth in *S. cerevisiae* (Giaever et al., 2002), *C. albicans* (Roemer et al., 2003; Gabriel et al., 2004), and *A. fumigatus* (Hu et al., 2007). Deletion of *GFA1* leads to a fungicidal terminal phenotype in both *A. fumigatus* (Hu et al., 2007) and *C. albicans* (Roemer et al., 2003). *GFA1* has long been sought as a promising target for the development of novel antifungal agents targeting the fungal cell wall, and multiple inhibitors have been described (Chittur and Griffith, 2002). The crystal structure for *GFA1* protein is also available to facilitate structure-based drug discovery (Teplyakov et al., 1999). However, most *GFA1* inhibitors possess reactive functionalities, which prevented their applications as clinical therapeutic agents (Chmara et al., 1998; Chittur and Griffith, 2002). Novel GFA1 inhibitors may also possess intrinsic target-based toxicity, as a human ortholog of *GFA1* has been reported (Smith et al., 1996).

SEC31

SEC31 encodes a component (p150) of the COPII coat of secretory pathway vesicles involved in endoplasmic reticulum-to-Golgi complex transport (Salama et al., 1997). The COPII coat consists of five principal components: Sar1p, Sec23p, Sec24p, Sce13p, and Sec31p (Yamasaki et al., 2006). The COPII coat complex buds vesicles from the endoplasmic reticulum membrane to transport newly synthesized proteins from the endoplasmic reticulum to the Golgi complex (Heidtman et al., 2003). *SEC31* has been shown to be essential for growth in *S. cerevisiae* (Salama et al., 2007), *C. albicans* (T. Roemer et al., unpublished data), and *A. fumigatus* (Hu et al., 2007). Depletion of *SEC31* caused a fungicidal terminal phenotype in *A. fumigatus* (Hu et al., 2007) and *C. albicans* (Roemer et al., unpublished). Moreover, this gene has been validated as essential in vivo both in *A. fumigatus* and *C. albicans* (Hu et al., 2007; Roemer et al., unpublished). Although no specific inhibitors targeting *SEC31* have been discovered, a known inhibitor of the COPI complex, brefeldin A, has been extensively studied, and this suggests COPII complex inhibitors may be similarly identified (South et al., 2000).

RSC9

RSC9 is a gene encoding a component of the abundant RSC chromatin remodeling complex, which is composed of 15 subunits and plays a critical role in controlling gene expression (Cairns et al., 1996; Damelin et al., 2002). Rsc9p is a DNA-binding protein involved in the synthesis of rRNA and in transcriptional repression and activation of genes regulated by the target of the rapamycin (TOR) pathway (Damelin et al., 2002). *RSC9* has been shown to be essential for growth in *S. cerevisiae* (Winzeler et al., 1999), *C. albicans* (Roemer et al., unpublished), and *A. fumigatus* (Firon et al., 2003). A BLASTp search revealed that no clear homolog exists in humans (e=0.087) (B. Jiang et al., unpublished data).

TRR1

TRR1 encodes thioredoxin reductase 1, a peroxidase component that functions in response to oxidative stress (Hirt et al., 2002). Thioredoxin reductase, a flavoenzyme homodimer which binds flavin adenine dinucleotide and NADPH, reduces the oxidoreductase thioredoxin. It has been found that there are two distinct isoforms of thioredoxin reductase, the high-molecular-weight isoform and the low-molecular-weight isoform (Missall and Lodge, 2005). The high-molecular-weight isoform exists in mammals and some parasites, while the low-molecular-weight isoform is present in most bacteria, plants, and fungi (Hirt et al., 2002; Missall and Lodge, 2005). Although these two isoforms have similar biological and biochemical functions, they are very distinct in protein sequences and structures, suggesting the potential of *TRR1* as a specific antifungal target. Indeed, a BLASTp search with the predicted *A. fumigatus* Trr1p against sequences in GenBank failed to identify any human homolog (Jiang et al., unpublished). *TRR1* was shown experimentally to be essential for growth in *S. cerevisiae* (Winzeler et al., 1999), *C. albicans* (Roemer et al., unpublished), *C. neoformans* (Missall and Lodge, 2005), and *A. fumigatus* (Hu et al., 2007). The crystal structure of recombinant Trr1p of *S. cerevisiae* treated with hydrogen peroxide is available for structure-based drug screening (Oliveira et al., 2005).

AUR1

The *AUR1* gene in *S. cerevisiae* was first identified as a dominant resistant mutant to the antifungal agent aureobasidin A (LY295337) (Heidler and Radding, 1995). *AUR1* encodes inositol phosphorylceramide (IPC) synthase, which is essential for the biosynthesis of sphingolipid (Nagiec et al., 1997). Sphingolipids are found in eukaryotic membranes and contain a hydrophobic segment (ceramide) which is a long chain base, sphingosine in mammals and phytosphingosine in fungi, yeast, and plants (Heidler and Radding, 2000). The biosynthesis of ceramide is similar in all eukaryotes, but the addition of the polar head group to ceramide is a process which is highly divergent between fungi and mammals. In fungi, ceramide is converted to IPC by transfer of inositol phosphate from phosphatidylinositol to the 1-OH group of ceramide (Dickson and Lester, 1999). This process is catalyzed by phosphatidylinositol: ceramide transferase (IPC synthase), a membrane-bound enzyme encoded by *AUR1*. This essential step in fungal sphingolipid biosynthesis is not found in humans, making it an ideal antifungal target (Dickson and Lester, 1999; Sugimoto et al., 2004). *AUR1* has been experimentally validated as essential for growth in *S. cerevisiae* (Hashida-Okado et al., 1996; Winzeler et al., 1999), *C.*

albicans (Roemer et al., unpublished), and *A. fumigatus* (Hu et al., 2007). Importantly, *AUR1* is highly conserved among a number of other human pathogenic fungi, including *C. glabrata*, *C. krusei*, *Candida parapsilosis*, *Candida tropicalis*, and *C. neoformans* (Heidler and Radding, 2000).

Consistent with these findings, aureobasidin A (AbA) is a broad-spectrum antifungal compound which is potent against many pathogenic fungi, including *Candida* spp., *C. neoformans*, and some *Aspergillus* spp. (Takesako et al., 1991, 1993). AbA was first isolated from *Aureobasidium pullulans* R106 and is a cyclic depsipeptide which contains eight amino acids and a hydroxy acid (Takesako et al., 1991). The molecular target of AbA was identified as *AUR1* in *S. cerevisiae* (Heidler and Radding, 1995; Hashida-Okado et al., 1996), and AbA was found to be a specific inhibitor of IPC synthase, with a 50% inhibitory concentration (IC_{50}) of 0.2 nM (Nagiec et al., 1997). Later, two other IPC synthase inhibitors, khafrefungin and rustmicin, were identified (Mandala et al., 1997, 1998). Khafrefungin is a potent IPC synthase inhibitor with an IC_{50} of 0.6 nM and has fungicidal activity against *C. albicans, C. neoformans*, and *S. cerevisiae* (Mandala et al., 1997). Rustmicin is a macrolide antifungal agent that was isolated from fermentations of *Micromonospora chalcea* and named rustmicin for its activity against the wheat stem rust fungus (*Puccinia graminis*) (Takatsu et al., 1985). Rustmicin has potent fungicidal activity against clinically important human pathogens and is especially potent against *C. neoformans*, in which it inhibits growth and sphingolipid synthesis at concentrations of <1 ng/ml and inhibits the enzyme with an IC_{50} of 70 pM (Mandala et al., 1998).

GUA1

GUA1 encodes guanosine 5′-monophosphate synthase (GMPS), which catalyzes the amination of xanthine monophosphate to guanine monophosphate in the guanine branch of the purine biosynthesis pathway (Gardner and Woods, 1979; Dujardin et al., 1994). In *S. cerevisiae, GUA1* is a conditional essential gene whose function is not required in the presence of guanine, provided the cell is competent to convert extracellular guanine into GMP (Dujardin et al., 1994). Similarly, deletion of *GUA1* in *C. albicans* and *A. fumigatus* also led to guanine auxotrophy (Rodriguez-Suarez et al., 2007). However, it was demonstrated that GMPS activity was essential for the pathogenicity of *C. albicans* and *A. fumigatus*, suggesting that GUA1 could be explored as an antifungal drug target (Rodriguez-Suarez et al., 2007). Indeed, a GMPS inhibitor, designated ECC1385, was described which displayed in vitro activity against both *C. albicans* and *A. fumigatus* (Rodriguez-Suarez et al., 2007). Like *GUA1*, a few other genes required for prototrophy of intermediary metabolism function have also been implicated in microbial pathogenesis. For example, *A. fumigatus* mutants defective in orotidine 5′-phosphate decarboxylase (*pyrG*) and *p*-amino-benzoic acid biosynthesis (*pabaA*), as well as lysine biosynthesis (*lysF*), are avirulent in a murine model of IA (d'Enfert et al., 1996; Brown et al., 2000; Liebmann et al., 2004). Similarly, *C. albicans* mutants defective in adenine (*ade2* or *ade6*) or heme (*hem3*) biosynthesis, as well as *ura3* and *ura7* mutants, were also shown to be avirulent in a candidiasis model of infection (Kirsch and Whitney, 1991; Rodriguez-Suarez et al., 2007). Thus, these results suggest that conditional essential genes (i.e., *GUA1*) could be explored as potential antifungal drug targets.

CONCLUDING REMARKS

In this chapter we reviewed recent advances in identifying *A. fumigatus* essential genes which not only can serve as potential antifungal targets but also have begun to shed light on critical biological processes required for fungal growth. Currently, identification of *A. fumigatus* essential genes largely depends on the following four approaches: conventional gene deletion and disruption, parasexual genetics, RNAi knockdown, and conditional promoter replacement strategies. Each of these strategies has its own advantages and disadvantages which should be carefully considered, depending on the gene under investigation and the scale of the molecular genetics analysis undertaken by the experimenter. Completion of the *A. fumigatus* genome sequence, however, combined with current molecular genetic strategies and their inevitable refinements, has now made large-scale genetic analysis of *A. fumigatus* possible for the first time, thus expanding our knowledge of its biology, pathogenesis, and potential antifungal targets.

REFERENCES

Agrawal, N., P. V. Dasaradhi, A. Mohmmed, P. Malhotra, R. K. Bhatnagar, and S. K. Mukherjee. 2003. RNA interference: biology, mechanism, and applications. *Microbiol. Mol. Biol. Rev.* **67:**657–685.

Ahting, U., M. Thieffry, H. Engelhardt, R. Hegerl, W. Neupert, and S. Nussberger. 2001. Tom40, the pore-forming component of the protein-conducting TOM channel in the outer membrane of mitochondria. *J. Cell Biol.* **153:**1151–1160.

Akins, R. A. 2005. An update on antifungal targets and mechanisms of resistance in *Candida albicans*. *Med. Mycol.* **43:**285–318.

Aleksenko, A. Y., and A. J. Clutterbuck. 1995. Recombinational stability of replicating plasmids in *Aspergillus nidulans* during transformation, vegetative growth and sexual reproduction. *Curr. Genet.* **28:**87–93.

Aleksenko, A., I. Nikolaev, Y. Vinetski, and A. J. Clutterbuck. 1996. Gene expression from replicating plasmids in *Aspergillus nidulans*. *Mol. Gen. Genet.* **253**:242–246.

Amaar, Y. G., and M. M. Moore. 1998. Mapping of the nitrate-assimilation gene cluster (*crnA-niiA-niaD*) and characterization of the nitrite reductase gene (*niiA*) in the opportunistic fungal pathogen *Aspergillus fumigatus*. *Curr. Genet.* **33**:206–215.

Baker, K. P., A. Schaniel, D. Vestweber, and G. Schatz. 1990. A yeast mitochondrial outer membrane protein essential for protein import and cell viability. *Nature* **348**:605–609.

Baulcombe, D. 2001. RNA silencing. Diced defence. *Nature* **409**:295–296.

Beauvais, A., D. Maubon, S. Park, W. Morelle, M. Tanguy, M. Huerre, D. S. Perlin, and J. P. Latgé. 2005. Two α(1-3) glucan synthases with different functions in *Aspergillus fumigatus*. *Appl. Environ. Microbiol.* **71**:1531–1538.

Bhabhra, R., M. D. Miley, E. Mylonakis, D. Boettner, J. Fortwendel, J. C. Panepinto, M. Postow, J. C. Rhodes, and D. S. Askew. 2004. Disruption of the *Aspergillus fumigatus* gene encoding nucleolar protein *CgrA* impairs thermotolerant growth and reduces virulence. *Infect. Immun.* **72**:4731–4740.

Bok, J. W., D. Chung, S. A. Balajee, K. A. Marr, D. Andes, K. F. Nielsen, J. C. Frisvad, K. A. Kirby, and N. P. Keller. 2006. *GliZ*, a transcriptional regulator of gliotoxin biosynthesis, contributes to *Aspergillus fumigatus* virulence. *Infect. Immun.* **74**:6761–6768.

Brakhage, A. A. 2005. Systemic fungal infections caused by *Aspergillus* species: epidemiology, infection process and virulence determinants. *Curr. Drug Targets* **6**:875–886.

Brakhage, A. A., and K. Langfelder. 2002. Menacing mold: the molecular biology of *Aspergillus fumigatus*. *Annu. Rev. Microbiol.* **56**:433–455.

Bromley, M., C. Gordon, N. Rovira-Graells, and J. Oliver. 2006. The *Aspergillus fumigatus* cellobiohydrolase B (*cbhB*) promoter is tightly regulated and can be exploited for controlled protein expression and RNAi. *FEMS Microbiol. Lett.* **264**:246–254.

Brookman, J. L., and D. W. Denning. 2000. Molecular genetics in *Aspergillus fumigatus*. *Curr. Opin. Microbiol.* **3**:468–474.

Brown, J. S., A. Aufauvre-Brown, J. Brown, J. M. Jennings, H. Arst, Jr., and D. W. Holden. 2000. Signature-tagged and directed mutagenesis identify PABA synthetase as essential for *Aspergillus fumigatus* pathogenicity. *Mol. Microbiol.* **36**:1371–1380.

Bruno, V. M., S. Kalachikov, R. Subaran, C. J. Nobile, C. Kyratsous, and A. P. Mitchell. 2006. Control of the *C. albicans* cell wall damage response by transcriptional regulator Cas5. *PLoS Pathog.* **2**:e21.

Cairns, B. R., Y. Lorch, Y. Li, M. Zhang, L. Lacomis, H. Erdjument-Bromage, P. Tempst, J. Du, B. Laurent, and R. D. Kornberg. 1996. RSC, an essential, abundant chromatin-remodeling complex. *Cell* **87**:1249–1260.

Caplen, N. J., S. Parrish, F. Imani, A. Fire, and R. A. Morgan. 2001. Specific inhibition of gene expression by small double-stranded RNAs in invertebrate and vertebrate systems. *Proc. Natl. Acad. Sci. USA* **98**:9742–9747.

Charbonneau, C., I. Fournier, S. Dufresne, J. Barwicz, and P. Tancrède. 2001. The interactions of amphotericin B with various sterols in relation to its possible use in anticancer therapy. *Biophys. Chem.* **91**:125–133.

Chittur, S., and R. Griffith. 2002. Multisubstrate analogue inhibitors of glucosamine-6-phosphate synthase from *Candida albicans*. *Bioorg. Med. Chem. Lett.* **12**:2639–2642.

Chmara, H., S. Milewski, R. Andruszkiewicz, F. Mignini, and E. Borowski. 1998. Antibacterial action of dipeptides containing an inhibitor of glucosamine-6-phosphate isomerase. *Microbiology* **144**:1349–1358.

Clutterbuck, A. J. 1992. Sexual and parasexual genetics of *Aspergillus* species. *Biotechnology* **23**:3–18.

Colot, H. V., G. Park, G. E. Turner, C. Ringelberg, C. M. Crew, L. Litvinkova, R. L. Weiss, K. A. Borkovich, and J. C. Dunlap. 2006. A high-throughput gene knockout procedure for *Neurospora* reveals functions for multiple transcription factors. *Proc. Natl. Acad. Sci. USA* **103**:10352–10357.

Costa, S., and M. Nucci. 2001. Can we decrease amphotericin nephrotoxicity? *Curr. Opin. Crit. Care* **7**:379–383.

Cramer, R. A., Jr., M. P. Gamcsik, R. M. Brooking, L. K. Najvar, W. R. Kirkpatrick, T. F. Patterson, C. J. Balibar, J. R. Graybill, J. R. Perfect, S. N. Abraham, and W. J. Steinbach. 2006. Disruption of a nonribosomal peptide synthetase in *Aspergillus fumigatus* eliminates gliotoxin production. *Eukaryot. Cell* **5**:972–980.

Damelin, M., I. Simon, T. I. Moy, B. Wilson, S. Komili, P. Tempst, F. P. Roth, R. A. Young, B. R. Cairns, and P. A. Silver. 2002. The genome-wide localization of Rsc9, a component of the RSC chromatin-remodeling complex, changes in response to stress. *Mol. Cell* **9**:563–573.

da Silva Ferreira, M. E., M. R. Kress, M. Savoldi, M. H. Goldman, A. Härtl, T. Heinekamp, A. A. Brakhage, and G. H. Goldman. 2006. The *akuB*^{KU80} mutant deficient for nonhomologous end joining is a powerful tool for analyzing pathogenicity in *Aspergillus fumigatus*. *Eukaryot. Cell* **5**:207–211.

Davydenko, S. G., J. K. Juselius, T. Munder, E. Bogengruber, J. Jäntti, and S. Keränen. 2004. Screening for novel essential genes of *Saccharomyces cerevisiae* involved in protein secretion. *Yeast* **21**:463–471.

De Backer, M. D., B. Nelissen, M. Logghe, J. Viaene, I. Loonen, S. Vandoninck, R. de Hoogt, S. Dewaele, F. A. Simons, P. Verhasselt, G. Vanhoof, R. Contreras, and W. H. Luyten. 2001. An antisense-based functional genomics approach for identification of genes critical for growth of *Candida albicans*. *Nat. Biotechnol.* **19**:235–241.

d'Enfert, C. 1996. Selection of multiple disruption events in *Aspergillus fumigatus* using the orotidine-5′-decarboxylase gene, *pyrG*, as a unique transformation marker. *Curr. Genet.* **30**:76–82.

d'Enfert, C., M. Diaquin, A. Delit, N. Wuscher, J. P. Debeaupuis, M. Huerre, and J. P. Latgé. 1996. Attenuated virulence of uridine-uracil auxotrophs of *Aspergillus fumigatus*. *Infect. Immun.* **64**:4401–4405.

Denning, D. W. 1998. Invasive aspergillosis. *Clin. Infect. Dis.* **26**:781–803.

Dickson, R. C., and R. L. Lester. 1999. Metabolism and selected functions of sphingolipids in the yeast *Saccharomyces cerevisiae*. *Biochim. Biophys. Acta* **1438**:305–321.

Dujardin, G., M. Kermorgant, P. P. Slonimski, and H. Boucherie. 1994. Cloning and sequencing of the GMP synthetase-encoding gene of *Saccharomyces cerevisiae*. *Gene* **139**:127–132.

Felenbok, B., M. Flipphi, and I. Nikolaev. 2001. Ethanol catabolism in *Aspergillus nidulans*: a model system for studying gene regulation. *Prog. Nucleic Acid Res. Mol. Biol.* **69**:149–204.

Fire, A. Z. 2007. Gene silencing by double-stranded RNA. *Cell Death Differ.* **14**:1998–2012.

Firon, A., and C. d'Enfert. 2002. Identifying essential genes in fungal pathogens of humans. *Trends Microbiol.* **10**:456–462.

Firon, A., F. Villalba, R. Beffa, and C. D'Enfert. 2003. Identification of essential genes in the human fungal pathogen *Aspergillus fumigatus* by transposon mutagenesis. *Eukaryot. Cell* **2**:247–255.

Fortwendel, J. R., W. Zhao, R. Bhabhra, S. Park, D. S. Perlin, D. S. Askew, and J. C. Rhodes. 2005. A fungus-specific Ras homolog contributes to the hyphal growth and virulence of *Aspergillus fumigatus*. *Eukaryot. Cell* **4**:1982–1989.

Fournier, I., J. Barwicz, and P. Tancrède. 1998. The structuring effects of amphotericin B on pure and ergosterol- or cholesterol-containing dipalmitoylphosphatidylcholine bilayers: a differential scanning calorimetry study. *Biochim. Biophys. Acta* **1373**:76–86.

Gabriel, I., J. Olchowy, A. Stanislawska-Sachadyn, T. Mio, J. Kur, and S. Milewski. 2004. Phosphorylation of glucosamine-6-phosphate synthase is important but not essential for germination and mycelial growth of *Candida albicans*. *FEMS Microbiol. Lett.* 235:73–80.

Gal-Mor, O., and B. B. Finlay. 2006. Pathogenicity islands: a molecular toolbox for bacterial virulence. *Cell. Microbiol.* **8**:1707–1719.

Gardner, W. J., and R. A. Woods. 1979. Isolation and characterisation of guanine auxotrophs in *Saccharomyces cerevisiae*. *Can. J. Microbiol.* **25**:380–389.

Gerbaud, E., F. Tamion, C. Girault, K. Clabault, S. Lepretre, J. Leroy, and G. Bonmarchand. 2003. Persistent acute tubular toxicity after switch from conventional amphotericin B to liposomal amphotericin B (Ambisome). *J. Antimicrob. Chemother.* **51**:473–475.

Giaever, G., A. M. Chu, L. Ni, C. Connelly, L. Riles, S. Véronneau, S. Dow, A. Lucau-Danila, K. Anderson, B. André, A. P. Arkin, A. Astromoff, M. El-Bakkoury, R. Bangham, R. Benito, S. Brachat, S. Campanaro, M. Curtiss, K. Davis, A. Deutschbauer, K. D. Entian, P. Flaherty, F. Foury, D. J. Garfinkel, M. Gerstein, D. Gotte, U. Güldener, J. H. Hegemann, S. Hempel, Z. Herman, D. F. Jaramillo, D. E. Kelly, S. L. Kelly, P. Kötter, D. LaBonte, D. C. Lamb, N. Lan, H. Liang, H. Liao, L. Liu, C. Luo, M. Lussier, R. Mao, P. Menard, S. L. Ooi, J. L. Revuelta, C. J. Roberts, M. Rose, P. Ross-Macdonald, B. Scherens, G. Schimmack, B. Shafer, D. D. Shoemaker, S. Sookhai-Mahadeo, R. K. Storms, J. N. Strathern, G. Valle, M. Voet, G. Volckaert, C. Y. Wang, T. R. Ward, J. Wilhelmy, E. A. Winzeler, Y. Yang, G. Yen, E. Youngman, K. Yu, H. Bussey, J. D. Boeke, M. Snyder, P. Philippsen, R. W. Davis, and M. Johnston. 2002. Functional profiling of the *Saccharomyces cerevisiae* genome. *Nature* **418**:387–391.

Goffeau, A., B. G. Barrell, H. Bussey, R. W. Davis, B. Dujon, H. Feldmann, F. Galibert, J. D. Hoheisel, C. Jacq, M. Johnston, E. J. Louis, H. W. Mewes, Y. Murakami, P. Philippsen, H. Tettelin, and S. G. Oliver. 1996. Life with 6000 genes. *Science* **274**:563–567

Groll, A. H., and T. J. Walsh. 2001. Caspofungin: pharmacology, safety and therapeutic potential in superficial and invasive fungal infections. *Expert Opin. Investig. Drugs* **10**:1545–1558.

Haselbeck, R., D. Wall, B. Jiang, T. Ketela, J. Zyskind, H. Bussey, J. G. Foulkes, and T. Roemer. 2002. Comprehensive essential gene identification as a platform for novel anti-infective drug discovery. *Curr. Pharm. Des.* **8**:1155–1172.

Hashida-Okado, T., A. Ogawa, M. Endo, R. Yasumoto, K. Takesako, and I. Kato. 1996. *AUR1*, a novel gene conferring aureobasidin resistance on *Saccharomyces cerevisiae*: a study of defective morphologies in Aur1p-depleted cells. *Mol. Gen. Genet.* **251**:236–244.

Heidler, S. A., and J. A. Radding. 1995. The *AUR1* gene in *Saccharomyces cerevisiae* encodes dominant resistance to the antifungal agent aureobasidin A (LY295337). *Antimicrob. Agents Chemother.* **39**:2765–2769.

Heidler, S. A., and J. A. Radding. 2000. Inositol phosphoryl transferases from human pathogenic fungi. *Biochim. Biophys. Acta* **1500**: 147–152.

Heidtman, M., C. Z. Chen, R. N. Collins, and C. Barlowe. 2003. A role for Yip1p in COPII vesicle biogenesis. *J. Cell Biol.* **163**:57–69.

Henry, C., I. Mouyna, and J. P. Latgé. 2007. Testing the efficacy of RNA interference constructs in *Aspergillus fumigatus*. *Curr. Genet.* **51**:277–284.

Hirt, R. P., S. Müller, T. M. Embley, and G. H. Coombs. 2002. The diversity and evolution of thioredoxin reductase: new perspectives. *Trends Parasitol.* **18**:302–308.

Hope, W. W., S. Shoham, and T. J. Walsh. 2007. The pharmacology and clinical use of caspofungin. *Expert Opin. Drug Metab. Toxicol.* **3**:263–274.

Hu, W., S. Sillaots, S. Lemieux, J. Davison, S. Kauffman, A. Breton, A. Linteau, C. Xin, J. Bowman, J. Becker, B. Jiang, and T. Roemer. 2007. Essential gene identification and drug target prioritization in *Aspergillus fumigatus*. *PLoS Pathog.* **3**:e24.

Jiang, B., H. Bussey, and T. Roemer. 2002. Novel strategies in antifungal lead discovery. *Curr. Opin. Microbiol.* **5**:466–471.

Jones, T., N. A. Federspiel, H. Chibana, J. Dungan, S. Kalman, B. B. Magee, G. Newport, Y. R. Thorstenson, N. Agabian, P. T. Magee, R. W. Davis, and S. Scherer. 2004. The diploid genome sequence of *Candida albicans*. *Proc. Natl. Acad. Sci. USA* **101**:7329–7334.

Kalb, V. F., C. W. Woods, T. G. Turi, C. R. Dey, T. R. Sutter, and J. C. Loper. 1987. Primary structure of the P450 lanosterol demethylase gene from *Saccharomyces cerevisiae*. *DNA* **6**:529–557.

Kauffman, C. A. 2006. Clinical efficacy of new antifungal agents. *Curr. Opin. Microbiol.* **9**:483–488.

Khalaj, V., H. Eslami, M. Azizi, N. Rovira-Graells, and M. Bromley. 2007. Efficient downregulation of *alb1* gene using an *AMA1*-based episomal expression of RNAi construct in *Aspergillus fumigatus*. *FEMS Microbiol. Lett.* **270**:250–254.

Kirsch, D. R., and R. R. Whitney. 1991. Pathogenicity of *Candida albicans* auxotrophic mutants in experimental infections. *Infect. Immun.* **59**:3297–3300.

Krappmann, S., C. Sasse, and G. H. Braus. 2006. Gene targeting in *Aspergillus fumigatus* by homologous recombination is facilitated in a nonhomologous end-joining-deficient genetic background. *Eukaryot. Cell* **5**:212–215.

Kupfahl, C., T. Heinekamp, G. Geginat, T. Ruppert, A. Härtl, H. Hof, and A. A. Brakhage. 2006. Deletion of the *gliP* gene of *Aspergillus fumigatus* results in loss of gliotoxin production but has no effect on virulence of the fungus in a low-dose mouse infection model. *Mol. Microbiol.* **62**:292–302.

Lagorce, A., V. Le Berre-Anton, B. Aguilar-Uscanga, H. Martin-Yken, A. Dagkessamanskaia, and J. François. 2002. Involvement of *GFA1*, which encodes glutamine-fructose-6-phosphate amidotransferase, in the activation of the chitin synthesis pathway in response to cell-wall defects in *Saccharomyces cerevisiae*. *Eur. J. Biochem.* **269**: 1697–1707.

Lamarre, C., O. Ibrahim-Granet, C. Du, R. Calderone, and J. P. Latgé. 2007. Characterization of the *SKN7* ortholog of *Aspergillus fumigatus*. *Fungal Genet. Biol.* **44**:682–690.

Latgé, J. P. 1999. *Aspergillus fumigatus* and aspergillosis. *Clin. Microbiol. Rev.* **12**:310–350.

Latgé, J. P. 2001. The pathobiology of *Aspergillus fumigatus*. *Trends Microbiol.* **9**:382–389.

Liebmann, B., T. W. Mühleisen, M. Müller, M. Hecht, G. Weidner, A. Braun, M. Brock, and A. A. Brakhage. 2004. Deletion of the *Aspergillus fumigatus* lysine biosynthesis gene *lysF* encoding homoaconitase leads to attenuated virulence in a low-dose mouse infection model of invasive aspergillosis. *Arch. Microbiol.* **181**:378–383.

Liu, H., T. R. Cottrell, L. M. Pierini, W. E. Goldman, and T. L. Doering. 2002. RNA interference in the pathogenic fungus *Cryptococcus neoformans*. *Genetics* **160**:463–470.

Lupetti, A., R. Danesi, M. Campa, M. Del Tacca, and S. Kelly. 2002. Molecular basis of resistance to azole antifungals. *Trends Mol. Med.* **8**:76–81.

Mandala, S. M., R. A. Thornton, M. Rosenbach, J. Milligan, M. Garcia-Calvo, H. G. Bull, and M. B. Kurtz. 1997. Khafrefungin, a novel inhibitor of sphingolipid synthesis. *J. Biol. Chem.* **272**:32709–32714.

Mandala, S. M., R. A. Thornton, J. Milligan, M. Rosenbach, M. Garcia-Calvo, H. G. Bull, G. Harris, G. K. Abruzzo, A. M. Flattery, C. J. Gill, K. Bartizal, S. Dreikorn, and M. B. Kurtz. 1998. Rustmicin, a potent antifungal agent, inhibits sphingolipid synthesis at inositol phosphoceramide synthase. *J. Biol. Chem.* **273**:14942–14949.

Mann, P. A., R. M. Parmegiani, S. Q. Wei, C. A. Mendrick, X. Li, D. Loebenberg, B. DiDomenico, R. S. Hare, S. S. Walker, and P. M. McNicholas. 2003. Mutations in *Aspergillus fumigatus* resulting in reduced susceptibility to posaconazole appear to be restricted to a single amino acid in the cytochrome P450 14α-demethylase. *Antimicrob. Agents Chemother.* **47**:577–581.

Marr, K. A., R. A. Carter, F. Crippa, A. Wald, and L. Corey. 2002. Epidemiology and outcome of mould infections in hematopoietic stem cell transplant recipients. *Clin. Infect. Dis.* **34:**909–917.

McNeil, M. M., S. L. Nash, R. A. Hajjeh, M. A. Phelan, L. A. Conn, B. D. Plikaytis, and D. W. Warnock. 2001.Trends in mortality due to invasive mycotic diseases in the United States, 1980-1997. *Clin. Infect. Dis.* **33:**641–647.

Mellado, E., T. M. Diaz-Guerra, M. Cuenca-Estrella, and J. L. Rodriguez-Tudela. 2001. Identification of two different 14-α sterol demethylase-related genes (*cyp51A* and *cyp51B*) in *Aspergillus fumigatus* and other *Aspergillus* species. *J. Clin. Microbiol.* **39:**2431–2438.

Missall, T. A., and J. K. Lodge. 2005. Thioredoxin reductase is essential for viability in the fungal pathogen *Cryptococcus neoformans*. *Eukaryot. Cell* **4:**487–489.

Mnaimneh, S., A. P. Davierwala, J. Haynes, J. Moffat, W. T. Peng, W. Zhang, X. Yang, J. Pootoolal, G. Chua, A. Lopez, M. Trochesset, D. Morse, N. J. Krogan, S. L. Hiley, Z. Li, Q. Morris, J. Grigull, N. Mitsakakis, C. J. Roberts, J, F. Greenblatt, C. Boone, C. A. Kaiser, B. J. Andrews, and T. R. Hughes. 2004. Exploration of essential gene functions via titratable promoter alleles. *Cell* **118:**31–44.

Model, K., T. Prinz, T. Ruiz, M. Radermacher, T. Krimmer, W. Kühlbrandt, N. Pfanner, and C. Meisinger. 2002. Protein translocase of the outer mitochondrial membrane: role of import receptors in the structural organization of the TOM complex. *J. Mol. Biol.* **316:**657–666.

Mouyna, I., C. Henry, T. L. Doering, and J. P. Latgé. 2004. Gene silencing with RNA interference in the human pathogenic fungus *Aspergillus fumigatus*. *FEMS Microbiol. Lett.* **237:**317–324.

Muro-Pastor, M. I., R. Gonzalez, J. Strauss, F. Narendja, and C. Scazzocchio. 1999. The GATA factor *AreA* is essential for chromatin remodelling in a eukaryotic bidirectional promoter. *EMBO J.* **18:** 1584–1597.

Nagiec, M. M., E. E. Nagiec, J. A. Baltisberger, G. B. Wells, R. L. Lester, and R. C. Dickson. 1997. Sphingolipid synthesis as a target for antifungal drugs. Complementation of the inositol phosphorylceramide synthase defect in a mutant strain of *Saccharomyces cerevisiae* by the *AUR1* gene. *J. Biol. Chem.* **272:**9809–9817.

Nargang, F. E., K. P. Künkele, A. Mayer, R. G. Ritzel, W. Neupert, and R. Lill. 1995. 'Sheltered disruption' of *Neurospora crassa MOM22*, an essential component of the mitochondrial protein import complex. *EMBO J.* **14:**1099–1108.

Nierman, W. C., A. Pain, M. J. Anderson, J. R. Wortman, H. S. Kim, J. Arroyo, M. Berriman, K. Abe, D. B. Archer, C. Bermejo, J. Bennett, P. Bowyer, D. Chen, M. Collins, R. Coulsen, R. Davies, P. S. Dyer, M. Farman, N. Fedorova, N. Fedorova, T. V. Feldblyum, R. Fischer, N. Fosker, A. Fraser, J. L. García, M. J. García, A. Goble, G. H. Goldman, K. Gomi, S. Griffith-Jones, R. Gwilliam, B. Haas, H. Haas, D. Harris, H. Horiuchi, J. Huang, S. Humphray, J. Jiménez, N. Keller, H. Khouri, K. Kitamoto, T. Kobayashi, S. Konzack, R. Kulkarni, T. Kumagai, A. Lafon, J. P. Latgé, W. Li, A. Lord, C. Lu, W. H. Majoros, G. S. May, B. L. Miller, Y. Mohamoud, M. Molina, M. Monod, I. Mouyna, S. Mulligan, L. Murphy, S. O'Neil, I. Paulsen, M. A. Peñalva, M. Pertea, C. Price, B. L. Pritchard, M. A. Quail, E. Rabbinowitsch, N. Rawlins, M. A. Rajandream, U. Reichard, H. Renauld, G. D. Robson, S. Rodriguez de Córdoba, J. M. Rodríguez-Peña, C. M. Ronning, S. Rutter, S. L. Salzberg, M. Sanchez, J. C. Sánchez-Ferrero, D. Saunders, K. Seeger, R. Squares, S. Squares, M. Takeuchi, F. Tekaia, G. Turner, C. R. Vazquez de Aldana, J. Weidman, O. White, J. Woodward, J. H. Yu, V. Fraser, J. E. Galagan, K. Asai, M. Machida, N. Hall, B. Barrell, and D. W. Denning. 2005. Genomic sequence of the pathogenic and allergenic filamentous fungus *Aspergillus fumigatus*. *Nature* **438:**1151–1156.

Ninomiya, Y., K. Suzuki, C. Ishii, and H. Inoue. 2004. Highly efficient gene replacements in *Neurospora* strains deficient for nonhomologous end-joining. *Proc. Natl. Acad. Sci. USA* **101:**12248–12253.

Odds, F. C. 2005. Genomics, molecular targets and the discovery of antifungal drugs. *Rev. Iberoam. Micol.* **22:**229–237.

Oliveira, M. A., K. F. Discola, S. V. Alves, J. A. Barbosa, F. J. Medrano, L. E. Netto, and N. G. Guimarães. 2005. Crystallization and preliminary X-ray diffraction analysis of NADPH-dependent thioredoxin reductase I from *Saccharomyces cerevisiae*. *Acta Crystallogr. F* **61:**387–390.

Osmani, A. H., J. Davies, H. L. Liu, A. Nile, and S. A. Osmani. 2004. Systematic deletion and mitotic localization of the nuclear pore complex proteins of *Aspergillus nidulans*. *Mol. Biol. Cell* **17:**4946–4961.

Osmani, A. H., B. R. Oakley, and S. A. Osmani. 2006. Identification and analysis of essential *Aspergillus nidulans* genes using the heterokaryon rescue technique. *Nat. Protoc.* **1:**2517–2526.

Osmani, S. A., D. B. Engle, J. H. Doonan, and N. R. Morris. 1988. Spindle formation and chromatin condensation in cells blocked at interphase by mutation of a negative cell cycle control gene. *Cell* **52:**241–251.

Pathak, A., F. D. Pien, and L. Carvalho. 1998. Amphotericin B use in a community hospital, with special emphasis on side effects. *Clin. Infect. Dis.* **26:**334–338.

Perfect, J. R. 1996. Fungal virulence genes as targets for antifungal chemotherapy. *Antimicrob. Agents Chemother.* **40:**1577–1583.

Perlin, D. S. 2007. Resistance to echinocandin-class antifungal drugs. *Drug Resist. Update* **10:**121–130.

Pontecorvo, G., J. A. Roper, and E. Forbes. 1953. Genetic recombination without sexual reproduction in *Aspergillus niger*. *J. Gen. Microbiol.* **8:**198–210.

Punt, P. J., J. Strauss, R. Smit, J. R. Kinghorn, C. A. van den Hondel, and C. Scazzocchio. 1995. The intergenic region between the divergently transcribed *niiA* and *niaD* genes of *Aspergillus nidulans* contains multiple *NirA* binding sites which act bidirectionally. *Mol. Cell. Biol.* **15:**5688–5699.

Rapaport, D. 2005. How does the TOM complex mediate insertion of precursor proteins into the mitochondrial outer membrane? *J. Cell Biol.* **17:**419–423.

Rementeria, A., N. López-Molina, A. Ludwig, A. B. Vivanco, J. Bikandi, J. Pontón, and J. Garaizar. 2005. Genes and molecules involved in *Aspergillus fumigatus* virulence. *Rev. Iberoam. Micol.* **22:** 1–23.

Rodriguez-Suarez, R., D. Xu, K. Veillette, J. Davison, S. Sillaots, S. Kauffman, W. Hu, J. Bowman, N. Martel, S. Trosok, H. Wang, L. Zhang, L. Y. Huang, Y. Li, F. Rahkhoodaee, T. Ransom, D. Gauvin, C. Douglas, P. Youngman, J. Becker, B. Jiang, and T. Roemer. 2007. Mechanism-of-action determination of GMP synthase inhibitors and target validation in *Candida albicans* and *Aspergillus fumigatus*. *Chem. Biol.* **14:**1163–1175.

Roemer, T., B. Jiang, J. Davison, T. Ketela, K. Veillette, A. Breton, F. Tandia, A. Linteau, S. Sillaots, C. Marta, N. Martel, S. Veronneau, S. Lemieux, S. Kauffman, J. Becker, R. Storms, C. Boone, and H. Bussey. 2003. Large-scale essential gene identification in *Candida albicans* and applications to antifungal drug discovery. *Mol. Microbiol.* **50:**167–181.

Romano, J., G. Nimrod, N. Ben-Tal, Y. Shadkchan, K. Baruch, H. Sharon, and N. Osherov. 2006. Disruption of the *Aspergillus fumigatus ECM33* homologue results in rapid conidial germination, antifungal resistance and hypervirulence. *Microbiology* **152:**1919–1928.

Romero, B., G. Turner, I. Olivas, F. Laborda, and J. R. De Lucas. 2003.The *Aspergillus nidulans alcA* promoter drives tightly regulated conditional gene expression in *Aspergillus fumigatus* permitting validation of essential genes in this human pathogen. *Fungal Genet. Biol.* **40:**103–114.

Ronning, C. M., N. D. Fedorova, P. Bowyer, R. Coulson, G. Goldman, H. S. Kim, G. Turner, J. R. Wortman, J. Yu, M. J. Anderson, D. W. Denning, and W. C. Nierman. 2005. Genomics of *Aspergillus fumigatus. Rev. Iberoam. Micol.* **22:**223–228.

Ross-Macdonald, P., P. S. Coelho, T. Roemer, S. Agarwal, A. Kumar, R. Jansen, K. H. Cheung, A. Sheehan, D. Symoniatis, L. Umansky, M. Heidtman, F. K. Nelson, H. Iwasaki, K. Hager, M. Gerstein, P. Miller, G. S. Roeder, and M. Snyder. 1999. Large-scale analysis of the yeast genome by transposon tagging and gene disruption. *Nature* **402:**413–418.

Salama, N. R., J. S. Chuang, and R. W. Schekman. 1997. *Sec31* encodes an essential component of the COPII coat required for transport vesicle budding from the endoplasmic reticulum. *Mol. Biol. Cell* **8:**205–217.

Schrettl, M., E. Bignell, C. Kragl, C. Joechl, T. Rogers, H. N. Arst, Jr., K. Haynes, and H. Haas. 2004. Siderophore biosynthesis but not reductive iron assimilation is essential for *Aspergillus fumigatus* virulence. *J. Exp. Med.* **200:**1213–1219.

Sheppard, D. C., T. Doedt, L. Y. Chiang, H. S. Kim, D. Chen, W. C. Nierman, and S. G. Filler. 2005. The *Aspergillus fumigatus StuA* protein governs the up-regulation of a discrete transcriptional program during the acquisition of developmental competence. *Mol. Biol. Cell* **16:**5866–5879.

Sherman, E. L., R. D. Taylor, N. E. Go, and F. E. Nargang. 2006. Effect of mutations in Tom40 on stability of the translocase of the outer mitochondrial membrane (TOM) complex, assembly of Tom40, and import of mitochondrial preproteins. *J. Biol. Chem.* **281:**22554–22565.

Smith, R. J., S. Milewski, A. J. Brown, and G. W. Gooday. 1996. Isolation and characterization of the *GFA1* gene encoding the glutamine: fructose-6-phosphate amidotransferase of *Candida albicans. J. Bacteriol.* **178:**2320–2327.

Smith, V., K. N. Chou, D. Lashkari, D. Botstein, and P. O. Brown. 1996. Functional analysis of the genes of yeast chromosome V by genetic footprinting. *Science* **274:**2069–2074.

Som, T., and V. S. Kolaparthi. 1994. Developmental decisions in *Aspergillus nidulans* are modulated by *Ras* activity. *Mol. Cell. Biol.* **14:** 5333–5348.

South, S. T., K. A. Sacksteder, X. Li, Y. Liu, and S. J. Gould. 2000. Inhibitors of COPI and COPII do not block PEX3-mediated peroxisome synthesis. *J. Cell Biol.* **149:**1345–1360.

Spanakis, E. K., G. Aperis, and E. Mylonakis. 2006. New agents for the treatment of fungal infections: clinical efficacy and gaps in coverage. *Clin. Infect. Dis.* **43:**1060–1068.

Steinbach, W. J., R. A. Cramer, Jr., B. Z. Perfect, Y. G. Asfaw, T. C. Sauer, L. K. Najvar, W. R. Kirkpatrick, T. F. Patterson, D. K. Benjamin, Jr., J. Heitman, and J. R. Perfect. 2006. Calcineurin controls growth, morphology, and pathogenicity in *Aspergillus fumigatus. Eukaryot. Cell* **5:**1091–1103.

Stromnaes, O., and E. D. Garber. 1963. Heterocaryosis and the parasexual cycle in *Aspergillus fumigatus. Genetics* **48:**653–662.

Sugimoto, Y., H. Sakoh, and K. Yamada. 2004. IPC synthase as a useful target for antifungal drugs. *Curr. Drug Targets Infect. Disord.* **4:**311–322.

Takatsu, T., H. Nakayama, A. Shimazu, K. Furihata, K. Ikeda, K. Furihata, H. Seto, and N. Otake. 1985. Rustmicin, a new macrolide antibiotic active against wheat stem rust fungus. *J. Antibiot.* (Tokyo) **38:**1806–1809.

Takesako, K., K. Ikai, F. Haruna, M. Endo, K. Shimanaka, E. Sono, T. Nakamura, I. Kato, and H. Yamaguchi. 1991. Aureobasidins, new antifungal antibiotics. Taxonomy, fermentation, isolation, and properties. *J. Antibiot.* (Tokyo) **44:**919–924.

Takesako, K., H. Kuroda, T. Inoue, F. Haruna, Y. Yoshikawa, I. Kato, K. Uchida, T. Hiratani, and H. Yamaguchi. 1993. Biological properties of aureobasidin A, a cyclic depsipeptide antifungal antibiotic. *J. Antibiot.* (Tokyo) **46:**1414–1420.

Taylor, R. D., B. J. McHale, and F. E. Nargang. 2003. Characterization of *Neurospora crassa* Tom40-deficient mutants and effect of specific mutations on Tom40 assembly. *J. Biol. Chem.* **278:**765–775.

Tekaia, F., and J. P. Latgé. 2005. *Aspergillus fumigatus*: saprophyte or pathogen? *Curr. Opin. Microbiol.* **8:**385–392.

Teplyakov, A., G. Obmolova, M. A. Badet-Denisot, and B. Badet. 1999. The mechanism of sugar phosphate isomerization by glucosamine 6-phosphate synthase. *Protein Sci.* **8:**596–602.

Timberlake, W. E., and M. A. Marshall. 1988. Genetic regulation of development in *Aspergillus nidulans. Trends Genet.* **4:**162–169.

Tsai, H. F., M. H. Wheeler, Y. C. Chang, and K. J. Kwon-Chung. 1999. A developmentally regulated gene cluster involved in conidial pigment biosynthesis in *Aspergillus fumigatus. J. Bacteriol.* **181:** 6469–6477.

Uhl, M. A., M. Biery, N. Craig, and A. D. Johnson. 2003. Haploinsufficiency-based large-scale forward genetic analysis of filamentous growth in the diploid human fungal pathogen *C. albicans. EMBO J.* **22:**2668–2678.

Vogt, K., R. Bhabhra, J. C. Rhodes, and D. S. Askew. 2005. Doxycycline-regulated gene expression in the opportunistic fungal pathogen *Aspergillus fumigatus. BMC Microbiol.* **5:**1.

Watson, J. M., A. F. Fusaro, M. Wang, and P. M. Waterhouse. 2005. RNA silencing platforms in plants. *FEBS Lett.* **579:**5982–5987.

Winzeler, E. A., D. D. Shoemaker, A. Astromoff, H. Liang, K. Anderson, B. Andre, R. Bangham, R. Benito, J. D. Boeke, H. Bussey, A. M. Chu, C. Connelly, K. Davis, F. Dietrich, S. W. Dow, M. El Bakkoury, F. Foury, S. H. Friend, E. Gentalen, G. Giaever, J. H. Hegemann, T. Jones, M. Laub, H. Liao, N. Liebundguth, D. J. Lockhart, A. Lucau-Danila, M. Lussier, N. M'Rabet, P. Menard, M. Mittmann, C. Pai, C. Rebischung, J. L. Revuelta, L. Riles, C. J. Roberts, P. Ross-MacDonald, B. Scherens, M. Snyder, S. Sookhai-Mahadeo, R. K. Storms, S. Véronneau, M. Voet, G. Volckaert, T. R. Ward, R. Wysocki, G. S. Yen, K. Yu, K. Zimmermann, P. Philippsen, M. Johnston, and R. W. Davis. 1999. Functional characterization of the *S. cerevisiae* genome by gene deletion and parallel analysis. *Science* **285:**901–906.

Yamasaki, A., K. Tani, A. Yamamoto, N. Kitamura, and M. Komada. 2006. The Ca^{2+}-binding protein ALG-2 is recruited to endoplasmic reticulum exit sites by *Sec31A* and stabilizes the localization of *Sec31A. Mol. Biol. Cell* **17:**4876–4887.

Zamore, P. D., and N. Aronin. 2003. siRNAs knock down hepatitis. *Nat. Med.* **9:**266–267.

II. GROWTH AND SENSING, OR RESISTING ENVIRONMENTAL STRESS

Aspergillus fumigatus and Aspergillosis
Edited by J.-P. Latgé and W. J. Steinbach

Chapter 6

Aspects of Primary Carbon and Nitrogen Metabolism

SVEN KRAPPMANN

Species of the ubiquitous *Ascomycetes* genus *Aspergillus* account for a considerable number of cases of fungal diseases, and from a plethora of studies it has become evident that the pathogenicity of this genus is not linked to a single cellular attribute but that it should be considered as a multifactorial trait. Among the characteristics that make *Aspergillus* a pathogen, metabolism has to be taken into account, and it is the purpose of this chapter to briefly summarize the current knowledge on the role of nutritional versatility and characteristics of primary metabolism in the pathogenicity of this fungal genus. The focus will be on routes of carbon and nitrogen metabolism, on specific aspects of their regulation, and on insights specifically gained for *Aspergillus fumigatus*, by far the most pathogenic species among aspergilli and the main causative agent of aspergillosis.

ASPERGILLUS IS A PATHOGENIC SAPROPHYTE

Aspergillus spp. conidia, the products from asexual reproduction of this fungal genus, constitute a major part of the airborne fungal flora. Their small size and surface characteristics allow them to be easily distributed by currents of air or water and make *Aspergillus* a ubiquitous inhabitant of the ecosphere. It colonizes the soil or food, but its primary ecological niche is decaying vegetation, as it has been found in compost piles (Haines, 1995). There, complex polymeric substrates have to be degraded and their breakdown products taken up into the fungal cell, and the enzymatic equipment that is expressed and secreted by the fungus gives reference to this. Unusual substrates, such as bark, aromatic compounds, melanin, or even chicken feather keratin, can support growth of *A. fumigatus* (Nordstrom, 1974; Santos et al., 1996); interestingly, inspection of the *A. fumigatus* genome reflects this saprophytic life-style that is based on a plethora of hydrolytic enzymes able to degrade, for instance, soft plant material but not hard wood (Tekaia and Latgé, 2005). Glycosylhydrolase-encoding genes have been annotated that provide the capacity to degrade cellulose, hemicellulose, or pectin and all cell wall components of leaves, but no lignin-degrading activities appear to be present.

Beyond this, *Aspergillus* species are able to colonize a very special ecological niche, the immunocompromised, warm-blooded host (Latgé, 1999, 2001). The most common route of infection is pulmonary. Eventually, systemic infection via dissemination may occur, resulting in a fatal outcome. It is evident that this type of infection is based on the opportunistic exploitation of the environment that is proffered by the infected host. Accordingly, this growth mode may mirror the saprophytic life-style of *Aspergillus* to a certain extent. As a consequence, the pathogenicity of *Aspergillus* may be mainly based on proper colonization of the immunocompromised host as a kind of an ecological niche, making fungal growth a major virulence determinant (Paisley et al., 2005).

EXPLORING NUTRITIONAL VERSATILITY OF *ASPERGILLUS*

Fungi contribute significantly to the global recycling routes that result in turnover of elements like carbon or nitrogen, and aspergilli with their saprophytic life-style inhabit a prominent role in these processes. Genomic data from several *Aspergillus* species, among them *A. fumigatus*, have not revealed the existence of true virulence factors that would cause significant host damage; in fact, they clearly mirror the routine of degrading external polymers and assimilation of nutrients (Nierman et al., 2005a, 2005b; Tekaia and Latgé, 2005). From this point of view it has become obvious that the metabolic features and characteristics of *Asper-*

Sven Krappmann • Research Center for Infectious Diseases, Julius-Maximilians-University, Würzburg, Germany.

gillus contribute to its proliferation and survival within a susceptible host (Rhodes, 2006), and studying selected pathways and biochemical routes with respect to their relevance for pathogenicity has led to several valuable conclusions.

Sequencing efforts to analyze the genomic content of various aspergilli have opened enormous opportunities to reconstitute pathways and regulatory circuits involved in primary metabolism by in silico analysis. For *Aspergillus nidulans* this has been performed to deduce nutrient-sensing signaling cascades (Muthuvijayan and Marten, 2004), and detailed inspection of the *A. fumigatus* genome in that way may uncover specific metabolic features unique to this species. Other approaches of the post-genome era are definitely appropriate to support these analyses in order to assign functional denotations to the annotated gene loci. Transcriptional profiling accompanied by proteome studies is the method of choice, as it allows multiplex monitoring of expression patterns under diverse conditions, and virulence-specific traits may be deduced from comparative surveys.

The assessment of mutant strains impaired in biosynthetic routes or regulatory networks has evolved as a valuable and convincing approach to define the metabolic needs of the fungal pathogen. In bacterial pathogens, host-specific adaptation of cellular metabolism has been extensively scrutinized and still is (Boyce et al., 2004); in fungi, however, this research field has been touched to only a limited extent. One explanation for this may lie in the laborious task of mutant generation in *Aspergillus*, and recent advances on the molecular biology of this pathogen will definitely accelerate attempts to create defined strains by means of gene targeting (Brakhage and Langfelder, 2002; Krappmann, 2006). There are several biosynthesis pathways or metabolic systems that are specific for the fungal pathogen but not the host, such as biosynthesis of essential amino acids or transport of oligopeptides (Wiles et al., 2006; Schoberle and May, 2007), and these sections of primary metabolism may represent interesting candidates for the identification of targets for antifungal therapy.

When testing *A. fumigatus* mutant strains for virulence, the model used for aspergillosis has to be considered (Zaas and Steinbach, 2007). Most commonly, animals immunocompromised by administering doses of cyclophosphamide in combination with corticosteroids are used in infection experiments. These hosts have to be considered as leukopenic, and in this environment, fungal growth in the pulmonary location becomes the prime virulence determinant. Animals treated solely with corticosteroids retain neutrophils; therefore, specific mutants might display different virulence traits under these circumstances, as it has been demonstrated for the role of gliotoxin, a fungal metabolite with numerous effects on metazoan cells (Cramer et al., 2006; Kupfahl et al., 2006; Sugui et al., 2007; Spikes et al., 2008). Accordingly, the type of model and protocol of immunosuppression have to be taken into account when evaluating the virulence of metabolic mutants. This is of relevance when considering the phagosomal compartments of macrophages or neutrophils as distinct nutritional environments: transcriptional profiling studies on *Candida albicans* have revealed that the neutrophil phagosome is likely an amino acid-deficient niche, whereas the macrophage compartment is characterized by glucose deprivation (Rubin-Bejerano et al., 2003; Lorenz et al., 2004).

IMPORTANCE OF METABOLISM FOR PATHOGENICITY: WHAT AUXOTROPHS TELL US

The genome of *A. fumigatus* appears not to encode proper virulence factors such as a capsule or potent toxins. A variety of characteristics are accountable for the capability of *Aspergillus* to cause disease, among them thermophyly, environmental conditions, morphology, and its versatile physiology (Krappmann, 2007). Whereas the small size of its conidia and its pronounced thermotolerance support infection via inhalation and growth in general, respectively, the contributions of the environment and the fungal metabolism to pathogenicity are less obvious. Based on the rationale that aspergillosis depends to a high degree on the host's immune status and to a lesser extent on specific fungal determinants, Tekaia and Latgé (2005) emphasized the host and therefore the environmental component of aspergillosis as the major disease determinant.

Metabolism and nutritional versatility are a prerequisite for colonizing the ecological niche that is encountered by *Aspergillus* when causing disease (Fig. 1), and numerous studies mainly using auxotrophic mutant strains have supported this point of view. In a first study on auxotrophs of *A. nidulans* and their virulence capacities in a systemic murine infection model, mutants carrying mutations in four gene loci involved in the biosynthesis of adenine, pyridoxine, or *para*-aminobenzoic acid (PABA) were characterized as avirulent (Purnell, 1973). Especially the latter requirement is of interest, as PABA-requiring aspergilli have proven to be avirulent in several instances. Sandhu et al. (1976) demonstrated that intravenous injection of *A. fumigatus* strains that require this vitamin precursor did not result in systemic infection; *A. nidulans pabaA1* mutants were unable to cause disease in pulmonary infection models, but adding PABA to the drinking water of infected animals resulted in aspergillosis from the time point of supplementation (Tang et al., 1994). In vivo screening of a pool of *A.*

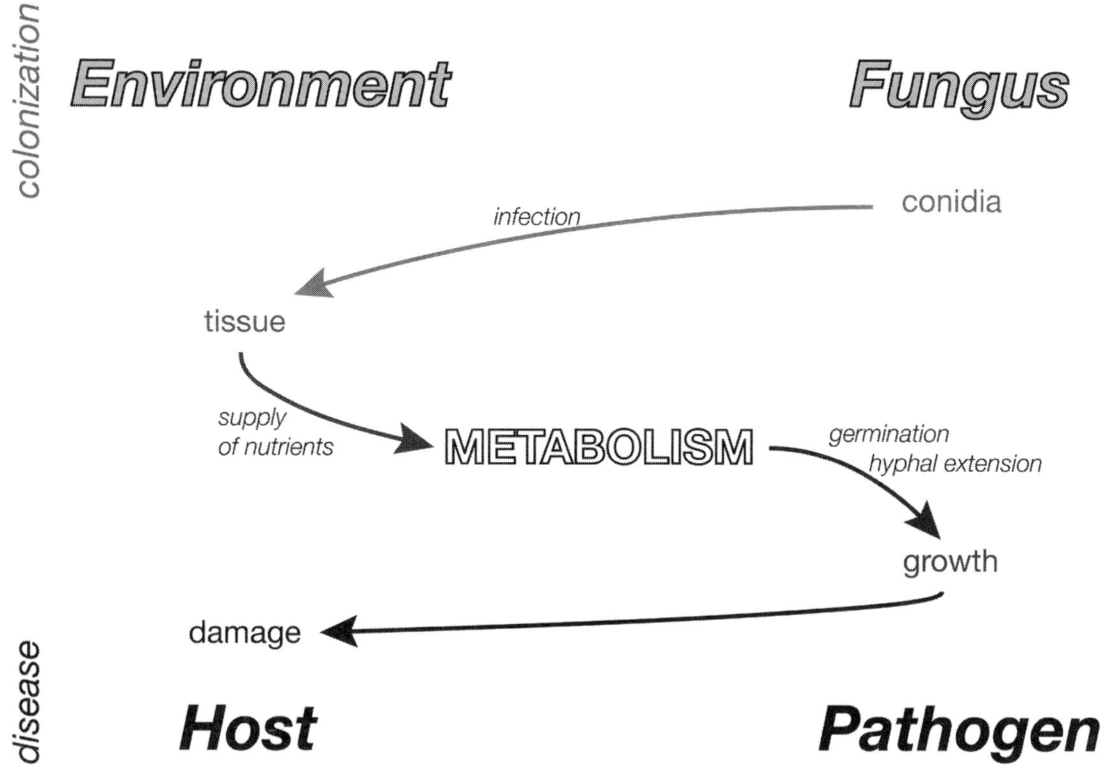

Figure 1. Metabolism as a virulence determinant of *Aspergillus*. This scheme illustrates the pathogen-host interaction as a variation of *Aspergillus* exploiting a specific environment. After infection by airborne conidia, the surrounding tissue has to be utilized to allow germination and hyphal growth, which eventually may result in a final outcome of disease. Nutritional versatility occupies a crucial role in this setting and therefore affects the aftermath of the pathogen-host encounter.

fumigatus mutants that had been randomly generated by signature-tagged mutagenesis resulted in identification and subsequent validation of the *pabaA* locus to be strictly required for germination as well as fungal proliferation inside the murine host (Brown et al., 2000).

Genes encoding enzymes of amino acid biosynthesis have attracted attention too, based on the fact that fungi are generally prototrophic for these building blocks of proteins, whereas animals require certain essential amino acids in their diet. Fungal biosynthesis of lysine is exceptional among the pathways resulting in amino acids, because in fungi the α-aminoadipate route is followed, in contrast to the situation in most other microorganisms or plants, where the diamino-pimelinic pathway is the rule (Garrad and Bhattacharjee, 1992; Nishida and Nishiyama, 2000; Zabriskie and Jackson, 2000). Accordingly, lysine-requiring *Aspergillus* mutants have been tested in infection models with conclusive results. An *A. nidulans lysA2* auxotroph displayed attenuated virulence in competitive infections of neutropenic mice (Tang et al., 1994), and an *A. fumigatus lysF*Δ deletant was almost avirulent in a murine low-dose infection model (Liebmann et al., 2004a).

Related to this are data from uridine monophosphate biosynthesis of *A. fumigatus* (d'Enfert et al., 1996). Testing a defined deletion mutant of the *pyrG* locus (which encodes orotidine-5′-monophosphate decarboxylase) that was auxotroph for uracil and uridine revealed that the corresponding conidia were unable to germinate in the lungs of infected mice. Accordingly, this uracil/uridine-requiring auxotroph was not virulent but regained virulence when the pyrimidine nucleoside was provided as a supplement in the drinking water.

In summary, these insights from studies using auxotrophic strains in established disease models give a hint towards the nutritional requirements faced by the fungal pathogen at the site of infection. Its primary metabolism has to adapt to the unique composition of this niche, and biosynthetic metabolic pathways specifically required for this are promising targets for antifungal therapy. However, detailed knowledge on the routes of primary metabolism necessary for aspergillosis is scarce, especially with respect to the pathways metabolizing fundamental elements, like carbon and nitrogen. But again, virulence studies of mutants deficient in enzymatic activities of biosynthetic cycles or impaired in regulatory circuits have recently revealed valuable information on the basic compounds that *Aspergillus* might be feeding from inside the infected host. Accordingly, it is also the purpose of this chapter to cover these aspects of primary nitrogen and carbon metabolism that appear to be relevant for the virulence of *A. fumigatus*.

SWEETS FOR THE SWEET: CARBON METABOLISM AND ITS REGULATION

As it is distinctive for their saprophytic life-style, aspergilli are able to utilize a wide spectrum of organic compounds and complex substrates. Especially with respect to carbon, a wide variety of substances can serve as sole source of this element, and moreover, complex mixtures and diverse polymeric substances can support growth. This was demonstrated, for instance, by a study on carbon compounds influencing the synthesis of an antibiotic from *A. fumigatus*, in which 29 carbon sources—poly-, oligo-, and monosaccharides as well as organic acids—in chemically defined media were tested as well as mixtures containing a secondary carbon source (Yang et al., 2003). This pronounced nutritional flexibility is reflected by the ample armamentarium for perception, degradation, uptake, and catabolism of organic substrates, which is complemented by regulatory networks that modulate expression of genes encoding pathway-specific enzymatic activities in accordance with the availability of nutrients. The ascomycete *A. nidulans* has served for decades as an established model system for studying carbon metabolism and its regulation (Hondmann and Visser, 1994; Ruijter and Visser, 1997; Hynes, 2007), and the essential facts deduced from research on this species will only be briefly summarized here. In contrast, detailed knowledge on carbon metabolism with respect to virulence of the pathogen *A. fumigatus* is evolving, and some of the recent highlight studies will be discussed in detail. By completion of the efforts to determine the genome sequence of an *A. fumigatus* reference strain and subsequent automated annotation (Denning et al., 2002; Mabey et al., 2004; Nierman et al., 2005b), loci putatively involved in carbon metabolism could be identified. For a comprehensive overview on deduced sequences the interested reader may refer to the KEGG pathway database (http://www.genome.ad.jp/kegg/pathway.html#carbohydrate), from which extensive lists of *A. fumigatus* candidate genes involved in carbohydrate metabolism may be retrieved.

As pointed out above, aspergilli express a wide array of hydrolytic enzymes with the capacity to degrade polymeric and oligomeric substrates, among them cellulose, pectin, xylan, and arabinan (Hondmann and Visser, 1994). Usually, the initial breakdown of polysaccharides results in the perception of signal molecules, like monomeric sugars or homo- and heterodisaccharides, which then serve as inducers of enzyme synthesis promoting rapid polymer degradation. Tight control of gene expression is executed in dependency of the substrates, as it was exemplified for the cellobiohydrolase-encoding gene *cbhB* (Bromley et al., 2006). As a logical consequence, signal transduction is required to adapt the cellular metabolism in accordance with the nutrients available. For instance, studies on fungal mitogen-activated protein kinases (MAPKs) (Xu, 2000; Roman et al., 2007), among them *A. fumigatus* MAPKs (May et al., 2005), resulted in valuable insights (Reyes et al., 2006), which are discussed in detail in chapter 13.

The uptake of oligomeric or monomeric degradation products is mediated by distinct uptake systems, and aspergilli express a large array of transporters and carriers (Diallinas, 2007). For instance, the genome of *A. nidulans* encodes at least 17 putative hexose transporters (Wei et al., 2004), and in silico analysis of the annotated *A. fumigatus* genome reveals 275 transporters of the major facilitator superfamily, to which sugar transporters belong (see http://www.membranetransport.org/). It is this high degree of redundancy that has hampered detailed analysis of particular transport systems that might represent fungus-specific targets.

The preferred source of carbon and energy is glucose, and studies on hexose metabolism of *Aspergillus* have a longstanding tradition in fungal research (Hondmann and Visser, 1994). Yet, other and more complex compounds must be utilized in the natural environment and, moreover, in the host, and virulence data from respective mutant strains have yielded valuable information on nutrient supply during infection. Hexoses are metabolized via glycolysis and the pentose phosphate pathway, whereas growth on fermentable carbon compounds such as acetate, fatty acids, ethanol, or amino acids requires the net formation of sugars by gluconeogenesis, based on the fact that these substrates generate intermediates of the tricarboxylic acid cycle (Hynes, 2007). As acetyl coenzyme A (CoA) is generated from these substrates, the glyoxalate bypass by the enzymes isocitrate lyase and malate synthase becomes active for the anaplerotic synthesis of oxaloacetate (Fig. 2A). Studies on the former enzyme of *A. fumigatus* have shed light on the question about the major carbon source during invasive aspergillosis: isocitrate lyase activity is present in conidia and is required to promote germination in the presence of C_2-generating carbon sources such as acetate (Ebel et al., 2006). Moreover, phagocytosis by macrophages induced isocitrate lyase expression in germlings, which suggests that lipids are metabolized under these circumstances. However, a mutant with the

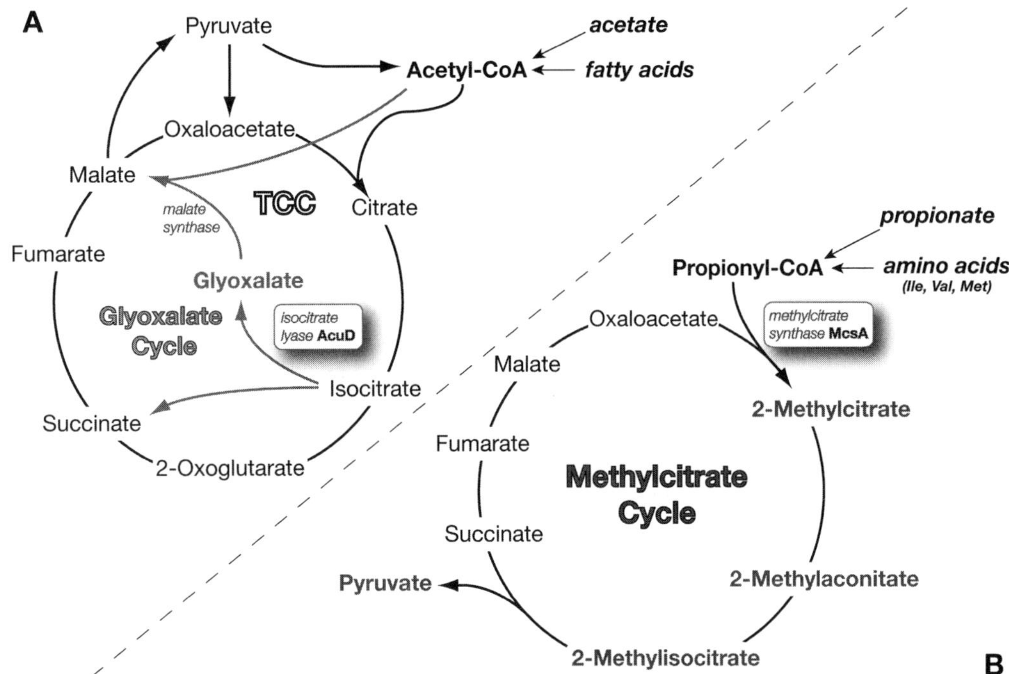

Figure 2. Routes of carbon metabolism to utilize fatty acids and amino acids. Various metabolic pathways are used by *Aspergillus* in utilizing carbon sources, including fatty acids and amino acids. (A) The former result in increased levels of acetyl-CoA and anaplerosis of the tricarbonic cycle (TCC) intermediate oxaloacetate via the glyoxalate bypass. The intermediate glyoxalate is formed by the action of an isocitrate lyase enzyme (AcuD; EC 4.1.3.1), which is dispensable in invasive aspergillosis (Schöbel et al., 2007). (B) The methylcitrate cycle feeds from propionyl-CoA, which derives from isoleucine, valine, or methionine as products from protein degradation. Its condensation with the TCC intermediate oxaloacetate is catalyzed by methylcitrate synthetase activity (McsA; EC 6.2.1.17), an enzyme that is essential for invasive aspergillosis (Ibrahim-Granet et al., 2008). Accordingly, *A. fumigatus* appears to utilize amino acids as a source for carbon and nitrogen.

acuD gene deleted was as virulent in a murine model of pulmonary aspergillosis as its wild-type progenitor (Schöbel et al., 2007). This finding indicates that the glyoxalate bypass is dispensable for invasive aspergillosis and that not lipids or fatty acids but alternative carbon sources are metabolized by the fungus during this disease. This hypothesis could be nicely substantiated by testing a methylcitrate synthase mutant of *A. fumigatus* for virulence. Amino acids generated by protein degradation may result in the formation of propionyl-CoA, which has to be metabolized by the methylcitrate cycle (Brock et al., 2000) (Fig. 2B). A mutant with the *mcsA* gene deleted, which encodes a key enzyme of this pathway (Maerker et al., 2005), displayed a pronounced attenuation in virulence when tested in a murine model for invasive aspergillosis (Ibrahim-Granet et al., 2008). One explanation for this phenotype is the accumulation of toxic propionyl-CoA, which indicates that amino acids, from which this intermediate is generated, serve as the carbon source during growth within the host.

As evident from the diversity of substrates that aspergilli can utilize, carbon catabolism has to be regulated in order to exploit nutritional sources with high efficiency. A well-studied paradigm for this is carbon catabolite repression in *A. nidulans* by the C2H2 zinc finger protein CreA, which ensures that the preferred carbon source glucose is selectively metabolized (Dowzer and Kelly, 1989, 1991; Ruijter and Visser, 1997). Carbon sources such as glucose, xylose, sucrose, and acetate strongly repress expression of genes required for utilization of less-favorable carbon sources, whereas glycerol, melibiose, lactose, and ethanol do not result in repression. The genes under carbon repression are classified into three major groups of utilization of less-favored C sources, the gluconeogenic and glyoxalate cycle, and secondary metabolism. A recent profiling study of *A. nidulans* elucidated the depth of the CreA-mediated transcriptome in response to two different carbon sources and revealed that carbon repression is a highly complex regulatory system affecting fungal metabolism over a wide range (Mogensen et al., 2006). Moreover, an extensive proteome study of *A. fumigatus* grown on ethanol or glucose, respectively, demonstrated changes in expression for enzymes involved in ethanol degradation, the glyoxalate cycle, or gluconeogenesis (Kniemeyer et al., 2006). However, the relevance of carbon catabolite repression for *Aspergillus* pathogenicity has not yet been studied in detail.

A particular regulatory system of *Aspergillus* that relates to carbon metabolism is the cyclic AMP (cAMP)/protein kinase A (PKA) signaling pathway (Oliver et al., 2002a), which has also been linked to pathogenicity of several fungi (Bölker, 1998; D'Souza and Heitman, 2001). Components of cAMP/PKA signal transduction have been targeted in several aspergilli, among them *A. fumigatus*, to find high diversity with respect to the G-protein-coupled receptor level but good conservation for the respective G proteins and modules of cAMP signaling (Lafon et al., 2006). Experimental studies in *A. fumigatus* on this pathway have aimed at the targeted deletion of genes encoding distinct components, such as the G protein α-subunit GpaB, the adenylyl cyclase AcyA, the catalytic subunit PkaC1, or the regulatory subunit PkaR (Oliver et al., 2002b; Liebmann et al., 2003; Liebmann et al., 2004b). Interestingly, levels of *pkaC* and *pkaR* transcripts were upregulated when *A. fumigatus* was cocultured with endothelial or pulmonary epithelial cells, which indicates that PKA signaling is triggered under these circumstances. Phenotypically, the corresponding mutants displayed poor conidiation and, except for the *gpaB*Δ deletant, reduced growth. When tested for virulence, all described PKA signaling mutants were severely impaired in pathogenicity, which can be attributed to poor growth characteristics. However, the significance of cAMP/PKA signaling for pathogenicity is validated by the *gpaB*Δ deletant, which is not impaired in growth but nevertheless is strongly attenuated in virulence. Further studies on the *A. fumigatus* cAMP/PKA pathway focused on the *pksP* gene, an established virulence determinant that is involved in synthesis of conidial pigment (Brakhage and Liebmann, 2005). In two mutants tested, *pkaC1*Δ and *gpaB*Δ, expression of a PksP-LacZ reporter was reduced to differing levels, which reflects GpaB-cAMP/PKA-dependent and -independent components acting on PksP expression.

From this brief inventory of the data on carbon metabolism and its regulation in *Aspergillus*, several conclusions may be drawn. It is evident that carbohydrate utilization affects growth and therefore virulence of aspergilli; hence, research on this particular aspect of fungal metabolism should be extended and intensified. Signal transduction pathways affecting the fungal response to varying environmental conditions are likely to contribute to its nutritional versatility, and identification of nutritional pathways supporting growth of *Aspergillus* in the ecological niche of a "susceptible host" could provide the capability to pinpoint fungus-specific virulence determinants and therefore targets of antifungal therapy. This is of course not limited to routes of carbon metabolism, as exemplified in the following section discussing the accumulated knowledge of nitrogen metabolism of *A. fumigatus* with respect to its pathogenicity.

MAKING BASIC BUILDING BLOCKS: NITROGEN METABOLISM AND REGULATORY CIRCUITS

The exploitation of nitrogen sources is essential for fungal growth, as fundamental compounds such as amino acids and nucleotides contain this macroelement. A variety of chemical substances can serve as sources of nitrogen for aspergilli, may they be simple ions like nitrate or ammonium, amino acids like histidine or proline, or complex, polymeric substrates such as proteins. Amino acids are of particular interest, as they may serve as nitrogen as well as carbon sources and represent the ultimate degradation products when the surrounding proteinaceous matrix has disintegrated. With respect to quality, substances can be distinguished as primary and secondary nitrogen sources, and ammonium together with the amino acids glutamate and glutamine represent the most suitable sources of nitrogen to support growth of *Aspergillus*. In contrast, substances like nitrate, purines, or amino acids other than glutamate and glutamine require conversion before feeding into the fungal metabolism; accordingly, they represent improper, secondary sources (Marzluf, 1997).

Of special interest are polymeric substrates that are present in the infected host tissue. The mammalian lung, which is the primary site of infection for *Aspergillus*, is rich in structural polypeptides and matrix proteins like elastin, collagen, laminin, and fibronectin, and their breakdown products represent nutritional sources of nitrogen. The amino acid proline is of importance as it constitutes a major portion and can serve as a nitrogen as well as a carbon source for aspergilli (Hull et al., 1989). Accordingly, uptake and metabolism of this amino acid are not only under carbon repression but also are additionally under so-called nitrogen catabolite repression (NCR). This wide domain regulatory system mediates fungal discrimination between rich and poor, or primary and secondary, nitrogen sources and has been studied extensively in the model organism *A. nidulans*, for instance, with respect to nitrate utilization (Caddick, 1994; Caddick et al., 1994). Central to the system is a GATA transcription factor, the *areA* gene product, which ensures that in the absence of primary N sources structural genes for utilization of alternatives are properly expressed (Caddick et al., 1986; Wilson and Arst, 1998). The requirement for *A. fumigatus* AreA in pulmonary disease was studied by creating loss-of function mutants via gene disruption as well as replacement (Hensel et al., 1998). Growth of such an *areA*$^-$ mutant was only supported by the presence of ammonium or glutamate and was up to wild-type levels when excess amounts were supplied. Interestingly, secretion of specific proteolytic activities was also impaired in the mutant. The *areA*$^-$ mutants were investigated with respect to virulence to reveal that AreA is not essential for fungal growth but contributes to fungal growth in the murine lung, as estimated from the fact that the onset of pulmonary disease was delayed. This finding was corroborated by competitive infections as well as by the fact that the *areA*$^-$ disruption mutant underwent significant reversion when passed through a cycle of infection. The latter result indicates that the source of nitrogen in the host's lung is not a primary one and that the nitrogen catabolite system with its master regulator AreA confers a selective advantage to support fungal growth in this environment.

Related to this, the nitrate assimilation cluster of *A. fumigatus* has been studied as a prime target of NCR. Extracellular nitrate is taken up by the CrnA transporter and subsequently converted to ammonium by the action of a nitrate and a nitrite reductase, NiaD and NiiA, respectively (Johnstone et al., 1990). A corresponding gene cluster comprising these three structural genes has been identified in *A. fumigatus* based on strong conservation and synteny to its *A. nidulans* counterpart (Amaar and Moore, 1998; Pain et al., 2004). Especially the intergenic region of the divergently transcribed genes *niiA* and *niaD* is of interest with respect to the binding and interplay of regulatory factors, among them AreA, and the high degree of conservation of this bidirectional promoter suggests similar regulatory patterns of nitrate assimilation for *A. nidulans* and *A. fumigatus*.

As it is evident for carbon source utilization, MAPK signal transduction also appears to influence nitrogen source sensing in *A. fumigatus*. Among the four MAPKs identified in the genome, the cellular function of the *sakA*-encoded one was studied initially by characterization of a deletion mutant (Xue et al., 2004). The *sakA*Δ strain displays sensitivity towards oxidative stress and, interestingly, seems to be impaired in nutrient sensing, as germination was accelerated in comparison to wild type when nitrate or nitrite was present as the source of nitrogen. Steady-state transcript levels of *sakA* increased specifically upon a shift to N- or C-depleted medium; however, no difference in germination was evident when different carbon sources were provided. Taken together, these data indicate that the SakA MAPK influences germination of *A. fumigatus* with respect to nitrogen source quality and may play a role in fungal starvation.

Starving *A. fumigatus* for nitrogen increases expression of a regulatory factor belonging to the Ras-related protein family of Rheb proteins (Panepinto et al., 2002). This RhbA protein is the sole Rheb homolog encoded by the *A. fumigatus* genome, and besides increased transcription under nitrogen, but not carbon starvation conditions, its transcript becomes upregulated when *A. fumigatus* is cultured in the presence of human endothelial cells, a situation that resembles conditions during angioinvasion (Rhodes et al., 2001). A corresponding deletion mutant was characterized to be viable but specifically retarded in growth when poor sources of nitrogen such as nitrate, histidine, or proline were provided (Panepinto et al., 2003). Moreover, the uptake of arginine was intensified in an *rhbA*Δ mutant, and this phenotype was evident only in the presence of the rich nitrogen source ammonium and not in the presence of nitrate or proline. Also, the deletant displayed abnormal development in submerged culture depending on the nitrogen source quality and was hypersensitive to rapamycin, the prime inhibitor of the target of rapamycin (TOR) kinase. Most interestingly, however, were the data on virulence of the deletion mutant as assessed in a leukopenic mouse model for pulmonary aspergillosis: survival of animals infected with the mutant was prolonged, accompanied by smaller lesion areas in the murine lungs, which indicates a reduced in vivo growth potential of the deletion mutant. Accordingly, this mutant study also linked sensing and utilization of nitrogen compounds to growth and therefore virulence of *A. fumigatus*.

Another recognized signal transduction cascade that regulates fungal nitrogen metabolism with respect to amino acid homeostasis is the cross-pathway control system, or the general control of amino acid biosynthesis (CPC/GC signaling) (Carsiotis and Jones, 1974; Carsiotis et al., 1974; Hinnebusch, 1986). CPC/GC signaling is a variation of signal transduction via phosphorylation of an initiation factor for translation, and this kind of pathway, which is highly conserved among the eukaryotic kingdom, is generally regarded as a sensor to perceive and counter conditions of environmental stress (Wek et al., 2006). In its core components (Fig. 3), this system includes a dimeric kinase, which perceives the stress condition to phosphorylate the α-subunit of the translation eukaryotic initiation factor 2 (eIF2) as its only known substrate, and a transcription factor to generate the cellular response by altering the transcriptional profile (Hinnebusch, 1997). In the case of CPC/GC signaling, the most prominent trigger is starvation for amino acids or imbalances in amino acid homeostasis, which is intracellularly mirrored by the accumulation of uncharged tRNA molecules. These bind to the sensor kinase, encoded by the *Aspergillus* gene *cpcC*, to prompt eIF2 phosphorylation at the Ser-51 residue of its α-subunit. As a result, translation initiation becomes decelerated to phase down cellular translation. For specific transcripts, however, eIF2α phosphorylation results in increased translation of the actual coding sequence, which is generally preceded by short regulatory upstream open reading frames (uORFs) in long transcript leader regions. Under conditions of amino acid starvation, the *Aspergillus* transcription factor CpcA becomes accordingly expressed at high levels as its mRNA comprises two uORFs (Hoffmann et al., 2001). It was dem-

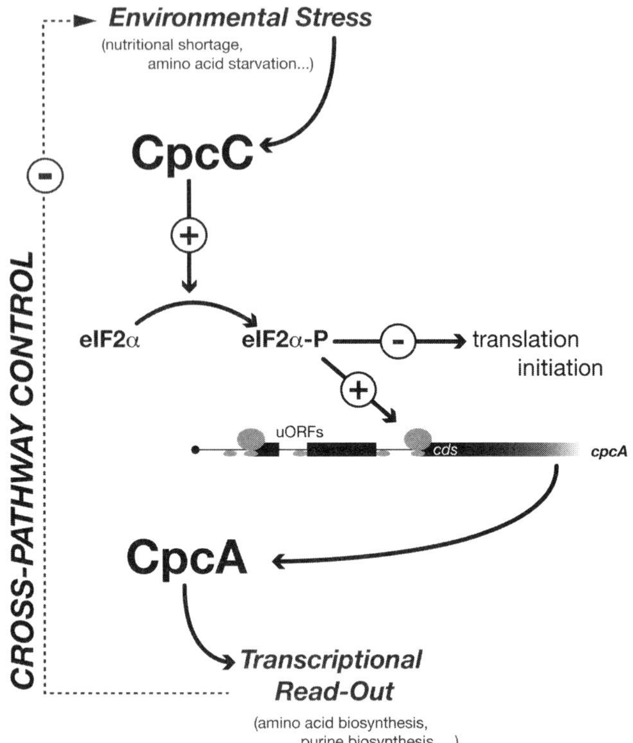

Figure 3. Cross-pathway control signaling in *Aspergillus*. The schematic outline shows the core components and regulatory effects in the eIF2α kinase signaling cascade of fungal cross-pathway control. Nutritional stress, such as amino acid starvation, is perceived by the kinase CpcC, which in turn phosphorylates the α-subunit of eIF2. This results in lower rates of translation initiation but elevated expression of the transcription factor CpcA, mediated by two uORFs on the *cpcA* transcript. Accordingly, transcriptional reprogramming occurs to counter the initial stress condition. The presence of the *cpcA* gene is required for full virulence of *A. fumigatus* (Krappmann et al., 2002), whereas a *cpcC*Δ deletant is as virulent as wild type (Sasse et al., 2008).

onstrated for Gcn4p, the yeast ortholog of CpcA, that this activator serves as a master regulator to modulate transcription of a wide array of genes, mainly involved in the biosynthesis of amino acids and purines but also in various other cellular processes (Natarajan et al., 2001). Transcriptional profiling data on the CpcA-directed transcriptome of *A. fumigatus* support this role as a wide domain effector. Moreover, a role of this transcription factor could be established in pulmonary aspergillosis: when leukopenic mice were infected with an *A. fumigatus* deletion strain ablated for CpcA, median survival times were increased, indicating significant attenuation of this mutant (Krappmann et al., 2004). This finding was further corroborated by competitive infection experiments with mixed inocula of the wild-type isolate and a *cpcA*Δ deletant. However, when a mutant with the CPC sensor kinase-encoding gene *cpcC* deleted was used for infection in this disease model, no attenuation was evident, which indicates that derepression of the CPC system does not occur in the murine lung (Sasse et al., 2008). By microscopic inspection of a suitable reporter strain expressing a functional GFP-CpcA fusion protein, this assumption became substantiated: no fluorescence was detected from conidia in bronchoalveolar lavage fluids, and phagocytic ingestion by alveolar macrophages did not trigger CpcA expression. This led to the conclusion that the murine lung is an environment supplying amino acids in sufficient amounts and that stress conditions triggering the CPC system are not present during infection. Yet, the actual role of the CpcA transcription factor during pulmonary aspergillosis remains to be resolved, as its basal expression level in a *cpcC*Δ deletant appears to support this disease.

A further issue that is worth elucidating is the influence of TOR kinase signaling in nitrogen sensing of *A. fumigatus* and its relevance for virulence. In fungi, growth and proliferation is controlled by nutrient availability, and in the model eukaryote *Saccharomyces cerevisiae*, sensing the quality of the nitrogen source is mediated by members of the TOR kinase family (Shamji et al., 2000). There, rich sources induce TOR kinase activity, and the AreA ortholog Gln3p is sequestered in the cytoplasm to suppress the NCR (Bertram et al., 2000). When cells are treated with the potent inhibitor of TOR kinases rapamycin, NCR is rapidly released and the cell cycle becomes arrested (Cardenas et al., 1999). Moreover, an influence of TOR on the general control has been validated in *S. cerevisiae* (Valenzuela et al., 2001). In *Aspergillus*, however, detailed knowledge on this conserved signaling pathway is just starting to evolve. Five components of the *A. nidulans* cascade were identified by a combination of approaches to find a minor influence on nitrogen metabolism (Fitzgibbon et al., 2005). Yet, an *A. fumigatus* mutant with the *rhbA* gene deleted displayed increased sensitivity towards rapamycin (Panepinto et al., 2003), as it appears to be the case in *S. cerevisiae*, a phenotype that suggests an overlap in nitrogen utilization and TOR-dependent signaling in this fungus. Straightforward analysis of the TOR pathway in *A. fumigatus* and assessment of its impact on virulence awaits experimental execution. One specific cellular process that is influenced by TOR signaling is autophagy (Klionsky, 2007; Onodera and Ohsumi, 2005), and the interested reader may refer to chapter 16 in this volume for more details on this cellular process with respect to *A. fumigatus* metabolism (Richie et al., 2007).

For the consumption of nitrogen-containing compounds a variety of regulatory systems have evolved in fungi, and it is obvious that aspergilli fall back upon some of these during growth in their natural habitat as well as in an infected individual. As is the case for carbon metabolism, the quantity and, more importantly, the quality of the nitrogen source is sensed to adapt the cellular metabolism accordingly. It will be of special in-

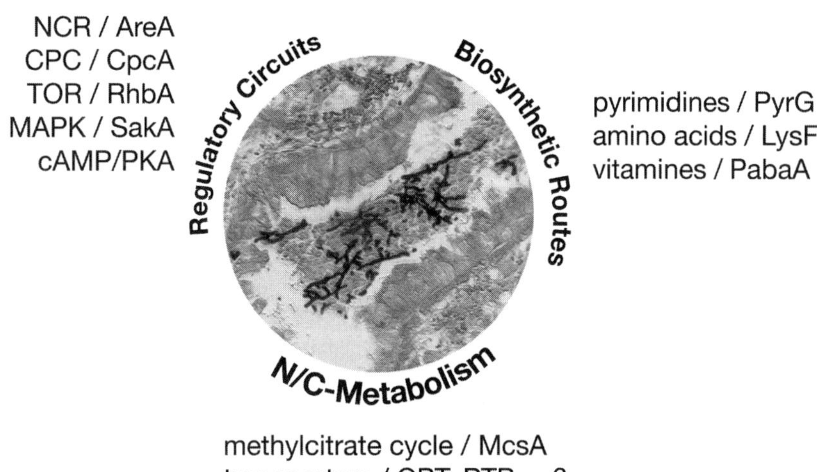

Figure 4. Aspects of primary carbon and nitrogen metabolism that influence pathogenicity of *Aspergillus*. Several biosynthetic routes have been identified as required for aspergillosis, and a variety of regulatory circuits have been analyzed in this respect, too. However, distinct aspects of basic C or N metabolism have been scrutinized only to a limited extent, and the role of nutritional transporters has not been addressed thoroughly. See text for further details.

terest to scrutinize these regulatory networks in more detail to obtain a brighter picture of the basic nutritional requirements of *Aspergillus* when it exploits an immunocompromised host.

CONCLUSIONS

Data from the briefly summarized available studies allow a rough and preliminary estimation of the nutritional situation faced by *Aspergillus* when entering and infecting a susceptible mammalian host (Fig. 4). The major nutritional source for *A. fumigatus* during pulmonary aspergillosis appears to be amino acids, which is in contrast to the situation for the commensal *C. albicans* or the pulmonary pathogen *Mycobacterium tuberculosis*, for which a supply of carbon is realized from products of lipid degradation (Lorenz and Fink, 2001; Munoz-Elias and McKinney, 2005). The environment encountered by *A. fumigatus* when causing pulmonary disease is likely to be rich in proteins; however, imbalances for particular amino acids may result from the bias in specific structural polymers constituting the surrounding tissue. Nutritional versatility of the "gourmand *Aspergillus*" (Wilson and Arst, 1998) impedes progress in defining the overlap between fungal metabolome and virulome. On the other hand, the distinct metabolic features of this fungal genus represent a hopeful perspective in pinpointing specific candidate pathways that may represent suitable targets for drug development. Substantial progress in this field of research has been achieved, but further analyses will have to be carried out in order to explore the nutritional environment during aspergillosis in more detail.

REFERENCES

Amaar, Y. G., and M. M. Moore. 1998. Mapping of the nitrate-assimilation gene cluster (*crnA-niiA-niaD*) and characterization of the nitrite reductase gene (*niiA*) in the opportunistic fungal pathogen *Aspergillus fumigatus*. *Curr. Genet.* **33:**206–215.

Bertram, P. G., J. H. Choi, J. Carvalho, W. Ai, C. Zeng, T. F. Chan, and X. F. Zheng. 2000. Tripartite regulation of Gln3p by TOR, Ure2p, and phosphatases. *J. Biol. Chem.* **275:**35727–35733.

Bölker, M. 1998. Sex and crime: heterotrimeric G proteins in fungal mating and pathogenesis. *Fungal Genet. Biol.* **25:**143-156.

Boyce, J. D., P. A. Cullen, and B. Adler. 2004. Genomic-scale analysis of bacterial gene and protein expression in the host. *Emerg. Infect. Dis.* **10:**1357–1362.

Brakhage, A. A., and K. Langfelder. 2002. Menacing mold: the molecular biology of *Aspergillus fumigatus*. *Annu. Rev. Microbiol.* **56:** 433–455.

Brakhage, A. A., and B. Liebmann. 2005. *Aspergillus fumigatus* conidial pigment and cAMP signal transduction: significance for virulence. *Med. Mycol.* **43**(Suppl. 1):S75–S82.

Brock, M., R. Fischer, D. Linder, and W. Buckel. 2000. Methylcitrate synthase from *Aspergillus nidulans*: implications for propionate as an antifungal agent. *Mol. Microbiol.* **35:**961–973.

Bromley, M., C. Gordon, N. Rovira-Graells, and J. Oliver. 2006. The *Aspergillus fumigatus* cellobiohydrolase B (*cbhB*) promoter is tightly regulated and can be exploited for controlled protein expression and RNAi. *FEMS Microbiol. Lett.* **264:**246–254.

Brown, J. S., A. Aufauvre-Brown, J. Brown, J. M. Jennings, H. Arst, Jr., and D. W. Holden. 2000. Signature-tagged and directed mutagenesis identify PABA synthetase as essential for *Aspergillus fumigatus* pathogenicity. *Mol. Microbiol.* **36:**1371–1380.

Caddick, M. X. 1994. Nitrogen metabolite repression. *Prog. Ind. Microbiol.* **29:**323–353.

Caddick, M. X., H. N. Arst, Jr., L. H. Taylor, R. I. Johnson, and A. G. Brownlee. 1986. Cloning of the regulatory gene *areA* mediating nitrogen metabolite repression in *Aspergillus nidulans*. *EMBO J.* **5**:1087–1090.

Caddick, M. X., D. Peters, and A. Platt. 1994. Nitrogen regulation in fungi. *Antonie Leeuwenhoek* **65**:169–177.

Cardenas, M. E., N. S. Cutler, M. C. Lorenz, C. J. Di Como, and J. Heitman. 1999. The TOR signaling cascade regulates gene expression in response to nutrients. *Genes Dev.* **13**:3271–3279.

Carsiotis, M., and R. F. Jones. 1974. Cross-pathway regulation: tryptophan-mediated control of histidine and arginine biosynthetic enzymes in *Neurospora crassa*. *J. Bacteriol.* **119**:889–892.

Carsiotis, M., R. F. Jones, and A. C. Wesseling. 1974. Cross-pathway regulation: histidine-mediated control of histidine, tryptophan, and arginine biosynthetic enzymes in *Neurospora crassa*. *J. Bacteriol.* **119**:893–898.

Cramer, R. A., Jr., M. P. Gamcsik, R. M. Brooking, L. K. Najvar, W. R. Kirkpatrick, T. F. Patterson, C. J. Balibar, J. R. Graybill, J. R. Perfect, S. N. Abraham, and W. J. Steinbach. 2006. Disruption of a nonribosomal peptide synthetase in *Aspergillus fumigatus* eliminates gliotoxin production. *Eukaryot. Cell* **5**:972–980.

d'Enfert, C., M. Diaquin, A. Delit, N. Wuscher, J. P. Debeaupuis, M. Huerre, and J.-P. Latgé. 1996. Attenuated virulence of uridine-uracil auxotrophs of *Aspergillus fumigatus*. *Infect. Immun.* **64**:4401–4405.

Denning, D. W., M. J. Anderson, G. Turner, J.-P. Latgé, and J. W. Bennett. 2002. Sequencing the *Aspergillus fumigatus* genome. *Lancet Infect. Dis.* **2**:251–253.

Diallinas, G. 2007. Aspergillus transporters, p. 301–320. *In* G. H. Goldman and S. A. Osmani (ed.), *The Aspergilli: Genomics, Medical Aspects, Biotechnology, and Research Methods*. CRC Press, Boca Raton, FL.

Dowzer, C. E., and J. M. Kelly. 1989. Cloning of the *creA* gene from *Aspergillus nidulans*: a gene involved in carbon catabolite repression. *Curr. Genet.* **15**:457–459.

Dowzer, C. E., and J. M. Kelly. 1991. Analysis of the *creA* gene, a regulator of carbon catabolite repression in *Aspergillus nidulans*. *Mol. Cell. Biol.* **11**:5701–5709.

D'Souza, C. A., and J. Heitman. 2001. Conserved cAMP signaling cascades regulate fungal development and virulence. *FEMS Microbiol. Rev.* **25**:349–364.

Ebel, F., M. Schwienbacher, J. Beyer, J. Heesemann, A. A. Brakhage, and M. Brock. 2006. Analysis of the regulation, expression, and localisation of the isocitrate lyase from *Aspergillus fumigatus*, a potential target for antifungal drug development. *Fungal Genet. Biol.* **43**:476–489.

Fitzgibbon, G. J., I. Y. Morozov, M. G. Jones, and M. X. Caddick. 2005. Genetic analysis of the TOR pathway in *Aspergillus nidulans*. *Eukaryot. Cell* **4**:1595–1598.

Garrad, R. C., and J. K. Bhattacharjee. 1992. Lysine biosynthesis in selected pathogenic fungi: characterization of lysine auxotrophs and the cloned *LYS1* gene of *Candida albicans*. *J. Bacteriol.* **174**:7379–7384.

Haines, J. 1995. *Aspergillus* in compost: straw man or fatal flaw. *Biocycle* **6**:32–35.

Hensel, M., H. N. Arst, Jr., A. Aufauvre-Brown, and D. W. Holden. 1998. The role of the *Aspergillus fumigatus areA* gene in invasive pulmonary aspergillosis. *Mol. Gen. Genet.* **258**:553–557.

Hinnebusch, A. G. 1986. The general control of amino acid biosynthetic genes in the yeast *Saccharomyces cerevisiae*. *CRC Crit. Rev. Biochem.* **21**:277–317.

Hinnebusch, A. G. 1997. Translational regulation of yeast GCN4. A window on factors that control initiator-tRNA binding to the ribosome. *J. Biol. Chem.* **272**:21661–21664.

Hoffmann, B., O. Valerius, M. Andermann, and G. H. Braus. 2001. Transcriptional autoregulation and inhibition of mRNA translation of amino acid regulator gene *cpcA* of filamentous fungus *Aspergillus nidulans*. *Mol. Biol. Cell* **12**:2846–2857.

Hondmann, D. H., and J. Visser. 1994. Carbon metabolism. *Prog. Ind. Microbiol.* **29**:61–139.

Hull, E. P., P. M. Green, H. N. Arst, Jr., and C. Scazzocchio. 1989. Cloning and physical characterization of the L-proline catabolism gene cluster of *Aspergillus nidulans*. *Mol. Microbiol.* **3**:553–559.

Hynes, M. J. 2007. Gluconeogenic carbon metabolism, p. 129–142. *In* G. H. Goldman and S. A. Osmani (ed.), *The Aspergilli: Genomics, Medical Aspects, Biotechnology, and Research Methods*. CRC Press, Boca Raton, FL.

Ibrahim-Granet, O., M. Dubourdeau, J.-P. Latgé, P. Ave, M. Huerre, A. A. Brakhage, and M. Brock. 2008. Methylcitrate synthase from *Aspergillus fumigatus* is essential for manifestation of invasive aspergillosis. *Cell. Microbiol.* **10**:134–148.

Johnstone, I. L., P. C. McCabe, P. Greaves, S. J. Gurr, G. E. Cole, M. A. Brow, S. E. Unkles, A. J. Clutterbuck, J. R. Kinghorn, and M. A. Innis. 1990. Isolation and characterisation of the *crnA-niiA-niaD* gene cluster for nitrate assimilation in *Aspergillus nidulans*. *Gene* **90**:181–192.

Klionsky, D. J. 2007. Autophagy: from phenomenology to molecular understanding in less than a decade. *Nat. Rev. Mol. Cell Biol.* **8**:931–937.

Kniemeyer, O., F. Lessing, O. Scheibner, C. Hertweck, and A. A. Brakhage. 2006. Optimisation of a 2-D gel electrophoresis protocol for the human-pathogenic fungus *Aspergillus fumigatus*. *Curr. Genet.* **49**:178–189.

Krappmann, S. 2006. Tools to study molecular mechanisms of *Aspergillus* pathogenicity. *Trends Microbiol.* **14**:356–364.

Krappmann, S. 2007. Pathogenicity determinants and allergens, p. 377–400. *In* G. H. Goldman and S. A. Osmani (ed.), *The Aspergilli: Genomics, Medical Aspects, Biotechnology, and Research Methods*. CRC Press, Boca Raton, FL.

Krappmann, S., E. M. Bignell, U. Reichard, T. Rogers, K. Haynes, and G. H. Braus. 2004. The *Aspergillus fumigatus* transcriptional activator CpcA contributes significantly to the virulence of this fungal pathogen. *Mol. Microbiol.* **52**:785–799.

Kupfahl, C., T. Heinekamp, G. Geginat, T. Ruppert, A. Hartl, H. Hof, and A. A. Brakhage. 2006. Deletion of the *gliP* gene of *Aspergillus fumigatus* results in loss of gliotoxin production but has no effect on virulence of the fungus in a low-dose mouse infection model. *Mol. Microbiol.* **62**:292–302.

Lafon, A., K. H. Han, J. A. Seo, J. H. Yu, and C. d'Enfert. 2006. G-protein and cAMP-mediated signaling in aspergilli: a genomic perspective. *Fungal Genet. Biol.* **43**:490–502.

Latgé, J.-P. 1999. *Aspergillus fumigatus* and aspergillosis. *Clin. Microbiol. Rev.* **12**:310–350.

Latgé, J.-P. 2001. The pathobiology of *Aspergillus fumigatus*. *Trends Microbiol.* **9**:382–389.

Liebmann, B., S. Gattung, B. Jahn, and A. A. Brakhage. 2003. cAMP signaling in *Aspergillus fumigatus* is involved in the regulation of the virulence gene *pksP* and in defense against killing by macrophages. *Mol. Genet. Genomics* **269**:420–435.

Liebmann, B., T. W. Mühleisen, M. Müller, M. Hecht, G. Weidner, A. Braun, M. Brock, and A. A. Brakhage. 2004a. Deletion of the *Aspergillus fumigatus* lysine biosynthesis gene *lysF* encoding homoaconitase leads to attenuated virulence in a low-dose mouse infection model of invasive aspergillosis. *Arch. Microbiol.* **181**:378–383.

Liebmann, B., M. Müller, A. Braun, and A. A. Brakhage. 2004b. The cyclic AMP-dependent protein kinase A network regulates development and virulence in *Aspergillus fumigatus*. *Infect. Immun.* **72**:5193–5203.

Lorenz, M. C., J. A. Bender, and G. R. Fink. 2004. Transcriptional response of *Candida albicans* upon internalization by macrophages. *Eukaryot. Cell* **3**:1076–1087.

Lorenz, M. C., and G. R. Fink. 2001. The glyoxylate cycle is required for fungal virulence. *Nature* **412**:83–86.

Mabey, J. E., M. J. Anderson, P. F. Giles, C. J. Miller, T. K. Attwood, N. W. Paton, E. Bornberg-Bauer, G. D. Robson, S. G. Oliver, and D. W. Denning. 2004. CADRE: the Central Aspergillus Data REpository. *Nucleic Acids Res.* **32:**D401–D405.

Maerker, C., M. Rohde, A. A. Brakhage, and M. Brock. 2005. Methylcitrate synthase from *Aspergillus fumigatus*. Propionyl-CoA affects polyketide synthesis, growth and morphology of conidia. *FEBS J.* **272:**3615–3630.

Marzluf, G. A. 1997. Genetic regulation of nitrogen metabolism in the fungi. *Microbiol. Mol. Biol. Rev.* **61:**17–32.

May, G. S., T. Xue, D. P. Kontoyiannis, and M. C. Gustin. 2005. Mitogen activated protein kinases of *Aspergillus fumigatus*. *Med. Mycol.* **43**(Suppl. 1)**:**S83–S86.

Mogensen, J., H. B. Nielsen, G. Hofmann, and J. Nielsen. 2006. Transcription analysis using high-density micro-arrays of *Aspergillus nidulans* wild-type and *creA* mutant during growth on glucose or ethanol. *Fungal Genet. Biol.* **43:**593–603.

Munoz-Elias, E. J., and J. D. McKinney. 2005. *Mycobacterium tuberculosis* isocitrate lyases 1 and 2 are jointly required for in vivo growth and virulence. *Nat. Med.* **11:**638–644.

Muthuvijayan, V., and M. R. Marten. 2004. In silico reconstruction of nutrient-sensing signal transduction pathways in *Aspergillus nidulans*. *In Silico Biol.* **4:**605–631.

Natarajan, K., M. R. Meyer, B. M. Jackson, D. Slade, C. Roberts, A. G. Hinnebusch, and M. J. Marton. 2001. Transcriptional profiling shows that Gcn4p is a master regulator of gene expression during amino acid starvation in yeast. *Mol. Cell. Biol.* **21:**4347–4368.

Nierman, W. C., G. May, H. S. Kim, M. J. Anderson, D. Chen, and D. W. Denning. 2005a. What the *Aspergillus* genomes have told us. *Med. Mycol.* **43**(Suppl. 1)**:**S3–S5.

Nierman, W. C., A. Pain, M. J. Anderson, J. R. Wortman, H. S. Kim, J. Arroyo, M. Berriman, K. Abe, D. B. Archer, C. Bermejo, J. Bennett, P. Bowyer, D. Chen, M. Collins, R. Coulsen, R. Davies, P. S. Dyer, M. Farman, N. Fedorova, N. Fedorova, T. V. Feldblyum, R. Fischer, N. Fosker, A. Fraser, J. L. Garcia, M. J. Garcia, A. Goble, G. H. Goldman, K. Gomi, S. Griffith-Jones, R. Gwilliam, B. Haas, H. Haas, D. Harris, H. Horiuchi, J. Huang, S. Humphray, J. Jimenez, N. Keller, H. Khouri, K. Kitamoto, T. Kobayashi, S. Konzack, R. Kulkarni, T. Kumagai, A. Lafon, J.-P. Latgé, W. Li, A. Lord, C. Lu, W. H. Majoros, G. S. May, B. L. Miller, Y. Mohamoud, M. Molina, M. Monod, I. Mouyna, S. Mulligan, L. Murphy, S. O'Neil, I. Paulsen, M. A. Penalva, M. Pertea, C. Price, B. L. Pritchard, M. A. Quail, E. Rabbinowitsch, N. Rawlins, M. A. Rajandream, U. Reichard, H. Renauld, G. D. Robson, S. Rodriguez de Cordoba, J. M. Rodriguez-Pena, C. M. Ronning, S. Rutter, S. L. Salzberg, M. Sanchez, J. C. Sanchez-Ferrero, D. Saunders, K. Seeger, R. Squares, S. Squares, M. Takeuchi, F. Tekaia, G. Turner, C. R. Vazquez de Aldana, J. Weidman, O. White, J. Woodward, J. H. Yu, C. Fraser, J. E. Galagan, K. Asai, M. Machida, N. Hall, B. Barrell, and D. W. Denning. 2005b. Genomic sequence of the pathogenic and allergenic filamentous fungus *Aspergillus fumigatus*. *Nature* **438:**1151–1156.

Nishida, H., and M. Nishiyama. 2000. What is characteristic of fungal lysine synthesis through the α-aminoadipate pathway? *J. Mol. Evol.* **51:**299–302.

Nordstrom, U. M. 1974. Bark degradation by *Aspergillus fumigatus*. Growth studies. *Can. J. Microbiol.* **20:**283–298.

Oliver, B. G., J. C. Panepinto, D. S. Askew, and J. C. Rhodes. 2002a. cAMP alteration of growth rate of *Aspergillus fumigatus* and *Aspergillus niger* is carbon-source dependent. *Microbiology* **148:**2627–2633.

Oliver, B. G., J. C. Panepinto, J. R. Fortwendel, D. L. Smith, D. S. Askew, and J. C. Rhodes. 2002b. Cloning and expression of *pkaC* and *pkaR*, the genes encoding the cAMP-dependent protein kinase of *Aspergillus fumigatus*. *Mycopathologia* **154:**85–91.

Onodera, J., and Y. Ohsumi. 2005. Autophagy is required for maintenance of amino acid levels and protein synthesis under nitrogen starvation. *J. Biol. Chem.* **280:**31582–31586.

Pain, A., J. Woodward, M. A. Quail, M. J. Anderson, R. Clark, M. Collins, N. Fosker, A. Fraser, D. Harris, N. Larke, L. Murphy, S. Humphray, S. O'Neil, M. Pertea, C. Price, E. Rabbinowitsch, M. A. Rajandream, S. Salzberg, D. Saunders, K. Seeger, S. Sharp, T. Warren, D. W. Denning, B. Barrell, and N. Hall. 2004. Insight into the genome of *Aspergillus fumigatus*: analysis of a 922 kb region encompassing the nitrate assimilation gene cluster. *Fungal Genet. Biol.* **41:**443–453.

Paisley, D., G. D. Robson, and D. W. Denning. 2005. Correlation between in vitro growth rate and in vivo virulence in *Aspergillus fumigatus*. *Med. Mycol.* **43:**397–401.

Panepinto, J. C., B. G. Oliver, T. W. Amlung, D. S. Askew, and J. C. Rhodes. 2002. Expression of the *Aspergillus fumigatus rheb* homologue, *rhbA*, is induced by nitrogen starvation. *Fungal Genet. Biol.* **36:**207–214.

Panepinto, J. C., B. G. Oliver, J. R. Fortwendel, D. L. Smith, D. S. Askew, and J. C. Rhodes. 2003. Deletion of the *Aspergillus fumigatus* gene encoding the Ras-related protein RhbA reduces virulence in a model of invasive pulmonary aspergillosis. *Infect. Immun.* **71:**2819–2826.

Purnell, D. M. 1973. The effects of specific auxotrophic mutations on the virulence of *Aspergillus nidulans* for mice. *Mycopathol. Mycol. Appl.* **50:**195–203.

Reyes, G., A. Romans, C. K. Nguyen, and G. S. May. 2006. Novel mitogen-activated protein kinase MpkC of *Aspergillus fumigatus* is required for utilization of polyalcohol sugars. *Eukaryot. Cell* **5:**1934–1940.

Rhodes, J. C. 2006. *Aspergillus fumigatus*: growth and virulence. *Med. Mycol.* **44**(Suppl. 1)**:**S77–S81.

Rhodes, J. C., B. G. Oliver, D. S. Askew, and T. W. Amlung. 2001. Identification of genes of *Aspergillus fumigatus* up-regulated during growth on endothelial cells. *Med. Mycol.* **39:**253–260.

Richie, D. L., K. K. Fuller, J. Fortwendel, M. D. Miley, J. W. McCarthy, M. Feldmesser, J. C. Rhodes, and D. S. Askew. 2007. Unexpected link between metal ion deficiency and autophagy in *Aspergillus fumigatus*. *Eukaryot. Cell* **6:**2437–2447.

Roman, E., D. M. Arana, C. Nombela, R. Alonso-Monge, and J. Pla. 2007. MAP kinase pathways as regulators of fungal virulence. *Trends Microbiol.* **15:**181–190.

Rubin-Bejerano, I., I. Fraser, P. Grisafi, and G. R. Fink. 2003. Phagocytosis by neutrophils induces an amino acid deprivation response in *Saccharomyces cerevisiae* and *Candida albicans*. *Proc. Natl. Acad. Sci. USA* **100:**11007–11012.

Ruijter, G. J., and J. Visser. 1997. Carbon repression in aspergilli. *FEMS Microbiol. Lett.* **151:**103–114.

Sandhu, D. K., R. S. Sandhu, Z. U. Khan, and V. N. Damodaran. 1976. Conditional virulence of a *p*-aminobenzoic acid-requiring mutant of *Aspergillus fumigatus*. *Infect. Immun.* **13:**527–532.

Santos, R., A. A. Firmino, C. M. de Sá, and C. R. Felix. 1996. Keratinolytic activity of *Aspergillus fumigatus* Fresenius. *Curr. Microbiol.* **33:**364–370.

Sasse, C., E. M. Bignell, M. Hasenberg, K. Haynes, M. Gunzer, G. H. Braus, and S. Krappmann. 2008. Basal expression of the *Aspergillus fumigatus* transcriptional activator CpcA is sufficient to support pulmonary aspergillosis. *Fungal Genet. Biol.* **45:**693–704.

Schöbel, F., O. Ibrahim-Granet, P. Ave, J.-P. Latgé, A. A. Brakhage, and M. Brock. 2007. *Aspergillus fumigatus* does not require fatty acid metabolism via isocitrate lyase for development of invasive aspergillosis. *Infect. Immun.* **75:**1237–1244.

Schoberle, T., and G. S. May. 2007. Fungal genomics: a tool to explore central metabolism of *Aspergillus fumigatus* and its role in virulence. *Adv. Genet.* **57:**263–283.

Shamji, A. F., F. G. Kuruvilla, and S. L. Schreiber. 2000. Partitioning the transcriptional program induced by rapamycin among the effectors of the Tor proteins. *Curr. Biol.* **10:**1574–1581.

Spikes, S., R. Xu, C. K. Nguyen, G. Chamilos, D. P. Kontoyiannis, R. H. Jacobson, D. E. Ejzykowicz, L. Y. Chiang, S. G. Filler, and G. S. May. 2008. Gliotoxin production in *Aspergillus fumigatus* contributes to host-specific differences in virulence. *J. Infect. Dis.* **197:** 479–486.

Sugui, J. A., J. Pardo, Y. C. Chang, K. A. Zarember, G. Nardone, E. M. Galvez, A. Mullbacher, J. I. Gallin, M. M. Simon, and K. J. Kwon-Chung. 2007. Gliotoxin is a virulence factor of *Aspergillus fumigatus*: *gliP* deletion attenuates virulence in mice immunosuppressed with hydrocortisone. *Eukaryot. Cell* **6:**1562–1569.

Tang, C. M., J. M. Smith, H. N. Arst, Jr., and D. W. Holden. 1994. Virulence studies of *Aspergillus nidulans* mutants requiring lysine or *p*-aminobenzoic acid in invasive pulmonary aspergillosis. *Infect. Immun.* **62:**5255–5260.

Tekaia, F., and J.-P. Latgé. 2005. *Aspergillus fumigatus*: saprophyte or pathogen? *Curr. Opin. Microbiol.* **8:**385–392.

Valenzuela, L., C. Aranda, and A. Gonzalez. 2001. TOR modulates GCN4-dependent expression of genes turned on by nitrogen limitation. *J. Bacteriol.* **183:**2331–2334.

Wei, H., K. Vienken, R. Weber, S. Bunting, N. Requena, and R. Fischer. 2004. A putative high affinity hexose transporter, *hxtA*, of *Aspergillus nidulans* is induced in vegetative hyphae upon starvation and in ascogenous hyphae during cleistothecium formation. *Fungal Genet. Biol.* **41:**148–156.

Wek, R. C., H. Y. Jiang, and T. G. Anthony. 2006. Coping with stress: eIF2 kinases and translational control. *Biochem. Soc. Trans.* **34:**7–11.

Wiles, A. M., F. Naider, and J. M. Becker. 2006. Transmembrane domain prediction and consensus sequence identification of the oligopeptide transport family. *Res. Microbiol.* **157:**395–406.

Wilson, R. A., and H. N. Arst, Jr. 1998. Mutational analysis of AREA, a transcriptional activator mediating nitrogen metabolite repression in *Aspergillus nidulans* and a member of the "streetwise" GATA family of transcription factors. *Microbiol. Mol. Biol. Rev.* **62:**586–596.

Xu, J. R. 2000. MAP kinases in fungal pathogens. *Fungal Genet. Biol.* **31:**137–152.

Xue, T., C. K. Nguyen, A. Romans, and G. S. May. 2004. A mitogen-activated protein kinase that senses nitrogen regulates conidial germination and growth in *Aspergillus fumigatus*. *Eukaryot. Cell* **3:** 557–560.

Yang, W., W. S. Kim, A. Fang, and A. L. Demain. 2003. Carbon and nitrogen source nutrition of fumagillin biosynthesis by *Aspergillus fumigatus*. *Curr. Microbiol.* **46:**275–279.

Zaas, A. K., and W. J. Steinbach. 2007. Mammalian models of aspergillosis, p. 401–412. *In* G. H. Goldman and S. A. Osmani (ed.), *The Aspergilli: Genomics, Medical Aspects, Biotechnology, and Research Methods*. CRC Press, Boca Raton, FL.

Zabriskie, T. M., and M. D. Jackson. 2000. Lysine biosynthesis and metabolism in fungi. *Nat. Prod. Rep.* **17:**85–97.

Aspergillus fumigatus and Aspergillosis
Edited by J.-P. Latgé and W. J. Steinbach

Chapter 7

Phosphlipases of *Aspergillus fumigatus*

Geoffrey D. Robson

In the environment, *Aspergillus fumigatus* is a thermotolerant saprophyte commonly isolated from soils, decaying plant material, and compost heaps (Campbell and McHardy, 1994; Haines, 1995; Beffa et al., 1998), and its genome encodes a broad repertoire of secretory hydrolases that enable it to utilize a diverse range of natural polymers in the environment (Nierman et al., 2005; Robson et al., 2005). *A. fumigatus* spores are common in the air, and it is estimated that on average, several hundred are inhaled per person every day (Hospenthal et al., 1998; Schmitt et al., 1990), where due to their small spore size they are able to reach the surfaces of the alveoli. The vast majority of diseases caused by *A. fumigatus* are caused by inhalation of spores, where in immunocompetent individuals they are normally removed by the innate immune system (Latgé, 2001; Hohl and Feldmesser, 2007). However, in immunocompromised individuals or in individuals with previous lung damage, spores germinate and grow in the lung, where they can cause inflammation and tissue damage, can invade the lung tissue and eventually escape into the bloodstream, and can infect other major organs. Small particles of a similar size to *A. fumigatus* spores have been shown to become coated in lung surfactant when inhaled (Morgenroth, 1998). Lung surfactant, which forms an essential coating on the lung surface that enables gaseous exchange, is composed of ca. 90% phospholipid with the remainder being composed largely of surfactant proteins (Turner et al., 1998; Veldhuizen et al., 1998). Spores entering and reaching the surface of the lung are therefore in a phospholipid-rich environment, and phospholipid degradation and metabolism are thus important in the subsequent colonization and invasion of the lung surface. Therefore, an understanding of the biology of the organism and its interaction with components of the lung surface is essential in understanding disease development.

PHOSPHOLIPIDS

Phospholipids and lipids are important biological cellular components that share a glycerol backbone esterified to saturated or unsaturated fatty acids. In the case of lipids, fatty acyl side chains are esterified to all three carbons of the glycerol backbone, forming triacylglycerol lipids that are hydrophobic high-energy storage molecules that are usually liquid at room temperature and accumulate in cells as lipid inclusion droplets. Phospholipids have fatty acyl side chains esterified to positions C-1 and C-2 of the glycerol backbone with a hydrophilic group esterified via a phosphoester bond at position C-3 (Fig. 1). Phospholipids are therefore amphiphilic molecules and are critical structural components of the membrane bilayers of cells. The most common groups esterified to position C-3 are choline, ethanolamine, serine, inositol, and glycerol and typically comprise over 90% of phospholipid membranes.

PHOSPHOLIPASES

Phospholipases are a group of esterases that are composed of two major categories, the acyl hydrolases and the phosphodiesterases. The acyl hydrolases consist of phospholipases A_1 and A_2 (PLA_1 and PLA_2, respectively), phospholipase B (PLB), and the lysophospholipases, while the phosphodiesterases consist of phospholipase C (PLC) and phospholipase D (PLD) (Waite, 1987). PLA_1 and PLA_2 remove the fatty acyl chain from position Sn1 and Sn2 of the glycerol backbone, respectively, while PLB removes both fatty acyl chains simultaneously but is also capable of removing the fatty acid chain from lysophospholipids and hence also has lysophospholipase activity. Diacylglycerides are formed by the removal of both the phosphate moiety and polar

Geoffrey D. Robson • Faculty of Life Sciences, 1800 Stopford Building, University of Manchester, Manchester M13 9PT, United Kingdom.

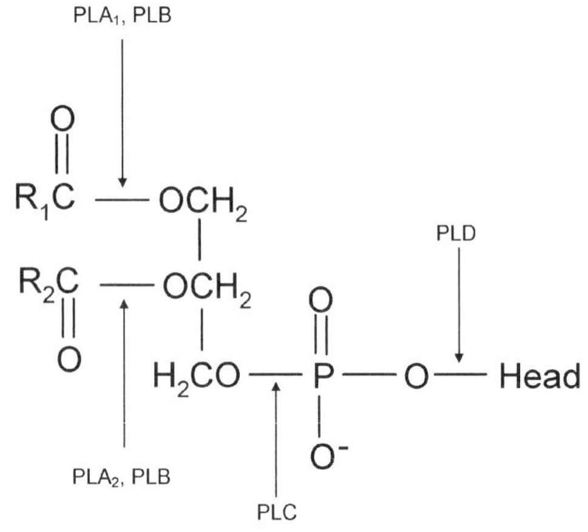

Figure 1. General structure of a phospholipid. R_1 and R_2 represent fatty acyl side chains and "Head" represents the polar head group (generally choline, ethanolamine, serine, inositol, or glycerol).

head group by the action of PLC, which cleaves the phosphodiester bond between the glycerol and phosphate group. Phosphatidic acid is formed by the action of PLD, which cleaves the phosphodiester bond between the polar head group and phosphate moiety. The phospholipid structure and the site of action of the phospholipases are shown in Fig. 1.

Investigations on the phospholipases of *Aspergillus* species have been very limited to date. Indirect evidence for extracellular phospholipase activity was demonstrated in *A. fumigatus* by identifying the accumulation of phospholipid breakdown products in cultures of *A. fumigatus* grown on lecithin using fast atom bombardment spectroscopy based on their mass ratios (Birch et al., 1996). On the basis of the specific degradation products found, it was predicted that *A. fumigatus* secretes multiple extracellular phospholipases, including PLA, PLB, PLC, and PLD. In a colorimetric assay, it was also confirmed that the presence of extracellular PLC activity was initially observed after 30 h of growth and accumulated in broth cultures up to 50 h.

In another study (unpublished data cited by Ghannoum, 2000), the production of phospholipases was investigated in five strains of *A. fumigatus* and three strains of *Aspergillus flavus*. After 3 days of growth, supernatants were concentrated and assayed for phospholipase activity by using a specific substrate radial diffusion assay capable of differentiating between PLA, PLB, and PLC. Phospholipase activity was detected in all strains regardless of temperature, and PLB was found to be the predominant phospholipase for both species and was shown to cause extensive cytolysis of cultured pneumocytes. More recently, the extracellular phospholipase activity was investigated in clinical and environmental isolates of *A. fumigatus*. In this survey the production of extracellular phospholipase of *A. fumigatus* collected from different centers worldwide was compared. All isolates showed extracellular phospholipase activity, including PLC and phospholipid acyl hydrolyse activity (PLA and/or PLB). The largest zone sizes in the diffusion assay were observed in clinical isolates of *A. fumigatus*, and these isolates produced more extracellular PLC than did the environmental isolates. In this study it was suggested that extracellular PLC activity rather than extracellular acyl hydrolase activity might be important in the pathogenicity of *A. fumigatus* (Birch et al., 2004), although the sample size for environmental isolates was small and geographically restricted. The genome of *A. fumigatus* is predicted to encode one PLA (secreted), three PLBs (two secreted), three PLCs (all secreted), and three PLDs (none predicted to be secreted), although little is known about their potential role in infection, survival, or growth in the lung environment. In addition, the genome encodes a number of intracellular patatin-like phospholipases of unknown function and a number of phosphoinositide-specific PLCs that are unrelated to the secreted PLCs and are involved in intracellular signaling. The properties of the predicted phospholipase proteins are shown in Table 1.

PLA_2

PLA_2 enzymes specifically remove the fatty acyl chain from position Sn2 from the glycerol backbone of phospholipids. The most widely distributed PLA_2s in fungi show the greatest homology to the mammalian cytosolic group IV PLAs and are phylogenetically more closely related to the mammalian cytosolic PLA_2 than fungal PLB enzymes, which also share homology with the mammalian cytosolic PLA_2s (Hong et al., 2005; Ghosh et al., 2006). They contain a single PLA2_B lysophospholipase domain (pfam01735), which is typically found in both PLA_2s and PLB and absent from the genomes of yeasts. Phylogenetically they fall broadly into three groups (Fig. 2): the first includes the *Aspergillus* species and contains an N-terminal secretory signal, and the other include two related PLAs encoded by *Ajellomyces capsulatus* and *Coccidiodes immitis*. *A. capsulatus* lacks an N-terminal secretory signal and is predicted to be cytosolic. The second group, which includes *Sclerotinia sclerotiorum*, *Trichoderma reesei*, and *Fusarium graminearum*, possesses either a signal peptide or signal anchor with the exception of *Chaetomium globosum* and *Neurospora crassa* PLA1. To date, only the PLA from *Aspergillus nidulans* has been functionally studied, and it was shown to be constitutively expressed in the presence of glucose, required calcium for maximal activity, and gave rise to no visible phenotype when deleted (Hong et al., 2005).

A third group of PLAs present in some fungi include that described from the mycorrhizal ascomycete

Table 1. Properties of predicted phospholipases from *A. fumigatus*

Phospholipase	Secretion signal[a]	Cleavage sequence	Predicted size (amino acids)	Predicted mass (Da)	Locus
afPLA	Yes (SP)	$ALG_{19}/L_{20}L$	807	89,359	afu2g11970
afPLB1	Yes (SP)	$VSG_{20}/A_{21}P$	613	66,202	afu4g08720
afPLB2	No		588	63,370	afu5g01340
afPLB3	Yes (SP)	$ATA_{16}/T_{17}P$	614	67,488	afu3g14680
afPLC-A	Yes (SA)	$AGA_{18}/A_{19}P$	415	45,872	afu3g01530
afPLC-B	Yes (SP)	$ASA_{18}/I_{19}P$	438	48,134	afu1g17590
afPLC-C	Yes (SP)	$AAA_{18}/A_{19}A$	442	48,532	afu7g04910
afPLD1	No		976	109,969	afu2g16520
afPLD2	No		1,807	204,858	afu3g05630
afPLD3	No		879	101,073	afu7g05580

[a] SA, predicted signal anchor; SP, predicted signal peptide.

Tuber borchi (Soragni et al., 2001), in which it may play a role during nutrient starvation and mycorrhizal development (Miozzi et al., 2005). However, orthologs are not universally conserved among the fungi and are unusual in that they contain a prokaryotic Phospholip_A2_3 (pfam09056) domain, are more closely related to the PLAs from the *Streptomycetes*, and have been classified as a group XIV PLA_2. While present in *Aspergillus oryzae*, its absence from the other sequenced *Aspergillus* genomes, including *A. fumigatus*, and the low degree of homology among the other fungal PLBs suggest this class of PLA was acquired at an evolutionarily early stage, possibly from a prokaryote, and was subsequently lost from many species.

A putative PLA_1 gene was cloned and the protein characterized from *A. oryzae* (Watanabe et al., 1999; Shiba et al., 2001), where it has been shown to release the fatty acid side chain from position Sn1, releasing lysophosphatidic acid. However, it appears unrelated to other PLA_1 eukaryotic sequences and shares homology to the lipase_3 domain containing triacylglcerol lipases. It is therefore uncertain as to whether this protein is a true PLA_1 or is a triacylglycerol monoacyl hydrolase with an ability to act on triacylglycerol and phospholipid.

PLB (Lysophospholipase)

Three genes have been identified in the genome of *A. fumigatus* that are predicted to encode PLBs; two (PLB1 and -3) are putatively secreted, while the third (PLB2) appears to be intracellular (Robson et al., 2005). All contain the catalytic triad Arg, Ser, and Asp typically found among the fungal PLBs and in human cytosolic PLA_2 (Pickard et al., 1996), and with the exception of PLB2 they are conserved among the other aspergilli sequenced to date. More recently, Shen et al. (2004) confirmed the cDNA sequence of PLB1 and PLB3 by rapid amplification of cDNA ends-PCR and demonstrated their upregulation when grown in the presence of phospholipid, whereas the expression of the intracellular PLB2 was unchanged. Phylogenetic analysis suggests that the three PLBs arose through gene duplication events, with the first duplication giving rise to PLB2 while the other underwent a second duplication, giving rise to PLB1 and PLB3 (Fig. 3). A more distantly related fourth group of secreted PLBs is present in *Aspergillus clavatus*, *A. oryzae*, *Aspergillus terreus*, and *A. nidulans* but is not present in *A. fumigatus*. Interestingly, while PLB2 from *A. fumigatus* lacks a secretory signal, the ortholog in *Neosartorya fischeri* contains a signal peptide, whereas *A. fumigatus* PLB3, which does contain a signal peptide, is lacking in the *N. fischeri* orthologs (Fig. 3).

The role of secreted PLB in the pathogenicity of *A. fumigatus* has yet to be evaluated; however, in *Candida albicans*, in which until recently extracellular phospholipase activity was thought to be restricted to two secreted PLBs (Leidich et al., 1998; Sugiyama et al., 1999), studies have demonstrated that PLB1 is produced during invasion of the gastrointestinal tract of mice and strains lacking PLB1 are significantly less virulent both in an intragastric infant mouse model and in an intravenous model of murine candidiasis and have an impaired ability to invade tissue (Ghannoum, 2000; Leidich et al., 1998), while reintroduction of the gene into the disrupted strain restored pathogenicity (Mukherjee et al., 2001). *C. albicans* is now known to contain five PLB members with a third, PLB5, containing a predicted glycosylphosphatidylinositol (GPI) anchor which in a knockout strain demonstrated reduced virulence and a reduced cell-associated PLB activity (Theiss et al., 2006). In the aspergilli, PLBs predicted to be secreted all contain signal peptides and lack a GPI anchor.

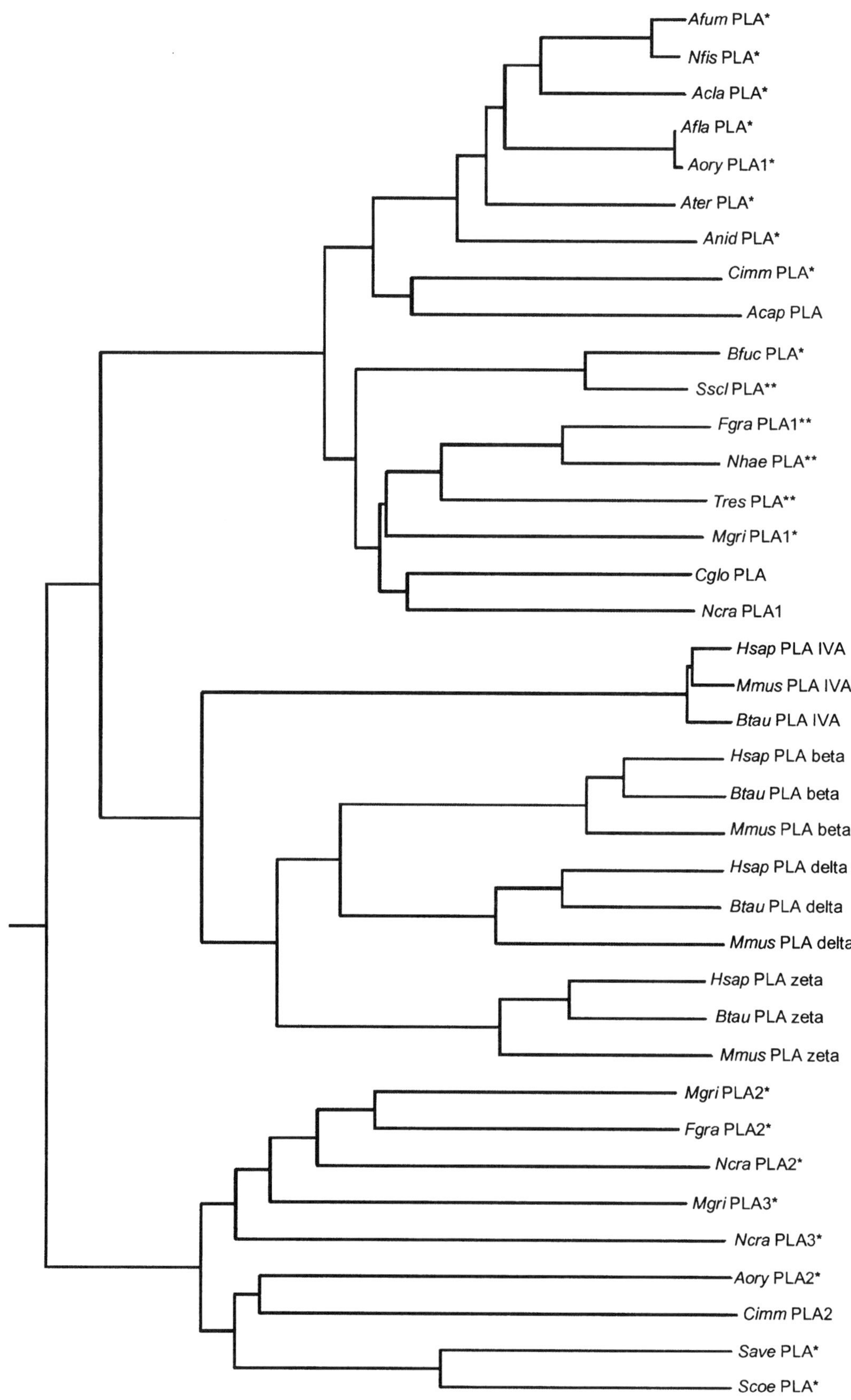
Afum PLA*
Nfis PLA*
Acla PLA*
Afla PLA*
Aory PLA1*
Ater PLA*
Anid PLA*
Cimm PLA*
Acap PLA
Bfuc PLA*
Sscl PLA**
Fgra PLA1**
Nhae PLA**
Tres PLA**
Mgri PLA1*
Cglo PLA
Ncra PLA1
Hsap PLA IVA
Mmus PLA IVA
Btau PLA IVA
Hsap PLA beta
Btau PLA beta
Mmus PLA beta
Hsap PLA delta
Btau PLA delta
Mmus PLA delta
Hsap PLA zeta
Btau PLA zeta
Mmus PLA zeta
Mgri PLA2*
Fgra PLA2*
Ncra PLA2*
Mgri PLA3*
Ncra PLA3*
Aory PLA2*
Cimm PLA2
Save PLA*
Scoe PLA*

In *Cryptococcus neoformans*, a single gene encoding a secreted PLB (*plb1*) has been cloned, targeted gene disruption has been shown to greatly reduce extracellular PLB activity, and the disrupted strain was less virulent than control strains in a mouse inhalation model and a rabbit meningitis model (Cox et al., 2001). More recently, the *plb1* gene was shown to encode an N-terminal leader peptide and C-terminal GPI anchor attachment motifs, suggesting PLB1 is GPI anchored. This was confirmed experimentally when PLB1 activity was demonstrated to be predominantly in the cell wall and was, in addition, found to play a role in maintaining cell wall integrity (Djordevic et al., 2005; Siafakas et al., 2007). In addition, inhibitors of *C. neoformans* PLB have been shown to have antifungal activity and to reduce brain load in a disseminated cryptococcosis mouse model, further implicating extracellular PLB as an important virulence component and leading to the suggestion that PLB may be an effective target for the development of novel antifungal compounds (Ganendren et al., 2004; Widmer et al., 2006). The activity of the secreted protein and its secretion in *C. neoformans* have been shown to be dependent on N-glycosylation (Chen et al., 2000; Turner et al., 2006), and a number of putative N-glycosylation sites have also been identified on the secreted PLBs in *A. fumigatus* (Shen et al., 2005). Extracellular PLB activity in both *C. neoformans* and *C. albicans* is therefore an important factor in the virulence of these organisms and additionally may also interfere with signaling mechanisms within the host cells (Shea et al., 2006). Currently, research into the role of the secreted PLBs in the pathobiology of *A. fumigatus* remains to be determined; while two genes are predicted to encode secreted PLBs and extracellular PLB activity has been demonstrated, orthologs are present throughout the aspergilli, and the only PLB which is present in *A. fumigatus* and the closely related *N. fischeri* and absent from the other aspergilli, PLB2, is predicted to be intracellular. Pathogenicity of gene knockout strains in a pulmonary animal model will be required to determine the significance of PLB activity in lung infection and disease progression.

PLC

The genome of *A. fumigatus* is predicted to encode three secretory PLC genes (one predicted to contain a secretory anchor signal with activity close to the inner cell, which may play not only an important nutritional role in lung colonization but may also through the release of the second messenger diacylglycerol interfere with cell signaling processes). In some bacterial species, including *Clostridium perfringens*, *Listeria monocytogenes*, and *Pseudomonas aeruginosa*, PLCs have been shown to be important pathogenicity factors, causing extensive cellular tissue damage (Titball, 1993; Songer, 1997). In the fungi, the presence of PLC genes is not universal but is largely restricted to the aspergilli along with some other *Ascomycetes*, including *T. reesei*, *F. graminearum*, and *M. grisea*, but is absent from *N. crassa* and the yeasts. Amongst the aspergilli sequenced to date, the number of PLC genes varies from three to five, whereas fungi outside of this group only have a single gene (Fig. 4). A recent analysis of *A. clavatus*, *A. flavus*, *A. nidulans*, *A. terreus*, *C. immitis*, *F. graminearum*, *N. fischeri*, and *T. reesei* with plant, bacterial, and *Tetrahymena thermophila* sequences revealed that the fungal PLCs were divided between two clades, one containing four groups (PLC-A to -D) and the second a single group (PLC-E). PLC appears to be absent in mammals (Tuckwell et al., 2006). All the fungal PLC genes contain a single phosphoesterase family domain (pfam 04185) and with the exception of the PLC-A group, which contains a signal anchor sequence, possess an N-terminal secretory signal sequence. It would appear likely from the phylogenetic analysis (Fig. 4) that two duplication events occurred, the first giving rise to PLC-A and PLC-B/C followed by a second duplication in PLC-B/C giving rise to PLB-B and PLC-C. It has been suggested that as the related PLC-D group is not universally found in all the aspergilli that this may have arisen from another duplication event and subsequently lost from some members of the aspergilli (Tuckwell et al., 2006). PLC-A to -D are more closely related to plant PLCs than members of the PLC-E group, which are more related to bacterial PLCs. PLC-E is absent in *A. fumigatus* but present in *A. terreus*, *A. oryzae*, and *A. flavus*, suggesting that this gene was lost from the genomes of the other aspergilli.

PLD

PLD acts mainly on choline-containing phospholipids and leads to the release of choline and phosphatidic

Figure 2. Phylogentic relationship of PLAs from fungi and other organisms. Fungi abbreviations: Acap, *Ajellomyces capsulatus*; Acla, *Aspergillus clavatus*; Afla, *Aspergillus flavus*; Afum, *Aspergillus fumigatus*; Anid, *Aspergillus nidulans*; Aory, *Aspergillus oryzae*; Ater, *Aspergillus terreus*; Bfuc, *Botryotinea fuckeliana*; Cimm, *Coccidiodes immitis*; Cglo, *Chaetomium globosum*; Fgra, *Fusarium graminearum*; Mgri, *Magnaporthe grisea*; Ncra, *Neurospora crassa*; Nfis, *Neosartorya fischeri*; Nhae, *Nectria haematococca*; Sscl, *Sclerotinia sclerotiorum*. Mammalian abbreviations: Btau, *Bos taurus*; Hsap, *Homo sapiens*; Mmus, *Mus musculus*. Bacteria abbreviations: Save, *Streptomyces avermitilis*; Scoe, *Streptomyces coelicolor*. The neighbor-joining tree was constructed using ClustalW, and predicted protein sequences were only selected from organisms for which the whole genome was sequenced and the annotation was publicly available. *, signal peptide predicted; **, signal anchor predicted.

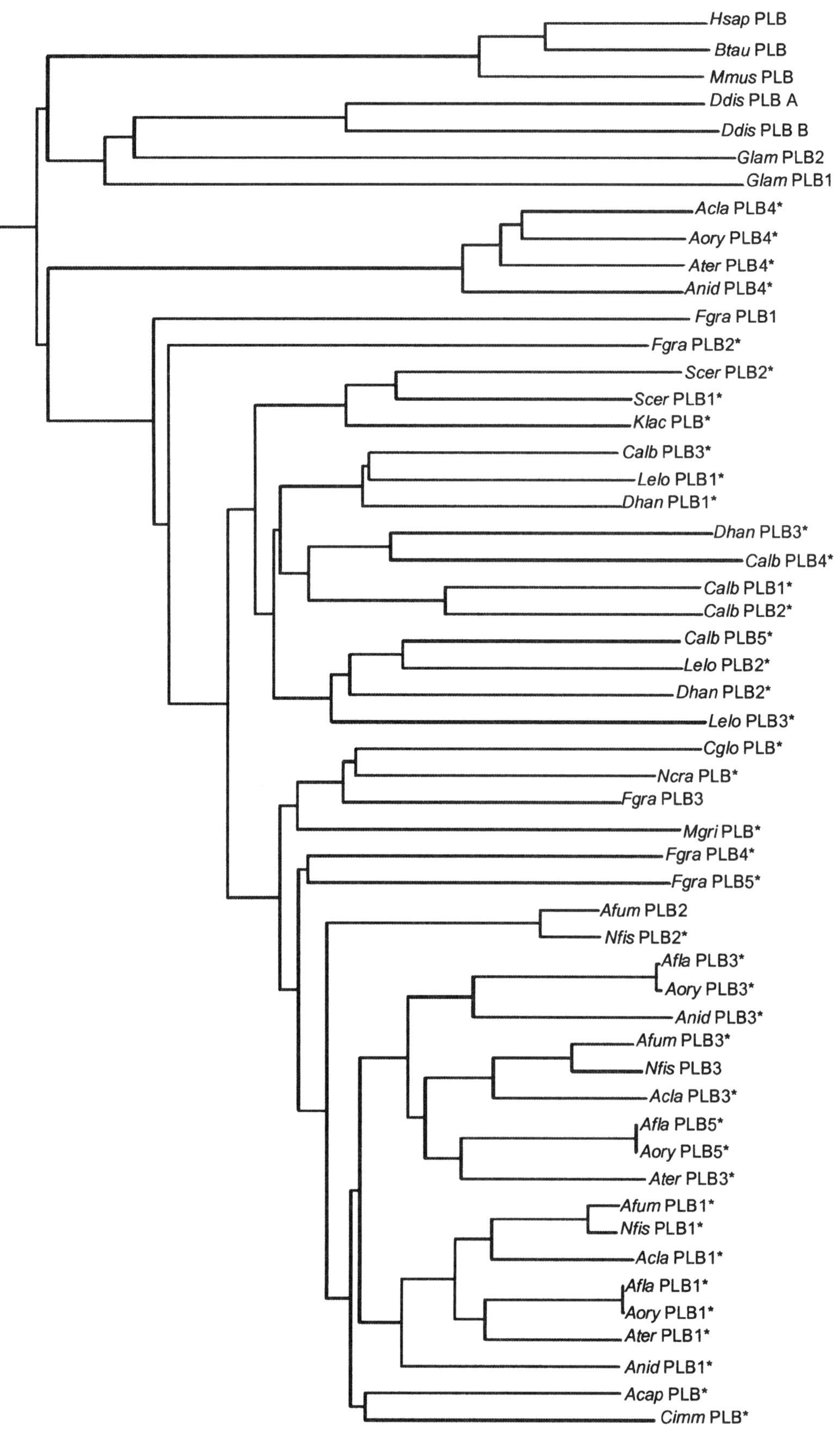
Hsap PLB
Btau PLB
Mmus PLB
Ddis PLB A
Ddis PLB B
Glam PLB2
Glam PLB1
Acla PLB4*
Aory PLB4*
Ater PLB4*
Anid PLB4*
Fgra PLB1
Fgra PLB2*
Scer PLB2*
Scer PLB1*
Klac PLB*
Calb PLB3*
Lelo PLB1*
Dhan PLB1*
Dhan PLB3*
Calb PLB4*
Calb PLB1*
Calb PLB2*
Calb PLB5*
Lelo PLB2*
Dhan PLB2*
Lelo PLB3*
Cglo PLB*
Ncra PLB*
Fgra PLB3
Mgri PLB*
Fgra PLB4*
Fgra PLB5*
Afum PLB2
Nfis PLB2*
Afla PLB3*
Aory PLB3*
Anid PLB3*
Afum PLB3*
Nfis PLB3
Acla PLB3*
Afla PLB5*
Aory PLB5*
Ater PLB3*
Afum PLB1*
Nfis PLB1*
Acla PLB1*
Afla PLB1*
Aory PLB1*
Ater PLB1*
Anid PLB1*
Acap PLB*
Cimm PLB*

acid as hydrolysis products. PLD is widely distributed across the eukaryotes, and PLD genes have been identified in mammals, plants, bacteria, and yeast (Brown et al., 2007). In mammalian cells, two isoforms of PLD have been identified, while up to three isoforms have been identified in plants (Exton, 1990; Hammond et al., 1997; Ueki et al., 1995). To date, most of the PLD activity described in fungi has been intracellular. PLD activity was identified in *C. albicans* and was associated with a membrane preparation (McLain and Dolan, 1997). Fetal bovine serum, a known inducer of dimorphic transition, was found to stimulate PLD activity, as does the addition of exogenous PLD. Moreover, the enzyme activity was capable of a transphosphatidylation reaction, a characteristic activity of PLD enzymes. Thus, intracellular PLD was shown to have a potential role in facilitating yeast-to-hyphal switching in *C. albicans*. PLD from *C. albicans* shares a 42% identity at the amino acid level with the PLD protein SPO14 from *S. cerevisiae* (Kanoh et al., 1998), which has been shown to be essential for sporulation (Ella et al., 1996; Waksman et al., 1996). More recently a PLD gene encoding an intracellular PLD was identified and disrupted in *A. nidulans*, although the mutant was found to display no visible phenotype (Hong et al., 2003).

The *A. fumigatus* genome encodes three PLD genes, all of which are predicted to encode nonsecretory proteins. However, PLD activity has been detected in the culture supernatant during the log phase growth of *A. fumigatus* which was fivefold higher in the presence of phospholipid (unpublished observation). As yet, no full-length cDNA sequences or sequenced proteins have been published, and therefore the predicted open reading frames have yet to be confirmed experimentally. Orthologs of the three PLD genes are also present in the other aspergilli and in *N. fischeri* with the exception of *A. terreus*, in which one gene appears to have been lost, and *A. nidulans*, in which an additional gene appears to have arisen through duplication. None of the predicted PLDs contains a signal sequence, and the role of these phospholipases in relation to phospholipid degradation in the lung remains unknown.

Patatin-Like Phospholipases

Another group of proteins with potential phospholipase activity found within the genomes of fungi are those containing patatin-like domains. Patatin is a large soluble storage protein found in the vacuoles of potato tubers and in the leaves of other plants that is activated in response to pathogen attack and environmental and oxidative stress (Prat et al., 1990; Dhondt et al., 2000, 2002; La Camera et al., 2005). The protein has esterase activity and possesses a Ser/Asp catalytic diad similar to that found in mammalian cytosolic PLA2 and has been found functionally to possess PLA2 activity, although it does not contain a Ca^{2+}-dependent binding domain (Dessen et al., 1999; Rydel et al., 2003).

The fungi also encode proteins (between four and eight) with a patatin-like domain and are widespread among yeasts, ascomycetes, and basidiomycetes. However, none appears to encode an N-terminal secretory signal and they are unlikely to play a direct role in extracellular phospholipid degradation, but as yet, their function in fungi remains to be determined.

Other Possible Functions of Fungal Extracellular Phospholipases

The limited evidence to date suggests that fungal phospholipases may be important nutritionally in lung colonization and may also play a role in pathogenicity by damaging host cell membranes. However, studies of phospholipases in other microorganisms, in particular bacteria, suggest that it is likely that phospholipases may have also other functions that facilitate virulence in addition to causing direct tissue damage to the host cell (Songer, 1997; Titball, 1993).

Many of the products generated by the effects of phospholipases on phospholipids are known to act as second messengers and mediators of signal transduction (Dennis et al., 1991; Serhan et al., 1996). For example, lysophospholipids produced by the action of PLA are reported to induce the activation of protein kinase C, which could interfere with normal cell signaling (Oishi et al., 1988; Songer, 1997). In addition, accumulation

Figure 3. Phylogentic relationship of PLBs from fungi and other organisms. Fungi abbreviations: Acap, *Ajellomyces capsulatus*; Acla, *Aspergillus clavatus*; Afla, *Aspergillus flavus*; Afum, *Aspergillus fumigatus*; Anid, *Aspergillus nidulans*; Aory, *Aspergillus oryzae*; Ater, *Aspergillus terreus*; Calb, *Candida albicans*; Cimm, *Coccidiodes immitis*; Cglo, *Chaetomium globosum*; Dhan, *Debaromyces hansenii*; Fgra, *Fusarium graminearum*; Klac, *Kluyveromyces lactis*; Lelo, *Lodderomyces elongisporus*; Mgri, *Magnaporthe grisea*; Ncra, *Neurospora crassa*; Nfis, *Neosartorya fischeri*; Scer, *Saccharomyces cerevisiae*. Mammalian abbreviations: Btau, *Bos taurus*; Hsap, *Homo sapiens*; Mmus, *Mus musculus*. Protist abbreviations: Ddis, *Dictyostelium discoideum*; Glam, *Giardia lamblia*. The neighbor-joining tree was constructed using ClustalW, and predicted protein sequences were only selected from organisms for which the whole genome was sequenced and the annotation was publicly available. *, signal peptide predicted; **, signal anchor predicted.

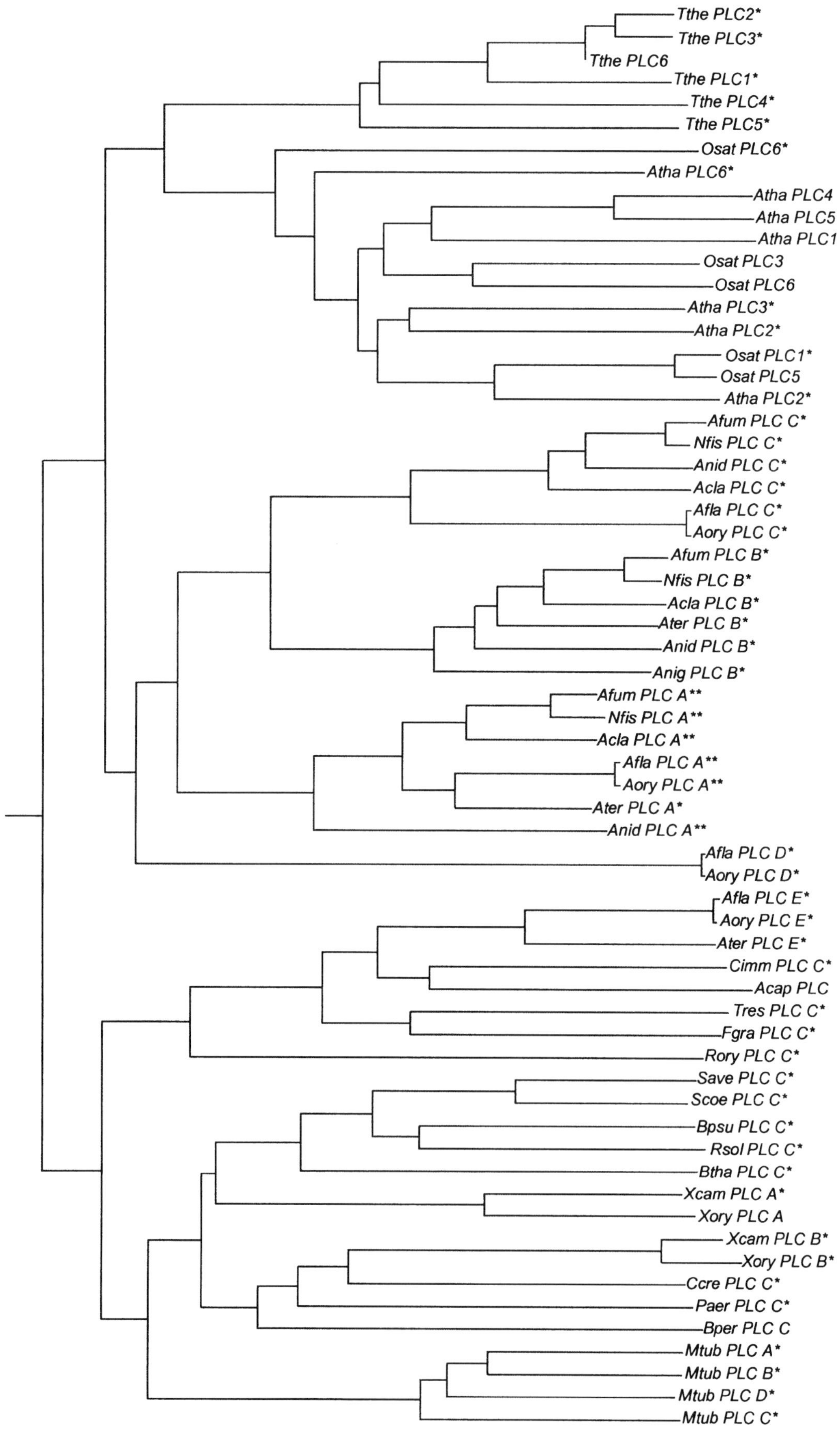
Tthe PLC2*
Tthe PLC3*
Tthe PLC6
Tthe PLC1*
Tthe PLC4*
Tthe PLC5*
Osat PLC6*
Atha PLC6*
Atha PLC4
Atha PLC5
Atha PLC1
Osat PLC3
Osat PLC6
Atha PLC3*
Atha PLC2*
Osat PLC1*
Osat PLC5
Atha PLC2*
Afum PLC C*
Nfis PLC C*
Anid PLC C*
Acla PLC C*
Afla PLC C*
Aory PLC C*
Afum PLC B*
Nfis PLC B*
Acla PLC B*
Ater PLC B*
Anid PLC B*
Anig PLC B*
Afum PLC A**
Nfis PLC A**
Acla PLC A**
Afla PLC A**
Aory PLC A**
Ater PLC A*
Anid PLC A**
Afla PLC D*
Aory PLC D*
Afla PLC E*
Aory PLC E*
Ater PLC E*
Cimm PLC C*
Acap PLC
Tres PLC C*
Fgra PLC C*
Rory PLC C*
Save PLC C*
Scoe PLC C*
Bpsu PLC C*
Rsol PLC C*
Btha PLC C*
Xcam PLC A*
Xory PLC A
Xcam PLC B*
Xory PLC B*
Ccre PLC C*
Paer PLC C*
Bper PLC C
Mtub PLC A*
Mtub PLC B*
Mtub PLC D*
Mtub PLC C*

of lysophospholipids is known to impair surfactant function (Hite et al., 2004). Phospholipases play an important role in the induction of cytokines in mammalian cells; for example, PLC activity from *C. perfringens* has been reported to induce the expression of interleukin-8 (IL-8) by endothelial cells (Bryant et al., 1993; Bunting et al., 1997). Phospholipase secretion by *Entamoeba histolytica* is known to stimulate epithelial cells to secrete IL-6 and IL-8 (Eckmann et al., 1995). Similarly, Kaplanski et al. (1995) reported that injury of endothelial cells by phospholipases following exposure to *Rickettsia conorii* led to the synthesis of IL-6 and IL-8.

The stimulation of cytokine production in response to lytic enzymes of microbial origin has also been demonstrated in vivo (May et al., 1996). The production of phospholipase is associated with candidal host injury (Leidich et al., 1998), so it is possible that the enzyme may directly or indirectly play a role in stimulating host cells to produce specific cytokines. One possible mechanism for the expression by epithelial cells of genes encoding IL-6 and IL-8 following microbial injury was suggested by Eckmann et al., 1995. They reported that *E. histolytica* injured epithelial cells and caused them to secrete IL-1 into the medium. This cytokine stimulated adjacent epithelial cells to produce IL-8 and IL-6. Microbial phospholipases have also been shown to be potent inflammatory agents, inducing the accumulation of inflammatory cells and plasma proteins and the release of various inflammatory mediators in vivo (Meyers and Berk, 1990; Wieland et al., 2002). In addition, microbial phospholipases are able to mobilize arachadonic acid and induce subsequent prostaglandin synthesis (Walker et al., 1990). PLD releases phosphatidic acid, which is a lipid mediator with a broad range of effects on cellular function, including cell signaling, growth proliferation, and stress responses (Wang et al., 2006). The action of PLC releases diacylglycerol, which is a potent activator of protein kinase C, leading to the activation of the protein kinase C pathway mediating numerous cellular processes (Nishizuka, 1995; Mellor and Parker, 1998). In the lung, treatment of cells with PLC or diacylglycerol has been shown to stimulate surfactant secretion by annexin A7-mediated fusion of lung lamellar bodies (Chander et al., 2007). When PLC was introduced into the lungs of rats it caused a rapid destruction of the alveolar lining cells, leading to interstitial pneumonia in 10 days (Ito et al., 1975). Thus, in addition to direct effects on cell and tissue integrity, extracellular phospholipases may also have indirect impacts by interfering with a number of host cell signaling pathways.

PHOSPHOLIPID METABOLISM

The combination of secreted phospholipases on the phospholipid-rich environment of the lung will generate a number of breakdown products which can be utilized as a carbon source for growth at the lung epithelium, principally fatty acids and glycerol. Fatty acid metabolism proceeds through the β-oxidation pathway (Kunau et al., 1995), which in yeast occurs exclusively in peroxisomes, while in filamentous fungi, like animals, β-oxidation of fatty acids occurs in both peroxisomes and mitochondria (Maggio-Hall and Keller, 2004). In fungi, acetyl coenzyme A (CoA) resulting from β-oxidation of fatty acids is directed through the glyoxylate shunt, which is essential for growth on C_2 compounds (Gainey et al., 1992; De Lucas et al., 1999). Moreover, isocitrate lyase, a key enzyme in the glyoxylate pathway, has been correlated with virulence in a number of microbial pathogens of humans and plants, including *C. albicans* (Lorenz and Fink, 2001; Idnurm and Howlett, 2002; Wang et al., 2003; Munoz-Elias and McKinney, 2005; Wall et al., 2005; Ramirez and Lorenz, 2007). In *C. albicans*, macrophage engulfment and subsequent nutrient starvation were accompanied by the induction of fatty acid metabolism and the glyoxylate cycle (Lorenz et al., 2004); this work followed an earlier study that implicated the glyoxylate pathway as essential for fungal virulence (Lorenz and Fink, 2001). However, while genes encoding fatty acid metabolism and the glyoxylate pathway were also upregulated during early stages of infection of a rabbit model with *C. neoformans*, strains in which isocitrate lyase had been knocked out showed no reduced pathogenicity (Rude et al., 2002). Moreover, a recent study in *A. fumigatus* in which the gene encoding

Figure 4. Phylogentic relationship of PLCs from fungi and other organisms. Fungi abbreviations: Acap, *Ajellomyces capsulatus*; Acla, *Aspergillus clavatus*; Afla, *Aspergillus flavus*; Afum, *Aspergillus fumigatus*; Anid, *Aspergillus nidulans*; Aory, *Aspergillus oryzae*; Ater, *Aspergillus terreus*; Cimm, *Coccidiodes immitis*; Fgra, *Fusarium graminearum*; Nfis, *Neosartorya fischeri*; Rory, *Rhizopus oryzae*; Tres, *Trichoderma reesei*. Plant abbreviations: Atha, *Arabidopsis thaliani*; Osat, *Oryza sativa*. Protist abbreviation: Tthe, *Tetrahymena thermophila*. Bacteria abbreviations: Bper, *Burkholderia mallei*; Bpsu, *Burkholderia psuedomallei*; Btha, *Burkholderia thailandensis*; Ccre, *Caulobacter crescentus*; Mtub, *Mycobacterium tuberculosis*; Paer, *Pseudomonas aeruginosa*; Rsol, *Ralstonia solanacearum*; Save, *Streptomyces avermitilis*; Scoe, *Streptomyces coelicolor*; Xcam, *Xanthomonas campestris*; Xory, *Xanthomonas oryzae*. The neighbor-joining tree was constructed using ClustalW, and predicted protein sequences were only selected from organisms for which the whole genome was sequenced and annotation was publicly available. *, signal peptide predicted; **, signal anchor predicted.

isocitrate lyase or malate synthase had been deleted showed no impairment in pathogenicity compared to the wild type in neutropenic mice in a model of invasive pulmonary aspergillosis (Olivas et al., 2008), and a separate study in which isocitrate lyase was deleted also showed no effect on pathogenicity or lung pathology in a mouse pulmonary model (Schöbel et al., 2007). These studies call into question the role of fatty acid metabolism in lung infection, although glycerol metabolism was unaffected and could represent a major source of carbon from the combined action of phospholipases on lung phospholipid. Recently, we demonstrated that growth on phosphatidylcholine leads to the upregulation of the β-oxidation pathway, glyoxylate pathway, and other genes involved in lipid metabolism and transport in *A. fumigatus* and *A. nidulans*, but that a higher proportion of genes associated with lipid metabolism were upregulated in *A. fumigatus* than in *A. nidulans*, suggesting that *A. fumigatus* may be better adapted to metabolizing phospholipids as a carbon source (unpublished observation).

CONCLUSIONS

Studies on the roles of extracellular phospholipases, while still at an early stage in *C. albicans* and *C. neoformans*, have clearly implicated their activities as important factors in the pathogenicity of these organisms in murine models, for cell adhesion, in tissue damage, and in generating lipid by-products that may interfere with host signaling pathways. By contrast, little is known about the role of extracellular phospholipases in the filamentous fungi in lung colonization and invasion. Genome studies have not revealed obvious differences in the predicted phospholipases between *A. fumigatus* and other aspergilli that would account for its higher level of pathogenicity. Nonetheless, the potential contribution of extracellular phospholipase activity, individually and collectively, is yet to be determined. In addition, the individual characteristics of the phospholipase proteins, such as specific activities and substrate specificities, as well as levels of gene expression and protein translation, may differ significantly between *A. fumigatus* and other aspergilli.

REFERENCES

Beffa, T., F. Staib, and J. Lott Fischer. 1998. Mycological control and surveillance of biological waste and compost. *Med. Mycol.* **36:**137–145.

Birch, M., G. Robson, D. Law, and D. W. Denning. 1996. Evidence of multiple extracellular phospholipase activities of *Aspergillus fumigatus*. *Infect. Immun.* **64:**751–755.

Birch, M., D. W. Denning, and G. Robson. 2004. A comparison of extracellular phospholipase activities in clinical and environmental *Aspergillus fumigatus* isolates. *Med. Mycol.* **42:**81–86.

Brown, H. A., L. G. Henage, A. M. Preininger, Y. Xiang, and J. H. Exton. 2007. Biochemical analysis of phospholipase D. *Methods Enzymol.* **434:**49–87.

Bryant, A. E., R. Bergstrom, G. A. Zimmerman, J. L. Salyer, H. R. Hill, R. K. Tweten, H. Sato, and D. L. Stevens. 1993. *Clostridium perfringens* invasiveness is enhanced by effects of theta toxin upon PMNL structure and function: the roles of leukocytotoxicity and expression of CD11/CD18 adherence glycoprotein. *FEMS Immunol. Med. Microbiol.* 7:321–336.

Bunting, M., D. E. Lorant, A. E. Bryant, G. A. Zimmerman, T. M. McIntyre, D. L. Stevens, and S. M. Prescott. 1997. Alpha toxin from *Clostridium perfringens* induces proinflammatory changes in endothelial cells. *J. Clin. Investig.* **100:**565–574.

Campbell, C. D., and W. I. McHardy. 1994. Scanning electron microscopy of the microbial colonization of composted tree bark. *Micron* **25:**253–255.

Chander, A., X.-L. Chen, and D. G. Naidu. 2007. Role for diacylglycerol in annexin A7-mediated fusion of lung lamellar bodies. *Biochim. Biophys. Acta* **1771:**1308–1318.

Chen, S. C., L. C. Wright, J. C. Golding, and T. C. Sorrell. 2000. Purification and characterization of secretory phospholipase B, lysophospholipase and lysophospholipase/transacylase from a virulent strain of the pathogenic fungus *Cryptococcus neoformans*. *Biochem. J.* **347:**431–439.

Cox, G. M., H. C. McDade, S. C. Chen, S. C. Tucker, M. Gottfredsson, L. C. Wright, T. C. Sorrell, S. D. Leidich, A. Casadevall, M. A. Ghannoum, and J. R. Perfect. 2001. Extracellular phospholipase activity is a virulence factor for *Cryptococcus neoformans*. *Mol. Microbiol.* 39:166–175.

De Lucas, J. R., A. I. Dominguez, S. Valenciano, G. Turner, and F. Laborda. 1999. The *acuH* gene of *Aspergillus nidulans*, required for growth on acetate and long-chain fatty acids, encodes a putative homologue of the mammalian carnitine/acyl-carnitine carrier. *Arch. Microbiol.* 171:386–396.

Dennis, E. A., S. G. Rhee, M. M. Billah, and Y. A. Hannun. 1991. Role of phospholipase in generating lipid second messengers in signal transduction. *FASEB J.* 5:2068–2077.

Derouin, F. 1994. Special issue on aspergillosis. *Pathol. Biol.* **42:**625–736.

Dessen, A., J. Tang, H. Schmidt, M. Stahl, J. D. Clark, J. Seehra, and W. S. Somers. 1999. Crystal structure of human cytosolic phospholipase A2 reveals a novel topology and catalytic mechanism. *Cell* 97:349–360.

Dhondt, S., P. Geoffroy, B. A. Stelmach, M. Legrand, and T. Heitz. 2000 Soluble phospholipase A2 activity is induced before oxylipin accumulation in tobacco mosaic virus-infected tobacco leaves and is contributed by patatin-like enzymes. *Plant J.* 23:431–440.

Dhondt, S., G. Gouzerh, A. Muller, M. Legrand, and T. Heitz. 2002. Spatio-temporal expression of patatin-like lipid acyl hydrolases and accumulation of jasmonates in elicitor-treated tobacco leaves are not affected by endogenous levels of salicylic acid. *Plant J.* 32:749–762.

Djordjevic, J. T., M. Del Poeta, T. C. Sorrel, K. M. Turner, and L. C. Wright. 2005. Secretion of cryptococcal phospholipase B1 PLB1 is regulated by a glycosylphosphatidylinositol GPI anchor. *Biochem. J.* 389:803–812.

Eckmann, L., S. L. Reed, R. Smith, and M. F. Kagnoff. 1995. *Entamoeba histolytica* trophozoites induce an inflammatory cytokine response by cultured human cells through the paracrine action of cytolytically released interleukin-1 alpha. *J. Clin. Investig.* **96:**1269–1279.

Ella, K. M., J. W. Dolan, C. Qi, and K. E. Meier. 1996. Characterization of *Saccharomyces cerevisiae* deficient in expression of phospholipase D. *Biochem. J.* **314:**15–19.

Exton, J. H. 1990. Signaling through phosphatidylcholine breakdown. *J. Biol. Chem.* **265:**1–4.

Gainey, L. D., I. F. Connerton, E. H. Lewis, G. Turner, and D. J. Balance. 1992. Characterization of the glyoxysomal isocitrate lyase genes of *Aspergillus nidulans* acuD and *Neurospora crassa* acu-3. *Curr. Genet.* **21:**43–47.

Ganendren, R., F. Widmer, V. Singha, C. Wilson, T. Sorrell, and L. Wright. 2004. In vitro antifungal activities of inhibitors of phospholipases from the fungal pathogen *Cryptococcus neoformans. Antimicrob. Agents Chemother.* **48:**1561–1569.

Ghannoum, M. A. 2000. Potential role of phospholipases in virulence and fungal pathogenesis. *Clin. Microbiol. Rev.* **13:**122–143.

Ghosh, M., D. E. Tucker, S. A. Burchett, and C. C. Leslie. 2006. Properties of the group IV phospholipase A_2 family. *Prog. Lipid Res.* **45:**487–510.

Haines, J. 1995. Aspergillus in compost: straw man or fatal flaw. *Biocycle* **6:**32–35.

Hammond, S. M., J. M. Jenco, S. Nakashima, K. Cadwallader, Q. Gu, S. Cook, Y. Nozawa, G. D. Prestwich, M. A. Frohman, and A. J. Morris. 1997. Characterization of two alternately spliced forms of phospholipase D1. Activation of the purified enzymes by phosphatidylinositol 4,5-bisphosphate, ADP-ribosylation factor, and Rho family monomeric GTP-binding proteins and protein kinase C-alpha. *J. Biol. Chem.* **272:**3860–3868.

Hite, R. D., M. C. Seeds, A. M. Safta, R. B. Jacinto, J. I. Gyves, D. A. Bass, and B. M. Waite. 2004. Lysophospholipid generation and phosphatidylglycerol depletion in phospholipase A_2-mediated surfactant dysfunction. *Am. J. Physiol. Lung Cell. Mol. Physiol.* **288:** L618–L624.

Hohl, T. M., and T. Feldmesser. 2007. *Aspergillus fumigatus*: principles of pathogenesis and host defense. *Eukaryot. Cell* **6:**1953–1963.

Hong, S., H. Horiuchi, and A. Ohta. 2003. Molecular cloning of a phospholipase D gene from *Aspergillus nidulans* and characterization of its deletion mutants. *FEMS Microbiol. Lett.* **224:**231–237.

Hong, S., H. Horiuchi, and A. Ohta. 2005. Identification and molecular cloning of a gene encoding phospholipase A2 plaA from *Aspergillus nidulans. Biochim. Biophys. Acta* **1735:**222–229.

Hospenthal, D. R., K. J. Kwon-Chung, and J. E. Bennett. 1998. Concentrations of airborne *Aspergillus* compared to the incidence of invasive aspergillosis: lack of correlation. *Med. Mycol.* **36:**165–168.

Idnurm, A., and B. J. Howlett. 2002. Isocitrate lyase is essential for pathogenicity of the fungus *Leptosphaeria maculans* to canola *Brassica napus. Eukaryot. Cell* **1:**719–724.

Ito, M., N. Takeuchi, T. Masuno, M. Kikui, and Y. Yamamura. 1975. Lung damage caused by phospholipase C and the changes in phospholipids in the rat lung. *Jpn. J. Exp. Med.* **45:**479–488.

Kanoh, H., S. Nakashima, Y. Zhao, Y. Sugiyama, Y. Kitajima, and Y. Nozawa. 1998. Molecular cloning of a gene encoding phospholipase D from the pathogenic and dimorphic fungus, *Candida albicans. Biochim. Biophys. Acta* **1398:**359–364.

Kaplanski, G., N. Teysseire, C. Farnarier, S. Kaplanski, J. C. Lissitzky, J. M. Durand, J. Soubeyrand, C. A. Dinarello, and P. Bongrand. 1995. IL-6 and IL-8 production from cultured human endothelial cells stimulated by infection with *Rickettsia conorii* via a cell-associated IL-1 alpha-dependent pathway. *J. Clin. Investig.* **96:** 2839–2844.

Kunau, W. H., V. Dommes, and H. Schulz. 1995. β-Oxidation of fatty acids in mitochondria, peroxisomes, and bacteria: a century of continued progress. *Prog. Lipid Res.* **34:**267–342.

La Camera, S., P. Geoffroy, H. Samaha, A. Ndiaye, G. Rahim, M. Legrand, and T. Heitz. 2005. A pathogen-inducible patatin-like lipid acyl hydrolase facilitates fungal and bacterial host colonization in *Arabidopsis. Plant J.* **44:**810–825.

Latgé, J.-P. 2001. The pathobiology of *Aspergillus fumigatus. Trends Microbiol.* **9:**382–389.

Leidich, S. D., A. S. Ibrahim, Y. Fu, A. Koul, C. Jessup, J. Vitullo, W. Fonzi, F. Mirbod, S. Nakashima, Y. Nozawa, and M. A. Ghannoum. 1998. Cloning and disruption of caPLB1, a phospholipase B gene involved in the pathogenicity of *Candida albicans. J. Biol. Chem.* **273:**26078–26086.

Lorenz, M. C., and G. R. Fink. 2001. The glyoxylate cycle is required for fungal virulence. *Nature* **412:**83–86.

Lorenz, M. C., J. A. Bender, and G. R. Fink. 2004. Transcriptional response of *Candida albicans* upon internalization by macrophages. *Eukaryot. Cell* **3:**1076–1087.

Maggio-Hall, L. A., and N. P. Keller. 2004. Mitochondrial beta-oxidation in *Aspergillus nidulans. Mol. Microbiol.* **54:**1173–1185.

May, A. K., R. G. Sawyer, T. Gleason, A. Whitworth, and T. L. Pruett. 1996. In vivo cytokine response to *Escherichia coli* alpha-hemolysin determined with genetically engineered hemolytic and nonhemolytic *E. coli* variants. *Infect. Immun.* **64:**2167–2171.

McLain, N., and J. W. Dolan. 1997. Phospholipase D activity is required for dimorphic transition in *Candida albicans. Microbiology* **143:**3521–3526.

Mellor, H., and P. J. Parker. 1998. The extended protein kinase C superfamily. *Biochem. J.* **332:**281–292.

Meyers, D. J., and R. S. Berk. 1990. Characterization of phospholipase C from *Pseudomonas aeruginosa* as a potent inflammatory agent. *Infect. Immun.* **58:**659–666.

Miozzi, L., R. Balestrini, A. Bolchi, M. Novero, O. Ottonello, and P. Bonfante. 2005. Phospholipase A2 up-regulation during mycorrhiza formation in *Tuber borchii. New Phytol.* **167:**229–238.

Morgenroth, K. 1988. *The Surfactant System of the Lungs.* Walter de Gruyter, Berlin, Germany.

Mukherjee, P. K., K. R. Seshan, S. D. Leidich, J. Chandra, G. T. Cole, and M. A. Ghannoum. 2001. Reintroduction of the PLB1 gene into *Candida albicans* restores virulence *in vivo. Microbiology* **147:** 2585–2597.

Muñoz-Elías, E. J., and J. D. McKinney. 2005. *Mycobacterium tuberculosis* isocitrate lyases 1 and 2 are jointly required for *in vivo* growth and virulence. *Nat. Med.* **11:**638–644.

Nierman, W. C., A. Pain, M. J. Anderson, J. R. Wortman, H. S. Kim, J. Arroyo, M. Berriman, K. Abe, D. B. Archer, C. Bermejo, J. Bennett, P. Bowyer, D. Chen, M. Collins, R. Coulsen, R. Davies, P. S. Dyer, M. Farman, N. Fedorova, N. Fedorova, T. V. Feldblyum, R. Fischer, N. Fosker, A. Fraser, J. L. García, M. J. García, A. Goble, G. H. Goldman, K. Gomi, S. Griffith-Jones, R. Gwilliam, B. Haas, H. Haas, D. Harris, H. Horiuchi, J. Huang, S. Humphray, J. Jiménez, N. Keller, H. Khouri, K. Kitamoto, T. Kobayashi, S. Konzack, R. Kulkarni, T. Kumagai, A. Lafon, J.-P. Latgé, W. Li, A. Lord, C. Lu, W. H. Majoros, G. S. May, B. L. Miller, Y. Mohamoud, M. Molina, M. Monod, I. Mouyna, S. Mulligan, L. Murphy, S. O'Neil, I. Paulsen, M. A. Peñalva, M. A. Pertea, C. Price, B. L. Pritchard, M. A. Quail, E. Rabbinowitsch, N. Rawlins, M. A. Rajandream, U. Reichard, H. Renauld, G. D. Robson, S. Rodriguez de Córdoba, J. M. Rodríguez-Peña, C. M. Ronning, S. Rutter, S. L. Salzberg, M. Sanchez, J. C. Sánchez-Ferrero, D. Saunders, K. Seeger, R. Squares, S. Squares, M. Takeuchi, F. Tekaia, G. Turner, C. R. Vazquez de Aldana, J. Weidman, O. White, J. Woodward, J. H. Yu, C. Fraser, J. E. Galagan, K. Asai, M. Machida, N. Hall, B. Barrell, and D. W. Denning. 2005. Genomic sequence of the pathogenic and allergenic filamentous fungus *Aspergillus fumigatus. Nature* **438:**1151–1156.

Nishizuka, Y. 1995. Protein kinase C and lipid signalling for sustained cellular responses. *FASEB J.* **9:**484–496.

Oishi, K., R. L. Raynor, P. A. Charp, and J. F. Kuo. 1988. Regulation of protein kinase C by lysophospholipids. *J. Biol. Chem.* **263:**6865–6871.

Olivas, I., M. Royuela, B. Romero, M. C. Monteiro, J. M. Mínguez, F. Laborda, and J. R. De Lucas. 2008. Ability to grow on lipids accounts for the fully virulent phenotype in neutropenic mice of

Aspergillus fumigatus null mutants in the key glyoxylate cycle enzymes. *Fungal Genet. Biol.* 45:45–60.

Pickard, T. R., X. G. Chiou, B. A. Strifler, M. R. DeFelippis, P. A. Hyslop, A. L. Tebbe, Y. K. Yee, L. J. Reynolds, E. A. Dennis, R. M. Kramer, and J. D. Sharp. 1996. Identification of essential residues for the catalytic function of 85-kDa cytosolic phospholipase A2. Probing the role of histidine, aspartic acid, cysteine, and arginine. *J. Biol. Chem.* **271:**19225–19231.

Prat, S., W. B. Frommer, R. Hofgen, M. Keil, J. Kossmann, M. Köster-Töpfer, X. J. Liu, B. Müller, H. Peña-Cortés, M. Rocha-Sosa, J. J. Sanchez-Serrano, U. Sonnewald, and L. Willmitzer. 1990. Gene expression during tuber development in potato plants. *FEBS Lett.* **268:** 334–338.

Ramírez, M. A., and L. C. Lorenz. 2007. Mutations in alternative carbon utilization pathways in *Candida albicans* attenuate virulence and confer pleiotropic phenotypes. *Eukaryot. Cell* **6:**280–290.

Robson, G. D., J. Huang, J. Wortman, and D. B. Archer. 2005. A preliminary analysis of the process of protein secretion and the diversity of putative secreted hydrolases encoded in *Aspergillus fumigatus*: insights from the genome. *Med. Mycol.* **43:**S41–S47.

Rude, T. H., D. L. Toffaletti, G. M. Cox, and J. R. Perfect. 2002. Relationship of the glyoxylate pathway to the pathogenesis of *Cryptococcus neoformans*. *Infect. Immun.* **70:**5684–5694.

Rydel, T. J., J. M. Williams, E. Krieger, F. Moshiri, W. C. Stallings, S. M. Brown, J. C. Pershing, J. P. Purcell, and M. F. Alibhai. 2003. The crystal structure, mutagenesis, and activity studies reveal that patatin is a lipid acyl hydrolase with a Ser-Asp catalytic dyad. *Biochemistry* **42:**6696–6708.

Schmitt, H. J., A. Blevins, K. Sobeck, and D. Armstrong. 1990. *Aspergillus* species from hospital air and from patients. *Mycoses* **33:** 539–541.

Schobel, F., O. Ibrahim-Granet, P. Ave, J.-P. Latgé, A. A. Brakhage, and M. Brock. 2007. *Aspergillus fumigatus* does not require fatty acid metabolism via isocitrate lyase for development of invasive aspergillosis. *Infect. Immun.* **75:**1237–1244.

Serhan, C. N., J. Z. Haeggstrom, and C. C. Leslie. 1996. Lipid mediator networks in cell signaling: update and impact of cytokines. *FASEB J.* **10:**1147–1158.

Shea, J. M., J. L. Henry, and M. Del Poeta. 2006. Lipid metabolism in *Cryptococcus neoformans*. *FEMS Yeast Res.* **6:**469–479.

Shen, D. K., A. D. Noodeh, A. Kazemi, R. Grillot, G. Robson, and J. F. Brugere. 2004. Characterisation and expression of phospholipases B from the opportunistic fungus *Aspergillus fumigatus*. *FEMS Microbiol. Lett.* **239:**87–93.

Shiba, Y., C. Ono, F. Fukui, I. Watanabe, N. Serizawa, K. Gomi, and H. Yoshikawa. 2001. High-level secretory production of phospholipase A1 by *Saccharomyces cerevisiae* and *Aspergillus oryzae*. *Biosci. Biotechnol. Biochem.* **65:**95–101.

Siafakas, A. R., T. C. Sorrell, L. C. Wright, C. Wilson, M. Larsen, R. Boadle, P. R. Williamson, and J. T. Djordjevic. 2007. Cell wall-linked cryptococcal phospholipase B1 is a source of secreted enzyme and a determinant of cell wall integrity. *J. Biol. Chem.* **282:**37508–37514.

Songer, J. G. 1997. Bacterial phospholipases and their role in virulence. *Trends Microbiol.* **5:**156–161.

Soragni, E., A. Bolchi, R. Balestrini, C. Gambaretto, R. Percudani, P. Bonfante, and S. Ottonello. 2001. A nutrient-regulated, dual localization phospholipase A2 in the symbiotic fungus *Tuber borchii*. *EMBO J.* **20:**5079–5090.

Sugiyama, Y., S. Nakashima, F. Mirbod, H. Kanoh, Y. Kitajima, M. A. Ghannoum, and Y. Nozawa. 1999. Molecular cloning of a second phospholipase B gene, caPLB2 from *Candida albicans*. *Med. Mycol.* **37:**61–67.

Theiss, S., G. Ishdorj, A. Brenot, M. Kretschmar, C. Y. Lan, T. Nichterlein, J. Hacker, S. Nigam, N. Agabian, and G. A. Köhler. 2006. Inactivation of the phospholipase B gene PLB5 in wild-type *Candida albicans* reduces cell-associated phospholipase A2 activity and attenuates virulence. *Int. J. Med. Microbiol.* **296:**405–420.

Titball, R. W. 1993. Bacterial phospholipases C. *Microbiol. Rev.* **57:** 347–366.

Tuckwell, D., S. E. Lavens, and M. Birch. 2006. Two families of extracellular phospholipase C genes are present in aspergilla. *Mycol. Res.* **110:**1140–1151.

Turner, K. M., L. C. Wright, T. C. Sorrell, J. T. Djordjevic, R. Veldhuizen, K. Nag, S. Orgeig, and F. Possmayer. 1998. The role of lipids in pulmonary surfactant. *Biochim. Biophys. Acta* **1408:**90–108.

Turner, K. M., L. C. Wright, T. C. Sorrell, and J. T. Djordjevic. 2006. N-linked glycosylation sites affect secretion of cryptococcal phospholipase B1, irrespective of glycosylphosphatidylinositol anchoring. *Biochim. Biophys. Acta* **1760:**1569–1579.

Ueki, J., S. Morioka, T. Komari, and T. Kumashiro. 1995. Purification and characterization of phospholipase D PLD from rice *Oryza sativa* L. and cloning of cDNA for PLD from rice and maize *Zea mays* L. *Plant Cell Physiol.* **36:**903–914.

Veldhuizen, R., K. Nag, S. Orgeig, and F. Possmayer. 1998. The role of lipids in pulmonary surfactant. *Biochim. Biophys. Acta* **1408:**90–108.

Waite, M. 1987. *The Phospholipases.* Plenum Press, New York, NY.

Waksman, M., Y. Eli, M. Liscovitch, and J. E. Gerst. 1996. Identification and characterization of a gene encoding phospholipase D activity in yeast. *J. Biol. Chem.* **271:**2361–2364.

Walker, T. S., J. S. Brown, C. S. Hoover, and D. A. Morgan. 1990. Endothelial prostaglandin secretion: effects of typhus rickettsiae. *J. Infect. Dis.* **162:**1136–1144.

Wall, D. M, P. S. Duffy, C. Dupont, J. F. Prescott, and W. G. Meijer. 2005. Isocitrate lyase activity is required for virulence of the intracellular pathogen *Rhodococcus equi*. *Infect. Immun.* **73:**6736–6741.

Wang, X. M., S. P. Devalah, W. H. Zhang, and R. Welti. 2006. Signaling functions of phosphatidic acid. *Prog. Lipid Res.* **45:**250–278.

Wang, Z. Y., C. R. Thornton, M. J. Kershaw, L. Debao, and N. J. Talbot. 2003. The glyoxylate cycle is required for temporal regulation of virulence by the plant pathogenic fungus *Magnaporthe grisea*. *Mol. Microbiol.* **47:**1601–1612.

Watanabe, I., R. Koishi, Y. Yao, T. Tsuji, and N. Serizawa. 1999. Molecular cloning and expression of the gene encoding a phospholipase A1 from *Aspergillus oryzae*. *Biosci. Biotechnol. Biochem.* **63:** 820–826.

Widmer, F., L. C. Wright, D. Obando, R. Handke, R. Ganendren, E. H. Ellis, and T. C. Sorrell. 2006. Hexadecylphosphocholine miltefosine has broad-spectrum fungicidal activity and is efficacious in a mouse model of cryptococcosis. *Antimicrob. Agents Chemother.* **50:**414–421.

Wieland, C. W., B. Siegmund, G. Senaldi, M. L. Vasil, C. A. Dinarello, and G. Fantuzzi. 2002. Pulmonary inflammation induced by *Pseudomonas aeruginosa* lipopolysaccharide, phospholipase C, and exotoxin A: role of interferon regulatory factor 1. *Infect. Immun.* **70:** 1352–1358.

Aspergillus fumigatus and Aspergillosis
Edited by J.-P. Latgé and W. J. Steinbach

Chapter 8

Aspergillus fumigatus Secreted Proteases

MICHEL MONOD, OLIVIER JOUSSON, AND UTZ REICHARD

Among various potential virulence factors, secreted proteolytic activity has attracted a lot of attention and has been intensively investigated in *Aspergillus fumigatus*. Like many other ascomycete fungi, *A. fumigatus* grows well in a medium containing protein as a sole nitrogen and carbon source and secretes both endo- and exoproteases (Reichard et al., 1990; Monod et al., 2005). Inspection of the *A. fumigatus* genome reveals more than 100 proteases. The aim of this review is to establish a catalog of the proteases secreted by *A. fumigatus*, to briefly describe their various properties, and to examine their biological functions, ranging from protein digestion into short peptides and assimilable amino acids to specific proteolysis during infection.

CLASSIFICATION OF PROTEASES

The term protease is synonymous with peptidase, proteolytic enzyme, and peptide hydrolase. The proteases include all enzymes that catalyze the cleavage of peptide bonds (CO-NH) of proteins, digesting them into peptides or free amino acids. The classification and the nomenclature of proteases can be found together with information about them in the *Handbook of Proteolytic Enzymes* (Barrett et al., 2004) and in the MEROPS database (Rawlings et al., 2006; accessible at http://merops.sanger.ac.uk/). Proteases are basically classified according to their catalytic mechanism and their active sites. Aspartic, cysteine, glutamic, metallo-serine, and threonine proteases as well as proteases with an unknown catalytic mechanism are recognized. Each protease has been assigned to a family representing a set of homologous enzymes (Table 1). These families are identified by a capital letter for the catalytic type of the protease they contain, together with a unique number. Subfamilies are labeled with a second capital letter (for instance, M28A and M28E in the serine proteases) (Table 1). Different families that are thought to have a common origin are grouped into a clan.

The proteases can be further divided into endoproteases (or endopeptidases) and exoproteases (or exopeptidases). Endoproteases cleave peptide bonds internally within a polypeptide. Exoproteases cleave peptide bonds only at the N or the C terminus of a polypeptide chain.

SECRETED PROTEASES

Most eukaryotic and prokaryotic secreted proteins are synthesized as precursors with a hydrophobic N-terminal extension of 15 to 30 amino acids (Fig. 1), known as the prepeptide or signal peptide (Blobel and Dobberstein, 1975). This signal peptide is necessary for entering the secretory pathway by enabling transport of the protein across the membrane of the endoplasmic reticulum (Pfeffer and Rothman, 1987), where it is subsequently cleaved by a signal peptidase (Milstein at al., 1972; Walker and Lively, 2004). Secretion into the environment may not necessarily be an obligate feature of proteases made with a signal peptide, as several of these proteases are vacuolar and remain intracellular. Other proteases with a signal peptide may be attached to the cell via a transmembrane domain or by a glycophosphatidylinositol (GPI) anchor.

Comparison of the amino acid residues of signal sequences allows the identification of three structurally distinct regions: a basic N-terminal region (the n-region), a central hydrophobic region of 8 to 12 amino acids (the h-region), and a usually more polar C-terminal region of 4 to 7 amino acids (the c-region) (von Heijne, 1985). Cleavage sites appear to conform to the following rules: (i) the residue in position −1 (from the cleavage site) must be small, i.e., either alanine, serine, glycine, cysteine, or threonine, and the residue in posi-

Michel Monod • Service de Dermatologie, Laboratoire de Mycologie, Centre Hospitalier Universitaire Vaudois, 1011 Lausanne, Switzerland. **Olivier Jousson** • Centre for Integrative Biology, University of Trento, 38100 Trento, Italy. **Utz Reichard** • Dept. of Medical Microbiology and National Reference Center for Systemic Mycoses, University Hospital of Goettingen, 37075 Goettingen, Germany.

Table 1. *A. fumigatus* proteases synthesized with a signal sequence and closely related proteases[a]

Family/subfamily[b]	Clan[b]	MERNUM [identifier][b]	Protease(s) as annotated[c]; gene(s) as published (reference)	Locus[c]	Protein NCBI accession no[d]	mRNA NCBI accession no[d]	GPI anchoring prediction[e] and general comments
A1	AA	MER004337 [A01.018]	Aspartic endopeptidase Pep2; *PEP2* (Reichard et al., 2000a)	AFUA_3G11400	XP_754479	XM_749386	(+) Vacuolar and possibly secreted (Reichard et al., 2000a); ortholog of *S. cerevisiae* proteinase A or saccharopepsin (MER00941)
A1	AA	MER001437 [01.026]	Aspartic endopeptidase Pep1/ aspergillopepsin F; *PEP1* (Reichard et al., 1995), PepF (Lee and Kolattukudy, 1995)	AFUA_5G13300	XP_753324	XM_748231	Main secreted aspartic protease (Reichard et al., 1994); ortholog of *A. niger* PepA or aspergillopepsin I (MER00919)
A1	AA	MER078983 [A01.077]	Aspartic-type endopeptidase (CtsD), putative; *CTSD* (Vickers et al., 2007)	AFUA_4G07040	XP_752122	XM_747029	(● and +) Secreted (Vickers et al., 2007)
A1	AA	MER082517 as nonpeptidase homolog	Aspartic-type endopeptidase, putative	AFUA_3G01220	XP_748443	XM_743350	(+) Putative secreted protease
A1	AA	MER071466 unassigned	Aspartic-type endopeptidase (OpsB), putative	AFUA_6G05350	XP_747542	XM_742449	(● and +) Putative secreted protease
A1	AA	MER107327 [A01.079]	Extracellular aspartic endopeptidase, putative	AFUA_2G15950	XP_755930	XM_750837	Putative secreted protease
# A1	AA	MER082513 as nonpeptidase homolog	Aspartic endopeptidase (AP1), putative	AFUA_6G03260	XP_747750	XM_742657	Similar to A1 secreted aspartic proteases, but no signal sequence
G1	GA	MER107323 [G01.002]	Aspergillopepsin, putative	AFUA_3G02970	XP_748619	XM_743526	Putative secreted protease; homologous to *A. niger* aspergillopepsin II (MER93102)
G1	GA	MER107322 unassigned	Aspergillopepsin, putative	AFUA_7G01200	XP_746812	XM_741719	Putative secreted protease; homologous to *A. niger* aspergillopepsin II (MER93102)
M12	MA	MER079660 unassigned	ADAM family of metalloprotease ADM-A; *ADM-A* (Lavens et al., 2005)	AFUA_6G14420	XP_751316	XM_746223	Putative membrane protease; domain structure similar to ADAM metalloproteases (Lavens et al., 2005); homologous to ADAM proteases and snake metallopeptidases
M12	MA	MER079661 unassigned	ADAM family of metalloprotease ADM-B; *ADM-B* (Lavens et al., 2005)	AFUA_4G11150	XP_751714	XM_746621	Putative membrane protease; domain structure similar to ADAM metalloproteases (Lavens et al., 2005); homologous to ADAM proteases and snake metallopeptidases; catalytic domain expressed in *E. coli*

M14	MC	MER082535 nonpeptidase homolog	Zinc carboxypeptidase, putative	AFUA_2G08790	XP_755213	XM_750120	Subfamily M14A nonpeptidase homolog, putative
M20A	MH	MER082533 [M20.002]	Vacuolar carboxypeptidase Cps1, putative	AFUA_3G07040	XP_754901	XM_749808	Closely related to vacuolar *S. cerevisiae* Gly-X carboxypeptidase (MER01269)
M20A	MH	MER049769 unassigned	Peptidase, putative	AFUA_6G06800	XP_750573	XM_745480	Putative carboxypeptidase
M20A	MH	MER082520 nonpeptidase homolog	WD repeat protein	AFUA_2G04360	XP_001481653	XM_001481603	Subfamily M20A nonpeptidase homolog, putative
M20D	MH	MER048266 unassigned	Amidohydrolase, putative	AFUA_1G11250	XP_752490	XM_747397	(+) Putative carboxypeptidase
M28A	MH	MER046068 [M28.001]	Aminopeptidase Y, putative; *LAP2* (Monod et al., 2005)	AFUA_3G00650	XP_748386	XM_743293	Secreted 50-kDa aminopeptidase; ortholog of *A. oryzae* Lap2 (Blinkowsky et al., 2000); closely related to *Streptomyces griseus* aminopeptidase S (MER02161)
M28A	MH	MER100678 [M28.001]	Aminopeptidase, putative	AFUA_2G00220	XP_749158	XM_744065	Putative vacuolar aminopeptidase, ortholog of *S. cerevisiae* aminopeptidase Y (MER01288)
M28E	MH	MER046067 [M28.006]	Aminopeptidase, putative; *LAP1* (Monod et al., 2005)	AFUA_4G04210	XP_746607	XM_741514	Secreted 35-kDa aminopeptidase; ortholog *of A. sojae* Lap1 (Chien et al, 2002); closely related to *Vibrio* aminopeptidase Ap1 (MER03354)
# M28	MH	MER050730 unassigned	Peptidase family M28 family mRNA	AFUA_1G05960	XP_750346	XM_745253	Putative membrane protein
M35	MA	MER002044 [M35.002]	Penicillolysin/deuterolysin metalloprotease, putative	AFUA_4G13750	XP_751456	XM_746363	Putative secreted endoprotease, homologous to *A. oryzae* neutral protease II (Tatsumi et al., 1991)
M35	MA	MER107321 [M35.002]	Metalloproteinase, putative	AFUA_4G02700	XP_746456	XM_741363	Putative secreted endoprotease, homologous to *A. oryzae* neutral protease II (Tatsumi et al., 1991)
M36	MA	MER001400 [M36.001]	Elastinolytic metalloproteinase Mep; *MEP* (Jaton-Ogay et al., 1994; Markaryan et al., 1994)	AFUA_8G07080	XP_747506	XM_742413	Ortholog of *A. oryzae* neutral protease I (MER06285)
M43	MA(M)	MER107325 unassigned	Metalloprotease MEP1	AFUA_1G07730	XP_750519	XM_745426	Putative secreted pappalysin-like endoprotease; homologous to *M. anisopliae* Mep1 (MER11122) and *C. podosaii* Mep1 (Hung et al., 2005)

Continued on following page

Table 1. *Continued*

Family/subfamily[b]	Clan[b]	MERNUM [identifier][b]	Protease(s) as annotated[c]; gene(s) as published (reference)	Locus[c]	Protein NCBI accession no[d]	mRNA NCBI accession no[d]	GPI anchoring prediction[e] and general comments
S8A	SB	MER003475 [S08.052]	Autophagic serine protease Alp; *ALP2* (Reichard et al., 2000b)	AFUA_5G09210	XP_753718	XM_748625	Vacuolar; ortholog of *S. cerevisiae* proteinase B or cerevisin (MER00356)
S8A	SB	MER000343 [S08.053]	Alkaline serine protease Alp1; *ALP1* (Jaton-Ogay et al., 1992; Reichard et al., 1990)	AFUA_4G11800	XP_751651	XM_746558	Secreted alkaline protease
S8B	SB	MER032524 [S08.070]	Pheromone processing endoprotease KexB	AFUA_4G12970	XP_751534	XM_746441	Putative ortholog of *S. cerevisiae* Kex2 (MER00364)
# S8A	SB	MER049758 as nonpeptidase homolog	Alkaline serine protease (PR1)/allergen F18-like	AFUA_7G04930	XP_749017	XM_743924	No signal sequence; 38% identical (146 of 383 amino acids) identical to Alp1 (MER00343)
S9B	SC	MER004383 [S09.008]	Extracellular dipeptidyl-peptidase Dpp4; *DPPIV* (Beauvais et al., 1997b)	AFUA_4G09320	XP_751893	XM_746800	DppIV, secreted X-prolyl peptidase
# S9B	SC	MER032528 [S09.006]	Pheromone maturation dipeptidyl aminopeptidase DapB	AFUA_3G07850	XP_754828	XM_749735	Putative dipeptidyl peptidase with signal anchor; no signal peptidase cleavage site; putative ortholog of *S. cerevisiae* DpapB (MER00405), vacuolar membrane prolyl oligopeptidase
S9 (not assigned to subfamily)	SC	MER000263 [S09.012]	Secreted dipeptidyl-peptidase DppV; *DPPV* (Beauvais et al., 1997a)	AFUA_2G09030	XP_755237	XM_750144	DppV, secreted X-alanyl peptidase
# S9 (not assigned to subfamily)	SC	MER077504 [S09.012]	Oligopeptidase family protein	AFUA_8G04730	XP_747279	XM_742186	No signal sequence; 42% identical (297 of 707 amino acids) to secreted DppV (MER00263)
S10	SC	MER079359 [S10.016]	Carboxypeptidase S1, putative; *CP1* (Zaugg et al., 2007)	AFUA_5G07330	XP_753901	XM_748808	Characterized as recombinant protease (Zaugg et al., 2008)
S10	SC	MER079360 [S01.016]	Carboxypeptidase S1, putative; *CP2* (Zaugg et al., 2007)	AFUA_8G04120	XP_747219	XM_742126	Characterized as recombinant protease (Zaugg et al., 2008)
S10	SC	MER079361 [S10.001]	Carboxypeptidase CpyA/Prc1, putative; *CP3* (Zaugg et al., 2008)	AFUA_6G13540	XP_751230	XM_746137	Putative ortholog of vacuolar *S. cerevisiae* CpY (MER02010)
S10	SC	MER079362 [S10.014]	Serine carboxypeptidase (CpdS), putative; *CP4* (Zaugg et al., 2008)	AFUA_6G00310	XP_731524	XM_726431	Closely related to *A. saitoi* and *A. niger* CpD-I (MER27994) (Zaugg et al., 2008)
S10	SC	MER079363 [S10.006]	Pheromone processing carboxypeptidase (Sxa2), putative; *CP5* (Zaugg et al., 2008)	AFUA_2G03510	XP_749486	XM_744393	Closely related to *A. niger* CpD-II (PepF) (MER00415) (Zaugg et al., 2008)

S10	SC	MER082516 [S10.008]	Carboxypeptidase S1, putative; *CP6* (Zaugg et al., 2008)	AFUA_5G01200	XP_748222	XM_743129	Closely related to *Penicillium janthinellum* CpS (MER00412)
S10	SC	MER082530 unassigned	Carboxypeptidase Y, putative (Cp7 in Fig. 4)	AFUA_3G12210	XP_754400	XM_749307	(+) Putative carboxypeptidase
S10	SC	MER032646 unassigned	Carboxypeptidase Y, putative (Cp8 in Fig. 4)	AFUA_5G14610	XP_753196	XM_748103	Putative carboxypeptidase
S10	SC	MER032643 unassigned	Serine carboxypeptidase, putative (Cp9 in Fig. 4)	AFUA_4G07270	XP_752099	XM747.006	Putative carboxypeptidase
S10	SC	MER082525 [S10.007]	Pheromone processing carboxypeptidase KexA	AFUA_1G08940	XP_752261	XM_747168	Putative ortholog of *S. cerevisiae* KexA (MER00413)
S26	SF	MER049786 [S26.010]	Signal peptidase I	AFUA_3G12840	XP_754337	XM_749244	Signal peptidase complex 21-kDa component
S28	SC	MER064064 as nonpeptidase homolog	Serine peptidase, putative	AFUA_2G01250	XP_749261	XM_744168	Putative ortholog of *A. niger* extracellular prolyl endopeptidase (Edens et al., 2005)
S28	SC	MER049751 unassigned	Serine peptidase, family S28, putative	AFUA_4G03790	XP_746566	XM_741473	Putative secreted exopeptidase
S28	SC	MER049791 unassigned	Extracellular serine carboxypeptidase, putative	AFUA_2G17330	XP_756068	XM_750975	Putative secreted exopeptidase
S33	SC	MER046012 unassigned	α/β Fold family hydrolase, putative	AFUA_1G11400	XP_752505	XM_747412	Putative secreted peptidase
S53	SB	MER049772 [S53.007]	Alkaline serine protease AorO, putative; *SEDA* (Reichard et al., 2006)	AFUA_6G10250	XP_750914	XM_745821	(+) Secreted acid endoprotease (Reichard et al., 2006), ortholog of *A. oryzae* aorsin (Lee et al., 2003)
S53	SB	MER044525 [S53.010]	Tripeptidyl-peptidase (TppA), putative; *SEDB* (Reichard et al., 2006)	AFUA_4G03490	XP_746536	XM_741443	Secreted tripeptidyl-peptidase
S53	SB	MER044603 [S53.010]	Tripeptidyl-peptidase SED3; *SEDC* (Reichard et al., 2006)	AFUA_3G08930	XP_754723	XM_749630	(+) Secreted tripeptidyl-peptidase
S53	SB	MER044526 [S53.010]	Tripeptidyl-peptidase A; *SEDD* (Reichard et al., 2006)	AFUA_4G14000	XP_751432	XM_746339	Secreted tripeptidyl-peptidase
# S53	SB	MER107324 unassigned	Serine protease, putative; *SEDE* (Reichard et al., 2006).	AFUA_7G06220	XP_748888	XM_743795	Putative serine protease

[a] Signal sequences were predicted using the program SignalP 3.0, accessible at http://www.cbs.dtu.dk/services/SignalP/. Other proteases related to secreted members of the same family are included and are labeled by a # in the left margin.
[b] Clans, families, protease numbers, and identifiers are from the *Handbook of Proteolytic Enzymes* (Barrett et al., 2004) and the MEROPS database (Rawling et al., 2004), accessible at http://merops.sanger.ac.uk/. With the same identifier are proteins which display a particular kind of peptidase activity and are closely related in sequence.
[c] Gene annotation and locus from the *A. fumigatus* Af293 genome (accessible at www.tigr.org/tdb/e2k1/afu1).
[d] Accessible at http://www.ncbi.nlm.nih.gov/.
[e] GPI anchoring prediction (indicated in parentheses): +, prediction using GPI-Som (Fankhauser and Mäser, 2005; accessible at http://gpi.unibe.ch/); •, prediction using BigPI Fungal Predictor (Eisenhaber et. al., 2004; accessible at http://mendel.imp.ac.at/gpi/fungi_server.html). The GPI-SOM program is less specific but more sensitive than the BigPI Fungal Predictor. However, with whichever prediction method was used, GPI anchoring can be proven only by an experimental approach. Lack of a symbol indicates that no GPI anchor was predicted by either program.

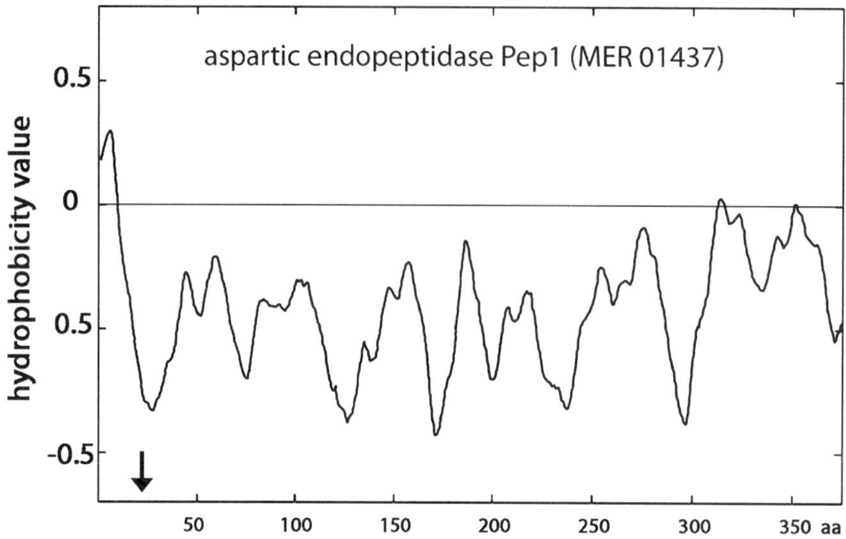

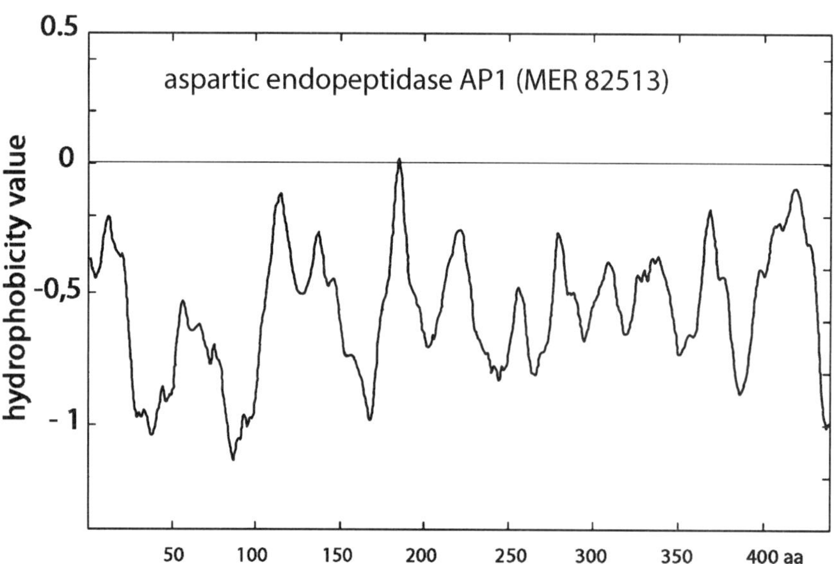

Figure 1. Hydrophobicity plots of two pepsin-like proteases, Pep1 (MER001437) and AP1 (MER082513) (A1 family) (Table 1). Pep1 is a secreted enzyme with a hydrophobic signal sequence. In contrast, AP1 has no hydrophobic signal sequence at its beginning. For the plot, the program TopPred, available at http://bioweb.pasteur.fr/seqanal/interfaces/toppred.html, was used. The Pep1 signal peptidase cleavage site is indicated by a vertical arrow.

tion −3 must also be small or hydrophobic but not aromatic, i.e., either leucine, isoleucine, or valine; (ii) acceptable signal peptidase cleavage sites should reside in a window of 4 to 10 amino acids downstream from the C-terminal amino acid of the hydrophobic core (von Heijne, 1983, 1984, 1985, 1986). Signal sequences can be predicted using the program SignalP 3.0 (available at http://www.cbs.dtu.dk/services/SignalP/) (Nielsen et al., 1997; Bendtsen et al., 2004).

Many, but not all, secreted proteases are synthesized as precursors in a preproprotein form. The propeptide (30 to 250 amino acids in length), which is an intermediate sequence between the signal sequence and the N terminus of the mature enzyme, has been found to be essential and specific in assisting with correct folding and secretion (Fabre et al., 1991; Eder and Fersht, 1995; Fukuda et al., 1996). Upon completion of folding, the propeptide is removed by an autoproteolytic or an exogenous proteolytic reaction to generate the active enzyme (Togni et al., 1996; Newport and Agabian, 1997; Marie-Claire et al., 1998; Zaugg et al., 2008).

A total of 111 proteases and 26 nonpeptidase homologs have been recorded in the MEROPS peptidase database for the *A. fumigatus* genome (as of January 2008). Roughly, proteases constitute 1% of the total *A. fumigatus* proteins (about 10,000). A detailed analysis performed with all recorded proteases revealed that 46 possess a signal sequence. A listing of these enzymes is given in Table 1. The distribution of the 14 families containing *A. fumigatus* secreted proteases varies among the organism kingdoms (Table 2). G1, M35, and M36 are typical fungal families.

PRODUCTION OF *ASPERGILLUS* SECRETED PROTEASES

Several proteases are simultaneously secreted by *A. fumigatus* and other *Aspergillus* species under growth conditions promoting their production (Reichard et al., 1990; Monod et al., 1993b). Purification of an individual protease requires several steps and is often a laborious process. However, numerous *Aspergillus* secreted enzymes can be individually and successfully produced as recombinant proteins using various expression systems. The availability of recombinant proteases has assisted many studies in the fields of secreted protease biochemistry, crystallography, and clinical microbiology. Using a reverse genetics approach (from gene to protein), it has also been possible to synthesize proteases which were revealed by genome sequencing but which have not yet been discovered and remain putative so far. In that manner, new secreted *Aspergillus* proteolytic activities have been discovered (Lavens et al., 2005; Reichard et al., 2006; Vickers et al., 2007; Zaugg et al., 2008).

Table 2. Distribution of protease families among taxonomic kingdoms

Protease family	Presence of protease family in:					
	Archea	Bacteria	Fungi	Protozoans	Plants	Animals
A1	−	−	+	+	+	+
G1	−	+ (few)	+	−	−	−
M12	−	+	+	+	−	+
M20	+	+	+	−	+	+
M28	+	+	+	−	+	+
M35	−	+ (few)	+	−	−	−
M36	−	+ (few)	+	−	−	−
M43	+	+	+	−	−	+
S8A	+	+	+	+	+	+
S9	+	+	+	+	+	+
S10	−	+	+	−	+	+
S28	−	−	+	−	+	+
S33	+	+	+	+	+	+
S53	+	+	+	−	−	+

Escherichia coli was used to produce the *A. fumigatus* aspartic protease CtsD (Vickers et al., 2007), a member of the desintegrin and metalloprotease family ADM-B (Lavens et al., 2005), alkaline protease Alp1 (Moser et al., 1994), *Aspergillus oryzae* deuterolysin (Fushimi et al., 1999), and *Aspergillus niger* aspergillopepsin II (Huang et al., 2000). Bacterial expression systems are quite easy to handle, but folding of the protein to obtain active enzyme has to be performed in vitro. Alternatively, numerous recombinant *A. fumigatus* secreted proteases were obtained using the yeast *Pichia pastoris* as an expression system (Beauvais et al., 1997a, 1997b; Monod et al., 2005; Reichard et al., 2006; Sarfati et al., 2006; Zaugg et al., 2008). The yield, which depended on the protease, varied from 1 to 200 μg/ml of yeast culture supernatant. *Aspergillus* proteases have also been produced by gene overexpression in *Aspergillus* (Edens et al., 2005) and in *Fusarium* spp. (Blinkovsky et al., 1999).

To produce a recombinant secreted protein such as a protease by using *P. pastoris*, the procedure consists of cloning the cDNA encoding the protein of interest downstream of a signal sequence under the control of the tightly regulated alcohol oxidase gene (*AOX1*) promoter in a *P. pastoris* expression vector. In general, the *P. pastoris* acid phosphatase gene (*PHO1*) signal sequence or the α-factor signal peptide sequence are used for entering the secretory pathway of the yeast (Higgins and Cregg, 1998). The construct which carries a gene for selection after transformation of *P. pastoris*, in addition to the cloned coding sequence of interest, is inserted into the *P. pastoris* genome at the *AOX1* locus via homologous recombination. Selected transformants are screened for protease production after induction of their encoding gene in a medium containing methanol. To be secreted, a protease must be synthesized with its propeptide when one exists (Beggah et al., 2000).

ENDOPROTEASES SECRETED BY *A. FUMIGATUS*

A. fumigatus secretes three major endoproteases when the fungus is cultivated in the presence of protein as the sole nitrogen source: an aspartic protease (Pep1; A1 family) (Reichard et al., 1994), a metalloprotease (Mep; M36 family) (Monod et al., 1993b), and a subtilisin (Alp1; S8 family) referred to as alkaline protease (Reichard et al., 1990, Monod et al., 1991). The expression of genes encoding these three proteases is completely inhibited in the presence of free amino acids or small peptides in the growth medium. *alp mep* knockout mutants were completely devoid of detectable neutral extracellular proteolytic activity in vitro, while single *alp* and *mep* mutants revealed 30 and 70% of the proteolytic activity of the wild-type strain, respectively (Monod et al., 1993a; Jaton-Ogay et al., 1994). However, the *A. fumigatus* genome contains multiple genes encoding other endoproteases which were either found to be secreted in minor amounts in culture supernatants or which remain hypothetical (Table 1). Some of these enzymes are similar to proteases which have been found to be secreted in substantial amounts by other *Aspergillus* species and have been characterized (for instance, G1 and M35 proteases).

Aspartic Proteases (A1 Family)

Aspergillus secreted proteases of the pepsin family have a molecular mass of 35 to 40 kDa and are synthesized as precursors in the form of preproproteins with a 30- to 40-amino-acid-long propeptide. Like all aspartic proteases, they are active at acidic pH and inhibited by pepstatin (Berka et al., 1990, 1993; Reichard et al., 1994, 2000a; Vickers et al., 2007).

Genes encoding aspartic proteases of the pepsin-like protease family are multiple in *Aspergillus* spp. (Fig. 2). *A. fumigatus, Aspergillus nidulans*, and *Aspergillus terreus* contain seven, seven, and six members of this family, respectively. *A. fumigatus* Pep1 closely resembles the orthologous aspartic proteases PepA (or aspergillopepsin I) and PepO, which are secreted by *A. niger* var. *awamori* and *A. oryzae*, respectively (Berka et al., 1990, 1993) (Fig. 2). The genes encoding *A. fumigatus* Pep1, *A. niger* var. *awamori* PepA, and *A. oryzae* PepO have the same exon-intron structures, with four exons and

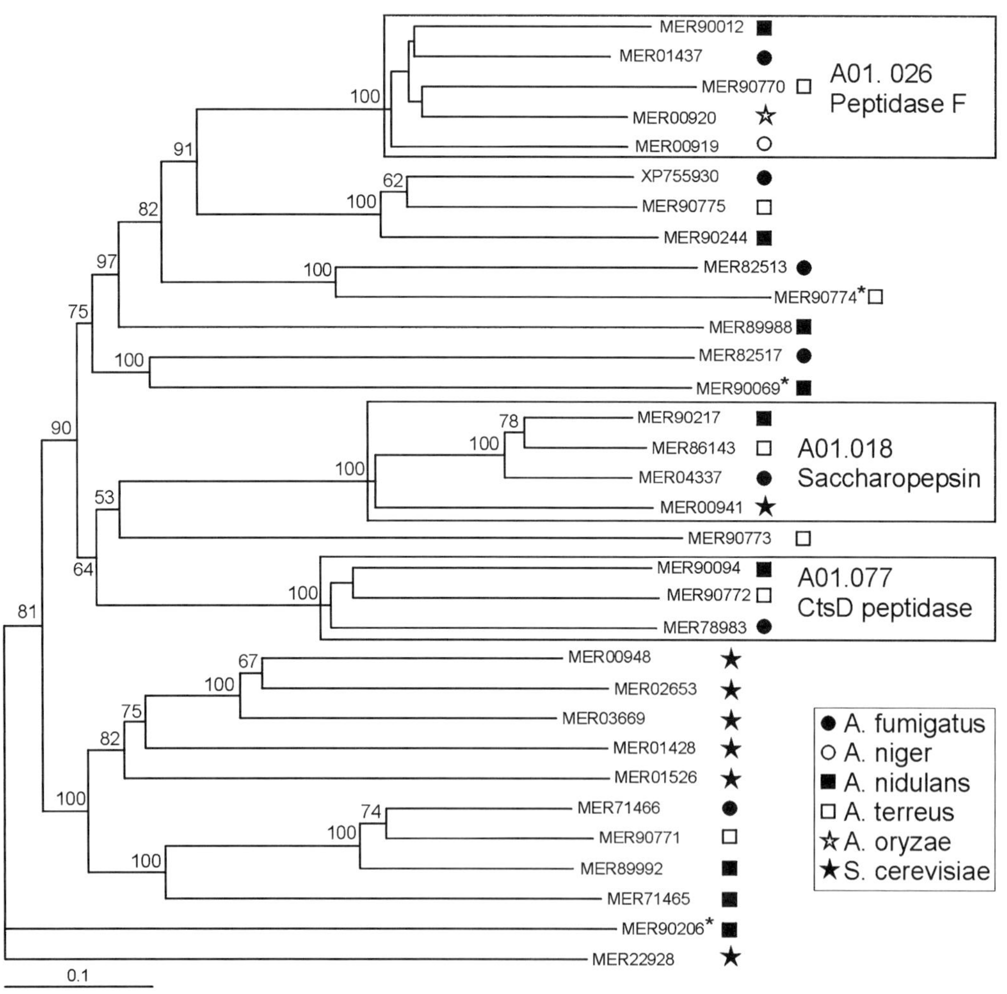

Figure 2. Phylogenetic relationships of aspartic proteases (A1 family) from *A. fumigatus, A. nidulans, A. terreus*, and *S. cerevisiae*. Sequences of the well-characterized *A. niger* PepA (MER00919) and *A. oryzae* PepO (MER00920) were also included. Phylogenetic analyses were performed as previously described by Zaugg et al. (2008). Three well-supported monophyletic groups correspond to MEROPS identifiers A01.026, A01.018, and A01.077. The other proteases are not assigned to specific identifiers. Nonprotease homologs in *A. nidulans* and *A. terreus* are indicated by an asterisk.

three introns in homologous positions (Reichard et al., 1995).

A second aspartic protease (Pep2) was found in the cell wall fraction of *A. fumigatus* (Reichard et al., 2000b) but was secreted when recombinantly produced in *P. pastoris*. However, Pep2 is also vacuolar and is the ortholog of saccharopepsin or *Saccharomyces cerevisiae* proteinase A (Ammerer et al., 1986; Winther et al., 2004). A third pepsin (CtsD) was recently characterized as a recombinant enzyme and found to be secreted by *A. fumigatus* in small amounts (Vickers et al., 2007). The low level of secretion of CtsD may reflect a predominant location of this protease elsewhere than in the supernatant, as potential GPI anchoring is predicted (Table 1). In accordance, two membrane-associated acid protease activities were previously characterized (Piechura et al., 1990).

Four proteases of the pepsin family identified in the *A. fumigatus* genome are still hypothetical and remain to be characterized. One of these proteases or CtsD could be an aspartic protease, which has been detected by immunofluorescence with anti-Pep1 polyclonal antibodies on submerged mycelium, on the tips of growing aerial hyphae, and on developing conidiophores of *A. fumigatus pep1* mutants (Reichard et al., 1996) (Fig. 3). This protease seemed to be tightly linked to the cell wall and was found in all tested aspergilli.

Glutamic Proteases (G1 Family)

A. fumigatus possesses two DNA sequences encoding protease homologs to *A. niger* aspergillopepsin II (Takahashi, 2004). This enzyme has also been termed proctase A or *Aspergillus* proteinase A. Neither of these *A. fumigatus* proteases has ever been characterized on the protein level, and they remain hypothetical. *A. niger* aspergillopepsin II is a two-chain protein of 212 residues composed of a 39-residue light chain and a 173-residue heavy chain bound noncovalently to each other. Both chains originate from a single precursor of 282 amino acids. This enzyme is an acidic protease with an optimum pH of 2.0 which is, however, resistant to pepstatin. It was therefore previously considered a non-pepsin-type acid protease and was classified with the aspartic proteases in family A4 (clan AX), which only contains proteases of fungal origin (Barett et al., 2004). Identification of a catalytic dyad formed by residues Glu and Gln (Yabuki et al., 2004) led to the creation of the new class of glutamic proteases.

Metallo-Endoproteases (M12, M35, M36, and M43 Families)

Genes encoding secreted metallo-endoproteases of four different families (M12, M35, M36, and M43) are present in the *A. fumigatus* genome. These proteases (several still hypothetical) belong to the same clan (MA), which gathers proteases with the specific HEXXH motif. More precisely, the HEXXH motif is found in the sequence Xaa-Xbb-Xcc-His-Glu-Xbb-Xbb-His-Xbb-Xdd, where Xaa is hydrophobic or Thr, Xbb is uncharged, Xcc is any amino acid except Pro, and Xdd is hydrophobic (Jongeneel et al., 1989; Barrett et al., 2004). The two His residues together with a third residue towards the C terminus of the protein are zinc ligands. The third residue varies depending on the family. It is a His in the M12 and M43 families and an Asp in the M35 family (see MEROPS database or Barrett et al., 2004). It remains to be formally identified in the M36 family. However, the sequence containing the third zinc ligand in thermolysin (M4 family) (van den Burg and Eijsink, 2004), which is a Glu, is partially conserved in

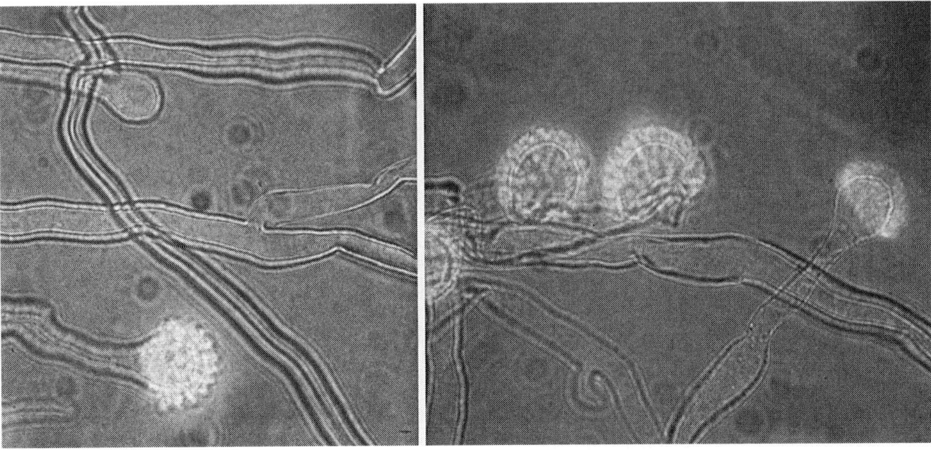

Figure 3. Immunofluorescence reaction of aspartic protease antigen at the surface of hyphae and conidiophores of *A. fumigatus*. Indirect immunofluorescence was achieved with anti-Pep1 rabbit antiserum.

A. fumigatus Mep (see below) (Sirakova et al., 1994) as well in all other recently discovered M36 proteases (Jousson et al., 2004b). Chelators inactivate MA proteases by removal of the ion Zn^{2+}. However, like thermolysin *A. fumigatus* Mep (M36 family) and *A. oryzae* deuterolisin (M35 family), it can be reactivated by the addition of Zn^{2+} or Co^{2+} (Markaryan et al., 1994; Doi et al., 2003). The Glu residue in the HEXXH motif has a catalytic function. With the exception of the MA clan specific signature, members of a given MA family have no or low amino acid sequence similarity with members of another MA family.

A. fumigatus Mep is secreted concomitantly with subtilisin Alp1 (see below) when the fungus is grown in a medium containing protein as sole nitrogen source. The mature protease was first isolated from *alp1* mutant cultures and is an unglycosylated protein of 40 kDa (Monod et al., 1993b). *A. fumigatus* Mep is the first member of the M36 family (fungalysin) for which the amino acid sequence was identified. It is synthesized as a precursor in the form of a preproprotein with a 226-amino-acid-long propeptide (Jaton-Ogay et al., 1994; Sirakova et al., 1994). Its amino acid sequence is 78% identical and 87% similar to that of *A. oryzae* neutral protease I (Nakadai et al., 1973a; Doumas et al., 1999). Mep cleaves collagen but does not cleave elastin-Congo red (Monod et al., 1993a, 1993b). However, elastinolytic activity was reported based on the detection of radioactivity released after incubation of the enzyme with [^{3}H]elastin (Markaryan et al., 1994). Mep substrate specificity is similar to that of thermolysin, which has a preference for a hydrophobic amino acid at the S'_1 position in the site of cleavage (Markaryan et al., 1994; Kollattukudy and Sirakova, 2004; van den Burg and Eijsink, 2004). The optimum pH of activity is between 7.0 and 8.0. This neutral protease is totally inhibited by phosphoramidon and chelating agents such as EDTA and 1-10 phenanthroline and is resistant to phenylmethylsulfonyl fluoride, antipain, leupeptin, chymostatin, pepstatin, and iodoacetamide (Monod, 1993b). While there is only one fungalysin in *A. fumigatus*, a second homolog (MER032459; BAC78815; published only in the MEROPS database) was found in *A. oryzae* in addition to neutral proteinase I. Multiple proteases of this family are secreted by dermatophytes under conditions promoting secreted proteolytic activity (Jousson et al., 2004a).

Genes encoding two metalloproteases of the M35 family and one of the M43 family have been identified in the *A. fumigatus* genome. However, no translation product has been detected so far in *A. fumigatus* culture supernatants. The two metalloproteases of the M35 family are homologous to neutral protease II of *A. oryzae* and *Aspergillus sojae* (Sekine, 1972; Nakadai et al., 1973b; Tatsumi et al., 1991). Neutral protease II was found to be highly active on basic nuclear proteins such as salmine, clupein, and histones (Sekine, 1973). This enzyme is specific for paired dibasic amino acid residues at the P_2-P_1 positions, but other amino acid residues at the P_1 position were recorded in histone H4 (Doi et al., 2003). The M43 family contains a metalloprotease secreted by the insect pathogen *Metarhizium anisopliae* (Mep1; MER011122; unpublished data) and a metalloprotease of *Coccidioides posadasii* (Mep1) which cleaves an immunodominant cell surface antigen and thereby renders the fungus less detectable by the host immune system (Hung et al., 2005).

Two ADAM metalloproteases (ADM-A and ADM-B) were found to be encoded by the *A. fumigatus* genome (Lavens et al., 2005). (The acronym ADAM designates members of *a d*esintegrin *a*nd *m*etalloprotease family [Wolfsberg et al., 1995]). The ADAM metalloproteases, several of which are found in humans (Wolfsberg and White, 2004), belong to the M12 family, which also contains numerous snake (*Crotalidae* and *Viperidae*) venom metallopeptidases (Barrett et al., 2004). ADM-A, ADM-B, and human ADAMs have a similar domain structure, with a signal peptide, a propeptide, the metalloprotease domain, a desintegrin domain, a transmembrane domain, and a cytoplasmic domain. Therefore, these proteases are not secreted but are attached to the plasma membrane. The ADM-B catalytic domain was produced as a recombinant protein using *E. coli* as an expression system and shown to be active on casein and albumin, but not on gelatine, attesting to its narrow substrate specificity (Lavens et al., 2005).

Serine Endoproteases (S8, S53, and S28 Families)

Two subtilisins (S8 family; Alp1 and Alp2) and one endoprotease of the sedolisin family (SedA; S53) have been characterized. The subtilisin Alp1 (formerly Alp) was isolated and characterized in the 1990s by several groups (Reichard et al., 1990; Monod et al., 1991; Jaton-Ogay et al., 1992; Larcher et al., 1992; Frosco et al., 1992, Kolattukudy et al., 1993). This enzyme has a molecular mass of 33 kDa and cleaves elastin as well as collagen. Alp1, like other subtilisins, very efficiently cleaves the synthetic substrate *N*-Suc-Ala-Ala-Pro-Phe–*p*-nitroanilide (pNA) (Larcher et al., 1992). It has optimal activity between pH 7.0 and 9.0 and is totally inhibited by phenylmethylsulfonyl fluoride, antipain, and chymostatin but not by inhibitors of other serine proteases, such as *N*-α-*p*-tosyl-L-lysine chloromethyl ketone and tosylsulfonyl phenylalanyl chloromethyl ketone (Reichard et al., 1990; Monod et al., 1991; Larcher et al., 1992).

A second subtilisin (Alp2), which is vacuolar, was also found in the cell wall fraction of *A. fumigatus*-like

Pep2 (Reichard et al., 2000b). Alp2 is the ortholog of cerevisin or *S. cerevisiae* proteinase B (Moehle et al., 1987; Jones and Naik, 2004) and *A. niger* PepC (Frederick et al., 1993). Alp1 and Alp2, like other secreted fungal and bacterial subtilisins, are synthesized as preproproteins with a propeptide of approximately 120 amino acids.

Targeted disruption of *ALP1* did not result in differences of morphology or growth of the fungus (Monod et al., 1993b). In contrast, *alp2* mutants showed a slightly reduced speed of growth, conidiophores of smaller size, and a significant decrease in number of conidia, but without specific block in the genesis of the conidiophore phialides and conidia (Reichard et al., 2000b). The starvation pattern of conidiophore development is explained by a possible function of Alp2 as vacuolar protease responsible for the general supply of amino acids for the synthesis of developmental proteins, like proteinase B in *S. cerevisiae*.

Two other genes coding for subtilisins were found in the *A. fumigatus* genome. One gene (XM_746441) codes for a protein which revealed extensive sequence homology to *S. cerevisiae* Kex2 (Julius et al., 1984), whereas the deduced amino acid sequence of the presumed open reading frames of the other gene (*ALP3*; XM_746441) suggested an intracellular location of the protein, as no signal peptide could be identified.

The *A. fumigatus* endoprotease SedA (Reichard et al., 2006), which belongs to the S53 family (sedolisin), is the ortholog of aorsin from *A. oryzae* (Lee et al., 2003). Only a 25-kDa degradation product of SedA, but no full-length translation product, has so far been detected in *A. fumigatus* culture supernatants (Reichard et al., 2006).

From its deduced amino acid sequence, one protease (064064) of the S28 family which contains exopeptidases appears to be closely related to the recently isolated and characterized *A. niger* acid prolylendopeptidase (64% identity) (Edens et al., 2005). This protease very efficiently degrades gliadins in gluten, which are proline rich and are highly resistant to proteolytic degradation within the gastrointestinal tract (Stepniak et al., 2006). Gliadin proline-rich peptides are the cause of the uncontrolled immune response in celiac disease, a human gastrointestinal disorder.

EXOPEPTIDASES SECRETED BY *A. FUMIGATUS*

Aminopeptidase, dipeptidyl-peptidase, tripeptidyl-peptidase, and carboxypeptidase activities were detected at different pHs in culture supernatants of *A. fumigatus* grown in various media. While several native aminopeptidases and carboxypeptidases were isolated and characterized from *A. oryzae* and *A. sojae* extracts (Nakadai et al., 1972a, 1972b, 1972c, 1973c, 1973d, 1973e, 1973f), most *A. fumigatus* exoproteases were characterized as recombinant enzymes. Only native *A. fumigatus* DppIV and DppV were isolated and characterized prior to the use of recombinant enzymes (Kobayashi et al., 1993; Beauvais et al., 1997b). Secreted aminopeptidases are metalloproteases of the M28 family which have two Zn^{2+}-binding sites (see MEROPS database). Secreted dipeptidyl-peptidases and carboxypeptidases (S9 and S10 families, respectively) are serine proteases with a Ser, Asp, and His catalytic triad. In tripeptidyl-peptidases of the sedolisin family (S53), residues of the catalytic site are Glu, Asp, Asp, and Ser (Wlodawer, 2003).

Aminopeptidases (M28 Family)

The *A. fumigatus* genome contains four genes encoding aminopeptidases of the M28 family with signal sequences. Two secreted aminopeptidases, orthologous to the previously characterized *A. oryzae* Lap1 and Lap2, were characterized as recombinant enzymes (Monod et al., 2005). From its deduced amino acid sequence, the third aminopeptidase (XP_749158) is closely related to the *S. cerevisiae* aminopeptidase Y (Nikai and Yasuhara, 2004) and is probably vacuolar. The fourth protease (MER050730) is much larger than the three others; it is synthesized as a 965-amino-acid polypeptidic chain with nine *trans*-membrane domains (one being at the N terminus of the protein) and is apparently a membrane protein.

Both *A. fumigatus* Lap1 and Lap2 have been called leucine aminopeptidases, like *Aspergillus* spp. orthologs, because of their preference for leucine-7-amido-4-methylcoumarin (Leu-AMC) as a substrate. However, these enzymes are able to remove any amino acid from the N terminus of a peptide, provided that a proline is not in the second position (Table 3). Lap1 and Lap2 are active between pH 6.5 and 10.5 with a broad optimum peak between pH 7.0 and 9.0 (Monod et al., 2005). Lap1 structurally belongs to the M28E subfamily, as does *Vibrio* spp. secreted aminopeptidase (Chevrier and D'Orchymont, 2004). Lap2 structurally belongs to the M28A subfamily, like the putative vacuolar aminopeptidase XP_749158, the *S. cerevisiae* aminopeptidase Y (Nikai and Yasuhara, 2004), and the *Streptomyces griseus* secreted aminopeptidase (Awad, 2004). Members of the M28A and M28E subfamilies share low sequence similarities. However, the amino acid sequences of two Zn^{2+}-binding sites are conserved. As observed in other fungal secreted aminopeptidases, *A. fumigatus* Lap1 and Lap2 were found to be sensitive to different ions. Like *S. griseus* aminopeptidase, Lap2 activity is highly enhanced by Co^{2+} (Monod et al., 2005).

Table 3. pNA release from various pNA substrates by different mixes of *A. fumigatus* exopeptidases[a]

A. fumigatus peptidases	Buffer (pH)	pNA release from exopeptidase mix						
		Ala-pNA	Gly-Pro-pNA	Ala-Ala-pNA	Phe-Pro-Ala-pNA	Ala-Ala-Phe-pNA	Ala-Ala-Pro-pNA	Ala-Ala-Pro-Leu-pNA
Lap1/2	Tris, 50 mM (7.4)	+	−	+	−	+	−	−
DppIV	Tris, 50 mM (7.4)	−	+	+	−	−	−	−
DppV	Tris, 50 mM (7.4)	−	−	+	−	−	−	−
SedB; SedC; SedD	Citrate, 50 mM (5.0–5.5)	−	−	−	+	+	−	−
Lap1/2+ DppIV	Tris, 50 mM (7.5)	+	+	+	+	+	+	+
Lap1/2+ SedB/C	Tris, 50 mM (6.8)	+	−	+	+	+	−	+
DppIV+ SedB/C	Tris, 50 mM (6.8)	−	+	+	+	+	−	−

[a] Optimal pH values are given for each enzyme mix.

Dipeptidyl-Peptidases (S9 Family)

A. fumigatus DppIV and DppV are serine proteases of the S9 family (Monod and Beauvais, 2004). Both enzymes are glycoproteins of approximately 90 kDa with about 10 kDa of N-linked carbohydrates (Beauvais et al., 1997a, 1997b). They are active between pH 6.5 and 10.5 with a broad optimum peak between pH 7.0 and 9.0. DppIV efficiently removes X-Pro and X-Ala from chromogenic substrates (e.g., X-Pro-AMC, X-Pro-pNA, X-Ala-AMC, and X-Ala-pNA) (Table 3) and X-Pro from active polypepeptides {e.g., neuropeptide Y; [des-Arg[1]] bradykinin, glucagon-like peptide-1-(7-36) amide, and substance P} (Beauvais et al., 1997b; Grouzman et al., 2002). *A. fumigatus* DppIV is homologous to human DppIV (CD26; MEROPS S09.003), which has the same substrate specificity. Both enzymes are inhibited by Lys-[Z(NO_2)]-pyrolidide and lys-[Z(NO_2)]-thiazolidide. Because it can be produced in large amounts as recombinant protease, *A. fumigatus* DppIV has been used instead of pig and rabbit DppIV to cleave proinflammatory peptides in animal models of rhinosinusitis and asthma (Grouzman et al., 2002; Landis et al., 2007).

DppV releases mainly X-Ala, but also His-Ser and Ser-Tyr, dipeptides from the N-terminal extremity of peptides (Beauvais, 1997a), but it is not capable of removing N-terminal Gly-Pro (Table 3). This protein was identified as one of the two major antigens used in the serodiagnosis of aspergillosis (Biguet et al., 1967) and was unfortunately called chymotrypsic antigen because of its ability to degrade, like chymotrypsin, the chromogenic substrate *N*-acetyl-phenylalanine naphthyl ester. Amino acids and tripeptides are not removed by either DppIV or DppV (Table 3).

Two other hypothetical secreted proteases of the S9 family were revealed in the *A. fumigatus* genome. The first (MER032528/XP_754828) has a signal anchor but no signal peptidase cleavage site. It is 39% identical to DppIV. This protease is most likely the ortholog of the *S. cerevisiae* dipeptidyl-peptidase B (41% identity), which is localized in the vacuolar membrane and also hydrolyzes X-Pro and X-Ala peptides (Misumi and Ikehara, 2004). The second (MER077504/XP747279), which has no signal sequence, is 42% identical to DppV.

Tripeptidyl-Peptidases (S53 Family)

Three *A. fumigatus* secreted tripeptidyl-peptidases, called SedB, SedC, and SedD, were characterized as recombinant enzymes (Reichard et al., 2006). These tripeptidyl-peptidases belong to the S53 family, which contains, in addition, the secreted acidic nonaspartic endoprotease SedA and a fifth hypothetical protease (MER107324/XP_748888) lacking a signal sequence.

SedB, SedC, and SedD very efficiently release pNA when Phe-Pro-Ala-pNA or Ala-Ala-Phe-pNA are used as substrates (Table 3). None of these three enzymes demonstrated activity towards either Ala-Ala-Pro-pNA or mono-, di-, and tetrapeptide pNA substrates. SedB and SedC were active between pH 3.0 and 7.0 with an optimum at pH 6.0, whereas SedD was active between pH 1.5 and 6.0 with an optimum at pH 5.0.

A. fumigatus tripeptidyl-peptidase activity was detected in hemoglobin liquid medium culture supernatants with Phe-Pro-Ala-pNA as a substrate at pH 5.0 (Reichard et al., 2006). At this pH, secreted *A. fumigatus* leucine aminopeptidases and DppIV activities are not detectable, and thus, interference by these enzymes was excluded. Western blot analysis unambiguously detected the presence of glycosylated SedB, SedC, and SedD. Deglycosylated native and heterologously expressed enzymes from *P. pastoris* had the same electrophoretic mobility (Reichard et al., 2006). The apparent molecular mass of each deglycosylated protein was lower than that of its predicted molecular mass. The N-terminal amino acid sequences obtained by Edman degradation showed the existence of a prosequence for each of these enzymes.

Carboxypeptidases (S10 Family)

The *A. fumigatus* genome contains 10 genes encoding serine carboxypeptidases of the S10 family with a signal sequence. Two of these carboxypeptidases, AfuCp1 and AfuCp2, are characterized as recombinant proteases (Zaugg et al., 2008). Both enzymes are glycoproteins of approximately 90 kDa, with polypeptidic chains of 538 and 595 amino acids, respectively. They very efficiently hydrolyze *N*-(2-furanacryloyl)-L-phenylalanyl-L-phenylalanine at 30°C between pH 4.0 and 8.0 with an optimum at pH 4.5. *Aspergillus* secreted acidic serine carboxypeptidases were first characterized in *A. niger* and *A. oryzae* (Svendsen and Dal Degan, 1998; Blinkovsky et al., 1999). These enzymes are similar to the well-characterized *S. cerevisiae* vacuolar carboxypeptidase Y (SceCpY), with a conserved catalytic triad consisting of Ser, Asp, and His residues (Ser^{240}, Asp^{459}, and His^{517} for SceCpY) (Mortensen et al., 2004).

Genes encoding carboxypeptidases were expanded to form families with a variable number of members in *Aspergillus* spp. A phylogenetic analysis including the *A. fumigatus* secreted carboxypeptidases and other biochemically well-characterized fungal carboxypeptidases of the S10 family produced a robust tree consisting of three main clades presenting a good correlation with MEROPS identifiers (Fig. 4). The most derived clade included AfuCp1, AfuCp2, and AfuCp6 (Zaugg et al., 2008) with *Penicillium janthinellum* and *A. oryzae* S1 carboxypeptidases. This clade also contains two *Trichophyton rubrum* carboxypeptidases which are membrane associated and likely GPI anchored (Zaugg et al., 2008). A second clade included all sequences closely related to *S. cerevisiae* CPY, which are AfuCp3 and other putative fungal vacuolar carboxypeptidases. The basal clade contained AfuCp4 and AfuCp5 and *A. niger* CpD1 and CpD2, respectively. The *A. fumigatus* genome also contains three unassigned carboxypeptidases, AfuCp7 to AfuCp9, which show little similarity with the other members of the S10 family.

A putative ortholog of *S. cerevisiae* Kex1 was also detected in the *A. fumigatus* genome. This enzyme is a vacuolar S10 carboxypeptidase which is involved in processing of killer toxin and α-factor precursors by removing Lys and Arg residues at their C terminus (Dmochowska et al., 1987; Latchinian-Sadek and Thomas, 1993).

Other Putative Exoproteases (M20, S28, and S33 Families)

The *A. fumigatus* genome contains genes encoding exoproteases of the M20, S28, and S33 families, but these enzymes remain hypothetical (Table 1). Among nine *A. fumigatus* M20 proteases, four display a signal peptide. The M20 family contains carboxypeptidases, dipeptidases, and one aminopeptidase, the bacterial peptidase T which acts on tripeptide substrates (Barrett et al., 2004). From its deduced amino acid sequence, one *A. fumigatus* M20 peptidase (XP_001481653) appears to be closely related to the vacuolar *S. cerevisiae* Gly-X carboxypeptidase (Suarez-Rendueles and Bordallo, 2004). This protease releases C-terminal residues from N-blocked peptides as long as glycine precedes the last position. In addition to *A. niger* acid prolyl-endopeptidase, the S28 family contains exopeptidases that hydrolyze prolyl bonds (Barrett et al., 2004).

PROTEIN DIGESTION AND ASSIMILATION OF PROTEOLYSIS PRODUCTS

The growth of a fungus in a protein medium depends on the synergistic action of secreted endo- and exoproteases, because large peptides cannot be used as nutrients. Only amino acids and short peptides can be assimilated via membrane transporters. In addition to amino acid permeases, *A. fumigatus* possesses seven transporters of the proton-dependent oligopeptide transporter (POT) family and eight transporters of the oligopeptide transporter (OPT) family (see the Transporter database at www.membranetransport.org/). Well-characterized members of the POT family are the di- and tripeptide transporters Ptr2 from *S. cerevisiae* and *Candida albicans* (Perry et al., 1994; Basrai et al.,

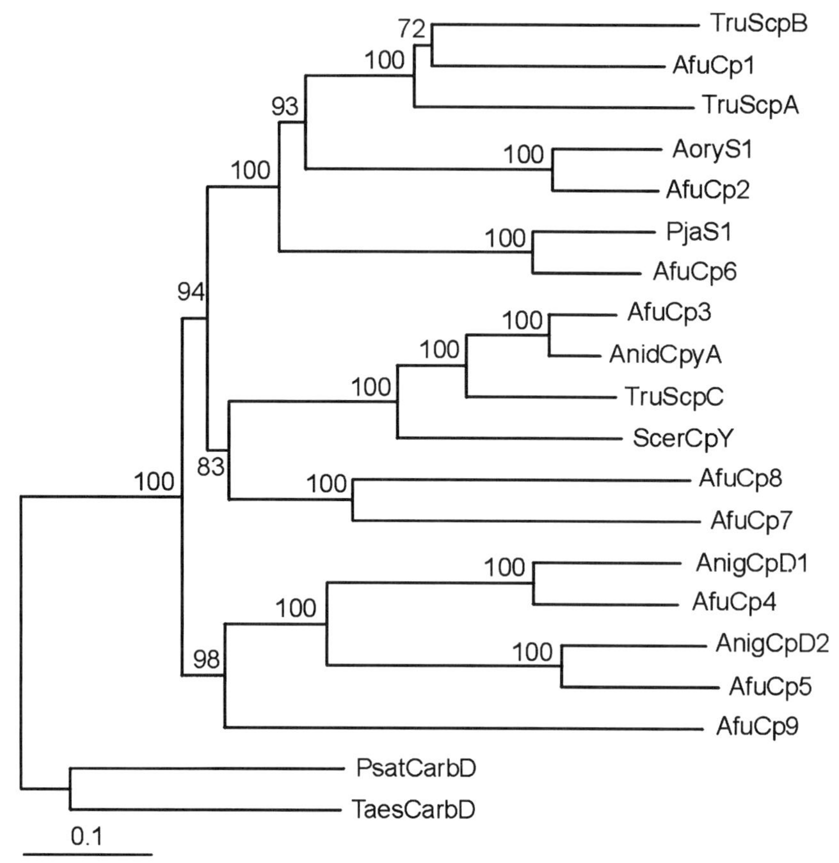

Figure 4. Phylogenetic relationships of *A. fumigatus* carboxypeptidases with other biochemically well-characterized carboxypeptidases of the S10 family. The tree is rooted with plant carboxypeptidases of the S10.005 subfamily. Phylogenetic analyses were performed as previously described (Zaugg et al., 2008). Abbreviations: Tru, *Trichophyton rubrum*; Afu, *A. fumigatus*; Aory, *A. oryzae*; Anid, *A. nidulans*; Anig, *A. niger*; Pja, *Penicillium janthinellum*; Scer, *S. cerevisiae*; Psat, *Pisum sativum*; Taes, *Triticum aestivum*. MEROPS identifiers are as follows: AfuCp1, MER079359; AfuCp2, MER079360; AfuCp3, MER079361; AfuCp4, MER079362; AfuCp5, MER079363; AfuCp6, MER082516; AfuCp7, MER082530; AfuCp8, MER032646; AfuCp9, MER032643; AoryS1, MER016549; AnidCpyA, MER090176, AnigCpD1, MER027994; AnigCpD2, MER000415; PjaS1, MER000412; ScerCpY, MER002010; TruScpB, MER079400; TruScpC, MER079401. TruScpA (Zaugg et al., 2008) is not registered in MEROPS.

1995). POTs are found in species from all taxonomic kingdoms. OPTs are limited to fungi and plants, allow the uptake of tetra- and pentapeptides, and were recently the object of intensive investigation in *S. cerevisiae, C. albicans*, and *Arabidopsis thaliana* (Hauser et al., 2001; Koh et al., 2002; Lubkowitz et al., 1997; Reuss and Morschhäuser, 2006). None of the *Aspergillus* OPTs and POTs have been characterized.

Protein digestion into amino acids and short peptides by *Aspergillus* spp. has been investigated extensively by the food fermentation industry. From a physiological point of view, *Aspergillus* secreted proteases can be classified as either acidic, neutral, or basic. Endoproteolysis at neutral and basic pH occurs with the dominant proteases Alp1 and Mep (Monod et al., 1993a, 1993b; Jaton-Ogay et al., 1994). Subsequently, Laps and DppIV synergistically digest large peptides into amino acids and X-pro dipeptides. Laps degrade peptides from their N terminus; however, X-Pro acts as a stop sequence, and in a complementary manner, these X-Pro sequences are removed by DppIV, which allows Laps access to the following residues. Synergistic action of *A. oryzae* Lap and DppIV at pH 7.5 leads to complete digestion of the Ala-Pro-Gly-Asp-Arg-Ile-Tyr-Val-His-Pro-Phe peptide into amino acids and X-Pro dipeptides (Byun et al., 2001). Likewise, pNA is released from Ala-Ala-Pro-Leu-pNA substrate by a mix of Lap and DppIV (Table 3). Serine carboxypeptidases can also digest large peptides at neutral pH.

Endoproteolysis at acidic pH may be achieved by the action of Pep and SedA. Amino acids and short peptides can be generated by the tripeptidyl-peptidases of the sedolisin family S53 (SedB, SedC, and SedD) and serine carboxypeptidases of the S10 family, which are active between pH 3 and 8 with an optimum between pH 4 and 5. Sed2 was shown to be able to bypass Pro residues by degrading large peptides from their N terminus when these Pro residues were in the second position (Reichard et al., 2006). Although substrate specificity of Tpps of the sedolisin family needs further investigation, these enzymes appear to be active when the amino acid in the P1 or P′1 position (amino acids in positions 3 and 4 from the N terminus of the substrate peptide) is not a proline.

REDUCTION OF CYSTEINE DISULFIDE BRIDGES PRECEDING PROTEOLYTIC ACTIVITY

Both *A. fumigatus* and *A. oryzae*, like dermatophytes, are able to grow in medium containing hard keratin as the sole source of nitrogen and carbon. Hard keratin grains are totally digested after 20 to 40 days at 30°C (Léchenne et al., 2007). However, fungal secreted proteases are incapable of dissolving compact keratinous tissues by themselves. Efficient protein degradation of hard keratin by hydrolytic enzymes has to be accompanied by simultaneous reduction of cysteine disulfide bridges. For instance, efficient in vitro degradation of the hair structure of keratin azure by either dermatophyte or *A. fumigatus* subtilisins was only possible in the presence of a reducing agent, such as 1% β-mercaptoethanol or dithiothreitol (Jousson et al., 2004b).

Filamentous fungi were shown to excrete sulfite as a reducing agent (Kunert, 1972, 1976, 2000). In the presence of sulfite, disulfide bonds of the keratin substrate are directly cleaved to cysteine and *S*-sulfocysteine. *S*-Sulfocysteine was detected in the culture supernatant when *A. fumigatus* was grown in a medium containing hard keratin as sole nitrogen source, thereby confirming sulfite secretion (Léchenne et al., 2007). Genes encoding *A. fumigatus* and dermatophyte sulfite efflux pumps (*SSU1*) were cloned and expressed in *S. cerevisiae* (Léchenne et al., 2007). These transporters, like the *S. cerevisiae* sulfite transporter Ssu1, belong to the tellurite resistance/dicarboxylate transporter (TDT) family, which also includes the *E. coli* tellurite efflux pump and the *Schizosaccharomyces pombe* malate transporter encoded by the genes *TEHA* and *MAEI*, respectively (Walter et al., 1991; Grobler et al., 1995). However, the expression of *A. fumigatus SSU1* is relatively low in comparison to that of *SSU1* in dermatophytes, which are specialized in keratin degradation (Léchenne et al., 2007). Apparently, the growth of *A. fumigatus* in keratin does not depend on Ssu1 activity, since *A. fumigatus ssu1* mutants grow well in a keratin medium. The existence of another sulfite efflux pump cannot be excluded in *A. fumigatus*. It is also possible that small amounts of sulfite can leave the mycelium by a route other than by sulfite transporters.

Although both *A. fumigatus* and *A. oryzae* are able to grow in keratin medium, only *A. oryzae* is repeatedly found as an infectious agent in onychomycoses (Fig. 5). *A. fumigatus* is often isolated from nail samples, but generally as a contaminant. In the particular case of onychomycosis, it is evident that secreted proteolytic activity of the fungus is essential for its parasitic development. However, fungal development in nails could rely on high-level expression of *SSU1* and therefore on the ability to secrete large amounts of sulfite.

SECRETED PROTEASES AND VIRULENCE OF *A. FUMIGATUS*

It is tempting to postulate that proteolytic activity is needed for invasion of host tissues during invasive aspergillosis. Various experimental results support this point of view. (i) Endo- and exoproteases were shown to be secreted in vivo. Indirect immunofluorescence studies revealed that protease-specific antisera labeled mycelia in the lungs of either patients or experimentally infected mice (Reichard et al., 1990, 1994; Moutaouakil et al., 1993; Markarian et al., 1994) (Fig. 6). (ii) Sera from patients with aspergilloma contain Mep, Pep, and DppV antibodies, which can be utilized to discriminate between aspergilloma patients and healthy controls (Biguet et al., 1967; Monod et al., 1993b; Beauvais et al., 1997a; Sarfati et al., 2006). (iii) Elastinolytic activity of *A. fumigatus* has also been correlated with the ability of

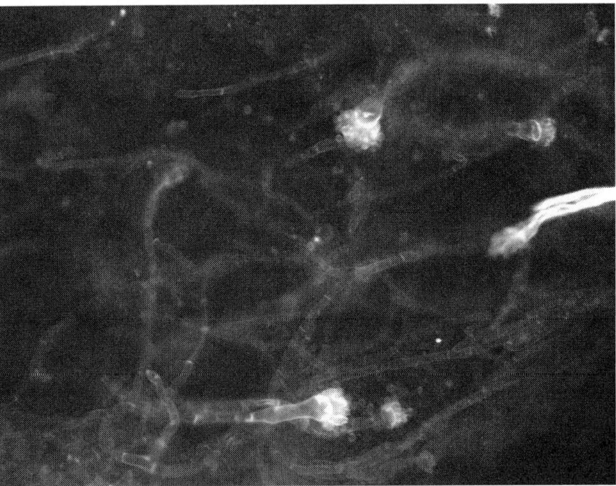

Figure 5. Direct mycological examination of an infected nail, showing *A. oryzae* hyphae and conidiophores. Nail scrapings were examined in a dissolving solution containing a fluorochrome (Monod et al., 1989). In situ identification of the fungus was performed using a PCR and sequencing method (Monod et al., 2006).

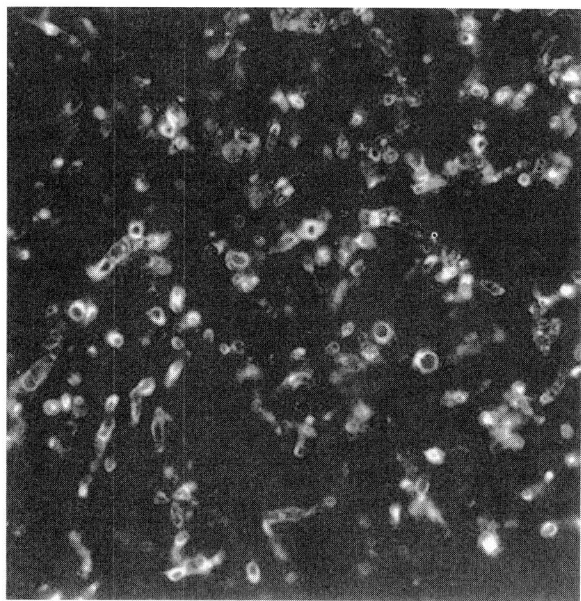

Figure 6. Reactivity of mycelia in the lung of an immunocompromised mouse infected with *A. fumigatus*. Indirect immunofluorescence was achieved with anti-Alp1 rabbit antiserum.

several wild-type strains to cause invasive aspergillosis in mice (Kothary et al., 1984). (iv) Alp was able to disrupt the actin fiber cytoskeleton of lung pneumocytes grown in culture (Kogan et al., 2004). (v) A not-yet-identified protease at the surface of conidia is able to degrade complement component C3 to low-molecular-mass fragments, which could play a role in resistance of the conidia to host phagocytosis (Sturtevant and Latgé, 1992).

However, other results render the role of secreted proteases in the virulence of *A. fumigatus* questionable. (i) Vessel wall elastinolysis was not observed in tissues obtained at autopsy from a patient who died from invasive aspergillosis (Denning et al., 1992). (ii) No difference in pathogenicity was observed between the wild-type strains and single *alp*, *mep*, or *pep* mutants and double *alp mep* mutants (the latter being totally deficient in proteolytic activity at neutral pH in vitro) (Monod et al., 1993a; Tang et al., 1992, 1993; Jaton-Ogay et al., 1994; Reichard et al., 1997). Mortality curves of wild-type and mutant strains were not statistically different, and histopathological studies of lungs from infected mice showed a similar extent of mycelial growth. (iii) In the same sense, the fungal virulence could not be attenuated by monoclonal antibodies inhibiting Alp1 (Frosco et al., 1994).

Although it can be stated that individual proteases such as Alp1, Mep, and Pep1 are not essential for tissue invasion, it cannot be ruled out that secreted proteases do not contribute to the establishment of invasive aspergillosis. Because the activities of several secreted proteases are redundant, any defect in one secreted individual protease may likely be compensated by the activities of other proteases. It is also possible that yet-uncharacterized proteases (for instance, members of the M12, M43, and S28 families) are produced during infection by induction through specific host factors. However, many fungi possess a battery of proteases allowing the degradation of proteins into amino acids and short peptides and are able to grow in a medium containing protein as the sole nitrogen source. Therefore, it is reasonable to concede that secreted proteolytic activity is not sufficient to explain the pathologic behavior of *A. fumigatus*, and it is necessary to search elsewhere for other factors to elucidate the development of the fungus in the human host. For instance, in contrast to many other molds, *A. fumigatus* grows rapidly at 37°C on rich and minimal media and therefore may be capable of finding a particular ecological niche in the bronchi.

CONCLUSION

A. fumigatus, like other *Aspergillus* species, secretes various endo- and exoproteases which belong to several families and which have redundant activities. However, virulence of *A. fumigatus* and other *Aspergillus* species led to an expectation of finding particular secreted proteases. In contrast, one can state that *Aspergillus* secreted proteases are similar to those of other ascomycetes, such as *Fusarium* spp., *Magnaporthe grisea*, and dermatophytes. Furthermore, proteases found in other ascomycetes were not found in *A. fumigatus* and other *Aspergillus* species. Genes encoding trypsin and chymotrypsin-like enzymes characterized in *Fusarium* spp. and *M. anisopliae* (Rypniewski et al., 1993; St. Leger et al., 1996; Screen and St. Leger, 2000) as well as metallocarboxypeptidases of the M14A subfamily characterized in *M. anisopliae* and dermatophytes (Joshi and St. Leger, 1999; Zaugg et al., 2008) are not present in *Aspergillus* spp.

A. fumigatus also does not possess specific large gene families encoding secreted proteases. Aspartic proteases of the A1 family and carboxypeptidases of the S10 family are multiple in many fungi. The emergence of multigenic families is most frequently due to ancient gene duplication processes allowing organisms to better adapt to different environmental conditions, and marked differences occur from one pathogenic species to another. For instance, dermatophytes, which are highly specialized in hard keratin degradation, secrete multiple endoproteases, such as subtilisins and fungalysins (Jousson et al., 2004a, 2004b). Another example is the human pathogenic yeast *C. albicans*, which expresses different genes encoding closely related secreted aspartic proteases in reaction to specific environmental conditions during infection (Staib et al., 2000; Hube and Naglik, 2001). The absence of specific large gene families encoding secreted proteases may well be a feature of nonspecialized opportunistic fungi which are saprophytic in the environment and degrade proteins into amino acids and short peptides in the process of recycling soil nitrogen.

Acknowledgments. We thank Massimo Lurati, Bertrand Favre, Peter Staib, and Eric Grouzman for critical reading of the manuscript. This work was supported by the Swiss National Foundation for Scientific Research, grant 3100-105313/1.

REFERENCES

Ammerer, G., C. P. Hunter, J. H. Rothman, G. C. Saari, L. A. Valls, and T. H. Stevens. 1986. *PEP4* gene of *Saccharomyces cerevisiae* encodes proteinase A, a vacuolar enzyme required for processing of vacuolar precursors. *Mol. Cell. Biol.* **6:**2490–2499.

Awad, W. M. 2004. *Streptomyces griseus* aminopeptidase, p. 957–959. *In* A. J. Barrett, N. D. Rawlings, and J. F. Woessner (ed.), *Handbook of Proteolytic Enzymes*, 2nd ed. Elsevier, London, United Kingdom.

Barrett, A. J., N. D. Rawlings, and J. F. Woessner. 2004. *Handbook of Proteolytic Enzymes*, 2nd ed. Elsevier Academic Press, London, United Kingdom.

Basrai, M. A., M. A. Lubkowitz, J. R. Perry, D. Miller, E. Krainer, F. Naider, and J. M. Becker. 1995. Cloning of a *Candida albicans* peptide transport gene. *Microbiology* **141:**1147–1156.

Beauvais, A., M. Monod, J. P. Debeaupuis, M. Diaquin, H. Kobayashi, and J. P. Latgé. 1997a. Biochemical and antigenic characterization of a new dipeptidyl-peptidase isolated from *Aspergillus fumigatus*. *J. Biol. Chem.* **272:**6238–6244.

Beauvais, A., M. Monod, J. Wyniger, J. P. Debeaupuis, E. Grouzmann, N. Brakch, J. Svab, A. G. Hovanessian, and J. P. Latgé. 1997b. Dipeptidyl-peptidase IV secreted by *Aspergillus fumigatus*, a fungus pathogenic to humans. *Infect. Immun.* **65:**3042–3047.

Beggah, S., B. Lechenne, U. Reichard, S. Foundling, and M. Monod. 2000. Intra- and intermolecular events direct the propeptide-mediated maturation of the *Candida albicans* secreted aspartic proteinase Sap1p. *Microbiology* **146:**2765–2773.

Bendtsen, J. D., H. Nielsen, G. von Heijne, and S. Brunak. 2004. Improved prediction of signal peptides: SignalP 3.0. *J. Mol. Biol.* **340:**783–795.

Berka, R. M., M. Ward, L. J. Wilson, K. J. Hayenha, K. H. Kodama, L. P. Carlomango, and S. A. Thompson. 1990. Molecular cloning and deletion of the gene encoding aspergillopepsin A from *Aspergillus awamori*. *Gene* **86:**153–162.

Berka, R. M., C. L. Carmona, K. J. Hayenga, S. A. Thompson, and M. Ward. 1993. Isolation and characterization of the *Aspergillus oryzae* gene encoding aspergillopepsin O. *Gene* **125:**195–198.

Biguet, J., J. Tran Van Ky, and S. Andrieu. 1967. Identification d'une activité chymotrypsique au niveau de fractions remarquables d'*Aspergillus fumigatus*. Répercussions sur le diagnostic immunologique de l'aspergillose. *Rev. Immunol. Paris* **31:**317–328.

Blinkovsky, A. M., T. Byun, K. M. Brown, and E. J. Golightly. 1999. Purification, characterization, and heterologous expression in *Fusarium venenatum* of a novel serine carboxypeptidase from *Aspergillus oryzae*. *Appl. Environ. Microbiol.* **65:**3298–3303.

Blinkovsky, A. M., T. Byun, K. M. Brown, E. J. Golightly, and A. V. Klotz. 2000. A non-specific aminopeptidase from *Aspergillus*. *Biochim. Biophys. Acta* **1480:**171–181.

Blobel, G., and B. Dobberstein. 1975. Transfer to proteins across membranes. II. Reconstitution of functional rough microsomes from heterologous components. *J. Cell Biol.* **67:**852–862.

Byun, T., L. Kofod, and A. Blinkovsky. 2001. Synergistic action of an X-prolyl dipeptidyl aminopeptidase and a non-specific aminopeptidase in protein hydrolysis. *J. Agric. Food Chem.* **49:**2061–2063.

Chevrier, B., and H. D'Orchymont. 2004. *Vibrio* aminopeptidase, p. 963–965. *In* A. J. Barrett, N. D. Rawlings, and J. F. Woessner (ed.), *Handbook of Proteolytic Enzymes*, 2nd ed. Elsevier, London, United Kingdom.

Chien, H. C., S. H. Lin, S. H. Chao, C. C. Chen, W. C. Wang, C. Y. Shaw, Y. C. Tsai, H. Y. Hu, and W. H. Hsu. 2002. Purification, characterisation, and genetic analysis of a leucine aminopepeptidase from *Aspergillus sojae*. *Biochim. Biophys. Acta* **1576:**119–126.

Denning, D. W, P. N. Ward, L. E. Fenelon, and E. W. Benbow. 1992. Lack of vessel wall elastolysis in human invasive aspergillosis. *Infect. Immun.* **60:**5153–5156.

Dmochowska, A., D. Dignard, D. Henning, D. Y. Thomas, and H. Bussey. 1987. Yeast KEX1 gene encodes a putative protease with a carboxypeptidase B-like function involved in killer toxin and alpha-factor precursor processing. *Cell* **50:**573–584.

Doi, Y., B. R. Lee, M. Ikeguchi, Y. Ohoba, T. Ikoma, S. Tero-Kubota, S. Yamauchi, K. Takahashi, and E. Ichishima. 2003. Substrate specificities of deuterolysin from *Aspergillus oryzae* and electron paramagnetic resonance measurement of cobalt-substituted deuterolysin. *Biosci. Biotechnol. Biochem.* **67:**264–270.

Doumas, A., R. Crameri, B. Léchenne, and M. Monod. 1999. Cloning of the gene encoding neutral protease I of the koji mold *Aspergillus oryzae* and its expression in *Pichia pastoris*. *J. Food Mycol.* **2:**271–279.

Edens, L., P. Dekker, R. van der Hoeven, F. Deen, A. de Roos, and R. Floris. 2005. Extracellular prolyl endoprotease from *A. niger* and its use in the debittering of protein hydrolysates. *J. Agric. Food Chem.* **53:**7950–7957.

Eder, J., and A. R. Fersht. 1995. Pro-sequence-assisted protein folding. *Mol. Microbiol.* **16:**609–614.

Eisenhaber, B., G. Schneider, M. Wildpaner, and F. Eisenhaber. 2004. A sensitive predictor for potential GPI lipid modification sites in fungal protein sequences and its application to genome-wide studies for *Aspergillus nidulans, Candida albicans, Neurospora crassa, Saccharomyces cerevisiae*, and *Schizosaccharomyces pombe*. *J. Mol.Biol.* **337:**243–253.

Fabre, E., J. M. Nicaud, M. C. Lopez, and C. Gaillardin. 1991. Role of the proregion in the production and secretion of the *Yarrowia lipolytica* alkaline extracellular protease. *J. Biol. Chem.* **266:**3782–3790.

Fankhauser, N., and P. Mäser. 2005. Identification of GPI anchor attachment signals by a Kohonen self-organizing map. *Bioinformatics* **21:**1846–1852.

Frederick, G. D., P. Rombouts, and F. P. Buxton. 1993. Cloning and characterization of *pepC*, a gene encoding a serine protease from *Aspergillus niger*. *Gene* **125:**57–64.

Frosco, M. B., T. Chase, and J. D. MacMillan. 1992. Purification and properties of the elastase from *Aspergillus fumigatus*. *Infect. Immun.* **60:**728–734.

Frosco, M. B., T. Chase, and J. D. MacMillan. 1994. The effect of elastase-specific monoclonal and polyclonal antibodies on the virulence of *Aspergillus fumigatus* in immunocompromised mice. *Mycopathologia* **125:**65–76.

Fukuda, R., K. Umebayashi, H. Horiuchi, A. Ohta, and M. Takagi. 1996. Degradation of *Rhizopus niveus* aspartic proteinase-I with mutated prosequences occurs in the endoplasmic reticulum of *Saccharomyces cerevisiae*. *J. Biol. Chem.* **271:**14252–14255.

Fushimi, N., C. E. Ee, T. Nakajima, and E. Ichishima. 1999. Aspzincin, a family of metalloendopeptidases with a new zinc-binding motif. Identification of new zinc-binding sites (His[128], His[132], and Asp[164]) and three catalytically crucial residues (Glu[129], Asp[143], and Tyr[106]) of deuterolysin from *Aspergillus oryzae* by site-directed mutagenesis. *J. Biol. Chem.* **274**:24195–24201.

Grobler, J., F. Bauer, R. E. Subden, and H. J. J. vanVuuren. 1995. The *mae1* gene of *Schizosaccharomyces pombe* encodes a permease for malate and other C4 dicarboxylic acids. *Yeast* **11**:1485–1491.

Grouzmann, E., M. Monod, B. Landis, S. Wilk, N. Brakch, K. Nicoucar, R. Giger, D. Malis, I. Szalay-Quinodoz, C. Cavadas, D. R. Morel, and J. S. Lacroix. 2002. Loss of dipeptidylpeptidase IV activity in chronic rhinosinusitis contributes to the neurogenic inflammation induced by substance P in the nasal mucosa. *FASEB J.* **16**: 1132–1134.

Hauser, M., V. Narita, A. M. Donhardt, F. Naider, and J. M. Becker. 2001. Multiplicity and regulation of genes encoding peptide transporters in *Saccharomyces cerevisiae*. *Mol. Membr. Biol.* **18**:105–112.

Higgins, D. R., and J. M. Cregg. 1998. *Pichia Protocols*. Humana Press, Totowa, NJ.

Huang, X. P., N. Kagami, H. Inoue, M. Kojima, T. Kimura, O. Makabe, K. Suzuki, and K. Takahashi. 2000. Identification of a glutamic acid and an aspartic acid residue essential for catalytic activity of aspergillopepsin II, a non-pepsin type acid proteinase. *J. Biol. Chem.* **275**:26607–26614.

Hube, B., and J. Naglik. 2001. *Candida albicans* proteinases: resolving the mystery of a gene family. *Microbiology* **147**:1997–2005.

Hung, C. Y., K. R. Seshan, J. J. Yu, R. Schaller, J. Xue, V. Basrur, M. J. Gardner, and G.T. Cole. 2005. A metalloproteinase of *Coccidioides posadasii* contributes to evasion of host detection. *Infect. Immun.* **73**:6689–6703.

Jaton-Ogay, K., M. Suter, R. Crameri, R. Falchetto, A. Fatih, and M. Monod. 1992. Nucleotide sequence of a genomic and a cDNA clone encoding an extracellular alkaline protease of *Aspergillus fumigatus*. *FEMS Microbiol. Lett.* **71**:163–168.

Jaton-Ogay, K., S. Paris, M. Huerre, M. Quadroni, R. Falchetto, G. Togni, J. P. Latgé, and M. Monod. 1994. Cloning and disruption of the gene encoding an extracellular metalloprotease of *Aspergillus fumigatus*. *Mol. Microbiol.* **14**:917–928.

Jones, E. W., and R. R. Naik. 2004. Cerevisin, p.1824–1827. *In* A. J. Barrett, N. D. Rawlings, and J. F. Woessner (ed.), *Handbook of Proteolytic Enzymes*, 2nd ed. Elsevier, London, United Kingdom.

Jongeneel, C. V., J. Bouvier, and A. Bairoch. 1989. A unique signature identifies a family of zinc-dependent metallopeptidases. *FEBS Lett.* **242**:211–214.

Joshi, L., and R. J. St. Leger. 1999. Cloning, expression, and substrate specificity of MeCPA, a zinc carboxypeptidase that is secreted into infected tissues by the fungal entomopathogen *Metarhizium anisopliae*. *J. Biol. Chem.* **274**:9803–9811.

Jousson, O., B. Léchenne, O. Bontems, S. Capoccia, B. Mignon, J. Barblan, M. Quadroni, and M. Monod. 2004a. Multiplication of an ancestral gene encoding secreted fungalysin preceded species differentiation in the dermatophytes *Trichophyton* and *Microsporum*. *Microbiology* **150**:301–310.

Jousson, O., B. Lechenne, O. Bontems, B. Mignon, U. Reichard, J. Barblan, M. Quadroni, and M. Monod. 2004b. Secreted subtilisin gene family in *Trichophyton rubrum*. *Gene* **339**:79-88.

Julius, D., A. Brake, L. Blair, R. Kunisawa, and J. Thorner. 1984. Isolation of the putative structural gene for the lysine-arginine-cleaving endopeptidase required for processing of yeast prepro-α-factor. *Cell* **37**:1075–1089.

Kobayashi, H., J. P. Debeaupuis, J. P. Bouchara, and J. P. Latgé. 1993. An 88-kilodalton antigen secreted by *Aspergillus fumigatus*. *Infect. Immun.* **61**:4767–4771.

Kogan, T. V., J. Jadoun, L. Mittelman, K. Hirschberg, and N. Osherov. 2004. Involvement of secreted *Aspergillus fumigatus* proteases in disruption of the actin fiber cytoskeleton and loss of focal adhesion sites in infected A549 lung pneumocytes. *J. Infect. Dis.* **189**: 1965–1973.

Koh, S., A. M. Wiles, J. S. Sharp, F. R. Naider, J. M. Becker, and G. Stacey. 2002. An oligopeptide transporter gene family in *Arabidopsis*. *Plant Physiol.* **128**:21–29.

Kolattukudy, P. E., J. D. Lee, L. M. Rogers, P. Zimmerman, S. Ceselski, B. Fox, B. Stein, and E. A. Copelan. 1993. Evidence for possible involvement of an elastolytic serine protease in aspergillosis. *Infect. Immun.* **61**:2357–2368.

Kolattukudy, P. E., and T. D. Sirakova. 2004. Fungalysin, p. 793–794. *In* A. J. Barrett, N. D. Rawlings, and J. F. Woessner (ed.), *Handbook of Proteolytic Enzymes*, 2nd ed. Elsevier, London, United Kingdom.

Kothary, M. H., T. Chase, and J. D. MacMillan. 1984. Correlation of elastase production by some strains of *Aspergillus fumigatus* with ability to cause pulmonary invasive aspergillosis in mice. *Infect. Immun.* **43**:320–325.

Kunert, J. 1972. Keratin decomposition by dermatophytes: evidence of the sulphitolysis of the protein. *Experientia* **28**:1025–1026.

Kunert, J. 1976. Keratin decomposition by dermatophytes. II. Presence of S-sulphocysteine and cysteic acid in soluble decomposition products. *Z. Allg. Mikrobiol.* **16**:97–105.

Kunert, J. 2000. Physiology of keratinophilic fungi, p. 77–85. *In* R. K. S. Kushwaha and J. Guarro (ed.), *Biology of Dermatophytes and Other Keratinophilic Fungi*. Revista Iberoamericana de Micología, Bilbao, Spain.

Landis, B. N., E. Grouzmann, M. Monod, N. Busso, F. Petak, A. Spiliopoulos, J. H. Robert, I. Szalay-Quinodoz, D. R. Morel, and J. S. Lacroix. 2008. Implication of dipeptidylpeptidase IV activity in human bronchial inflammation and in bronchoconstriction evaluated in anesthetized rabbits. *Respiration* **75**:89–97.

Larcher, G., J. P. Bouchara, V. Annaix, F. Symoens, D. Chabasse, and G. Tronchin. 1992. Purification and characterization of a fibrinogenolytic serine proteinase from *Aspergillus fumigatus* culture filtrate. *FEBS Lett.* **308**:65–69.

Latchinian-Sadek L., and D. Y. Thomas. 1993. Expression, purification, and characterization of the yeast *KEX1* gene product, a polypeptide precursor processing carboxypeptidase. *J. Biol. Chem.* **268**: 534–540.

Lavens, S. E., N. Rovira-Graells, M. Birch, and D. Tuckwell. 2005. ADAMs are present in fungi: identification of two novel ADAM genes in *Aspergillus fumigatus*. *FEMS Microbiol. Lett.* **248**:23–30.

Léchenne, B., U. Reichard, C. Zaugg, M. Fratti, J. Kunert, O. Boulat, and M. Monod. 2007. Sulphite efflux pumps in *Aspergillus fumigatus* and dermatophytes. *Microbiology* **153**:905–913.

Lee, B. R., M. Furukawa, K. Yamashita, Y. Kanasugi, C. Kawabata, K. Hirano, K. Ando, and E. Ichishima. 2003. Aorsin, a novel serine proteinase with trypsin-like specificity at acidic pH. *Biochem. J.* **371**: 541–548.

Lee, J. D., and P. E. Kolattukudy. 1995. Molecular cloning of the cDNA and gene for an elastinolytic aspartic proteinase from *Aspergillus fumigatus* and evidence of its secretion by the fungus during invasion of the host lung. *Infect. Immun.* **63**:3796–3803.

Lubkowitz, M. A., L. Hauser, M. Breslav, F. Naider, and J. M. Becker. 1997. An oligopeptide transport gene from *Candida albicans*. *Microbiology* **143**:387–396.

Marie-Claire, C., B. P. Roques, and A. Beaumont. 1998. Intramolecular processing of prothermolysin. *J. Biol. Chem.* **273**:5697–5701.

Markaryan, A., I. Morozova, H. Yu, and P. E. Kolattukudy. 1994. Purification and characterization of an elastinolytic metalloprotease from *Aspergillus fumigatus* and immunoelectron microscopic evidence of secretion of this enzyme by the fungus invading the murine lung. *Infect. Immun.* **62**:2149–2157.

Milstein, C., G. G. Brownlee, T. M. Harrison, and M. B. Mathews. 1972. A possible precursor of immunoglobulin light chains. *Nat. New Biol.* **239**:117–120.

Misumi, Y., and Y. Ikehara. 2004. Dipeptidyl-peptidases A and B, p. 1910–1911. *In* A. J. Barrett, N. D. Rawlings, and J. F. Woessner (ed.), *Handbook of Proteolytic Enzymes*, 2nd ed. Elsevier, London, United Kingdom.

Moehle, C. M., M. W. Aynardi, M. R. Kolodny, F. J. Park, and E. W. Jones. 1987. Protease B of *Saccharomyces cerevisiae*: isolation and regulation of the PRB1 structural gene. *Genetics* **115**:255–263.

Monod, M., F. Baudraz-Rosselet, A. A. Ramelet, and E. Frenk. 1989. Direct mycological examination in dermatology: a comparison of different methods. *Dermatologica* **179**:183–186.

Monod, M., G. Togni, L. Rahalison, and E. Frenk. 1991. Isolation and characterisation of an extracellular alkaline protease of *Aspergillus fumigatus*. *J. Med. Microbiol.* **35**:23–28.

Monod, M., S. Paris, J. Sarfati, K. Jaton-Ogay, P. Ave, and J. P. Latgé, 1993a. Virulence of alkaline protease-deficient mutants of *Aspergillus fumigatus*. *FEMS Microbiol. Lett.* **106**:39–46.

Monod, M., S. Paris, D. Sanglard, K. Jaton-Ogay, J. Bille, and J. P. Latgé. 1993b. Isolation and characterization of a secreted metalloprotease of *Aspergillus fumigatus*. *Infect. Immun.* **61**:4099–4104.

Monod, M. M., and A. Beauvais. 2004. Dipeptidyl-peptidases IV and V of *Aspergillus*, p. 1911–1913. *In* A. J. Barrett, N. D. Rawlings, and J. F. Woessner (ed.), *Handbook of Proteolytic Enzymes*, 2nd ed. Elsevier, London, United Kingdom.

Monod, M., B. Lechenne, O. Jousson, D. Grand, C. Zaugg, R. Stocklin, and E. Grouzmann. 2005. Aminopeptidases and dipeptidylpeptidases secreted by the dermatophyte *Trichophyton rubrum*. *Microbiology* **151**:145–155.

Monod, M., O. Bontems, C. Zaugg, B. Léchenne, M. Fratti, and R. Panizzon. 2006. Fast and reliable PCR/sequencing/RFLP assay for identification of fungi in onychomycoses. *J. Med. Microbiol.* **55**: 1211–1216.

Mortensen, U. H., K. Olesen, and K. Breddam. 2004. Serine carboxypeptidase C including carboxypeptidase Y, p. 1919–1923. *In* A. J. Barrett, N. D. Rawlings, and J. F. Woessner (ed.), *Handbook of Proteolytic Enzymes*, 2nd ed. Elsevier, London, United Kingdom.

Moser, M., G. Menz, K. Blaser, and R. Crameri. 1994. Recombinant expression and antigenic properties of a 32-kilodalton extracellular alkaline protease, representing a possible virulence factor from *Aspergillus fumigatus*. *Infect. Immun.* **62**:936–942.

Moutaouakil, M., M. Monod, M. C. Prevost, J. P. Bouchara, and J. P. Latgé. 1993. Identification of the 33-kDa alkaline protease of *Aspergillus fumigatus* in vitro and in vivo. *J. Med. Microbiol.* **39**:393–399.

Nakadai, T., S. Nasuno, and N. Iguchi. 1972a. Purification and properties of acid carboxypeptidase I from *Aspergillus oryzae*. *Agric. Biol. Chem.* **36**:1343–1352.

Nakadai, T., S. Nasuno, and N. Iguchi. 1972b. Purification and properties of acid carboxypeptidase II from *Aspergillus oryzae*. *Agric. Biol. Chem.* **36**:1473–1480.

Nakadai, T., S. Nasuno, and N. Iguchi. 1972c. Purification and properties of acid carboxypeptidase III from *Aspergillus oryzae*. *Agric. Biol. Chem.* **36**:1481–1488.

Nakadai, T., S. Nasuno, and N. Iguchi. 1973a. Purification and properties of neutral proteinase I from *Aspergillus oryzae*. *Agric. Biol. Chem.* **37**:2695–2701.

Nakadai, T., S. Nasuno, and N. Iguchi. 1973b. Purification and properties of neutral proteinase II from *Aspergillus oryzae*. *Agric. Biol. Chem.* **37**:2703–2708.

Nakadai, T., S. Nasuno, and N. Iguchi. 1973c. Purification and properties of leucine aminopeptidase I from *Aspergillus oryzae*. *Agric. Biol. Chem.* **37**:757–765.

Nakadai, T., S. Nasuno, and N. Iguchi. 1973d. Purification and properties of leucine aminopeptidase II from *Aspergillus oryzae*. *Agric. Biol. Chem.* **37**:767–774.

Nakadai, T., S. Nasuno, and N. Iguchi. 1973e. Purification and properties of leucine aminopeptidase III in *Aspergillus oryzae*. *Agric. Biol. Chem.* **37**:775–782.

Nakadai, T., S. Nasuno, and N. Iguchi. 1973f. Purification and properties of acid carboxypeptidase IV from *Aspergillus oryzae*. *Agric. Biol. Chem.* **37**:1237–1251.

Newport, G., and N. Agabian. 1997. *KEX2* influences *Candida albicans* proteinase secretion and hyphal formation. *J. Biol. Chem.* **272**: 28954–28961.

Nielsen, H., J. Engelbrecht, S. Brunak, and G. von Heijne 1997. Identification of prokaryotic and eukaryotic signal peptides and prediction of their cleavage sites. *Protein Eng.* **10**:1–6.

Nikai, T., and T. Yasuhara. 2004. Aminopeptidase Y, p. 956–957. *In* A. J. Barrett, N. D. Rawlings, and J. F. Woessner (ed.), *Handbook of Proteolytic Enzymes*, 2nd ed. Elsevier, London, United Kingdom.

Perry, J. R., M. A. Basrai, H. Y. Steiner, F. Naider, and J. M. Becker. 1994. Isolation and characterization of a *Saccharomyces cerevisiae* peptide transport gene. *Mol. Cell. Biol.* **14**:104–115.

Pfeffer, S. R., and J. E. Rothman. 1987. Biosynthetic protein transport and sorting by the endoplasmic reticulum and Golgi. *Annu. Rev. Biochem.* **56**:829–852.

Piechura, J. E., V. P. Kurup, and L. J. Daft. 1990. Isolation and immunochemical characterization of fractions from membranes of *Aspergillus fumigatus* with protease activity. *Can. J. Microbiol.* **36**:33–41.

Rawlings, N. D, F. R. Morton, and A. J. Barrett. 2006. MEROPS: the peptidase database. *Nucleic Acids Res.* **34**:D270–D272.

Reichard, U., S. Büttner, H. Eiffert, F. Staib, and R. Rüchel. 1990. Purification and characterisation of an extracellular serine proteinase from *Aspergillus fumigatus* and its detection in tissue. *J. Med. Microbiol.* **33**:243–251.

Reichard, U., H. Eiffert, and R. Rüchel. 1994. Purification and characterization of an extracellular aspartic proteinase from *Aspergillus fumigatus*. *J. Med. Vet. Mycol.* **32**:427–436.

Reichard, U., M. Monod, and R. Rüchel. 1995. Molecular cloning and sequencing of the gene encoding an extracellular aspartic proteinase from *Aspergillus fumigatus*. *FEMS Microbiol. Lett.* **130**:69–74.

Reichard, U., M. Monod, and R. Rüchel. 1996. Expression pattern of aspartic proteinase antigens in aspergilli. *Mycoses* **39**:99–101.

Reichard, U., M. Monod, F. Odds, and R. Rüchel. 1997. Virulence of an aspergillopepsin-deficient mutant of *Aspergillus fumigatus* and evidence for another aspartic proteinase linked to the fungal cell wall. *J. Med. Vet. Mycol.* **35**:189–195.

Reichard, U., G. T. Cole, R. Rüchel, and M. Monod. 2000a. Molecular cloning and targeted deletion of *PEP2* which encodes a novel aspartic proteinase from *Aspergillus fumigatus*. *Int. J. Med. Microbiol.* **290**:85–96.

Reichard, U., G. T. Cole, T. W. Hill, R. Rüchel, and M. Monod. 2000b. Molecular characterization and influence on fungal development of ALP2, a novel serine proteinase from *Aspergillus fumigatus*. *Int. J. Med. Microbiol.* **290**:549–558.

Reichard, U., B. Léchenne, A.R. Asif, F. Streit, E. Grouzmann, O. Jousson, and M. Monod. 2006. Sedolisins, as new class of secreted proteases from *Aspergillus fumigatus* with endoprotease or tripeptidyl-peptidase activity at acidic pH. *Appl. Environ. Microbiol.* **72**:1739–1748.

Reuss, O., and J. Morschhäuser. 2006. A family of oligopeptide transporters is required for growth of *Candida albicans* on proteins. *Mol. Microbiol.* **60**:795-812. (Erratum, **62**:916.)

Rypniewski, W. R., S. Hastrup, C. Betzel, M. Dauter, Z. Dauter, G. Papendorf, S. Branner, and K. S. Wilson. 1993. The sequence and X-ray structure of the trypsin from *Fusarium oxysporum*. *Protein Eng.* **6**:341–348.

Sarfati, J., M. Monod, P. Recco, A. Sulahian, C. Pinel, E. Candolfi, T. Fontaine, J. P. Debeaupuis, M. Tabouret, and J. P. Latgé. 2006. Recombinant antigens as diagnostic markers for aspergillosis. *Diagn. Microbiol. Infect. Dis.* **55**:279–291.

Screen, S. E., and R. J. St. Leger. 2000. Cloning, expression, and substrate specificity of a fungal chymotrypsin. Evidence for lateral gene transfer from an actinomycete bacterium. *J. Biol. Chem.* **275:**6689–6694.

Sekine, H. 1972. Neutral proteinases I and II of *Aspergillus sojae*. Isolation in homogeneous form. *Agric. Biol. Chem.* **36:**198–206.

Sekine, H. 1973. Neutral proteinases II of *Aspergillus sojae*: an enzyme specifically active on protamine and histone. *Agric. Biol. Chem.* **37:** 1765–1767.

Sirakova, T. D., A. Markaryan, and P. E. Kolattukudy. 1994. Molecular cloning and sequencing of the cDNA and gene for a novel elastinolytic metalloproteinase from *Aspergillus fumigatus* and its expression in *Escherichia coli*. *Infect. Immun.* **62:**4208–4218.

Staib, P., M. Kretschmar, T. Nichterlein, H. Hof, and J. Morschhäuser. 2000. Differential activation of a *Candida albicans* virulence gene family during infection. *Proc. Natl. Acad. Sci. USA* **97:**6102–6107.

Stepniak, D., L. Spaenij-Dekking, C. Mitea, M. Moester, A. de Ru, R. Baak-Pablo, P. van Veelen, L. Edens, and F. Koning. 2006. Highly efficient gluten degradation with a newly identified prolyl endoprotease: implications for celiac disease. *Am. J. Physiol. Gastrointest. Liver Physiol.* **291:**G621–G629.

St. Leger, R. J., L. Joshi, M. J. Bidochka, N. W. Rizzo, and D. W. Roberts. 1996. Biochemical characterization and ultrastructural localization of two extracellular trypsins produced by *Metarhizium anisopliae* in infected insect cuticles. *Appl. Environ. Microbiol.* **62:** 1257–1264.

Sturtevant, J., and J. P. Latgé. 1992. Interactions between conidia of *Aspergillus fumigatus* and human complement component C3. *Infect. Immun.* **60:**1913–1918.

Suarez-Rendueles, P., and J. Bordallo. 2004. Gly-X carboxypeptidase, p. 956–957. *In* A. J. Barrett, N. D. Rawlings, and J. F. Woessner (ed.), *Handbook of Proteolytic Enzymes*, 2nd ed. Elsevier, London, United Kingdom.

Svendsen, I., and F. Dal Degan. 1998. The amino acid sequences of carboxypeptidases I and II from *Aspergillus niger* and their stability in the presence of divalent cations. *Biochim. Biophys. Acta* **1387:** 369–377.

Takahashi, K. 2004. Aspergillopepsin II, p. 221–224. *In* A. J. Barrett, N. D. Rawlings, and J. F. Woessner (ed.), *Handbook of Proteolytic Enzymes*, 2nd ed. Elsevier, London, United Kingdom.

Tang, C. M., J. Cohen, and D. W. Holden. 1992. An *Aspergillus fumigatus* alkaline protease mutant constructed by gene disruption is deficient in extracellular elastase activity. *Mol. Microbiol.* **6:**1663–1671.

Tang, C. M., J., Cohen, T., Krausz, S., Van Noorden, and D. W. Holden. 1993. The alkaline protease of *Aspergillus fumigatus* is not a virulence determinant in two murine models of invasive pulmonary aspergillosis. *Infect. Immun.* **61:**1650–1656.

Tatsumi, H., S. Murakami, R. F. Tsuji, Y. Ishida, K. Murakami, A. Masaki, H. Kawabe, H. Arimura, E. Nakano, and H. Motai. 1991. Cloning and expression in yeast of a cDNA clone encoding *Aspergillus oryzae* neutral protease II, a unique metalloprotease. *Mol. Gen. Genet.* **228:**97–103.

Togni, G., D. Sanglard, M. Quadroni, S.I. Foundling, and M. Monod. 1996. Acid proteinase secreted by *Candida tropicalis*: functional analysis of preproregion cleavages in *C. tropicalis* and *Saccharomyces cerevisiae*. *Microbiology* **142:**493–503.

van den Burg, B., and V. Eijsink. 2004. Thermolysin and related *Bacillus* metallopeptidases, p. 374–387. *In* A. J. Barrett, N. D. Rawlings, and J. F. Woessner (ed.), *Handbook of Proteolytic Enzymes*, 2nd ed. Elsevier, London, United Kingdom.

Vickers, I., E. P. Reeves, K. A. Kavanagh, and S. Doyle. 2007. Isolation, activity and immunological characterisation of a secreted aspartic protease, CtsD, from *Aspergillus fumigatus*. *Protein Expr. Purif.* **53:**216–224.

von Heijne, G. 1983. Patterns of amino acids near signal-sequence cleavage sites. *Eur. J. Biochem.* **133:**17–21.

von Heijne, G. 1984. How signal sequences maintain cleavage specificity. *J. Mol. Biol.* **173:**243–251.

von Heijne, G. 1985. Signal sequences. The limits of variation. *J. Mol. Biol.* **184:**99–105.

von Heijne, G. 1986. A new method for predicting signal sequence cleavage sites. *Nucleic Acids Res.* **14:**4683–4690.

Walker, S. J., and M. O. Lively. 2004. Signal peptidase (eukaryote), p. 1991–1997. *In* A. J. Barrett, N. D. Rawlings, and J. F. Woessner (ed.), *Handbook of Proteolytic Enzymes*, 2nd ed. Elsevier, London, United Kingdom.

Walter, E. G., J. H. Weiner, and D. E. Taylor. 1991. Nucleotide sequence and overexpression of the tellurite-resistance determinant from the IncHII plasmid pHH1508a. *Gene* **101:**1–7.

Winther, J. R., L. Phylip, and J. Kay. 2004. Saccharopepsin, p. 87–90. In A. J. Barrett, N. D. Rawlings, and J. F. Woessner (ed.), *Handbook of Proteolytic Enzymes*, 2nd ed. Elsevier, London, United Kingdom.

Wlodawer, A., M. Li, A. Gustchina, H. Oyama, B. M. Dunn, and K. Oda. 2003. Structural and enzymatic properties of the sedolisin family of serine-carboxyl peptidases. *Acta Biochim. Pol.* **50:**81–102.

Wolfsberg, T. G., P. Primakoff, D. G. Myles, and J. M. White. 1995. ADAM, a novel family of membrane proteins containing a disintegrin and metalloprotease domain: multipotential functions in cell-cell and cell-matrix interactions. *J. Cell Biol.* **131:**275–278.

Wolfsberg, T. G., and J. M. White. 2004. ADAM metalloproteinases, p. 709–714. *In* A. J. Barrett, N. D. Rawlings, and J. F. Woessner (ed.), *Handbook of Proteolytic Enzymes*, 2nd ed. Elsevier, London, United Kingdom.

Yabuki, Y., K. Kubota, M. Kojima, H. Inoue, and K. Takahashi. 2004. Identification of a glutamine residue essential for catalytic activity of aspergilloglutamic peptidase by site-directed mutagenesis. *FEBS Lett.* **569:**161–164.

Zaugg, C., O. Jousson, B. Léchenne, P. Staib, and M. Monod. 25 January 2008 posting date. *Trichophyton rubrum* secreted and membrane-associated carboxypeptidases. *Int. J. Med. Microbiol.* [Epub ahead of print.]

Aspergillus fumigatus and Aspergillosis
Edited by J.-P. Latgé and W. J. Steinbach

Chapter 9

Cations (Zn, Fe)

José Antonio Calera and Hubertus Haas

Iron and zinc are the most abundant transition metals in cells. Although present only in trace amounts, both metals play critical roles in a wide variety of biochemical processes. Thus, iron is present in hemoproteins, iron-sulfur proteins, and other ferroproteins, most of which are oxidoreductases in which the oxidation state of iron interchanges between the ferrous (Fe^{2+}) and ferric (Fe^{3+}) state (except certain iron-sulfur proteins, ferritin, and transferrin). In contrast, Zn^{2+} is a cofactor that never changes its oxidation state and is required by more than 300 enzymes belonging to six major functional classes, as well as by many other nonenzymatic proteins. Therefore, iron and zinc are essential micronutrients for growth and development of all organisms including microbes, although excess amounts of these metals are toxic for cells.

During saprophytic growth *Aspergillus fumigatus* usually has access to sufficient amounts of iron and zinc. In contrast, availability of both iron and zinc in hosts is usually kept low enough to inhibit growth of most microorganisms. The concentration of free iron within a mammalian host is extremely low, since virtually all iron is associated with proteins. In addition, the low-iron environment is maintained in particular by the iron-binding proteins transferrin and lactoferrin (Papanikolaou and Pantopoulos, 2005). Thus, even though the total amount of iron in the human body is about 4 g (Coleman, 1992), the concentration of free iron in plasma is as low as 10^{-12} μM (Bullen et al., 2006). Similarly, the concentration of free zinc is also very low, since nearly all zinc is tightly bound to (i) enzymes, which use zinc as a cofactor for their catalytic or cocatalytic activities; (ii) transcription factors that require zinc for their structural stability; (iii) metallothioneins (MTs), which contribute to both zinc storage and detoxification; and (iv) plasma zinc-binding proteins, such as α_2-macroglobulin and albumin, the main zinc carriers in plasma. Hence, the concentration of free zinc in plasma is about 10^{-4} μM (Magneson et al., 1987), even though the total zinc concentration in plasma is about 15 μM (Tapiero and Tew, 2003). However, the minimal concentration of both iron and zinc required to ensure optimal growth of *A. fumigatus* is approximately 1.0 μM (unpublished data). Therefore, the estimated concentrations of free iron and free zinc in plasma are about 10^{12}- and 10^4-fold, respectively, short of the minimal amounts required by *A. fumigatus*. Consequently, *A. fumigatus*, as other pathogens, must be able to extract iron and zinc from host proteins. Recently, mechanisms for high-affinity uptake of iron and control of zinc acquisition have been identified and proven to be essential for pathogenic growth of *A. fumigatus* (Moreno et al., 2007b; Schrettl et al., 2004a). As *A. fumigatus* is not an obligate pathogen, these mechanisms have probably evolved for saprophytic growth under conditions of shortage of these metals. Possibly, these mechanisms have been adapted due to environmental selective pressures imposed by natural predators, such as fungivory amoebas and nematodes, as suggested for the origin of virulence factors in *Cryptococcus* spp. (Steenbergen and Casadevall, 2003).

Microbial pathogens' requirements for iron and zinc are the basis of an elaborate mammalian defense system against microbial infection which relies upon withholding mechanisms for iron and zinc to deny access to these metals for invading microbes. Thus, infection and inflammation trigger a systemic, acute-phase response that includes hypoferremia and hypozincemia (Moshage, 1997). The main proinflammatory cytokine responsible for inducing this status is interleukin-6 (IL-6). This cytokine induces the hepatic production of hepcidin, an iron regulatory hormone, in response to inflammation. Hepcidin inhibits iron release by macrophages and absorption of dietary iron from the intestine and thereby produces hypoferremia (Ganz and Nemeth, 2006). Additionally, IL-6 also induces hypo-

José Antonio Calera • Instituto de Microbiología-Bioquímica, Centro mixto CSIC/USAL, Dept. de Microbiología y Genética (Universidad de Salamanca), Plaza Doctores de la Reina s/n, 37007 Salamanca, Spain. **Hubertus Haas** • Biocenter, Division of Molecular Biology, Innsbruck Medical University, Fritz-Pregl-Str. 3, A-6020 Innsbruck, Austria.

zincemia through enhancing expression of MTs and the Zip14 zinc transporter in hepatocytes which, in turn, results in an increased hepatic zinc accumulation (Liuzzi et al., 2005; Schroeder and Cousins, 1990). In addition, hypozincemia is induced by secretion of calprotectin by neutrophils, monocytes, endothelial cells, and epidermal cells in response to infection. The zinc-binding protein calprotectin inhibits microbial growth by competition for zinc (Striz and Trebichavsky, 2004).

In summary, iron and zinc supply play critical roles in the pathogenicity of *A. fumigatus*. In this chapter we have compiled the current knowledge about *A. fumigatus* mechanisms involved in maintaining iron and zinc homeostasis and their impact on pathogenicity.

ZINC HOMEOSTASIS

Due to the enormous diversity of biochemical processes for which zinc is essential, all organisms must maintain permanently an adequate intracellular concentration of zinc to support cell growth regardless of the ambient zinc availability. To accomplish this feat, cells have evolved different classes of proteins that function coordinately to prevent inhibition of cell growth either by zinc starvation or zinc intoxication. These zinc homeostasis-maintaining proteins include the following.

1. *Zinc transporters of the ZIP family.* The zinc transporters of the ZIP family (Zrt-, Irt-like proteins) have eight predicted transmembrane domains and similar predicted topologies, with the N and C termini located towards the extracellular surface of the plasma membrane or towards the lumen of the organellar membrane. ZIP transporters pump zinc and/or other metal ions from the extracellular space or organellar lumen into the cytoplasm (Gaither and Eide, 2001).

2. *Zinc transporters of the CDF family.* The fungal proteins of the CDF (cation diffusion facilitator) family usually have six predicted transmembrane domains, and their N and C termini are located towards the cytosolic surface of the organellar membrane. CDF proteins transport zinc and/or other metal ions from the cytoplasm into the lumen of intracellular organelles (e.g., the vacuole plays a role in zinc storage and/or detoxification) (Gaither and Eide, 2001).

3. *Zinc-chelating MTs.* MTs are low-molecular-mass proteins with high cysteine content that lack histidine and aromatic amino acids. Their structural characteristics allow potent metal binding (particularly Cu, Cd, and Zn) and redox capabilities. MTs play a role in zinc homeostasis, particularly in detoxification and regulation of metabolism via zinc donation, sequestration, and/or redox control (Coyle et al., 2002; Maret, 2006).

4. *Transcription factors.* Two classes of transcription factors controlling the expression of a broad range of zinc-regulated genes at the transcriptional level can be distinguished: (i) zinc-responsive transcriptional activators that induce gene expression under zinc-limiting conditions, and (ii) other regulatory factors that might interact with the zinc-responsive transcriptional activator to modulate gene expression in response to environmental factors other than zinc availability but which, nevertheless, may influence zinc availability, e.g., pH.

Among all organisms, the molecular mechanisms for maintaining zinc homeostasis have been best characterized in *Saccharomyces cerevisiae* (Eide, 2006; Rutherford and Bird, 2004), which undoubtedly facilitates the investigation of zinc homeostasis in *A. fumigatus*. In this yeast, all classes of proteins involved in maintaining zinc homeostasis mentioned above are present: four ZIP family members (Ztr1 to -3 and Yke4) (Eide, 2006; Kumanovics et al., 2006), four CDF transporters (Zrc1, Co1, Msc2, and Zrg17), one zinc-thionein (Csr5) (Pagani et al., 2007), and one zinc-responsive transcriptional activator (Zap1) (Rutherford and Bird, 2004). Additional transporters possibly contributing to zinc uptake are Fet4, which is also involved in low-affinity uptake of iron and copper (Waters and Eide, 2002), and Pho84, a high-affinity phosphate transporter that also plays a role in manganese homeostasis (Jensen et al., 2003). In spite of the detailed knowledge about zinc and pH regulation in this yeast, connections have not been characterized beyond a physical interaction between Zap1 and the pH regulatory transcription factor Rim101, which was identified in a large-scale two-hybrid screen but not confirmed otherwise (Uetz et al., 2000).

Investigations of zinc homeostasis-maintaining mechanisms in *Aspergillus* have been initiated recently (Moreno et al., 2007a, 2007b; Vicentefranqueira et al., 2005). Thus, the knowledge about this aspect of *A. fumigatus* physiology is limited. In contrast to *S. cerevisiae*, however, several lines of evidence suggest that the ambient pH largely affects zinc homeostasis (see below). *A. fumigatus* is able to grow at similar rates over a broad range of pH values, whereas *S. cerevisiae* is adapted to acidic conditions rather than neutral or alkaline environments, in which it actually grows poorly (Caddick et al., 1986; Hong et al., 1999; Serrano et al., 2006). In *A. fumigatus*, the zinc supply appears to be of particular importance under neutral or alkaline conditions (unpublished data). Moreover, the immunodominant antigen ASPND1 of *Aspergillus nidulans* and its ortholog Aspf2 of *A. fumigatus* are produced only under neutral and zinc-limiting conditions, which suggests a relationship between zinc regulation, pH control, and pathogenicity in *A. fumigatus* (Calera et al., 1997; Segurado et al., 1999). Therefore, *A. fumigatus* represents an attractive model organism to study the link between zinc homeostasis, pH homeostasis, and their connection to virulence.

COMPONENTS OF THE SYSTEM THAT GOVERNS ZINC HOMEOSTASIS

Inspection of the genomic sequence of *A. fumigatus* revealed homologs to all classes of proteins involved in maintaining zinc homeostasis in *S. cerevisiae* except to MTs: eight members of the ZIP family (ZrfA to -H), four transporters belonging to the CDF family, and a zinc-responsive transcriptional activator (ZafA) (Table 1 and Fig. 1). Moreover, *A. fumigatus* has orthologs to Fet4 and Pho84 of *S. cerevisiae*, which could also contribute to zinc uptake. Surprisingly, in spite of the presence of MTs in all animal phyla and most plants and fungi, *A. fumigatus* does not appear to possess MT-encoding genes. However, MTs are small metal-binding proteins whose encoding genes may have passed unnoticed during the genome annotation process. For example, the MT Mmt1 of *Magnaporthe grisea* was not identified during the annotation process of its genome (Tucker et al., 2004).

At present, only the functions of ZrfA, ZrfB, and ZafA have been investigated in some detail, while deciphering the functions of other genes awaits further investigations. The intronless genes *zrfA* and *zrfB* encode proteins of 359 and 353 amino acids, respectively. They are likely embedded within the plasma membrane through eight transmembrane domains and function as zinc transporters, similar to other proteins of the ZIP family. ZrfA and ZrfB show identities of 54.5% and 37.6% with proteins Zrt1 and Zrt2 of *S. cerevisiae*, respectively. Nevertheless, each gene improved growth of a yeast *zrt1Δ zrt2Δ* strain to the same extent under severe zinc-limiting conditions when expressed under the control of the same promoter (Vicentefranqueira et al., 2005). Therefore, it is likely that ZrfA and ZrfB exhibit a similar zinc transport activity regardless of their affinity for zinc binding and uptake.

Fungal growth is severely affected by zinc availability, which is in turn influenced by pH, as solubility of zinc decreases gradually around neutrality (Barrow, 1993). Consistently, the *A. fumigatus* zinc regulon (genes whose expression is influenced by zinc) is se-

Table 1. Selected genes of *A. fumigatus* whose expression is or might be regulated by ZafA

Code	Gene[a]	Structural feature and/or function of protein
Afu1g01550	*zrfA*[1]	ZIP transporter
Afu2g03860	*zrfB*[1]	ZIP transporter
Afu4g09560	*zrfC*[1]	ZIP transporter
Afu6g00470	*zrfD*[2]	ZIP transporter
Afu8g04010	*zrfE*[2]	ZIP transporter
Afu2g08740	*zrfF*[1]	ZIP transporter
Afu2g01460	*zrfG*[2]	ZIP transporter
Afu2g12050	*zrfH*[2]	ZIP transporter
Afu7g06570	*zrcA*[1]	CDF transporter
Afu2g14570	*cotA*[2]	CDF transporter
Afu6g14170	*mscA*[2]	CDF transporter
Afu1g12090	*zrgA*[3]	CDF transporter
Afu1g10080	*zafA*[1]	C2H2 transcription factor
Afu4g14640	*fetD*[2]	Similar to Fet4 (low-affinity iron transporter) of *S. cerevisiae*
Afu4g03610	*phoA*[2]	Similar to Pho84 (high-affinity phosphate transporter) of *S. cerevisiae*
Afu1g15960	*glrA*[2]	Similar to Glr1 (glutathione reductase) of *S. cerevisiae*
Afu8g06080	*fhpA*[2]	Similar to flavohemoproteins
Afu5g13640	*trxE*[2]	Similar to thioredoxins
Afu2g14960	*trxF*[2]	Similar to thioredoxins
Afu8g07130	*prxA*[2]	Similar to peroxiredoxins
Afu6g02280	*prxC*[2]	Allergen Aspf3 similar to peroxiredoxins
Afu5g09240	*sodA*[1]	Cu,Zn-SOD[b]
Afu1g11640	*sodB*[2]	Cu,Zn-SOD
Afu1g14550	*sodC*[1]	Allergen Mn-SOD
Afu6g07210	*sodE*[2]	Fe-SOD
Afu4g09580	*aspf2*[1]	Immunodominant antigen Aspf2

[a] Superscript numbers indicate the following: 1, the gene harbors a ZR motif in its promoter region and expression of the gene is upregulated by ZafA in zinc-limiting medium (Moreno et al., 2007b; unpublished data); 2, the gene harbors a ZR-like motif in its promoter region; 3, the gene lacks a ZR motif in the promoter region.
[b] SOD, superoxide dismutase.

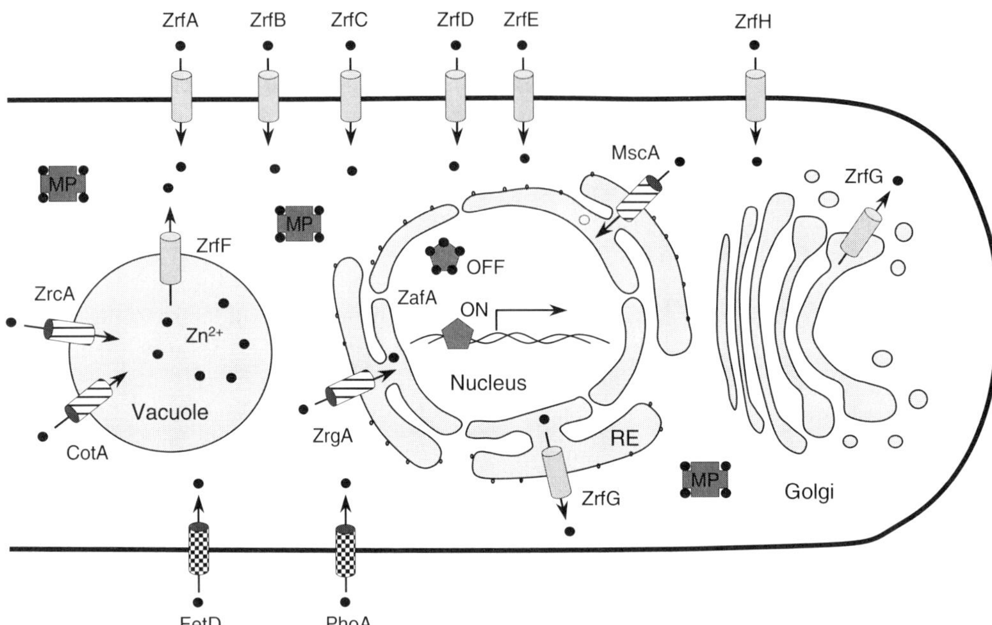

Figure 1. Schematic representation of proteins putatively involved in maintaining zinc homeostasis in *A. fumigatus*. Each protein is represented in its more likely subcellular location according to what is currently known about its correspondent yeast ortholog. Zinc transporters of the ZIP and CDF families are represented by gray cylinders and dashed cylinders, respectively. Other proteins that also may be involved in zinc homeostasis are depicted with a dotted pattern. The zinc-responsive transcriptional activator ZafA is represented by a dark gray pentagon in either an active or inactive state (saturated with zinc ions). MP, metalloproteins. Zinc ions are represented by small black circles.

verely influenced by environmental pH (unpublished data) (Moreno et al., 2007a; Vicentefranqueira et al., 2005). Thus, the study of zinc-regulated gene expression in *A. fumigatus* requires control of the pH of the culture, as this affects zinc availability. To accomplish this, the expression analysis of zinc-regulated genes should be performed using culture either containing or lacking zinc, which allows physiological control of the culture pH without the use of any buffer system, such as the SDA and SDN media. SDA contains ammonia and SDN contains nitrate as the nitrogen source, resulting in acidic and neutral conditions, respectively. Notably, zinc-lacking media contain residual amounts of zinc (2.3 ± 1.9 μM in SDA; 0.18 ± 0.1 μM in SDN), which initially support fungal growth but result in zinc starvation (Moreno et al., 2007a). Our studies showed that (i) the transcription of *zrfB* is turned off at a concentration 50-fold higher than *zrfA* transcription during acidic conditions (the initial concentration of zinc to repress the transcription of *zrfA* is 2.0 μM in SDA, or about 4.0 μM including the residual zinc), that is, the expression of *zrfA* and *zrfB* is differentially downregulated by the environmental concentration of zinc; (ii) neutral pH downregulates the expression of *zrfA* slightly and that of *zrfB* greatly under zinc-limiting conditions, that is, *zrfA* and *zrfB* reach their highest levels of expression in acidic media. Consistent with their maximal expression under acidic conditions, deletion of both genes largely impairs growth of *A. fumigatus* under acidic, zinc-limiting conditions but not neutral, zinc-limiting conditions. Consequently, another zinc uptake system must exist in addition to ZrfA and -B to support growth of *A. fumigatus* under neutral or alkaline conditions. Indeed, the main zinc transporter responsible for zinc uptake under neutral or slightly alkaline culture conditions is ZrfC (unpublished data). Interestingly, the genes encoding ZrfC and the immunodominant antigen Aspf2 are divergently transcribed from a common promoter region, which indicates involvement of both ZrfC and Aspf2 in the same metabolic process and stresses the possibility that both play a role in virulence, an aspect currently under active investigation.

In the promoter region of several zinc-regulated *A. fumigatus* genes (Table 1) we identified two different DNA motifs that play a role in transcriptional regulation: (i) zinc response motifs (5′-YYYCAAGGTNYBY-3′) recognized by the zinc-responsive transcriptional activator ZafA (unpublished data) (Moreno et al., 2007b) and (ii) pH response motifs (5′-GCCARG-3′) recognized by the *A. fumigatus* PacC protein, an ortholog of the *A. nidulans* PacC protein, which is involved in regulation of gene expression in response to changes in the environmental pH (Peñalva and Arst, 2004). Moreover, the particular arrangement of these two sequence motifs in the promoter regions appears to determine whether transcription is upregulated simultaneously in response to both the concentration of zinc and alkaline pH (unpublished data). As all these genes are expressed only during zinc starvation, zinc regulation mediated by ZafA clearly prevails over pH regulation by PacC. Therefore, pH regulation appears to only modulate the expression level (unpublished data).

In *A. fumigatus*, zinc regulation is mediated by the transcription factor ZafA (Moreno et al., 2007b). ZafA has four typical zinc fingers of the C2H2 class (designated ZnF3 to -6) matching the zinc finger consensus motif $X_2CX_{2-4}CX_{12}HX_{3-5}H$ and two "atypical" zinc fingers (ZnF1 and -2) containing a characteristic tryptophan residue at the fifth position that defines the $X_2CXWX_{2-10}C$-type Cys2 module. Most cysteine and histidine residues (83%) located in the first 357 amino acids of ZafA are clustered within four regions that may correspond to putative zinc-binding domains 1 to 4 (ZBD1 to -4). ZafA also contains a putative nuclear localization signal, LKRHMRT, similar to the transcriptional activators of the PacC/Rim101p and GLI/ZIC families (Fernández-Martínez et al., 2003). ZafA does not show significant overall similarity to any human protein. Apart from the zinc finger region, ZafA displays no significant similarity to Zap1 from *S. cerevisiae*, the only zinc-responsive transcriptional activator that has been finely characterized to date (Bird et al., 2000, 2003, 2006a, 2006b; Evans-Galea et al., 2003; Herbig et al., 2005; Zhao et al., 1998; Zhao and Eide, 1997). ZafA Zap1 exhibit two important structural differences: (i) ZafA appears to lack the zinc fingers, which are involved in both zinc sensing and repressing the activity of Zap1 (Herbig et al., 2005); (ii) ZafA has a predicted nuclear localization signal associated within its C-terminal zinc finger, while Zap1 has a potential nuclear localization signal located outside the zinc finger region (Zhao and Eide, 1997). Nevertheless, the ZafA protein also functions as a zinc-responsive transcriptional activator in the background provided by *S. cerevisiae*.

In *A. fumigatus*, the *zafA* gene is only expressed under zinc-limiting but not iron- or copper-limiting conditions. The only divalent metal ion that mimics the effects of zinc on expression of *zrfA*, *zrfB*, and *zafA* is Cd^{2+}, that is, Cd^{2+} represses expression of these genes (Moreno et al., 2007b; Vicentefranqueira et al., 2005), which underscores the similarity of cadmium and zinc.

ZafA is essential for germination and subsequent hyphal growth of *A. fumigatus* under zinc-limiting conditions. Consistently, ZafA is required to activate transcription of several genes under zinc-limiting conditions, including *zrfA* and *zrfB* (Moreno et al., 2007b). Moreover, ZafA might also regulate transcription of other metabolic processes, e.g., detoxification of antimicrobial reactive oxygen (ROS) and reactive nitrogen (RNS) species (Table 1) (unpublished data). ZafA is also required to reach maximal growth capability under neutral zinc-replete conditions, which indicates that even under these circumstances a minimal amount of ZafA is required, probably for activating expression of certain genes involved in zinc uptake (e.g., ZrfC).

HOST-FUNGUS INTERACTION, ZINC AVAILABILITY, AND VIRULENCE

Zinc is an essential micronutrient. However, zinc availability in living tissues is kept constant at levels low enough to restrict microbial growth. Several pieces of evidence indicate that the amount of free zinc available in the microhabitats provided by the host is very scarce: (i) sera of immunocompetent patients suffering from different forms of aspergillosis contain antibodies against the immunodominant antigen Aspf2 (Banerjee et al., 1998; Calera et al., 1997; López-Medrano et al., 1996), which is only synthesized by *A. fumigatus* during zinc-limiting growth (Segurado et al., 1999); (ii) a *zafA*Δ strain is not able to grow within the lungs of immunosuppressed mice and is nonpathogenic (Moreno et al., 2007b); and (iii) the concentration of total zinc in fetal bovine serum, which provides metal ions in both a concentration and physical state quite similar to that present in host tissues, is too low to allow optimal growth of a *zafA*Δ mutant (Moreno et al., 2007b). Therefore, ZafA-mediated regulation of zinc uptake is an essential attribute of the pathogenicity of *A. fumigatus* despite its dispensability during zinc-replete saprophytic growth.

It is well-known that infection triggers the release of the proinflammatory cytokine IL-1 by the host's macrophages and monocytes which, in turn, upregulates a broad spectrum of systemic acute-phase responses involved in host defense, including the release of glucocorticoids (GCs) and synthesis of IL-6 (Dinarello, 1988; Webster et al., 2002). Both IL-1 and IL-6 can synergistically activate the release of GCs, which also contribute to stimulate the synthesis of several proteins involved in the liver acute-phase response (Beltramini et al., 2004; Moshage, 1997; Schroeder and Cousins, 1990; Yeager et al., 2004). Among these proteins there are transporters that mediate uptake of zinc from plasma (Cousins et al., 2006; Liuzzi et al., 2005) and MTs that bind and store zinc (Beltramini et al., 2004; Coyle et al., 2002; Schroeder and Cousins, 1990). These proteins enhance the redistribution of labile zinc (zinc that is only loosely bound to proteins and can easily be exchanged between different binding sites) from plasma to several organs, mainly to the liver and central nervous system, where zinc accumulates, resulting in plasma zinc deficiency or hypozincemia.

Hypozincemia triggers a reprogramming of the immune system by increasing myelopoiesis at the expense of lymphopoiesis in the bone marrow in an attempt to protect the first line of immune defense at all cost (Fraker and King, 2004). In addition, hypozincemia also contributes to inhibit microbial growth through limiting access of pathogens to host zinc. However, the lowered zinc concentrations in plasma resulting from the acute-phase response might still be sufficient for the growth of many pathogens unless extremely zinc-limiting microenvironments are created around the infected areas. Indeed, during the acute-phase response IL-1 and GCs stimulate the cells of the innate immune system that first respond to the infection (neutrophils and monocytes) to secrete the antimicrobial, zinc-chelating protein calprotectin (Striz and Trebichavsky, 2004).

On the other hand, GCs at physiological concentrations modulate the inflammatory response by downregulating the expression of IL-1 and IL-6 to prevent tissue damage from overactivity of the host inflammatory response (Yeager et al., 2004), whereas increased doses of GCs result in a total anti-inflammatory response characterized by inhibition of activated macrophages, reduction of the number of circulating monocytes, and inhibition of neutrophil function, including phagocytosis and the production of antimicrobial ROS and RNS (Fang, 2004; Webster et al., 2002). Hence, patients subjected to long-term anti-inflammatory therapies induced with corticosteroids may undergo a chronic hypozincemia-like status. However, in the long run these treatments cause mobilization of Zn^{2+} from MTs upon cysteine oxidation (Haase and Rink, 2007; Kroncke, 2007; Maret, 2006), which results in an increased labile zinc pool within cells that impairs the overall immune function and the innate resistance to infection, leading to an increased risk for infectious diseases (Fischer Walker and Black, 2004). In addition, these long anti-inflammatory treatments also create environmental conditions that favor growth of *A. fumigatus*. Indeed, increased doses of GCs stimulate the growth rate of *A. fumigatus* under laboratory culture conditions (Ng et al., 1994), and oxidative and nitrosative stresses increase the concentration of free Zn^{2+} within phagocytes (Kroncke, 2007; Spahl et al., 2003) that can be readily used by *A. fumigatus* to grow. Thus, it is very likely that the avirulence of a *zafA*Δ null mutant in immunosuppressed mice primarily resides in the inability of this mutant to obtain zinc from host tissue, leading to an eventual arrest of conidial germination and/or elongation of hyphae (Color Plate 4). Moreover, ZafA may also modulate the expression of genes that play a role in protecting fungal cells against the ROS and/or RNS produced by those residual phagocytes that remain active in immunosuppressed individuals, because ZafA-binding motifs have been found in the promoter regions of *A. fumigatus* genes encoding glutathione reductase, flavohemoglobin, Fe-superoxide dismutase, thioredoxins, peroxiredoxins, and Cu/Zn-superoxide dismutases (Table 1). Consistently, expression of the gene encoding the peroxiredoxin Tsa1 is regulated by Zap1 in *S. cerevisiae* (Wu et al., 2007).

In summary, the lung tissue of immunocompetent individuals can be considered an extremely zinc-limiting environment efficiently protected by the innate immune system, which can only be rarely colonized by *A. fumigatus*. In contrast, GC-treated, immunosuppressed patients provide a less zinc-limiting environment virtually unprotected by the innate immune system and can be readily colonized by wild-type strains of *A. fumigatus*. In addition, zinc redistribution following immunosuppression may also explain the pathophysiology of *Aspergillus* infections. For example, after dissemination of pulmonary aspergillosis, both brain and liver are among the organs most frequently colonized by *A. fumigatus* (Kleinschmidt-DeMasters, 2002). Strikingly, the brain is the organ of the body with the highest concentration of zinc, which has been estimated to be around 150 μM in healthy humans—about 10-fold higher than the total zinc concentration in plasma (Takeda et al., 2001). Finally, the importance of the zinc supply for the virulence of *A. fumigatus* might explain why the mortality rate in a rodent model of invasive pulmonary aspergillosis can be reduced by the administration of EDTA, which is a strong zinc chelator, together with standard antifungal chemotherapy (Hachem et al., 2006).

IRON ACQUISITION SYSTEMS

In fungi, four different mechanisms for iron uptake have been characterized at the molecular level: (i) siderophore-mediated Fe^{3+} uptake, (ii) reductive iron assimilation (RIA), (iii) heme uptake, and (iv) low-affinity Fe^{2+} uptake. Most fungal species employ more than one of these systems in parallel. Fungal iron metabolism has been studied in greatest detail in the fungal prototype *S. cerevisiae*. In contrast to most fungi, this yeast is not able to synthesize siderophores (low-molecular-mass, Fe^{3+}-specific chelators), although it can utilize iron bound to siderophores produced by other microbial species (xenosiderophores) and employs an iron regulatory mechanism that is different from that used by most other fungi. Thus, the rich literature on iron homeostatic mechanisms in *S. cerevisiae* is not sufficient for comprehensive understanding of fungal iron metabolism. At the molecular level, biosynthesis, regulation, and physiological function of siderophores are probably best understood in *Aspergillus* spp. Both the saprobe *A. nidulans* and the pathogen *A. fumigatus* secrete the siderophore triacetylfusarinine C (TAFC) for iron acquisition and utilize the siderophore ferricrocin (FC) for hyphal iron storage. For conidial iron storage, *A. nidulans* employs FC and *A. fumigatus* uses the FC derivative hydroxyferricrocin. Both *Aspergillus* species possess a low-affinity iron uptake system, and *A. fumigatus* but not *A. nidulans* employs RIA. Both *Aspergillus* species lack orthologs to the endoplasmic reticulum-localized heme oxygenase Hmx1 that is present, for example, in *Candida albicans* and *S. cerevisiae* (Protchenko and Philpott, 2003; Santos et al., 2003) and, consistently, are not able to use heme as an iron source (Eisendle et al., 2003; Schrettl et al., 2004a). Figure 2 schematically compares the systems for uptake and storage of iron from *S. cerevisiae* and *A. fumigatus*. In Table 2 siderophore production of selected fungal species is given. Explanations follow in the text.

Siderophores and Their Biosynthesis

Siderophores are designed to form tight complexes with Fe^{3+} to overcome the problem of low bioavailability by solubilization. Production of these chelators is found among bacteria, fungi, and plants. Siderophores can be classified into three main groups depending on the chemical nature of the moieties donating the oxygen ligands for Fe^{3+} (Miethke and Marahiel, 2007): (i) aryl caps (catecholates and phenolates), (ii) carboxylates, and (iii) hydroxamates. With the exception of the carboxylate rhizoferrin produced by certain zygomycetes, all fungal siderophores identified so far are hydroxamates (van der Helm and Winkelmann, 1994). Fungal hydroxamates are derived from the nonproteinogenic amino acid ornithine and different acyl groups and can be grouped into four structural families: (i) rhodothorulic acid, (ii) fusarinines, (iii) coprogens, and (iv) ferrichromes. A detailed description of the chemistry of hydroxamates has been presented previously (van der Helm and Winkelmann, 1994). The *Aspergillus* siderophores TAFC and FC belong to the fusarinines and ferrichromes, respectively (Fig. 3A). TAFC is a cyclic tripeptide consisting of three N^5-*cis*-anydromevalonyl-N^5-hydroxyornithine residues linked by ester bonds in a head-to-tail fashion. FC is a cyclic hexapeptide with the structure glycine-serine-glycine-(N^5-actyl-N^5-hydroxyornithine)$_3$ (Fig. 3A). TAFC and FC form 1:1 complexes with Fe^{3+}. Depending on the pH (Miethke and Marahiel, 2007), the iron-binding constant for siderophores can reach $>10^{30}$ M, permitting microbial extraction of iron even from stainless steel.

A schematic view of the proposed siderophore biosynthetic pathway of *Aspergillus* spp. is given in Fig. 3B. The first committed step is N^5-hydroxylation of ornithine catalyzed by ornithine-N^5-monooxygenase, termed SidA in *A. nidulans* and *A. fumigatus* (Eisendle et al., 2003; Schrettl et al., 2004a). Genes encoding this enzyme have been functionally characterized also from

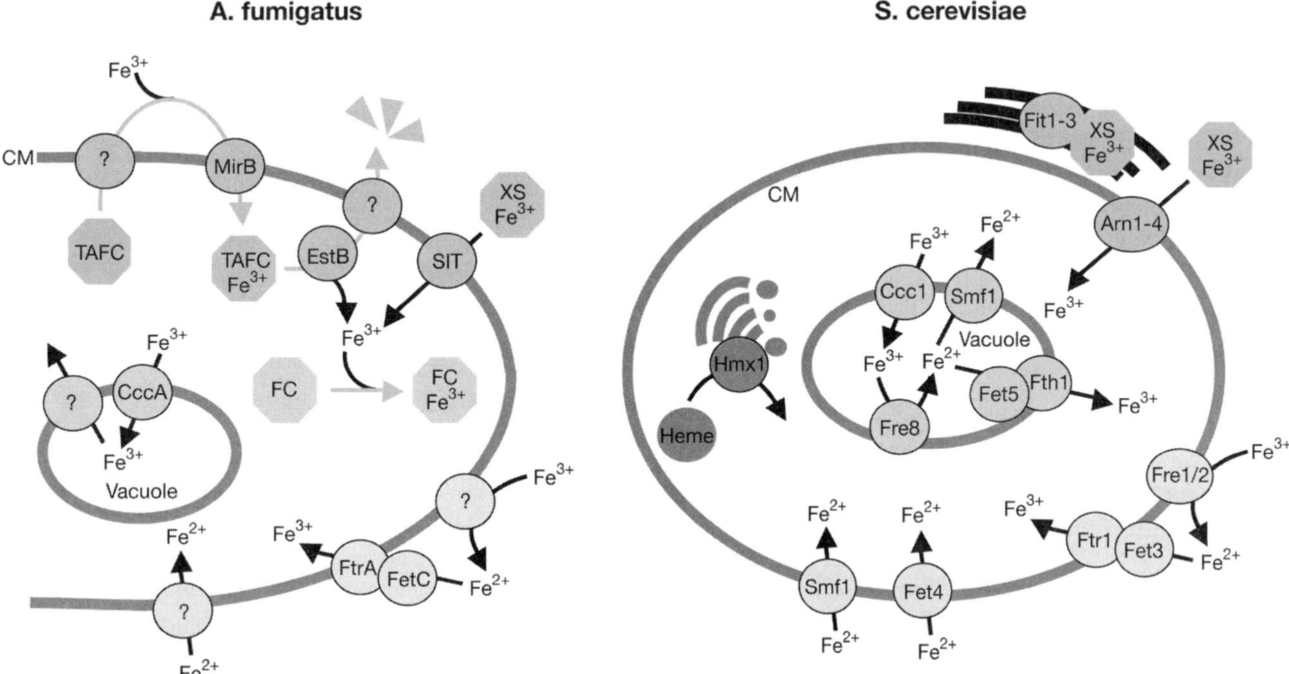

Figure 2. Comparison of systems for uptake and storage in *S. cerevisiae* and *A. fumigatus*. CM, cytoplasmic membrane; XS, xenosiderophore.

Table 2. Siderophore production by selected fungi

Fungal species	Siderophore(s)[a]			Reference(s)
	Excreted	Hyphal	Conidial	
Ascomycota, Pezizomycotina				
Alternaria brassicicola	*N*-Dimethylcoprogen	FC	ND	Oide et al., 2006, 2007
Aspergillus fumigatus	TAFC, fusarinine C	FC	Hydroxy-ferricrocin	Schrettl et al., 2007
Aspergillus nidulans	TAFC, fusarinine C	FC	FC	Charlang et al., 1981; Oberegger et al., 2001
Cochliobolus heterostrophus	Coprogen, neocoprogen I, isoneocoprogen I, neocoprogen II	FC	ND	Oide et al., 2006, 2007
Epichloe festucae	Novel fusarinine siderophore	FC	ND	Johnson et al., 2007
Fusarium graminearum	TAFC	FC	ND	Oide et al., 2006, 2007
Magnaporthae grisea	Coprogen, coprogen B, N^2-methylcoprogen, N^2-methylcoprogen B	FC	ND	Antelo et al., 2006; Hof et al., 2007
Neurospora crassa	Coprogen	FC	FC	Matzanke et al., 1987
Ascomycota, Saccharomycotina				
Candida albicans	—	—	—	Schrettl et al., 2004b
Saccharomyces cerevisiae	—	—	—	Neilands, 1995
Ascomycota, Taphrinomycotina				
Schizosaccharomyces pombe	Ferrichrome	Ferrichrome	ND	Schrettl et al., 2004b
Basidiomycota				
Cryptococcus neoformans	—	—	—	Howard, 2004
Rhodotorula pilimanae	Rhodotorulic acid	Rhodotorulic acid	ND	Muller et al., 1985
Ustilago maydis	Ferrichrome, ferrichrome A	Ferrichrome	ND	Leong and Winkelmann, 1998

[a] —, no siderophore production; ND, not determined.

Aspergillus oryzae (*dffA*), *Fusarium graminearum* (*sid1*), and *Ustilago maydis* (*sid1*) (Greenshields et al., 2007; Mei et al., 1993; Yamada et al., 2003). Fungal ornithine-N^5-monooxygenases show significant similarity at the protein level to bacterial siderophore biosynthetic ornithine-N^5-monooxygenases and lysine-N^6-monooxygenases. Inactivation of SidA blocks synthesis of all siderophores. Notably, the respective *A. nidulans* mutant is barely able to grow unless supplemented with siderophores or large amounts of Fe^{2+} due to the lack of another high-affinity iron acquisition system (Eisendle et al., 2003). In contrast, lack of siderophore biosynthesis affects axenic growth of *A. fumigatus* only partly under iron-depleted conditions and barely under iron-replete conditions, as *A. fumigatus* also employs RIA.

In the second siderophore biosynthetic step (Fig. 3B), the hydroxamate group is formed by transfer of an acyl group from acyl-coenzyme A derivatives to N^5-hydroxyornithine. Here, the pathway for different siderophores splits due to the choice of the acyl group, with acetyl for FC and *cis*-anhydromevalonyl for TAFC. In *A. fumigatus*, SidF is probably the N^5-hydroxyornithine:*cis*-anhydromevalonyl coenzyme A-N^5-transacylase, because its inactivation disrupts TAFC biosynthesis (Schrettl et al., 2007) and because it is homologous to the N^6-hydroxylysine:acetyl coenzyme A-N^6-transacetylase IucB, which is involved in synthesis of the siderophore aerobactin in *Escherichia coli* (de Lorenzo et al., 1986). Fungal transacylases involved in FC biosynthesis have not been identified so far.

In the third siderophore biosynthetic step, the hydroxamates are covalently linked via ester (TAFC) or peptide (FC) bonds accomplished by nonribosomal peptide synthetases (NRPSs) (Fig. 3B). NRPSs are large multifunctional enzymes with a modular structure and synthesize peptides, independent of the ribosome, by formation of peptide or ester bonds between proteinogenic and nonproteinogenic precursors (Mootz et al., 2002). NRPSs are found exclusively in bacteria and fungi and these enzymes are involved in the biosynthesis of a wide range of secondary metabolites, as this "thiotemplate" mechanism enormously expands the structural diversity. The involvement of NRPSs in their

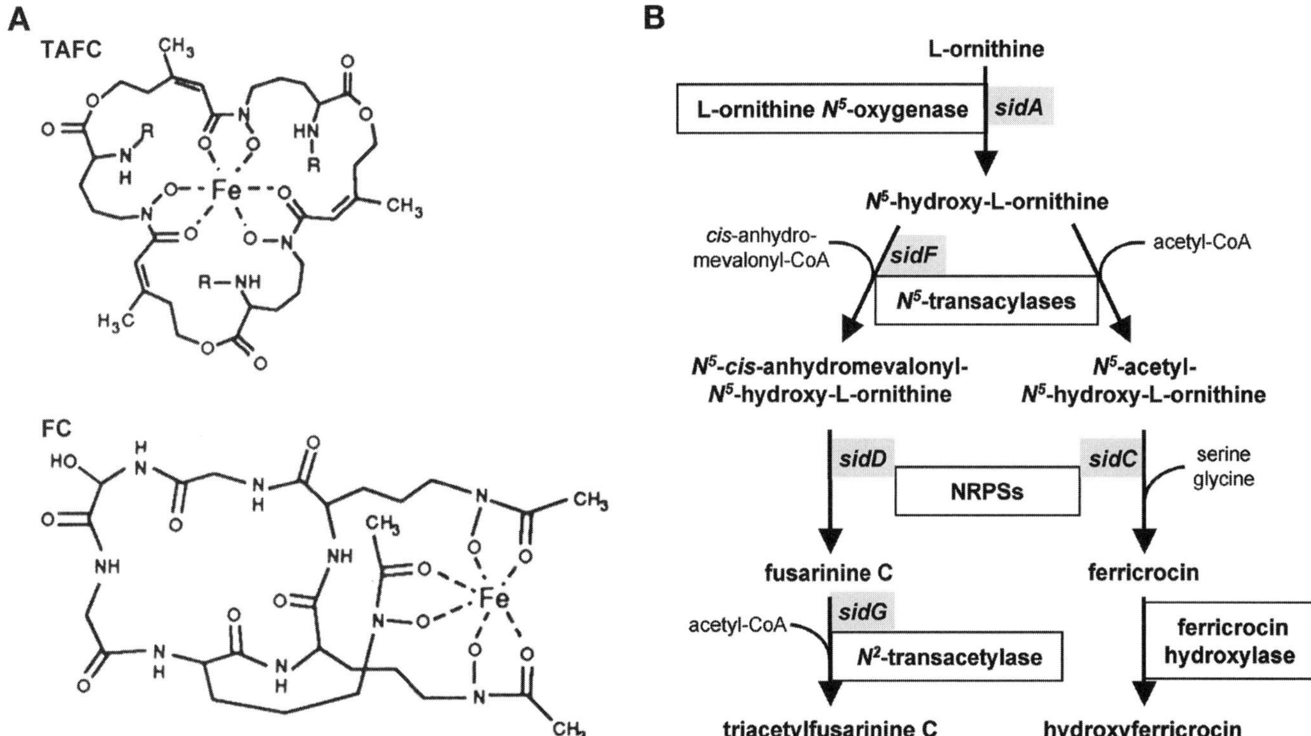

Figure 3. Structures of TAFC and FC (A) and the proposed biosynthetic pathway (B). Enzymatic activities are boxed, and encoding enzymes are shaded in gray. CoA, coenzyme A.

biosynthesis might account for siderophores being sometimes ranked among secondary metabolites; our preference is to call this type of metabolite a natural product, since iron uptake and storage is a primary metabolic function. NRPSs have a modular structure. One full module harbors all the catalytic units for incorporation of one amino acid (or amino acid-like) residue: an adenylation domain for substrate specificity and activation, a peptidyl carrier domain for attachment of the activated substrate, and a condensation domain for bond formation. Linear NRPSs contain one module for every substrate to be incorporated into the peptide, while in iterative NRPSs one or more modules are used repeatedly (Mootz et al., 2002). Fungal siderophore NRPSs appear to belong to the latter class. In *A. fumigatus* the NRPSs required for biosynthesis of TAFC and FC are SidD and SidC, respectively (Schrettl et al., 2007). The modular organization of SidC is similar, but not identical, to that of other NRPSs involved in biosynthesis of ferrichromes (Table 3). All these enzymes contain at least three full modules plus a differing number of additional incomplete modules, indicating flexibility in the enzymatic machinery for synthesis of similar or even identical peptides. The fusarinine C NRPS SidD is closely related to NRPSs involved in biosynthesis of coprogens, as it contains only one full adenylation domain (Table 3).

The identification and inactivation of these transacylases and NRPSs allowed dissection of the roles of intracellular and extracellular siderophores. In *A. fumigatus* as well as in the phytopathogenic fungi *Alternaria brassicicola, Cochliobolus heterostrophus, Cochliobolus miyabeanus*, and *F. graminearum*, loss of extracellular siderophores due to inactivation of the respective NRPSs or transacylase completely inhibits growth in the presence of Fe^{2+} chelators such as bathophenanthroline disulfonate (Oide et al., 2006; Schrettl et al., 2007), indicating absolute dependence on RIA, the alternative high-affinity iron uptake system present in these species. During iron limitation, such mutants display a decrease in growth rate, asexual sporulation, and oxidative stress resistance (Oide et al., 2006; Schrettl et al., 2007). These data demonstrate that RIA cannot fully compensate the loss of siderophore-mediated iron. The increased sensitivity to oxidative stress might be explained by the dependence of several oxidative stress-detoxifying enzymes on iron; for example, catalases and peroxidases require heme as cofactor. The consequences of loss of intracellular FC are discussed later in this chapter.

As the acyl carrier domain of fatty acid and polyketide synthetases, the peptidyl carrier domain of NRPSs requires covalent attachment of the cofactor phosphopantetheine, conducted by 4′-

Table 3. Characterized fungal siderophore NRPSs

NRPS group and name	Siderophore	Fungal species	Modular organization[a]	Reference(s)
Ferrichrome NRPSs				
Sid2	Ferrichrome	*U. maydis*	ATCATCATCTC*	Eichhorn et al., 2006; Yuan et al., 2001
Sib2	Ferrichrome	*S. pombe*	ATCTCA*TCATCTCTC	Schwecke et al., 2006
SidC	Ferricrocin	*A. nidulans*	ATCATCATCTCTC	Eisendle et al., 2003
SidC	Ferricrocin	*A. fumigatus*	ATCATCATCTCTC	Schrettl et al., 2007
Nps1	Ferricrocin	*C. heterostrophus*	ATCATCATCATCTCTC	Oide et al., 2007
Nps1	Ferricrocin	*F. graminearum*	ATCATCTCATCTCTC	Oide et al., 2007; Tobiasen et al., 2007
Ssm1	Ferricrocin	*M. grisea*	ATCATCTCATCTCTC	Hof et al., 2007
Nrps9	Ferricrocin	*E. festucae*	Unpublished	Johnson et al., 2007
Fer3	Ferrichrome A[b]	*U. maydis*	ATCATCATCTCTC	Eichhorn et al., 2006; Schwecke et al., 2006
Fso1	Ferrichrome A[b]	*O. olearius*	ATCATCATCTCTC	Schwecke et al., 2006; Welzel et al., 2005
Coprogen NRPSs				
Nps6	*N*-Dimethylcoprogen	*A. brassicicola*	ATCA*TT	Oide et al., 2006
Nps6	Coprogen, neocoprogen I, isoneocoprogen I, neocoprogen II	*C. heterostrophus*	ATCA*TTC	Oide et al., 2006
Nps6	Coprogen	*N. crassa*	ATA*CT	Oide et al., 2006
Fusarinine NRPSs				
SidD	Fusarinine C	*A. fumigatus*	ATCA*TC	Schrettl et al., 2007
Nrps2/SidF	Novel fusarinine	*E. festucae*	Unpublished	Johnson et al., 2007
Nps6	Fusarinine C	*F. graminearum*	ATCA*TC	Oide et al., 2006

[a] A, T, and C denote an adenylation domain, peptidyl carrier domain, and condensation domain, respectively. *, mutated inactive domain.
[b] Specificity predicted by genomic localization, expression analysis, and molecular modeling.

phosphopantetheine transferase. In contrast to most bacteria (Finking et al., 2002), *A. nidulans* possesses only a single enzyme for activation of polyketide synthetases and NRPSs, termed NpgA (Oberegger et al., 2003). Loss of NpgA blocks synthesis of all polyketides and nonribosomal peptides, including TAFC and FC (Oberegger et al., 2003). NpgA renders *A. nidulans* auxotrophic for siderophores and lysine, the biosynthesis of which also requires 4′-phosphopantetheinylation (Oberegger et al., 2003). *A. fumigatus* possesses a single NpgA ortholog (data not shown).

Modification of the NRPS products leads to further siderophore variants (Fig. 3B). The fusarinine C–acetyl coenzyme A–N^2-transacetylase SidG forms TAFC from fusarinine C in *A. fumigatus* (Schrettl et al., 2007). SidG-deficient strains lack TAFC and produce the ultimate precursor fusarinine C as major siderophore. Such mutants are phenotypically indistinguishable from the wild type, suggesting that fusarinine C can fully compensate the loss of TAFC. Among fungi, the possession of SidG orthologs appears to be indicative for the potential of TAFC synthesis. The enzyme catalyzing hydroxylation of FC (Fig. 3B) has not been identified so far.

Little is known regarding the subcellular localization of siderophore biosynthesis and the mechanism of siderophore excretion. However, the synthesis and storage of intracellular iron-free siderophores must be precisely controlled, as these molecules are potentially harmful based on their ability to chelate cellular iron.

Siderophore Uptake

In fungi, siderophore-iron chelates are usually taken up through transporters of the SIT (*s*iderophore-*i*ron *t*ransporter) subfamily of the major facilitator superfamily, which likely function as proton symporters energized by the plasma membrane potential (Pao et al., 1998; Winkelmann, 2001).

Fungal siderophore uptake has been studied in greatest detail in *S. cerevisiae*. This yeast expresses four different siderophore transporters that differ in substrate specificity (alternative gene names are indicated): Sit1p/Arn3p, Arn1p, Taf1p/Arn2p, and Enb1p/Arn4p (Heymann et al., 1999, 2000a, 2000b; Lesuisse et al., 1998; Yun et al., 2000a, 2000b). Two of these transporters are highly specific: Enb1p/Arn4p for the bacterial catecholate siderophore enterobactin and Taf1p/

Arn2p for TAFC. The other two transporters show broad and overlapping specificity: Arn1p transports coprogen and a wide range of ferrichromes. Sit1p/Arn3p exhibits the broadest range of substrate specificity by recognizing the bacterial hydroxamate ferrioxamine B, coprogen, and a variety of ferrichromes lacking anhydromevalonic acid. The K_m values for the uptake of ferrichrome-bound iron were determined to be 0.9 and 2.3 μM for Arn1p and Sit1p/Arn3p, respectively (Yun et al., 2000a). Substrate specificities of siderophore transporters from several fungi have been determined by measuring growth stimulation or uptake of radiolabeled Fe^{3+} in the presence of a particular siderophore after heterologous expression in an *S. cerevisiae* mutant lacking siderophore transport and RIA (Ardon et al., 2001; Haas et al., 2003; Heymann et al., 2002; Park et al., 2006; Pelletier et al., 2003), e.g., enterobactin and TAFC for *A. nidulans* MirA and MirB, respectively.

A search of the genome sequences revealed 7 putative siderophore transporters in *A. fumigatus* and 10 in *A. nidulans* (Fig. 4). Phylogenetic analysis shows that all *S. cerevisiae* transporters are more similar to each other than to permeases from *Aspergillus* spp., indicating that these transporters arose after the split of *S. cerevisiae* and *Aspergillus* spp. Consequently, the specificity of transporters from *Aspergillus* species cannot be predicted on the basis of sequence similarity to the *S. cerevisiae* transporters. In contrast, the transporters of the two *Aspergillus* species mostly form clusters of orthologs (Fig. 4). The possession of multiple siderophore transporters most likely reflects the ability to utilize multiple siderophore types, including xenosiderophores. For example, *A. nidulans* is able to take up the bacterial catecholate enterobactin via MirA (Haas et al., 2003). Iron chelated by siderophores represents an energized form of iron due to its high solubility. In certain niches siderophore iron is very abundant: in soil, siderophore contents have been reported to range between 2 and 279 nM, with ferrichrome, FC, and bacterial ferrioxamines being most prominent (Essen et al., 2006; Powell et al., 1980). Consequently, utilization of xenosiderophores saves energy. Moreover, uptake of xenosiderophores might mirror microbial competition for iron. From this view, the best strategy for a microorganism would be the ability to use multiple xenosiderophores and to produce a siderophore not utilizable by other species. In this respect it is interesting that iron-free TAFC inhibits growth of numerous bacterial species (Anke, 1977).

Most fungal species possess at least one siderophore transporter-encoding gene (data not shown). Interestingly, siderophore transporters constitute 1 of only 17 protein families that are unique to fungi and are not present in prokaryotes or other eukaryotes (Hsiang and Baillie, 2005). Consequently, these transporters might allow specific drug delivery to fungi during infection by a "Trojan horse" approach, whereby known antimicrobial agents would be covalently attached to siderophores. Synthesis of siderophore-drug conjugates that are active against bacteria and *C. albicans* have been described (Roosenberg et al., 2000). There are also several natural examples for a Trojan horse approach that take advantage of bacterial siderophore transporters (Miethke and Marahiel, 2007); for example, the *Klebsiella pneumoniae* peptide microcin MccE492m mimics a siderophore, which enhances its translocation by outer membrane siderophore receptors into the periplasmic space of enterobacterial species, where it exerts its bactericidal activity (Destoumieux-Garzon et al., 2006). Notably, the bacterial and fungal uptake systems for siderophores are structurally different (Ratledge and Dover, 2000).

In *S. cerevisiae*, some of the siderophore transporters undergo precise changes in cellular location controlled in part by the concentration of their specific substrate (Froissard et al., 2007; Kim et al., 2007). In the absence of their substrate, Arn1p and Sit1p/Arn3p are located in endosomes and targeted to the vacuolar degradation process. The presence of their substrates causes sorting of both transporters from the endosomal compartments to the plasma membrane. In contrast to Arn1p and Sit1p/Arn3p, Enb1p/Arn4p is directly targeted to the plasma membrane independent of substrate availability. Thus, different siderophore transporters behave differently in terms of intracellular trafficking in *S. cerevisiae*.

Also, the fate of structurally different siderophores appears to differ after internalization. In *S. cerevisiae*, ferrichrome has been reported to accumulate in the cytoplasm (Moore et al., 2003) while ferrioxamine B is probably compartmentalized to the vacuole (Froissard et al., 2007). In contrast, coprogens and ferrichromes are reexcreted in intact form in *Neurospora crassa* and *Ustilago sphaerogena* (Emery, 1971; van der Helm and Winkelmann, 1994). In *A. nidulans* and *A. fumigatus* (Kragl et al., 2007; Oberegger et al., 2001), the ester bonds of TAFC are hydrolyzed after cellular uptake and the resulting fusarinine moieties are excreted. The cleavage of TAFC is reminiscent of utilization of ester bond-containing enterobactin in *E. coli*, which also involves a hydrolytic step, carried out by the esterase Fes (Brickman and McIntosh, 1992). Recently, the gene encoding the TAFC-hydrolyzing enzyme EstB was identified in *A. fumigatus* (Kragl et al., 2007). EstB displays limited similarity to *E. coli* Fes and is located in the cytoplasm. Despite the presence of an alternative uncharacterized TAFC degradation mechanism in *A. fumigatus*, lack of EstB reduces the rate of TAFC hydrolysis, decreases the transfer rate of iron from TAFC to FC, delays iron sensing, and decreases the growth rate during iron limitation. Taken together, EstB is not essential for but optimizes TAFC-mediated iron uptake. In agreement with

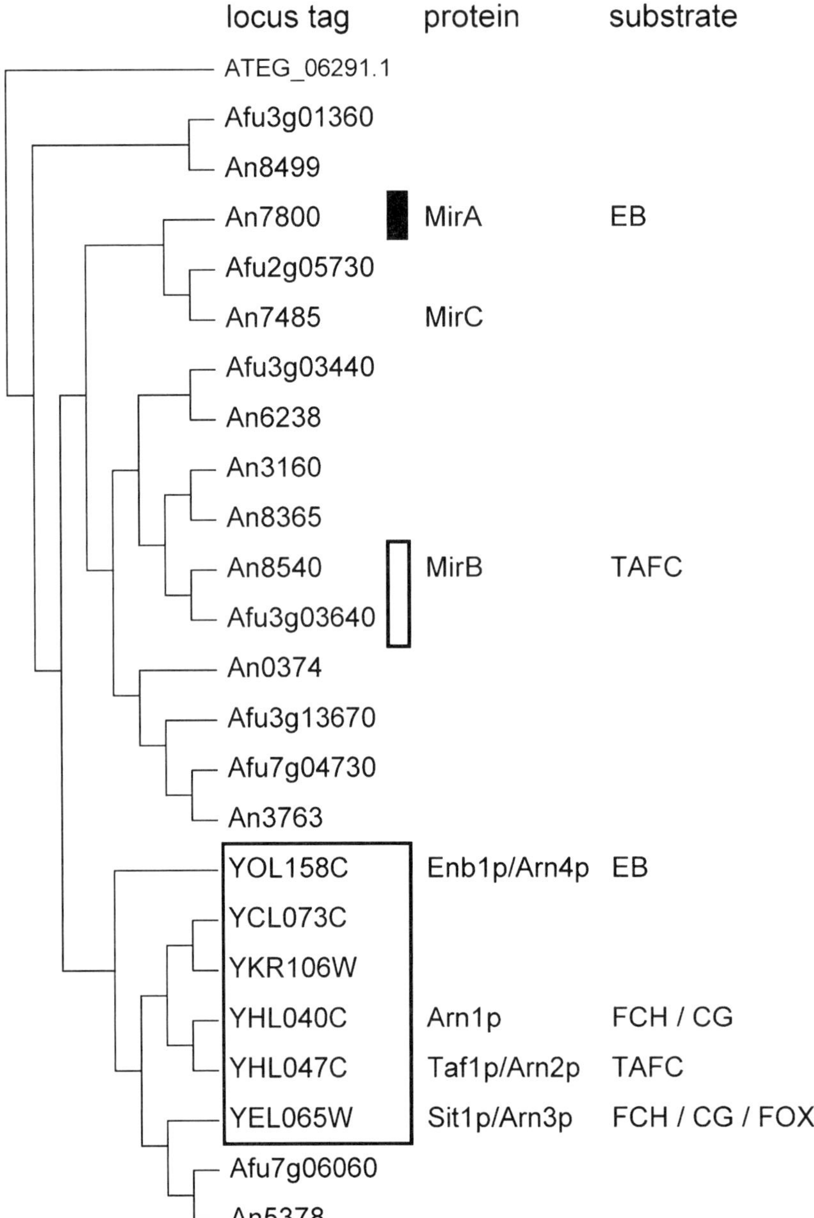

Figure 4. Phylogenetic analysis of siderophore transporters from *A. fumigatus* (Afu), *A. nidulans* (An), and *S. cerevisiae* (Y). Available gene names and substrate specificities are indicated. CG, coprogen; EB, enterobactin; FCH, ferrichromes; FOX, ferrioxamines. *A. terreus* ATEG_06291.1 is a major facilitator not belonging to the SIT family and served as an outgroup. *S. cerevisiae* transporters are boxed. Permeases likely to be involved in transport of TAFC are marked by an open bar.

its dispensability, fungal species utilizing but not producing TAFC lack EstB orthologs, e.g., *S. cerevisiae* and *C. albicans*.

In fungi that possess intracellular siderophores, iron is transferred from the internalized extracellular siderophores to the intracellular siderophores (Eisendle et al., 2006; Kragl et al., 2007). Release of iron from internalized extracellular and intracellular siderophores has been suggested to be accomplished through a reductive mechanism (De Luca and Wood, 2000). As an alternative for the uptake of siderophore-iron complexes, siderophore-chelated iron can also be utilized via RIA, at least in *S. cerevisiae* (Yun et al., 2001).

Genomic Organization of Genes Encoding Components of the Siderophore System

In filamentous fungi, genes encoding components of common pathways are often organized in gene clusters defined by coregulation, which facilitates molecular analysis of pathways. The genes encoding ornithine-N^5-monooxygenase and ferrichrome-type NRPS are bidirectionally transcribed from a common promoter region in *U. maydis, Schizosaccharomyces pombe, N. crassa, Aureobasidium pullulans, F. graminearum*, and *Omphalotus olearius* (Haas, 2003; Schrettl et al., 2004b; Tobiasen et al., 2007; Yuan et al., 2001), but not in *A. nidulans* and *A. fumigatus* (Eisendle et al., 2003; Schrettl et al., 2007; Welzel et al., 2005). Additionally, the *Aspergillus pullulans* cluster contains a gene encoding an ATP-binding cassette (ABC) transporter and the *O. olearius* cluster contains a transacylase. Microarray-based gene expression profiling identified a gene cluster encoding putative new components of the siderophore system of *U. maydis*, including the putative ferrichrome A NRPS (Fer3), enoyl-coenzyme A hydratase (Fer4), transacylase (Fer5), ABC transporter (Fer6), and a siderophore transporter (Fer7) (Eichhorn et al., 2006). In *A. fumigatus*, the genes encoding TAFC esterase EstB, fusarinine C–acetyl coenzyme A–N^2-transacetylase SidG, an ABC transporter, and a siderophore transporter are clustered together (Kragl et al., 2007). A second siderophore-related gene cluster of *A. fumigatus* includes the genes encoding SidD and SidF (H. Haas, unpublished data).

RIA

RIA is a two-step process that begins with the extracellular reduction of Fe^{3+} to Fe^{2+} at the plasma membrane followed by high-affinity uptake of Fe^{2+}. RIA has been studied in greatest detail in *S. cerevisiae* (Kosman, 2003). In this yeast, reduction of Fe^3 at the plasma membrane is mediated by the iron-regulated paralogous metalloreductases Fre1p, Fre2p, Fre3p, and Fre4p (Dancis et al., 1992; Georgatsou and Alexandraki, 1994; Yun et al., 2001). Fre1p and Fre2p have additionally been shown to facilitate copper uptake (Georgatsou et al., 1997; Hassett and Kosman, 1995). The Fre proteins share significant similarity with the gp91^{phox} subunit of cytochrome b_{558}, the human phagocyte respiratory burst oxidase (Finegold et al., 1996; Rotrosen et al., 1992). Substrates for the reductive iron assimilatory system include iron salts, low-affinity iron chelates as ferric citrate, and siderophores. Fre1p appears to constitute the major activity for reduction of iron salts and low-affinity iron chelates, because *FRE1* deletion mutants are unable to grow on iron-limited medium (Dancis et al., 1990, 1992). In *A. fumigatus* respective metalloreductases have not been functionally characterized yet. Notably, fungi that do not employ RIA possess such genes, e.g., *freA* in *A. nidulans* (Oberegger et al., 2002a). Nevertheless, *freA* is likely involved in iron metabolism, possibly intracellular reduction of iron, because its expression is repressed by iron.

In *S. cerevisiae*, high-affinity uptake of Fe^{2+} is promoted by a protein complex consisting of the multicopper ferroxidase Fet3p and the iron permease Ftr1p (Askwith et al., 1994; De Silva et al., 1995; Stearman et al., 1996). This bipartite complex operates with an apparent K_m of 0.2 μM. The Fe^{2+} to be transported is first oxidized by Fet3p and then transported into the cytosol by Ftr1p. Fet3p shows remarkable similarity to various multicopper oxidases, e.g., laccases, ascorbate oxidase, ceruloplasmin, and hephaestin. All these proteins catalyze copper-dependent oxidation of substrates followed by reduction of dioxygen to water. The copper requirement of Fet3p is underscored by the fact that defects in copper metabolism also impair iron homeostasis (Dancis, 1998; Lin et al., 1997). Notably, Fet3p also contributes to copper resistance due to its additional activity as a copper oxidase (Shi et al., 2003; Stoj et al., 2007). Assembly of the Fet3p-Ftr1p complex is required for trafficking to the plasma membrane. In addition to transcriptional regulation (see below), Fet3p-Ftr1p-mediated iron uptake is subject to posttranslational regulation (Felice et al., 2005): high levels of iron trigger internalization with subsequent degradation in the vacuole of the Fet3p-Ftr1p complex, ensuring a fast response to avoid iron toxicity.

Most fungal species possess orthologs to Ftr1p and Fet3p, suggesting the ability for RIA. The presence of RIA has been experimentally confirmed in *A. fumigatus, F. graminearum, C. albicans, S. pombe, C. neoformans*, and *U. maydis* (Eichhorn et al., 2006; Labbe et al., 2007; Lian et al., 2005; Park et al., 2007; Ramanan and Wang, 2000; Schrettl et al., 2004a). Except for *S. cerevisiae* and *C. albicans*, the genes encoding orthologs of Fet3p and Ftr1p are adjacent and divergently transcribed in all of these fungi.

Inactivation of FtrA, the Ftr1p ortholog of *A. fumigatus*, blocks RIA but does not affect the growth of the respective mutants due to compensation by increased siderophore biosynthesis (Schrettl et al., 2004a). *A. nidulans* lacks orthologs to Ftr1p and Fet3p and, consistently, does not employ RIA, as loss of the siderophore system blocks high-affinity iron uptake in *A. nidulans* (Eisendle et al., 2003).

Low-Affinity Fe^{2+} Uptake

At the molecular level, direct Fe^{2+} uptake has been studied exclusively in *S. cerevisiae*. Here, Fe^{2+} is taken up directly by Fet4p with an apparent K_m of approximately 30 μM (Dix et al., 1994; Eide et al., 1992). This low-affinity system is not specific for Fe^{2+} but transports also other metals, e.g., copper and zinc (Hassett et al., 2000; Waters and Eide, 2002). Moreover, the natural resistance-associated macrophage protein (NRAMP) family member Smf1p and fluid-phase endocytosis followed by mobilization of iron from the vacuole contribute to the iron supply for *S. cerevisiae* (Chen et al., 1999; Li et al., 2001). Supplementation with large amounts of Fe^{2+}, but not Fe^{3+}, partly rescues growth of *A. nidulans* and *A. fumigatus* mutants lacking high-affinity iron acquisition systems, indicating low-affinity Fe^{2+} uptake (Eisendle et al., 2003; Schrettl et al., 2004a). Both *Aspergillus* species possess Smf1 homologs; *A. fumigatus* but not *A. nidulans* possesses a Fet4p ortholog.

FUNGAL IRON STORAGE

In order to ensure a steady supply of iron, cells need to store iron. In animals, plants, and bacteria, iron is stored as ferritin, phytoferritin, and bacterioferritin, respectively (Andrews, 1998). With exception of zygomycetes, fungi appear to lack ferritin-like molecules (Matzanke, 1994). Two different mechanisms have been described in fungi, vacuolar and siderophore-mediated iron storage.

Vacuolar Iron Storage

In *S. cerevisiae*, Ccc1p mediates transport of iron into the vacuole (Li et al., 2001), whereas export of iron from the vacuole is supported by the Smf1p paralog Smf3p and the bipartite Fet5p-Fth1p protein complex, which is paralogous to the Fet3p-Ftr1p complex (Portnoy et al., 2000; Urbanowski and Piper, 1999). The vacuolar membrane-localized metalloreductase Fre7p supplies Fe^{2+} to both vacuolar efflux systems (Singh et al., 2007). Consistent with roles in mobilization of vacuolar iron stores, deficiency in Smf3p or the Fet5p-Fth1p complex causes signs of iron starvation. In contrast, lack of Ccc1p causes sensitivity to iron. Identification of genes encoding Ccc1p orthologs in *A. nidulans* (Eisendle et al., 2006) and *A. fumigatus* (M. Eisendle and H. Haas, unpublished data), which are downregulated during iron starvation, suggests vacuolar iron storage also in these siderophore-producing fungi. However, it is not clear if and how iron can be exported from the vacuole, because both *Aspergillus* species lack orthologs to *S. cerevisiae* Fet5p and Fth1p.

Siderophore-Mediated Iron Storage

Consistent with a role in iron storage, FC accumulation increases during intracellular iron excess due to a shift from low to high iron availability or due to derepression of iron uptake by inactivation of the iron regulator SreA (see below) in *A. nidulans* (Eisendle et al., 2006; Oberegger et al., 2001) and *A. fumigatus* (M. Schrettl and H. Haas, unpublished data). FC constitutes only about 5% of the total hyphal iron content during iron-replete growth but reaches up to 64% during intracellular iron excess in *A. nidulans* (Eisendle et al., 2006). The latter is visible even to the naked eye, because hyphae accumulating large amounts of FC turn orange, which is the color of the FC-iron chelate (Eisendle et al., 2006; Oberegger et al., 2001). Siderophore-chelated iron is not available for Fenton-Haber Weiss chemistry, which might be the rationale for increased FC iron storage under conditions of oxidative stress (Eisendle et al., 2006). In agreement, lack of FC increases the labile iron pool and decreases the resistance to oxidative stress in *A. nidulans* and *A. fumigatus* (Eisendle et al., 2006; Schrettl et al., 2007). Notably, also iron starvation increases hyphal FC accumulation, but in this case in the iron-free form as shown in *A. nidulans* and *A. fumigatus* (Eisendle et al., 2006; Schrettl et al., 2007). This might represent a protective mechanism to cope with incoming iron. Alternatively, FC might optimize intracellular iron transfer. In agreement with the latter possibility, inactivation of FC biosynthesis decreases the growth rate during iron limitation, increases iron uptake resulting in a twofold increase in iron content (Eisendle et al., 2006), and impairs asexual conidiation in *A. nidulans* and *A. fumigatus* (Eisendle et al., 2006; Schrettl et al., 2007). Furthermore, FC deficiency blocks sexual development in *A. nidulans*, *Gibberella zeae* (anamorph *F. graminearum*), and *C. heterostrophus* (Eisendle et al., 2006; Oide et al., 2007).

Inactivation of FC biosynthesis reduces the conidial iron content by 34 and 76% in *A. nidulans* and *A. fumigatus*, respectively (Eisendle et al., 2006; Schrettl et al., 2007), underlining the importance of FC and hydroxyferricrocin in conidial iron storage. Moreover, in both *Aspergillus* species, loss of FC biosynthesis decreases resistance to hydrogen peroxide of conidia due to a decrease in the iron-dependent catalase CatA activity, as shown in *A. fumigatus* (Eisendle et al., 2006; Schrettl et al., 2007). The conidial siderophore storage is also an important germination factor, as inactivation of FC biosynthesis or loss of conidial FC by treatment with a high salt concentration delays germination of conidia during iron limitation in *A. fumigatus* and *A. nidulans* (Charlang et al., 1981; Eisendle et al., 2006; Horowitz et al., 1976; Schrettl et al., 2007).

REGULATION OF IRON METABOLISM

Maintaining the balance between iron deficiency and iron toxicity requires fine-tuned regulatory mechanisms. Two different strategies to regulate iron acquisition at the transcriptional level have been identified in fungi: activation via Aft1p-like proteins during iron starvation, and repression via iron-responsive GATA factors (IRGFs) under iron-replete conditions. Recent studies revealed that these factors not only regulate iron acquisition but are also involved in downregulation of iron-dependent metabolic pathways to reduce the cellular requirement for iron during periods of iron starvation. Iron regulatory mechanisms of *S. cerevisiae* and *A. nidulans* are schematically compared in Fig. 5.

In *S. cerevisiae*, transcriptional control of iron homeostasis is mediated by Aft1p and to a lesser degree by its paralog Aft2p, which together activate their targets in response to iron starvation (Bird, 2007; Yamaguchi-Iwai et al., 1995). Targets include genes involved in RIA, low-affinity Fe^{2+} uptake, siderophore uptake, mobilization of vacuolar iron stores, and heme degradation. Another Aft1p target is the gene encoding Cth2p, which posttranscriptionally downregulates iron-consuming pathways. Cth2p decreases the transcript stability of a large number of transcripts encoding proteins that either bind iron or that are involved in iron-dependent processes, including heme and iron-sulfur cluster biosynthesis, mitochondrial respiration, sterol metabolism, and fatty acid metabolism (Puig et al., 2005). Orthologs to Aft1p and Cth2p are found in some *Saccharomycotina* species, but most other fungi including *Aspergillus* spp. appear to lack these regulators (data not shown).

The majority of fungal species appears to employ iron regulation via IRGFs. The first characterized member of this group was Urbs1 from *U. maydis*, followed by orthologs of *A. nidulans* (SreA), *Penicillium chrysogenum* (SreP), *N. crassa* (Sre), *S. pombe* (Fep1), *C. albicans* (Sfu1), *C. neoformans* (Cir1), and *Pichia pastoris* (Fep1) (Haas et al., 1997, 1999; Jung et al., 2006; Lan et al., 2004; Miele et al., 2007; Pelletier et al., 2002, 2003; Zhou et al., 1998). Iron regulation of *A. fumigatus* resembles that of *A. nidulans* (M. Schrettl and H. Haas, unpublished data). GATA factors are a group of transcription factors characterized by conserved zinc finger motifs that recognize the core sequence 5′-GATA-3′. Most fungi contain several GATA factors involved in various regulatory circuits, e.g., nitrogen metabolism, sexual development, light response, and circadian rhythmicity (Scazzocchio, 2000). The fungal IRGFs are distinguished from all other GATA factors by the presence of two zinc fingers flanking a highly conserved cysteine-rich region. Under iron-replete conditions, IRGFs repress transcription of their targets, including genes encoding components of RIA as well as the siderophore system (Lan et al., 2004; Oberegger et al., 2001, 2002b; Pelletier et al., 2003). At least in *S. pombe*, the IRGF functions by recruiting a cosuppressor protein complex that causes chromatin-mediated transcriptional repression (Smith and Johnson, 2000; Znaidi et al., 2004). IRGF transcript levels are constitutive in *U. maydis, S. pombe, N. crassa, C. albicans, C. neoformans, P. pastoris*, and *S. pombe*. In contrast, iron represses the transcription of the gene encoding the IRGF SreA in *A. nidulans* (Haas et al., 1999) and *A. fumigatus* (Schrettl and Haas, unpublished), adding another layer of regulation. IRGFs probably sense the cellular iron status via direct binding of this metal by the cysteine-rich region, as suggested by studies in *S. pombe* and *N. crassa* (Harrison and Marzluf, 2002; Labbe et al., 2007). Inactivation of SreA leads to derepression of biosynthesis and uptake of siderophores, which causes increased iron accumulation and increased oxidative stress in *A. nidulans* (Haas et al., 1999; Oberegger et al., 2001) and *A. fumigatus* (Schrettl and Haas, unpublished).

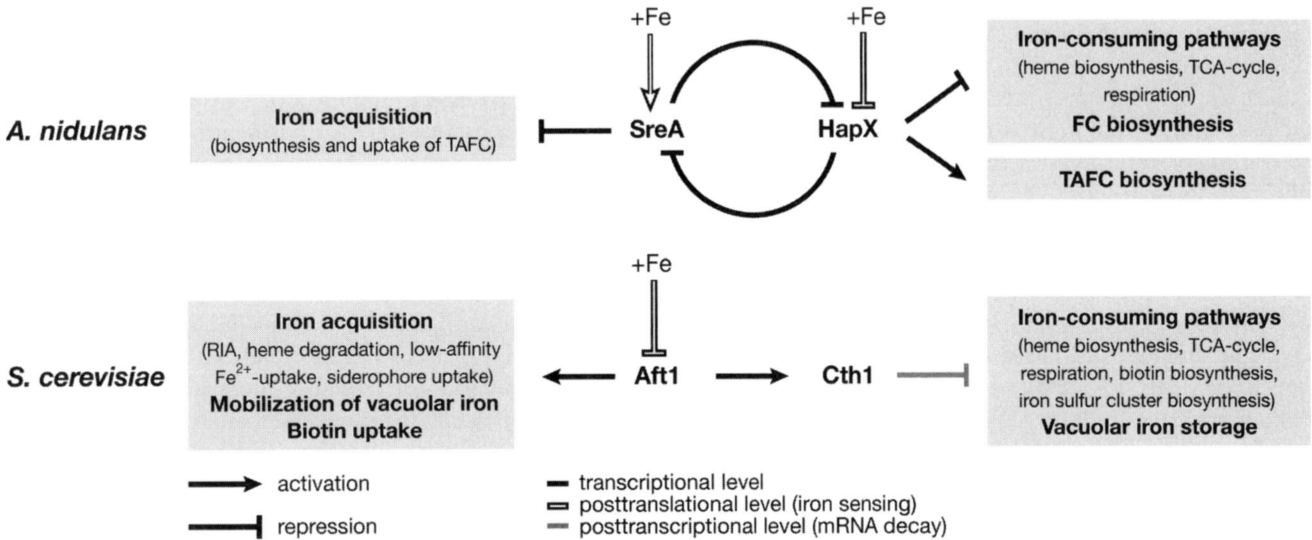

Figure 5. Schematic comparison of iron regulatory mechanisms of *A. nidulans* and *S. cerevisiae*.

In *A. nidulans* (Hortschansky et al., 2007; Oberegger et al., 2002a) and *A. fumigatus* (Schrettl and Haas, unpublished) iron starvation downregulates genes encoding iron-dependent proteins (e.g., homoaconitase, aconitase, and cytochrome *c*) and iron-consuming pathways (e.g., heme biosynthesis) via physical interaction of HapX with the CCAAT-binding complex. The CCAAT-binding complex is conserved in all eukaryotes and suggested to be involved in regulation of about 30% of the genes. In *A. nidulans* it consists of three subunits, HapB, HapC, and HapE, which are responsible for DNA binding. In contrast to HapX, the function of which is confined to iron-depleted conditions, the CCAAT-binding complex has a general role independent of the iron status in *A. nidulans*; for example, it positively regulates utilization of nitrogen and carbon sources (e.g., acetamide, formamide, γ-aminobutyrate, and starch) and production of the secondary metabolite penicillin (Brakhage et al., 1999). Consistently, expression of the genes encoding the subunits of the CCAAT-binding complex is not subject to iron regulation. Interaction of the CCAAT-binding complex and HapX is controlled at both transcriptional and posttranslational levels (Hortschansky et al., 2007): iron represses expression of *hapX* via SreA and destroys the physical interaction of HapX with the CCAAT-binding complex. HapX and SreA form a regulatory loop by mutual transcriptional control, because HapX in turn represses the expression of *sreA* during iron starvation. Synthetic lethality resulting from inactivation of both SreA and HapX provides further evidence for a tight interplay of these two regulators. HapX/CCAAT-binding complex-mediated control also affects siderophore biosynthesis: inactivation of HapX or of the CCAAT-binding complex increases FC accumulation and decreases TAFC production (Hortschansky et al., 2007). It is unclear how this complex acts both negatively and positively. HapX-CCAAT-binding complex-mediated downregulation seems to be conserved in *S. pombe* and *C. albicans* (Labbe et al., 2007). In *S. cerevisiae* interaction of Hap4p with the CCAAT-binding complex is required for transcriptional activation of target genes involved in oxidative phosphorylation in response to growth on nonfermentable carbon sources (McNabb et al., 1995). Hap4p displays similarity to *A. nidulans* HapX in an N-terminal 16-amino-acid region, indicating a common origin of these functionally different regulatory mechanisms.

The ambient pH has a wide influence on iron acquisition: (i) iron solubility is significantly elevated under acidic conditions because of stabilization of Fe^{2+}, which prevents autooxidation and formation of insoluble ferric oxyhydroxides; (ii) due to competition from protons for the Fe^{3+}-binding sites, the affinity constants of siderophores for iron decrease under acidic conditions, which makes the siderophore system a less efficient iron uptake tool at acidic pH; (iii) uptake of siderophores likely requires cotransport with protons, providing another rationale for cross talk between siderophore metabolism and pH sensing. In agreement, biosynthesis and uptake of TAFC are elevated under alkaline conditions in *A. nidulans* and are mediated at least in part by the pH regulatory transcription factor PacC (Eisendle et al., 2004). Similarly, expression of genes encoding siderophore transporters and components of the reductive iron assimilatory system are induced by alkaline pH in *S. cerevisiae* and *C. albicans*, in part mediated by the respective PacC ortholog (Bensen et al., 2004; Lamb et al., 2001). Consistently, iron has been shown to be a limiting factor for growth of *S. cerevisiae* in alkaline environments (Serrano et al., 2004).

Additionally, oxidative stress upregulates FC accumulation in *A. nidulans* (Eisendle et al., 2006), and formation of hydroxyferricrocin is developmentally regulated in *A. fumigatus* (Schrettl et al., 2007). In *S. cerevisiae* and *C. neoformans*, oxygen availability has been found to influence expression of genes involved in iron acquisition (Bird, 2007; Chang et al., 2007; Kaplan et al., 2006). To prevent iron toxicity, aberrant expression of iron uptake and heme recycling is additionally limited by RNA surveillance pathways in *S. cerevisiae* (Lee et al., 2005).

IRON ACQUISITION AND VIRULENCE

There are multiple examples for involvement of siderophores in bacterial pathogenicity (Ratledge and Dover, 2000). In contrast to most fungi, bacteria do not employ intracellular siderophores but use ferritin and bacterioferritin for iron storage (Smith, 2004). The complete block of siderophore biosynthesis of *A. fumigatus* by inactivation of SidA (Fig. 3B) prevents initiation of mammalian infection in a mouse model for pulmonary aspergillosis (Hissen et al., 2005; Schrettl et al., 2004a). In contrast, inhibition of RIA by inactivation of FtrA is inconsequential for virulence (Schrettl et al., 2004a). In agreement, siderophore-mediated removal of iron from transferrin has been suggested to be crucial for survival of *A. fumigatus* in serum (Hissen et al., 2004), and siderophore-deficient *A. fumigatus* mutants are not able to grow on blood agar plates (Schrettl et al., 2004a). Abrogation of extracellular siderophore biosynthesis by inactivation of either SidD or SidF causes partial attenuation of virulence, indicating that RIA can partly compensate for the loss of siderophore-mediated iron uptake during infection (Schrettl et al., 2007). Compared to SidF deficiency, SidD inactivation causes higher sensitivity to iron depletion and oxidative stress and results in a stronger attenuation of virulence (Schrettl et al., 2007). These differences suggest that abrogation of TAFC biosynthesis at different steps of the pathway has different consequences which might be related to accumulation of different pathway intermediates with differing consequences for fungal metabolism. *A. fumigatus* mutants producing fusarinine C instead of TAFC due to a deficiency in SidG have wild-type virulence, indicating that fusarinine C can satisfactorily replace TAFC as a siderophore not only in axenic growth but also during pathogenic growth. Blocking of FC biosynthesis by inactivation of SidC also caused partial attenuation of *A. fumigatus* virulence, which might be attributed to the role of FC in promoting germination and resisting oxidative stress. FC deficiency downregulates conidial catalase CatA activity and oxidative stress resistance (Schrettl et al., 2007), but this does not appear to be the reason, at least not the only one, for the attenuation of *A. fumigatus* mutant strains lacking SidA or SidC, because CatA deficiency does not affect virulence (Paris et al., 2003). The partial rescue of virulence of the avirulent SidA-deficient *A. fumigatus* mutant following reconstitution of the conidial hydroxyferricrocin content (which is possible by supplementation of the sporulation medium with FC in the absence of de novo synthesis of both TAFC and FC) indicates the importance of the conidial siderophore during the initial phase of infection. Consistent with the role of siderophores in virulence, *A. fumigatus sidD* was found to be the most highly expressed *A. fumigatus* NRPS-encoding gene following incubation with macrophages (Cramer et al., 2006), and genes encoding *A. fumigatus* siderophore biosynthetic enzymes (SidC, SidD, SidF, and SidG) are significantly induced at the level of gene expression at an early stage of infection in neutropenic mice (Schrettl et al., 2007).

The *A. nidulans* pH regulator PacC acts positively on expression of "alkaline" genes and negatively on "acidic" genes. The attenuation of *A. nidulans* virulence caused by PacC deficiency (Bignell et al., 2005) might, at least partly, be due to impairment of siderophore-mediated iron uptake, because the "alkaline" genes include those coding for biosynthesis and uptake of siderophore (Eisendle et al., 2004),

The microbial quest for iron is the basis for an elaborate mammalian defense system against microbial infection, which relies upon extensive iron-withholding mechanisms to effectively deny access to iron for invading microbes (Fluckinger et al., 2004; Weinberg, 1999; Weiss, 2002). Upon inflammation, iron is withheld by macrophages, leaving extracellular fluids iron depleted, causing so-called anemia of chronic disease. The crucial role of siderophores in virulence is reflected by the fact that mammals possess at least two proteins, the lipocalins Lcn2 and Lcn1, that sequester siderophores (Fluckinger et al., 2004; Goetz et al., 2002). Lcn2, and also siderocalin or NGAL, protects mammals from infection with *E. coli* by sequestering the siderophore enterobactin and preventing iron acquisition (Flo et al., 2004). However, glycosylation of enterobactin inhibits recognition by Lcn2 and allows evasion of the native immune system, which represents an excellent example of host-pathogen coevolution (Fischbach et al., 2006). Lcn2 binds its substrates with high affinity (K_d of about 4×10^{-10}) but has a rather narrow substrate specificity by recognizing only particular catecholate but not hydroxamate siderophores (Goetz et al., 2002). Lcn1 has a broader substrate specificity, recognizing a variety of bacterial and fungal catecholate and hydroxamate siderophores, and therefore has bacteriostatic and fungistatic activities but binds its substrates with lower affinity (K_d of about 1×10^{-5} to 5×10^{-5}) (Fluckinger et al., 2004).

In agreement with iron playing an important role in the pathophysiology of *A. fumigatus*, increased bone marrow iron stores represent an independent risk factor for invasive aspergillosis in high-risk patients (Kontoyiannis et al., 2007; Zarember et al., 2007). Moreover, polymorphonuclear leukocytes inhibit growth of *A. fumigatus* by lactoferrin-mediated iron depletion (Kontoyiannis et al., 2007; Zarember et al., 2007). Consistently, the chelator EDTA (which binds iron among other metals) as an adjunct was found to improve the effectiveness of other antifungal agents in a rodent model for invasive pulmonary aspergillosis (Hachem et al., 2006). Recently, iron chelation therapy, using deferasirox, was shown to protect mice from mucormycosis (Ibrahim et al., 2007). In contrast, deferoxamine B, a *Streptomyces* spp. siderophore widely used for treatment of iron and aluminum overload, serves as a xenosiderophore for *Rhizopus* spp. and increases the risk for mucormycosis (Boelaert et al., 1993).

In summary, siderophore production is crucial for virulence of *A. fumigatus*. However, the role of individual iron uptake mechanisms in virulence largely depends on the pathogen-host system, because there are siderophore-lacking animal pathogenic fungi, e.g., *C. albicans* and *C. neoformans*. In *C. albicans*, RIA is crucial for systemic virulence and, remarkably, the xenosiderophore permease Sit1p/Arn1p is required for epithelial invasion (Heymann et al., 2002; Ramanan and Wang, 2000).

CONCLUSIONS AND OUTLOOK

Fungal requirements for zinc and iron could potentially open up perspectives for the development of novel antifungal treatments. For instance, since humans do not possess a ZafA ortholog, the ZafA protein might constitute a target for the development of chemotherapeutic agents that selectively interfere with fungal zinc homeostasis. Similarly, blocking of siderophore transporters or siderophore biosynthetic enzymes appears to be particularly promising because the involved proteins are not present in mammals.

Acknowledgments. J.A.C. acknowledges support from the Spanish Ministry of Education and Science and from the Junta de Castilla y León (Spain). H.H. acknowledges support from Austrian Science Foundation (FWF), Tyrolean Science Foundation (TWF), and Austrian National Bank (OENB).

REFERENCES

Andrews, S. C. 1998. Iron storage in bacteria. *Adv. Microb. Physiol.* **40:**281–351.

Anke, H. 1977. Desferritriacetylfusigen, an antibiotic from *Aspergillus deflectus*. *J. Antibiot.* **30:**125–128.

Antelo, L., C. Hof, K. Welzel, K. Eisfeld, O. Sterner, and H. Anke. 2006. Siderophores produced by *Magnaporthe grisea* in the presence and absence of iron. *Z. Naturforsch. Teil C* **61:**461–464.

Ardon, O., H. Bussey, C. Philpott, D. M. Ward, S. Davis-Kaplan, S. Verroneau, B. Jiang, and J. Kaplan. 2001. Identification of a *Candida albicans* ferrichrome transporter and its characterization by expression in *Saccharomyces cerevisiae*. *J. Biol. Chem.* **276:**43049–43055.

Askwith, C., D. Eide, A. Van Ho, P. S. Bernard, L. Li, S. Davis-Kaplan, D. M. Sipe, and J. Kaplan. 1994. The FET3 gene of *S. cerevisiae* encodes a multicopper oxidase required for ferrous iron uptake. *Cell* **76:**403–410.

Banerjee, B., P. A. Greenberger, J. N. Fink, and V. P. Kurup. 1998. Immunological characterization of Asp f 2, a major allergen from *Aspergillus fumigatus* associated with allergic bronchopulmonary aspergillosis. *Infect. Immun.* **66:**5175–5182.

Barrow, N. J. 1993. Mechanisms of reaction of zinc with soil and soil components, p. 15–27. *In* A. D. Robson (ed.), *Zinc in Soils and Plants*. Kluwer Academic Publishers, Dordrecht, The Netherlands.

Beltramini, M., P. Zambenedetti, W. Wittkowski, and P. Zatta. 2004. Effects of steroid hormones on the Zn, Cu and MTI/II levels in the mouse brain. *Brain Res.* **1013:**134–141.

Bensen, E. S., S. J. Martin, M. Li, J. Berman, and D. A. Davis. 2004. Transcriptional profiling in *Candida albicans* reveals new adaptive responses to extracellular pH and functions for Rim101p. *Mol. Microbiol.* **54:**1335–1351.

Bignell, E., S. Negrete-Urtasun, A. M. Calcagno, K. Haynes, H. N. Arst, Jr., and T. Rogers. 2005. The Aspergillus pH-responsive transcription factor PacC regulates virulence. *Mol. Microbiol.* **55:**1072–1084.

Bird, A. J. 2007. Metallosensors, the ups and downs of gene regulation. *Adv. Microb. Physiol.* **53:**231–267.

Bird, A., M. V. Evans-Galea, E. Blankman, H. Zhao, H. Luo, D. R. Winge, and D. J. Eide. 2000. Mapping the DNA binding domain of the Zap1 zinc-responsive transcriptional activator. *J. Biol. Chem.* **275:**16160–16166.

Bird, A. J., K. McCall, M. Kramer, E. Blankman, D. R. Winge, and D. J. Eide. 2003. Zinc fingers can act as Zn^{2+} sensors to regulate transcriptional activation domain function. *EMBO J.* **22:**5137–5146.

Bird, A. J., S. Swierczek, W. Qiao, D. J. Eide, and D. R. Winge. 2006a. Zinc metalloregulation of the zinc finger pair domain. *J. Biol. Chem.* **281:**25326–25335.

Bird, A. J., H. Zhao, H. Luo, L. T. Jensen, C. Srinivasan, M. Evans-Galea, D. R. Winge, and D. J. Eide. 2000b. A dual role for zinc fingers in both DNA binding and zinc sensing by the Zap1 transcriptional activator. *EMBO J.* **19:**3704–3713.

Boelaert, J. R., M. de Locht, J. Van Cutsem, V. Kerrels, B. Cantinieaux, A. Verdonck, H. W. Van Landuyt, and Y. J. Schneider. 1993. Mucormycosis during deferoxamine therapy is a siderophore-mediated infection. In vitro and in vivo animal studies. *J. Clin. Investig.* **91:**1979–1986.

Brakhage, A. A., A. Andrianopoulos, M. Kato, S. Steidl, M. A. Davis, N. Tsukagoshi, and M. J. Hynes. 1999. HAP-Like CCAAT-binding complexes in filamentous fungi: implications for biotechnology. *Fungal Genet. Biol.* **27:**243–252.

Brickman, T. J., and M. A. McIntosh. 1992. Overexpression and purification of ferric enterobactin esterase from *Escherichia coli*. Demonstration of enzymatic hydrolysis of enterobactin and its iron complex. *J. Biol. Chem.* **267:**12350–12355.

Bullen, J. J., H. J. Rogers, P. B. Spalding, and C. G. Ward. 2006. Natural resistance, iron and infection: a challenge for clinical medicine. *J. Med. Microbiol.* **55:**251–258.

Caddick, M. X., A. G. Brownlee, and H. N. Arst, Jr. 1986. Regulation of gene expression by pH of the growth medium in *Aspergillus nidulans*. *Mol. Gen. Genet.* **203:**346–353.

Calera, J. A., M. C. Ovejero, R. Lopez-Medrano, M. Segurado, P. Puente, and F. Leal. 1997. Characterization of the *Aspergillus nidulans aspnd1* gene demonstrates that the ASPND1 antigen, which it encodes, and several *Aspergillus fumigatus* immunodominant antigens belong to the same family. *Infect. Immun.* **65:**1335–1344.

Chang, Y. C., C. M. Bien, H. Lee, P. J. Espenshade, and K. J. Kwon-Chung. 2007. Sre1p, a regulator of oxygen sensing and sterol homeostasis, is required for virulence in *Cryptococcus neoformans*. *Mol. Microbiol.* **64:**614–629.

Charlang, G., B. Ng, N. H. Horowitz, and R. M. Horowitz. 1981. Cellular and extracellular siderophores of *Aspergillus nidulans* and *Penicillium chrysogenum*. *Mol. Cell. Biol.* **1:**94–100.

Chen, X. Z., J. B. Peng, A. Cohen, H. Nelson, N. Nelson, and M. A. Hediger. 1999. Yeast SMF1 mediates H^+-coupled iron uptake with concomitant uncoupled cation currents. *J. Biol. Chem.* **274:**35089–35094.

Coleman, J. E. 1992. Zinc proteins: enzymes, storage proteins, transcription factors, and replication proteins. *Annu. Rev. Biochem.* **61:** 897–946.

Cousins, R. J., J. P. Liuzzi, and L. A. Lichten. 2006. Mammalian zinc transport, trafficking, and signals. *J. Biol. Chem.* **281:**24085–24089.

Coyle, P., J. C. Philcox, L. C. Carey, and A. M. Rofe. 2002. Metallothionein: the multipurpose protein. *Cell. Mol. Life Sci.* **59:**627–647.

Cramer, R. A., Jr., J. E. Stajich, Y. Yamanaka, F. S. Dietrich, W. J. Steinbach, and J. R. Perfect. 2006. Phylogenomic analysis of non-ribosomal peptide synthetases in the genus *Aspergillus*. *Gene* **383:** 24–32.

Dancis, A. 1998. Genetic analysis of iron uptake in the yeast *Saccharomyces cerevisiae*. *J. Pediatr.* **132:**S24–S29.

Dancis, A., R. D. Klausner, A. G. Hinnebusch, and J. G. Barriocanal. 1990. Genetic evidence that ferric reductase is required for iron uptake in *Saccharomyces cerevisiae*. *Mol. Cell. Biol.* **10:**2294–2301.

Dancis, A., D. G. Roman, G. J. Anderson, A. G. Hinnebusch, and R. D. Klausner. 1992. Ferric reductase of *Saccharomyces cerevisiae*: molecular characterization, role in iron uptake, and transcriptional control by iron. *Proc. Natl. Acad. Sci. USA* **89:**3869–3873.

de Lorenzo, V., A. Bindereif, B. H. Paw, and J. B. Neilands. 1986. Aerobactin biosynthesis and transport genes of plasmid ColV-K30 in *Escherichia coli* K-12. *J. Bacteriol.* **165:**570–578.

De Luca, N. G., and P. M. Wood. 2000. Iron uptake by fungi: contrasted mechanisms with internal or external reduction. *Adv. Microb. Physiol.* **43:**39–74.

De Silva, D. M., C. C. Askwith, D. Eide, and J. Kaplan. 1995. The FET3 gene product required for high affinity iron transport in yeast is a cell surface ferroxidase. *J. Biol. Chem.* **270:**1098–1101.

Destoumieux-Garzon, D., J. Peduzzi, X. Thomas, C. Djediat, and S. Rebuffat. 2006. Parasitism of iron-siderophore receptors of *Escherichia coli* by the siderophore-peptide microcin E492m and its unmodified counterpart. *Biometals* **19:**181–191.

Dinarello, C. A. 1988. Biology of interleukin 1. *FASEB J.* **2:**108–115.

Dix, D. R., J. T. Bridgham, M. A. Broderius, C. A. Byersdorfer, and D. J. Eide. 1994. The FET4 gene encodes the low affinity Fe(II) transport protein of *Saccharomyces cerevisiae*. *J. Biol. Chem.* **269:** 26092–26099.

Eichhorn, H., F. Lessing, B. Winterberg, J. Schirawski, J. Kamper, P. Muller, and R. Kahmann. 2006. A ferroxidation/permeation iron uptake system is required for virulence in *Ustilago maydis*. *Plant Cell* **18:**3332–3345.

Eide, D. J. 2006. Zinc transporters and the cellular trafficking of zinc. *Biochim. Biophys. Acta* **1763:**711–722.

Eide, D., S. Davis-Kaplan, I. Jordan, D. Sipe, and J. Kaplan. 1992. Regulation of iron uptake in *Saccharomyces cerevisiae*. The ferrireductase and Fe(II) transporter are regulated independently. *J. Biol. Chem.* **267:**20774–20781.

Eisendle, M., H. Oberegger, R. Buttinger, P. Illmer, and H. Haas. 2004. Biosynthesis and uptake of siderophores is controlled by the PacC-mediated ambient-pH regulatory system in *Aspergillus nidulans*. *Eukaryot. Cell* **3:**561–563.

Eisendle, M., H. Oberegger, I. Zadra, and H. Haas. 2003. The siderophore system is essential for viability of *Aspergillus nidulans*: functional analysis of two genes encoding L-ornithine N5-monooxygenase (sidA) and a non-ribosomal peptide synthetase (sidC). *Mol. Microbiol.* **49:**359–375.

Eisendle, M., M. Schrettl, C. Kragl, D. Muller, P. Illmer, and H. Haas. 2006. The intracellular siderophore ferricrocin is involved in iron storage, oxidative-stress resistance, germination, and sexual development in *Aspergillus nidulans*. *Eukaryot. Cell* **5:**1596–1603.

Emery, T. 1971. Role of ferrichrome as a ferric ionophore in *Ustilago sphaerogena*. *Biochemistry* **10:**1483–1488.

Essen, S. A., D. Bylund, S. J. Holmstrom, M. Moberg, and U. S. Lundstrom. 2006. Quantification of hydroxamate siderophores in soil solutions of podzolic soil profiles in Sweden. *Biometals* **19:**269–282.

Evans-Galea, M. V., E. Blankman, D. G. Myszka, A. J. Bird, D. J. Eide, and D. R. Winge. 2003. Two of the five zinc fingers in the Zap1 transcription factor DNA binding domain dominate site-specific DNA binding. *Biochemistry* **42:**1053–1061.

Fang, F. C. 2004. Antimicrobial reactive oxygen and nitrogen species: concepts and controversies. *Nat. Rev. Microbiol.* **2:**820–832.

Felice, M. R., I. De Domenico, L. Li, D. M. Ward, B. Bartok, G. Musci, and J. Kaplan. 2005. Post-transcriptional regulation of the yeast high affinity iron transport system. *J. Biol. Chem.* **280:**22181–22190.

Fernández-Martínez, J., C. V. Brown, E. Diez, J. Tilburn, H. N. Arst, Jr., M. A. Peñalva, and E. A. Espeso. 2003. Overlap of nuclear localisation signal and specific DNA-binding residues within the zinc finger domain of PacC. *J. Mol. Biol.* **334:**667–684.

Finegold, A. A., K. P. Shatwell, A. W. Segal, R. D. Klausner, and A. Dancis. 1996. Intramembrane bis-heme motif for transmembrane electron transport conserved in a yeast iron reductase and the human NADPH oxidase. *J. Biol. Chem.* **271:**31021–31024.

Finking, R., J. Solsbacher, D. Konz, M. Schobert, A. Schafer, D. Jahn, and M. A. Marahiel. 2002. Characterization of a new type of phosphopantetheinyl transferase for fatty acid and siderophore synthesis in Pseudomonas aeruginosa. *J. Biol. Chem.* **277:**50293–50302.

Fischbach, M. A., H. Lin, L. Zhou, Y. Yu, R. J. Abergel, D. R. Liu, K. N. Raymond, B. L. Wanner, R. K. Strong, C. T. Walsh, A. Aderem, and K. D. Smith. 2006. The pathogen-associated iroA gene cluster mediates bacterial evasion of lipocalin 2. *Proc. Natl. Acad. Sci. USA* **103:**16502–16507.

Fischer Walker, C., and R. E. Black. 2004. Zinc and the risk for infectious disease. *Annu. Rev. Nutr.* **24:**255–275.

Flo, T. H., K. D. Smith, S. Sato, D. J. Rodriguez, M. A. Holmes, R. K. Strong, S. Akira, and A. Aderem. 2004. Lipocalin 2 mediates an innate immune response to bacterial infection by sequestrating iron. *Nature* **432:**917–921.

Fluckinger, M., H. Haas, P. Merschak, B. J. Glasgow, and B. Redl. 2004. Human tear lipocalin exhibits antimicrobial activity by scavenging microbial siderophores. *Antimicrob. Agents Chemother.* **48:** 3367–3372.

Fraker, P. J., and L. E. King. 2004. Reprogramming of the immune system during zinc deficiency. *Annu. Rev. Nutr.* **24:**277–298.

Froissard, M., N. Belgareh-Touze, M. Dias, N. Buisson, J. M. Camadro, R. Haguenauer-Tsapis, and E. Lesuisse. 2007. Trafficking of siderophore transporters in *Saccharomyces cerevisiae* and intracellular fate of ferrioxamine B conjugates. *Traffic* **8:**1601–1616.

Gaither, L. A., and D. J. Eide. 2001. Eukaryotic zinc transporters and their regulation. *Biometals* **14:**251–270.

Ganz, T., and E. Nemeth. 2006. Iron imports. IV. Hepcidin and regulation of body iron metabolism. *Am. J. Physiol. Gastrointest. Liver Physiol.* **290:**G199–G203.

Georgatsou, E., and D. Alexandraki. 1994. Two distinctly regulated genes are required for ferric reduction, the first step of iron uptake in *Saccharomyces cerevisiae*. *Mol. Cell. Biol.* **14:**3065–3073.

Georgatsou, E., L. A. Mavrogiannis, G. S. Fragiadakis, and D. Alexandraki. 1997. The yeast Fre1p/Fre2p cupric reductases facilitate copper uptake and are regulated by the copper-modulated Mac1p activator. *J. Biol. Chem.* **272:**13786–13792.

Goetz, D. H., M. A. Holmes, N. Borregaard, M. E. Bluhm, K. N. Raymond, and R. K. Strong. 2002. The neutrophil lipocalin NGAL is a bacteriostatic agent that interferes with siderophore-mediated iron acquisition. *Mol. Cell* **10:**1033–1043.

Greenshields, D., L. Guosheng, J. Feng, G. Selvaraj, and Y. Wei. 2007. The siderophore biosynthetic gene SID1, but not the ferroxidase gene FET3, is required for full *Fusarium graminearum* virulence. *Mol. Plant Pathol.* **8:**411–421.

Haas, H. 2003. Molecular genetics of fungal siderophore biosynthesis and uptake: the role of siderophores in iron uptake and storage. *Appl. Microbiol. Biotechnol* **62:**316–330.

Haas, H., K. Angermayr, and G. Stoffler. 1997. Molecular analysis of a *Penicillium chrysogenum* GATA factor encoding gene (sreP) exhibiting significant homology to the *Ustilago maydis urbs1* gene. *Gene* **184:**33–37.

Haas, H., M. Schoeser, E. Lesuisse, J. F. Ernst, W. Parson, B. Abt, G. Winkelmann, and H. Oberegger. 2003. Characterization of the *Aspergillus nidulans* transporters for the siderophores enterobactin and triacetylfusarinine C. *Biochem. J.* **371:**505–513.

Haas, H., I. Zadra, G. Stoffler, and K. Angermayr. 1999. The *Aspergillus nidulans* GATA factor SREA is involved in regulation of siderophore biosynthesis and control of iron uptake. *J. Biol. Chem.* **274:**4613–4619.

Haase, H., and L. Rink. 2007. Signal transduction in monocytes: the role of zinc ions. *Biometals* **20:**579–585.

Hachem, R., P. Bahna, H. Hanna, L. C. Stephens, and I. Raad. 2006. EDTA as an adjunct antifungal agent for invasive pulmonary aspergillosis in a rodent model. *Antimicrob. Agents Chemother.* **50:**1823–1827.

Harrison, K. A., and G. A. Marzluf. 2002. Characterization of DNA binding and the cysteine rich region of SRE, a GATA factor in *Neurospora crassa* involved in siderophore synthesis. *Biochemistry* **41:** 15288–15295.

Hassett, R., D. R. Dix, D. J. Eide, and D. J. Kosman. 2000. The Fe(II) permease Fet4p functions as a low affinity copper transporter and supports normal copper trafficking in *Saccharomyces cerevisiae*. *Biochem. J.* **351:**477–484.

Hassett, R., and D. J. Kosman. 1995. Evidence for Cu(II) reduction as a component of copper uptake by *Saccharomyces cerevisiae*. *J. Biol. Chem.* **270:**128–134.

Herbig, A., A. J. Bird, S. Swierczek, K. McCall, M. Mooney, C. Y. Wu, D. R. Winge, and D. J. Eide. 2005. Zap1 activation domain 1 and its role in controlling gene expression in response to cellular zinc status. *Mol. Microbiol.* **57:**834–846.

Heymann, P., J. F. Ernst, and G. Winkelmann. 1999. Identification of a fungal triacetylfusarinine C siderophore transport gene (TAF1) in *Saccharomyces cerevisiae* as a member of the major facilitator superfamily. *Biometals* **12:**301–306.

Heymann, P., J. F. Ernst, and G. Winkelmann. 2000a. A gene of the major facilitator superfamily encodes a transporter for enterobactin (Enb1p) in *Saccharomyces cerevisiae*. *Biometals* **13:**65–72.

Heymann, P., J. F. Ernst, and G. Winkelmann. 2000b. Identification and substrate specificity of a ferrichrome-type siderophore transporter (Arn1p) in *Saccharomyces cerevisiae*. *FEMS Microbiol. Lett.* **186:**221–227.

Heymann, P., M. Gerads, M. Schaller, F. Dromer, G. Winkelmann, and J. F. Ernst. 2002. The siderophore iron transporter of *Candida albicans* (Sit1p/Arn1p) mediates uptake of ferrichrome-type siderophores and is required for epithelial invasion. *Infect. Immun.* **70:**5246–5255.

Hissen, A. H., J. M. Chow, L. J. Pinto, and M. M. Moore. 2004. Survival of *Aspergillus fumigatus* in serum involves removal of iron from transferrin: the role of siderophores. *Infect. Immun.* **72:**1402–1408.

Hissen, A. H., A. N. Wan, M. L. Warwas, L. J. Pinto, and M. M. Moore. 2005. The *Aspergillus fumigatus* siderophore biosynthetic gene *sidA*, encoding L-ornithine N^5-oxygenase, is required for virulence. *Infect. Immun.* **73:**5493–5503.

Hof, C., K. Eisfeld, K. Welzel, L. Antelo, A. J. Foster, and H. Anke. 2007. Ferricrocin synthesis in *Magnaporthae grisaea* and its role in pathogenicity in rice. *Mol. Plant Pathol.* **8:**163–172.

Hong, S. K., S. B. Han, M. Snyder, and E. Y. Choi. 1999. *SHC1*, a high pH inducible gene required for growth at alkaline pH in *Saccharomyces cerevisiae*. *Biochem. Biophys. Res. Commun.* **255:**116–122.

Horowitz, N. H., G. Charlang, G. Horn, and N. P. Williams. 1976. Isolation and identification of the conidial germination factor of *Neurospora crassa*. *J. Bacteriol.* **127:**135–140.

Hortschansky, P., M. Eisendle, Q. Al-Abdallah, A. D. Schmidt, S. Bergmann, M. Thon, O. Kniemeyer, B. Abt, B. Seeber, E. R. Werner, M. Kato, A. A. Brakhage, and H. Haas. 2007. Interaction of HapX with the CCAAT-binding complex-a novel mechanism of gene regulation by iron. *EMBO J.* **26:**3157–3168.

Howard, D. H. 2004. Iron gathering by zoopathogenic fungi. *FEMS Immunol. Med. Microbiol.* **40:**95–100.

Hsiang, T., and D. L. Baillie. 2005. Comparison of the yeast proteome to other fungal genomes to find core fungal genes. *J. Mol. Evol.* **60:**475–483.

Ibrahim, A. S., T. Gebermariam, Y. Fu, L. Lin, M. I. Husseiny, S. W. French, J. Schwartz, C. D. Skory, J. E. Edwards, Jr., and B. J. Spellberg. 2007. The iron chelator deferasirox protects mice from mucormycosis through iron starvation. *J. Clin. Investig.* **117:**2649–2657.

Jensen, L. T., M. Ajua-Alemanji, and V. C. Culotta. 2003. The *Saccharomyces cerevisiae* high affinity phosphate transporter encoded by *PHO84* also functions in manganese homeostasis. *J. Biol. Chem.* **278:**42036–42040.

Johnson, R., C. Voisey, L. Johnson, J. Pratt, D. Fleetwood, A. Khan, and G. Bryan. 2007. Distribution of NRPS gene families within the *Neotyphodium/Epichloe* complex. *Fungal Genet. Biol.* **44:**1180–1190.

Jung, W. H., A. Sham, R. White, and J. W. Kronstad. 2006. Iron regulation of the major virulence factors in the AIDS-associated pathogen *Cryptococcus neoformans*. *PLoS Biol.* **4:**e410.

Kaplan, J., D. McVey Ward, R. J. Crisp, and C. C. Philpott. 2006. Iron-dependent metabolic remodeling in *S. cerevisiae*. *Biochim. Biophys. Acta* **1763:**646–651.

Kim, Y., Y. Deng, and C. C. Philpott. 2007. GGA2- and ubiquitin-dependent trafficking of Arn1, the ferrichrome transporter of *Saccharomyces cerevisiae*. *Mol. Biol. Cell* **18:**1790–1802.

Kleinschmidt-DeMasters, B. K. 2002. Central nervous system aspergillosis: a 20-year retrospective series. *Hum. Pathol.* **33:**116–124.

Kontoyiannis, D. P., G. Chamilos, R. E. Lewis, S. Giralt, J. Cortes, I. I. Raad, J. T. Manning, and X. Han. 2007. Increased bone marrow iron stores is an independent risk factor for invasive aspergillosis in patients with high-risk hematologic malignancies and recipients of allogeneic hematopoietic stem cell transplantation. *Cancer* **110:**1303–1306.

Kosman, D. J. 2003. Molecular mechanisms of iron uptake in fungi. *Mol. Microbiol.* **47:**1185–1197.

Kragl, C., M. Schrettl, B. Abt, B. Sarg, H. H. Lindner, and H. Haas. 2007. EstB-mediated hydrolysis of the siderophore triacetylfusarinine C optimizes iron uptake of *Aspergillus fumigatus*. *Eukaryot. Cell* **6:**1278–1285.

Kroncke, K. D. 2007. Cellular stress and intracellular zinc dyshomeostasis. *Arch. Biochem. Biophys.* **463:**183–187.

Kumanovics, A., K. E. Poruk, K. A. Osborn, D. M. Ward, and J. Kaplan. 2006. *YKE4* (YIL023C) encodes a bidirectional zinc transporter in the endoplasmic reticulum of *Saccharomyces cerevisiae*. *J. Biol. Chem.* **281:**22566–22574.

Labbe, S., B. Pelletier, and A. Mercier. 2007. Iron homeostasis in the fission yeast *Schizosaccharomyces pombe*. *Biometals* **20:**523–537.

Lamb, T. M., W. Xu, A. Diamond, and A. P. Mitchell. 2001. Alkaline response genes of *Saccharomyces cerevisiae* and their relationship to the RIM101 pathway. *J. Biol. Chem.* **276:**1850–1856.

Lan, C. Y., G. Rodarte, L. A. Murillo, T. Jones, R. W. Davis, J. Dungan, G. Newport, and N. Agabian. 2004. Regulatory networks affected by iron availability in *Candida albicans*. *Mol. Microbiol.* **53:**1451–1469.

Lee, A., A. K. Henras, and G. Chanfreau. 2005. Multiple RNA surveillance pathways limit aberrant expression of iron uptake mRNAs and prevent iron toxicity in *S. cerevisiae*. *Mol. Cell* **19:**39–51.

Leong, S. A., and G. Winkelmann. 1998. Molecular biology of iron transport in fungi. *Metal Ions Biol. Syst.* **35:**147–186.

Lesuisse, E., M. Simon-Casteras, and P. Labbe. 1998. Siderophore-mediated iron uptake in *Saccharomyces cerevisiae*: the SIT1 gene encodes a ferrioxamine B permease that belongs to the major facilitator superfamily. *Microbiology* **144:**3455–3462.

Li, L., O. S. Chen, D. McVey Ward, and J. Kaplan. 2001. CCC1 is a transporter that mediates vacuolar iron storage in yeast. *J. Biol. Chem.* **276:**29515–29519.

Lian, T., M. I. Simmer, C. A. D'Souza, B. R. Steen, S. D. Zuyderduyn, S. J. Jones, M. A. Marra, and J. W. Kronstad. 2005. Iron-regulated transcription and capsule formation in the fungal pathogen *Cryptococcus neoformans*. *Mol. Microbiol.* **55:**1452–1472.

Lin, S. J., R. A. Pufahl, A. Dancis, T. V. O'Halloran, and V. C. Culotta. 1997. A role for the *Saccharomyces cerevisiae* ATX1 gene in copper trafficking and iron transport. *J. Biol. Chem.* **272:**9215–9220.

Liuzzi, J. P., L. A. Lichten, S. Rivera, R. K. Blanchard, T. B. Aydemir, M. D. Knutson, T. Ganz, and R. J. Cousins. 2005. Interleukin-6 regulates the zinc transporter Zip14 in liver and contributes to the hypozincemia of the acute-phase response. *Proc. Natl. Acad. Sci. USA* **102:**6843–6848.

López-Medrano, R., M. C. Ovejero, J. A. Calera, P. Puente, and F. Leal. 1996. Immunoblotting patterns in the serodiagnosis of aspergilloma: antibody response to the 90kDa *Aspergillus fumigatus* antigen. *Eur. J. Clin. Microbiol. Infect. Dis.* **15:**146–152.

Magneson, G. R., J. M. Puvathingal, and W. J. Ray, Jr. 1987. The concentrations of free Mg^{2+} and free Zn^{2+} in equine blood plasma. *J. Biol. Chem.* **262:**11140–11148.

Maret, W. 2006. Zinc coordination environments in proteins as redox sensors and signal transducers. *Antioxid. Redox Signal.* **8:**1419–1441.

Matzanke, B. F. 1994. Iron storage in fungi, p. 179–213. *In* G. Winkelmann and D. R. Winge (ed.), *Metal Ions in Fungi.* Marcel Dekker, Inc., New York, NY.

Matzanke, B. F., E. Bill, A. X. Trautwein, and G. Winkelmann. 1987. Role of siderophores in iron storage in spores of *Neurospora crassa* and *Aspergillus ochraceus. J. Bacteriol.* **169:**5873–5876.

McNabb, D. S., Y. Xing, and L. Guarente. 1995. Cloning of yeast HAP5: a novel subunit of a heterotrimeric complex required for CCAAT binding. *Genes Dev.* **9:**47–58.

Mei, B., A. D. Budde, and S. A. Leong. 1993. *sid1*, a gene initiating siderophore biosynthesis in *Ustilago maydis*: molecular characterization, regulation by iron, and role in phytopathogenicity. *Proc. Natl. Acad. Sci. USA* **90:**903–907.

Miele, R., D. Barra, and M. C. Bonaccorsi di Patti. 2007. A GATA-type transcription factor regulates expression of the high-affinity iron uptake system in the methylotrophic yeast *Pichia pastoris. Arch. Biochem. Biophys.* **465:**172–179.

Miethke, M., and M. A. Marahiel. 2007. Siderophore-based iron acquisition and pathogen control. *Microbiol. Mol. Biol. Rev.* **71:**413–451.

Moore, R. E., Y. Kim, and C. C. Philpott. 2003. The mechanism of ferrichrome transport through Arn1p and its metabolism in *Saccharomyces cerevisiae. Proc. Natl. Acad. Sci. USA* **100:**5664–5669.

Mootz, H. D., D. Schwarzer, and M. A. Marahiel. 2002. Ways of assembling complex natural products on modular nonribosomal peptide synthetases. *Chembiochem* **3:**490–504.

Moreno, M. A., J. Amich, R. Vicentefranqueira, F. Leal, and J. A. Calera. 2007a. Culture conditions for zinc- and pH-regulated gene expression studies in *Aspergillus fumigatus. Int. Microbiol.* **10:**187–192.

Moreno, M. A., O. Ibrahim-Granet, R. Vicentefranqueira, J. Amich, P. Ave, F. Leal, J. P. Latgé, and J. A. Calera. 2007b. The regulation of zinc homeostasis by the ZafA transcriptional activator is essential for *Aspergillus fumigatus* virulence. *Mol. Microbiol.* **64:**1182–1197.

Moshage, H. 1997. Cytokines and the hepatic acute phase response. *J. Pathol.* **181:**257–266.

Muller, G., S. J. Barclay, and K. N. Raymond. 1985. The mechanism and specificity of iron transport in *Rhodotorula pilimanae* probed by synthetic analogs of rhodotorulic acid. *J. Biol. Chem.* **260:**13916–13920.

Neilands, J. B. 1995. Siderophores: structure and function of microbial iron transport compounds. *J. Biol. Chem.* **270:**26723–26726.

Ng, T. T., G. D. Robson, and D. W. Denning. 1994. Hydrocortisone-enhanced growth of *Aspergillus* spp.: implications for pathogenesis. *Microbiology* **140:**2475–2479.

Oberegger, H., M. Eisendle, M. Schrettl, S. Graessle, and H. Haas. 2003. 4′-Phosphopantetheinyl transferase-encoding npgA is essential for siderophore biosynthesis in *Aspergillus nidulans. Curr. Genet.* **44:**211–215.

Oberegger, H., M. Schoeser, I. Zadra, B. Abt, and H. Haas. 2001. SREA is involved in regulation of siderophore biosynthesis, utilization and uptake in *Aspergillus nidulans. Mol. Microbiol.* **41:**1077–1089.

Oberegger, H., M. Schoeser, I. Zadra, M. Schrettl, W. Parson, and H. Haas. 2002a. Regulation of *freA, acoA, lysF*, and *cycA* expression by iron availability in *Aspergillus nidulans. Appl. Environ. Microbiol.* **68:**5769–5772.

Oberegger, H., I. Zadra, M. Schoeser, B. Abt, W. Parson, and H. Haas. 2002b. Identification of members of the *Aspergillus nidulans* SREA regulon: genes involved in siderophore biosynthesis and utilization. *Biochem. Soc. Trans.* **30:**781–783.

Oide, S., S. B. Krasnoff, D. M. Gibson, and B. G. Turgeon. 2007. Intracellular siderophores are essential for ascomycete sexual development in heterothallic *Cochliobolus heterostrophus* and homothallic *Gibberella zeae. Eukaryot. Cell* **6:**1339–1353.

Oide, S., W. Moeder, H. Haas, S. Krasnoff, D. Gibson, K. Yoshioka, and B. G. Turgeon. 2006. NPS6, encoding a non-ribosomal peptide synthetase involved in siderophore-mediated iron metabolism, is a conserved virulence determinant of plant pathogenic ascomycetes. *Plant Cell* **18:**2836–2853.

Pagani, A., L. Villarreal, M. Capdevila, and S. Atrian. 2007. The *Saccharomyces cerevisiae* Crs5 metallothionein metal-binding abilities and its role in the response to zinc overload. *Mol. Microbiol.* **63:**256–269.

Pao, S. S., I. T. Paulsen, and M. H. Saier, Jr. 1998. Major facilitator superfamily. *Microbiol. Mol. Biol. Rev.* **62:**1–34.

Papanikolaou, G., and K. Pantopoulos. 2005. Iron metabolism and toxicity. *Toxicol. Appl. Pharmacol.* **202:**199–211.

Paris, S., D. Wysong, J. P. Debeaupuis, K. Shibuya, B. Philippe, R. D. Diamond, and J. P. Latgé. 2003. Catalases of *Aspergillus fumigatus. Infect. Immun.* **71:**3551–3562.

Park, Y. S., J. H. Kim, J. H. Cho, H. I. Chang, S. W. Kim, H. D. Paik, C. W. Kang, T. H. Kim, H. C. Sung, and C. W. Yun. 2007. Physical and functional interaction of FgFtr1-FgFet1 and FgFtr2-FgFet2 is required for iron uptake in *Fusarium graminearum. Biochem. J.* **408:**97–104.

Park, Y. S., T. H. Kim, H. I. Chang, H. C. Sung, and C. W. Yun. 2006. Cellular iron utilization is regulated by putative siderophore transporter FgSit1 not by free iron transporter in *Fusarium graminearum. Biochem. Biophys. Res. Commun.* **345:**1634–1642.

Pelletier, B., J. Beaudoin, Y. Mukai, and S. Labbe. 2002. Fep1, an iron sensor regulating iron transporter gene expression in *Schizosaccharomyces pombe. J. Biol. Chem.* **277:**22950–22958.

Pelletier, B., J. Beaudoin, C. C. Philpott, and S. Labbe. 2003. Fep1 represses expression of the fission yeast *Schizosaccharomyces pombe* siderophore-iron transport system. *Nucleic Acids Res.* **31:**4332–4344.

Peñalva, M. A., and H. N. Arst, Jr. 2004. Recent advances in the characterization of ambient pH regulation of gene expression in filamentous fungi and yeasts. *Annu. Rev. Microbiol.* **58:**425–451.

Portnoy, M. E., X. F. Liu, and V. C. Culotta. 2000. *Saccharomyces cerevisiae* expresses three functionally distinct homologues of the Nramp family of metal transporters. *Mol. Cell. Biol.* **20:**7893–7902.

Powell, P. E., G. R. Cline, C. P. P. Reid, and P. J. Szanizlo. 1980. Occurrance of hydroxamate siderophore iron in soils. *Nature* **287:**833–834.

Protchenko, O., and C. C. Philpott. 2003. Regulation of intracellular heme levels by HMX1, a homologue of heme oxygenase, in *Saccharomyces cerevisiae. J. Biol. Chem.* **278:**36582–36587.

Puig, S., E. Askeland, and D. J. Thiele. 2005. Coordinated remodeling of cellular metabolism during iron deficiency through targeted mRNA degradation. *Cell* **120:**99–110.

Ramanan, N., and Y. Wang. 2000. A high-affinity iron permease essential for *Candida albicans* virulence. *Science* **288:**1062–1064.

Ratledge, C., and L. G. Dover. 2000. Iron metabolism in pathogenic bacteria. *Annu. Rev. Microbiol.* **54:**881–941.

Roosenberg, J. M., Y. M. Lin, Y. Lu, and M. J. Miller. 2000. Studies and syntheses of siderophores, microbial iron chelators, and analogs as potential drug delivery agents. *Curr. Med. Chem.* **7:**159–197.

Rotrosen, D., C. L. Yeung, T. L. Leto, H. L. Malech, and C. H. Kwong. 1992. Cytochrome b_{558}: the flavin-binding component of the phagocyte NADPH oxidase. *Science* **256:**1459–1462.

Rutherford, J. C., and A. J. Bird. 2004. Metal-responsive transcription factors that regulate iron, zinc, and copper homeostasis in eukaryotic cells. *Eukaryot. Cell* **3:**1–13.

Santos, R., N. Buisson, S. Knight, A. Dancis, J. M. Camadro, and E. Lesuisse. 2003. Haemin uptake and use as an iron source by *Candida albicans*: role of CaHMX1-encoded haem oxygenase. *Microbiology* **149:**579–588.

Scazzocchio, C. 2000. The fungal GATA factors. *Curr. Opin. Microbiol.* 3:126–131.

Schrettl, M., E. Bignell, C. Kragl, C. Joechl, T. Rogers, H. N. Arst, Jr., K. Haynes, and H. Haas. 2004a. Siderophore biosynthesis but not reductive iron assimilation is essential for *Aspergillus fumigatus* virulence. *J. Exp. Med.* **200:**1213–1219.

Schrettl, M., E. Bignell, C. Kragl, Y. Sabiha, O. Loss, M. Eisendle, A. Wallner, H. N. Arst, K. Haynes, and H. Haas. 2007. Distinct roles for intra- and extracellular siderophores during *Aspergillus fumigatus* infection. *PLoS Pathog.* 3:e128.

Schrettl, M., G. Winkelmann, and H. Haas. 2004b. Ferrichrome in *Schizosaccharomyces pombe*: an iron transport and iron storage compound. *Biometals* **17:**647–654.

Schroeder, J. J., and R. J. Cousins. 1990. Interleukin 6 regulates metallothionein gene expression and zinc metabolism in hepatocyte monolayer cultures. *Proc Natl Acad Sci USA* **87:**3137–3141.

Schwecke, T., K. Gottling, P. Durek, I. Duenas, N. F. Kaufer, S. Zock-Emmenthal, E. Staub, T. Neuhof, R. Dieckmann, and H. von Dohren. 2006. Nonribosomal peptide synthesis in *Schizosaccharomyces pombe* and the architectures of ferrichrome-type siderophore synthetases in fungi. *ChemBioChem* **7:**612–622.

Segurado, M., R. Lopez-Aragon, J. A. Calera, J. M. Fernandez-Abalos, and F. Leal. 1999. Zinc-regulated biosynthesis of immunodominant antigens from *Aspergillus* spp. *Infect. Immun.* **67:**2377–2382.

Serrano, R., D. Bernal, E. Simon, and J. Arino. 2004. Copper and iron are the limiting factors for growth of the yeast *Saccharomyces cerevisiae* in an alkaline environment. *J. Biol. Chem.* **279:**19698–19704.

Serrano, R., H. Martin, A. Casamayor, and J. Arino. 2006. Signaling alkaline pH stress in the yeast *Saccharomyces cerevisiae* through the Wsc1 cell surface sensor and the Slt2 MAPK pathway. *J. Biol. Chem.* **281:**39785–39795.

Shi, X., C. Stoj, A. Romeo, D. J. Kosman, and Z. Zhu. 2003. Fre1p Cu^{2+} reduction and Fet3p Cu^{1+} oxidation modulate copper toxicity in *Saccharomyces cerevisiae*. *J. Biol. Chem.* **278:**50309–50315.

Singh, A., N. Kaur, and D. J. Kosman. 2007. The metalloreductase Fre6p in Fe-efflux from the yeast vacuole. *J. Biol. Chem.* **282:**28619–28626.

Smith, J. L. 2004. The physiological role of ferritin-like compounds in bacteria. *Crit. Rev. Microbiol.* **30:**173–185.

Smith, R. L., and A. D. Johnson. 2000. Turning genes off by Ssn6-Tup1: a conserved system of transcriptional repression in eukaryotes. *Trends Biochem. Sci.* **25:**325–330.

Spahl, D. U., D. Berendji-Grun, C. V. Suschek, V. Kolb-Bachofen, and K. D. Kroncke. 2003. Regulation of zinc homeostasis by inducible NO synthase-derived NO: nuclear metallothionein translocation and intranuclear Zn^{2+} release. *Proc. Natl. Acad. Sci. USA* **100:**13952–13957.

Stearman, R., D. S. Yuan, Y. Yamaguchi-Iwai, R. D. Klausner, and A. Dancis. 1996. A permease-oxidase complex involved in high-affinity iron uptake in yeast. Science **271:**1552–1557.

Steenbergen, J. N., and A. Casadevall. 2003. The origin and maintenance of virulence for the human pathogenic fungus *Cryptococcus neoformans*. *Microbes Infect.* **5:**667–675.

Stoj, C. S., A. J. Augustine, E. I. Solomon, and D. J. Kosman. 2007. Structure-function analysis of the cuprous oxidase activity in Fet3p from *Saccharomyces cerevisiae*. *J. Biol. Chem.* **282:**7862–7868.

Striz, I., and I. Trebichavsky. 2004. Calprotectin: a pleiotropic molecule in acute and chronic inflammation. *Physiol. Res.* **53:**245–253.

Takeda, A., A. Minami, S. Takefuta, M. Tochigi, and N. Oku. 2001. Zinc homeostasis in the brain of adult rats fed zinc-deficient diet. *J. Neurosci. Res.* **63:**447–452.

Tapiero, H., and K. D. Tew. 2003. Trace elements in human physiology and pathology: zinc and metallothioneins. *Biomed. Pharmacother.* **57:**399–411.

Tobiasen, C., J. Aahman, K. S. Ravnholt, M. J. Bjerrum, M. N. Grell, and H. Giese. 2007. Nonribosomal peptide synthetase (NPS) genes in *Fusarium graminearum*, *F. culmorum* and *F. pseudograminearium* and identification of NPS2 as the producer of ferricrocin. *Curr. Genet.* **51:**43–58.

Tucker, S. L., C. R. Thornton, K. Tasker, C. Jacob, G. Giles, M. Egan, and N. J. Talbot. 2004. A fungal metallothionein is required for pathogenicity of *Magnaporthe grisea*. *Plant Cell* **16:**1575–1588.

Uetz, P., L. Giot, G. Cagney, T. A. Mansfield, R. S. Judson, J. R. Knight, D. Lockshon, V. Narayan, M. Srinivasan, P. Pochart, A. Qureshi-Emili, Y. Li, B. Godwin, D. Conover, T. Kalbfleisch, G. Vijayadamodar, M. Yang, M. Johnston, S. Fields, and J. M. Rothberg. 2000. A comprehensive analysis of protein-protein interactions in *Saccharomyces cerevisiae*. *Nature* **403:**623–627.

Urbanowski, J. L., and R. C. Piper. 1999. The iron transporter Fth1p forms a complex with the Fet5 iron oxidase and resides on the vacuolar membrane. *J. Biol. Chem.* **274:**38061–38070.

van der Helm, D., and G. Winkelmann. 1994. Hydroxamates and polycarbonates as iron transport agents (siderophores) in fungi, p. 39–148. *In* G. Winkelmann and D. R. Winge (ed.), *Metal Ions in Fungi*. Marcel Dekker, Inc., New York, NY.

Vicentefranqueira, R., M. A. Moreno, F. Leal, and J. A. Calera. 2005. The *zrfA* and *zrfB* genes of *Aspergillus fumigatus* encode the zinc transporter proteins of a zinc uptake system induced in an acid, zinc-depleted environment. *Eukaryot. Cell* **4:**837–848.

Waters, B. M., and D. J. Eide. 2002. Combinatorial control of yeast *FET4* gene expression by iron, zinc, and oxygen. *J. Biol. Chem.* **277:**33749–33757.

Webster, J. I., L. Tonelli, and E. M. Sternberg. 2002. Neuroendocrine regulation of immunity. *Annu. Rev. Immunol.* **20:**125–163.

Weinberg, E. D. 1999. The role of iron in protozoan and fungal infectious diseases. *J. Eukaryot. Microbiol.* **46:**231–238.

Weiss, G. 2002. Iron and immunity: a double-edged sword. *Eur. J. Clin. Investig.* **32**(Suppl. 1):70–78.

Welzel, K., K. Eisfeld, L. Antelo, T. Anke, and H. Anke. 2005. Characterization of the ferrichrome A biosynthetic gene cluster in the homobasidiomycete *Omphalotus olearius*. *FEMS Microbiol. Lett.* **249:**157–163.

Winkelmann, G. 2001. Siderophore transport in fungi, p. 463–480. *In* G. Winkelmann (ed.), *Microbial Transport Systems*. Wiley-VCH, Weinheim, Germany.

Wu, C. Y., A. J. Bird, D. R. Winge, and D. J. Eide. 2007. Regulation of the yeast *TSA1* peroxiredoxin by *ZAP1* is an adaptive response to the oxidative stress of zinc deficiency. *J. Biol. Chem.* **282:**2184–2195.

Yamada, O., S. Na Nan, T. Akao, M. Tominaga, H. Watanabe, T. Satoh, H. Enei, and O. Akita. 2003. *dffA* gene from *Aspergillus oryzae* encodes L-ornithine N5-oxygenase and is indispensable for deferriferrichrysin biosynthesis. *J. Biosci. Bioeng.* **95:**82–88.

Yamaguchi-Iwai, Y., A. Dancis, and R. D. Klausner. 1995. AFT1: a mediator of iron regulated transcriptional control in *Saccharomyces cerevisiae*. *EMBO J.* **14:**1231–1239.

Yeager, M. P., P. M. Guyre, and A. U. Munck. 2004. Glucocorticoid regulation of the inflammatory response to injury. *Acta Anaesthesiol. Scand.* **48:**799–813.

Yuan, W. M., G. D. Gentil, A. D. Budde, and S. A. Leong. 2001. Characterization of the *Ustilago maydis sid2* gene, encoding a multidomain peptide synthetase in the ferrichrome biosynthetic gene cluster. *J. Bacteriol.* **183:**4040–4051.

Yun, C. W., M. Bauler, R. E. Moore, P. E. Klebba, and C. C. Philpott. 2001. The role of the FRE family of plasma membrane reductases in the uptake of siderophore-iron in *Saccharomyces cerevisiae*. *J. Biol. Chem.* **276:**10218–10223.

Yun, C. W., T. Ferea, J. Rashford, O. Ardon, P. O. Brown, D. Botstein, J. Kaplan, and C. C. Philpott. 2000a. Desferrioxaminemediated iron uptake in *Saccharomyces cerevisiae*. Evidence for two pathways of iron uptake. *J. Biol. Chem.* 275:10709–10715.

Yun, C. W., J. S. Tiedeman, R. E. Moore, and C. C. Philpott. 2000b. Siderophore-iron uptake in *Saccharomyces cerevisiae*. Identification of ferrichrome and fusarinine transporters. *J. Biol. Chem.* 275: 16354–16359.

Zarember, K. A., J. A. Sugui, Y. C. Chang, K. J. Kwon-Chung, and J. I. Gallin. 2007. Human polymorphonuclear leukocytes inhibit *Aspergillus fumigatus* conidial growth by lactoferrin-mediated iron depletion. *J. Immunol.* 178:6367–6373.

Zhao, H., E. Butler, J. Rodgers, T. Spizzo, S. Duesterhoeft, and D. Eide. 1998. Regulation of zinc homeostasis in yeast by binding of the *ZAP1* transcriptional activator to zinc-responsive promoter elements. *J. Biol. Chem.* 273:28713–28720.

Zhao, H., and D. J. Eide. 1997. Zap1p, a metalloregulatory protein involved in zinc-responsive transcriptional regulation in *Saccharomyces cerevisiae*. *Mol. Cell. Biol.* 17:5044–5052.

Zhou, L. W., H. Haas, and G. A. Marzluf. 1998. Isolation and characterization of a new gene, sre, which encodes a GATA-type regulatory protein that controls iron transport in *Neurospora crassa*. *Mol. Gen. Genet.* 259:532–540.

Znaidi, S., B. Pelletier, Y. Mukai, and S. Labbe. 2004. The *Schizosaccharomyces pombe* corepressor Tup11 interacts with the iron-responsive transcription factor Fep1. *J. Biol. Chem.* 279:9462–9474.

Aspergillus fumigatus and Aspergillosis
Edited by J.-P. Latgé and W. J. Steinbach

Chapter 10

Conidial Germination in *Aspergillus fumigatus*

NIR OSHEROV

The asexual spore or conidium is critical in the life cycle of many filamentous fungi. The spore is the primary means of dispersal in the environment and serves as a safe house for the fungal genome when nutrients are limited or the environment threatens it. Indeed, the spores of some fungal species can remain viable after desiccation for several decades (Gams and Stalpers, 1994). This chapter discusses the physiological and biochemical aspects of conidial germination in *Aspergillus fumigatus* and the regulatory pathways used to activate the process. Our definition of conidial germination takes into account only those events that occur during the first few hours and does not include later responses, such as the initiation of nuclear division and hyphal growth.

Remarkably, we still know very little about the mechanism of conidial germination in *A. fumigatus*. Most of our insights into fungal conidial germination come from studies in the related *Aspergillus nidulans* as well as in *Neurospora crassa*, and these will be described. The physiological processes taking place during early germination include the uptake of water and isotropic swelling and changes in the properties of the cell wall. Biochemical changes include the rapid initiation of respiration, protein and RNA synthesis, and the breakdown of trehalose, followed later by DNA synthesis. Several signaling pathways, including the cyclic AMP/protein kinase A (cAMP/PKA), RAS, and mitogen-activated protein kinase (MAPK) pathways, appear to be involved in transmitting the germination signal.

The pathogenesis of *A. fumigatus* is intimately related to conidial inhalation and germination in the alveoli. By achieving a molecular understanding of this process, it may be possible to develop novel germination inhibitors that block infection at its outset.

APPROACHES AND LIMITATIONS TO ANALYSIS OF CONIDIAL GERMINATION

Conidial germination appears to be a regulated process that responds to environmental stimuli by a signaling cascade. Therefore, it should be amenable to genetic and biochemical inquiry. For example, conditional mutants unable to germinate can be generated via random mutagenesis and the responsible genes analyzed. Biochemical analysis can identify the earliest biochemical changes that take place during germination. The pathways regulating these changes can then be dissected, working upstream to identify the initial sensors and activators. Transcriptomics and proteomics can identify genes that are preferentially active and transcripts and proteins that are preferentially accumulated during germination, and some of these may be directly involved in controlling the process. Bioinformatics can generate lists of candidate conidial germination genes based on sequence homology

In view of our advanced ability to perform molecular manipulations in several species of fungi that serve as model organisms, it is somewhat surprising that the basic molecular steps regulating early conidial germination have not yet been clearly defined. Why? One plausible explanation for the difficulties involved in dissecting conidial germination at the molecular level is that it may be controlled by multiple sensors and pathways, each one sensitive to a particular environmental stimulus, working in a specific combination. This greatly complicates genetic dissection. Moreover, it is difficult to differentiate between the true early signaling events and the explosion of essential metabolic and housekeeping activities that occurs almost at the very beginning of this process. It is also difficult to differentiate genetically between early signal transduction events and early essential housekeeping activities. Genetic damage in either will yield a germination-deficient mutant, but only the former will be instructive.

PHYSIOLOGY OF *ASPERGILLUS* CONIDIAL GERMINATION

Conidial germination of *A. fumigatus* is initiated by the presence of water, carbon, phosphate, and nitrate

Nir Osherov • Dept. of Human Microbiology, Sackler School of Medicine, Tel-Aviv University, Ramat Aviv, Tel-Aviv 69978, Israel.

sources (Table 1), and oxygen (Taubitz et al., 2007). The fact that optimal and efficient germination occurs only when all the main nutrient sources are available suggests that *A. fumigatus* conidia can sense and combine multiple external inputs and respond accordingly. It also raises the possibility that different sensors may activate independent signaling pathways to initiate germination (see "Candidate Gene Approach" below).

A. fumigatus is unable to germinate under anaerobic conditions or in the presence of respiratory chain inhibitors. Active mitochondria can be detected during early conidial swelling, indicating that respiration is an early event in germination (Taubitz et al., 2007).

A. fumigatus conidia apparently contain a germination self-inhibitor. High concentrations of *A. fumigatus* conidia germinate less readily than low concentrations (Fig. 1). Self-inhibition has the obvious advantage of preventing rapid germination of all spores at the same time and place, which ensures survival under fluctuating environmental conditions and stimulates dissemination in nature. This phenomenon has been described in many fungi, including *Aspergillus niger* and *Penicillium paneum* (Barrios-Gonzáles et al., 1989; Chitarra et al., 2004). It is a reversible process which is alleviated by the extensive washing of the conidia (Macko et al., 1976). The inhibitory compounds involved in this process in *Aspergillus* have not been identified.

The earliest observable changes in conidial morphology, occurring approximately 1 h after the addition of rich medium, are swelling, nuclear decondensation, and adhesion (Fig. 2). Swelling or isotrophic growth involves both water uptake (rehydration) and wall growth (Griffin, 1994; Van Etten et al., 1983). Nuclear decondensation occurs in preparation for DNA replication (~2 h) and mitosis (~4 h) (Harris, 1999; Momany and Taylor, 2000). Adhesion takes place both between conidia (forming conidial clumps) and the substrate (Tronchin et al., 1995). Many of the physical aspects of *A. fumigatus* conidia have not been comprehensively determined, including their rate of swelling and adherence in different media, their stability and resistance to various environmental insults (temperature, humidity, UV light, acid, etc.), and their ability to aerosolize in comparison to other *Aspergillus* species. Also missing are additional cell markers (biochemical activities, reporter genes, or expression of tagged proteins) that could parse and define the very early steps of conidial germination.

Microscopic examination shows that the conidial cytoplasm in *A. fumigatus* contains a very high density of ribosomes, mitochondria, glycogen deposits, and at least two types of lipid storage inclusions. The plasma membrane shows numerous small infoldings visible with transmission electron microscopy or freeze-etching electron microscopy (Ghiorse and Edwards, 1973). The conidial cell wall is composed of three layers: a convoluted melanin-rich outer layer, a light inner layer, and a dark innermost layer tightly associated with the plasma membrane. The outer layer is hydrophobic and contains a family of small proteins known collectively as hydrophobins (Paris et al., 2003). It is shed during germination, revealing a smooth surface (Ma et al., 2006; Rohde et al., 2002; Tronchin et al., 1997). Conidial coat shedding can also be visualized with monoclonal antibodies generated against the cell wall, or lectins such as peanut agglutinin (Momany et al., 2004; Tronchin et al., 1995).

BIOCHEMICAL CHANGES DURING EARLY GERMINATION

Within 15 to 30 min of adding an appropriate carbon source, conidia initiate the process of RNA and protein synthesis. This has been shown in both *A. nidulans* and *N. crassa*, but not in *A. fumigatus* (Bainbridge,

Table 1. *A. fumigatus* conidial germination in different media

Medium	% germination[a]
Double-distilled water	0
1% glucose	46 ± 9
12 mM KH_2PO_4	28 ± 2
7 mM KCl + 4 mM $MgSO_4$ + 70 mM $NaNO_3$	4.4 ± 3
1% glucose + 12 mM KH_2PO_4	71 ± 6
1% glucose + 7 mM KCl + 4 mM $MgSO_4$ + 70 mM $NaNO_3$	44 ± 7
12 mM KPO_4 + 7 mM KCl + 4 mM $MgSO_4$ + 70 mM $NaNO_3$	21 ± 3
Minimal medium (MM)[b]	81 ± 4
YAG[c]	100

[a] Percent conidia of *A. fumigatus* AF293 exhibiting one or more germ tubes after an incubation of 16 h at 37°C in liquid medium at a concentration of 10^4 conidia/ml.
[b] Minimal medium contains 70 mM $NaNO_3$, 1% (wt/vol) glucose, 12 mM KPO_4 (pH 6.8), 4 mM $MgSO_4$, and 7 mM KCl, supplemented with trace elements and vitamins.
[c] YAG rich medium contains 0.5% (wt/vol) yeast extract, 1% (wt/vol) glucose, and 10 mM $MgCl_2$, supplemented with trace elements and vitamins.

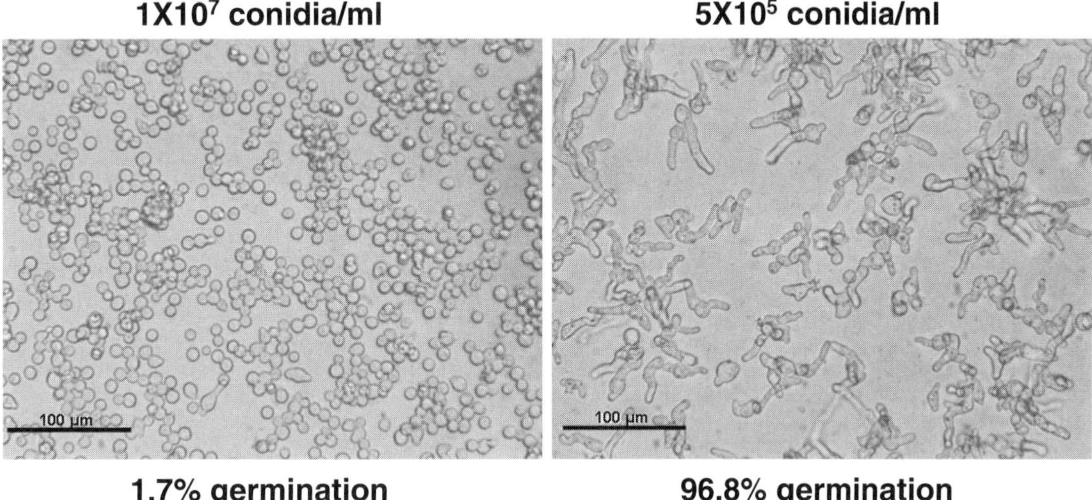

Figure 1. Germination of *A. fumigatus* conidia is inhibited at high conidial concentrations. *A. fumigatus* conidia at high (1×10^7 conidia/ml) and low (5×10^5 conidia/ml) concentrations were incubated for 8 h at 37°C in YAG rich medium, and the percentage of germinating conidia with emergent hyphal tubes was counted ($n = 300$).

1971; Hollomon, 1970; Mirkes, 1974; Schmit and Brody, 1976). RNA synthesis does not appear to be essential for early germination to proceed: the addition of RNA polymerase inhibitors fails to block conidial swelling in *A. nidulans* and *N. crassa*. However, definitive studies with RNA polymerase conditional mutants have not been carried out. In contrast, protein synthesis is an absolute prerequisite for conidial germination. Conidia of *A. fumigatus, A. nidulans*, and *N. crassa* display no signs of germination in the presence of cycloheximide, an inhibitor of protein synthesis (Fortwendel et al., 2004; Inoue and Ishikawa, 1970; Osherov and May, 2000; Schmit and Brody, 1976). *A. nidulans* and *N. crassa* mutants that are temperature sensitive for protein synthesis fail to germinate at the restrictive temperature, showing no signs of swelling, nuclear decondensation, or adhesion (Loo et al., 1981; Osherov and May, 2000). Germination resumes normally if the conidia are returned to the permissive temperature.

Polysome assembly is another early event in germination, at least in *N. crassa*. In dry resting *N. crassa* conidia, most of the ribosomes are freely dispersed. Less than 3% sediment as polysomes. Within 15 min of suspending the conidia in medium, however, 20 to 40% of the ribosomes are assembled into actively synthesizing polysomes (Mirkes, 1974; Mirkes and McCalley, 1976).

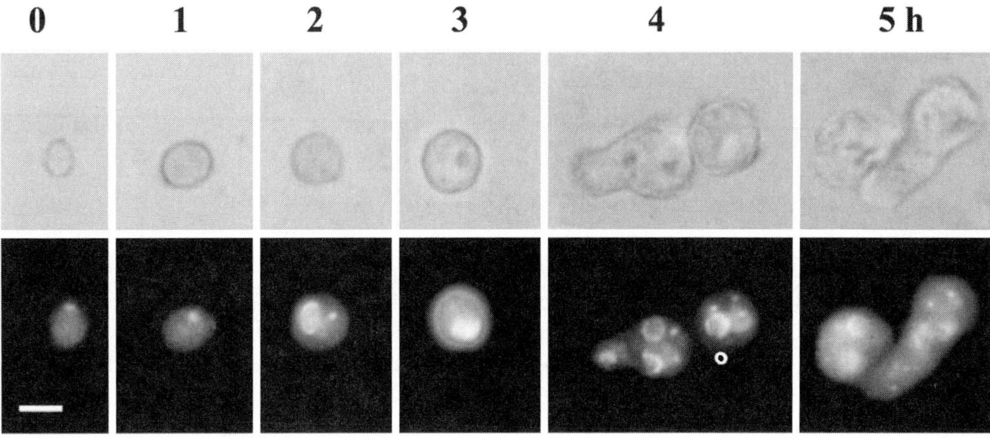

Figure 2. Germination of *A. fumigatus* conidia. Cells were fixed on glass coverslips and visualized following calcofluor (cell wall) and 4′,6′-diamidino-2-phenylindole (nuclear) staining. The morphological and biochemical changes that occur during the first 2 h are the subject of this review.

These results suggest that dormant conidia contain a preexisting pool of mRNA and ribosomes, primed for activation and translation in the presence of nutrients (see "Genetic Analysis of Conidial Germination" below).

The conidia of numerous fungi, including *N. crassa* and *A. nidulans*, contain high levels of polyols, including glycerol, mannitol, and the disaccharide trehalose. Upon induction of germination, the trehalose and mannitol pools are rapidly degraded and a glycerol pool is transiently accumulated (d'Enfert and Fontaine, 1997; Witteveen and Visser, 1995). Trehalose breakdown is not essential for germination to proceed. Deletion of the *A. nidulans treB* and *tpsA* genes, responsible for conidial trehalose mobilization and trehalose biosynthesis, respectively, did not substantially affect germination (d'Enfert et al., 1999; Fillinger et al., 2001). Mutant *tpsA* conidia show increased sensitivity to moderate stress conditions and rapid loss of viability upon storage, suggesting that intracellular trehalose is important for the acquisition of stress tolerance and for conidial stability. An analysis of trehalose breakdown and the pathways controlling it has provided insights into the signaling pathways controlling germination (see "Candidate Gene Approach" below).

GENETIC ANALYSIS OF CONIDIAL GERMINATION

The genetic analysis of spore germination has been applied successfully in several microorganisms, including *Bacillus subtilis, Bacillus cereus*, and *Dictyostelium discoideum* (Moir, 2006; Setlow, 2003; Xu et al., 2004). Isolation of germination-deficient *Bacillus* mutants has identified key carbon-sensing putative receptors, ion transporters, and cortex-lysing enzymes (Moir and Smith, 1990). Although genetic analysis is not feasible in *A. fumigatus*, because it lacks a known sexual cycle, it can be carried out in closely related sexually reproducing species such as *A. nidulans* (Osherov and May, 2000). The aim of the genetic approach is to isolate conditional germination mutants and clone the responsible gene. First, freshly harvested spores undergo saturating mutagenesis and an expression cycle in which the mutagenized conidia are allowed to germinate and sporulate. This step guarantees that the conidia subsequently used in the screening contain mutant versions of the protein. Mutagenized conidia are replica plated and a visual screen is carried out to identify temperature-sensitive nongerminating mutants. Enrichment can be used before the screen to ensure that the large number of "general" temperature-sensitive mutants that are often found are excluded. For example, Osherov and May (2000) used nystatin to selectively kill germinating conidia incubated at the restrictive temperature, thereby enriching for spores that did not germinate. It is also important to differentiate between (i) those mutants that, following a shift to the restrictive temperature, are specifically blocked in germination alone (i.e., mutated in a germination-specific gene) and those that are blocked at all stages of growth (i.e., mutated in essential housekeeping genes) and (ii) those mutants that are blocked in germination on a specific carbon source but that germinate normally on others (suggesting a mutation in a specific "carbon receptor").

Genetic analysis of conidial germination in *N. crassa* and *A. nidulans* has met with only limited success (Osherov and May, 2000; Schmit and Brody, 1976) (unpublished data). In *A. nidulans*, 12 spore germination-deficient (*sgd*) heat-sensitive mutants were identified from 12,000 colonies screened. Complementation analysis showed that the 12 *sgd* mutants defined eight genes, of which two (*sgdA* and *sgdB*) were identified multiple times, indicating that the screen had approached saturation. None of the *sgd* genes was germination specific, as they were necessary for both germination and vegetative growth. Five of them (*sgdA to -E*) were cloned by complementation. Three of these five genes encoded proteins involved in translational initiation (*sgdA*) and elongation (*sgdB* and *-C*), and a fourth (*sgdE*) encoded a protein involved in protein stabilization and folding. Thus, out of a bewildering array of early biochemical events, an unbiased genetic approach was able to specifically highlight a central role for translation in conidial germination. The results of this genetic screen suggest that conidial germination may be controlled by signaling pathways that directly activate protein translation. Translational initiation in eukaryotes is primarily regulated by phosphorylation of the α subunit of eukaryotic initiation factor 2 (eIF2α) and the eIF4E-binding proteins (Sunnerhagen, 2007). This phosphorylation is regulated by a number of regulatory pathways in budding yeast and other organisms: the TOR pathway (Inoki et al., 2005), the stress-activated MAPKs (Swaminathan et al., 2006), the cAMP/PKA pathway (Ashe et al., 2000), and the general amino acid control pathway (Hinnebusch, 2005). Why were these pathways not identified in the genetic screen? A plausible explanation is that because several independent pathways regulate translational initiation, inactivation of a single pathway is not sufficient to completely obstruct the process and, with this, to block germination. It would be interesting to identify and clone the homologs of these genes in *A. fumigatus* or *A. nidulans* and assess their involvement, singly and in combination, in conidial germination.

GENOMIC APPROACHES TO ANALYZING CONIDIAL GERMINATION

In *N. crassa* and *Aspergillus oryzae*, dormant conidia contain high levels of free ribosomes which, in the presence of a carbon source, associate with mRNA within 15 min to form polysomes (Horikoshi et al., 1965; Mirkes, 1974). These results suggest that dormant conidia contain a preexisting pool of mRNA and ribosomes, primed for rapid activation and translation in the presence of nutrients. The identity of the conidium-specific transcripts could prove instructive, possibly revealing which proteins need to be translated at the very beginning of germination. Pioneering studies by Timberlake and coworkers showed that about 300 such genes exist in *A. nidulans*. Most of the clones were isolated but their sequences were not identified (Timberlake, 1980; Zimmermann et al., 1980). Approximately 80% of these genes are organized in clusters (Orr and Timberlake, 1982). One of these, the *spoC1* cluster, has been studied in detail. It contains 14 conidium-specific genes of unknown function. Deletion of the entire region or of specific genes has no phenotypic effect (Aramayo et al., 1989; Stephens et al., 1999).

We used suppressive subtractive hybridization to identify and clone 12 transcripts (*c*onidia-*e*nriched *t*ranscripts *cetA* to *-L*) that are stored in dormant *A. nidulans* conidia and disappear during germination (Osherov et al., 2002). Based on sequence homology, one of these genes, *cetA*, is weakly similar to plant thaumatin-like proteins, which exhibit glucanase activity and antifungal properties (Grenier at al., 1999), and four genes encode metabolic enzymes involved in the synthesis of glucose, carbohydrates, nucleic acids, and amino acid precursors. The functions of the rest are unknown. Three *cetA*-like homologs are found in *A. fumigatus* (Afu3g09690, Afu8g01710, and Afu3g00510), and although none is stored in dormant conidia, two are expressed during early germination (Greenstein et al., 2006). *A. nidulans* contains another *cetA*-like gene, *calA* (AN7619.2), whose transcript is not present in dormant conidia but only during germination (Greenstein et al., 2006). Both *cetA* and *calA* encode small secreted proteins that accumulate in the spore cell wall and surrounding matrix during early germination (Belaish et al., 2008; Greenstein et al., 2006). The *cetA* and *calA* genes are synthetically lethal: whereas deletion of either one results in no obvious phenotype, deletion of both genes is lethal, resulting in inhibition of germination at an early stage, preceding nuclear decondensation and mitosis. Most of the double mutant conidia collapse and lyse during the first few hours of germination and show profound defects in their cell wall architecture (Fig. 3). Based on their sequence homology to plant thaumatin-like proteins with glucanase activity, the CETA and CALA proteins may have a similar activity in *A. nidulans* during germination, perhaps acting as cell wall-softening agents by binding or hydrolyzing conidial β-1,3-glucans.

Several recent studies have analyzed the "conidial transcriptome" in various fungi during dormancy and

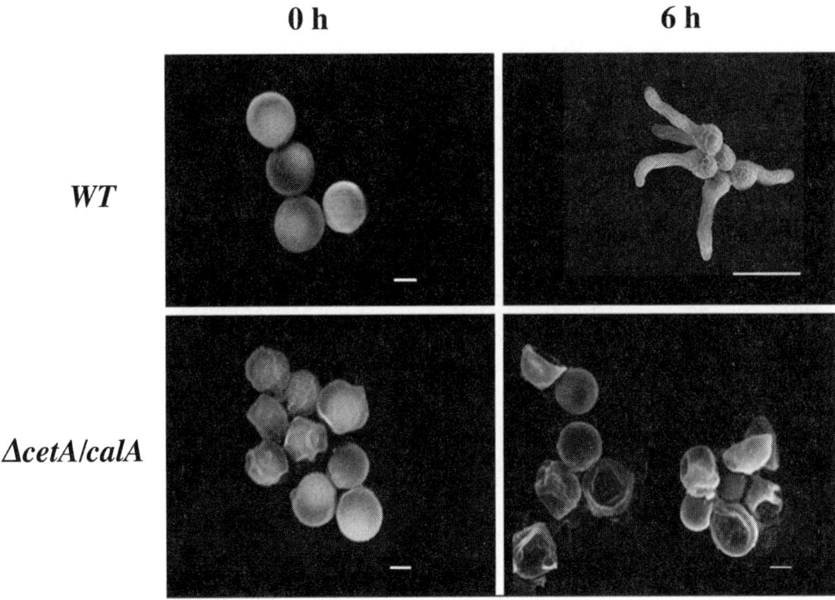

Figure 3. The *cetA/calA*-K/O1 double mutant has a defective cell wall and autolyses during germination. Dormant wild-type (WT) control and *cetA/calA*-K/O1 mutant conidia (0 h) and conidia that were allowed to germinate in liquid minimal medium for 6 h at 37°C (6 h) were fixed and analyzed by scanning electron microscopy.

germination with microarrays. These fungi include *Fusarium oxysporum* (Deng et al., 2006), *N. crassa* (Kasuga et al. 2005), and *Trychophytum rubrum* (Liu et al., 2007). Analysis has also been undertaken in *A. nidulans* and *A. fumigatus* but is still unpublished (M. Momany, personal communication). Some generalizations can be made regarding the genes identified in these fungi. (i) There is only a handful of truly conidium-specific transcripts in each of these organisms. (ii) Most have not been assigned a function, and almost none have been studied. (iii) The few annotated genes among them suggest that they are primarily involved in glucose and fatty acid metabolism, nutrient uptake, defense against oxidative stress, and cell wall remodeling. (iv) Each fungal species exhibits primarily its own unique set of conidiation-specific transcripts, with only a few common genes among the species.

An analysis of the changes that occur in the conidial proteome during germination, although now feasible, has not yet been carried out. Likewise, a genome-wide screen to identify germination-deficient mutants will be informative once such collections are prepared. A genome-wide screen for ascospore germination-deficient mutants in *Saccharomyces cerevisiae* identified 158 postgermination-defective strains of which two, a glucose-regulated pump and a mannosyltransferase, may play a specific role in germination (Deutschbauer et al. 2002).

CANDIDATE GENE APPROACH

The "candidate gene" approach has looked at signaling pathways with a high probability of involvement in conidial germination. They include the cAMP/PKA, RAS, MAPK, HOG-histidine kinase, and calcium-calcineurin signaling pathways. Evidence for and against their involvement in conidial germination is described below.

MAPK signaling in *A. fumigatus* is comprehensively described in chapter 13 and will not be dealt with in this chapter. Here will be specifically mentioned key points of MAPK signaling related to germination that are not covered in chapter 13.

cAMP Signaling

cAMP signaling in *A. fumigatus* is comprehensively described in chapter 13. Our discussion here will be limited to analyzing the findings that link cAMP signaling to conidial germination in *Aspergillus* species. Taking into account that cAMP signaling controls many of the events that are linked to the resumption of fungal growth (Santangelo et al., 2006), one would expect that gene deletions in the cAMP pathway would lead to a block or delay in germination for positively acting elements and precocious or rapid germination for negatively acting elements. Experimental results seem to bear out this simple prediction: deletion of the *A. nidulans ganB* gene (which encodes the G protein Gα subunit), *pkaA* (encoding the PKA catalytic subunit), or *cyaA* (encoding adenylate cyclase) resulted in delayed germination and trehalose breakdown (Chang et al., 2004; Fillinger et al., 2001; Lafon et al., 2005; Shimizu and Keller, 2001). However, these mutants also displayed extensive morphological defects, including reduced hyphal growth and changes in colony morphology and sporulation, highlighting the pleiotropic function of the cAMP/PKA pathway. Because this pathway plays a central role throughout the fungal life cycle, it is possible that the delay in germination and trehalose breakdown may simply mirror a general loss of viability. However, two important findings with gain-of-function mutants suggest that this is not the case: deletion of *A. nidulans rgsA*, which downregulates *ganB* G-protein activity, or constitutive activation of *ganB* by point mutation resulted in conidial germination in the absence of a carbon source (Chang et al., 2004; Lafon et al., 2005). These results show that constitutive activation of the cAMP/PKA pathway bypasses the need for glucose as a signal and results in precocious germination.

Investigation into the cAMP/PKA pathway homologs in *A. fumigatus* and their involvement in conidial germination has been limited. Deletion of *pkaC*, which encodes the PKA catalytic subunit *pkaR*, the PKA regulatory subunit, *acyA* adenylate cyclase, and *gpaB*, the G-protein α-subunit-encoding gene, resulted in reduced sporulation, slow growth (except *gpaB*), and lowered germination rates (Liebmann et al., 2003, 2004; Zhao et al., 2006). However, the phenotype was not correlated to the rate of trehalose breakdown, and gain-of-function mutants were not generated as they were in *A. nidulans*.

The putative glucose receptor has not been identified in *Aspergillus*. In yeast, extracellular glucose is sensed by the G-protein-coupled receptor Gpr1p, which subsequently activates the PKA pathway (Rolland et al. 2002; Santangelo, 2006). However, the *A. nidulans* and *A. fumigatus* genomes do not contain a similar gene. *A. nidulans* contains nine putative G-protein-coupled receptors (*gprA* to *-L*). Six of them have been disrupted, but only three (*gprA*, *-B*, and *-D*) resulted in a clear mutant phenotype (Han et al., 2004); only the deletion of *gprD* was described. Strains with *gprD* deleted were primarily affected in uncontrolled activation of sexual development. Although germination was also delayed, this was most likely an indirect effect.

It is also worth noting a lesson from *S. cerevisiae*: glucose sensing in this fungus employs a whole range of pathways and mechanisms, including direct activation of glycolytic enzymes by glucose intermediary metabolites, activation through the hexose transporters Rgt2p and Snf3p, and activation of the PKA pathway as described above (reviewed by Rolland et al., 2002). Similarly, carbon sensing in *Aspergillus* species may employ multiple sensors and pathways to induce germination.

RAS Signaling

Homologs of the RAS family of GTPase proteins affect morphology and virulence in several pathogenic fungi (Lengeler et al., 2000). Deletion of components of the *ras2*-signaling pathway in *S. cerevisiae* blocks ascospore germination, supporting a conserved role for RAS proteins in the germination of fungal spores (Herman and Rine, 1997).

In both *A. nidulans* and *A. fumigatus*, overexpression of a dominant negative form of *rasA* causes a delay in germination, whereas overexpression of dominant active *rasA* causes initiation of germination, including spore swelling, adhesion, and nuclear decondensation in the absence of a carbon source (Fortwendel et al., 2004; Osherov and May, 2000; Som and Kolaparthi, 1994). This suggests that activation of RAS bypasses the need for glucose as a germination trigger. RAS subsequently activates a downstream signaling pathway that initiates early germination. The downstream targets of RAS in the filamentous fungi remain unknown.

There is apparently no cross talk between RAS and the cAMP/PKA pathway during conidial germination; epistatic analysis in *A. nidulans* suggests that *rasA* and the cAMP/PKA pathway activate germination separately, through different signaling cascades (Fillinger et al., 2001). This contrasts with *S. cerevisiae*, where the cAMP/PKA pathway is regulated by RAS proteins.

A. fumigatus contains another RAS-like gene, *rasB*, more closely related to *S. cerevisiae* Rsr1p, which is involved in bud site selection. Results after deletion of the *A. fumigatus rasB* gene suggest that it is primarily involved in hyphal tip maintenance and polarized growth (Fortwendel et al., 2004).

Histidine Kinase Signaling

In fungi, histidine kinase (HK) signaling plays a vital role in response to changes in osmolarity and oxidative stress (Santos and Shiozaki, 2001). Two-component signaling systems are composed of an HK sensor and a response regulator (RR) protein. Signaling involves a phospho-relay from the histidine of the kinase sensor to the aspartic acid of the RR. High osmolarity or oxidative stress downregulate this activity, leading to the activation of a MAPK-signaling module and the induction of genes responsible for adaptation (Santos and Shiozaki, 2001). The *A. fumigatus* genome contains approximately 15 putative HK genes and two RRs. Only two of the HKs, *fos-1* and *tcsB*, have been studied. Deletion of *fos-1* and *tcsB* resulted in no obvious phenotype, suggesting that there is some redundancy in this family (Du et al., 2006; Pott et al., 2000).

Interestingly, deletion of the two RRs in *A. nidulans, sskA* and *srrA*, resulted in reduced conidiophore density and pigmentation and poor viability during storage at 4°C in distilled water. These results indicate that the HK-signaling pathway is required for full sporulation and spore viability in *A. nidulans*.

Calcium Signaling

Although calcium signaling plays a clear role in fungal morphogenesis and development, its direct involvement in the signaling events taking place during early conidial germination is debatable. In *Aspergillus*, calcium signaling has been mainly analyzed in *A. nidulans* (Kahl and Means, 2003). Cytosolic calcium levels increase rapidly in response to various stimuli, including osmotic shock, mechanical perturbation, and intriguingly, activation of the PKA pathway (Bencina et al., 2005; Nelson et al., 2004). However, changes that occur in cytosolic Ca^{2+} concentrations during early germination have not been reported. Ca^{2+}-dependent signaling takes place primarily through the binding of calcium to calmodulin (CaM). In *A. nidulans* CaM can subsequently activate three Ca^{2+}/CaM-dependent kinases (*CMKA* to *-C*) and one phosphatase, calcineurin A (*CnA*). *CMKA* to *-C* are involved in cell cycle progression and play a role in cell cycle reentry during early germination. Deletion of *CMKA* or *CMKB* is lethal, leading to G_2 or G_1 arrest, respectively, and a complete block in germination (Dayton and Means, 1996; Joseph and Means, 2000). Interestingly, expression of constitutively active *CMKA* also prevents entry into G_1 and blocks conidial germination (Dayton et al., 1997). *CMKC* is not essential, but deletion of this gene also delays germination and G_1 entry (Joseph and Means, 2000). Taken together, the data suggest a role for *CMKA* to *-C* in cell cycle initiation, a relatively late stage of conidial germination.

Calcineurin is a serine/threonine-specific phosphatase heterodimer consisting of the catalytic subunit A and the Ca^{2+}/calmodulin-binding subunit B. Calcineurin is activated by binding of Ca^{2+}/CaM to the catalytic subunit, leading to the dephosphorylation and activation of downstream transcription factors. In *A. fumigatus*, deletion of *calA*, encoding the catalytic subunit of calcineurin, resulted in severe defects in growth extension, branching, and conidial architecture (da Silva Ferreira et al., 2006, 2007; Steinbach et al., 2006). Approximately half of the Δ*calA* conidia were abnormally large and some were tear shaped and lacked a nucleus. Mutant conidia formed long chains that were difficult to disrupt because of the presence of thick junctions between the cells. The surface morphology of the conidia was smooth and lacked the rodlet-containing protrusions seen in wild-type conidia. These results suggest that in *A. fumigatus* calcineurin plays an important role in both hyphal growth and conidiogenesis, but not in the control of early conidial germination.

CONIDIAL GERMINATION AND THE CELL WALL

The cell wall serves as a dynamic and flexible interface between the fungal cell and its environment. During conidial germination, the cell wall of *A. fumigatus* changes dramatically: the outer melanin-rich layer is remodeled and a new inner layer is produced (Bernard and Latgé, 2001; Latgé et al., 2005).

The chemical composition of the conidial cell wall is slightly different from the mycelial cell wall. It does not contain galactosamine, contains low levels of chitin (~2 versus 13% in mycelia) and α-1,3-glucan (19 versus 42% in mycelia), and has high levels of galactomannan (~10 to 15% versus 1 to 2% in mycelia) (Maubon et al., 2006). Recently, progress has been made in the analysis of cell wall-associated proteins (CWPs) and their contribution to conidial properties and to germination. The *A. fumigatus* genome contains 80 genes encoding putative glycosylphosphatidylinositol (GPI)-anchored CWPs (Nierman et al., 2005). Five of these were positively identified in the plasma membrane fraction of the *A. fumigatus* mycelium (Bruneau et al., 2001). A proteomic analysis of *A. fumigatus* conidial CWPs identified 26 proteins. Of these, 12 had a signal for secretion and one (RodAp) also contained a GPI anchor motif (Asif et al., 2006). The contribution of known and recently reported CWPs to conidial germination is discussed below.

Hydrophobins

The outermost cell wall layer of *Aspergillus* conidia is characterized by the presence of interwoven fascicles of clustered proteinaceous microfibrils called rodlets. The rodlets are primarily composed of hydrophobins: small (14- to 16-kDa) cysteine-rich hydrophobic proteins. They are abundantly expressed during conidiogenesis. The conidia of *A. fumigatus* contain two hydrophobins, RodAp and RodBp. Deletion of *rodA* resulted in the disappearance of the conidial rodlet microfibrils and the conidia's physiochemical properties changed markedly. They showed decreased hydrophobicity, increased clumping, and reduced aerosolization. Conidial adhesion to collagen- or albumin-coated surfaces was reduced (Thau et al., 1994). Deletion of *rodB* had no phenotypic effect, but deletion of both *rodA* and *rodB* resulted in a further smoothening of the conidial cell wall (Paris et al., 2003).

Taken together, these results indicate that the primary role of the hydrophobins in *A. fumigatus* is to keep the conidia dry and enable them to form an efficient aerosol. However, they do not appear to play any direct role in conidial germination.

Proteins Involved in Cell Wall Biosynthesis and CWPs Can Affect the Rate of Conidial Germination

Recently, several proteins involved in cell wall biosynthesis were implicated in controlling the speed of conidial germination in *A. fumigatus*. Deletion of AfuEcm33 (a plasma membrane protein which influences cell wall biosynthesis), Afu6g08990 (a CWP of unknown function), *ags3* (α-1,3-glucan synthase), or *chsG*/*chsE* (chitin synthases) led to rapid conidial germination and hyphal emergence relative to the wild type (Levdansky et al., 2007; Maubon et al., 2006; Mellado et al., 2003; Romano et al., 2006; Beauvais et al., 2005; Chabane et al., 2006) (Fig. 4). Deletion of Af*PigA*, which catalyzes the addition of the GPI anchor to CWPs, also resulted in rapid germination (Li et al., 2007). These results suggest that differences in the constitution of the cell wall can affect the rate of germination. One possibility is that these mutations weaken the cell wall, enabling uncoating, swelling, and nutrient uptake to proceed more rapidly. Interestingly, conidia of the *chsG*/*chsE* double mutant germinated in the presence of water alone, suggesting that the glucose-sensing apparatus can be circumvented by changes in the cell wall.

Conidial Adhesion

Germinating conidia markedly increase their adhesive properties. *A. fumigatus* conidia adhere to extracellular matrix proteins such as fibronectin, collagen, and laminins that are found in the host tissue. To date, no true adhesin has been cloned and identified in *A. fumigatus*. Two putative adhesins, a 72-kDa cell wall laminin-binding protein and a 32-kDa sialic acid-binding lectin, have been isolated from resting *A. fumigatus* conidia, but their identities remain unknown (Tronchin et al., 1997, 2002). Deletion of the *rodA* hydrophobin gene decreased adherence to collagen but not to laminin or fibrinogen (Thau et al., 1994). Deletion of the repeat-containing CWP Afu3g08990 resulted in decreased adhesion to A549 cell-derived extracellular matrix but not to laminin or polystyrene (Levdansky et al., 2007). Still, it is not clear in either case whether these genes encode true adhesins or if deletion resulted in nonspecific changes to the conidial surface that reduced adherence.

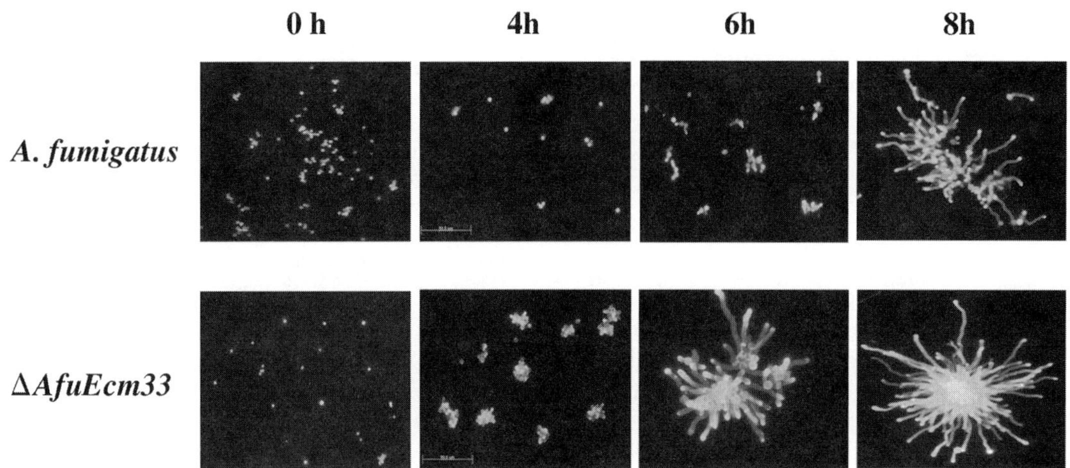

Figure 4. Disruption of AfuEcm33 results in rapid conidial germination. AF293 wild-type and AfuEcm33-disrupted cells were fixed on glass coverslips and visualized following calcofluor (cell wall) staining. Note the early germination and cell-cell clumping in the mutant strain (lower panel) relative to the control AF293 wild-type strain (upper panel).

In *A. fumigatus*, conidial adherence is also mediated by the presence of cell surface sialic acid. Sialic acids are negatively charged sugars that are covalently attached to glycosylated CWPs. Interestingly, pathogenic *Aspergillus* species such as *A. fumigatus* have greater amounts of sialic acids on their surface than their nonpathogenic counterparts, *Aspergillus ornatus*, *Aspergillus wentii*, and *Aspergillus auricomus* (Wasylnka et al., 2001). Enzymatic removal of conidial sialic acid in nongerminating conidia substantially reduced both adhesion to fibronectin and uptake by macrophage and lung epithelial cell lines (Warwas et al., 2007).

Recently, Beauvais et al. (2007) showed that an extracellular matrix surrounds and glues the hyphae of *A. fumigatus* growing on solid agar. This fungal extracellular matrix is composed of galactomannan, α-1,3-glucans, monosaccharides and polyols, melanin, and proteins, including major antigens and hydrophobins. It would be of great interest to analyze the formation of this matrix during conidial germination and early hy-

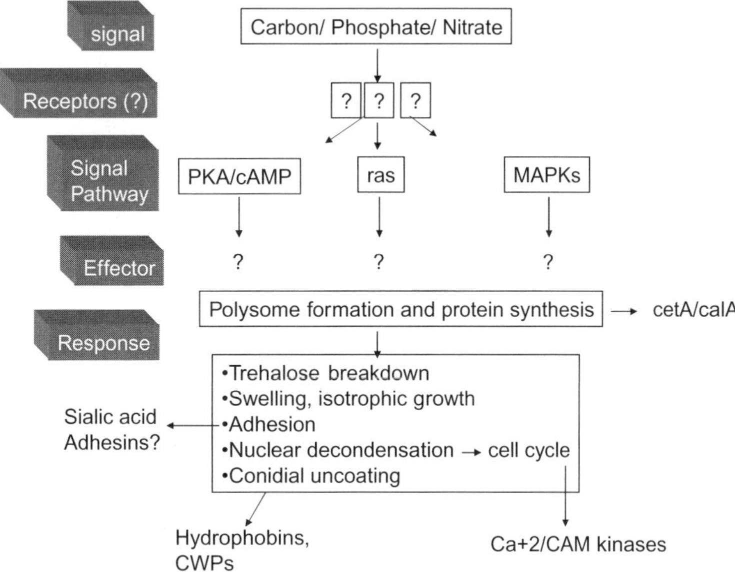

Figure 5. Tentative model of conidial germination in *A. fumigatus*. A carbon, phosphate, and nitrate source activates the cAMP/PKA, RAS, and MAPK pathways. The main bottleneck controlling conidial germination is the initiation of protein synthesis. Rapid assembly of polysomes onto prestored mRNA is followed by translation of key enzymes and proteins necessary for tighter adhesion, metabolic activation, conidial uncoating, nuclear decondensation, and isotrophic growth.

phal growth. If such a matrix were secreted during early growth, it might assist in conidial adhesion and possibly protect the spore during uncoating and early hyphal growth.

SUMMARY AND CONCLUSIONS

It is quite clear that conidial germination in *A. fumigatus* is a complex process. We are still collecting the pieces of the puzzle and do not yet have a clear understanding of how they fit together. Although the currently available data create only a fractured mosaic, several generalizations can be made (Fig. 5). The carbon source acts as a signal to activate the cAMP/PKA, RAS, and MAPK pathways. These pathways somehow connect to a host of biochemical and physiological responses, of which the synthesis of new proteins seems to be the most crucial for initiation of germination.

Exciting findings are emerging with respect to the identity of conidium-specific proteins that are involved in early germination. Genomic studies promise to reveal the identity of all conidium-specific transcripts and proteins. Only a systematic deletion of these genes will identify their contribution to germination. Subsequently, a detailed analysis of the transcriptional and translational control of some of these genes may reveal how they are regulated by the above-described pathways, as well as other signal transduction pathways.

Acknowledgments. I thank Haim Sharon and Jacob Romano for their help in the preparation of the figures and Gregory May for critical reading of the manuscript.

REFERENCES

Aramayo, R., T. H. Adams, and W. E. Timberlake. 1989. A large cluster of highly expressed genes is dispensable for growth and development in *Aspergillus nidulans*. *Genetics* **122:**65–71.

Ashe, M. P., S. K. De Long, and A. B. Sachs. 2000. Glucose depletion rapidly inhibits translation initiation in yeast. *Mol. Biol. Cell* **11:** 833–848.

Asif, A. R., M. Oellerich, V. W. Amstrong, B. Riemenschneider, M. Monod, and U. Reichard. 2006. Proteome of conidial surface associated proteins of *Aspergillus fumigatus* reflecting potential vaccine candidates and allergens. *J. Proteome Res.* **5:**954–962.

Bainbridge, B. W. 1971. Macromolecular composition and nuclear division during spore germination in *Aspergillus nidulans*. *J. Gen. Microbiol.* **66:**319–325.

Barrios-Gonzáles, J., C. Martinez, A. Aguilera, and M. Raimbault. 1989. Germination of concentrated suspensions of spores from *Aspergillus niger*. *Biotechnol. Lett.* **11:**551–554.

Beauvais, A., D. Maubon, S. Park, W. Morelle, M. Tanguy, M. Huerre, D. S. Perlin, and J. P. Latgé. 2005. Two α(1-3) glucan synthases with different functions in *Aspergillus fumigatus*. *Appl. Environ. Microbiol.* **71:**1531–1538.

Beauvais, A., C. Schmidt, S. Guadagnini, P. Roux, E. Perret, C. Henry, S. Paris, A. Mallet, M. C. Prevost, and J. P. Latgé. 2007. An extracellular matrix glues together the aerial-grown hyphae of *Aspergillus fumigatus*. *Cell. Microbiol.* **9:**1588–1600.

Belaish, R., H. Sharon, E. Levdansky, S. Greenstein, Y. Shadkchan, and N. Osherov. 2008. The *Aspergillus nidulans cetA* and *calA* genes are involved in conidial germination and cell wall morphogenesis. *Fungal Genet. Biol.* **45:**232–242.

Bencina, M., M. Legisa, and N. D. Read. 2005. Cross-talk between cAMP and calcium signaling in *Aspergillus niger*. *Mol. Microbiol.* **56:** 268–281.

Bernard, M., and J. P. Latgé. 2001. *Aspergillus fumigatus* cell wall: composition and biosynthesis. *Med. Mycol.* **39**(Suppl. 10)**:**9–17.

Bruneau, J. M., T. Magnin, E. Tagat, R. Legrand, M. Bernard, M. Diaquin, C. Fudali, and J. P. Latgé. 2001. Proteome analysis of *Aspergillus fumigatus* identifies glycosylphosphatidylinositol-anchored proteins associated to the cell wall biosynthesis. *Electrophoresis* **22:**2812–2823.

Chabane, S., J. Sarfati, O. Ibrahim-Granet, C. Du, C. Schmidt, I. Mouyna, M. C. Prevost, R. Calderone, and J. P. Latgé. 2006. Glycosylphosphatidylinositol-anchored Ecm33p influences conidial cell wall biosynthesis in *Aspergillus fumigatus*. *Appl. Environ. Microbiol.* **72:**3259–3267.

Chang, M. H., K. S. Chae, D. M. Han, and K. Y. Jahng. 2004. The *GanB* Gα-protein negatively regulates asexual sporulation and plays a positive role in conidial germination in *Aspergillus nidulans*. *Genetics* **167:**1305–1315.

Chitarra, G. S., T. Abee, F. M. Rombouts, M. A. Posthumus, and J. Dijksterhuis. 2004. Germination of *Penicillium paneum* conidia is regulated by 1-octen-3-ol, a volatile self-inhibitor. *Appl. Environ. Microbiol.* **70:**2823–2829.

da Silva Ferreira, M. E., T. Heinekamp, A. Hartl, A. A. Brakhage, C. P. Semighini, S. D. Harris, M. Savoldi, P. F. de Gouvea, M. H. Goldman, and G. H. Goldman. 2007. Functional characterization of the *Aspergillus fumigatus* calcineurin. *Fungal Genet. Biol.* **44:** 219–230.

da Silva Ferreira, M. E., M. R. Kress, M. Savoldi, M. H. Goldman, A. Hartl, T. Heinekamp, A. A. Brakhage, and G. H. Goldman. 2006. The *akuB*^{KU80} mutant deficient for nonhomologous end joining is a powerful tool for analyzing pathogenicity in *Aspergillus fumigatus*. *Eukaryot. Cell* **5:**207–211.

Dayton, J. S., and A. R. Means. 1996. Ca^{2+}/calmodulin-dependent kinase is essential for both growth and nuclear division in *Aspergillus nidulans*. *Mol. Biol. Cell* **7:**1511–1519.

Dayton, J. S., M. Sumi, N. N. Nanthakumar, and A. R. Means. 1997. Expression of a constitutively active Ca^{2+}/calmodulin-dependent kinase in *Aspergillus nidulans* spores prevents germination and entry into the cell cycle. *J. Biol. Chem.* **272:**3223–3230.

d'Enfert, C., B. M. Bonini, P. D. Zapella, T. Fontaine, A. M. da Silva, and H. F. Terenzi. 1999. Neutral trehalases catalyse intracellular trehalose breakdown in the filamentous fungi *Aspergillus nidulans* and *Neurospora crassa*. *Mol. Microbiol.* **32:**471–483.

d'Enfert, C., and T. Fontaine. 1997. Molecular characterization of the *Aspergillus nidulans treA* gene encoding an acid trehalase required for growth on trehalose. *Mol. Microbiol.* **24:**203–216.

Deng, Y., H. Dong, Q. Jin, C. Dai, Y. Fang, S. Liang, K. Wang, J. Shao, Y. Lou, W. Shi, D. J. Vakalounakis, and D. Li. 2006. Analysis of expressed sequence tag data and gene expression profiles involved in conidial germination of *Fusarium oxysporum*. *Appl. Environ. Microbiol.* **72:**1667–1671.

Deutschbauer, A. M., R. M. Williams, A. M. Chu, and R. W. Davis. 2002. Parallel phenotypic analysis of sporulation and postgermination growth in *Saccharomyces cerevisiae*. *Proc. Natl. Acad. Sci. USA* **99:**15530–15535.

Du, C., J. Sarfati, J. P. Latgé, and R. Calderone. 2006. The role of the *sakA* (Hog1) and *tcsB* (*sln1*) genes in the oxidant adaptation of *Aspergillus fumigatus*. *Med. Mycol.* **44:**211–218.

Fillinger, S., M. K. Chaveroche, P. van Dijck, R. de Vries, G. Ruijter, J. Thevelein, and C. d'Enfert. 2001. Trehalose is required for the acquisition of tolerance to a variety of stresses in the filamentous fungus *Aspergillus nidulans*. *Microbiology* **147**:1851–1862.

Fortwendel, J. R., J. C. Panepinto, A. E. Seitz, D. S. Askew, and J. C. Rhodes. 2004. *Aspergillus fumigatus rasA* and *rasB* regulate the timing and morphology of asexual development. *Fungal Genet. Biol.* **41**:129–139.

Gams, W., and J. A. Stalpers. 1994. Has the prehistoric ice-man contributed to the preservation of living fungal spores? *FEMS Microbiol. Lett.* **120**:9–10.

Gancedo, J. M. 1998. Yeast carbon catabolite repression. *Microbiol. Mol. Biol. Rev.* **62**:334–361.

Ghiorse, W. C., and M. R. Edwards. 1973. Ultrastructure of *Aspergillus fumigatus* conidia development and maturation. *Protoplasma* **76**:49–59.

Greenstein, S., Y. Shadkchan, J. Jadoun, C. Sharon, S. Markovich, and N. Osherov. 2006. Analysis of the *Aspergillus nidulans* thaumatin-like *cetA* gene and evidence for transcriptional repression of *pyr4* expression in the *cetA*-disrupted strain. *Fungal Genet. Biol.* **43**:42–53.

Grenier, J., C. Potvin, J. Trudel, and A. Asselin. 1999. Some thaumatin-like proteins hydrolyse polymeric beta-1,3-glucans. *Plant J.* **19**:473–480.

Griffin, D. H. 1994. *Fungal Physiology*, 2nd ed. Wiley-Liss, New York, NY.

Han, K. H., J. A. Seo, and J. H. Yu. 2004. A putative G protein-coupled receptor negatively controls sexual development in *Aspergillus nidulans*. *Mol. Microbiol.* **51**:1333–1345.

Harris, S. D. 1999. Morphogenesis is coordinated with nuclear division in germinating *Aspergillus nidulans* conidiospores. *Microbiology* **145**:2747–2756.

Herman, P. K, and J. Rine. 1997. Yeast spore germination: a requirement for Ras protein activity during re-entry into the cell cycle. *EMBO J.* **16**:6171–6181.

Hinnebusch, A. G. 2005. Translational regulation of GCN4 and the general amino acid control of yeast. *Annu. Rev. Microbiol.* **59**:407–450.

Hollomon, D. W. 1970. Ribonucleic acid synthesis during fungal spore germination. *J. Gen. Microbiol.* **62**:75–87.

Horikoshi, K., Y. Okitaka, and K. Ikeda. 1965. Ribosomes in dormant and germinating conidia of *Aspergillus oryzae*. *Agric. Biol. Chem.* **29**:724–727.

Inoki, K., H. Ouyang, Y. Li, and K. L. Guan. 2005. Signaling by target of rapamycin proteins in cell growth control. *Microbiol. Mol. Biol. Rev.* **69**:79–100.

Inoue, H., and T. Ishikawa. 1970. Macromolecule synthesis and germination of conidia in temperature sensitive mutants of *Neurospora crassa*. *Jpn. J. Genet.* **45**:357–369.

Joseph, J. D., and A. R. Means. 2000. Identification and characterization of two Ca^{2+}/CaM-dependent protein kinases required for normal nuclear division in *Aspergillus nidulans*. *J. Biol. Chem.* **275**: 38230–38238.

Kahl, C. R., and A. R. Means. 2003. Regulation of cell cycle progression by calcium/calmodulin-dependent pathways. *Endocr. Rev.* **24**:719–736.

Kasuga, T., J. P. Townsend, C. Tian, L. B. Gilbert, G. Mannhaupt, J. W. Taylor, and N. L. Glass. 2005. Long-oligomer microarray profiling in *Neurospora crassa* reveals the transcriptional program underlying biochemical and physiological events of conidial germination. *Nucleic Acids Res.* **33**:6469–6485.

Lafon, A., J. A. Seom, K. H. Han, J. H. Yu, and C. d'Enfert. 2005. The heterotrimeric G-protein *GanBα-SfaDβ-GpgAγ* is a carbon source sensor involved in early cAMP-dependent germination in *Aspergillus nidulans*. *Genetics* **171**:71–80.

Latgé, J. P., I. Mouyna, F. Tekaia, A. Beauvais, J. P. Debeaupuis, and W. Nierman. 2005. Specific molecular features in the organization and biosynthesis of the cell wall of *Aspergillus fumigatus*. *Med. Mycol.* 43(Suppl. 1):S15–S22.

Lengeler, K. B., R. C. Davidson, C. D'souza, T. Harashima, W. C. Shen, P. Wang, X. Pan, M. Waugh, and J. Heitman. 2000. Signal transduction cascades regulating fungal development and virulence. *Microbiol. Mol. Biol. Rev.* **64**:746–785.

Levdansky, E., J. Romano, Y. Shadkchan, H. Sharon, K. J. Verstrepen, G. R. Fink, and N. Osherov. 2007. Coding tandem repeats generate diversity in *Aspergillus fumigatus* genes. *Eukaryot. Cell* **6**:1380–1391.

Li, H., H. Zhou, Y. Luo, H. Ouyang, H. Hu, and C. Jin. 2007. Glycosylphosphatidylinositol (GPI) anchor is required in *Aspergillus fumigatus* for morphogenesis and virulence. *Mol. Microbiol.* **64**:1014–1027.

Liebmann, B., S. Gattung, B. Jahn, and A. A. Brakhage. 2003. cAMP signaling in *Aspergillus fumigatus* is involved in the regulation of the virulence gene *pksP* and in defense against killing by macrophages. *Mol. Genet. Genomics* **269**:420–435.

Liebmann, B., M. Muller, A. Braun, and A. A. Brakhage. 2004. The cyclic AMP-dependent protein kinase a network regulates development and virulence in *Aspergillus fumigatus*. *Infect. Immun.* **72**: 5193–5203.

Liu, T., Q. Zhang, L. Wang, L. Yu, W. Leng, J. Yang, L. Chen, J. Peng, L. Ma, J. Dong, X. Xu, Y. Xue, Y. Zhu, W. Zhang, L. Yang, W. Li, L. Sun, Z. Wan, G. Ding, F. Yu, K. Tu, Z. Qian, R. Li, Y. Shen, Y. Li, and Q. Jin. 2007. The use of global transcriptional analysis to reveal the biological and cellular events involved in distinct development phases of *Trichophyton rubrum* conidial germination. *BMC Genomics* **8**:100–109.

Loo, M. W., N. S. Schricker, and P. J. Russell. 1981. Heat-sensitive mutant strain of *Neurospora crassa*, 4M(t), conditionally defective in 25S ribosomal ribonucleic acid production. *Mol. Cell. Biol.* **1**: 199–207.

Ma, H., L. A. Snook, C. Tian, S. G. Kaminskyj, and T. E. Dahms. 2006. Fungal surface remodelling visualized by atomic force microscopy. *Mycol. Res.* **110**:879–886.

Macko, V., R. C. Staples, Z. Yaniv, and R. R. Granados. 1976. Self-inhibitors of fungal spore germination, p. 73–100. *In* D. J. Weber and W. M. Hess (ed.), *The Fungal Spore*. John Wiley, New York, NY.

Maubon, D., S. Park, M. Tanguy, M. Huerre, C. Schmitt, M. C. Prevost, D. S. Perlin, J. P. Latgé, and A. Beauvais. 2006. *AGS3*, an α(1-3)glucan synthase gene family member of *Aspergillus fumigatus*, modulates mycelium growth in the lung of experimentally infected mice. *Fungal Genet. Biol.* **43**:366–375.

Mellado, E., G. Dubreucq, P. Mol, J. Sarfati, S. Paris, M. Diaquin, D. W. Holden, J. L. Rodriguez-Tudela, and J. P. Latgé. 2003. Cell wall biogenesis in a double chitin synthase mutant ($chsG^-/chsE^-$) of *Aspergillus fumigatus*. *Fungal Genet. Biol.* **38**:98–109.

Mirkes, P. E. 1974. Polysomes, ribonucleic acid, and protein synthesis during germination of *Neurospora crassa* conidia. *J. Bacteriol.* **117**: 196–202.

Mirkes, P. E, and B. McCalley. 1976. Synthesis of polyadenylic acid-containing ribonucleic acid during the germination of *Neurospora crassa* conidia. *J. Bacteriol.* **125**:174–180.

Moir, A. 2006. How do spores germinate? *J. Appl. Microbiol.* **101**: 526–530.

Moir, A., and D. A. Smith. 1990. The genetics of bacterial spore germination. *Annu. Rev. Microbiol.* **44**:531–553.

Momany, M., R. Lindsey, T. W. Hill, E. A. Richardson, C. Momany, M. Pedreira, G. M. Guest, J. F. Fisher, R. B. Hessler, and K. A. Roberts. 2004. The *Aspergillus fumigatus* cell wall is organized in domains that are remodelled during polarity establishment. *Microbiology* **150**:3261–3268.

Momany, M., and I. Taylor. 2000. Landmarks in the early duplication cycles of *Aspergillus fumigatus* and *Aspergillus nidulans*: polarity, germ tube emergence and septation. *Microbiology* **146:**3279–3284.

Neirman, W. C., et al. 2005. Genomic sequence of the pathogenic and allergenic filamentous fungus *Aspergillus fumigatus*. *Nature* **438:**1151–1156.

Nelson, G., O. Kozlova-Zwinderman, A. J. Collis, M. R. Knight, J. R. Fincham, C. P. Stanger, A. Renwick, J. G. Hessing, P. J. Punt, C. A. van den Hondel, and N. D. Read. 2004. Calcium measurement in living filamentous fungi expressing codon-optimized aequorin. *Mol. Microbiol.* **52:**1437–1450.

Orr, W. C., and W. E. Timberlake. 1982. Clustering of spore-specific genes in *Aspergillus nidulans*. *Proc. Natl. Acad. Sci. USA* **79:**5976–5980.

Osherov, N., J. Mathew, A. Romans, and G. S. May. 2002. Identification of conidial-enriched transcripts in *Aspergillus nidulans* using suppression subtractive hybridization. *Fungal Genet. Biol.* **37:**197–204.

Osherov, N., and G. S. May. 2000. Conidial germination in *Aspergillus nidulans* requires RAS signaling and protein synthesis. *Genetics* **155:**647–656.

Paris, S., J. P. Debeaupuis, R. Crameri, M. Carey, F. Charles, M. C. Prevost, C. Schmitt, B. Philippe, and J. P. Latgé. 2003. Conidial hydrophobins of *Aspergillus fumigatus*. *Appl. Environ. Microbiol.* **69:**1581–1588.

Pott, G. B., T. K. Miller, J. A. Bartlett, J. S. Palas, and C. P. Selitrennikoff. 2000. The isolation of *FOS-1*, a gene encoding a putative two-component histidine kinase from *Aspergillus fumigatus*. *Fungal Genet. Biol.* **31:**55–67.

Rohde, M., M. Schwienbacher, T. Nikolaus, J. Heesemann, and F. Ebel. 2002. Detection of early phase specific surface appendages during germination of *Aspergillus fumigatus* conidia. *FEMS Microbiol. Lett.* **206:**99–105.

Rolland, F., J. Winderickx, and J. M. Thevelein. 2002. Glucose-sensing and -signalling mechanisms in yeast. *FEMS Yeast Res.* **2:**183–201.

Romano, J., G. Nimrod, N. Ben-Tal, Y. Shadkchan, K. Baruch, H. Sharon, and N. Osherov. 2006. Disruption of the *Aspergillus fumigatus ECM33* homologue results in rapid conidial germination, antifungal resistance and hypervirulence. *Microbiology* **152:**1919–1928.

Santangelo, G. M. 2006. Glucose signaling in *Saccharomyces cerevisiae*. *Microbiol. Mol. Biol. Rev.* **70:**253–282.

Santos, J. L, and K. Shiozaki. 2001. Fungal histidine kinases. *Sci. STKE* **98:**RE1.

Schmit, J. C., and S. Brody. 1976. Biochemical genetics of *Neurospora crassa* conidial germination. *Bacteriol. Rev.* **40:**1–41.

Setlow, P. 2003. Spore germination. *Curr. Opin. Microbiol.* **6:**550–556.

Shimizu, K., and N. P. Keller. 2001. Genetic involvement of a cAMP-dependent protein kinase in a G protein signaling pathway regulating morphological and chemical transitions in *Aspergillus nidulans*. *Genetics* **157:**591–600.

Som, T., and V. S. Kolaparthi. 1994. Developmental decisions in *Aspergillus nidulans* are modulated by Ras activity. *Mol. Cell. Biol.* **14:**5333–5348.

Steinbach, W. J., R. A. Cramer, Jr., B. Z. Perfect, Y. G. Asfaw, T. C. Sauer, L. K. Najvar, W. R. Kirkpatrick, T. F. Patterson, D. K. Benjamin, Jr., J. Heitman, and J. R. Perfect. 2006. Calcineurin controls growth, morphology, and pathogenicity in *Aspergillus fumigatus*. *Eukaryot. Cell* **5:**1091–1103.

Stephens, K. E., K. Y. Miller, and B. L. Miller. 1999. Functional analysis of DNA sequences required for conidium-specific expression of the *SpoI-C1C* gene of *Aspergillus nidulans*. *Fungal Genet. Biol.* **27:**231–242.

Sunnerhagen, P. 2007. Cytoplasmatic post-transcriptional regulation and intracellular signalling. *Mol. Genet. Genomics* **277:**341–355.

Swaminathan, S., T. Masek, C. Molin, M. Pospisek, and P. Sunnerhagen. 2006. Rck2 is required for reprogramming of ribosomes during oxidative stress. *Mol. Biol. Cell* **17:**1472–1482.

Taubitz, A., B. Bauer, J. Heesemann, and F. Ebel. 2007. Role of respiration in the germination process of the pathogenic mold *Aspergillus fumigatus*. *Curr. Microbiol.* **54:**354–360.

Thau, N., M. Monod, B. Crestani, C. Rolland, G. Tronchin, J. P. Latgé, and S. Paris. 1994. Rodletless mutants of *Aspergillus fumigatus*. *Infect. Immun.* **62:**4380–4388.

Timberlake, W. E. 1980. Developmental gene regulation in *Aspergillus nidulans*. *Dev. Biol.* **78:**497–510.

Tronchin, G., J. P. Bouchara, M. Ferron, G. Larcher, and D. Chabasse. 1995. Cell surface properties of *Aspergillus fumigatus* conidia: correlation between adherence, agglutination, and rearrangements of the cell wall. *Can. J. Microbiol.* **41:**714–721.

Tronchin, G., K. Esnault, G. Renier, R. Filmon, D. Chabasse, and J. P. Bouchara. 1997. Expression and identification of a laminin-binding protein in *Aspergillus fumigatus* conidia. *Infect. Immun.* **65:**9–15.

Tronchin, G., K. Esnault, M. Sanchez, G. Larcher, A. Marot-Leblond, and J. P. Bouchara. 2002. Purification and partial characterization of a 32-kilodalton sialic acid-specific lectin from *Aspergillus fumigatus*. *Infect. Immun.* **70:**6891–6895.

van Etten, J. L., K. R. Dahlbaerg, and G. M. Russo. 1983. Fungal spore germination, p. 235–266. *In* J. E. Smith (ed.), *Fungal Differentiation: a Contemporary Synthesis*. Marcel Dekker, New York, NY.

Warwas, M. L., J. N. Watson, A. J. Bennet, and M. M. Moore. 2007. Structure and role of sialic acids on the surface of *Aspergillus fumigatus* conidiospores. *Glycobiology* **17:**401–410.

Wasylnka, J. A., M. I. Simmer, and M. M. Moore. 2001. Differences in sialic acid density in pathogenic and non-pathogenic *Aspergillus* species. *Microbiology* **147:**869–877.

Witteveen, C. F., and J. Visser. 1995. Polyol pools in *Aspergillus niger*. *FEMS Microbiol. Lett.* **134:**57–62.

Xu, Q., M. Ibarra, D. Mahadeo, C. Shaw, E. Huang, A. Kuspa, D. Cotter, and G. Shaulski. 2004. Transcriptional transitions during *Dictyostelium* spore germination. *Eukaryot. Cell* **3:**1101–1110.

Zhao, W., J. C. Panepinto, J. R. Fortwendel, L. Fox, B. G. Oliver, D. S. Askew, and J. C. Rhodes. 2006. Deletion of the regulatory subunit of protein kinase A in *Aspergillus fumigatus* alters morphology, sensitivity to oxidative damage, and virulence. *Infect. Immun.* **74:**4865–4874.

Zimmermann, C. R., W. C. Orr, R. F. Leclerc, E. C. Barnard, and W. E. Timberlake. 1980. Molecular cloning and selection of genes regulated in *Aspergillus* development. *Cell* **21:**709–715.

Aspergillus fumigatus and Aspergillosis
Edited by J.-P. Latgé and W. J. Steinbach
©2009 ASM Press, Washington, DC

Chapter 11

Growth Polarity

MICHELLE MOMANY AND YAINITZA HERNÁNDEZ-RODRÍGUEZ

ASPERGILLOSIS AND POLAR GROWTH

All filamentous fungi use highly polar extension, adding new materials to the tips of hyphae and branches, to explore their environments in search of nutrients. Highly polar growth allows *Aspergillus fumigatus* to explore and invade blood vessels and tissues, resulting in the necrosis characteristic of invasive aspergillosis (Denning and Stevens, 1990; Latgé, 2001).

The small (3-μm) conidia of *A. fumigatus* are ubiquitous in the environment and are frequently inhaled (Beffa et al., 1998; Latgé, 2001; Pitt, 1994). In individuals with competent immune systems, alveolar macrophages and neutrophils internalize and destroy conidia before they cause disease. Conidia can also be internalized by cells which do not destroy them. Endothelial and epithelial cells have been shown to take up conidia (Lopes Bezerra and Filler, 2004; Paris et al., 1997; Wasylnka and Moore, 2002). Conidia within epithelial lung cells eventually germinate and grow. Their hyphae penetrate the cell from the inside and escape to the extracellular space (Wasylnka and Moore, 2002).

A. fumigatus conidia that survive to germinate eventually elaborate hyphae that invade blood vessels and from there disseminate to other sites in the host (Denning et al., 1992; Latgé, 1999) and continue to extend by polar growth, forming the filamentous, branching network common in histological sections from invasive aspergillosis patients.

A. FUMIGATUS POLAR GROWTH IN VITRO

Most of the more detailed knowledge we have of polar growth in *A. fumigatus* has come from in vitro characterization of early development in both *A. fumigatus* and *Aspergillus nidulans*. One thing that is clear from in vitro studies is that early development is characterized by predictable, synchronous switches between isotropic (round) and polar (tubular) morphologies. Dormant conidia are round with a uniform size and nuclei arrested in interphase (Bergen and Morris, 1983; Harris, 1997; Robinow and Canten, 1969). When *A. fumigatus* conidia are inoculated into medium containing a carbon source, they break dormancy and begin synchronous nuclear division and morphological development. Nuclear division and morphological development remain roughly synchronous for several nuclear divisions, at about 12 h in rich medium at 37°C (Fig. 1) (Momany and Taylor, 2000; A. Breakspear and M. Momany, unpublished data). After breaking dormancy conidia expand isotropically for approximately 4 h at 37°C. They polarize and send out the first germ tube by 6 h. Germ tubes continue polar growth to become hyphae, with the first septa forming near the bases of the hyphae by 10 h. At about the same time, the first branches emerge on the apical sides of septa. Hyphae continue polar growth, extending and branching to give rise to visible colonies by 24 h (Fig. 2). Within 36 h conidiophores develop. These asexual reproductive structures include several isotropic cell layers and produce isotropic dormant conidia, beginning the whole cycle again.

MECHANISMS OF FUNGAL POLARITY

Saccharomyces cerevisiae undergoes polar growth, although the growth is less dramatic than that seen in filamentous fungi (Pruyne and Bretscher, 2000). Unbudded yeast cells expand isotropically, switching to polar growth with bud emergence. For bud expansion, growth is once more isotropic. In a broad overview, polar growth in *S. cerevisiae* can be thought of as having three steps: (i) cortical markers specify the position for polar growth; (ii) signaling proteins such as the Rho GTPase

Michelle Momany and Yainitza Hernández-Rodríguez • Dept. of Plant Biology, University of Georgia, Athens, GA 30602.

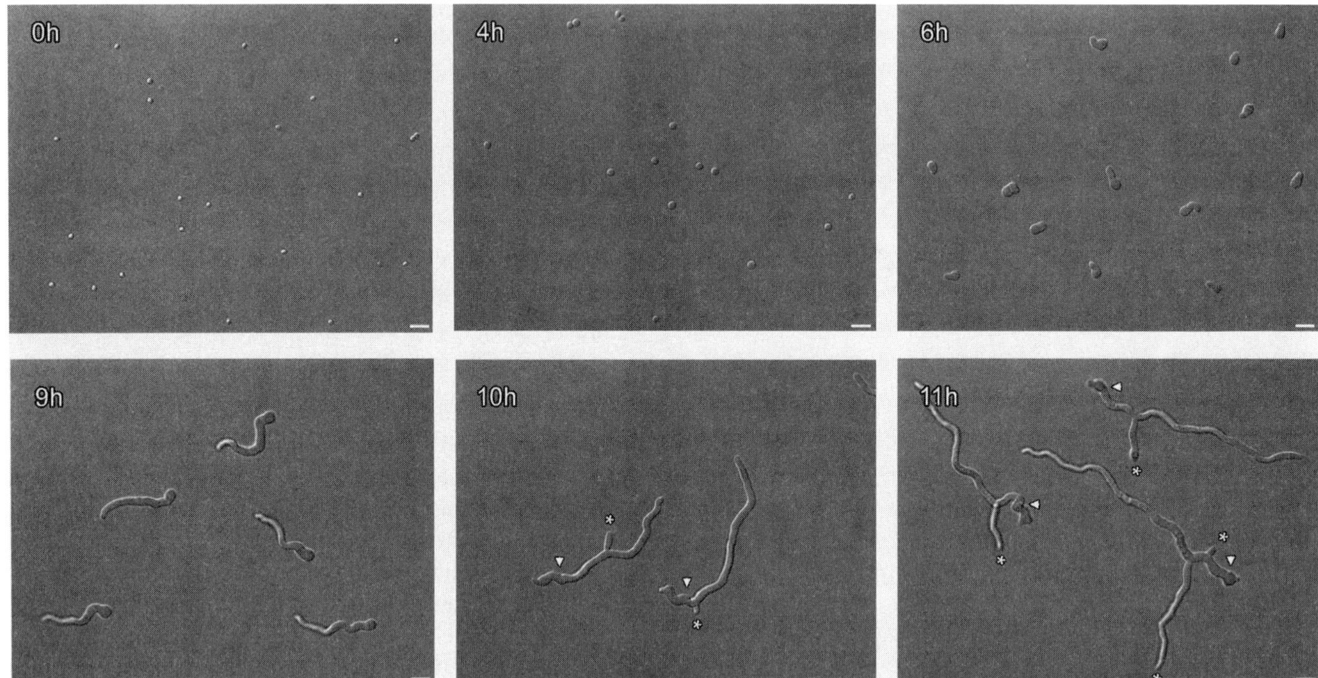

Figure 1. *A. fumigatus* conidia were inoculated into complete medium and incubated at 37°C. Samples were fixed at the time points indicated. Arrowheads indicate septa. Asterisks indicate branches. Bars, 10 μm. (Images courtesy of Andrew Breakspear.)

Cdc42 and its associated proteins relay this information to the morphogenetic machinery; (iii) the morphogenetic machinery deposits new cellular material in the appropriate area. The morphogenetic machinery includes everything needed to make and direct new cellular material: the cytoskeleton, secretory system, and cell wall biosynthetic machinery. Filamentous fungi generally lack homologs of the *S. cerevisiae* cortical markers, suggesting that cortical markers are very divergent in different fungi or that filamentous fungi use a different system for establishing positional information (Harris, 2006; Harris and Momany, 2004). One intriguing possibility is that the polarity axis in filamentous fungi is established in a stochastic manner and then becomes

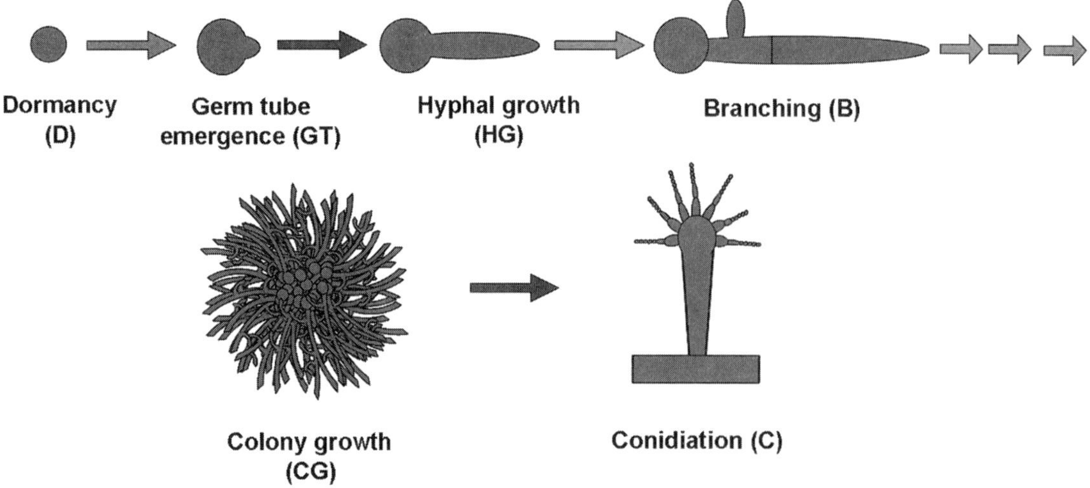

Figure 2. Diagrammatic representation of *A. fumigatus* growth stages. See text for details.

stabilized (Harris, 2006). In contrast to cortical markers, Rho GTPases and their associated proteins and the morphogenetic machinery proteins are highly conserved among fungi.

Despite sharing core machinery, important differences between polar growth in filamentous fungi and yeasts are beginning to emerge. A recent review of polar growth in filamentous fungi highlighted several of the unique features, most notably a role for microtubules in polar growth of filamentous fungi and the importance of the Spitzenkorpefer, a collection of vesicles at the tip that appears to be analogous to the polarisome (Harris, 2006). In the last few years, several *A. fumigatus* mutations have been described that affect the isotropic-to-polar switch at germ tube emergence or the ability to maintain polar growth once it is established in the hypha. These polarity-related mutations fall into one of two categories. They are either mutations in signaling genes or in cell wall genes. It should be noted that because filamentous fungi are characterized by polar extension, in many cases it is difficult to say whether the inability to properly polarize results from a polarity defect or is a consequence of a more general growth defect.

POLARITY-RELATED SIGNALING GENES

Calcineurin A is a serine/threonine-specific protein phosphatase involved in the antigenic response through calcium signaling. Deletion of the calcineurin A catalytic subunit, *cnaA*, or the Ca^{2+}-calmodulin binding subunit, *calA*, results in conidia with reduced surface rodlets and attenuated virulence. Deletion of *calA* also results in polarized conidia (a teardrop rather than round shape) that are unable to germinate (da Silva Ferreira et al., 2007; Steinbach et al., 2006).

rasA and *rasB* encode GTPases involved in morphogenesis and signaling. Mutation of *rasA* causes reduced polarization and other morphological abnormalities. Deletion of *rasB* results in delayed polarization and a failure to properly maintain polar growth seen as dichotomous branching. Deletion of *rasB* also results in a loss of virulence (Fortwendel et al., 2004, 2005).

mpkA encodes a mitogen-activated protein kinase involved in cell wall integrity and signaling (Du et al., 2006; Valiante et al., 2007). Deletion of *mpkA* results in broad, hyperpolar hyphae (hyperbranching) with no reduction in virulence.

Protein kinase A (PKA) is involved in cyclic AMP signaling, which is important for development and virulence. *pkaC1* encodes the catalytic subunit of PKA, and deletion of *pkaC1* results in a delay of polarization (Liebmann et al., 2004). Deletion of the regulatory subunit *pkaR* also results in delayed polarization along with broader germ tubes and hyphae (Zhao et al., 2006). Deletion of either PKA subunit also causes a loss of virulence.

POLARITY-RELATED CELL WALL GENES

The conidial cell wall is composed of chitin, galactomannan, α-1,3-glucan, and β-1,3-glucan (Latgé et al., 2005). *A. fumigatus* has seven different chitin synthase genes (Af*chsA*, Af*chsB*, Af*chsC*, Af*chsD*, Af*chsE*, Af*chsF*, and Af*chsG*). Strains with a deletion of any single chitin synthase gene do not show a phenotype. However, Δ*chsG* Δ*chsE* double deletion conidia show improper polarization, taking on a teardrop shape, and have less chitin and more α-1,3-glucans in their walls. Interestingly, the Δ*chsG* Δ*chsE* mutant germinates early but shows no reduction in virulence (Aufauvre-Brown et al., 1997; Mellado et al., 1996, 2003).

ags1, *ags2*, and *ags3* encode α-1,3-glucan synthases. Ags1 localizes to the cell wall periphery, while Ags2 localization is intracellular. Deletion of *ags1* results in reduced α-1,3-glucan in the cell wall. Deletion of *ags1* and *ags2* results in tip splitting, basically a failure to maintain polarity. Deletion of *ags1* and *ags2* does not affect virulence (Beauvais et al., 2005). Deletion of *ags3* results in early germination, an increase in melanin in the conidial cell wall, and increased virulence (Maubon et al., 2006).

gel1 and *gel2* encode glycosylphosphatidyinositol (GPI)-anchored β-1,3-glucanosyltranferases involved in cell wall biosynthesis. Though deletion of *gel1* shows no phenotype, deletion of *gel2* or both *gel1* and *gel2* results in hyperpolarized phialides on conidiophores which produce conidia without melanin. Virulence of these mutants is reduced (Mouyna et al., 2005).

ecm33 encodes a GPI-anchored protein involved in fungal morphology. Its deletion results in conidia with an increased diameter, defective separation, and increased chitin content (Chabane et al., 2006; Romano et al., 2006). Interestingly, Δ*ecm33* shows an early switch to polar growth (germination) and increased virulence.

Af*pig-a* encodes the catalytic subunit of the GPI–*N*-acetylglucosaminyl-transferase complex, which is involved in biosynthesis of GPI-anchored proteins. Deletion of Af*pig-a* results in swollen conidia, an early switch to polar growth (germination), cell lysis, and early conidiation. Interestingly, in this case early germination and conidiation don't result in increased virulence, but rather in decreased virulence (Li et al., 2007).

Af*pmt1* encodes a protein mannosyl transferase that adds mannose to serine or threonine residues of target proteins. In systems where targets have been identified, they are mostly cell wall or secreted proteins. De-

Table 1. Polarity-related genes in *A. fumigatus*[a]

Related function and gene(s)	Protein	Defects upon deletion	Deletion fully virulent?	Reference(s)
Signaling-related genes				
cnaA	Calcineurin A (Ser/Thr protein phosphatase) catalytic subunit	Conidia lack surface rodlets; abnormal hyphae; reduced colony growth and conidiation	No[b]	Steinbach et al., 2006
calA	Calcineurin A (Ser/Thr protein phosphatase) Ca^{2+}/calmodulin-binding unit	Conidia have reduced surface rodlets, some with abnormal polarized morphology (teardrop) and some unable to polarize; hyperpolarized hyphae (hyperbranching); reduced colony growth	No[b]	da Silva Ferreira et al., 2007
rasA	Ras GTPase	Dominant-negative mutants have reduced polarization; dominant-active mutants have reduced conidiation and abnormal conidiophores	NT	Fortwendel et al., 2004
rasB	Ras GTPase	Delayed polarization, failure to maintain polarized hyphae (dichotomous branching) and reduced colony growth	No[b]	Fortwendel et al., 2005
mpkA	MAP kinase	Thick hyperpolarized (hyperbranching) hyphae and reduced colony growth	Yes	Valiante et al., 2007
pkaC1	PKA catalytic subunit, cAMP signaling pathway	Delayed polarization and reduced colony growth and conidiation	No[b]	Liebmann et al., 2004
pkaR	PKA regulatory subunit	Reduced and delayed polarization, broader germ tubes and hyphae, reduced colony growth and conidiation, and abnormal conidiophores	No[b]	Zhao et al., 2006
Cell wall-related genes				
chsG and *chsE*	Chitin synthases	Double mutant has polarized conidia (teardrop); early switch to polar growth; reduced colony growth	Yes	Mellado et al., 2003
ags1 and *ags2*	α-1,3-Glucan synthase	Double deletion does not maintain polarized growth integrity (dichotomous branching); abnormal conidiophores and poor conidiation	Yes	Beauvais et al., 2005
ags3	α-1,3-Glucan synthase	Thicker conidial cell wall; increased melanin; early switch to polar growth	Yes, hypervirulent	Maubon et al., 2006
gel1 and *gel2*	GPI-anchored β-1,3-glucanosyl transferase	*gel2* or double deletion; no melanin in conidia; germination requires special medium; slow colony growth rate; some hyperpolarized phialides	No[b]	Mouyna et al., 2005
ecm33/*sps2*	GPI-anchored protein involved in fungal morphology	Increased conidial diameter; early switch to polar growth	Yes, hypervirulent	Chabane et al., 2006; Romano et al., 2006
Af*pig-a*	GPI-anchored protein; biosynthesis of GPI anchors	Swollen conidia; early switch to polar growth, cell lysis and early conidiation	No[b]	Li et al., 2007
Af*pmt1*	PMT O-mannosyl-transferases	Thin conidial cell wall; early switch to polar growth; reduced colony growth and poor conidiation at 42°C	Yes	Zhou et al., 2007

[a] NT, not tested; MAP, mitogen-activated protein; cAMP, cyclic AMP.
[b] Attenuated or no virulence.

letion of Af*pmt1* results in conidia that switch early to polar growth (early germination) and have thin walls. Strains with *Afpmt1* deleted are fully virulent (Zhou et al., 2007).

CONCLUSIONS

Because polar extension is critical to robust growth of filamentous fungi, one could predict that it is critical for virulence in invasive aspergillosis. Bearing in mind the caveats that it is often difficult to distinguish polarity defects from general growth defects and that the number of polarity-related genes so far examined is small, the recent literature supports the idea that polarity is needed for virulence. In most cases the deletion of polarity-related signaling genes leads to reduced or late polarization as measured by germ tube emergence. Deletion or mutation of *calA, rasA, rasB, pkaC1*, and *pkaR* all resulted in delayed or reduced germination and reduced virulence (Table 1). The only polarity-related signaling gene whose mutation did not lead to reduced polar growth was *mpkA*, which showed hyperpolarization. Consistent with the idea that polar growth is critical for virulence, Δ*mpkA* showed no reduction in virulence.

In most cases the deletion of polarity-related cell wall genes leads to early polarization as measured by germ tube emergence. Deletion of *chsG* and *chsE* or of *ags3*, *ecm33*, Af*pig-a*, or Af*pmt1* leads to early germ tube emergence (Table 1). With the exception of Δ-Af*pig-a*, these mutants were fully virulent. Indeed, two of the deletions with early germ tube emergence, Δ*ags3* and Δ*ecm33*, were more virulent than wild type. Δ-Af*pig-a*, the deletion strain that showed early germ tube emergence along with a decrease in virulence, also showed cell lysis, perhaps offsetting the advantages of early polarization.

At this point it is still difficult to say how much of the correlation between polarity and virulence is simply a correlation between rapid, robust growth of the pathogen and increased virulence. A better understanding of the mechanisms underlying polar growth of *A. fumigatus* in vitro and in vivo might allow the contributions of polarity and of robust growth to the organism's pathogenesis to be separated.

REFERENCES

Aufauvre-Brown, A., E. Mellado, N. A. R. Gow, and D. W. Holden. 1997. *Aspergillus fumigatus chsE*: a gene related to *CHS3* of *Saccharomyces cerevisiae* and important for hyphal growth and conidiophore development but not pathogenicity. *Fungal Genet. Biol.* **21:** 141–152.

Beauvais, A., D. Maubon, S. Park, W. Morelle, M. Tanguy, M. Huerre, D. S. Perlin, and J. P. Latgé. 2005. Two α(1-3) glucan synthases with different functions in *Aspergillus fumigatus. Appl. Environ. Microbiol.* **71:**1531–1538.

Beffa, T., F. Staib, J. Lott Fischer, P. F. Lyon, P. Gumowski, O. E. Marfenina, S. Dunoyer-Geindre, F. Georgen, R. Roch-Susuki, L. Gallaz, et al. 1998. Mycological control and surveillance of biological waste and compost. *Med. Mycol.* **36**(Suppl. 1):137–145.

Bergen, L. G., and N. R. Morris. 1983. Kinetics of the nuclear division cycle of *Aspergillus nidulans. J. Bacteriol.* **156:**155–160.

Chabane, S., J. Sarfati, O. Ibrahim-Granet, C. Du, C. Schmidt, I. Mouyna, M. C. Prevost, R. Calderone, and J. P. Latgé. 2006. Glycosylphosphatidylinositol-anchored Ecm33p influences conidial cell wall biosynthesis in *Aspergillus fumigatus. Appl. Environ. Microbiol.* **72:**3259–3267.

da Silva Ferreira, M. E., T. Heinekamp, A. Hartl, A. A. Brakhage, C. P. Semighini, S. D. Harris, M. Savoldi, P. F. de Gouvea, M. H. de Souza Goldman, and G. H. Goldman. 2007. Functional characterization of the *Aspergillus fumigatus* calcineurin. *Fungal Genet. Biol.* **44:**219–230.

Denning, D. W., and D. A. Stevens. 1990. Antifungal and surgical treatment of invasive aspergillosis: review of 2,121 published cases. *Rev. Infect. Dis.* **12:**1147–1200.

Denning, D. W., P. N. Ward, L. E. Fenelon, and E. W. Benbow. 1992. Lack of vessel wall elastolysis in human invasive pulmonary aspergillosis. *Infect. Immun.* **60:**5153–5156.

Du, C., J. Sarfati, J. P. Latgé, and R. Calderone. 2006. The role of the sakA (Hog1) and tcsB (sln1) genes in the oxidant adaptation of *Aspergillus fumigatus. Med. Mycol.* **44:**211–218.

Fortwendel, J. R., J. C. Panepinto, A. E. Seitz, D. S. Askew, and J. C. Rhodes. 2004. *Aspergillus fumigatus rasA* and *rasB* regulate the timing and morphology of asexual development. *Fungal Genet. Biol.* **41:**129–139.

Fortwendel, J. R., W. Zhao, R. Bhabhra, S. Park, D. S. Perlin, D. S. Askew, and J. C. Rhodes. 2005. A fungus-specific Ras homolog contributes to the hyphal growth and virulence of *Aspergillus fumigatus. Eukaryot. Cell* **4:**1982–1989.

Harris, S. D. 1997. The duplication cycle in *Aspergillus nidulans. Fungal Genet. Biol.* **22:**1–12.

Harris, S. D. 2006. Cell polarity in filamentous fungi: shaping the mold. *Int. Rev. Cytol.* **251:**41–77.

Harris, S. D., and M. Momany. 2004. Polarity in filamentous fungi: moving beyond the yeast paradigm. *Fungal Genet. Biol.* **41:**391–400.

Latgé, J. P. 1999. *Aspergillus fumigatus* and aspergillosis. *Clin. Microbiol. Rev.* **12:**310–350.

Latgé, J. P. 2001. The pathobiology of *Aspergillus fumigatus. Trends Microbiol.* **9:**382–389.

Latgé, J. P., I. Mouyna, F. Tekaia, A. Beauvais, J. P. Debeaupuis, and W. Nierman. 2005. Specific molecular features in the organization and biosynthesis of the cell wall of *Aspergillus fumigatus. Med. Mycol.* **43**(Suppl. 1):S15–S22.

Li, H., H. Zhou, Y. Luo, H. Ouyang, H. Hu, and C. Jin. 2007. Glycosylphosphatidylinositol (GPI) anchor is required in *Aspergillus fumigatus* for morphogenesis and virulence. *Mol. Microbiol.* **64:**1014–1027.

Liebmann, B., M. Muller, A. Braun, and A. A. Brakhage. 2004. The cyclic AMP-dependent protein kinase A network regulates development and virulence in *Aspergillus fumigatus. Infect. Immun.* **72:** 5193–5203.

Lopes Bezerra, L. M., and S. G. Filler. 2004. Interactions of *Aspergillus fumigatus* with endothelial cells: internalization, injury, and stimulation of tissue factor activity. *Blood* **103:**2143–2149.

Maubon, D., S. Park, M. Tanguy, M. Huerre, C. Schmitt, M. C. Prevost, D. S. Perlin, J. P. Latgé, and A. Beauvais. 2006. AGS3, an α(1-3)glucan synthase gene family member of *Aspergillus fumigatus*, modulates mycelium growth in the lung of experimentally infected mice. *Fungal Genet. Biol.* **43:**366–375.

Mellado, E., A. Aufauvre-Brown, N. A. R. Gow, and D. W. Holden. 1996. The *Aspergillus fumigatus chsC* and *chsG* genes encode class III chitin synthases with different functions. *Mol. Microbiol.* **20:** 667–679.

Mellado, E., G. Dubreucq, P. Mol, J. Sarfati, S. Paris, M. Diaquin, D. W. Holden, J. L. Rodriguez-Tudela, and J. P. Latgé. 2003. Cell wall biogenesis in a double chitin synthase mutant ($chsG^-/chsE^-$) of *Aspergillus fumigatus*. *Fungal Genet. Biol.* **38:**98–109.

Momany, M., and I. Taylor. 2000. Landmarks in the early duplication cycles of *Aspergillus fumigatus* and *Aspergillus nidulans*: polarity, germ tube emergence and septation. *Microbiology* **146:**3279–3284.

Mouyna, I., W. Morelle, M. Vai, M. Monod, B. Lechenne, T. Fontaine, A. Beauvais, J. Sarfati, M. C. Prevost, C. Henry, et al. 2005. Deletion of GEL2 encoding for a β(1-3)glucanosyltransferase affects morphogenesis and virulence in *Aspergillus fumigatus*. *Mol. Microbiol.* **56:**1675–1688.

Paris, S., E. Boisvieux-Ulrich, B. Crestani, O. Houcine, D. Taramelli, L. Lombardi, and J. P. Latgé. 1997. Internalization of *Aspergillus fumigatus* conidia by epithelial and endothelial cells. *Infect. Immun.* **65:**1510–1514.

Pitt, J. I. 1994. The current role of *Aspergillus* and *Penicillium* in human and animal health. *J. Med. Vet. Mycol.* **32**(Suppl. 1)**:**17–32.

Pruyne, D., and A. Bretscher. 2000. Polarization of cell growth in yeast. I. Establishment and maintenance of polarity states. *J. Cell Sci.* **113:**365–375.

Robinow, C. F., and C. E. Canten. 1969. Mitosis in *Aspergillus nidulans*. *J. Cell Sci.* **5:**403–431.

Romano, J., G. Nimrod, N. Ben-Tal, Y. Shadkchan, K. Baruch, H. Sharon, and N. Osherov. 2006. Disruption of the *Aspergillus fumigatus* ECM33 homologue results in rapid conidial germination, antifungal resistance and hypervirulence. *Microbiology* **152:**1919–1928.

Steinbach, W. J., R. A. Cramer, Jr., B. Z. Perfect, Y. G. Asfaw, T. C. Sauer, L. K. Najvar, W. R. Kirkpatrick, T. F. Patterson, D. K. Benjamin, Jr., J. Heitman, et al. 2006. Calcineurin controls growth, morphology, and pathogenicity in *Aspergillus fumigatus*. *Eukaryot. Cell* **5:**1091–1103.

Valiante, V., T. Heinekamp, R. Jain, A. Hartl, and A. A. Brakhage. 2007. The mitogen-activated protein kinase MpkA of *Aspergillus fumigatus* regulates cell wall signaling and oxidative stress response. *Fungal Genet. Biol.* **45:**618–627.

Wasylnka, J. A., and M. M. Moore. 2002. Uptake of *Aspergillus fumigatus* conidia by phagocytic and nonphagocytic cells in vitro: quantitation using strains expressing green fluorescent protein. *Infect. Immun.* **70:**3156–3163.

Zhao, W., J. C. Panepinto, J. R. Fortwendel, L. Fox, B. G. Oliver, D. S. Askew, and J. C. Rhodes. 2006. Deletion of the regulatory subunit of protein kinase A in *Aspergillus fumigatus* alters morphology, sensitivity to oxidative damage, and virulence. *Infect. Immun.* **74:**4865–4874.

Zhou, H., H. Hu, L. Zhang, R. Li, H. Ouyang, J. Ming, and C. Jin. 2007. O-Mannosyltransferase 1 in *Aspergillus fumigatus* (AfPmt1p) is crucial for cell wall integrity and conidium morphology, especially at an elevated temperature. *Eukaryot. Cell* **6:**2260–2268.

Aspergillus fumigatus and Aspergillosis
Edited by J.-P. Latgé and W. J. Steinbach

Chapter 12

Biofilm Formation in *Aspergillus fumigatus*

ANNE BEAUVAIS AND FRANK-MICHAEL MÜLLER

The first step of pathobiological development of *Aspergillus fumigatus* following the inhalation of airborne conidia is the colonization of the upper respiratory tract. In allergic bronchopulmonary aspergillosis, *A. fumigatus* develops on the bronchial epithelial cells; in aspergilloma, a ball of tightly associated hyphae is formed in a preexisting pulmonary cavity or in chronically obstructed paranasal sinuses; in invasive pulmonary aspergillosis, the most severe *Aspergillus* disease, the first step is the invasion of the alveolar epithelia. Common to all these developments is that the fungus encounters an aerial and static environment quite similar to the conditions found by the fungus during growth in a petri dish in vitro. In contrast, all studies undertaken to date with this fungal pathogen in vitro were performed with mycelia grown under shaken and submerged conditions, in flasks or in fermentors.

Biofilms are defined as a community of microorganisms that are attached to a surface and embedded in an extracellular polysaccharidic matrix. A biofilm is obtained by the development of many cells close together at the same time and in the same environment. Can a colony mat of *A. fumigatus* grown in vitro under static and aerial conditions fit this definition?

A major problem with *A. fumigatus* infections is the poor efficiency of antifungal compounds that are active in vivo. Growing as a multicellular community helps to colonize the substratum and resist external aggressions. This is well-known with bacterial and yeast biofilms, or flocculating yeasts. Established bacterial biofilms are 10 to 1,000 times more resistant to antimicrobials than equivalent planktonic cells and are also highly resistant to phagocytosis, making them difficult to eradicate from living hosts (Jefferson, 2004). Similarly, biofilms formed by different *Candida* species show an increase of the MIC_{50} of up to 10-fold to almost all antifungals, particularly azole drugs (d'Enfert, 2006). Flocs of *Saccharomyces cerevisiae* that are composed of yeast cells embedded in a matrix are resistant to fluconazole (A. Beauvais, unpublished results). Is an aerial colony more resistant to antifungals than planktonic cells?

Although bacterial biofilms have been known and studied for 100 years, and yeast biofilms for >10 years, no studies have been undertaken until recently to analyze the presence of a biofilm in *A. fumigatus* (O'Toole et al., 2000; Hawser and Douglas, 1994). The first studies on the development of filamentous fungi under static and aerial conditions were published in the early 1970s (Righelato, 1979). In some *Neurospora crassa* mutants, the biomass specific growth rate in submerged culture was less than the hyphal length specific growth rate determined on a monospore colony (Steele and Trinci, 1975). In contrast, a highly branched mutant of *Aspergillus nidulans* had a lower colony radial growth rate and a smaller peripheral growth zone than the wild type but an identical specific growth rate in submerged culture (Righelato, 1979). The morphology of the colony analyzed in these studies was, however, different from the biofilm analyzed by Beauvais et al. (2007). A colony is formed from one spore, whereas a biofilm originates from many cells. The structure of the mycelial mat is different in both cases. In a colony of any filamentous fungus the growth ceases at the center and the growth of the colony is due to the mycelia at the outer annulus (Prosser, 1994). In the biofilm, active growth was seen on the entire surface of the petri dish because a high conidial inoculum was used (Beauvais et al., 2007).

The understanding of the physiological mechanisms governing the cohesion between hyphae on solid medium is poor. Only the role of hydrophobicity during this process was shown in a study about the formation of aerial hyphae of the filamentous fungus *Schizophillum commune* (Wösten et al., 1999). The authors showed that hydrophobin SC3 is secreted at the surface of the cell wall and self-assembles as a rodlet layer, conferring hydrophobicity to these hyphae (Wösten et al., 1999). All these data suggest that the growth of the fungus under submerged or aerial conditions is different.

Anne Beauvais • Unité des Aspergillus, Institut Pasteur, 25, Rue du Dr Roux, 75015 Paris, France. **Frank-Michael Müller** • Päd. Pneumologie & Speciale Infektiologie, Zentrum für Kinder-u. Jugendmedizin, INF 153, D-69120 Heidelberg, Germany.

Another aspect of aerial hyphal network formation is that such a mat can serve as a substratum for the development of other microorganisms, such as bacteria. In cystic fibrosis patients, *A. fumigatus* is often found in close relationship with *Pseudomonas aeruginosa* in the bronchial airways. Mixed fungal and bacterial communities have been poorly studied, although the interactions between the two species can totally modify the behavior of each partner. Some bacteria can enhance the development of the fungus by secreting compounds that increase the virulence of the fungus, but some bacteria can secrete toxins to kill the fungus.

Such considerations have led to the analysis of putative biofilms produced during mycelial growth of *A. fumigatus* (i) under aerial and static conditions in a petri dish in a rich medium, (ii) on polystyrene in tissue culture plates in cell culture medium, or (iii) on bronchial epithelial cells in tissue culture plates in cell culture medium. These three conditions have in common a start from a pool of conidia and a static condition of growth. The resistance of the putative biofilm against antifungal drugs, in particular, has also been investigated, since aspergillomas are highly resistant to antifungal drugs and surgery is often the only way to eradicate the fungus.

The first studies of *A. fumigatus* growing from a pool of conidia as a biofilm, in vitro on agar medium or on polystyrene, were only published recently (Beauvais et al., 2007; Mowat et al., 2007) or soon will be submitted for publication (M. Seidler et al., unpublished results). The results of these studies are summarized in this chapter.

FORMATION, GROWTH, AND STRUCTURE OF *A. FUMIGATUS* BIOFILMS

Static and Aerial Conditions

The growth of *A. fumigatus* is usually performed in protein hydrolysate, glucose-rich medium in agar or in liquid. Fundamental differences exist between the two conditions: in agar medium, access to nutrients occurs only at the lower part of the mat, whereas the hyphae at the surface of the agar have a preferential access to air. Under these conditions, the fungus has to develop a system to drive the nutrients to the surface of the mat and avoid desiccation. Accordingly, many differences have been observed between mycelia growing under liquid submerged shaken conditions (ShS) or under agar aerial static ones (StA). First, the mycelial growth was higher under StA conditions, started much earlier, and lasted longer than under shaken conditions (Fig. 1) (Beauvais et al., 2007). In bacteria or yeast biofilms the general structure is reversed: the lower layer is attached to the surface (catheter) and has poor access to nutrients. In contrast to *A. fumigatus*, bacterial and yeast biofilms grow more slowly than in planktonic cultures because of the limited availability of nutrients (Lewis, 2001; Baillie and Douglas, 1998).

Morphological changes between the aerial and shaken conditions also occur. Until recently, investigations of the structure, growth, morphology, and development of biofilms were difficult because most techniques were inadequate. For electron microscopy, the techniques required sample dehydration and consequently the loss of many characteristics, such as the ultrastructure of the extracellular matrix (ECM). Recent advances in cryo-scanning electron microscopy and in confocal scanning laser microscopy now allow an extremely accurate analysis of the structure of a hydrated biofilm in three dimensions (3D). Under aerial conditions, a thick undissociable network of hyphae is produced, and this is not observed in shaken cultures. Under liquid unshaken conditions, the germ tubes reach the surface and develop the same aerial network of hyphae as under static aerial conditions (Beauvais et al., unpublished). The StA hydrophobic, strong, and nondissociable network of hyphae is difficult to observe under non-confocal microscopy because of the presence of

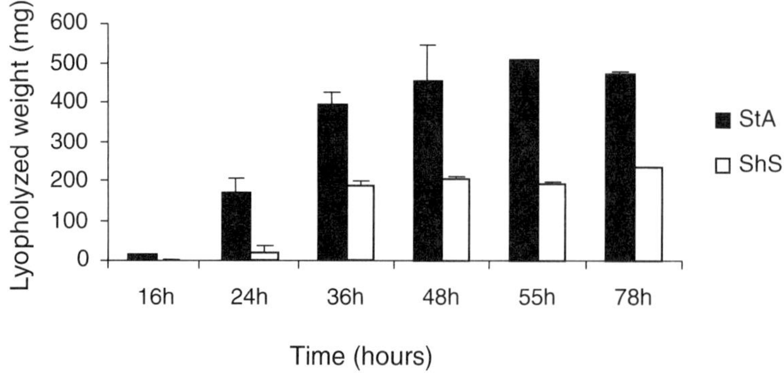

Figure 1. Growth of StA and ShS mycelia in 20 ml of culture medium containing 3% glucose–2% yeast extract, from 16 to 78 h (Beauvais et al., 2007).

a "glue" at the surface of the mycelium mat, whereas the ShS mycelium develops observable hydrophilic hyphae (Beauvais et al., 2007). Confocal microscopy shows that the StA mycelium develops an aerial stage which reaches 25 to 30 μm in thickness after 16 h and 100 to 150 μm after 24 h of growth. The development is vertical from the beginning of growth, explaining the depth of the network and the presence of channels between groups of hyphae (Color Plate 5). The extracellular material surrounding the hyphae is easily observed by cryo-scanning electron microscopy after 20 h of growth (Beauvais et al., 2007). This material increases in thickness with the age of the culture (Fig. 2). This matrix is probably necessary to avoid desiccation of the fungus. In a *Candida albicans* biofilm, the emergence of the ECM characterizes the intermediate development phase, whereas in *Cryptococcus neoformans*, the exopolymeric material is observed during the early maturation phase of the biofilm (Chandra et al., 2001; Martinez and Casadevall, 2006). Transmission electron microscopy studies of *A. fumigatus* biofilms have confirmed that inside the network, all hyphae are bound together by an electron-dense fibrillar matrix. Air channels with few ECM are also observed (Fig. 3). In contrast, with growth time disregarded, the ShS mycelium was totally devoid of any matrix and the hyphal surface was smooth (Fig. 4) (Beauvais et al., 2007). For all these reasons, the aerial network of hyphae developed under static conditions can be called a biofilm, as it is a community of hyphal cells embedded in an ECM.

All hyphae of the StA mat of a biofilm are alive and have the same morphology from the agar base to the upper layer in contact with the air and from the periphery to the inner center of the biofilm (Beauvais et al., 2007). As mentioned above, this is different from the growth of a colony originating from a single conidium, since biofilm hyphae have the same stage of differentiation at the periphery and are in a more central position inside the hyphal mat (Color Plate 5C). This is also different from *Candida* and bacterial biofilms. *Candida* biofilms are composed of a 3D structure consisting of yeast microcolonies produced in the early phase and hyphae and pseudohyphae produced in the later phases (Ramage et al., 2005). The presence of dormant cells inside the biofilm was also demonstrated (Boucherit et al., 2007). In bacterial biofilms, cells located in the upper regions are metabolically active and have a normal size, whereas cells unmeshed within the matrix are dormant and smaller in size (Anwar et al., 1992). Such morphological dimorphism does not exist in *A. fumigatus*.

The cell wall of *A. fumigatus* is different when the fungus is grown under static, aerial, or shaken conditions. The cell wall of *A. fumigatus* is mainly composed of polysaccharides: β-1,3-glucans, chitin, α-1,3-glucans, galactomannan, and a galactosaminogalactan polymer. β-1,3-Glucans represent 32% of the StA cell wall and only 21% of the ShS cell wall. This was associated with a similar decrease in the galactosaminogalactan polymer in the StA cell wall. No differences were observed in the other polysaccharides. The hyphae embedded in a biofilm have a slightly more rigid cell wall than in the shaken submerged medium, since the ratio of alkali-soluble and alkali-insoluble polymers increases from 0.3 to 0.4 for the StA and ShS cell wall, respectively (Beauvais et al., unpublished). Modifications of the cell wall during biofilm formation are not unique to *A. fumigatus*.

Figure 2. Cryo-scanning electron microscopy images of an *A. fumigatus* biofilm (StA; 30 h of growth) (Beauvais et al., 2007).

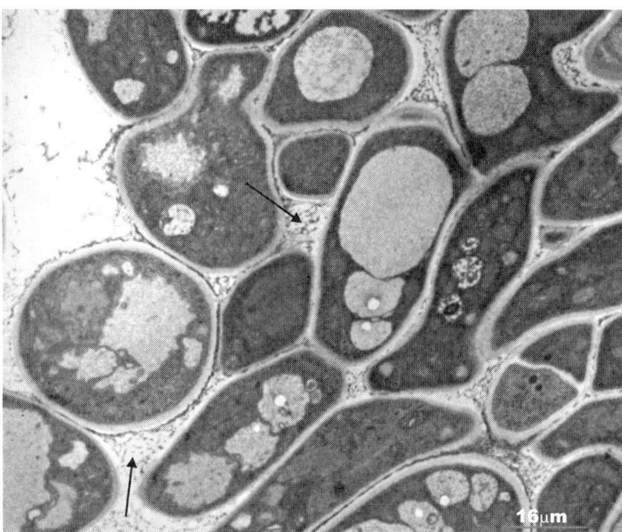

Figure 3. Ultrastructure of an *A. fumigatus* biofilm (StA; 24 h of growth). High magnification of the network of hyphae shows the electron-dense ECM on the surface of the cell wall (arrows) (Beauvais et al., 2007).

The cell walls of *C. albicans* biofilms are two times thicker than planktonic cells and contain about 40% more β-1,3-glucans (Nett et al., 2007).

Static and Submerged Conditions

The development of *A. fumigatus* on polystyrene on microtiter plates is quite similar to yeast and bacterial biofilms, since the lower layer is attached to the abiotic surface and the fungus is submerged in the medium. However, the culture conditions are also static. The conidial seeding density has critical importance for the metabolic activity and the biomass of the biofilm, and particularly for the depth of the resultant biofilm (Mowat et al., 2007). Biofilms formed at lower concentrations than 10^5 conidia/ml are thinner and have longer mycelial frameworks which are more easily removed by mechanical forces. At 10^5 conidia/ml, the growth kinetics starts with a lag phase (until 10 h), then an exponential rise in the depth of the biofilm (until 18 h), which reaches a plateau during the stationary developmental phase after 30 h of culture. As in agar-rich medium, the structural complexity of the biofilm increases after 16 h of culture (Mowat et al., 2007). After attachment of the conidia, the biofilm begins to mature and forms a hyphal network. This network spreads in every direction. When the culture matures for more than 72 h, the fruiting bodies can grow out off the medium surface (Seidler, unpublished).

For a better understanding of biofilm formation in chronic *A. fumigatus* infections (e.g., aspergilloma and allergic bronchopulmonary aspergillosis), an *A. fumigatus* biofilm was produced on human bronchial epithelial cells (16HBE). An *A. fumigatus* conidial suspension was added to confluent 16HBE14o- cells to let the fungus adhere to the cell surface (Seidler et al., 2006b). Biofilm production was evidenced by an increase of the dry weight after 48 h in comparison to a nonbiofilm control. By scanning electron microscopy the surface displays a highly coordinated network of hyphal structures (Fig. 5). The covering of the hyphal structures and in particular ending flow tubes is an indicator of biofilm production. Parallel packed hyphae strengthen the structure in one direction, while crossing hyphae further stabilize the structure. A self-produced matrix is seen between the hyphae. For confocal scanning laser microscopy, the polysaccharides in the cell wall were stained green by the fluorescence of concanavalin A (ConA)-Alexa Fluor 488. The cytoplast of the mold stained red with the fluorescent FUN-1 cell stain. ConA-positive material was observed surrounding conidia, hyphae, or free float-

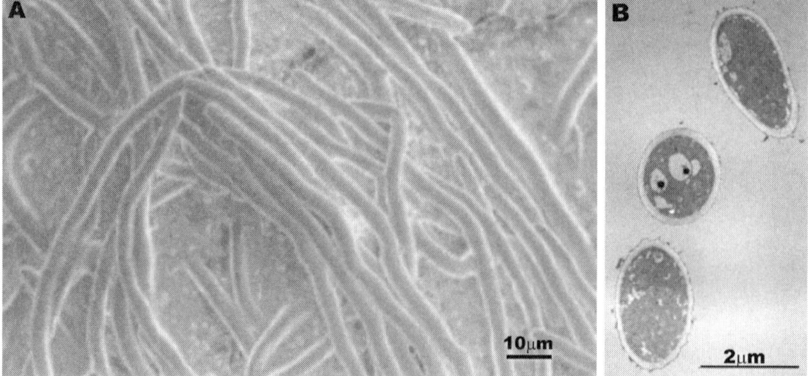

Figure 4. Cryo-scanning (A) and transmission (B) electron microscopy images of ShS mycelium (30 h of growth) (Beauvais et al., 2007).

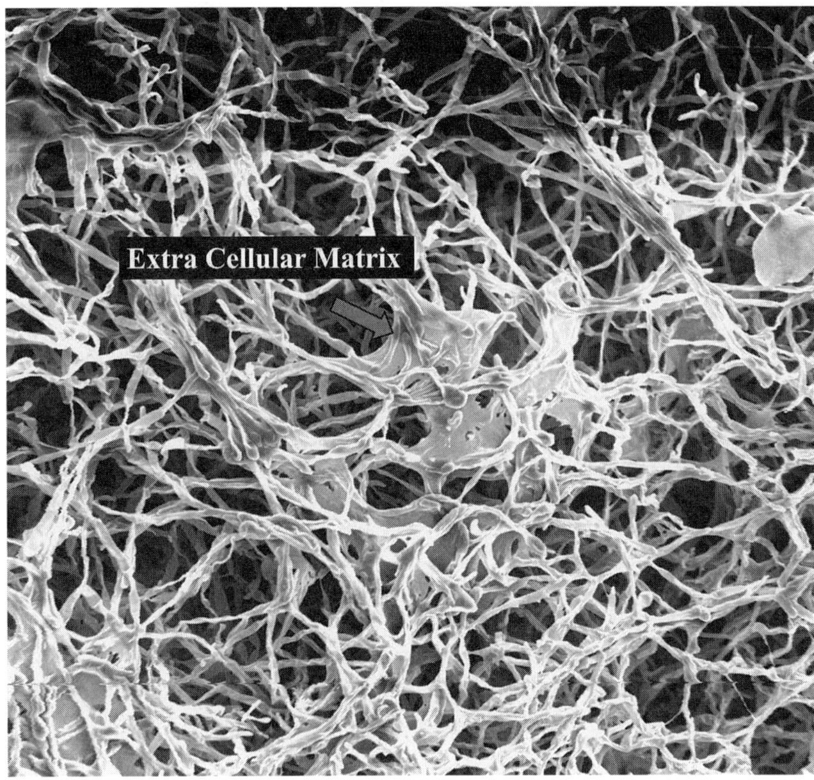

Figure 5. Scanning electron microscopy images of an *A. fumigatus* biofilm on bronchial epithelial cells (Zeiss Novascan, Jena, Germany), showing the presence of ECM (arrow) (Seidler, unpublished data).

ing without contacting the mold. A 3D reconstruction of *A. fumigatus* on epithelial cells demonstrated the viability of the cells after 72 h of coculturing. Similar to bacterial species, *A. fumigatus* biofilm formation on epithelial cells proceeds through an early phase (12 h), an intermediate phase characterized by hyphal development (until 48 h), and a maturation phase, which is associated with the production of the extracellular ConA-positive material that is heterogeneously distributed in the culture (until 72 h), as on polystyrene (Color Plate 6) (Seidler, unpublished). Compared to a *C. albicans* biofilm at 48 h, a biofilm of *A. fumigatus* requires more time to fully mature.

Under all in vitro conditions, the development of an *A. fumigatus* biofilm has three steps in common: (i) the early phase, including the lag phase, during which the germ tubes are formed and adhere to the surface; (ii) the intermediate phase, including the linear growth phase, during which parallel-packed hyphae grow vertically to increase the depth of the network; and (iv) the maturation phase, including the stationary phase, during which crossing hyphae stabilize the biofilm and there is formation of an ECM which increases during the stationary phase.

How is the biofilm produced in vivo? The abundant mucus secreted by cystic fibrosis patients favors colonization by aerial opportunistic pathogens. Among them, *A. fumigatus* and *P. aeruginosa* are often found in close association in these patients (Ritz et al., 2005). Preliminary assays show the occurrence of important cross talk between these two microbes. The results of this cross talk favor *P. aeruginosa*, since in vitro, after 24 h at 37°C, *P. aeruginosa* forms a biofilm on *A. fumigatus* and kills the fungus (Color Plate 7) (Gastebois and Latgé, 2008). The molecular response of *A. fumigatus* biofilms to bacteria is under study.

ANTIFUNGAL DRUG RESISTANCE IN *A. FUMIGATUS* BIOFILMS

It has been known for many years that biofilms protect microorganisms from antimicrobial agents. In addition to evolutionarily developed antifungal drug resistance, temporary drug resistance is observed inside yeast biofilms. Several studies have reported that biofilm production aids in the development of antimicrobial drug resistance in *C. albicans* and bacteria. Persister cells, interactions of the drugs with polysaccharides of the ECM, and the ECM acting as a barrier can contribute to reduced antifungal drug susceptibility. Mature *Can-*

dida biofilms are resistant to amphotericin B (AMB) and other antifungals (Seidler et al., 2006a).

In *A. fumigatus*, Beauvais et al. found that colonies were more resistant to polyenes but not to azoles and echinocandins compared to shaken, submerged mycelia (Beauvais et al., 2007). Under different biofilm formation conditions, Seidler et al. (unpublished data) and Mowat et al. (2007) showed that the MICs of AMB and liposomal AMB and also of voriconazole (VRC), posaconazole, itraconazole (ITC), caspofungin (CSP), and micafungin are higher when *A. fumigatus* grows in a biofilm (Seidler et al., 2006b; Mowat et al., 2007). The MICs of the polyenes and the azoles are increased at least threefold when *A. fumigatus* is grown as a biofilm. For the echinocandins, eightfold-higher MICs are observed with *A. fumigatus* biofilms on HBE cells (Seidler et al., unpublished). Mowat et al. (2007) reported that ITC and CSP were ineffective against multicellular structures, exhibiting over 1,000-fold more resistance than their planktonic counterparts (Mowat et al., 2007). A high variation in the concentrations of the different products was seen. This may have been a result of the method used, since Mowat et al. (2007) reported that the effects of the antifungal drugs on multicellular structures, which are no longer actively dividing, could only be quantified accurately in a modified XTT assay (Antachopoulos et al., 2008; Van Vianen et al., 2006). In the study by Mowat et al. (2007), CSP demonstrated poor overall activity against adherent multicellular *A. fumigatus* cells. ITC was ineffective against adherent multicellular populations of *A. fumigatus*. Overall, AMB was the most effective antifungal drug against *A. fumigatus* biofilms at the lowest concentrations, followed by VRC, CSP, and ITC. Both AMB and VRC had the ability to reduce cellular viability by over 50% at relatively low concentrations (Mowat et al., 2007).

However, how does an *A. fumigatus* biofilm resist antifungal drugs? Many antimicrobial drugs have an absolute requirement for cell growth in order to kill. The slow growth of a bacterial biofilm is a major factor in the increased resistance to drugs (Lewis, 2001). However, in an *A. fumigatus* biofilm, the growth is faster than in a shaken culture (Beauvais et al., 2007).

Very little is known so far about the mechanisms involved in the development of antifungal resistance in *A. fumigatus* biofilms. It only has been demonstrated that the presence of the ECM delays penetration of polyenes into the cells (Beauvais et al., 2007). Mechanisms of antifungal drug resistance have been better studied in *C. albicans* and bacterial biofilms (d'Enfert, 2006; Lewis, 2001). Persister cells look to be an important key for the resistance to antimicrobial drugs. In bacterial biofilms, the majority of cells are killed when treated with a bactericidal antibiotic, but persisters survive and are ultimately responsible for the high level of resistance of the biofilm (Lewis, 2001). Expression profiling of RNA from isolated persister cells indicates that these cells are probably dormant (Lewis, 2007). Similarly, in *C. albicans*, in the presence of AMB, a certain number of cells become dormant and renew their proliferation when the drug is removed (Boucherit et al., 2007). In *A. fumigatus*, observation under transmission electron microscopy of the biofilm after AMB treatment showed some perfectly alive cells embedded in many dead cells (Fig. 6) (Beauvais, unpublished). Withdrawal of the drug allows these cells to start growing and to reform the mat. The nature of these cells is still unknown, but the lack of methods for the isolation of persister or dormant cells, particularly in a mat of highly bound hyphae, makes their study difficult. So far, many multidrug resistance efflux pumps and major facilitator superfamilies that transport antimicrobials have been identified in *A. fumigatus*, and transcriptome studies of *A. fumigatus* growing under planktonic or biofilm conditions are under way to identify biofilm-specific genes that could be associated with antifungal drug resistance.

COMPOSITION AND ROLE OF THE ECM

Composition of the ECM

The ECM produced by *A. fumigatus* growing in rich-agar medium is the only matrix for which the composition has been completely established (Table 1) (Beauvais et al., 2007).

Hexoses are the main component of the ECM. They are represented by 25% polysaccharides and 68% monosaccharides (Beauvais et al., 2007). The polysaccharides are galactomannans and α-1,3-glucans, and glucose is, as in *C. albicans* biofilms, the predominant monosaccharide. Glucose can be seen as a nutrient sliding along the ECM and is able to provide nutrients to the cells more distally located from the agar surface. Galactomannan and α-1,3-glucan are also cell wall components (Maubon et al., 2006). However, these two polysaccharides are specific to the biofilm, since β-1,3-glucan and hexosamines (chitin and polygalactosamine), which are also cell wall carbohydrates, are not found in the biofilm matrix (Beauvais et al., 2007). Compared to the ShS cell wall, α-1,3-glucans are not localized in the biofilm, since they are only observed at the surface of the cell wall and not in the distinct intermediate layer as in the cell wall of the shaken mycelium (Fig. 7). The galactomannan present in the matrix has a fibrillar structure (Beauvais et al., 2007). Polysaccharides are also the main components of bacterial and yeast biofilms. In the bacterium *Staphylococcus epidermidis*, the biofilm matrix is mainly composed of a linear β-1,6-glucosaminoglycan (Mack et al., 1996). The composi-

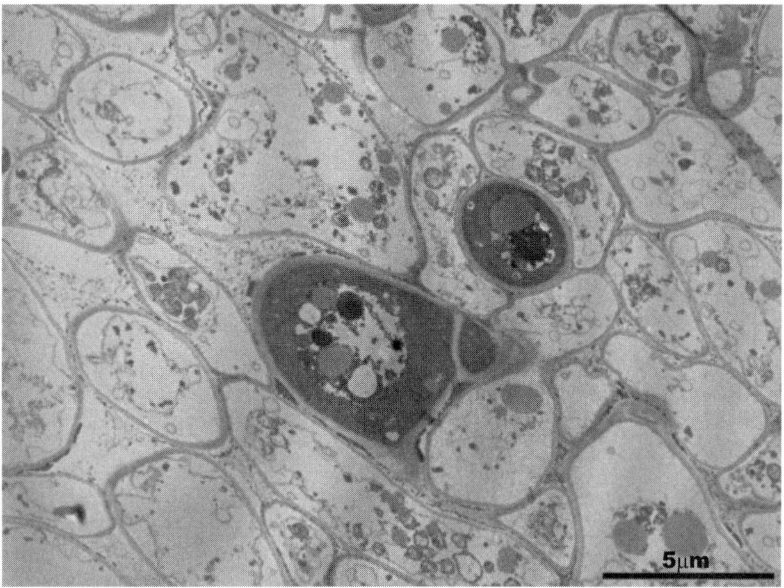

Figure 6. Transmission electron microscopy image of an *A. fumigatus* biofilm (StA) after AMB treatment. Note the presence of persister-like cells (Beauvais, unpublished).

tion of the matrix in a *C. albicans* biofilm is still unknown, except that it contains 41% carbohydrates, including β-1,3-glucan and 16% glucose, which is the most abundant monosaccharide in the biofilm (Baillie and Douglas, 2000; Nett et al., 2007).

Melanin, as shown by transmission electron microscopy and confirmed chemically, represents the electron-dense material present in the ECM (Beauvais et al., 2007). Its nature is still unknown, but preliminary high-performance liquid chromatography experiments have shown that it is not DHN-melanin (S. Dadachova and A. Casadevall, personal communication).

Proteins (2%), including the major secreted antigens recognized by aspergilloma patient sera and hydrophobins, are also part of the matrix of *A. fumigatus* (Beauvais et al., 2007). The hydrophobicity is a main characteristic of *A. fumigatus* biofilms. Only two hydrophobins, RodBp and RodEp, are specifically expressed in *A. fumigatus* biofilms. RodBp is a typical class I glycophosphatidylinositol-anchored hydrophobin characterized by eight conserved cysteine residues and the conserved spacing of hydrophilic and hydrophobic regions (Paris et al., 1993). RodBp is 50% similar to other *A. fumigatus* class I hydrophobins. RodEp is not a typical hydrophobin; RodEp has no signal peptide but has a hydrophilic N-terminal region, 11 cysteine residues, and a long amino acid end after the last cysteine, and RodEp has no sequence homology to the other known hydrophobins (Beauvais et al., 2007).

Table 1. Composition of the extracellular matrix of *A. fumigatus* biofilm

Component	
Polyols	74% glucose
	5% mannitol
	5% glycerol
	3% trehalose
Polysaccharides	Galactomannan
	α-1,3-glucan
Melanin	
Proteins	Antigens
	Hydrophobins: RodB, RodE

Role of the ECM

The role of the ECM in the formation of an *A. fumigatus* biofilm and its resistance to antifungal drugs is still unknown. However, in *C. albicans*, the matrix polysaccharide binds fluconazole and when the mature biofilm is exposed to fluconazole together with β-1,3-glucanase in nontoxic concentrations, the biofilm is rapidly eliminated (Nett et al., 2007). This result suggests that the β-1,3-glucan of the matrix protects the yeast from external aggressions. A similar role can be foreseen for the galactomannan and α-1,3-glucans present in the *A. fumigatus* matrix. Moreover, α-1,3-glucans are known to play a role in virulence of pathogenic fungi by protecting the fungus against host defense reactions (Beauvais et al., 2006). In the bacterium *S. epidermidis*, the biofilm matrix glucosaminoglycan is referred to as the polysaccharide intercellular adhesin (Mack et al., 1996). In an *A. fumigatus* biofilm, α-1,3-glucans can also be considered as an adhesin, since preliminary results

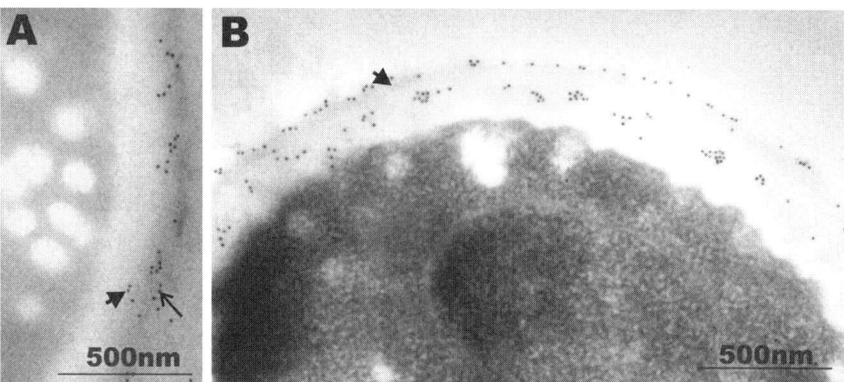

Figure 7. Immunolabeling of the ECM (thin arrow) and the cell wall (thick arrow) of StA (A) and ShS (B) hyphae with anti-α-1,3-glucan antibodies (Beauvais et al., 2007).

have shown that they interfere with the agglutination of germinated *A. fumigatus* conidia (T. Fontaine, personal communication).

The matrix plays an important role in bacterial biofilms for adherence to an abiotic surface. For example, in *S. epidermidis*, mutants in which the *ICA* locus encoding the production of the matrix β-1,6-glucosaminoglycan was deleted showed negligible biofilm formation (McKenney et al., 1998). Cellulose biosynthesis is a prerequisite for the attachment and colonization of plants by *Salmonella enterica* (Barak et al., 2007). In *A. fumigatus*, adherence to a substratum is not required for formation of a biofilm, since under liquid static conditions *A. fumigatus* develops a biofilm at the air surface of the medium (Beauvais et al., unpublished).

A cell integrity pathway could also mediate the interactions and the formation of the matrix between hyphae. In *C. albicans* during growth on a solid surface, Mkc1p, the cell integrity mitogen-activated protein kinase, is overexpressed and contributes to the biofilm structure, since an *mkc1*-null mutant produces an abnormal fluconazole-sensitive biofilm (Kumamoto, 2005). In *A. fumigatus*, four different mitogen-activated protein kinases were identified in the genome: MpkA, MpkB, MpkC, and SakA (Afu4g13720, Afu6g12820, Afu5g09100, and Afu1g12940, respectively). Some of them have a role in determining the structure of a colony, but the association with biofilm formation has not been investigated (Du et al., 2006; Valiante et al., 2008).

In other fungi, hydrophobins present in the ECM play a structural role in aerial mycelium development by lowering the surface tension of water enclosing the biofilm. This has been demonstrated for SC3 of *S. commune* and the streptofactin of *Streptomyces tendae* (Wösten et al., 1999; Richter et al., 1998). Application of streptofactin restored the capacity to produce aerial mycelium in a *Streptomyces* Blad (*bld*) mutant which was defective in the hydrophobic peptide SapB (Richter et al., 1998; Tillotson et al., 1998). In *Ustilago maydis*, the *REP1* gene encodes a pre-pro-protein cleaved in different hydrophobic repeats known as repellents. Eight were identified in cell walls of aerial hyphae, and the disruption of REP1 results in a reduction of aerial formation of hyphae (Wösten et al., 1996). In *A. fumigatus*, a *rodB rodE* double mutant has been constructed in an effort to understand the role of hydrophobins in biofilms (Beauvais et al., unpublished).

In conclusion, many studies remain to be done to understand the role of the ECM in *A. fumigatus*. The release of the genome annotation, the progress with the transcriptome and proteome, and the availability of strains adapted to improve the efficacy of the transformation in this fungus are now among many tools available for such studies.

CONCLUSION

A. fumigatus can produce in vitro an extracellular hydrophobic matrix with typical characteristics of a biofilm under all static conditions tested: on agar medium, polystyrene, or epithelial cells. Under static conditions the mycelial growth is greater than under shaken submerged conditions. The ECM is composed of galactomannan, α-1,3-glucans, monosaccharides and polyols, melanin, and proteins, including major antigens and hydrophobins. All antifungal drugs are significantly less effective when *A. fumigatus* is grown under biofilm versus planktonic conditions. However, the role of a modified cell wall, the role of the ECM, particularly α-1,3-glucan and hydrophobins, and the eventual role of multidrug resistance efflux pumps in the establishment of biofilms in vivo and in vitro and their resistance to antifungal drugs remain to be established. Future work is required to demonstrate in vivo growth of *A. fumigatus* biofilms in patients. Different strategies will be undertaken. (i) Transcriptome and proteome studies will allow a more detailed knowledge of the genes regulated in biofilm formation and the molecular mechanisms causing antifungal resistance in *Aspergillus* biofilms. (ii) The construction of mutants impaired in static growth or sensitive to antifungal drugs under static conditions will be of great benefit. (iii) Ultrastructure analysis and immunolabeling of aspergilloma and mouse invasive aspergillosis lesions with anti-galactomannan and anti-α-1,3-glucan antibodies under transmission electron microscopy will eventually allow observation of biofilms and evaluation of the composition of an ECM in vivo.

REFERENCES

Antachopoulos, C., J. Meletiadis, T. Sein, and E. Roilides. 2008. Comparative in vitro pharmacodynamics of caspofungin, micafungin, and anidulafungin against *Aspergillus* germinated and nongerminated conidia. *Antimicrob. Agents Chemother.* **52:**321–328.

Anwar, H., J. L. Strap, and J. W. Costerton. 1992. Establishment of aging biofilms: possible mechanism of bacterial resistance to antimicrobial therapy. *Antimicrob. Agents Chemother.* **36:**1347–1351.

Baillie, G. S., and L. J. Douglas. 1998. Effect of growth rate on resistance of *Candida albicans* biofilms to antifungal agents. *J. Antimicrob. Chemother.* **42:**1900–1905.

Baillie, G. S., and L. J. Douglas. 2000. Matrix polymers of *Candida* biofilms and their possible role in biofilm resistance to antifungal agents. *J. Antimicrob. Chemother.* **46:**397–403.

Barak, J. D., C. E. Jahn, D. L. Gibson, and A. O. Charkowski. 2007. The role of cellulose and O-antigen capsule in the colonization of plants by *Salmonella enterica*. *Mol. Plant Microbe Interact.* **20:** 1083–1091.

Beauvais, A., D. S. Perlin, and J. P. Latgé. 2006. Role of α(1-3) glucan in *Aspergillus fumigatus* and other human fungal pathogens, p. 269–288. *In* G. M. Gadd, S. C. Watkinson, and P. Dyer (ed.), *Fungi in the Environment*. Cambridge University Press, Cambridge, United Kingdom.

Beauvais, A., C. Schmidt, S. Guadagnini, P. Roux, E. Perret, C. Henry, S. Paris, A. Mallet, M. C. Prevost, and J. P. Latgé. 2007. An extracellular matrix glues together the aerial grown hyphae of *Aspergillus fumigatus*. *Cell. Microbiol.* **9:**1588–1600.

Boucherit, Z., O. Seksek, and J. Bolard. 2007. Dormancy of *Candida albicans* cells in the presence of the polyene antibiotic amphotericin B: simple demonstration by flow cytometry. *Med. Mycol.* **45:**525–533.

Chandra, J., D. M. Kuhn, P. K. Mukherjee, L. L. Hoyer, T. McCormick, and M. A. Ghannoum. 2001. Biofilm formation by the fungal pathogen *Candida albicans*: development, architecture, and drug resistance. *J. Bacteriol.* **183:**5385–5394.

D'Enfert, C. 2006. Biofilms and their role in the resistance of pathogenic *Candida* to antifungal agents. *Curr. Drug Targets* **7:**465–470.

Du, C., J. Sarfati, J. P. Latgé, and R. Calderone. 2006. The role of the sakA (Hog1) and tcsB (sln1) genes in the oxidant adaptation of *Aspergillus fumigatus*. *Med. Mycol.* **44:**211–218.

Gastebois, A., and J. P. Latgé. 2008. Bactéries et champignons pathogènes de l'homme: amis-ennemis. *Biofutur* **284:**22–27.

Hawser, S. P., and L. J. Douglas. 1994. Biofilm formation by *Candida* species on the surface of catheter materials in vitro. *Infect. Immun.* **62:**915–921.

Jefferson, K. K. 2004. What drives bacteria to produce a biofilm? *FEMS Microbiol. Lett.* **236:**163–173.

Kumamoto, C. A. 2005. A contact-activated kinase signals *Candida albicans* invasive growth and biofilm development. *Proc. Natl. Acad. Sci. USA* **102:**5576–5581.

Lewis, K. 2001. Riddle of biofilm resistance. *Antimicrob. Agents Chemother.* **45:**999–1007.

Lewis, K. 2007. Persister cells, dormancy and infectious disease. *Nat. Rev. Microbiol.* **5:**48–56.

Mack, D., W. Fischer, A. Krokotsch, K. Leopold, R. Hartmann, H. Egge, and R. Laufs. 1996. The intercellular adhesin involved in biofilm accumulation of *Staphylococcus epidermidis* is a linear β-1,6-linked glucosaminoglycan: purification and structural analysis. *J. Bacteriol.* **178:**175–183.

Martinez, L. R., and A. Casadevall. 2006. Susceptibility of *Cryptococcus neoformans* biofilms to antifungal agents in vitro. *Antimicrob. Agents Chemother.* **50:**1021–1033.

Maubon, D., S. Park, M. Tanguy, M. Huerre, C. Schmitt, M. C. Prévost, D. S. Perlin, J. P. Latgé, and A. Beauvais. 2006. *AGS3*, an α(1-3)glucan synthase gene family member of *Aspergillus fumigatus*, modulates mycelium growth in the lung of experimentally infected mice. *Fungal Genet. Biol.* **43:**366–375.

McKenney, D., J. Hübner, E. Muller, Y. Wang, D. A. Goldmann, and G. B. Pier. 1998. The *ica* locus of *Staphylococcus epidermidis* encodes production of the capsular polysaccharide/adhesin. *Infect. Immun.* **66:**4711–4720.

Mikkelsen L., S. Sarrocco, M. Lübeck, and D. F. Jensen. 2003. Expression of the red fluorescent protein DsRed-Express in filamentous ascomycete fungi. *FEMS Microbiol. Lett.* **223:**135–139.

Mowat, E., J. Butcher, S. Lang, C. Williams, and G. Ramage. 2007. Development of a simple model for studying the effects of antifungal agents on multicellular communities of *Aspergillus fumigatus*. *J. Med. Microbiol.* **56:**1205–1212.

Nett, J., L. Lincoln, K. Marchillo, and D. Andes. 2007. Beta-1,3 glucan as a test for central venous catheter biofilm infection. *J. Infect. Dis.* **195:**1705–1712.

O'Toole, G., H. B. Kaplan, and K. Kolter. 2000. Biofilm formation as microbial development. *Annu. Rev. Microbiol.* **54:**49–79.

Paris, S., M. Monod, M. Diaquin, B. Lamy, L. K. Arruda, J. P. Punt, and J. P. Latgé. 1993. A transformant of *Aspergillus fumigatus* deficient in the antigenic cytotoxin ASPF1. *FEMS Microbiol. Lett.* **111:** 31–36.

Prosser, J. I. 1994. Kinetics of filamentous growth and branching, p. 301–317. *In* N. A. R. Gow and G. M. Gadd (ed.), *The Growing Fungus*. Chapman and Hall, London, United Kingdom.

Ramage, G., S. P. Saville, D. P. Thomas, and J. L. Lopez-Ribot. 2005. *Candida* biofilms: an update. *Eukaryot. Cell* **4:**633–638.

Richter, M., J. M. Willey, R. Süssmuth, G. Jung, and H. P. Hiedler. 1998. Streptofactin, a novel biosurfactant with aerial mycelium inducing activity from *Streptomyces tendae* Tü 901-8c. *FEMS Microbiol. Lett.* **163:**165–171.

Righelato, R. C. 1979. The kinetics of mycelial growth, p. 385–401. *In* J. H. Burnett and A. P. J. Trinci (ed.), *Fungal Wall and Hyphal Growth*. Cambridge University Press, Cambridge, United Kingdom.

Ritz, N., R. A. Ammann, C. Casaulta-Aebischer, F. Schoeni-Affolter, and M. H. Schoeni. 2005. Risk factors for allergic bronchopulmonary aspergillosis and sensitisation to *Aspergillus fumigatus* in patients with cystic fibrosis. *Eur. J. Pediatr.* **164:**577–582.

Seidler, M., S. Salvenmoser, F.-M. Müller. 2006a. In vitro effects of micafungin against *Candida* biofilms on polystyrene and central venous catheter sections. *Int. J. Antimicrob. Agents* **28:**568–573.

Seidler, M., S. Salvenmoser, and F.-M. Müller. 2006b. *Aspergillus fumigatus* biofilm formation on polystyrene and bronchial epithelial cells. *Abstr. 46th Intersci. Conf. Antimicrob. Agents Chemother.*, abstr. M-1766.

Steele, G. C., and P. J. Trinci. 1975. Morphology and growth kinetics of hyphae of differentiated and undifferentiated mycelia of *Neurospora crassa*. *J. Gen. Microbiol.* **91:**362–368.

Tillotson, R. D., H. A. Wosten, M. Richter, and J. M. Willey. 1998. A surface active protein involved in aerial hyphae formation in the filamentous fungus *Schizophillum commune* restores the capacity of a bald mutant of the filamentous bacterium *Streptomyces coelicolor* to erect aerial structures. *Mol. Microbiol.* **30:**595–602.

Valiante, V., T. Heinekamp, R. Jain, A. Härtl, and A. A. Brakhage. 2008. The mitogen-activated protein kinase MpkA of *Aspergillus fumigatus* regulates cell wall signaling and oxidative stress response. *Fungal Genet. Biol.* **45:**618–627.

van Vianen, W., S. de Marie, M. T. ten Kate, and R. A. Mathot. 2006. Caspofungin: antifungal activity in vitro, pharmacokinetics, and effects on fungal load and animal survival in neutropenic rats with invasive pulmonary aspergillosis. *J. Antimicrob. Chemother.* **57:**732–740.

Wösten, H. A., R. Bohlmann, C. Eckerskorn, F. Lottspeich, M. Bolker, and R. Kahmann. 1996. A novel class of small amphipathic peptides affect aerial hyphal growth and surface hydrophobicity in *Ustilago maydis*. *EMBO J.* **15:**4274–4281.

Wösten, H. A., M. A. van Wetter, L. G. Lugones, H. C. van der Mei, H. J. Busscher and J. G. Wessels. 1999. How a fungus escapes the water to grow into the air. *Curr. Biol.* **9:**85–88.

Aspergillus fumigatus and Aspergillosis
Edited by J.-P. Latgé and W. J. Steinbach

Chapter 13

Signal Transduction

GREGORY S. MAY AND TAYLOR SCHOBERLE

Microorganisms must adapt to changing environments in order to maintain a constant intracellular state that is conducive to life. As a result, organisms have evolved elaborate regulatory systems that control cellular responses to changes in the environment. Environmental changes that can alter cellular physiology to which microorganisms must respond to maintain cellular homeostasis include nutrient availability, pH, temperature, osmotic stress, and other microorganisms. In microbial eukaryotes there are several homeostatic systems that contribute to maintaining a constant intracellular environment. These include the small GTPase proteins of the Ras superfamily of proteins, mitogen-activated protein (MAP) kinase pathways, cyclic AMP (cAMP)-regulated protein kinase (PKA), and the tripartite G-protein signaling pathways. In this chapter we will discuss the current status of our understanding of these regulatory systems in *Aspergillus fumigatus* and their role in regulating the physiology of this fungus. We will also discuss how these regulatory networks contribute to pathogenesis in *A. fumigatus*.

SMALL GTPase PROTEIN SIGNALING IN *A. FUMIGATUS* AND VIRULENCE

Small GTPase proteins, like the Ras family members, exist in either the GTP-bound active state or the GDP-bound inactive state. In the GTP-bound state the GTPase protein can bind to a target protein, resulting in activation of a specific response pathway. Thus, the Ras family of GTPases functions as molecular switches, turning signaling networks on and off. The switch between the two states, active and inactive, is facilitated by proteins that are guanine nucleotide exchange factors, or GEFs, and GTPase-activating proteins, or GAPs (Fig. 1). Currently there have been no studies of GEF or GAP functions in *A. fumigatus*. The functions of three Ras or Ras-related GTPases have been studied in *A. fumigatus*, including RasA, RasB, and RhbA (Fortwendel et al., 2004, 2005; Panepinto et al., 2002, 2003).

There are two Ras genes, *rasA* and *rasB*, in *A. fumigatus*. Interestingly, the presence of two Ras genes is a common feature in the filamentous fungal genomes, as there are orthologous genes in many other species (Fortwendel et al., 2004, 2005). A distinguishing feature of the pairs of Ras proteins in the filamentous fungi is a 20-amino-acid insertion just upstream of the third GDP/GTP-binding site of the polypeptide RasB. Conservation of this insertion suggests that RasB may have novel functions and interacting proteins that are specific to filamentous fungi. A mutational analysis of *rasA* and *rasB* was undertaken in an attempt to elucidate each gene's functions in cellular morphogenesis in *A. fumigatus* (Fortwendel et al., 2004). Both dominant activating and dominant negative mutations were made for each of the genes. The growth and development of conidia were investigated for strains carrying a single copy of the mutant form of the *rasA* or *rasB* genes expressed from their endogenous promoter. Expression of dominant activating mutants of *rasA* or *rasB* had no significant effect on conidial germination. In contrast, expression of dominant negative mutants of *rasA* and *rasB* did alter conidial germination. The dominant negative mutant *rasA* strain had a much reduced rate of conidial germination overall, while the dominant negative *rasB* mutant strain exhibited a delay in conidial germination, but once germination began it proceeded at the wild-type rate. These results are consistent with studies previously done in *Aspergillus nidulans* (Fillinger et al., 2002; Osherov and May, 2000; Som and Kolaparthi, 1994).

A. fumigatus is a multicellular fungus that elaborates specialized cell types to complete the asexual life cycle that ends with the formation of abundant conidia.

Gregory S. May and Taylor Schoberle • Division of Pathology and Laboratory Medicine, Dept. of Laboratory Medicine, Research, The University of Texas, M. D. Anderson Cancer Center, Houston, TX 77030.

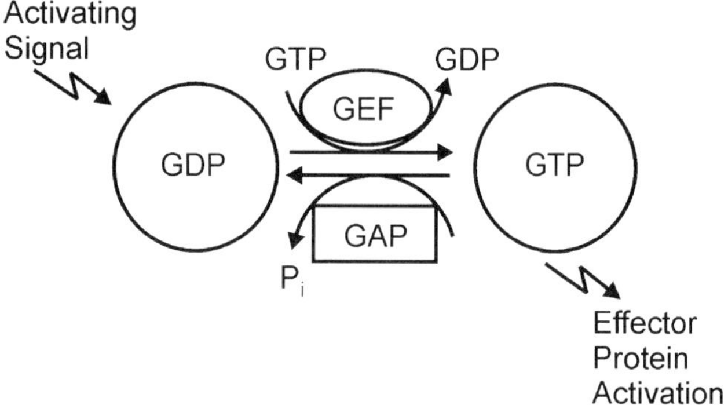

Figure 1. Model for the small G-protein activation and inactivation cycle. An activating signal leads to the exchange of GTP for GDP on the G-protein, catalyzed by GEF, producing the active GTP-bound G-protein. Inactivation of the GTP-bound G-protein is catalyzed by GAP, which increases the rate of hydrolysis of GTP to GDP.

This developmental program has been studied extensively in *A. nidulans*, and a considerable body of knowledge exists about the process and its regulation (Adams et al., 1998; Mah and Yu, 2006). Briefly, hyphae at an air-liquid interface will initiate the developmental program that leads to the formation of specialized aerial hyphae, foot cells, whose apex swells to form a vesicle. On the surface of the vesicle a single row of phialides is formed, following a nuclear division, and then chains of conidia form by budding from the end of the phialide. The conidial development process was normal in the dominant negative mutant *rasA* strain. In contrast, in the dominant active mutant *rasA* strain, conidial development was greatly delayed and reduced. This was seen on plates as the formation of a fluffy white colony, indicative of reduced conidiation. Microscopic examination determined that fewer conidiophores were formed, and those present lacked the vesicle and had reduced numbers of phialides. These results suggest that RasA signaling may negatively regulate conidial development and particularly the formation of conidiophores. Like the dominant negative *rasA* mutant, conidiophore development was normal in the dominant negative *rasB* mutant. In contrast, the dominant negative *rasB* mutant exhibited reduced growth on plates and the hyphae in submerged culture were structurally abnormal, displaying apical hyphal branching. The dominant active *rasB* mutant also impacted conidiophore morphology but was distinct from the morphological defects seen in the dominant active *rasA* mutant. The conidiophores formed in the dominant active *rasB* mutant were small with no vesicle and formed longer phialides that were correctly positioned, in contrast to those formed in the *rasA* mutant. Another phenotype of the dominant negative *rasB* mutant was that this strain formed conidiophores in submerged culture and expressed *brlA*, a primary transcriptional activator of conidial development, continuously in submerged culture, whereas *brlA* is not expressed in submerged culture with the wild type. This suggests that *rasB* activity represses *brlA* expression in submerged cultures.

A *rasB* deletion mutant was made and found to have even more pronounced hyphal growth defects than the dominant negative *rasB* strain (Fortwendel et al., 2004). The deletion mutant had a much reduced growth rate on plates, and the hyphae in liquid culture displayed a high frequency of apical branch formation as well as increased lateral branch formation. As reported for the dominant negative *rasB* mutant, the *rasB* deletion mutant displayed delayed conidial germination, but the delay was of increased time to germination. The results for the dominant negative *rasB* allele and the deletion allele suggest that RasB is needed for efficient conidial germination. Interestingly, a *rasB* gene lacking the 20-amino-acid insertion was unable to complement the *rasB* deletion mutant, suggesting that this region of the polypeptide is required for RasB activity. When the *rasB* deletion mutant was tested for virulence in a neutropenic mouse model of pulmonary aspergillosis, it had a greatly reduced virulence. The mutant also produced lower fungal burden in the lungs of infected animals, and the lesions produced were smaller. Given the severe growth defects displayed by the *rasB* deletion mutant, it is not surprising that there are reduced virulence and growth effects in this model.

The Rehb family of small GTPase proteins is another member of the Ras superfamily of proteins, and they share a common set of amino acid differences (Panepinto et al., 2002). In *A. fumigatus* there is a single *rehb* gene, *rhbA*, that was initially identified as a transcript shown to increase in the fungus when grown in the presence of human endothelial cells in culture (Panepinto et al., 2002; Rhodes et al., 2001). The *rhbA* complementary DNA clone was able to complement the canavanine-sensitive phenotype of an *rhb1* mutant of the budding yeast *Saccharomyces cerevisiae* (Panepinto et al., 2002). The *rhbA* transcript is detectable by reverse transcription-PCR (RT-PCR) throughout the life cycle of the fungus. Using the RT-PCR assay the author also showed that the amount of *rhbA* mRNA increased under nitrogen starvation conditions but not during carbon source starvation. Similar results were obtained by Northern blot hybridization.

Deletion of *rhbA* resulted in a strain that had no significant growth difference from the wild type when grown in the presence of ammonium as the sole nitrogen source (Panepinto et al., 2003). In contrast, when nitrate, histidine, or proline was used as the sole nitrogen source, the *rhbA* deletion mutant had statistically significant slower growth than the wild type, with the slowest growth on proline-containing medium. Given the apparent role for RhbA in sensing nitrogen and regulating growth, the authors tested the effect of deletion of *rhbA* on arginine uptake, as this is one of the defects of the *rhb1* mutant in *S. cerevisiae*. As had been previously reported for the *S. cerevisiae rhb1* mutant, the *rhbA* deletion mutant display increased arginine uptake (Urano et al., 2000). The target of rapamycin (TOR) pathway is involved in regulating genes in response to nitrogen availability in both fission and budding yeast (Cardenas et al., 1999; Kawai et al., 2001). Thus, the authors investigated the effect of rapamycin on the growth of the *rhbA* deletion mutant strain and found that the mutant had increased sensitivity to the drug. This is likely explained by the fact that two pathways regulating nitrogen utilization are affected, one by the *rhbA* deletion and the other by inhibition of TOR with rapamycin. A curious phenotype of the *rhbA* deletion mutant strain is that the mutant was able to complete the asexual developmental conidiation program in submerged culture when grown in rich medium or ammonium-containing medium, but not when the nonpreferred nitrogen sources proline or nitrate were used. Since there has been a connection between starvation stress and *brlA* expression in *A. nidulans* (Skromne et al., 1995), perhaps there is a stronger link to nutrient sensing and conidial development in *A. fumigatus* that leads to precocious conidiation in submerged culture that is not seen in *A. nidulans*.

Given the link between nitrogen utilization and *rhbA* function, it was logical to assume that *rhbA* plays an important role during infectious growth. The authors tested this possibility in a neutropenic murine model of pulmonary aspergillosis (Panepinto et al., 2003). The *rhbA* deletion mutant had reduced virulence in this model, and the size of lesions produced by the mutant in the lung was reduced by about 75% relative to the wild type. Thus, it appears the *rhbA* function is necessary for efficient growth in an animal host and that nitrogen utilization plays an important role in virulence.

cAMP-DEPENDENT PROTEIN KINASE PATHWAY

cAMP-dependent PKA pathways are used to regulate a number of cellular responses to environmental changes, particularly in response to glucose utilization (Fillinger et al., 2002; Lafon et al., 2005; Oliver et al., 2002a). In the microbial eukaryotes the PKA pathway has also been linked to morphogenetic changes and stress response (Fillinger et al., 2002; Lafon et al., 2005; Oliver et al., 2002a). PKA is a heterotetramer composed of two regulatory subunits and two catalytic subunits. The regulatory subunits bind cAMP, producing a conformational change that allows the catalytic subunits to autophosphorylate one another, leading to full protein kinase activity (Fig. 2). Phosphorylation of downstream target proteins then leads to both physiological changes resulting from a combination of changes in gene expression and also enzyme activities that result in a homeostatic response in the cell. Thus, the PKA pathway plays a critical role in regulating cellular responses to environmental changes that ensure maintenance of a constant cellular environment.

In *A. fumigatus* the PKA pathway has been extensively studied (Brakhage and Liebmann, 2005; Liebmann et al., 2003; Oliver et al., 2002a, 2002b). Research on the PKA pathway in *A. fumigatus* was stimulated by two observations. The first was that the mRNAs for the regulatory and catalytic subunits increase in response to exposure to human cells in culture (Oliver et al., 2002b; Rhodes et al., 2001). Second was the finding that the polyketide synthase PKSP, a known virulence trait, is regulated in part by the PKA pathway. Deletion mutants have been made for genes encoding the catalytic subunit, *pkaC*, the regulatory subunit, *pkaR*, the adenylate cyclase, *acyA*, and *gpaB*, the G-protein α-subunit (Liebmann et al., 2003, 2004; Zhao

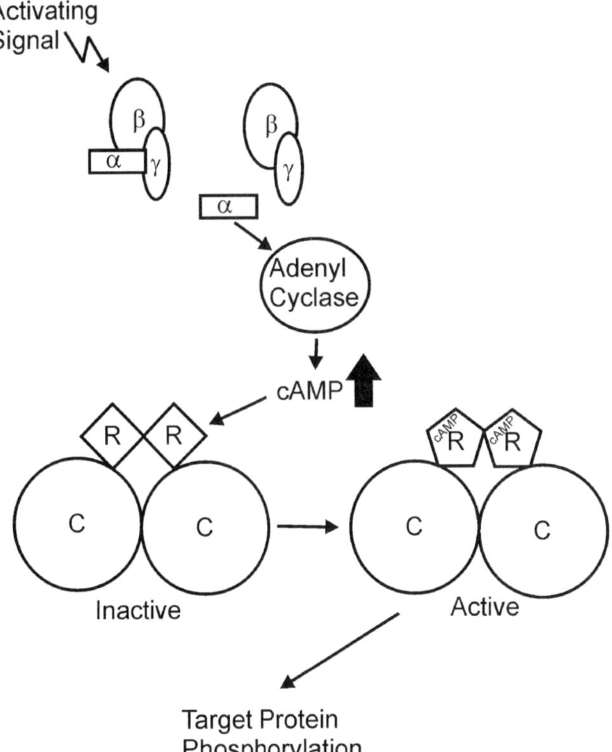

Figure 2. Model for regulation of PKA. An activating signal leads to dissociation of the tripartite G-protein into the α-subunit and the β-γ dimer. The free α-subunit stimulates adenyl cyclase, leading to increased cAMP levels. cAMP binds to the regulatory subunits (R) of the PKA tetramer, leading to a conformational change. The now-active protein kinase catalytic subunits (C) can then phosphorylate downstream target proteins.

et al., 2006). The α-subunits of heterotrimeric G-proteins either stimulate or inhibit adenylate cyclase activity, thus affecting cellular levels of cAMP (Liebmann et al., 2003). These deletion mutants provide a rich source of experimental material to further investigate how the PKA pathway contributes to virulence and what role it plays in regulating the growth of *A. fumigatus*, hyphal morphogenesis, and conidial development.

Deletion of the adenlyate cyclase gene, *acyA*, results in a strain whose growth on minimal medium is greatly diminished relative to the parental strain. In addition, the *acyA* deletion mutant fails to form conidia and produces smaller conidiophores with poorly defined vesicles containing few phialides. Interestingly, the addition of dibutyryl-cAMP partially rescued both the growth and conidiation phenotypes. These results demonstrate the requirement for cAMP, both for robust hyphal growth and conidial development, presumably acting through the action of cAMP as a regulator of the PKA pathway. In contrast to the dramatic phenotypic effects of deleting *acyA*, deletion of *gpaB*, the G-protein α-subunit gene, resulted in more subtle phenotypic effects. Growth and conidiation of the *gpaB* deletion mutant was barely affected on minimal medium and not at all on malt extract medium, and the modest effects seen on minimal medium were corrected for by the addition of dibutyryl-cAMP. In comparison to the parental strains, the *acyA* and the *gpaB* deletion mutants were approximately three times more susceptible to killing by monocyte-derived macrophages, though killing of the *acyA* deletion mutant required long incubation times, presumably reflecting a delay in conidial germination.

Although the previous experiments implicated the PKA pathway with respect to hyphal growth and conidial development, they are incomplete. It is possible that the phenotypes seen in *acyA* and *gpaB* deletion mutants result from effects on cAMP levels that influence pathways other than those mediated through PKA. Additional experiments addressed this potential through the construction of *pkaC* and *pkaR* deletion mutants that directly impacted the PKA pathway (Liebmann et al., 2004; Zhao et al., 2006). The *pkaC* deletion mutant had a reduced growth rate compared to the parental strains (Liebmann et al., 2004). In addition, conidiation in the *pkaC* deletion mutant was reduced by about 90% of that of the parental strain, and the conidia of the deletion mutant germinated more slowly. Like the phenotypes seen for the *pkaC* deletion mutant, the *pkaR* deletion mutant grew at only about 50% the rate of the parental strain and formed abnormal hyphae and condiophores (Zhao et al., 2006). In addition, the conidial germination in the *pkaR* deletion mutant was very slow, achieving about 50% germination in 11 h, compared to about 4.5 h for the parental strain. The conidia of the *pkaR* deletion mutant were more susceptible to killing by treatment with hydrogen peroxide, menadione, paraquat, and diamide, whereas the hyphae were only more susceptible to paraquat and diamide.

When deletion mutants for genes that function in the PKA pathway were tested for virulence in a murine model of invasive pulmonary aspergillosis, they uniformly had reduced virulence. The *pkaC* deletion mutant was effectively avirulent in the model used, while the *gpaB* deletion mutant was somewhat more virulent, killing approximately 25% of the mice (Liebmann et al., 2004). In contrast, the *pkaR* deletion mutant was even more virulent, in that only 50% of the animals survived the 2-week duration of the experiment (Zhao et al., 2006). One caveat to these experiments is that many of the mutants tested germinated poorly and had restricted hyphal growth in vitro, so the lack of virulence may simply be a reflection of these growth characteristics. This is certainly the case for both the *pkaC* and *pkaR* deletion mutants, because they have the most severe germination and growth defects, whereas this is probably not sufficient to explain the reduced virulence of the *gpaB* mutant, since it has near-wild-type germination and growth in vitro.

MAPKs OF *A. FUMIGATUS*

Mitogen-activated protein kinases (MAPKs) play a pivotal role in the regulation of fungal cell physiology in response to nutritional status and environmental stresses (e.g., hypertonic shock, heat shock, oxidative stress, and reactive nitrogen species). MAPKs are also involved in fungal cell morphogenesis and asexual and sexual development (May et al., 2005; Paoletti et al., 2007; Reyes et al., 2006; Xue et al., 2004). MAPKs function in a cascade of kinases, sending signals from one molecule to another, thus amplifying these signals. Essentially, an intracellular signal created from upstream activators responding to a receptor-ligand interaction will activate a MAP kinase kinase kinase, which phosphorylates a serine and a threonine residue in a conserved amino-terminal domain of the MAP kinase kinase, which in turn phosphorylates a threonine and a tyrosine located in the activation loop in the conserved kinase domain on the MAPK (Fig. 3). This MAPK sends a signal to a downstream target, often a transcription factor that will in turn adjust gene expression to meet the requirements of the environment (Bussink and Osmani, 1999; May et al., 2005; Valiante et al., 2008). Thus, defects in one component of a MAPK signaling pathway could decrease the fitness and virulence of the fungus. Studies of plant pathogenic fungi have shown a clear role of the MAPK signaling pathway in virulence. There is an agricultural antifungal drug, fludioxonil, that targets a MAPK signaling pathway (Kojima et al., 2004;

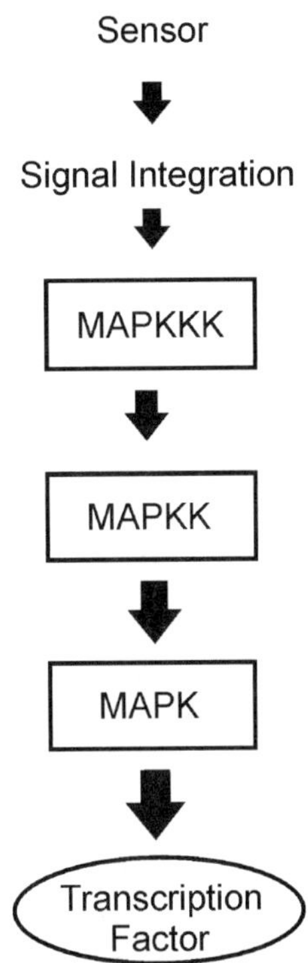

Figure 3. Model for the MAPK cascade. An intracellular signal created from upstream activators responding to a receptor-ligand interaction will activate a MAP kinase kinase kinase (MAPKKK), which phosphorylates a MAP kinase kinase (MAPKK), which in turn phosphorylates the MAPK. This MAPK sends a signal to a downstream target, often a transcription factor that will in turn adjust gene expression to meet the requirements of the environment.

May et al., 2005; Reyes et al., 2006). These MAPK genes are highly conserved through all eukaryotes. In *A. fumigatus*, there are four genes encoding MAPKs: *mpkA, mpkB, mpkC*, and *sakA/hogA* (May et al., 2005; Reyes et al., 2006; Valiante et al., 2008).

MpkA

Many studies have focused on the importance of *A. fumigatus* cell wall structure and composition, because these aspects have great potential as targets for antifungal drugs (Bruneau et al., 2001). Although a number of genes involved in cell wall composition have been isolated from *A. fumigatus*, the knowledge regarding specific signaling pathways that regulate these genes is still limited (Beauvais and Latgé, 2001). One particular gene, *mpkA*, has been studied in numerous species, such as *A. nidulans, Schizosaccharomyces pombe, Saccharomyces cerevisiae*, and *Candida albicans* (Bussink and Osmani, 1999; Valiante et al., 2008). It was demonstrated that deletion of *mpkA* from *A. fumigatus* resulted in a severe growth deficiency on minimal medium with glucose as the sole carbon source. Upon further investigation, it was shown that the Δ*mpkA* mutant formed thicker, less-elongated hyphae that were more branched than the wild-type strain, suggesting a role for *mpkA* in hyphal growth and filamentation. Although Δ*mpkA* colony morphology is different from the wild type, normal conidiophores and conidia are formed by the mutant. Conidial germination of the Δ*mpkA* mutant does not seem to be delayed in Sabouraud liquid medium compared to the wild type (Valiante et al., 2008).

In addition to being involved in cell wall composition, MpkA is believed to be involved in a signaling pathway that regulates responses to cell wall damage. There are several compounds used to test the integrity of mutant cell walls. Calcofluor white and Congo red are fluorochromes that afflict damage by binding to chitin and β-1,3-glucan, respectively, known components of fungal cell walls (Ram and Klis, 2006; Valiante et al., 2008). Gluconex is used in transformations to digest fungal cell walls because of its strong β-1,3, β-1,6-glucanase activity. Caffeine and sodium dodecyl sulfate (SDS) are also known to inflict cell wall damage. It is believed that sensitivity to any one of these compounds gives evidence of a defect in the cell wall. The Δ*mpkA* mutant shows a clear growth defect at various temperatures in response to caffeine, SDS, Calcofluor white, and Congo red. When *Aspergillus* hyphae are exposed to different damaging agents, such as gluconex, caffeine, or SDS, expression of *mpkA* dramatically increases. These results suggest that *mpkA* plays an important role in cell wall metabolism and integrity (Valiante et al., 2008).

When testing *A. fumigatus* mutants, one often finds that altered morphology due to impairment of a signaling pathway leads to attenuated virulence (da Silva Ferreira et al., 2006; Fortwendel et al., 2005; Zhao et al., 2006). This does not seem to be the case with *mpkA*. Compared to wild-type strains, Δ*mpkA* mutants show no significant difference in mortality when tested in low-dose murine infection models (Valiante et al., 2008). Although *mpkA* functions within a signaling pathway responsible for cell wall development and damage response, this gene does not seem to be necessary for the survival and virulence of *A. fumigatus*. The continued isolation and analysis of genes involved in cell wall biosynthesis of *A. fumigatus* will help scientists gain clues into the synthesis of cell wall components and its response to environmental stress.

MpkB

No studies have been done on *mpkB* of *A. fumigatus*; however, this gene has been studied in other species. Genes in other species of fungi that have high homology to *mpkB* of *A. fumigatus* have been shown to play roles in pheromone signaling and mating (May et al., 2005; Paoletti et al., 2007). In fungi, there are two ways to reproduce sexually: by self-fertilization and by outcrossing. Outcrossing occurs when nuclei that are supplied by two different partners of compatible mating types fuse and undergo meiosis. By convention, there are two mating types seen in species capable of sexual reproduction (Dyer and Paoletti, 2005; Paoletti et al., 2007; Poggeler, 2002). In *S. cerevisiae*, these mating types are termed *MAT*a and *MAT*α (Gustin et al., 1998). With regard to filamentous fungi, the mating types are identified as *MAT*-1 (or *MAT*α) and *MAT*-2 (*MAT*-HMG). Research over the years has divulged that identity of mating type is conferred by mating-type genes (*MAT*) residing at a MAT locus. This MAT locus contains idiomorphs, or highly dissimilar portions of DNA, in isolates of opposite mating type. *MAT*-1 contains a conserved α-domain at the *MAT* locus, while *MAT*-2 contains a conserved high-mobility group sequence (HMG) at the MAT locus (Dyer and Paoletti, 2005; Paoletti et al., 2007; Poggeler, 2002).

In homothallic species, such as *A. nidulans*, both mating types are present within each organism. Heterothallic (obligate outcrossing) species, such as *A. fumigatus*, only have one mating type within each strain. In heterothallic species, sexual reproduction can only occur when isolates of compatible mating types are present (Dyer and Paoletti, 2005; Paoletti et al., 2007). Interestingly, although the majority of *Aspergillus* species are not known to reproduce sexually, both mating-type sequences have been isolated from several asexual fungi, including *A. fumigatus* (Arie et al., 2000; Dyer and Paoletti, 2005; Pitt et al., 2000; Poggeler, 2002; Raper and Fennell, 1965; Yun et al., 2000). In self-fertilization ("selfing"), reproduction involves fusion of nuclei within one individual (Dyer and Paoletti, 2005; Paoletti et al., 2007). This is predominantly seen in homothallic species, such as *A. nidulans*, where both mating types are present within each organism. Self-fertilization has been seen in fungi, flowering plants, and even some animal taxa (Barriere and Felix, 2005; Murtagh et al., 2000; Nasrallah et al., 2004; Shimizu et al., 2004). This phenomenon is thought to serve as a reproductive assurance for organisms that find themselves without mating partners (Murtagh et al., 2000; Nasrallah et al., 2004; Shimizu et al., 2004). Self-fertilization still remains unclear to scientists, even though many genetic controls and signaling pathways involved in outcrossing are well understood (Casselton, 2002; Hiscock and Kues, 1999; Lengeler et al., 2000).

MpkB of *A. fumigatus* is closely related to the Fus3p and Kss1p MAP kinases of *S. cerevisiae*, as well as MpkB of *A. nidulans* (May et al., 2005; Paoletti et al., 2007). It has been shown in *A. nidulans* that knocking out *mpkB* leads to sterility with regard to sexual reproduction. Therefore, it can be predicted that *mpkB* of *A. fumigatus* plays a role in sexual reproduction and pheromone responses. Although *A. fumigatus* is not known to have a sexual cycle, several genes correlated with sexual reproduction in other fungi have been identified in *A. fumigatus* through genome searches. This suggests that there may be a possibility for sexual reproduction that has yet to be determined (Dyer et al., 2003; Dyer and Paoletti, 2005; May et al., 2005). Another possibility is that pathogenic development of *A. fumigatus* could be linked to a mating pathway, as seen in *Ustilago maydis* (May et al., 2005). In any case, it would be useful to study the function of MpkB in *A. fumigatus*.

MpkC

The protein sequence of MpkC is most closely related to Hog1p, a stress-activated protein kinase, in *S. cerevisiae*. Deletion of *mpkC* in *A. fumigatus* results in poor growth with polyalcohol sugars as the sole carbon source. Conidial germination of *mpkC* deletion mutants has been measured on minimal medium (MM) with sorbitol compared to MM with glucose. These studies revealed that a significantly reduced number of Δ*mpkC* mutants germinated in the presence of sorbitol compared to the wild type (Af293), while both strains showed equal germination in glucose medium. Even when Δ*mpkC* mutants were germinated overnight in MM with glucose, growth was significantly impaired when mycelia were transferred to MM with sorbitol. This suggests that MpkC is required for effective conidial germination and hyphal growth on certain carbon sources (Reyes et al., 2006).

When mycelia from Af293 are grown in MM with glucose and shifted to MM with sorbitol, RNA levels of *mpkC* increase in abundance. Interestingly, *mpkC* expression is so low at basal levels that it is virtually undetectable by Northern blot analysis. It has also been reported that *mpkC* mRNA levels could not be detected in *A. nidulans*. When exposed to oxidative stress (17 mM H_2O_2), *mpkC* mRNA levels exhibited a rapid and prolonged increase (Reyes et al., 2006). When Δ*mpkC* mutants were grown on plates containing various amounts of H_2O_2, there was no significant change in growth compared to Af293 (unpublished data). This raises the question as to why *mpkC* mRNA has such a significant response to H_2O_2. This suggests that MpkC has an overlapping function with other genes when exposed to oxidative stress. Studies on virulence of Δ*mpkC* mutants in *A. fumigatus* have not been done, as the function of this gene is not yet fully understood. If this MAPK proves to be important to virulence, this signaling pathway could be a target for antifungal therapy.

SakA

There is a subfamily of MAPK genes called stress-activated protein kinases. These protein kinases specifically convey environmental stress signals. Members of this subfamily include Hog1p from *S. cerevisiae* and Spc1p/Sty1p from *S. pombe*, as well as SakA in *A. nidulans* (Kawasaki et al., 2002; May et al., 2005). There are numerous forms of stress, such as hyperosmolarity, heat shock, UV irradiation, and oxidative stress, which elicit responses from mammalian and fission yeast stress-activated protein kinases, whereas high-osmolarity stress is the main activator of Hog1p of *S. cerevisiae* (Kawasaki et al., 2002). SakA of *A. fumigatus* shows similar responses to stress conditions. Deletion of *sakA* results in growth arrest of germlings in response to a shift to high-osmolarity medium. Interestingly, conidia grown continuously under the same hypertonic stress will germinate and grow hyphae, although the rate is slower than the parent strain. This suggests that the signaling pathway that *sakA* is involved in is not active during metabolically dormant periods or may be used for other purposes (May et al., 2005; Xue et al., 2004).

When Af293 (wild type) and Δ*sakA* are grown on complete medium (CM) and MM, there is a difference in conidiation. Both strains grow similarly on CM, but Δ*sakA* germinates at a faster rate than Af293 on MM. The presence of yeast extract, peptone, and tryptone is the major difference between MM and CM. All of these compounds are sources of reduced nitrogen. Although there is no difference in germination between the wild-type strain and Δ*sakA* on different carbon sources, there seems to be a difference in relation to nitrogen content. On MM containing nonpreferred nitrogen sources, such as sodium nitrate or sodium nitrite, Af293 conidia germinate more slowly than Δ*sakA* conidia. When exposed to reduced-N sources, such as ammonium chloride or proline, both strains, wild type and Δ*sakA*, germinate equally well. From these results, one could hypothesize that SakA functions to negatively regulate conidial germination in response to less-preferred nitrogen sources. In addition to these studies, it has been shown that transcription of *sakA* increases robustly when hyphae are shifted from rich medium to MM lacking either a nitrogen or carbon source (Xue et al., 2004). Although complete studies have not been done on the role of *sakA* in virulence of *A. fumigatus*, it would be interesting to see if there is an effect. It is also worth noting that Δ*sakA* mutants have shown a loss of virulence in other fungal species (Alonso-Monge et al., 1999; Bahn et al., 2006).

CONCLUSIONS

As demonstrated through numerous studies over a wide variety of species, MAPKs are involved within signaling pathways responsible for individual maintenance and integrity for a range of environmental and nutritional stresses. These MAPK signaling pathways could prove to be potential targets for antifungals, given their central role in governing fundamental homeostatic systems regulating fungal cellular physiology. Although scientists have considerable understanding of MAPK signaling pathways in a number of yeast and filamentous fungi, there is still much to be learned about their roles in medically important fungi, such as *A. fumigatus*. Work in this area of study may not only expand our general knowledge of how MAPKs function in the regulation of fungal physiology, but also they may lead to the development of new and effective classes of antifungal drugs.

In summary, major regulatory signaling pathways that regulate fungal cell physiology and contribute to robust hyphal growth are good candidates for pathways that might be exploited in the development of a novel mechanism to inhibit fungal growth in an animal host. It is significant that the body of research in this area for *A. fumigatus* is limited. To fully grasp the importance of protein kinases in regulating fungal growth and morphogenesis, a more concerted effort is needed. The development of improved methods and strains for targeting gene deletions should facilitate the study of these important pathways in *A. fumigatus*, an important and increasingly common opportunistic human pathogen (da Silva Ferreira et al., 2006; Krappmann et al., 2006).

REFERENCES

Adams, T. H., J. K. Wieser, and J. H. Yu. 1998. Asexual sporulation in *Aspergillus nidulans*. *Microbiol. Mol. Biol. Rev.* **62:**35–54.

Alonso-Monge, R., F. Navarro-Garcia, G. Molero, R. Diez-Orejas, M. Gustin, J. Pla, M. Sanchez, and C. Nombela. 1999. Role of the mitogen-activated protein kinase Hog1p in morphogenesis and virulence of *Candida albicans*. *J. Bacteriol.* **181:**3058–3068.

Arie, T., I. Kaneko, T. Yoshida, M. Noguchi, Y. Nomura, and I. Yamaguchi. 2000. Mating-type genes from asexual phytopathogenic ascomycetes *Fusarium oxysporum* and *Alternaria alternata*. *Mol. Plant Microbe Interact.* **13:**1330–1339.

Bahn, Y. S., K. Kojima, G. M. Cox, and J. Heitman. 2006. A unique fungal two-component system regulates stress responses, drug sensitivity, sexual development, and virulence of *Cryptococcus neoformans*. *Mol. Biol. Cell* **17:**3122–3135.

Barriere, A., and M. A. Felix. 2005. High local genetic diversity and low outcrossing rate in *Caenorhabditis elegans* natural populations. *Curr. Biol.* **15:**1176–1184.

Beauvais, A., and J. P. Latgé. 2001. Membrane and cell wall targets in *Aspergillus fumigatus*. *Drug Resist. Update* **4:**38–49.

Brakhage, A. A., and B. Liebmann. 2005. *Aspergillus fumigatus* conidial pigment and cAMP signal transduction: significance for virulence. *Med. Mycol.* **43**(Suppl. 1)**:**S75–S82.

Bruneau, J. M., T. Magnin, E. Tagat, R. Legrand, M. Bernard, M. Diaquin, C. Fudali, and J. P. Latgé. 2001. Proteome analysis of *Aspergillus fumigatus* identifies glycosylphosphatidylinositol-anchored proteins associated to the cell wall biosynthesis. *Electrophoresis* **22:**2812–2823.

Bussink, H. J., and S. A. Osmani. 1999. A mitogen-activated protein kinase (MPKA) is involved in polarized growth in the filamentous fungus, *Aspergillus nidulans*. *FEMS Microbiol. Lett.* **173:**117–125.

Cardenas, M. E., N. S. Cutler, M. C. Lorenz, C. J. Di Como, and J. Heitman. 1999. The TOR signaling cascade regulates gene expression in response to nutrients. *Genes Dev.* **13:**3271–3279.

Casselton, L. A. 2002. Mate recognition in fungi. *Heredity* **88:**142–147.

da Silva Ferreira, M. E., M. R. Kress, M. Savoldi, M. H. Goldman, A. Hartl, T. Heinekamp, A. A. Brakhage, and G. H. Goldman. 2006. The *akuB*[KU80] mutant deficient for nonhomologous end joining is a powerful tool for analyzing pathogenicity in *Aspergillus fumigatus*. *Eukaryot. Cell* **5:**207–211.

Dyer, P. S., M. Paoletti, and D. B. Archer. 2003. Genomics reveals sexual secrets of *Aspergillus*. *Microbiology* **149:**2301–2303.

Dyer, P. S., and M. Paoletti. 2005. Reproduction in *Aspergillus fumigatus*: sexuality in a supposedly asexual species? *Med. Mycol.* **43**(Suppl. 1)**:**S7–S14.

Fillinger, S., M. K. Chaveroche, K. Shimizu, N. Keller, and C. d'Enfert. 2002. cAMP and Ras signalling independently control spore germination in the filamentous fungus *Aspergillus nidulans*. *Mol. Microbiol.* **44:**1001–1016.

Fortwendel, J. R., J. C. Panepinto, A. E. Seitz, D. S. Askew, and J. C. Rhodes. 2004. *Aspergillus fumigatus* rasA and rasB regulate the timing and morphology of asexual development. *Fungal Genet. Biol.* **41:**129–139.

Fortwendel, J. R., W. Zhao, R. Bhabhra, S. Park, D. S. Perlin, D. S. Askew, and J. C. Rhodes. 2005. A fungus-specific Ras homolog contributes to the hyphal growth and virulence of *Aspergillus fumigatus*. *Eukaryot. Cell* **4:**1982–1989.

Gustin, M. C., J. Albertyn, M. Alexander, and K. Davenport. 1998. MAP kinase pathways in the yeast *Saccharomyces cerevisiae*. *Microbiol. Mol. Biol. Rev.* **62:**1264–1300.

Hiscock, S. J., and U. Kues. 1999. Cellular and molecular mechanisms of sexual incompatibility in plants and fungi. *Int. Rev. Cytol.* **193:**165–295.

Kawai, M., A. Nakashima, M. Ueno, T. Ushimaru, K. Aiba, H. Doi, and M. Uritani. 2001. Fission yeast Tor1 functions in response to various stresses including nitrogen starvation, high osmolarity, and high temperature. *Curr. Genet.* **39:**166–174.

Kawasaki, L., O. Sanchez, K. Shiozaki, and J. Aguirre. 2002. SakA MAP kinase is involved in stress signal transduction, sexual development and spore viability in *Aspergillus nidulans*. *Mol. Microbiol.* **45:**1153–1163.

Kojima, K., Y. Takano, A. Yoshimi, C. Tanaka, T. Kikuchi, and T. Okuno. 2004. Fungicide activity through activation of a fungal signalling pathway. *Mol. Microbiol.* **53:**1785–1796.

Krappmann, S., C. Sasse, and G. H. Braus. 2006. Gene targeting in *Aspergillus fumigatus* by homologous recombination is facilitated in a nonhomologous end-joining-deficient genetic background. *Eukaryot. Cell* **5:**212–215.

Lafon, A., J. A. Seo, K. H. Han, J. H. Yu, and C. d'Enfert. 2005. The heterotrimeric G-protein GanBα-SfaDβ-GpgAγ is a carbon source sensor involved in early cAMP-dependent germination in *Aspergillus nidulans*. *Genetics* **171:**71–80.

Lengeler, K. B., R. C. Davidson, C. D'Souza, T. Harashima, W. C. Shen, P. Wang, X. Pan, M. Waugh, and J. Heitman. 2000. Signal transduction cascades regulating fungal development and virulence. *Microbiol. Mol. Biol. Rev.* **64:**746–785.

Liebmann, B., S. Gattung, B. Jahn, and A. A. Brakhage. 2003. cAMP signaling in *Aspergillus fumigatus* is involved in the regulation of the virulence gene *pksP* and in defense against killing by macrophages. *Mol. Genet. Genomics* **269:**420–435.

Liebmann, B., M. Muller, A. Braun, and A. A. Brakhage. 2004. The cyclic AMP-dependent protein kinase A network regulates development and virulence in *Aspergillus fumigatus*. *Infect. Immun.* **72:**5193–5203.

Mah, J. H., and J. H. Yu. 2006. Upstream and downstream regulation of asexual development in *Aspergillus fumigatus*. *Eukaryot. Cell* **5:**1585–1595.

May, G. S., T. Xue, D. P. Kontoyiannis, and M. C. Gustin. 2005. Mitogen activated protein kinases of *Aspergillus fumigatus*. *Med. Mycol.* **43**(Suppl. 1)**:**S83–S86.

Murtagh, G. J., P. S. Dyer, and P. D. Crittenden. 2000. Sex and the single lichen. *Nature* **404:**564.

Nasrallah, M. E., P. Liu, S. Sherman-Broyles, N. A. Boggs, and J. B. Nasrallah. 2004. Natural variation in expression of self-incompatibility in *Arabidopsis thaliana*: implications for the evolution of selfing. *Proc. Natl. Acad. Sci. USA* **101:**16070–16074.

Oliver, B. G., J. C. Panepinto, D. S. Askew, and J. C. Rhodes. 2002a. cAMP alteration of growth rate of *Aspergillus fumigatus* and *Aspergillus niger* is carbon-source dependent. *Microbiology* **148:**2627–2633.

Oliver, B. G., J. C. Panepinto, J. R. Fortwendel, D. L. Smith, D. S. Askew, and J. C. Rhodes. 2002b. Cloning and expression of pkaC and pkaR, the genes encoding the cAMP-dependent protein kinase of *Aspergillus fumigatus*. *Mycopathologia* **154:**85–91.

Osherov, N., and G. May. 2000. Conidial germination in *Aspergillus nidulans* requires RAS signaling and protein synthesis. *Genetics* **155:**647–656.

Panepinto, J. C., B. G. Oliver, T. W. Amlung, D. S. Askew, and J. C. Rhodes. 2002. Expression of the *Aspergillus fumigatus rheb* homologue, *rhbA*, is induced by nitrogen starvation. *Fungal Genet. Biol.* **36:**207–214.

Panepinto, J. C., B. G. Oliver, J. R. Fortwendel, D. L. Smith, D. S. Askew, and J. C. Rhodes. 2003. Deletion of the *Aspergillus fumigatus* gene encoding the Ras-related protein RhbA reduces virulence in a model of invasive pulmonary aspergillosis. *Infect. Immun.* **71:**2819–2826.

Paoletti, M., F. A. Seymour, M. J. Alcocer, N. Kaur, A. M. Calvo, D. B. Archer, and P. S. Dyer. 2007. Mating type and the genetic basis of self-fertility in the model fungus *Aspergillus nidulans*. *Curr. Biol.* **17:**1384–1389.

Pitt, J. I., R. A. Samson, and J. C. Frisvad. 2000. Nomenclature of *Penicillium* and *Aspergillus* and their teleomorphs, p. 3–82. *In* R. Samson and J. Pitt (ed.), *Integration of Modern Taxonomic Methods for* Penicillium *and* Aspergillus *Classification*. Hardwood Academic Publishers, Amsterdam, The Netherlands.

Poggeler, S. 2002. Genomic evidence for mating abilities in the asexual pathogen *Aspergillus fumigatus*. *Curr. Genet.* **42:**153–160.

Ram, A. F., and F. M. Klis. 2006. Identification of fungal cell wall mutants using susceptibility assays based on Calcofluor white and Congo red. *Nat. Protoc.* **1:**2253–2256.

Raper, K. B., and D. I. Fennell. 1965. *The Genus* Aspergillus. Williams & Wilkins, Baltimore, MD.

Reyes, G., A. Romans, C. K. Nguyen, and G. S. May. 2006. Novel mitogen-activated protein kinase MpkC of *Aspergillus fumigatus* is required for utilization of polyalcohol sugars. *Eukaryot. Cell* **5:**1934–1940.

Rhodes, J. C., B. G. Oliver, D. S. Askew, and T. W. Amlung. 2001. Identification of genes of *Aspergillus fumigatus* up-regulated during growth on endothelial cells. *Med. Mycol.* **39:**253–260.

Shimizu, K. K., J. M. Cork, A. L. Caicedo, C. A. Mays, R. C. Moore, K. M. Olsen, S. Ruzsa, G. Coop, C. D. Bustamante, P. Awadalla, and M. D. Purugganan. 2004. Darwinian selection on a selfing locus. *Science* **306:**2081–2084.

Skromne, I., O. Sanchez, and J. Aguirre. 1995. Starvation stress modulates the expression of the *Aspergillus nidulans brlA* regulatory gene. *Microbiology* **141:**21–28.

Som, T., and V. S. Kolaparthi. 1994. Developmental decisions in *Aspergillus nidulans* are modulated by Ras activity. *Mol. Cell. Biol.* **14:**5333–5348.

Urano, J., A. P. Tabancay, W. Yang, and F. Tamanoi. 2000. The *Saccharomyces cerevisiae* Rheb G-protein is involved in regulating canavanine resistance and arginine uptake. *J. Biol. Chem.* **275:** 11198–11206.

Valiante, V., T. Heinekamp, R. Jain, A. Härtl, and A. A. Brakhage. 2008. The mitogen-activated protein kinase MpkA of *Aspergillus fumigatus* regulates cell wall signaling and oxidative stress response. *Fungal Genet. Biol.* **45:**618–627.

Xue, T., C. K. Nguyen, A. Romans, and G. S. May. 2004. A mitogen-activated protein kinase that senses nitrogen regulates conidial germination and growth in *Aspergillus fumigatus. Eukaryot. Cell* **3:** 557–560.

Yun, S. H., T. Arie, I. Kaneko, O. C. Yoder, and B. G. Turgeon. 2000. Molecular organization of mating type loci in heterothallic, homothallic, and asexual *Gibberella/Fusarium* species. *Fungal Genet. Biol.* **31:**7–20.

Zhao, W., J. C. Panepinto, J. R. Fortwendel, L. Fox, B. G. Oliver, D. S. Askew, and J. C. Rhodes. 2006. Deletion of the regulatory subunit of protein kinase A in *Aspergillus fumigatus* alters morphology, sensitivity to oxidative damage, and virulence. *Infect. Immun.* **74:**4865–4874.

Aspergillus fumigatus and Aspergillosis
Edited by J.-P. Latgé and W. J. Steinbach

Chapter 14

Cell Wall of *Aspergillus fumigatus*: a Dynamic Structure

ISABELLE MOUYNA AND THIERRY FONTAINE

The cell wall is an essential extracellular organelle that accounts for 20 to 40% of the cellular dry weight of most fungi (Latgé, 2007). The cell wall is permanently in contact with the external fungal environment. It is essential in resisting osmotic pressure, and it allows the fungal cell to withstand environmental aggressions. Because of the pathogenicity of *Aspergillus fumigatus*, cell wall composition and chemical organization have been studied to identify fungal molecules involved in host-pathogen interactions, resistance to microbial killing mechanisms, and immune defenses. In addition, because of its essentiality for fungal life, the biochemical pathways responsible for its biosynthesis are drug targets.

GENERAL ORGANIZATION AND COMPOSITION OF THE CELL WALL

The cell wall structure varies with the fungal morphotype and culture conditions. Figure 1 shows the cell wall of a resting conidium, a germinating conidium, and a hypha. Transmission electron microscopy shows that the conidium has a two-layer cell wall: a translucent inner layer and an electron-dense pigmented outer layer. On the surface is a rodlet layer that confers hydrophobicity to the conidium. Scanning electron microscopy by atomic force microscopy shows dramatic changes of the cell surface structures after 2 h of germination. The echinulate morphology and the hydrophobic layer of the resting conidia are progressively lost before the swelling, and the cell surface layer is changed into a layer of amorphous material (Dague et al., 2007; Rhode et al., 2002; Tronchin et al., 1995). Swelling is associated with the emergence of β-1,3-glucan to the cell surface that is accessible to dectin-1 or anti-β-1,3-glucan monoclonal antibody (Gersuk et al., 2006). The swelling of conidia requires both a cell wall plasticity and the formation of a new cell wall inner layer during the isodiametric growth that will emerge as the new hypha cell wall (Fig. 1). These morphogenetic events require coordinated hydrolases and synthase activities that are not yet understood.

In static aerial growth, in contrast to liquid growth, the mycelium forms a highly hydrophobic compact colony with a water-repellent surface (Beauvais et al., 2007). Electron microscopy studies show the presence of a single electron-translucent layer under shaken liquid conditions. In contrast, a thin electron-dense layer and an extracellular material cover the hyphal cell wall surface of mycelia grown under aerial conditions.

Polysaccharides represent the major part of the fungal cell wall and are responsible for its physical properties, such as rigidity and plasticity, necessary to maintain the cell morphology. Their chemical organization and interaction are responsible for the separation of two fractions discriminated by their solubility in sodium hydroxide solution. The biochemical approaches used to determine this were based on specific polysaccharide solubilization with recombinant enzymes to isolate soluble oligosaccharides by liquid chromatography and characterize them by chemical methods developed for carbohydrate chemistry. Six different polysaccharides have been observed in the cell wall of *A. fumigatus* mycelia: β-1,3-glucan, chitin, galactomannan, α-1,3-glucan, β-1,3/1,4-glucan, and a polymer of galactosaminogalactan (Table 1). Most of them are present both in mycelia and conidia. No β-1,6-glucans, which have been identified in yeasts, have been detected in *A. fumigatus*.

CHITIN AND β-1,3-GLUCANS: ESSENTIAL COMPONENTS OF THE ALKALI-INSOLUBLE FRACTION

β-1,3-Glucan and chitin have only been observed in the alkali-insoluble fraction and are described to

Isabelle Mouyna and Thierry Fontaine • Unité des Aspergillus, Institut Pasteur, 25 Rue du Dr Roux, 75015 Paris, France.

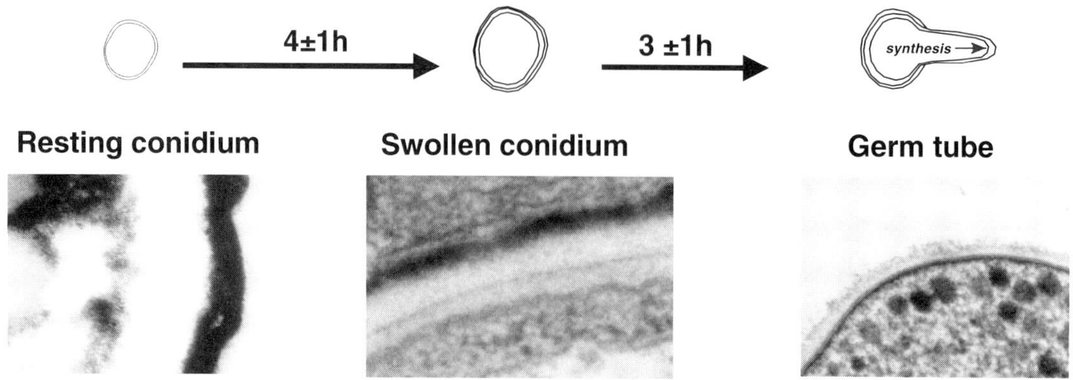

Figure 1. Cell cycle of *A. fumigatus*. Transmission electron micrographs show resting conidia, swelling conidia, and a germ tube of *A. fumigatus*.

be responsible for the frame of the cell wall. β-1,3-Glucan, the main alkali-insoluble component, is highly branched, with β-1,6 linkages (4% of branch points) constituting a three-dimensional network with a large number of side chains and ramifications. Other polysaccharides, such as chitin, galactomannan, and β-1,3/1,4-glucan, are cross-linked to the branched β-1,3/1,6-glucan network at the nonreducing ends of side chains (Fig. 2) (Fontaine et al., 2000). The β-1,3/1,4-glucan is a linear polymer representing 10% of the total amount of β-glucans of the cell wall and is composed by polymerization of the repeat unit, -3-βGlc1-4βGlc-1, chemical structure similar to lichen β-glucans. Chitin, a polymer of *N*-acetylglucosamine, is cross-linked to β-1,3-glucan through a β-1,4 linkage. The chitin/branched β-1,3-glucan complex is common to yeast and filamentous fungi. The strong insolubility of the chitin chain is essential for the alkali insolubility of β-glucan of the cell wall (Hartland et al., 1994).

Table 1. Polysaccharidic composition of the *A. fumigatus* mycelium and conidial cell wall[a]

Polysaccharide	Mycelium		Conidia	
	AI	AS	AI	AS
β-1,3-Glucan	30	—	38	5
β-1,3/1,4-Glucan	3	—	ND	ND
Chitin	13	—	1.7	0.3
Chitosan	4	—	3.9	0.2
α-1,3-Glucan	—	42	—	14
Galactosaminogalactan	4	2.3	—	—
Mannose	2	0.9	16	9
Galactose	3	0.5	10	4

[a]The composition was obtained after SDS–β-mercaptoethanol extraction on crude cell preparations (Maubon et al., 2006). Mycelia were produced in Brian medium. Conidia were produced in malt solid medium. AI, alkali insoluble; AS, alkali soluble; —, not detected; ND, not determined.

α-GLUCAN, THE MAIN COMPONENT OF THE ALKALI-SOLUBLE FRACTION

The alkali-soluble fraction of cell wall polysaccharides is mainly composed of α-1,3-glucan. This type of polymer, containing some α-1,4 linkages, was described in yeast and filamentous fungi a long time ago; nigeran, a hot-water-soluble α-glucan from *Aspergillus niger*, has alternative 1,3 and 1,4 linkages (Barker et al., 1957; Bobbitt et al., 1980), and pseudonigeran is an unbranched α-1,3-glucan containing 1 to 2% α-1,4 linkages (Horisberger et al., 1972). Cell wall α-glucan produced by *A. fumigatus* has a pseudonigeran structure containing around 1% α-1,4 linkages (W. Morelle, personal communication) and is also similar to the α-glucan of *Schizosaccharomyces pombe* (Grün et al., 2005). No nigeran has been found in *A. fumigatus* mycelia or conidia (Bobitt and Nordin, 1978; A. Beauvais, personal communication). In *A. fumigatus*, α-1,3-glucan is the major cell wall polysaccharide in conidia and mycelia and is only soluble in sodium hydroxide. Immunolabeling with anti-α-1,3-glucan antibodies reveals that α-1,3-glucans are not present in the inner layer of the cell wall but are mainly localized at the outer layer of the cell wall (Beauvais et al., 2007).

GALACTOMANNAN, A UNIQUE FUNGAL POLYSACCHARIDE

The polysaccharide galactomannan is present in both alkali-insoluble and alkali-soluble fractions. In contrast to the yeast cell wall, where mannans are N- or O-linked to proteins, mannoproteins represent a minor part of the *A. fumigatus* cell wall mannans. Galactomannan was first isolated from the culture medium and found in serum of patients with aspergillosis infection.

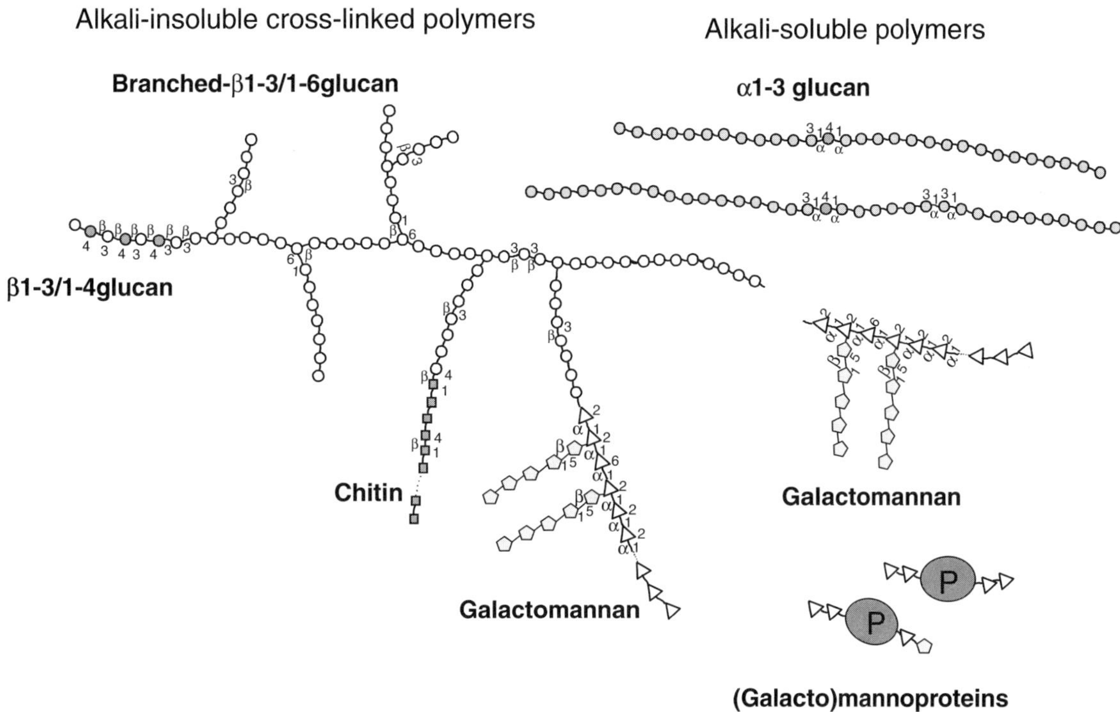

Figure 2. Schematic representation of alkali-insoluble and alkali-soluble polymers of the mycelial cell wall of *A. fumigatus*.

Galactomannan is constituted by a linear chain of α-mannoside residues where small side chains of one to five galactofuranoses are linked (Latgé et al., 1994). It is represented as a comb with mannan repeat units in Fig. 3.

The production of galactomannan and the ratio of galactofuranose on mannose residues depend on the culture medium. *A. fumigatus* produces a higher amount of galactomannan in a protein-rich medium than in a synthetic medium (T. Fontaine, unpublished results). In addition to the soluble form, the galactomannan is cross-linked to the nonreducing end of β-1,3-glucan chains of the alkali-insoluble fraction, indicating that it is involved in the cell wall organization of mycelia (Fig. 2) (Fontaine et al., 2000). Recently, galactomannan has also been described to be membrane bound through a glycosylphosphatidylinositol (GPI) anchor in the mycelium. This lipid anchor is a ceramide composed of C18:0-phytosphyngosine with a 2OH-C24:0 fatty acid (Costachel et al., 2005). This lipogalactomannan, with the same polysaccharide structure, is the first GPI-anchored fungal polysaccharide and may represent the galactomannan circulating in the culture medium, since *A. fumigatus* contains PI-phospholipase C (Bruneau et al., 2001). Moreover, the lipogalactomannan is also present in conidia (Fontaine, unpublished). The lipogalactomannan was purified by liquid chromatography on a column of hydrophobic interactions. However, the same chromatographic experiments with the galactomannan released from a cell wall extract showed that this cell wall galactomannan contains a lipid anchor (Fontaine, unpublished). These data suggest that the lipogalactomannan is also associated with the cell wall and could be involved in the plasma membrane-cell wall interactions.

-[-2Manα1-2Manα1-6Manα1-2Manα1-]-

|6 |3

Galfβ Galfβ

|5 |5

$[\text{Galf}\beta]_{1-2}$ $[\text{Galf}\beta]_{1-2}$

|5 |5

Galfβ Galfβ

Figure 3. Structure of the repeating unit of the galactomannan produced by *A. fumigatus* (Latgé et al., 1994).

GALACTOSAMINOGALACTAN, A SECRETED POLYSACCHARIDE

Galactosaminogalactan is a polymer composed of *N*-acetyl-galactosamine and galactopyranose. This polymer is produced by *Aspergillus* species (Bardalaye and Nordin, 1976; Takada et al., 1981). In *A. fumigatus*, this polymer is mainly secreted and recovered in the culture medium but has also been found in both alkali-insoluble and alkali-soluble fractions of mycelium cell wall (Fontaine et al., 2000) (Table 2). Its presence in mycelium cell wall seems to be due to its water insolubility, suggesting that the galactosaminogalactan has no function in the cell wall organization. Galactosaminogalactan is absent from the conidial cell wall.

PROTEINS OF THE MYCELIAL CELL WALL IN *A. FUMIGATUS*

The fungal cell wall contains many antigenic proteins and enzymes. The majority of these proteins are secreted proteins transiently found in the cell wall before they are secreted in the extracellular environment (Klis, 1994). These proteins can be easily removed by boiling in sodium dodecyl sulfate (SDS)–β-mercaptoethanol. In *A. fumigatus*, it has been estimated that an SDS–β-mercaptoethanol extraction released about 3 to 5% of proteins from a crude cell wall preparation (dry weight). Among the noncovalent cell wall-associated protein, no proteins with a carbohydrate-binding domain, as described for *Blastomyces dermatitidis*, were found in *A. fumigatus* (Brandhorst and Klein, 2000). Two types of cell wall-associated antigenic proteins can be found: (i) nonglycosylated or glycosylated proteins that do not contain galactofuranose residues, e.g., catalase 1 (Afu3g02270), RNase H (Afu5g02330), dipeptidylpeptidase DPPV (Afu2g09030), and AFMP2 (Afu2g05150), which have been previously isolated from the culture medium, indicating that they are transiently cell wall associated (Bernard et al., 2002; Chong et al., 2004); (ii) galactofuranose-containing glycosylated proteins. This galactofuran moiety is responsible for the reactivity of specific monoclonal antibodies (Morelle et al., 2005).

Table 2. Enzymes associated with synthesis of the mycelium and conidial cell wall in *A. fumigatus*

Group and enzyme name (locus)	Phenotype of mutant
β-1,3-Glucan synthases	
FKS1 (Afu6g12400)	Essential
RHO1 (Afu6g06900)	Essential
Chitin synthases (total, 8)	
CHSA (Afu2g01870), CHSB (Afu4g04180), CHSC (Afu5g00760), CHSD (Afu1g12600)	No
CHSG (Afu3g14420)	Reduced growth rate, increased hyphal branching
CHSF (Afu8g05630)	No
CHSE (Afu2g13440), CHSE[prime] (Afu2g13430)	Swollen hyphae, altered conidiation
α-1,3-Glucan synthases (total, 3)	
AGS1 (Afu3g00910)	Altered conidiation, hyphae are more branched
AGS2 (Afu2g11270)	Altered conidiation, hyphae are more branched
AGS3 (Afu1g15440)	More virulent
β-1,3-Glucan elongases (total, 7)	
GEL1 (Afu2g01170)	No
GEL2 (Afu6g11390)	Reduced growth, abnormal conidiogenesis
GEL4 (Afu2g05340)	
β-1,3-Glucanosyltransferase	
BGT1 (Afu1g11460)	No
β-1,3-Glucanases (total, >20)	
ENGL1 (AAF13033)	No
Fungal/bacterial chitinase	
CHIB (Afu8g01410)	No
Fungal/plant chitinases	
CHIa1, CHIa2, CHIa3, CHIa4, CHIa5	No
Melanin biosynthetic enzymes	
ALB1/PKS (Afu2g17600)	White conidia, reduced virulence
ARP1,2 (Afu2g17580, Afu2g17560)	Reddish-pink conidia
ABR1,2 (Afu2g17540, Afu2g17530)	Brown conidia
AYG1 (Afu2g17550)	Yellow-green condia
Hydrophobins (total, 7)	
RodA (Afu8g07270)	Conidia are not hydrophobic
RodB (Afu1g17250)	No

These results show that the *A. fumigatus* cell wall is a major antigen reservoir: patient sera contain high levels of antibodies directed against the *Aspergillus* mycelium wall (Hearn et al., 1991). Indeed, O-linked or N-linked glycans of *A. fumigatus* glycoproteins contain the galactofuranose epitope (Goto, 2007; Haido et al., 1998; Leitao et al., 2003; Morelle et al., 2005; Yuen et al., 2001). In addition to these antigenic proteins, the cell wall contains other soluble proteins. Enzyme activities have also been isolated from cell wall autolysate, including exo- and endo-β-glucanases (Fontaine et al., 1997a, 1997b) and transglucosidases (Hartland et al., 1996; Mouyna et al., 1998) able to modify cell wall β-1,3-glucan chains (see below). In contrast to yeast, in which GPI proteins are covalently linked to β-1,3-glucan via a β-1,6 linkage through GPI remnants or Pir protein directly to β-1,3-glucan through a glutamine residue (Ecker et al., 2006), no proteins covalently bound to cell wall polysaccharides have been found in *A. fumigatus*. Moreover, no β-1,6-glucan in the *A. fumigatus* cell wall has been detected.

A proteomic approach of cell wall-associated proteins showed that the major GPI-anchored protein, an acid phosphatase (PhoAp; Afu1g03570), is strongly associated with the cell wall β-1,3-glucans (Bernard et al., 2002). However, the covalent linkage was not established by carbohydrate chemical analysis. Moreover, in medium repressing *PHOA* expression, there is no effect on growth and cell wall organization, suggesting that PhoAp is not involved in cell wall stability. In addition, comparative genomic analyses have found that no gene coding for cell wall covalently linked protein orthologs of yeast cells, such as Flop, CWP, Tirp, or agglutinin, has been found in the *A. fumigatus* genome.

SPECIFICITIES OF THE CELL WALL OF CONIDIA OF *A. FUMIGATUS*

Rodlet Layer

A thin layer of regularly arranged rodlets covers the surface of aerial fungal conidia. This structure favors air buoyancy and dispersion of conidia in air (Beever and Dempsey, 1978). Hydrophobins are the proteins responsible for the rodlet structure and for the conidial surface properties. In *A. fumigatus*, two small proteins have been isolated from conidia, RodAp (16 kDa; Afu8g07270) and RodBp (14 kDa; Afu1g17250), with 44% similarity between them. Both proteins present a similar organization: after the hydrophobic signal peptide, there is a neutral-to-hydrophobic domain and a hydrophobic central core, followed by a small hydrophilic region (Paris et al., 2003). RodAp and RodBp are GPI-anchored proteins. RodAp is removed by HF, suggesting it is bound to polysaccharide through a phosphodiester bond (V. Kumar, unpublished results). In spite of these similarities, only RodAp is essential to the formation of a rodlet structure on the conidial surface. Moreover, conidia of a RodA mutant are easily wetted (Thau et al., 1994), are not dispersed in the air, and are less hydrophobic (Girardin et al., 1999). Conidial adhesion to mammalian extracellular matrix components is weakly altered in the RODA mutant (Paris et al., 2003), but the RODA mutant is less resistant to killing in mouse alveolar macrophages, showing that the rodlet layer or RodAp protects the conidia against the defense mechanism of the host (Paris et al., 2003).

Presence of Sialic Acid at the Cell Surface of Conidia

In addition to the rodlet layer, studies of *A. fumigatus*-host interaction have allowed scientists to identify ligands of human receptors involved in immune host defense. Sialic acid has been found to play an important role during viral or bacterial infection. Several fungal human pathogens express sialic acid on their cell surface (Alviano et al., 1999). *A. fumigatus* conidia are recognized by *Sambucus nigra* agglutinin, a specific lectin of *N*-acetyl-neuraminic acid (NeuAc) linked to galactose through an α-2,6 linkage. Further analyses by high-performance liquid chromatography and mass spectrometry have confirmed the presence of NeuAc on the cell surface of conidia (Warwas et al., 2007). Digestion of conidia with neuraminidase showed that NeuAc residues help the conidial adhesion to extracellular matrix proteins, such as fibronectin (Wasylnka et al., 2001). However, the orthologous gene of sialyl transferase has not yet been identified in the *A. fumigatus* genome.

Melanin Layer

In contrast to the mycelium cell wall, conidia show an echinulate surface morphology and a large electron-dense outer layer due to the presence of pigment (Fig. 1). This melanin layer has been described as a virulence factor for fungal pathogenic species. In *A. fumigatus*, the gray-green cell surface of conidia is due to the synthesis of melanin composed of macromolecules formed by the oxidative polymerization of phenolic compounds. The type of fungal melanin found in the conidial cell wall is 1,8-dihydroxynaphthalene-melanin (DHN-melanin). Genetic and biochemical investigations have shown that biosynthesis of the conidial DHN-melanin in *A. fumigatus* requires a six-gene cluster that includes the genes *ALB1/PKS* (Agu2g17600), *AYG1* (Afu2g17550), *ARP2* (Afu2g17560), *ARP1* (Afu2g17580), *ABR1* (Afu2g17540), and *ABR2* (Afu2g17530) (Tsai et al., 1998, 1999) (Fig. 4). In *A. fumigatus*, the disruption of

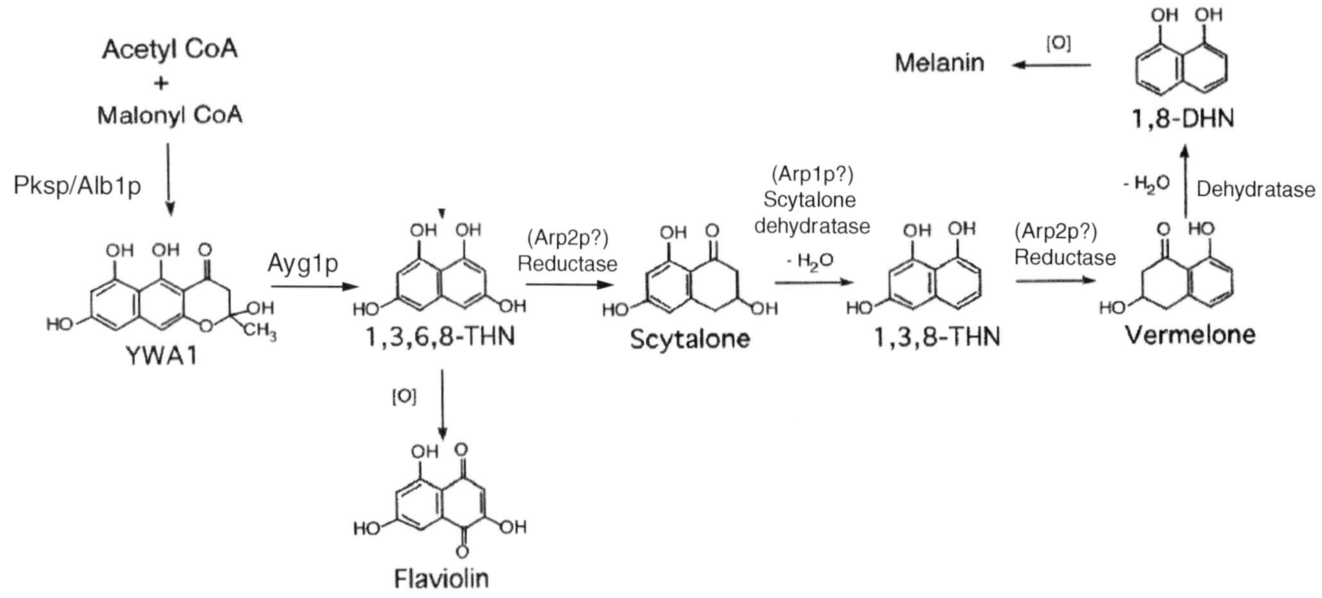

Figure 4. Schematic melanin biosynthetic pathway in *A. fumigatus* (modified from Tsai et al., 1998, 1999, 2001; Langfelder et al., 2003). YWA1, heptaketide naphthopyrone; 1,3,6,8-THN, 1,3,6,8-tetrahydronaphtalene; 1,3,8-THN, 1,3,8-trihydronaphthalene; 1,8-DHN, 1,8-dihydronaphthalene; [O], oxidation step.

PKS/ALB1 genes encoding the polyketide synthase that is involved in the first step of DHN-melanin synthesis is responsible for the formation of a pigmentless white mutant (Langfelder et al., 1998, 2003; Tsai et al., 1998). PKSp is a single large polypeptide with four active site domains, a β-keto-synthase, acyl transferase, acyl carrier, and Claisen-type cyclase domain. Instead of pentaketide-tetrahydroxynaphthalene, which is found in numerous fungi, in *A. fumigatus* the Pksp/Alb1p product is a heptaketide-naphthopyrone (*YWA1*) (Fig. 4), suggesting that AfAlb1p is a naphthopyrone synthase (Fujii et al., 2000; Watanabe et al., 2000). Scanning electron microscopy studies revealed that an *ALB1/PKS* disruptant produced nearly smooth, white conidia. In contrast, no change in conidia hydrophobicity, the water-repellent phenotype, was observed (Tsai et al., 1998). Nevertheless, melanin-defective mutants are more sensitive to hydrogen peroxide or sodium hypochlorite than the wild type and show a reduced virulence in the mouse model due to an increase of phagocytosis and of conidial sensitivity to reactive oxygen species (ROS) (Jahn et al., 1997, 2000). DHN-melanin is able to quench ROS derived from human granulocytes, which are important in eliminating fungal conidia (Latgé, 1999).

The second biosynthetic step for DHN-melanin in *A. fumigatus*, which requires Ayg1p, converts naphthapyrone to 1,3,6,8-tetrahydroxy-naphthalene (1,3,6,8-THN) to follow the biosynthetic pathway of 1,8-DHN (Tsai et al., 2001). The third step involved the reduction of 1,3,6,8-THN to scytalone, which is putatively catalyzed by the reductase enzyme encoded by *ARP2* (Tsai et al., 1999). Arp1p, encoded by *ARP1* and similar to a scytalone dehydratase, catalyzes the dehydration of scytalone to 1,3,8-trihydroxynaphthalene (1,3,8-THN). Deletion of this gene results in *A. fumigatus* colonies with pink conidia (Tsai et al., 1998). *ABR1* and *ABR2* genes code for putative multicopper oxidase and laccase, which catalyze the oxidation of various polyphenol compounds. Disruption of the *ABR1* or *ABR2* gene resulted, as described for other genes of the cluster, in an alteration of the conidial color phenotype but not of the virulence or ROS susceptibility (Tsai et al., 1999; Sugareva et al., 2006).

POLYSACCHARIDE BIOSYNTHESIS

β-1,3-Glucan Synthesis

β-1,3-Glucans are synthesized by a plasma membrane-bound glucan synthase complex which uses UDP-glucose as a substrate and extrudes linear β-1,3-glucan chains through the membrane into the periplasmic space (Douglas, 2001). The protein complex contains at least two proteins, a putative catalytic subunit encoded by the gene *FKS* with a mass of >200 kDa and a regulatory subunit encoded by the *RHO1* gene with a 20-kDa protein. The reaction adds 1 mol of glucose for every mole of UDP-glucose hydrolyzed, producing chains of increasing length. In *A. fumigatus*, 1,500 glucose residues are produced per chain (Beauvais et al., 1993, 2001). The number of genes encoding β-1,3-glucan synthases and the essentiality of each individual gene vary according to the fungal species.

In contrast to yeast, a single *FKS1* ortholog is present in the genome of *A. fumigatus* (Afu6g12400). It is essential, as shown in a diploid background after haploidization following transformation (Firon et al., 2002) or by RNA interference (Mouyna et al., 2004), indicating the essential function of β-1,3-glucan in the cell wall organization. *FKS* proteins are thus an obvious target for antifungal compounds. Three general structural families of known natural product inhibitors of β-1,3-glucan are known (Nyfeler and Keller-Schierlein, 1974; Onishi et al., 2000; Traxler et al., 1977). Among them, the echinocandins, such as caspofungin, micafungin, and anidulafungin, are now in clinical use (Kahn et al., 2006; Morrison, 2006).

Hydropathy analysis of the large protein encoded by all *FKS* family members predicts a localization within the plasma membrane, with as many as 16 transmembrane helices. A central hydrophilic domain of about 580 amino acids displays a remarkable degree of identity (>80%) among all known *FKS* protein sequences (Douglas, 2001). It has been proposed that this region is located on the cytoplasmic face of the plasma membrane and must have some essential, conserved function. Two aspartate residues at positions D392 and D441 in this region have been recognized as essential for function of the glucan synthase in yeast (Douglas, 2001).

β-1,3-Glucan synthase is regulated by Rhop-GTPase. Rhop-GTPase is regulated by switching between a GDP-bound inactive state and a GTP-bound active state with conformational changes (Wei et al., 1997). After synthesis in the endoplasmic reticulum (ER), Rho1p is geranylgeranylated, a modification required for Rho1p attachment to the membrane and intracellular traffic. In yeast, geranylgeranylated Rho1p and Fks1p are transported to the plasma membrane as an inactive complex through the classical secretory pathway. Rho1p is activated on its arrival at the plasma membrane by Rom2p, the GDP/GTP exchange factor of Rho1p that is only localized in the plasma membrane. The activation and the movement of Fks1p on the plasma membrane are required for proper cell wall β-1,3-glucan localization (Abe et al., 2003; Inoue et al., 1999). The role of Rho1-GTP in the regulation of β-1,3-glucan synthesis has also been shown in *A. fumigatus* (Afu6g06900) (Beauvais et al., 2001). In *A. fumigatus*, four *RHO* genes have been cloned. However, no experiments using recombinant proteins or with conditional mutants have identified which of the four Rho proteins identified in the *A. fumigatus* genome is the regulatory subunit of the glucan synthase of *A. fumigatus*. The role of guanine nucleotide exchange factors might be interchangeable, however, in contrast to yeast, as the unique ortholog of *ROM2* is not essential in *A. fumigatus* (Hu et al., 2007).

The gap in our knowledge of the β-1,3-glucan synthase mechanism raises many questions. In the absence of purification to homogeneity of the enzyme, the first provocative question is whether Fks1p is the true catalytic subunit of the glucan synthase.

Chitin Synthesis

The chitin synthases that are responsible for the synthesis of linear chains of β-1,4-*N*-acetylgucosamine from the substrate UDP-*N*-acetylglucosamine are a family of integral membrane proteins with molecular masses of 100 to 130 kDa (Roncero, 2002). The largest gene families of chitin synthases are found in filamentous fungi. Eight genes are found in *A. fumigatus*. The presence of the highest number of genes in filamentous ascomycetes is correlated with a higher chitin content in the mold cell wall. In contrast to β-1,3-glucan synthase inhibitor, no chitin inhibitor is used in the clinical setting. The peptide nucleotide antibiotics polyoxins and nikkomycins are strong competitive inhibitors of chitin synthase and are substrate analogs of UDP-N-GlcNAc (Munro and Gow, 2001). Polyoxin and nikkomycin are highly active in vitro but only poorly active in vivo, and none of the inhibitors of chitin synthesis has been launched yet in clinical practice. Six families of chitin synthases have been identified. Biochemical data have suggested that classes I to III are zymogenic, that is, they are stimulated in vitro by trypsin, whereas classes IV to VI are nonzymogenic. Three are specific to filamentous fungi (classes III, V, and VI) (Bowen et al., 1992). The significance of each of these six *CHS* classes is indeed not understood, as mutations affecting members of a common family do not always result in a similar phenotype. Two groups of mutants have been identified; however, the first group has reduced chitin content with normal in vitro chitin synthase activity. The second group is affected in enzyme activity but has a normal cell wall chitin content. The motif QR-RRW, present in all *CHS* genes from different fungi, has been proposed as a catalytic domain, since mutation in this domain results in a loss of chitin synthase activity (Cos et al., 1998).

Analysis of the common domains among the eight chitin synthases of *A. fumigatus* has shown two clusters of three and four genes and a singleton which are associated with chitin synthesis (Latgé et al., 2005). This bioinformatic analysis of the two chitin synthase families in *A. fumigatus* showed that most of the motifs are found in both families but their localization, including the localization of the QRRW motif in each family, is different, suggesting that the UDP-GlcNAc-binding domains inside each protein lead to different outcomes (Latgé et al., 2005). Two genes in the *CHS* families of *A. fumigatus* and other molds have a consensus domain that is homologous to kinesin or myosin motor-like domains (Takeshita et al., 2005). It has been shown in *Aspergillus nidulans* that this myosin motor-like domain of class V localizes near actin structures at the hyphal tips and septation sites. It binds to actin, and its binding is necessary for chitin synthase activity.

In contrast to *Candida albicans*, none of the *CHS* genes of *A. fumigatus* are essential (Mellado et al., 1995, 2003; C. Jimenez and C. Roncero, personal communication). Mutants with the most altered phenotype result only from inactivation of *CHSE* (Afu2g13440), CHSE′ (Afu2g13430), and *CHSG* (Afu3g14420) genes belonging to classes III and V (Aufauvre-Brown et al., 1997; Mellado et al., 1996). The phenotypes of these mutants include a reduction in hyphal growth, periodic swellings along the lengths of hyphae, and a block of conidiation which is partially restored by growth in the presence of an osmotic stabilizer. A double *CHSE*/*CHSG* disruption mutant has been obtained, and the phenotype of the double mutant is only additive (Mellado et al., 2003); the cell wall still contains chitin (at half the concentration of the parental strain) that becomes partly alkali soluble, suggesting a decrease of chain length or an altered cross-linked β-glucan acceptor. The double mutant still displays zymogenic and nonzymogenic chitin synthase activities.

Although relatedness and divergent domain structures can be correlated with function, expression and mutagenesis studies in fungi have not been very informative in this area. It remains to be seen whether there is a relationship between a specific fungal *CHS* and the structure of the chitin polymer in *A. fumigatus* at a cellular location, since it was shown in *C. albicans*. Moreover, in contrast to yeast, the regulation of *CHS* activity has not been studied in *A. fumigatus*.

Synthesis of α-1,3-Glucans

In *A. fumigatus*, three orthologs of the putative α-glucan synthase of *S. pombe* have been identified. Agsp contains two main regions with homologies to two consensus sequences, an amylase-like domain and a starch/glycogen synthase-like domain with three consensus UDP-glucose-binding regions (Beauvais et al., 2005; Hochstenbach et al., 1998; Katayama et al., 1999; Maubon et al., 2006). All three of the *A. fumigatus* genes are expressed during vegetative growth. In contrast to *S. pombe*, in which *AGS1* is an essential gene, none of the *A. fumigatus* genes is essential. Both Af*AGS1* (Afu3g00910) and Af*AGS2* (Afu2g11270) single mutants produce a slightly altered hyphal growth with excessive hyphal branching and dichotomous apices (Beauvais et al., 2005). Both mutants also present a reduced conidiation associated with altered phialides on the *Aspergillus* head. However, these two proteins are not redundant, since Ags1p is localized at the septum of the germ tube, whereas Ags2p has an intracellular localization. Only the *AGS1* mutant produced a cell wall with a reduction of 50% of α-1,3-glucan content, which was not compensated by the presence of the *AGS2* gene (Beauvais et al., 2005). In contrast, the *AGS3* (Afu1g15440) mutant does not present a cell wall defect, morphological alteration, or a reduction of conidiation because of the high compensatory expression of *AGS1* (Maubon et al., 2006). In an experimental mouse model of invasive aspergillosis, the *AGS3* mutant is more virulent than the wild type, and this is correlated with an increased melanin content of the conidial cell wall and a better resistance to ROS (Maubon et al., 2006). In *A. fumigatus* α-1,3-glucans seem associated with the melanin layer deposition of the conidial cell wall (Maubon et al., 2006). In agreement with α-1,3-glucan localization in the cell wall of *A. fumigatus*, these data suggest that α-1,3-glucans interact with several cell wall compounds, have a cementing function in the cell wall, and limit the interaction of other compounds with the external environment.

Interestingly, in the double mutant of chitin synthase genes *CHSE*/*CHSG* of *A. fumigatus*, the defect of chitin is partly compensated by an increase of α-1,3-glucan content in the cell wall (Mellado et al., 2003). The compensatory phenomenon of the cell wall defect has also been observed in *A. niger*, in which the *AGS1* gene is induced in response to cell wall stress (Damveld et al., 2005), suggesting that α-glucan synthesis as well as chitin synthesis is upregulated to prevent the cell wall defect.

None of the *AGS*-encoded proteins has ever been purified and characterized in vitro to confirm that they are able to synthesize α-1,3-glucan in the absence of any cofactor or accessory protein.

In *Histoplasma capsulatum*, another gene encoding a putative α-1,4-amylase is essential for α-1,3-glucan production (Marion et al., 2006). In *A. niger*, GPI-anchored α-glucanotransferase, producing new α-1,4-glucosidic bonds on maltooligosaccharides, has been recently described (van der Kaaij et al., 2007). This transglucosidase activity is able to use nigero-oligosaccharides as an acceptor. Homologs of such enzymes are present in other fungal species with α-glucans in their cell wall, but not in yeast species lacking cell wall α-glucans. Biosynthesis of fungal α-glucans and their cell wall deposition require several enzymatic steps. The relative similarity with glycogen/starch synthase and/or hydrolase suggest that maltooligosaccharides might serve as a primer for α-1,3-glucan chain elongation (Marion et al., 2006; Vos et al., 2007). The biosynthesis in vitro of α-1,3-glucan remains a mystery.

BIOSYNTHESIS OF GALACTOMANNAN

The biosynthesis of the galactomannan- and galactofuranose-containing molecules (glycoproteins or glycolipids) remains unknown. Galactomannan seems to be secreted to the plasma membrane with a GPI anchor. At this stage, the galactomannan could be transferred to a β-1,3-glucan chain, as has been proposed for some yeast GPI-anchored proteins (Klis, 1994), or the galactomannan could be released in the culture medium by a specific glycosidase. The galactomannan structure is very different from that of yeast or mold N- or O-mannans isolated from glycoproteins. A comparative genomic analysis indicated that orthologs of most yeast mannosyltransferase genes can be found in the genome of *A. fumigatus*: *OCH1*, or the mannosyltransferase complex of *MNN9, VAN1*, and *ANP1* (Fontaine et al., 2004; Nierman et al., 2005). A novel α-mannosyltransferase activity has been detected in mycelium lysate of *A. fumigatus* (Fontaine et al., 2004). This activity transfers a mannose residue from GDP-Man to a GlcN-inositolphospholipid, where the lipid could be a diacylglycerol or a ceramide (Fontaine, unpublished), suggesting that this activity could be involved in lipogalactomannan biosynthesis. The elongation of the α-mannoside chain needs specific mannosyltransferase activities. The addition of galactofuranose residues to the mannan core to form a side chain also requires specific activities of galactofuranosyl transferase.

These enzymes use UDP-galactofuranose as the donor. A UDP-Gal mutase able to catalyze the conversion of UDP-galactopyranose to UDP-galactofuranose has been described for *A. fumigatus* (Bakker et al., 2005). Only one gene encoding this protein is present in the *A. fumigatus* genome (Afu3g12690), indicating that it is essential for the addition of galactofuranose residues on carbohydrate structures.

Recently, studies done in Routier's laboratory and in our laboratory showed that the deletion of this gene induces the absence of galactofuranose residue on any carbohydrate structure (Schmalhorst et al., 2008; C. Lamarre, unpublished data). Although galactofuranose is not an essential component, this mutant shows a growth defect, indicating that galactofuranose is involved in cell wall structure (Damveld et al., 2008; Schmalhorst et al., 2008).

POLYSACCHARIDE HYDROLYSIS

Conidial swelling and hyphal branching are morphogenetic events that require cell wall plasticity, which depends upon the activities of a range of hydrolytic enzymes found intimately associated with the fungal cell wall. Since glucan and chitin are the prominent components of the fungal cell wall, it is obvious that most of the fungal cell wall hydrolases characterized to date have chitinase or glucanase activity.

A survey of the *A. fumigatus* genome, using the protein sequences of ChiA1p (Afu5g03760) and ChiB1p (Afu8g01410) from this organism, reveals that *A. fumigatus* contains 18 homologous chitinase genes (Adams, 2004). Chitinases are divided into 13 fungal/bacterial chitinases found in bacteria and 5 fungal/plant chitinases, which are similar to chitinases from plants. Only one gene encoding a chitinase has been disrupted. Disruption of the gene encoding the ChiB1p chitinase of *A. fumigatus* had no effect on growth or morphogenesis in these organisms (Jaques et al., 2003). In *A. fumigatus*, five fungal/plant-type chitinase genes are present in the genome, and only one is a GPI-anchored protein. The single mutants and the multiple mutants have been obtained and do not show any phenotype, suggesting that these proteins are not involved in morphogenesis (C. Lamarre, personal communication). It is possible that related enzymes in *A. fumigatus* compensate for the loss of chitinase, since most of the chitinase genes are expressed during swelling of conidia. Alternatively, these secreted enzymes may have no morphogenetic roles. Instead, they may contribute to the digestion and utilization of exogenous chitin as a source of organic nutrients for energy and biosynthesis.

β-1,3-Glucan-hydrolyzing enzymes have also been investigated, since β-1,3-glucan is an essential cell wall polysaccharide and the formation of numerous nonreducing ends is necessary for the activity of β-1,3-glucanosyltransferase. Several glucanolytic activities have been detected in the cell wall autolysates of *A. fumigatus*. Fontaine et al. (1997a) detected monomeric and dimeric exo-1,3-glucanases with molecular masses of 82 and 230 kDa, respectively. An endo-β-1,3-glucanase with a molecular mass of 74 kDa was also detected (Fontaine et al., 1997b); disruption of the *ENGL1* gene encoding this enzyme did not lead to a phenotype distinct from that of the parental strain (Mouyna et al., 2002), suggesting that this endo-β-1,3-glucanase activity does not play a morphogenetic role. However, although only one enzyme has been analyzed, the genome of *A. fumigatus* contains >20 putative exo- and endo-β-glucanases which have to be studied before a role for endo-β-1,3-glucanase in fungal growth can be demonstrated. If genes encoding a putative α-1,3-

glucanase are present in the genome of *A. fumigatus*, none of them has been characterized.

To date, no study has shown that the cell separation requires the use of glycosylhydrolases. β-1,3-Glucan and chitin are synthesized at the apex as a linear polysaccharide, and the maintenance of the cell wall plastic rigidification of the cell wall, responsible for protection against osmotic pressure, results from cross-linking between these individual polysaccharides.

β-1,3-GLUCAN BRANCHING AND CROSS-LINKING ENZYMES

The chronological steps in the synthesis of the core structural polysaccharides in the fungal cell wall are depicted in Fig. 5. β-1,3-Glucan chains produced by the β-1,3-glucan synthase complex remain unorganized and alkali soluble until covalent linkages occur between β-1,3-glucans and other cell wall components. The transglycosidases responsible for these two major steps are being eagerly investigated. To date, only two β-1,3-glucanosyltransferases have been identified biochemically.

The first β-1,3-glucanosyltransferase of *A. fumigatus* (Bgl2p/Bgt1p; Afu1g11460) present in both yeasts and molds cleaves laminaribiose from the reducing end of linear β-1,3-glucans and transfers the remaining glucan to the nonreducing end of another β-1,3-glucan acceptor with a β-1,6 linkage (Goldman et al., 1995, Mouyna et al., 1998). Null mutants, however, do not display a phenotype different from the wild-type parental strain. Since the gene is present as a single copy in the *A. fumigatus* genome, the absence of phenotype for the null mutant suggests that this enzyme does not play a major role in cell wall morphogenesis and, in particular, in cross-linking of cell wall polysaccharides, a result expected since this enzyme requires a free reducing end to display its activity.

A second β-1,3-glucanosyltransferase, isolated originally from an autolysate of the cell wall of *A. fumigatus*, has also been characterized (Hartland et al., 1996; Mouyna et al., 2000a). This enzyme splits internally a β-1,3-glucan chain and transfers the newly generated reducing end to the nonreducing end of another β-1,3-glucan molecule. The generation of a new β-1,3 linkage between the acceptor and donor molecules results in the elongation of β-1,3-glucan chains. The predicted amino acid sequence of *GEL1* encoding this enzymatic activity is homologous to several yeast protein gene families, such as the *GAS* family in *Saccharomyces cerevisiae* and *PHR* in *C. albicans* (Muhlschlegel and Fonzi, 1997; Popolo and Vai, 1999; Ragni et al., 2007; Saporito-Irwin et al., 1995). In yeast, five *GAS* genes are present in the genome. In *A. fumigatus*, *GEL1* belongs to a family with seven members.

Sequence comparison and hydrophobic cluster analysis showed that all these proteins belong to family

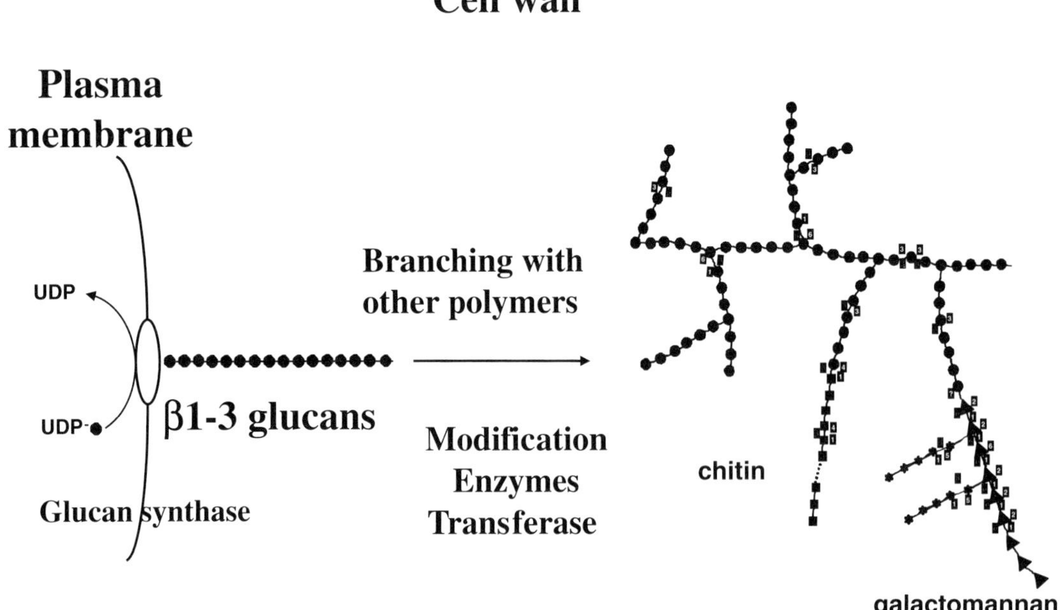

Figure 5. Temporal events in the biosynthesis of the structural β-1,3-glucan-chitin core. For simplification, only glucan synthase is represented, but chitin synthase is active at the same time and at the same location as glucan synthase. Linear chains of glucan are modified by both branching and cross-linking enzymes.

72 of the glycosylhydrolases and that two very conserved glutamate residues are involved in the active site of this family (Mouyna et al., 2000b). The C-terminal region has a variable length and is very dissimilar among the family of proteins, due to a domain containing a motif of six cysteine residues. This domain has been classified as CBM43 in the carbohydrate-binding modules (CBM) database of CaZy, since it was shown to act as an autonomous noncatalytic module for binding to laminarin in the protein of olive pollen Ole10 (Barral et al., 2005). Among the seven genes in *A. fumigatus*, only *GEL1* (Afu2g01170), *GEL2* (Afu6g11390), and *GEL4* (Afu2g05340) are expressed during spore germination and vegetative growth (A. Gastebois, personal communication). Disruption of *GEL1* did not result in a phenotype, whereas a Δ*gel2* mutant and the double mutant Δ*gel1* Δ*gel2* exhibit slower growth, abnormal conidiogenesis, and an altered cell wall composition. The reduced growth rate results in lower virulence of the Δ*gel1* Δ*gel2* mutant in a murine model of invasive aspergillosis (Mouyna et al., 2005). The Gelp family proteins, like their yeast orthologs, are glycosylated proteins attached to the membrane through a GPI anchor. The GPI anchoring of this glucanosyltransferase is in agreement with the function of this protein. As a result of anchoring to the plasma membrane, the active site of this enzyme faces the cell wall. This organization is consistent with its biological function, since this enzyme modifies the linear nascent β-1,3-glucan extruded from the plasma membrane into the cell wall space. Because of their putative role in cell wall biosynthesis, an analysis of genes coding for GPI-anchored proteins in *A. fumigatus* was undertaken. Prediction analysis by bioinformatics algorithms for GPI-anchored proteins (big-PI Fungal predictor; http://mendel.imp.ac.at/gpi/fungi_server.html) has been carried out. GPI-anchored proteins common to all fungi were especially researched, since GPI-anchored proteins with a putative role in cross-linking would be common to all fungi. Among the 81 genes coding for putative GPI-anchored proteins identified in *A. fumigatus*, only six families of GPI proteins in *A. fumigatus* were homologous to yeast GPI proteins: *SPS2, GAS, DFG, PLB, CRH*, and *YPS* (Bernard et al., 2002). Genomic data are in agreement with a proteomic study of GPI-anchored proteins of *A. fumigatus* (Bruneau et al., 2001). Five of these families were classified as membrane-bound GPI proteins in yeast (Caro et al., 1997; Hamada et al., 1998). Four of the GPI-bound proteins (*SPS2, GAS, DFG*, and *CRH*) are involved in cell wall construction, as SPS2 (*ECM33*; Afu4g06820), *DFG5*/*DCW1*, *CRH1*/*CRH2*, and *GEL*/*GAS* mutants have defects in the cell wall that are associated with a reduced growth phenotype (Cabib et al., 2007; Chabane et al., 2006; Kitagaki et al., 2002; Mouyna et al., 2000b, 2005; Rodriguez-Pena et al., 2000). However, the in vitro transglycosidase activity of all of them, except Gelp/Gasp, remains to be determined.

It has been shown that the cross-linking between β-1,3-glucans and chitin is essential for the formation of a resistant fibrillar skeletal component in *Ascomycetes*. No transglycosidase has been identified bioinformatically or biochemically which can achieve the branching of β-1,3-glucans and the subsequent cross-linking of chitin and β-1,3-glucan.

PERSPECTIVES

The fungal cell wall of *A. fumigatus* is a complex organization of macromolecules resulting from several biosynthetic pathways: protein synthesis in the ER, biosynthesis of mannan and glycans in the ER and Golgi complex, and synthesis of homopolysaccharide at the plasma membrane. Molecular biology methods and biochemical approaches have allowed researchers to identify genes and proteins involved in polysaccharide synthesis. However, the knowledge of fungal cell wall organization and biosynthesis remains limited and raises several questions:

(i) Genes encoding polysaccharide synthase activities have been characterized, but none of these enzyme activities has been demonstrated in vitro, especially since it is impossible to produce them as recombinant proteins because of their multiple transmembrane domains. Knowledge on their mode of action remains very limited. In *A. fumigatus*, as in other fungi, polysaccharides can be separated into two groups, the fibrillar alkali-insoluble and the alkali-soluble α-1,3-glucans that can play a cement cohesive function between cell wall polysaccharides. The kinetics of biosynthesis of each polysaccharide are unknown, but preliminary data suggest that the polysaccharides are not synthesized together or at the same cellular location.

(ii) GPI-anchored proteins have a role in the postsynthesis modifications responsible for the formation of a resistant skeleton that occurs in the cell wall space. A few proteins, such as Gelp, Crhp, and Ecm33p, identified in yeast and filamentous fungi, have been identified as involved in cell wall organization. Even when the enzymatic activity is known, biochemical data are insufficient for an understanding of their enzymatic and biological functions.

(iii) The branching of β-1,3-glucan chains and the cross-linking between chitin and β-1,3-glucan remain the most important postsynthesis modifications. In spite of the essential role of the cell wall for fungi and the number of research teams working in this area, no gene or protein responsible for these modifications has been identified in *A. fumigatus* or in other fungi.

(iv) The membrane-cell wall interactions seem essential for anchoring cell wall polysaccharides and for the cooperation between transmembrane synthase and GPI-anchored transglycosidase activities. Due to the presence of its lipid anchor, the lipogalactomannan from *A. fumigatus* could be involved in the plasma membrane-cell wall connection. The presence of membrane proteins linking the cytoskeleton and the cell wall polysaccharide through the plasma membrane is an interesting concept that has not been demonstrated biochemically.

REFERENCES

Abe, M., H. Qadota, A. Hirata, and Y. Ohya. 2003. Lack of GTP-bound Rho1p in secretory vesicles of *Saccharomyces cerevisiae*. *J. Cell Biol.* **162:**85–97.

Adams, D. J. 2004. Fungal cell wall chitinases and glucanases. *Microbiology* **150:**2029–2035.

Alviano, C. S., L. R. Travassos, and R. Schauer. 1999. Sialic acids in fungi: a minireview. *Glycoconj. J.* **16:**545–554.

Aufauvre-Brown, A., E. Mellado, N. A. R. Gow, and D. W. Holden. 1997. *Aspergillus fumigatus chsE:* a gene related to *CHS3* of *Saccharomyces cerevisiae* and important for hyphal growth and conidiophore development but not pathogenicity. *Fungal Genet. Biol.* **21:** 141–152.

Bakker, H., B. Kleczka, R. Gerardy-Schahn, and F. H. Routier. 2005. Identification and partial characterization of two eukaryotic UDP-galactopyranose mutases. *J. Biol. Chem.* **386:**657–661.

Bardalaye, P. C., and J. H. Nordin. 1976. Galactoaminogalactan from cell walls of *Aspergillus niger*. *J. Bacteriol.* **125:**655–669.

Barker, S. A., E. J. Bourne, D. M. O'Mant, and M. Stacey. 1957. Studies of *Aspergillus niger*. Part VI. The separation and structures of oligosaccharides from nigeran. *J. Chem. Soc.* **1957:**2448–2454.

Barral, P., C. Suarez, E. Batanero, C. Alfonso, D. Alche Jde, M. I. Rodriguez-Garcia, M. Villalba, G. Rivas, and R. Rodriguez. 2005. An olive pollen protein with allergenic activity, Ole e 10, defines a novel family of carbohydrate-binding modules and is potentially implicated in pollen germination. *Biochem. J.* **390:**77–84.

Beauvais, A., R. Drake, K. Ng, M. Diaquin, and J. P. Latgé. 1993. Characterization of the 1,3-beta-glucan synthase of *Aspergillus fumigatus*. *J. Gen. Microbiol.* **39:**3071–3078.

Beauvais, A., J. M. Bruneau, P. C. Mol, M. J. Buitrago, R. Legrand, and J. P. Latgé. 2001. Glucan synthase complex of *Aspergillus fumigatus*. *J. Bacteriol.* **183:**2273–2279.

Beauvais, A., D. Maubon, S. Park, W. Morelle, M. Tanguy, M. Huerre, D. S. Perlin, and J. P. Latgé. 2005. Two α(1-3) glucan synthases with different functions in *Aspergillus fumigatus*. *Appl. Environ. Microbiol.* **71:**1531–1538.

Beauvais, A., C. Schmidt, S. Guadagnini, P. Roux, E. Perret, C. Henry, S. Paris, A. Mallet, M. C. Prevost, and J. P. Latgé. 2007. An extracellular matrix glues together the aerial-grown hyphae of *Aspergillus fumigatus*. *Cell. Microbiol.* **9:**1588–1600.

Beever, R. E., and G. P. Dempsey. 1978. Function of rodlets on the surface of fungal spores. *Nature* **272:**608–610.

Bernard, M., I. Mouyna, G. Dubreucq, J. P. Debeaupuis, T. Fontaine, C. Vorgias, C. Fuglsang, and J. P. Latgé. 2002. Characterization of a cell-wall acid phosphatase (PhoAp) in *Aspergillus fumigatus*. *Microbiology* **148:**2819–2829.

Bobbitt, T. F., and J. H. Nordin. 1978. Hyphal nigeran as a potential phylogenetic marker for *Aspergillus* and *Penicillium* species. *Mycologia* **70:**1201–1211.

Bobbitt, T. F., J. H. Nordin, D. Gagnaire, and M. Vincendon. 1980. NMR investigation of specifically labeled [^{13}C]nigeran. *Carbohydr. Res.* **81:**177–181.

Bowen, A. R., J. L. Chen-Wu, M. Momany, R. Young, P. J. Szaniszlo, and P. W. Robbins. 1992. Classification of fungal chitin synthases. *Proc. Natl. Acad. Sci. USA* **89:**519–523.

Brandhorst, T., and B. Klein. 2000. Cell wall biogenesis of *Blastomyces dermatitidis*. Evidence for a novel mechanism of cell surface localization of a virulence-associated adhesin via extracellular release and reassociation with cell wall chitin. *J. Biol. Chem.* **275:** 7925–7934.

Bruneau, J. M., T. Magnin, E. Tagat, R. Legrand, M. Bernard, M. Diaquin, C. Fudali, and J. P. Latgé. 2001. Proteome analysis of *Aspergillus fumigatus* identifies glycosylphosphatidylinositol-anchored proteins associated to the cell wall biosynthesis. *Electrophoresis* **22:**2812–2823.

Cabib, E., N. Blanco, C. Grau, J. M. Rodriguez-Pena, and J. Arroyo. 2007. Crh1p and Crh2p are required for the cross-linking of chitin to β(1-6)glucan in the *Saccharomyces cerevisiae* cell wall. *Mol. Microbiol.* **63:**921–935.

Caro, L. H., H. Tettelin, J. H. Vossen, A. F. Ram, H. van den Ende, and F. M. Klis. 1997. In silicio identification of glycosyl-phosphatidylinositol-anchored plasma-membrane and cell wall proteins of *Saccharomyces cerevisiae*. *Yeast* **13:**1477–1489.

Chabane, S., J. Sarfati, O. Ibrahim-Granet, C. Du, C. Schmidt, I. Mouyna, M. C. Prevost, R. Calderone, and J. P. Latgé. 2006. Glycosylphosphatidylinositol-anchored Ecm33p influences conidial cell wall biosynthesis in *Aspergillus fumigatus*. *Appl. Environ. Microbiol.* **72:**3259–3267.

Chong, K. T. K., P. C. Y. Woo, S. K. P., S. K. P. Lau, Y. Huang, and K. Y. Yuen. 2004. *AFMP2* encodes a novel immunogenic protein of the antigenic mannoprotein superfamily in *Aspergillus fumigatus*. *J. Clin. Microbiol.* **42:**2287–2291.

Cos, T., R. A. Ford, J. A. Trilla, A. Duran, E. Cabib, and C. Roncero. 1998. Molecular analysis of Chs3p participation in chitin synthase III activity. *Eur. J. Biochem.* **256:**419–426.

Costachel, C., B. Coddeville, J. P. Latgé, and T. Fontaine. 2005. Glycosylphosphatidylinositol-anchored fungal polysaccharide in *Aspergillus fumigatus*. *J. Biol. Chem.* **280:**39835–39842.

Dague, E., D. Alsteens, J. P. Latgé, and Y. Dufrene. 2007. High-resolution cell surface dynamics of germinating *Aspergillus fumigatus* conidia. *Biophys. J.* **94:**656–660.

Damveld, R. A., A. Franke, M. Arentshorst, P. J. Punt, F. M. Klis, C. A. M. J. J. Van den Hondel, and A. F. J. Ram. 2008. A novel screening method for cell wall mutants in *Aspergillus niger* identifies UDP-galactopyranose mutase as an important protein in fungal cell wall biosynthesis. *Genetics* **178:**873–881.

Damveld, R. A., P. A. vanKuyk, M. Arentshorst, F. M. Klis, C. A. van den Hondel, and A. F. Ram. 2005. Expression of agsA, one of five 1,3-alpha-D-glucan synthase-encoding genes in *Aspergillus niger*, is induced in response to cell wall stress. *Fungal Genet. Biol.* **42:** 165–177.

Douglas, C. M. 2001. Fungal β(1,3)-D-glucan synthesis. *Med. Mycol.* **39**(Suppl. 1)**:**55–66.

Ecker, M., R. Deutzmann, L. Lehle, V. Mrsa, and W. Tanner. 2006. Pir proteins of *Saccharomyces cerevisiae* are attached to β(1-3)glucan by a new protein-carbohydrate linkage. *J. Biol. Chem.* **281:** 11523–11529.

Firon, A., A. Beauvais, J. P. Latgé, E. Couvé, M. C. Grosjean-Cournoyer, and C. d'Enfert. 2002. Characterization of essential genes by parasexual genetics in the human fungal pathogen *Aspergillus fumigatus*: impact of genomic rearrangements associated with electroporation of DNA. *Genetics* **161:**1077–1087.

Fontaine, T., R. P. Hartland, M. Diaquin, C. Simenel, and J. P. Latgé. 1997a. Differential patterns of activity displayed by two exo-β-1,3 glucanases associated with the *Aspergillus fumigatus* cell wall. *J. Bacteriol.* **179:**3154–3163.

Fontaine, T., R. Hartland, A. Beauvais, M. Diaquin, and J. P. Latgé. 1997b. Purification and characterization of an endo-1,3-β-glucanase from *Aspergillus fumigatus. Eur. J. Biochem.* **243:**315–321.

Fontaine, T., C. Simenel, G. Dubreucq, O. Adam, M. Delepierre, J. Lemoine, C. E. Vorgias, M. Diaquin, and J. P. Latgé. 2000. Molecular organization of the alkali-insoluble fraction of *Aspergillus fumigatus* cell wall. *J. Biol. Chem.* **275:**27594–27607.

Fontaine, T., T. K. Smith, A. Crossman, J. S. Brimacombe, J. P. Latgé, and M. A. Ferguson. 2004. *In vitro* biosynthesis of glycosylphosphatidylinositol in *Aspergillus fumigatus. Biochemistry* **43:**15267–15275.

Fujii, I., Y. Mori, A. Watanabe, Y. Kubo, G. Tsuji, and Y. Ebizuka. 2000. Enzymatic synthesis of 1,3,6,8-tetrahydroxynaphthalene solely from malonyl coenzyme A by a fungal iterative type I polyketide synthase PKS1. *Biochem. J.* **39:**8853–8858.

Gersuk, G. M., D. M. Underhill, L. Zhu, and K. A. Marr. 2006. Dectin-1 and TLRs permit macrophages to distinguish between different *Aspergillus fumigatus* cellular states. *J. Immunol.* **176:**3717–3724.

Girardin, H., S. Paris, J. Rault, M. N. Bellon-Fontaine, and J. P. Latgé. 1999. The role of the rodlet structure on the physicochemical properties of *Aspergillus* conidia. *Appl. Environ. Microbiol.* **29:**364–369.

Goldman, R. C., P. A. Sullivan, D. Zakula, and J. O. Capobianco. 1995. Kinetics of β-1,3 glucan interaction at the donor and acceptor sites of the fungal glucosyltransferase encoded by the *BGL2* gene. *Eur. J. Biochem.* **227:**372–378.

Goto, M. 2007. Protein O-glycosylation in fungi: diverse structures and multiple functions. *Biosci. Biotechnol. Biochem.* **71:**1415–1427.

Grün, C. H., F. Hochstenbach, B. M. Humbel, A. J. Verkleij, J. H. Sietsma, F. M. Klis, J. P. Kamerling, and J. F. G. Vliegenthart. 2005. The structure of cell wall α-glucan from fission yeast. *Glycobiology* **15:**245–257.

Haido, R. M. T., M. H. Silva, R. Ejzemberg, E. A. Leitao, V. M. Hearn, G. V. Evans, and E. Barreto Bergter. 1998. Analysis of peptidogalactomannans from the mycelial surface of *Aspergillus fumigatus. Med. Mycol.* **36:**313–321.

Hamada, K., S. Fukuchi, M. Arisawa, M. Baba, and K. Kitada. 1998. Screening for glycosylphosphatidylinositol (GPI)-dependent cell wall proteins in *Saccharomyces cerevisiae. Mol. Gen. Genet.* **258:** 53–59.

Hartland, R. P., C. A. Vermeulen, F. M. Klis, J. H. Sietsma, and J. G. Wessels. 1994. The linkage of (1-3)-β-glucan to chitin during cell wall assembly in *Saccharomyces cerevisiae. Yeast* **10:**1591–1599.

Hartland, R. P., T. Fontaine, J. P. Debeaupuis, C. Simenel, M. Delepierre, and J. P. Latgé. 1996. A novel β-(1-3)-glucanosyltransferase from the cell wall of *Aspergillus fumigatus. J. Biol. Chem.* **271:** 26843–26849.

Hearn, V. M., J. P. Latgé, and M. C. Prevost. 1991. Immunolocalization of *Aspergillus fumigatus* mycelial antigens. *J. Med. Vet. Mycol.* **29:**73–81.

Hochstenbach, F., F. M. Klis, H. Van Den Ende, E. Van Donselaar, P. J. Peters, and R. D. Klausner. 1998. Identification of a putative alpha-glucan synthase essential for cell wall construction and morphogenesis in fission yeast. *Proc. Natl. Acad. Sci. USA* **95:**9161–9166.

Horisberger, M., B. A. Lewis, and F. Smith. 1972. Structure of a (1→3)-β-D-glucan (pseudonigeran) of *Aspergillus niger* NNRL 326 cell wall. *Carbohydr. Res.* **23:**183–188.

Hu, W., S. Sillaots, S. Lemieux, J. Davison, S. Kauffman, A. Breton, A. Linteau, C. Xin, J. Bowman, J. Becker, B. Jiang, and T. Roemer. 2007. Essential gene identification and drug target prioritization in *Aspergillus fumigatus. PLoS Pathog.* **3:**1–14.

Inoue, S. B., H. Qadota, M. Arisawa, T. Watanabe, and Y. Ohya. 1999. Prenylation of Rho1p is required for activation of yeast 1,3-beta-glucan synthase. *J. Biol. Chem.* **274:**38119–38124.

Jahn, B., A. Koch, A. Schmidt, G. Wanner, H. Gehringer, S. Bhakdi, and A. A. Brakhage. 1997. Isolation and characterization of a pigmentless-conidium mutant of *Aspergillus fumigatus* with altered conidial surface and reduced virulence. *Infect. Immun.* **65:**5110–5117.

Jahn, B., F. Boukhallouk, J. Lotz, K. Langfelder, G. Wanner, and A. A. Brakhage. 2000. Interaction of human phagocytes with pigmentless *Aspergillus* conidia. *Infect. Immun.* **68:**3736–3739.

Jaques, A. K., T. Fukamizo, D. Hall, R. C. Barton, G. M. Escott, T. Parkinson, C. A. Hitchcock, and D. J. Adams. 2003. Disruption of the gene encoding the ChiB1 chitinase of *Aspergillus fumigatus* and characterization of a recombinant gene product. *Microbiology* **149:** 2931–2939.

Kahn, J. N., M.-J. Hsu, F. Racine, R. Giacobbe, and M. Motyl. 2006. Caspofungin susceptibility in *Aspergillus* and non-*Aspergillus* molds: inhibition of glucan synthase and reduction of β-D-1,3 glucan levels in culture. *Antimicrob. Agents Chemother.* **50:**2214–2216.

Katayama, S., D. Hirata, M. Arellano, P. Pérez, and T. Toda. 1999. Fission yeast α-glucan synthase *mok1* requires the actin cytoskeleton to localize the sites of growth and plays an essential role in cell morphogenesis downstream of protein kinase C function. *J. Cell Biol.* **144:**1173–1186.

Kitagaki, H., H. Wu., H. Shimoi, and K. Ito. 2002. Two homologous genes, *DCW1* (YKL046c) and *DFG5*, are essential for cell growth and encode glycosylphosphatidylinositol (GPI)-anchored membrane proteins required for cell wall biogenesis in *Saccharomyces cerevisiae. Mol. Microbiol.* **46:**1011–1022.

Klis, F. M. 1994. Review: cell wall assembly in yeast. *Yeast* **10:**851–869.

Langfelder, K., B. Jahn, H. Gehringer, A. Schmidt, G. Wanner, and A. A. Brakhage. 1998. Identification of a polyketide synthase gene (*pksP*) of *Aspergillus fumigatus* involved in conidial pigment biosynthesis and virulence. *Med. Microbiol. Immunol.* **187:**79–89.

Langfelder, K., M. Streibel, B. Jahn, G. Haase, and A. A. Brakhage. 2003. Biosynthesis of fungal melanins and their importance for human pathogenic fungi. *Fungal Genet. Biol.* **38:**143–158.

Latgé, J. 1999. *Aspergillus fumigatus* and aspergillosis. *Clin. Microbiol. Rev.* **12:**310-350.

Latgé, J. P. 2007. The cell wall: a carbohydrate armour for the fungal cell. *Mol. Microbiol.* **66:**279–290.

Latgé, J. P., H. Kobayashi, J. P. Debeaupuis, M. Diaquin, J. Sarfati, J. M. Wieruszeski, E. Parra, and B. Fournet. 1994. Chemical and immunological characterization of the galactomannan secreted by *Aspergillus fumigatus. Infect. Immun.* **62:**5424–5433.

Latgé, J. P., I. Mouyna, F. Tekaia, A. Beauvais, J. P. Debeaupuis, and W. Nierman. 2005. Specific molecular features in the organization and biosynthesis of the cell wall of *Aspergillus fumigatus. Med. Mycol.* 43(Suppl. 1):S15–S22.

Leitao, E. A., V. C. Bittencourt, R. M. Haido, A. P. Valente, J. Peter-Katalinic, M. Letzel, L. M. de Souza, and E. Barreto-Bergter. 2003. Beta-galactofuranose-containing O-linked oligosaccharides present in the cell wall peptidogalactomannan of *Aspergillus fumigatus* contain immunodominant epitopes. *Glycobiology* **13:**681–692.

Marion, C. L., C. A. Rappleye, J. T. Engle, and W. E. Goldman. 2006. An alpha-(1,4)-amylase is essential for alpha-(1,3)-glucan production and virulence in *Histoplasma capsulatum. Mol. Microbiol.* **62:** 970–983.

Maubon, D., S. Park, M. Tanguy, M. Huerre, C. Schmitt, M. C. Prevost, D. S. Perlin, J. P. Latgé, and A. Beauvais. 2006. *AGS3*, an α(1-3)glucan synthase gene family member of *Aspergillus fumigatus*, modulates mycelium growth in the lung of experimentally infected mice. *Fungal Genet. Biol.* **43:**366–375.

Mellado, E., A. Aufauvre-Brown, C. A. Specht, P. W. Robbins, and D. W. Holden. 1995. A multigene family related to chitin synthase genes of yeast in the opportunistic pathogen *Aspergillus fumigatus*. *Mol. Gen. Genet.* 246:353–359.

Mellado, E., A. Aufauvre-Brown, N. A. R. Gow, and D. W. Holden. 1996. The *Aspergillus fumigatus* chsC and chsG genes encode class III chitin synthases with different functions. *Mol. Microbiol.* 20: 667–679.

Mellado, E., G. Dubreucq, P. Mol, J. Sarfati, S. Paris, M. Diaquin, D. W. Holden, J. L. Rodriguez-Tudela, and J. P. Latgé. 2003. Cell wall biogenesis in a double chitin synthase mutant ($chsG^-/chsE^-$) of *Aspergillus fumigatus*. *Fungal Genet. Biol.* 38:98–109.

Morelle, W., M. Bernard, J. P. Debeaupuis, M. Buitrago, M. Tabouret, and J. P. Latgé. 2005. Galactomannoproteins of *Aspergillus fumigatus*. *Eukaryot. Cell* 4:1308–1316.

Morrison, V. A. 2006. Echinocandin antifungals: review and update. *Expert Rev. Anti Infect. Ther.* 4:325–342.

Mouyna, I., R. P. Hartland, T. Fontaine, M. Diaquin, and J. P. Latgé. 1998. A β(1-3) glucanosyltransferase isolated from the cell wall of *Aspergillus fumigatus* is an homolog of the yeast Bgl2p. *Microbiology* 144:3171–3180.

Mouyna, I., T. Fontaine, M. Vai, M. Monod, W. A. Fonzi, M. Diaquin, L. Popolo, R. P. Hartland, and J. P. Latgé. 2000a. GPI-anchored glucanosyltransferases play an active role in the biosynthesis of the fungal cell wall. *J. Biol. Chem.* 275:14882–14889.

Mouyna, I., M. Monod, T. Fontaine, B. Henrissat, B. Léchenne, and J. P. Latgé. 2000b. Identification of the catalytic residues of the first family of β(1-3)glucanosyltransferases identified in fungi. *Biochem. J.* 347:741–747.

Mouyna, I., J. Sarfati, P. Recco, T. Fontaine, B. Henrissat, and J. P. Latgé. 2002. Molecular characterization of a cell wall-associated β(1-3) endoglucanase of *Aspergillus fumigatus*. *Med. Mycol.* 40: 455–464.

Mouyna, I., C. Henry, T. L. Doering, and J. P. Latgé. 2004. Gene silencing with RNA interference in the human pathogenic fungus *Aspergillus fumigatus*. *FEMS Microbiol. Lett.* 237:317–324.

Mouyna, I., W. Morelle, M. Vai, M. Monod, B. Lechenne, T. Fontaine, A. Beauvais, J. Sarfati, M. C. Prevost, C. Henry, and J. P. Latgé. 2005. Deletion of *GEL2* encoding for a β(1-3)glucanosyltransferase affects morphogenesis and virulence in *Aspergillus fumigatus*. *Mol. Microbiol.* 56:1675–1688.

Mühlschlegel, F. A., and W. A. Fonzi. 1997. *PHR2* of *Candida albicans* encodes a functional homolog of the pH-regulated gene *PHR1* with an inverted pattern of pH-dependent expression. *Mol. Cell. Biol.* 17:5960–5967.

Munro, C. A., and N. A. Gow. 2001. Chitin synthesis in human pathogenic fungi. *Med. Mycol.* 39(Suppl. 1):41–53.

Nierman, W. C., A. Pain, M. J. Anderson, J. R. Wortman, H. S. Kim, J. Arroyo, M. Berriman, K. Abe, D. B. Archer, C. Bermejo, J. Bennett, P. Bowyer, D. Chen, M. Collins, R. Coulsen, R. Davies, P. S. Dyer, M. Farman, N. Fedorova, T. V. Feldblyum, R. Fischer, N. Fosker, A. Fraser, J. L. Garcia, M. J. Garcia, A. Goble, G. H. Goldman, K. Gomi, S. Griffith-Jones, R. Gwilliam, B. Haas, H. Haas, D. Harris, H. Horiuchi, J. Huang, S. Humphray, J. Jimenez, N. Keller, H. Khouri, K. Kitamoto, T. Kobayashi, S. Konzack, R. Kulkarni, T. Kumagai, A. Lafon, J. P. Latgé, W. Li, A. Lord, C. Lu, W. H. Majoros, G. S. May, B. L. Miller, Y. Mohamoud, M. Molina, M. Monod, I. Mouyna, S. Mulligan, L. Murphy, S. O'Neil, I. Paulsen, M. A. Penalva, M. Pertea, C. Price, B. L. Pritchard, M. A. Quail, E. Rabbinowitsch, N. Rawlins, M. A. Rajandream, U. Reichard, H. Renauld, G. D. Robson, S. Rodriguez de Cordoba, J. M. Rodriguez-Pena, C. M. Ronning, S. Rutter, S. L. Salzberg, M. Sanchez, J. C. Sanchez-Ferrero, D. Saunders, K. Seeger, R. Squares, S. Squares, M. Takeuchi, F. Tekaia, G. Turner, C. R. Vazquez de Aldana, J. Weidman, O. White, J. Woodward, J. H. Yu, C. Fraser, J. E. Galagan, K. Asai, M. Machida, N. Hall, B. Barrell, and D. W. Denning. 2005. Genomic sequence of the pathogenic and allergenic filamentous fungus *Aspergillus fumigatus*. *Nature* 438:1151–1156.

Nyfeler, R., and W. Keller-Schierlein. 1974. Metabolites of microorganisms. 143. Echinocandin B, a novel polypeptide-antibiotic from *Aspergillus nidulans* var. *echinulatus*: isolation and structural components. *Helv. Chim. Acta* 57:2459–2477.

Onishi, J., M. Meinz, J. Thompson, J. Curotto, S. Dreikorn, M. Rosenbach, C. Douglas, G. Abruzzo, A. Flattery, L. Kong, et al. 2000. Discovery of novel antifungal (1-3)-β-D-glucan synthase inhibitors. *Antimicrob. Agents Chemother.* 44:368–377.

Paris, S., J. P. Debeaupuis, R. Crameri, M. Carey, F. Charlès, M. C. Prévost, C. Schmitt, B. Philippe, and J. P. Latgé. 2003. Conidial hydrophobins of *Aspergillus fumigatus*. *Appl. Environ. Microbiol.* 69:1581–1588.

Popolo, L., and M. Vai. 1999. The Gas1 glycoprotein, a putative wall polymer cross-linker. *Biochim. Biophys. Acta* 1426:385–400.

Ragni, E., T. Fontaine, C. Gissi, J. P. Latgé, and L. Popolo. 2007. The Gas family of proteins of *Saccharomyces cerevisiae*: characterization and evolutionary analysis. *Yeast* 24:297–308.

Rhode, M., M. Schwienbacher, T. Nikolaus, J. Heesemann, and F. Ebel. 2002. Detection of early phase specific surface appendages during germination of *Aspergillus fumigatus* conidia. *FEMS Microbiol. Lett.* 206:99–105.

Rodriguez-Pena, J. M., V. J. Cid, J. Arroyo, and C. Nombela. 2000. A novel family of cell wall related proteins regulated differently during the yeast life cycle. *Mol. Cell. Biol.* 20:3245–3255.

Roncero, C. 2002. The genetic complexity of chitin synthesis in fungi. *Curr. Genet.* 41:367–378.

Saporito-Irwin, S. M., C. E. Birse, P. S. Sypherd, and W. A. Fonzi. 1995. *PHR1*, a pH-regulated gene of *Candida albicans*, is required for morphogenesis. *Mol. Cell. Biol.* 15:601–613.

Schmalhorst, P. S., S. Krappmann, W. Vervecken, M. Rohde, M. Müller, G. H. Braus, R. Contreras, A. Braun, H. Bakker, and F. H. Routier. 2008. Contribution of galactofuranose to the virulence of the opportunistic pathogen *Aspergillus fumigatus*. *Eukaryot. Cell* 7: 1268–1277.

Sugareva, V., A. Hartl, M. Brock, K. Hubner, M. Rohde, T. Heinekamp, and A. A. Brakhage. 2006. Characterisation of the laccase-encoding gene abr2 of the dihydroxynaphthalene-like melanin gene cluster of *Aspergillus fumigatus*. *Arch. Microbiol.* 186:345–355.

Takada, H., Y. Araki, and E. Ito. 1981. Structure of polygalactosamine produced by *Aspergillus parasiticus*. *J. Biochem.* 89:1265–1274.

Takeshita, N., A. Ohta, and H. Horiuchi. 2005. CsmA, a class V chitin synthase with a myosin motor-like domain, is localized through direct interaction with the actin cytoskeleton in *Aspergillus nidulans*. *Mol. Biol. Cell* 16:1961–1970.

Thau, N., M. Monod, B. Crestani, C. Roland, G. Tronchin, J. P. Latgé, and S. Paris. 1994. Rodletless mutants of *Aspergillus fumigatus*. *Infect. Immun.* 62:4380–4388.

Traxler, P., J. Gruner, and J. A. Auden. 1977. Papulacandins, a new family of antibiotics with antifungal activity. I. Fermentation, isolation, chemical and biological characterization of papulacandins A, B, C, D and E. *J. Antibiot.* (Tokyo) 30:289–296.

Tronchin, G., J. P. Bouchara, M. Ferron, G. Larcher, and D. Chabasse. 1995. Cell surface properties of *Aspergillus fumigatus* conidia—correlation between adherence, agglutination, and rearrangements of the cell wall. *Can. J. Microbiol.* 41:714–721.

Tsai, H. F., Y. C. Chang, R. G. Washburn, M. H. Wheeler, and K. J. Kwon-Chung. 1998. The developmentally regulated *alb1* gene of *Aspergillus fumigatus*: its role in modulation of conidial morphology and virulence. *J. Bacteriol.* 180:3031–3038.

Tsai, H. F., M. H. Wheeler, Y. C. Chang, and K. J. Kwon-Chung. 1999. A developmentally regulated gene cluster involved in conidial pigment biosynthesis in *Aspergillus fumigatus*. *J. Bacteriol.* **181:** 6469–6477.

Tsai, H. F., I. Fujii, A. Watanabe, M. H. Wheeler, Y. C. Chang, Y. Yasuoka, Y. Ebizuka, and K. J. Kwon-Chung. 2001. Pentaketide melanin biosynthesis in *Aspergillus fumigatus* requires chain-length shortening of a heptaketide precursor. *J. Biol. Chem.* **276:**29292–29298.

Van der Kaaij, R. M., X. L. Yuan, A. Franken, A. F. J. Ram, P. J. Punt, M. J. E. C. van der Maarel, and L. Dijkhuizen. 2007. Two novel, putatively cell wall-associated and glycosylphosphatidylinositol-anchored α-glucanotransferase enzymes of *Aspergillus niger*. *Eukaryot. Cell* **6:**1178–1188.

Vos, A., N. Dekker, B. Distel, J. A. M. Leunissen, and F. Hochstenbach. 2007. Role of the synthase domain of Ags1p in cell wall α-glucan biosynthesis in fission yeast. *J. Biol. Chem.* **282:**18969–18979.

Warwas, M. L., J. N. Watson, A. J. Bennet, and M. M. Moore. 2007. Structure and role of sialic acids on the surface of *Aspergillus fumigatus* conidiospores. *Glycobiology* **17:**401–410.

Wasylnka, J. A., M. I. Simmer, and M. M. Moore. 2001. Differences in sialic acid density in pathogenic and non-pathogenic *Aspergillus* species. *Microbiology* **147:**869–877.

Watanabe, H., T. Fujii, H. F. Tsai, D. Chang, K. J. Kwon-Chung, and Y. Ebizuka. 2000. *Aspergillus fumigatus alb1* encodes naphthopyrone synthase when expressed in *Aspergillus oryzae*. *FEMS Microbiol. Lett.* **192:**39–44.

Wei, Y., Y. Zhang, U. Derewenda, X. Liu, W. Minor, R. K. Nakamoto, A. V. Somlyo, A. P. Somlyo, and Z. S. Derewenda. 1997. Crystal structure of RhoA-GDP and its functional implications. *Nat. Struct. Biol.* **4:**699–703.

Yuen, K. Y., C. M. Chan, K. M. Chan, P. C. Woo, X. Y. Che, A. S. Leung, and L. Cao. 2001. Characterization of *AFMP1*: a novel target for serodiagnosis of aspergillosis. *J. Clin. Microbiol.* **39:**3830–3837.

Aspergillus fumigatus and Aspergillosis
Edited by J.-P. Latgé and W. J. Steinbach

Chapter 15

Genetic Regulation of *Aspergillus* Secondary Metabolites and Their Role in Fungal Pathogenesis

ROBERT A. CRAMER, JR., E. KEATS SHWAB, AND NANCY P. KELLER

Humankind has benefited enormously from the production of small-molecular-weight metabolites by numerous filamentous fungi, including the genus *Aspergillus*. Indeed, one has to look no further than the discovery of penicillin to recognize the impact these natural products can and have had on humankind (Hewitt, 1967; Stollerman, 1993). Ironically, with respect to opportunistic mycoses, our ability to successfully perform many solid organ transplants is in part due to the discovery of the immunosuppressive fungal secondary metabolite cyclosporine (Cohen et al., 1984; Dummer et al., 1983; Perfect and Durack, 1985). Since these metabolites have often been found to be dispensable for in vitro growth and reproduction of the fungi that produce them, they are often referred to as "secondary metabolites," differentiating them from essential metabolites produced by primary metabolism (Bennett and Bentley, 1989; Keller et al., 2005).

Recent research on secondary metabolites in filamentous fungi has revealed their importance in the general biology and fitness of fungi (Kim et al., 2007; Lee et al., 2005; Rohlfs et al., 2007). Thus, secondary metabolites are anything but "secondary" in importance to the biology of the fungi that produce them. With respect to fungal pathogenesis and virulence, fungal secondary metabolites are primary virulence factors in several fungus-plant interactions (Osbourn, 2001). However, our knowledge of fungal secondary metabolites and their potential role in human mycoses is limited.

The purpose of this chapter is to explore our current state of knowledge on the production of secondary metabolites by the opportunistic human pathogen *Aspergillus fumigatus*. In particular, we will discuss the potential role of secondary metabolites in the fungus-host interaction and how secondary metabolite production is regulated at the molecular level.

SECONDARY METABOLITES PRODUCED BY *A. FUMIGATUS*

Like most filamentous fungi, *A. fumigatus* produces a wide variety of secondary metabolites. Surveys of the whole genome sequence of *A. fumigatus* strain AF293 have revealed numerous nonribosomal peptide synthetase (NRPS) and polyketide synthetase genes that are responsible in part for secondary metabolite production in filamentous fungi (Cramer et al., 2006b; Nierman et al., 2005). Several types of secondary metabolites have been isolated from *A. fumigatus* in vitro cultures, including fumitremorgins, verruculogen, fumigaclavines, fumagillin, helvolic acid, and gliotoxin, among other, less-studied compounds (Larsen et al., 2007) (Fig. 1). Importantly, gliotoxin has been detected in the serum of patients with invasive pulmonary aspergillosis (IPA) and in experimental murine models of IPA (Lewis et al., 2005a). Discussion of the most frequently studied *A. fumigatus* secondary metabolites and their potential role in aspergillosis follows.

Fumitremorgins and Verruculogen

Aspergillus and *Penicillium* species produce a group of secondary metabolites collectively referred to as tremorgenic mycotoxins due to their ability to cause neuromuscular tremors (Gallagher and Latch, 1977; Suzuki et al., 1984; Yamazaki et al., 1980, 1983; Yamazaki and Suzuki, 1986). Fumitremorgins, from *A. fumigatus*, are prenylated indole alkaloids derived from the amino acids tryptophan and proline. Maiya et al. (2006) recently identified a gene cluster responsible for fumitremorgin biosynthesis in *A. fumigatus*. Overexpression of the NRPS *ftmA* in *A. fumigatus* AF293.1 and *Aspergillus nidulans* resulted in accumulation of brevianamide F, the

Robert A. Cramer, Jr. • Dept. of Veterinary Molecular Biology, Montana State University–Bozeman, Bozeman, MT 59718. E. Keats Shwab and Nancy P. Keller • Dept. of Plant Pathology and Dept. of Medical Microbiology and Immunology, University of Wisconsin–Madison, Madison, WI 53706.

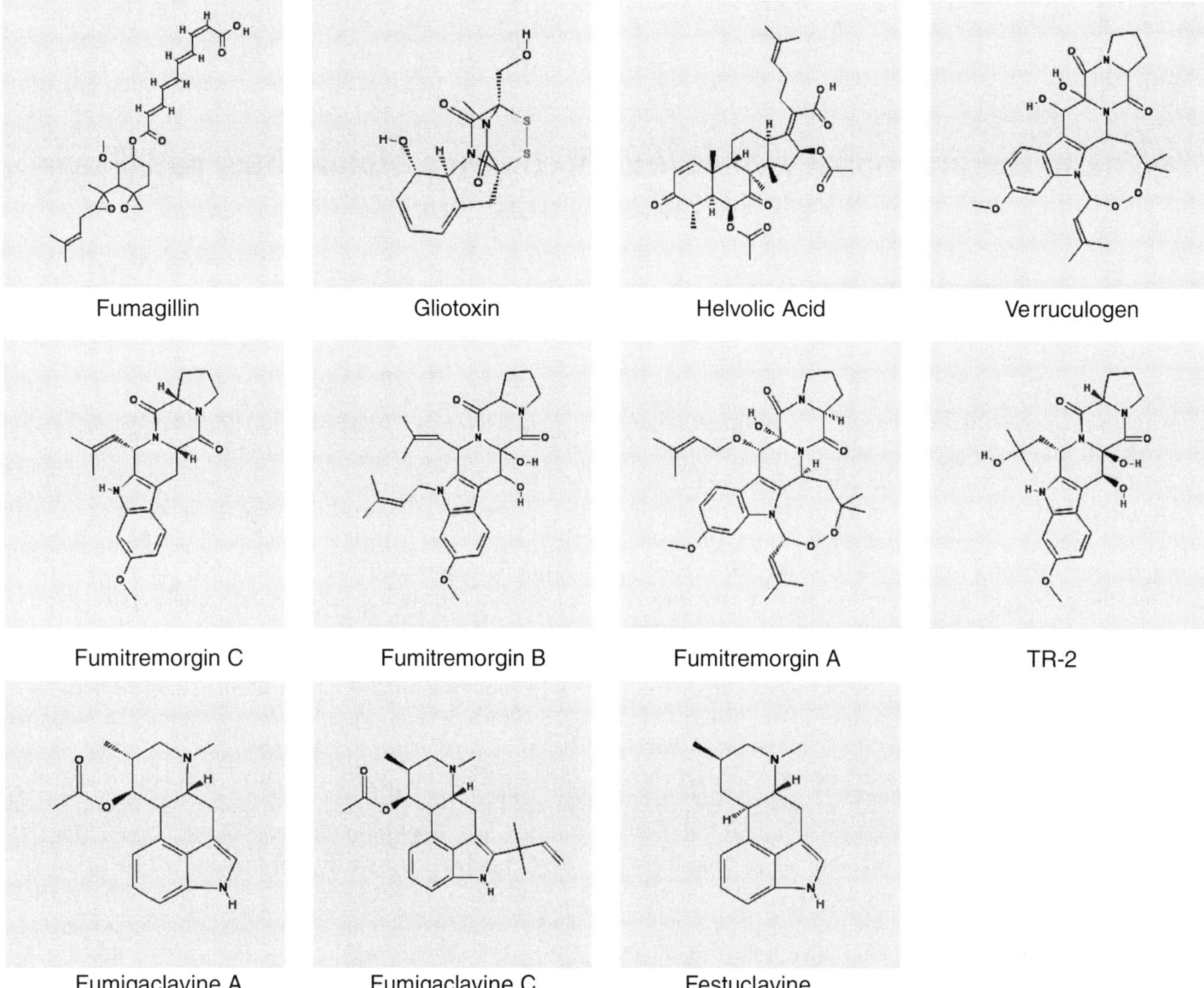

Figure 1. Secondary metabolites produced by *A. fumigatus*. Chemical structures were obtained from the PubChem project (http://pubchem.ncbi.nlm.nih.gov/).

likely precursor of fumitremorgin biosynthesis. Interestingly, the authors could not detect fumitremorgin or brevianamide F biosynthesis in *A. fumigatus* strain AF293 under conditions reported to induce fumitremorgin biosynthesis. Thus, a Δ*ftmA* mutant strain in the AF293 background proved inconclusive in definitively confirming *ftmA*'s role in fumitremorgin biosynthesis. This result highlights an important finding with respect to secondary metabolite production in filamentous fungi, i.e., the presence of a specific gene cluster and secondary metabolite in one species does not always lead to the production of the associated secondary metabolite in different species or strains that also contain the biosynthetic gene cluster. However, the results reported by Maiya et al. strongly suggest that *ftmA* is indeed involved in fumitremorgin biosynthesis.

The potential role of fumitremorgins in aspergillosis has not been examined. Fumitremorgin C is a potent inhibitor of the ABC-type transporter ABCG2, which is a breast cancer resistance protein (Allen et al., 2002; Garimella et al., 2004; Rabindran et al., 2000; van Loevezijn et al., 2001). Thus, it may be possible that fumitremorgins affect immune effector cell or epithelial and/or endothelial cell functions at sites of fungal infections. To date, fumitremorgins have not been detected in vivo in patients with IPA or in experimental murine models. In addition, fumitremorgins have been implicated in causing symptoms associated with wood trimmer's disease (Land et al., 1987, 1993).

Another tremorgenic mycotoxin produced by *A. fumigatus*, verruculogen, is a lipophilic secondary metabolite that can cross the blood-brain barrier and affect neurotransmitters, potentially via inhibition of calcium-activated potassium channels (Bradford et al., 1990; Knaus et al., 1994). The tremorgenic mycotoxin TR-2, also produced by *A. fumigatus*, has been implicated as a biosynthetic intermediary in the biosynthesis of verruculogen (Willingale et al., 1983). Verruculogen has recently been implicated in altering airway epithelium properties, including a decrease in transepithelial electrical resistance and polarization (Khoufache et al., 2007). Verruculogen was detected in both conidial and hyphal extracts, and thus it is possible that verruculogen affects the conidia-airway epithelial cell interaction and contributes to the ability of *A. fumigatus* to colonize the respiratory tract.

Fumigaclavines

Conidia of *A. fumigatus* have been found to contain four ergot alkaloids: fumigaclavine C, festuclavine, fumigaclavine A, and fumigaclavine B (listed in order of abundance, according to Panaccione and Coyle [2005]). In addition, culture broths of *A. fumigatus* are found to contain ergot alkaloids, indicating that these alkaloids are likely secreted during hyphal growth (Cole et al., 1977; Spilsbury and Wilkinson, 1961). The fumigaclavine biosynthetic gene cluster has been identified by two independent groups in the genome sequence of *A. fumigatus* strain AF293 (Coyle and Panaccione, 2005; Unsold and Li, 2005). In distinct but complementary methods, the dimethylallyltyptophan synthase gene, which is responsible for the prenylation of L-tryptophan in the biosynthesis of ergot alkaloids, was functionally characterized. Unsold and Li heterologously overproduced the dimethylallyltryptophan synthase enzyme named *FgaPT2* in *Saccharomyces cerevisiae* and confirmed that it catalyzed the conversion of tryptophan to dimethylallyltryptophan (Unsold and Li, 2005). In a conventional gene disruption approach, Coyle and Panaccione created a gene disruption in the dimethylallyltryptophan synthase, which they called *dmaW*, and showed that Δ*dmaA* strains no longer produced any of the known ergot alkaloids produced by *A. fumigatus* (Coyle and Panaccione, 2005).

The potential role of ergot alkaloids in the aspergillosis disease complex has not been examined. However, creation of ergot alkaloid-deficient *A. fumigatus* strains now makes addressing this question possible. The presence of ergot alkaloids on *A. fumigatus* conidia may influence the initial interaction with the airway epithelium and perhaps alter the initial immune response via alteration of cytokine signaling and/or alteration of alveolar macrophage functions. Intraperitoneal injection of fumigaclavine C into mice treated with concanavalin A (to induce liver injury) demonstrated that fumigaclavine C inhibited lymphocyte activation, proliferation, and adhesion to extracellular matrices (Coyle and Panaccione, 2005). In addition, tumor necrosis factor alpha production, an important cytokine in immune responses to invasive aspergillosis, was markedly decreased in mice treated with fumigaclavine C. Thus, it is possible that ergot alkaloids affect immune responses and alter the outcome of various forms of aspergillosis.

Fumagillin

Fumagillin, a sesquiterpene produced by *A. fumigatus*, is a potent angiogenesis inhibitor and has been extensively explored as an anticancer agent (Ingber et al., 1990; Sin et al., 1997). However, fumagillin has cytotoxic side effects that complicate its use in patients. Fumagillin has been shown to directly inhibit endothelial cell proliferation and ciliary function of human respiratory epithelium and to cause genotoxic aneuploidy in cultured human lymphocytes (Amitani et al., 1995; Griffith et al., 1998; Ingber et al., 1990; Stanimirovic et al., 2007). Due to its genotoxicity, several analogs of fumagillin have been developed and are currently being tested for their anticancer properties. In addition, fumagillin is one of the few potent antimicrosporidia compounds available. Currently, the only registered treatment available to control *Nosema* diseases in honey bees is the use of fumagillin (Bailey, 1953; Hartwig and Przelecka, 1971; Katznelson and Jamieson, 1952; Whittington and Winston, 2003).

The molecular target of fumagillin has been identified as methionine aminopeptidase 2 (MetAP-2), a metallopeptidase that catalyzes the removal of initiator methionine residues from nascent polypeptide chains (Sin et al., 1997). Methionine aminopeptidases are essential for both prokaryotic and eukaryotic growth and consequently are considered excellent drug targets (Chang et al., 1989; Li and Chang, 1995). In mammals, MetAP-2 inhibition has been shown to inhibit endothelial cell proliferation (Bernier et al., 2005; Griffith et al., 1997).

Currently, the role of fumagillin in the aspergillosis disease complex is unknown. Angioinvasion during pulmonary aspergillosis is characterized by hyphal invasion of the endothelial lining of blood vessels (Lopes Bezerra and Filler, 2004), and thus fumagillin may potentially play a role in stimulating and altering endothelial or other immune effector cell functions.

Helvolic Acid

Helvolic acid belongs to a small group of naturally occurring steroidal antibiotics called fusidines (Amitani et al., 1995; Okuda et al., 1964, 1966; Williams, 1952). Helvolic acid and fusidine steroidal antibiotics are synthesized from oxidosqualene, which can be cyclized into a variety of polycyclic triterpenes, including protostadienol, the precursor of helvolic acid and fusidine (Abe et al., 2001; Kawaguchi et al., 1973). Like many other *A. fumigatus*-produced secondary metabolites, the potential role of helvolic acid in aspergillosis is unknown. At relatively high concentrations (10 μg/ml), helvolic acid was shown to inhibit superoxide production in rat alveolar macrophages (Mitchell et al., 1997). To date, helvolic acid has not been detected in vivo during IPA; thus, it is questionable whether the effects observed at the high concentration, which would be detectable in vivo, are physiologically relevant. Given the close link between steroid and helvolic acid biosynthesis, the ability to create a helvolic acid-deficient strain of *A. fumigatus* without affecting important aspects of steroid biosynthesis may be a challenge.

Gliotoxin

Gliotoxin is the most-studied secondary metabolite produced by *A. fumigatus*. A member of the epipolythiodioxopiperazine (ETP) class of secondary metabolites, gliotoxin was first isolated as a "toxic" substance from a *Trichoderma* species by Weindling and Emerson in 1936 (Weindling and Emerson, 1936). However, the toxin gets its name from an isolation in 1943 by Johnson et al. (1943) from *Gliocladium fimbriatum*. Gliotoxin has been reported to be produced by *Aspergillus*, *Penicillium*, and *Trichoderma* spp., which has led to the hypothesis that it is a basic component of these fungi's defense mechanisms in their soil-borne ecological niches (Macdonald and Slater, 1975). In addition, it has been reported that specific isolates of *Candida albicans* produce gliotoxin (Shah and Larsen, 1991; Shah et al., 1995), but this result was recently suggested to be inaccurate (Kupfahl et al., 2007; Shah and Larsen, 1991; Shah et al., 1995). Recent studies have demonstrated that other *Aspergillus* species also produce gliotoxin, but the frequency and quantity are significantly decreased compared to production by *A. fumigatus*, and no gliotoxin biosynthetic genes have been found in these fungi (Kupfahl et al., 2008; Lewis et al., 2005b). This either implicates an alternative gliotoxin biosynthetic pathway in these fungi or brings into question the true identity of these compounds (Kupfahl et al., 2008).

Contamination of tissue culture medium by *A. fumigatus* in 1984 instigated a series of experiments aimed at deducing gliotoxin's role in the pathogenesis of invasive aspergillosis (Eichner and Mullbacher, 1984; Mullbacher and Eichner, 1984). These early studies led to the observation that gliotoxin inhibited T- and B-cell proliferation after antigen stimulation, ameliorated the function of cytotoxic T cells produced in mixed lymphocyte reactions, and significantly reduced the phagocytosis of particles by macrophages (Eichner et al., 1986; Mullbacher and Eichner, 1984; Mullbacher et al., 1985, 1987). Gliotoxin was shown to have selective activity against mature cells of hematopoietic origin, as progenitor cells were less sensitive to the toxin (Mullbacher et al., 1987). In addition, injection of gliotoxin into irradiated mice demonstrated that immunocompromised mice were more susceptible to the lethal effects of gliotoxin than immunocompetent animals (Sutton et al., 1994).

Later studies began to explore the mechanisms by which gliotoxin exerted its immunosuppressive effects. The ETP class of secondary metabolites is characterized by an internal disulfide bridge, which has been shown to be essential for gliotoxin's biological activities (Rodriguez and Carrasco, 1992; Trown and Bilello, 1972). This disulfide bridge can cause significant damage to proteins and cells via two distinct mechanisms. First, it can conjugate with susceptible thiol residues on proteins and lead to their inactivation. Second, the redox cycling between the dithiol (reduced) and disulfide (oxidized) forms of the toxin can generate toxic reactive oxygen species that are harmful to cells (Bernardo et al., 2003; Waring et al., 1994, 1995). No consensus exists as to which mechanism is primarily responsible for gliotoxin's biological activities. Evidence suggests that both redox cycling and direct inactivation of proteins via conjugation to susceptible thiol residues are likely important (Hurne et al., 2002).

Several protein targets of gliotoxin have now been identified and help explain the observations that gliotoxin induces apoptosis in mammalian cells at concentrations below 10 μM and necrosis at concentrations above that threshold (Beaver and Waring, 1994). Pahl et al. (1996) demonstrated that gliotoxin inhibits the important regulatory protein nuclear factor κB (NF-κB). NF-κB is an important regulator of granulocyte apoptosis, and thus gliotoxin's inhibition of NF-κB likely explains, in part, its immunosuppressive activity (Ward et al., 1999). Kroll et al. (1999) showed that gliotoxin targeted proteolytic activities of the proteasome. Gliotoxin was found to inhibit proteasome-mediated degradation of IκBα, which consequently suppresses NF-κB activation (Kroll et al., 1999). Pardo et al. (2006) recently discovered that the mitochondrial protein Bak was required for gliotoxin-induced apoptosis. Interestingly, Bak-deficient mice were more resistant to invasive aspergillosis than wild-type mice. This surprising and intriguing result suggests a role for apoptosis in resistance and susceptibility to *A. fumigatus* and warrants further examination.

Neutrophils are critical components of resistance to *A. fumigatus* infections, and gliotoxin has been shown to inhibit the assembly and function of the NADPH oxidase complex in neutrophils (Nishida et al., 2005;

Tsunawaki et al., 2004; Yoshida et al., 2000). Surprisingly, a recent study suggested that gliotoxin failed to induce apoptosis in neutrophils but still inhibited polymorphonuclear cell (PMN) phagocytosis and reactive oxygen species generation (Orciuolo et al., 2007). However, in combination with methylprednisolone, gliotoxin actually increased reactive oxygen species generation. Thus, in patients receiving corticosteroid therapy, the presence of gliotoxin may exacerbate the disease by increasing levels of PMN-mediated inflammation (Orciuolo et al., 2007). Importantly, these studies were conducted with gliotoxin levels consistent with levels found in IPA patients. Finally, gliotoxin has been shown to induce significant levels of apoptosis in monocytes and impair antigen presentation by T cells (Stanzani et al., 2005).

Taken together, the results summarized in the preceding paragraphs clearly demonstrate that gliotoxin is a strong immunosuppressant and has the potential to play an important role in the pathophysiology of IPA. However, arguments against the role of gliotoxin in IPA include the fact that species of *Aspergillus* that have not been reported to produce gliotoxin also cause IPA, and importantly, patients that acquire IPA infections are already severely immunocompromised. In addition, since gliotoxin is not found in or associated with *A. fumigatus* conidia, and at least in vitro is not detectable in culture for 24 h, it seems gliotoxin does not play a role in the establishment of disease.

Most recently, creation of *A. fumigatus* mutants deficient in gliotoxin production were created to genetically address the role of gliotoxin in IPA. This was possible because the genome sequence of *A. fumigatus* led to the identification of a gene cluster with strong similarity to a gene cluster responsible for the production of the ETP sirodesmin PL in the plant pathogen *Leptosphaeria maculans* (Gardiner and Howlett, 2005; Gardiner et al., 2005b; Nierman et al., 2005). One of the genes in the putative gliotoxin cluster, *gliP*, encodes an NRPS predicted to catalyze the formation of the diketopiperazine scaffold (Gardiner and Howlett, 2005), a mechanism recently chemically confirmed (Balibar and Walsh, 2006).

Four separate groups have now confirmed the role of the *gliP*-containing gene cluster in gliotoxin biosynthesis (Table 1) and have examined the respective mutant's ability to cause disease in distinct murine models of IPA (Bok et al., 2006a; Cramer et al., 2006a; Kupfahl et al., 2006; Sugui et al., 2007a). Three of the laboratories created mutations in the NRPS *gliP* coding sequence, while the other laboratory mutated the transcriptional regulator *gliZ*. Table 1 summarizes the experimental methods and results from these four studies. Although different background *A. fumigatus* strains, mouse strains, and inoculum delivery methods were utilized, the three studies that utilized cyclophosphamide and cortisone acetate as immunosuppressive agents to establish IPA found no statistical difference in mortality between wild-type gliotoxin-producing strains and gliotoxin-deficient mutant strains (Bok et al., 2006a; Cramer et al., 2006a; Kupfahl et al., 2006). However, all three studies did reveal possible roles of gliotoxin in host damage at the cellular level. For example, a decrease in PMN apoptotic cell death was found from supernatants of the gliotoxin mutant (Bok et al., 2006a). These results suggest that gliotoxin is not required for establishment or persistence of disease in severely immunocompromised hosts but may contribute to host cell damage.

In contrast to the above studies, when cortisone acetate was utilized as the sole immunosuppressive agent in an experimental model of IPA, *gliP* mutant strains in the B5233 background were significantly less pathogenic (Sugui et al., 2007a). Although the majority of patients affected with IPA are neutropenic and severely immunocompromised, IPA occurs more frequently in the postengraftment period following stem cell transplantation, when onset of acute or chronic graft-versus-host disease occurs (Fukuda et al., 2004). Graft-versus-host disease patients who are no longer neutropenic typically receive corticosteroid therapies to minimize inflammation, which puts them at risk for fungal infections (Grow et al., 2002; Marr et al., 2002; Ribaud et al., 1999). The results of Sugui et al. (2007a) also support the recent results of Orciuolo et al. (2007), which suggested that the effects of gliotoxin on host PMNs are exacerbated in patients receiving corticosteroid therapies leading to increased inflammation.

These studies nicely illustrate the importance of the role and status of the host immune system in determining the outcome of IPA infections. However, much work remains to be done in elucidating gliotoxin's role in aspergillosis. While these four distinct but complementary studies illustrated gliotoxin's role in different murine models of IPA by measuring mortality and effects on specific cells of the immune system, the immune response(s) induced or inhibited by the presence or absence of gliotoxin in these models in vivo has not been extensively explored. Unfortunately, the use of relatively nonspecific chemotherapies for creation of immunosuppression makes mechanistic examination of the host response to gliotoxin difficult, if not impossible, in these murine models. Finally, no studies to date have explored the role of gliotoxin in allergic bronchopulmonary aspergillosis. Given the potent immunosuppressive activity of this toxin, further studies are needed to uncover its precise role in the pathophysiology of the aspergillosis disease complex.

Table 1. Pathogenesis studies of gliotoxin-deficient *A. fumigatus* strains in murine models of IPA

Study	*A. fumigatus* strain(s)	Mutation	Mouse strain(s), gender, and weight range	Immunosuppression regimen	Inoculum delivery	Result
Cramer et al. (2006)	AF293.1	3.7-kb internal fragment of *gliP* replaced with *pyrG*	Outbred ICR males, 19–21 g	Cyclophosphamide (250 mg/kg on day −2, 200 mg/kg on day +3); cortisone acetate (250 mg/kg on day −1 and day +3)	Inhalational aerosol, 10^9 CFU/ml for 1 h	No change in murine mortality
Kupfahl et al. (2006)	ATCCΔ*akuB*^{KU80}, KU80Δ*pyrG*	*gliP* replaced from start codon to stop codon with *ble* or *pyrG*	BALB/c females, 16–19 g	Cyclophosphamide (150 mg/kg on day −4 and every third day after); cortisone acetate (200 mg/kg on day −1)	Intranasal, 3×10^4 conidia delivered in 25 μl for strains derived from CEA17; 6×10^4 conidia for strains derived from ATCC 46645	No change in murine mortality
Bok et al. (2006)	AF293.1	*gliZ* replaced from start codon to stop codon with *pyrG*	Outbred Swiss ICR mice, gender not reported, 24–27 g	Cyclophosphamide (150 mg/kg on days −4, −1, and +3; cortisone acetate (200 mg/kg on day 0)	Intranasal, 2.5×10^5 conidia delivered in 50 μl	20% attenuation of mortality, but not statistically significant
Sugui et al. (2007)	B-5233	*gliP* replaced from start codon to stop codon with *hygB*	BALB/c and 129/Sv males 10–11 weeks of age (weight not reported)	Hydrocortisone acetate (2 mg in 100 μl PBS on days −4, −2, 0, and +4)	Intranasal, 5×10^6 conidia delivered in 20 μl	75% attenuation of mortality in 129/Sv mice; no difference in mortality in BALB/c mice but 6-day difference in time to reach date of 50% mortality

REGULATION OF SECONDARY METABOLISM IN *A. FUMIGATUS* AND OTHER ASPERGILLI

Gliotoxin and other secondary metabolites are not constitutively produced by aspergilli. Production of secondary metabolites is an energetically costly process involving many enzymatic steps. Presumably, production is beneficial to the fungus only under conditions when it would be most advantageous to the organism. As such, there are many diverse factors involved in the regulation of secondary metabolite production (Fig. 2). These include the developmental stage of the fungus and general environmental factors, such as carbon and nitrogen sources, temperature, light, and pH, that act to regulate secondary metabolite biosynthetic gene expression by affecting molecular signaling pathways. These pathways may influence general fungal physiology and regulate secondary metabolism on a global scale or may target specific metabolites. Often, such general signaling pathways ultimately regulate secondary metabolite production by controlling the expression or activity of transcription factors that specifically regulate only the genes required for production of a single secondary metabolite.

As inferred earlier, the genes devoted to production of a fungal secondary metabolite are typically arranged contiguously in a cluster (Keller and Hohn, 1997). Secondary metabolite gene clusters are found in the majority of filamentous fungi and may range from only a few to more than 20 genes. The reason this clustering is maintained by fungi is not fully understood. It has been suggested that these clusters may result from horizontal transfer from prokaryotes, but only the cluster responsible for biosynthesis of penicillin shows evidence for this (Landan et al., 1990). Instead, it is assumed that clustering confers some selective advantage to the fungus, and it is likely that this advantage may be related to efficiency of gene regulation and secondary metabolite production, as will be discussed below.

Signaling through Environmental Cues

In fungi, signals generated in response to the environment are typically relayed through DNA-binding

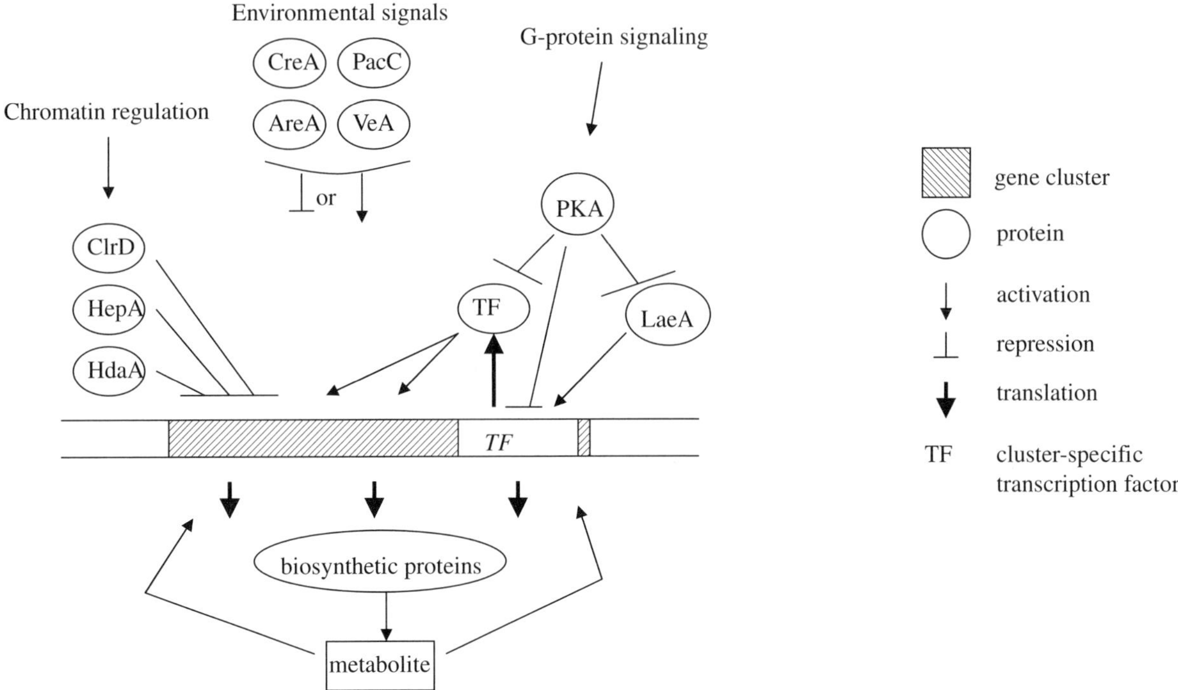

Figure 2. Regulation of a hypothetical secondary metabolite gene cluster. This figure illustrates modes of regulation of production for a hypothetical secondary metabolite based on known regulatory mechanisms for gliotoxin, aflatoxin, sterigmatocystin, and others. A cluster-specific regulatory gene located within the cluster is activated by LaeA and in turn activates expression of the rest of the cluster genes, leading to biosynthesis of the metabolite product, which then may act to promote expression of the cluster in a positive feedback mechanism. Cluster expression is repressed by the chromatin-modifying proteins ClrD, HepA, and HdaA. G-protein signaling activates PKA, which suppresses metabolite production by inhibiting expression of *laeA* and the in-cluster transcription factor gene and also by inactivation of its protein product via phosphorylation. Environmental signals, such as carbon and nitrogen sources, pH, and light, may alter production of this metabolite through the actions of proteins such as CreA, PacC, AreA, and VeA.

Cys2His2 zinc-finger proteins, including CreA for carbon signaling (Dowzer and Kelly, 1989), AreA for nitrogen signaling (Hynes, 1975), and PacC for pH signaling (Tilburn et al., 1995). These proteins may have either positive or negative regulatory effects on metabolite production. For example, penicillin production in the model organism *A. nidulans* is positively regulated by CreA and negatively regulated by PacC (Martin, 2000).

Many other environmental factors also affect fungal secondary metabolism. The production of gliotoxin by *A. fumigatus* is strongly influenced by temperature and environmental oxygen content (Belkacemi et al., 1999; Watanabe et al., 2004). This is also true of aflatoxin production in *Aspergillus parasiticus* and *Aspergillus flavus* (Clevstrom et al., 1983; Ellis et al., 1994; O'Brian et al., 2007; Schindler et al., 1967). Several studies have also implicated a role for light in aflatoxin production in *A. parasiticus* (Bennett et al., 1978, 1981). VeA, a light-regulated protein (Stinnett et al., 2007) whose mechanism of action is not yet known, positively regulates production of several *Aspergillus* secondary metabolites, including aflatoxin (Duran et al., 2007; Kato et al., 2003). Interestingly, although deletion of *veA* results in decreased production of both sterigmatocystin and penicillin in *A. nidulans*, overexpression of *veA* also decreases penicillin production (Sprote and Brakhage, 2007). VeA also couples morphogenesis with secondary metabolism (Duran et al., 2007; Kato et al., 2003), a general phenomenon in filamentous fungi (Calvo et al., 2002; Yu and Keller, 2005).

G-Protein Coupling of Development and Secondary Metabolism

Insight into signaling pathways linking growth, development, and secondary metabolism came from a key study in *A. nidulans* showing heterotrimeric G-protein activation simultaneously inhibits sporulation and production of the mycotoxin sterigmatocystin while promoting vegetative growth (Hicks et al., 1997). This regulation is conserved in the aflatoxin-producing species *A. flavus* and *A. parasiticus* (reviewed by Brodhagen and Keller, 2006, and Yu and Keller, 2005). G-protein repression of both sterigmatocystin and aflatoxin is mediated through protein kinase A (PKA) activity, implicating the importance of cyclic AMP levels (Roze et al., 2004; Shimizu and Keller, 2001). PKA has been shown to regulate sterigmatocystin production both by preventing expression of a transcription factor necessary for expression of sterigmatocystin biosynthesis genes and by posttranscriptional inactivation (via phosphorylation) of the same transcription factor (Shimizu and Keller, 2001; Shimizu et al., 2003). In contrast to sterigmatocystin, production of penicillin by *A. nidulans* is activated rather than repressed by G-protein signaling, as is trichothecene production by *Fusarium sporotrichioides*, indicating that different secondary metabolites may respond differently to developmental signals (Tag et al., 2000). This is also illustrated by the deletion of *rgsA*, a negative regulator of G-protein signaling, which results in greatly reduced sterigmatocystin production by *A. nidulans* but marked overproduction of pigments (Han et al., 2004). Moreover, PhLP, a positive regulator of G-protein signaling, promotes rather than inhibits sterigmatocystin production by *A. nidulans*, indicating that multiple G-protein signaling pathways may have opposing effects on secondary metabolism (Seo and Yu, 2006).

Lipid Signaling Molecules

Hormone-like signaling molecules known as oxylipins also contribute to regulation of secondary metabolites (Tsitsigiannis and Keller, 2007). These oxygenated lipid molecules mediate the balance of sexual and asexual spore production in aspergilli and are produced by fatty acid oxygenases encoded by *ppo* genes (Tsitsigiannis et al., 2004). The number of *ppo* genes varies among *Aspergillus* species. In *A. nidulans*, deletion of the genes *ppoA* and *ppo*C results in decreased sterigmatocystin production but increased penicillin production, indicating an additional link between development and secondary metabolism (Tsitsigiannis and Keller, 2006). This latter result suggests oxylipin effects may be mediated through the G-protein pathways described above.

Coregulation of Clusters

Transcription factors

In general, the genes in a secondary metabolite cluster are coregulated, with transcription of all the genes being activated or repressed simultaneously and independently of genes immediately outside the cluster. This type of regulation appears to be dependent on the presence of a pathway-specific factor and the chromosomal organization of the biosynthetic genes. Most cluster genes encode enzymes that catalyze steps in the biosynthetic process, but frequently clusters will also include a gene encoding a transcription factor necessary for expression of the other cluster genes (Hoffmeister and Keller, 2007). *gliZ*, mentioned earlier, specifically regulates the gliotoxin cluster of *A. fumigatus* in this manner (Bok et al., 2006a).

The best-studied pathway-specific regulator is AflR, found in the aflatoxin clusters of *A. flavus* and *A. parasiticus* as well as the sterigmatocystin cluster of *A. nidulans* (Brown et al., 1996; Chang et al., 1993; Fernandes et al., 1998; Woloshuk et al., 1994; Yu et al., 1996).

Elimination of *aflR* results in loss of transcription of aflatoxin and sterigmatocystin biosynthetic genes and subsequent reduction in aflatoxin and sterigmatocystin production (Yu et al., 1996). This regulation is mainly limited to genes within the cluster, though several *aflR*-regulated genes were recently identified elsewhere in the genome of *A. parasiticus* (Price et al., 2006). Although AflR is required for aflatoxin cluster activation, research indicates that its presence alone is not sufficient for toxin production to occur. For example, *aflR* expression in *A. nidulans* is still present in several sterigmatocystin-deficient strains (Butchko et al., 1999). In a mutated strain of *A. parasiticus* deficient in secondary metabolite production and deficient in expression of aflatoxin cluster genes, including *aflR*, ectopic expression of *aflR* did not effect an increase in production of aflatoxin, despite confirmation of the presence of the AflR protein via Western blot analysis (Kale et al., 2007). Both of these studies support the concept that other proteins and metabolites are required for metabolite production.

Both *gliZ* and *aflR* encode $Zn(II)_2Cys_6$ zinc binuclear proteins that activate transcription by binding to palindromic DNA sequences in the promoters of regulated cluster genes (Ehrlich et al., 1999; Fernandes et al., 1998; Fox et al., 2008; Payne and Brown, 1998). Zinc binuclear proteins are the most common type of in-cluster pathway regulator and have only been identified in fungi (Todd and Andrianopoulos, 1997). Other in-cluster regulatory proteins include Cys2His2 zinc finger proteins, such as Tri6 for trichothecene production by *Fusarium sporotrichiodes* (Proctor et al., 1995), and ankyrin-repeat proteins, such as ToxE for production of HC toxin by *Cochliobolus carbonum* (Pedley and Walton, 2001), among others. Positive-acting regulators that are present outside of clusters have also been identified, such as the PENR1 transcriptional complex required for penicillin production in a number of fungal species (Brakhage et al., 1999; Litzka et al., 1998). The aflatoxin and sterigmatocystin clusters also contain the gene *aflJ*, the product of which appears to interact with AflR to produce optimal expression of the cluster genes, although low levels of expression are observed in the absence of *aflJ* in *A. parasiticus* (Chang et al., 2000; Meyers et al., 1998). Its function has not yet been examined in any other aspergilli.

In the case of the *A. fumigatus* gliotoxin cluster, deletion of the NRPS gene *gliP*, encoding an enzyme essential for gliotoxin production, results in downregulation of all genes in the cluster. However, addition of exogenous gliotoxin restores expression of the cluster, indicating that the toxin itself acts as a regulator of cluster expression in a positive feedback mechanism, rather than indicating a specific regulatory role for *gliP* (Cramer et al., 2006a). Another example of such a feedback mechanism is provided by the fungus *Leptosphaeria maculans*, which produces sirodesmin, a toxin closely related to gliotoxin. Production of sirodesmin is increased by deletion of the sirodesmin transporter gene *sirA*, located within the sirodesmin cluster. However, deletion of this gene also results in decreased tolerance of the fungus to sirodesmin, indicating that SirA contributes to self-protection of *L. maculans* against its toxin (Gardiner et al., 2005a).

Chromatin regulation

A novel mechanism of global secondary metabolite regulation was identified when the protein LaeA was discovered through complementation of an *A. nidulans* mutant deficient in secondary metabolism (Bok and Keller, 2004). Deletion of *laeA* has been shown to result in loss of production of multiple secondary metabolites in numerous *Aspergillus* species (Bok and Keller, 2004; Keller et al., 2006; Perrin et al., 2007; S. P. Kale, L. Milde, M. K. Trapp, N. P. Keller, and J. W. Bok, unpublished results). Detailed comparisons of gene expression patterns in wild-type and Δ*laeA* strains of *A. fumigatus* were made using microarray technology (Perrin et al., 2007). A total of 943 genes were found to be differentially regulated in the Δ*laeA* mutant. Of the 943 genes, 102 of these genes were identified as part of 13 secondary metabolite clusters, and all 13 clusters were down-regulated. These include clusters known to be involved in production of pigments, fumitremorigens, festuclavine, elymoclavine, fumigaclavines, ergot alkaloids, and gliotoxin. Interestingly, 7 of these 13 clusters were found to be located within 300 kb of a telomere, and approximately 25% of all LaeA-regulated genes were located in these regions, compared with only 16% of all *A. fumigatus* genes. Subtelomeric regions of fungal chromosomes are found to be highly variable and often contain genes involved in niche specialization, which is in agreement with a role for secondary metabolites in adaptation to specific environments (Fairhead and Dujon, 2006; Perrin et al., 2007; Rehmeyer et al., 2006). The *laeA* deletion strains demonstrated a reduction in virulence in murine models of IPA (Bok et al., 2005; Sugui et al., 2007b), likely due in part to the decrease in multiple toxins.

In all clusters examined, regulation by LaeA is spatially limited to the genes within the cluster and does not extend to genes immediately adjacent (Bok and Keller, 2004; Bok et al., 2006b, 2006c; Perrin et al., 2007). Moreover, placement of a copy of *aflR* at a location outside the sterigmatocystin cluster remediates the loss of sterigmatocystin production in Δ*laeA* mutants, whereas placement of an extra copy within the cluster does not (Bok et al., 2006c). Placement of the primary metabolic gene *argB* within the sterigmatocystin cluster also results in its down-regulation in the Δ*laeA* mutant

(Bok et al., 2006c). These accumulated data implicate a role for LaeA in chromatin regulation of cluster expression (Turner et al., 2005). Interestingly, LaeA appears to be a methyltransferase, as mutation of an *S*-adenosyl methionine-binding site results in a Δ*laeA* phenotype (Bok et al., 2006c). LaeA contains some motifs similar to histone methyltransferases, enzymes that are known to play an important role in epigenetic gene regulation. A number of chromatin-modifying proteins have been found to influence production of secondary metabolites in *Aspergillus* species. Deletion of *hdaA*, a major histone deacetylase gene, in *A. nidulans* results in significant overproduction of both sterigmatocystin and penicillin, as well as increased expression levels of genes in the corresponding clusters (Shwab et al., 2007), and preliminary data suggest the same for *A. fumigatus hdaA* mutants (I. Lee and N. P. Keller, unpublished data). Additionally, the spread of histone acetylation in the promoter areas of genes in the aflatoxin cluster of the plant pathogen *A. parasiticus* has been shown to correlate closely with the onset of gene expression in this cluster (Roze et al., 2007). Histone methylation also appears to play a role in secondary metabolite regulation, as deletion of the histone methyltransferase gene *clrD* of *A. nidulans* (homolog of human SUV39H1) or *hepA* (a homolog of heterochromatin-associated proteins) results in increased expression of sterigmatocystin cluster genes and a corresponding increase in sterigmatocystin production (Y. Reyes-Dominguez, J. W. Bok, C. Scazzocchio, E. K. Shwab, N. P. Keller, and J. Strauss, unpublished data). Mutation of *hdaA* also results in partial remediation of the loss of metabolite production in Δ*laeA* mutants (Shwab et al., 2007). The prevalence of secondary metabolite clusters in subtelomeric areas also supports the possibility of epigenetic regulation, as such regulation is commonly associated with certain subtelomeric genes in a wide variety of eukaryotes (De Las Penas et al., 2003; Domergue et al., 2005; Freitas-Junior et al., 2005; Halme et al., 2004; Horn and Barry, 2005; Ralph and Scherf, 2005; Robyr et al., 2002).

The identification of epigenetic regulatory processes for fungal secondary metabolism may be a significant step towards determining the reason that natural selection has favored the clustering of secondary metabolite genes. As histone modification and chromatin remodeling are processes that affect specific localized regions of DNA, the clustering of genes would presumably facilitate their regulation by these means. The varied environmental factors and molecular signals discussed above may alter secondary metabolite production ultimately by effecting reorganization of chromatin structure at secondary metabolite gene cluster loci, enabling all of the genes in a cluster to be activated or inactivated simultaneously in response to environmental conditions. However, the extent to which chromatin modification is involved in fungal secondary metabolite regulation is not yet known, and it remains to be seen if epigenetic regulation alone is sufficient to explain the occurrence of secondary metabolite gene clusters or if other evolutionary factors also play a role, as seems likely.

CONCLUSIONS

Fungal secondary metabolites are important biomolecules with significant biological activities. *A. fumigatus* and other aspergilli are no exception when it comes to production of bioactive secondary metabolites. For some secondary metabolites like gliotoxin, which are abundantly produced and detectable in patient serum, future studies undoubtedly will definitively uncover their role in aspergillosis. However, we still do not know or understand whether specific low-abundance or undetectable secondary metabolites are produced in the microenvironments found at sites of infection in the lung, where they may influence or alter local immune responses. As genomics and molecular biology techniques allow us to uncover more regulatory and biosynthetic pathways for specific secondary metabolites produced by *A. fumigatus* and related aspergilli, answers to these questions will be forthcoming.

REFERENCES

Abe, I., K. Naito, Y. Takagi, and H. Noguchi. 2001. Molecular cloning, expression, and site-directed mutations of oxidosqualene cyclase from *Cephalosporium caerulens*. *Biochim. Biophys. Acta* **1522:** 67–73.

Allen, J. D., A. van Loevezijn, J. M. Lakhai, M. van der Valk, O. van Tellingen, G. Reid, J. H. Schellens, G. J. Koomen, and A. H. Schinkel. 2002. Potent and specific inhibition of the breast cancer resistance protein multidrug transporter in vitro and in mouse intestine by a novel analogue of fumitremorgin C. *Mol. Cancer Ther.* **1:**417–425.

Amitani, R., G. Taylor, E. N. Elezis, C. Llewellyn-Jones, J. Mitchell, F. Kuze, P. J. Cole, and R. Wilson. 1995. Purification and characterization of factors produced by *Aspergillus fumigatus* which affect human ciliated respiratory epithelium. *Infect. Immun.* **63:**3266–3271.

Bailey, L. 1953. Effect of fumagillin upon *Nosema apis* (Zander). *Nature* **171:**212–213.

Balibar, C. J., and C. T. Walsh. 2006. GliP, a multimodular nonribosomal peptide synthetase in *Aspergillus fumigatus*, makes the diketopiperazine scaffold of gliotoxin. *Biochemistry* **45:**15029–15038.

Beaver, J. P., and P. Waring. 1994. Lack of correlation between early intracellular calcium ion rises and the onset of apoptosis in thymocytes. *Immunol. Cell. Biol.* **72:**489–499.

Belkacemi, L., R. C. Barton, V. Hopwood, and E. G. Evans. 1999. Determination of optimum growth conditions for gliotoxin production by *Aspergillus fumigatus* and development of a novel method for gliotoxin detection. *Med. Mycol.* **37:**227–233.

Bennett, J. W., F. A. Fernholz, and L. S. Lee. 1978. Effect of light on aflatoxins, anthraquinones, and sclerotia in *Aspergillus flavus* and *A. parasiticus*. *Mycologia* **70:**104–116.

Bennett, J. W., J. J. Dunn, and C. I. Goldsman. 1981. Influence of white light on production of aflatoxins and anthraquinones in *Aspergillus parasiticus*. *Appl. Environ. Microbiol.* **41:**488–491.

Bennett, J. W., and R. Bentley. 1989. What's in a name? Microbial secondary metabolism. *Adv. Appl. Microbiol.* **34:**1–28.

Bernardo, P. H., N. Brasch, C. L. Chai, and P. Waring. 2003. A novel redox mechanism for the glutathione-dependent reversible uptake of a fungal toxin in cells. *J. Biol. Chem.* **278:**46549–46555.

Bernier, S. G., N. Taghizadeh, C. D. Thompson, W. F. Westlin, and G. Hannig. 2005. Methionine aminopeptidases type I and type II are essential to control cell proliferation. *J. Cell. Biochem.* **95:**1191–1203.

Bok, J. W., and N. P. Keller. 2004. LaeA, a regulator of secondary metabolism in *Aspergillus* spp. *Eukaryot. Cell* **3:**527–535.

Bok, J. W., S. A. Balajee, K. A. Marr, D. Andes, K. F. Nielsen, J. C. Frisvad, and N. P. Keller. 2005. LaeA, a regulator of morphogenetic fungal virulence factors. *Eukaryot. Cell* **4:**1574–1582.

Bok, J. W., D. Chung, S. A. Balajee, K. A. Marr, D. Andes, K. F. Nielsen, J. C. Frisvad, K. A. Kirby, and N. P. Keller. 2006a. GliZ, a transcriptional regulator of gliotoxin biosynthesis, contributes to *Aspergillus fumigatus* virulence. *Infect. Immun.* **74:**6761–6768.

Bok, J. W., D. Hoffmeister, L. A. Maggio-Hall, R. Murillo, J. D. Glasner, and N. P. Keller. 2006b. Genomic mining for *Aspergillus* natural products. *Chem. Biol.* **13:**31–37.

Bok, J. W., D. Noordermeer, S. P. Kale, and N. P. Keller. 2006c. Secondary metabolic gene cluster silencing in *Aspergillus nidulans*. *Mol. Microbiol.* **61:**1636–1645.

Bradford, H. F., P. J. Norris, and C. C. Smith. 1990. Changes in transmitter release patterns in vitro induced by tremorgenic mycotoxins. *J. Environ. Pathol. Toxicol. Oncol.* **10:**17–30.

Brakhage, A. A., A. Andrianopoulos, M. Kato, S. Steidl, M. A. Davis, N. Tsukagoshi, and M. J. Hynes. 1999. HAP-like CCAAT-binding complexes in filamentous fungi: implications for biotechnology. *Fungal Genet. Biol.* **27:**243–252.

Brodhagen, M., and N. P. Keller. 2006. Signaling pathways connecting mycotoxin production and sporulation. *Mol. Plant Pathol.* **7:**285–301.

Brown, D. W., J. H. Yu, H. S. Kelkar, M. Fernandes, T. C. Nesbitt, N. P. Keller, T. H. Adams, and T. J. Leonard. 1996. Twenty-five coregulated transcripts define a sterigmatocystin gene cluster in *Aspergillus nidulans*. *Proc. Natl. Acad. Sci. USA* **93:**1418–1422.

Butchko, R. A., T. H. Adams, and N. P. Keller. 1999. *Aspergillus nidulans* mutants defective in *stc* gene cluster regulation. *Genetics* **153:**715–720.

Calvo, A. M., R. A. Wilson, J. W. Bok, and N. P. Keller. 2002. Relationship between secondary metabolism and fungal development. *Microbiol. Mol. Biol. Rev.* **66:**447–459.

Chang, P. K., J. W. Cary, D. Bhatnagar, T. E. Cleveland, J. W. Bennett, J. E. Linz, C. P. Woloshuk, and G. A. Payne. 1993. Cloning of the *Aspergillus parasiticus apa-2* gene associated with the regulation of aflatoxin biosynthesis. *Appl. Environ. Microbiol.* **59:**3273–3279.

Chang, P. K., J. Yu, D. Bhatnagar, and T. E. Cleveland. 2000. Characterization of the *Aspergillus parasiticus* major nitrogen regulatory gene, *areA*. *Biochim. Biophys. Acta* **1491:**263–266.

Chang, S. Y., E. C. McGary, and S. Chang. 1989. Methionine aminopeptidase gene of *Escherichia coli* is essential for cell growth. *J. Bacteriol.* **171:**4071–4072.

Clevstrom, G., H. Ljunggren, S. Tegelstrom, and K. Tideman. 1983. Production of aflatoxin by an *Aspergillus flavus* isolate cultured under a limited oxygen supply. *Appl. Environ. Microbiol.* **46:**400–405.

Cohen, D. J., R. Loertscher, M. F. Rubin, N. L. Tilney, C. B. Carpenter, and T. B. Strom. 1984. Cyclosporine: a new immunosuppressive agent for organ transplantation. *Ann. Intern. Med.* **101:**667–682.

Cole, R. J., J. W. Kirksey, J. W. Dorner, D. M. Wilson, J. C. Johnson, Jr., A. N. Johnson, D. M. Bedell, J. P. Springer, K. K. Chexal, J. C. Clardy, and R. H. Cox. 1977. Mycotoxins produced by *Aspergillus fumigatus* species isolated from molded silage. *J. Agric. Food Chem.* **25:**826–830.

Coyle, C. M., and D. G. Panaccione. 2005. An ergot alkaloid biosynthesis gene and clustered hypothetical genes from *Aspergillus fumigatus*. *Appl. Environ. Microbiol.* **71:**3112–3118.

Cramer, R. A., Jr., M. P. Gamcsik, R. M. Brooking, L. K. Najvar, W. R. Kirkpatrick, T. F. Patterson, C. J. Balibar, J. R. Graybill, J. R. Perfect, S. N. Abraham, and W. J. Steinbach. 2006a. Disruption of a nonribosomal peptide synthetase in *Aspergillus fumigatus* eliminates gliotoxin production. *Eukaryot. Cell* **5:**972–980.

Cramer, R. A., Jr., J. E. Stajich, Y. Yamanaka, F. S. Dietrich, W. J. Steinbach, and J. R. Perfect. 2006b. Phylogenomic analysis of nonribosomal peptide synthetases in the genus *Aspergillus*. *Gene* **383:**24–32.

De Las Penas, A., S. J. Pan, I. Castano, J. Alder, R. Cregg, and B. P. Cormack. 2003. Virulence-related surface glycoproteins in the yeast pathogen *Candida glabrata* are encoded in subtelomeric clusters and subject to RAP1- and SIR-dependent transcriptional silencing. *Genes Dev.* **17:**2245–2258.

Domergue, R., I. Castano, A. De Las Penas, M. Zupancic, V. Lockatell, J. R. Hebel, D. Johnson, and B. P. Cormack. 2005. Nicotinic acid limitation regulates silencing of *Candida* adhesins during UTI. *Science* **308:**866–870.

Dowzer, C. E., and J. M. Kelly. 1989. Cloning of the *creA* gene from *Aspergillus nidulans*: a gene involved in carbon catabolite repression. *Curr. Genet.* **15:**457–459.

Dummer, J. S., A. Hardy, A. Poorsattar, and M. Ho. 1983. Early infections in kidney, heart, and liver transplant recipients on cyclosporine. *Transplantation* **36:**259–267.

Duran, R. M., J. W. Cary, and A. M. Calvo. 2007. Production of cyclopiazonic acid, aflatrem, and aflatoxin by *Aspergillus flavus* is regulated by *veA*, a gene necessary for sclerotial formation. *Appl. Microbiol. Biotechnol.* **73:**1158–1168.

Ehrlich, K. C., B. G. Montalbano, and J. W. Cary. 1999. Binding of the C_6-zinc cluster protein, AflR, to the promoters of aflatoxin pathway biosynthesis genes in *Aspergillus parasiticus*. *Gene* **230:**249–257.

Eichner, R. D., and A. Mullbacher. 1984. Hypothesis: fungal toxins are involved in aspergillosis and AIDS. *Aust. J. Exp. Biol. Med. Sci.* **62:**479–484.

Eichner, R. D., M. Al Salami, P. R. Wood, and A. Mullbacher. 1986. The effect of gliotoxin upon macrophage function. *Int. J. Immunopharmacol.* **8:**789–797.

Ellis, W. O., J. P. Smith, B. K. Simpson, H. Ramaswamy, and G. Doyon. 1994. Growth of and aflatoxin production by *Aspergillus flavus* in peanuts stored under modified atmosphere packaging (MAP) conditions. *Int. J. Food Microbiol.* **22:**173–187.

Fairhead, C., and B. Dujon. 2006. Structure of *Kluyveromyces lactis* subtelomeres: duplications and gene content. *FEMS Yeast Res.* **6:**428–441.

Fernandes, M., N. P. Keller, and T. H. Adams. 1998. Sequence-specific binding by *Aspergillus nidulans* AflR, a C_6 zinc cluster protein regulating mycotoxin biosynthesis. *Mol. Microbiol.* **28:**1355–1365.

Fox, E. M., D. M. Gardiner, N. P. Keller, and B. J. Howlett. 2008. A $Zn(II)_2Cys_6$ DNA binding protein regulates the sirodesmin PL biosynthetic gene cluster in *Leptosphaeria maculans*. *Fungal Genet. Biol.* **45:**671–682.

Freitas-Junior, L. H., R. Hernandez-Rivas, S. A. Ralph, D. Montiel-Condado, O. K. Ruvalcaba-Salazar, A. P. Rojas-Meza, L. Mancio-Silva, R. J. Leal-Silvestre, A. M. Gontijo, S. Shorte, and A. Scherf.

2005. Telomeric heterochromatin propagation and histone acetylation control mutually exclusive expression of antigenic variation genes in malaria parasites. *Cell* **121**:25–36.

Fukuda, T., M. Boeckh, K. A. Guthrie, D. K. Mattson, S. Owens, A. Wald, B. M. Sandmaier, L. Corey, R. F. Storb, and K. A. Marr. 2004. Invasive aspergillosis before allogeneic hematopoietic stem cell transplantation: 10-year experience at a single transplant center. *Biol. Blood Marrow Transplant.* **10**:494–503.

Gallagher, R. T., and G. C. Latch. 1977. Production of the tremorgenic mycotoxins verruculogen and fumitremorgin B by *Penicillium piscarium* Westling. *Appl. Environ. Microbiol.* **33**:730–731.

Gardiner, D. M., and B. J. Howlett. 2005. Bioinformatic and expression analysis of the putative gliotoxin biosynthetic gene cluster of *Aspergillus fumigatus*. *FEMS Microbiol. Lett.* **248**:241–248.

Gardiner, D. M., R. S. Jarvis, and B. J. Howlett. 2005a. The ABC transporter gene in the sirodesmin biosynthetic gene cluster of *Leptosphaeria maculans* is not essential for sirodesmin production but facilitates self-protection. *Fungal Genet. Biol.* **42**:257–263.

Gardiner, D. M., P. Waring, and B. J. Howlett. 2005b. The epipolythiodioxopiperazine (ETP) class of fungal toxins: distribution, mode of action, functions and biosynthesis. *Microbiology* **151**:1021–1032.

Garimella, T. S., D. D. Ross, and K. S. Bauer. 2004. Liquid chromatography method for the quantitation of the breast cancer resistance protein ABCG2 inhibitor fumitremorgin C and its chemical analogues in mouse plasma and tissues. *J. Chromatogr. B* **807**:203–208.

Griffith, E. C., Z. Su, B. E. Turk, S. Chen, Y. H. Chang, Z. Wu, K. Biemann, and J. O. Liu. 1997. Methionine aminopeptidase (type 2) is the common target for angiogenesis inhibitors AGM-1470 and ovalicin. *Chem. Biol.* **4**:461–471.

Griffith, E. C., Z. Su, S. Niwayama, C. A. Ramsay, Y. H. Chang, and J. O. Liu. 1998. Molecular recognition of angiogenesis inhibitors fumagillin and ovalicin by methionine aminopeptidase 2. *Proc. Natl. Acad. Sci. USA* **95**:15183–15188.

Grow, W. B., J. S. Moreb, D. Roque, K. Manion, H. Leather, V. Reddy, S. A. Khan, K. J. Finiewicz, H. Nguyen, C. J. Clancy, P. S. Mehta, and J. R. Wingard. 2002. Late onset of invasive *Aspergillus* infection in bone marrow transplant patients at a university hospital. *Bone Marrow Transplant.* **29**:15–19.

Halme, A., S. Bumgarner, C. Styles, and G. R. Fink. 2004. Genetic and epigenetic regulation of the FLO gene family generates cell-surface variation in yeast. *Cell* **116**:405–415.

Han, K. H., J. A. Seo, and J. H. Yu. 2004. Regulators of G-protein signalling in *Aspergillus nidulans*: RgsA downregulates stress response and stimulates asexual sporulation through attenuation of GanB (Gα) signalling. *Mol. Microbiol.* **53**:529–540.

Hartwig, A., and A. Przelecka. 1971. Nucleic acids in intestine of *Apis mellifica* infected with *Nosema apis* and treated with fumagillin DCH: cytochemical and autoradiographic studies. *J. Invertebr. Pathol.* **18**:331–336.

Hewitt, W. L. 1967. Penicillin—historical impact on infection control. *Ann. N. Y. Acad. Sci.* **145**:212–215.

Hicks, J. K., J. H. Yu, N. P. Keller, and T. H. Adams. 1997. *Aspergillus* sporulation and mycotoxin production both require inactivation of the FadA G alpha protein-dependent signaling pathway. *EMBO J.* **16**:4916–4923.

Hoffmeister, D., and N. P. Keller. 2007. Natural products of filamentous fungi: enzymes, genes, and their regulation. *Nat. Prod. Rep.* **24**:393–416.

Horn, D., and J. D. Barry. 2005. The central roles of telomeres and subtelomeres in antigenic variation in African trypanosomes. *Chromosome Res.* **13**:525–533.

Hurne, A. M., C. L. Chai, K. Moerman, and P. Waring. 2002. Influx of calcium through a redox-sensitive plasma membrane channel in thymocytes causes early necrotic cell death induced by the epipolythiodioxopiperazine toxins. *J. Biol. Chem.* **277**:31631–31638.

Hynes, M. J. 1975. Studies on the role of the *areA* gene in the regulation of nitrogen catabolism in *Aspergillus nidulans*. *Aust. J. Biol. Sci.* **28**:301–313.

Ingber, D., T. Fujita, S. Kishimoto, K. Sudo, T. Kanamaru, H. Brem, and J. Folkman. 1990. Synthetic analogues of fumagillin that inhibit angiogenesis and suppress tumour growth. *Nature* **348**:555–557.

Johnson, J. R., W. F. Bruce, and J. D. Dutcher. 1943. Gliotoxin, the antibiotic principle of *Gliocladium fimbriatum*. Production, physical and biological properties. *J. Am. Chem. Soc.* **65**:2005–2009.

Kale, S. P., J. W. Cary, N. Hollis, J. R. Wilkinson, D. Bhatnagar, J. Yu, T. E. Cleveland, and J. W. Bennett. 2007. Analysis of aflatoxin regulatory factors in serial transfer-induced non-aflatoxigenic *Aspergillus parasiticus*. *Food Addit. Contam.* **24**:1061–1069.

Kato, N., W. Brooks, and A. M. Calvo. 2003. The expression of sterigmatocystin and penicillin genes in *Aspergillus nidulans* is controlled by *veA*, a gene required for sexual development. *Eukaryot. Cell* **2**:1178–1186.

Katznelson, H., and C. A. Jamieson. 1952. Control of nosema disease of honeybees with fumagillin. *Science* **115**:70–71.

Kawaguchi, A., H. Kobayashi, and S. Okuda. 1973. Cyclization of 2,3-oxidosqualene with microsomal fraction of *Cephalosporium caerulens*. *Chem. Pharm. Bull.* **21**:577–583.

Keller, N., J. Bok, D. Chung, R. M. Perrin, and E. K. Shwab. 2006. LaeA, a global regulator of *Aspergillus* toxins. *Med. Mycol.* **44**(Suppl.):83–85.

Keller, N. P., and T. M. Hohn. 1997. Metabolic pathway gene clusters in filamentous fungi. *Fungal Genet. Biol.* **21**:17–29.

Keller, N. P., G. Turner, and J. W. Bennett. 2005. Fungal secondary metabolism: from biochemistry to genomics. *Nat. Rev. Microbiol.* **3**:937–947.

Khoufache, K., O. Puel, N. Loiseau, M. Delaforge, D. Rivollet, A. Coste, C. Cordonnier, E. Escudier, F. Botterel, and S. Bretagne. 2007. Verruculogen associated with *Aspergillus fumigatus* hyphae and conidia modifies the electrophysiological properties of human nasal epithelial cells. *BMC Microbiol.* **7**:5.

Kim, K. H., Y. R. Cho, M. I. Rota, R. A. Cramer, Jr., and C. B. Lawrence. 2007. Functional analysis of the *Alternaria brassicicola* non-ribosomal peptide synthetase gene *AbNPS2* reveals a role in conidial cell wall construction. *Mol. Plant Pathol.* **8**:23–39.

Knaus, H. G., O. B. McManus, S. H. Lee, W. A. Schmalhofer, M. Garcia-Calvo, L. M. Helms, M. Sanchez, K. Giangiacomo, J. P. Reuben, A. B. Smith III, et al. 1994. Tremorgenic indole alkaloids potently inhibit smooth muscle high-conductance calcium-activated potassium channels. *Biochemistry* **33**:5819–5828.

Kroll, M., F. Arenzana-Seisdedos, F. Bachelerie, D. Thomas, B. Friguet, and M. Conconi. 1999. The secondary fungal metabolite gliotoxin targets proteolytic activities of the proteasome. *Chem. Biol.* **6**:689–698.

Kupfahl, C., T. Heinekamp, G. Geginat, T. Ruppert, A. Hartl, H. Hof, and A. A. Brakhage. 2006. Deletion of the *gliP* gene of *Aspergillus fumigatus* results in loss of gliotoxin production but has no effect on virulence of the fungus in a low-dose mouse infection model. *Mol. Microbiol.* **62**:292–302.

Kupfahl, C., A. Michalka, C. Lass-Florl, G. Fischer, G. Haase, T. Ruppert, G. Geginat, and H. Hof. 2008. Gliotoxin production by clinical and environmental *Aspergillus fumigatus* strains. *Int. J. Med. Microbiol.* **298**:319–327.

Kupfahl, C., T. Ruppert, A. Dietz, G. Geginat, and H. Hof. 2007. *Candida* species fail to produce the immunosuppressive secondary metabolite gliotoxin in vitro. *FEMS Yeast Res.* **7**:986–992.

Land, C. J., K. Hult, R. Fuchs, S. Hagelberg, and H. Lundstrom. 1987. Tremorgenic mycotoxins from *Aspergillus fumigatus* as a possible occupational health problem in sawmills. *Appl. Environ. Microbiol.* **53**:787–790.

Land, C. J., H. Lundstrom, and S. Werner. 1993. Production of tremorgenic mycotoxins by isolates of *Aspergillus fumigatus* from sawmills in Sweden. *Mycopathologia* **124**:87–93.

Landan, G., G. Cohen, Y. Aharonowitz, Y. Shuali, D. Graur, and D. Shiffman. 1990. Evolution of isopenicillin N synthase genes may have involved horizontal gene transfer. *Mol. Biol. Evol.* 7:399–406.

Larsen, T. O., J. Smedsgaard, K. F. Nielsen, M. A. Hansen, R. A. Samson, and J. C. Frisvad. 2007. Production of mycotoxins by *Aspergillus lentulus* and other medically important and closely related species in section *Fumigati*. *Med. Mycol.* 45:225–232.

Lee, B. N., S. Kroken, D. Y. T. Chou, B. Robbertse, O. C. Yoder, and B. G. Turgeon. 2005. Functional analysis of all nonribosomal peptide synthetases in *Cochliobolus heterostrophus* reveals a factor, NPS6, involved in virulence and resistance to oxidative stress. *Eukaryot. Cell* 4:545–555.

Lewis, R. E., N. P. Wiederhold, J. Chi, X. Y. Han, K. V. Komanduri, D. P. Kontoyiannis, and R. A. Prince. 2005a. Detection of gliotoxin in experimental and human aspergillosis. *Infect. Immun.* 73:635–637.

Lewis, R. E., N. P. Wiederhold, M. S. Lionakis, R. A. Prince, and D. P. Kontoyiannis. 2005b. Frequency and species distribution of gliotoxin-producing *Aspergillus* isolates recovered from patients at a tertiary-care cancer center. *J. Clin. Microbiol.* 43:6120–6122.

Li, X., and Y. H. Chang. 1995. Amino-terminal protein processing in *Saccharomyces cerevisiae* is an essential function that requires two distinct methionine aminopeptidases. *Proc. Natl. Acad. Sci. USA* 92: 12357–12361.

Litzka, O., P. Papagiannopolous, M. A. Davis, M. J. Hynes, and A. A. Brakhage. 1998. The penicillin regulator PENR1 of *Aspergillus nidulans* is a HAP-like transcriptional complex. *Eur. J. Biochem.* 251:758–767.

Lopes Bezerra, L. M., and S. G. Filler. 2004. Interactions of *Aspergillus fumigatus* with endothelial cells: internalization, injury, and stimulation of tissue factor activity. *Blood* 103:2143–2149.

Macdonald, J. C., and G. P. Slater. 1975. Biosynthesis of gliotoxin and mycelianamide. *Can. J. Biochem.* 53:475–478.

Maiya, S., A. Grundmann, S. M. Li, and G. Turner. 2006. The fumitremorgin gene cluster of *Aspergillus fumigatus*: identification of a gene encoding brevianamide F synthetase. *ChemBioChem* 7:1062–1069.

Marr, K. A., R. A. Carter, M. Boeckh, P. Martin, and L. Corey. 2002. Invasive aspergillosis in allogeneic stem cell transplant recipients: changes in epidemiology and risk factors. *Blood* 100:4358–4366.

Martin, J. F. 2000. Molecular control of expression of penicillin biosynthesis genes in fungi: regulatory proteins interact with a bidirectional promoter region. *J. Bacteriol.* 182:2355–2362.

Meyers, D. M., G. Obrian, W. L. Du, D. Bhatnagar, and G. A. Payne. 1998. Characterization of *aflJ*, a gene required for conversion of pathway intermediates to aflatoxin. *Appl. Environ. Microbiol.* 64: 3713–3717.

Mitchell, C. G., J. Slight, and K. Donaldson. 1997. Diffusible component from the spore surface of the fungus *Aspergillus fumigatus* which inhibits the macrophage oxidative burst is distinct from gliotoxin and other hyphal toxins. *Thorax* 52:796–801.

Mullbacher, A., and R. D. Eichner. 1984. Immunosuppression in vitro by a metabolite of a human pathogenic fungus. *Proc. Natl. Acad. Sci. USA* 81:3835–3837.

Mullbacher, A., P. Waring, and R. D. Eichner. 1985. Identification of an agent in cultures of *Aspergillus fumigatus* displaying antiphagocytic and immunomodulating activity in vitro. *J. Gen. Microbiol.* 131:1251–1258.

Mullbacher, A., D. Hume, A. W. Braithwaite, P. Waring, and R. D. Eichner. 1987. Selective resistance of bone marrow-derived hemopoietic progenitor cells to gliotoxin. *Proc. Natl. Acad. Sci. USA* 84: 3822–3825.

Nierman, W. C., A. Pain, M. J. Anderson, J. R. Wortman, H. S. Kim, J. Arroyo, M. Berriman, K. Abe, D. B. Archer, C. Bermejo, J. Bennett, P. Bowyer, D. Chen, M. Collins, R. Coulsen, R. Davies, P. S. Dyer, M. Farman, N. Fedorova, T. V. Feldblum, R. Fischer, N. Fosker, A. Fraser, J. L. Garcia, M. J. Garcia, A. Goble, G. H. Goldman, K. Gomi, S. Griffith-Jones, R. Gwilliam, B. Haas, H. Haas, D. Harris, H. Horiuchi, J. Huang, S. Humphray, J. Jimenez, N. Keller, H. Khouri, K. Kitamoto, T. Kobayashi, S. Konzack, R. Kulkarni, T. Kumagai, A. Lafton, J. P. Latgé, W. Li, A. Lord, C. Lu, W. H. Majoros, G. S. May, B. L. Miller, Y. Mohamoud, M. Molina, M. Monod, I. Mouyna, S. Mulligan, L. Murphy, S. O'Neil, I. Paulsen, M. A. Penalva, M. Pertea, C. Price, B. L. Pritchard, M. A. Quail, E. Rabbinowitsch, N. Rawlins, M. A. Rajandream, U. Reichard, H. Renauld, G. D. Robson, S. Rodriguez de Cordoba, J. M. Rodriguez-Pena, C. M. Ronning, S. Rutter, S. L. Salzberg, M. Sanchez, J. C. Sanchez-Ferrero, D. Saunders, K. Seeger, R. Squares, S. Squares, M. Takeuchi, F. Tekaia, G. Turner, C. R. Vazquez de Aldana, J. Weidman, O. White, J. Woodward, J. H. Yu, C. Fraser, J. E. Galagan, K. Asai, M. Machida, N. Hall, B. Barrell, and D. W. Denning. 2005. Genomic sequence of the pathogenic and allergenic filamentous fungus *Aspergillus fumigatus*. *Nature* 438:1151–1156.

Nishida, S., L. S. Yoshida, T. Shimoyama, H. Nunoi, T. Kobayashi, and S. Tsunawaki. 2005. Fungal metabolite gliotoxin targets flavocytochrome b_{558} in the activation of the human neutrophil NADPH oxidase. *Infect. Immun.* 73:235–244.

O'Brian, G. R., D. R. Georgianna, J. R. Wilkinson, J. Yu, H. K. Abbas, D. Bhatnagar, T. E. Cleveland, W. Nierman, and G. A. Payne. 2007. The effect of elevated temperature on gene transcription and aflatoxin biosynthesis. *Mycologia* 99:232–239.

Okuda, S., S. Iwasaki, K. Tsuda, Y. Sano, T. Hata, S. Udagawa, Y. Nakayama, and H. Yamaguchi. 1964. The structure of helvolic acid. *Chem. Pharm. Bull.* (Tokyo) 12:121–124.

Okuda, S., Y. Nakayama, and K. Tsuda. 1966. Studies on microbial products. I. Helvolic acid and related compounds. I. 7-desacetoxyhelvolic acid and helvolinic acid. *Chem. Pharm. Bull.* (Tokyo) 14:436–441.

Orciuolo, E., M. Stanzani, M. Canestraro, S. Galimberti, G. Carulli, R. Lewis, M. Petrini, and K. V. Komanduri. 2007. Effects of *Aspergillus fumigatus* gliotoxin and methylprednisolone on human neutrophils: implications for the pathogenesis of invasive aspergillosis. *J. Leukoc. Biol.* 82:839–848.

Osbourn, A. E. 2001. Tox-boxes, fungal secondary metabolites, and plant disease. *Proc. Natl. Acad. Sci. USA* 98:14187–14188.

Pahl, H. L., B. Krauss, K. Schulze-Osthoff, T. Decker, E. B. Traenckner, M. Vogt, C. Myers, T. Parks, P. Warring, A. Muhlbacher, A. P. Czernilofsky, and P. A. Baeuerle. 1996. The immunosuppressive fungal metabolite gliotoxin specifically inhibits transcription factor NF-κB. *J. Exp. Med.* 183:1829–1840.

Panaccione, D. G., and C. M. Coyle. 2005. Abundant respirable ergot alkaloids from the common airborne fungus *Aspergillus fumigatus*. *Appl. Environ. Microbiol.* 71:3106–3111.

Pardo, J., C. Urban, E. M. Galvez, P. G. Ekert, U. Muller, J. Kwon-Chung, M. Lobigs, A. Mullbacher, R. Wallich, C. Borner, and M. M. Simon. 2006. The mitochondrial protein Bak is pivotal for gliotoxin-induced apoptosis and a critical host factor of *Aspergillus fumigatus* virulence in mice. *J. Cell Biol.* 174:509–519.

Payne, G. A., and M. P. Brown. 1998. Genetics and physiology of aflatoxin biosynthesis. *Annu. Rev. Phytopathol.* 36:329–362.

Pedley, K. F., and J. D. Walton. 2001. Regulation of cyclic peptide biosynthesis in a plant pathogenic fungus by a novel transcription factor. *Proc. Natl. Acad. Sci. USA* 98:14174–14179.

Perfect, J. R., and D. T. Durack. 1985. Effects of cyclosporine in experimental cryptococcal meningitis. *Infect. Immun.* 50:22–26.

Perrin, R., N. Federova, J.-W. Bok, J. Wortman, R. A. Cramer, Jr., H. Kim, W. Nierman, and N. Keller. 2007. Transcriptional regulation of chemical diversity in *Aspergillus fumigatus* by LaeA. *PLoS Pathog.* 3:e50.

Price, M. S., J. Yu, W. C. Nierman, H. S. Kim, B. Pritchard, C. A. Jacobus, D. Bhatnagar, T. E. Cleveland, and G. A. Payne. 2006. The aflatoxin pathway regulator AflR induces gene transcription inside and outside of the aflatoxin biosynthetic cluster. *FEMS Microbiol. Lett.* 255:275–279.

Proctor, R. H., T. M. Hohn, S. P. McCormick, and A. E. Desjardins. 1995. *tri6* encodes an unusual zinc finger protein involved in regulation of trichothecene biosynthesis in *Fusarium sporotrichioides*. *Appl. Environ. Microbiol.* **61**:1923–1930.

Rabindran, S. K., D. D. Ross, L. A. Doyle, W. Yang, and L. M. Greenberger. 2000. Fumitremorgin C reverses multidrug resistance in cells transfected with the breast cancer resistance protein. *Cancer Res.* **60**:47–50.

Ralph, S. A., and A. Scherf. 2005. The epigenetic control of antigenic variation in *Plasmodium falciparum*. *Curr. Opin. Microbiol.* **8**:434–440.

Rehmeyer, C., W. Li, M. Kusaba, Y. S. Kim, D. Brown, C. Staben, R. Dean, and M. Farman. 2006. Organization of chromosome ends in the rice blast fungus, *Magnaporthe oryzae*. *Nucleic Acids Res.* **34**: 4685–4701.

Ribaud, P., C. Chastang, J. P. Latgé, L. Baffroy-Lafitte, N. Parquet, A. Devergie, H. Esperou, F. Selimi, V. Rocha, H. Esperou, F. Selimi, V. Rocha, F. Derouin, G. Socie, and E. Gluckman. 1999. Survival and prognostic factors of invasive aspergillosis after allogeneic bone marrow transplantation. *Clin. Infect. Dis.* **28**:322–330.

Robyr, D., Y. Suka, I. Xenarios, S. K. Kurdistani, A. Wang, N. Suka, and M. Grunstein. 2002. Microarray deacetylation maps determine genome-wide functions for yeast histone deacetylases. *Cell* **109**: 437–446.

Rodriguez, P. L., and L. Carrasco. 1992. Gliotoxin: inhibitor of poliovirus RNA synthesis that blocks the viral RNA polymerase 3Dpol. *J. Virol.* **66**:1971–1976.

Rohlfs, M., M. Albert, N. P. Keller, and F. Kempken. 2007. Secondary chemicals protect mould from fungivory. *Biol. Lett.* **3**:523–525.

Roze, L. V., R. M. Beaudry, N. P. Keller, and J. E. Linz. 2004. Regulation of aflatoxin synthesis by FadA/cAMP/protein kinase A signaling in *Aspergillus parasiticus*. *Mycopathologia* **158**:219–232.

Roze, L. V., R. M. Beaudry, A. E. Arthur, A. M. Calvo, and J. E. Linz. 2007. *Aspergillus* volatiles regulate aflatoxin synthesis and asexual sporulation in *Aspergillus parasiticus*. *Appl. Environ. Microbiol.* **73**:7268–7276.

Schindler, A. F., J. G. Palmer, and W. V. Eisenberg. 1967. Aflatoxin production by *Aspergillus flavus* as related to various temperatures. *Appl. Microbiol.* **15**:1006–1009.

Seo, J. A., and J. H. Yu. 2006. The phosducin-like protein PhnA is required for $G\beta\gamma$-mediated signaling for vegetative growth, developmental control, and toxin biosynthesis in *Aspergillus nidulans*. *Eukaryot. Cell* **5**:400–410.

Shah, D. T., and B. Larsen. 1991. Clinical isolates of yeast produce a gliotoxin-like substance. *Mycopathologia* **116**:203–208.

Shah, D. T., D. D. Glover, and B. Larsen. 1995. In situ mycotoxin production by *Candida albicans* in women with vaginitis. *Gynecol. Obstet. Investig.* **39**:67–69.

Shimizu, K., and N. P. Keller. 2001. Genetic involvement of a cAMP-dependent protein kinase in a G protein signaling pathway regulating morphological and chemical transitions in *Aspergillus nidulans*. *Genetics* **157**:591–600.

Shimizu, K., J. K. Hicks, T. P. Huang, and N. P. Keller. 2003. Pka, Ras and RGS protein interactions regulate activity of AflR, a $Zn(II)_2Cys_6$ transcription factor in *Aspergillus nidulans*. *Genetics* **165**:1095–1104.

Shwab, E. K., J. W. Bok, M. Tribus, J. Galehr, S. Graessle, and N. P. Keller. 2007. Histone deacetylase activity regulates chemical diversity in *Aspergillus*. *Eukaryot. Cell* **6**:1656–1664.

Sin, N., L. Meng, M. Q. Wang, J. J. Wen, W. G. Bornmann, and C. M. Crews. 1997. The anti-angiogenic agent fumagillin covalently binds and inhibits the methionine aminopeptidase, MetAP-2. *Proc. Natl. Acad. Sci. USA* **94**:6099–6103.

Spilsbury, J. F., and S. Wilkinson. 1961. The isolation of festuclavine and two new clavine alkaloids from *Aspergillus fumigatus* Fres. *J. Chem. Soc.* **5**:2085–2091.

Sprote, P., and A. A. Brakhage. 2007. The light-dependent regulator velvet A of *Aspergillus nidulans* acts as a repressor of the penicillin biosynthesis. *Arch. Microbiol.* **188**:69–79.

Stanimirovic, Z., J. Stevanovic, V. Bajic, and I. Radovic. 2007. Evaluation of genotoxic effects of fumagillin by cytogenetic tests in vivo. *Mutat. Res.* **628**:1–10.

Stanzani, M., E. Orciuolo, R. Lewis, D. P. Kontoyiannis, S. L. Martins, L. S. St. John, and K. V. Komanduri. 2005. *Aspergillus fumigatus* suppresses the human cellular immune response via gliotoxin-mediated apoptosis of monocytes. *Blood* **105**:2258–2265.

Stinnett, S. M., E. A. Espeso, L. Cobeno, L. Araujo-Bazan, and A. M. Calvo. 2007. *Aspergillus nidulans* VeA subcellular localization is dependent on the importin alpha carrier and on light. *Mol. Microbiol.* **63**:242–255.

Stollerman, G. H. 1993. The global impact of penicillin: then and now. *Mt. Sinai J. Med.* **60**:112–119.

Sugui, J. A., J. Pardo, Y. C. Chang, K. A. Zarember, G. Nardone, E. M. Galvez, A. Mullbacher, J. I. Gallin, M. M. Simon, and K. J. Kwon-Chung. 2007a. Gliotoxin is a virulence factor of *Aspergillus fumigatus: gliP* deletion attenuates virulence in mice immunosuppressed with hydrocortisone. *Eukaryot. Cell* **6**:1562–1569.

Sugui, J. A., J. Pardo, Y. C. Chang, A. Müllbacher, K. A. Zarember, E. M. Galvez, L. Brinster, P. Zerfas, J. I. Gallin, M. M. Simon, and K. J. Kwon-Chung. 2007b. Role of *laeA* in the regulation of *alb1*, *gliP*, conidial morphology, and virulence in *Aspergillus fumigatus*. *Eukaryot. Cell* **6**:1552–1561.

Sutton, P., N. R. Newcombe, P. Waring, and A. Mullbacher. 1994. In vivo immunosuppressive activity of gliotoxin, a metabolite produced by human pathogenic fungi. *Infect. Immun.* **62**:1192–1198.

Suzuki, S., K. Kikkawa, and M. Yamazaki. 1984. Abnormal behavioral effects elicited by a neurotropic mycotoxin, fumitremorgin A in mice. *J. Pharmacobiodyn.* **7**:935–942.

Tag, A., J. Hicks, G. Garifullina, C. Ake, Jr., T. D. Phillips, M. Beremand, and N. Keller. 2000. G-protein signalling mediates differential production of toxic secondary metabolites. *Mol. Microbiol.* **38**:658–665.

Tilburn, J., S. Sarkar, D. A. Widdick, E. A. Espeso, M. Orejas, J. Mungroo, M. A. Penalva, and H. N. Arst, Jr. 1995. The *Aspergillus* PacC zinc finger transcription factor mediates regulation of both acid- and alkaline-expressed genes by ambient pH. *EMBO J.* **14**: 779–790.

Todd, R. B., and A. Andrianopoulos. 1997. Evolution of a fungal regulatory gene family: the $Zn(II)_2Cys_6$ binuclear cluster DNA binding motif. *Fungal Genet. Biol.* **21**:388–405.

Trown, P. W., and J. A. Bilello. 1972. Mechanism of action of gliotoxin: elimination of activity by sulfhydryl compounds. *Antimicrob. Agents Chemother.* **2**:261–266.

Tsitsigiannis, D. I., T. M. Kowieski, R. Zarnowski, and N. P. Keller. 2004. Endogenous lipogenic regulators of spore balance in *Aspergillus nidulans*. *Eukaryot. Cell* **3**:1398–1411.

Tsitsigiannis, D. I., and N. P. Keller. 2006. Oxylipins act as determinants of natural product biosynthesis and seed colonization in *Aspergillus nidulans*. *Mol. Microbiol.* **59**:882–892.

Tsitsigiannis, D. I., and N. P. Keller. 2007. Oxylipins as developmental and host-fungal communication signals. *Trends Microbiol.* **15**: 109–118.

Tsunawaki, S., L. S. Yoshida, S. Nishida, T. Kobayashi, and T. Shimoyama. 2004. Fungal metabolite gliotoxin inhibits assembly of the human respiratory burst NADPH oxidase. *Infect. Immun.* **72**:3373–3382.

Unsold, I. A., and S. M. Li. 2005. Overproduction, purification and characterization of FgaPT2, a dimethylallyltryptophan synthase from *Aspergillus fumigatus*. *Microbiology* **151**:1499–1505.

van Loevezijn, A., J. D. Allen, A. H. Schinkel, and G. J. Koomen. 2001. Inhibition of BCRP-mediated drug efflux by fumitremorgin-type indolyl diketopiperazines. *Bioorg. Med. Chem. Lett.* **11:**29–32.

Ward, C., E. R. Chilvers, M. F. Lawson, J. G. Pryde, S. Fujihara, S. N. Farrow, C. Haslett, and A. G. Rossi. 1999. NF-κB activation is a critical regulator of human granulocyte apoptosis in vitro. *J. Biol. Chem.* **274:**4309–4318.

Waring, P., N. Newcombe, M. Edel, Q. H. Lin, H. Jiang, A. Sjaarda, T. Piva, and A. Mullbacher. 1994. Cellular uptake and release of the immunomodulating fungal toxin gliotoxin. *Toxicon* **32:**491–504.

Waring, P., A. Sjaarda, and Q. H. Lin. 1995. Gliotoxin inactivates alcohol dehydrogenase by either covalent modification or free radical damage mediated by redox cycling. *Biochem. Pharmacol.* **49:** 1195–1201.

Watanabe, A., K. Kamei, T. Sekine, M. Waku, K. Nishimura, M. Miyaji, K. Tatsumi, and T. Kuriyama. 2004. Effect of aeration on gliotoxin production by *Aspergillus fumigatus* in its culture filtrate. *Mycopathologia* **157:**245–254.

Weindling, R., and O. H. Emerson. 1936. The isolation of a toxic substance from the culture filtrate of *Trichoderma. Phytopathology* **31:**991–1003.

Whittington, R., and M. L. Winston. 2003. Effects of *Nosema bombi* and its treatment fumagillin on bumble bee (*Bombus occidentalis*) colonies. *J. Invertebr. Pathol.* **84:**54–58.

Williams, T. I. 1952. Some chemical properties of helvolic acid. *Biochem. J.* **51:**538–542.

Willingale, J., K. P. Perera, and P. G. Mantle. 1983. An intermediary role for the tremorgenic mycotoxin TR-2 in the biosynthesis of verruculogen. *Biochem. J.* **214:**991–993.

Woloshuk, C. P., K. R. Foutz, J. F. Brewer, D. Bhatnagar, T. E. Cleveland, and G. A. Payne. 1994. Molecular characterization of *aflR*, a regulatory locus for aflatoxin biosynthesis. *Appl. Environ. Microbiol.* **60:**2408–2414.

Yamazaki, M., H. Fujimoto, and T. Kawasaki. 1980. Chemistry of tremorogenic metabolites. I. Fumitremorgin A from *Aspergillus fumigatus. Chem. Pharm. Bull.* (Tokyo) **28:**245–254.

Yamazaki, M., S. Suzuki, and N. Ozaki. 1983. Biochemical investigation on the abnormal behaviors induced by fumitremorgin A, a tremorgenic mycotoxin to mice. *J. Pharmacobiodyn.* **6:**748–751.

Yamazaki, M., and S. Suzuki. 1986. Toxicology of tremorgenic mycotoxins, fumitremorgin A and B. *Dev. Toxicol. Environ. Sci.* **12:** 273–282.

Yoshida, L. S., S. Abe, and S. Tsunawaki. 2000. Fungal gliotoxin targets the onset of superoxide-generating NADPH oxidase of human neutrophils. *Biochem. Biophys. Res. Commun.* **268:**716–723.

Yu, J. H., R. A. Butchko, M. Fernandes, N. P. Keller, T. J. Leonard, and T. H. Adams. 1996. Conservation of structure and function of the aflatoxin regulatory gene *aflR* from *Aspergillus nidulans* and *A. flavus. Curr. Genet.* **29:**549–555.

Yu, J. H., and N. Keller. 2005. Regulation of secondary metabolism in filamentous fungi. *Annu. Rev. Phytopathol.* **43:**437–458.

Aspergillus fumigatus and Aspergillosis
Edited by J.-P. Latgé and W. J. Steinbach

Chapter 16

Aspergillus fumigatus: Survival and Death under Stress

DAVID S. ASKEW AND JUDITH C. RHODES

Aspergillus fumigatus is responsible for the vast majority of aspergillosis cases, yet we still do not fully understand why this species has become the predominant mold pathogen. Environmental estimates do not support the idea that disease association is simply a reflection of environmental prevalence, so it is generally believed that *A. fumigatus* has evolved some unique properties that allow it to cause opportunistic infections more readily than other common molds. *A. fumigatus* is normally found in the inhospitable environment of soil and decaying organic debris, where it must compete with other microorganisms to survive. *A. fumigatus* conidia also encounter a hostile environment when they enter the lung, and their subsequent survival depends on their intrinsic ability to adapt to this stress. Could the factors involved in the survival of the organism in the harsh environment of decaying vegetation also contribute to its success as an opportunistic pathogen? Here, we summarize some of the remarkable attributes of this fungus that are important for survival in the environment and highlight their potential for overlapping functions in the host (Casadevall and Pirofski, 2007).

THERMAL STRESS

In nature, *A. fumigatus* resides primarily in decaying vegetation, where it contributes to the degradation of organic debris and the recycling of carbon and nitrogen in the environment (Latgé, 1999). Composting is the controlled manipulation of this process, with the goal of transforming raw organic materials into humic substances that are suitable for the support of plant growth (Epstein, 1997). The composting reaction is highly dependent on microbial activity, a substantial proportion of which is derived from *A. fumigatus* (Beffa et al., 1998; Epstein, 1997). In the presence of adequate aeration, moisture, and nutrients, the early stage of composting is associated with the rapid growth of multiple species of bacteria and fungi. Heat is an inevitable byproduct of this intense microbial activity, resulting in a progressive increase in temperature, if the compost pile is sufficiently insulated to retain the thermal energy (de Bertoldi et al., 1983). Since high temperature facilitates the decomposition process, composting systems are often designed to generate some heat. The process of thermophilic composting can last from days to months, depending on the nature of the particular system, but can be divided into three general stages (Epstein, 1997; Trautmann et al., 2003). The initial phase is characterized by the proliferation of mesophilic bacteria and fungi that thrive at temperatures up to about 40°C. These organisms grow poorly at higher temperatures, so they become progressively less abundant as the temperature rises and are eventually replaced by thermotolerant species at temperatures above 40°C. However, when microbial growth eventually slows and the temperature declines, a second mesophilic phase will follow, characterized by protracted growth of mesophilic bacteria and fungi, as well as colonization by invertebrates. This latter period facilitates the breakdown of complex organic polymers that are resistant to degradation, such as lignin and cellulose. As a thermotolerant fungus, *A. fumigatus* is well-equipped for growth on diverse substrates throughout the compost cycle, including the thermophilic phase (Beffa et al., 1998; Epstein, 1997). This remarkable thermotolerance allows the fungus to survive when the compost achieves 60°C, which is considered the maximum for any eukaryote (Tansey and Brock, 1972).

Thermotolerance Is Linked to Virulence

The ability to thrive at mammalian body temperature is a requirement of all human pathogens. In bacteria, temperature plays an important role in modulating

David S. Askew and Judith C. Rhodes • Dept. of Pathology and Laboratory Medicine, University of Cincinnati College of Medicine, Cincinnati, OH 45267-0529.

the expression of gene products that influence virulence (Gophna and Ron, 2003; Konkel and Tilly, 2000). This raises the possibility that genes that evolved to support the growth of *A. fumigatus* in self-heating compost may also contribute to the aggressiveness of the fungus in the human host. These genes are of interest because they may represent targets for the development of new antifungal drugs or improved diagnostic methods. To date, four studies have provided insight into the molecular basis of thermotolerance in *A. fumigatus.*

Temperature-Regulated Gene Expression in *A. fumigatus*

A global perspective on temperature-regulated gene expression in *A. fumigatus* has been obtained using a whole-genome expression array. This experimental approach was designed to compare global gene expression patterns following a shift from 30°C (representing tropical soil) to 37°C (representing the mammalian body temperature) or to 48°C (representing thermophilic composting) (Machida et al., 2005; Nierman et al., 2005). The analysis revealed temperature-dependent expression of a distinct set of genes, 323 of which showed higher expression at 48°C than at 37°C and 135 of which had higher expression at 37°C than at 48°C. The first category was enriched for mRNAs encoding heat shock proteins, which are likely to play important roles in the ability of *A. fumigatus* to survive the thermophilic phase of composting. These clusters contained surprisingly few of the genes involved in the general stress response of *Saccharomyces cerevisiae*, suggesting that the thermotolerance of *A. fumigatus* has limited overlap with this other, well-characterized yeast stress response. Nevertheless, the identification of temperature-regulated mRNAs provides an important foundation for future studies into what this fungus requires for growth in the host.

Proteins are the ultimate effectors of cell phenotype, and eukaryotic protein levels do not always correlate with mRNA abundance (Greenbaum et al., 2003; Gygi et al., 1999; Washburn et al., 2003), so a comprehensive understanding of the thermotolerant phenotype of *A. fumigatus* will eventually require a determination of which of these thermally regulated mRNAs are actively translated into protein. Structural features in the 5′ untranslated region of mRNA play a major role in translational regulation (Meijer and Thomas, 2002; Mignone et al., 2002), the most well-characterized of which includes short upstream open reading frames (uORFs) in the 5′ untranslated region (Vilela and McCarthy, 2003). uORFs are frequently involved in cellular processes that control cell growth, particularly under conditions of environmental stress (Hoffmann et al., 2001; van den Brink et al., 2000; Vilela et al., 1998; Vilela and McCarthy, 2003). The genome sequence of *A. fumigatus* suggests that there is widespread translational regulation by uORFs in this organism (Nierman et al., 2005). At least one established virulence determinant in *A. fumigatus*, the cross-pathway control transcription factor CpcA (Krappmann et al., 2004), is subject to uORF-mediated translational regulation (Hoffmann et al., 2001), suggesting that the predominance of uORFs in *A. fumigatus* may indeed have relevance to the virulence of this organism.

THTA Is Required for Thermotolerance of *A. fumigatus*

Since little is known of the genetic basis of thermotolerance, a chemical mutagenesis strategy has been employed to identify *A. fumigatus* genes that are required for growth at 48°C but not at 42°C (Chang et al., 2004). DNA from a cosmid library was used to complement the resulting temperature-sensitive mutants, leading to the identification of *THTA*, encoding a protein of unknown function. Subsequent deletion of *THTA* in wild-type *A. fumigatus* by homologous recombination confirmed that *THTA* is a thermotolerance gene required for growth at 48°C. However, the gene was dispensable for virulence in a mouse model of aspergillosis, indicating that the contribution of *THTA* to high-temperature growth is not required in vivo. The precise function of the gene product encoded by *THTA*, and also the mechanism by which it contributes to thermotolerance, is presently unclear. Predicted homologs can be identified in the genomes of less thermotolerant *Aspergillus* species, such as *A. nidulans, A. niger, A. terreus*, and *A. clavatus*, and transfecting the *A. fumigatus THTA* gene into *A. nidulans* failed to increase the thermotolerance of that species (Chang et al., 2004). Together, these findings indicate that although *THTA* is likely to contribute to the success of *A. fumigatus* in thermophilic composting, it is not sufficient for thermotolerance, presumably because of the complex polygenic nature of the thermotolerant phenotype (Bhabhra and Askew, 2005).

CgrA Is Required for Thermotolerance of *A. fumigatus*

When *A. fumigatus* conidia enter the lung, host defenses are mobilized to eradicate the conidia before they can germinate into invasive hyphae. *A. fumigatus* conidia have the highest germination rate of the three most commonly isolated pathogenic species of *Aspergillus* at 37°C (Araujo and Rodrigues, 2004), and analysis of growth rate variability among clinical isolates has revealed that a faster growth rate correlates with increased virulence in animal models (Paisley et al., 2005). Thus, if the immune system experiences a decline in functional activity, the rapidly germinating conidia are able to outpace residual host defenses, resulting in extensive tissue damage and dissemination of the fungus. Further compounding this problem is the fact that the growth of *A. fumigatus* is enhanced by the same immunosuppressive corticosteroid treatments that predispose to *A. fumigatus* infections in the first place (Ng et al., 1994). This high rate of growth places considerable demand on the translational machinery of *A. fumigatus*, necessitating an increase in ribosome production in proportion to the demand for new proteins. The rapid growth of *A. fumigatus* at 37°C requires CgrA, a nucleolar protein involved in ribosome biogenesis (Bhabhra et al., 2004; Moy et al., 2002). A mutant of *A. fumigatus* that lacks CgrA grows normally at temperatures below 25°C but becomes increasingly growth impaired as the temperature exceeds 25°C and is unable to grow at 48°C (Bhabhra et al., 2004). Since the ribosome biogenesis defect caused by loss of CgrA was not temperature dependent (Bhabhra et al., 2008), it appears that the demand for ribosomes is higher at 37°C than it is at 25°C, presumably reflecting a faster growth rate and increased synthesis of thermoprotective proteins. CgrA was required for wild-type virulence in a mouse model of invasive aspergillosis but was dispensable for virulence in a low-temperature *Drosophila melanogaster* model (Bhabhra et al., 2004). These findings suggest that CgrA contributes to pathogenesis by providing sufficient protein synthetic capacity to support rapid growth in the mammalian host, and they underscore the importance of ribosome biogenesis to the thermotolerant phenotype of this organism.

O-Glycosylation Is Required for Thermotolerance

O-Glycosylation is an important posttranslational modification, catalyzed by a family of highly conserved O-mannosyltransferases (Ernst and Prill, 2001). The protein O-mannosyltransferase Pmt1 has been shown to contribute to thermotolerance in *A. fumigatus* (Zhou et al., 2007). A Δ*pmt1* mutant grew normally at 37°C but became increasingly growth impaired at temperatures above 42°C. This was attributed to a temperature-sensitive defect in cell wall integrity, since thermotolerance could be restored by growth on osmotically stabilized medium. These findings underscore the importance of O-glycosylation to the maintenance of cell wall integrity during thermotolerant growth.

NUTRIENT SENSING

Before any organism can reprogram its metabolism to correspond to the growth milieu, the quantity and quality of available nutrients must be determined. In *A. fumigatus*, two of the signaling pathways that are involved in nitrogen and carbon sensing are the Rheb/TOR and protein kinase A (PKA) pathways.

Rheb/TOR Senses Nitrogen and Influences Virulence

Ras family proteins are small GTPases that function as molecular switches in many different pathways (Reuther and Der, 2000). The related protein Rheb is a small lipid-anchored membrane protein that was originally identified as the *R*as *h*omolog *e*nriched in *b*rain. However, Rheb is now known to be present in many, if not all, mammalian cell types and to be essential for coordinating growth with nutrient availability (Uritani et al., 2006). The activity of most Ras proteins is controlled by GTP binding, which is regulated by the activity of GTPase-activating proteins (GAPs) and guanine exchange factors (GEFs). Rheb also has GTPase activity and shuttles between a GDP-bound form and a GTP-bound form. However, there is evidence that Rheb exists predominantly in its GTP-bound form, and although a Rheb GAP has been identified there is as yet no recognized Rheb GEF (Aspuria and Tamanoi, 2004). In addition to controlling its activity through GTP binding, Rheb is also regulated at the transcriptional level. Current evidence suggests that Rheb is an upstream regulator of the TOR kinase signaling pathway, which is essential for translating growth cues into the accumulation of cell mass (Arsham and Neufeld, 2006; Martin and Hall, 2005; Patel et al., 2003; Schmelzle and Hall, 2000; Wullschleger et al., 2006). This key link between nutrient sensing and growth has been shown to involve nitrogen sensing in both fission and budding yeasts (Mach et al., 2000; Urano et al., 2000, 2006; Weisman et al., 2007). Indeed, in lower eukaryotes, activated Rheb is a positive regulator of processes controlled by the TOR1 complex, processes that include the activation of ribosome biogenesis, translation, and the transcription of stress factors and the inhibition of autophagy. In *Drosophila*, these TOR-dependent growth regulatory networks are positively regulated by Rheb, leading to a smaller cell size in Rheb-deficient mutants (Patel et al., 2003). Many of the downstream genes in the Rheb/TOR pathway have been identified in *A. nidulans*; however, their characterization has not been pursued in any other *Aspergillus* species (Fitzgibbon et al., 2005).

In *A. fumigatus*, the Rheb protein RhbA has been shown to function in both nitrogen sensing and growth control. A Rheb deletion mutant of *A. fumigatus*, Δ*rhbA*, shares phenotypic features in common with the *rhb1* mutant of *Saccharomyces cerevisiae* and the *Rhb1* mutant of *Schizosaccharomyces pombe* (Mach et al., 2000; Panepinto et al., 2003; Urano et al., 2000). The expression of *A. fumigatus rhbA* is regulated by nitrogen availability; nitrogen, but not carbon, starvation results in rapid accumulation of steady-state *rhbA* mRNA within 30 min (Panepinto et al., 2002). The growth of the *rhbA* deletion mutant is indistinguishable from that of wild-type *A. fumigatus* when grown on rich nitrogen medium such as peptone or ammonium tartrate. However, the Δ*rhbA* mutant is growth impaired when it is forced to use nitrogen sources that require induction of nitrogen catabolism pathways, such as nitrate or proline (Panepinto et al., 2003). This suggests a central role for RhbA in the adaptive response to a suboptimal nitrogen source. The ability to sense and rapidly adjust to a secondary nitrogen source is likely to be of considerable benefit in the complex environment of decaying plant material. Similarly, balancing growth rate in proportion to nitrogen availability may also be relevant in the host. For example, deletion of the *areA* gene, encoding a transcription factor required for the expression of genes involved in the utilization of secondary nitrogen sources, attenuates the virulence of *A. fumigatus* (Hensel et al., 1998). Increased *rhbA* mRNA abundance has been reported in *A. fumigatus* cocultured with endothelial cells in vitro or during growth in vivo in a mouse model (Rhodes et al., 2001; Zhang et al., 2005). This suggests that increased *rhbA* mRNA levels are part of the normal adaptive response of *A. fumigatus* to the host environment, which was confirmed by demonstrating that a Δ*rhbA* mutant had attenuated virulence in a mouse model of aspergillosis (Panepinto et al., 2003).

PKA Senses Carbon Availability and Influences Virulence

The cyclic AMP (cAMP)-dependent PKA is a conserved kinase in all eukaryotes, and the PKA pathway has been shown to be involved in nutrient sensing, regulation of cell proliferation, carbon storage, and stress response (D'Souza et al., 2001; Thevelein et al., 2000; Wilson and Roach, 2002). The PKA holoenzyme complex is made up of a homodimer of regulatory subunits bound to two catalytic subunits. Although the heterotetramer is inactive, elevated cAMP levels in the cell result in activation of the kinase; cAMP binds to the regulatory subunits, inducing a conformational change that results in liberation of the catalytic subunits. The free catalytic subunits autophosphorylate a key serine residue and, thus, become active. At this point the catalytic subunits may phosphorylate serines and threonines found in target proteins that contain the motif RRXS/T. PKA activity is regulated by controlling the level of cAMP in the cell, in addition to modulating the localization of the holoenzyme complex at different intracellular sites (Griffioen and Thevelein, 2002). The control of PKA localization in higher eukaryotes is accomplished by the binding of regulatory subunits in the holoenzyme complex to proteins called A kinase anchor proteins (Colledge and Scott, 1999). Although A kinase anchor proteins have not been demonstrated in lower eukaryotes, including *Aspergillus*, compartmentalization still appears to play a significant role in regulation of PKA activity. Indeed, proper localization of the regulatory subunit in yeast is required for optimal viability during stress, such as glucose starvation.

Although the data on carbon sensing by PKA in *A. fumigatus* are limited, there is ample precedent for regulation of PKA activity by glucose in *S. cerevisiae*. In yeast, addition of fresh glucose to a glucose-depleted culture results in a G-protein-dependent increase in adenyl cyclase activity, which is followed by an increase in PKA activity (Thevelein et al., 2000; Wilson and Roach, 2002). Multiple pathways are then regulated by phosphorylation by PKA, either directly or indirectly. In *A. fumigatus*, the response of the PKA pathway to the addition of exogenous cAMP and the basal level of PKA activity have been shown to be dependent upon the carbon source on which the fungus is grown (Oliver et al., 2002). Such carbon source-dependent regulation of PKA activity may be of particular benefit to the organism in the milieu of compost, where plant polysaccharides are abundant. Indeed, the ability to alter growth rate based not only on abundance but also on quality of carbon source is an important factor in promoting survival of *A. fumigatus* in the environment. In the absence of sufficient glucose, the PKA pathway would be downregulated, signaling for a reduced growth rate. The reduced growth rate on glycerol when compared with that on glucose correlates with the lower measured PKA activity in hyphae grown on the sugar alcohol (Oliver et al., 2002). In turn, the reduced growth on glycerol correlates with the decreased growth rate and lower PKA activity seen when the PKA catalytic subunit 1 (C1) is deleted. Disruption of the PKA signaling pathway, by deleting either the regulatory or C1 catalytic subunits, has been shown to be detrimental not just to growth in vitro but also to mouse virulence (Liebmann et al., 2003, 2004b; Zhao et al., 2006). Since transcript levels for both of these genes are increased when the fungus is grown in the presence of pulmonary epithelial cells, the requirement for an intact PKA pathway for growth in the environment also applies in the host (Oliver et al., 2001).

STARVATION RESPONSES

The ability of *A. fumigatus* to thrive on decaying plant and animal debris requires an enzymatic machinery that can efficiently degrade this material into component molecules (Beffa et al., 1998; Robson et al., 2005; Tekaia and Latgé, 2005). This must be accomplished in the presence of numerous other microorganisms that are also competing for access to the same substrates, making nutrient limitation an important stress encountered by *A. fumigatus* in nature. Accordingly, *A. fumigatus* has evolved metabolic capabilities that promote its survival when faced with nutrient insufficiency. Some of these functions are also required to support growth in the host, including iron and zinc acquisition (Hissen et al., 2005; Moreno et al., 2007; Schrettl et al., 2007), folate biosynthesis (Brown et al., 2000), uridine-uracil biosynthesis (D'Enfert et al., 1996), lysine biosynthesis (Liebmann et al., 2004a), and propionate metabolism (Ibrahim-Granet et al., 2008; Maerker et al., 2005).

Autophagy Allows Conidiation of *A. fumigatus* during Nutrient Limitation

Autophagy is a highly conserved stress response in eukaryotes that helps organisms survive periods of nutrient deprivation by activating a process of limited intracellular digestion to fuel essential cellular functions (Klionsky, 2007; Mizushima, 2007). Autophagic degradation is triggered primarily by nutrient starvation, particularly nitrogen deficiency, resulting in the expansion of "isolation membranes" in the cytoplasm that eventually coalesce and sequester cytoplasmic material into an enclosed sac called the autophagosome. The autophagosome is delivered to the vacuole, thus releasing the contents for degradation. The component macromolecules are then returned to the cytoplasm via vacuolar membrane permeases, thereby providing a source of recycled building blocks to support vital functions until a new external supply can be found. At least 30 autophagy (ATG) genes have been implicated in autophagy in *S. cerevisiae* (Xie and Klionsky, 2007). The core autophagy machinery is also conserved in *A. fumigatus*, although some differences may exist between genera (Meijer et al., 2007).

In nature, starvation for nitrogen is one of the major signals that triggers sporulation in *Aspergillus* spp., allowing the fungus to produce conidia that are resistant to adverse environmental conditions (Adams et al., 1998). Sporulation requires the construction of new morphological structures and the production of copious quantities of nuclei. However, when sporulation is triggered by nitrogen deficiency, the lack of sufficient nitrogen poses a significant challenge to the ability of the organism to complete this energetically costly developmental program. Recent evidence has demonstrated that autophagy plays a major role in facilitating this process (Richie and Askew, 2007; Richie et al., 2007a). An autophagy-deficient strain of *A. fumigatus* was constructed by deleting the gene encoding Atg1, a serine/threonine kinase that is required for an early step in autophagy in other species. The *A. fumigatus* Δ*atg1* mutant conidiates poorly on nitrogen-limiting medium but conidiates normally if sufficient nitrogen is present. This suggests that autophagy facilitates conidiation by recycling intracellular materials to maintain nitrogen levels above the threshold required to support the developmental program (Richie and Askew, 2007; Richie et al., 2007a).

Autophagy Enables Growth of *A. fumigatus* during Nutrient Limitation

A. fumigatus is able to sustain a limited amount of growth at the colony edge when starved for nutrients (Robson, 1999). This starvation response is an important feature of filamentous growth that allows a starved colony to explore the environment for new sources of nutrients. The Atg1 kinase is required for this process, suggesting that *A. fumigatus* relies upon autophagy for this foraging response (Richie et al., 2007a). Interestingly, autophagy is also required for the growth of *A. fumigatus* under conditions of metal ion deficiency, even in the presence of abundant carbon and nitrogen, suggesting that autophagic degradation can be used as a mechanism to release metals from preexisting metal-associated proteins when needed. Taken together, these findings demonstrate a vital role for autophagy in supporting the growth and survival of *A. fumigatus* under starvation conditions, a function that is likely to be indispensable to the organism during periods of fluctuating nutrient availability in the environment. However, since autophagy was dispensable for the virulence of *A. fumigatus* in a mouse model of aspergillosis, it appears that the fungus is capable of adjusting its metabolism to use mammalian tissues as a food source, without the need for additional nutritional support from autophagy.

OXIDATIVE STRESS

Reactive oxygen species are generated during the course of the compost process via photolysis of humic substances and from photosensitizing toxins secreted by other microbes (Ehrenshaft et al., 1998; Paul et al., 2004). Encounters with such environmental toxins have led to the development of systems in *A. fumigatus* that equip the fungus for inimical encounters with host-generated toxic oxygen species. The ability of *A. fumigatus* to defend itself against oxidative stress requires an armamentarium of antioxidants (Chauhan et al., 2006). These proteins include catalases, superoxide dismutases, and thioredoxins, all of which may play significant roles in the pathogenesis of aspergillosis (Hamilton et al., 1996; Paris et al., 2003; Thon et al., 2007). The sensing of oxidative stress, and coordinating an appropriate response, is carried out by a number of different pathways that have unique and overlapping functions (Moye-Rowley, 2003). For example, the mitogen-activated protein kinase MpkA negatively regulates the survival of *A. fumigatus* after exposure to oxidative damage, whereas the PKA pathway appears to regulate the survival mechanism positively (Liebmann et al., 2003, 2004b; Valiante et al., 2008; Zhao et al., 2006). Transcription factors such as Skn7 and AfYap1 are downstream effectors of oxidative stress response pathways, resulting in the enhanced expression of antioxidant defenses (Lamarre et al., 2007; Lessing et al., 2007). Deletion of either of these transcription factors creates mutants that have markedly increased susceptibility to oxidative damage, although the panel of oxidants that cause the damage does not completely overlap. The roles that these antioxidant factors play have been well described under laboratory conditions, and we can only assume that they display a similar function in the compost. The importance of the antioxidant response in the compost and in the host continues to be a source of controversy, due to the difficulty of demonstrating a clear role for genes encoding catalase, MpkA, SKN7, Yap1, and others in pathogenesis in mouse models (Lamarre et al., 2007; Lessing et al., 2007; Paris et al., 2003; Valiante et al., 2008).

PKA Regulates Oxidative Stress Responses and Virulence

The PKA pathway has been studied in a number of fungal pathogens of animals and plants (D'Souza et al., 2001). Transcriptional regulation of genes controlled by the PKA pathway has been shown to play an important role in the virulence of *Cryptococcus, Candida, Ustilago*, and *Magnaporthe* species. Studies of PKA in *Aspergillus* species have been reported for *A. niger, A. nidulans*, and *A. fumigatus*. *A. niger* is an industrially important organism in the production of citric acid. Control of citric acid production during fermentation requires the sensing of sucrose concentration and PKA-regulated phosphorylation of phosphofructokinase, which then directs carbon flux through glycolysis to produce citric acid (Bencina et al., 1997; Gradisnik-Grapulin and Legisa, 1997). Stress factors, such as temperature and hyper- and hypo-osmolarity, have been shown to increase mRNA levels for the catalytic subunit of PKA in *A. niger*, suggesting a role for PKA in divergent stress responses. In *A. nidulans*, PKA has been shown to function in the control of morphological development and the regulation of secondary metabolism (Ni et al., 2005). Overexpression of the catalytic subunit (*pkaA*) leads to a reduction in conidiation, whereas deletion of *pkaA* resulted in limited radial growth and an increased production of conidia. Deletion of the second catalytic subunit (*pkaB*) did not have an obvious phenotype, but *pkaB* overexpression resulted in enhanced hyphal proliferation and ability to rescue the growth defects of the Δ*pkaA* strain and reduced tolerance to oxidative damage.

In *A. fumigatus*, the ortholog of *pkaA, pkaC1*, is required for wild-type growth and conidiation, germination, and expression of the polyketide synthase gene *pksP* (Liebmann et al., 2003, 2004b). Analysis of upstream members of the pathway indicates that deletion of adenyl cyclase (*acyA*) results in markedly reduced conidiation and growth, whereas deletion of the G-protein α-subunit (*gpaB*) results in decreased conidiation but does not influence growth rate. The fitness of the mutants in the PKA pathway in the compost heap may be adversely affected by their markedly reduced conidiation, as this would limit their ability to spread and recolonize other substrates in the area. The downregulation of *pksP* transcription could also lead to decreased melanin production, which could increase the susceptibility of the mutants to reactive oxygen species, which are scavenged by fungal melanins (Jacobson, 2000). The reduced virulence of mutants with *gpaB* or *pkaC1* deleted indicates the importance of this pathway in the host and emphasizes the link between environmental and host fitness.

Deletion of the regulatory subunit of PKA in *A. fumigatus* also results in decreased growth and delayed germination and conidiation, along with morphological abnormalities in the condiophores (Zhao et al., 2006). The Δ*pkaR* mutants were tested for susceptibility to killing by oxidative agents that generate different reactive oxygen species. Whereas Δ*pkaR* mutant conidia were hypersensitive to killing by hydrogen peroxide, paraquat, menadione, and diamide, the Δ*pkaR* hyphae only showed increased susceptibility to diamide and paraquat. Taken together, these findings suggest that the PKA pathway is of key importance in mediating resistance to oxidative damage in wild-type *A. fumigatus*. This could be mediated by a number of processes shown to be downstream of PKA in *Aspergillus* or in other organisms. For example, Leibmann et al. have shown that expression of the polyketide synthase, *pksP*, in *A. fumigatus* is positively controlled by both *pkaC1* and *gpaB* (Liebmann et al., 2003). Deletion of *pksP* has been shown to result in white conidia, which lack melanin, and are more susceptible to phagocyte killing (Brakhage and Liebmann, 2005; Jahn et al., 2002). White conidia would be expected to be more susceptible to UV irradiation from sunlight, as well as to oxidative damage, because of the loss of the reactive oxygen-scavenging properties of melanins. Likewise, PKA is known to be involved in regulating the cell cycle, often working through release of cell cycle checkpoints (Searle and Sanchez, 2004). Loss of this level of control would be expected to reduce overall fitness by interfering with the ability to repair DNA damage, as well as preventing the organism from linking growth and cell cycle progression to adequate nutrient availability. Therefore, PKA signaling plays an important role in numerous processes that would contribute to the ability of *A. fumigatus* to survive lethal challenges to its growth in the environment.

PROGRAMMED CELL DEATH AND STRESS RESPONSES

Programmed cell death (PCD) is a mechanism of cell removal that eliminates excess or damaged cells and is a universally conserved stress response among metazoans (Elmore, 2007). Until recently, the conventional wisdom argued that apoptosis was unique to metazoan species. However, within the last few years it has become apparent that lower eukaryotes possess divergent homologs of the PCD machinery and that these pathways contribute to death in response to adverse environmental conditions (Frohlich et al., 2007). Most of what is known about apoptosis in lower eukaryotes has come from studies in yeast. Yeast cells undergoing apoptosis exhibit the same markers that have been well-described in mammals, such as membrane exposure of phosphatidylserine, chromatin condensation, DNA degradation, nuclear fragmentation, and cytochrome *c* release. A diverse array of adverse conditions that disrupt cell homeostasis trigger yeast apoptosis, similar to what has been reported in mammalian apoptosis (Frohlich et al., 2007). The biological role of yeast apoptosis is incompletely understood. This is primarily because it is counterintuitive to consider that suicide would be beneficial to a unicellular organism. However, evidence suggests that yeast apoptosis may strengthen a community of cells by eliminating the weaker members so that a source of nutrients can be recycled for the rest of the population (Fabrizio et al., 2004; Herker et al., 2004). For example, yeast cells at the center of an aging colony are under nutrient stress and undergo apoptotic-like death, whereas those at the periphery do not. The removal of cells from the central dying zone reduces the growth of cells at the colony periphery, suggesting that apoptosis in the center benefits the colony by providing nutrients to support the healthy peripheral cells (Vachova and Palkova, 2005). It is somewhat easier to conceive of a reason why PCD would be of benefit to a filamentous fungus like *A. fumigatus*. The mycelium of *A. fumigatus* is composed of a network of interconnected hyphae that show spatial compartmentalization (Vinck et al., 2005). The ability to partition function in the fungal colony is analogous to the segregation of cellular activities that characterize the multicellular tissues of higher eukaryotes. Since PCD is essential to maintain homeostasis in metazoan tissues, this raises the possibility that fungal PCD may enhance the fitness of the mycelium by sacrificing damaged biomass for the benefit of the organism as a unit, particularly under conditions of stress. Further knowledge of PCD in fungi may lead to the development of novel strategies to manipulate the pathway for therapeutic gain.

Metacaspase Function in *A. fumigatus* Enables Growth during ER Stress

Two major types of PCD have been distinguished: caspase-dependent apoptosis (type I PCD) and autophagy-dependent PCD (type II PCD) (Galluzzi et al., 2007) (Fig. 1). Type I PCD is the most studied of the two pathways and is of considerable interest as a source of novel targets for the manipulation of PCD in human disease (Fernandez-Luna, 2007; Fulda, 2007; Jana and Paliwal, 2007). The aspartate-specific cysteine proteases that comprise the caspase superfamily are central to apoptosis. Multiple members of the caspase family exist in metazoans, and their sequential activation during apoptosis triggers a caspase cascade that culminates in proteolytic cleavage of downstream substrates and cell destruction (Elmore, 2007). The discovery of a single metacaspase in yeast, Yca1p, provided the first evidence that fungal PCD is mediated by divergent members of the caspase superfamily (Madeo et al., 2002). Yca1p is classified as a metacaspase, which is one of three divisions within the caspase superfamily: the caspases of metazoans, the paracaspases of metazoans and *Dictyostelium*, and the metacaspases of plants, fungi, and protozoa (Uren et al., 2000). The yeast apoptotic response can be either Yca1 dependent (Bettiga et al., 2004; Flower et al., 2005; Herker et al., 2004; Khan et al., 2005; Madeo et al., 2002; Mazzoni et al., 2005; Reiter et al., 2005; Silva et al., 2005; Wadskog et al., 2004; Weinberger et al., 2005) or Yca1 independent (Fahrenkrog et al., 2004; Hauptmann et al., 2006; Maeta et al., 2005; Wissing et al., 2004; Wysocki and Kron, 2004), similar to what has been reported in mammals.

Ultrastructural and biochemical changes that are characteristic of apoptosis have also been reported in filamentous fungi, including *Mucor racemosus* treated with lovastatin (Roze and Linz, 1998), *A. nidulans* treated with farnesol, the antifungal protein PAF, or phytosphingosine (Cheng et al., 2003; Leiter et al., 2005; Semighini et al., 2006), or *A. fumigatus* exposed to amphotericin B, oxidative stress, or stationary phase stress (Mousavi and Robson, 2003, 2004). Some of these conditions may be encountered in the environment, so it is intriguing to speculate that they are elaborated by competing organisms to induce PCD in susceptible fungi. Organisms with increasing complexity

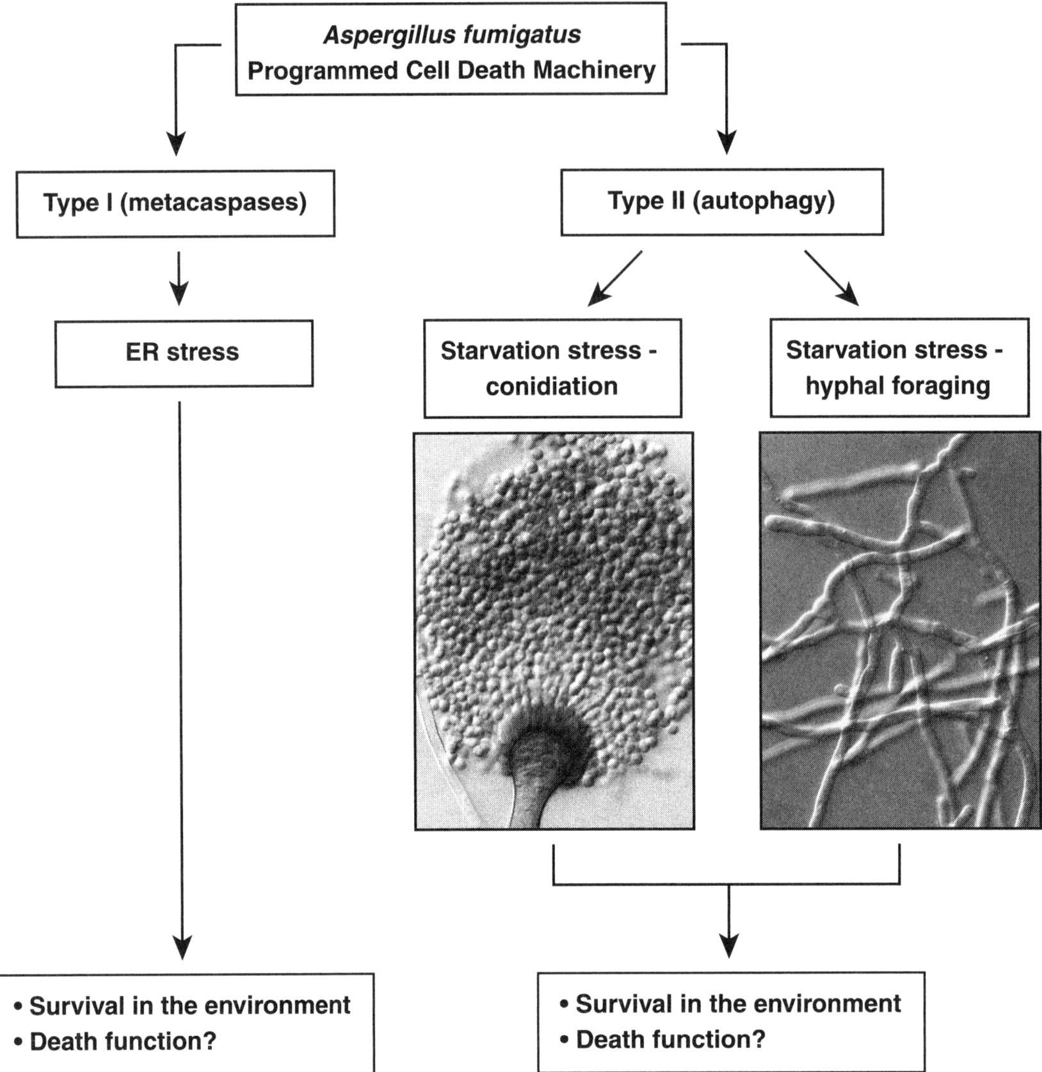

Figure 1. The PCD machinery in *A. fumigatus*. In higher eukaryotes, two types of PCD have been described: type I (caspase dependent) and type II (autophagy dependent). Current evidence suggests that the *A. fumigatus* metacaspases have a prosurvival function that is active under conditions of ER stress. Autophagy also provides a survival advantage when *A. fumigatus* is under stress, allowing for conidiophore development when external nitrogen sources are limiting, as well as providing a mechanism to fuel the foraging of hyphae into unexplored substrate. Each of these functions is likely to play an important role in the survival of the fungus in the competitive environment of compost. However, it remains to be determined whether these functions also have death-promoting activities under certain other conditions.

have multiple caspase members (Aravind et al., 2001), two of which have been identified in the *A. fumigatus* genome, *casA* and *casB* (Fedorova et al., 2005). These metacaspases are required for the externalization of phosphatidylserine that occurs during stationary phase stress in *A. fumigatus*, suggesting that this well-known marker of apoptotic death is metacaspase dependent in *A. fumigatus* (Richie et al., 2007b). However, the metacaspases are not required for death mediated by agents that trigger apoptotic-like death in yeast or filamentous fungi, suggesting that there may be redundant pathways of cell death that can be activated under these conditions.

In higher eukaryotes, prolonged ER stress activates caspase-dependent apoptosis. This is in striking contrast to *A. fumigatus*, where the metacaspases have a protective role under conditions of ER stress rather than a death-promoting function (Richie et al., 2007b). One possible explanation for this difference is that fungal metacaspases initially evolved to act in a stress response pathway that benefits the organism during adverse environmental conditions. With the advent of multicellularity, these functions may have diverged to include PCD, which also provides a benefit to the organism by deleting irreversibly damaged cells in a multicellular tissue. This idea is consistent with increasing evidence linking mammalian caspases to functions that are not directly involved in cell death (Garrido and Kroemer, 2004; Schwerk and Schulze-Osthoff, 2003; Zeuner et al., 1999).

Is Autophagy an Alternative Form of PCD in *A. fumigatus*?

Autophagy is an alternative form of PCD in higher eukaryotes (Bursch, 2001), sometimes referred to as type II PCD. Although caspase-dependent and autophagy-dependent PCD are distinct pathways, there is evidence that they have complementary roles in cell death signaling and may also utilize some common regulatory mechanisms (Maiuri et al., 2007). Recent evidence supports a role for autophagy-dependent PCD in the virulence of two eukaryotic pathogens, the filamentous fungus *Magnaporthe grisea* (Liu et al., 2007; Veneault-Fourrey et al., 2006) and the parasitic protozoan *Leishmania major* (Besteiro et al., 2006). In each of these organisms, autophagy is used as a remodeling tool to produce a morphological form that is required for infection. Although autophagy is required to support some important functions during starvation responses in *A. fumigatus* (Richie et al., 2007a), we do not yet know whether autophagy is involved in PCD in this organism.

Heterokaryon incompatibility (HI) is a type of nonself recognition that triggers PCD when fungal hyphae of unlike genotypes fuse (Dementhon et al., 2003, 2004; Pinan-Lucarre et al., 2003). The process has not been studied in *A. fumigatus*, but it is thought to be ubiquitous among filamentous ascomycetes (Glass and Dementhon, 2006), and many of the genes that have been implicated in HI are present in the *A. fumigatus* genome (Fedorova et al., 2005). Autophagy is induced during HI, raising the possibility that it might play a role in the PCD that occurs during HI. However, studies in *Podospora anserina* have shown that autophagy actually protects the fungus against death rather than promoting it (Pinan-Lucarre et al., 2007). Thus, although it is clear that *A. fumigatus* has orthologs of many of the proteins implicated in metazoan PCD pathways, including heterokaryon incompatibility (Fedorova et al., 2005), much more research is needed to determine the exact nature and function of such pathways in this fungus.

CONCLUSION

Implicit in the search for classical virulence factors in *A. fumigatus* is the goal of improving therapeutic options through the specific targeting of these factors. Unfortunately, no single feature has been identified that can adequately explain the success of *A. fumigatus* as the predominant opportunistic mold pathogen, supporting the concept that virulence is multifactorial. The apparent ease with which *A. fumigatus* can adapt to growth in the environment of an immunocompromised host illustrates the remarkable versatility of this fungus for growth on diverse substrates. This is consistent with a multifactorial model of "dual-use" virulence, in which *A. fumigatus* has evolved an extensive armamentarium of defenses and metabolic capabilities that facilitate growth in both organic debris and a living host. Since *A. fumigatus* must acquire nutrients from host tissues throughout the infection, further understanding of the pathways that support this metabolic versatility may offer novel strategies for therapeutic intervention.

REFERENCES

Adams, T. H., J. K. Wieser, and J.-H. Yu. 1998. Asexual sporulation in *Aspergillus nidulans. Microbiol. Mol. Biol. Rev.* **62:**35–54.

Araujo, R., and A. G. Rodrigues. 2004. Variability of germinative potential among pathogenic species of *Aspergillus. J. Clin. Microbiol.* **42:**4335–4337.

Aravind, L., V. M. Dixit, and E. V. Koonin. 2001. Apoptotic molecular machinery: vastly increased complexity in vertebrates revealed by genome comparisons. *Science* **291:**1279–1284.

Arsham, A. M., and T. P. Neufeld. 2006. Thinking globally and acting locally with TOR. *Curr. Opin. Cell Biol.* **18:**589–597.

Aspuria, P.-J., and F. Tamanoi. 2004. The Rheb family of GTP-binding proteins. *Cell. Signal.* **16:**1105–1112.

Beffa, T., F. Staib, J. Lott Fischer, P. F. Lyon, P. Gumowski, O. E. Marfenina, S. Dunoyer-Geindre, F. Georgen, R. Roch-Susuki, L. Gallaz, and J. P. Latgé. 1998. Mycological control and surveillance of biological waste and compost. *Med. Mycol.* **36**(Suppl. 1)**:**137–145.

Bencina, M., H. Panneman, G. Ruijter, M. Legisa, and J. Visser. 1997. Characterization and overexpression of the *Aspergillus niger* gene encoding the cAMP-dependent protein kinase catalytic subunit. *Microbiology* **143:**1211–1220.

Besteiro, S., R. A. Williams, L. S. Morrison, G. H. Coombs, and J. C. Mottram. 2006. Endosome sorting and autophagy are essential for differentiation and virulence of *Leishmania major. J. Biol. Chem.* **281:**11384–11396.

Bettiga, M., L. Calzari, I. Orlandi, L. Alberghina, and M. Vai. 2004. Involvement of the yeast metacaspase Yca1 in *ubp10*Δ-programmed cell death. *FEMS Yeast Res.* **5:**141–147.

Bhabhra, R., and D. S. Askew. 2005. Thermotolerance and virulence of *Aspergillus fumigatus*: role of the fungal nucleolus. *Med. Mycol.* **43**(Suppl. 1)**:**S87–S93.

Bhabhra, R., M. D. Miley, E. Mylonakis, D. Boettner, J. Fortwendel, J. C. Panepinto, M. Postow, J. C. Rhodes, and D. S. Askew. 2004. Disruption of the *Aspergillus fumigatus* gene encoding nucleolar protein CgrA impairs thermotolerant growth and reduces virulence. *Infect. Immun.* **72:**4731–4740.

Bhabhra, R., D. Richie, H. Kim, W. Nierman, J. Fortwendel, J. Aris, J. Rhodes, and D. Askew. 2008. Impaired ribosome biogenesis disrupts the integration between morphogenesis and nuclear duplication during the germination of *Aspergillus fumigatus. Eukaryot. Cell* **7:**575–583.

Brakhage, A. A., and B. Liebmann. 2005. Aspergillus fumigatus conidial pigment and cAMP signal transduction: significance for virulence. *Med. Mycol.* 43(Suppl. 1):75–82.

Brown, J. S., A. Aufauvre-Brown, J. Brown, J. M. Jennings, H. Arst, Jr., and D. W. Holden. 2000. Signature-tagged and directed mutagenesis identify PABA synthetase as essential for *Aspergillus fumigatus* pathogenicity. *Mol. Microbiol.* **36:**1371–1380.

Bursch, W. 2001. The autophagosomal-lysosomal compartment in programmed cell death. *Cell Death Differ.* **8:**569–581.

Casadevall, A., and L. Pirofski. 2007. Accidental virulence, cryptic pathogenesis, Martians, lost hosts, and the pathogenicity of environmental microbes. *Eukaryot. Cell* **6:**2169–2174.

Chang, Y. C., H. F. Tsai, M. Karos, and K. J. Kwon-Chung. 2004. THTA, a thermotolerance gene of *Aspergillus fumigatus*. *Fungal Genet. Biol.* **41:**888–896.

Chauhan, N., J.-P. Latgé, and R. Calderone. 2006. Signalling and oxidant adaptation in *Candida albicans* and *Aspergillus fumigatus*. *Nat. Rev. Microbiol.* **4:**435–444.

Cheng, J., T. S. Park, L. C. Chio, A. S. Fischl, and X. S. Ye. 2003. Induction of apoptosis by sphingoid long-chain bases in *Aspergillus nidulans*. *Mol. Cell. Biol.* **23:**163–177.

Colledge, M., and J. D. Scott. 1999. AKAPs: from structure to function. *Trends Cell. Biol.* **9:**216–221.

de Bertoldi, M., G. Vallini, and A. Pera. 1983. The biology of composting: a review. *Waste Manage. Res.* **1:**157–176.

Dementhon, K., M. Paoletti, B. Pinan-Lucarre, N. Loubradou-Bourges, M. Sabourin, S. J. Saupe, and C. Clave. 2003. Rapamycin mimics the incompatibility reaction in the fungus *Podospora anserina*. *Eukaryot. Cell* **2:**238–246.

Dementhon, K., S. J. Saupe, and C. Clave. 2004. Characterization of IDI-4, a bZIP transcription factor inducing autophagy and cell death in the fungus *Podospora anserina*. *Mol. Microbiol.* **53:**1625–1640.

D'Enfert, C., M. Diaquin, A. Delit, N. Wuscher, J. P. Debeaupuis, M. Huerre, and J. P. Latgé. 1996. Attenuated virulence of uridine-uracil auxotrophs of *Aspergillus fumigatus*. *Infect. Immun.* **64:**4401–4405.

D'Souza, C., J. Alspaugh, C. Yue, T. Harashima, G. Cox, J. Perfect, and J. Heitman. 2001. Cyclic AMP-dependent protein kinase controls virulence of the fungal pathogen *Cryptococcus neoformans*. *Mol. Cell. Biol.* **21:**3179–3191.

Ehrenshaft, M., A. E. Jenns, K. R. Chung, and M. E. Daub. 1998. SOR1, a gene required for photosensitizer and singlet oxygen resistance in *Cercospora* fungi, is highly conserved in divergent organisms. *Mol. Cell* **1:**603–609.

Elmore, S. 2007. Apoptosis: a review of programmed cell death. *Toxicol. Pathol.* **35:**495–516.

Epstein, E. 1997. *The Science of Composting*. Technomic Publishing, Lancaster, PA.

Ernst, J. F., and S. K.-H. Prill. 2001. O-glycosylation. *Med. Mycol.* 39(Suppl. 1):67–74.

Fabrizio, P., L. Battistella, R. Vardavas, C. Gattazzo, L. L. Liou, A. Diaspro, J. W. Dossen, E. B. Gralla, and V. D. Longo. 2004. Superoxide is a mediator of an altruistic aging program in *Saccharomyces cerevisiae*. *J. Cell Biol.* **166:**1055–1067.

Fahrenkrog, B., U. Sauder, and U. Aebi. 2004. The *S. cerevisiae* HtrA-like protein Nma111p is a nuclear serine protease that mediates yeast apoptosis. *J. Cell Sci.* **117:**115–126.

Fedorova, N. D., J. H. Badger, G. D. Robson, J. R. Wortman, and W. C. Nierman. 2005. Comparative analysis of programmed cell death pathways in filamentous fungi. *BMC Genomics* **6:**177.

Fernandez-Luna, J. L. 2007. Apoptosis regulators as targets for cancer therapy. *Clin. Transl. Oncol.* **9:**555–562.

Fitzgibbon, G. J., I. Y. Morozov, M. G. Jones, and M. X. Caddick. 2005. Genetic analysis of the TOR pathway in *Aspergillus nidulans*. *Eukaryot. Cell* **4:**1595–1598.

Flower, T. R., L. S. Chesnokova, C. A. Froelich, C. Dixon, and S. N. Witt. 2005. Heat shock prevents alpha-synuclein-induced apoptosis in a yeast model of Parkinson's disease. *J. Mol. Biol.* **351:**1081–1100.

Frohlich, K. U., H. Fussi, and C. Ruckenstuhl. 2007. Yeast apoptosis: from genes to pathways. *Semin. Cancer Biol.* **17:**112–121.

Fulda, S. 2007. Inhibitor of apoptosis proteins as targets for anticancer therapy. *Expert Rev. Anticancer Ther.* **7:**1255–1264.

Galluzzi, L., M. C. Maiuri, I. Vitale, H. Zischka, M. Castedo, L. Zitvogel, and G. Kroemer. 2007. Cell death modalities: classification and pathophysiological implications. *Cell Death Differ.* **14:**1237–1243.

Garrido, C., and G. Kroemer. 2004. Life's smile, death's grin: vital functions of apoptosis-executing proteins. *Curr. Opin. Cell Biol.* **16:**639–646.

Glass, N. L., and K. Dementhon. 2006. Non-self recognition and programmed cell death in filamentous fungi. *Curr. Opin. Microbiol.* **9:**553–558.

Gophna, U., and E. Z. Ron. 2003. Virulence and the heat shock response. *Int. J. Med. Microbiol.* **292:**453–461.

Gradisnik-Grapulin, M., and M. Legisa. 1997. A spontaneous change in the cAMP level in *Aspergillus niger* is influenced by the sucrose concentration and light. *Appl. Environ. Microbiol.* **63:**2844–2849.

Greenbaum, D., C. Colangelo, K. Williams, and M. Gerstein. 2003. Comparing protein abundance and mRNA expression levels on a genomic scale. *Genome Biol.* **4:**117.

Griffioen, G., and J. M. Thevelein. 2002. Molecular mechanisms controlling the localisation of protein kinase A. *Curr. Genet.* **41:**199–207.

Gygi, S. P., Y. Rochon, B. R. Franza, and R. Aebersold. 1999. Correlation between protein and mRNA abundance in yeast. *Mol. Cell. Biol.* **19:**1720–1730.

Hamilton, A. J., M. D. Holdom, and L. Jeavons. 1996. Expression of the Cu,Zn superoxide dismutase of *Aspergillus fumigatus* as determined by immunochemistry and immunoelectron microscopy. *FEMS Immunol. Med. Microbiol.* **14:**95–102.

Hauptmann, P., C. Riel, L. A. Kunz-Schughart, K. U. Frohlich, F. Madeo, and L. Lehle. 2006. Defects in N-glycosylation induce apoptosis in yeast. *Mol. Microbiol.* **59:**765–778.

Hensel, M., H. N. Arst, Jr., A. Aufauvre-Brown, and D. W. Holden. 1998. The role of the *Aspergillus fumigatus areA* gene in invasive pulmonary aspergillosis. *Mol. Gen. Genet.* **258:**553–557.

Herker, E., H. Jungwirth, K. A. Lehmann, C. Maldener, K. U. Frohlich, S. Wissing, S. Buttner, M. Fehr, S. Sigrist, and F. Madeo. 2004. Chronological aging leads to apoptosis in yeast. *J. Cell Biol.* **164:**501–507.

Hissen, A. H., A. N. Wan, M. L. Warwas, L. J. Pinto, and M. M. Moore. 2005. The *Aspergillus fumigatus* siderophore biosynthetic gene *sidA*, encoding L-ornithine N^5-oxygenase, is required for virulence. *Infect. Immun.* **73:**5493–5503.

Hoffmann, B., O. Valerius, M. Andermann, and G. H. Braus. 2001. Transcriptional autoregulation and inhibition of mRNA translation of amino acid regulator gene *cpcA* of filamentous fungus *Aspergillus nidulans*. *Mol. Biol. Cell* **12:**2846–2857.

Ibrahim-Granet, O., M. Dubourdeau, J. P. Latgé, P. Ave, M. Huerre, A. A. Brakhage, and M. Brock. 2008. Methylcitrate synthase from *Aspergillus fumigatus* is essential for manifestation of invasive aspergillosis. *Cell. Microbiol.* **10:**134–148.

Jacobson, E. S. 2000. Pathogenic roles for fungal melanins. *Clin. Microbiol. Rev.* **13:**708–717.

Jahn, B., K. Langfelder, U. Schneider, C. Schindel, and A. A. Brakhage. 2002. PKSP-dependent reduction of phagolysosomal fusion and intracellular kill of *Aspergillus fumigatus* conidia by human monocyte-derived macrophages. *Cell. Microbiol.* **4:**793–803.

Jana, S., and J. Paliwal. 2007. Apoptosis: potential therapeutic targets for new drug discovery. *Curr. Med. Chem.* **14:**2369–2379.

Khan, M. A., P. B. Chock, and E. R. Stadtman. 2005. Knockout of caspase-like gene, YCA1, abrogates apoptosis and elevates oxidized proteins in *Saccharomyces cerevisiae*. *Proc. Natl. Acad. Sci. USA* **102:** 17326–17331.

Klionsky, D. J. 2007. Autophagy: from phenomenology to molecular understanding in less than a decade. *Nat. Rev. Mol. Cell Biol.* **8:** 931–937.

Konkel, M. E., and K. Tilly. 2000. Temperature-regulated expression of bacterial virulence genes. *Microbes Infect.* **2:**157–166.

Krappmann, S., E. M. Bignell, U. Reichard, T. Rogers, K. Haynes, and G. H. Braus. 2004. The *Aspergillus fumigatus* transcriptional activator CpcA contributes significantly to the virulence of this fungal pathogen. *Mol. Microbiol.* **52:**785–799.

Lamarre, C., O. Ibrahim-Granet, C. Du, R. Calderone, and J.-P. Latgé. 2007. Characterization of the SKN7 ortholog of *Aspergillus fumigatus*. *Fungal Genet. Biol.* **44:**682–690.

Latgé, J.-P. 1999. *Aspergillus fumigatus* and aspergillosis. *Clin. Microbiol. Rev.* **12:**310–350.

Leiter, E., H. Szappanos, C. Oberparleiter, L. Kaiserer, L. Csernoch, T. Pusztahelyi, T. Emri, I. Pocsi, W. Salvenmoser, and F. Marx. 2005. Antifungal protein PAF severely affects the integrity of the plasma membrane of *Aspergillus nidulans* and induces an apoptosis-like phenotype. *Antimicrob. Agents Chemother.* **49:**2445–2453.

Lessing, F., O. Kniemeyer, I. Wozniok, J. Loeffler, O. Kurzai, A. Haertl, and A. A. Brakhage. 2007. The *Aspergillus fumigatus* transcriptional regulator AfYap1 represents the major regulator for defense against reactive oxygen intermediates but is dispensable for pathogenicity in an intranasal mouse infection model. *Eukaryot. Cell* **6:**2290–2302.

Liebmann, B., S. Gattung, B. Jahn, and A. A. Brakhage. 2003. cAMP signaling in *Aspergillus fumigatus* is involved in the regulation of the virulence gene *pksP* and in defense against macrophage killing by macrophages. *Mol. Genet. Genom.* **269:**420–435.

Liebmann, B., T. W. Muhleisen, M. Muller, M. Hecht, G. Weidner, A. Braun, M. Brock, and A. A. Brakhage. 2004a. Deletion of the *Aspergillus fumigatus* lysine biosynthesis gene *lysF* encoding homoaconitase leads to attenuated virulence in a low-dose model mouse infection model of invasive aspergillosis. *Arch. Microbiol.* **181:**378–383.

Liebmann, B., M. Muller, A. Braun, and A. A. Brakhage. 2004b. The cyclic AMP-dependent protein kinase A network regulates development and virulence in *Aspergillus fumigatus*. *Infect. Immun.* **72:** 5193–5203.

Liu, X.-H., J.-P. Lu, L. Zhang, B. Dong, H. Min, and F.-C. Lin. 2007. Involvement of a *Magnaporthe grisea* serine/threonine kinase gene, MgATG1, in appressorium turgor and pathogenesis. *Eukaryot. Cell* **6:**997–1005.

Mach, K. E., K. A. Furge, and C. F. Albright. 2000. Loss of Rhb1, a Rheb-related GTPase in fission yeast, causes growth arrest with a terminal phenotype similar to that caused by nitrogen starvation. *Genetics* **155:**611–622.

Machida, M., K. Asai, M. Sano, T. Tanaka, T. Kumagai, G. Terai, K. Kusumoto, T. Arima, O. Akita, Y. Kashiwagi, K. Abe, K. Gomi, H. Horiuchi, K. Kitamoto, T. Kobayashi, M. Takeuchi, D. W. Denning, J. E. Galagan, W. C. Nierman, J. Yu, D. B. Archer, J. W. Bennett, D. Bhatnagar, T. E. Cleveland, N. D. Fedorova, O. Gotoh, H. Horikawa, A. Hosoyama, M. Ichinomiya, R. Igarashi, K. Iwashita, P. R. Juvvadi, M. Kato, Y. Kato, T. Kin, A. Kokubun, H. Maeda, N. Maeyama, J. Maruyama, H. Nagasaki, T. Nakajima, K. Oda, K. Okada, I. Paulsen, K. Sakamoto, T. Sawano, M. Takahashi, K. Takase, Y. Terabayashi, J. R. Wortman, O. Yamada, Y. Yamagata, H. Anazawa, Y. Hata, Y. Koide, T. Komori, Y. Koyama, T. Minetoki, S. Suharnan, A. Tanaka, K. Isono, S. Kuhara, N. Ogasawara, and H. Kikuchi. 2005. Genome sequencing and analysis of *Aspergillus oryzae*. *Nature* **438:**1157–1161.

Madeo, F., E. Herker, C. Maldener, S. Wissing, S. Lachelt, M. Herlan, M. Fehr, K. Lauber, S. J. Sigrist, S. Wesselborg, and K. U. Frohlich. 2002. A caspase-related protease regulates apoptosis in yeast. *Mol. Cell* **9:**911–917.

Maerker, C., M. Rohde, A. A. Brakhage, and M. Brock. 2005. Methylcitrate synthase from *Aspergillus fumigatus*. Propionyl-CoA affects polyketide synthesis, growth and morphology of conidia. *FEBS J.* **272:**3615–3630.

Maeta, K., S. Izawa, and Y. Inoue. 2005. Methylglyoxal, a metabolite derived from glycolysis, functions as a signal initiator of the high osmolarity glycerol-mitogen-activated protein kinase cascade and calcineurin/Crz1-mediated pathway in *Saccharomyces cerevisiae*. *J. Biol. Chem.* **280:**253–260.

Maiuri, M. C., E. Zalckvar, A. Kimchi, and G. Kroemer. 2007. Self-eating and self-killing: crosstalk between autophagy and apoptosis. *Nat. Rev. Mol. Cell Biol.* **8:**741–752.

Martin, D., and M. N. Hall. 2005. The expanding TOR signaling network. *Curr. Opin. Cell Biol.* **17:**158–166.

Mazzoni, C., E. Herker, V. Palermo, H. Jungwirth, T. Eisenberg, F. Madeo, and C. Falcone. 2005. Yeast caspase 1 links messenger RNA stability to apoptosis in yeast. *EMBO Rep.* **6:**1076–1081.

Meijer, H. A., and A. A. Thomas. 2002. Control of eukaryotic protein synthesis by upstream open reading frames in the 5′-untranslated region of an mRNA. *Biochem. J.* **367:**1–11.

Meijer, W. H., I. J. van der Klei, M. Veenhuis, and J. A. Kiel. 2007. ATG genes involved in non-selective autophagy are conserved from yeast to man, but the selective Cvt and pexophagy pathways also require organism-specific genes. *Autophagy* **3:**106–116.

Mignone, F., C. Gissi, S. Liuni, and G. Pesole. 2002. Untranslated regions of mRNAs. *Genome Biol.* **3:**reviews0004.0001–0004.0010.

Mizushima, N. 2007. Autophagy: process and function. *Genes Dev.* **21:**2861–2873.

Moreno, M. A., O. Ibrahim-Granet, R. Vicentefranqueira, J. Amich, P. Ave, F. Leal, J. P. Latgé, and J. A. Calera. 2007. The regulation of zinc homeostasis by the ZafA transcriptional activator is essential for *Aspergillus fumigatus* virulence. *Mol. Microbiol.* **64:**1182–1197.

Mousavi, S. A., and G. D. Robson. 2003. Entry into the stationary phase is associated with a rapid loss of viability and an apoptotic-like phenotype in the opportunistic pathogen *Aspergillus fumigatus*. *Fungal Genet. Biol.* **39:**221–229.

Mousavi, S. A., and G. D. Robson. 2004. Oxidative and amphotericin B-mediated cell death in the opportunistic pathogen *Aspergillus fumigatus* is associated with an apoptotic-like phenotype. *Microbiology* **150:**1937–1945.

Moy, T. I., D. Boettner, J. C. Rhodes, P. A. Silver, and D. S. Askew. 2002. Identification of a role for *Saccharomyces cerevisiae* Cgr1p in pre-rRNA processing and 60S ribosome subunit synthesis. *Microbiology* **148:**1081–1090.

Moye-Rowley, W. S. 2003. Regulation of the transcriptional response to oxidative stress in fungi: similarities and differences. *Eukaryot. Cell* **2:**381–389.

Ng, T. T., G. D. Robson, and D. W. Denning. 1994. Hydrocortisone-enhanced growth of *Aspergillus* spp.: implications for pathogenesis. *Microbiology* **140:**2475–2479.

Ni, W., S. Rierson, J. Seo, and J. Yu. 2005. The *pkaB* gene encoding the secondary protein kinase A catalytic subunit has a synthetic lethal interaction with *pkaA* and plays overlapping and opposite roles in *Aspergillus nidulans*. *Eukaryot. Cell* **4:**1465–1476.

**Nierman, W. C., A. Pain, M. J. Anderson, J. R. Wortman, H. S. Kim, J. Arroyo, M. Berriman, K. Abe, D. B. Archer, C. Bermejo, J. Bennett, P. Bowyer, D. Chen, M. Collins, R. Coulsen, R. Davies, P. S. Dyer, M. Farman, N. Fedorova, T. V. Feldblyum, R. Fischer, N. Fosker, A. Fraser, J. L. Garcia, M. J. Garcia, A. Goble, G. H. Goldman, K. Gomi, S. Griffith-Jones, R. Gwilliam, B. Haas, H. Haas, D. Harris, H. Horiuchi, J. Huang, S. Humphray, J. Jimenez, N. Keller, H. Khouri, K. Kitamoto, T. Kobayashi, S. Konzack, R. Kulkarni, T. Kumagai, A. Lafon, J. P. Latgé, W. Li, A. Lord, C. Lu, W. H. Majoros, G. S. May, B. L. Miller, Y. Mohamoud, M. Molina, M. Monod, I. Mouyna, S. Mulligan, L. Murphy, S. O'Neil, I. Paulsen, M. A. Penalva, M. Pertea, C. Price, B. L. Pritchard, M. A. Quail, E. Rabbinowitsch, N. Rawlins, M. A. Rajandream, U. Reichard, H. Renauld, G. D. Robson, S. Rodriguez de Cordoba, J. M. Rodriguez-Pena, C. M. Ronning, S. Rutter, S. L. Salzberg, M. Sanchez, J. C. Sanchez-Ferrero, D. Saunders, K. Seeger, R. Squares, S. Squares, M. Takeuchi, F. Tekaia, G. Turner, C. R. Vazquez de Aldana, J. Weidman, O. White, J. Woodward, J. H. Yu, C. Fraser, J. E. Galagan, K. Asai, M. Machida, N. Hall, B. Barrell, and D. W. Denning. 2005. Genomic sequence of the pathogenic and allergenic filamentous fungus *Aspergillus fumigatus*. *Nature* 438:1151–1156.

Oliver, B. G., J. C. Panepinto, D. S. Askew, and J. C. Rhodes. 2002. cAMP alteration of growth rate of *Aspergillus fumigatus* and *Aspergillus niger* is carbon-source dependent. *Microbiology* 148:2627–2633.

Oliver, B. G., J. C. Panepinto, J. R. Fortwendel, D. L. Smith, D. S. Askew, and J. C. Rhodes. 2001. Cloning and expression of pkaC and pkaR, the genes encoding the cAMP-dependent protein kinase of *Aspergillus fumigatus*. *Mycopathologia* 154:85–91.

Paisley, D., G. D. Robson, and D. W. Denning. 2005. Correlation between in vitro growth rate and in vivo virulence in *Aspergillus fumigatus*. *Med. Mycol.* 43:397–401.

Panepinto, J. C., B. G. Oliver, T. W. Amlung, D. S. Askew, and J. C. Rhodes. 2002. Expression of the *Aspergillus fumigatus rheb* homologue, *rhbA*, is induced by nitrogen starvation. *Fungal Genet. Biol.* 36:207–214.

Panepinto, J. C., B. G. Oliver, J. R. Fortwendel, D. L. H. Smith, D. S. Askew, and J. C. Rhodes. 2003. Deletion of the *Aspergillus fumigatus* gene encoding the Ras-related protein RhbA reduces virulence in a model of invasive pulmonary aspergillosis. *Infect. Immun.* 71: 2819–2826.

Paris, S., D. Wysong, J.-P. Debeaupuis, K. Shibuya, B. Philippe, R. D. Diamond, and J.-P. Latgé. 2003. Catalases of *Aspergillus fumigatus*. *Infect. Immun.* 71:3551–3562.

Patel, P. H., N. Thapar, L. Guo, M. Martinez, J. Maris, C.-L. Gau, J. A. Lengyei, and F. Tamanoi. 2003. *Drosophila* Rheb GTPase is required for cell cycle progression and cell growth. *J. Cell Sci.* 116: 3601–3610.

Paul, A., S. Hackbarth, R. D. Vogt, B. Roder, B. K. Burnison, and C. E. Steinberg. 2004. Photogeneration of singlet oxygen by humic substances: comparison of humic substances of aquatic and terrestrial origin. *Photochem. Photobiol. Sci.* 3:273–280.

Pinan-Lucarre, B., M. Paoletti, and C. Clave. 2007. Cell death by incompatibility in the fungus *Podospora*. *Semin. Cancer Biol.* 17: 101–111.

Pinan-Lucarre, B., M. Paoletti, K. Dementhon, B. Coulary-Salin, and C. Clave. 2003. Autophagy is induced during cell death by incompatibility and is essential for differentiation in the filamentous fungus *Podospora anserina*. *Mol. Microbiol.* 47:321–333.

Reiter, J., E. Herker, F. Madeo, and M. J. Schmitt. 2005. Viral killer toxins induce caspase-mediated apoptosis in yeast. *J. Cell Biol.* 168: 353–358.

Reuther, G. W., and C. J. Der. 2000. The Ras branch of small GTPases: Ras family members don't fall far from the tree. *Curr. Opin. Cell Biol.* 12:157–165.

Rhodes, J. C., B. G. Oliver, D. S. Askew, and T. W. Amlung. 2001. Identification of genes of *Aspergillus fumigatus* up-regulated during growth on endothelial cells. *Med. Mycol.* 39:253–260.

Richie, D., and D. Askew. 2007. Autophagy: a role in metal ion homeostasis? *Autophagy* 4:115–117.

Richie, D., K. Fuller, J. Fortwendel, M. Miley, J. McCarthy, M. Feldmesser, J. Rhodes, and D. Askew. 2007a. Unexpected link between metal ion deficiency and autophagy in *Aspergillus fumigatus*. *Eukaryot. Cell* 6:2437–2447.

Richie, D., M. Miley, R. Bhabhra, G. Robson, J. Rhodes, and D. Askew. 2007b. The *Aspergillus fumigatus* metacaspases CasA and CasB facilitate growth under conditions of endoplasmic reticulum stress. *Mol. Microbiol.* 63:591–604.

Robson, G. 1999. Hyphal cell biology, p. 164–184. *In* R. Oliver and M. Schweizer (ed.), *Molecular Fungal Biology*. Cambridge University Press, Oxford, United Kingdom.

Robson, G. D., J. Huang, J. Wortman, and D. B. Archer. 2005. A preliminary analysis of the process of protein secretion and the diversity of putative secreted hydrolases encoded in *Aspergillus fumigatus*: insights from the genome. *Med. Mycol.* 43(Suppl. 1):S41–S47.

Roze, L. V., and J. E. Linz. 1998. Lovastatin triggers an apoptosis-like cell death process in the fungus *Mucor racemosus*. *Fungal Genet. Biol.* 25:119–133.

Schmelzle, T., and M. N. Hall. 2000. TOR, a central controller of cell growth. *Cell* 103:253–262.

Schrettl, M., E. Bignell, C. Kragl, Y. Sabiha, O. Loss, M. Eisendle, A. Wallner, H. N. Arst, Jr., K. Haynes, and H. Haas. 2007. Distinct roles for intra- and extracellular siderophores during *Aspergillus fumigatus* infection. *PLoS Pathog.* 3:1195–1207.

Schwerk, C., and K. Schulze-Osthoff. 2003. Non-apoptotic functions of caspases in cellular proliferation and differentiation. *Biochem. Pharmacol.* 66:1453–1458.

Searle, J., and Y. Sanchez. 2004. Stopped for repairs: a new role for nutrient sensing pathways. *Cell Cycle* 3:865–868.

Semighini, C. P., J. M. Hornby, R. Dumitru, K. W. Nickerson, and S. D. Harris. 2006. Farnesol-induced apoptosis in *Aspergillus nidulans* reveals a possible mechanism for antagonistic interactions between fungi. *Mol. Microbiol.* 59:753–764.

Silva, R. D., R. Sotoca, B. Johansson, P. Ludovico, F. Sansonetty, M. T. Silva, J. M. Peinado, and M. Corte-Real. 2005. Hyperosmotic stress induces metacaspase- and mitochondria-dependent apoptosis in *Saccharomyces cerevisiae*. *Mol. Microbiol.* 58:824–834.

Tansey, M. R., and T. D. Brock. 1972. The upper temperature limit for eukaryotic organisms. *Proc. Natl. Acad. Sci. USA* 69:2426–2428.

Tekaia, F., and J. P. Latgé. 2005. *Aspergillus fumigatus*: saprophyte or pathogen? *Curr. Opin. Microbiol.* 8:385–392.

Thevelein, J. M., L. Cauwenberg, S. Colombo, J. H. DeWinde, M. Donation, F. Dumortier, L. Kraakman, K. Lemaire, P. Ma, D. Nauwelaers, F. Rolland, A. Teunissen, P. Van Dijck, M. Versele, S. Wera, and J. Winderickx. 2000. Nutrient-induced signal transduction through the protein kinase A pathway and its role in the control of metabolism, stress resistance, and growth in yeast. *Enzyme Microb. Technol.* 26:819–825.

Thon, M., Q. Al-Abdallah, P. Hortschansky, and A. A. Brakhage. 2007. The thioredoxin system of the filamentous fungus *Aspergillus nidulans*: impact on development and oxidative stress response. *J. Biol. Chem.* 282:27259–27269.

Trautmann, N. M., T. Richard, and M. E. Krasny. 2003. *Monitoring Compost pH*. Cornell University Composting Resources, Ithaca, NY.

Urano, J., A. P. Tabancay, W. Yang, and F. Tamanoi. 2000. The *Saccharomyces cerevisiae* Rheb G-protein is involved in regulating canavanine resistance and arginine uptake. *J. Biol. Chem.* 275: 11198–11206.

Uren, A. G., K. O'Rourke, L. A. Aravind, M. T. Pisabarro, S. Seshagiri, E. V. Koonin, and V. M. Dixit. 2000. Identification of paracaspases and metacaspases: two ancient families of caspase-like proteins, one of which plays a key role in MALT lymphoma. *Mol. Cell* 6:961–967.

Uritani, M., H. Hidaka, Y. Hotta, M. Ueno, T. Ushimaru, and T. Toda. 2006. Fission yeast Tor2 links nitrogen signals to cell proliferation and acts downstream of the Rheb GTPase. *Genes Cells* 11: 1367–1379.

Vachova, L., and Z. Palkova. 2005. Physiological regulation of yeast cell death in multicellular colonies is triggered by ammonia. *J. Cell Biol.* 169:711–717.

Valiante, V., T. Heinekamp, R. Jain, A. Hartl, and A. A. Brakhage. 2008. The mitogen-activated protein kinase MpkA of *Aspergillus fumigatus* regulates cell wall signaling and oxidative stress response. *Fungal Genet. Biol.* **45:**618–627.

van den Brink, J. M., P. J. Punt, R. F. van Gorcom, and C. A. van den Hondel. 2000. Regulation of expression of the *Aspergillus niger* benzoate para-hydroxylase cytochrome P450 system. *Mol. Gen. Genet.* **263:**601–609.

Veneault-Fourrey, C., M. Barooah, M. Egan, G. Wakley, and N. J. Talbot. 2006. Autophagic fungal cell death is necessary for infection by the rice blast fungus. *Science* **312:**580–583.

Vilela, C., B. Linz, C. Rodrigues-Pousada, and J. E. McCarthy. 1998. The yeast transcription factor genes YAP1 and YAP2 are subject to differential control at the levels of both translation and mRNA stability. *Nucleic Acids Res.* **26:**1150–1159.

Vilela, C., and J. E. McCarthy. 2003. Regulation of fungal gene expression via short open reading frames in the mRNA 5′ untranslated region. *Mol. Microbiol.* **49:**859–867.

Vinck, A., M. Terlou, W. R. Pestman, E. P. Martens, A. F. Ram, C. A. van den Hondel, and H. A. Wosten. 2005. Hyphal differentiation in the exploring mycelium of *Aspergillus niger*. *Mol. Microbiol.* **58:** 693–699.

Wadskog, I., C. Maldener, A. Proksch, F. Madeo, and L. Adler. 2004. Yeast lacking the SRO7/SOP1-encoded tumor suppressor homologue show increased susceptibility to apoptosis-like cell death on exposure to NaCl stress. *Mol. Biol. Cell* **15:**1436–1444.

Washburn, M. P., A. Koller, G. Oshiro, R. R. Ulaszek, D. Plouffe, C. Deciu, E. Winzeler, and J. R. Yates III. 2003. Protein pathway and complex clustering of correlated mRNA and protein expression analyses in Saccharomyces cerevisiae. *Proc. Natl. Acad. Sci. USA* **100:**3107–3112.

Weinberger, M., L. Ramachandran, L. Feng, K. Sharma, X. Sun, M. Marchetti, J. A. Huberman, and W. C. Burhans. 2005. Apoptosis in budding yeast caused by defects in initiation of DNA replication. *J. Cell Sci.* **118:**3543–3553.

Weisman, R., I. Roitburg, M. Schonbrun, R. Harari, and M. Kupiec. 2007. Opposite effects of Tor1 and Tor2 on nitrogen starvation responses in fission yeast. *Genetics* **175:**1153–1162.

Wilson, W. A., and P. J. Roach. 2002. Nutrient-regulated protein kinases in budding yeast. *Cell* **111:**155–158.

Wissing, S., P. Ludovico, E. Herker, S. Buttner, S. M. Engelhardt, T. Decker, A. Link, A. Proksch, F. Rodrigues, M. Corte-Real, K. U. Frohlich, J. Manns, C. Cande, S. J. Sigrist, G. Kroemer, and F. Madeo. 2004. An AIF orthologue regulates apoptosis in yeast. *J. Cell Biol.* **166:**969–974.

Wullschleger, S., R. Loewith, and M. N. Hall. 2006. TOR signaling growth and metabolism. *Cell* **124:**471–484.

Wysocki, R., and S. J. Kron. 2004. Yeast cell death during DNA damage arrest is independent of caspase or reactive oxygen species. *J. Cell Biol.* **166:**311–316.

Xie, Z., and D. J. Klionsky. 2007. Autophagosome formation: core machinery and adaptations. *Nat. Cell Biol.* **9:**1102–1109.

Zeuner, A., A. Eramo, C. Peschle, and R. De Maria. 1999. Caspase activation without death. *Cell Death Differ.* **6:**1075–1080.

Zhang, L., M. Wang, R. Lui, and R. Calderone. 2005. Expression of *Aspergillus fumigatus* virulence-related genes detected in vitro and in vivo with competitive RT-PCR. *Mycopathologia* **160:**201–206.

Zhao, W., J. Panepinto, J. Fortwendel, L. Fox, B. Oliver, D. Askew, and J. Rhodes. 2006. Deletion of the regulatory subunit of protein kinase A in *Aspergillus fumigatus* alters morphology, sensitivity to oxidative damage, and virulence. *Infect. Immun.* **74:**4865–4874.

Zhou, H., H. Hu, L. Zhang, R. Li, H. Ouyang, J. Ming, and C. Jin. 2007. *O*-Mannosyltransferase 1 in *Aspergillus fumigatus* (AfPmt1p) is crucial for cell wall integrity and conidium morphology, especially at an elevated temperature. *Eukaryot. Cell* **6:**2260–2268.

III. IMMUNE DEFENSE AGAINST *ASPERGILLUS*

Aspergillus fumigatus and Aspergillosis
Edited by J.-P. Latgé and W. J. Steinbach
©2009 ASM Press, Washington, DC

Chapter 17

Reactive Oxygen Intermediates, pH, and Calcium

ELAINE BIGNELL

Immune cells of mice and humans can show us how to kill *Aspergillus fumigatus* effectively, since phagocytes from healthy humans launch potent chemical assaults to eliminate *A. fumigatus* spores. This is no mean feat, given that the fungal spore is designed by nature to withstand extended periods of time under unfavorable conditions. Laden with a genetically encoded protective arsenal, including pigment, immunotoxin, antioxidants, and hydrolytic enzymes, and under physical cover of a hardy cell wall, the spore is tasked with maintaining continued survival of the species. For aspergilli this involves spore germination and growth of a mycelium before generation and dispersal of new progeny.

Reactive oxygen intermediates (ROIs), pH exacerbations, and calcium fluxes can act together to effect fungal killing at the level of the phagocyte; therefore, resilience in the face of rapid environmental alterations is necessary to promote fungal survival within the susceptible host. In addition to effecting fungal killing, variations in this ionic cocktail can orchestrate cellular responses and mediate damage at the level of both host and pathogen. Thus, the benefits of understanding this complex interplay include acquiring the means to emulate or enhance innate microbicidal activity or to block *A. fumigatus* adaptation to host-imposed stresses, thereby preventing colonization. With such goals in mind the scientific community has pursued an understanding of innate immune defenses against *A. fumigatus* spores and hyphae.

REACTIVE OXYGEN INTERMEDIATES

Reduction is the usual route of cellular oxygen metabolism, ROIs being the intermediate products of sequential oxygen reduction, resulting from a series of four electron acceptance reactions, that ultimately yields water (Fig. 1). Partially reduced forms of oxygen, namely superoxide, hydrogen peroxide, and hydroxyl radical, and reactive products of these with halides and amides are unstable and thus highly reactive. Like superoxide, hydrogen peroxide can lead to production of other reactive species; for this reason it is often also referred to as an ROI.

Oxygen radicals have diverse effects on cells, causing cellular damage via oxidation of protein, lipid, and carbohydrate, as well as by oxidative alteration of DNA molecules (Yu, 1994). The threat posed by ROIs to cellular structures and functions is an unavoidable consequence of anaerobic respiration; thus, aerobic organisms have evolved mechanisms to neutralize ROIs resulting from normal physiological processes. Conversely, a growing body of evidence supports a role for ROIs in cellular signaling in plant (Mittler, 2002), mammalian (Franklin et. al., 2006), and fungal cells (Greene et al., 2002; Takemoto et al., 2007). Appropriate modulation of ROI homeostasis is a prerequisite for successful ROI-mediated signaling such that low levels of ROIs are maintained for mediation of cell signaling and excess ROIs are immediately quenched.

Higher-order modulation of ROI production is required for defense against pathogens in plant and animal cells. Phagocytosis by mammalian neutrophils and macrophages is accompanied by nonmitochondrial oxygen consumption, known as the respiratory burst, which is catalyzed by the NADPH-oxidase enzyme complex (Babior, 2004) (Fig. 2). Spatial segregation of membrane-bound, cytoplasmic, and granule-associated NADPH-oxidase components prevents constitutive activity of the enzyme, thereby limiting host cell damage under unstimulated conditions. Phagocytosis creates a membrane-bound, microbe-containing intracellular vesicle called the phagosome which simultaneously brings NADPH components at the cellular boundary (flavocytochrome b_{558} and a heterodimer comprised of gp91^{phox} and p22^{phox}) into close proximity with cytoplasmic compo-

Elaine Bignell • Dept. of Microbiology, Centre for Molecular Microbiology and Infection, Imperial College London, Armstrong Road, London SW7 2AZ, United Kingdom.

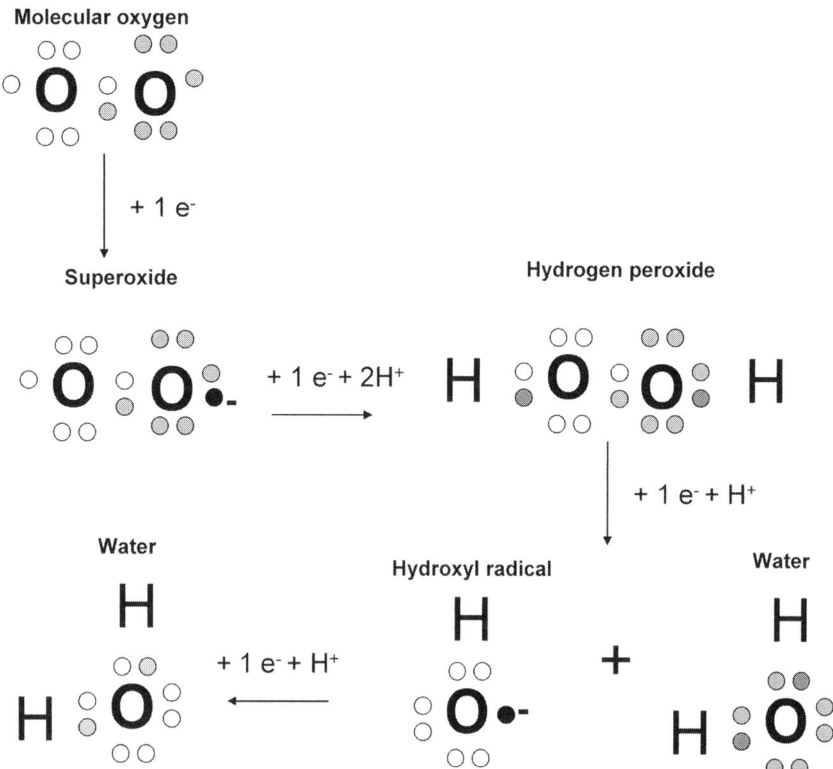

Figure 1. Four-step reduction of molecular oxygen and resulting ROIs. The Lewis dot diagram depicts the stepwise reduction of O_2. Complete reduction of molecular oxygen to water involves the addition of four protons and four electrons. The hydroxyl radical is the strongest ROI, being capable of indiscriminate oxidation of organic compounds such as biological macromolecules.

nents (p67^{phox}, p47^{phox}, p40^{phox}, and p21^{rac}-GTP), thus partitioning the newly formed "killing chamber" from the rest of the cell. Assembly of the membrane-associated NADPH-oxidase complex creates a channel for the efficient transfer of electrons from cytosolic NADPH to dissolved molecular oxygen at the cell surface (and/or in the forming phagolysosome), generating the highly unstable anion superoxide. Dismutation of superoxide generates hydrogen peroxide, a progenitor of hypochlorous acid (HOCl, via myeloperoxidase [MPO]-mediated halide oxidation), singlet oxygen, ozone, and hydroxyl radicals, the latter two by largely speculative mechanisms (Fig. 2). Also required is the delivery of microbicidal enzymes to the phagosome, agents which are stored in subcellular organelles called granules. Thus, coincident with, or soon after, the phagocytic event preformed cytoplasmic granules fuse with the phagosomal membrane, discharging microbicidal enzymes and polypeptides. Granules contain a variety of proteins, including MPO and flavocytochrome b_{558}, as well as protein-degrading enzymes known as proteases, the differential action of which against various microorganisms has been demonstrated (Appelberg, 2007; Borregaard et al., 2007).

The respiratory burst provides a primary and generalized phagocytic defense mechanism against infecting microorganisms, and chronic granulomatous disorder (CGD) is an inherited genetic disorder of NADPH-oxidase that renders patients susceptible to bacterial and fungal infections. Underlying genetic lesions, which abrogate or greatly reduce activity of the enzyme in both macrophages and neutrophils, can occur in four of the five NADPH-oxidase subunits, p22^{phox}, p40^{phox}, p47^{phox}, or gp91^{phox} (Segal et al., 2000). Seminal studies by Schaffner and colleagues (1982) discerned distinct roles for macrophages and neutrophils in protection against *A. fumigatus* spores and hyphae, respectively. The high incidence of aspergillosis in leukopenic, agranulocytic, and CGD patients indicates the gravity of this cellular collaboration in the defense against *Aspergillus* infection.

Role of ROIs in Macrophage-Mediated *A. fumigatus* Killing

Numerous studies have addressed the kinetics of *A. fumigatus* spore killing by cultured murine and human

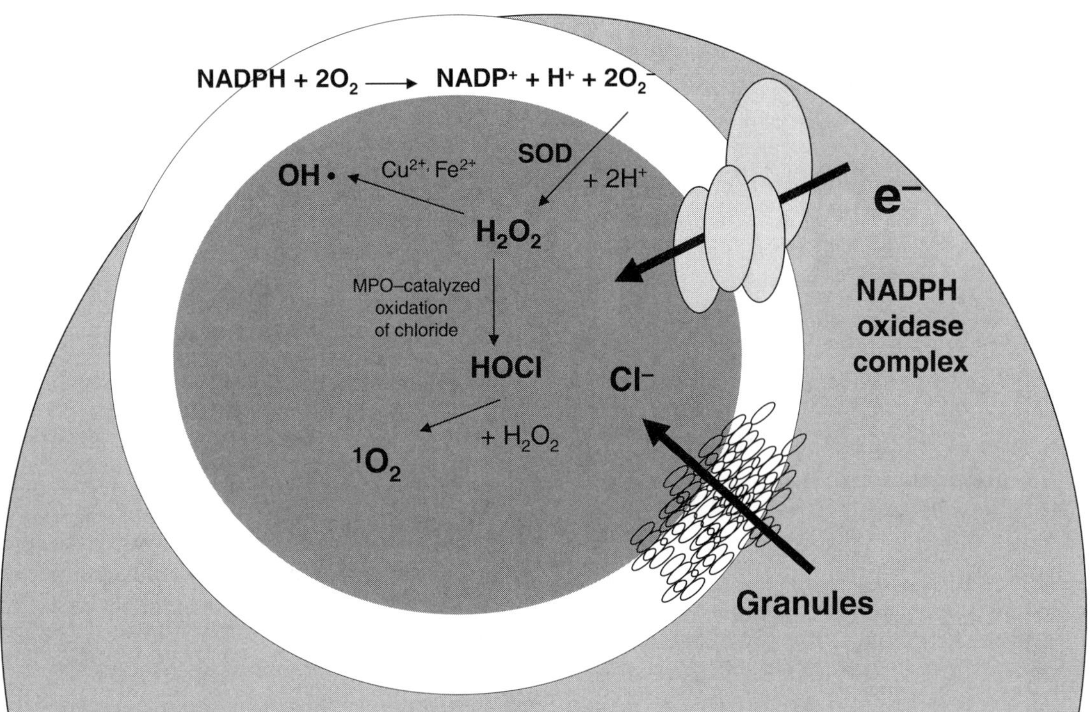

Figure 2. ROI production during phagocytosis. Shown is a schematic view of an *A. fumigatus* spore during phagocytosis and concomitant generation of ROIs. The spatially segregated NADPH-oxidase molecular components gp91^{phox}, p22^{phox}, p67^{phox}, p47^{phox}, p40^{phox}, and p21^{rac}-GTP (represented by ovals) assemble into an active NADPH-oxidase complex in the phagosomal membrane of activated phagocytes. Transfer of electrons from cytosolic NADPH to dissolved molecular oxygen at the cell surface (and/or in the forming phagolysosome) generates the highly unstable superoxide anion (O_2^-). Dismutation of superoxide (via superoxide dismutase [SOD]) generates hydrogen peroxide (H_2O_2), a progenitor of HOCl (via MPO-mediated halide oxidation), singlet oxygen ($^1O_2$), and hydroxyl radicals ($OH^•$). Microbicidal enzymes are delivered to the phagosome from subcellular organelles called granules, coincident with, or soon after, the phagocytic event.

macrophages or macrophages directly isolated from mice, rabbits, or humans (Levitz et al., 1986; Ibrahim-Granet et al., 2003; Philippe et al., 2003). Inherent variability among such studies is unavoidable due to variations among host cell origin and experimental conditions; however, certain widely accepted conclusions have emerged, among them that macrophages efficiently phagocytose, and kill, *A. fumigatus* conidia. Further, macrophage-dependent *A. fumigatus* killing requires ROIs (Schaffner et al., 1982; Philippe et al., 2003; Perkhofer et al., 2007).

Using a luminol peroxidase assay, Philippe et al. observed NADPH-oxidase-dependent ROI production at 3 h postphagocytosis in murine alveolar macrophages. Maximal ROI production could be correlated with spore swelling within the phagolysosome and heightened *A. fumigatus* killing. In agreement with earlier findings (Schaffner et al., 1982; Waldorf et al., 1984) macrophages from corticosteroid-treated mice displayed partial reductions in killing, and this could now be correlated with a marked reduction in extracellular ROI production. Accordingly, macrophages from a p47$^{phox-/-}$ murine model of CGD were completely unable to prevent germination of phagocytosed conidia, indicating the importance of NADPH-oxidase catalysis for the oxidative killing of *A. fumigatus* spores. The established proof of absolute ROI requirement does not, however, establish a direct role for ROIs in *A. fumigatus* killing.

If ROIs acted directly to kill *A. fumigatus* spores during infection, it would be reasonable to hypothesize that *A. fumigatus* mutants having heightened sensitivity to ROIs in vitro would be more susceptible to killing by macrophages. Such a prediction would be correct for null mutants of the polyketide synthase PksP/Alb1, which catalyzes the first step of dihydroxynaphthalene-like melanin biosynthesis in *A. fumigatus*. Pigmentless conidia, lacking PksP, have heightened sensitivity to oxidative stress in vitro (Jahn et al., 1997) relative to the wild type, are more susceptible to macrophage-mediated damage (Jahn et al., 2002), and are less virulent in a murine model of infection (Jahn et al., 1997). Currently it is not possible to causatively link lack of pigmentation to the aforementioned phenotypes, as morphological effects on conidia, resulting from the absence of PksP, may affect physical properties of the spores. One can, however, conclude that mutational modulation of *A. fumigatus* susceptibility to macrophage-mediated killing is achievable and, furthermore, that the physiological impact of such manipulations can be extrapolated to measurable reductions in virulence.

The removal of distinct antioxidant functions more simplistically tests the role of ROIs in *A. fumigatus* survival during infection. Among those characterized in *A. fumigatus*, only the catalases have thus far been subject to close scrutiny. Catalase is a major antioxidant defense component (Yu, 1994; Hamilton and Holdom, 1999) that catalyzes hydrogen peroxide decomposition, yielding water. The *A. fumigatus* genome encodes three such catalase enzymes, a spore-specific catalase, CatA (Afu6g03890), and two mycelial catalases, Cat1 (Afu3g02270) and Cat2 (Afu8g01670). *A. fumigatus* mutants singly and multiply deficient in these enzymes have been characterized (Paris et al., 2003). CatA deficiency can be correlated with hydrogen peroxide sensitivity in vitro, but no phenotype for any of the characterized mutants is observable in macrophage killing, polymorphonuclear cell (PMN) damage assays, or mammalian models of virulence. Cat2 expression is, however, upregulated during murine infection relative to laboratory culture (McDonagh et al., 2008). The initial impetus for investigating the role of catalases was derived from the observation that catalase-negative organisms rarely infect CGD patients, leading to the hypothesis that these organisms generate enough hydrogen peroxide (in the absence of intrinsic neutralizing activity) to catalyze MPO-mediated halogenation, even in the absence of the respiratory burst, and thereby self-destruct. In the idealized setting of neutrophil killing in vitro, hydrogen peroxide effectively kills fungal hyphae, and neutrophil-mediated damage is blockable by commercial catalase (Diamond and Clark, 1982). However, the physiological relevance of such findings is difficult to judge. Notwithstanding obstructions posed to virulence testing by the aberrant inflammatory response of $p47^{phox-/-}$ mice (Bignell et al., 2005), catalase-negative *A. nidulans* mutants remain virulent in this CGD model of infection (Chang et al., 1998). The finding that CatA cannot defend against the cell-mediated respiratory burst ex vivo or during murine infection (Paris et al., 2003) suggests either, as concluded by the authors, that the main *A. fumigatus*-damaging ROI is not hydrogen peroxide or that redundancy among *A. fumigatus* ROI-scavenging mechanisms exists.

A global proteomic approach to assessing the *A. fumigatus* cellular responses to oxidative stress identified 28 proteins having altered migration or intensity following hydrogen peroxide exposure (Lessing et al., 2007). This cohort included *A. fumigatus* homologs of several proteins dependent upon the *Saccharomyces cerevisiae* metabolic stress response protein Yap1p. Deletion of an *A. fumigatus* homolog, AfYap1p, leads to hydrogen peroxide and menadione sensitivity relative to the wild type in vitro. Thus, AfYap1p protects against peroxide and superoxide anions, respectively, in vitro. However, AfYap1 deletion did not promote attenuation of virulence in neutropenic mice and was inconsequential for ROI production or hyphal damage by PMNs. It therefore appears that one cannot universally assign attenuated virulence to strains having hypersensitivity to oxidative stresses in vitro, a conclusion also reached by Lamarre et al. upon observing increased peroxide sensitivity (but wild-type virulence) of AfSkn7p null mutants lacking a transcription factor contributing to the oxidative stress response (Lamarre et al., 2007). On the other hand, loss of a virulence-modulating α-1,3-glucan synthase activity, Ags3, speeds germination in vitro and the mutant has increased melanin content, heightened resistance to hydrogen peroxide in vitro, and a more rapid onset of invasive tissue infection in mice (Maubon et al., 2006). Based upon this observation, and notwithstanding the usual considerations with pleiotropism of phenotype, the fact that acquired resistance to stresses normally able to kill conidia can be correlated with increases in virulence reinforces what is known about the important role such host defenses perform. This apparent inability to equate loss of ROI scavengers with attenuation of *A. fumigatus* virulence may simply mean the real agents of ROI quenching have yet to be identified, or that there are multiple enzymes performing such a role and multiple gene deletions will be required in order to reach firm conclusions. It also remains possible that current murine models of infection cannot appropriately address phagocytic *A. fumigatus* killing, particularly given the recently observed dichotomy of virulence phenotypes, for example, in *A. fumigatus* gliotoxin biosynthetic mutants in various murine models of *Aspergillus* infection (Bok et al., 2006; Cramer et al., 2006; Kupfahl et al., 2006; Spikes et al., 2008).

In summary then, controversy surrounds the precise role of ROIs in macrophage-mediated *A. fumigatus* killing. The notion that ROI generation directly damages *A. fumigatus* spores and hyphae is supported by a considerable body of experimental evidence generated by multiple complementary approaches. However, investigations of macrophage killing mechanisms have received less attention than those of PMNs or neutrophils.

Role of ROIs in Neutrophil-Mediated *A. fumigatus* Killing

Superoxide production by phagocytic cells was first demonstrated by Babior et al. (1973) in PMNs. The search for such an antimicrobial mechanism had been prompted by several observations (Curnutte, 2004). First, dramatic increases in neutrophil oxygen uptake had been observed during phagocytosis of multiple microorganisms (Sbarra and Karnovsky, 1959). Second, the occurrence of superoxide dismutase enzymes, which provide a cellular defense against the harmful effects of superoxide (Yu, 1994), was evident in aerobic but not in anaerobic cells. Finally, MPO-catalyzed peroxidation of halides failed to fully account for the microbicidal activity of leukocytes, as evidenced by the largely asymptomatic presentation of congenital MPO deficiency in humans (Lehrer and Cline, 1969), suggesting an alternative mechanism of oxygen-dependent microbial killing. Strong support for a model of superoxide-directed microbial damage was soon provided by studies in CGD patients (Curnutte et al., 1974), and the inability of CGD neutrophils to kill infecting microorganisms conclusively linked the respiratory burst with microbial killing, as well as presenting a powerful tool with which to probe NADPH-oxidase function, from both microbicidal and immunomodulatory perspectives.

Initial studies on phagocyte killing of fungi were centered predominantly on the role of the neutrophil, given the predominance of neutropenia among susceptible patients. Probing the importance of PMN-mediated killing in defense against primary (*Coccidioides, Histoplasma, Paracoccidioides, Blastomyces*, and *Sporothrix* species) and opportunistic (*Candida, Mucoraceae, Aspergillus, Petriellidium*, and *Trichosporon* species) fungal pathogens revealed a clear correlation between resistance to PMN-mediated killing in vitro and primary fungal pathogenesis (Schaffner et al., 1986), the dimorphic primary fungal pathogens tested being much less susceptible to hydrogen peroxide than the opportunists. Susceptibility to HOCl was indistinguishable between classes, however, and catalase activity was broadly similar among all species tested, so catalase activity levels could not be correlated with susceptibility or resistance to hydrogen peroxide or PMNs.

Reconstitution of H_2O_2-MPO-halide systems in vitro can kill *A. fumigatus* spores and hyphae (Diamond and Clark, 1982). Such "cell-free" studies enjoy the advantage that the action of specific inhibitors of killing can be tested in the absence of concomitant effects on cellular metabolism, which can distort findings in whole-cell assays. For example, sodium azide and sodium cyanide inhibit MPO but may affect cellular respiration also. Diamond and Clark (1982) devised a means to kill *A. fumigatus* hyphae in vitro by combining MPO and halides, supplying hydrogen peroxide directly or from in vitro syntheses, for example, by using glucose- and glucose oxidase-generating systems. Sodium azide was found to block hyphal damage in this system, whereas superoxide dismutase did not. Addition of catalase, however, completely eliminated hyphal damage. Such studies established the relative importance of several potential fungicidal products of neutrophils, assigning primary importance to the action of both MPO and hydrogen peroxide rather than to superoxide, which was concluded to be important solely in providing a source of hydrogen peroxide for downstream reactions.

Reeves and colleagues concluded, from eloquent and extensive biochemical analyses of phagocytosis, that the role of the NADPH oxidase is to create an environment in which the phagocytic granule enzymes are functional (Reeves et al., 2002). Their conclusions are supported by the observation that knockout mice, lacking the neutral proteases cathepsin G and elastase, do not kill microbes despite normal ROI production and halogenation. The passage of electrons (and resulting differential charge) across the wall of the phagocytic vacuole during the respiratory burst is compensated by the inward passage of chloride anions originating from the granules, and the pH of the phagocytic vacuole is regulated by the exchange by vacuolar Na^+ for cytoplasmic H^+, along with a massive flux of potassium cations into the vacuole. These ion fluxes, Reeves et al. hypothesized, and the consequential pH changes serve to promote microbial killing and digestion by optimizing conditions for the action of the enzymes released from the cytoplasmic granules.

Although there is no direct evidence to support similar biochemical activity during killing of *A. fumigatus*, indirect support for this model can be gleaned from studies of *A. fumigatus* infection using knockout mice. The relative contributions of MPO and NADPH-oxidase to the host defense against *A. fumigatus* have been assessed by direct comparison of $MPO^{-/-}$ and X-linked ($gp91^{phox-/-}$) $CGD^{-/-}$ mice (Aratani et al., 2002). Whereas HOCl production was found to be comparable in neutrophils of wild-type and $CGD^{-/-}$ mice, neutrophils from $MPO^{-/-}$ mice failed to produce HOCl, and superoxide production was negligible in neutrophils from $CGD^{-/-}$ mice. Neutrophils originating from mice with both deficiencies produced neither HOCl nor MPO. MPO is therefore essential for HOCl production, and since $MPO^{-/-}$ mice were found to be resistant to *A. fumigatus* infection (despite delayed clearance of *A. fumigatus* spores), MPO-catalyzed HOCl generation is dispensable for murine defense against *A. fumigatus* infection. Furthermore, MPO deficiency was not additive with $gp91^{phox}$ deficiency in terms of severity of disease; thus, it seems that MPO, when functional, depends upon integrity of NADPH-oxidase to exert its effect.

To recap, the respiratory burst generates hydrogen peroxide, superoxide, and various oxidation products, including hydroxyl radical, singlet oxygen, and HOCl. Combining MPO with halides and hydrogen peroxide in vitro is microbicidal, but MPO deficiency does not increase susceptibility to *A. fumigatus* in mice or in human beings. How, then, do neutrophils kill *A. fumigatus*? Mice deficient in one or both of the granulocyte enzymes elastase and cathepsin G are susceptible to infection with *A. fumigatus* in an intravenous model of murine infection (Tkalcevic et al., 2000). Thus, knockout mice demonstrate the importance of granule proteases for *Aspergillus* killing, suggesting that, in line with the Reeves findings (Reeves et al., 2002), both proteases and ROIs are important for killing.

pH

pH is a negative logarithmic measure of hydrogen ion activity, devised in 1909 by the Danish biochemist Søren Peter Lauritz Sørensen to conveniently express the widely varying H^+ concentrations encountered in aqueous solution (Sorensen, 1909). Such is the convenience of this scale for assignation of acidity (pH 1 to 6.9) and alkalinity (pH 7.1 to 14) to aqueous solutions that the actual ionic effect parameterized by the scale can be often overlooked. For example, a pH of 2 equates approximately to a hydrogen ion concentration of 0.01 M, and pH 13 equates to a hydrogen ion concentration of 0.0000000000001 M. In all circumstances where metabolism is active, there is a steady flux of protons into and out of the cytoplasmic proton pool. Therefore, in addition to ROIs, aerobic cellular metabolism generates protons which must be removed from the cytoplasm to ensure optimal activity of major metabolic pathways. In fungi the H^+-ATPase serves this function, thereby regulating intracellular pH, maintaining ion balance, and generating an electrochemical proton gradient which in turn drives an array of secondary transport systems (Serrano, 1988). Large deviations of intracellular pH must be avoided to maintain control of pH-sensitive physiological processes, and cellular homeostatic mechanisms collaborate to ensure that even large perturbations of extracellular pH impact minimally on the intracellular pH of the cell. Some cellular functions, however, must respond to exacerbations of extracellular pH, and clear benefits can be attributed to the possession of a means to modulate gene expression according to the ambient environment, promoting rapid adaptation to environmental change and energy efficiency with respect to protein synthesis, transport, and export.

Fungal Adaptation to Environmental pH Flux

A conserved fungal regulatory mechanism which acts independently of intracellular homeostasis to regulate acid- and alkaline-expressed genes is dependent upon the action of PacC/Rim101p fungal transcription factors, which direct appropriate cellular responses to changes in the environment (Arst and Penalva, 2003). In *Aspergillus nidulans* PacC activates alkaline-expressed genes, such as alkaline protease (Tilburn et al., 1995), following two successive proteolytic processing reactions occurring in the cytoplasm of the cell and in response to elevated extracellular pH, thereby permitting nuclear entry of the transcription factor (Diez et al., 2002; Hervas-Aguilar et al., 2007, Fernadez-Martinez et al., 2003). Six proteins, PalA, PalB, PalC, PalF, PalH, and PalI, participate in ambient pH signaling to promote PacC proteolysis at alkaline pH (Fig. 3). The notion that these proteins participate in a pathway upstream of PacC originated from classical genetic studies which revealed epistasis of the *pal* null allele phenotype to constitutivity of PacC processing (Caddick et al., 1986) and also from identical null phenotypes among all but one of the pathway components, PalI (Negrete-Urtasun et al., 1999). Much progress has been made in recent years in characterizing and ordering the molecular steps involved in ambient pH signal transmission, and it is now clear that spatial segregation of the pH signaling complex components serves to prevent PacC processing under acidic environmental conditions (Peñalva et al., 2008). pH signal transmission requires formation of, and communication between, two distinct molecular complexes: one comprising PalI, PalH, and PalF and located at the plasma membrane of the cell, and the second associated with components of ESCRT (*e*ndosomal *s*orting *c*omplex *r*equired for *t*ransport) complexes of the endosomal membranes and involving PalA, PalB, and PacC (Fig. 3). The ESCRT pathway carries transmembrane proteins from the outer membrane of the cell into the interior of multivesicular bodies (for a review, see Williams and Urbe, 2007).

Stoichiometry with respect to cellular ratios of PalH and PalI, observed using green fluorescent protein-tagged alleles driven from regulatable promoters, is required to ensure plasma membrane versus vacuolar localization of PalH and is the basis for the current assumption that PalI assists PalH localization to the plasma membrane in a nonessential capacity (Calcagno-Pizarelli et al., 2007). This would explain the merely partial acidity mimicry of the phenotype associated with *A. nidulans palI* null alleles. Mutational and coimmunoprecipitation analyses established the requirement for an interaction between the carboxy terminus of the 7-transmembrane domain protein PalH with the arrestin-like protein PalF in order for pH sensing to occur (Herranz et al., 2005). Eukaryotic cells modulate signaling protein populations in the plasma membrane by trafficking proteins between the plasma membrane and the lysosome or vacuole. Downregulation of membrane protein populations can be achieved by delivery of internalized receptors to the vacuole for degradation; con-

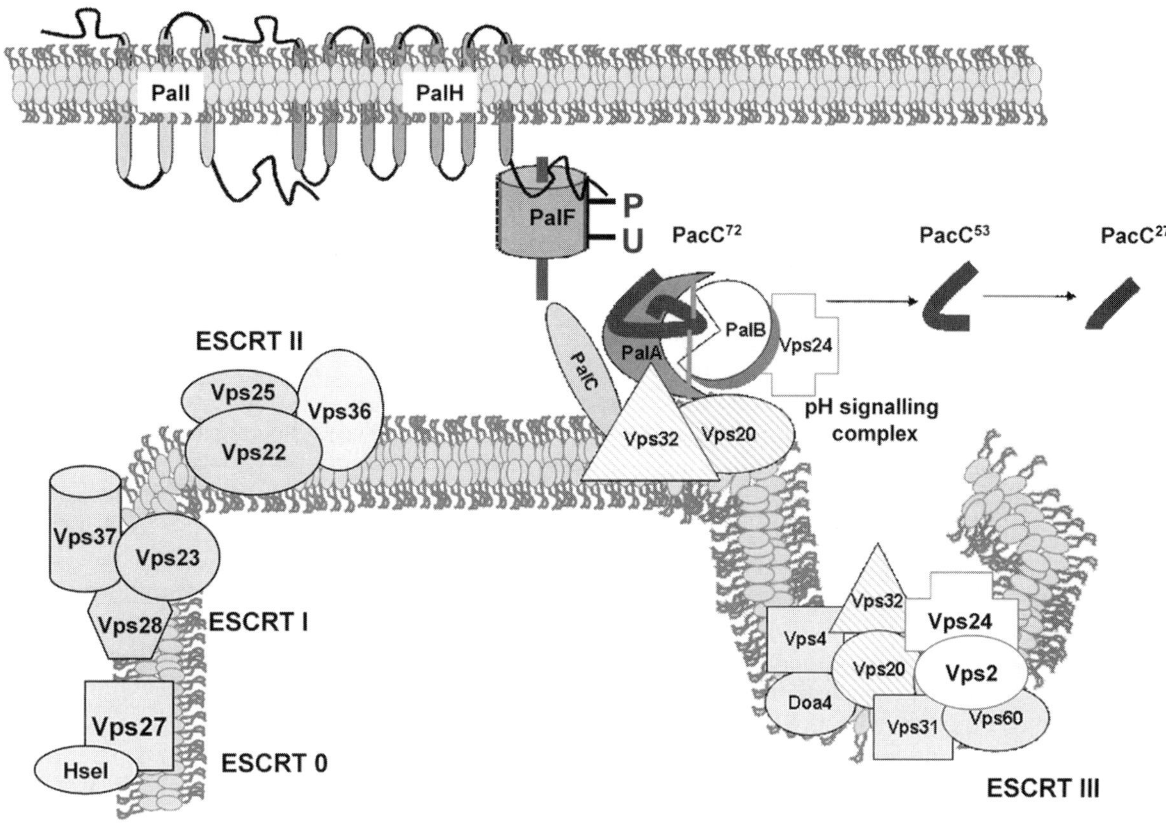

Figure 3. Schematic representation of plasma and endosomal membrane associations with pH signaling and ESCRT complexes. Molecules having demonstrated roles in pH signaling in *S. cerevisae* are shaded dark gray. Those leading to partially constitutive pH signaling in *S. cerevisiae* are shaded light gray. Links between ESCRTs are not shown.

versely, delivery to endosomal ESCRT complexes is thought to facilitate signaling (Williams and Urbe, 2007), and accumulating evidence supports a role for ESCRT complexes I, II, and III in fungal pH signal transduction. All components of ESCRT-I and ESCRT-II and some ESCRT-III components (Fig. 3) are required for pH signaling in *S. cerevisiae* (Boysen and Mitchell, 2006). In *A. nidulans* PalA recognizes the short amino acid motif YPXL/I, two residues of which are present in PacC, flanking the pH signaling protease (which is likely the calpain-like protease PalB) cleavage site (Vincent et al., 2003). Thus, the current model views PalA as a scaffold for the PacC processing machinery, which is tethered via Vps32 interaction to the endosomal membrane of the cell (Fig. 3).

pH Adaptation and Fungal Virulence

Due to the existence of sexual and parasexual stages in the *A. nidulans* life cycle, this organism serves as a model among members of the genus *Aspergillus*. Decades of classical genetic manipulations, mutagenic screens, meiotic and mitotic analyses, and more recently genomic sequencing have provided a detailed physical map of the *A. nidulans* genome, a wealth of auxotrophic markers for genetic selection, and an extensive strain collection for the *Aspergillus* community (Todd et al., 2007). Comparatively high efficiencies of transformation and homologous integration have added to the ease with which genetic analyses can be performed. *A. nidulans*, despite rarely infecting human beings, closely mirrors *A. fumigatus* infection kinetics in neutropenic mice (Tang et al., 1994) and completes the tool set required for rapid hypothesis testing of pathogenetic mechanisms. Interrogation of pH signaling during *Aspergillus* infection provides an excellent example of the power of such modeling, since the ability to test virulence of a proteolysis-resistant *pacC* allele hinged upon introduction of a single nucleotide substitution into the *pacC* open reading frame (Bignell et al., 2005). UV mutagenesis in *A. nidulans* already provided an appropriate allele for such testing, whereas construction of the equivalent *A. fumigatus* mutant by available molecular genetic methodologies would have been almost impossible. A panel of prototrophic *A. nidulans* strains was constructed to probe the role of alkaline adaptation in virulence. Among them, a *palB*$^{-}$ null allele, a mutant with a frameshift mutation in the initiating methionine codon of the *pacC* gene (a *pacC*$^{-}$ null allele), and the aforementioned processing-resistant *pacC* allele (*pacC*$^{+/-}$ 209) allow a test for mere absence of PacC processing, in addition to complete absence of PacC protein. Mutational truncation of the PacC protein, achieved by introduction of a stop codon, permitted analysis of constitutive PacC processing during virulence (*pacC*^{c} 14 allele). Color Plate 8A shows the physiological effects of such *A. nidulans* pH regulatory mutations on growth on simple pH-buffered medium and the corresponding effects on fungal virulence (Color Plate 8B), relative to complemented strains. Despite suffering no overall effects on growth rate and conidiation on rich medium in vitro, mutations preventing PacC proteolysis resulted in complete attenuation of virulence in neutropenic mice, and mutations promoting constitutive PacC processing led to earlier tissue invasion and accelerated mortality. Therefore, it is possible to conclude that *A. nidulans* PacC is required for virulence. Further, PacC must be present and proteolytically cleaved in response to Pal-mediated pH signaling for virulence, and constitutive processing of PacC can facilitate tissue invasion and pathogenesis of *A. nidulans* infection (Bignell et al., 2005). Extrapolation of these findings to the major fungal pathogen *A. fumigatus* is fully expected (and actively sought by us) in view of the high level of conservation of pH signaling components between the two species and given the importance of pH signaling to virulence in multiple other animal and plant pathogens (Peñalva et al., 2008).

pH and Phagocytosis

Many investigators have monitored phagolysosomal pH during microbial internalization and killing (Dri et al., 2002; Ibrahim-Granet et al., 2003; Jahn et al., 2002; Reeves et al., 2002; Segal et al., 1981). Ibrahim-Granet et al. observed a drop in phagosomal pH during *A. fumigatus* killing which was essential for microbicidal activity based upon bafilomycin-mediated inhibition. Segal et al. (1981) noted from early studies that the respiratory burst is accompanied by a large disturbance in the pH of the neutrophil phagocytic vacuole, which rises from 6 to 8, and based upon defective alkalinization among CGD neutrophils proposed this as a basis for killing activity. An attempt to test this hypothesis using acid- and alkaline-nonresponsive *A. nidulans* pH mutants was thwarted by exuberant neutrophil recruitment to p47$^{pho-/-}$ murine lungs (Bignell et al., 2005), which were equally susceptible to infection by all strains tested, including an auxotrophic *p*-amino benzoic acid synthetase mutant which was unable to grow in vivo. Histological analysis revealed no growth advantage (or deficit in fungal burden) associated with any mutant tested, a strong indication that environmental pH is not the dominant physiological challenge in this murine model.

With reference to Color Plate 8A, and considering the naturally high pH tolerance demonstrated by *Aspergillus* species, one can comfortably conclude that pH alteration, over the time scale required to kill fungal spores and/or hyphae, could not directly account for killing. A rather broader consideration of conditions pertaining to the vacuole must be undertaken. Returning to the Reeves et al. neutrophil study, amounts of superoxide are predicted to be enormous, in the region of 1 to 4 M, and concentrations of granule proteins are very high, at an estimated 500 mg/ml. pH is between 7.4 and 8.0, and the anionic charge inside the vacuole is offset by massive influx of potassium ions through the vacuolar membrane, rendering the vesicle grossly hypertonic. Analysis of combinatorial stress on *A. fumigatus* survival has been minimal, although lithium and potassium ions can be toxic to *Aspergillus* at high concentrations and alkaline pH (O. Loss, personal communication).

The pH of the phagosome is believed to be important for appropriate intracellular trafficking and is a means by which some microbes can subvert host attack. *Histoplasma capsulatum* induces sustained local pH change within late-stage phagolysosomes, altering the pH of the phagocytic vacuole as a survival strategy (Eissenberg et al., 1993). This presumably allows the yeast cells to avoid killing by lysosomal acid hydrolases and still acquire iron from transferrin, which would be half-saturated at pH 6.5 (Newman et al., 2006). There is no evidence to support any such hijacking of intracellular trafficking mechanisms following *A. fumigatus* phagocytosis; indeed, microscopy using immunolabeled markers reveals that *A. fumigatus* phagolysosomes fuse with both early and late endosomes, culminating in lysosomal acidification, an absolute requirement for killing (Ibrahim-Granet et al., 2003). Increased phagolysosomal fusion has been observed in human monocyte-derived macrophages following infection with the polyketide synthase mutant Δ*pksP*, however, so modulation of intracellular trafficking can be induced genetically, and this can be extrapolated to enhanced killing, which is significant given that variations on this experimental theme might ultimately have therapeutic significance.

CALCIUM

The divalent calcium cation can rapidly bind to, and precipitate, both organic and inorganic anions. The insolubility of the resulting calcium salts is thought, in evolutionary terms, to have driven rejection of calcium from the cellular cytoplasm when life became land based and eukaryotic cells evolved some 2 million years ago (Williams, 2006). High intracellular calcium is incompatible with life, and significant increases in cytoplasmic calcium are cytotoxic to all cells. Calcium is therefore another tightly regulated ion within the intracellular environment, and the steep concentration gradient between extracellular and intracellular compartments requires constant maintenance. This gradient, coupled with the ability of calcium to interact with many biological molecules, underlies an evolved cellular signaling mechanism of incredible sophistication.

Eukaryotic cells regulate a variety of cellular processes using calcium-elicited signals, including cell cycle progression, cation homeostasis, and morphogenesis. Cells have evolved mechanisms for maintaining low levels of calcium in the cytoplasm, including regulated entry, active removal, and intracellular sequestration to the vacuole or endoplasmic reticulum (Clapham, 2007). Calcium-mediated signaling via calmodulin and calcineurin is critical for the regulation of stress responses in model and pathogenic fungi (Kraus and Heitman, 2003), and exposure of *S. cerevisiae* to a number of extracellular stresses triggers a rise in cytoplasmic calcium. Transient elevation of cytoplasmic calcium concentrations can result from influxes of extracellular calcium or from release of calcium from internal stores, such as the vacuole, and results in the activation of calcineurin. Calcineurin is a heterodimer of catalytic and regulatory subunits which, in *S. cerevisiae*, directs the phosphorylation-dependent activation of the transcription factor Crz1p, permitting nuclear entry and activation of stress-responsive genes via binding to calcineurin-dependent responsive elements in the promoters of relevant genes (Cyert, 2003). From a therapeutic standpoint calcineurin is interesting because calcineurin inhibitors demonstrate fungicidal synergy with azole antifungals (Heitman, 2005; Steinbach et al., 2007).

Deletion of the *A. fumigatus* calcineurin catalytic subunit, CnaA, is achievable (da Silva Ferreira et al., 2006, 2007; Steinbach et al., 2006) but highly deleterious to the cell, preventing growth and filamentation. Δ*cnaA* hyphae are compact, dense, and extremely blunted, and accordingly an almost complete attenuation of virulence in neutropenic mice is observed.

At the time of writing no detailed analyses of calcium signaling in *A. fumigatus* have been published. Current annotation of the *A. fumigatus* genome suggests the presence of multiple channels and transporters of likely relevance, but to date only a single calcium ATPase has been cloned and characterized to any degree (Soriani et al., 2005). Concerned as this chapter is with immune defenses against *A. fumigatus*, the most relevant considerations of calcium signaling relate to stress adaptation. With respect to *A. fumigatus* virulence, distinct roles for calcineurin-mediated stress adaptation and morphogenesis are discernable mutationally through deletion of CrzA, the homolog of *S. cerevisiae* Crz1p (Soriani et al., 2008). *crz*A deletion in *A. fumigatus* leads to reductions in ion tolerance, in the absence of any drastic effect on growth. In this genetic background a possible role for calcineurin-dependent stress adaptation in virulence is discernable with reduced virulence, relative to a parental isolate, in neutropenic mice.

Calcium Changes within Neutrophil Phagosomes

In neutrophils, changes in cytosolic calcium concentration play an important role in cell activation. A rise in intracellular calcium is not required for uptake of microorganisms through phagocytosis, but it is necessary for killing of ingested prey (Wilsson et al., 1996). This requirement has been attributed to assembly of the NADPH-oxidase complex and the docking and fusion process leading to phagolysosome formation. Local rises in cytoplasmic calcium in the periphagosomal area have been attributed to either release of calcium from intracellular stores or an opening of channels in the plasma membrane, or both. Based on the fact that the content of the phagosome constitutes a part of the previously extracellular milieu (having a calcium concentration above 1 mM), a plausible hypothesis is one which sees calcium released from the phagosome and into the cytoplasm to create the local calcium signal required for phagosome processing and phagolysosomal fusion. Intraphagosomal calcium measurement using fura-2 zymosan particles (Lundqvist-Gustafsson et al., 2000) identified a lowering of phagosomal calcium to levels matching the periphagosomal area as an early event in phagosome processing, and those authors reported a decrease in calcium concentration from millimolar to nanomolar levels, occurring within a few minutes of closure of plasma membrane invagination.

Studies of *Salmonella enterica* serovar Typhimurium (Garcia Vescovi et al., 1997) suggest that the phagosome is a calcium-limited environment, and in *H. capsulatum* major differences in calcium dependence of yeast and mycelial growth forms are observed with respect to concentrations required to support growth. Yeast forms of *H. capsulatum* are capable of growing under calcium deprivation, secreting a 7.8-kDa calcium-binding protein which is not produced by mycelia and is essential for virulence. Calcium limitation within the phagosome can therefore be overcome by yeast forms of the organism, facilitating intracellular survival and pathogenesis (Sebghati et al., 2000).

ROIs, pH, AND CALCIUM

With respect to host defense against *A. fumigatus*, ROIs, pH, and calcium serve differential, and likely complementary, roles on the part of both the host and the pathogen. From a host perspective, the respiratory burst is of primary importance, although the precise relevance of ROIs within the phagocytic vacuole remains to be determined. pH and calcium fluxes are further parameters which can crucially affect manifestation of an adequate host defense.

It is important to qualify the rationale behind studies of microbicidal activity carefully. Healthy alveolar macrophages and PMNs kill *A. fumigatus* spores and hyphae highly efficiently; therefore, *A. fumigatus* gene deletions resulting in heightened susceptibility to killing merely represent a further weakening of a somewhat defenseless microbe. From a therapeutic viewpoint, loss of function in such genes may disable *A. fumigatus* sufficiently to allow residual immunity to cope with spores and hyphae. Presumably during invasive disease, surmountable host defenses are the relevant challenge, so relevant studies of therapeutic value might also focus upon factors weakening *A. fumigatus* under compromised host defense conditions. Murine modeling does this routinely through immunosuppression, but cellular assays, for the most part, have not. A further consideration is the effect of combinatorial stress on survival of *A. fumigatus*; for example, in *S. cerevisiae* calcium is required for growth under conditions of hypertonic shock. Under circumstances in which *A. fumigatus* escapes from the phagocytic vacuole, appropriate adaptation to the impaired phagolysosomal environment is an important aspect of survival in the host. The field is poised to interrogate microarray analyses of *A. fumigatus* stress adaptation, in in vitro, ex vivo, and in vivo studies, and these promise to illuminate both host and fungal mechanisms contributing to survival and killing. To underestimate the importance of phagocyte studies would be perilous, given those first findings of Schaffner et. al. in distinguishing between opportunistic and primary fungal pathogens. Such studies raise the menacing, but formal, possibility that primary pathogenesis may lie only a few evolutionary steps away for *A. fumigatus*, greatly reinforcing the importance of mastering the dynamics of this host-pathogen interaction.

REFERENCES

Appelberg, R. 2007. Neutrophils and intracellular pathogens: beyond phagocytosis and killing. *Trends Microbiol.* **15:**87–92.

Aratani, Y., F. Kura, H. Watanabe, H. Akagawa, Y. Takano, K. Suzuki, M. C. Dinauer, N. Maeda, and H. Koyama. 2002. Relative contributions of myeloperoxidase and NADPH-oxidase to the early host defense against pulmonary infections with Candida albicans and Aspergillus fumigatus. *Med. Mycol.* **40:**557–563.

Arst, H. N., and M. A. Penalva. 2003. pH regulation in Aspergillus and parallels with higher eukaryotic regulatory systems. *Trends Genet.* **19:**224–231.

Babior, B. M. 2004. NADPH oxidase. *Curr. Opin. Immunol.* **16:**42–47.

Babior, B. M., R. S. Kipnes, and J. T. Curnutte. 1973. Biological defense mechanisms. The production by leukocytes of superoxide, a potential bactericidal agent. *J. Clin. Investig.* **52:**741–744.

Bignell, E., S. Negrete-Urtasun, A. M. Calcagno, H. N. Arst, Jr., T. Rogers, and K. Haynes. 2005. Virulence comparisons of *Aspergillus nidulans* mutants are confounded by the inflammatory response of $p47^{phox-/-}$ mice. *Infect. Immun.* **73:**5204–5207.

Bignell, E., S. Negrete-Urtasun, A. M. Calcagno, K. Haynes, H. N. Arst, Jr., and T. Rogers. 2005. The Aspergillus pH-responsive transcription factor PacC regulates virulence. *Mol. Microbiol.* **55:**1072–1084.

Bok, J. W., D. Chung, S. A. Balajee, K. A. Marr, D. Andes, K. F. Nielsen, J. C. Frisvad, K. A. Kirby, and N. P. Keller. 2006. GliZ, a transcriptional regulator of gliotoxin biosynthesis, contributes to *Aspergillus fumigatus* virulence. *Infect. Immun.* **74:**6761–6768.

Borregaard, N., O. E. Sorensen, and K. Theilgaard-Monch. 2007. Neutrophil granules: a library of innate immunity proteins. *Trends Immunol.* **28:**340–345.

Boysen, J. H., and A. P. Mitchell. 2006. Control of Bro1-domain protein Rim20 localization by external pH, ESCRT machinery, and the Saccharomyces cerevisiae Rim101 pathway. *Mol. Biol. Cell* **17:** 1344–1353.

Caddick, M. X., A. G. Brownlee, and H. N. Arst, Jr. 1986. Regulation of gene expression by pH of the growth medium in Aspergillus nidulans. *Mol. Gen. Genet.* **203:**346–353.

Calcagno-Pizarelli, A. M., S. Negrete-Urtasun, S. H. Denison, J. D. Rudnicka, H. J. Bussink, T. Munera-Huertas, L. Stanton, A. Hervas-Aguilar, E. A. Espeso, J. Tilburn, H. N. Arst, Jr., and M. A. Penalva. 2007. Establishment of the ambient pH signaling complex in *Aspergillus nidulans*: PalI assists plasma membrane localization of PalH. *Eukaryot. Cell* **6:**2365–2375.

Chang, Y. C., B. H. Segal, S. M. Holland, G. F. Miller, and K. J. Kwon-Chung. 1998. Virulence of catalase-deficient Aspergillus nidulans in $p47^{phox-/-}$ mice. Implications for fungal pathogenicity and host defense in chronic granulomatous disease. *J. Clin. Investig.* **101:**1843–1850.

Clapham, D. E. 2007. Calcium signaling. *Cell* **131:**1047–1058.

Cramer, R. A., Jr., M. P. Gamcsik, R. M. Brooking, L. K. Najvar, W. R. Kirkpatrick, T. F. Patterson, C. J. Balibar, J. R. Graybill, J. R. Perfect, S. N. Abraham, and W. J. Steinbach. 2006. Disruption of a nonribosomal peptide synthetase in *Aspergillus fumigatus* eliminates gliotoxin production. *Eukaryot. Cell* **5:**972–980.

Curnutte, J. T. 2004. Superoxide production by phagocytic leukocytes: the scientific legacy of Bernard Babior. *J. Clin. Investig.* **114:** 1054–1057.

Curnutte, J. T., D. M. Whitten, and B. M. Babior. 1974. Defective superoxide production by granulocytes from patients with chronic granulomatous disease. *N. Engl. J. Med.* **290:**593–597.

Cyert, M. S. 2003. Calcineurin signalling in Saccharomyces cerevisiae: how yeast go crazy in response to stress. *Biochem. Biophys. Res. Commun.* **311:**1143–1150.

da Silva Ferreira, M. E., T. Heinekamp, A. Hartl, A. A. Brakhage, C. P. Semighini, S. D. Harris, M. Savoldi, P. F. de Gouvea, M. H. de Souza Goldman, and G. H. Goldman. 2007. Functional characterization of the Aspergillus fumigatus calcineurin. *Fungal Genet. Biol.* **44:**219–230.

da Silva Ferreira, M. E., M. R. Kress, M. Savoldi, M. H. Goldman, A. Hartl, T. Heinekamp, A. A. Brakhage, and G. H. Goldman. 2006. The *akuB*^{KU80} mutant deficient for nonhomologous end joining is a powerful tool for analyzing pathogenicity in *Aspergillus fumigatus. Eukaryot. Cell* 5:207–211.

Diamond, R. D., and R. A. Clark. 1982. Damage to *Aspergillus fumigatus* and *Rhizopus oryzae* hyphae by oxidative and nonoxidative microbicidal products of human neutrophils in vitro. *Infect. Immun.* 38:487–495.

Diez, E., J. Alvaro, E. A. Espeso, L. Rainbow, T. Suarez, J. Tilburn, H. N. Arst, Jr., and M. A. Penalva. 2002. Activation of the Aspergillus PacC zinc finger transcription factor requires two proteolytic steps. *EMBO J.* 21:1350–1359.

Dri, P., G. Presani, S. Perticarari, L. Alberi, M. Prodan, and E. Decleva. 2002. Measurement of phagosomal pH of normal and CGD-like human neutrophils by dual fluorescence flow cytometry. *Cytometry* 48:159–166.

Eissenberg, L. G., W. E. Goldman, and P. H. Schlesinger. 1993. Histoplasma capsulatum modulates the acidification of phagolysosomes. *J. Exp. Med.* 177:1605–1611.

Fernandez-Martinez, J., C. V. Brown, E. Diez, J. Tilburn, H. N. Arst, Jr., M. A. Penalva, and E. A. Espeso. 2003. Overlap of nuclear localisation signal and specific DNA-binding residues within the zinc finger domain of PacC. *J. Mol. Biol.* 334:667–684.

Franklin, R. A., O. G. Rodriguez-Mora, M. M. Lahair, and J. A. McCubrey. 2006. Activation of the calcium/calmodulin-dependent protein kinases as a consequence of oxidative stress. *Antioxid. Redox Signal.* 8:1807–1817.

Greene, V., H. Cao, F. A. Schanne, and D. C. Bartelt. 2002. Oxidative stress-induced calcium signalling in Aspergillus nidulans. *Cell. Signal.* 14:437–443.

Hamilton, A. J., and M. D. Holdom. 1999. Antioxidant systems in the pathogenic fungi of man and their role in virulence. *Med. Mycol.* 37:375–389.

Heitman, J. 2005. Cell biology. A fungal Achilles' heel. *Science* 309: 2175–2176.

Herranz, S., J. M. Rodriguez, H. J. Bussink, J. C. Sanchez-Ferrero, H. N. Arst, Jr., M. A. Penalva, and O. Vincent. 2005. Arrestin-related proteins mediate pH signaling in fungi. *Proc. Natl. Acad. Sci. USA* 102:12141–12146.

Hervás-Aguilar, A., J. M. Rodríguez, J. Tilburn, H. N. Arst, Jr., and M. A. Peñalva. 2007. Evidence for the direct involvement of the proteasome in the proteolytic processing of the Aspergillus nidulans zinc finger transcription factor PacC. *J. Biol. Chem.* 282:34735–34747.

Ibrahim-Granet, O., B. Philippe, H. Boleti, E. Boisvieux-Ulrich, D. Grenet, M. Stern, and J. P. Latgé. 2003. Phagocytosis and intracellular fate of *Aspergillus fumigatus* conidia in alveolar macrophages. *Infect. Immun.* 71:891–903.

Jahn, B., A. Koch, A. Schmidt, G. Wanner, H. Gehringer, S. Bhakdi, and A. A. Brakhage. 1997. Isolation and characterization of a pigmentless-conidium mutant of *Aspergillus fumigatus* with altered conidial surface and reduced virulence. *Infect. Immun.* 65:5110–5117.

Jahn, B., K. Langfelder, U. Schneider, C. Schindel, and A. A. Brakhage. 2002. PKSP-dependent reduction of phagolysosome fusion and intracellular kill of Aspergillus fumigatus conidia by human monocyte-derived macrophages. *Cell. Microbiol.* 4:793–803.

Kraus, P. R., and J. Heitman. 2003. Coping with stress: calmodulin and calcineurin in model and pathogenic fungi. *Biochem. Biophys. Res. Commun.* 311:1151–1157.

Kupfahl, C., T. Heinekamp, G. Geginat, T. Ruppert, A. Hartl, H. Hof, and A. A. Brakhage. 2006. Deletion of the gliP gene of Aspergillus fumigatus results in loss of gliotoxin production but has no effect on virulence of the fungus in a low-dose mouse infection model. *Mol. Microbiol.* 62:292–302.

Lamarre, C., O. Ibrahim-Granet, C. Du, R. Calderone, and J. P. Latgé. 2007. Characterization of the SKN7 ortholog of Aspergillus fumigatus. *Fungal Genet. Biol.* 44:682–690.

Lehrer, R. I., and M. J. Cline. 1969. Leukocyte myeloperoxidase deficiency and disseminated candidiasis: the role of myeloperoxidase in resistance to Candida infection. *J. Clin. Investig.* 48:1478–1488.

Lessing, F., O. Kniemeyer, I. Wozniok, J. Loeffler, O. Kurzai, A. Haertl, and A. A. Brakhage. 2007. The *Aspergillus fumigatus* transcriptional regulator AfYap1 represents the major regulator for defense against reactive oxygen intermediates but is dispensable for pathogenicity in an intranasal mouse infection model. *Eukaryot. Cell* 6:2290–2302.

Levitz, S. M., M. E. Selsted, T. Ganz, R. I. Lehrer, and R. D. Diamond. 1986. In vitro killing of spores and hyphae of Aspergillus fumigatus and Rhizopus oryzae by rabbit neutrophil cationic peptides and bronchoalveolar macrophages. *J. Infect. Dis.* 154:483–489.

Lundqvist-Gustafsson, H., M. Gustafsson, and C. Dahlgren. 2000. Dynamic Ca^{2+} changes in neutrophil phagosomes: a source for intracellular Ca^{2+} during phagolysosome formation? *Cell. Calcium* 27:353–362.

Maubon, D., S. Park, M. Tanguy, M. Huerre, C. Schmitt, M. C. Prevost, D. S. Perlin, J. P. Latgé, and A. Beauvais. 2006. AGS3, an α(1-3)glucan synthase gene family member of Aspergillus fumigatus, modulates mycelium growth in the lung of experientally infected mice. *Fungal Genet. Biol.* 43:366–375.

McDonagh, A., N. D. Fedorova, J. Crabtree, J., Y. Yu, S. Kim, D. Chen, O. Loss, T. Cairns, G. H. Goldman, D. Armstrong-James, K. Haynes, H. Haas, M. Schrettl, G. May, W. C. Nierman, and E. Bignell. 2008. Sub-telomere directed gene expression during initiation of invasive aspergillosis. *PLoS Pathogens* September.

Mittler, R. 2002. Oxidative stress, antioxidants and stress tolerance. *Trends Plant Sci.* 7:405–410.

Negrete-Urtasun, S., W. Reiter, E. Diez, S. H. Denison, J. Tilburn, E. A. Espeso, M. A. Penalva, and H. N. Arst, Jr. 1999. Ambient pH signal transduction in Aspergillus: completion of gene characterization. *Mol. Microbiol.* 33:994–1003.

Newman, S. L., L. Gootee, J. Hilty, and R. E. Morris. 2006. Human macrophages do not require phagosome acidification to mediate fungistatic/fungicidal activity against Histoplasma capsulatum. *J. Immunol.* 176:1806–1813.

Paris, S., D. Wysong, J. P. Debeaupuis, K. Shibuya, B. Philippe, R. D. Diamond, and J. P. Latgé. 2003. Catalases of *Aspergillus fumigatus. Infect. Immun.* 71:3551–3562.

Peñalva, M. A., J. Tilburn, E. Bignell, and H. N. Arst, Jr. 2008. Ambient pH gene regulation in fungi: making connections. *Trends Microbiol.* 16:291–300.

Perkhofer, S., C. Speth, M. P. Dierich, and C. Lass-Florl. 2007. In vitro determination of phagocytosis and intracellular killing of Aspergillus species by mononuclear phagocytes. *Mycopathologia* 163: 303–307.

Philippe, B., O. Ibrahim-Granet, M. C. Prevost, M. A. Gougerot-Pocidalo, M. Sanchez Perez, A. Van der Meeren, and J. P. Latgé. 2003. Killing of *Aspergillus fumigatus* by alveolar macrophages is mediated by reactive oxidant intermediates. *Infect. Immun.* 71: 3034–3042.

Reeves, E. P., H. Lu, H. L. Jacobs, C. G. Messina, S. Bolsover, G. Gabella, E. O. Potma, A. Warley, J. Roes, and A. W. Segal. 2002. Killing activity of neutrophils is mediated through activation of proteases by K^+ flux. *Nature* 416:291–297.

Sbarra, A. J., and M. L. Karnovsky. 1959. The biochemical basis of phagocytosis. I. Metabolic changes during the ingestion of particles by polymorphonuclear leukocytes. *J. Biol. Chem.* 234:1355–1362.

Schaffner, A., C. E. Davis, T. Schaffner, M. Markert, H. Douglas, and A. I. Braude. 1986. In vitro susceptibility of fungi to killing by neutrophil granulocytes discriminates between primary pathogenicity and opportunism. *J. Clin. Investig.* 78:511–524.

Schaffner, A., H. Douglas, and A. Braude. 1982. Selective protection against conidia by mononuclear and against mycelia by polymorphonuclear phagocytes in resistance to Aspergillus. Observations on these two lines of defense in vivo and in vitro with human and mouse phagocytes. *J. Clin. Investig.* **69**:617–631.

Sebghati, T. S., J. T. Engle, and W. E. Goldman. 2000. Intracellular parasitism by Histoplasma capsulatum: fungal virulence and calcium dependence. *Science* **290**:1368–1372.

Segal, A. W., M. Geisow, R. Garcia, A. Harper, and R. Miller. 1981. The respiratory burst of phagocytic cells is associated with a rise in vacuolar pH. *Nature* **290**:406–409.

Segal, B. H., T. L. Leto, J. I. Gallin, H. L. Malech, and S. M. Holland. 2000. Genetic, biochemical, and clinical features of chronic granulomatous disease. *Medicine* (Baltimore) **79**:170–200.

Serrano, R. 1988. Structure and function of proton translocating ATPase in plasma membranes of plants and fungi. *Biochim. Biophys. Acta* **947**:1–28.

Sorensen, O. E. 1909. Enzymstudien. II: Mitteilung. Über die Messung und die Bedeutung der Wasserstoffionenkoncentration bei enzymatischen Prozessen. *Biochem. Zeitsch.* **1909**:131–304.

Soriani, F. M., I. Malavazi, M. E. da Silva Ferreira, M. Savoldi, M. R. Von Zeska Kress, M. H. de Souza Goldman, O. Loss, E. Bignell, and G. H. Goldman. 2008. Functional characterization of the *Aspergillus fumigatus* CRZ1 homologue, CrzA. *Mol. Micro.* **67**:1274–1291.

Soriani, F. M., V. P. Martins, T. Magnani, V. G. Tudella, C. Curti, and S. A. Uyemura. 2005. A PMR1-like calcium ATPase of Aspergillus fumigatus: cloning, identification and functional expression in S. cerevisiae. *Yeast* **22**:813–824.

Spikes, S., R. Xu, C. K. Nguyen, G. Chamilos, D. P. Kontoyiannis, R. H. Jacobson, D. E. Ejzykowicz, L. Y. Chiang, S. G. Filler, and G. S. May. 2008. Gliotoxin production in Aspergillus fumigatus contributes to host-specific differences in virulence. *J. Infect. Dis.* **197**:479–486.

Steinbach, W. J., R. A. Cramer, Jr., B. Z. Perfect, Y. G. Asfaw, T. C. Sauer, L. K. Najvar, W. R. Kirkpatrick, T. F. Patterson, D. K. Benjamin, Jr., J. Heitman, and J. R. Perfect. 2006. Calcineurin controls growth, morphology, and pathogenicity in *Aspergillus fumigatus. Eukaryot. Cell* **5**:1091–1103.

Steinbach, W. J., J. L. Reedy, R. A. Cramer, Jr., J. R. Perfect, and J. Heitman. 2007. Harnessing calcineurin as a novel anti-infective agent against invasive fungal infections. *Nat. Rev. Microbiol.* **5**:418–430.

Takemoto, D., A. Tanaka, and B. Scott. 2007. NADPH oxidases in fungi: diverse roles of reactive oxygen species in fungal cellular differentiation. *Fungal Genet. Biol.* **44**:1065–1076.

Tang, C. M., J. M. Smith, H. N. Arst, Jr., and D. W. Holden. 1994. Virulence studies of *Aspergillus nidulans* mutants requiring lysine or *p*-aminobenzoic acid in invasive pulmonary aspergillosis. *Infect. Immun.* **62**:5255–5260.

Tilburn, J., S. Sarkar, D. A. Widdick, E. A. Espeso, M. Orejas, J. Mungroo, M. A. Penalva, and H. N. Arst, Jr. 1995. The Aspergillus PacC zinc finger transcription factor mediates regulation of both acid- and alkaline-expressed genes by ambient pH. *EMBO J.* **14**: 779–790.

Tkalcevic, J., M. Novelli, M. Phylactides, J. P. Iredale, A. W. Segal, and J. Roes. 2000. Impaired immunity and enhanced resistance to endotoxin in the absence of neutrophil elastase and cathepsin G. *Immunity* **12**:201–210.

Todd, R. B., M. A. Davis, and M. J. Hynes. 2007. Genetic manipulation of Aspergillus nidulans: heterokaryons and diploids for dominance, complementation and haploidization analyses. *Nat. Protoc.* **2**:822–830.

Vescovi, E. G., Y. Ayala, E. A. Di Cera, and E. A. Groisman. 1997. Characterisation of the bacterial sensor protein PhoQ. *J. Biol. Chem.* **272**:1440–1443.

Vincent, O., L. Rainbow, J. Tilburn, H. N. Arst, Jr., and M. A. Penalva. 2003. YPXL/I is a protein interaction motif recognized by *Aspergillus* PalA and its human homologue, AIP1/Alix. *Mol. Cell. Biol.* **23**:1647–1655.

Waldorf, A. R., S. M. Levitz, and R. D. Diamond. 1984. In vivo bronchoalveolar macrophage defense against Rhizopus oryzae and Aspergillus fumigatus. *J. Infect. Dis.* **150**:752–760.

Williams, R. J. 2006. The evolution of calcium biochemistry. *Biochim. Biophys. Acta* **1763**:1139–1146.

Williams, R. L., and S. Urbe. 2007. The emerging shape of the ESCRT machinery. *Nat. Rev. Mol. Cell Biol.* **8**:355–368.

Wilsson, A., H. Lundqvist, M. Gustafsson, and O. Stendahl. 1996. Killing of phagocytosed Staphylococcus aureus by human neutrophils requires intracellular free calcium. *J. Leukoc. Biol.* **59**:902–907.

Yu, B. P. 1994. Cellular defenses against damage from reactive oxygen species. *Physiol. Rev.* **74**:139–162.

Aspergillus fumigatus and Aspergillosis
Edited by J.-P. Latgé and W. J. Steinbach

Chapter 18

Innate Defense against *Aspergillus*: the Phagocyte

MICHEL CHIGNARD

PHAGOCYTES

Macrophages and neutrophils are essential components of the innate immune response of the host to infection by microorganisms (Greenberg and Grinstein, 2002; Aderem, 2003). Macrophages have three main functions: (i) to take up and kill pathogens by phagocytosis (Aderem and Underhill, 1999; Greenberg and Grinstein, 2002; Stuart and Ezekowitz, 2005); (ii) to generate a large array of biologically active molecules, including cytokines, chemokines, and lipid mediators (Streiter et al., 2002), that orchestrate the recruitment of other phagocytes, such as monocytes and neutrophils; and (iii) to present antigens to lymphocytes. Neutrophils also play a key role in the innate immune system, as demonstrated with neutropenia (such as that occurring after chemotherapy), which increases susceptibility to infections (Witko-Sarsat et al., 2000; Nathan, 2006). Neutrophils are ideally suited to the elimination of pathogenic microbes, as they have large stores of proteolytic enzymes and express NADPH oxidase, which rapidly produces reactive oxygen species (ROS) that degrade and kill pathogens (Shepherd, 1986; Sibille and Reynolds, 1990; Hampton et al., 1998). Patients with chronic granulomatous disease (CGD) clearly illustrate the importance of NADPH oxidase and ROS in the process of microbial killing (Umeki, 1994; Roos et al., 2003; El-Benna et al., 2005). These individuals have defects in the NADPH oxidase enzyme complex and are highly susceptible to infections caused by bacteria and fungi, including *Aspergillus fumigatus*. The oxygen-independent mechanisms involve various proteins, including in particular three serine proteinases, elastase, cathepsin G, and proteinase 3 (Hiemstra, 2002; Pham, 2006; Wiedow and Meyer-Hoffert, 2005; Korkmaz et al., 2008), that have well-characterized antimicrobial activities, particularly against bacteria. There are currently grounds for suggesting that granule proteinases play a major role in the killing process, whereas ROS facilitate this killing but are not essential (Segal, 2005).

However, neutrophils are a double-edged sword in that they play important roles in both host defense and inflammation, which has long been recognized to be deleterious, particularly when it concerns the lung (Weiss, 1989; Smith, 1994). During the activation process, neutrophils degranulate, releasing the contents of their granules and producing ROS, both of which are released into the extracellular medium (Pham, 2006). These toxic products damage host tissues and are responsible for the detrimental effects of neutrophils (Moraes et al., 2006).

Activated neutrophils were recently shown to extrude a web of fibers composed of chromatin and DNA, referred to as neutrophil extracellular traps (NETs). NETs are formed during a novel ROS-dependent death program in neutrophils (Fuchs et al., 2007). These structures provide a high local concentration of antimicrobial components that bind and kill bacteria extracellularly. Elastase is present in NETs and probably plays a key role, together with defensins, histones (Brinkmann et al., 2004; Brinkmann and Zychlinsky, 2007), and the long pentraxin 3 (PTX3) (Jaillon et al., 2007). It has even been suggested that the delivery of neutrophil proteinases into NETs may prevent them from diffusing away and damaging nearby tissue. NETs may capture and kill *A. fumigatus*, as it has been shown that NETs kill both the yeast form and hyphal cells of *Candida albicans* (Urban et al., 2006).

THE INNATE IMMUNE SYSTEM

The innate immune system is the first line of host defense against pathogens (Medzhitov and Janeway, 2000; Akira et al., 2006). Pathogen recognition is based on a limited repertoire of germ line-encoded receptors

Michel Chignard • Unité de Défense innée et Inflammation, Institut Pasteur, and Unité 874, INSERM, 75015 Paris, France.

called pattern recognition receptors (PRR), of two types, expressed on the surface of certain cells or soluble in biological fluids. PRR recognize pathogen-associated molecular patterns (PAMP), highly invariant structures shared by large groups of microorganisms and essential for their survival or pathogenicity (Medzhitov and Janeway, 1997; Medzhitov, 2007). The archetypal examples of PAMP are lipopolysaccharides from gram-negative bacteria, lipopeptides from gram-positive bacteria, double-stranded RNA from viruses, and glucans from fungi.

Cellular PRR belong to various functional and structural groups, including Toll-like receptors (TLR), nucleotide-binding oligomerization domain proteins (NOD), NOD-like receptors, scavenger receptors, lectin receptors (such as dectin-1), the double-stranded RNA helicases, and G-protein-coupled receptors for formyl peptides (Gordon, 2002; Werts et al., 2006; Martinon et al., 2007). Soluble PRR are also diverse and include collectins (mannose-binding lectin, surfactant proteins, and C1q), ficolin, and PTX (Romani, 2004).

Macrophages and neutrophils express many of the receptors of the innate immune system for pathogen recognition and the triggering of defense mechanisms (Taylor et al., 2005). These receptors are involved in the phagocytosis process or in inducing the synthesis of enzymes (e.g., Cox2, nitric oxide [NO] synthase), mediators (e.g., cytokines, chemokines), and adhesion molecules (e.g., integrins, selectins), all of which contribute to the eradication of invasive pathogens. Indeed, macrophages (Smith et al., 2005) and neutrophils (Burg and Pillinger, 2001; Kobayashi et al., 2005) constitute the bulwark of the innate immune system.

ASPERGILLOSIS AND PHAGOCYTES

A. fumigatus produces abundant small conidia that readily form aerosols. These airborne conidia are small enough to reach the alveoli during breathing, where alveolar macrophages engulf and destroy them with or without the assistance of recruited neutrophils. However, in immunocompromised patients, inhaled conidia germinate to produce hyphae that invade the parenchyma and, possibly, other tissues through hematogenous dissemination. The development of these hyphae leads to the tissue destruction and respiratory failure typical of invasive pulmonary aspergillosis (IPA). Neutropenia is the most important risk factor, closely followed by transplantation, particularly lung and bone marrow transplantation (Segal and Walsh, 2006). The importance of transplantation as a risk factor is related to various immune defects, including neutropenia induced during the preengraftment phase and the use of antirejection treatments, such as corticosteroids.

Macrophages and Phagocytosis

One of the first prominent demonstrations of the role of macrophages in IPA was published in 1982 by Schaffner et al. However, the very first study listed in PubMed describing such a role in an in vivo model of infection was published more than a decade earlier (Gernez-Rieux et al., 1967). The role of neutrophils was recognized a few years later (Olenchock et al., 1979). In their study, Schaffner et al. (1982) brilliantly described a first line of defense based on macrophages and directed against spores and a second line of defense involving the neutrophil protecting against the hyphal form of *A. fumigatus*. The authors concluded that, "The host, thus, can call upon two independent phagocytic cell lines that form graded defense systems against *Aspergillus*. These lines of defense function in the absence of a specific immune response, which seems superfluous in the control and elimination of this fungus." They described a very efficient "military" organization of host defenses, with macrophages in place on the battleground and responsible for destroying the invading enemies, the conidia. If the macrophages are outnumbered (Philippe et al., 2003), or unsuccessful for any other reason, their enemies have time to deploy strategic weapons, in this case by allowing the conidia to germinate and produce hyphae. The macrophages then send messages (inflammatory mediators) to call for reinforcements in the form of neutrophils. These new "troops" are armed with specific deadly weapons (proteases and ROS) for the precise destruction of hyphae.

When conidia arrive in the airways, the initial response is thus the phagocytosis and intracellular killing of conidia by alveolar macrophages. Using alveolar macrophages from p47*phox*$^{-/-}$ and iNOS$^{-/-}$ mice, Philippe et al. (2003) clearly showed that the killing of conidia is not dependent on NO production, confirming results obtained in a previous study (Michaliszyn et al., 1995). Instead, the killing of conidia is specifically associated with ROS synthesis. However, this mechanism is not consistent with the findings of Morgenstern et al. (1997), who found, with another type of targeted gene disruption of the NADPH oxidase complex (working with mice with X-linked CGD generated by disruption of the gp91^{phox} subunit of the NADPH oxidase [X-CGD mice]), that alveolar macrophages from wild-type and X-CGD mice killed conidia equally efficiently. According to Philippe et al. (2003), the rate of killing was almost eight times higher for the wild-type than for p47*phox*$^{-/-}$ macrophages, whereas phagocytosis was similarly effective in both types of macrophages. Alveolar macrophages from human CGD patients are also unable to kill conidia (B. Philippe et al., unpublished data). Following engulfment, the phagocytic process follows a classical scheme, with fusion of the phagosome with early and late endosomes and maturation of the phagolysosome (Ibrahim-Granet et al., 2003). Molecules other than ROS within the phagolysosome, such as cationic peptides (Levitz et al., 1986), proteases (Rodriguez et al., 1997), and chitinases (Boot et al., 1995), may also play a role. Once engulfed, the conidia swell within the phagosome, this step being an apparent prerequisite for the killing of conidia (Levitz et al., 1986; Philippe et al. 2003). Swelling always precedes the germination of conidia, but none of the conidia subjected to phagocytosis produces germ tubes (Philippe et al., 2003), illustrating the efficacy of this mechanism. By contrast, alveolar macrophages from cortisone acetate-treated mice display impaired killing, with the production of germ tubes within the phagolysosome, ending in an outgrowth that disrupts the membrane and results in the death of the alveolar macrophage (Waldorf et al., 1984; Philippe et al., 2003). This result is consistent with the known inhibitory effect of corticosteroids on ROS production (Russo-Marie, 1992; Dandona et al., 1999).

Macrophages internalize microorganisms via different types of receptors expressed on their surface. These receptors bind PAMP directly or via opsonins. *A. fumigatus* is recognized directly, via its carbohydrates, by DC-SIGN and dectin-1. Only one group has reported the role of DC-SIGN in the binding and internalization of conidia (Serrano-Gomez et al., 2004), whereas the involvement of dectin-1, a well-known β-1,3-glucan receptor, has attracted more attention. Nonetheless, only one report has clearly demonstrated a role for dectin-1 in the phagocytosis of *A. fumigatus* (Luther et al., 2007). By contrast, a larger number of reports have provided evidence of a well-characterized role of dectin-1 in the induction of an inflammatory response in terms of cytokine and chemokine production (Hohl et al., 2005; Steele et al., 2005; Gersuk et al., 2006). All reports to date have indicated that dectin-1 does not recognize resting conidia but instead recognizes swollen and germinating conidia, consistent with the presence of β-1,3-glucan on the surface of swollen and germinating conidia. The other important finding is the dependence of inflammatory responses on both dectin-1 and TLR2 signaling, the two pathways being additive, at least for tumor necrosis factor alpha (TNF-α) synthesis (Underhill, 2007). PTX3 functions as an opsonin for the dectin-1-dependent internalization of zymosan by macrophages (Diniz et al., 2004). This function is probably involved in the phagocytosis of *A. fumigatus*, as PTX3 binds to conidia (Garlanda et al., 2002). Moreover, alveolar macrophages from PTX3-deficient mice display defective recognition of conidia. These mice are highly susceptible to IPA even without immunosuppression (Garlanda et al., 2002). A similar situation possibly takes place for molecules of the collectin family, such as the surfactant proteins A and D. Thus, both proteins enhance the killing of conidia by alveolar macrophages (Madan et al., 1997, 2005).

Macrophages and Inflammation

As indicated above, the activation of alveolar macrophages via dectin-1 and TLR2 induces the synthesis of various proinflammatory cytokines and chemokines, including TNF-α and macrophage inflammatory protein 2 (Hohl et al., 2005). In normal human monocytes, *A. fumigatus* triggers differential regulation of the expression of 1,827 genes ($P < 0.05$), many of which encode cytokines and chemokines involved in host defense (Cortez et al., 2006). The second most important function of macrophages is the recruitment of circulating neutrophils to the site of infection, via the production of cytokines and chemokines. Classically, this inflammatory function of macrophages follows a chronological pattern: (i) recognition of the pathogen, (ii) induction of intracellular signaling pathways, and finally (iii) expression of different genes of the innate immune response (Luther and Ebel, 2006; Netea et al., 2006; Zelante et al., 2007; Chignard et al., 2007). The synthesis of a large array of these molecules is partly dependent on the activation of MyD88, an adaptor protein downstream from the TLR. Thus, alveolar macrophages (Hohl et al., 2005) and peritoneal macrophages (Mambula et al., 2002; Gersuk et al., 2006) from Myd88$^{-/-}$ mice produce significantly less TNF-α than control wild-type macrophages when challenged with conidia. It has been inferred that the MyD88-independent production of TNF-α is probably due to activation of the cells through dectin-1. Indeed, TLR/MyD88-induced responses are dispensable, as it has been shown that another intracellular molecule, Card9, controls dectin-1-mediated cytokine production and innate antifungal immunity (Gross et al., 2006). Nonetheless, TNF-α synthesis by the lung in vivo is, paradoxically, almost abolished in MyD88-deficient mice infected intranasally with *A. fumigatus* (Bellocchio et al., 2004a). Unfortunately, with the exception of TNF-α, it is not known how the synthesis of other innate defense mediators is affected in the absence of MyD88. Activation of the transcription factor NF-κB is presumably affected, probably at least reducing the synthesis of these other molecules. However, there is another gap in our knowledge, as paradoxically, MyD88$^{-/-}$ mice are no more susceptible to *A. fumigatus* than wild-type mice (Bellochio et al., 2004a; Dubourdeau et al., 2006; Chignard et al., 2007).

The TLR involved in the recognition of *A. fumigatus* remains to be identified. Experiments on transfected HEK cells expressing TLR1 to -10 suggested that only TLR2 and TLR4 recognized *A. fumigatus* (Meier et al., 2003). These findings have been confirmed for bone marrow-derived (Hohl et al, 2005), peritoneal (Mambula et al., 2002; Netea et al., 2003; Gersuk et al., 2006), and alveolar (Balloy et al., 2005a) macrophages collected from knockout mice. It has been suggested that TLR2 may recognize both conidia and hyphae, whereas TLR4 recognizes only conidia (Netea et al., 2003). In any case, we can infer from these findings that the host recognizes at least two different TLR ligands.

Neutrophils and Phagocytosis

Alveolar macrophages are important for the host defense against *A. fumigatus*, as demonstrated in a murine model of IPA (Fig. 1). The depletion of cells of this type results in greater outgrowth of the fungus in the lung and in lower levels of animal survival. However, this is clearly observed only when mice have been rendered neutropenic (Chignard et al., 2007). Clinicians knew a long time ago that neutrophils play an important role in combating *A. fumigatus* infection. Indeed, neutropenic patients with aplastic anemia (Weinberger et al., 1992) or chemotherapy-induced neutropenia (Gerson et al., 1984; Bodey and Vartivarian, 1989; Walsh and Dixon, 1989; Wiley et al., 1990) were observed to be at risk of acquiring IPA. As early as 1970, it was observed that neutrophils ingested conidia in vitro (Lehrer and Jan, 1970), although the conidia remained viable for 3 h. This seems to be characteristic of resting conidia, their resistance to the fungicidal mechanisms of neutrophils possibly being due to both their failure to stimulate optimal ROS production and their resistance to neutrophil oxidants. By contrast, swollen and germinated conidia show no such resistance (Levitz and Diamond, 1985). Their phagocytosis is enhanced by complement (Sturtevant and Latgé, 1992) and possibly also by characteristic lung molecules, such as surfactant proteins A and D (Madan et al., 1997). Along the same lines, the uptake of conidia by neutrophils is enhanced in the presence of mannan-binding lectin. Consistent with this, in a murine model of IPA, recombinant human mannan-binding lectin-treated mice had a survival rate of 80%, whereas no untreated mice with IPA survived (Kaur et al., 2007). An unexpected new mechanism was recently discovered in studies of PTX3. This molecule is stored in specific granules, and at least some is transferred to NETs on its release. Jaillon et al. (2007) showed that neutrophils from PTX3 knockout mice present significantly lower levels of conidial phagocytosis than wild-type mice. Similarly, the conidicidal activity of neutrophils from these mice is much weaker than that of neutrophils from wild-type mice and is increased by the addition of recombinant PTX3. Finally, platelets may increase neutrophil-mediated killing as they damage *A. fumigatus* hyphae (Christin et al., 1998). Gliotoxin, a sulfur-containing antibiotic produced by *A. fumigatus* (Mullbacher et al., 1985), inhibits phagocytosis and decreases ROS generation by neutrophils (Tsunawaki et al., 2004; Orciuolo et al., 2007; Comera et al., 2007). A genetically engineered mutant that does not produce gliotoxin has a pathogenicity in neutropenic mice similar to that of the wild type, whereas its virulence is attenuated in corticosteroid-treated mice

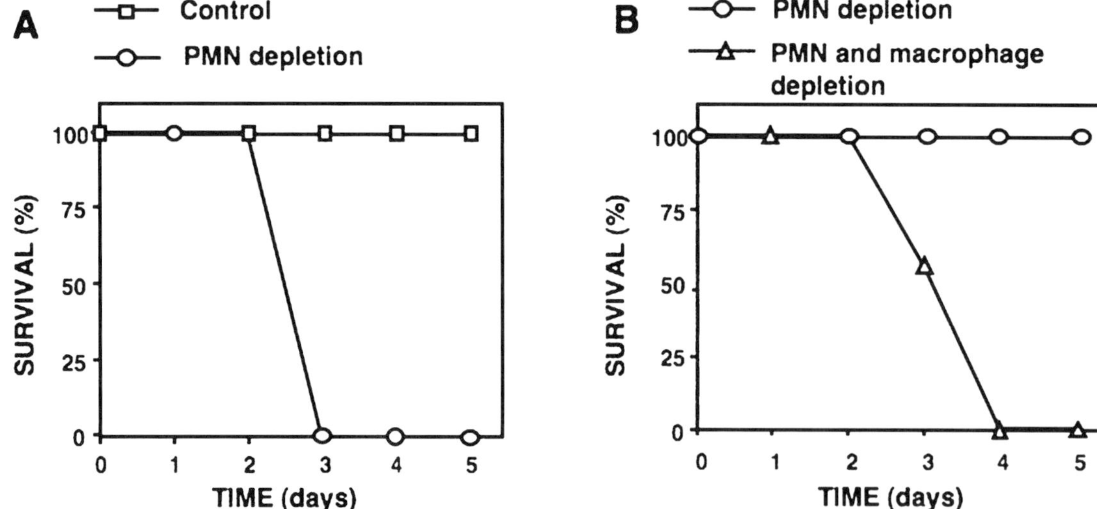

Figure 1. Role of neutrophils and alveolar macrophages in experimental IPA. Mice were infected intratracheally with 10^7 (A) or 10^6 (B) conidia of *A. fumigatus*. Mice were depleted of neutrophils and macrophages by treatment with vinblastine (Balloy et al., 2005b) and chlodronate (Chignard et al., 2007), respectively. (From Chignard et al. [2007] with permission of the publisher.)

(Cramer et al., 2006; Kupfahl et al., 2006, Sugui et al., 2007; Spikes et al., 2008). These results are reminiscent of other experimental (Balloy et al. 2005b; Stephens-Romero et al., 2005) and clinical (Stergiopoulou et al., 2007) reports showing that several features of IPA pathogenesis differ according to whether immunosuppression is the result of corticosteroid treatment or chemotherapy (Chignard et al., 2007).

Neutrophils and Inflammation

It remains a matter of debate whether the impact of neutrophils on IPA results from their phagocytic properties or their ability to degranulate and to kill *A. fumigatus* from the outside, with concomitant damage to the nearby epithelium. As mentioned earlier, Schaffner et al. (1982) showed that the role of neutropenia in vivo becomes evident only after conidia escape from the killing mediated by alveolar macrophages and begin to produce mycelia. They concluded that alveolar macrophages control the conidia responsible for establishing infection, whereas neutrophils kill hyphae. Consistent with this hypothesis, bronchoalveloar lavage samples (BAL) collected from mice 3 h after infection contain alveolar macrophages with engulfed conidia (Color Plate 9A). By contrast, BAL collected 48 h after infection contain hyphae with adherent neutrophils (Color Plate 9B). Nonetheless, although neutrophils have been shown to attach to hyphae, spread on their surfaces, and degranulate and damage hyphae (Diamond et al., 1978), they have also been shown to engulf conidia (Lehrer and Jan, 1970; Sturtevant and Latgé, 1992; Madan et al., 1997; Feldmesser, 2006). It remains possible that neutrophils carry out both activities, at different times in the infection process. Indeed, neutrophils reach the site of infection within 3 to 6 h (Fig. 2), at a time at which the conidia have probably not yet germinated, as deduced from the stable chitin concentrations measured in whole lung (Fig. 2) and consistent with the known pattern of development of hyphae, which takes between 6 and 8 h, as observed in vitro (Bernard and Latgé, 2001; Latgé, 2001). Thus, neutrophils arriving rapidly at the infection site encounter conidia. They clearly continue to migrate for a few days after the start of infection, as their numbers in airspaces continue to increase (Fig. 2). They are therefore still present when the conidia have swollen and germinated, expressing an initial morphotype, known as a germling, followed by a second morphotype, the hyphae (Luther and Ebel, 2006), as reflected by increases in chitin concentration (Fig. 2). Thus, neutrophils arriving at later times mostly encounter hyphae, as shown in Color Plate 9B. Along similar lines, Bonnett et al. (2006) recently showed that the early recruitment of neutrophils inhibits germination through the release of ROS. They concluded that the destruction of hyphae by neutrophils described by Schaffner et al. (1982) was probably of secondary importance in protecting the lung following exposure to large numbers of conidia.

Neutrophils have several weapons for killing *A. fumigatus*, at early or late time points. ROS are a key weapon for the killing of *A. fumigatus* by neutrophils, as demonstrated by cases of severe IPA in patients with CGD (Gallin et al., 1983; Mamishi et al., 2007). This finding has been confirmed experimentally, using X-CGD mice. X-CGD mice were more susceptible to pul-

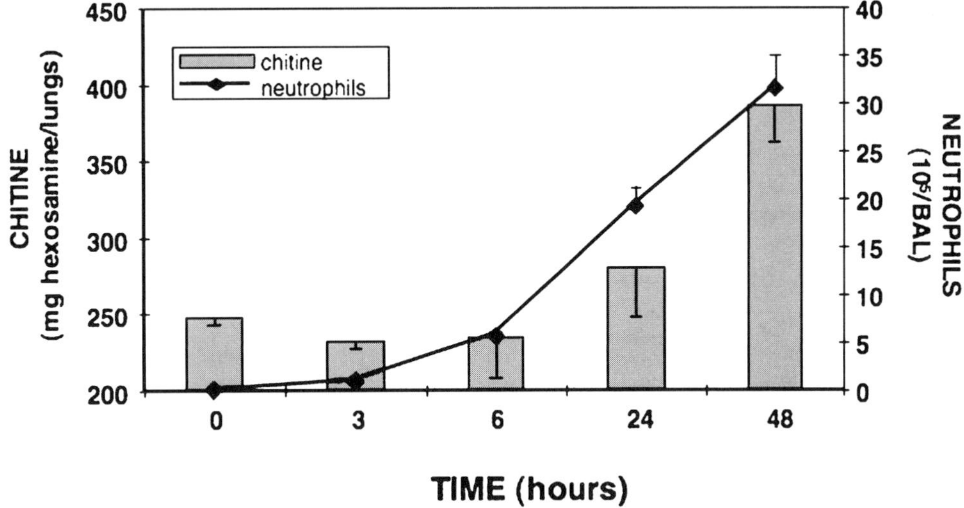

Figure 2. Neutrophil recruitment and chitin concentration in the lung in experimental IPA. Mice were infected intratracheally with 10^7 conidia of *A. fumigatus* and BAL was obtained at various time points. Neutrophils were counted and chitin quantified. (Images courtesy of V. Balloy et al. [unpublished data].)

monary infection than normal mice following exposure to conidia (Morgenstern et al., 1997). X-CGD mice displayed higher mortality levels, associated with 10 to 100 times higher fungal outgrowth rates (Aratani et al., 2002). Another type of animal model has shed light on the impact of neutrophil serine proteases on the risk of acquiring IPA. Mice lacking the granule serine proteases elastase and/or cathepsin G are susceptible to *A. fumigatus* infection, despite displaying normal neutrophil development and recruitment (Tkalcevic et al., 2000). Iron sequestration by neutrophil lactoferrin is another important host defense mechanism, as this secondary granule protein arrests the growth of *A. fumigatus* conidia in vitro (Zarember et al., 2007). Cationic peptides, also stored in neutrophilic granules, participate in the killing process (Levitz et al., 1986).

As for alveolar macrophages, pathogen recognition is the first important step. Neutrophils express various TLRs but have no TLR3 (Hayashi et al., 2003). Some of these TLRs mediate recognition and trigger conidicidal and hyphal damage responses in a morphotype-specific manner (Bellocchio et al., 2004b). Nonetheless, MyD88$^{-/-}$, TLR2$^{-/-}$ (Balloy et al., 2005a), and TLR4$^{-/-}$ (Bellocchio et al., 2004a) mice are more susceptible to IPA than wild-type mice when neutropenic. By contrast, immunocompetent mice are not susceptible to *A. fumigatus*, regardless of their status for MyD88, TLR2, or TLR4. Thus, TLR2 and/or TLR4 is essential only in the absence of neutrophils, and these molecules are not required in healthy hosts (Chignard et al., 2007). It has been suggested that the TLRs expressed by neutrophils are potentially less important than other PRR in vivo and that dectin-1 expression by neutrophils (Taylor et al., 2002) may play a role (Chignard et al., 2007).

CONCLUSION

Alveolar macrophages and neutrophils are clearly of prime importance for combating inhaled *A. fumigatus*. Indeed, if for any reason these cells are absent or incapacitated, the fungus grows and causes a life-threatening infection. A recent study showed how these phagocytes function in vitro in the ingestion of *A. fumigatus*, based on a model of the specific environment of lung alveoli, in a two-dimensional space (Behnsen et al., 2007). In vivo, the two-dimensional movements of alveolar macrophages and neutrophils can be observed at the surface of the respiratory epithelium (Sibille and Reynolds, 1990). Surprisingly, little attention has been paid to this third cell type present in airspaces, although most incoming airborne conidia clearly encounter epithelial cells. Indeed, it is now thought that these cells play an active role in innate defenses against pathogens (Message and Johnston, 2004; Diamond et al., 2000), although the interaction between *A. fumigatus* and epithelial cells leads to the production of comparatively modest levels of interleukin-6 and interleukin-8 (Zhang et al., 2005). Other potential consequences of this interaction have been described. Several studies have described the effects of *A. fumigatus*-derived proteinases on respiratory epithelial cells, including cell detachment (Robinson et al., 1990) and cytokine production (Tomee et al., 1997; Borger et al., 1999; Kogan et al., 2004). The uptake of conidia by lung epithelial cells, including type II pneumocytes in particular, has also been observed (Paris et al., 1997; Wasylnka and Moore, 2002). These conidia survive and are able to germinate (Wasylnka and Moore, 2003) and block apoptosis of pulmonary epithelial cells (Berkova et al., 2006). It has been suggested that epithelial cells may constitute a reservoir, contributing to virulence of this microorganism (Filler and Sheppard, 2006) (see also chapter 19 by Hope and Filler).

Epithelial cells and neutrophils are both able to express defensins. These molecules are small cationic peptides with antimicrobial activity. Human neutrophils contain large amounts of three α-defensins (HNP-1 to HNP-3) and smaller amounts of a fourth defensin, HNP-4. β-Defensins are mostly produced by epithelial cells, and their synthesis is induced by cell surface TLR or cytoplasmic NOD (Semple et al., 2006; Lehrer, 2007). Thus, unlike neutrophil α-defensins, they do not appear to reside in granules. The inhibition of antimicrobial peptide activity or gene expression may increase susceptibility to infections (Laube et al., 2006), confirming the involvement of these molecules. The most frequently described antimicrobial activity mostly concerns effects on bacteria and, more rarely, on fungi. One of the most comprehensive reviews in recent years (Jenssen et al., 2006) devoted a paragraph to the antifungal peptide activity, but few studies have been performed, and those that have been carried out were mostly concerned with *C. albicans*. Only two groups of studies on *A. fumigatus* have been reported, but they concern two peptides of insect origin: cecropin (De Lucca et al., 1998) and tenecin-3 (Lee et al., 1999). SLPI, a proteinase inhibitor mostly produced by epithelial cells, is another peptide with antimicrobial activity against *A. fumigatus* (Tomee et al., 1997). Another peptide that potently inhibits the growth of *A. fumigatus* has been isolated from human amniotic fluid. This peptide has a partial sequence that is 100% identical to the N-terminal sequence of ubiquitin (Kim et al., 2007). These peptides may serve as templates in the search for new antibiotics active against *A. fumigatus*.

REFERENCES

Aderem, A. 2003. Phagocytosis and the inflammatory response. *J. Infect. Dis.* 187(Suppl. 2):S340–S345.

Aderem, A., and D. M. Underhill. 1999. Mechanisms of phagocytosis in macrophages. *Annu. Rev. Immunol.* **17:**593–623.

Akira, S., S. Uematsu, and O. Takeuchi. 2006. Pathogen recognition and innate immunity. *Cell* **124:**783–801.

Aratani, Y., F. Kura, H. Watanabe, H. Akagawa, Y. Takano, K. Suzuki, M. C. Dinauer, N. Maeda, and H. Koyama. 2002. Relative contributions of myeloperoxidase and NADPH-oxidase to the early host defense against pulmonary infections with *Candida albicans* and *Aspergillus fumigatus*. *Med. Mycol.* **40:**557–563.

Balloy, V., M. Si-Tahar, O. Takeuchi, B. Philippe, M. A. Nahori, M. Tanguy, M. Huerre, S. Akira, J. P. Latgé, and M. Chignard. 2005a. Involvement of toll-like receptor 2 in experimental invasive pulmonary aspergillosis. *Infect. Immun.* **73:**5420–5425.

Balloy, V., M. Huerre, J. P. Latgé, and M. Chignard. 2005b. Differences in patterns of infection and inflammation for corticosteroid treatment and chemotherapy in experimental invasive pulmonary aspergillosis. *Infect. Immun.* **73:**494–503.

Behnsen, J., P. Narang, M. Hasenberg, F. Gunzer, U. Bilitewski, N. Klippel, M. Rohde, M. Brock, A. A. Brakhage, and M. Gunzer. 2007. Environmental dimensionality controls the interaction of phagocytes with the pathogenic fungi *Aspergillus fumigatus* and *Candida albicans*. *PLoS Pathog.* **3:**e13.

Bellocchio, S., C. Montagnoli, S. Bozza, R. Gaziano, G. Rossi, S. S. Mambula, A. Vecchi, A. Mantovani, S. M. Levitz, and L. Romani. 2004a. The contribution of the Toll-like/IL-1 receptor superfamily to innate and adaptive immunity to fungal pathogens *in vivo*. *J. Immunol.* **172:**3059–3069.

Bellocchio, S., S. Moretti, K. Perruccio, F. Fallarino, S. Bozza, C. Montagnoli, P. Mosci, G. B. Lipford, L. Pitzurra, and L. Romani. 2004b. TLRs govern neutrophil activity in aspergillosis. *J. Immunol.* **173:**7406–7415.

Berkova, N., S. Lair-Fulleringer, F. Féménia, D. Huet, M. C. Wagner, K. Gorna, F. Tournier, O. Ibrahim-Granet, J. Guillot, R. Chermette, P. Boireau, and J. P. Latgé. 2006. *Aspergillus fumigatus* conidia inhibit tumour necrosis factor- or staurosporine-induced apoptosis in epithelial cells. *Int. Immunol.* **18:**139–150.

Bernard, M., and J. P. Latgé. 2001. *Aspergillus fumigatus* cell wall: composition and biosynthesis. *Med. Mycol.* **39**(Suppl. 1)**:**9–17.

Bodey, G. P., and S. Vartivarian. 1989. Aspergillosis. *Eur. J. Clin. Microbiol. Infect. Dis.* **8:**413–437.

Bonnett, C. R., E. J. Cornish, A. G. Harmsen, and J. B. Burritt. 2006. Early neutrophil recruitment and aggregation in the murine lung inhibit germination of *Aspergillus fumigatus* conidia. *Infect. Immun.* **74:**6528–6539.

Boot, R. G., G. H. Renkema, A. Strijland, A. J. van Zonneveld, and J. M. Aerts. 1995. Cloning of a cDNA encoding chitotriosidase, a human chitinase produced by macrophages. *J. Biol. Chem.* **270:**26252–26256.

Borger, P., G. H. Koëter, J. A. Timmerman, E. Vellenga, J. F. Tomee, and H. F. Kauffman. 1999. Proteases from *Aspergillus fumigatus* induce interleukin (IL)-6 and IL-8 production in airway epithelial cell lines by transcriptional mechanisms. *J. Infect. Dis.* **180:**1267–1274.

Brinkmann, V., U. Reichard, C. Goosmann, B. Fauler, Y. Uhlemann, D. S. Weiss, Y. Weinrauch, and A. Zychlinsky. 2004. Neutrophil extracellular traps kill bacteria. *Science* **303:**1532–1535.

Brinkmann, V., and A. Zychlinsky. 2007. Beneficial suicide: why neutrophils die to make NETs. *Nat. Rev. Microbiol.* **5:**577–582.

Burg, N. D., and M. H. Pillinger. 2001. The neutrophil: function and regulation in innate and humoral immunity. *Clin. Immunol.* **99:**7–17.

Chignard, M., V. Balloy, J. M. Sallenave, and M. Si-Tahar. 2007. Role of Toll-like receptors in lung innate defense against invasive aspergillosis. Distinct impact in immunocompetent and immunocompromized hosts. *Clin. Immunol.* **124:**238–243.

Christin, L., D. R. Wysong, T. Meshulam, R. Hastey, E. R. Simons, and R. D. Diamond. 1998. Human platelets damage *Aspergillus fumigatus* hyphae and may supplement killing by neutrophils. *Infect. Immun.* **66:**1181–1189.

Comera, C., K. Andre, J. Laffitte, X. Collet, P. Galtier, and I. Maridonneau-Parini. 2007. Gliotoxin from *Aspergillus fumigatus* affects phagocytosis and the organization of the actin cytoskeleton by distinct signalling pathways in human neutrophils. *Microbes Infect.* **9:**47–54.

Cortez, K. J., C. A. Lyman, S. Kottilil, H. S. Kim, E. Roilides, J. Yang, B. Fullmer, R. Lempicki, and T. J. Walsh. 2006. Functional genomics of innate host defense molecules in normal human monocytes in response to *Aspergillus fumigatus*. *Infect. Immun.* **74:**2353–2365.

Cramer, R. A., Jr., M. P. Gamcsik, R. M. Brooking, L. K. Najvar, W. R. Kirkpatrick, T. F. Patterson, C. J. Balibar, J. R. Graybill, J. R. Perfect, S. N. Abraham, and W. J. Steinbach. 2006. Disruption of a nonribosomal peptide synthetase in *Aspergillus fumigatus* eliminates gliotoxin production. *Eukaryot. Cell* **5:**972–980.

Dandona, P., P. Mohanty, W. Hamouda, A. Aljada, Y. Kumbkarni, and R. Garg. 1999. Effect of dexamethasone on reactive oxygen species generation by leukocytes and plasma interleukin-10 concentrations: a pharmacodynamic study. *Clin. Pharmacol. Ther.* **66:**58–65.

De Lucca, A. J., J. M. Bland, T. J. Jacks, C. Grimm, and T. J. Walsh. 1998. Fungicidal and binding properties of the natural peptides cecropin B and dermaseptin. *Med. Mycol.* **36:**291–298.

Diamond, R. D., R. Krzesicki, B. Epstein, and W. Jao. 1978. Damage to hyphal forms of fungi by human leukocytes *in vitro*. A possible host defense mechanism in aspergillosis and mucormycosis. *Am. J. Pathol.* **91:**313–328.

Diamond, G., D. Legarda, and L. K. Ryan. 2000. The innate immune response of the respiratory epithelium. *Immunol. Rev.* **173:**27–38.

Diniz, S. N., R. Nomizo, P. S. Cisalpino, M. M. Teixeira, G. D. Brown, A. Mantovani, S. Gordon, L. F. Reis, and A. A. Dias. 2004. PTX3 function as an opsonin for the dectin-1-dependent internalization of zymosan by macrophages. *J. Leukoc. Biol.* **75:**649–656.

Dubourdeau, M., R. Athman, V. Balloy, M. Huerre, M. Chignard, D. J. Philpott, J. P. Latgé, and O. Ibrahim-Granet. 2006. *Aspergillus fumigatus* induces innate immune responses in alveolar macrophages through the MAPK pathway independently of TLR2 and TLR4. *J. Immunol.* **177:**3994–4001.

El-Benna, J., P. M. Dang, M. A. Gougerot-Pocidalo, and C. Elbim. 2005. Phagocyte NADPH oxidase: a multicomponent enzyme essential for host defenses. *Arch. Immunol. Ther. Exp.* **53:**199–206.

Feldmesser, M. 2006. Role of neutrophils in invasive aspergillosis. *Infect. Immun.* **74:**6514–6516.

Filler, S. G., and D. C. Sheppard. 2006. Fungal invasion of normally non-phagocytic host cells. *PLoS Pathog.* **2:**e129.

Fuchs, T. A., U. Abed, C. Goosmann, R. Hurwitz, I. Schulze, V. Wahn, Y. Weinrauch, V. Brinkmann, and A. Zychlinsky. 2007. Novel cell death program leads to neutrophil extracellular traps. *J. Cell Biol.* **176:**231–241.

Gallin, J. I., E. S. Buescher, B. E. Seligmann, J. Nath, T. Gaither, and P. Katz. 1983. NIH conference. Recent advances in chronic granulomatous disease. *Ann. Intern. Med.* **99:**657–674.

Garlanda, C., E. Hirsch, S. Bozza, A. Salustri, M. De Acetis, R. Nota, A. Maccagno, F. Riva, B. Bottazzi, G. Peri, A. Doni, L. Vago, M. Botto, R. De Santis, P. Carminati, G. Siracusa, F. Altruda, A. Vecchi, L. Romani, and A. Mantovani. 2002. Non-redundant role of the long pentraxin PTX3 in anti-fungal innate immune response. *Nature* **420:**182–186.

Gernez-Rieux, C., C. Voisin, C. Aerts, F. Wattel, and B. Gosselin. 1967. Experimental aspergillosis in the guinea pig. Dynamic study of the role of alveolar macrophages in the defense of the respiratory tract, after massive inhalation of *Aspergillus fumigatus* spores. *Rev. Tuberc. Pneumol.* (Paris) **31:**705–725. (In French.)

Gerson, S. L., G. H. Talbot, S. Hurwitz, B. L. Strom, E. J. Lusk, and P. A. Cassileth. 1984. Prolonged granulocytopenia: the major risk factor for invasive pulmonary aspergillosis in patients with acute leukemia. *Ann. Intern. Med.* **100:**345–351.

Gersuk, G. M., D. M. Underhill, L. Zhu, and K. A. Marr. 2006. Dectin-1 and TLRs permit macrophages to distinguish between different *Aspergillus fumigatus* cellular states. *J. Immunol.* **176:**3717–3724.

Gordon, S. 2002. Pattern recognition receptors: doubling up for the innate immune response. *Cell* **111:**927–930.

Greenberg, S., and S. Grinstein. 2002. Phagocytosis and innate immunity. *Curr. Opin. Immunol.* **14:**136–145.

Gross, O., A. Gewies, K. Finger, M. Schafer, T. Sparwasser, C. Peschel, I. Forster, and J. Ruland. 2006. Card9 controls a non-TLR signalling pathway for innate anti-fungal immunity. *Nature* **442:** 651–656.

Hampton, M. B., A. J. Kettle, and C. C. Winterbourn. 1998. Inside the neutrophil phagosome: oxidants, myeloperoxidase, and bacterial killing. *Blood* **92:**3007–3017.

Hayashi, F., T. K. Means, and A. D. Luster. 2003. Toll-like receptors stimulate human neutrophil function. *Blood* **102:**2660–2669.

Hiemstra, P. S. 2002. Novel roles of protease inhibitors in infection and inflammation. *Biochem. Soc. Trans.* **30:**116–120.

Hohl, T. M., H. L. Van Epps, A. Rivera, L. A. Morgan, P. L. Chen, M. Feldmesser, and E. G. Pamer. 2005. *Aspergillus fumigatus* triggers inflammatory responses by stage-specific beta-glucan display. *PLoS Pathog.* **1:**e30.

Ibrahim-Granet, O., B. Philippe, H. Boleti, E. Boisvieux-Ulrich, D. Grenet, M. Stern, and J. P. Latgé. 2003. Phagocytosis and intracellular fate of *Aspergillus fumigatus* conidia in alveolar macrophages. *Infect. Immun.* **71:**891–903.

Jaillon, S., G. Peri, Y. Delneste, I. Fremaux, A. Doni, F. Moalli, C. Garlanda, L. Romani, H. Gascan, S. Bellocchio, S. Bozza, M.A. Cassatella, P. Jeannin, and A. Mantovani. 2007. The humoral pattern recognition receptor PTX3 is stored in neutrophil granules and localizes in extracellular traps. *J. Exp. Med.* **204:**793–804.

Jenssen, H., P. Hamill, and R. E. Hancock. 2006. Peptide antimicrobial agents. *Clin. Microbiol. Rev.* **19:**491–511.

Kaur, S., V. K. Gupta, S. Thiel, P. U. Sarma, and T. Madan. 2007. Protective role of mannan-binding lectin in a murine model of invasive pulmonary aspergillosis. *Clin. Exp. Immunol.* **148:**382–389.

Kim, J. Y., S. Y. Lee, S. C. Park, S. Y. Shin, S. J. Choi, Y. Park, and K. S. Hahm. 2007. Purification and antimicrobial activity studies of the N-terminal fragment of ubiquitin from human amniotic fluid. *Biochim. Biophys. Acta* **1774:**1221–1226.

Kobayashi, S. D., J. M. Voyich, C. Burlak, and F. R. DeLeo. 2005. Neutrophils in the innate immune response. *Arch. Immunol. Ther. Exp.* **53:**505–517.

Kogan, T. V., J. Jadoun, L. Mittelman, K. Hirschberg, and N. Osherov. 2004. Involvement of secreted *Aspergillus fumigatus* proteases in disruption of the actin fiber cytoskeleton and loss of focal adhesion sites in infected A549 lung pneumocytes. *J. Infect. Dis.* **189:** 1965–1973.

Korkmaz, B., T. Moreau, and F. Gauthier. 2008. Neutrophil elastase, proteinase 3 and cathepsin G: physicochemical properties, activity and physiopathological functions. *Biochimie* **90:**227–242.

Kupfahl, C., T. Heinekamp, G. Geginat, T. Ruppert, A. Hartl, H. Hof, and A. A. Brakhage. 2006. Deletion of the *gliP* gene of *Aspergillus fumigatus* results in loss of gliotoxin production but has no effect on virulence of the fungus in a low-dose mouse infection model. *Mol. Microbiol.* **62:**292–302.

Latgé, J. P. 2001. The pathobiology of *Aspergillus fumigatus*. *Trends Microbiol.* **9:**382–389.

Laube, D. M., S. Yim, L. K. Ryan, K. O. Kisich, and G. Diamond. 2006. Antimicrobial peptides in the airway. *Curr. Top. Microbiol. Immunol.* **306:**153–182.

Lee, Y. T., D. H. Kim, J. Y. Suh, J. H. Chung, B. L. Lee, Y. Lee, and B. S. Choi. 1999. Structural characteristics of tenecin 3, an insect antifungal protein. *Biochem. Mol. Biol. Int.* **47:**369–376.

Lehrer, R. I. 2007. Multispecific myeloid defensins. *Curr. Opin. Hematol.* **14:**16–21.

Lehrer, R. I., and R. G. Jan. 1970. Interaction of *Aspergillus fumigatus* spores with human leukocytes and serum. *Infect. Immun.* **1:**345–350.

Levitz, S. M., and R. D. Diamond. 1985. Mechanisms of resistance of *Aspergillus fumigatus* conidia to killing by neutrophils *in vitro*. *J. Infect. Dis.* **152:**33–42.

Levitz, S. M., M. E. Selsted, T. Ganz, R. I. Lehrer, and R. D. Diamond. 1986. *In vitro* killing of spores and hyphae of *Aspergillus fumigatus* and *Rhizopus oryzae* by rabbit neutrophil cationic peptides and bronchoalveolar macrophages. *J. Infect. Dis.* **154:**483–489.

Luther, K., and K. Ebel. 2006. Toll-like receptors: recent advances, open questions and implications for aspergillosis control. *Med. Mycol.* **44:**S219–S227.

Luther, K., A. Torosantucci, A. A. Brakhage, J. Heesemann, and F. Ebel. 2007. Phagocytosis of *Aspergillus fumigatus* conidia by murine macrophages involves recognition by the dectin-1 beta-glucan receptor and Toll-like receptor 2. *Cell. Microbiol.* **9:**368–381.

Madan, T., P. Eggleton, U. Kishore, P. Strong, S. S. Aggrawal, P. U. Sarma, and K. B. Reid. 1997. Binding of pulmonary surfactant proteins A and D to *Aspergillus fumigatus* conidia enhances phagocytosis and killing by human neutrophils and alveolar macrophages. *Infect. Immun.* **65:**3171–3179.

Madan, T., S. Kaur, S. Saxena, M. Singh, U. Kishore, S. Thiel, K. B. Reid, and P. U. Sarma. 2005. Role of collectins in innate immunity against aspergillosis. *Med. Mycol.* **43**(Suppl. l):S155–S163.

Mambula, S. S., K. Sau, P. Henneke, D. T. Golenbock, and S. M. Levitz. 2002. Toll-like receptor (TLR) signaling in response to *Aspergillus fumigatus*. *J. Biol. Chem.* **277:**39320–39326.

Mamishi, S., N. Parvaneh, A. Salavati, S. Abdollahzadeh, and M. Yeganeh. 2007. Invasive aspergillosis in chronic granulomatous disease: report of 7 cases. *Eur. J. Pediatr.* **166:**83–84.

Martinon, F., O. Gaide, V. Pétrilli, A. Mayor, and J. Tschopp. 2007. NALP inflammasomes: a central role in innate immunity. *Semin. Immunopathol.* **29:**213–229.

Medzhitov, R. 2007. Recognition of microorganisms and activation of the immune response. *Nature* **449:**819–826.

Medzhitov, R. and C. A. Janeway, Jr. 1997. Innate immunity: impact on the adaptive immune response. *Curr. Opin. Immunol.* **9:**4–9.

Medzhitov, R., and C. Janeway, Jr. 2000. Innate immunity. *N. Engl. J. Med.* **343:**338–344.

Meier, A., C. J. Kirschning, T. Nikolaus, H. Wagner, J. Heesemann, and F. Ebel. 2003. Toll-like receptor (TLR) 2 and TLR4 are essential for Aspergillus-induced activation of murine macrophages. *Cell. Microbiol.* **5:**561–570.

Message, S. D., and S. L. Johnston. 2004. Host defense function of the airway epithelium in health and disease: clinical background. *J. Leukoc. Biol.* **75:**5–17.

Michaliszyn, E., S. Senechal, P. Martel, and L. de Repentigny. 1995. Lack of involvement of nitric oxide in killing of *Aspergillus fumigatus* conidia by pulmonary alveolar macrophages. *Infect. Immun.* **63:**2075–2078.

Moraes, T. J., J. H. Zurawska, and G. P. Downey. 2006. Neutrophil granule contents in the pathogenesis of lung injury. *Curr. Opin. Hematol.* **13:**21–27.

Morgenstern, D. E., V. Gifford, M. A. Li, C. M. Doerschuk, and M. C. Dinauer. 1997. Absence of respiratory burst in X-linked chronic granulomatous disease mice leads to abnormalities in both host defense and inflammatory response to *Aspergillus fumigatus*. *J. Exp. Med.* **185:**207–218.

Mullbacher, A., P. Waring, and R. D. Eichner. 1985. Identification of an agent in cultures of *Aspergillus fumigatus* displaying antiphagocytic and immunomodulating activity *in vitro*. *J. Gen. Microbiol.* **131:**1251–1258.

Nathan, C. 2006. Neutrophils and immunity: challenges and opportunities. *Nat. Rev. Immunol.* **6:**173–182.

Netea, M. G., G. Ferwerda, C. A. van der Graaf, J. W. Van der Meer, and B. J. Kullberg. 2006. Recognition of fungal pathogens by toll-like receptors. *Curr. Pharm. Des.* **12:**4195–4201.

Netea, M. G., A. Warris, J. W. Van der Meer, M. J. Fenton, T. J. Verver-Janssen, L. E. Jacobs, T. Andresen, P. E. Verweij, and B. J. Kullberg. 2003. *Aspergillus fumigatus* evades immune recognition during germination through loss of toll-like receptor-4-mediated signal transduction. *J. Infect. Dis.* **188:**320–326.

Olenchock, S. A., M. S. Mentnech, J. C. Mull, M. E. Gladish, F. H. Green, and P. C. Manor. 1979. Complement, polymorphonuclear leukocytes and platelets in acute experimental respiratory reactions to *Aspergillus*. *Comp. Immunol. Microbiol. Infect. Dis.* **2:**113–124.

Orciuolo, E., M. Stanzani, M. Canestraro, S. Galimberti, G. Carulli, R. Lewis, M. Petrini, and K. V. Komanduri. 2007. Effects of *Aspergillus fumigatus* gliotoxin and methylprednisolone on human neutrophils: implications for the pathogenesis of invasive aspergillosis. *J. Leukoc. Biol.* **82:**839–848.

Paris, S., E. Boisvieux-Ulrich, B. Crestani, O. Houcine, D. Taramelli, L. Lombardi, and J. P. Latgé. 1997. Internalization of *Aspergillus fumigatus* conidia by epithelial and endothelial cells. *Infect. Immun.* **65:**1510–1514.

Pham, C. T. 2006. Neutrophil serine proteases: specific regulators of inflammation. *Nat. Rev. Immunol.* **6:**541–550.

Philippe, B., O. Ibrahim-Granet, M. C. Prevost, M. A. Gougerot-Pocidalo, M. Sanchez Perez, A. Van der Meeren, and J. P. Latgé. 2003. Killing of *Aspergillus fumigatus* by alveolar macrophages is mediated by reactive oxidant intermediates. *Infect. Immun.* **71:** 3034–3042.

Robinson, B. W., T. J. Venaille, A. H. Mendis, and R. McAleer. 1990. Allergens as proteases: an *Aspergillus fumigatus* proteinase directly induces human epithelial cell detachment. *J. Allergy Clin. Immunol.* **86:**726–731.

Rodriguez, E., F. Boudard, M. Mallie, J. M. Bastide, and M. Bastide. 1997. Murine macrophage elastolytic activity induced by *Aspergillus fumigatus* strains *in vitro*: evidence of the expression of two macrophage-induced protease genes. *Can. J. Microbiol.* **43:**649–657.

Romani, L. 2004. Immunity to fungal infections. *Nat. Rev. Immunol.* **4:**1–23.

Roos, D., R. van Bruggen, and C. Meischl. 2003. Oxidative killing of microbes by neutrophils. *Microbes Infect.* **5:**1307–1315.

Russo-Marie, F. 1992. Macrophages and the glucocorticoids. *J. Neuroimmunol.* **40:**281–286.

Schaffner, A., H. Douglas, and A. Braude. 1982. Selective protection against conidia by mononuclear and against mycelia by polymorphonuclear phagocytes in resistance to *Aspergillus*. Observations on these two lines of defense *in vivo* and *in vitro* with human and mouse phagocytes. *J. Clin. Investig.* **69:**617–631.

Segal, A. W. 2005. How neutrophils kill microbes. *Annu. Rev. Immunol.* **23:**197–223.

Segal, B. H., and T. J. Walsh. 2006. Current approaches to diagnosis and treatment of invasive aspergillosis. *Am. J. Respir. Crit. Care Med.* **173:**707–717.

Semple, C. A., P. Gautier, K. Taylor, and J. R. Dorin. 2006. The changing of the guard: molecular diversity and rapid evolution of beta-defensins. *Mol. Divers.* **10:**575–584.

Serrano-Gomez, D., A. Dominguez-Soto, J. Ancochea, J. A. Jimenez-Heffernan, J. A. Leal, and A. L. Corbi. 2004. Dendritic cell-specific intercellular adhesion molecule 3-grabbing nonintegrin mediates binding and internalization of *Aspergillus fumigatus* conidia by dendritic cells and macrophages. *J. Immunol.* **173:**5635–5643.

Shepherd, V. L. 1986. The role of the respiratory burst of phagocytes in host defense. *Semin. Respir. Infect.* **1:**99–106.

Sibille, Y., and H. Y. Reynolds. 1990. Macrophages and polymorphonuclear neutrophils in lung defense and injury. *Am. Rev. Respir. Dis.* **141:**471–501.

Smith, J. A. 1994. Neutrophils, host defense, and inflammation: a double-edged sword. *J. Leukoc. Biol.* **56:**672–686.

Smith, P. D, C. Ochsenbauer-Jambor, and L. E. Smythies. 2005. Intestinal macrophages: unique effector cells of the innate immune system. *Immunol. Rev.* **206:**149–159.

Spikes, S., R. Xu, C. K. Nguyen, G. Chamilos, D. P. Kontoyiannis, R. H. Jacobson, D. E. Ejzykowicz, L. Y. Chiang, S. G. Filler, and G. S. May. 2008. Gliotoxin production in *Aspergillus fumigatus* contributes to host-specific differences in virulence. *J. Infect. Dis.* **197:** 479–486.

Steele, C., R. R. Rapaka, A. Metz, S. M. Pop, D. L. Williams, S. Gordon, J. K. Kolls, and G. D. Brown. 2005. The beta-glucan receptor dectin-1 recognizes specific morphologies of *Aspergillus fumigatus*. *PLoS Pathog.* **1:**e42.

Stephens-Romero, S. D., A. J. Mednick, and M. Feldmesser. 2005. The pathogenesis of fatal outcome in murine pulmonary aspergillosis depends on the neutrophil depletion strategy. *Infect. Immun.* **73:**114–125.

Stergiopoulou, T., J. Meletiadis, E. Roilides, D. E. Kleiner, R. Schaufele, M. Roden, S. Harrington, L. Dad, B. Segal, and T. J. Walsh. 2007. Host-dependent patterns of tissue injury in invasive pulmonary aspergillosis. *Am. J. Clin. Pathol.* **127:**349–355.

Strieter, R. M., J. A. Belperio, and M. P. Keane. 2002. Cytokines in innate host defense in the lung. *J. Clin. Investig.* **109:**699–705.

Stuart, L. M., and R. A. Ezekowitz. 2005. Phagocytosis: elegant complexity. *Immunity* **22:**539–550.

Sturtevant, J., and J. P. Latgé. 1992. Participation of complement in the phagocytosis of the conidia of *Aspergillus fumigatus* by human polymorphonuclear cells. *J. Infect. Dis.* **166:**580–586.

Sugui, J. A., J. Pardo, Y. C. Chang, K. A. Zarember, G. Nardone, E. M. Galvez, A. Mullbacher, J. I. Gallin, M. M. Simon, and K. J. Kwon-Chung. 2007. Gliotoxin is a virulence factor of *Aspergillus fumigatus: gliP* deletion attenuates virulence in mice immunosuppressed with hydrocortisone. *Eukaryot. Cell* **6:**1562–1569.

Taylor, P. R., G. D. Brown, D. M. Reid, J. A. Willment, L. Martinez-Pomares, S. Gordon, and S. Y. Wong. 2002. The beta-glucan receptor, dectin-1, is predominantly expressed on the surface of cells of the monocyte/macrophage and neutrophil lineages. *J. Immunol.* **169:**3876–3882.

Taylor, P. R., L. Martinez-Pomares, M. Stacey, H. H. Lin, G. D. Brown, and S. Gordon. 2005. Macrophage receptors and immune recognition. *Annu. Rev. Immunol.* **23:**901–944.

Tkalcevic, J., M. Novelli, M. Phylactides, J. P. Iredale, A. W. Segal, and J. Roes. 2000. Impaired immunity and enhanced resistance to endotoxin in the absence of neutrophil elastase and cathepsin G. *Immunity* **12:**201–210.

Tomee, J. F., P. S. Hiemstra, R. Heinzel-Wieland, and H. F. Kauffman. 1997. Antileukoprotease: an endogenous protein in the innate mucosal defense against fungi. *J. Infect. Dis.* **176:**740–747.

Tomee, J. F., A. T. Wierenga, P. S. Hiemstra, and H. K. Kauffman. 1997. Proteases from *Aspergillus fumigatus* induce release of proinflammatory cytokines and cell detachment in airway epithelial cell lines. *J. Infect. Dis.* **176:**300–303.

Tsunawaki, S., L. S. Yoshida, S. Nishida, T. Kobayashi, and T. Shimoyama. 2004. Fungal metabolite gliotoxin inhibits assembly of the human respiratory burst NADPH oxidase. *Infect. Immun.*72:3373–3382.

Umeki, S. 1994. Mechanisms for the activation/electron transfer of neutrophil NADPH-oxidase complex and molecular pathology of chronic granulomatous disease. *Ann. Hematol.* 68:267–277.

Underhill, D. M. 2007. Collaboration between the innate immune receptors dectin-1, TLRs, and NODs. *Immunol. Rev.* 219:75–87.

Urban, C. F., U. Reichard, V. Brinkmann, and A. Zychlinsky. 2006. Neutrophil extracellular traps capture and kill *Candida albicans* yeast and hyphal forms. *Cell. Microbiol.* 8:668–676.

Waldorf, A. R., S. M. Levitz, and R. D. Diamond. 1984. *In vivo* bronchoalveolar macrophage defense against *Rhizopus oryzae* and *Aspergillus fumigatus*. *J. Infect. Dis.* 150:752–760.

Walsh, T. J., and D. M. Dixon. 1989. Nosocomial aspergillosis: environmental microbiology, hospital epidemiology, diagnosis and treatment. *Eur. J. Epidemiol.* 5:131–142.

Wasylnka, J. A., and M. M. Moore. 2002. Uptake of *Aspergillus fumigatus* conidia by phagocytic and nonphagocytic cells in vitro: quantitation using strains expressing green fluorescent protein. *Infect. Immun.*70:3156–3163.

Wasylnka, J. A., and M. M. Moore. 2003. *Aspergillus fumigatus* conidia survive and germinate in acidic organelles of A549 epithelial cells. *J. Cell Sci.* 116:1579–1587.

Weinberger, M., I. Elattar, D. Marshall, S. M. Steinberg, R. L. Redner, N. S. Young, and P. Pizzo. 1992. Patterns of infection in patients with aplastic anemia and the emergence of *Aspergillus* as a major cause of death. *Medicine* 71:24–43.

Weiss, S. J. 1989. Tissue destruction by neutrophils. *N. Engl. J. Med.* 320:365–376.

Werts, C., S. E. Girardin, and D. J. Philpott. 2006. TIR, CARD and PYRIN: three domains for an antimicrobial triad. *Cell Death Differ.* 13:798–815.

Wiedow, O., and U. Meyer-Hoffert. 2005. Neutrophil serine proteases: potential key regulators of cell signalling during inflammation. *J. Intern. Med.* 257:319–328.

Wiley, J. M., N. Smith, B. G. Leventhal, M. L. Graham, L. C. Strauss, C. A. Hurwitz, J. Modlin, D. Mellits, R. Baumgardner, B. J. Corden, and C. I. Civin. 1990. Invasive fungal disease in pediatric acute leukemia patients with fever and neutropenia during induction chemotherapy: a multivariate analysis of risk factors. *J. Clin. Oncol.* 8: 280–286.

Witko-Sarsat, V., P. Rieu, B. Descamps-Latscha, P. Lesavre, and L. Halbwachs-Mecarelli. 2000. Neutrophils: molecules, functions and pathophysiological aspects. *Lab. Investig.* 80:617–653.

Zarember, K. A., J. A. Sugui, Y. C. Chang, K. J. Kwon-Chung, and J. I. Gallin. 2007. Human polymorphonuclear leukocytes inhibit *Aspergillus fumigatus* conidial growth by lactoferrin-mediated iron depletion. *J. Immunol.* 178:6367–6373.

Zelante, T., C. Montagnoli, S. Bozza, R. Gaziano, S. Bellocchio, P. Bonifazi, S. Moretti, F. Fallarino, P. Puccetti, and L. Romani. 2007. Receptors and pathways in innate antifungal immunity: the implication for tolerance and immunity to fungi. *Adv. Exp. Med. Biol.* 590:209–221.

Zhang, Z., R. Liu, J. A. Noordhoek, and H. F. Kauffman. 2005. Interaction of airway epithelial cells (A549) with spores and mycelium of *Aspergillus fumigatus*. *J. Infect.* 51:375–382.

Aspergillus fumigatus and Aspergillosis
Edited by J.-P. Latgé and W. J. Steinbach

Chapter 19

Interactions of *Aspergillus* with the Mucosa

WILLIAM W. HOPE AND SCOTT G. FILLER

Conidia of *Aspergillus* spp. are approximately 2 μm in diameter. Because of this small size, the conidia can be deposited in the nasal passages, pulmonary airways, and alveoli when they are inhaled. Although some inhaled conidia interact with macrophages and dendritic cells, other conidia must interact with epithelial cells that line the nasal and pulmonary mucosa. These interactions have the potential to significantly influence the host response to *Aspergillus* and determine whether invasive or allergic disease develops.

NORMAL SINUS MUCOSA AND FUNCTION

Structure and Function

A normally functioning mucosa is critical to the prevention of microbial colonization and infection. The sinuses are normally sterile (despite being in contact with the nonsterile nasal mucosa). The nasal passages and paranasal sinuses (consisting of maxillary, ethmoid, frontal, and sphenoid sinuses) are continuously exposed to a wide range of microorganisms, including fungal genera such as *Aspergillus, Penicillium, Cladosporium, Alternaria*, and *Aureobasidium* (Buzina et al., 2003). The mucosal lining of the nasal passages and paranasal sinus represents a critical barrier to the prevention of disease; this barrier consists of (i) a pseudostratified ciliated columnar epithelium, (ii) goblet cells and submucosal glands, and (iii) a mucus layer.

Epithelium

The nasal passages and paranasal sinuses are lined with a pseudostratified columnar ciliated epithelium (also referred to as a modified respiratory epithelium). Individual cells are connected by tight junctions which ensure the overall integrity of the epithelial layer and separation of sinus contents from underlying structures, such as the basement membrane (Ali and Pearson, 2007). These tight junctions are formed from transmembrane adhesion proteins (occludins and claudins) and intracellular proteins such as ZO-1, ZO-2, and ZO-3. Tight junctions limit access to paracellular transport, and their disruption (even temporarily) means that allergens may translocate across the mucosa and reach underlying dendritic cells, where an inflammatory reaction may be initiated (Ami and Pearson, 2007; Reed and Kita, 2004; Robinson et al., 2001). In addition to their barrier role, epithelial cells probably have a crucial role in early immunological signaling events which subsequently result in coordinated host responses.

Goblet Cells and Submucosal Glands

Goblet cells are fairly evenly located throughout the sinuses at a density of approximately 5,700 to 11,000 cells/mm^3 (Ali and Pearson, 2007). Goblet cells secrete mucin, which is critical to mucosal defenses (see below). The maxillary sinus mucosa is thinner, with a much higher number of goblet cells. Normal sinus mucosa exhibits scattered submucosal tubuloacinar cells, but no true glandular layer. There is a significantly higher number of seromucosal glands in the nose compared with the sinuses.

Mucus Layer

The mucus layer serves as a selective barrier between the extracellular milieu of the sinuses and the underlying epithelial cell layer. The mucus layer primarily serves to protect the epithelium but also enables hydration, lubrication, and transportation of foreign material.

William W. Hope • School of Translational Medicine, The University of Manchester, Manchester, United Kingdom. **Scott G. Filler** • Division of Infectious Diseases, Dept. of Medicine, Los Angeles Biomedical Research Institute at Harbor-UCLA Medical Center, Torrance, CA 90502, and The David Geffen School of Medicine at UCLA, Los Angeles, CA 90502.

Respiratory mucus is thick enough to enable inhaled particles to be efficiently trapped and transported but thin enough to prevent mucus retention, which may interfere with respiratory function and lead to infection. The gel-like properties of mucous are primarily related to the content of high-molecular-weight glycopeptides called mucins, which are stored in vesicles within Golgi cells and released by exocytosis. Mucins are large molecules with a molecular mass in the range of 1×10^6 to 50×10^6 Da (Ali and Pearson, 2007).

Mucins can be classified into secretory and membrane-associated forms. The membrane forms contain a hydrophobic membrane-spanning domain which serves to anchor the molecule to the apical cell membrane. Mucus gel is produced on the cellular surface following secretion (exocytosis) of mucin from the Golgi vesicles and constitutes the viscous element of sinus secretions. At the cell surface, mucins rapidly absorb water, resulting in a marked increase in volume (Ali and Pearson, 2007). The mucin molecule is a linear and flexible amino acid chain composed of subunits joined by disulfide bonds. Approximately 90% of the molecular weight of mucin consists of carbohydrate in the form of oligosaccharide chains attached to a central protein core. The wide array of oligosaccharide side chains may facilitate broad binding specificity for pathogenic microorganisms (Ali and Pearson, 2007).

Mucus secreted into the nasal and paranasal sinuses is drained by the mucociliary transport system. The mucus layer consists of two layers: the sol phase, in which the cilia recover from their active beat, and the gel phase, which is a more outer viscous layer that is transported by the ciliary beat. Inhaled foreign particulate matter is trapped within the mucus layer and swept away under the action of cilia. The mucus layer in the sinus is replaced two to three times per hour, and mucus transit times have been estimated to be 4.6 to 12.3 mm/min (Ali and Pearson, 2007).

Other Factors

The normal sterility of the sinuses suggests that in addition to the physical barrier conferred by the epithelial cell layer and mucus, a variety of immunological mechanisms are likely to be operational and important. Dendritic cells resident immediately beneath the mucosa respond to any initial breach by invading pathogens. A number of additional immunological mechanisms are also likely to be important, including collectins (surfactant protein A and D [Ooi et al., 2007b; Woodworth et al., 2007] and long pentraxin 3 [Baruah et al., 2007]), as well as defensins (Ooi et al., 2007a) and secreted immunoglobulin A antibody.

IN VITRO MODELS OF NASAL MUCOSA AND SINUSES

A number of in vitro models have been used to study the interaction between *Aspergillus* and the nasal mucosa. In each of these models, primary cells obtained directly from patients are used: there are no secondary cell lines available for study. Perhaps the most straightforward approach is to propagate ciliated epithelial cells obtained from nasal epithelial swabs (collected with cytology brushes) or nasal biopsies in tissue culture plates; these preparations can be used to study the production of cytokines or other morphological changes following exposure to *Aspergillus* spp. (Ooi et al., 2007a, 2007b; Shin et al., 2006). A more sophisticated model consists of growing nasal epithelial cells to confluence on collagen-containing membranes before enabling an air-liquid interface to form; this model can be used to assess the integrity of the epithelium (measured in terms of the transmembrane resistance) following direct exposure to *Aspergillus* spp. or culture filtrates (Botterel et al., 2002; Khoufache et al., 2007). Ciliary motility has been quantified by mounting ciliated epithelial cells on a slide which can then be visualized with a phase contrast microscope, and ciliary beat frequency is analyzed through the fluctuations in transmitted light (Amitani et al, 1995a; Cody et al., 1997).

A. FUMIGATUS MOLECULES INTERACTING WITH NASAL EPITHELIAL CELLS AND THE SINUS MUCOSA

The mechanisms of fungal colonization, invasion, and the initiation of allergic manifestations remain poorly elucidated. *Aspergillus* produces many cytotoxic proteins with allergenic properties and a wide range of secreted proteases. Although multiple putative virulence factors have been proposed in *Aspergillus* spp., none of them has been investigated using null mutants to determine their contribution to rhinosinusitis disease.

Proteases

Proteases have been implicated in the pathogenicity of *Aspergillus fumigatus* (Kheradmand et al., 2002; Markaryan et al., 1994; Reichard et al., 2000; Tomee et al., 1997). This fungus produces metalloproteases (e.g., Aspf5), aspartic proteases (e.g., Aspf10), and serine proteases (e.g., Aspf13 and Aspf18). These proteases may act in the allergic inflammatory cascade by (i) degrading epithelial cells to permit the passage of the allergen through the mucosal barrier, (ii) cleavage of receptors on the surface of innate immune cells, (iii) cleavage of receptors on T and B cells, (iv) activation of mast cells via protease-activated receptors (PARs), and (v) alteration of the protease-antiprotease balance of the mucosal epithelium (Donnelly et al., 2006). Purified proteases from *Aspergillus* cell filtrates lead to production of the proinflammatory cytokines interleukin-6 (IL-6) and IL-8 in a time- and dose-dependent manner, probably via interaction with PARs on epithelial cells (Kauffmann et al., 2000).

PARs are 7-transmembrane G-protein-coupled receptors that are stimulated by serine proteases and activated by cleavage of an N-terminal domain. The extracellular protease cleaves the amino acids at a specific site in the extracellular N terminus to expose a new N-terminal ligand; this new ligand then binds to another site on the same molecule, thus activating the receptor. The proteolytic activation is irreversible. Once cleaved, the receptor is degraded by lysosomes. Activated PARs lead to G-coupled protein signaling, which increases phospholipase C levels and intracellular calcium concentrations, as well as generation of heightened transcription of mitogen-activated protein kinase and nuclear factor κB. PARs are present on epithelial cells, mast cells, eosinophils, neutrophils, macrophage-monocytes, lymphocytes, smooth muscle cells, endothelium, fibroblasts, and neurons (Reed and Kita, 2004).

Protease-induced epithelial damage may facilitate the translocation of fungal antigens across the mucosa, and the liberation of proinflammatory cytokines may induce an inflammatory reaction within nasal tissues, leading to allergic manifestations following fungal exposure. Such a model appears tenable for the initiation of invasive disease and has also been proposed to account for the relationship between fungal exposure and allergic syndromes, such as asthma and allergic fungal sinusitis (Reed and Kita, 2004).

Secondary Metabolites

Aspergillus spp. produce a vast array of secondary metabolites that are potentially important in the initiation of invasive and allergic disease. Crude culture filtrates placed on human nasal epithelial cultures cause cellular damage as manifested by (i) decreased transepithelial resistance and (ii) decreased transepithelial potential differences; these electrophysiological changes may be important in the pathogenesis by disrupting normal mucosal structure and function. Recently, Khoufache and colleagues (2007) attempted to characterize and identify the specific component which may be responsible for these changes. The secondary metabolites gliotoxin, fumagillin, and helvolic acid did not reproduce the findings induced by whole filtrates (at the relevant concentrations and in a similar manner). Interestingly, verruculogen, a tremorgenic mycotoxin, replicates the changes induced by whole filtrates, modifies the electrophysiological properties of human nasal epithelial cells, and can be detected in both conidial and hyphal suspensions, and thus is important at the earliest point of contact and has been implicated in the pathogenesis of invasive aspergillosis.

ALLERGIC AND INVASIVE SYNDROMES RELATED TO MUCOSAL DYSFUNCTION

Acute invasive *Aspergillus* sinusitis was initially described in patients with profound and prolonged neutropenia (Talbot et al., 1991). As well as the major defects in systemic immunity which are associated with this syndrome, abnormalities of the nasal mucosa in bone marrow transplant patients have been documented (Cordonnier et al., 1996) and may result in decreased clearance of inhaled conidia and provide the opportunity for colonization of the sinuses and initiation of invasive disease.

Sinus aspergillomas have been linked to root canal filling materials which contain metals such as zinc. While these metals may serve as growth promoters for *Aspergillus*, they may also lead to disruption of local mucosal function and provide a nidus from which *Aspergillus* is able to become established and grow (Hope et al., 2005).

Perhaps the most obvious connection with mucosal dysfunction and *Aspergillus*-related disease is seen in cystic fibrosis. Cystic fibrosis, a lethal genetic disease, is characterized by altered viscoelastic properties of mucus; mucus may be 30 to 60 times more viscous than usual, which leads to mucus retention, reduced mucociliary clearance, and nasal polyposis. Approximately one-third of patients with cystic fibrosis have a positive fungal isolate from samples taken at endoscopic sinus surgery (Flume et al., 1994; Wise et al., 2005). While invasive disease in this population is highly unusual, allergic manifestations of fungal exposure are relatively common and well-recognized.

INTERACTIONS OF *ASPERGILLUS* SPP. WITH PULMONARY EPITHELIAL CELLS

Epithelial Cell Defense Mechanisms against *Aspergillus* spp.

The respiratory epithelium has multiple mechanisms of defense against *Aspergillus* that function in the absence of leukocytes. One mechanism is the presence of cilia, which prevent *Aspergillus* conidia from adhering to the epithelial cells and transport the conidia out of the airways (Paris et al., 1997). Interestingly, *A. fumigatus* culture filtrates inhibit ciliary beat frequency in tracheal explants. Gliotoxin is one factor within culture filtrates that inhibits ciliary beat frequency, although other toxins may also contribute to this inhibition (Amitani et al., 1995b). Whether intact *A. fumigatus* cells inhibit ciliary function in vivo has not yet been determined.

Pulmonary epithelial cells also express small cationic antimicrobial peptides, especially human β-defensin 2 (Bals, 2000; Duits et al., 2003; Harder et al., 2000; MacRedmond et al., 2005; Mendez-Samperio et al., 2006; Wang et al., 2003). These defensins have direct antimicrobial activity and are also chemotactic for dendritic cells and T lymphocytes. Although the activity of most human defensins against *Aspergillus* species has not been reported, they have significant activity against other fungi, such as *Candida* species (Joly et al., 2004).

Another pulmonary epithelial cell defense mechanism against *Aspergillus* is the production of surfactant protein D. In a mouse model of allergic bronchopulmonary aspergillosis, antigens of *A. fumigatus* induced the production of IL-4 and IL-13. These type 2 cytokines in turn stimulate type II alveolar cells to synthesize surfactant protein D (Haczku et al., 2006). This protein prevents activation of Th2 cells and thereby modulates the allergic response (Haczku et al., 2006). Other proteins produced by pulmonary epithelial cells, such as surfactant protein A, may also downregulate the allergic response to *A. fumigatus* antigens (Scanlon et al., 2005).

Adherence to Pulmonary Epithelial Cells

Surfactant protein D also inhibits the binding of *A. fumigatus* conidia to pulmonary epithelial cells in vitro (Yang et al., 2000). However, it is highly probable that conidia do adhere to and invade pulmonary epithelial cells during the initiation of invasive pulmonary aspergillosis (Fig. 1). The majority of studies of the interactions of *A. fumigatus* conidia with pulmonary epithelial cells in vitro have used the A549 type II alveolar epithelial cell line. *A. fumigatus* conidia adhere avidly to these cells (DeHart et al., 1997; Hope et al., 2007; Paris et al., 1997). At present the adhesin(s) expressed by *A. fumigatus* conidia and their epithelial cell ligands are unknown. However, it is known that the hydrophobic conidial protein RodA is dispensable for conidial adherence to pulmonary epithelial cells (Thau et al., 1994). In addition, it has been reported that conidial adherence to A549 cells is increased when these host cells are stimulated with gamma interferon, suggesting that at least one of the epithelial cell ligands is inducible (Bromley and Donaldson, 1996).

Epithelial Cell Invasion and Damage

After *Aspergillus* conidia adhere to pulmonary epithelial cells, they invade these cells by inducing their own endocytosis. Both conidia and germlings of *A. fumigatus* are endocytosed by pulmonary epithelial cells in vitro (DeHart et al., 1997; Hope et al., 2007; Paris et al., 1997; Wasylnka and Moore, 2003). This process is passive on the part of the organism, as even killed conidia are endocytosed (Kogan et al., 2004). Endocytosis of *A. fumigatus* conidia requires intact epithelial cell microfilaments and microtubules, and the conidia traffic to acidic phagosomes that contain the lysosomal marker LAMP-1 (Wasylnka and Moore, 2002, 2003). *A. fumigatus* conidia and hyphae also invade other normally nonphagocytic host cells, such as endothelial cells, by inducing their own endocytosis (Lopes-Bezerra and Filler, 2004; Paris et al., 1997; Wasylnka et al., 2002). This endocytic process resembles the endocytosis of bacteria by epithelial cells, which is a receptor-mediated process (Pizarro-Cerda and Cossart, 2006). Efforts are currently being made to identify the epithelial receptor(s) and *A. fumigatus* ligand(s) that mediate the endocytosis of this organism.

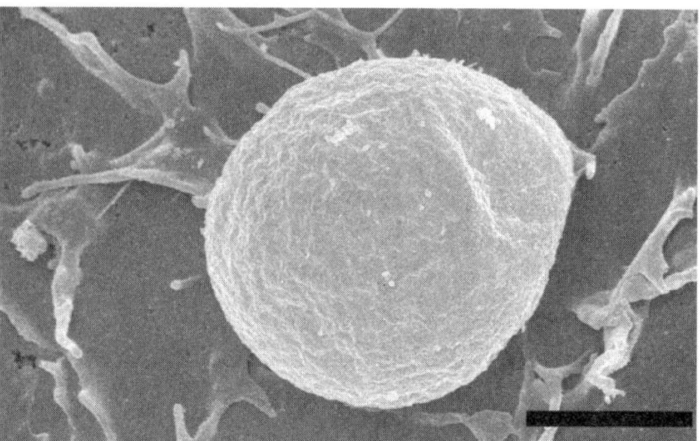

Figure 1. Scanning electron micrograph showing the interaction of an *A. fumigatus* conidium with a type II pneumocyte (A549 cell). Magnification, ×11,000. Bar, 3 μm.

Adherence to and invasion of the pulmonary epithelium induces responses in both the fungus and the host cell. It has been reported that conidia that have been endocytosed by A549 cells have delayed germination compared to conidia incubated in the absence of these cells (Wasylnka and Moore, 2002). In addition, mRNA levels of *pkaR* and *pkaC*, which encode regulatory and catalytic subunits of protein kinase A, respectively, are higher when *A. fumigatus* cells are grown in the presence of alveolar epithelial cells compared to their absence (Oliver et al., 2002).

Kogan et al. (2004) found that the endocytosis of intact *A. fumigatus* conidia by A549 pulmonary epithelial cells resulted in loss of actin stress fibers and the formation of membrane blebs. These responses were observed after only 30 min of contact between the conidia and A549 cells. Subsequently, there was disruption of focal adhesion sites which was associated with detachment of the epithelial cells from the basement membrane and loss of viability. This epithelial cell damage response was also induced by *A. fumigatus* culture filtrates and could be inhibited by adding a serine protease inhibitor to the medium. Moreover, this response was not induced by an alkaline serine protease-deficient strain of *A. fumigatus*. The authors concluded that serine proteases are essential for inducing epithelial cell damage in vitro.

Hope et al. (2007) constructed a bilayer model of the alveolus in which A549 cells were grown on top of human pulmonary endothelial cells. They found that after *A. fumigatus* conidia were endocytosed by the epithelial cells, they subsequently germinated and penetrated through both the epithelial cells and the adjacent endothelial cells (Color Plate 10 and Fig. 2). However, they observed that both the epithelial and endothelial cell monolayers remained intact for at least 14 h after infection. Using a chromium release assay, Filler and Kamai also determined that *A. fumigatus* conidia do not cause detectable damage to A549 cells for the first 8 h of infection (unpublished data). Why Kogan et al. detected *A. fumigatus*-induced epithelial cell injury after such a short incubation period is unclear, but it may have been due to differences in culture conditions, inocula, and the strain of *A. fumigatus* used.

Interestingly, live *A. fumigatus* conidia also inhibit the activation of caspase 3 and apoptosis induced by staurosporine or TNF in both A549 cells and human tracheal epithelial cells (Berkova et al., 2006). How this inhibition of epithelial cell apoptosis influences the pathogenesis of invasive pulmonary aspergillosis is uncertain at present.

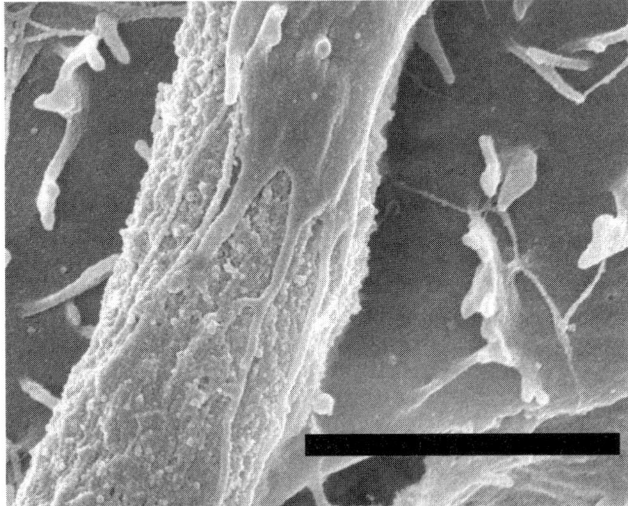

Figure 2. Scanning electron micrograph of an *A. fumigatus* hypha emerging from an endothelial cell and penetrating into the vascular lumen. Magnification, ×18,000. Bar, 2 μm.

Epithelial Cell Stimulation

Exposure to both viable *A. fumigatus* cells and *A. fumigatus* culture filtrates stimulates a proinflammatory response in pulmonary epithelial cells in vitro, as manifested by secretion of IL-6, IL-8, and monocyte chemoattractant protein 1 (Borger et al., 1999; Tomee et al., 1997; Zhang et al., 2005). Pulmonary epithelial cells of mice exposed to *A. fumigatus* antigens are also stimulated to express increased TNF, IL-1α, and intercellular adhesion molecule 1 (Chu et al., 1996). Epithelial cell stimulation by culture filtrates of *A. fumigatus* is induced by serine proteases, which induce activation of the transcription factor NF-κB in A549 cells. These results suggest that epithelial cell stimulation may be mediated via a PAR type 2 mechanism (Borger et al., 1999; Tomee et al., 1997). Whether intact *A. fumigatus* cells also stimulate epithelial cells via a similar mechanism has not yet been determined. However, it is likely that the secretion of proinflammatory cytokines and expression of leukocyte adhesion molecules by the infected pulmonary epithelial cells contribute to the recruitment of leukocytes to foci of epithelial cell invasion.

REFERENCES

Ali, M. S., and J. P. Pearson. 2007. Upper airway mucin gene expression: a review. *Laryngoscope* **117:**932–938.

Amitani, R., T. Murayama, R. Nawada, W. J. Lee, A. Niimi, K. Suzuki, E. Tanaka, and F. Kuze. 1995a. *Aspergillus* culture filtrates and sputum sols from patients with pulmonary aspergillosis cause damage to human respiratory ciliated epithelium in vitro. *Eur. Respir. J.* **8:**1681–1687.

Amitani, R., G. Taylor, E. N. Elezis, C. Llewellyn-Jones, J. Mitchell, F. Kuze, P. J. Cole, and R. Wilson. 1995b. Purification and characterization of factors produced by *Aspergillus fumigatus* which affect human ciliated respiratory epithelium. *Infect. Immun.* **63:**3266–3271.

Bals, R. 2000. Epithelial antimicrobial peptides in host defense against infection. *Respir. Res.* **1:**141–150.

Baruah, P., M. Trimarchi, I. E. Dumitriu, G. Dellantonio, C. Doglioni, P. Rovere-Querini, M. Bussi, and A. A. Manfredi. 2007. Innate responses to *Aspergillus*: role of C1q and pentraxin 3 in nasal polyposis. *Am. J. Rhinol.* **21:**224–230.

Berkova, N., S. Lair-Fulleringer, F. Femenia, D. Huet, M. C. Wagner, K. Gorna, F. Tournier, O. Ibrahim-Granet, J. Guillot, R. Chermette, P. Boireau, and J. P. Latgé. 2006. *Aspergillus fumigatus* conidia inhibit tumour necrosis factor- or staurosporine-induced apoptosis in epithelial cells. *Int. Immunol.* **18:**139–150.

Borger, P., G. H. Koeter, J. A. Timmerman, E. Vellenga, J. F. Tomee, and H. F. Kauffman. 1999. Proteases from *Aspergillus fumigatus* induce interleukin (IL)-6 and IL-8 production in airway epithelial cell lines by transcriptional mechanisms. *J. Infect. Dis.* **180:**1267–1274.

Botterel, F., C. Cordonnier, V. Barbier, L. Wingerstmann, M. Liance, A. Coste, E. Escudier, and S. Bretagne. 2002. *Aspergillus fumigatus* causes in vitro electrophysiological and morphological modifications in human nasal epithelial cells. *Histol. Histopathol.* **17:**1095–1101.

Bromley, I. M., and K. Donaldson. 1996. Binding of *Aspergillus fumigatus* spores to lung epithelial cells and basement membrane proteins: relevance to the asthmatic lung. *Thorax* **51:**1203–1209.

Buzina, W., H. Braun, K. Freudenschuss, A. Lackner, W. Habermann, and H. Stammberger. 2003. Fungal biodiversity: as found in nasal mucus. *Med. Mycol.* **41:**149–161.

Chu, H. W., J. M. Wang, M. Boutet, and M. Laviolette. 1996. Tumor necrosis factor-alpha and interleukin-1 alpha expression in a murine model of allergic bronchopulmonary aspergillosis. *Lab. Anim. Sci.* **46:**42–47.

Cody, D. T., II, T. V. McCaffrey, G. Roberts, and E. B. Kern. 1997. Effects of *Aspergillus fumigatus* and *Alternaria alternata* on human ciliated epithelium in vitro. *Laryngoscope* **107:**1511–1514.

Cordonnier, C., L. Gilain, F. Ricolfi, L. Deforges, F. Girard-Pipau, F. Poron, M. C. Millepied, and E. Escudier. 1996. Acquired ciliary abnormalities of nasal mucosa in marrow recipients. *Bone Marrow Transplant.* **17:**611–616.

DeHart, D. J., D. E. Agwu, N. C. Julian, and R. G. Washburn. 1997. Binding and germination of *Aspergillus fumigatus* conidia on cultured A549 pneumocytes. *J. Infect. Dis.* **175:**146–150.

Donnelly, S., J. P. Dalton, and A. Loukas. 2006. Proteases in helminth- and allergen-induced inflammatory responses. *Chem. Immunol. Allergy* **90:**45–64.

Duits, L. A., P. H. Nibbering, E. van Strijen, J. B. Vos, S. P. Mannesse-Lazeroms, M. A. van Sterkenburg, and P. S. Hiemstra. 2003. Rhinovirus increases human beta-defensin-2 and -3 mRNA expression in cultured bronchial epithelial cells. *FEMS Immunol. Med. Microbiol.* **38:**59–64.

Flume, P. A., T. M. Egan, L. J. Paradowski, F. C. Detterbeck, J. T. Thompson, and J. R. Yankaskas. 1994. Infectious complications of lung transplantation. Impact of cystic fibrosis. *Am. J. Respir. Crit. Care Med.* **149:**1601–1607.

Haczku, A., Y. Cao, G. Vass, S. Kierstein, P. Nath, E. N. Atochina-Vasserman, S. T. Scanlon, L. Li, D. E. Griswold, K. F. Chung, F. R. Poulain, S. Hawgood, M. F. Beers, and E. C. Crouch. 2006. IL-4 and IL-13 form a negative feedback circuit with surfactant protein-D in the allergic airway response. *J. Immunol.* **176:**3557–3565.

Harder, J., U. Meyer-Hoffert, L. M. Teran, L. Schwichtenberg, J. Bartels, S. Maune, and J. M. Schroder. 2000. Mucoid *Pseudomonas aeruginosa*, TNF-alpha, and IL-1beta, but not IL-6, induce human beta-defensin-2 in respiratory epithelia. *Am. J. Respir. Cell Mol. Biol.* **22:**714–721.

Hope, W. W., M. J. Kruhlak, C. A. Lyman, R. Petraitiene, V. Petraitis, A. Francesconi, M. Kasai, D. Mickiene, T. Sein, J. Peter, A. M. Kelaher, J. E. Hughes, M. P. Cotton, C. J. Cotten, J. Bacher, S. Tripathi, L. Bermudez, T. K. Maugel, P. M. Zerfas, J. R. Wingard, G. L. Drusano, and T. J. Walsh. 2007. Pathogenesis of *Aspergillus fumigatus* and the kinetics of galactomannan in an in vitro model of early invasive pulmonary aspergillosis: implications for antifungal therapy. *J. Infect. Dis.* **195:**455–466.

Hope, W. W., T. J. Walsh, and D. W. Denning. 2005. The invasive and saprophytic syndromes due to Aspergillus spp. *Med. Mycol.* **43**(Suppl. 1)**:**S207–S238.

Joly, S., C. Maze, P. B. McCray, Jr., and J. M. Guthmiller. 2004. Human beta-defensins 2 and 3 demonstrate strain-selective activity against oral microorganisms. *J. Clin. Microbiol.* **42:**1024–1029.

Kauffman, H. F., J. F. Tomee, M. A. van de Riet, A. J. Timmerman, and P. Borger. 2000. Protease-dependent activation of epithelial cells by fungal allergens leads to morphologic changes and cytokine production. *J. Allergy Clin. Immunol.* **105:**1185–1193.

Kheradmand, F., A. Kiss, J. Xu, S. H. Lee, P. E. Kolattukudy, and D. B. Corry. 2002. A protease-activated pathway underlying Th cell type 2 activation and allergic lung disease. *J. Immunol.* **169:**5904–5911.

Khoufache, K., O. Puel, N. Loiseau, M. Delaforge, D. Rivollet, A. Coste, C. Cordonnier, E. Escudier, F. Botterel, and S. Bretagne. 2007. Verruculogen associated with *Aspergillus fumigatus* hyphae and conidia modifies the electrophysiological properties of human nasal epithelial cells. *BMC Microbiol.* **7:**5.

Kogan, T. V., J. Jadoun, L. Mittelman, K. Hirschberg, and N. Osherov. 2004. Involvement of secreted *Aspergillus fumigatus* proteases in disruption of the actin fiber cytoskeleton and loss of focal adhesion sites in infected A549 lung pneumocytes. *J. Infect. Dis.* **189:**1965–1973.

Lopes-Bezerra, L. M., and S. G. Filler. 2004. Interactions of *Aspergillus fumigatus* with endothelial cells: internalization, injury, and stimulation of tissue factor activity. *Blood* **103:**2143–2149.

MacRedmond, R., C. Greene, C. C. Taggart, N. McElvaney, and S. O'Neill. 2005. Respiratory epithelial cells require Toll-like receptor 4 for induction of human beta-defensin 2 by lipopolysaccharide. *Respir. Res.* **6:**116.

Markaryan, A., I. Morozova, H. Yu, and P. E. Kolattukudy. 1994. Purification and characterization of an elastinolytic metalloprotease from *Aspergillus fumigatus* and immunoelectron microscopic evidence of secretion of this enzyme by the fungus invading the murine lung. *Infect. Immun.* **62:**2149–2157.

Mendez-Samperio, P., E. Miranda, and A. Trejo. 2006. *Mycobacterium bovis* bacillus Calmette-Guerin (BCG) stimulates human beta-defensin-2 gene transcription in human epithelial cells. *Cell. Immunol.* **239:**61–66.

Oliver, B. G., J. C. Panepinto, J. R. Fortwendel, D. L. Smith, D. S. Askew, and J. C. Rhodes. 2002. Cloning and expression of pkaC and pkaR, the genes encoding the cAMP-dependent protein kinase of *Aspergillus fumigatus*. *Mycopathologia* **154:**85–91.

Ooi, E. H., P. J. Wormald, A. S. Carney, C. L. James, and L. W. Tan. 2007a. Fungal allergens induce cathelicidin LL-37 expression in chronic rhinosinusitis patients in a nasal explant model. *Am. J. Rhinol.* **21:**367–372.

Ooi, E. H., P. J. Wormald, A. S. Carney, C. L. James, and L. W. Tan. 2007b. Surfactant protein D expression in chronic rhinosinusitis patients and immune responses in vitro to *Aspergillus* and *Alternaria* in a nasal explant model. *Laryngoscope* **117:**51–57.

Paris, S., E. Boisvieux-Ulrich, B. Crestani, O. Houcine, D. Taramelli, L. Lombardi, and J. P. Latgé. 1997. Internalization of *Aspergillus fumigatus* conidia by epithelial and endothelial cells. *Infect. Immun.* **65:**1510–1514.

Pizarro-Cerda, J., and P. Cossart. 2006. Bacterial adhesion and entry into host cells. *Cell* **124:**715–727.

Reed, C. E., and H. Kita. 2004. The role of protease activation of inflammation in allergic respiratory diseases. *J. Allergy Clin. Immunol.* **114:**997–1008.

Reichard, U., G. T. Cole, T. W. Hill, R. Ruchel, and M. Monod. 2000. Molecular characterization and influence on fungal development of ALP2, a novel serine proteinase from *Aspergillus fumigatus*. *Int. J. Med. Microbiol.* **290:**549–558.

Robinson, C., S. F. Baker, and D. R. Garrod. 2001. Peptidase allergens, occludin and claudins. Do their interactions facilitate the development of hypersensitivity reactions at mucosal surfaces? *Clin. Exp. Allergy* **31:**186–192.

Scanlon, S. T., T. Milovanova, S. Kierstein, Y. Cao, E. N. Atochina, Y. Tomer, S. J. Russo, M. F. Beers, and A. Haczku. 2005. Surfactant protein-A inhibits *Aspergillus fumigatus*-induced allergic T-cell responses. *Respir. Res.* **6:**97.

Shin, S. H., Y. H. Lee, and C. H. Jeon. 2006. Protease-dependent activation of nasal polyp epithelial cells by airborne fungi leads to migration of eosinophils and neutrophils. *Acta Otolaryngol.* **126:**1286–1294.

Talbot, G. H., A. Huang, and M. Provencher. 1991. Invasive aspergillus rhinosinusitis in patients with acute leukemia. *Rev. Infect. Dis.* **13:**219–232.

Thau, N., M. Monod, B. Crestani, C. Rolland, G. Tronchin, J. P. Latgé, and S. Paris. 1994. Rodletless mutants of *Aspergillus fumigatus*. *Infect. Immun.* **62:**4380–4388.

Tomee, J. F., A. T. Wierenga, P. S. Hiemstra, and H. K. Kauffman. 1997. Proteases from *Aspergillus fumigatus* induce release of proinflammatory cytokines and cell detachment in airway epithelial cell lines. *J. Infect. Dis.* **176:**300–303.

Wang, X., Z. Zhang, J. P. Louboutin, C. Moser, D. J. Weiner, and J. M. Wilson. 2003. Airway epithelia regulate expression of human beta-defensin 2 through Toll-like receptor 2. *FASEB J.* **17:**1727–1729.

Wasylnka, J. A., and M. M. Moore. 2003. *Aspergillus fumigatus* conidia survive and germinate in acidic organelles of A549 epithelial cells. *J. Cell Sci.* **116:**1579–1587.

Wasylnka, J. A., and M. M. Moore. 2002. Uptake of *Aspergillus fumigatus* conidia by phagocytic and nonphagocytic cells in vitro: quantitation using strains expressing green fluorescent protein. *Infect. Immun.* **70:**3156–3163.

Wise, S. K., T. T. Kingdom, L. McKean, J. M. DelGaudio, and G. Venkatraman. 2005. Presence of fungus in sinus cultures of cystic fibrosis patients. *Am. J. Rhinol.* **19:**47–51.

Woodworth, B. A., J. G. Neal, D. Newton, K. Joseph, A. P. Kaplan, J. E. Baatz, and R. J. Schlosser. 2007. Surfactant protein A and D in human sinus mucosa: a preliminary report. *ORL J. Otorhinolaryngol. Relat. Spec.* **69:**57–60.

Yang, Z., S. M. Jaeckisch, and C. G. Mitchell. 2000. Enhanced binding of *Aspergillus fumigatus* spores to A549 epithelial cells and extracellular matrix proteins by a component from the spore surface and inhibition by rat lung lavage fluid. *Thorax* **55:**579–584.

Zhang, Z., R. Liu, J. A. Noordhoek, and H. F. Kauffman. 2005. Interaction of airway epithelial cells (A549) with spores and mycelium of *Aspergillus fumigatus*. *J. Infect.* **51:**375–382.

Aspergillus fumigatus and Aspergillosis
Edited by J.-P. Latgé and W. J. Steinbach
©2009 ASM Press, Washington, DC

Chapter 20

Dendritic Cells in *Aspergillus* Infection and Allergy

LUIGINA ROMANI

The inherent resistance to diseases caused by *Aspergillus fumigatus* suggests the occurrence of regulatory mechanisms that provide the host with adequate defense without necessarily eliminating the fungus or causing unacceptable levels of host damage. As a matter of fact, immunocompetent and nonatopic subjects are relatively resistant to *A. fumigatus* diseases, and disease occurs in the setting of host damage (Latgé, 1999; Marr et al., 2002). Most of the inhaled conidia are eliminated by exclusion mechanisms, which include physical barriers such as mucus and cilia as well as a variety of mediators of the collectin (Garlanda et al., 2002; Madan et al., 2005; McCormack et al., 2002) and cytokine (Phadke and Mehrad, 2005) families. Effector mechanisms of the innate immune system, such as resident alveolar macrophages and polymorphonuclear neutrophils (PMN), have long been recognized as major host defenses against aspergillosis (Latgé, 1999; Walsh et al., 2005). PMN are the predominant immune cells in the acute stage of the infection and are essential in initiation and execution of the acute inflammatory response and subsequent resolution of the infection. However, despite extensive fungal growth, pulmonary pathology is reduced under conditions of PMN deficiency, in both humans and mice (Balloy et al., 2005; Marr et al., 2002), a finding suggesting that PMN may act as double-edged swords, as the excessive release of oxidants and proteases may be responsible for injury to organs and fungal sepsis (Bellocchio et al., 2004b).

IMMUNITY TO *ASPERGILLUS*: AN EVOLVING FUNCTION

Innate and adaptive immune responses act to generate the most effective form of immunity for protection against *A. fumigatus* (Hohl and Feldmesser, 2007; Montagnoli et al., 2006a; Walsh et al., 2005). The decision of how to respond is still primarily determined by interactions between the fungus and cells of the innate immune system, but the actions of T cells will feed back into this dynamic equilibrium to regulate antifungal effector functions and the balance between proinflammatory and anti-inflammatory signals. Generation of a dominant Th1 response driven by interleukin-12 (IL-12) is essentially required for the expression of protective immunity to the fungus. Through the production of the signature cytokine gamma interferon (IFN-γ), the activation of Th1 cells is instrumental in the optimal activation of phagocytes at sites of infection (Romani, 2004). The inflammatory allergic manifestations that follow contact with or inhalation of *Aspergillus* spp. all constitute compelling evidence for the pathogenic role of T-cell dysreactivity in fungal diseases. Allergy is an overzealous Th2 response to environmental airborne allergens. Moreover, and consistent with the inflammatory state, patients with pulmonary aspergillosis are genetically low producers of both transforming growth factor β (TGF-β) and IL-10 (Sambatakou et al., 2006). While the importance of Th1/Th2 dysregulation in the outcome of invasive aspergillosis is undisputed, recent evidence points to a further complexity of cross-regulatory innate and adaptive Th pathways that operate from the state of continuous sensitization to ubiquitous airborne fungi to infections and fungal diseases. In this regard, a role for "inflammatory T cells," or Th17 cells, producing IL-17, as a link between T-cell inflammation and granulocytic influx has been observed in allergic airway inflammation (Romani and Puccetti, 2007; Romani and Puccetti, in press; Romani, 2008).

THE BIPOLAR NATURE OF INFLAMMATION

A major role of the immune system, innate and adaptive, is to maintain tissue homeostasis by stopping

Luigina Romani • Microbiology Section, Dept. of Experimental Medicine and Biochemical Sciences, and Fondazione "Istituto di Ricovero e Cura per le Biotecnologie Trapiantologiche," Perugia, Italy.

and/or preventing cell damage caused by invading pathogens and promoting tissue repair following formation of a lesion. The control of the immune response is pivotal for preventing damage. Prolonged inflammation is a hallmark of a wide range of chronic diseases and autoimmune conditions (Gutcher and Becher, 2007). Although the inflammatory response to fungi may serve to limit infection, an overzealous or heightened inflammatory response may contribute to pathogenicity. This conceptual principle is best exemplified by the occurrence of severe fungal infections in patients with immune reconstitution syndrome, an entity characterized by localized and systemic inflammatory reactions of varying degrees that have both beneficial and noxious features during an invasive mycosis (Miceli et al., 2007; Singh and Perfect, 2007). The association of persistent inflammation with intractable *Aspergillus* infection is common in nonneutropenic patients after allogeneic hematopoietic stem cell transplantation (HSCT) (Ortega et al., 2006) as well as in those with allergic fungal diseases (Schubert, 2006). Additionally, a high incidence of fungal infections and sensitization to *Aspergillus* spp. has been described in hyper-immunoglobulin E syndrome, in which increased levels of proinflammatory gene transcripts have recently been described (Holland et al., 2007). Therefore, paradoxically, an increased inflammatory innate response may predispose to either fungal infections or dysregulated immune responses to the fungus. As a matter of fact, the status of innate host immunity also may contribute significantly to the histological patterns associated with fungal infections (Marr et al., 2002). Thus, although host immunity is crucial in the eradication of infection, immunological recovery can also be detrimental and may contribute towards worsening disease expression.

Most prominently, in patients with chronic granulomatous disease (CGD), a hyperinflammatory phenotype and defective fungus (typically *A. fumigatus*) clearance have long been known to benefit each other (Segal et al., 2000). In experimental CGD, an intrinsic, genetically determined failure to control inflammation to sterile fungal components determines the animal's inability to resolve an actual infection with the fungus (Romani et al., 2008). The above observations highlight a truly bipolar nature of the inflammatory process in infection. Early inflammation prevents or limits infection, but an uncontrolled response may eventually oppose disease eradication. A main implication of these findings is that, at least in specific clinical settings, it is an exaggerated inflammatory response that likely compromises a patient's ability to eradicate infection and not an "intrinsic" susceptibility to infection that determines the state of chronic or intractable disease. Because *Aspergillus*, through the subversion of host inflammatory pathways (Moretti et al., 2008), may promote inflammatory responses, an inflammatory vicious circle seems to be at work, the manipulation of which may offer strategies to control or prevent exacerbations of *Aspergillus* infections and diseases.

There have been tremendous advances in the field of innate immunity over the last decade, including the elucidation of the toll-like receptors (TLR), which are activated by recognition of fungus-associated molecular patterns (see also chapters 21 and 22). A compelling body of evidence indicates that hosts rely upon TLR and C-type lectins, such as dectin-1, to sense the presence of and respond to *Aspergillus* ligands (Bellocchio et al., 2004a; Braedel et al., 2004; Bretz et al., 2008; Chignard et al., 2007). Nonspecific activation by various microbial stimuli can lead to important antimicrobial effects but can also result in inflammation and injury because of release of inflammatory mediators, such as cytokines, reactive oxygen species, and nitric oxide (Trinchieri and Sher, 2007). Thus, TLR activation itself is a double-edged sword. By hyperinduction of proinflammatory cytokines, by facilitating tissue damage, or by impairing protective immunity, TLR might also promote the pathogenesis of infections. Another function of the innate immunity that is emerging is its role in sterile inflammation, that is, inflammation caused by endogenous TLR ligands. Recent studies have provided new insights into the host pathogenic determinants responsible for inflammatory pathology in *A. fumigatus* infections (Moretti et al., 2008) and diseases (Grohmann et al., 2007).

This chapter will highlight how the remarkable functional plasticity of dendritic cells (DC) in response to the fungus may accommodate the activation of different mechanisms of immunity and can be exploited for the deliberate targeting of cells and pathways of protective cell-mediated immunity and for the identification of candidate fungal vaccines.

DENDRITIC CELLS

The DC system, first discovered by Ralph Steinman and Zanvil Cohn in 1973 (Steinman and Cohn, 1973), comprises a network of different subpopulations. DC show a unique functional duality during their development, designed to ultimately provide secondary lymphoid tissues with useful information about the antigenic composition in the periphery. They are strategically located at epithelial barriers that often serve as major portals of pathogen entry. There are specialized features of DC that contribute to their capacity to control T-cell recognition and responsiveness and, in turn, either prevent or generate disease. DC avidly internalize nonopsonized pathogens by macropinocytosis, phagocytosis, or through C-type lectins and mannose receptors (MR) or complexes of antibody and microbial antigen via receptors for the Fc portion of immunoglobulins (FcR). By virtue of their high phagocytic and endocytic capacities, DC constitutively internalize samples of their antigenic microenvironment, which in the event of infection will also include microbial antigens. After antigen capture in the presence of maturation signals associated with inflammation or infection, activated DC undergo a complex maturation process. In infections, DC become activated and mobilized by either direct recognition of pathogen-derived ligands by TLR or indirectly through receptors for inflammatory cytokines and chemokines and T-cell products. In vivo this process is paralleled by migration of DC to T-cell-rich areas of lymphoid organs, sites for the generation of immunity and tolerance. Here, the ability of DC to influence the pattern of cytokines secreted by T cells represents a critical function which can profoundly influence the final outcome of the immune response to a pathogen. Several factors appear to influence the ability of DC to polarize T-cell cytokine responses. These include the following.

DC Subsets

DC can develop along two pathways, myeloid DC (also called conventional DC) and plasmacytoid DC (pDC). DC subsets differ in their phenotype, microenvironmental localization, migration potential, pattern recognition receptor expression, responsiveness to microbes, and capacity to induce and regulate distinct arms of the innate and adaptive immune systems (Pulendran, 2004; Steinman 2007; Steinman and Banchereau, 2007). Overall, the ability of a given DC subset to respond with flexible activating programs to the different stimuli as well as the ability of different subsets to convert into each other (Diebold et al., 2003; Zuniga et al., 2004) confer unexpected plasticity to the DC system. In this regard, as further discussed below, through their ability to produce type I IFNs, pDC indisputably drive protective antiviral inflammation but have also been implicated in the induction and exacerbation of the inflammatory processes associated with autoimmunity and allergy (Colonna et al., 2004). The bipolar functions of pDC, and of the associated type I IFN response, appear to be an intrinsic ability of the immune system to coactivate cytostatic mechanisms and induce the death of pathogenic T cells and polarization of T cells towards a Treg-cell phenotype. Tryptophan catabolism (see below) fulfills all the requirements to effect these functions (Grohmann et al., 2003; Mellor and Munn, 2004).

Dynamics of DC Migration to the Lymphoid Organs

Early after initiation of the immune response, large numbers of recently stimulated DC actively secreting IL-12 and entering into the T-cell areas may preferentially induce Th1-cell priming. In contrast, at a later time, when DC influx decreases and the surviving DC in T-cell areas have downregulated their secretion of IL-12, preferential priming for Th2 and Treg may occur (Lanzavecchia and Sallusto, 2001).

Nature of the Maturation Stimuli

TLR function as sensors of microbial infection and play a critical role in the induction of innate and adaptive immune responses. TLR-mediated DC activation and maturation lead to upregulation of major histocompatibility complex and costimulatory molecules, as well as production of cytokines and soluble factors. However, DC maturation can also occur in response to signals from newly activated $CD4^+$ T cells independently of innate priming. T-cell-driven DC maturation takes place both in *cis* and in *trans*, affecting all DC in the microenvironment, irrespective of antigen specificity (Reis e Sousa, 2006).

Microenvironmental Factors

Cytokine production by DC can be influenced by mediators released in the DC microenvironment (Pulendran, 2004). DC and T cells can also interact through the formation of an immunological synapse in which T-cell receptors and costimulatory molecules congregate in a central area surrounded by a ring of adhesion molecules. Sustained signaling via these synaptic interactions is required in order for the T cell to enter the first cell division cycle. It is believed that antigen presentation by DC to T cells may involve the concerted action of multiple subsets of DC, each of which contributes a different facet to this process through either an antigen-processing pathway that is direct (antigen-loaded DC that migrate to the secondary lymphoid organ could undergo apoptosis or necrosis and be phagocytosed by resident DC) or one that is indirect (migrating DC could actively release antigen-bearing vesicle exosomes derived from the DC lysosomal compartment, which could then be captured by resident DC) (Pulendran, 2004). In contrast, antigen capture in the absence of activation stimuli, as seen in "steady-state" migration, may lead to the induction of T-cell tolerance as a result of antigen presentation by immature DC in the absence of costimulation (Steinman, 2007). Therefore, DC are central in the early decision-making mechanisms that result in a given type of immune response and determine the balance between immunopathology and protective immunity generated by host-microbe interactions.

DC AT THE HOST-FUNGUS INTERFACE

Studies in vivo suggest that DC have the ability to internalize *Aspergillus* at sites of the infection (Bozza et al., 2002b). It is known that DC of the respiratory tract are specialized for uptake and processing but not for antigen presentation, the latter requiring cytokine maturation signals that are encountered after migration to regional lymph nodes. In aspergillosis, DC present in the alveolar spaces phagocytose conidia, translocate to the space below, within the alveolar septal wall, and reach the draining lymph nodes, where fungus-pulsed DC instruct local development of antifungal Th reactivity (Fig. 1). It seems, therefore, that DC are uniquely able to decode the fungus-associated information at the host-fungus interface and translate it in qualitatively different

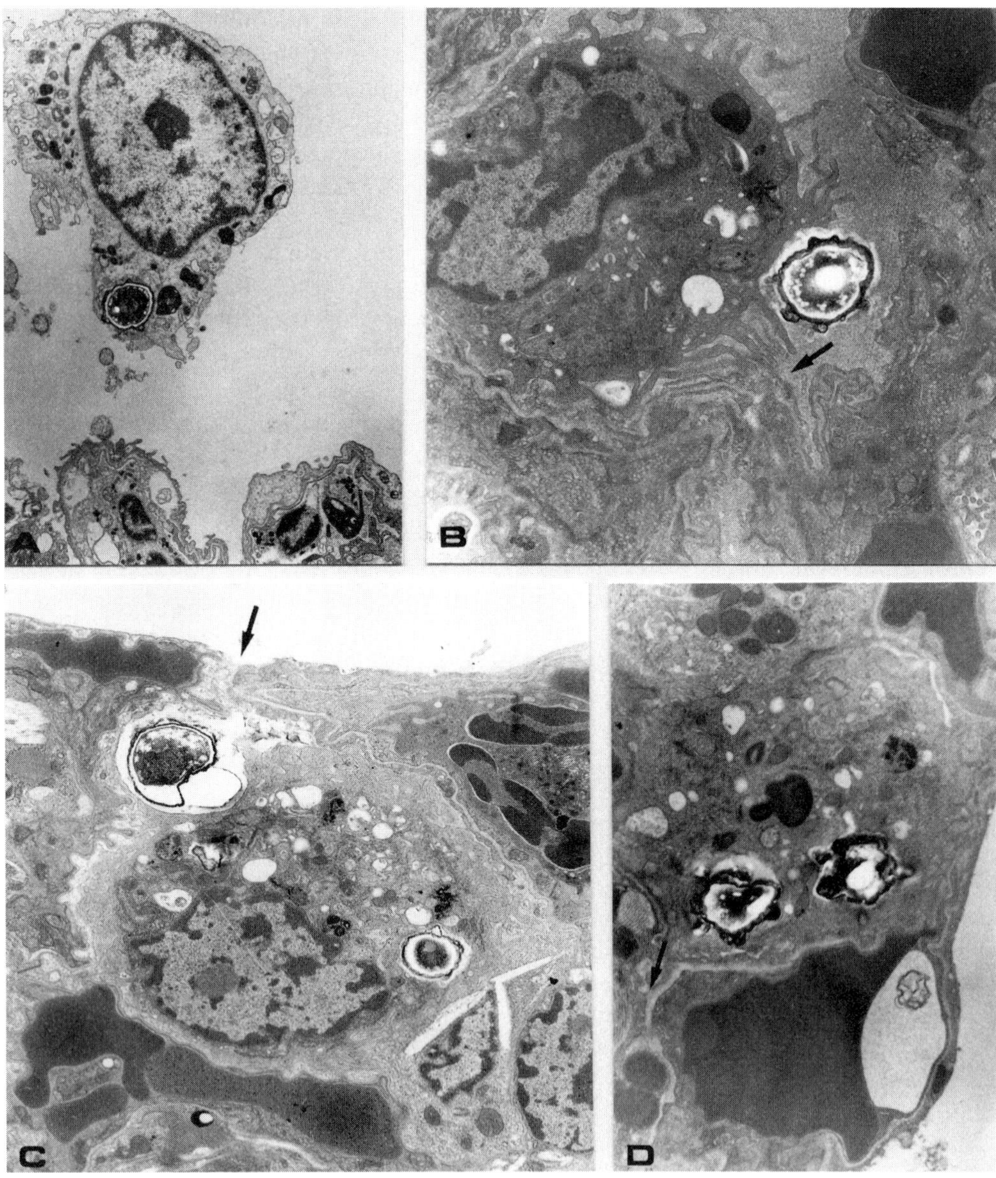

Figure 1. Transmission electron microscopy images of lungs of mice with *A. fumigatus* infection. Mice were intratracheally injected with viable *Aspergillus* conidia 2 h before being processed for transmission electron microscopy. (A) Conidia are internalized by phagocytic cells with characteristics of DC morphology, as judged by the numerous cytoplasmic extensions and abundant cytoplasm present in the alveolar spaces. Magnification, ×12,000. (B) Through emission of pseudopods, DC engulf conidia and make contact with the epithelial barrier (arrow). Magnification, ×8,000. (C) DC with engulfed conidia and free conidia migrate through invaginated epithelial cells (arrow). Magnification, ×8,000. (D) DC with engulfed conidia are present within the alveolar septal wall. Magnification, ×12,000. Reproduced from Bozza et al. (2002b) with permission of the publisher.

Th immune responses in vivo (Bozza et al., 2002b, 2003, 2004).

In vitro, the engagement of distinct receptors by different fungal morphotypes translates into downstream signaling events, ultimately regulating fungus survival, DC activation, and Th differentiation (Behnsen et al., 2007; Bellocchio et al., 2004a; Bozza et al., 2002b, 2003, 2004; Grazziutti et al., 2001; Romani et al., 2002; Serrano-Gomez et al., 2004). The fungus has exploited common pathways for entry into DC, which include a lectin-like pathway that includes an MR with galactomannan specificity, DC-SIGN and partly CR3 for the unicellular form and opsono-dependent pathways for the filamentous form, which occurs by a more conventional, zipper-type phagocytosis and involves the cooperative action of FcγR II and III and CR3. Transmission electronic microscopy indicated that internalization of conidia occurred predominantly by coiling phagocytosis, characterized by the presence of overlapping bilateral pseudopods that led to a pseudopodal stack before transforming into a phagosome wall. In contrast, entry of hyphae occurred by a more conventional zipper-type phagocytosis, characterized by the presence of symmetrical pseudopods which strictly followed the contour of the hyphae before fusion. The fate of the different forms of the fungus inside cells appeared to be quite different. Two hours after the exposure, numerous conidia were found inside DC with no evidence of conidial destruction, as opposed to hyphae that were rapidly degraded once inside cells (Fig. 2). As killing of conidia would seem to be a necessary prerequisite to obtain efficient antigen presentation, it can be postulated that either a small number of conidia are actually degraded by mature DC, thus allowing their antigen processing and presentation, or alternatively, antigens are processed and regurgitated by other infected phagocytes and then transferred to DC for presentation.

Entry of *Aspergillus* conidia through MR and dectin-1 results in the production of proinflammatory cytokines, including IL-12, and upregulation of costimulatory molecules and histocompatibility class II antigens. IL-12 and IL-23 production by DC require the MyD88 pathway, with the implication of distinct TLR involvement (Romani, 2004; Zelante et al., 2007). In contrast, coligation of CR3 with FcγR, as in the phagocytosis of hyphae, resulted in the production of IL-4/IL-10 and upregulation of costimulatory molecules and histocompatibility class II antigens (Bozza et al., 2002b). The production of IL-10 was largely MyD88 independent (Bellocchio et al., 2004a). Therefore, TLR collaborate with other innate immune receptors in the activation of DC against the fungus through MyD88-dependent and -independent pathways (Moretti et al., 2008). It is of interest that TLR gene expression on DC can be affected upon fungal exposure in a morphotype-dependent manner (Bozza et al., 2004) and that the TLR9 agonist CpG-ODN can convert an *Aspergillus* allergen to a potential protective antigen (Bozza et al., 2002a), suggesting the potential for TLR agonists to act upon the degree of flexibility of the immune recognition pathways to antigens and allergens.

DC SUBSETS ACTIVATE DISTINCT Th CELL RESPONSES

Fungus-pulsed DC activate different $CD4^+$ Th cells upon adoptive transfer into immunocompetent mice (Bozza et al., 2002a, 2002b, 2003, 2004; Shao et al., 2005). Adoptive transfer of purified ex vivo DC pulsed with conidia or hyphae resulted in priming of $CD4^+$ T cells for Th1 or Th2 cytokine production, respectively. The ability of fungus-pulsed DC to prime for Th1 and Th2 cell activation upon adoptive transfer in vivo correlated with the occurrence of resistance and susceptibility to the infection (Bozza et al., 2002a, 2002b, 2003, 2004). Antifungal protective immunity in vivo was also observed upon adoptive transfer of ex vivo DC transfected with fungal RNA. The efficacy was restricted to DC transfected with RNA from conidia and not those transfected with hyphal RNA (Bozza et al., 2003). It is of interest that conidial RNA, more efficiently than live fungi, concurrently activated IL-10-producing Tregs. These findings expand upon the vaccination potential for DC in fungal infections.

More recent data indicate that conventional DC or pDC activate distinct T-cell priming in response to the fungus in vitro and also in vivo upon adoptive transfer in hematopoietic stem cell-transplanted mice with aspergillosis (see below).

CONVENTIONAL $CD11c^+$ DC CONTRIBUTE TO Th17 ACTIVATION

The fact that when *A. fumigatus* spores are delivered directly to the lung, the inflammatory response is dependent upon the germination-dependent expression of fungal β-glucans that bind to the dectin-1 receptor (Hohl et al., 2005), suggests a role for this receptor in the inflammatory response to the fungus. It is interesting, in this regard, that the dectin-1 recognition of fungal β-glucan mediated the activation of inflammatory Th17 in *Candida albicans* infection (Leibundgut-Landmann et al., 2007). It has recently been shown that Th17 cells are responsible for various organ-related autoimmune diseases (Bettelli et al., 2007; Dong, 2006), and both IL-23 and the Th17 pathway correlate with disease severity and immunopathology in diverse infec-

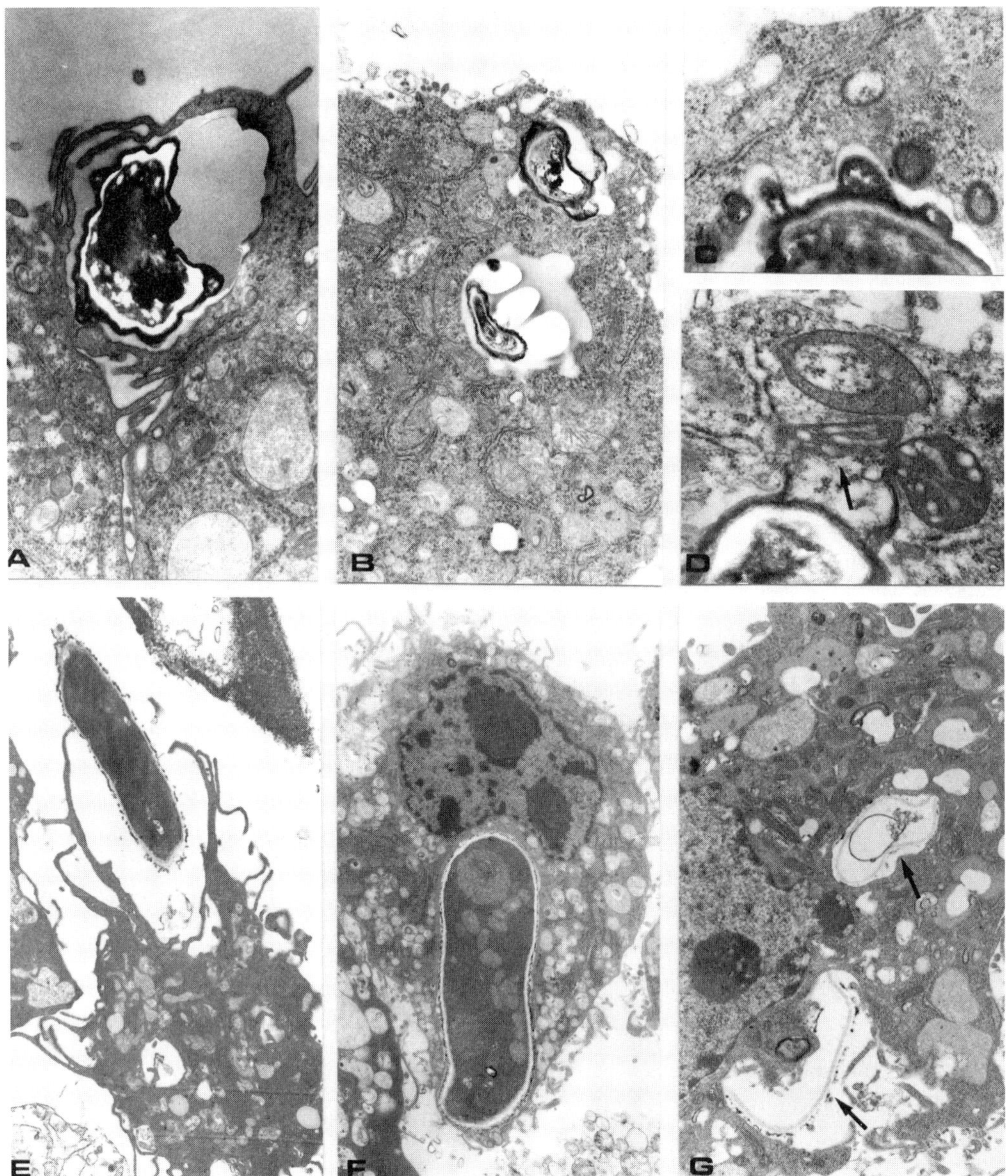

Figure 2. Transmission electron microscopy images of phagocytosis of *A. fumigatus* by DC. Fetal skin-derived murine DC were incubated with live unopsonized *A. fumigatus* conidia (A to D) or hyphae (E to G) for 1 h (A and E) or 3 h (B, C, D, F, and G) before processing for transmission electron microscopy. (A) Conidial engulfment through coiling phagocytosis. Magnification, ×20,000. (B) Conidia inside the cells 3 h later. Magnification, ×12,000. (C and D) Conidia are emanating thick projections (C; magnification, ×30,000) through which they make contact with mitochondria (D, arrow; magnification, ×35,000). (E and F) Hyphal uptake through zipper-type phagocytosis at 1 h after infection (E; magnification, ×8,000) and inside the cells (F; magnification, ×8,000). (G) Hyphae in partially degraded forms at 3 h after exposure (arrows). Magnification, ×8,000. Reproduced from Bozza et al. (2002b) with permission from the publisher.

tions (Hunter, 2005; Langrish et al., 2004; Trinchieri et al., 2003). The IL-23/Th17 developmental pathway acted as a negative regulator of the Th1-mediated immune resistance to the fungus and played an inflammatory role previously attributed to uncontrolled Th1 cell responses. Both inflammation and infection were exacerbated by a heightened Th17 response against *A. fumigatus*. These new findings provide a molecular connection between the failure to resolve inflammation and lack of antifungal immune resistance and point to strategies for immune therapy of fungal infections that attempt to limit inflammation to stimulate an effective immune response. As a matter of fact, IL-17 neutralization increased fungal clearance, ameliorated inflammatory pathology, and restored protective Th1 antifungal resistance (Zelante et al., 2007).

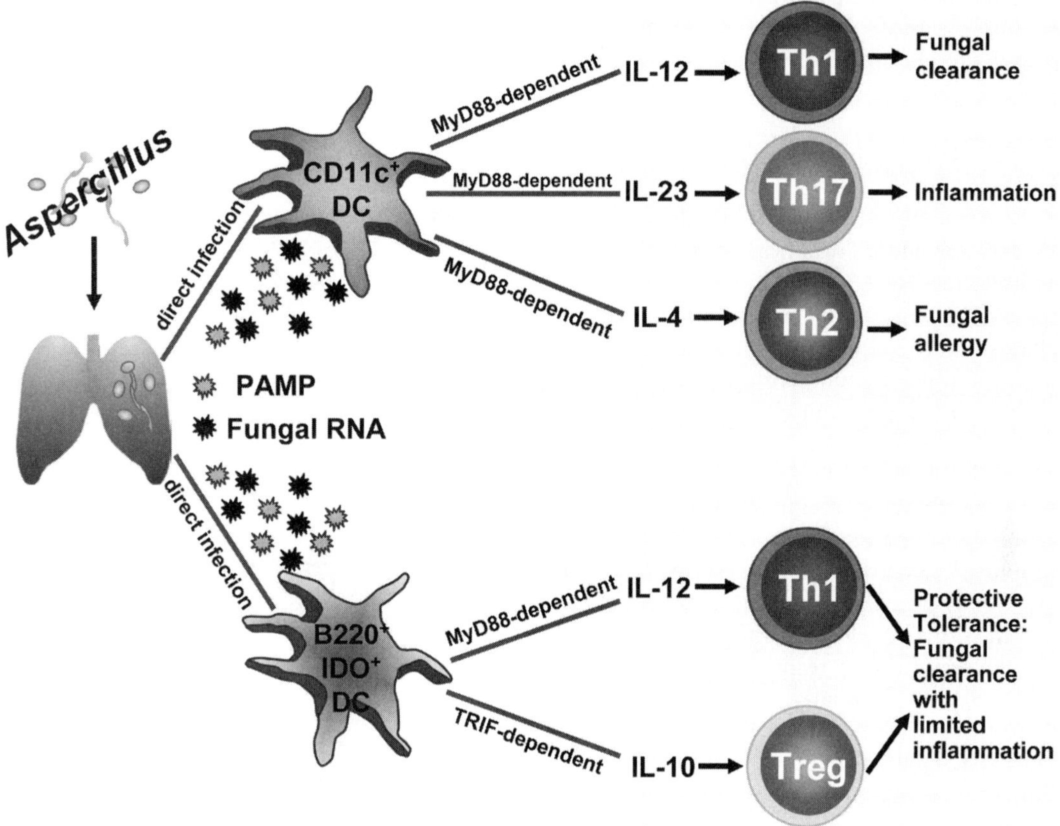

Figure 3. Encounter of DC subsets with *A. fumigatus* leads to a distinct outcome. Under steady-state conditions, in the absence of accompanying danger signals in the lung, inhaled conidia and small hyphal fragments are picked up by lung DC, which take the cargo antigen to draining lymph nodes, where the close interaction with naïve Th cells results in the activation of distinct Th cell responses (Bozza et al., 2002b). Conventional CD11c$^+$ DC and CD11c$^+$ B200$^+$ IDO$^+$ pDC sense fungi in a morphotype-dependent manner through the engagement of distinct receptors (Bozza et al., 2002b; Romani et al., 2006). This translates into downstream signaling events that differentially affect cytokine production and Th cell activation. PAMP, pathogen-associated molecular pattern; MyD88, *Drosophila melanogaster* myeloid differentiation primary response gene 88; TRIF, Toll/IL-1R domain-containing adaptor inducing IFN-β.

IL-12 and IL-23 are members of a family of proinflammatory heterodimeric cytokines that share a common p40 subunit linked to the IL-12p35 chain or the IL-23p19 chain. IL-23 functions through a receptor complex composed of the IL-12Rβ1 subunit and a unique component, the IL-23R chain (Bettelli et al., 2007; Langrish et al., 2004). While both cytokines can induce IFN-γ expression in CD4$^+$ T cells, IL-23, although not involved in Th17 differentiation, plays an important role in maintaining Th17 effector function (Bettelli et al., 2007). IL-23 acted as a molecular connection between uncontrolled fungal growth and inflammation, being produced by inflammatory, conventional DC in response to *Aspergillus* conidia, in a MyD88-dependent manner and counterregulating IL-12p70 production. Thus, conventional DC, due to the simultaneous production of IL-12 and IL-23, contribute to both Th1 and Th17 cell activation in response to the fungus (Fig. 3). This finding is consistent with the notion that distinct signal transduction pathways contribute to DC activation (Pulendran, 2004).

The finding that IL-23 and IL-17, by means of negative regulation of tryptophan catabolism (Zelante et al., 2007), promote inflammation while subverting protective antifungal immunity may serve to accommodate the paradoxical association of chronic inflammatory responses with intractable forms of fungal infections where fungal persistence occurs in the face of an ongoing inflammation. Because both IL-17A and IL-17F may contribute to the expression of airway inflammation and pulmonary hyperreactivity, free soluble IL-17A is increased in asthma (Linden et al., 2005), and allergic cellular and humoral responses are suppressed in IL-17-deficient mice (Nakae et al., 2002), these findings indicate that the Th17 pathway may also be involved in fungus-associated allergic lung diseases.

pDC CONTRIBUTE TO Tregs ACTIVATION

$CD11c^+$ $B220^+$ pDC produced IL-10 in response to the fungus and activated Tregs both in vitro and in vivo (Romani and Puccetti, 2006; Romani et al., 2006). In addition to efficient control of pathogens, tight regulatory mechanisms are required in order to balance protective immunity and immunopathology both in infection and allergy (Montagnoli et al., 2006b). To limit the pathologic consequences of an excessive inflammatory cell-mediated reaction, Tregs, capable of fine-tuning protective antimicrobial immunity in order to minimize harmful immune pathology, have become an integral component of the immune response. Tregs serve to restrain exuberant immune reactivity, which in many chronic infections benefits the host by limiting tissue damage. However, the Treg responses may handicap the efficacy of protective immunity.

Besides ensuring self-tolerance, different types of Tregs actively participate in immune responses. Naturally occurring $CD4^+$ $CD25^+$ Tregs (nTregs), expressing the Treg lineage-specific factor forkhead box P3 (FoxP3), originate in the thymus and survive in the periphery as natural regulators, whereas inducible (or adaptive) Treg (iTreg) cells develop from naive $CD4^+$ T cells in the periphery. Peripheral development of $Foxp3^+$ Tregs represents a mechanism that helps broaden the Treg repertoire in specialized anatomical sites (Belkaid, 2007). Recent studies have also revealed a reciprocal relationship between the development of $FoxP3^+$ Tregs and effector T cells, so that naïve $CD4^+$ T cells differentiate into $Foxp3^+$ Tregs in the presence of TGF-β or into Th17 cells in the presence of TGF-β and IL-6. Thus, T-cell activation in the presence of innate stimuli diverts iTreg generation to Th17 generation (Bettelli et al., 2007).

Tregs with tolerogenic activity have been described in fungal infections (Romani and Puccetti, 2006). Fungal growth, inflammatory immunity, and tolerance to *A. fumigatus* were all controlled by the coordinate activation of nTregs—limiting early inflammation at the sites of infection—and pathogen-induced iTregs, which regulated the expression of adaptive Th immunity in secondary lymphoid organs. Early in infection, inflammation was controlled by the expansion, activation, and local recruitment of nTregs suppressing neutrophils through the combined actions of IL-10 and cytotoxic T-lymphocyte antigen 4 (CTLA-4) acting on the enzyme indoleamine 2,3-dioxygenase (IDO) (see below). Late in infection, and similarly in allergy, the inflammatory immunity was modulated by iTregs, which acted through activation of IDO in pDC, inhibited Th2 cells, and prevented allergy to the fungus. Collectively, these observations suggest that the capacity of Tregs to inhibit aspects of innate and adaptive immunity is pivotal in their regulatory function and further support the concept of "protective tolerance" to fungi, implying that a host's immune defense may be adequate for protection without necessarily eliminating fungal pathogens—which would impair immune memory—or causing an unacceptable level of tissue damage (Romani and Puccetti, 2006).

THE TRYPTOPHAN METABOLIC PATHWAY PIVOTALLY CONTRIBUTES TO ACTIVATION OF TOLEROGENIC DC

The inflammatory/anti-inflammatory state of DC is strictly controlled by the metabolic pathway involved in tryptophan catabolism and mediated by IDO (Grohmann et al., 2003; Mellor and Munn, 2004). IDO has a complex role in immunoregulation in infection, pregnancy, autoimmunity, transplantation, and neoplasia (Grohmann et al., 2003; Mellor and Munn, 2004). There is an increasing appreciation of the unifying role that the immunosuppressive pathway of tryptophan catabolism mediated by the enzyme IDO may have in promoting tolerance under a variety of physiopathologic conditions (Grohmann et al., 2003; Mellor and Munn, 2004; Puccetti and Grohmann, 2007). In experimental aspergillosis, IDO blockade greatly exacerbated infections and allergy to the fungus, as a result of deregulated innate and adaptive immune responses caused by the impaired activation and functioning of suppressor $CD4^+$ $CD25^+$ Tregs producing IL-10 (Montagnoli et al., 2006b). IDO-expressing DC are regarded as regulatory DC specialized to cause antigen-specific deletional tolerance or otherwise negatively regulating responding T cells. Recently, however, the tolerogenic potential of otherwise-immunogenic DC could be induced by exposure to tryptophan catabolites, and this resulted in TGF-β-dependent conversion of naïve $CD4^+$ cells into $FoxP3^+$ Tregs (Fallarino et al., 2006). In addition, while capable of inducing the Foxp3-encoding gene transcriptionally, tryptophan catabolites were also found to suppress the gene encoding retinoid-related orphan receptor gamma t (RORγt), the Th17 lineage specification factor (De Luca et al., 2007). Thus, regulation of iTregs/Th17 balance at the host-pathogen interface further emphasizes the pivotal role of tryptophan catabolism and IDO in tolerogenesis and prevention of autoimmune inflammation and allergy.

Type 1 IFN is required for functional IDO enzymatic activity in DC (Grohmann et al., 2003; Puccetti and Grohmann, 2007). Activation of IDO in CD11c$^+$ B220$^+$ pDC in response to the fungus occurred through TLR9 and correlated with IL-10 production and Treg activation (Romani et al., 2006). Interestingly, thymosin α1 (Tα1), a naturally occurring thymic peptide, known to modulate human pDC functions through TLR9 (Romani et al., 2004), affected mobilization and tolerization of human and murine pDC by activating the IDO-dependent pathway (Romani et al., 2006; Romani and Puccetti, 2006). IDO activation by Tα1 required TLR9 and type I IFN receptor signaling and resulted in IL-10 production and Treg development and tolerization. These data suggest that Tα1 may act on the development program of DC precursors, which results in the promotion of IDO-expressing pDC. In transfer experiments, functionally distinct subsets of differentiated DC were required for priming and tolerance to the fungus or alloantigens (see below). In contrast, Tα1-primed DC fulfilled multiple requirements, including the induction of T helper type 1 immunity within a regulatory environment (Romani et al., 2006). Thus, instructive immunotherapy with Tα1 targeting IDO-competent DC could allow for a balanced control of inflammation and tolerance.

It is intriguing that fungi have exploited IDO manipulation as a means to induce or subvert the tolerogenic program of pDC. Regulation of IDO activity in pDC occurred in a morphotype-dependent manner and, interestingly, in an opposite manner for *Candida* and *Aspergillus*. IDO activity was promoted by *Candida* hyphae and by *Aspergillus* resting conidia and was inhibited by *Aspergillus* swollen conidia or hyphae. The implication is that *Candida* hyphae, by promoting tolerance, contribute to commensalism and eventually to immunoevasion, while swollen *Aspergillus* conidia promote a host inflammatory response by subverting tolerance (Romani and Puccetti, 2006).

IDO$^+$ pDC PREVENT FUNGAL ALLERGY

Modulation of tryptophan catabolism represents a general mechanism of action of Tregs expressing surface CTLA-4 (Fallarino et al., 2003), and different cell types respond to CTLA-4 engagement of CD80 receptor molecules with the activation of IDO, including conventional and pDC, CD4$^+$ T cells, and neutrophils (Puccetti and Grohmann, 2007). The IDO mechanism has revealed an unexpected potential in the control of inflammation, allergy, and allergic airway inflammation, all conditions in which pDC could have a protective function (Puccetti and Grohmann, 2007). IDO expression is paradoxically upregulated in patients with allergy (Hayashi et al., 2004; von Bubnoff et al., 2004) or autoimmune inflammation (Kwidzinski et al., 2005), a finding suggesting the occurrence of a homeostatic mechanism to halt ongoing inflammation. In experimental allergic aspergillosis, the level of inflammation and IFN-γ in the early stage of contact with the fungus set the subsequent adaptive stage by conditioning the IDO-dependent tolerogenic program of DC and the subsequent activation and expansion of tolerogenic Tregs, preventing allergy to the fungus (Montagnoli et al., 2006b). Therefore, regulatory mechanisms operating in the control of inflammation and allergy to the fungus are different but interdependent, as the level of the inflammatory response early in infection may impact on susceptibility to allergy under conditions of continuous exposure to the fungus. Early nTregs, by affecting IFN-γ production, indirectly exert a fine control over the induction of late tolerogenic iTregs. Thus, a unifying mechanism linking nTregs to tolerogenic iTregs via IDO appears to be at work in response to the fungus and is consistent with the revisited "hygiene hypothesis" of allergy in infections, that is, an early reduction in microbial burden may predispose to allergy (Wills-Karp et al., 2001).

Recent data have confirmed the protective role of the IDO/Tregs axis in fungal allergy (Grohmann et al., 2007). Thanks to reverse signaling, indicating a two-way communication between cells or cell types via a single pair of transducing molecules whereby information actually flows in both directions (Puccetti and Grohmann, 2007), Tregs may condition DC via costimulatory ligands expressed by DC. Reverse signaling involves Tregs expressing CTLA-4, and glucocorticoid-induced tumor necrosis factor receptor (GITR) is mediated by at least two coreceptor pairs, CTLA-4 with CD80 and GITR with GITR ligand (GITRL) and results in transcriptional activation of type I IFN genes (in pDC) and *Indo* (in conventional and pDC). Reverse signaling is an effector mechanism of Treg function, a means of self-propagation by Tregs, and participates in the pharmacologic induction of Tregs (Puccetti and Grohmann, 2007). In aspergillosis, reverse signaling cooperates with TLR signaling to mediate protective responses while optimally balancing inflammation and tolerance (Grohmann et al., 2007; Montagnoli et al., 2006b).

TARGETING IDO-MEDIATED IMMUNE HOMEOSTASIS IN FUNGAL ALLERGY WITH CORTICOSTEROIDS

Allergic bronchopulmonary aspergillosis (ABPA) is a Th2-sustained allergic condition of the lung that is responsive to corticosteroid treatment (Judson and Stevens, 2001; Virnig and Bush, 2007). Experimental models of the disease have been used to demonstrate a pivotal role for Treg cells, pDC, and tryptophan catabolism in protecting mice from allergic airway inflammation (Grohmann et al., 2007; Montagnoli et al., 2006b). A recent study showed the IDO-dependent effects of dexamethasone on the hypersensitivity response to *Aspergillus* antigens in the murine lung (Grohmann et al., 2007). Consistent with the responsiveness of ABPA to corticosteroid treatment (Judson and Stevens, 2001; Virnig and Bush, 2007), the Th2-allergic phenotype was greatly attenuated by dexamethasone, which enhanced production of IL-10 and enhanced *Foxp3* transcripts, both markers of protective Treg activity in *Aspergillus* allergy. The data demonstrate that dexamethasone downregulates exacerbation of Th2 responses in ABPA by inhibiting the expansion and activation of Th2 cells, and it upregulates the expression of *Foxp3* via mechanisms that require tryptophan catabolism (Grohmann et al., 2007). In this model, modulation of tryptophan catabolism via the GITR and GITRL inhibited Th2-cell responses and allergy and induced the expression of $FoxP3^+$ Tregs through mechanisms dependent on IDO induction by components of the noncanonical NF-κB signaling pathway. Thus, induction of IDO could be an important mechanism underlying the anti-inflammatory action of corticosteroids (Puccetti and Grohmann, 2007). The data suggested that IDO modulation by noncanonical NF-κB is essential for the maintenance of TLR-driven immune homeostasis in the airways. It is conceivable that fungal aeroantigen is prevented from initiating airway inflammation by the integrity and antimicrobial defense of the epithelium, in an environment in which TLR9-driven induction of IDO and consequent inhibition of Th2 cells will contrast with the onset of allergic inflammation. Asymptomatic atopy associated with increased IDO activity and IL-10 production in seasonal allergen exposure has been described (von Bubnoff et al., 2004). Clinical trials of TLR9-based immunotherapy are ongoing, and the investigators have suggested that, "TLR9 ligands may revolutionize the treatment of allergic diseases" (Hessel et al., 2005). As the effects of treatments with antifungal agents on symptoms and clinical findings in patients with allergic fungal diseases are controversial (Denning et al., 2006; Judson and Stevens, 2001), targeting IDO-mediated immune homeostasis in fungal diseases is a promising option.

EXPLOITING DC AS FUNGAL VACCINES IN TRANSPLANTATION

DC are now being exploited to improve vaccine efficacy (Steinman, 2008). Pathogen-pulsed DC or RNA-transfected DC acted, indeed, as a potent fungal vaccine in experimental HSCT (Bozza et al., 2003), a model in which autologous reconstitution of host stem cells is greatly reduced to the benefit of a long-term, donor-type chimerism in more than 95% of the mice and low incidence of graft-versus-host disease. Protection was associated with myeloid and T-cell recovery, the activation of $CD4^+$ Th1 lymphocytes, and the concomitant IL-10-driven Tregs. Our studies have expanded upon these earlier observations and suggest that DC may contribute to the educational program of reconstituting T cells in HSCT and also that active vaccination can represent an option in the immunotherapy of fungal infections in transplantation.

Over recent years experimental models have shown that it is possible to exploit the mechanisms that normally maintain immune homeostasis and tolerance to self-antigens to induce tolerance to alloantigens (Martinic and von Herrath, 2006; Waldmann and Cobbold, 2004). Like natural tolerance, transplantation tolerance is achieved through control of T-cell reactivity by central and peripheral mechanisms of tolerance. Although as yet it has not been possible to define a single cell surface marker that is unique to Treg cells, a growing body of evidence has pointed to the conclusion that regulatory $CD4^+$ $CD25^+$ T cells constitutively expressing FoxP3 maintain transplantation tolerance (Sakaguchi, 2005), and these cells can have indirect allospecificity for donor antigens (Graca et al., 2006). The interplay between Treg and antigen-responsive T cells is modulated by DC. Type I IFN-producing pDC are implicated in the induction and maintenance of tolerance and contribute to engraftment facilitation and prevention of graft-versus-host disease in HSCT. Expansion of pDC is contingent upon the hematopoietic growth factor FLT3L (D'Amico and Wu, 2003; Weigel et al., 2002), which acts on FLT3-expressing myeloid and nonmyeloid hematopoietic precursors (D'Amico and Wu, 2003). As tolerance is also one major requirement of a successful immune response to fungi (Romani and Puccetti, 2006), tolerogenic DC, including pDC, may be pivotal in the generation of some form of dominant regulation that ultimately controls inflammation, pathogen immunity, and tolerance in transplant recipients (Montagnoli et al., 2008). Because host DC function is impaired during the immediate period posttransplantation, the administration of donor DC may be useful for the educational program of recovering T cells.

Distinct DC subsets activate specialized antifungal effector and regulatory functions upon adoptive transfer in experimental HSCT (Fig. 4). Specialization and complementarity in priming and tolerization by the different DC subsets, either murine or human, were observed. FLT3 ligand-derived DC (mainly $B220^+$ IDO^+ pDC) or Tα1-conditioned DC fulfilled the requirement for Th1/Treg antifungal priming, an activity to which the combined action of several DC populations contributed (Ro-

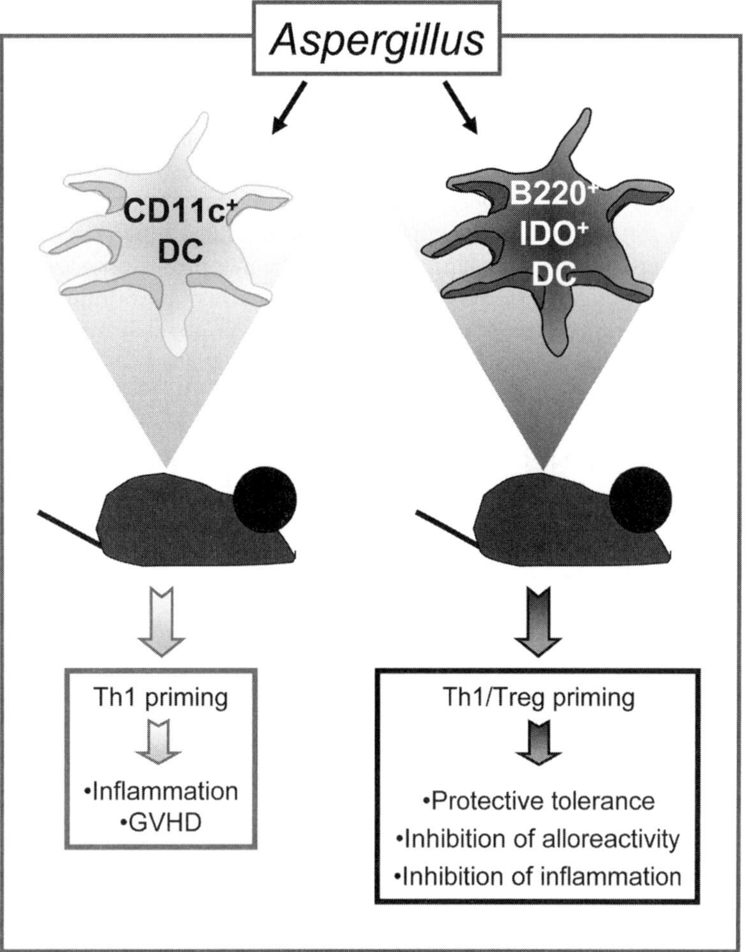

Figure 4. Exploiting DC for transplantation tolerance and concomitant pathogen clearance in hematopoietic transplantation. The figure shows the relative contributions of murine DC subsets to antifungal priming and alloantigen tolerization upon adoptive transfer in vivo in HSCT mice with aspergillosis (Romani et al., 2006). Specialization and complementarity in priming and tolerization by the different DC subsets is shown. Whereas CD11c$^+$ DC activate Th1 cells and concomitant inflammatory toxicity, CD11c$^+$ B220$^+$ IDO$^+$ DC fulfilled the requirement for (i) Th1/Treg antifungal priming, (ii) tolerization toward alloantigens, and (iii) diversion from alloantigen-specific to antigen-specific T-cell responses in the presence of donor T lymphocytes. Interestingly, Tα1, known to modulate human pDC functions through TLR9, affected mobilization and tolerization of pDC by activating the IDO-dependent pathway, and this resulted in Treg development and tolerization (Romani et al., 2006). Thus, transplantation tolerance and concomitant pathogen clearance can be achieved through the therapeutic induction of antigen-specific Tregs via instructive immunotherapy with pathogen- or TLR-conditioned donor DC. GVHD, graft-versus-host disease.

mani et al., 2006). In contrast, conventional CD11c$^+$ DC, through the production of IL-12, contribute to Th1 activation but also to immunotoxicity, which includes the promotion of inflammation and graft-versus-host disease. Whether the contribution of conventional DC to inflammatory pathology can be ascribed to the activation of IL-23-driven Th17 responses is presently unknown, but it is highly suspected.

It is of interest that human DC subsets also produce different patterns of cytokines in response to *Aspergillus* and activate distinct Th cell responses in vitro (Romani et al., 2006). Conventional DC produce mainly IL-12 and activate IFN-γ-producing cells, whereas pDC mainly produce IFN-α/IL-12/IL-10 and activate Th cells producing IFN-γ and IL-10. Cytokine production by each DC subset was severely impaired in HSCT, along with the ability to phagocytose conidia and to upregulate costimulatory molecule expression. Antigen-specific proliferation was induced by conventional DC, more than pDC, from healthy donors but not by DC from HSCT patients (Romani et al., 2006). These results suggest that human pDC are fully competent at inducing IFN-γ-/IL-10-producing cells in response to the fungus in vitro and are defective at early stages after HSCT.

It is intriguing that antifungal Treg cells concomitantly suppressed inflammation and alloreactivity (Montagnoli et al., 2008; Romani and Puccetti, 2006). Several studies have addressed the effect that infections have on transplantation tolerance, and the overall view is that both prior and concurrent exposure to pathogens can prevent tolerance induction. However, much less attention has been paid to the effect that pathogen-directed tolerance based on active T-cell regulation might have on tolerance to donor antigens. Because of the cross-reactivity in the T-cell repertoire between antimicrobial, environmental, and transplantation antigens (Mason, 1998; Pantenburg et al., 2002), our results raise the intriguing possibility that pathogen-conditioned DC could be potential reagents to promote donor-specific transplantation through the induction of $CD4^+$ $CD25^+$ Treg cells with indirect antidonor allospecificity. Strategies to generate human $CD4^+$ $CD25^+$ T-cell lines with indirect allospecificity for therapeutic use for the induction of donor-specific transplantation tolerance have recently been described (Jiang et al., 2006). Thus, transplantation tolerance and concomitant pathogen clearance could be achieved through the therapeutic induction of antigen-specific Tregs via instructive immunotherapy with pathogen- or TLR-conditioned donor DC (Fig. 4).

CONCLUSIONS AND PERSPECTIVES: EXPLOITATION OF DC FOR DEVELOPMENT OF FUNGAL VACCINES AND ANTI-INFLAMMATORY STRATEGIES

Developments in DC biology are providing opportunities for improved strategies for the prevention and management of fungal diseases in immunocompromised patients. The ultimate challenge will be to design fungal vaccines capable of inducing optimal immune responses by targeting specific receptors on DC (Serrano-Gomez et al., 2007). This will require, however, further studies aimed at elucidating the convergence and divergence of pathways of immune protection elicited in infections or upon vaccination. At a time when vaccination of stem cell transplant recipients is highly recommended (Ljungman et al., 2005), developing new strategies to expand and modulate the functions of distinct DC associated with specific regulation of host immunity may provide novel immune-based therapies for HSCT (Steinman and Pope, 2002). Within the instructive model of DC-mediated regulation of the Th repertoire, it is conceivable that an improved understanding of the pathogen-DC interaction will allow the potential use of pathogen- or TLR-conditioned DC for the induction of patient-tailored $CD4^+$ $CD25^+$ Tregs with indirect antidonor allospecificity. In addition, the new CD4 T-cell subset of Th17 T cells has filled many gaps in our understanding of how immune responses to the fungus are regulated. The new finding has provided a molecular connection between the failure to resolve inflammation and lack of antifungal immune resistance and point to strategies for immune therapy of fungal infections that attempt to limit inflammation to stimulate an effective immune response. Thus, inhibition of RORγt activity, by alleviating the symptoms associated with the Th17-dependent inflammatory autoimmune diseases, may potentially represent a novel strategy for the prevention of inflammatory immunity and allergy to *Aspergillus*. In this regard, IDO^+ DC and tryptophan metabolites may prove to be potent regulators capable of taming overzealous or heightened inflammatory host responses, to the benefit of pathogen eradication and host survival.

Acknowledgments. We thank Cristina Massi Benedetti for editorial assistance. This study was supported by the Specific Targeted Research Project "MANASP" (LSHE-CT-2006), contract number 037899 (FP6).

REFERENCES

Balloy, V., M. Huerre, J. P. Latgé, and M. Chignard. 2005. Differences in patterns of infection and inflammation for corticosteroid treatment and chemotherapy in experimental invasive pulmonary aspergillosis. *Infect. Immun.* 73:494–503.

Behnsen, J., P. Narang, M. Hasenberg, F. Gunzer, U. Bilitewski, N. Klippel, M. Rohde, M. Brock, A. A. Brakhage, and M. Gunzer. 2007. Environmental dimensionality controls the interaction of phagocytes with the pathogenic fungi *Aspergillus fumigatus* and *Candida albicans*. *PLoS Pathog.* 3:e13.

Belkaid, Y. 2007. Regulatory T cells and infection: a dangerous necessity. *Nat. Rev. Immunol.* 7:875–888.

Bellocchio, S., C. Montagnoli, S. Bozza, R. Gaziano, G. Rossi, S. S. Mambula, A. Vecchi, A. Mantovani, S. M. Levitz, and L. Romani. 2004a. The contribution of the Toll-Like/IL-1 receptor superfamily to innate and adaptive immunity to fungal pathogens in vivo. *J. Immunol.* 172:3059–3069.

Bellocchio, S., S. Moretti, K. Perruccio, F. Fallarino, S. Bozza, C. Montagnoli, P. Mosci, G. B. Lipford, L. Pitzurra, and L. Romani. 2004b. TLRs govern neutrophil activity in aspergillosis. *J. Immunol.* 173:7406–7415.

Bettelli, E., M. Oukka, and V. K. Kuchroo. 2007. T_H-17 cells in the circle of immunity and autoimmunity. *Nat. Immunol.* 8:345–350.

Bozza, S., R. Gaziano, G. B. Lipford, C. Montagnoli, A. Bacci, P. Di Francesco, V. P. Kurup, H. Wagner, and L. Romani. 2002a. Vaccination of mice against invasive aspergillosis with recombinant *Aspergillus* proteins and CpG oligodeoxynucleotides as adjuvants. *Microbes Infect.* 4:1281–1290.

Bozza, S., R. Gaziano, A. Spreca, A. Bacci, C. Montagnoli, P. di Francesco, and L. Romani. 2002b. Dendritic cells transport conidia and hyphae of *Aspergillus fumigatus* from the airways to the draining lymph nodes and initiate disparate Th responses to the fungus. *J. Immunol.* 168:1362–1371.

Bozza, S., C. Montagnoli, R. Gaziano, G. Rossi, G. Nkwanyuo, S. Bellocchio, and L. Romani. 2004. Dendritic cell-based vaccination against opportunistic fungi. *Vaccine* 22:857–864.

Bozza, S., K. Perruccio, C. Montagnoli, R. Gaziano, S. Bellocchio, E. Burchielli, G. Nkwanyuo, L. Pitzurra, A. Velardi, and L. Romani.

2003. A dendritic cell vaccine against invasive aspergillosis in allogeneic hematopoietic transplantation. *Blood* **102:**3807–3814.

Braedel, S., M. Radsak, H. Einsele, J. P. Latgé, A. Michan, J. Loeffler, Z. Haddad, U. Grigoleit, H. Schild, and H. Hebart. 2004. *Aspergillus fumigatus* antigens activate innate immune cells via toll-like receptors 2 and 4. *Br. J. Haematol.* **125:**392–399.

Bretz, C., G. Gersuk, S. Knoblaugh, N. Chaudhary, J. Randolph-Habecker, R. Hackman, J. Staab, and K. A. Marr. 2008. MyD88-signaling contributes to early pulmonary responses to *Aspergillus fumigatus. Infect. Immun.* **76:**952–958.

Chignard, M., V. Balloy, J. M. Sallenave, and M. Si-Tahar. 2007. Role of Toll-like receptors in lung innate defense against invasive aspergillosis. Distinct impact in immunocompetent and immunocompromized hosts. *Clin. Immunol.* **124:**238–243.

Colonna, M., G. Trinchieri, and Y. J. Liu. 2004. Plasmacytoid dendritic cells in immunity. *Nat. Immunol.* **5:**1219–1226.

D'Amico, A., and L. Wu. 2003. The early progenitors of mouse dendritic cells and plasmacytoid predendritic cells are within the bone marrow hemopoietic precursors expressing Flt3. *J. Exp. Med.* **198:** 293–303.

De Luca, A., C. Montagnoli, T. Zelante, P. Bonifazi, S. Bozza, S. Moretti, C. D'Angelo, C. Vacca, L. Boon, F. Bistoni, P. Puccetti, F. Fallarino, and L. Romani. 2007. Functional yet balanced reactivity to *Candida albicans* requires TRIF, MyD88, and IDO-dependent inhibition of Rorc. *J. Immunol.* **179:**5999–6008.

Denning, D. W., B. R. O'Driscoll, C. M. Hogaboam, P. Bowyer, and R. M. Niven. 2006. The link between fungi and severe asthma: a summary of the evidence. *Eur. Respir. J.* **27:**615–626.

Diebold, S. S., M. Montoya, H. Unger, L. Alexopoulou, P. Roy, L. E. Haswell, A. Al-Shamkhani, R. Flavell, P. Borrow, and C. Reis e Sousa. 2003. Viral infection switches non-plasmacytoid dendritic cells into high interferon producers. *Nature* **424:**324–328.

Dong, C. 2006. Diversification of T-helper-cell lineages: finding the family root of IL-17-producing cells. *Nat. Rev. Immunol.* **6:**329–333.

Fallarino, F., U. Grohmann, K. W. Hwang, C. Orabona, C. Vacca, R. Bianchi, M. L. Belladonna, M. C. Fioretti, M. L. Alegre, and P. Puccetti. 2003. Modulation of tryptophan catabolism by regulatory T cells. *Nat. Immunol.* **4:**1206–1212.

Fallarino, F., U. Grohmann, S. You, B. C. McGrath, D. R. Cavener, C. Vacca, C. Orabona, R. Bianchi, M. L. Belladonna, C. Volpi, P. Santamaria, M. C. Fioretti, and P. Puccetti. 2006. The combined effects of tryptophan starvation and tryptophan catabolites down-regulate T cell receptor zeta-chain and induce a regulatory phenotype in naive T cells. *J. Immunol.* **176:**6752–6761.

Garlanda, C., E. Hirsch, S. Bozza, A. Salustri, M. De Acetis, R. Nota, A. Maccagno, F. Riva, B. Bottazzi, G. Peri, A. Doni, L. Vago, M. Botto, R. De Santis, P. Carminati, G. Siracusa, F. Altruda, A. Vecchi, L. Romani, and A. Mantovani. 2002. Non-redundant role of the long pentraxin PTX3 in anti-fungal innate immune response. *Nature* **420:**182–186.

Graca, L., B. Silva-Santos, and A. Coutinho. 2006. The blind-spot of regulatory T cells. *Eur. J. Immunol.* **36:**802–805.

Grazziutti, M., D. Przepiorka, J. H. Rex, I. Braunschweig, S. Vadhan-Raj, and C. A. Savary. 2001. Dendritic cell-mediated stimulation of the in vitro lymphocyte response to *Aspergillus. Bone Marrow Transplant.* **27:**647–652.

Grohmann, U., F. Fallarino, and P. Puccetti. 2003. Tolerance, DCs and tryptophan: much ado about IDO. *Trends Immunol.* **24:**242–248.

Grohmann, U., C. Volpi, F. Fallarino, S. Bozza, R. Bianchi, C. Vacca, C. Orabona, M. L. Belladonna, E. Ayroldi, G. Nocentini, L. Boon, F. Bistoni, M. C. Fioretti, L. Romani, C. Riccardi, and P. Puccetti. 2007. Reverse signaling through GITR ligand enables dexamethasone to activate IDO in allergy. *Nat. Med.* **13:**579–586.

Gutcher, I., and B. Becher. 2007. APC-derived cytokines and T cell polarization in autoimmune inflammation. *J. Clin. Investig.* **117:** 1119–1127.

Hayashi, T., L. Beck, C. Rossetto, X. Gong, O. Takikawa, K. Takabayashi, D. H. Broide, D. A. Carson, and E. Raz. 2004. Inhibition of experimental asthma by indoleamine 2,3-dioxygenase. *J. Clin. Investig.* **114:**270–279.

Hessel, E. M., M. Chu, J. O. Lizcano, B. Chang, N. Herman, S. A. Kell, M. Wills-Karp, and R. L. Coffman. 2005. Immunostimulatory oligonucleotides block allergic airway inflammation by inhibiting Th2 cell activation and IgE-mediated cytokine induction. *J. Exp. Med.* **202:**1563–1573.

Hohl, T. M., and M. Feldmesser. 2007. *Aspergillus fumigatus*: principles of pathogenesis and host defense. *Eukaryot. Cell* **6:**1953–1963.

Hohl, T. M., H. L. Van Epps, A. Rivera, L. A. Morgan, P. L. Chen, M. Feldmesser, and E. G. Pamer. 2005. *Aspergillus fumigatus* triggers inflammatory responses by stage-specific beta-glucan display. *PLoS Pathog.* **1:**e30.

Holland, S. M., F. R. DeLeo, H. Z. Elloumi, A. P. Hsu, G. Uzel, N. Brodsky, A. F. Freeman, A. Demidowich, J. Davis, M. L. Turner, V. L. Anderson, D. N. Darnell, P. A. Welch, D. B. Kuhns, D. M. Frucht, H. L. Malech, J. I. Gallin, S. D. Kobayashi, A. R. Whitney, J. M. Voyich, J. M. Musser, C. Woellner, A. A. Schaffer, J. M. Puck, and B. Grimbacher. 2007. STAT3 mutations in the hyper-IgE syndrome. *N. Engl. J. Med.* **357:**1608–1619.

Hunter, C. A. 2005. New IL-12-family members: IL-23 and IL-27, cytokines with divergent functions. *Nat. Rev. Immunol.* **5:**521–531.

Jiang, S., J. Tsang, D. S. Game, S. Stevenson, G. Lombardi, and R. I. Lechler. 2006. Generation and expansion of human $CD4^+$ $CD25^+$ regulatory T cells with indirect allospecificity: potential reagents to promote donor-specific transplantation tolerance. *Transplantation* **82:**1738–1743.

Judson, M. A., and D. A. Stevens. 2001. Current pharmacotherapy of allergic bronchopulmonary aspergillosis. *Expert Opin. Pharmacother.* **2:**1065–1071.

Kwidzinski, E., J. Bunse, O. Aktas, D. Richter, L. Mutlu, F. Zipp, R. Nitsch, and I. Bechmann. 2005. Indolamine 2,3-dioxygenase is expressed in the CNS and down-regulates autoimmune inflammation. *FASEB J.* **19:**1347–1349.

Langrish, C. L., B. S. McKenzie, N. J. Wilson, R. de Waal Malefyt, R. A. Kastelein, and D. J. Cua. 2004. IL-12 and IL-23: master regulators of innate and adaptive immunity. *Immunol. Rev.* **202:**96–105.

Lanzavecchia, A., and F. Sallusto. 2001. Regulation of T cell immunity by dendritic cells. *Cell* **106:**263–266.

Latgé, J. P. 1999. *Aspergillus fumigatus* and aspergillosis. *Clin. Microbiol. Rev.* **12:**310–350.

Leibundgut-Landmann, S., O. Gross, M. J. Robinson, F. Osorio, E. C. Slack, S. V. Tsoni, E. Schweighoffer, V. Tybulewicz, G. D. Brown, J. Ruland, and C. Reis e Sousa. 2007. Syk- and CARD9-dependent coupling of innate immunity to the induction of T helper cells that produce interleukin 17. *Nat. Immunol.* **8:**630–638.

Linden, A., M. Laan, and G. P. Anderson. 2005. Neutrophils, interleukin-17A and lung disease. *Eur. Respir. J.* **25:**159–172.

Ljungman, P., D. Engelhard, R. de la Camara, H. Einsele, A. Locasciulli, R. Martino, P. Ribaud, K. Ward, and C. Cordonnier. 2005. Vaccination of stem cell transplant recipients: recommendations of the Infectious Diseases Working Party of the EBMT. *Bone Marrow Transplant.* **35:**737–746.

Madan, T., S. Kaur, S. Saxena, M. Singh, U. Kishore, S. Thiel, K. B. Reid, and P. U. Sarma. 2005. Role of collectins in innate immunity against aspergillosis. *Med. Mycol.* **43**(Suppl. 1):S155–S163.

Marr, K. A., T. Patterson, and D. Denning. 2002. Aspergillosis. Pathogenesis, clinical manifestations, and therapy. *Infect. Dis. Clin. North Am.* **16:**875–894.

Martinic, M. M., and M. G. von Herrath. 2006. Control of graft-versus-host disease by regulatory T cells: which level of antigen specificity? *Eur. J. Immunol.* **36:**2299–2303.

Mason, D. 1998. A very high level of crossreactivity is an essential feature of the T-cell receptor. *Immunol. Today* **19:**395–404.

McCormack, F. X., and J. A. Whitsett. 2002. The pulmonary collectins, SP-A and SP-D, orchestrate innate immunity in the lung. *J. Clin. Investig.* **109:**707–712.

Mellor, A. L., and D. H. Munn. 2004. IDO expression by dendritic cells: tolerance and tryptophan catabolism. *Nat. Rev. Immunol.* **4:** 762–774.

Miceli, M. H., J. Maertens, K. Buve, M. Grazziutti, G. Woods, M. Rahman, B. Barlogie, and E. J. Anaissie. 2007. Immune reconstitution inflammatory syndrome in cancer patients with pulmonary aspergillosis recovering from neutropenia: proof of principle, description, and clinical and research implications. *Cancer* **110:**112–120.

Montagnoli, C., S. Bozza, R. Gaziano, T. Zelante, P. Bonifazi, S. Moretti, S. Bellocchio, L. Pitzurra, and L. Romani. 2006a. Immunity and tolerance to *Aspergillus fumigatus*. *Novartis Found. Symp.* **279:** 66–77.

Montagnoli, C., F. Fallarino, R. Gaziano, S. Bozza, S. Bellocchio, T. Zelante, W. P. Kurup, L. Pitzurra, P. Puccetti, and L. Romani. 2006b. Immunity and tolerance to *Aspergillus* involve functionally distinct regulatory T cells and tryptophan catabolism. *J. Immunol.* **176:**1712–1723.

Montagnoli, C., K. Perruccio, S. Bozza, P. Bonifazi, T. Zelante, A. De Luca, S. Moretti, C. D'Angelo, F. Bistoni, M. Martelli, F. Aversa, A. Velardi, and L. Romani. 2008. Provision of antifungal immunity and concomitant alloantigen tolerization by conditioned dendritic cells in experimental hematopoietic transplantation. *Blood Cells Mol. Dis.* **40:**55–62.

Moretti, S., S. Bellocchio, P. Bonifazi, S. Bozza, T. Zelante, F. Bistoni, and L. Romani. 2008. The contribution of PARs to inflammation and immunity to fungi. *Mucosal Immunol.* **1:**156–168.

Nakae, S., Y. Komiyama, A. Nambu, K. Sudo, M. Iwase, I. Homma, K. Sekikawa, M. Asano, and Y. Iwakura. 2002. Antigen-specific T cell sensitization is impaired in IL-17-deficient mice, causing suppression of allergic cellular and humoral responses. *Immunity* **17:** 375–387.

Ortega, M., M. Rovira, X. Filella, J. A. Martinez, M. Almela, J. Puig, E. Carreras, and J. Mensa. 2006. Prospective evaluation of procalcitonin in adults with non-neutropenic fever after allogeneic hematopoietic stem cell transplantation. *Bone Marrow Transplant.* **37:** 499–502.

Pantenburg, B., F. Heinzel, L. Das, P. S. Heeger, and A. Valujskikh. 2002. T cells primed by *Leishmania major* infection cross-react with alloantigens and alter the course of allograft rejection. *J. Immunol.* **169:**3686–3693.

Phadke, A. P., and B. Mehrad. 2005. Cytokines in host defense against *Aspergillus*: recent advances. *Med. Mycol.* **43**(Suppl. 1):S173–S176.

Puccetti, P., and U. Grohmann. 2007. IDO and regulatory T cells: a role for reverse signalling and non-canonical NF-κB activation. *Nat. Rev. Immunol.* **7:**817–823.

Pulendran, B. 2004. Modulating vaccine responses with dendritic cells and Toll-like receptors. *Immunol. Rev.* **199:**227–250.

Reis e Sousa, C. 2006. Dendritic cells in a mature age. *Nat. Rev. Immunol.* **6:**476–483.

Romani, L. 2008. Cell mediated immunity to fungi: a reassessment. *Med. Mycol.* **12:**1–15.

Romani, L. 2004. Immunity to fungal infections. *Nat. Rev. Immunol.* **4:**1–23.

Romani, L., F. Bistoni, R. Gaziano, S. Bozza, C. Montagnoli, K. Perruccio, L. Pitzurra, S. Bellocchio, A. Velardi, G. Rasi, P. Di Francesco, and E. Garaci. 2004. Thymosin alpha 1 activates dendritic cells for antifungal Th1 resistance through toll-like receptor signaling. *Blood* **103:**4232–4239.

Romani, L., F. Bistoni, K. Perruccio, C. Montagnoli, R. Gaziano, S. Bozza, P. Bonifazi, G. Bistoni, G. Rasi, A. Velardi, F. Fallarino, E. Garaci, and P. Puccetti. 2006. Thymosin α1 activates dendritic cell tryptophan catabolism and establishes a regulatory environment for balance of inflammation and tolerance. *Blood* **108:**2265–2274.

Romani, L., F. Bistoni, and P. Puccetti. 2002. Fungi, dendritic cells and receptors: a host perspective of fungal virulence. *Trends Microbiol.* **10:**508–514.

Romani, L., F. Fallarino, A. De Luca, C. Montagnoli, C. D'Angelo, T. Zelante, C. Vacca, F. Bistoni, M. C. Fioretti, U. Grohmann, B. H. Segal, and P. Puccetti. 2008. Defective tryptophan catabolism underlies inflammation in mouse chronic granulomatous disease. *Nature* **451:**211–215.

Romani, L., and P. Puccetti. 2007. Controlling pathogenic inflammation to fungi. *Expert Rev. Anti Infect. Ther.* **5:**1007–1017.

Romani, L., and P. Puccetti. Immune regulation and tolerance to fungi in the lungs and skin. *Chem. Immunol.*, in press.

Romani, L., and P. Puccetti. 2006. Protective tolerance to fungi: the role of IL-10 and tryptophan catabolism. *Trends Microbiol.* **14:**183–189.

Sakaguchi, S. 2005. Naturally arising Foxp3-expressing $CD25^+CD4^+$ regulatory T cells in immunological tolerance to self and non-self. *Nat. Immunol.* **6:**345–352.

Sambatakou, H., V. Pravica, I. V. Hutchinson, and D. W. Denning. 2006. Cytokine profiling of pulmonary aspergillosis. *Int. J. Immunogenet.* **33:**297–302.

Schubert, M. S. 2006. Allergic fungal sinusitis. *Clin. Rev. Allergy Immunol.* **30:**205–216.

Segal, B. H., T. L. Leto, J. I. Gallin, H. L. Malech, and S. M. Holland. 2000. Genetic, biochemical, and clinical features of chronic granulomatous disease. *Medicine* (Baltimore) **79:**170–200.

Serrano-Gomez, D., A. Dominguez-Soto, J. Ancochea, J. A. Jimenez-Heffernan, J. A. Leal, and A. L. Corbi. 2004. Dendritic cell-specific intercellular adhesion molecule 3-grabbing nonintegrin mediates binding and internalization of *Aspergillus fumigatus* conidia by dendritic cells and macrophages. *J. Immunol.* **173:**5635–5643.

Serrano-Gomez, D., R. T. Martinez-Nunez, E. Sierra-Filardi, N. Izquierdo, M. Colmenares, J. Pla, L. Rivas, J. Martinez-Picado, J. Jimenez-Barbero, J. L. Alonso-Lebrero, S. Gonzalez, and A. L. Corbi. 2007. AM3 modulates dendritic cell pathogen recognition capabilities by targeting DC-SIGN. *Antimicrob. Agents Chemother.* **51:**2313–2323.

Shao, C., J. Qu, L. He, Y. Zhang, J. Wang, H. Zhou, Y. Wang, and X. Liu. 2005. Dendritic cells transduced with an adenovirus vector encoding interleukin-12 are a potent vaccine for invasive pulmonary aspergillosis. *Genes Immun.* **6:**103–114.

Singh, N., and J. R. Perfect. 2007. Immune reconstitution syndrome associated with opportunistic mycoses. *Lancet Infect. Dis.* **7:**395–401.

Steinman, R. M. 2008. Dendritic cells and vaccines. *Baylor Univ. Med. Ctr. Proc.* **21:**3–8.

Steinman, R. M. 2007. Lasker Basic Medical Research Award. Dendritic cells: versatile controllers of the immune system. *Nat. Med.* **13:**1155–1159.

Steinman, R. M., and J. Banchereau. 2007. Taking dendritic cells into medicine. *Nature* **449:**419–426.

Steinman, R. M., and Z. A. Cohn. 1973. Identification of a novel cell type in peripheral lymphoid organs of mice. I. Morphology, quantitation, tissue distribution. *J. Exp. Med.* **137:**1142–1162.

Steinman, R. M., and M. Pope. 2002. Exploiting dendritic cells to improve vaccine efficacy. *J. Clin. Investig.* **109:**1519–1526.

Trinchieri, G., S. Pflanz, and R. A. Kastelein. 2003. The IL-12 family of heterodimeric cytokines: new players in the regulation of T cell responses. *Immunity* **19:**641–644.

Trinchieri, G., and A. Sher. 2007. Cooperation of Toll-like receptor signals in innate immune defence. *Nat. Rev. Immunol.* 7:179–190.

Virnig, C., and R. K. Bush. 2007. Allergic bronchopulmonary aspergillosis: a US perspective. *Curr. Opin. Pulm. Med.* 13:67–71.

von Bubnoff, D., R. Fimmers, M. Bogdanow, H. Matz, S. Koch, and T. Bieber. 2004. Asymptomatic atopy is associated with increased indoleamine 2,3-dioxygenase activity and interleukin-10 production during seasonal allergen exposure. *Clin. Exp. Allergy* 34:1056–1063.

Waldmann, H., and S. Cobbold. 2004. Exploiting tolerance processes in transplantation. *Science* 305:209–212.

Walsh, T. J., E. Roilides, K. Cortez, S. Kottilil, J. Bailey, and C. A. Lyman. 2005. Control, immunoregulation, and expression of innate pulmonary host defenses against *Aspergillus fumigatus*. *Med. Mycol.* 43(Suppl. 1):S165–S172.

Weigel, B. J., N. Nath, P. A. Taylor, A. Panoskaltsis-Mortari, W. Chen, A. M. Krieg, K. Brasel, and B. R. Blazar. 2002. Comparative analysis of murine marrow-derived dendritic cells generated by Flt3L or GM-CSF/IL-4 and matured with immune stimulatory agents on the in vivo induction of antileukemia responses. *Blood* 100:4169–4176.

Wills-Karp, M., J. Santeliz, and C. L. Karp. 2001. The germless theory of allergic disease: revisiting the hygiene hypothesis. *Nat. Rev. Immunol.* 1:69–75.

Zelante, T., A. De Luca, P. Bonifazi, C. Montagnoli, S. Bozza, S. Moretti, M. L. Belladonna, C. Vacca, C. Conte, P. Mosci, F. Bistoni, P. Puccetti, R. A. Kastelein, M. Kopf, and L. Romani. 2007. IL-23 and the Th17 pathway promote inflammation and impair antifungal immune resistance. *Eur. J. Immunol.* 37:2695–2706.

Zuniga, E. I., D. B. McGavern, J. L. Pruneda-Paz, C. Teng, and M. B. Oldstone. 2004. Bone marrow plasmacytoid dendritic cells can differentiate into myeloid dendritic cells upon virus infection. *Nat. Immunol.* 5:1227–1234.

Aspergillus fumigatus and Aspergillosis
Edited by J.-P. Latgé and W. J. Steinbach
©2009 ASM Press, Washington, DC

Chapter 21

CD4$^+$ T-Cell Responses to *Aspergillus fumigatus*

AMARILIZ RIVERA AND ERIC G. PAMER

Aspergillus fumigatus is a ubiquitous fungus that sporulates abundantly, producing small conidia that can reach the lung alveoli. For the most part, inhalation of *A. fumigatus* spores has no adverse consequences but, in the context of severe immunosuppression, invasive aspergillosis (IA) can ensue (Hohl and Feldmesser, 2007; Latgé, 1999). Mortality rates from IA remain quite high, even when antifungal therapy is administered (Latgé, 1999). Although T cells were originally considered to be dispensable in antifungal defense, they have been increasingly recognized as important mediators of protection against IA (Bellocchio et al., 2005). Studies in mice and humans suggest a protective role for Th1 gamma interferon (IFN-γ)-producing CD4$^+$ T lymphocytes against invasive disease (Fig. 1) (Cenci et al., 1997; Hebart et al., 2002). Moreover, several studies have demonstrated a beneficial impact against IA of adoptively transferred, *A. fumigatus*-specific T cells (Cenci et al., 2000; Perruccio et al., 2005).

In addition to causing invasive fungal disease in immunosuppressed individuals, exposure to *A. fumigatus* spores can lead to allergic responses in susceptible individuals (Latgé, 1999). The generation of allergic responses is driven by CD4$^+$ T lymphocytes that secrete the Th2 cytokines interleukin-4 (IL-4) and IL-13 (Fig. 1) (Gibson, 2006; Moss, 2005). In addition to contributing to allergic disease, Th2-biased CD4$^+$ responses are also detrimental in the setting of IA (Cenci et al., 1997, 1999). The in vivo factors that determine whether Th1 or Th2 responses are elicited upon exposure to *A. fumigatus* antigens are currently unknown. A detailed understanding of the factors that contribute to *A. fumigatus*-specific CD4$^+$ T-cell activation and differentiation will not only further the development of fungus-specific, T-cell-based therapies for IA but also will lead to improved therapies against allergic diseases caused by fungi. In this chapter, we will review recent advances in our understanding of *A. fumigatus*-specific CD4$^+$ T-cell activation and differentiation.

ALLERGIC RESPONSES TO *A. FUMIGATUS*

Role of CD4$^+$ T Lymphocytes

It is estimated that humans inhale an average of several hundred *A. fumigatus* conidia every day (Hohl and Feldmesser, 2007; Latgé, 1999). Repeated exposure to fungal antigens is inconsequential in most individuals, due primarily to the rapid induction of antimicrobial defenses by innate immune cells of the airways. In some susceptible individuals, however, a variety of allergic responses to *A. fumigatus*, including allergic rhinitis, allergic sinusitis, and allergic asthma, can ensue. Allergic pulmonary responses to *A. fumigatus* are particularly prevalent in individuals suffering from atopic asthma or cystic fibrosis and are referred to as allergic bronchopulmonary aspergillosis (ABPA) (Gibson, 2006; Moss, 2005). ABPA is characterized by eosinophilic infiltration of the airways, increased immunoglobulin E (IgE) production, airway remodeling, and enhanced Th2 cytokine production (Gibson, 2006; Moss, 2005). It has been shown that CD4$^+$ T cells are crucial mediators of the asthmatic features of ABPA (Corry et al., 1998). Mice devoid of lymphocytes, due to a disruption of the recombinase-activating gene (RAG$^{-/-}$), were resistant to *A. fumigatus*-induced asthma (Corry et al., 1998). Moreover, adoptive transfer of purified CD4$^+$ T cells reversed the asthma-resistant phenotype of RAG$^{-/-}$ mice (Corry et al., 1998). Interestingly, transferred T cells mediated airway inflammation and eosinophilia even in the absence of B cells and IgE production, implicating T cells directly in the pathogenesis of ABPA (Corry et al., 1998).

The cytokines IL-4 and IL-13 are produced by Th2 CD4$^+$ T cells and are key components of the CD4$^+$ T-cell-driven airway hyperresponsiveness that occurs in asthma and ABPA (Cohn et al., 2004; Gibson, 2006) (Fig. 1). IL-4 and IL-13 promote IgE production and eosinophil recruitment (Lewis, 2002; Wynn, 2003).

Amariliz Rivera • Immunology Program, Memorial Sloan Kettering Cancer Center, New York, NY 10021. **Eric G. Pamer** • Infectious Disease Service, Immunology Program, Memorial Sloan Kettering Cancer Center, New York, NY 10021.

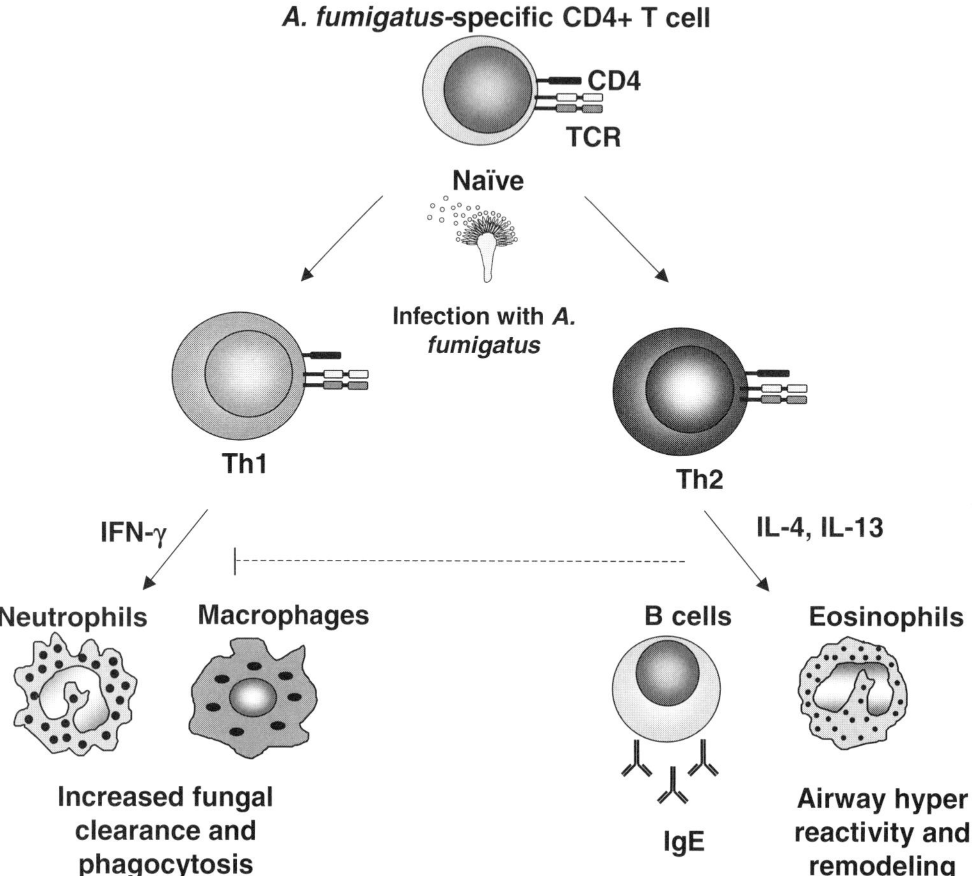

Figure 1. After infection with *A. fumigatus* conidia, *A. fumigatus*-specific $CD4^+$ T cells can differentiate along the Th1 or Th2 lineage. $CD4^+$ T cells that differentiate into Th1 cells acquire the capacity to make IFN-γ. Exposure of macrophages and neutrophils to IFN-γ leads to the production of antimicrobial products that promote clearance of *A. fumigatus*. Th1-mediated enhancement of antifungal responses provides protection from IA. *A. fumigatus*-specific T cells that differentiate into Th2 are central to the pathogenesis of ABPA. Through the production of IL-4 and IL-13, Th2 T cells promote eosinophil infiltration of the airways and the production of IgE antibodies. This $CD4^+$ Th2-mediated response leads to airway hyperreactivity. Th2 responses also contribute to the progression of IA by limiting Th1 responses.

Neutralization of IL-4 and IL-13 can prevent airway hyperresponsiveness, thus demonstrating their central role in asthmatic responses (Blease et al., 2001a, 2001b; Corry et al., 1996; Grunig et al., 1998). It has also been shown that IL-4 and IL-13 neutralization can improve the outcome of IA, providing further evidence for the notion that Th2 responses are detrimental for host defense against IA (Cenci et al., 1997, 1998, 1999). The factors that determine whether protective or detrimental responses are elicited in response to *A. fumigatus* exposure are currently unclear. Several studies suggest that the nature of the antigen, metabolic activity of the spores, and innate cell recognition can all influence $CD4^+$ T-cell differentiation in response to *A. fumigatus*.

A. fumigatus Allergens

Several dozen proteins from *A. fumigatus* have been identified as human allergens (Banerjee and Kurup, 2003). To isolate these allergens, expression cDNA libraries were screened with serum IgE from ABPA patients. Some of these allergens have been characterized in detail and have been found to possess a variety of enzymatic activities (Banerjee and Kurup, 2003). *A. fumigatus* allergens include, but are not limited to, β-glucanases, amylases, and proteases (Banerjee and Kurup, 2003). Of particular interest are allergens with protease activity. It has been shown that intranasal delivery of a single purified *A. fumigatus*-derived protease together with the model antigen ovalbumin can induce airway hyperresponsiveness in mice (Kheradmand et al., 2002). The key role played by the proteolytic activity was demonstrated by the absence of allergic airway responses in mice challenged with the same allergen but whose enzymatic activity was inhibited prior to delivery (Kheradmand et al., 2002). Interestingly, the most common human allergen, Der p 1, is a protease and has also been shown to require enzymatic activity to be allergenic (Kikuchi et al., 2006). Thus, it is likely that one of the important factors contributing to *A. fumigatus* allergenic potential is the presence of many enzymatically active proteins, particularly within inhaled fungal spores.

Human T-Cell Responses to *A. fumigatus* Antigens

Although many studies of *A. fumigatus* allergens have focused on humoral responses, T-cell recognition of fungal antigens is known to be central to pathogenesis. Several studies with human peripheral blood lymphocytes have demonstrated reactivity of $CD4^+$ T cells against several *A. fumigatus* allergens, including Asp f 1, Asp f 2, and Asp f 16 (Chauhan et al., 1996; Ramadan et al., 2005b; Rathore et al., 2001). A detailed analysis of responsiveness to these allergens identified amino acids (aa) 46 to 65 and 106 to 125 of Asp f 1, aa 54 to 74 of Asp f 2, and aa 174 to 182 of Asp f 16 to contain immunodominant peptides recognized by human T cells (Chauhan et al., 1996). The majority of Asp f 1- and Asp f 2-specific clones secreted IL-4 and IL-5 with minimal IFN-γ production, thus demonstrating the predominant Th2 phenotype of responding T cells (Chauhan et al., 1996; Rathore et al., 2001). On the other hand, Asp f 16-specific $CD4^+$ T-cell clones produced high amounts of IFN-γ and displayed cytotoxic activities against *A. fumigatus* hyphae and conidia in vitro (Ramadan et al., 2005b). Healthy individuals and patients surviving IA also mounted Th1-biased $CD4^+$ T-cell responses against the nonallergenic antigens 90-kDa catalase and dipeptidylpeptidase V (Hebart et al., 2002). These findings provide functional evidence for the protective role of Th1 $CD4^+$ T-cell responses and suggest that the Th1 versus Th2 bias of $CD4^+$ T-cell responses to *A. fumigatus* is determined, at least in part, by the antigen specificity of responding T cells.

$CD4^+$ T-cell responses to Asp f 1 were found to be restricted by HLA-DR2 and HLA-DR5, while responses to Asp f 16 were restricted by HLA-DRB1-0301 (Chauhan et al., 1996; Ramadan et al., 2005b). Genotype analysis of HLA expression frequencies among ABPA patients and controls found a significant correlation between the HLA-DR2/DR5 genotype and likelihood of developing ABPA (Chauhan et al., 1997). The strong association of these HLA-DR alleles with ABPA demonstrates that development of allergic responses to *A. fumigatus* are at least partly genetically determined. Although $CD8^+$ T-cell responses against *A. fumigatus* have not been implicated in allergic disease, and thus have not been the focus of extensive investigation, human $CD8^+$ T-cell clones specific for Asp f 16 have been isolated (Ramadan et al., 2005a). These $CD8^+$ T-cell clones were restricted by the HLA B*3501 class I molecule and displayed in vitro cytotoxic activity against *A. fumigatus* (Ramadan et al., 2005a). While these studies suggest that $CD8^+$ T cells might contribute to the pulmonary immune defense against *A. fumigatus*, further studies are necessary to determine the magnitude of their contribution.

Mouse T-Cell Responses to *A. fumigatus* Allergens

Several *A. fumigatus* proteins identified as human allergens also elicit immune responses in mice. Among these, humoral and cellular responses to Asp f 1, Asp f 2, Asp f 3, and Asp f 16 have been documented in a variety of mouse strains (Banerjee and Kurup, 2003; Rivera et al., 2005; Svirshchevskaya et al., 2002). Detailed analysis of T-cell responses to *A. fumigatus* antigens identified aa 155 to 167 of Aspf 1 and aa 6 to 71 and aa 235 to 249 of Asp f 2 as immunodominant $CD4^+$ T-cell epitopes in mice of the *H-2*^{d} haplotype (Svirshchevskaya et al., 2000, 2002). $CD4^+$ T-cell responses to Asp f 2 and Asp f 3 were detected in mice exposed to a single dose of live *A. fumigatus* spores, demonstrating that repeated exposure to inhaled spores is not required to elicit responses against these allergens (Rivera et al., 2005; Svirshchevskaya et al., 2002). Th1 responses against both allergens dominated the $CD4^+$ T-cell response after intratracheal challenge of mice with live spores, while metabolically inactive spores induced a Th2-biased response (Rivera et al., 2005). These findings suggest that the metabolic activity of *A. fumigatus* spores influences the allergic potential of this fungus, with live, metabolically active spores promoting Th1 responses and inactive spores inducing Th2 responses. The utility of the mouse model to investigate *A. fumigatus*-specific T-cell responses and T-cell differentiation is supported by the findings that human and mouse $CD4^+$ T cells respond to the same fungal antigens and that both mice and humans, depending on the circumstances, can mount predominantly Th1 or Th2 responses to this fungus (Fig. 1). Thus, the mouse model of respiratory challenge with live *A. fumigatus* spores will continue to contribute to the development and validation of novel therapeutic approaches to treat allergic responses to fungal infections.

DCs AND INITIATION OF T-CELL RESPONSES TO *A. FUMIGATUS*

Pulmonary DCs and Acquisition of *A. fumigatus* Spores In Vivo

The activation of $CD4^+$ T lymphocytes is critically dependent on antigen processing and peptide presentation by major histocompatibility complex class II

(MHC-II) molecules expressed on specialized antigen-presenting cells. Although macrophages, B cells, and dendritic cells (DCs) can all express MHC-II molecules, DCs are the principal cell population that activates naïve $CD4^+$ T cells. Many different DC subsets have been identified on the basis of differential expression of cell surface molecules, but the major subsets are classified as myeloid DCs (expressing CD11b), lymphoid DCs (expressing CD8α), and plasmacytoid DCs (expressing B220 and secreting large amounts of IFN-α and IFN-β) (Iwasaki, 2007). In the respiratory tract, DC populations have been subdivided according to their location as conducting airway DCs ($CD11c^+$ $CD11b^+$ $MHC\text{-}II^+$ $DEC205^{low}$), interstitial DCs ($CD11c^+$ $CD11b^{hi}$ $MHC\text{-}II^{hi}$ $DEC205^+$), and alveolar DCs ($CD11c^+$ $CD11b^+$ $MHC\text{-}II^+$ $CD205^+$ $Gr\text{-}1^+$ $F4/80^+$) (Hammad and Lambrecht, 2007; Iwasaki, 2007). At steady state, all pulmonary DCs display an immature phenotype, with low-level expression of the costimulatory molecules CD80, CD86, and CD40, and constitutively migrate from the lung to the lung-draining lymph nodes (mediastinal lymph nodes [MLN]). After antigen uptake, DCs migrate from the lung to the MLN and mature into competent antigen-presenting cells by upregulating the expression of CD40, CD80, CD86, and MHC-II (Hammad and Lambrecht, 2007; Iwasaki, 2007).

Upon inhalation, antigen uptake by pulmonary DCs is quite rapid. The intratracheal delivery of fluorescein isothiocyanate (FITC)-labeled *A. fumigatus* conidia or hyphae resulted in the uptake of fungal antigens by pulmonary DCs. $FITC^+$ $CD11c^+$ pulmonary DCs were detectable in the lungs of infected mice as early as 3 h postinjection (Bozza et al., 2002). By 6 h postchallenge, $FITC^+$ $CD11c^+$ DCs were detected in the lung-draining lymph nodes and the spleen (Bozza et al., 2002). These observations suggest that pulmonary DCs can transport *A. fumigatus* conidia or hyphae to secondary lymphoid organs, where they can initiate fungus-specific T-cell responses. It remains unclear whether $CD4^+$ T-cell priming is elicited directly, by the migratory DCs, or by lymph node-resident DCs that take up the antigen from the migratory population. It is also possible that lymph node-resident DCs acquire *A. fumigatus*-derived antigens from the afferent lymph and thus function as the main antigen-presenting cells priming *A. fumigatus*-specific $CD4^+$ T cells. The phenotype of DCs responsible for *A. fumigatus*-specific T-cell priming in the MLN is currently unknown. The timing of antigen uptake and migration reported for *A. fumigatus* is similar to that observed for other pathogens and model antigens (Cook and Bottomly, 2007; de Heer et al., 2005).

DC Migration and $CD4^+$ T-Cell Activation

There is evidence that migration of DCs in the steady state leads to presentation of self-antigens or innocuous molecules in a tolerogenic fashion (de Heer et al., 2005). It has also been documented that antigen presentation by pulmonary DCs primes Th2-biased $CD4^+$ T-cell responses (de Heer et al., 2005). Although antigen inhalation, in the absence of innate immune receptor signaling, is considered to lead primarily to either antigenic tolerance or Th2-biased responses, under certain circumstances, immune defense against many pathogens requires recruitment of Th1 $CD4^+$ T cells to the lung. Th1 $CD4^+$ T-cell responses are protective against a wide range of pathogens, and DCs can specify the differentiation of T-cell responses to optimally combat the challenging microorganism. To direct the differentiation of $CD4^+$ T-cell responses, DCs rely on a spectrum of cell surface receptors to decode the nature of the invading organism. DCs utilize both secreted and cell surface receptors to recognize *A. fumigatus* and can discriminate between fungal conidia and hyphae. One of the most studied mechanisms of pathogen recognition employed by DCs and other innate cells is through the family of Toll-like receptors (TLRs). TLRs recognize conserved molecular components of microorganisms. Triggering of TLRs on DCs leads to the production of inflammatory mediators that can instruct $CD4^+$ Th1 differentiation (Takeda et al., 2003). TLR2 and TLR4 recognize *A. fumigatus* conidia and mediate the production of the inflammatory cytokines tumor necrosis factor (TNF) and IL-12 (Fig. 2) (Balloy et al., 2005; Mambula et al., 2002; Meier et al., 2003). For binding and uptake of *A. fumigatus* conidia, DCs also utilize the mannose receptor, DC-SIGN, and DEC-205, while CR3 and FcγR receptors are used for recognition of hyphae (Bozza et al., 2002; Hohl and Feldmesser, 2007; Serrano-Gomez et al., 2004). The secreted molecule pentraxin 3 also binds conidia and has been shown to contribute to the induction of Th1 responses to *A. fumigatus* (Garlanda et al., 2002, 2005). More recently, the transmembrane C-type lectin receptor dectin-1 was found to recognize β-1,3-glucan exposed on germinating *A. fumigatus* conidia and to trigger the secretion of the inflammatory mediators TNF and IL-12 (Fig. 2) (Gersuk et al., 2006; Hohl et al., 2005; Steele et al., 2005). Together these receptors allow DCs that contact *A. fumigatus* to mature and secrete proinflammatory cytokines that subsequently promote $CD4^+$ T-cell priming and differentiation (Fig. 2).

Triggering of innate receptors upon pathogen exposure leads not only to the production of inflammatory cytokines but also induces expression of CCR7 on the surface of DCs (Cook and Bottomly, 2007). CCR7 is a chemokine receptor required for cellular entry into T-cell zones of secondary lymphoid organs. The acquisition of CCR7 expression by DCs in the lung facilitates their entry into the MLN. Several studies have demonstrated that migration of DCs from the lung occurs pri-

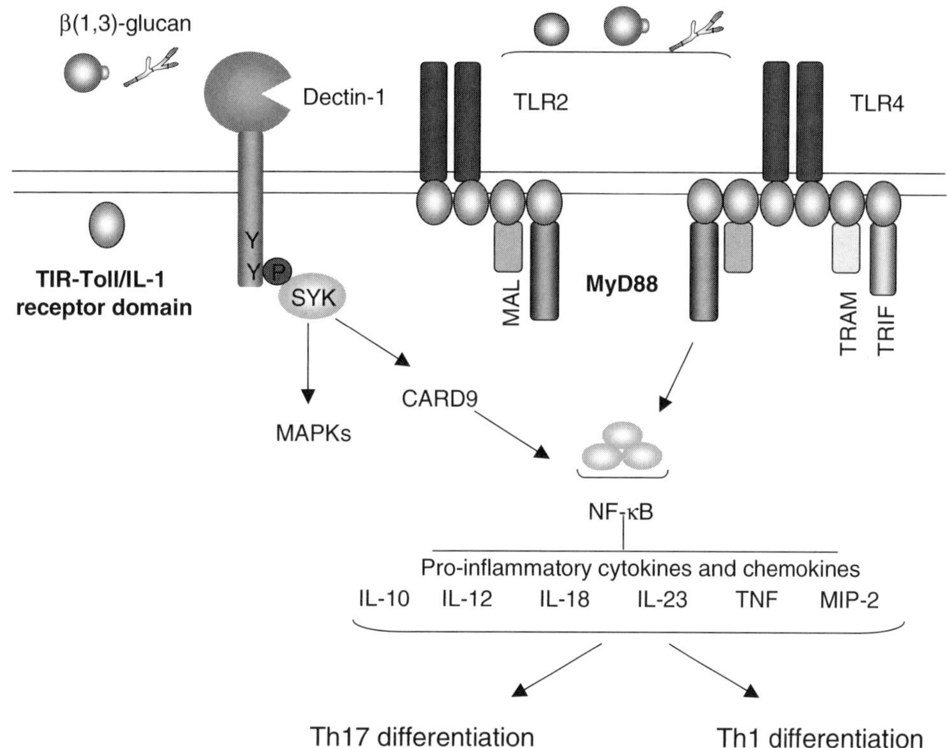

Figure 2. *A. fumigatus* conidia and hyphae are recognized by DCs via the TLR family and by the C-type lectin dectin-1. During the process of germination, *A. fumigatus* exposes β-1,3-glucans as well as unidentified molecules recognized by TLRs. Recognition of β-1,3-glucans by dectin-1 triggers mitogen-activated protein kinases (MAPKs) as well as Syk kinase and CARD9-dependent signals that ultimately lead to NF-κB activation and the production of inflammatory cytokines and chemokines. Both TLR2 and TLR4 mediate recognition of *A. fumigatus*. Upon engagement, TLR signals are transmitted by adaptor proteins that interact with the receptors via Toll/IL-1 receptor domains (TIR). The adaptor proteins MyD88 and MyD88-adaptor-like (MAL) mediate signals from TLR2 and TLR4. TLR4 signals are delivered by MAL-MyD88 as well as by TRIF-related adaptor molecule (TRAM) and TIR-domain containing adaptor protein inducing IFN-β (TRIF). Together, dectin-1- and TLR-derived signals result in the activation of DCs in response to *A. fumigatus* infection and influence CD4+ T-cell differentiation.

marily in a CCR7-dependent fashion and that this migration is required for the activation of naïve CD4+ T cells (Hintzen et al., 2006; Jakubzick et al., 2006). In addition to activating resident lung DCs and promoting their migration, with antigen, to draining lymph nodes, infection triggers the recruitment of DC precursors from the bloodstream into the lung. Recruitment of DC precursors from the blood to the lung can be mediated in a CCR2- and CCR6-dependent manner and is necessary to replenish DC populations in the lung during inflammation. Recruitment of DCs to the lung via CCR6 was recently shown to be important for protection against *A. fumigatus* in a mouse model of IA (Phadke et al., 2007). CCR6-deficient mice were more susceptible to infection and had higher mortality than wild-type control mice (Phadke et al., 2007). CCR2 is a chemokine receptor that is expressed on a subset of circulating monocytes and binds monocyte chemoattractant protein 1 (MCP-1), MCP-3, and MCP-5. CCR2-deficient mice are highly susceptible to a number of microbial pathogens and have been shown to suffer from more severe airway inflammation after sensitization with *A. fumigatus* antigens (Blease et al., 2000; Peters et al., 2001; Sato et al., 2000; Serbina et al., 2003). Although recruitment of DCs or DC precursors to the lungs of infected mice has been shown to occur in a CCR2- and/or CCR6-dependent manner, it remains to be determined how these populations and their recruitment impact CD4+ T-cell activation in response to *A. fumigatus* infection (Grayson et al., 2007; Jakubzick et al., 2006; Ravindran et al., 2007; Robays et al., 2007).

KINETICS OF *A. FUMIGATUS*-SPECIFIC CD4+ T-CELL ACTIVATION IN VIVO

Initiation of CD4+ T-Cell Proliferative Responses

The frequencies of CD4+ T cells specific for a particular antigen are quite low in naïve animals (Moon et al., 2007). Priming of CD4$^+$ T cells by activated DCs presenting their cognate antigen results in vigorous T-cell proliferation and expansion of antigen-specific clones that can mediate protection against infection. Pulmonary challenge with live *A. fumigatus* spores elicits robust CD4$^+$ T-cell proliferation in the draining MLN (Rivera et al., 2005, 2006). Although DCs containing *A. fumigatus* conidia are detected in the MLN as early as 6 h after infection, CD4$^+$ T-cell proliferation occurs considerably later, with polyclonal *A. fumigatus*-specific CD4$^+$ T-cell responses being detectable 72 h after inoculation (Bozza et al., 2002; Rivera et al., 2005). The delay between the arrival of infected DCs in the lymph node and T-cell proliferation is likely due, at least in part, to the need for degradation of spores, antigen processing, and expression of MHC-II–peptide complexes on the surface of DCs. The timing of antigen presentation appears to depend on the nature of the antigen encountered, as purified protein antigens are processed and presented earlier than antigens derived from whole organisms (Ravindran et al., 2007). In a model of *Salmonella* spp. infection, it was recently shown that secreted antigens were more rapidly presented than cell-associated proteins following bacterial infection (Ravindran et al., 2007). Thus, the kinetics of CD4$^+$ T-cell proliferation are likely influenced by availability of antigens for processing and presentation. Regardless of the efficiency and rapidity of antigen presentation, however, most studies have demonstrated that in vivo CD4$^+$ T-cell proliferation occurs approximately 72 h after inoculation of antigen-expressing microbial pathogens (Catron et al., 2006; Jelley-Gibbs et al., 2005; McSorley et al., 2002; Rivera et al., 2005).

Visualization of In Vivo CD4$^+$ T-Cell Proliferation

Although studies of endogenous, polyclonal CD4$^+$ T-cell responses to *A. fumigatus* have provided valuable information regarding the general kinetics of CD4$^+$ T-cell activation and have demonstrated the contribution of CD4$^+$ T cells in immunity to this fungal pathogen, they have not provided a full picture of CD4$^+$ T-cell responses to *A. fumigatus*. To better understand fungus-specific CD4$^+$ T-cell activation in vivo, a T-cell receptor transgenic (TCR-tg) mouse strain specific for *A. fumigatus* was generated (Rivera et al., 2006). These *A. fumigatus*-specific TCR-tg mice (Af 3.16) served as a source of fungus-specific, monoclonal, naïve CD4$^+$ T cells which, upon transfer into recipient mice, enabled direct visualization of in vivo CD4$^+$ T-cell activation, expansion, differentiation, and trafficking following fungal infection (Rivera et al., 2006).

Proliferation of adoptively transferred, *A. fumigatus*-specific T cells was determined by carboxy fluorescein diacetate succinimidyl ester (CFSE) dilution. Exposure of naïve T cells to CFSE, a fluorescent dye, prior to adoptive transfer labels these cells and, as T cells proliferate in the recipient, they become progressively less fluorescent, a process that can be readily quantified by flow cytometry. Thus, the fluorescence intensity correlates inversely with the number of divisions the labeled cell has undergone. By using this strategy, CFSE-labeled, naive Af3.16 TCR-tg cells were transferred into recipient mice which were then intratracheally infected with *A. fumigatus* spores, and T-cell proliferation was measured at various times afterwards. Two days after infection, Af3.16 cells were present in the MLN and remained highly CFSE fluorescent, indicating that T-cell proliferation had not occurred at this early time point (Rivera et al., 2006). On the other hand, between 3 and 4 days after infection, Af3.16 TCR-tg CD4$^+$ T cells underwent extensive proliferation in the MLN and the spleen, such that by 4 days after challenge, *A. fumigatus*-specific T cells had undergone at least six rounds of proliferation (Rivera et al., 2006). As the delivery of *A. fumigatus* conidia to the MLN and the spleen has been documented to occur with similar kinetics, it is possible that *A. fumigatus*-specific CD4$^+$ T cells are primed at both sites. Alternatively, CD4$^+$ T cells primed in the MLN may exit the lymph node and circulate to the spleen. Further studies will be necessary to distinguish between these possibilities.

CD4$^+$ T-Cell Expansion and Contraction

The total number of *A. fumigatus*-specific CD4$^+$ T cells present in the MLN peaked 7 days after infection and then declined. The finding that the frequency and absolute number of Af3.16 T cells peak approximately 7 days after inoculation with *A. fumigatus* conidia is consistent with results obtained when polyclonal, endogenous T-cell responses to *A. fumigatus* are quantified (Rivera et al., 2005, 2006). Vigorous proliferation of *A. fumigatus*-specific CD4$^+$ T cells after infection resulted in greatly increased frequencies of antigen-specific T cells in the MLN at the peak of the response. Similar to what has been determined in other systems in which TCR-transgenic T cells are transferred into recipient mice, the number of adoptively transferred Af3.16 TCR-tg CD4$^+$ T cells influenced their in vivo expansion potential (Badovinac et al., 2007; Hataye et al., 2006). Thus, adoptive transfer of 10^3 or 10^4 naïve Af3.16 TCR-tg CD4$^+$ T cells resulted in a 200-fold expansion, while transfer of 10^6 T cells resulted in markedly attenuated, less-than-10-fold in vivo expansion (Rivera et al., 2006). Altogether, these findings demonstrate that, despite the production of several immunosuppressive mycotoxins with the capacity to inhibit T-cell responses in vitro, in vivo CD4$^+$ T-cell responses to *A. fumigatus* are robust and, in terms of kinetics and magnitude, are similar to

CD4$^+$ T-cell responses to viral and bacterial pathogens (Catron et al., 2006; Jelley-Gibbs et al., 2005; Pahl et al., 1996; Rivera et al., 2006; Roman et al., 2002).

Contraction of Af3.16 TCR-tg T-cell populations was evident by 10 days postinfection and, between 10 and 14 days following infection, the number of Af3.16 TCR-tg cells present in the MLN steadily declined (Rivera et al., 2006). The contraction of *A. fumigatus*-specific CD4$^+$ T cells was evident even though viable *A. fumigatus* conidia were still present in the lungs of infected mice (unpublished observations). How T-cell responses are terminated despite continued infection and the persistence of antigen remains an unresolved issue. A detailed analysis of in vivo antigen presentation to CD8$^+$ T cells following a *Listeria monocytogenes* infection demonstrated that effective antigen presentation was terminated prior to bacterial clearance (Wong and Pamer, 2003). In this setting, primed CD8$^+$ T cells that acquired effector mechanisms mediated the elimination of antigen-presenting cells, effectively terminating antigen presentation (Wong and Pamer, 2003). It remains to be determined whether antigen presentation to CD4$^+$ T cells during fungal infection is similarly restricted. After extensive contraction, Af3.16 CD4$^+$ T cells could still be detected in the MLN 42 days after the initial challenge with *A. fumigatus*, indicating that a memory T-cell population had been established (unpublished observations).

INNATE RECOGNITION OF *A. FUMIGATUS* AND CD4$^+$ T-CELL ACTIVATION AND DIFFERENTIATION

TLRs and *A. fumigatus*-Specific CD4$^+$ T-Cell Activation

As discussed earlier, TLRs are potent mediators of DC activation, and studies with model antigens have demonstrated a requirement for TLR-mediated activation of DCs to obtain optimal CD4$^+$ T-cell activation and differentiation (Layland et al., 2005; Muraille et al., 2003; Pasare and Medzhitov, 2004; Schnare et al., 2001). In vitro and in vivo studies have consistently documented that exposure of DCs and other innate cells to live *A. fumigatus* spores leads to the secretion of inflammatory mediators that promote Th1 CD4$^+$ T-cell differentiation (Brieland et al., 2001; Gafa et al., 2006, 2007). IL-12, IL-18, and IFN-γ are produced in response to *A. fumigatus* and have been found to contribute to Th1 differentiation (Brieland et al., 2001). Thus, although allergic responses to *A. fumigatus* can be triggered in some circumstances, Th1 CD4$^+$ T-cell responses are dominant after intratracheal exposure to live *A. fumigatus* (Cenci et al., 2000; Rivera et al., 2005, 2006). Studies assessing CD4$^+$ T-cell responses to live and inactivated spores are in agreement with these observations, in that live spores induced predominant Th1 responses (Rivera et al., 2005). On the other hand, inactive spores triggered Th2 responses (Rivera et al., 2005).

Since several studies have demonstrated that *A. fumigatus* can trigger TLR2 and TLR4 signaling, experiments were performed to assess the impact of TLR-derived signals on the activation and differentiation of *A. fumigatus*-specific CD4$^+$ T cells. Most TLRs signal through the MyD88 adaptor molecule and, thus, MyD88-deficient mice have markedly impaired TLR-mediated responses (Takeda et al., 2003). Indeed, MyD88-deficient mice succumb rapidly to many infections, primarily due to impaired innate immune responses that mediate early inhibition of pathogen replication and survival (de Veer et al., 2003; Seki et al., 2002; Takeuchi et al., 2000). Although MyD88-deficient mice are quite resistant to respiratory tract infection with *A. fumigatus* spores, MyD88-deficient mice are more susceptible to *A. fumigatus* infection than control mice when both groups are treated with immunosuppressive agents (Balloy et al., 2005; Bellocchio et al., 2004). More recent studies found a defect in early cytokine secretion by innate cells from unmanipulated MyD88-deficient mice infected with *A. fumigatus* (Bretz et al., 2007). The defects in early recognition and clearance of *A. fumigatus* conidia observed in the latter study normalized by 72 h postinfection and did not result in increased mortality (Bretz et al., 2007). Studies aimed at assessing the contribution of TLR/MyD88 to CD4$^+$ T-cell activation have been undertaken in several models with varying results (Fremond et al., 2004; Kursar et al., 2004; Layland et al., 2005; Way et al., 2003; Zhou et al., 2005). For responses to *A. fumigatus*, CD4$^+$ T-cell proliferation in the MLN was not affected by the absence of TLR-mediated recognition of live *A. fumigatus* conidia (Rivera et al., 2006). *A. fumigatus*-specific CD4$^+$ T cells accumulated in the MLN and were recruited to airways independent of TLR/MyD88-mediated signals, demonstrating a limited contribution of TLR/MyD88 signaling to the generation of fungus-specific T-cell responses (Rivera et al., 2006).

Innate Receptors and *A. fumigatus*-Specific CD4$^+$ T-Cell Differentiation

Since TLR/MyD88-mediated signals have been shown to influence not only CD4$^+$ T-cell activation but also differentiation, the impact of MyD88 deficiency on *A. fumigatus*-specific CD4$^+$ T-cell differentiation has also been assessed (Layland et al., 2005; Muraille et al., 2003; Pasare and Medzhitov, 2004; Schnare et al., 2001). Although *A. fumigatus*-specific CD4$^+$ T-cell proliferation was not altered by the absence of MyD88-mediated signals, Th1 differentiation was affected (Rivera et al., 2006). *A. fumigatus*-specific CD4$^+$ T cells primed in MyD88-deficient mice differentiated into Th1 cells but displayed a diminished capacity to produce IFN-γ at the site of infection. The induction of T-bet in the MLN was diminished in the absence of TLR/MyD88-mediated signals (Rivera et al., 2006). In contrast, T-bet expression and IFN-γ production upon trafficking to the airways was not altered by the absence of TLR/MyD88-mediated signals (Rivera et al., 2006). Overall, these studies demonstrated the relatively limited contribution of TLR/MyD88 on the activation and trafficking of fungus-specific CD4$^+$ T cells and revealed a multistage differentiation process where MyD88 signals contribute to Th1 differentiation in the lymph node but not at the site of infection.

DC activation in response to *A. fumigatus* is not limited to signals mediated by MyD88, and other receptors also contribute to innate immune activation. Among these receptors, dectin-1 is a likely candidate to influence CD4$^+$ T-cell activation following infection with *A. fumigatus*. It was recently shown that dectin-1 specifically recognizes β-1,3-glucans displayed by germinating *A. fumigatus* conidia (Gersuk et al., 2006; Hohl et al., 2005; Steele et al., 2005). Dectin-1 has been shown to signal through Syk and CARD9 and mediates immunity against other fungal pathogens, such as *Candida albicans* and *Pneumocystis carinii* (Leibund-Gut-Landmann et al., 2007; Saijo et al., 2007; Taylor et al., 2007). Moreover, it was recently demonstrated that selective in vitro activation of DCs by dectin-1 promoted the activation and differentiation of CD4$^+$ T cells towards the Th1 and Th17 phenotype (Leibund-Gut-Landmann et al., 2007). The selective recognition of live *A. fumigatus* spores by dectin-1 can also provide an explanation for the altered Th1 differentiation observed in mice challenged with inactive spores. It is possible that the recognition of germinating conidia by dectin-1 leads to the production of proinflammatory cytokines that promote Th1 differentiation. As inactivated spores likely do not induce the same level of dectin-1-mediated proinflammatory signals, a potential outcome might be the predominant Th2 response. This interpretation suggests an important role for dectin-1 in instructing Th1 T helper differentiation, but further studies will be necessary to determine whether this is indeed the case. Thus, it is likely that the limited contribution of TLR/MyD88 to *A. fumigatus*-specific CD4$^+$ T-cell activation is attributable to an important role of other pathogen recognition receptors, in particular dectin-1, in directing *A. fumigatus*-specific responses. Further studies, however, are necessary to provide direct experimental evidence for the contribution of dectin-1 to *A. fumigatus*-specific CD4$^+$ T-cell activation and differentiation in vivo.

RECRUITMENT OF *A. FUMIGATUS*-SPECIFIC CD4$^+$ T CELLS TO THE INFECTED LUNG

After priming and proliferation in secondary lymphoid organs, antigen-specific CD4$^+$ T cells traffic to the site of infection, where they activate effector functions and mediate microbial clearance. In order to initially respond to cognate antigen, naïve CD4$^+$ T cells circulate within peripheral lymphoid organs and scan DCs for the presence of specific peptide–MHC-II complexes. Naïve T cells enter the lymph nodes from the blood by expressing the CD62L surface molecule, which interacts specifically with polysaccharides expressed on the surface of high endothelial venules, and traffic into the T-cell zone following a gradient of the CCL21 chemokine, which is detected by lymphocyte-expressed CCR7 (Cyster, 2005). Upon activation, CD4$^+$ T cells lose expression of both CCR7 and CD62L and, thus, effector T cells, upon departure, do not traffic back to lymph nodes. Exit of lymphocytes from the lymph nodes occurs through the efferent lymph and is mediated by the interaction of sphingosine 1-phosphate receptor 1 ($S1P_1$) along a gradient of S1P (Cyster, 2005). Thus, activated CD4$^+$ T cells downregulate CCR7 and CD62L and follow an S1P gradient to exit the MLN through the efferent lymph, where they then enter the blood and subsequently circulate to the inflamed lung. Activated *A. fumigatus*-specific CD4$^+$ T cells leave the MLN and can be readily detected in the airways and lung tissue of infected mice by 6 days postinfection (Rivera et al., 2006). The number of *A. fumigatus*-specific CD4$^+$ T cells recruited to the airways peaks at 7 days postinfection, and airway T cells display an activated phenotype, with high levels of CD44 expression and low levels of CD62L expression (Rivera et al., 2006). Fungus-specific CD4$^+$ T cells that enter the airways produce the Th1 cytokines TNF and IFN-γ. After the peak of the T-cell response, the number of *A. fumigatus*-specific CD4$^+$ T cells present in the airways steadily declines (Rivera et al., 2006). By 30 days postchallenge, *A. fumigatus*-specific CD4$^+$ T cells are no longer detectable in the airways of infected mice (unpublished observations). Our studies with Af3.16 TCR-tg cells demonstrated that entry to the airways was restricted to CD4$^+$ T cells that underwent many rounds of proliferation, as measured by complete dilution of CFSE in cells recovered from the airways of *A. fumigatus*-infected mice (Rivera et al., 2006). The kinetics of Af3.16 TCR-tg CD4$^+$ T cell entry to the airways corresponded to previous observations with endogenous, polyclonal populations (Rivera et al., 2005).

Adoptive transfer studies with specific CD4$^+$ T cells suggest that trafficking from MLN to the airways of *A. fumigatus*-infected mice promotes the expression of effector functions by activated CD4$^+$ T cells. A much greater proportion of *A. fumigatus*-specific CD4$^+$ T cells in the airways of infected mice produce IFN-γ than in the MLN (Rivera et al., 2006). Analysis of IFN-γ production by intracellular cytokine staining revealed that antigen-specific cells acquire the capacity to produce this cytokine after undergoing many rounds of proliferation. A comparison of IFN-γ production by Af3.16 present in the MLN with Af3.16 recovered from the airways of infected mice demonstrated that, even though highly proliferating CD4$^+$ T cells are present in both the MLN and the airways, CD4$^+$ T cells in the lung have a greater capacity to produce IFN-γ (Rivera et al., 2006). Increased IFN-γ production by airway-derived, *A. fumigatus*-specific CD4$^+$ T cells was accompanied by increased expression of the Th1-specific transcription factor T-bet (Rivera et al., 2006). These results suggest that activation and differentiation of *A. fumigatus*-specific T cells occur in distinct stages, with complete differentiation into Th1 T cells occurring at the site of infection (i.e., the lung) rather than at the site of T-cell priming (i.e., MLN). The stimuli that promote T-cell differentiation at the site of infection remain incompletely defined.

REGULATION OF CD4$^+$ T-CELL RESPONSES TO *A. FUMIGATUS*

CD4$^+$ CD25$^+$ Natural Treg-Mediated Regulation

Since the airways are continuously exposed to inhaled antigens, most of which are harmless, mechanisms to avoid overly robust inflammatory responses are critical for the maintenance of optimal pulmonary function. Moreover, it is also necessary to control responses elicited by pathogenic invasion once the organism has been successfully eliminated. The termination of antigen-specific responses is crucial for the prevention of unnecessary collateral damage to the infected organ. To this end, the immune system has evolved several regulatory mechanisms that promote tolerance to innocuous proteins or actively regulate harmful responses. As mentioned earlier, one of the mechanisms employed in the lung and other mucosal surfaces involves the induction of antigen-specific tolerance by immature DCs (Cook and Bottomly, 2007; Langlois and Legge, 2007; Steinman et al., 2003). In this setting, antigen presentation in the absence of costimulation and innate immunity-mediated inflammation leads to long-lasting antigen-specific unresponsiveness. Another mechanism of regulation involves the selection of a specific regulatory T-cell lineage to actively constrain harmful immune responses (Sakaguchi, 2004). Recent studies have provided evidence for the role of regulatory CD4$^+$ CD25$^+$ T cells (Tregs) in maintaining peripheral immune tolerance. Tregs represent a unique lineage of CD4$^+$ T cells that are selected in the thymus and require the transcription factor Foxp3 for differentiation (Fontenot and Rudensky, 2005). Several elegant studies have demonstrated the importance of Tregs in maintaining tolerance throughout the life span of the host (Kim et al., 2007; Sakaguchi, 2004). While the potential involvement of Tregs in controlling pathogen-specific adaptive immune responses is just beginning to be examined in detail, it is apparent that the role of Tregs in responses to infectious organisms will vary according to the pathogen in question (Belkaid, 2007). Although it is believed that Tregs are mostly specific for self-antigens, it has been shown that Tregs can be specific for pathogen-derived antigens and actively respond to infection (Suffia et al., 2006). The latter studies were performed in a mouse model of *Leishmania major*, and antigen-specific Treg responses facilitated the persistence of parasites (Suffia et al., 2006). It remains to be established whether Tregs recognize other pathogens in an antigen-specific manner.

The complex role of Tregs in regulating immune responses to *A. fumigatus* has been recently documented. Distinct contributions by Tregs were observed at early and late stages following *A. fumigatus* infection and during allergic responses to fungus-derived antigens (Montagnoli et al., 2006). At early times following the intranasal delivery of *A. fumigatus* spores, Tregs were activated in a B7/CD28-dependent manner and suppressed neutrophil function through the actions of IL-10 and CTLA-4 (Montagnoli et al., 2006). The early inhibition of neutrophil function was primarily achieved through the induction of the immunosuppressive enzyme indoleamine 2,3-dioxygenase (IDO). IDO induction contributed to the expansion of Tregs at later stages, and those Tregs inhibited Th2 CD4$^+$ T-cell responses in allergic responses to *A. fumigatus* (Montagnoli et al., 2006). Similar to what has been observed in many other studies with *A. fumigatus*, the invasive potential of the fungus shaped the ensuing immune response; in this setting, swollen but not resting conidia counteracted Treg activity. The involvement of IDO in regulating immune responses to *A. fumigatus* was also observed in studies assessing the impact of thymosin α-1 (Tα1) on fungal immunity (Romani et al., 2006). Tα1 is a naturally occurring thymic peptide with immunoregulatory properties that is currently being used therapeutically against viral infections like hepatitis C and human immunodeficiency virus infection (Romani et al., 2007). In responses to *A. fumigatus*, Tα1 influenced T-cell priming and tolerance through TLR9-mediated induction of IDO and DC-dependent regulatory functions (Romani et al., 2006).

Role of Th17 T Cells

Another way of controlling CD4$^+$ T-cell responses is through the cross-regulation of different T helper lineages. Over the past 2 decades, extensive studies have documented the reciprocal regulation of Th1 and Th2 lineages, demonstrating that cytokines not only instruct a particular lineage but also inhibit the alternate pathway. In addition to Th1 and Th2 lineages, more recent studies have identified another T helper lineage characterized by the production of IL-17 (Weaver et al., 2007). The role of Th17 in immune responses and the requirements for their differentiation have been the focus of intense study over the past several years. Thus far, Th17 CD4$^+$ T cells have been found to play a pathogenic role in autoimmune diseases, and their differentiation is instructed by IL-6 and transforming growth factor β and also influenced by IL-23 (Weaver et al., 2007). Although the role of Th17 CD4$^+$ T-cell responses in infection is just starting to be elucidated, it seems that Th17 CD4$^+$ T-cell responses may be particularly important in immunity to fungi (Huang et al., 2004; Rudner et al., 2007). IL-17 facilitates neutrophil recruitment at sites of infection and was found to mediate protection from systemic *Candida* infection (Huang et al., 2004). Moreover, dectin-1-mediated activation of DCs was recently found to direct CD4$^+$ T-cell differentiation along the Th17 pathway (Leibund-Gut-Landmann et al., 2007). In vitro stimulation of DCs with *A. fumigatus* not only induced the production of IL-12, as expected, but also of IL-23, suggesting the potential involvement of Th17 CD4$^+$ T cells in immunity to *A. fumigatus* (Gafa et al., 2006). Indeed, recent in vivo studies have identified a surprising role for IL-23/IL-17 in immune responses to this fungus. IL-23-deficient mice or mice treated with anti-IL-17 antibodies had a lower fungal burden, reduced inflammation, and diminished neutrophil influx in the lungs following *A. fumigatus* infection (Zelante et al., 2007). Moreover, neutralization of these cytokines in vivo resulted in increased numbers of IFN-γ-producing T cells in the MLN, indicating that IL-23/IL-17 responses inhibit protective Th1 CD4$^+$ T-cell responses and thus promote susceptibility to *A. fumigatus* (Zelante et al., 2007).

As can be appreciated from the experimental evidence, the regulation of antifungal CD4$^+$ T-cell responses is complex, with multiple, potentially parallel mechanisms contributing to the control of in vivo infection with *A. fumigatus.*

CD4$^+$ T-CELL-MEDIATED PROTECTION FROM INVASIVE FUNGAL DISEASE

Involvement of T Cells in Protection from IA

The importance of adaptive immune responses against *A. fumigatus* was underappreciated for many years due, in part, to early experiments that failed to identify an increased susceptibility to *A. fumigatus* infection in lymphocyte-deficient mice (nude or SCID) (Schaffner et al., 1982; Williams et al., 1981). Even though lymphocytes appear, in some circumstances, to be dispensable for defense against *A. fumigatus*, several clinical and experimental observations indicate that CD4$^+$ T lymphocytes do participate in immunity against aspergillosis and can be harnessed for therapeutic purposes (Bellocchio et al., 2005; Perruccio et al., 2004). Although neutropenia remains the main risk factor for IA, epidemiological studies of IA in bone marrow transplant recipients have also identified the occurrence of "late-onset" IA in patients with adequate or normal neutrophil counts (Jantunen et al., 1997; Wald et al., 1997). These observations suggest that other factors, in addition to neutropenia, can contribute to susceptibility to IA. The increased susceptibility to IA observed in patients undergoing immunosuppressive treatment to prevent graft-versus-host disease supports the involvement of T-cell-mediated immunity against *A. fumigatus* (Jantunen et al., 1997). Moreover, recent studies assessing the risk factors underlying susceptibility to IA in large patient cohorts identified lymphopenia as a previously underappreciated risk factor (Fukuda et al., 2003). The observation that the production of Th1 cytokines like IFN-γ correlates with increased survival from IA in humans and in mouse models of IA is also indicative of the beneficial role of CD4$^+$ T cells in the immune response to *A. fumigatus* (Bozza et al., 2003; Cenci et al., 1997, 1998; Hebart et al., 2002).

Vaccination Strategies

Several studies in mice have demonstrated the beneficial impact of vaccination against aspergillosis (Fig. 3) (Bozza et al., 2003; Cenci et al., 2000; Ito and Lyons, 2002; Ito et al., 2006). Prior intranasal infection with *A. fumigatus* conidia or the intranasal delivery of crude preparations of *A. fumigatus* antigens conferred significant protection against invasive disease in mice with cyclophosphamide-induced immunosuppression (Cenci et al., 2000). Moreover, it was found that intranasal delivery of *A. fumigatus* antigens greatly enhanced survival of mice from invasive fungal disease after corticosteroid-induced immunosuppression, while subcutaneous vaccination with the same antigen preparation conferred complete protection from IA (Ito and Lyons, 2002; Ito

Vaccination

Intranasal delivery of *A. fumigatus* antigens

Subcutaneous delivery of *A. fumigatus* antigens or rAsp f 3

Intravenous transfer of DCs presenting *A. fumigatus* antigens

infection

Immunosuppressed after vaccination

Protection from IA

Adoptive transfer

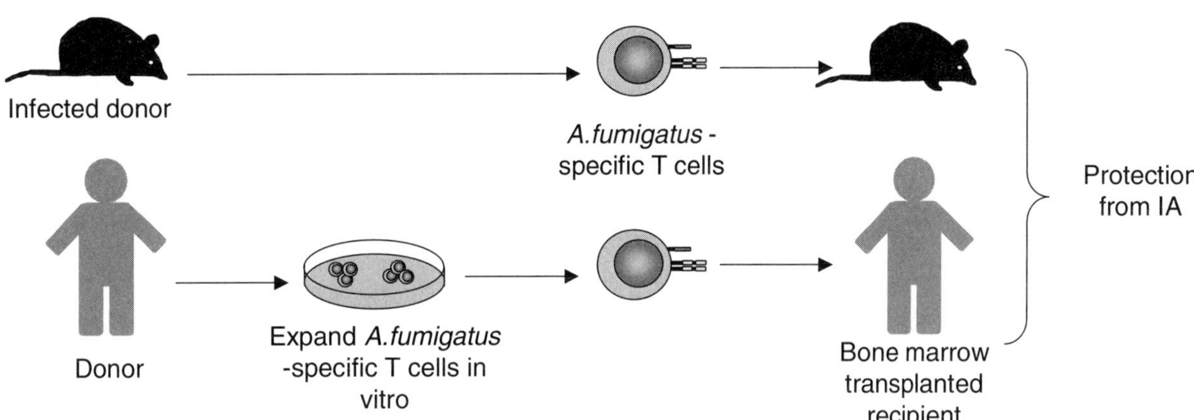

Figure 3. The potential use of *A. fumigatus*-specific CD4+ T cells as therapeutics for the treatment of IA has been demonstrated by various approaches. Potential strategies include vaccination and adoptive transfer of antigen-specific T cells. Vaccination of animals via intranasal or subcutaneous delivery of *A. fumigatus* antigens leads to the development of T-cell responses capable of conferring protection from IA. The systemic delivery of DCs modified to present *A. fumigatus* antigens can also lead to the induction of protective T-cell responses in treated animals. *A. fumigatus*-specific CD4+ T cells generated in vivo in infected donor mice can confer protection against IA in recipient animals that have undergone allogeneic bone marrow transplantation. Similarly, human-derived, *A. fumigatus*-specific CD4+ T cells can be recovered from donors and expanded in vitro with a variety of *A. fumigatus* antigens. Expanded *A. fumigatus*-specific T cells have been transferred into bone marrow transplant recipients.

et al., 2006). Analysis of sera obtained from protectively vaccinated mice identified Asp f 3 as the principal recognized antigen (Ito et al., 2006). Furthermore, vaccination by subcutaneous delivery of purified recombinant Asp f 3 in adjuvant provided significant protection against IA.

The protection achieved with Asp f 3 alone, or truncated versions of the protein in which IgE-binding domains were removed, was comparable to that obtained with crude antigen preparations, suggesting that Asp f 3-based vaccines might be clinically beneficial (Ito et al., 2006). Other DC-based vaccination strategies have also been successfully employed to prevent IA in systemically infected mice (intravenous delivery of conidia) or after allogeneic bone marrow transplantation (Fig. 3) (Bozza et al., 2003, 2004; Montagnoli et al., 2008; Perruccio et al., 2004). In these studies, mouse-derived as well as human-derived DCs were cultured with live conidia or transfected with *A. fumigatus* conidial RNA (Bozza et al., 2003). Such treatments induced maturation of the DCs and were accompanied by the secretion of IL-12 and other Th1-inducing cytokines. Adoptive transfer of DCs early after stem cell transplantation conferred protection against IA for more than 60 days after infection with live *A. fumigatus* conidia (Bozza et al., 2003). Protected mice had accelerated recovery of myeloid and lymphoid cells and increased frequencies of IFN-γ-secreting CD4+ T cells in the lung, suggesting that protective, pathogen-specific responses have beneficial effects that extend beyond defense against a single pathogen (Bozza et al., 2003, 2004).

Adoptive Antigen-Specific T-Cell Therapies

The protective role of CD4$^+$ T cells, as conferred by some of the vaccination strategies, is increasingly accepted as clinically relevant. Adoptive transfer of polyclonal CD4$^+$ T cells from the spleens of mice immunized by prior infection confers significant protection to recipient mice (Fig. 3) (Cenci et al., 2000). Along similar lines, adoptive transfer of Th1 CD4$^+$ T cells to mice undergoing allogeneic bone marrow transplantation conferred significant protection from invasive fungal disease, with prolonged survival compared to control mice (Bozza et al., 2003). The beneficial impact of transferring CD4$^+$ Th1 T cells as a preventive and potentially therapeutic approach to combat invasive fungal disease in bone marrow transplant recipients has been demonstrated (Perruccio et al., 2005). In such studies, donor-derived *A. fumigatus*-specific CD4$^+$ T-cell clones were extensively expanded in vitro after culture in the presence of heat-inactivated conidia (Perruccio et al., 2005). Prior to transfer, T-cell clones were screened for alloreactivity against host cells, and only nonalloreactive clones were transferred. Adoptive transfer of *A. fumigatus*-specific clones did not cause graft-versus-host disease and mediated a reduction of serum galactomannan antigenemia in patients with IA (Perruccio et al., 2005). Moreover, 9 of 10 patients receiving pathogen-specific clones survived IA and maintained high frequencies of antigen-specific T cells for the duration of the clinical trial (Perruccio et al., 2005). An improved method for expansion of *A. fumigatus*-specific CD4$^+$ T cells from the peripheral blood of healthy donors has been recently developed (Beck et al., 2006). In the latter study, T-cell clones induced damage to *A. fumigatus* hyphae and also significantly enhanced the capacity of human neutrophils to damage the fungus (Beck et al., 2006). Altogether, these studies provide experimental evidence supporting the use of vaccination and therapeutic T-cell transfer as a means to combat invasive fungal disease in patients with increased risk for IA.

CONCLUDING REMARKS

The studies summarized in this chapter provide extensive experimental data demonstrating the participation of adaptive immune responses, in particular by CD4$^+$ T lymphocytes, in antifungal immunity. Much remains to be learned about the regulation of CD4$^+$ T cells in antifungal defense, the factors that trigger their recruitment and differentiation, and how they mediate their protective functions. Future studies in humans and mice will undoubtedly provide surprising answers to these questions and will further the development of novel therapeutic approaches against IA.

REFERENCES

Badovinac, V. P., J. S. Haring, and J. T. Harty. 2007. Initial T cell receptor transgenic cell precursor frequency dictates critical aspects of the CD8$^+$ T cell response to infection. *Immunity* **26:**827–841.

Balloy, V., M. Si-Tahar, O. Takeuchi, B. Philippe, M. A. Nahori, M. Tanguy, M. Huerre, S. Akira, J. P. Latgé, and M. Chignard. 2005. Involvement of toll-like receptor 2 in experimental invasive pulmonary aspergillosis. *Infect. Immun.* **73:**5420–5425.

Banerjee, B., and V. P. Kurup. 2003. Molecular biology of *Aspergillus* allergens. *Front. Biosci.* **8:**S128–S139.

Beck, O., M. S. Topp, U. Koehl, E. Roilides, M. Simitsopoulou, M. Hanisch, J. Sarfati, J. P. Latgé, T. Klingebiel, H. Einsele, and T. Lehrnbecher. 2006. Generation of highly purified and functionally active human TH1 cells against *Aspergillus fumigatus*. *Blood* **107:** 2562–2569.

Belkaid, Y. 2007. Regulatory T cells and infection: a dangerous necessity. *Nat. Rev. Immunol.* **7:**875–888.

Bellocchio, S., S. Bozza, C. Montagnoli, K. Perruccio, R. Gaziano, L. Pitzurra, and L. Romani. 2005. Immunity to *Aspergillus fumigatus*: the basis for immunotherapy and vaccination. *Med. Mycol.* **43**(Suppl. 1)**:**S181–S188.

Bellocchio, S., C. Montagnoli, S. Bozza, R. Gaziano, G. Rossi, S. S. Mambula, A. Vecchi, A. Mantovani, S. M. Levitz, and L. Romani. 2004. The contribution of the Toll-like/IL-1 receptor superfamily to innate and adaptive immunity to fungal pathogens in vivo. *J. Immunol.* **172:**3059–3069.

Blease, K., C. Jakubzick, J. M. Schuh, B. H. Joshi, R. K. Puri, and C. M. Hogaboam. 2001a. IL-13 fusion cytotoxin ameliorates chronic fungal-induced allergic airway disease in mice. *J. Immunol.* **167:**6583–6592.

Blease, K., C. Jakubzick, J. Westwick, N. Lukacs, S. L. Kunkel, and C. M. Hogaboam. 2001b. Therapeutic effect of IL-13 immunoneutralization during chronic experimental fungal asthma. *J. Immunol.* **166:**5219–5224.

Blease, K., B. Mehrad, T. J. Standiford, N. W. Lukacs, J. Gosling, L. Boring, I. F. Charo, S. L. Kunkel, and C. M. Hogaboam. 2000. Enhanced pulmonary allergic responses to *Aspergillus* in CCR2$^{-/-}$ mice. *J. Immunol.* **165:**2603–2611.

Bozza, S., R. Gaziano, A. Spreca, A. Bacci, C. Montagnoli, P. di Francesco, and L. Romani. 2002. Dendritic cells transport conidia and hyphae of *Aspergillus fumigatus* from the airways to the draining lymph nodes and initiate disparate Th responses to the fungus. *J. Immunol.* **168:**1362–1371.

Bozza, S., C. Montagnoli, R. Gaziano, G. Rossi, G. Nkwanyuo, S. Bellocchio, and L. Romani. 2004. Dendritic cell-based vaccination against opportunistic fungi. *Vaccine* **22:**857–864.

Bozza, S., K. Perruccio, C. Montagnoli, R. Gaziano, S. Bellocchio, E. Burchielli, G. Nkwanyuo, L. Pitzurra, A. Velardi, and L. Romani. 2003. A dendritic cell vaccine against invasive aspergillosis in allogeneic hematopoietic transplantation. *Blood* **102:**3807–3814.

Bretz, C., G. Gersuk, S. Knoblaugh, N. Chaudhary, J. Randolph-Habecker, R. C. Hackman, J. Staab, and K. A. Marr. 2007. MyD88 signaling contributes to early pulmonary responses to *Aspergillus fumigatus*. *Infect. Immun.* **76:**952–958.

Brieland, J. K., C. Jackson, F. Menzel, D. Loebenberg, A. Cacciapuoti, J. Halpern, S. Hurst, T. Muchamuel, R. Debets, R. Kastelein, T. Churakova, J. Abrams, R Hare, and A. O'Garra. 2001. Cytokine networking in lungs of immunocompetent mice in response to inhaled *Aspergillus fumigatus*. *Infect. Immun.* 69:1554–1560.

Catron, D. M., L. K. Rusch, J. Hataye, A. A. Itano, and M. K. Jenkins. 2006. CD4+ T cells that enter the draining lymph nodes after antigen injection participate in the primary response and become central-memory cells. *J. Exp. Med.* **203:**1045–1054.

Cenci, E., A. Mencacci, A. Bacci, F. Bistoni, V. P. Kurup, and L. Romani. 2000. T cell vaccination in mice with invasive pulmonary aspergillosis. *J. Immunol.* **165:**381–388.

Cenci, E., A. Mencacci, G. Del Sero, A. Bacci, C. Montagnoli, C. F. d'Ostiani, P. Mosci, M. Bachmann, F. Bistoni, M. Kopf, and L. Romani. 1999. Interleukin-4 causes susceptibility to invasive pulmonary aspergillosis through suppression of protective type I responses. *J. Infect. Dis.* **180:**1957–1968.

Cenci, E., A. Mencacci, C. Fe d'Ostiani, G. Del Sero, P. Mosci, C. Montagnoli, A. Bacci, and L. Romani. 1998. Cytokine- and T helper-dependent lung mucosal immunity in mice with invasive pulmonary aspergillosis. *J. Infect. Dis.* **178:**1750–1760.

Cenci, E., S. Perito, K. H. Enssle, P. Mosci, J. P. Latgé, L. Romani, and F. Bistoni. 1997. Th1 and Th2 cytokines in mice with invasive aspergillosis. *Infect. Immun.* **65:**564–570.

Chauhan, B., A. Knutsen, P. S. Hutcheson, R. G. Slavin, and C. J. Bellone. 1996. T cell subsets, epitope mapping, and HLA-restriction in patients with allergic bronchopulmonary aspergillosis. *J. Clin. Investig.* 97:2324–2331.

Chauhan, B., L. Santiago, D. A. Kirschmann, V. Hauptfeld, A. P. Knutsen, P. S. Hutcheson, S. L. Woulfe, R. G. Slavin, H. J. Schwartz, and C. J. Bellone. 1997. The association of HLA-DR alleles and T cell activation with allergic bronchopulmonary aspergillosis. *J. Immunol.* **159:**4072–4076.

Cohn, L., J. A. Elias, and G. L. Chupp. 2004. Asthma: mechanisms of disease persistence and progression. *Annu. Rev. Immunol.* **22:** 789–815.

Cook, D. N., and K. Bottomly. 2007. Innate immune control of pulmonary dendritic cell trafficking. *Proc. Am. Thorac. Soc.* **4:**234–239.

Corry, D. B., H. G. Folkesson, M. L. Warnock, D. J. Erle, M. A. Matthay, J. P. Wiener-Kronish, and R. M. Locksley. 1996. Interleukin 4, but not interleukin 5 or eosinophils, is required in a murine model of acute airway hyperreactivity. *J. Exp. Med.* **183:**109–117.

Corry, D. B., G. Grunig, H. Hadeiba, V. P. Kurup, M. L. Warnock, D. Sheppard, D. M. Rennick, and R. M. Locksley. 1998. Requirements for allergen-induced airway hyperreactivity in T and B cell-deficient mice. *Mol. Med.* **4:**344–355.

Cyster, J. G. 2005. Chemokines, sphingosine-1-phosphate, and cell migration in secondary lymphoid organs. *Annu. Rev. Immunol.* **23:** 127–159.

de Heer, H. J., H. Hammad, M. Kool, and B. N. Lambrecht. 2005. Dendritic cell subsets and immune regulation in the lung. *Semin. Immunol.* **17:**295–303.

de Veer, M. J., J. M. Curtis, T. M. Baldwin, J. A. DiDonato, A. Sexton, M. J. McConville, E. Handman, and L. Schofield. 2003. MyD88 is essential for clearance of Leishmania major: possible role for lipophosphoglycan and Toll-like receptor 2 signaling. *Eur. J. Immunol.* **33:**2822–2831.

Fontenot, J. D., and A. Y. Rudensky. 2005. A well adapted regulatory contrivance: regulatory T cell development and the forkhead family transcription factor Foxp3. *Nat. Immunol.* **6:**331–337.

Fremond, C. M., V. Yeremeev, D. M. Nicolle, M. Jacobs, V. F. Quesniaux, and B. Ryffel. 2004. Fatal *Mycobacterium tuberculosis* infection despite adaptive immune response in the absence of MyD88. *J. Clin. Investig.* **114:**1790–1799.

Fukuda, T., M. Boeckh, R. A. Carter, B. M. Sandmaier, M. B. Maris, D. G. Maloney, P. J. Martin, R. F. Storb, and K. A. Marr. 2003. Risks and outcomes of invasive fungal infections in recipients of allogeneic hematopoietic stem cell transplants after nonmyeloablative conditioning. *Blood* **102:**827–833.

Gafa, V., R. Lande, M. C. Gagliardi, M. Severa, E. Giacomini, M. E. Remoli, R. Nisini, C. Ramoni, P. Di Francesco, D. Aldebert, R. Grillot, and E. M. Coccia. 2006. Human dendritic cells following *Aspergillus fumigatus* infection express the CCR7 receptor and a differential pattern of interleukin-12 (IL-12), IL-23, and IL-27 cytokines, which lead to a Th1 response. *Infect. Immun.* **74:**1480–1489.

Gafa, V., M. E. Remoli, E. Giacomini, M. C. Gagliardi, R. Lande, M. Severa, R. Grillot, and E. M. Coccia. 2007. In vitro infection of human dendritic cells by *Aspergillus fumigatus* conidia triggers the secretion of chemokines for neutrophil and Th1 lymphocyte recruitment. *Microbes Infect.* 9:971–980.

Garlanda, C., B. Bottazzi, A. Bastone, and A. Mantovani. 2005. Pentraxins at the crossroads between innate immunity, inflammation, matrix deposition, and female fertility. *Annu. Rev. Immunol.* **23:** 337–366.

Garlanda, C., E. Hirsch, S. Bozza, A. Salustri, M. De Acetis, R. Nota, A. Maccagno, F. Riva, B. Bottazzi, G. Peri, A. Doni. L. Vago, M. Botto, R. De Santis, P. Carminati, G. Siracusa, F. Altruda, A. Vecchi, L. Romani, and A. Mantovani. 2002. Non-redundant role of the long pentraxin PTX3 in anti-fungal innate immune response. *Nature* **420:**182–186.

Gersuk, G. M., D. M. Underhill, L. Zhu, and K. A. Marr. 2006. Dectin-1 and TLRs permit macrophages to distinguish between different *Aspergillus fumigatus* cellular states. *J. Immunol.* **176:**3717–3724.

Gibson, P. G. 2006. Allergic bronchopulmonary aspergillosis. *Semin. Respir. Crit. Care Med.* **27:**185–191.

Grayson, M. H., M. S. Ramos, M. M. Rohlfing, R. Kitchens, H. D. Wang, A. Gould, E. Agapov, and M. J. Holtzman. 2007. Controls for lung dendritic cell maturation and migration during respiratory viral infection. *J. Immunol.* **179:**1438–1448.

Grunig, G., M. Warnock, A. E. Wakil, R. Venkayya, F. Brombacher, D. M. Rennick, D. Sheppard, M. Mohrs, D. D. Donaldson, R. M. Locksley, and D. B. Corry. 1998. Requirement for IL-13 independently of IL-4 in experimental asthma. *Science* **282:**2261–2263.

Hammad, H., and B. N. Lambrecht. 2007. Lung dendritic cell migration. *Adv. Immunol.* **93:**265–278.

Hataye, J., J. J. Moon, A. Khoruts, C. Reilly, and M. K. Jenkins. 2006. Naive and memory CD4+ T cell survival controlled by clonal abundance. *Science* **312:**114–116.

Hebart, H., C. Bollinger, P. Fisch, J. Sarfati, C. Meisner, M. Baur, J. Loeffler, M. Monod, J. P. Latgé, and H. Einsele. 2002. Analysis of T-cell responses to *Aspergillus fumigatus* antigens in healthy individuals and patients with hematologic malignancies. *Blood* **100:** 4521–4528.

Hintzen, G., L. Ohl, M. L. del Rio, J. I. Rodriguez-Barbosa, O. Pabst, J. R. Kocks, J. Krege, S. Hardtke, and R. Forster. 2006. Induction of tolerance to innocuous inhaled antigen relies on a CCR7-dependent dendritic cell-mediated antigen transport to the bronchial lymph node. *J. Immunol.* **177:**7346–7354.

Hohl, T. M., and M. Feldmesser. 2007. *Aspergillus fumigatus:* principles of pathogenesis and host defense. *Eukaryot. Cell* **6:**1953–1963.

Hohl, T. M., H. L. Van Epps, A. Rivera, L. A. Morgan, P. L. Chen, M. Feldmesser, and E. G. Pamer. 2005. *Aspergillus fumigatus* triggers inflammatory responses by stage-specific beta-glucan display. *PLoS Pathog.* **1:**e30.

Huang, W., L. Na, P. L. Fidel, and P. Schwarzenberger. 2004. Requirement of interleukin-17A for systemic anti-*Candida albicans* host defense in mice. *J. Infect. Dis.* **190:**624–631.

Ito, J. I., and J. M. Lyons. 2002. Vaccination of corticosteroid immunosuppressed mice against invasive pulmonary aspergillosis. *J. Infect. Dis.* **186:**869–871.

Ito, J. I., J. M. Lyons, T. B. Hong, D. Tamae, Y. K. Liu, S. P. Wilczynski, and M. Kalkum. 2006. Vaccinations with recombinant variants of *Aspergillus fumigatus* allergen Asp f 3 protect mice against invasive aspergillosis. *Infect. Immun.* 74:5075–5084.

Iwasaki, A. 2007. Mucosal dendritic cells. *Annu. Rev. Immunol.* 25: 381–418.

Jakubzick, C., F. Tacke, J. Llodra, N. van Rooijen, and G. J. Randolph. 2006. Modulation of dendritic cell trafficking to and from the airways. *J. Immunol.* 176:3578–3584.

Jantunen, E., P. Ruutu, L. Niskanen, L. Volin, T. Parkkali, P. Koukila-Kahkola, and T. Ruutu. 1997. Incidence and risk factors for invasive fungal infections in allogeneic BMT recipients. *Bone Marrow Transplant.* 19:801–808.

Jelley-Gibbs, D. M., D. M. Brown, J. P. Dibble, L. Haynes, S. M. Eaton, and S. L. Swain. 2005. Unexpected prolonged presentation of influenza antigens promotes CD4 T cell memory generation. *J. Exp. Med.* 202:697–706.

Kheradmand, F., A. Kiss, J. Xu, S. H. Lee, P. E. Kolattukudy, and D. B. Corry. 2002. A protease-activated pathway underlying Th cell type 2 activation and allergic lung disease. *J. Immunol.* 169:5904–5911.

Kikuchi, Y., T. Takai, T. Kuhara, M. Ota, T. Kato, H. Hatanaka, S. Ichikawa, T. Tokura, H. Akiba, K. Mitsuishi, S. Ikeda, K. Okumura, and H. Ogawa. 2006. Crucial commitment of proteolytic activity of a purified recombinant major house dust mite allergen Der p1 to sensitization toward IgE and IgG responses. *J. Immunol.* 177:1609–1617.

Kim, J. M., J. P. Rasmussen, and A. Y. Rudensky. 2007. Regulatory T cells prevent catastrophic autoimmunity throughout the lifespan of mice. *Nat. Immunol.* 8:191–197.

Kursar, M., H. W. Mittrucker, M. Koch, A. Kohler, M. Herma, and S. H. Kaufmann. 2004. Protective T cell response against intracellular pathogens in the absence of Toll-like receptor signaling via myeloid differentiation factor 88. *Int. Immunol.* 16:415–421.

Langlois, R. A., and K. L. Legge. 2007. Respiratory dendritic cells: mediators of tolerance and immunity. *Immunol. Res.* 39:128–145.

Latgé, J. P. 1999. *Aspergillus fumigatus* and aspergillosis. *Clin. Microbiol. Rev.* 12:310–350.

Layland, L. E., H. Wagner, and C. U. da Costa. 2005. Lack of antigen-specific Th1 response alters granuloma formation and composition in *Schistosoma mansoni*-infected MyD88$^{-/-}$ mice. *Eur. J. Immunol.* 35:3248–3257.

Leibund Gut-Landmann, S., O. Gross, M. J. Robinson, F. Osorio, E. C. Slack, S. V. Tsoni, E. Schweighoffer, V. Tybulewicz, G. D. Brown, J. Ruland, and C. Reis e Sousa. 2007. Syk- and CARD9-dependent coupling of innate immunity to the induction of T helper cells that produce interleukin 17. *Nat. Immunol.* 8:630–638.

Lewis, D. B. 2002. Allergy immunotherapy and inhibition of Th2 immune responses: a sufficient strategy? *Curr. Opin. Immunol.* 14: 644–651.

Mambula, S. S., K. Sau, P. Henneke, D. T. Golenbock, and S. M. Levitz. 2002. Toll-like receptor (TLR) signaling in response to *Aspergillus fumigatus*. *J. Biol. Chem.* 277:39320–39326.

McSorley, S. J., S. Asch, M. Costalonga, R. L. Reinhardt, and M. K. Jenkins. 2002. Tracking salmonella-specific CD4 T cells in vivo reveals a local mucosal response to a disseminated infection. *Immunity* 16:365–377.

Meier, A., C. J. Kirschning, T. Nikolaus, H. Wagner, J. Heesemann, and F. Ebel. 2003. Toll-like receptor (TLR) 2 and TLR4 are essential for *Aspergillus*-induced activation of murine macrophages. *Cell. Microbiol.* 5:561–570.

Montagnoli, C., F. Fallarino, R. Gaziano, S. Bozza, S. Bellocchio, T. Zelante, W. P. Kurup, L. Pitzurra, P. Puccetti, and L. Romani. 2006. Immunity and tolerance to *Aspergillus* involve functionally distinct regulatory T cells and tryptophan catabolism. *J. Immunol.* 176:1712–1723.

Montagnoli, C., K. Perruccio, S. Bozza, P. Bonifazi, T. Zelante, A. De Luca, S. Moretti, C. D'Angelo, F. Bistoni, M. Martelli, F. Aversa, A Velardi, and L. Romani. 2008. Provision of antifungal immunity and concomitant alloantigen tolerization by conditioned dendritic cells in experimental hematopoietic transplantation. *Blood Cells Mol. Dis.* 40:55–62.

Moon, J. J., H. H. Chu, M. Pepper, S. J. McSorley, S. C. Jameson, R. M. Kedl, and M. K. Jenkins. 2007. Naive CD4$^+$ T cell frequency varies for different epitopes and predicts repertoire diversity and response magnitude. *Immunity* 27:203–213.

Moss, R. B. 2005. Pathophysiology and immunology of allergic bronchopulmonary aspergillosis. *Med. Mycol.* 43(Suppl. 1):S203–S206.

Muraille, E., C. De Trez, M. Brait, P. De Baetselier, O. Leo, and Y. Carlier. 2003. Genetically resistant mice lacking MyD88-adapter protein display a high susceptibility to *Leishmania major* infection associated with a polarized Th2 response. *J. Immunol.* 170:4237–4241.

Pahl, H. L., B. Krauss, K. Schulze-Osthoff, T. Decker, E. B. Traenckner, M. Vogt, C. Myers, T. Parks, P. Warring, A. Muhlbacher, A. P. Czernilofsky, and P. A. Baeuerle. 1996. The immunosuppressive fungal metabolite gliotoxin specifically inhibits transcription factor NF-κB. *J. Exp. Med.* 183:1829–1840.

Pasare, C., and R. Medzhitov. 2004. Toll-dependent control mechanisms of CD4 T cell activation. *Immunity* 21:733–741.

Perruccio, K., S. Bozza, C. Montagnoli, S. Bellocchio, F. Aversa, M. Martelli, F. Bistoni, A. Velardi, and L. Romani. 2004. Prospects for dendritic cell vaccination against fungal infections in hematopoietic transplantation. *Blood Cells Mol. Dis.* 33:248–255.

Perruccio, K., A. Tosti, E. Burchielli, F. Topini, L. Ruggeri, A. Carotti, M. Capanni, E. Urbani, A. Mancusi, F. Aversa, M. F. Martelli, L. Romani, and A. Velardi. 2005. Transferring functional immune responses to pathogens after haploidentical hematopoietic transplantation. *Blood* 106:4397–4406.

Peters, W., H. M. Scott, H. F. Chambers, J. L. Flynn, I. F. Charo, and J. D. Ernst. 2001. Chemokine receptor 2 serves an early and essential role in resistance to *Mycobacterium tuberculosis*. *Proc. Natl. Acad. Sci. USA* 98:7958–7963.

Phadke, A. P., G. Akangire, S. J. Park, S. A. Lira, and B. Mehrad. 2007. The role of CC chemokine receptor 6 in host defense in a model of invasive pulmonary aspergillosis. *Am. J. Respir. Crit. Care Med.* 175:1165–1172.

Ramadan, G., B. Davies, V. P. Kurup, and C. A. Keever-Taylor. 2005a. Generation of cytotoxic T cell responses directed to human leucocyte antigen class I restricted epitopes from the *Aspergillus* f16 allergen. *Clin. Exp. Immunol.* 140:81–91.

Ramadan, G., B. Davies, V. P. Kurup, and C. A. Keever-Taylor. 2005b. Generation of Th1 T cell responses directed to a HLA Class II restricted epitope from the *Aspergillus* f16 allergen. *Clin. Exp. Immunol.* 139:257–267.

Rathore, V. B., B. Johnson, J. N. Fink, K. J. Kelly, P. A. Greenberger, and V. P. Kurup. 2001. T cell proliferation and cytokine secretion to T cell epitopes of Asp f 2 in ABPA patients. *Clin. Immunol.* 100: 228–235.

Ravindran, R., L. Rusch, A. Itano, M. K. Jenkins, and S. J. McSorley. 2007. CCR6-dependent recruitment of blood phagocytes is necessary for rapid CD4 T cell responses to local bacterial infection. *Proc. Natl. Acad. Sci. USA* 104:12075–12080.

Rivera, A., G. Ro, H. L. Van Epps, T. Simpson, I. Leiner, D. B. Sant'Angelo, and E. G. Pamer. 2006. Innate immune activation and CD4$^+$ T cell priming during respiratory fungal infection. *Immunity* 25:665–675.

Rivera, A., H. L. Van Epps, T. M. Hohl, G. Rizzuto, and E. G. Pamer. 2005. Distinct CD4$^+$-T-cell responses to live and heat-inactivated *Aspergillus fumigatus* conidia. *Infect. Immun.* 73:7170–7179.

Robays, L. J., T. Maes, S. Lebecque, S. A. Lira, W. A. Kuziel, G. G. Brusselle, G. F. Joos, and K. V. Vermaelen. 2007. Chemokine receptor CCR2 but not CCR5 or CCR6 mediates the increase in pulmonary dendritic cells during allergic airway inflammation. *J. Immunol.* 178:5305–5311.

Roman, E., E. Miller, A. Harmsen, J. Wiley, U. H. Von Andrian, G. Huston, and S. L. Swain. 2002. CD4 effector T cell subsets in the response to influenza: heterogeneity, migration, and function. *J. Exp. Med.* **196:**957–968.

Romani, L., F. Bistoni, C. Montagnoli, R. Gaziano, S. Bozza, P. Bonifazi, T. Zelante, S. Moretti, G. Rasi, E. Garaci, and P. Puccetti. 2007. Thymosin α1: an endogenous regulator of inflammation, immunity, and tolerance. *Ann. N. Y. Acad. Sci.* **1112:**326–338.

Romani, L., F. Bistoni, K. Perruccio, C. Montagnoli, R. Gaziano, S. Bozza, P. Bonifazi, G. Bistoni, G. Rasi, A. Velardi, F. Fallarino, E. Garaci, and P. Puccetti. 2006. Thymosin α1 activates dendritic cell tryptophan catabolism and establishes a regulatory environment for balance of inflammation and tolerance. *Blood* **108:**2265–2274.

Rudner, X. L., K. I. Happel, E. A. Young, and J. E. Shellito. 2007. Interleukin-23 (IL-23)–IL-17 cytokine axis in murine *Pneumocystis carinii* infection. *Infect. Immun.* **75:**3055–3061.

Saijo, S., N. Fujikado, T. Furuta, S. H. Chung, H. Kotaki, K. Seki, K. Sudo, S. Akira, Y. Adachi, N. Ohno, T. Kinjo, K. Nakamura, K. Kawakami, and Y. Iwakura. 2007. Dectin-1 is required for host defense against *Pneumocystis carinii* but not against *Candida albicans. Nat. Immunol.* **8:**39–46.

Sakaguchi, S. 2004. Naturally arising CD4+ regulatory T cells for immunologic self-tolerance and negative control of immune responses. *Annu. Rev. Immunol.* **22:**531–562.

Sato, N., S. K. Ahuja, M. Quinones, V. Kostecki, R. L. Reddick, P. C. Melby, W. A. Kuziel, and S. S. Ahuja. 2000. CC chemokine receptor (CCR)2 is required for langerhans cell migration and localization of T helper cell type 1 (Th1)-inducing dendritic cells. Absence of CCR2 shifts the *Leishmania major*-resistant phenotype to a susceptible state dominated by Th2 cytokines, B cell outgrowth, and sustained neutrophilic inflammation. *J. Exp. Med.* **192:**205–218.

Schaffner, A., H. Douglas, and A. Braude. 1982. Selective protection against conidia by mononuclear and against mycelia by polymorphonuclear phagocytes in resistance to *Aspergillus*. Observations on these two lines of defense in vivo and in vitro with human and mouse phagocytes. *J. Clin. Investig.* **69:**617–631.

Schnare, M., G. M. Barton, A. C. Holt, K. Takeda, S. Akira, and R. Medzhitov. 2001. Toll-like receptors control activation of adaptive immune responses. *Nat. Immunol.* **2:**947–950.

Seki, E., H. Tsutsui, N. M. Tsuji, N. Hayashi, K. Adachi, H. Nakano, S. Futatsugi-Yumikura, O. Takeuchi, K. Hoshino, S. Akira, J. Fujimoto, and K. Nakanishi. 2002. Critical roles of myeloid differentiation factor 88-dependent proinflammatory cytokine release in early phase clearance of *Listeria monocytogenes* in mice. *J. Immunol.* **169:**3863–3868.

Serbina, N. V., T. P. Salazar-Mather, C. A. Biron, W. A. Kuziel, and E. G. Pamer. 2003. TNF/iNOS-producing dendritic cells mediate innate immune defense against bacterial infection. *Immunity* **19:**59–70.

Serrano-Gomez, D., A. Dominguez-Soto, J. Ancochea, J. A. Jimenez-Heffernan, J. A. Leal, and A. L. Corbi. 2004. Dendritic cell-specific intercellular adhesion molecule 3-grabbing nonintegrin mediates binding and internalization of *Aspergillus fumigatus* conidia by dendritic cells and macrophages. *J. Immunol.* **173:**5635–5643.

Steele, C., R. R. Rapaka, A. Metz, S. M. Pop, D. L. Williams, S. Gordon, J. K. Kolls, and G. D. Brown. 2005. The beta-glucan receptor dectin-1 recognizes specific morphologies of *Aspergillus fumigatus. PLoS Pathog.* **1:**e42.

Steinman, R. M., D. Hawiger, and M. C. Nussenzweig. 2003. Tolerogenic dendritic cells. *Annu. Rev. Immunol.* **21:**685–711.

Suffia, I. J., S. K. Reckling, C. A. Piccirillo, R. S. Goldszmid, and Y. Belkaid. 2006. Infected site-restricted Foxp3+ natural regulatory T cells are specific for microbial antigens. *J. Exp. Med.* **203:**777–788.

Svirshchevskaya, E., E. Frolova, L. Alekseeva, O. Kotzareva, and V. P. Kurup. 2000. Intravenous injection of major and cryptic peptide epitopes of ribotoxin, Asp f 1 inhibits T cell response induced by crude *Aspergillus fumigatus* antigens in mice. *Peptides* **21:**1–8.

Svirshchevskaya, E. V., L. Alekseeva, A. Marchenko, N. Viskova, T. M. Andronova, S. V Benevolenskii, and V. P. Kurup. 2002. Immune response modulation by recombinant peptides expressed in virus-like particles. *Clin. Exp. Immunol.* **127:**199–205.

Takeda, K., T. Kaisho, and S. Akira. 2003. Toll-like receptors. *Annu. Rev. Immunol.* **21:**335–376.

Takeuchi, O., K. Hoshino, and S. Akira. 2000. Cutting edge: TLR2-deficient and MyD88-deficient mice are highly susceptible to *Staphylococcus aureus* infection. *J. Immunol.* **165:**5392–5396.

Taylor, P. R., S. V. Tsoni, J. A. Willment, K. M. Dennehy, M. Rosas, H. Findon, K. Haynes, C. Steele, M. Botto, S. Gordon, and G. D. Brown. 2007. Dectin-1 is required for beta-glucan recognition and control of fungal infection. *Nat. Immunol.* **8:**31–38.

Wald, A., W. Leisenring, J. A. van Burik, and R. A. Bowden. 1997. Epidemiology of *Aspergillus* infections in a large cohort of patients undergoing bone marrow transplantation. *J. Infect. Dis.* **175:**1459–1466.

Way, S. S., T. R. Kollmann, A. M. Hajjar, and C. B. Wilson. 2003. Cutting edge: protective cell-mediated immunity to *Listeria monocytogenes* in the absence of myeloid differentiation factor 88. *J. Immunol.* **171:**533–537.

Weaver, C. T., R. D. Hatton, P. R. Mangan, and L. E. Harrington. 2007. IL-17 family cytokines and the expanding diversity of effector T cell lineages. *Annu. Rev. Immunol.* **25:**821–852.

Williams, D. M., M. H. Weiner, and D. J. Drutz. 1981. Immunologic studies of disseminated infection with *Aspergillus fumigatus* in the nude mouse. *J. Infect. Dis.* **143:**726–733.

Wong, P., and E. G. Pamer. 2003. Feedback regulation of pathogen-specific T cell priming. *Immunity* **18:**499–511.

Wynn, T.A. 2003. IL-13 effector functions. *Annu. Rev. Immunol.* **21:**425–456.

Zelante, T., A. De Luca, P. Bonifazi, C. Montagnoli, S. Bozza, S. Moretti, M. L. Belladonna, C. Vacca, C. Conte, P. Mosci, F. Bistoni, P. Puccetti, R. A. Kastelein, M. Kopf, and L. Romani. 2007. IL-23 and the Th17 pathway promote inflammation and impair antifungal immune resistance. *Eur. J. Immunol.* **37:**2695–2706.

Zhou, S., E. A. Kurt-Jones, L. Mandell, A. Cerny, M. Chan, D. T. Golenbock, and R. W. Finberg. 2005. MyD88 is critical for the development of innate and adaptive immunity during acute lymphocytic choriomeningitis virus infection. *Eur. J. Immunol.* **35:**822–830.

Aspergillus fumigatus and Aspergillosis
Edited by J.-P. Latgé and W. J. Steinbach
©2009 ASM Press, Washington, DC

Chapter 22

Innate Recognition of *Aspergillus fumigatus* by the Mammalian Immune System

Lisa M. Graham and Gordon D. Brown

The innate arm of the immune system is responsible for the initial recognition and elimination of pathogens. This evolutionarily ancient form of defense is mediated primarily by phagocytic cells, such as macrophages, dendritic cells (DCs), and neutrophils, which bear germ line-encoded pattern recognition receptors (PRRs) (Medzhitov and Janeway, 2000). PRRs recognize specific molecular structures found in microbes which have been termed pathogen-associated molecular patterns (PAMPs) (Medzhitov and Janeway, 2000). As these structures are generally not found in the host, it enables the discrimination of nonself from self. PAMPs are often essential for the microbe's survival and are therefore highly conserved, decreasing the possibility of escape via mutation. PAMPs are also conserved among groups of pathogens, such as β-glucans, which are found in the cell walls of several fungi, and this allows a limited number of PRRs to recognize a wide range of pathogens.

Recognition can be direct, mediated by PRRs that are expressed on the cell surface, in intracellular vesicles, or in the cytoplasm, or recognition can be indirect, whereby soluble PRRs in bodily fluids coat or opsonize the pathogen, allowing recognition via surface-expressed phagocyte receptors, "the opsonic receptors." Recognition by PRRs results in numerous cellular responses triggered by a variety of intracellular signaling cascades, including microbial uptake by phagocytosis, induction of microbial killing machinery (such as the respiratory burst), and cytokine and chemokine production (Medzhitov, 2007).

Mounting the correct immune response, which is mediated in part through the production of specific cytokines, is pivotal for the successful control of fungal infections, and this control is influenced by the immunological status of the host. In immunocompetent hosts, who do not succumb to *Aspergillus* infections, protective antifungal responses are characterized by the development of a Th1 cytokine profile, including tumor necrosis factor alpha (TNF-α), interleukin-12 (IL-12), IL-8, and gamma interferon (IFN-γ). The pathogen is then rapidly eliminated through the activatory effect of these cytokines on phagocyte antimicrobial activities. In immunocompromised hosts, however, this protective response cannot be mounted, increasing susceptibility to infection and the possibility of developing invasive aspergillosis (IA) (Latgé, 2001).

PAMPs SPECIFIC TO *ASPERGILLUS FUMIGATUS*

PRRs appear to primarily recognize fungal cell wall components of *A. fumigatus*, which may only be displayed with different growth morphologies (Hohl et al., 2005; Steele et al., 2005). Initially, the pathogen enters the host airways as resting conidia, which are rapidly recognized and ingested by alveolar macrophages but do not elicit inflammatory cytokine production (Latgé, 2001). The conidia are covered by a layer of hydrophobic proteins known as the rodlet layer and a dense pigmented layer, which are suspected of being a means of protection from host recognition (Latgé, 2001). It is possible that the immune system limits pulmonary damage by ignoring these dormant conidia and limiting inflammatory responses to metabolically active conidia.

Following inhalation, spores begin to germinate, inducing an inflammatory response which rapidly recruits neutrophils to the site of infection. Germination of the conidia begins with conidial swelling, progresses to germ tube formation, and ultimately results in branched, septate vegetative hyphae (Latgé, 2001). The hyphal cell wall contains a core β-1,3-glucan chain with β-1,3/1,6 branches and chitin, galactomannan, and linear β-1,3/1,4-glucan linked to the nonreducing end of the β-1,3-glucan chain (see earlier chapters for a detailed descrip-

Lisa M. Graham and Gordon D. Brown • Institute of Infectious Disease and Molecular Medicine, Division of Immunology, University of Cape Town, Observatory, 7925, South Africa.

tion of the *Aspergillus* cell wall) (Beauvais and Latgé, 2001). Morphological changes expose these various structures, which then can act as PAMPs and are recognized by specific PRRs (Fig. 1).

Although many PRRs for *Aspergillus* have been identified, as described in detail below, their ligands remain undefined. β-1,3-Glucan has been shown to be the ligand for dectin-1 (Hohl et al., 2005; Steele et al., 2005; Gersuk et al., 2006), while DC-SIGN and PTX3 recognize conidial galactomannan (Garlanda et al., 2002; Serrano-Gomez et al., 2004). Surfactant proteins have also been shown to bind conidial galactomannan and possibly β-1,6-glucan (Allen et al., 2001). Chitin, β-1,4-glucan, and β-1,6-glucan are also likely to act as PAMPs.

PRRs WHICH RECOGNIZE *A. FUMIGATUS*

Numerous receptors have been identified which recognize *A. fumigatus*. These PRRs have been shown to play various roles in the host response to the fungus by activating complement and opsonizing the pathogen, for example, or inducing intracellular signaling leading to cytokine production. We describe here the PRRs which have been shown to recognize *Aspergillus* and detail what is known about their role in the control of this pathogen.

Toll-Like Receptors

Toll-like receptors (TLRs) are a conserved family of type I transmembrane proteins which were originally identified by their homology with Toll, a protein that is essential for the anti-*Aspergillus* response in *Drosphilia melanogaster* flies (Lemaitre et al., 1996). TLRs are composed of an extracellular domain containing between 19 and 25 leucine-rich repeat motifs and a signaling cytoplasmic Toll/Il-1R (TIR) homology domain (Akira et al., 2006). TLRs are expressed in numerous cell types, including DCs, macrophages, T cells, B cells, and epithelial cells, and can either be expressed on the cell surface or on the membranes of intracellular organelles, where they act as PRRs (Akira et al., 2006). All members of the TLR family, except TLR3, signal via an association with the main adaptor molecule, MyD88, which also contains a TIR domain. The association of this adaptor with the TLRs is thought to take place via homophilic interactions of the TIR domains, initiating signaling cascades that culminate in proinflammatory cytokine and chemokine production (O'Neill, 2006). MyD88 recruits IRAK-4 to TLRs via interaction of the molecules' death domains. IRAK-4 phosphorylates IRAK-1, which associates with TRAF-6 and initiates IκB degradation, which culminates in NF-κB activation and proinflammatory cytokine and chemokine production (O'Neill, 2006).

Other adaptor molecules besides MyD88 have also been implicated in TLR signaling. TIR domain-containing adaptor protein (TIRAP) is an adaptor protein involved specifically in bridging TLR4 and TLR2 to MyD88, resulting in NF-κB activation (O'Neill, 2006). TLR4 and TLR3 are additionally able to signal via a MyD88-independent pathway, employing TIR-related adaptor protein inducing interferon (Trif). Signaling via this adaptor results in activation of interferon regulatory factor 3 and induces IFN-β production. TLR4, but not TLR3, associates with Trif by means of the bridging molecule, Trif-related adaptor molecule (Tram) (O'Neill, 2006). These various signaling pathways are employed in response to the different PAMPs recognized by the TLRs. For example, TLR2 is able to recognize bacterial lipoproteins, TLR4 recognizes lipopolysaccharide, TLR3 recognizes double-stranded RNA, TLR5 recognizes flagellin, and TLR9 recognizes CpG motifs of bacterial DNA.

The discovery of the TLRs was of great importance, as it revealed the mechanism of innate pathogen recognition and intracellular signaling. These receptors play an important role in the control of most pathogens, and there is accumulating evidence that TLRs, in particular TLR2, TLR4, and their adaptor MyD88, also play a role in defense against *A. fumigatus*. In vitro studies, however, have shown somewhat contradictory data as to whether TLR2 and/or TLR4 is involved in this defense. Initially, whole human blood stimulated with *Aspergillus* hyphae displayed increased TNF-α, IL-1β, and IL-6 production in a TLR4- and CD14-dependent manner (Wang et al., 2001). Subsequently, another study showed that macrophages deficient in TLR2 or MyD88, but not those deficient in TLR4, showed decreased TNF-α production when stimulated with different forms of *Aspergillus* (Mambula et al., 2002). Soon thereafter, Meier et al. (2003) and Netea et al. (2003) showed that both receptors were in fact involved in anti-*Aspergillus* defense. Specifically, TLR2 recognized both conidia and hyphae, while TLR4 recognized only conidia. Recognition of conidia resulted in proinflammatory cytokine production, but hyphae resulted in IL-10 production, via TLR2 signaling (Netea et al., 2003). TLR2 and TLR4 have additionally been shown to promote the oxidative antifungal pathways in neutrophils, albeit via different mechanisms (Bellocchio et al., 2004b). Interestingly, TLR2-deficient alveolar macrophages have been shown to display only slightly reduced cytokine production when stimulated with conidia, indicating alternative pathways for defense against *Aspergillus* infection in these cells (Hohl et al., 2005; Steele et al., 2005).

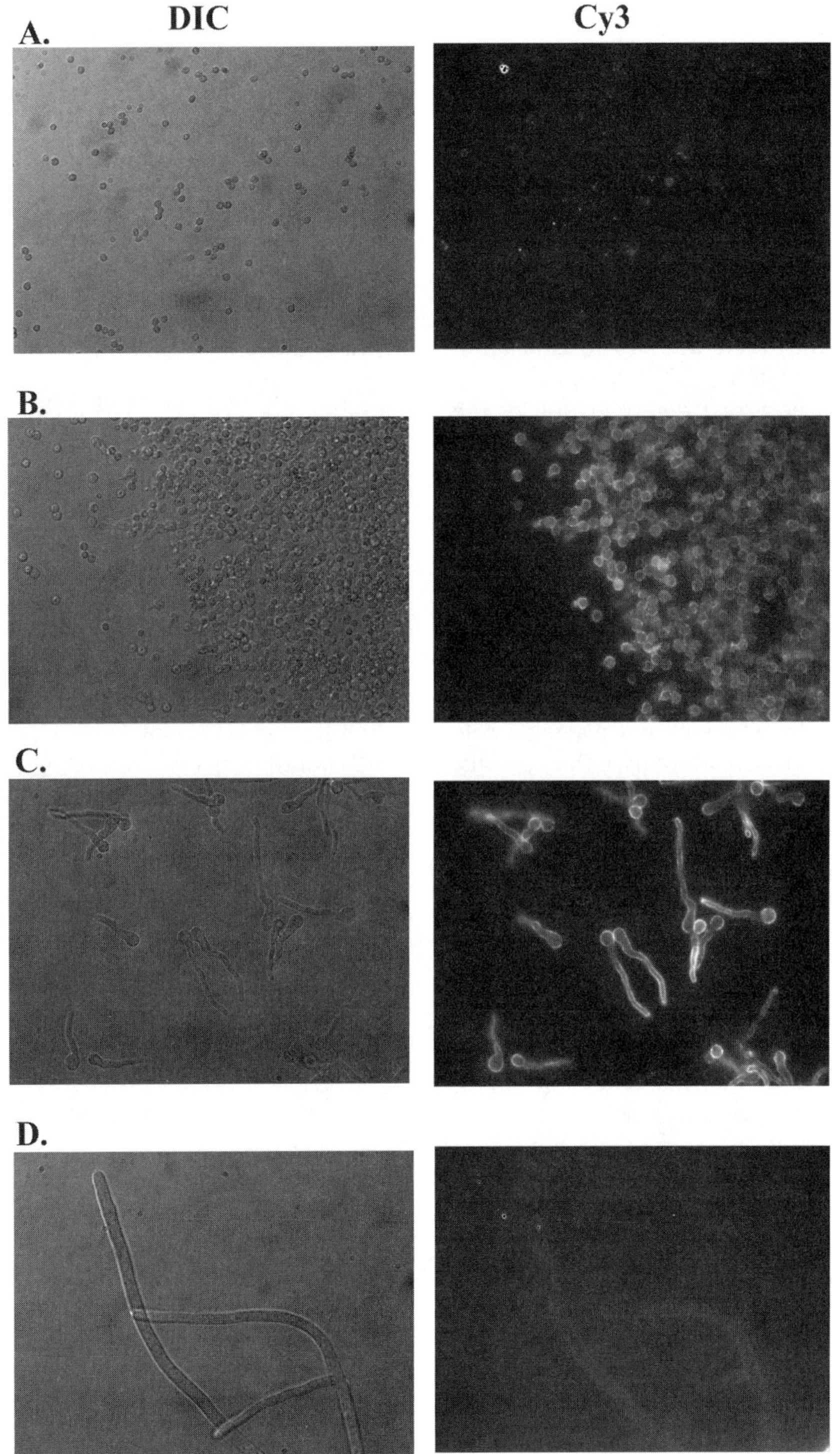

Figure 1. β-Glucan is displayed on the cell wall during specific growth morphologies of *A. fumigatus*. Soluble dectin-1 is able to bind to exposed β-glucan of swollen conidia (B) and germ tubes (C) but not to resting conidia (A) or hyphae (D). DIC, differential interference contrast images; Cy3, fluorescent image of soluble dectin-1 binding. Reprinted from Steele et al. (2005) with permission of the publisher.

The role of TLRs in immunity to *Aspergillus* has also been studied extensively in vivo. In the largest study so far, immunosuppressed mice lacking various TLRs or MyD88 were analyzed for susceptibility following infection with *Aspergillus* conidia (Bellocchio et al., 2004a). TLR2-, TLR4-, and MyD88-deficient mice were found to have higher lung fungal burdens than wild-type animals, but only the TLR4- and MyD88-deficient mice showed decreased survival (Bellocchio et al., 2004a). In a subsequent study, TLR2 deficiency was also shown to result in higher respiratory distress and fungal burdens than wild-type animals, resulting in reduced levels of TNF-α, IL-12, and macrophage inhibitory protein 2α (MIP-2α), but in this study there was also an association with decreased survival (Balloy et al., 2005).

It is important to note that determination of the role of the TLRs in these in vivo studies required the use of immunosuppressed mice. Immunocompetent TLR2, TLR4, and MyD88 knockout mice infected with *Aspergillus* at a dose able to kill immunosuppressed mice are resistant to infection, irrespective of the mode of infection (Dubourdeau et al., 2006). Although this suggests that the TLRs are redundant in healthy animals, there is evidence that signals initiated from this pathway are involved during the initial control of infection (Bretz et al., 2008). Additionally, although not playing a role in priming the Th1 differentiation of naïve T cells, MyD88 signaling has been shown to enhance IFN-γ production of antigen-specific cells in the lung (Rivera et al., 2006).

There is also clinical evidence for a role of the TLRs in the control of IA. Immunosuppressed patients who developed IA after hematopoetic stem cell transplantation were evaluated for single-nucleotide polymorphisms, and an association was found with a single TLR1 polymorphism and with polymorphisms in both TLR1 and TLR6. No association was found with polymorphisms of TLR4. The fact that TLR2 forms heterodimers with TLR1 or TLR6 in order to induce a proinflammatory response (Ozinsky et al., 2000) strengthens the proposal that TLR2, but not TLR4, is involved in defense against *Aspergillus*. Thus, the TLRs are clearly involved in immunity to *Aspergillus*, but in vivo they only appear to play an essential role when the host is immunocompromised.

C-Type Lectins

The C-type lectins are a large superfamily of proteins that all have at least one C-type lectin domain (CTLD, previously termed a carbohydrate recognition domain) and have diverse functions in both immunity and homeostasis. This family has been divided into 17 groups based on the organization of their CTLDs and phylogeny (Drickamer and Fadden, 2002; Zelensky and Gready, 2005). The C-type lectins may be either cell surface or soluble receptors and can loosely be defined as either classical or nonclassical (Drickamer and Fadden, 2002). Classical C-type lectins contain conserved amino acids in their CTLDs, enabling them to bind carbohydrate ligands in a calcium-dependent manner. Nonclassical C-type lectins (also called lectin-like receptors) do not have these conserved residues and typically bind noncarbohydrate ligands, such as proteins. However, as discussed below, some nonclassical C-type lectins, such as dectin-1, can recognize sugars. Both classical and nonclassical C-type lectins have been implicated in anti-*Aspergillus* immunity, including DC-SIGN, the collectins, and dectin-1 (Fig. 2).

DC-SIGN

DC-specific ICAM-3-grabbing nonintegrin (DC-SIGN; CD209) is a transmembrane C-type lectin which is expressed primarily by DCs but is also found on macrophage subsets and endothelium (Koppel et al., 2005). Structurally, DC-SIGN contains a single CTLD, a stalk region consisting of seven repeats which allows tetramerization, and a cytoplasmic domain with internalization and recycling motifs (Fig. 2) (Geijtenbeek et al., 2000). The CTLD of DC-SIGN contains a conserved EPN motif which allows recognition of high-mannose glycans and fucose-containing Lewis^x antigens (Koppel et al., 2005). DC-SIGN plays an important role as a cell adhesion molecule by binding ICAM-3 on resting T cells and ICAM-2 on endothelial cells (Koppel et al., 2005). Binding to ICAM-2 allows emigration of DCs from the blood and migration across the endothelium, while binding to ICAM-3 is important to initiate and stabilize contact with resting T cells. DC-SIGN also binds a range of pathogens, including viruses such as human immunodeficiency virus, Ebola virus, hepatitis C virus, and Dengue virus, as well as *Mycobacterium tuberculosis, Helicobacter pylori*, and *Shistosoma mansoni* soluble egg antigen (Koppel et al., 2005) and fungi such as *Candida albicans* (Taylor et al., 2004) and *A. fumigatus* (Serrano-Gomez et al., 2004, 2005). Pathogen binding results in multimerization of DC-SIGN (Serrano-Gomez et al., 2008), which in turn plays a role in targeting receptor-antigen complexes to late endosomes and lysosomes, resulting in processing and presentation of the antigen to $CD4^+$ T cells (Engering et al., 2002).

Recognition of *A. fumigatus* conidial galactomannan by DC-SIGN results in internalization of the fungus in DCs and macrophages (Serrano-Gomez et al., 2004, 2005). In DCs, this internalization triggers IL-10 production, without an increase in IL-12, implying a skewing towards Th2 polarization (Serrano-Gomez et al., 2004). This polarization is thought to occur for many DC-SIGN-associated pathogens, resulting in impaired

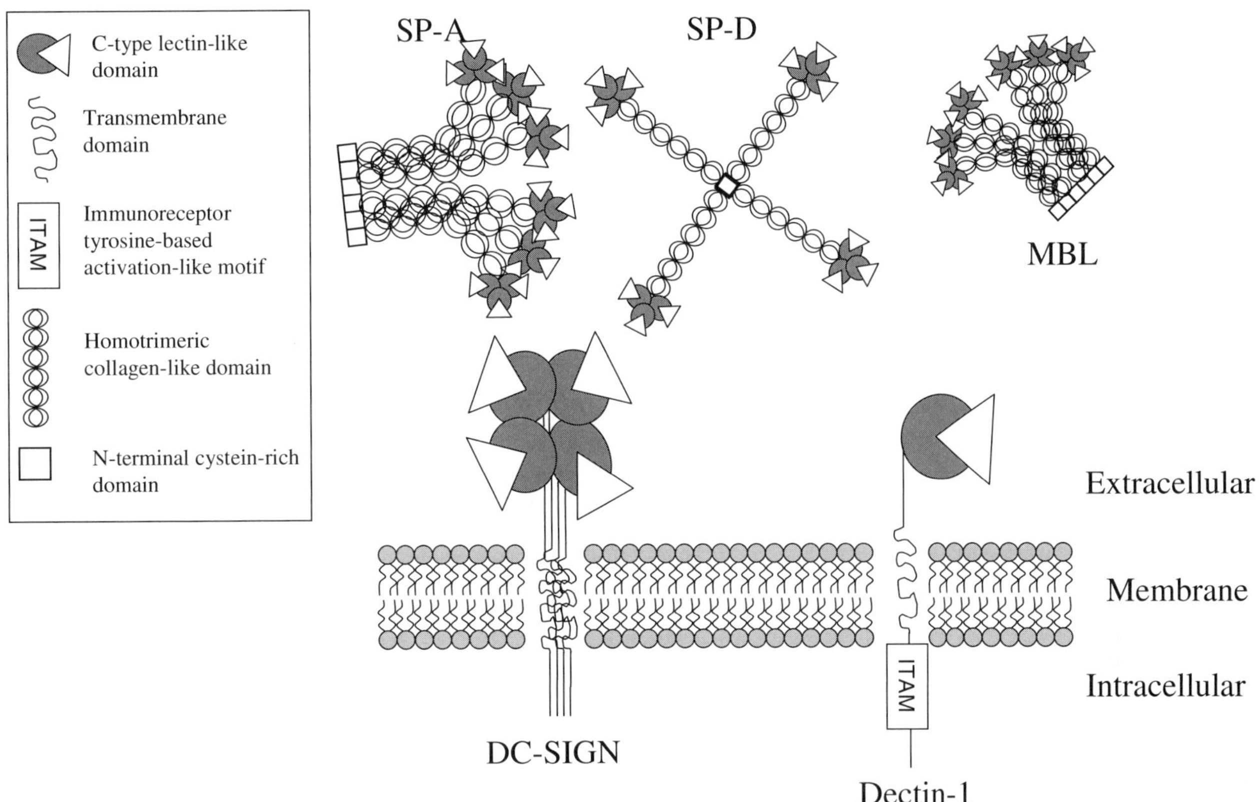

Figure 2. Cartoon representation of C-type lectin receptors which recognize *A. fumigatus* (not drawn to scale).

pathogen clearance and establishment of chronic infections. Indeed, this skewing can be explained by the ability of DC-SIGN to initiate signaling via ERK1/2 and Akt, which promotes the production of IL-10 but inhibits the production of IL-12 (Caparros et al., 2006). As DC-SIGN expression has been shown to be IL-4 dependent, it has been speculated that establishment of a Th2 milieu might increase DC-SIGN and exacerbate the disease.

The Collectins: SP-A, SP-D, and MBL

The collectins are a group of soluble PRRs composed of a short N-terminal cysteine-rich domain, a collagen-like domain, a stalk region, and a C-terminal CTLD. Collectins multimerize into homotrimers, which then associate to form complex tertiary structures. Mannose-binding lectin (MBL) and surfactant protein A (SP-A) assemble into hexamers of homotrimers, forming bouquet-like structures, whereas SP-D is formed by the arrangement of homotrimers into cruciform-shaped tetramers, which may oligomerize into a wheel structure (Fig. 2) (Holmskov et al., 2003). The CTLDs of collectins recognize terminal mannose, L-fucose, glucose, *N*-acetyl-mannosamine (ManNAc), and *N*-acetylglucosamine (GlcNAc) in a calcium-dependent fashion and are clustered to achieve high-avidity binding to microbial surfaces (Holmskov et al., 2003).

MBL is produced by hepatocytes in the liver and is secreted into the serum, where it recognizes a number of pathogens, including *Staphylococcus aureus*, β-hemolytic group A streptococci, *C. albicans*, and *A. fumigatus* (Neth et al., 2000). Binding of pathogens by MBL results in activation of a serine protease cascade (MBL-associated serine proteases) and opsonization of the pathogen by deposition of complement components C4b and C3b (Holmskov et al., 2003; Thiel, 2007). This antibody-independent activation of the classical complement cascade, known as the lectin pathway, is however not the only mode of MBL functioning. MBL is also able to bind directly to phagocytic receptors, including C1q receptor, calreticulin, and complement receptor 1 (Holmskov et al., 2003). Gene polymorphisms are often found in humans and can lead to reduced serum levels of functional MBL and increased susceptibility to infections, including infection with *Aspergillus* (Crosdale et al., 2001).

In mice, MBL-A deficiency did not increase susceptibility to *Aspergillus* infection, but there was an effect on airway hyperresponsiveness in a model of chronic fungal asthma (Hogaboam et al., 2004). Interestingly, the Th2 response generally observed in the asthma model was markedly decreased in MBL-A$^{-/-}$ mice. In immunosuppressed mice, however, treatment with recombinant MBL was shown to increase survival in a model of invasive pulmonary aspergillosis by 80% (Kaur et al., 2007). The treated mice demonstrated increased production of TNF-α and IL-1α and decreased levels of IL-10, which led to higher levels of IFN-γ and a protective Th1 response. In vitro, the addition of recombinant MBL and MBL-free serum, but not MBL alone, was shown to increase conidial uptake and induction of the oxidative burst (Kaur et al., 2007). This demonstrates the cooperative role of MBL and serum components in strengthening the host defense to *Aspergillus* infection.

SP-A and SP-D are produced by alveolar type II and nonciliated bronchial epithelial cells and are secreted into the alveolar space (Kishore et al., 2006). These collectins were originally identified by their function in reducing surface tension at the air-liquid interface of the lung but have since been shown to also function as PRRs, critical for innate defense against inhaled pathogens (Madan et al., 1997). Unlike MBL, SP-A and -D do not activate complement but may act as opsonins and have been shown to bind to a number of soluble and surface receptors, including C1q receptor, glycoprotein 340, CD14, calreticulin, and signal regulating protein α (Kishore et al., 2006).

SP-A and SP-D are involved in both allergic and invasive aspergillosis. These collectins bind and agglutinate conidia, in a β-1,6-glucan-dependent manner, at least for SP-D (Allen et al., 2001). Opsonization with these collectins can enhance fungal uptake and killing by macrophages and neutrophils, but not in all species, such as the rat (Madan et al., 1997; Allen et al., 1999). In immunosuppressed mice, intranasal treatment with recombinant SP-D or SP-A was shown to increase survival by 80% and 20%, respectively, in a model of invasive pulmonary aspergillosis, highlighting the potential therapeutic use of these PRRs (Madan et al., 2001b). SP-A and SP-D have additionally been shown to have a direct fungistatic effect and can inhibit the germination and growth of clinical isolates of *Aspergillus* in an as-yet-undefined manner (Kishore et al., 2006). However, to date, there have been no infection studies performed in surfactant knockout mice, possibly due to the alterations in surfactant homeostasis seen in SP-D null mice which result in accumulation of surfactant lipids and foamy macrophages in the alveolar space (Kishore et al., 2006).

Surfactants have also been considered as a treatment for allergic disorders caused by *A. fumigatus*, such as allergic bronchopulmonary aspergillosis. Allergic bronchopulmonary aspergillosis is characterized by a Th2 response, including elevated specific IgE and IgG, blood and lung eosinophilia and pulmonary inflammation, and hyperreactivity (Kurup and Grunig, 2002). Treatment with SP-A and SP-D has been shown to shift this response towards a Th1 profile, alleviating the disease (Madan et al., 2001a; Strong et al., 2002; Erpenbeck et al., 2006). Surfactant knockout mice have also been used to study the role of SP-A and SP-D in allergy. These mice naturally show a bias to a Th2 immune response, with hypereosinophilia, increased IL-5 and IL-13, and lowered IFN-γ in the lungs, which can be reversed by treatment of the respective gene-deficient mice with SP-A or SP-D (Madan et al., 2005). In SP-D knockouts, sensitization with *Aspergillus* antigens rendered the mice more susceptible to pulmonary hypersensitivity, while SP-A knockouts were mostly resistant. However, when treated with exogenous surfactant, SP-D knockout mice recovered, whereas SP-A knockout mice developed severe eosinophilia and lung damage.

Dectin-1

Dectin-1 is a transmembrane protein that belongs to the nonclassical C-type lectin receptors and is expressed on a range of cells, including DCs, macrophages, monocytes, neutrophils, and a subset of T cells (Taylor et al., 2002). The receptor contains one CTLD, a stalk, and a transmembrane region and an intracellular tail which contains an immunoreceptor tyrosine-based activation (ITAM)-like motif (Fig. 2) (Ariizumi et al., 2000). Dectin-1 recognizes soluble and particulate β-1,3-glucan in a calcium-independent manner and has been shown to be the major receptor for these carbohydrates on leukocytes (Brown and Gordon, 2001; Brown et al., 2002; Palma et al., 2006; Adams et al., 2008). Ligand binding by dectin-1 results in intracellular signaling via its ITAM-like motif, inducing a variety of cellular responses, including phagocytosis, induction of the respiratory burst, and cytokine and chemokine production (Brown et al., 2003). Signaling takes place, at least in part, via an unusual interaction with spleen tyrosine kinase (Syk) and downstream signaling through the adaptor caspase recruitment domain 9 (CARD9), and many of these responses involve collaborative signaling from the TLRs (Hohl et al., 2005; Rogers et al., 2005; Gross et al., 2006; Luther et al., 2007; Dennehy et al., 2008).

Dectin-1 has been shown to play an important role in antifungal immunity and can recognize a range of fungal species, including *Candida, Pneumocystis, Coccidoides, Saccharomyces,* and *Aspergillus.* However, dectin-1 cannot recognize all morphologies of *A. fumigatus*, being dependent on recognizing β-glucan exposed on swollen conidia and early germlings (Fig. 1) (Hohl et al., 2005; Steele et al., 2005; Gersuk et al., 2006).

Dectin-1 can mediate the uptake of this organism, an activity which is enhanced by TLR2 (Luther et al., 2007). Recognition of *Aspergillus* by dectin-1 also leads to the activation of transcription factors, including NF-κB and activator protein 1, and results in the production of numerous cytokines and chemokines, including TNF-α, IL-6, IL-18, IL-10, IL-1α, IL-1β, MIP-2, MIP-1α, granulocyte colony-stimulating factor (G-CSF), and granulocyte-macrophage CSF (GM-CSF) (Steele et al., 2005; Taylor et al., 2007; Mezger et al., 2008; Toyotome et al., 2008). Furthermore, recognition of *A. fumigatus* germ tubes stimulates the production of reactive oxygen in human monocyte-derived macrophages and murine macrophages, which is partly mediated by dectin-1 (Gersuk et al., 2006). As these inflammatory responses are not triggered by resting conidia, it has been suggested that dectin-1-mediated defense towards *Aspergillus* is tailored against the metabolically active growth stages, helping to reduce unnecessary tissue damage (Hohl et al., 2005).

The role of dectin-1 in anti-*Aspergillus* immunity has not yet been extensively examined in vivo; however, blockage of dectin-1 in immunocompetent mice, by using a soluble inhibitor, was shown to reduce the production of inflammatory cytokines and chemokines, including TNF-α, KC, MIP-1α, IL-6, and GM-CSF, leading to defective cellular recruitment and higher fungal burdens in the lung (Steele et al., 2005). Dectin-1 knockout mice have recently become available (Saijo et al., 2007; Taylor et al., 2007), and their study will be important for further elucidating the role of this receptor in *Aspergillus*-related diseases.

PENTRAXINS

The pentraxins are a highly conserved family of soluble proteins which are subdivided into two subclasses depending on their length and structure. The classical short pentraxins include C-reactive protein (CRP) and serum amyloid P (SAP) and are composed of monomers that assemble into pentameric structures (Bottazzi et al., 2006). CRP and SAP are produced in the liver in response to inflammatory signals and recognize a wide range of ligands in a calcium-dependent manner. The long pentraxin 3 (PTX3) is similar to the short pentraxins in that it contains the C-terminal pentraxin domain and forms multimers, but it differs in having an unrelated long N-terminal domain (Bottazzi et al., 1997). PTX3 also differs in terms of expression, ligand binding, and gene organization, as described below.

PTX3

PTX3 is produced and released by a wide variety of cell types in response to inflammatory signals, including myeloid DCs, mononuclear phagocytes, fibroblasts, endothelial and epithelial cells, and neutrophils (Garlanda et al., 2005). PTX3 has a diverse array of functions, including clearance of apoptotic cells, extracellular matrix architecture, and female fertility (Garlanda et al., 2005). PTX also acts as a PRR and can recognize and opsonize a range of pathogens, including *Salmonella enterica* serotype Typhimurium, *Pseudomonas aeruginosa*, and *A. fumigatus* (Garlanda et al., 2002).

PTX3 recognizes *A. fumigatus* via binding to live or heat-inactivated conidia in a galactomannan-dependent manner but is unable to recognize *Aspergillus* hyphae (Garlanda et al., 2002). This recognition facilitates the phagocytosis of conidia by alveolar macrophages, indicating the presence of a receptor for PTX3. PTX3 can also act as an opsonin and cause particle aggregation, resulting in increased phagocytosis by macrophages in a dectin-1-dependent manner (Diniz et al., 2004). *Aspergillus* conidia can induce the production of PTX3 in mononuclear phagocytes, DCs, bronchoalveolar lavage fluid, and in plasma. In neutrophils, PTX3 is preformed and stored in the specific granules, and neutrophil-derived PTX3 is important for the conidial uptake and killing (Jaillon et al., 2007). Upon stimulation with *A. fumigatus* conidia, neutrophils release PTX3, which becomes associated with neutrophil extracellular traps, antimicrobial structures which are formed through a novel cell death program (Brinkmann et al., 2004; Fuchs et al., 2007). Although conidia were found to associate with PTX3 located in the neutrophil extracellular traps, the functional relevance of this interaction has not yet been demonstrated.

In a murine model of invasive pulmonary aspergillosis using immunocompetent animals, PTX3$^{-/-}$ mice had increased susceptibility to infection compared to wild-type animals, due to defective recognition of conidia (Garlanda et al., 2002). Knockout mice showed massive inflammation in the lungs, the presence of hyphae, and many extracellular conidia, while wild-type mice demonstrated moderate inflammation with few intracellular conidia. Additionally, levels of IFN-γ and IL-12 production were decreased in the lungs of knockout mice, while IL-4 production was increased, showing a defect in the development of an antifungal Th1 response and a skewing towards a Th2 response. This study represents the only example so far in immunocompetent animals of a PRR defect which leads to increased susceptibility to infection with *Aspergillus*.

PTX3 has been examined as a potential therapeutic for IA. In murine models of bone marrow transplantation and of cytomegalovirus infection, treatment with PTX3 caused an increased resistance to *A. fumigatus* infection which was characterized by increased survival, reduced fungal growth, and a decrease in lung pathology (Gaziano et al., 2004; Bozza et al., 2006). Administration of PTX3 was also shown to enhance efficacy when used in conjunction with suboptimal doses of known antifungal drugs (Gaziano et al., 2004).

CRP

CRP is a soluble acute-phase protein in humans that is produced in the liver and rapidly detectable in the serum in response to the proinflammatory cytokine IL-6 (Garlanda et al., 2005). CRP was first identified by its ability to bind *Streptococcus pneumoniae*, but it has since been shown to recognize diverse pathogens, including bacteria, protozoa, and fungi including *C. albicans* and *A. fumigatus* (Jensen et al., 1986; Szalai, 2002). CRP can bind C1q and activate the classical complement pathway, and it can also directly bind the Fcγ receptor, both of which facilitate phagocytosis, making CRP an important opsonin in the defense against pathogens (Szalai, 2002). Indeed, CRP has been shown to bind to metabolically active *A. fumigatus* conidia and enhance complement-independent phagocytosis by human neutrophils (Richardson et al., 1991). To date, no infection studies have been performed in CRP knockout mice, and although transgenic mice expressing human or rabbit CRP are available, they have not yet been studied in the context of *Aspergillus* infection (Garlanda et al., 2005).

COMPLEMENT

The complement system plays an essential role in the defense against pathogens (for a complete review on the complement system, see Carroll, 1998). *A. fumigatus* activates complement through both the lectin and alternative pathway in naive hosts, with resting conidia activating the alternative pathway and swollen conidia and hyphae activating both the lectin pathway (such as MBL, discussed above) and alternative pathways (Kozel et al., 1989; Neth et al., 2000). Deposition of complement leads to enhanced recognition and uptake of the fungus by phagocytes (Sturtevant and Latgé, 1992). However, *A. fumigatus* conidia were recently shown to bind complement regulators, including factor H, factor H-like protein 1, C4-binding protein, and plasminogen, suggesting that this pathogen has mechanisms to avoid activation of the complement system (Behnsen et al., 2008; Vogl et al., 2008).

CONCLUSIONS

Early innate recognition of *A. fumigatus* is pivotal in mounting the correct defense against the pathogen and preventing disease. Recognition of the fungus takes place via PRRs, which bind to specific PAMPs present on the fungal cell wall. Although we are beginning to understand the protective mechanisms induced by many of these PRRs, the exact role these molecules in vivo remains largely undefined. In addition, it is important to remember that recognition involves multiple interactions with many PRRs, which together coordinates the resultant immune response. Despite the development of new antifungal drugs, the incidence of aspergillosis is increasing. Further investigation into these receptors and their underlying molecular mechanisms may provide a means to develop novel immunotherapeutic strategies to combat infection with this organism.

Acknowledgments. We thank the Wellcome Trust, Medical Research Council (South Africa), National Research Foundation (South Africa), University of Cape Town, German Academic Exchange Service, and the U.S. National Institutes of Health for funding. G.D.B. is a Wellcome Trust International Senior Fellow in South Africa.

REFERENCES

Adams, E. L., P. J. Rice, B. Graves, H. E. Ensley, H. Yu, G. D. Brown, S. Gordon, M. A. Monteiro, E. Papp-Szabo, D. W. Lowman, T. D. Power, M. F. Wempe, and D. L. Williams. 2008. Differential high affinity interaction of dectin-1 with natural or synthetic glucans is dependent upon primary structure and is influenced by polymer chain length and side chain branching. *J. Pharmacol. Exp. Ther.* **325:** 115–123.

Akira, S., S. Uematsu, and O. Takeuchi. 2006. Pathogen recognition and innate immunity. *Cell* **124:**783–801.

Allen, M. J., R. Harbeck, B. Smith, D. R. Voelker, and R. J. Mason. 1999. Binding of rat and human surfactant proteins A and D to *Aspergillus fumigatus* conidia. *Infect. Immun.* **67:**4563–4569.

Allen, M. J., D. R. Voelker, and R. J. Mason. 2001. Interactions of surfactant proteins A and D with *Saccharomyces cerevisiae* and *Aspergillus fumigatus*. *Infect. Immun.* **69:**2037–2044.

Ariizumi, K., G. L. Shen, S. Shikano, S. Xu, R. Ritter III, T. Kumamoto, D. Edelbaum, A. Morita, P. R. Bergstresser, and A. Takashima. 2000. Identification of a novel, dendritic cell-associated molecule, dectin-1, by subtractive cDNA cloning. *J. Biol. Chem.* **275:** 20157–20167.

Balloy, V., M. Si-Tahar, O. Takeuchi, B. Philippe, M. A. Nahori, M. Tanguy, M. Huerre, S. Akira, J. P. Latgé, and M. Chignard. 2005. Involvement of toll-like receptor 2 in experimental invasive pulmonary aspergillosis. *Infect. Immun.* **73:**5420–5425.

Beauvais, A., and J. P. Latgé. 2001. Membrane and cell wall targets in *Aspergillus fumigatus*. *Drug Resist. Update* **4:**38–49.

Behnsen, J., A. Hartmann, J. Schmaler, A. Gehrke, A. A. Brakhage, and P. F. Zipfel. 2008. The opportunistic human pathogenic fungus *Aspergillus fumigatus* evades the host complement system. *Infect. Immun.* **76:**820–827.

Bellocchio, S., C. Montagnoli, S. Bozza, R. Gaziano, G. Rossi, S. S. Mambula, A. Vecchi, A. Mantovani, S. M. Levitz, and L. Romani. 2004a. The contribution of the Toll-like/IL-1 receptor superfamily to innate and adaptive immunity to fungal pathogens in vivo. *J. Immunol.* **172:**3059–3069.

Bellocchio, S., S. Moretti, K. Perruccio, F. Fallarino, S. Bozza, C. Montagnoli, P. Mosci, G. B. Lipford, L. Pitzurra, and L. Romani. 2004b. TLRs govern neutrophil activity in aspergillosis. *J. Immunol.* **173:**7406–7415.

Bottazzi, B., C. Garlanda, G. Salvatori, P. Jeannin, A. Manfredi, and A. Mantovani. 2006. Pentraxins as a key component of innate immunity. *Curr. Opin. Immunol.* **18:**10–15.

Bottazzi, B., V. Vouret-Craviari, A. Bastone, L. De Gioia, C. Matteucci, G. Peri, F. Spreafico, M. Pausa, C. D'Ettorre, E. Gianazza, A. Tagliabue, M. Salmona, F. Tedesco, M. Introna, and A. Mantovani. 1997. Multimer formation and ligand recognition by the long pentraxin PTX3. Similarities and differences with the short pentraxins C-reactive protein and serum amyloid P component. *J. Biol. Chem.* **272:**32817–32823.

Bozza, S., F. Bistoni, R. Gaziano, L. Pitzurra, T. Zelante, P. Bonifazi, K. Perruccio, S. Bellocchio, M. Neri, A. M. Iorio, G. Salvatori, R. De Santis, M. Calvitti, A. Doni, C. Garlanda, A. Mantovani, and L. Romani. 2006. Pentraxin 3 protects from MCMV infection and reactivation through TLR sensing pathways leading to IRF3 activation. *Blood* **108:**3387–3396.

Bretz, C., G. Gersuk, S. Knoblaugh, N. Chaudhary, J. Randolph-Habecker, R. Hackman, J. Staab, and K. A. Marr. 2008. Myd88 signaling contributes to early pulmonary responses to *Aspergillus fumigatus*. *Infect. Immun.* **76:**952–958.

Brinkmann, V., U. Reichard, C. Goosmann, B. Fauler, Y. Uhlemann, D. S. Weiss, Y. Weinrauch, and A. Zychlinsky. 2004. Neutrophil extracellular traps kill bacteria. *Science* **303:**1532–1535.

Brown, G. D., and S. Gordon. 2001. Immune recognition. A new receptor for beta-glucans. *Nature* **413:**36–37.

Brown, G. D., J. Herre, D. L. Williams, J. A. Willment, A. S. Marshall, and S. Gordon. 2003. Dectin-1 mediates the biological effects of beta-glucans. *J. Exp. Med.* **197:**1119–1124.

Brown, G. D., P. R. Taylor, D. M. Reid, J. A. Willment, D. L. Williams, L. Martinez-Pomares, S. Y. Wong, and S. Gordon. 2002. Dectin-1 is a major beta-glucan receptor on macrophages. *J. Exp. Med.* **196:**407–412.

Caparros, E., P. Munoz, E. Sierra-Filardi, D. Serrano-Gomez, A. Puig-Kroger, J. L. Rodriguez-Fernandez, M. Mellado, J. Sancho, M. Zubiaur, and A. L. Corbi. 2006. DC-SIGN ligation on dendritic cells results in ERK and PI3K activation and modulates cytokine production. *Blood* **107:**3950–3958.

Carroll, M. C. 1998. The role of complement and complement receptors in induction and regulation of immunity. *Annu. Rev. Immunol.* **16:**545–568.

Crosdale, D. J., K. V. Poulton, W. E. Ollier, W. Thomson, and D. W. Denning. 2001. Mannose-binding lectin gene polymorphisms as a susceptibility factor for chronic necrotizing pulmonary aspergillosis. *J. Infect. Dis.* **184:**653–656.

Dennehy, K. M., G. Ferwerda, I. Faro-Trindade, E. Pyz, J. A. Willment, P. R. Taylor, A. Kerrigan, S. V. Tsoni, S. Gordon, F. Meyer-Wentrup, G. J. Adema, B. J. Kullberg, E. Schweighoffer, V. Tybulewicz, H. M. Mora-Montes, N. A. Gow, D. L. Williams, M. G. Netea, and G. D. Brown. 2008. Syk kinase is required for collaborative cytokine production induced through Dectin-1 and Toll-like receptors. *Eur. J. Immunol.* **38:**500–506.

Diniz, S. N., R. Nomizo, P. S. Cisalpino, M. M. Teixeira, G. D. Brown, A. Mantovani, S. Gordon, L. F. Reis, and A. A. Dias. 2004. PTX3 function as an opsonin for the dectin-1-dependent internalization of zymosan by macrophages. *J. Leukoc. Biol.* **75:**649–656.

Drickamer, K., and A. J. Fadden. 2002. Genomic analysis of C-type lectins. *Biochem. Soc. Symp.* **69:**59–72.

Dubourdeau, M., R. Athman, V. Balloy, M. Huerre, M. Chignard, D. J. Philpott, J. P. Latgé, and O. Ibrahim-Granet. 2006. *Aspergillus fumigatus* induces innate immune responses in alveolar macrophages through the MAPK pathway independently of TLR2 and TLR4. *J. Immunol.* **177:**3994–4001.

Engering, A., T. B. Geijtenbeek, S. J. van Vliet, M. Wijers, E. van Liempt, N. Demaurex, A. Lanzavecchia, J. Fransen, C. G. Figdor, V. Piguet, and Y. van Kooyk. 2002. The dendritic cell-specific adhesion receptor DC-SIGN internalizes antigen for presentation to T cells. *J. Immunol.* **168:**2118–2126.

Erpenbeck, V. J., M. Ziegert, D. Cavalet-Blanco, C. Martin, R. Baelder, T. Glaab, A. Braun, W. Steinhilber, B. Luettig, S. Uhlig, H. G. Hoymann, N. Krug, and J. M. Hohlfeld. 2006. Surfactant protein D inhibits early airway response in *Aspergillus fumigatus*-sensitized mice. *Clin. Exp. Allergy* **36:**930–940.

Fuchs, T. A., U. Abed, C. Goosmann, R. Hurwitz, I. Schulze, V. Wahn, Y. Weinrauch, V. Brinkmann, and A. Zychlinsky. 2007. Novel cell death program leads to neutrophil extracellular traps. *J. Cell Biol.* **176:**231–241.

Garlanda, C., B. Bottazzi, A. Bastone, and A. Mantovani. 2005. Pentraxins at the crossroads between innate immunity, inflammation, matrix deposition, and female fertility. *Annu. Rev. Immunol.* **23:**337–366.

Garlanda, C., E. Hirsch, S. Bozza, A. Salustri, M. De Acetis, R. Nota, A. Maccagno, F. Riva, B. Bottazzi, G. Peri, A. Doni, L. Vago, M. Botto, R. De Santis, P. Carminati, G. Siracusa, F. Altruda, A. Vecchi, L. Romani, and A. Mantovani. 2002. Non-redundant role of the long pentraxin PTX3 in anti-fungal innate immune response. *Nature* **420:**182–186.

Gaziano, R., S. Bozza, S. Bellocchio, K. Perruccio, C. Montagnoli, L. Pitzurra, G. Salvatori, R. De Santis, P. Carminati, A. Mantovani, and L. Romani. 2004. Anti-*Aspergillus fumigatus* efficacy of pentraxin 3 alone and in combination with antifungals. *Antimicrob. Agents Chemother.* **48:**4414–4421.

Geijtenbeek, T. B., R. Torensma, S. J. van Vliet, G. C. van Duijnhoven, G. J. Adema, Y. van Kooyk, and C. G. Figdor. 2000. Identification of DC-SIGN, a novel dendritic cell-specific ICAM-3 receptor that supports primary immune responses. *Cell* **100:**575–585.

Gersuk, G. M., D. M. Underhill, L. Zhu, and K. A. Marr. 2006. Dectin-1 and TLRs permit macrophages to distinguish between different *Aspergillus fumigatus* cellular states. *J. Immunol.* **176:**3717–3724.

Gross, O., A. Gewies, K. Finger, M. Schafer, T. Sparwasser, C. Peschel, I. Forster, and J. Ruland. 2006. Card9 controls a non-TLR signalling pathway for innate anti-fungal immunity. *Nature* **442:**651–656.

Hogaboam, C. M., K. Takahashi, R. A. Ezekowitz, S. L. Kunkel, and J. M. Schuh. 2004. Mannose-binding lectin deficiency alters the development of fungal asthma: effects on airway response, inflammation, and cytokine profile. *J. Leukoc. Biol.* **75:**805–814.

Hohl, T. M., H. L. Van Epps, A. Rivera, L. A. Morgan, P. L. Chen, M. Feldmesser, and E. G. Pamer. 2005. *Aspergillus fumigatus* triggers inflammatory responses by stage-specific beta-glucan display. *PLoS Pathog.* **1:**e30.

Holmskov, U., S. Thiel, and J. C. Jensenius. 2003. Collections and ficolins: humoral lectins of the innate immune defense. *Annu. Rev. Immunol.* **21:**547–578.

Jaillon, S., G. Peri, Y. Delneste, I. Fremaux, A. Doni, F. Moalli, C. Garlanda, L. Romani, H. Gascan, S. Bellocchio, S. Bozza, M. A. Cassatella, P. Jeannin, and A. Mantovani. 2007. The humoral pattern recognition receptor PTX3 is stored in neutrophil granules and localizes in extracellular traps. *J. Exp. Med.* **204:**793–804.

Jensen, T. D., H. Schonheyder, P. Andersen, and A. Stenderup. 1986. Binding of C-reactive protein to *Aspergillus fumigatus* fractions. *J. Med. Microbiol.* **21:**173–177.

Kaur, S., V. K. Gupta, S. Thiel, P. U. Sarma, and T. Madan. 2007. Protective role of mannan-binding lectin in a murine model of invasive pulmonary aspergillosis. *Clin. Exp. Immunol.* **148:**382–389.

Kishore, U., T. J. Greenhough, P. Waters, A. K. Shrive, R. Ghai, M. F. Kamran, A. L. Bernal, K. B. Reid, T. Madan, and T. Chakraborty. 2006. Surfactant proteins SP-A and SP-D: structure, function and receptors. *Mol. Immunol.* **43:**1293–1315.

Koppel, E. A., K. P. van Gisbergen, T. B. Geijtenbeek, and Y. van Kooyk. 2005. Distinct functions of DC-SIGN and its homologues L-SIGN (DC-SIGNR) and mSIGNR1 in pathogen recognition and immune regulation. *Cell. Microbiol.* 7:157–165.

Kozel, T. R., M. A. Wilson, T. P. Farrell, and S. M. Levitz. 1989. Activation of C3 and binding to *Aspergillus fumigatus* conidia and hyphae. *Infect. Immun.* 57:3412–3417.

Kurup, V. P., and G. Grunig. 2002. Animal models of allergic bronchopulmonary aspergillosis. *Mycopathologia* 153:165–177.

Latgé, J. P. 2001. The pathobiology of *Aspergillus fumigatus*. *Trends Microbiol.* 9:382–389.

Lemaitre, B., E. Nicolas, L. Michaut, J. M. Reichhart, and J. A. Hoffmann. 1996. The dorsoventral regulatory gene cassette spatzle/Toll/cactus controls the potent antifungal response in *Drosophila* adults. *Cell* 86:973–983.

Luther, K., A. Torosantucci, A. A. Brakhage, J. Heesemann, and F. Ebel. 2007. Phagocytosis of *Aspergillus fumigatus* conidia by murine macrophages involves recognition by the dectin-1 beta-glucan receptor and Toll-like receptor 2. *Cell. Microbiol.* 9:368–381.

Madan, T., P. Eggleton, U. Kishore, P. Strong, S. S. Aggrawal, P. U. Sarma, and K. B. Reid. 1997. Binding of pulmonary surfactant proteins A and D to *Aspergillus fumigatus* conidia enhances phagocytosis and killing by human neutrophils and alveolar macrophages. *Infect. Immun.* 65:3171–3179.

Madan, T., U. Kishore, M. Singh, P. Strong, H. Clark, E. M. Hussain, K. B. Reid, and P. U. Sarma. 2001a. Surfactant proteins A and D protect mice against pulmonary hypersensitivity induced by *Aspergillus fumigatus* antigens and allergens. *J. Clin. Investig.* 107:467–475.

Madan, T., U. Kishore, M. Singh, P. Strong, E. M. Hussain, K. B. Reid, and P. U. Sarma. 2001b. Protective role of lung surfactant protein D in a murine model of invasive pulmonary aspergillosis. *Infect. Immun.* 69:2728–2731.

Madan, T., K. B. Reid, M. Singh, P. U. Sarma, and U. Kishore. 2005. Susceptibility of mice genetically deficient in the surfactant protein (SP)-A or SP-D gene to pulmonary hypersensitivity induced by antigens and allergens of *Aspergillus fumigatus*. *J. Immunol.* 174:6943–6954.

Mambula, S. S., K. Sau, P. Henneke, D. T. Golenbock, and S. M. Levitz. 2002. Toll-like receptor (TLR) signaling in response to *Aspergillus fumigatus*. *J. Biol. Chem.* 277:39320–39326.

Medzhitov, R. 2007. Recognition of microorganisms and activation of the immune response. *Nature* 449:819–826.

Medzhitov, R., and C. Janeway, Jr. 2000. Innate immune recognition: mechanisms and pathways. *Immunol. Rev.* 173:89–97.

Meier, A., C. J. Kirschning, T. Nikolaus, H. Wagner, J. Heesemann, and F. Ebel. 2003. Toll-like receptor (TLR) 2 and TLR4 are essential for *Aspergillus*-induced activation of murine macrophages. *Cell. Microbiol.* 5:561–570.

Mezger, M., S. Kneitz, I. Wozniok, O. Kurzai, H. Einsele, and J. Loeffler. 2008. Proinflammatory response of immature human dendritic cells is mediated by dectin-1 after exposure to *Aspergillus fumigatus* germ tubes. *J. Infect. Dis.* 197:924–931.

Netea, M. G., A. Warris, J. W. Van der Meer, M. J. Fenton, T. J. Verver-Janssen, L. E. Jacobs, T. Andresen, P. E. Verweij, and B. J. Kullberg. 2003. *Aspergillus fumigatus* evades immune recognition during germination through loss of toll-like receptor-4-mediated signal transduction. *J. Infect. Dis.* 188:320–326.

Neth, O., D. L. Jack, A. W. Dodds, H. Holzel, N. J. Klein, and M. W. Turner. 2000. Mannose-binding lectin binds to a range of clinically relevant microorganisms and promotes complement deposition. *Infect. Immun.* 68:688–693.

O'Neill, L. A. 2006. How Toll-like receptors signal: what we know and what we don't know. *Curr. Opin. Immunol.* 18:3–9.

Ozinsky, A., D. M. Underhill, J. D. Fontenot, A. M. Hajjar, K. D. Smith, C. B. Wilson, L. Schroeder, and A. Aderem. 2000. The repertoire for pattern recognition of pathogens by the innate immune system is defined by cooperation between toll-like receptors. *Proc. Natl. Acad. Sci. USA* 97:13766–13771.

Palma, A. S., T. Feizi, Y. Zhang, M. S. Stoll, A. M. Lawson, E. Diaz-Rodriguez, M. A. Campanero-Rhodes, J. Costa, S. Gordon, G. D. Brown, and W. Chai. 2006. Ligands for the beta-glucan receptor, Dectin-1, assigned using "designer" microarrays of oligosaccharide probes (neoglycolipids) generated from glucan polysaccharides. *J. Biol. Chem.* 281:5771–5779.

Richardson, M. D., G. S. Shankland, and C. A. Gray. 1991. Opsonizing activity of C-reactive protein in phagocytosis of *Aspergillus fumigatus* conidia by human neutrophils. *Mycoses* 34:141-143.

Rivera, A., G. Ro, H. L. Van Epps, T. Simpson, I. Leiner, D. B. Sant'Angelo, and E. G. Pamer. 2006. Innate immune activation and $CD4^+$ T cell priming during respiratory fungal infection. *Immunity* 25:665–675.

Rogers, N. C., E. C. Slack, A. D. Edwards, M. A. Nolte, O. Schulz, E. Schweighoffer, D. L. Williams, S. Gordon, V. L. Tybulewicz, G. D. Brown, and C. Reis e Sousa. 2005. Syk-dependent cytokine induction by Dectin-1 reveals a novel pattern recognition pathway for C type lectins. *Immunity* 22:507–517.

Saijo, S., N. Fujikado, T. Furuta, S. H. Chung, H. Kotaki, K. Seki, K. Sudo, S. Akira, Y. Adachi, N. Ohno, T. Kinjo, K. Nakamura, K. Kawakami, and Y. Iwakura. 2007. Dectin-1 is required for host defense against *Pneumocystis carinii* but not against *Candida albicans*. *Nat. Immunol.* 8:39–46.

Serrano-Gomez, D., A. Dominguez-Soto, J. Ancochea, J. A. Jimenez-Heffernan, J. A. Leal, and A. L. Corbi. 2004. Dendritic cell-specific intercellular adhesion molecule 3-grabbing nonintegrin mediates binding and internalization of *Aspergillus fumigatus* conidia by dendritic cells and macrophages. *J. Immunol.* 173:5635–5643.

Serrano-Gomez, D., J. A. Leal, and A. L. Corbi. 2005. DC-SIGN mediates the binding of Aspergillus fumigatus and keratinophylic fungi by human dendritic cells. *Immunobiology* 210:175–183.

Serrano-Gomez, D., E. Sierra-Filardi, R. T. Martinez-Nunez, E. Caparros, R. Delgado, M. A. Munoz-Fernandez, M. A. Abad, J. Jimenez-Barbero, M. Leal, and A. L. Corbi. 2008. Structural requirements for multimerization of the pathogen receptor dendritic cell-specific ICAM3-grabbing non-integrin (CD209) on the cell surface. *J. Biol. Chem.* 283:3889–3903.

Steele, C., R. R. Rapaka, A. Metz, S. M. Pop, D. L. Williams, S. Gordon, J. K. Kolls, and G. D. Brown. 2005. The beta-glucan receptor dectin-1 recognizes specific morphologies of *Aspergillus fumigatus*. *PLoS Pathog.* 1:e42.

Strong, P., K. B. Reid, and H. Clark. 2002. Intranasal delivery of a truncated recombinant human SP-D is effective at down-regulating allergic hypersensitivity in mice sensitized to allergens of *Aspergillus fumigatus*. *Clin. Exp. Immunol.* 130:19–24.

Sturtevant, J., and J. P. Latgé. 1992. Participation of complement in the phagocytosis of the conidia of *Aspergillus fumigatus* by human polymorphonuclear cells. *J. Infect. Dis.* 166:580–586.

Szalai, A. J. 2002. The antimicrobial activity of C-reactive protein. *Microbes Infect.* 4:201–205.

Taylor, P. R., G. D. Brown, J. Herre, D. L. Williams, J. A. Willment, and S. Gordon. 2004. The role of SIGNR1 and the beta-glucan receptor (dectin-1) in the nonopsonic recognition of yeast by specific macrophages. *J. Immunol.* 172:1157–1162.

Taylor, P. R., G. D. Brown, D. M. Reid, J. A. Willment, L. Martinez-Pomares, S. Gordon, and S. Y. Wong. 2002. The beta-glucan receptor, dectin-1, is predominantly expressed on the surface of cells of the monocyte/macrophage and neutrophil lineages. *J. Immunol.* 169:3876–3882.

Taylor, P. R., S. V. Tsoni, J. A. Willment, K. M. Dennehy, M. Rosas, H. Findon, K. Haynes, C. Steele, M. Botto, S. Gordon, and G. D.

Brown. 2007. Dectin-1 is required for beta-glucan recognition and control of fungal infection. *Nat. Immunol.* **8:**31–38.

Thiel, S. 2007. Complement activating soluble pattern recognition molecules with collagen-like regions, mannan-binding lectin, ficolins and associated proteins. *Mol. Immunol.* **44:**3875–3888.

Toyotome, T., Y. Adachi, A. Watanabe, E. Ochiai, N. Ohno, and K. Kamei. 2008. Activator protein 1 is triggered by *Aspergillus fumigatus* beta-glucans surface-exposed during specific growth stages. *Microb. Pathog.* **44:**141–150.

Vogl, G., I. Lesiak, D. B. Jensen, S. Perkhofer, R. Eck, C. Speth, C. Lass-Florl, P. F. Zipfel, A. M. Blom, M. P. Dierich, and R. Wurzner. 2008. Immune evasion by acquisition of complement inhibitors: the mould *Aspergillus* binds both factor H and C4b binding protein. *Mol. Immunol.* **45:**1485–1493.

Wang, J. E., A. Warris, E. A. Ellingsen, P. F. Jorgensen, T. H. Flo, T. Espevik, R. Solberg, P. E. Verweij, and A. O. Aasen. 2001. Involvement of CD14 and toll-like receptors in activation of human monocytes by *Aspergillus fumigatus* hyphae. *Infect. Immun.* **69:**2402–2406.

Zelensky, A. N., and J. E. Gready. 2005. The C-type lectin-like domain superfamily. *FEBS J.* **272:**6179–6217.

IV. THE SPECTRUM OF DISEASE

Aspergillus fumigatus and Aspergillosis
Edited by J.-P. Latgé and W. J. Steinbach

Chapter 23

Invasive Pulmonary Aspergillosis

Aimee K. Zaas and Barbara D. Alexander

Invasive fungal infections are frequently reported in the literature. This is due not only to the development of a clear case definition, improved diagnostic methods, and reporting, but also to an ever-enlarging at-risk population (Ascioglu et al., 2002). Surveillance data indicate that over the past several decades there has been an increasing incidence of invasive fungal infections due to *Aspergillus* and that *Aspergillus* remains the most common mold associated with invasive disease (Marr et al., 2002; Warnock, 2007). The mode of transmission of the pathogen and the burden of the epidemiologic exposure, as well as the patient's overall net state of immunosuppression, all play a role in determining the pattern of disease. In this chapter, we review the spectrum of invasive pulmonary aspergillosis (IPA), including its epidemiology, clinical and radiographic presentation, outcomes, and unique situations related to IPA. Intensive discussions of new and emerging diagnostics, therapeutic agents, and therapeutic strategies can be found in other chapters.

EPIDEMIOLOGY

Aspergillus is a ubiquitous hyalohyphomycete (a mold with nonpigmented, regularly septate hyphae) found in soil, dust, compost, rotted plants, and other organic debris, including foods and spices (Marr et al., 2002). Over 200 species are known, although only a few have been reported as pathogenic to humans. The more commonly reported human pathogens include *A. fumigatus, A. flavus, A. niger*, and *A. terreus*. Of these, *A. fumigatus* is the most common species to cause invasive disease, and *A. flavus* is the second most commonly reported. Although *A. terreus* is less common, it is resistant to amphotericin B and has historically been associated with an exceptionally high mortality (Hachem et al., 2004; Iwen et al., 1998; Steinbach et al., 2004; Tritz and Woods, 1993). Infections with other *Aspergillus* species, such as *A. clavatus* and *A. nidulans*, are increasingly reported (Marr et al., 2002).

Aspergillus grows best at 37°C, forming hyaline hyphae with asexual reproduction conidia that give each species a distinct colony color. Conidia are easily aerosolized, and when small airborne conidia (2 to 3 μm for *A. fumigatus*) are inhaled, they can settle deep in the lung, where colonization and a variety of clinical syndromes may develop. The type of host plays a role in the clinical spectrum of disease, as the host's immune response and the ability of *Aspergillus* to invade and destroy tissue determine the clinical presentation. In immunocompromised hosts, invasive disease may develop, either as locally invasive tracheobronchial disease, IPA, invasive sinusitis, or dissemination to extrapulmonary sites (Marr et al., 2002; Soubani and Chandrasekar, 2002).

Persons at highest risk of development of disease include those with hematologic malignancies, recipients of hematopoietic stem cell transplantation (HSCT; more so with allogeneic than autologous), and recipients of solid organ transplants (SOT) (Cornillet et al., 2006; Gavalda and Roman, 2007; Marr et al., 2002). A detailed discussion of diseases in these special populations can be found in chapters 37 to 39. Other at-risk groups include persons who require chronic corticosteroid therapy and those with primary immunodeficiencies (e.g., chronic granulomatous disease), have been hospitalized for a prolonged time in an intensive care unit, have poorly controlled diabetes mellitus, or have AIDS (Almyroudis et al., 2005; Groll et al., 1996; Vandewoude et al., 2006).

The cumulative incidence of IA for two of the highest populations at risk, HSCT and SOT recipients, has been reported from a recent multicenter study in the United States (Morgan et al., 2005). In the HSCT population, *Aspergillus* now exceeds *Candida* as the most

Aimee K. Zaas • Dept. of Medicine/Infectious Diseases, Duke University Medical Center, DUMC Box 3355, Durham, NC 27710. Barbara D. Alexander • Dept. of Medicine/Infectious Diseases and Dept. of Pathology, Duke University Medical Center, DUMC Box 3035, Durham, NC 27710.

common invasive fungal pathogen, and the cumulative incidence is higher at 12 months in patients with allogeneic unrelated donors (3.9%) than in those with allogeneic HLA-mismatched (3.2%), allogeneic HLA-matched (2.3%), or autologous (0.5%) donors (Morgan et al., 2005; Pappas et al., 2006). The rates were similar for myeloablative and nonmyeloablative conditioning regimens. Mortality at 3 months was 53.8% for autologous transplants versus 84.6% for allogeneic transplants with unrelated donors. In the SOT population, the cumulative incidence of IA at 12 months was 2.4% for lung, 0.8% for heart, 0.3% for liver, and 0.1% for kidney transplant recipients. Mortality at 3 months ranged from 20% for lung recipients to 66.7% for heart and kidney recipients.

In addition to factors associated with the development of IA (Table 1), retrospective studies have also evaluated risk factors for death in patients with proven or probable IA (including pulmonary, central nervous system, and disseminated disease) following HSCT (PMID 16511759 and 17243056). Notably, poor lung function prior to HSCT, receipt of an HLA-mismatched allograft, neutropenia, pleural effusions, receipt of >2 mg/kg corticosteroids in the 2 months preceding diagnosis, disseminated disease, active graft-versus-host disease (GVHD), and diagnosis of IA more than 40 days after transplant were associated with death. One study found administration of voriconazole to be protective, as well as the increased use of nonmyeloablative conditioning regimens and peripheral blood stem cells (Upton et al., 2007). Additionally, probability of survival at 90 days after diagnosis improved to 45% in patients diagnosed between 2002 and 2004 compared to those diagnosed earlier. Other studies have noted the poor outcome for IA, independent of the timing of IA post-transplant; survival is approximately 30% at 6 months and 20% at 12 months after the diagnosis of infection.

Table 1. Risk factors for IPA[a]

Risk factor
Cancer and chemotherapy
Genetic background (polymorphisms)
Immunosuppressive drugs
Corticosteroids
Diabetes mellitus
HSCT
Older age at transplant
Underlying disease
Other than chronic myelogenous leukemia in chronic phase
Multiple myeloma
Type of HSCT
Receipt of T-cell-depleted or CD34-selected transplant
Unrelated or HLA-mismatched peripheral blood SCT
Prolonged neutropenia
CMV disease
GVHD
Grade II to IV GVHD
Steroids for GVHD
Respiratory virus infection
SOT
Pulmonary colonization with *Aspergillus*
Acute rejection
Thrombocytopenia (liver transplant)
Primary immunodeficiencies (e.g., chronic granulomatous disease)
AIDS
Prolonged hospitalization in an intensive care unit
Acute renal failure, hemodialysis, or peritoneal dialysis
Endotracheal intubation or mechanical ventilation

[a]Sources: Cordonnier et al., 2006; Herbrecht et al., 2002; Maertens et al., 2002, 2004; Upton et al., 2007.

PATHOPHYSIOLOGY

It is well established that IPA arises from a failure of pulmonary defenses. *A. fumigatus* conidia are continuously inhaled by humans but rarely have any adverse effects, as they are eliminated efficiently by the immune response. Pulmonary alveolar macrophages can ingest inhaled *A. fumigatus* conidia and inhibit germination. If the macrophages are unable to clear the inhaled conidia, the conidia germinate into hyphae. The host must then rely on polymorphonuclear leukocytes (PMNs) to defend against continued fungal growth. Notably, animal models and reports from human studies are beginning to question the longstanding dogma of initial macrophage defense against conidia with subsequent PMN defense against hyphae (Stephens-Romero et al., 2005; Stergiopoulou et al., 2007). However, it is certain that the pattern of pulmonary injury and the subsequent outcome following disease acquisition are directly related to the immune status of the host (Ibrahim-Granet et al., 2003; Philippe et al., 2003). Numerous cytokines and chemokines are involved in host defense against aspergillosis (Phadke and Mehrad, 2005), with increasing appreciation for the role of adaptive immunity in response to this disease as well (Mehrad et al., 1999, 2002; Morrison et al., 2003; Park et al., 2006; Phadke et al., 2007). Once germinated, *A. fumigatus* hyphae grow and invade along vascular planes, causing tissue destruction, pulmonary hemorrhage, and microvascular obstruction. Given the significant role of the innate immune system for host defense against development of IPA, risk factors for acquiring this disease reflect significant compromise in one or more elements of defense, as is seen in HSCT or SOT.

Despite similar degrees of exogenous immunosuppression among at-risk patients impairing the immune clearance of *A. fumigatus* conidia and hyphae, only a subset of at-risk patients will develop IA. As such, there is a growing interest in host genetic differences that contribute to risk of IA. Functional genetic polymorphisms in any component of host defenses may be capable of altering the balance towards establishment of IA. Polymorphisms in Toll-like receptor 4, the interleukin-10 promoter (Sainz et al., 2007; Seo et al., 2005), and mannose-binding lectin (Kaur et al., 2007) have been evaluated for a role in acquisition of IA. While two studies demonstrated a protective role for the functional interleukin-10 promoter polymorphism 1082A, these evaluations need further validation prior to being used for clinical risk prediction or direction of prophylactic algorithms.

Animal studies using a rabbit model of IPA revealed a distinct pattern of response to invasive fungal infection on histopathology, depending on whether immunosuppression was secondary to neutropenia or was induced with cyclosporine plus methylprednisolone (Berenguer et al., 1995). The pattern noted in the neutropenic rabbits was one of scarce mononuclear infiltrate with coagulative necrosis and intra-alveolar hemorrhage, whereas in rabbits treated with cyclosporine plus methylprednisolone the histology consisted of a neutrophilic and monocytic inflammatory necrosis and scant hemorrhage. In humans, neutropenic patients and recipients of HSCT complicated by GVHD can have angioinvasion by *Aspergillus* hyphae and intra-alveolar hemorrhage with scant neutrophilic and monocytic infiltrate seen more frequently than in nonneutropenic patients with aspergillosis. The HSCT group also has a higher incidence of coagulative necrosis (eosinophilic inflammatory infiltrate with loss of cell membrane but preservation of pulmonary architecture), while the nonneutropenic group shows inflammatory necrosis (neutrophilic and monocytic infiltrate with necrosis of pulmonary parenchyma) and granuloma formation on histology (Stergiopoulou et al., 2007).

The histolopathologic pattern found in patients with profound neutropenia and IA underscores the importance of neutrophils as a first line of defense, as they prevent proliferation and angioinvasion by hyphae. In nonneutropenic patients, the host immune response is able to contain the hyphae and produce a necrotic inflammatory picture. Although the recovery of white blood cells is important in controlling IA, rapid recovery of the white blood cell count (with or without granulocyte colony-stimulating factor) after profound neutropenia in patients with hematologic malignancy following chemotherapy has led to complications in some patients (Takuma et al., 2002; Todeschini et al., 1999). Pulmonary hemorrhage with hemoptysis, pneumothorax, and respiratory failure, characterized by hypoxia and worsening infiltrates, were reported in a small study (Todeschini et al., 1999). The authors postulated that, along with the angioinvasiveness of *Aspergillus*, the necrotic inflammatory response, secondary to functional PMNs interacting with mold and cellular debris, may stimulate cell metabolism and proteolytic enzymes to a degree which can injure the pulmonary vasculature.

CLINICAL PRESENTATION

IPA most often presents with fever, cough which may be dry or productive, and dyspnea. Pleuritic chest pain and hemoptysis may be present, as can altered mental status and respiratory failure. Early diagnosis is often complicated by the paucity of clinical signs and symptoms in severely immune-compromised persons. IPA is characterized as being more invasive than chronic necrotizing aspergillosis, as it often includes invasion of small vessels with hemorrhage and/or infarction and the possibility of dissemination (Marr et al., 2002; Soubani and Chandrasekar, 2002; Trullas et al., 2005).

While certain findings are often seen in IPA (fever, cough, abnormal chest imaging), the presentation of IPA is highly dependent on host risk factors for disease development. Classically, IPA is considered in a patient with hematologic malignancy or HSCT who presents with fever despite broad-spectrum antibacterial therapy. Current diagnostic algorithms then involve chest imaging (X-ray, computed tomography [CT]), serologic testing (galactomannan enzyme immunoassay), and bronchoscopy with bronchoalveolar lavage. Biopsy is often not possible due to bleeding diathesis. The presentation of IPA in other hosts (i.e., SOT recipients or persons receiving high-dose corticosteroid therapy) tends to be less fulminant than that seen in HSCT recipients or persons with hematologic malignancies (Iversen et al., 2007; Sole et al., 2005; Trof et al., 2007; Bulpa et al., 2007).

Pulmonary Immune Reconstitution Inflammatory Syndrome

It is widely accepted that successful therapy of IA requires immune system recovery. However, the potential detrimental effects of an immune response to infection recently have emerged as an area of intense investigation (Miceli et al., 2007). Clinically, patients with IPA often experience worsening of both respiratory status and radiologic imaging as the immune system recovers following chemotherapy or engraftment of an HSCT. The availability of molecular markers of aspergillosis (galactomannan) allowed investigators to follow 19 patients with IA during the time of neutrophil recovery. Galactomannan antigenemia declined, although the clinico-radiographic picture worsened as the neutrophil count increased. The majority of patients (16 of 19) survived without any changes in antifungal therapy, and in the three deaths, autopsy revealed no evidence of IPA.

These findings led to the definition of pulmonary immune reconstitution inflammatory syndrome as the new onset of or worsening of clinical and radiologic pulmonary findings consistent with an infectious or inflammatory pulmonary condition temporally related to neutrophil recovery with evidence of microbiologic response (a decrease of 50% in serum galactomannan index titers on two consecutive tests within 4 days of each other) despite no change in antifungal therapy. The definition also required absence of new extrapulmonary lesions of aspergillosis (e.g., new skin lesions, as described above) and/or other processes, such as newly acquired infection, failure of treatment of a known infection, or medication side effects. Comprehensive management strategies for pulmonary immune reconstitution inflammatory syndrome are not well-defined but include supportive care and possibly corticosteroids in case of severe pulmonary decline.

RADIOGRAPHIC PRESENTATION

Radiologically, alveolar infiltrates, either bilateral or diffuse, nodules, cavitation, and pleural effusion can be present. The pathophysiology of infection directly translates into the radiographic findings. Macronodules are the most common manifestation of IPA in neutropenic patients (Austin et al., 1996; Greene et al., 2007). Review of the baseline chest CT findings from 235 patients with IPA, the majority of whom had hematologic malignancies as the underlying diagnosis, revealed that at presentation most patients (94%) had one or more macronodules (Greene et al., 2007). In patients with neutropenia and IPA, the CT scan may show a nodule surrounded by ground glass attenuation, the classic halo sign. The halo sign is related to the presence of hemorrhage surrounding a central necrotic nodule. Although the halo sign is not specific for IPA, initiation of treatment on the basis of the CT halo sign alone is associated with a better prognosis. In studies comprised mostly of neutropenic patients, the halo sign has been reported to be present in 61 to 82% of early CT scans (Greene et al., 2007; Horger et al., 2005a, 2005b). Unfortunately, this lesion is transitory, and by the first week, three-fourths of the CT halo signs disappear. With the recovery of the neutrophil count, an air crescent sign (representing early cavitation) may be seen (Greene et al., 2007; Horger et al., 2005a, 2005b). The air crescent sign is due to hemorrhagic infarction associated with lung sequestra resulting in a crescentic area of cavitation. It is considered to be highly suggestive for IPA (Caillot et al., 2001), but its utility in diagnosis is limited in that it is a late sign, appearing during convalescence. The hypodense core sign is more useful for early diagnosis. It has low sensitivity but high specificity for IPA in neutropenic patients (Horger et al., 2005a). It represents differences in density between pulmonary necrosis following infarction ("hypodense" sign) and surrounding hemorrhage and inflammation.

The early diagnosis of IPA is critical to successful treatment, and chest CT clearly has a central role. CT is a noninvasive method of investigation, and the results of CT scans are available immediately. However, much less is known about the CT appearance of IPA in nonmaligancy risk groups. The radiographic manifestations of IPA are very likely to differ for these patients; for example, the crescent sign is generally associated with the recovery of neutrophil counts and its value, therefore, is limited in the evaluation of nonneutropenic populations (Miceli et al., 2007).

DIAGNOSIS

Diagnosis of IPA relies on a combination of clinical, radiographic, microbiologic, serologic, and histologic parameters. A high index of suspicion is typically required, particularly in the highly immunosuppressed, who may have a paucity of clinical symptoms, and uncommon risk groups. When IPA is suspected, evaluation includes high-resolution chest CT, evaluation of sputum with potassium hydroxide microscopic examination and fungal culture, and potentially, measurement of serum galactomannan or β-D-glucan. While bronchoscopy with bronchoalveolar lavage and biopsy is ideal, often the clinical status of the patient precludes obtaining a biopsy, due to thrombocytopenia or other coagulation abnormalities. Increasingly, metabolic imaging, such as FDG-PET scanning, is employed to evaluate metabolic activity within lesions seen on the chest CT and to perhaps guide clinicians to the most informative area for biopsy (Mahfouz et al., 2005). As the lungs are not a sterile site, recovery of *Aspergillus* species from respiratory specimens is suggestive, but not diagnostic, of IA. Investigators have explored the predictive value of finding *Aspergillus* species in sputum cultures from various hosts and have noted that the positive predictive value increases from 1% in a host with connective tissue disease or cystic fibrosis to upwards of 64% in an allogeneic HSCT recipient or neutropenic host. Patients with hematologic malignancies (50% of patients had a sputum culture in cases of proven or probable IPA) were also considered a group for which identification of *Aspergillus* species in culture could not be considered contamination. In this particular retrospective study, intermediate risk groups included allogeneic HSCT recipients, SOT recipients, persons with solid organ malignancies, diabetes, chronic corticosteroid use, or underlying lung disease (Perfect et al., 2001). These data underscore the concept that the immune status of the host is critical in interpreting data with regards to the diagnosis of pulmonary aspergillosis.

TREATMENT

Therapy for IA is discussed in depth in other chapters. Table 2 highlights the major therapeutic trials for IA. Notably, a majority of patients in these trials had invasive disease limited to the lungs, but cases of central nervous system and disseminated disease are included as well. Management of IA involves several different strategies, including prophylactic therapy, empiric antifungal therapy, preemptive therapy, and treatment of known IA (Segal et al., 2007). Due to both disease and patient-specific factors, diagnosis of IA is often complicated. Despite the inherent difficulty in establishing a definitive diagnosis of IA, interpretation of clinical trial data investigating new antifungal agents or new therapeutic strategies depends on accurate diagnoses. Clinical trials can ensure the inclusion of true cases of IA by utilizing the European Organization for the Research and Treatment of Cancer—Mycoses Study Group criteria for diagnosis of IA (Ascioglu et al., 2002). These definitions are not meant to be used as a clinical tool for individual physicians outside of a clinical trial setting; however, in addition to their utility in standardizing clinical trial data, they serve as a guide for clinicians involved in the care of patients with suspected IA. Notably, these definitions apply best to patients with hematologic malignancies or those who are post-HSCT and are less applicable to SOT recipients and other risk groups.

CONCLUSIONS

IPA remains the most common *Aspergillus* syndrome in the immune-compromised host. Risk groups are expanding beyond the traditional groups of hematologic malignancy patients and HSCT recipients to include rheumatologic patients on chronic corticosteroid therapy and patients with a prolonged intensive care unit admission. Despite considerable advances over the past decade, particularly in the diagnostic (Maertens et al., 2002, 2004) and therapeutic arenas, mortality from IPA remains unacceptably high, particularly in patients who experience delay in recovery of the immune system (Upton et al., 2007). Important issues that remain to be addressed include elucidation of the optimal monitoring strategy (diagnostic test of choice, frequency, specimen tested), the role of combination antifungal therapy in the treatment of the disease, and the role of novel adjunctive therapies in disease management.

Table 2. Major antifungal trials for treatment of IA (includes disseminated disease)[a]

Reference	Agent(s)	Trial details	Trial outcome(s)	Additional key issues
Herbrecht et al., 2002	Voriconazole (6 mg/kg, q12 × 2 doses, then 4 mg/kg q12) vs. AmB (1–1.5 mg/kg q24)	277 patients, mostly HM or HSCT, randomized double blind	Success rate 52.8% with voriconazole vs. 31.6% with AmB (95% CI, 10.4–32.9)	Overall survival higher in voriconazole group, with fewer adverse events
Denning et al., 2002	Voriconazole salvage therapy	116 patients, mostly HM or HSCT, refractory to/intolerant of OLAT	Clinical response in 48%, complete response in 14%	Pretreated patients; open label
Kartsonis et al., 2005	Caspofungin compassionate use	48 patients refractory to/intolerant of AmB	Favorable response in 44% (20/45)	Pretreated patients; open label
Maertens et al., 2006	Caspofungin alone or in combination	90 patients refractory to or intolerant of OLAT, 83 MITT	Favorable response in 37/83 (45%) (32/50 pulmonary and 3/13 disseminated)	Only two discontinuations of caspofungin for adverse events
Walsh et al., 2007	Posaconazole salvage therapy	107 patients receiving posaconazole compared to 86 contemporaneous controls	Favorable response, 42% POS vs. 26% controls (OR, 4.06; 95% CI, 1.5–11.04; $P = 0.006$)	Patients previously received amphotericin products, not voriconazole or an echinocandin
Chandrasekar and Ito, 2005	ABLC, primary and salvage	101 HSCT, 101 HM, 109 SOT recipients; primary or salvage; mono- or combination therapy	65% favorable response overall (44% CR, 21% stabilized)	Data from the CLEAR registry; minimal renal toxicity

[a]Abbreviations: AmB, amphotericin B deoxycholate; ABLC, amphotericin B lipid complex; CI, confidence interval; CLEAR, Collaborative Exchange of Antifungal Research; CR, complete response; HM, hematologic malignancy; MITT, modified intent to treat population; OLAT, other licensed antifungal therapy; OR, odds ratio; POS, posaconazole.

REFERENCES

Almyroudis, N. G., S. M. Holland, and B. H. Segal. 2005. Invasive aspergillosis in primary immunodeficiencies. *Med. Mycol.* **43**(Suppl. 1):S247–59.

Ascioglu, S., J. H. Rex, B. de Pauw, J. E. Bennett, J. Bille, F. Crokaert, D. W. Denning, J. P. Donnelly, J. E. Edwards, Z. Erjavec, D. Fiere, O. Lortholary, J. Maertens, J. F. Meis, T. F. Patterson, J. Ritter, D. Selleslag, P. M. Shah, D. A. Stevens, and T. J. Walsh. 2002. Defining opportunistic invasive fungal infections in immunocompromised patients with cancer and hematopoietic stem cell transplants: an international consensus. *Clin. Infect. Dis.* **34**:7–14.

Austin, J. H., N. L. Muller, P. J. Friedman, D. M. Hansell, D. P. Naidich, M. Remy-Jardin, W. R. Webb, and E. A. Zerhouni. 1996. Glossary of terms for CT of the lungs: recommendations of the Nomenclature Committee of the Fleischner Society. *Radiology* **200**: 327–331.

Berenguer, J., M. C. Allende, J. W. Lee, K. Garrett, C. Lyman, N. M. Ali, J. Bacher, P. A. Pizzo, and T. J. Walsh. 1995. Pathogenesis of pulmonary aspergillosis. Granulocytopenia versus cyclosporine and methylprednisolone-induced immunosuppression. *Am. J. Respir. Crit. Care Med.* **152**:1079–1086.

Bulpa, P., A. Dive, and Y. Sibille. 2007. Invasive pulmonary aspergillosis in patients with chronic obstructive pulmonary disease. *Eur. Respir. J.* **30**:782–800.

Caillot, D., J. F. Couaillier, A. Bernard, O. Casasnovas, D. W. Denning, L. Mannone, J. Lopez, G. Couillault, F. Piard, O. Vagner, and H. Guy. 2001. Increasing volume and changing characteristics of invasive pulmonary aspergillosis on sequential thoracic computed tomography scans in patients with neutropenia. *J. Clin. Oncol.* **19**: 253–259.

Chandrasekar, P. H., and J. I. Ito. 2005. Amphotericin B lipid complex in the management of invasive aspergillosis in immunocompromised patients. *Clin. Infect. Dis.* **40**(Suppl. 6):S392–S400.

Cordonnier, C., P. Ribaud, R. Herbrecht, N. Milpied, D. Valteau-Couanet, C. Morgan, and A. Wade. 2006. Prognostic factors for death due to invasive aspergillosis after hematopoietic stem cell transplantation: a 1-year retrospective study of consecutive patients at French transplantation centers. *Clin. Infect. Dis.* **42**:955–963.

Cornillet, A., C. Camus, S. Nimubona, V. Gandemer, P. Tattevin, C. Belleguic, S. Chevrier, C. Meunier, C. Lebert, M. Aupee, S. Caulet-Maugendre, M. Faucheux, B. Lelong, E. Leray, C. Guiguen, and J. P. Gangneux. 2006. Comparison of epidemiological, clinical, and biological features of invasive aspergillosis in neutropenic and non-neutropenic patients: a 6-year survey. *Clin. Infect. Dis.* **43**:577–584.

Denning, D. W., P. Ribaud, N. Milpied, D. Caillot, R. Herbrecht, E. Thiel, A. Haas, M. Ruhnke, and H. Lode. 2002. Efficacy and safety of voriconazole in the treatment of acute invasive aspergillosis. *Clin. Infect. Dis.* **34**:563–571.

Gavalda, J., and A. Roman. 2007. Infection in lung transplantation. *Enferm. Infecc. Microbiol. Clin.* **25**:639–650. (In French.)

Greene, R. E., H. T. Schlamm, J. W. Oestmann, P. Stark, C. Durand, O. Lortholary, J. R. Wingard, R. Herbrecht, P. Ribaud, T. F. Patterson, P. F. Troke, D. W. Denning, J. E. Bennett, B. E. de Pauw, and R. H. Rubin. 2007. Imaging findings in acute invasive pulmonary aspergillosis: clinical significance of the halo sign. *Clin. Infect. Dis.* **44**:373–379.

Groll, A. H., P. M. Shah, C. Mentzel, M. Schneider, G. Just-Nuebling, and K. Huebner. 1996. Trends in the postmortem epidemiology of invasive fungal infections at a university hospital. *J. Infect.* **33**:23–32.

Hachem, R. Y., D. P. Kontoyiannis, M. R. Boktour, C. Afif, C. Cooksley, G. P. Bodey, I. Chatzinikolaou, C. Perego, H. M. Kantarjian, and I. I. Raad. 2004. Aspergillus terreus: an emerging amphotericin B-resistant opportunistic mold in patients with hematologic malignancies. *Cancer* **101**:1594–1600.

Herbrecht, R., D. W. Denning, T. F. Patterson, J. E. Bennett, R. E. Greene, J. W. Oestmann, W. V. Kern, K. A. Marr, P. Ribaud, O. Lortholary, R. Sylvester, R. H. Rubin, J. R. Wingard, P. Stark, C. Durand, D. Caillot, E. Thiel, P. H. Chandrasekar, M. R. Hodges, H. T. Schlamm, P. F. Troke, and B. de Pauw. 2002. Voriconazole versus amphotericin B for primary therapy of invasive aspergillosis. *N. Engl. J. Med.* **347**:408–415.

Horger, M., H. Einsele, U. Schumacher, M. Wehrmann, H. Hebart, C. Lengerke, R. Vonthein, C. D. Claussen, and C. Pfannenberg. 2005a. Invasive pulmonary aspergillosis: frequency and meaning of the "hypodense sign" on unenhanced CT. *Br. J. Radiol.* **78**:697–703.

Horger, M., H. Hebart, H. Einsele, C. Lengerke, C. D. Claussen, R. Vonthein, and C. Pfannenberg. 2005b. Initial CT manifestations of invasive pulmonary aspergillosis in 45 non-HIV immunocompromised patients: association with patient outcome? *Eur. J. Radiol.* **55**: 437–444.

Ibrahim-Granet, O., B. Philippe, H. Boleti, E. Boisvieux-Ulrich, D. Grenet, M. Stern, and J. P. Latgé. 2003. Phagocytosis and intracellular fate of *Aspergillus fumigatus* conidia in alveolar macrophages. *Infect. Immun.* **71**:891–903.

Iversen, M., C. M. Burton, S. Vand, L. Skovfoged, J. Carlsen, N. Milman, C. B. Andersen, M. Rasmussen, and M. Tvede. 2007. Aspergillus infection in lung transplant patients: incidence and prognosis. *Eur. J. Clin. Microbiol. Infect. Dis.* **26**:879–886.

Iwen, P. C., M. E. Rupp, A. N. Langnas, E. C. Reed, and S. H. Hinrichs. 1998. Invasive pulmonary aspergillosis due to Aspergillus terreus: 12-year experience and review of the literature. *Clin. Infect. Dis.* **26**:1092–1097.

Kartsonis, N. A., A. J. Saah, C. Joy Lipka, A. F. Taylor, and C. A. Sable. 2005. Salvage therapy with caspofungin for invasive aspergillosis: results from the caspofungin compassionate use study. *J. Infect.* **50**:196–205.

Kaur, S., V. K. Gupta, S. Thiel, P. U. Sarma, and T. Madan. 2007. Protective role of mannan-binding lectin in a murine model of invasive pulmonary aspergillosis. *Clin. Exp. Immunol.* **148**:382–389.

Maertens, J., A. Glasmacher, R. Herbrecht, A. Thiebaut, C. Cordonnier, B. H. Segal, J. Killar, A. Taylor, N. Kartsonis, T. F. Patterson, M. Aoun, D. Caillot, and C. Sable. 2006. Multicenter, noncomparative study of caspofungin in combination with other antifungals as salvage therapy in adults with invasive aspergillosis. *Cancer* **107**: 2888–2897.

Maertens, J., K. Theunissen, E. Verbeken, K. Lagrou, J. Verhaegen, M. Boogaerts, and J. V. Eldere. 2004. Prospective clinical evaluation of lower cut-offs for galactomannan detection in adult neutropenic cancer patients and haematological stem cell transplant recipients. *Br. J. Haematol.* **126**:852–860.

Maertens, J., J. Van Eldere, J. Verhaegen, E. Verbeken, J. Verschakelen, and M. Boogaerts. 2002. Use of circulating galactomannan screening for early diagnosis of invasive aspergillosis in allogeneic stem cell transplant recipients. *J. Infect. Dis.* **186**:1297–1306.

Mahfouz, T., M. H. Miceli, F. Saghafifar, S. Stroud, L. Jones-Jackson, R. Walker, M. L. Grazziutti, G. Purnell, A. Fassas, G. Tricot, B. Barlogie, and E. Anaissie. 2005. ^{18}F-fluorodeoxyglucose positron emission tomography contributes to the diagnosis and management of infections in patients with multiple myeloma: a study of 165 infectious episodes. *J. Clin. Oncol.* **23**:7857–7863.

Marr, K. A., T. Patterson, and D. Denning. 2002. Aspergillosis: pathogenesis, clinical manifestations, and therapy. *Infect. Dis. Clin. North Am.* **16**:875–894.

Mehrad, B., R. M. Strieter, and T. J. Standiford. 1999. Role of TNF-alpha in pulmonary host defense in murine invasive aspergillosis. *J. Immunol.* **162**:1633–1640.

Mehrad, B., M. Wiekowski, B. E. Morrison, S. C. Chen, E. C. Coronel, D. J. Manfra, and S. A. Lira. 2002. Transient lung-specific expression of the chemokine KC improves outcome in invasive aspergillosis. *Am. J. Respir. Crit. Care Med.* **166:**1263–1268.

Miceli, M. H., J. Maertens, K. Buve, M. Grazziutti, G. Woods, M. Rahman, B. Barlogie, and E. J. Anaissie. 2007. Immune reconstitution inflammatory syndrome in cancer patients with pulmonary aspergillosis recovering from neutropenia: proof of principle, description, and clinical and research implications. *Cancer* **110:**112–120.

Morgan, J., K. A. Wannemuehler, K. A. Marr, S. Hadley, D. P. Kontoyiannis, T. J. Walsh, S. K. Fridkin, P. G. Pappas, and D. W. Warnock. 2005. Incidence of invasive aspergillosis following hematopoietic stem cell and solid organ transplantation: interim results of a prospective multicenter surveillance program. *Med. Mycol.* **43**(Suppl. 1):S49–S58.

Morrison, B. E., S. J. Park, J. M. Mooney, and B. Mehrad. 2003. Chemokine-mediated recruitment of NK cells is a critical host defense mechanism in invasive aspergillosis. *J. Clin. Investig.* **112:** 1862–1870.

Park, S. J., M. T. Wiekowski, S. A. Lira, and B. Mehrad. 2006. Neutrophils regulate airway responses in a model of fungal allergic airways disease. *J. Immunol.* **176:**2538–2545.

Pappas, P. G., D. Andes, M. Schuster, S. Hadley, J. Rabkin, R. M. Merion, C. A. Kauffman, C. Huckabee, G. A. Cloud, W. E. Dismukes, and A. W. Karchmer. 2006. Invasive fungal infections in low-risk liver transplant recipients: a multi-center prospective observation study. *Am. J. Transplant.* **6:**386–391.

Perfect, J. R., G. M. Cox, J. Y. Lee, C. A. Kauffman, L. de Repentigny, S. W. Chapman, V. A. Morrison, P. Pappas, J. W. Hiemenz, and D. A. Stevens. 2001. The impact of culture isolation of Aspergillus species: a hospital-based survey of aspergillosis. *Clin. Infect. Dis.* **33:**1824–1833.

Phadke, A. P., G. Akangire, S. J. Park, S. A. Lira, and B. Mehrad. 2007. The role of CC chemokine receptor 6 in host defense in a model of invasive pulmonary aspergillosis. *Am. J. Respir. Crit. Care Med.* **175:**1165–1172.

Phadke, A. P., and B. Mehrad. 2005. Cytokines in host defense against Aspergillus: recent advances. *Med. Mycol.* **43**(Suppl. 1):S173–S176.

Philippe, B., O. Ibrahim-Granet, M. C. Prevost, M. A. Gougerot-Pocidalo, M. Sanchez Perez, A. Van der Meeren, and J. P. Latgé. 2003. Killing of *Aspergillus fumigatus* by alveolar macrophages is mediated by reactive oxidant intermediates. *Infect. Immun.* **71:** 3034–3042.

Sainz, J., L. Hassan, E. Perez, A. Romero, A. Moratalla, E. Lopez-Fernandez, S. Oyonarte, and M. Jurado. 2007. Interleukin-10 promoter polymorphism as risk factor to develop invasive pulmonary aspergillosis. *Immunol. Lett.* **109:**76–82.

Segal, B. H., N. G. Almyroudis, M. Battiwalla, R. Herbrecht, J. R. Perfect, T. J. Walsh, and J. R. Wingard. 2007. Prevention and early treatment of invasive fungal infection in patients with cancer and neutropenia and in stem cell transplant recipients in the era of newer broad-spectrum antifungal agents and diagnostic adjuncts. *Clin. Infect. Dis.* **44:**402–409.

Seo, K. W., D. H. Kim, S. K. Sohn, N. Y. Lee, H. H. Chang, S. W. Kim, S. B. Jeon, J. H. Baek, J. G. Kim, J. S. Suh, and K. B. Lee. 2005. Protective role of interleukin-10 promoter gene polymorphism in the pathogenesis of invasive pulmonary aspergillosis after allogeneic stem cell transplantation. *Bone Marrow Transplant.* **36:** 1089–1095.

Sole, A., P. Morant, M. Salavert, J. Peman, and P. Morales. 2005. Aspergillus infections in lung transplant recipients: risk factors and outcome. *Clin. Microbiol. Infect.* **11:**359–365.

Soubani, A. O., and P. H. Chandrasekar. 2002. The clinical spectrum of pulmonary aspergillosis. *Chest* **121:**1988–1999.

Steinbach, W. J., D. K. Benjamin, Jr., D. P. Kontoyiannis, J. R. Perfect, I. Lutsar, K. A. Marr, M. S. Lionakis, H. A. Torres, H. Jafri, and T. J. Walsh. 2004. Infections due to Aspergillus terreus: a multicenter retrospective analysis of 83 cases. *Clin. Infect. Dis.* **39:**192–198.

Stephens-Romero, S. D., A. J. Mednick, and M. Feldmesser. 2005. The pathogenesis of fatal outcome in murine pulmonary aspergillosis depends on the neutrophil depletion strategy. *Infect. Immun.* **73:**114–125.

Stergiopoulou, T., J. Meletiadis, E. Roilides, D. E. Kleiner, R. Schaufele, M. Roden, S. Harrington, L. Dad, B. Segal, and T. J. Walsh. 2007. Host-dependent patterns of tissue injury in invasive pulmonary aspergillosis. *Am. J. Clin. Pathol.* **127:**349–355.

Takuma, T., K. Okada, Y. Uchida, A. Yamagata, and Y. Sawae. 2002. Invasive pulmonary aspergillosis resulting in respiratory failure during neutrophil recovery from postchemotherapy neutropenia in three patients with acute leukaemia. *J. Intern. Med.* **252:**173–177.

Todeschini, G., C. Murari, R. Bonesi, G. Pizzolo, G. Verlato, C. Tecchio, V. Meneghini, M. Franchini, C. Giuffrida, G. Perona, and P. Bellavite. 1999. Invasive aspergillosis in neutropenic patients: rapid neutrophil recovery is a risk factor for severe pulmonary complications. *Eur. J. Clin. Investig.* **29:**453–457.

Tritz, D. M., and G. L. Woods. 1993. Fatal disseminated infection with Aspergillus terreus in immunocompromised hosts. *Clin. Infect. Dis.* **16:**118–122.

Trof, R. J., A. Beishuizen, Y. J. Debets-Ossenkopp, A. R. Girbes, and A. B. Groeneveld. 2007. Management of invasive pulmonary aspergillosis in non-neutropenic critically ill patients. *Intensive Care Med.* **33:**1694–1703.

Trullas, J. C., C. Cervera, N. Benito, J. P. de la Bellacasa, C. Agusti, M. Rovira, A. Mas, M. Navasa, F. Cofan, M. J. Ricart, F. Perez-Villa, and A. Moreno. 2005. Invasive pulmonary aspergillosis in solid organ and bone marrow transplant recipients. *Transplant. Proc.* **37:**4091–4093.

Upton, A., K. A. Kirby, P. Carpenter, M. Boeckh, and K. A. Marr. 2007. Invasive aspergillosis following hematopoietic cell transplantation: outcomes and prognostic factors associated with mortality. *Clin. Infect. Dis.* **44:**531–540.

Vandewoude, K. H., S. I. Blot, P. Depuydt, D. Benoit, W. Temmerman, F. Colardyn, and D. Vogelaers. 2006. Clinical relevance of Aspergillus isolation from respiratory tract samples in critically ill patients. *Crit. Care* **10:**R31.

Walsh, T. J., I. Raad, T. F. Patterson, P. Chandrasekar, G. R. Donowitz, R. Graybill, R. E. Greene, R. Hachem, S. Hadley, R. Herbrecht, A. Langston, A. Louie, P. Ribaud, B. H. Segal, D. A. Stevens, J. A. van Burik, C. S. White, G. Corcoran, J. Gogate, G. Krishna, L. Pedicone, C. Hardalo, and J. R. Perfect. 2007. Treatment of invasive aspergillosis with posaconazole in patients who are refractory to or intolerant of conventional therapy: an externally controlled trial. *Clin. Infect. Dis.* **44:**2–12.

Warnock, D. W. 2007. Trends in the epidemiology of invasive fungal infections. *Nippon Ishinkin Gakkai Zasshi* **48:**1–12.

Aspergillus fumigatus and Aspergillosis
Edited by J.-P. Latgé and W. J. Steinbach

Chapter 24

Aspergillus Sinusitis and Cerebral Aspergillosis

STEFAN SCHWARTZ AND MARKUS RUHNKE

SINUSITIS

Introduction

Aspergillus species are the most frequently identified pathogens in patients with fungal sinusitis (Parikh et al., 2004; Willinger et al., 2003). Aspergillosis of the paranasal sinuses virtually always represents an airborne disease, which is acquired by inhalation of conidia. Occasionally, the disease may occur as a complication after invasive procedures, such as transphenoidal surgery (Batra et al., 2005). In addition, aspergillosis of the maxillary sinuses has been reported in association with dental procedures, such as endodontic treatment (Beck-Mannagetta et al., 1983; Giardino et al., 2006; Odell and Pertl, 1995a). In these patients, root canal overfilling with translocation of sealer material into the maxillary sinus is a frequent finding. Interestingly, experimental data demonstrate that zinc, potentially released from the sealer material, promotes growth of *Aspergillus* species (Mensi et al., 2004; Odell and Pertl, 1995b; Willinger et al., 1996).

Rhinosinusitis caused by *Aspergillus* species was first described more than a century ago (Fig. 1), but a proposal for a comprehensive classification of fungal sinusitis that considered clinical, radiological, and histopathological characteristics was not published until 1997 (deShazo et al., 1997; Schubert, 1885). The primary feature, which allows a distinction of various forms of fungal sinusitis, is the absence (noninvasive sinusitis) or the presence (invasive sinusitis) of invasion by fungal elements and tissue necrosis. *Aspergillus* infections of the paranasal sinuses can be grouped into five main subtypes. The invasive forms are acute sinusitis (fulminant), chronic sinusitis (indolent), and chronic granulomatous sinusitis, whereas the noninvasive forms are fungus ball (aspergilloma) and allergic fungal sinusitis (Table 1).

Despite this classification of *Aspergillus* sinusitis into at least five subtypes, available epidemiological data about the incidence and prevalence of these entities are limited. In one of the largest published series, data on 86 patients with histopathologically proven fungus-related disease of the nasal sinuses were analyzed (Driemel et al., 2007). Invasive fungal sinusitis was seen in 22 individuals (11 males) with a mean age of 57 years (22 to 84 years). Of these, 41% had immunocompromising conditions, including diabetes mellitus (three patients), various malignancies (five patients), and bacterial endocarditis (one patient). A fungus ball was diagnosed in 60 patients (26 males) who had a mean age of 54 years (22 to 88 years). An immunocompromising condition was seen in only 15% (9/60) of these patients, including diabetes mellitus (two patients), solid tumor with combined chemotherapy and radiation therapy (four patients), myocarditis (one patient), and chronic hepatitis (two patients). An allergic fungal sinusitis was recorded in only four patients, and these patients had a lower mean age of 43 years (17 to 63 years) compared to all other patients.

Interestingly, other reports on acute invasive fungal sinusitis found this entity almost exclusively in severely immunocompromised patients, particularly in individuals with hematological malignancies, such as acute leukemia or after hematopoietic stem cell transplantation (Drakos et al., 1993; Gillespie et al., 2000; Kennedy et al., 1997; Talbot et al., 1991). Finally, there is allergic fungal sinusitis, which is not within the scope of this chapter.

Noninvasive *Aspergillus* Sinusitis

Acute rhinosinusitis in adults is mostly caused by bacterial or viral pathogens. In chronic and recurrent forms of noninvasive rhinosinusitis, the causative path-

Stefan Schwartz • Medizinische Klinik III, Charité–Universitätsmedizin Berlin, Campus Benjamin Franklin, 12200 Berlin, Germany.
Markus Ruhnke • Medizinische Klinik und Poliklinik II, Charité–Universitätsmedizin Berlin, Campus Mitte, 10117 Berlin, Germany.

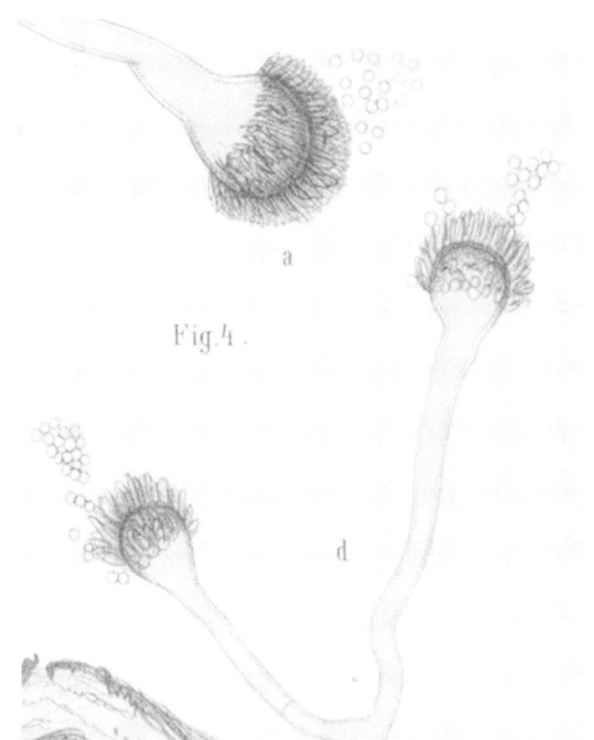

Figure 1. Drawings from the first description of fungal rhinosinusitis, showing a fungal morphology compatible with *A. fumigatus* (Schubert, 1885). Reprinted with permission of Blackwell Publishing.

ogens may be fungi (Rosenfeld et al., 2007). The presenting symptoms are usually nonspecific and may result in a delayed diagnosis. However, in isolated sphenoid sinusitis, approximately 20% of the disease could be caused by fungus balls, with *Aspergillus* species as the predominant pathogen (Friedman et al., 2005). In up to 60% of cases the diagnosis of a fungus ball may not be established by mycological culture but only by histopathological examination (Pagella et al., 2007).

Classifications

The noninvasive forms of *Aspergillus* sinusitis, occurring almost exclusively in immunocompetent hosts, are mostly categorized into allergic sinusitis and sinus fungus ball or mycetoma (deShazo et al., 1997; Karci et al., 2001). However, different entities have been proposed in other publications. In a prospective study from India, three types of paranasal sinus aspergillosis were described which were called either chronic invasive, noninvasive (fungus ball), and noninvasive destructive (Panda et al., 2004). Adjuvant chemotherapy was employed in noninvasive destructive and chronic invasive disease.

Diagnosis

The majority of patients with allergic fungal sinusitis suffer from chronic sinusitis, nasal polyps, asthma, and atopy (deShazo and Swain, 1995). The hallmark of allergic fungal sinusitis is the presence of "allergic mucin" within the sinuses, which is frequently multilayered and contains cellular debris, eosinophils, Charcot-Leyden crystals, and only a few fungal elements (Katzenstein et al., 1983). The second noninvasive form of *Aspergillus* sinusitis, sinus mycetoma, should preferentially be termed sinus fungus ball or aspergilloma. In a study from Turkey performed between 1993 and 1997 there were 27 cases of fungal sinusitis, 22 of which were noninvasive forms and 5 of which were invasive forms (Karci et al., 2001). Eleven patients were diagnosed with mycetoma, nine with allergic fungal sinusitis, three with acute fulminant sinusitis, and two with chronic indolent sinusitis, while two patients were not included in any of the four subgroups of sinusitis. In all mycetoma cases the fungal pathogen was an *Aspergillus* species.

Patients with sinus aspergilloma typically present with facial pain, nasal obstruction, rhinorrhea, and fetid odor (cacosmia) (Pagella et al., 2007). Radiological evaluation most frequently exhibits unilateral maxillary sinus involvement, but any sinus or multiple sinuses may be affected (Pagella et al., 2007). In the majority of patients with sinus aspergilloma, computed tomography (CT) scans disclose heterogenous opacifications in affected sinuses, including microcalcifications or metal-dense material (Klossek et al., 1997). These radiation-dense areas are due to enrichment of calcium salts and formation of fungal concrements (Stammberger et al., 1984). Demonstration of filamentous fungi within the fungus ball has a >90% sensitivity in establishing the diagnosis, whereas fungal cultures are much less sensitive (~30%) in this particular subtype of fungal sinusitis (Klossek et al., 1997). Thus, due to the low sensitivity of mycological culture, histopathological examination for fungal elements should be always performed to diagnose fungal sinusitis. It is largely unknown what factors, apart from allergy, contribute to formation of *Aspergillus* sinusitis in immunocompetent individuals. Recent data from a nonimmunocompromised rabbit model indicate that impaired sinus aeration subsequent to inoculation of *Aspergillus* conidia is a major factor which promotes fungal sinusitis (Dufour et al., 2005).

Despite the absent tissue invasion of fungi in allergic fungal sinusitis due to *Aspergillus* species and sinus aspergilloma, involvement of neighboring structures into the inflammatory process may develop in these subtypes of fungal sinusitis and occasionally require extensive surgery (Liu et al., 2004). Allergic *Aspergillus* sinusitis or sinus aspergilloma may be accompanied by orbital and even intracranial extension causing propto-

Table 1. Clinicopathological subtypes of *Aspergillus* sinusitis[a]

Sinusitis subtype	Clinical features	Immunosuppression	Histopathology	Treatment
Noninvasive				
Allergic	Chronic sinusitis, nasal polyps, frequently atopy	–	"Allergic mucin" with eosinophils, Charcot-Leyden crystals, but few hyphae; no soft tissue invasion	Debridement, sinus aeration, steroids
Fungus ball (aspergilloma[b] or mycetoma)	Chronic sinus symptoms, nasal polyps, sinus calcifications, occasionally atopy	–	Fungal ball with dense accumulation of hyphae, may contain concrements; no tissue invasion	Debridement, sinus aeration
Invasive				
Acute (fulminant)	Fever, facial pain, nasal discharge or congestion, epistasis, periorbital swelling, rapid progression	+++	Invasion of mucosa, submucosa, bone, and vessels with extensive tissue necrosis	Debridement/resection should be attempted, early initiation of systemic antifungal therapy (voriconazole preferred)
Chronic	Chronic sinus symptoms, often misdiagnosed as inflammatory pseudotumor, association with orbital apex syndrome	+	Sparse chronic inflammatory infiltration with vascular invasion of fungal elements, dense accumulation of hyphae	Debridement/resection should be attempted, early initiation of systemic antifungal therapy (voriconazole preferred)
Granulomatous	Chronic, slowly progressive sinusitis associated with proptosis	–	Florid granulomatous inflammation without tissue necrosis but frequently extending beyond the confines of sinuses	Debridement/resection should be attempted, initiation of systemic antifungal therapy is suggested

[a] Modified according to deShazo et al., 1997.
[b] The original description of aspergilloma was recognized as a lung disease with a fungus ball that colonized in a healed lung scar or abscess cavity from a previous disease.

sis, diplopia, visual loss, or cranial nerve palsy (Carter et al., 1999; Pagella et al., 2007). Bone erosion may be found in a number of individuals with allergic fungal sinusitis or sinus aspergilloma, which likely results from chronic inflammation and mass expansion rather than from fungal tissue invasion. Any bony sinus wall may be affected, but a preference for the lamina papyracea has been repeatedly observed (Ghegan et al., 2006; Nussenbaum et al., 2001; Pagella et al., 2007).

In a series reported by Liu et al., the authors described their experience in 21 immunocompetent patients with an average age of 25 years (range, 9 to 46 years) and a male/female ratio of 3.75:1 (Liu et al., 2004). All patients had a history of chronic sinusitis with imaging findings of disease involving multiple sinuses. Fifteen patients had nasal polyposis, eight had erosion of bone demonstrated with CT scans, eight had disease with intracranial extension, and six had disease that involved the lamina papyracea.

Due to the extension of the inflammatory process with bony erosion in a subset of immunocompetent patients with noninvasive fungal sinusitis, some authors have created the terms "destructive noninvasive paranasal sinus aspergillosis" and "erosive fungal sinusitis" to characterize this disease as intermediate between aspergilloma, allergic, and chronic invasive fungal sinsusitis, but this has not been generally accepted as a distinct disease entity (Rowe-Jones and Moore-Gillon, 1994; Uri et al., 2003).

Therapy

Therapy of noninvasive forms of *Aspergillus* sinusitis consists of surgical removal of allergic mucin or the fungus ball and aeration of sinuses under endoscopic guidance. Using this approach, the majority of patients with sinus aspergilloma experience long-term remissions and do not require any additional treatment (Klossek et al., 1997). However, patients with allergic forms of *Aspergillus* sinusitis frequently experience a relapse of their disease. Irrigation with isotonic saline or topical or systemic corticosteroid therapy, as well as immunomodulatory therapy, might prevent repeat mucus impaction and suppress the inflammatory response in these patients (Schubert, 2004).

The limited nature of the infection does not require systemic antifungal therapy, which is of unproven benefit in patients with noninvasive forms of *Aspergillus* sinusitis. However, corticosteroid therapy may reduce local inflammation and is associated with a lower relapse rate in patients with allergic fungal sinusitis (Schubert, 2004). In a report on 21 patients with allergic fungal sinusitis, all underwent transnasal and/or transmaxillary endoscopic approaches for debridement and irrigation, 6 patients underwent orbital decompression, and 3 patients underwent a bifrontal craniotomy for removal of intracranial extradural disease. None of the patients developed a cerebrospinal fluid leakage. Postoperatively, 1 patient was treated with amphotericin B and the other 20 were treated with a short course of corticosteroids (Liu et al., 2004). Karci et al. reported on a series of 27 patients with fungal sinusitis, including 22 patients with noninvasive forms of *Aspergillus* sinusitis, all of whom were treated with endoscopic sinus surgery (Karci et al., 2001). The infection recurred in two patients with allergic fungal sinusitis and in another patient with chronic invasive sinusitis within 20 months.

Acute (Fulminant) Invasive *Aspergillus* Sinusitis

Fulminant or acute invasive *Aspergillus* sinusitis was first described as a distinct disease entity in 1980 (McGill et al., 1980). This aggressive form of *Aspergillus* sinusitis is characterized by an abrupt onset with rapid progression and a tendency of destructive invasion into neighboring structures. It occurs almost exclusively in severely immunocompromised hosts, including patients with profound neutropenia (e.g., acute leukemia, aplastic anemia, and postchemotherapy), AIDS patients, or patients after hematopoietic stem cell transplantation (Drakos et al., 1993; McGill et al., 1980; Mylonakis et al., 1997; Schwartz and Thiel, 1997; Viollier et al., 1986). Remarkably, acute invasive *Aspergillus* sinusitis is less common than invasive pulmonary aspergillosis, with a frequency of sinus infections of only 5% compared to a frequency of pulmonary infections of ≥56% among immunocompromised patients with invasive aspergillosis (Herbrecht et al., 2002; Patterson et al., 2000). However, invasive aspergillosis of the lungs and sinuses may coexist in individual patients (Pauksens and Oberg, 2006).

The reported incidence of invasive fungal sinusitis in patients with hematopoietic stem cell transplantation varied between 1.7 and 2.6% in two institutions between 1983 and 1993 (Drakos et al., 1993; Kennedy et al., 1997). Kennedy et al. reported that survival from invasive fungal sinusitis was not affected by patient age, white blood cell count at onset, dose and type of antifungal therapy, or extent of surgical resection. Their study concluded that only intracranial and/or orbital involvement is predictive for a poor outcome. A significant proportion of patients (50%) did not recover from invasive fungal sinusitis despite neutrophil recovery after transplantation. In this series, 61% of patients who succumbed to the infection had undergone extensive surgical procedures, versus 55% of those who resolved the infection. In contrast, Gillespie et al. concluded that complete surgical resection with negative margins and the reversal of neutropenia appear to be critical factors for survival in patients with invasive fungal sinusitis (Gillespie et al., 1998).

Microbiology and pathology

In the majority of patients with acute invasive fungal sinusitis, the infecting fungus is *Aspergillus flavus* (Iwen et al., 1997; Talbot et al., 1991; Viollier et al., 1986). Kennedy et al. reported that *A. flavus* (n = 9), *Aspergillus fumigatus* (n = 3), and unspecified *Aspergillus* spp. (n = 2) were isolated from 26 bone marrow transplantation patients with invasive fungal sinusitis (Kennedy et al., 1997). Drakos et al. reported on 11 patients with invasive fungal sinusitis that occurred among 423 consecutive patients with bone marrow transplantation, and they found *A. flavus* in 7 patients and *Aspergillus quadrilineatus* in 1 patient (Drakos et al., 1993). However, most data on invasive fungal sinusitis were published more than 10 years ago, and the current epidemiology of *A. flavus* in this setting is largely unknown. It is unclear why *A. flavus* and not *A. fumigatus* is the predominant mold found in acute invasive fungal sinusitis. Interestingly, conidia of *A. flavus* are somewhat larger than those of *A. fumigatus* (8 versus 3.5 μm), which may favor their retention in the upper airway tract (Klich, 2002). In acute invasive sinusitis, histopathological examination typically shows fungal invasion of mucosa, submucosa, bone, and vessels, with extensive tissue necrosis (McGill et al., 1980; Taxy, 2006). Tissue invasion with neutrophils is usually pronounced but may be less marked or absent in patients with persistent neutropenia. Recently, the important role of neutrophils as a defense mechanism against aspergillosis of the paranasal sinuses has been studied in a mouse model in which neutrophils are depleted using anti-Gr-1 monoclonal antibody (Rodriguez et al., 2007). The presence of neutrophils was essential in protecting the sinuses against acute *Aspergillus* infection and in clearing of established hyphal masses.

Diagnosis

Continuous invasion of fungi into neighboring structures is a common and threatening complication, and it could result in orbital cellulitis, retinitis, palate destruction, or brain abscess formation (Fig. 2 to 5 and Color Plates 11 to 14). Symptoms that should prompt an immediate diagnostic evaluation in at-risk patients include fever, facial pain, nasal discharge or congestion, epistaxis, and periorbital swelling (Gillespie et al., 1998; Iwen et al., 1997; Talbot et al., 1991). CT or magnetic

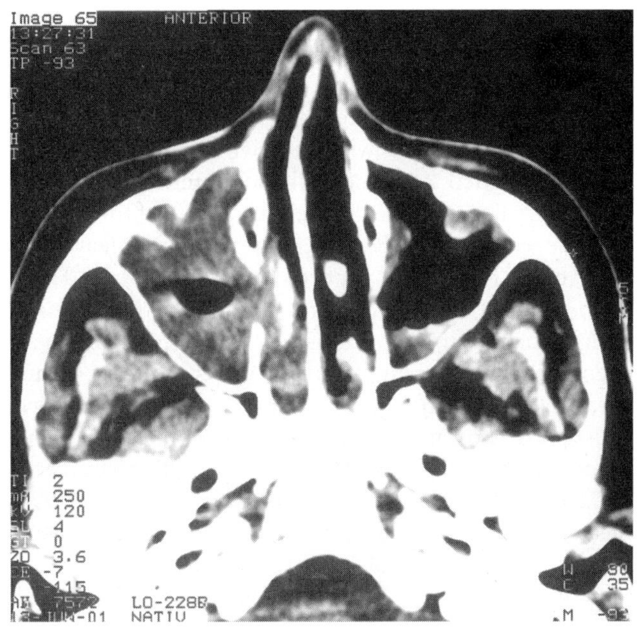

Figure 2. Acute invasive sinusitis (*A. fumigatus*) with orbital cellulitis in a patient after hematopoietic stem cell transplantation (see also Color Plate 11). CT scan showing sinusitis with destruction of the medial wall of the right maxillary sinus.

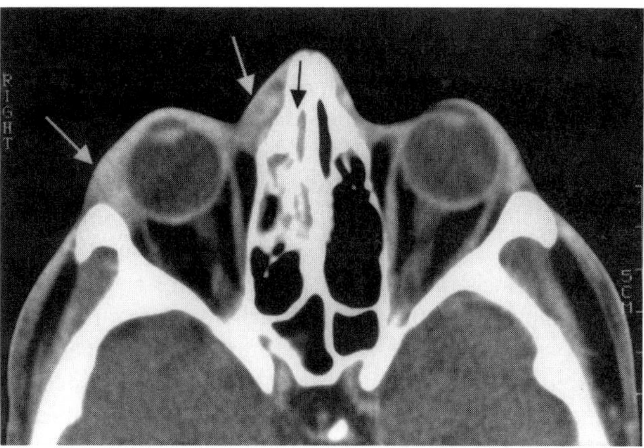

Figure 4. Acute invasive sinusitis with orbital cellulitis caused by *Aspergillus* spp., resulting in palate destruction in a patient with acute leukemia (see also Color Plate 13). The CT scan shows opacification of the right ethmoidal sinus (black arrow) and thickening of periorbital soft tissue (white arrows). Reprinted from the *New England Journal of Medicine* (Schwartz and Thiel, 1997) with permission of the Massachusetts Medical Society.

resonance imaging scans allow early detection of inflammatory soft tissue edema, bone destruction, or invasion into adjacent structures and could guide the subsequent diagnostic approach and surgical management (Aribandi et al., 2007; DelGaudio et al., 2003). A proven diagnosis of acute invasive *Aspergillus* sinusitis requires tissue biopsy, but this carries the unwanted risk of hemorrhage in patients with thrombocytopenia and may require general anesthesia. Rigid nasal endoscopy has been advocated in at-risk patients with specific symptoms and may identify mucosal discoloration, crusts, or ulcerations and could guide a targeted biopsy (Gillespie et al., 1998). Once a biopsy is available, conventional processing of specimens is time-consuming and could cause a

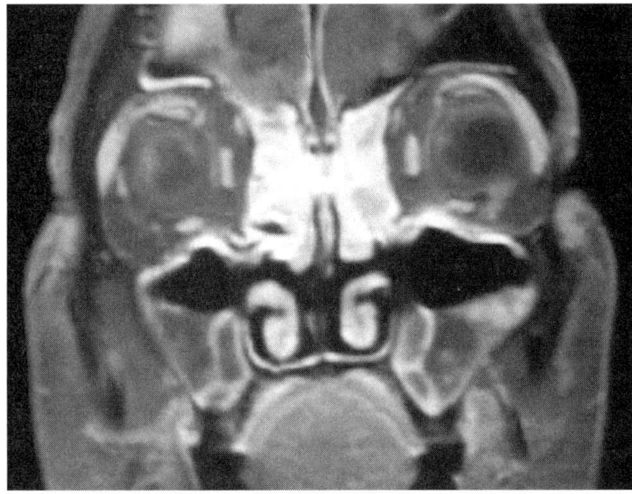

Figure 3. Acute invasive fungal sinusitis (*A. fumigatus*) resulting in fungal retinitis in a patient after hematopoietic stem cell transplantation (see also Color Plate 12). Magnetic resonance imaging scan (T1-weighted, Gd-enhanced imaging) showing sinusitis of the ethmoidal sinuses with extension into the orbital cavities.

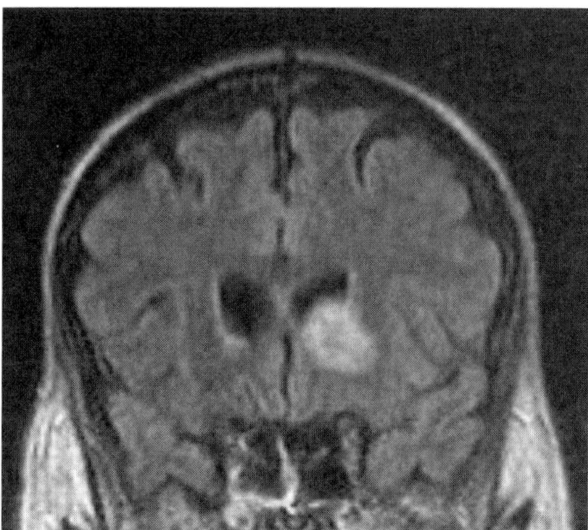

Figure 5. Fungal brain abscess caused by hematogenous spread from invasive pulmonary aspergillosis in a patient after hematopoietic stem cell transplantation. Brain section and histological photographs were kindly provided by A. Harder, Berlin, Germany (see also Color Plate 14). The magnetic resonance imaging scan (fluid-attenuated inversion recovery imaging) shows a brain abscess within the left-sided region of the nucleus caudatus.

delay of a firm diagnosis. Recently, frozen section biopsies were compared to permanent tissue sections from 20 patients with invasive mold sinusitis. Evaluation of frozen section biopsies identified with a sensitivity of 84% and specificity of 100% the presence of an invasive fungal infection. Furthermore, frozen section analysis correctly separated *Aspergillus* and non-*Aspergillus* cases (Ghadiali et al., 2007). Using this technique, a more rapid diagnosis may facilitate initiation of an appropriate therapy and could also guide the extent of resection during surgery. Nasal swab cultures have an acceptable sensitivity but are of limited value due to a low specificity in predicting invasive *Aspergillus* sinusitis (Viollier et al., 1986). Remarkably, cultures from nasal flushings in healthy volunteers frequently show fungal growth (>90%), including *Aspergillus* species (Buzina et al., 2003).

Detection and identification of fungi from fungus balls of the maxillary sinus by PCR amplification with universal fungal primers for 28S rDNA and amplicon identification by hybridization with species-specific probes as well as sequencing, including *Aspergillus* species such as *A. fumigatus, A. flavus, A. niger, A. terreus*, and *A. glaucus*, is a promising diagnostic approach. In one study, 112 samples were obtained from patients with histologically proven fungal infections. Eighty-one samples were paraffin-embedded tissue sections of the maxillary sinus and 31 samples were fresh biopsies. Fungal DNA was detected in all fresh samples but in only 71 (87.7%) paraffin-embedded tissue samples. Sequence analysis was the most sensitive technique, as positive results could be obtained for 28 (90.3%) fresh samples using this method compared to 24 (77.4%) samples using the hybridization technique and only 16 (51.6%) samples by culture (Willinger et al., 2003).

Therapy

When symptoms and signs of invasive *Aspergillus* sinusitis are present, surgery should be performed to obtain material for histopathological evaluation and to aggressively debride the devitalized tissue, which may support fungal growth. Thereafter, broad antifungal treatment should be initiated immediately, even before availability of histopathological studies confirming tissue invasion (deShazo et al., 1997). Complete resection, including gross resection, was associated with improved survival and should be attempted (Gillespie et al., 1998).

For antifungal therapy, amphotericin B has been the standard treatment in the past and may continue to be an alternative under resource-limited conditions (Stevens et al., 2000). Alternatively, itraconazole has been used successfully either alone or in combination therapy with amphotericin B in some studies (Panda et al., 2008; Rowe-Jones and Freedman, 1994; Streppel et al., 1999). However, the response to amphotericin B is limited, resulting in cure rates or rates of lasting remissions of around only 30% (Talbot et al., 1991; Viollier et al., 1986). Using liposomal amphotericin B as salvage therapy in seven patients with invasive sinonasal aspergillosis who failed conventional amphotericin B, Weber and Lopez-Berestein (1987) reported cure of this disease in five patients. Conversely, fatality rates of at least 50% have been repeatedly reported (Drakos et al., 1993; Iwen et al., 1997; Talbot et al., 1991). Newer agents, such as caspofungin, micafungin, or voriconazole, have been described to treat effectively either as monotherapy or as combination (salvage) therapy of acute invasive *Aspergillus* sinusitis in immunocompromised patients (Baumann et al., 2007; Myoken et al., 2006; Said et al., 2005; Tsiodras et al., 2004). These case reports demonstrate that effective salvage therapy with newer agents is feasible in individual cases even without surgical intervention. In general, because data from randomized, prospective trials are lacking for this rare indication, antifungal therapy for acute invasive *Aspergillus* sinusitis should be similar to treatment strategies for invasive pulmonary aspergillosis, which favor first-line therapy with voriconazole (Walsh et al., 2008).

Chronic Invasive *Aspergillus* Sinusitis

Patients with chronic invasive *Aspergillus* sinusitis suffer from concurrent conditions carrying a low level of immunosuppression (e.g., poorly controlled diabetes or chronic steroid therapy). Chronic invasive fungal sinusitis can be distinguished from the other two forms of invasive fungal sinusitis by its chronic course, dense accumulation of hyphae resembling a mycetoma, and association with the orbital apex syndrome, diabetes mellitus, and corticosteroid treatment (deShazo et al., 1997). The orbital apex syndrome is characterized by decreasing vision and ocular immobility resulting from an orbital mass. This condition may be misdiagnosed as an inflammatory pseudotumor, and corticosteroid therapy may be initiated before appropriate orbital exploration and biopsy are performed. The optimal treatment strategy is not well-defined, but because of its poor prognosis, chronic invasive *Aspergillus* sinusitis should be treated similar to acute invasive *Aspergilllus* sinusitis, with broad-spectrum antifungal agents.

Chronic Granulomatous Invasive *Aspergillus* Sinusitis

Chronic granulomatous sinusitis is a syndrome of slowly progressive sinusitis associated with proptosis that has also been termed indolent fungal sinusitis or primary paranasal granuloma. Florid granulomatous inflammation is the histological hallmark of this condition (deShazo et al., 1997; Miloshev et al., 1966). According to one definition, primary paranasal *Aspergillus* granuloma is a slowly progressive chronic infection of the sinuses extending beyond the confines of the sinus (Currens et al., 2002). It has been reported only in patients from the Sudan, India, and single cases from Saudi Arabia and the United States (Alrajhi et al., 2001; Gumaa et al., 1992; Milosev et al., 1969; Miloshev et al., 1966; Panda et al., 2004; Yagi et al., 1999). Microscopically, it differs from chronic invasive fungal sinusitis in that there are pseudotubercles containing giant cells, histiocytes, lymphocytes, plasma cells, newly formed capillaries, eosinophils, and *Aspergillus* fungal elements (Currens et al., 2002). Dawlatly et al. suggested that in view of the geographical similarities between northern Sudan and Saudi Arabia, some of the granulomatous inflammatory conditions occurring in Saudi Arabia, for which no definite etiologic agent has been ascribed, may fall into this category (Dawlatly et al., 1988).

The reports outside of Africa or the Indian subcontinent (e.g., the United States) suggest that primary paranasal *Aspergillus* granuloma affects almost exclusively African Americans (Currens et al., 2002). Whether this reflects climatic conditions and/or a genetic predisposition is unknown. Patients appear to be immunocompetent and are infected almost exclusively with *A. flavus* (Alrajhi et al., 2001; Gumaa et al., 1992; Hedayati et al., 2007; Yagi et al., 1999). Interestingly, bone erosion is a common finding, and tissue destruction occurs as a result of expansion of the mass rather than vascular invasion (Yagi et al., 1999). Most individuals present with a unilateral proptosis (Milosev et al., 1969). Marked regression generally occurs following surgical intervention designed to produce adequate aeration of the sinuses. However, the recurrence rate is high (about 80%) and some evidence suggests that the use of antifungal drugs may offer benefit, but the optimal treatment is unclear (Yagi et al., 1999).

CEREBRAL ASPERGILLOSIS

Introduction

Before the advent of modern medical mycology, central nervous system (CNS) aspergillosis was recognized as a rare medical condition. The first description of CNS aspergillosis was reported more than 100 years ago and only 33 well-documented cases were published up to 1969 (Mukoyama et al., 1969; Oppe, 1897). In recent years, CNS involvement has been reported in 14 to 42% of patients with invasive aspergillosis and acute leukemia or allogeneic hematopoietic stem cell transplantation (Jantunen et al., 2003; Pagano et al., 1996). *Aspergillus* species have been identified as the most frequent causative microorganism in patients with brain abscess and hematopoietic stem cell or solid organ transplantation (Baddley et al., 2002; Hagensee et al., 1994). In a large-scale autopsy data analysis from Japan, brain or meningeal involvement was found in only 3.3% of patients with invasive aspergillosis (Yamazaki et al., 1999). However, a 20% frequency of cerebral involvement in patients with invasive aspergillosis was reported from a multinational autopsy survey with a focus on fungal infections in cancer patients, suggesting that CNS infections are a frequent terminal event in this particular patient group (Bodey et al., 1992).

Until recently, patients with cerebral aspergillosis virtually always experienced a fatal outcome, and data from patients surviving this devastating infection were only available from anecdotal case reports. In patients with cerebral aspergillosis, symptoms tend to progress rapidly, leading to death within days if untreated (Walsh et al., 1985). Literature-based surveys, covering the era of antifungal therapy limited to amphotericin B and itraconazole, indicated that patients with cerebral aspergillosis carry a particularly poor prognosis with a mortality rate approaching 100% (Denning, 1996; Lin et al., 2001). Despite a growing awareness with more frequent use of preemptive antifungal therapy in the absence of a confirmed diagnosis, the mortality continued to be 100% in patients treated between 1993 and 1999 with amphotericin B or itraconazole for cerebral aspergillosis, which suggests that limited CNS penetration of these antifungal drugs contributes to their lack of efficacy (Schwartz et al., 2007b).

Clinical Presentation

Cerebral aspergillosis almost exclusively occurs in severely immunocompromised patients, including patients with hematopoietic malignancies or aplastic anemia, hematopoietic stem cell or solid organ transplantation recipients, patients with long-term drug-induced immunosuppression (e.g., corticosteroid therapy), or patients with inherited defects of phagocytes, such as chronic granulomatous disease (Alsultan et al., 2006; Baddley et al., 2002; Bodey and Glann, 1993; Pagano et al., 1996; Rodriguez et al., 1999). However, cerebral aspergillosis has also been observed in patients with an absent or a less severe degree of immunosuppression (e.g., diabetes or a short course of corticosteroids) (Apostolidis et al., 2001; Kim et al., 1993; Leroy et al., 2006; Mikolich et al., 1996).

In the majority of patients with cerebral aspergillosis, the CNS is invaded by hematogenous spread from primary sites of infection such as the lungs (Fig. 5 and Color Plate 14) (Boes et al., 1994). In addition, cerebral aspergillosis may also arise by continuous invasion from adjacent anatomical sites, such as the paranasal sinuses (Fig. 6) (Saah et al., 1994). Less frequently, cerebral as-

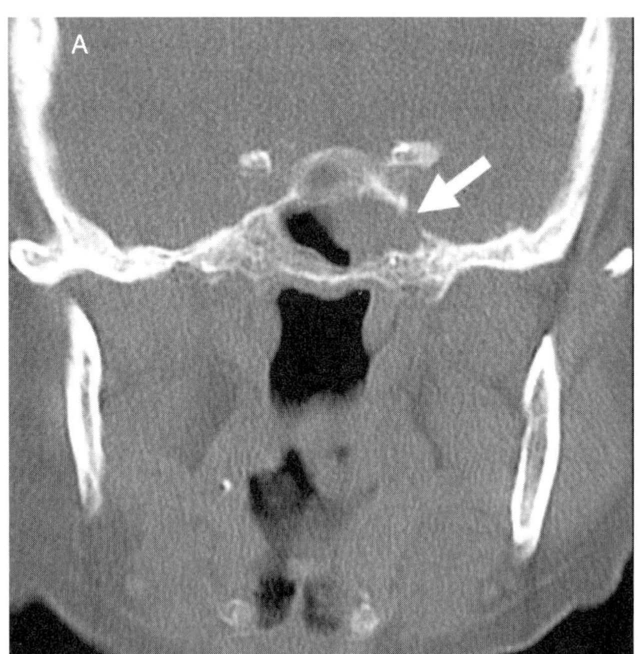

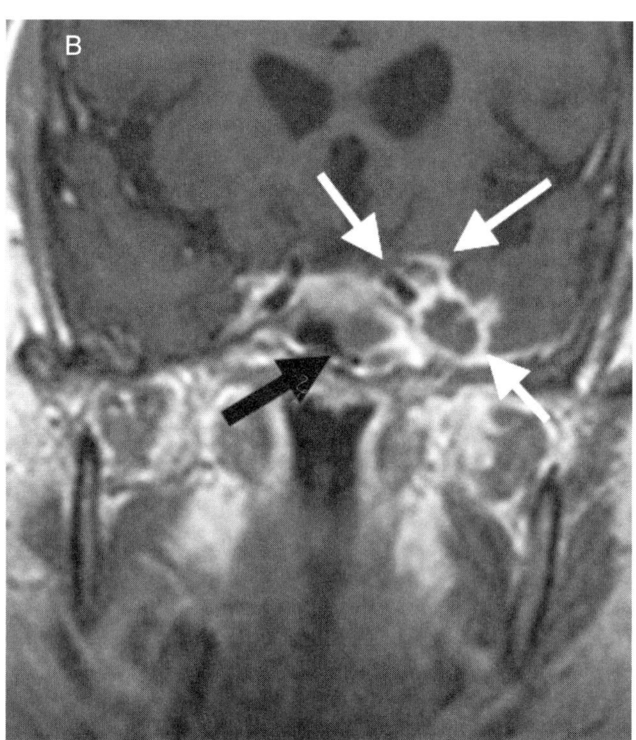

Figure 6. Acute invasive sinusitis caused by *A. niger*, with continuous intracranial invasion resulting in a brain abscess in a hematopoietic stem cell recipient. Radiographs were generously provided by C. P. Heussel, Heidelberg, Germany. (A) CT scan demonstrating bone destruction (arrow) of the lateral sphenoidal sinus. (B) Magnetic resonance imaging scan (T1-weighted, Gd-enhanced imaging) disclosing sphenoidal infiltration (black arrow) and adjacent temporal lobe abscess formation (white arrows).

pergillosis cases following neurosurgical procedures or vascular intervention, or in association with fungal endocarditis, have been reported (Darras-Joly et al., 1996; Endo et al., 2002; Kim et al., 1993; Scherer et al., 2005).

Brain abscess formation is the most frequent finding in patients with cerebral aspergillosis (Pagano et al., 1996). Septic embolism or vascular thrombosis due to angioinvasion of fungi may cause cerebral infarction, and a loss of vessel integrity could promote cerebral hemorrhage (Boes et al., 1994; Torre-Cisneros et al., 1993). Although angioinvasion is a common finding in invasive aspergillosis, true mycotic aneurysms due to *Aspergillus* species have been infrequently observed and may cause vascular thrombosis, septic embolism, or fatal hemorrhage (Ho and Deruytter, 2004; Hurst et al., 2001). Invasive aspergillosis causing granuloma formation has been reported in mostly nonimmunocompromised patients from the Sudan and Indian subcontinent or in patients with chronic granulomatous disease, and cerebral involvement has been repeatedly reported in these patients (Alsultan et al., 2006; Dubey et al., 2005; Hotchi et al., 1980; Milosev et al., 1969; Murthy et al., 2000). Overt meningitis due to *Aspergillus* species is uncommon and is difficult to diagnose unless careful diagnostic workup, including testing for *Aspergillus* DNA or galactomannan, is applied (Lammens et al., 1992; Verweij et al., 1999). Meningism or meningeal involvement may also be noted in patients with cerebral aspergillosis and brain abscess formation, which could be due to involvement of adjacent cerebral tissue (Boes et al., 1994; Schwartz et al., 1997). Thus, these peculiarities of cerebral aspergillosis do not represent distinct subentities of the disease, as they may occur simultaneously or at different time points during the course of the disease (Kleinschmidt-DeMasters, 2002; Kurino et al., 1994).

Clinical symptoms in cerebral aspergillosis are variable and nonspecific. Fever, refractory to broad-spectrum antibiotics, is noted in the majority of patients with cerebral aspergillosis but may be caused by other opportunistic CNS infections (Maschke et al., 1999). An altered mental status, focal neurological deficits, and seizures are the most common symptoms reported in patients with cerebral aspergillosis (Boes et al., 1994; Hagensee et al., 1994; Jantunen et al., 2003; Walsh et al., 1985). Due to the lack of specificity, any of these symptoms in patients at risk should prompt a rapid diagnostic workup allowing an early initiation of antifungal therapy.

Pharmacological Considerations

The CNS is a sanctuary site, and drug penetration into the brain parenchyma is limited by various barriers. The blood-brain barrier is mainly represented by endothelial cells of the brain capillaries, which are connected with tight junctions, thus limiting penetration of substances into brain parenchyma. The blood-cerebrospinal fluid (CSF) barrier comprises mainly the choroidal and arachnoidal epithelium, which is fenestrated and faces the CSF. Although substances may enter more easily into the CSF, further penetration into brain tissue is limited by the ependyma, a single layer of epithelial cells which covers the ventricles (de Boer and Gaillard, 2007). In general, substances may cross the blood-brain barrier by transcytosis using active transport systems or diffusion. Paracellular diffusion of hydrophilic substances across the blood-brain barrier is limited, but lipophilic compounds smaller than 400 to 600 Da may freely enter the brain tissue through transcellular diffusion. Various transport systems are functionally active at the blood-brain barrier and maintain brain homeostasis. The energy-dependent efflux pump P-glycoprotein is highly expressed in human cerebral cortex microvessels, targets a variety of hydrophilic compounds, and may therefore counteract uptake of lipophilic drugs into the CNS (Virgintino et al., 2002). Physicochemical properties, such as the molecular mass and lipophilicity, as well as various pharmacokinetic parameters, such as the blood drug levels, half-life, and degree of protein binding, have an impact on CNS penetration of antifungal drugs. Furthermore, the cerebral blood flow, a disruption of the blood-brain barrier, and targeting of transport systems at the blood-brain barrier may also have an impact on CNS levels of antifungal drugs.

Amphotericin B exhibits limited penetration into the CNS. Experimental data in a rabbit model with intravenous challenge with *Candida albicans* revealed mean amphotericin B concentrations of only 0.033 μg/ml in the CSF and mean brain tissue concentrations of amphotericin B of up to 1.99 μg/g after treatment with amphotericin B deoxycholate or various lipid preparations of amphotericin B (Table 2). Interestingly, brain

Table 2. CSF and brain tissue levels of selected antifungal drugs

Drug	Mass (Da)	CSF level (μg/ml) (data source)[a]	Brain tissue level (μg/g) (data source)[a]	Reference(s)
Micafungin	1,292	< 1 (A) 0.007–0.017 (H)	< 1–~4.5[b] (A)	Hope et al., 2008 Okugawa et al., 2007
Caspofungin	1,213	No published data	≤0.08,[c] ≤0.164[c] (A) ≤0.2 T/P[d] ratio (A)	Hajdu et al., 1997; Stone et al., 2004 Hajdu et al., 1997
Anidulafungin	1,140	No published data	≤3.9 (A)	Groll et al., 2001a
Amphotericin B	924	≤0.033[c] (A) ≤ 0.11 (H) 0.4–0.9[g] (N)	≤1.99[e] (A) ≤0.5[f] (H) ≤0.7[f] (H)	Groll et al., 2000 Korfel et al., 1998; Collette et al., 1989 Baley et al., 1990; Collette et al., 1991
Posaconazole	708	Undetectable (A)	No published data	Perfect et al., 1996
Itraconazole	705	≤0.156[c] (A) ≤0.07 (H)	<0.3 (A)	Perfect and Durack, 1985; Heykants et al., 1987 Verweij et al., 1999; Heykants et al., 1987
Voriconazole	349	~ 0.5, 0.68–1.23[g] (A) 0.04–3.93 (H) 0.22–1[g] (H)	≤6.8, ~2 T/P[d] ratio (A) 11.8, 58.5 (H) 1.2–1.4 μg/g, 5.1μg/ml[h] (H)	Jezequel et al., 1995; Lutsar et al., 2003 Schwartz et al., 1997; Verweij et al., 1999 Lutsar et al., 2003 Elter et al., 2006; Schwartz et al., 2007a

[a] Data sources: A, animal model; H, human data; N, neonates.
[b] Mean value; dose dependent (0.25 to 16 mg/kg).
[c] Mean value.
[d] T/P, tissue/plasma drug concentration ratio.
[e] Brain tissue drug concentrations were higher after liposomal amphotericin B.
[f] Median value.
[g] Percent of blood level.
[h] Brain abscess material.

tissue concentrations were significantly higher in animals treated with liposomal amphotericin B compared to those treated with amphotericin B deoxycholate or any other lipid amphotericin B preparation (Groll et al., 2000). In humans, concentrations of amphotericin B in the CSF usually do not exceed 4% of corresponding serum drug concentrations (Groll et al., 1998). However, strikingly high CSF concentrations of amphotericin B relative to the serum values of 40 to 90% have been measured in a single study in neonates with invasive candidiasis, but absolute drug concentrations in the CSF were not reported (Baley et al., 1990). Treatment with amphotericin B deoxycholate and subsequent switch to liposomal amphotericin B in a patient with cryptococcal meningitis resulted in amphotericin B CSF levels of only 0.068 and 0.11 μg/ml, respectively (Table 2) (Korfel et al., 1998). Brain tissue concentrations of amphotericin B have been evaluated in necropsy specimens from patients treated with amphotericin B deoxycholate or liposomal amphotericin B. Using a methanolic extraction method and high-performance liquid chromatography, brain tissue amphotericin B concentrations ranging from 0.2 to 5.8 μg/g (median, 0.5 μg/g) and from undetectable to 1.6 μg/g (median, 0.7 μg/g) were measured in patients treated with amphotericin B deoxycholate and liposomal amphotericin B, respectively (Table 2). It is noteworthy that concentrations of amphotericin B measured by a bioassay in tissue homogenates were only 15 to 41% of values determined by high-performance liquid chromatography after methanolic extraction, indicating that relevant amounts of amphotericin B in tissue are biologically inactive (Collette et al., 1989, 1991).

The echinocandins are large molecules with a molecular mass well above 1,000 Da and are highly protein bound, making sufficient penetration into brain tissue unlikely (Denning, 2003). However, animal model data demonstrate that fungicidal echinocandin concentrations may be attained in the CNS. Doses of micafungin ranging from 0.5 to 2 mg/kg resulted in detectable brain tissue drug concentrations of up to 0.18 μg/g in healthy rabbits (Groll et al., 2001b). In a subsequent study, rabbits, challenged or not with *C. albicans* and treated with various doses of micafungin, showed dose-dependent increases of drug concentrations in brain tissue. In animals treated with a daily micafungin dose of 16 mg/kg, brain tissue concentrations of micafungin slightly exceeded 4 μg/g, while CSF drug concentrations remained below 1 μg/ml (Hope et al., 2008). Interestingly, the infectious challenge had no significant impact on CNS drug concentrations. In a single patient, successfully treated with 300 mg micafungin/day for recurrent CNS aspergillosis, CSF drug concentrations were 0.007 and 0.017 μg/ml, which was less than 0.2% of the corresponding plasma drug levels (Table 2) (Okugawa et al., 2007). Brain tissue pharmacokinetics of caspofungin have been evaluated in rodents. Mean caspofungin brain tissue levels in a rat model were ≤0.164 μg/g after a single dose of 2 mg/kg. Interestingly, caspofungin remained detectable in brain tissue for a prolonged period of time with a mean concentration of 0.022 μg/g measured 288 h after drug administration (Stone et al., 2004). Lower mean brain tissue levels of caspofungin (≤0.08 μg/g) have been reported from a murine model, but drug levels were determined after a single dose of only 1 mg/kg of caspofungin (Hajdu et al., 1997). Data on CNS pharmacokinetics of anidulafungin have been generated from a rabbit model with intravenous *C. albicans* challenge. With daily doses of anidulafungin at ≥0.5 mg/kg given for 10 days, brain tissue concentrations of anidulafungin were detectable and increased in a dose-dependent manner. In animals receiving 10 mg/kg anidulafungin/day, the mean brain tissue concentration of anidulafungin was 3.9 μg/g (Table 2) (Groll et al., 2001a). Data on caspofungin or anidulafungin CNS levels in humans have not yet been published.

Itraconazole CSF levels did not exceed 0.156 μg/ml in a study with healthy rabbits or rabbits with bacterial meningitis (Perfect and Durack, 1985). In humans, including two patients with cerebral aspergillosis, CSF levels of only 0.07 μg/ml or lower have been detected (Heykants et al., 1987; Imai et al., 1999; Verweij et al., 1999). Brain tissue concentrations of itraconazole in rodents remained below 0.3 μg/g, with a brain tissue to plasma drug concentration ratio of less than 0.3 after a single oral dose of 10 mg/kg itraconazole (Heykants et al., 1987). Thus, the CNS penetration of itraconazole is remarkably low, despite its marked lipophilicity favoring penetration across the blood-brain barrier through transcellular diffusion. More recent studies in rats showed a rapid decline of itraconazole brain tissue concentrations compared to those in plasma and liver tissue (half-lives of 0.4 h compared to 5 h), indicating a nonlinear efflux of this azole from brain tissue to the blood (Miyama et al., 1998). Further studies of this phenomenon in knockout mice with deficient expression of P-glycoprotein (mdr1a$^{-/-}$) demonstrated higher brain tissue levels of itraconazole in these mice compared to wild-type animals. Pretreatment with verapamil, a known inhibitor of P-glycoprotein, resulted in a significant increase in the brain tissue drug concentration relative to the plasma concentration of itraconazole, indicating that P-glycoprotein contributes to a relevant efflux of itraconazole from brain tissue to the blood (Miyama et al., 1998).

Very limited data are available on CNS pharmacokinetics properties of posaconazole. In an animal study evaluating the effects of posaconazole in experimental cryptococcal meningitis, posaconazole exhibited effects comparable to fluconazole treatment, but posaconazole CSF levels remained below the lower limit of detection (<0.05 μg/ml) (Perfect et al., 1996). Up to now, concentrations of posaconazole in brain tissue or human CSF have not been investigated.

Voriconazole is a derivative of fluconazole with a broadened spectrum of antifungal activity, and this new azole retains some beneficial pharmacokinetic properties of fluconazole, including excellent CNS penetration. In an early study, steady-state concentrations of voriconazole in CSF and brain tissue of guinea pigs were approximately half and double that measured in plasma, respectively (Jezequel et al., 1995). In a subsequent animal study, concentrations of voriconazole in CSF and plasma were found to be equivalent in healthy guinea pigs with a CSF/plasma ratio ranging from 0.68 to 1.23. Peak brain tissue concentrations were 6.8 μg/g measured at 15 min after a dose of 10 mg/kg. Thus, peak CSF concentrations of voriconazole were rapidly achieved, suggesting that voriconazole readily crosses the blood-brain barrier by transcellular diffusion (Lutsar et al., 2003). In humans, CSF concentrations of voriconazole ranging from 0.04 to 3.93 μg/ml were found, corresponding to a CSF/plasma ratio ranging from 0.22 to 1 (Lutsar et al., 2003; Schwartz et al., 1997; Verweij et al., 1999). Interestingly, high voriconazole concentrations of 11.8 and 58.5 μg/g were measured in postmortem brain tissue specimens from two patients with pulmonary aspergillosis in the absence of a CNS infection (Lutsar et al., 2003). Furthermore, penetration of voriconazole into brain abscess material has also been reported in two patients with CNS fungal disease. Voriconazole concentrations in abscess material ranged from 1.2 to 1.4 μg/g in a patient successfully treated for rhinocerebral aspergillosis and were 5.1 μg/ml and 1.4 μg/g in liquid abscess material and adjacent dura mater, respectively, in another patient with *Candida* meningoencephalitis (Table 2) (Elter et al., 2006; Schwartz et al., 2007a). Thus, fungicidal concentrations of voriconazole may be attained in CSF, as well as in infected and uninfected areas of the CNS, making this azole an ideal substance for treating patients with cerebral aspergillosis.

Treatment Options

Until recently, responses to antifungal therapy and survival in patients with cerebral aspergillosis were observed very infrequently and the vast majority of patients with this disease experienced a fatal outcome. In a retrospective survey evaluating 595 patients with invasive aspergillosis, 34 patients had CNS involvement. Antifungal treatment consisted mainly of amphotericin B deoxycholate or lipid amphotericin B formulations or itraconazole, either given as single agents, sequentially, or in combination. The overall response rate was 37%, but only 3 of 34 patients with cerebral involvement experienced a complete or partial treatment response (Patterson et al., 2000). Likewise, a mortality of 99% was found among 141 patients with cerebral aspergillosis in a literature-based analysis covering studies with at least four patients with invasive aspergillosis published before 1996 (Denning, 1996). A more recent retrospective analysis confirmed the negligible effects of amphotericin B in patients with cerebral aspergillosis. In this series of 17 patients with proven or probable cerebral aspergillosis, 15 patients were treated with amphotericin B deoxycholate or lipid preparations of amphotericin B. The mortality was 100% with a median survival of only 10 days after onset of symptoms or first radiological evidence of cerebral aspergillosis, which underscores that amphotericin B has no role in the treatment of this type of CNS fungal disease (Schwartz et al., 2007b). Survival of cerebral aspergillosis has been reported occasionally in patients treated with high doses (10 to 15 mg/kg/day) of liposomal amphotericin B, suggesting that enhanced CNS levels of amphotericin B may have contributed to the successful outcome (Coleman et al., 1995; Ng et al., 2000). However, meaningful clinical data to support this assumption are lacking, and in a large, randomized trial evaluating patients with invasive filamentous fungal diseases (97% of patients with invasive aspergillosis), initial treatment with high doses (10 mg/kg/day) of liposomal amphotericin B compared to standard dose therapy (3 mg/kg/day) showed equivalent treatment efficacy (Cornely et al., 2007).

A response to treatment with echinocandins has been occasionally reported in patients with cerebral aspergillosis. A diabetic patient with cerebral aspergillosis responded to micafungin (300 mg/day) but experienced a relapse after changeover to low dose (100 mg/day) itraconazole. Micafungin therapy (300 mg/day) was resumed, and the patient again achieved a response (Okugawa et al., 2007). Among 83 evaluable patients with invasive aspergillosis treated in the first clinical study exploring the use of caspofungin in patients refractory or intolerant to previous therapies, 6 patients had proven or possible CNS involvement. Of these, 2 (33%) had a favorable response to standard doses (day 1, 70 mg; day +2, 50 mg) of caspofungin (Maertens et al., 2004).

Although CNS penetration by itraconazole is poor, successful treatment with this azole has been reported occasionally in patients with cerebral aspergillosis (Imai et al., 1999; Mikolich et al., 1996; Sanchez et al., 1995; Saulsbury, 2001). The use of higher doses of itraconazole (800 mg/day) has been advocated in a retrospective analysis of 12 patients. In this analysis, three of four patients with high-dose (800 mg/day) itraconazole survived, whereas seven of eight patients with low-dose (400 mg/day) itraconazole died (Sanchez et al., 1995).

The conclusion from this report has been questioned, as all surviving patients in this analysis experienced recovery from neutropenia, whereas all other patients died with ongoing neutropenia (Verweij et al., 1996).

In an open-label clinical trial, treatment responses to oral posaconazole therapy (800 mg/day in two to four divided doses) were assessed in a subset of 39 patients with various types of CNS fungal diseases. All but one patient had an underlying immunocompromising condition, including human immunodeficiency virus infection ($n = 29$), hematologic malignancies ($n = 4$), solid organ transplantation ($n = 3$), diabetes mellitus ($n = 1$), and hypogammaglobulinemia ($n = 1$). The majority of patients had cryptococcal meningitis ($n = 29$), and four patients suffered from cerebral aspergillosis. Successful outcomes at the end of treatment were observed in 49% (19/39) of patients, and a single patient with cerebral aspergillosis experienced a partial treatment response. However, this latter patient died after termination of posaconazole therapy due to progression of cerebral aspergillosis (Pitisuttithum et al., 2005).

The efficacy of voriconazole in invasive aspergillosis has been investigated in a large, randomized trial demonstrating that this azole produced superior response and survival rates compared to amphotericin B deoxycholate (Herbrecht et al., 2002). Successful responses to voriconazole treatment with long-term survival were repeatedly reported in single patients with cerebral aspergillosis (Alsultan et al., 2006; Schwartz et al., 1997; Verweij et al., 1999). In a recent retrospective analysis of 81 patients with proven or probable cerebral aspergillosis, which represents the largest series of patients with this disease published to date, the rate of complete and partial responses to voriconazole treatment was 35% (Schwartz et al., 2005). It is noteworthy that the majority of patients (96%, 78 of 81) had received prior treatment with various antifungal drugs other than voriconazole, and the reason for changeover to voriconazole was efficacy failure of prior antifungal therapy in 62 patients. Twenty-five (31%) of 81 patients survived the infection for a median observation time of 390 days (Fig. 7). Of these, 15 had a reported median survival time after termination of voriconazole therapy of 237 days, suggesting that some long-term survivors were potentially cured from cerebral aspergillosis (Schwartz et al., 2005). In this patient series, various degrees of immunosuppression were present, which likely had an impact on the outcome. In 32 patients with cerebral aspergillosis as a complication of hematopoietic stem cell transplantation, including 30 patients with allogeneic transplants, the survival rate was significantly lower compared to all other patients. However, the survival rate in this high-risk group of patients with hematopoietic stem cell transplantation was still 22%, with a median survival time for those surviving the infection of 203 days.

Aggressive and early neurosurgical intervention in patients with cerebral aspergillosis has been advocated in a report on four pediatric patients with acute leukemia and who were successfully treated for cerebral aspergillosis by combined medical and neurosurgical management (Middelhof et al., 2005). In these four patients, all cerebral lesions were resected by a stereotactic approach or image-guided craniotomy, except for a stereotactic biopsy of only one lesion and resection of another lesion in a single patient. Adjunct voriconazole therapy was given to all four patients. Three patients survived for at least 2 years without recurrence of their

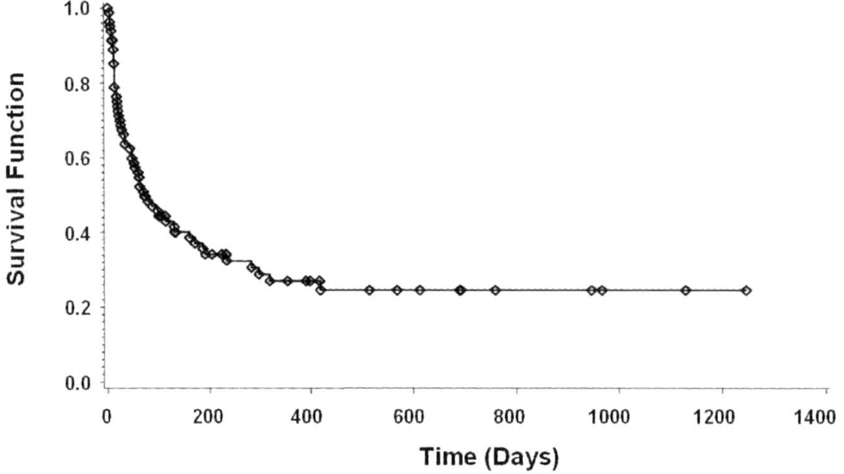

Figure 7. Survival curve for 81 patients with proven or probable cerebral aspergillosis treated with primary or salvage voriconazole therapy (Schwartz et al., 2005). This research was originally published in the journal *Blood* and is reprinted here with the permission of the American Society of Hematology.

cerebral infection, and the remaining patient died due to acute leukemia 4 months after successful combined medical and neurosurgical therapy of cerebral aspergillosis (Middelhof et al., 2005). In an earlier case report on a 2-year-old boy with acute leukemia and successful management of cerebral aspergillosis, another 25 previously published cases with successful outcomes were reviewed (Coleman et al., 1995). The medical therapy consisted mainly of amphotericin B or itraconazole. Remarkably, 19 of these 26 surviving patients underwent neurosurgical interventions, including multiple abscess resections, craniotomies, stereotactic drainage, and placement of intracavitary catheters. Thirty-one of 81 patients from the retrospective study on voriconazole in cerebral aspergillosis underwent neurosurgical management (Schwartz et al., 2005). A variety of interventions were performed, including craniotomy or abscess resection (n = 14), abscess drainage (n = 12), ventricular shunt (n = 4), and placement of an Ommaya reservoir (n = 1). In this recent patient series, the use of neurosurgery was significantly associated with an improved survival (Schwartz et al., 2005). However, the optimal neurosurgical approach in cerebral aspergillosis remains unclear. An abscess aspiration or resection through a stereotactic or neuronavigated approach may be regarded as the least invasive neurosurgical procedures and may allow removal of avital debris and reduction of the mass effects of a fungal brain abscess. Any invasive procedure carries a risk of severe hemorrhage, especially in patients with low platelet counts or otherwise-impaired coagulation, and the potential benefits of neurosurgery should carefully be weighed against the risks of invasive procedures in individual patients.

FUTURE PROSPECTS

Given the promising data on voriconazole treatment and neurosurgical management in patients with cerebral aspergillosis, overall treatment results for this threatening infection are still far from satisfying. Future studies should explore a more refined treatment approach, which may include higher drug doses, combinations of various antifungal agents, and sophisticated neurosurgical management guided by the individual pattern of cerebral involvement.

To investigate new treatment strategies in cerebral aspergillosis, a murine model was developed with cyclophosphamide-induced immunosuppression and intracerebral inoculation of *Aspergillus* conidia, producing a 100% mortality in untreated animals (Chiller et al., 2002). In this animal model, treatment with amphotericin B deoxycholate, itraconazole, or posaconazole resulted in a prolongation of survival (Chiller et al., 2003; Imai et al., 2004). Various combination therapies have been tested in this model. None of the tested treatment regimens have produced a cure in this immunocompromised host model, but a significant prolongation of animal survival over monotherapy was observed with a combination of amphotericin B lipid complex with voriconazole and liposomal amphotericin B with voriconazole (Clemons et al., 2005, 2006).

Preliminary clinical data on the effects of combination therapies in invasive aspergillosis are available from retrospective patient series, analysis of sequential patient cohorts, or prospectively collected patient series, from an open-label study and from a small randomized trial. These studies evaluated combination therapies of liposomal amphotericin B with caspofungin, voriconazole with caspofungin, and caspofungin in combination with any other mold-active agent (Caillot et al., 2007; Singh and Pursell, 2008). In particular, in sequential patient cohorts treated with voriconazole alone or in combination with caspofungin as salvage therapy after failure with amphotericin B deoxycholate treatment, a significant survival benefit was observed for the combination therapy cohort (Marr et al., 2004). In a recently published randomized trial, 30 patients were allocated to receive either liposomal amphotericin B at a daily dose of 10 mg/kg or a combination of standard doses (3 mg/kg/day) of liposomal amphotericin B with caspofungin (day 1, 70 mg; day +2, 50 mg). At the end of treatment complete or partial responses were observed more frequently in the combination group than in the high-dose single-agent group (67% versus 27%; P = 0.028) (Caillot et al., 2007). However, meaningful clinical data supporting the use of antifungal combination therapies in patients with cerebral aspergillosis are currently lacking, and this issue should be addressed in future studies.

REFERENCES

Alrajhi, A. A., M. Enani, Z. Mahasin, and K. Al Omran. 2001. Chronic invasive aspergillosis of the paranasal sinuses in immunocompetent hosts from Saudi Arabia. *Am. J. Trop. Med. Hyg.* **65:**83–86.

Alsultan, A., M. S. Williams, S. Lubner, and F. D. Goldman. 2006. Chronic granulomatous disease presenting with disseminated intracranial aspergillosis. *Pediatr. Blood Cancer* **47:**107–110.

Apostolidis, J., M. Tsandekidi, D. Kousiafes, M. Pagoni, C. Mitsouli, T. Karmiris, M. Bakiri, D. Karakasis, N. Harhalakis, and E. Nikiforakis. 2001. Short-course corticosteroid-induced pulmonary and apparent cerebral aspergillosis in a patient with idiopathic thrombocytopenic purpura. *Blood* **98:**2875–2877.

Aribandi, M., V. A. McCoy, and C. Bazan III. 2007. Imaging features of invasive and noninvasive fungal sinusitis: a review. *Radiographics* **27:**1283–1296.

Baddley, J. W., D. Salzman, and P. G. Pappas. 2002. Fungal brain abscess in transplant recipients: epidemiologic, microbiologic, and clinical features. *Clin. Transplant.* **16:**419–424.

Baley, J. E., C. Meyers, R. M. Kliegman, M. R. Jacobs, and J. L. Blumer. 1990. Pharmacokinetics, outcome of treatment, and toxic effects of amphotericin B and 5-fluorocytosine in neonates. *J. Pediatr.* 116:791–797.

Batra, P. S., M. J. Citardi, and D. C. Lanza. 2005. Isolated sphenoid sinusitis after transsphenoidal hypophysectomy. *Am. J. Rhinol.* **19:** 185–189.

Baumann, A., S. Zimmerli, R. Hausler, and M. Caversaccio. 2007. Invasive sphenoidal aspergillosis: successful treatment with sphenoidotomy and voriconazole. *J. Otorhinolaryngol. Relat. Spec.* **69:** 121–126.

Beck-Mannagetta, J., D. Necek, and M. Grasserbauer. 1983. Solitary aspergillosis of maxillary sinus, a complication of dental treatment. *Lancet* ii:1260.

Bodey, G., B. Bueltmann, W. Duguid, D. Gibbs, H. Hanak, M. Hotchi, G. Mall, P. Martino, F. Meunier, S. Milliken, et al. 1992. Fungal infections in cancer patients: an international autopsy survey. *Eur. J. Clin. Microbiol. Infect. Dis.* **11:**99–109.

Bodey, G. P., and A. S. Glann. 1993. Central nervous system aspergillosis following steroidal therapy for allergic bronchopulmonary aspergillosis. *Chest* **103:**299–301.

Boes, B., R. Bashir, C. Boes, F. Hahn, J. R. McConnell, and R. McComb. 1994. Central nervous system aspergillosis. Analysis of 26 patients. *J. Neuroimag.* **4:**123–129.

Buzina, W., H. Braun, K. Freudenschuss, A. Lackner, W. Habermann, and H. Stammberger. 2003. Fungal biodiversity—as found in nasal mucus. *Med. Mycol.* **41:**149–161.

Caillot, D., A. Thiebaut, R. Herbrecht, S. de Botton, A. Pigneux, F. Bernard, J. Larche, F. Monchecourt, S. Alfandari, and L. Mahi. 2007. Liposomal amphotericin B in combination with caspofungin for invasive aspergillosis in patients with hematologic malignancies: a randomized pilot study (Combistrat trial). *Cancer* **110:**2740–2746.

Carter, K. D., S. M. Graham, and K. M. Carpenter. 1999. Ophthalmic manifestations of allergic fungal sinusitis. *Am. J. Ophthalmol.* **127:** 189–195.

Chiller, T. M., J. C. Luque, R. A. Sobel, K. Farrokhshad, K. V. Clemons, and D. A. Stevens. 2002. Development of a murine model of cerebral aspergillosis. *J. Infect. Dis.* **186:**574–577.

Chiller, T. M., R. A. Sobel, J. C. Luque, K. V. Clemons, and D. A. Stevens. 2003. Efficacy of amphotericin B or itraconazole in a murine model of central nervous system *Aspergillus* infection. *Antimicrob. Agents Chemother.* **47:**813–815.

Clemons, K. V., M. Espiritu, R. Parmar, and D. A. Stevens. 2005. Comparative efficacies of conventional amphotericin B, liposomal amphotericin B (AmBisome), caspofungin, micafungin, and voriconazole alone and in combination against experimental murine central nervous system aspergillosis. *Antimicrob. Agents Chemother.* **49:** 4867–4875.

Clemons, K. V., R. Parmar, M. Martinez, and D. A. Stevens. 2006. Efficacy of Abelcet alone, or in combination therapy, against experimental central nervous system aspergillosis. *J. Antimicrob. Chemother.* **58:**466–469.

Coleman, J. M., G. G. Hogg, J. V. Rosenfeld, and K. D. Waters. 1995. Invasive central nervous system aspergillosis: cure with liposomal amphotericin B, itraconazole, and radical surgery. Case report and review of the literature. *Neurosurgery* **36:**858–863.

Collette, N., P. van der Auwera, A. P. Lopez, C. Heymans, and F. Meunier. 1989. Tissue concentrations and bioactivity of amphotericin B in cancer patients treated with amphotericin B-deoxycholate. *Antimicrob. Agents Chemother.* **33:**362–368.

Collette, N., P. van der Auwera, F. Meunier, C. Lambert, J. P. Sculier, and A. Coune. 1991. Tissue distribution and bioactivity of amphotericin B administered in liposomes to cancer patients. *J. Antimicrob. Chemother.* **27:**535–548.

Cornely, O. A., J. Maertens, M. Bresnik, R. Ebrahimi, A. J. Ullmann, E. Bouza, C. P. Heussel, O. Lortholary, C. Rieger, A. Boehme, M. Aoun, H. A. Horst, A. Thiebaut, M. Ruhnke, D. Reichert, N. Vianelli, S. W. Krause, E. Olavarria, and R. Herbrecht. 2007. Liposomal amphotericin B as initial therapy for invasive mold infection: a randomized trial comparing a high-loading dose regimen with standard dosing (AmBiLoad trial). *Clin. Infect. Dis.* **44:**1289–1297.

Currens, J., P. S. Hutcheson, R. G. Slavin, and M. J. Citardi. 2002. Primary paranasal *Aspergillus* granuloma: case report and review of the literature. *Am. J. Rhinol.* **16:**165–168.

Darras-Joly, C., B. Veber, J. P. Bedos, B. Gachot, B. Regnier, and M. Wolff. 1996. Nosocomial cerebral aspergillosis: a report of 3 cases. *Scand. J. Infect. Dis.* **28:**317–319.

Dawlatly, E. E., J. T. Anim, S. Sowayan, and A. Y. el Hassan. 1988. Primary paranasal *Aspergillus* granuloma in Saudi Arabia. *Trop. Geogr. Med.* **40:**247–250.

de Boer, A. G., and P. J. Gaillard. 2007. Drug targeting to the brain. *Annu. Rev. Pharmacol. Toxicol.* **47:**323–355.

DelGaudio, J. M., R. E. Swain, Jr., T. T. Kingdom, S. Muller, and P. A. Hudgins. 2003. Computed tomographic findings in patients with invasive fungal sinusitis. *Arch. Otolaryngol. Head Neck Surg.* **129:**236–240.

Denning, D. W. 1996. Therapeutic outcome in invasive aspergillosis. *Clin. Infect. Dis.* **23:**608–615.

Denning, D. W. 2003. Echinocandin antifungal drugs. *Lancet* **362:** 1142–1151.

deShazo, R. D., K. Chapin, and R. E. Swain. 1997. Fungal sinusitis. *N. Engl. J. Med.* **337:**254–259.

deShazo, R. D., and R. E. Swain. 1995. Diagnostic criteria for allergic fungal sinusitis. *J. Allergy Clin. Immunol.* **96:**24–35.

Drakos, P. E., A. Nagler, R. Or, E. Naparstek, J. Kapelushnik, D. Engelhard, G. Rahav, D. Ne'emean, and S. Slavin. 1993. Invasive fungal sinusitis in patients undergoing bone marrow transplantation. *Bone Marrow Transplant.* **12:**203–208.

Driemel, O., C. Wagner, S. Hurrass, U. Muller-Richter, T. Kuhnel, T. E. Reichert, and H. Kosmehl. 2007. Allergic fungal sinusitis, fungus ball and invasive sinonasal mycosis: three fungal-related diseases. *Mund. Kiefer Gesichtschir.* **11:**153–159. (In German.)

Dubey, A., R. V. Patwardhan, S. Sampth, V. Santosh, S. Kolluri, and A. Nanda. 2005. Intracranial fungal granuloma: analysis of 40 patients and review of the literature. *Surg. Neurol.* **63:**254–260.

Dufour, X., C. Kauffmann-Lacroix, J. M. Goujon, G. Grollier, M. H. Rodier, and J. M. Klossek. 2005. Experimental model of fungal sinusitis: a pilot study in rabbits. *Ann. Otol. Rhinol. Laryngol.* **114:** 167–172.

Elter, T., M. Sieniawski, A. Gossmann, C. Wickenhauser, U. Schroder, H. Seifert, J. Kuchta, J. Burhenne, K. D. Riedel, G. Fatkenheuer, and O. A. Cornely. 2006. Voriconazole brain tissue levels in rhinocerebral aspergillosis in a successfully treated young woman. *Int. J. Antimicrob. Agents* **28:**262–265.

Endo, T., T. Tominaga, H. Konno, and T. Yoshimoto. 2002. Fatal subarachnoid hemorrhage, with brainstem and cerebellar infarction, caused by *Aspergillus* infection after cerebral aneurysm surgery: case report. *Neurosurgery* **50:**1147–1150.

Friedman, A., P. S. Batra, S. Fakhri, M. J. Citardi, and D. C. Lanza. 2005. Isolated sphenoid sinus disease: etiology and management. *Otolaryngol. Head Neck Surg.* **133:**544–550.

Ghadiali, M. T., N. A. Deckard, U. Farooq, F. Astor, P. Robinson, and R. R. Casiano. 2007. Frozen-section biopsy analysis for acute invasive fungal rhinosinusitis. *Otolaryngol. Head Neck Surg.* **136:** 714–719.

Ghegan, M. D., F. S. Lee, and R. J. Schlosser. 2006. Incidence of skull base and orbital erosion in allergic fungal rhinosinusitis (AFRS) and non-AFRS. *Otolaryngol. Head Neck Surg.* **134:**592–595.

Giardino, L., F. Pontieri, E. Savoldi, and F. Tallarigo. 2006. *Aspergillus* mycetoma of the maxillary sinus secondary to overfilling of a root canal. *J. Endod.* **32:**692–694.

Gillespie, M. B., D. M. Huchton, and B. W. O'Malley. 2000. Role of middle turbinate biopsy in the diagnosis of fulminant invasive fungal rhinosinusitis. *Laryngoscope* **110:**1832–1836.

Gillespie, M. B., B. W. O'Malley, Jr., and H. W. Francis. 1998. An approach to fulminant invasive fungal rhinosinusitis in the immunocompromised host. *Arch. Otolaryngol. Head Neck Surg.* **124:**520–526.

Groll, A. H., N. Giri, V. Petraitis, R. Petraitiene, M. Candelario, J. S. Bacher, S. C. Piscitelli, and T. J. Walsh. 2000. Comparative efficacy and distribution of lipid formulations of amphotericin B in experimental *Candida albicans* infection of the central nervous system. *J. Infect. Dis.* **182:**274–282.

Groll, A. H., D. Mickiene, R. Petraitiene, V. Petraitis, C. A. Lyman, J. S. Bacher, S. C. Piscitelli, and T. J. Walsh. 2001a. Pharmacokinetic and pharmacodynamic modeling of anidulafungin (LY303366): reappraisal of its efficacy in neutropenic animal models of opportunistic mycoses using optimal plasma sampling. *Antimicrob. Agents Chemother.* **45:**2845–2855.

Groll, A. H., D. Mickiene, V. Petraitis, R. Petraitiene, K. H. Ibrahim, S. C. Piscitelli, I. Bekersky, and T. J. Walsh. 2001b. Compartmental pharmacokinetics and tissue distribution of the antifungal echinocandin lipopeptide micafungin (FK463) in rabbits. *Antimicrob. Agents Chemother.* **45:**3322–3327.

Groll, A. H., S. C. Piscitelli, and T. J. Walsh. 1998. Clinical pharmacology of systemic antifungal agents: a comprehensive review of agents in clinical use, current investigational compounds, and putative targets for antifungal drug development. *Adv. Pharmacol.* **44:** 343–500.

Gumaa, S. A., E. S. Mahgoub, and R. J. Hay. 1992. Post-operative responses of paranasal *Aspergillus* granuloma to itraconazole. *Trans. R. Soc. Trop. Med. Hyg.* **86:**93–94.

Hagensee, M. E., J. E. Bauwens, B. Kjos, and R. A. Bowden. 1994. Brain abscess following marrow transplantation: experience at the Fred Hutchinson Cancer Research Center, 1984–1992. *Clin. Infect. Dis.* **19:**402–408.

Hajdu, R., R. Thompson, J. G. Sundelof, B. A. Pelak, F. A. Bouffard, J. F. Dropinski, and H. Kropp. 1997. Preliminary animal pharmacokinetics of the parenteral antifungal agent MK-0991 (L-743,872). *Antimicrob. Agents Chemother.* **41:**2339–2344.

Hedayati, M. T., A. C. Pasqualotto, P. A. Warn, P. Bowyer, and D. W. Denning. 2007. *Aspergillus flavus*: human pathogen, allergen and mycotoxin producer. *Microbiology* **153:**1677–1692.

Herbrecht, R., D. W. Denning, T. F. Patterson, J. E. Bennett, R. E. Greene, J. W. Oestmann, W. V. Kern, K. A. Marr, P. Ribaud, O. Lortholary, R. Sylvester, R. H. Rubin, J. R. Wingard, P. Stark, C. Durand, D. Caillot, E. Thiel, P. H. Chandrasekar, M. R. Hodges, H. T. Schlamm, P. F. Troke, and B. de Pauw. 2002. Voriconazole versus amphotericin B for primary therapy of invasive aspergillosis. *N. Engl. J. Med.* **347:**408–415.

Heykants, J., M. Michiels, W. Meuldermans, J. Monbaliu, K. Lavrijsen, A. van Peer, J. C. Levron, R. Woestenborghs, and G. Cauwenbergh. 1987. The pharmacokinetics of itraconazole in animals and man: an overview, p. 223–249. *In* R. A. Fromtling (ed.), *Recent Trends in the Discovery, Development and Evaluation of Antifungal Agents.* J.R. Prous Science, Barcelona, Spain.

Ho, C. L., and M. J. Deruytter. 2004. CNS aspergillosis with mycotic aneurysm, cerebral granuloma and infarction. *Acta Neurochir.* **146:** 851–856.

Hope, W. W., D. Mickiene, V. Petraitis, R. Petraitiene, A. M. Kelaher, J. E. Hughes, M. P. Cotton, J. Bacher, J. J. Keirns, D. Buell, G. Heresi, D. K. Benjamin, Jr., A. H. Groll, G. L. Drusano, and T. J. Walsh. 2008. The pharmacokinetics and pharmacodynamics of micafungin in experimental hematogenous *Candida* meningoencephalitis: implications for echinocandin therapy in neonates. *J. Infect. Dis.* **197:**163–171.

Hotchi, M., M. Fujiwara, S. Hata, and T. Nasu. 1980. Chronic granulomatous disease associated with peculiar *Aspergillus* lesions. Patho-anatomical report based on two autopsy cases and a brief review of all autopsy cases reported in Japan. *Virchows Arch. A* **387:** 1–15.

Hurst, R. W., A. Judkins, W. Bolger, A. Chu, and L. A. Loevner. 2001. Mycotic aneurysm and cerebral infarction resulting from fungal sinusitis: imaging and pathologic correlation. *Am. J. Neuroradiol.* **22:**858–863.

Imai, J. K., G. Singh, K. V. Clemons, and D. A. Stevens. 2004. Efficacy of posaconazole in a murine model of central nervous system aspergillosis. *Antimicrob. Agents Chemother.* **48:**4063–4066.

Imai, T., T. Yamamoto, S. Tanaka, M. Kashiwagi, S. Chiba, H. Matsumoto, and T. Uede. 1999. Successful treatment of cerebral aspergillosis with a high oral dose of itraconazole after excisional surgery. *Intern. Med.* **38:**829–832.

Iwen, P. C., M. E. Rupp, and S. H. Hinrichs. 1997. Invasive mold sinusitis: 17 cases in immunocompromised patients and review of the literature. *Clin. Infect. Dis.* **24:**1178–1184.

Jantunen, E., L. Volin, O. Salonen, A. Piilonen, T. Parkkali, V. J. Anttila, A. Paetau, and T. Ruutu. 2003. Central nervous system aspergillosis in allogeneic stem cell transplant recipients. *Bone Marrow Transplant.* **31:**191–196.

Jezequel, S. G., M. Clark, K. Evans, S. Roffey, P. Wastall, and R. Webster. 1995. UK-109,496, a novel, wide-spectrum triazole derivative for the treatment of fungal infections: disposition in animals, abstr. 126. *35th Intersci. Conf. Antimicrob. Agents Chemother.* American Society for Microbiology, Washington, DC.

Karci, B., D. Burhanoglu, T. Erdem, S. Hilmioglu, R. Inci, and A. Veral. 2001. Fungal infections of the paranasal sinuses. *Rev. Laryngol. Otol. Rhinol.* **122:**31–35.

Katzenstein, A. L., S. R. Sale, and P. A. Greenberger. 1983. Allergic *Aspergillus* sinusitis: a newly recognized form of sinusitis. *J. Allergy Clin. Immunol.* **72:**89–93.

Kennedy, C. A., G. L. Adams, J. P. Neglia, and G. S. Giebink. 1997. Impact of surgical treatment on paranasal fungal infections in bone marrow transplant patients. *Otolaryngol. Head Neck Surg.* **116:** 610–616.

Kim, D. G., S. C. Hong, H. J. Kim, J. G. Chi, M. H. Han, K. S. Choi, and D. H. Han. 1993. Cerebral aspergillosis in immunologically competent patients. *Surg. Neurol.* **40:**326–331.

Kleinschmidt-DeMasters, B. K. 2002. Central nervous system aspergillosis: a 20-year retrospective series. *Hum. Pathol.* **33:**116–124.

Klich, M. A. 2002. *Identification of Common* Aspergillus *Species*, 1st ed. Centraalbureau voor Schimmelcultures, Utrecht, The Netherlands.

Klossek, J. M., E. Serrano, L. Peloquin, J. Percodani, J. P. Fontanel, and J. J. Pessey. 1997. Functional endoscopic sinus surgery and 109 mycetomas of paranasal sinuses. *Laryngoscope* **107:**112–117.

Korfel, A., H. D. Menssen, S. Schwartz, and E. Thiel. 1998. Cryptococcosis in Hodgkin's disease: description of two cases and review of the literature. *Ann. Hematol.* **76:**283–286.

Kurino, M., J. Kuratsu, T. Yamaguchi, and Y. Ushio. 1994. Mycotic aneurysm accompanied by aspergillotic granuloma: a case report. *Surg. Neurol.* **42:**160–164.

Lammens, M., W. Robberecht, M. Waer, H. Carton, and R. Dom. 1992. Purulent meningitis due to aspergillosis in a patient with systemic lupus erythematosus. *Clin. Neurol. Neurosurg.* **94:**39–43.

Leroy, P., A. Smismans, and T. Seute. 2006. Invasive pulmonary and central nervous system aspergillosis after near-drowning of a child: case report and review of the literature. *Pediatrics* **118:**e509–e513.

Lin, S. J., J. Schranz, and S. M. Teutsch. 2001. Aspergillosis case-fatality rate: systematic review of the literature. *Clin. Infect. Dis.* **32:**358–366.

Liu, J. K., S. D. Schaefer, A. L. Moscatello, and W. T. Couldwell. 2004. Neurosurgical implications of allergic fungal sinusitis. *J. Neurosurg.* **100:**883–890.

Lutsar, I., S. Roffey, and P. Troke. 2003. Voriconazole concentrations in the cerebrospinal fluid and brain tissue of guinea pigs and immunocompromised patients. *Clin. Infect. Dis.* **37:**728–732.

Maertens, J., I. Raad, G. Petrikkos, M. Boogaerts, D. Selleslag, F. B. Petersen, C. A. Sable, N. A. Kartsonis, A. Ngai, A. Taylor, T. F. Patterson, D. W. Denning, and T. J. Walsh. 2004. Efficacy and safety of caspofungin for treatment of invasive aspergillosis in patients refractory to or intolerant of conventional antifungal therapy. *Clin. Infect. Dis.* **39:**1563–1571.

Marr, K. A., M. Boeckh, R. A. Carter, H. W. Kim, and L. Corey. 2004. Combination antifungal therapy for invasive aspergillosis. *Clin. Infect. Dis.* **39:**797–802.

Maschke, M., U. Dietrich, M. Prumbaum, O. Kastrup, B. Turowski, U. W. Schaefer, and H. C. Diener. 1999. Opportunistic CNS infection after bone marrow transplantation. *Bone Marrow Transplant.* **23:**1167–1176.

McGill, T. J., G. Simpson, and G. B. Healy. 1980. Fulminant aspergillosis of the nose and paranasal sinuses: a new clinical entity. *Laryngoscope* **90:**748–754.

Mensi, M., S. Salgarello, G. Pinsi, and M. Piccioni. 2004. Mycetoma of the maxillary sinus: endodontic and microbiological correlations. *Oral Surg. Oral Med. Oral Pathol. Oral Radiol. Endod.* **98:**119–123.

Middelhof, C. A., W. G. Loudon, M. D. Muhonen, C. Xavier, and C. S. Greene, Jr. 2005. Improved survival in central nervous system aspergillosis: a series of immunocompromised children with leukemia undergoing stereotactic resection of aspergillomas. Report of four cases. *J. Neurosurg.* **103:**374–378.

Mikolich, D. J., L. J. Kinsella, G. Skowron, J. Friedman, and A. M. Sugar. 1996. *Aspergillus* meningitis in an immunocompetent adult successfully treated with itraconazole. *Clin. Infect. Dis.* **23:**1318–1319.

Milosev, B., S. el Mahgoub, O. A. Aal, and A. M. el Hassan. 1969. Primary aspergilloma of paranasal sinuses in the Sudan. A review of seventeen cases. *Br. J. Surg.* **56:**132–137.

Miloshev, B., C. M. Davidson, J. C. Gentles, and A. T. Sandison. 1966. Aspergilloma of paranasal sinuses and orbit in northern Sudanese. *Lancet* **i:**746–747.

Miyama, T., H. Takanaga, H. Matsuo, K. Yamano, K. Yamamoto, T. Iga, M. Naito, T. Tsuruo, H. Ishizuka, Y. Kawahara, and Y. Sawada. 1998. P-glycoprotein-mediated transport of itraconazole across the blood-brain barrier. *Antimicrob. Agents Chemother.* **42:**1738–1744.

Mukoyama, M., K. Gimple, and C. M. Poser. 1969. Aspergillosis of the central nervous system. Report of a brain abscess due to *A. fumigatus* and review of the literature. *Neurology* **19:**967–974.

Murthy, J. M., C. Sundaram, V. S. Prasad, A. K. Purohit, S. Rammurti, and V. Laxmi. 2000. Aspergillosis of central nervous system: a study of 21 patients seen in a university hospital in south India. *J. Assoc. Physicians India* **48:**677–681.

Mylonakis, E., J. Rich, P. R. Skolnik, D. F. De Orchis, and T. Flanigan. 1997. Invasive *Aspergillus* sinusitis in patients with human immunodeficiency virus infection. Report of 2 cases and review. *Medicine* (Baltimore) **76:**249–255.

Myoken, Y., T. Sugata, Y. Fujita, M. Fujihara, K. Iwato, S. Y. Murayama, and Y. Mikami. 2006. Early diagnosis and successful management of atypical invasive *Aspergillus* sinusitis in a hematopoietic cell transplant patient: a case report. *J. Oral Maxillofac. Surg.* **64:**860–863.

Ng, A., N. Gadong, A. Kelsey, D. W. Denning, J. Leggate, and O. B. Eden. 2000. Successful treatment of *Aspergillus* brain abscess in a child with acute lymphoblastic leukemia. *Pediatr. Hematol. Oncol.* **17:**497–504.

Nussenbaum, B., B. F. Marple, and N. D. Schwade. 2001. Characteristics of bony erosion in allergic fungal rhinosinusitis. *Otolaryngol. Head Neck Surg.* **124:**150–154.

Odell, E., and C. Pertl. 1995a. Chronic sinusitis. *Br. Dent. J.* **179:**327.

Odell, E., and C. Pertl. 1995b. Zinc as a growth factor for *Aspergillus* sp. and the antifungal effects of root canal sealants. *Oral Surg. Oral Med. Oral Pathol. Oral Radiol. Endod.* **79:**82–87.

Okugawa, S., Y. Ota, K. Tatsuno, K. Tsukada, S. Kishino, and K. Koike. 2007. A case of invasive central nervous system aspergillosis treated with micafungin with monitoring of micafungin concentrations in the cerebrospinal fluid. *Scand. J. Infect. Dis.* **39:**344–346.

Oppe. 1897. Zur Kenntnis der Schimmelmykosen beim Menschen. *Zentralbl. Allg. Pathol. Anat.* **8:**301–306.

Pagano, L., P. Ricci, M. Montillo, A. Cenacchi, A. Nosari, A. Tonso, L. Cudillo, A. Chierichini, C. Savignano, M. Buelli, L. Melillo, E. O. La Barbera, S. Sica, S. Hohaus, A. Bonini, G. Bucaneve, A. Del Favero, et al. 1996. Localization of aspergillosis to the central nervous system among patients with acute leukemia: report of 14 cases. *Clin. Infect. Dis.* **23:**628–630.

Pagella, F., E. Matti, F. De Bernardi, L. Semino, C. Cavanna, P. Marone, C. Farina, and P. Castelnuovo. 2007. Paranasal sinus fungus ball: diagnosis and management. *Mycoses* **50:**451–456.

Panda, N. K., P. Balaji, A. Chakrabarti, S. C. Sharma, and C. E. Reddy. 2004. Paranasal sinus aspergillosis: its categorization to develop a treatment protocol. *Mycoses* **47:**277–283.

Panda, N. K., K. Saravanan, and A. Chakrabarti. 2008. Combination antifungal therapy for invasive aspergillosis: can it replace high-risk surgery at the skull base? *Am. J. Otolaryngol.* **29:**24–30.

Parikh, S. L., G. Venkatraman, and J. M. DelGaudio. 2004. Invasive fungal sinusitis: a 15-year review from a single institution. *Am. J. Rhinol.* **18:**75–81.

Patterson, T. F., W. R. Kirkpatrick, M. White, J. W. Hiemenz, J. R. Wingard, B. Dupont, M. G. Rinaldi, D. A. Stevens, J. R. Graybill, et al. 2000. Invasive aspergillosis. Disease spectrum, treatment practices, and outcomes. *Medicine* (Baltimore) **79:**250–260.

Pauksens, K., and G. Oberg. 2006. Concomitant invasive pulmonary aspergillosis and aspergillus sinusitis in a patient with acute leukaemia. *Acta Biomed.* **77**(Suppl. 4)**:**23–25.

Perfect, J. R., G. M. Cox, R. K. Dodge, and W. A. Schell. 1996. In vitro and in vivo efficacies of the azole SCH56592 against *Cryptococcus neoformans. Antimicrob. Agents Chemother.* **40:**1910–1913.

Perfect, J. R., and D. T. Durack. 1985. Penetration of imidazoles and triazoles into cerebrospinal fluid of rabbits. *J. Antimicrob. Chemother.* **16:**81–86.

Pitisuttithum, P., R. Negroni, J. R. Graybill, B. Bustamante, P. Pappas, S. Chapman, R. S. Hare, and C. J. Hardalo. 2005. Activity of posaconazole in the treatment of central nervous system fungal infections. *J. Antimicrob. Chemother.* **56:**745–755.

Rodriguez, D. L., C. A. Lopez, E. B. Cobos, A. J. Blanco, A. F. Fernandez, and L. F. Araujo. 1999. Invasive cerebral aspergillosis in a patient with aplastic anemia. Response to liposomal amphotericin and surgery. *Haematologica* **84:**758–759.

Rodriguez, T. E., N. R. Falkowski, J. R. Harkema, and G. B. Huffnagle. 2007. Role of neutrophils in preventing and resolving acute fungal sinusitis. *Infect. Immun.* **75:**5663–5668.

Rosenfeld, R. M., D. Andes, N. Bhattacharyya, D. Cheung, S. Eisenberg, T. G. Ganiats, A. Gelzer, D. Hamilos, R. C. Haydon III, P. A. Hudgins, S. Jones, H. J. Krouse, L. H. Lee, M. C. Mahoney, B. F. Marple, C. J. Mitchell, R. Nathan, R. N. Shiffman, T. L. Smith, and D. L. Witsell. 2007. Clinical practice guideline: adult sinusitis. *Otolaryngol. Head Neck Surg.* **137:**S1–S31.

Rowe-Jones, J. M., and A. R. Freedman. 1994. Adjuvant itraconazole in the treatment of destructive sphenoid aspergillosis. *Rhinology* **32:**203–207.

Rowe-Jones, J. M., and V. Moore-Gillon. 1994. Destructive noninvasive paranasal sinus aspergillosis: component of a spectrum of disease. *J. Otolaryngol.* **23:**92–96.

Saah, D., P. E. Drakos, J. Elidan, I. Braverman, R. Or, and A. Nagler. 1994. Rhinocerebral aspergillosis in patients undergoing bone marrow transplantation. *Ann. Otol. Rhinol. Laryngol.* **103:**306–310.

Said, T., M. R. Nampoory, M. P. Nair, M. Al Saleh, K. H. Al Haj, M. A. Halim, K. V. Johny, M. Samhan, and M. Al Mousawi. 2005.

Safety of caspofungin for treating invasive nasal sinus aspergillosis in a kidney transplant recipient. *Transplant. Proc.* 37:3038–3040.

Sanchez, C., E. Mauri, D. Dalmau, S. Quintana, A. Aparicio, and J. Garau. 1995. Treatment of cerebral aspergillosis with itraconazole: do high doses improve the prognosis? *Clin. Infect. Dis.* **21**:1485–1487.

Saulsbury, F. T. 2001. Successful treatment of aspergillus brain abscess with itraconazole and interferon-gamma in a patient with chronic granulomatous disease. *Clin. Infect. Dis.* **32**:E137–E139.

Scherer, M., H. G. Fieguth, T. Aybek, Z. Ujvari, A. Moritz, and G. Wimmer-Greinecker. 2005. Disseminated *Aspergillus fumigatus* infection with consecutive mitral valve endocarditis in a lung transplant recipient. *J. Heart Lung Transplant.* **24**:2297–2300.

Schubert, M. S. 2004. Allergic fungal sinusitis: pathogenesis and management strategies. *Drugs* **64**:363–374.

Schubert, P. 1885. Zur Casuistik der Aspergillusmykosen. *Dtsch. Arch. Klin. Med.* **36**:162–179.

Schwartz, S., J. Burhenne, A. Haisch, M. Ruhnke, E. Thiel, M. Brock, and S. Hammersen. 2007a. Fungicidal concentrations of voriconazole (VRC) in brain abscess and cerebrospinal fluid (CSF) in a patient (pt) with *Candida* meningoencephalitis, abstr. 440. *47th Intersci. Conf. Antimicrob. Agents Chemother.* American Society for Microbiology, Washington, DC.

Schwartz, S., D. Milatovic, and E. Thiel. 1997. Successful treatment of cerebral aspergillosis with a novel triazole (voriconazole) in a patient with acute leukaemia. *Br. J. Haematol.* **97**:663–665.

Schwartz, S., M. Ruhnke, P. Ribaud, L. Corey, T. Driscoll, O. A. Cornely, U. Schuler, I. Lutsar, P. Troke, and E. Thiel. 2005. Improved outcome in central nervous system aspergillosis, using voriconazole treatment. *Blood* **106**:2641–2645.

Schwartz, S., M. Ruhnke, P. Ribaud, E. Reed, P. Troke, and E. Thiel. 2007b. Poor efficacy of amphotericin B-based therapy in CNS aspergillosis. *Mycoses* **50**:196–200.

Schwartz, S., and E. Thiel. 1997. Images in clinical medicine. Palate destruction by *Aspergillus*. *N. Engl. J. Med.* **337**:241.

Singh, N., and K. J. Pursell. 2008. Combination therapeutic approaches for the management of invasive aspergillosis in organ transplant recipients. *Mycoses* **51**:99–108.

Stammberger, H., R. Jakse, and F. Beaufort. 1984. Aspergillosis of the paranasal sinuses: X-ray diagnosis, histopathology, and clinical aspects. *Ann. Otol. Rhinol. Laryngol.* **93**:251–256.

Stevens, D. A., V. L. Kan, M. A. Judson, V. A. Morrison, S. Dummer, D. W. Denning, J. E. Bennett, T. J. Walsh, T. F. Patterson, and G. A. Pankey. 2000. Practice guidelines for diseases caused by *Aspergillus*: Infectious Diseases Society of America. *Clin. Infect. Dis.* **30**:696–709.

Stone, J. A., X. Xu, G. A. Winchell, P. J. Deutsch, P. G. Pearson, E. M. Migoya, G. C. Mistry, L. Xi, A. Miller, P. Sandhu, R. Singh, F. deLuna, S. C. Dilzer, and K. C. Lasseter. 2004. Disposition of caspofungin: role of distribution in determining pharmacokinetics in plasma. *Antimicrob. Agents Chemother.* **48**:815–823.

Streppel, M., G. Bachmann, G. Arnold, M. Damm, and E. Stennert. 1999. Successful treatment of an invasive aspergillosis of the skull base and paranasal sinuses with liposomal amphotericin B and itraconazole. *Ann. Otol. Rhinol. Laryngol.* **108**:205-207.

Talbot, G. H., A. Huang, and M. Provencher. 1991. Invasive aspergillus rhinosinusitis in patients with acute leukemia. *Rev. Infect. Dis.* **13**:219–232.

Taxy, J. B. 2006. Paranasal fungal sinusitis: contributions of histopathology to diagnosis. A report of 60 cases and literature review. *Am. J. Surg. Pathol.* **30**:713–720.

Torre-Cisneros, J., O. L. Lopez, S. Kusne, A. J. Martinez, T. E. Starzl, R. L. Simmons, and M. Martin. 1993. CNS aspergillosis in organ transplantation: a clinicopathological study. *J. Neurol. Neurosurg. Psychiatry* **56**:188–193.

Tsiodras, S., R. Zafiropoulou, J. Giotakis, G. Imbrios, A. Antoniades, and E. K. Manesis. 2004. Deep sinus aspergillosis in a liver transplant recipient successfully treated with a combination of caspofungin and voriconazole. *Transplant. Infect. Dis.* **6**:37–40.

Uri, N., R. Cohen-Kerem, I. Elmalah, I. Doweck, and E. Greenberg. 2003. Classification of fungal sinusitis in immunocompetent patients. *Otolaryngol. Head Neck Surg.* **129**:372–378.

Verweij, P. E., K. Brinkman, H. P. Kremer, B. J. Kullberg, and J. F. Meis. 1999. *Aspergillus* meningitis: diagnosis by non-culture-based microbiological methods and management. *J. Clin. Microbiol.* **37**: 1186–1189.

Verweij, P. E., J. P. Donnelly, and J. F. Meis. 1996. High-dose itraconazole for the treatment of cerebral aspergillosis. *Clin. Infect. Dis.* **23**:1196–1197.

Viollier, A. F., D. E. Peterson, C. A. De Jongh, K. A. Newman, W. C. Gray, J. C. Sutherland, M. A. Moody, and S. C. Schimpff. 1986. *Aspergillus* sinusitis in cancer patients. *Cancer* **58**:366–371.

Virgintino, D., D. Robertson, M. Errede, V. Benagiano, F. Girolamo, E. Maiorano, L. Roncali, and M. Bertossi. 2002. Expression of P-glycoprotein in human cerebral cortex microvessels. *J. Histochem. Cytochem.* **50**:1671–1676.

Walsh, T. J., E. J. Anaissie, D. W. Denning, R. Herbrecht, D. P. Kontoyiannis, K. A. Marr, V. A. Morrison, B. H. Segal, W. J. Steinbach, D. A. Stevens, J. A. van Burik, J. R. Wingard, and T. F. Patterson. 2008. Treatment of aspergillosis: clinical practice guidelines of the Infectious Diseases Society of America. *Clin. Infect. Dis.* **46**:327–360.

Walsh, T. J., D. B. Hier, and L. R. Caplan. 1985. Aspergillosis of the central nervous system: clinicopathological analysis of 17 patients. *Ann. Neurol.* **18**:574–582.

Weber, R. S., and G. Lopez-Berestein. 1987. Treatment of invasive *Aspergillus* sinusitis with liposomal-amphotericin B. *Laryngoscope* **97**:937–941.

Willinger, B., J. Beck-Mannagetta, A. M. Hirschl, A. Makristathis, and M. L. Rotter. 1996. Influence of zinc oxide on *Aspergillus* species: a possible cause of local, non-invasive aspergillosis of the maxillary sinus. *Mycoses* **39**:361–366.

Willinger, B., A. Obradovic, B. Selitsch, J. Beck-Mannagetta, W. Buzina, H. Braun, P. Apfalter, A. M. Hirschl, A. Makristathis, and M. Rotter. 2003. Detection and identification of fungi from fungus balls of the maxillary sinus by molecular techniques. *J. Clin. Microbiol.* **41**:581–585.

Yagi, H. I., S. A. Gumaa, A. I. Shumo, N. Abdalla, and A. A. Gadir. 1999. Nasosinus aspergillosis in Sudanese patients: clinical features, pathology, diagnosis, and treatment. *J. Otolaryngol.* **28**:90–94.

Yamazaki, T., H. Kume, S. Murase, E. Yamashita, and M. Arisawa. 1999. Epidemiology of visceral mycoses: analysis of data in annual of the pathological autopsy cases in Japan. *J. Clin. Microbiol.* **37**: 1732–1738.

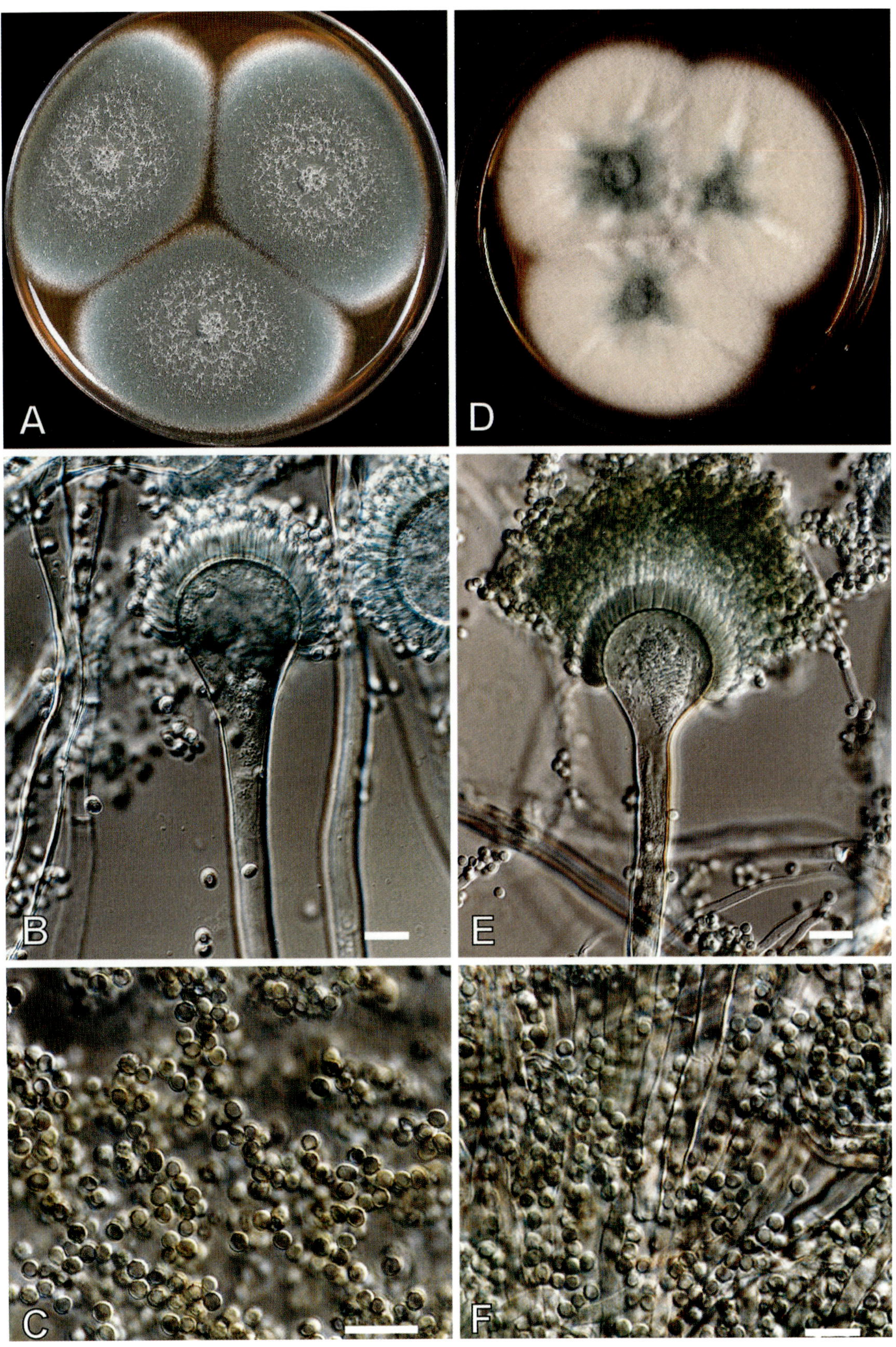

Color Plate 1 (chapter 2). (A to C) *A. fumigatus*: culture on MEA at 30°C for 7 days (A), conidiophores (B), and conidia (C). (D to F) *A. lentulus*: culture on MEA at 30°C for 7 days (D), conidiophores (E), and conidia (F). Bars, 10 μm.

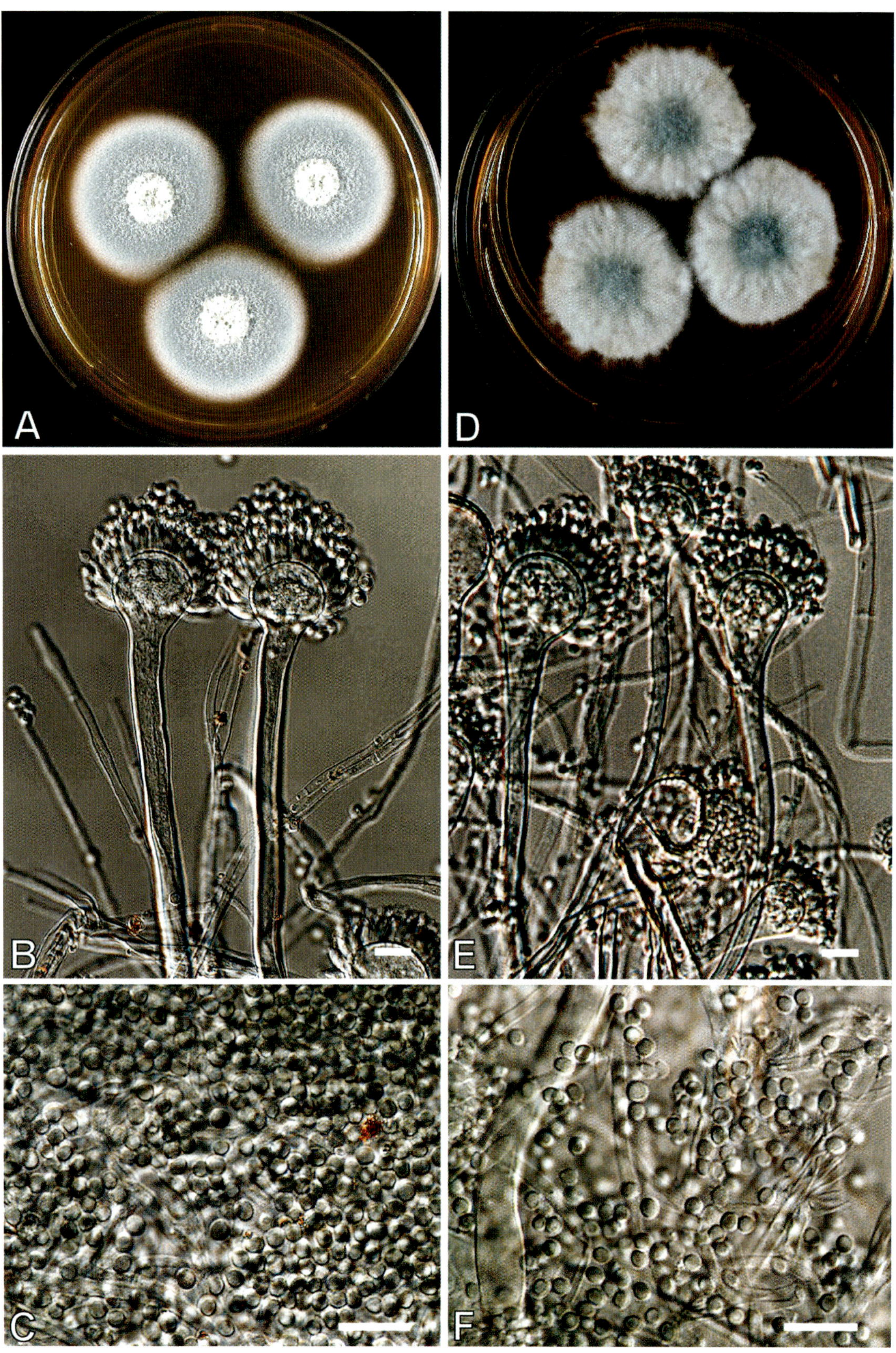

Color Plate 2 (chapter 2). (A to C) *A. fumisynnematus*: culture on MEA at 30°C for 7 days (A), conidiophores (B), and conidia (C). (D to F) *A. novofumigatus*: culture on MEA at 30°C for 7 days (D), conidiophores (E), and conidia (F). Bars, 10 μm.

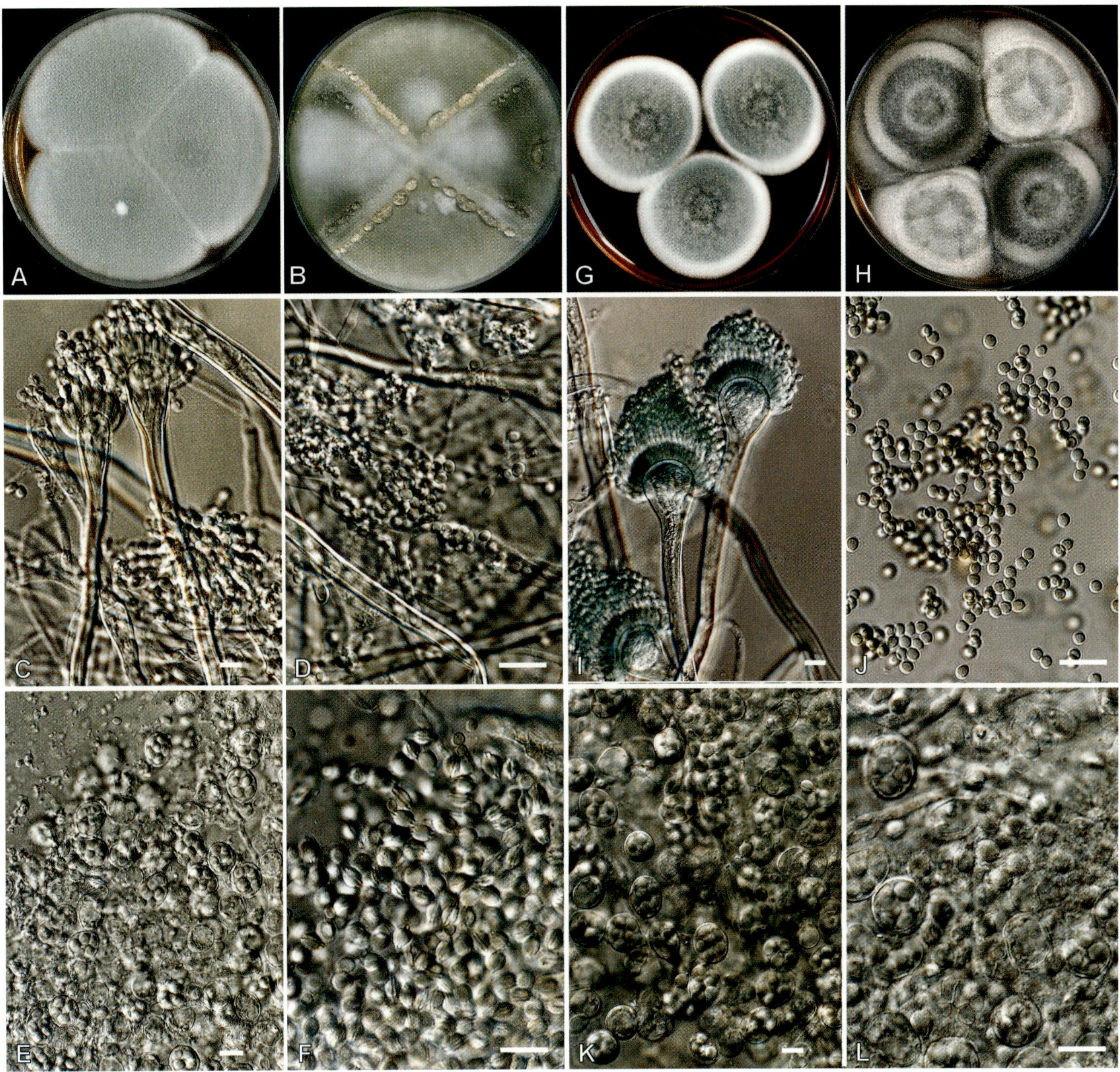

Color Plate 3 (chapter 2). (A to F) *N. fennelliae*: culture on MEA at 30°C for 7 days (A), crossing of two mating types at 37°C after 14 days (B), conidiophores (C), conidia (D), asci (E), and ascospores (F). (G to L) *N. udagawae*: culture on MEA at 30°C for 7 days (G), crossing of two mating types at 37°C after 14 days (H), conidiophores (I), conidia (J), asci (K), and ascospores (L). Bars, 10 μm.

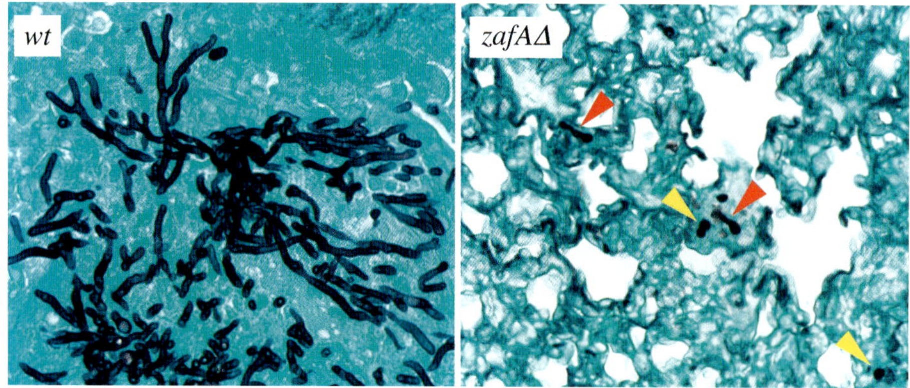

Color Plate 4 (chapter 9). Lung sections from immunosuppressed mice infected with either a wild-type or a *zafA*Δ mutant strain of *A. fumigatus* at 48 h postinfection. Conidia of the *zafA*Δ mutant did not germinate (yellow arrows) or exhibited very short germ tubes (red arrows), and no fungal mycelia spread through the adjacent tissue.

Color Plate 5 (chapter 12). Confocal and apotome microscopy images and a schematic representation of an *A. fumigatus* biofilm (StA) (Beauvais et al., 2007). (A1 to A4) Depth projections from a confocal image stack recorded from an StA culture after calcofluor white staining (24 h of growth; Plan-Apochromat, 20×/0.75 objective). Panels A1 to A3 are sections of a 24-h StA culture from the agar surface (in red) to the air surface (blue). A4 is a reconstitution of panels A1 to A3. (B1 and B2) 3D images of an StA culture (Plan-Neofluar; 40× objective, 0.8 numerical aperture), showing the vertical growth of the fungus (16 h of growth [B1]), followed by the appearance of extracellular material holding the hyphae together (20 h of growth [B2]). (C) Schematic representation of growth of an *A. fumigatus* biofilm compared to colony formation.

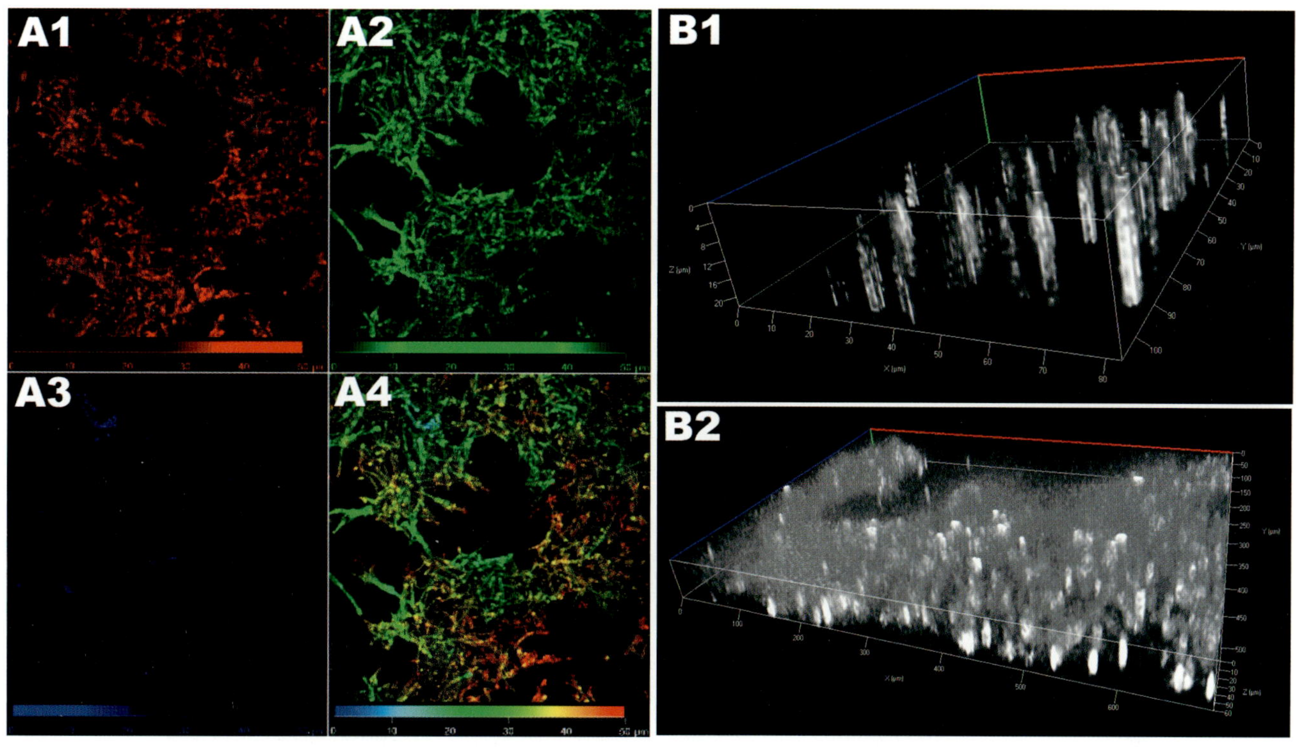

A1
A2
B1
A3
A4
B2

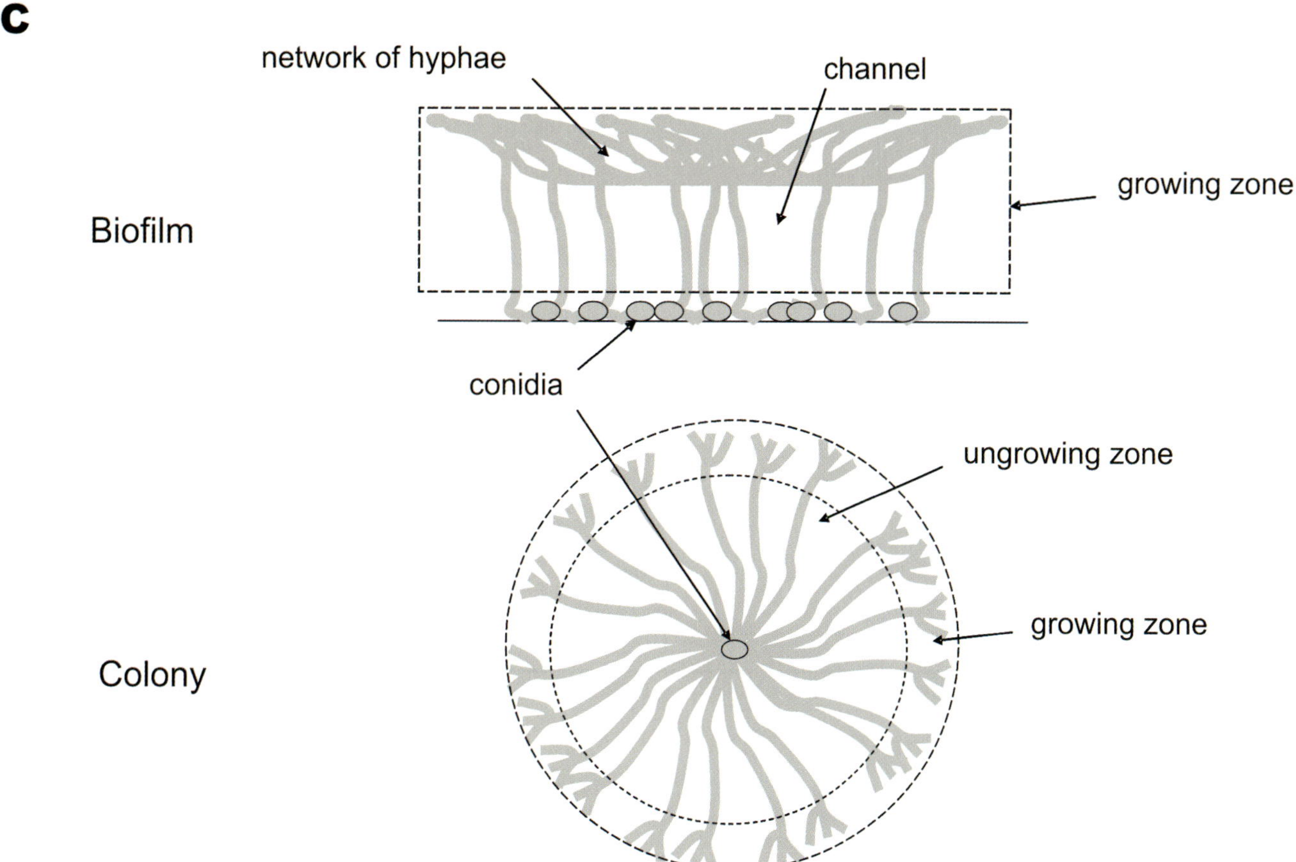

C
network of hyphae
channel
growing zone
Biofilm
conidia
ungrowing zone
growing zone
Colony

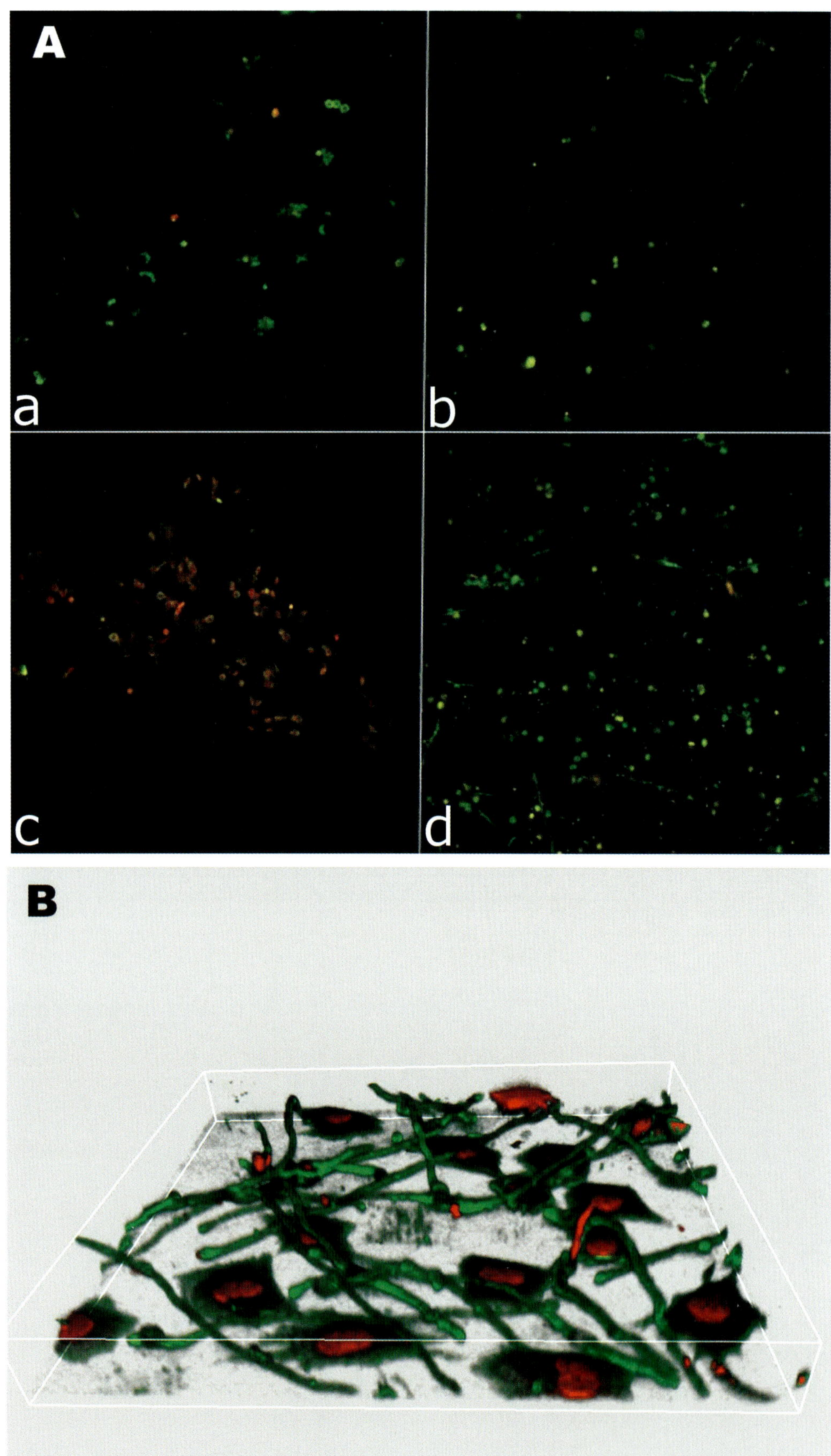
A
a
b
c
d
B

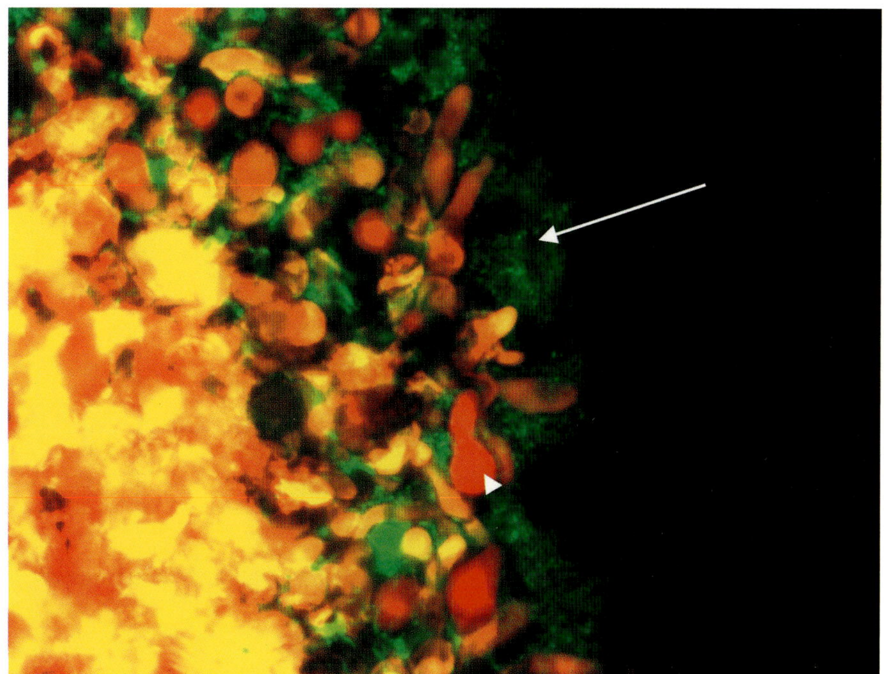

Color Plate 7 (chapter 12). Biofilm of *P. aeruginosa* on *A. fumigatus* mycelia. *P. aeruginosa* was labeled (arrow) with green fluorescent protein (kind gift of S. Garvis, Imperial College of Science Technology and Medicine, London, United Kingdom), and *A. fumigatus* (arrowhead) was stained with DS-Red (kind gift of A. Eberhard, CNRS Biomérieux, Lyon, France) (Mikkelsen et al., 2003). The yellow areas show the superposition of the green (green fluorescent protein) and the red (DS-Red) dyes (Gastebois and Latgé, 2008).

Color Plate 6 (chapter 12). Images from confocal scanning laser microscopy, with excitation at 543 nm (HeNe laser) and 488 nm (argon laser), beam splitting at 488 and 543 nm, and emission at 560 and 505 nm for FUN-1 (red) and ConA-Alexa Fluor 488 (green). (A) *A. fumigatus* coculture on bronchial epithelial cells after 1 day (a) or 2 days (c) with cells from healthy individuals (HBE) (a) or after 1 day (b) or 2 days (d) with cells from cystic fibrosis patients (d). (B) 3D image of an *A. fumigatus* biofilm after 3 days on HBE Green-stained polysaccharides (green arrows) and showing metabolically active sites (red arrow) (Seidler, unpublished data).

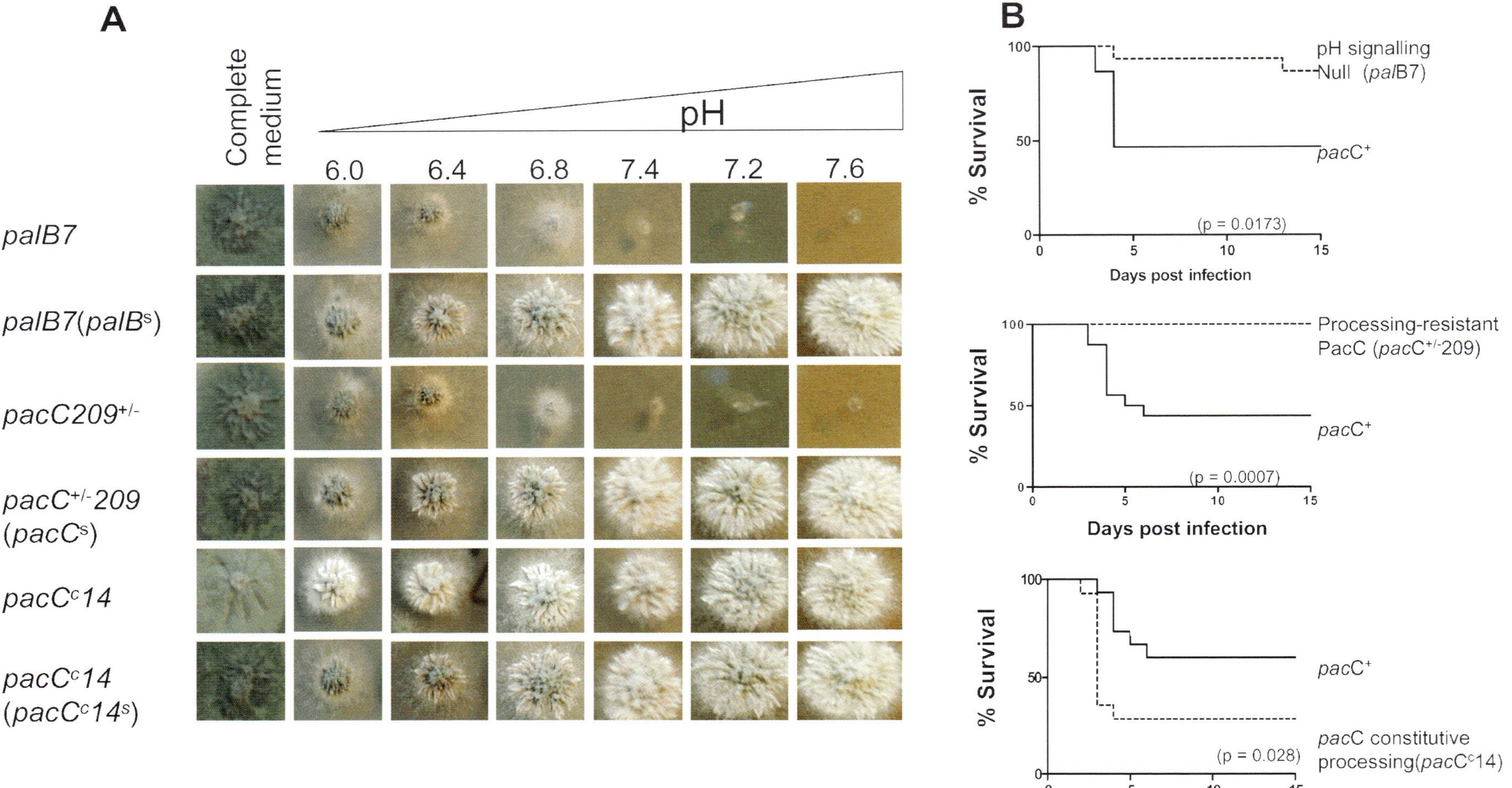

Color Plate 8 (chapter 17). Physiological consequences of pH signal abrogation in *A. nidulans.* (A) Mutational perturbation of pH signal transduction (*palB7*) and PacC processing (*pacC*$^{+/-}$ 209) prevent in vitro alkaline adaptation in *A. nidulans* relative to complemented (*palB*^{s} and *pacC*^{s}) strains. Constitutive PacC processing (*pacC*^{c} 14) does not affect in vitro alkaline adaptation. Prototrophic *A. nidulans* strains were grown on pH-buffered complete and minimal media containing ammonium tartrate and glucose. (B) Mutational perturbation of *A. nidulans* pH signal transduction (*palB7*) and PacC processing (*pacC*$^{+/-}$ 209) prevents murine aspergillosis, relative to complemented (*palB*^{s} and *pacC*^{s}) strains. Constitutive PacC processing (*pacC*^{c}14) increases *A. nidulans* virulence relative to an isogenic complemented strain (*pacC*^{c}14^{s}). Murine virulence testing was performed in neutropenic mice.

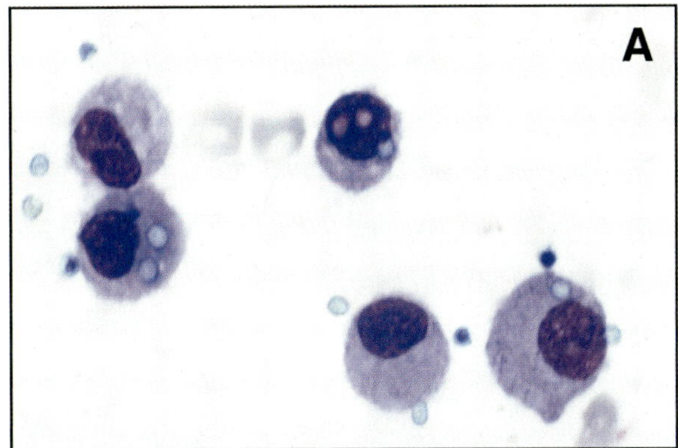

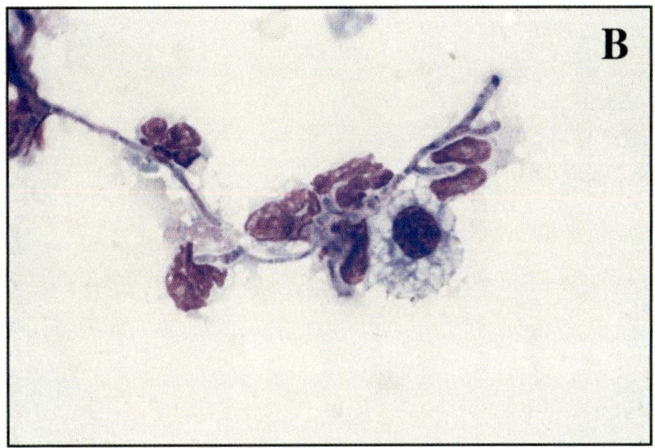

Color Plate 9 (chapter 18). Microscopy of neutrophils and alveolar macrophages in experimental IPA. Mice were infected intratracheally with 10^7 conidia of *A. fumigatus* and BAL was obtained 3 h (A) and 48 h (B) after infection. Macrophages collected at 3 h (A) have ingested resting conidia (light blue), whereas some conidia in the extracellular medium are swollen (dark blue). At 48 h, conidia have germinated and hyphae can be observed. In panel B, hyphae are surrounded by several neutrophils and one macrophage. (Images courtesy of V. Balloy et al. [unpublished data].)

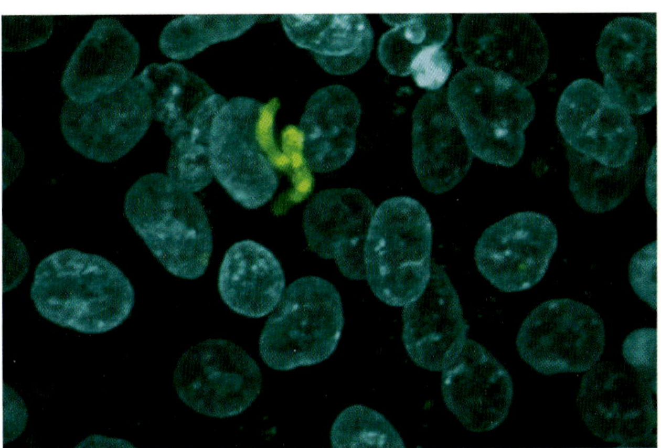

Color Plate 10 (chapter 19). Confocal micrograph showing the penetration of the endothelium by a hypha expressing green fluorescent protein in an in vitro model of early invasive pulmonary aspergillosis. The nuclei of the endothelial cells have been stained with 4′,6-diamidino-2-phenylindole.

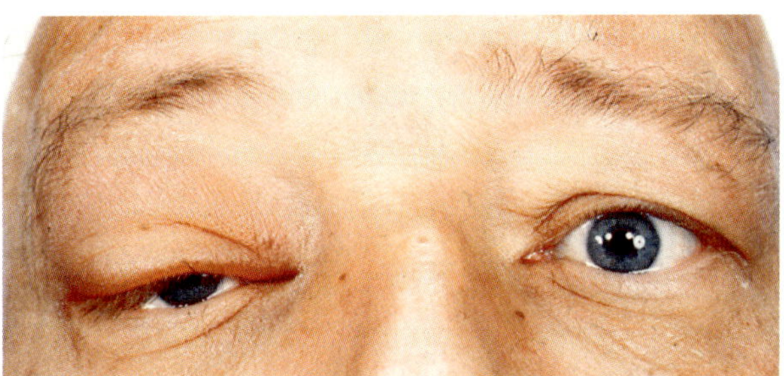

Color Plate 11 (chapter 24). Acute invasive sinusitis (*A. fumigatus*) with orbital cellulitis in a patient after hematopoietic stem cell transplantation (see also Fig. 2, chapter 24). Right-sided periorbital swelling indicating orbital cellulitis.

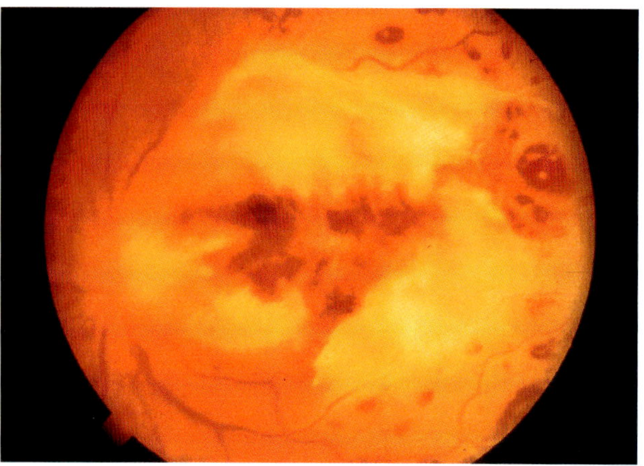

Color Plate 12 (chapter 24). Fungal retinitis with inflammatory retinal infiltration of the right eye in a patient with acute invasive fungal sinusitis (*A. fumigatus*) after hematopoietic stem cell transplantation (see also Fig. 3, chapter 24).

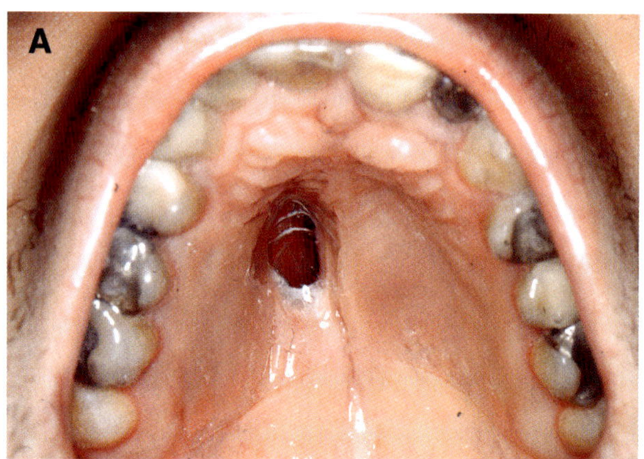

Color Plate 13 (chapter 24). Acute invasive sinusitis (*Aspergillus* sp.) with orbital cellulitis and palate destruction in a patient with leukemia (see also Fig. 4, chapter 24). (A) An ulcerative lesion of the hard palate progressed to a palatal cleft with recovery from neutropenia (translucent prosthesis in situ). (B) Hyphal tissue invasion in a surgical specimen (Grocott silver stain). (Reprinted from the *New England Journal of Medicine* [Schwartz and Thiel, 1997] with permission of the Massachusetts Medical Society.)

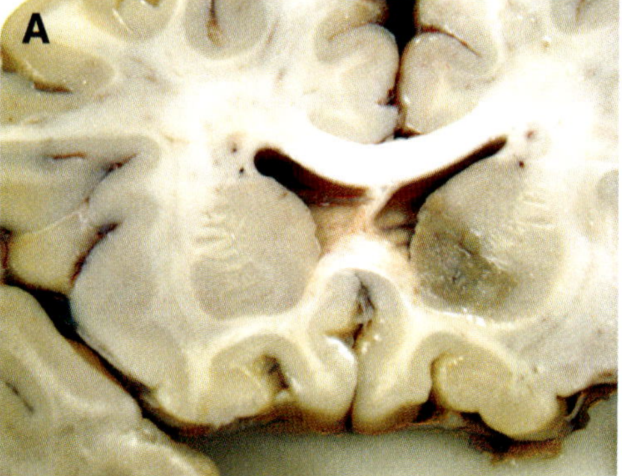

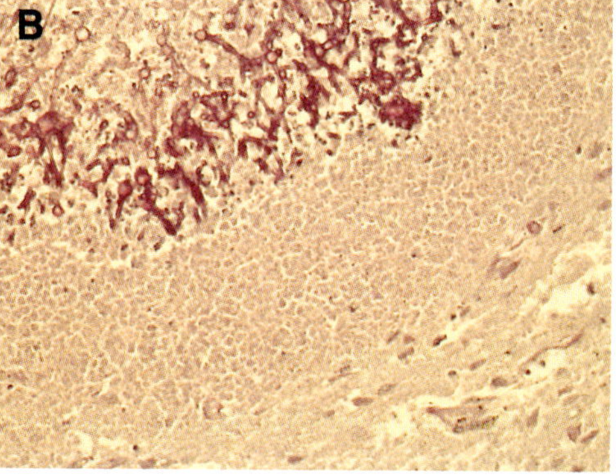

Color Plate 14 (chapter 24). Fungal brain abscess caused by hematogenous spread of invasive pulmonary aspergillosis in a hematopoietic stem cell transplant patient (see also Fig. 5, chapter 24). (A) Brain section view with brain abscess within the left-sided region of the nucleus caudatus. (B) Histological section demonstrating septate, dichotomous branching hyphae highly compatible with *Aspergillus* spp. and surrounding brain tissue necrosis (PAS stain).

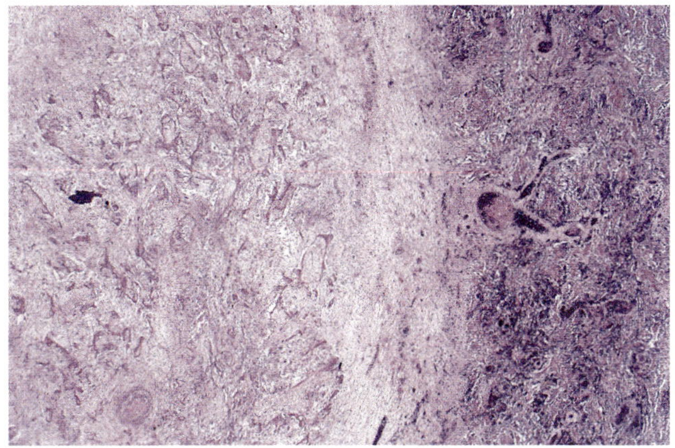

Color Plate 15 (chapter 27). Histopathology of the discrete nodule in IPA. Arcuate margin of the discrete nodule with coagulation necrosis on the left and acute hemorrhage on the right. This pathology mirrors the CT halo sign. (Hematoxylin and eosin stain; magnification, ×100.)

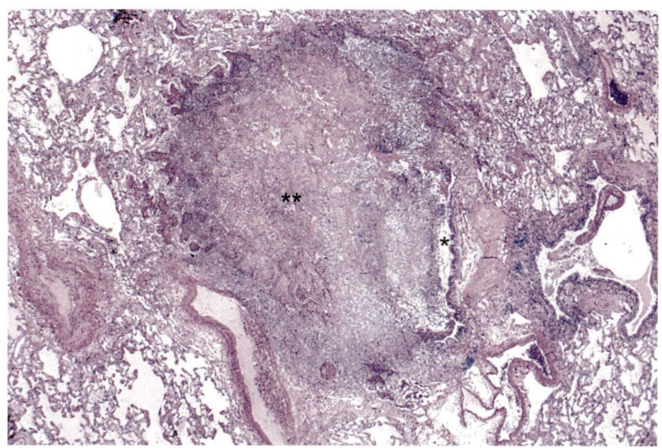

Color Plate 16 (chapter 27). Transitional lesion in IPA. A crescentic void (*) resulting from liquefaction necrosis is present at the periphery of a core of coagulation necrosis (**) in a discrete nodule. There are patent vessels at the periphery of the nodule. The transition lesion is the pathologic analog of the CT air crescent sign. (Hematoxylin and eosin stain; magnification, ×4.)

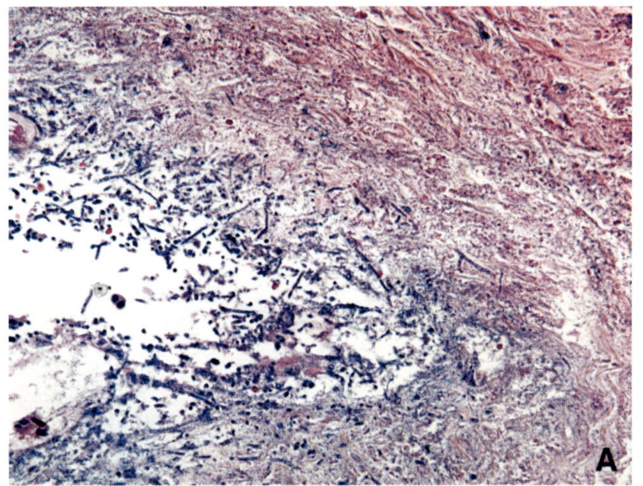

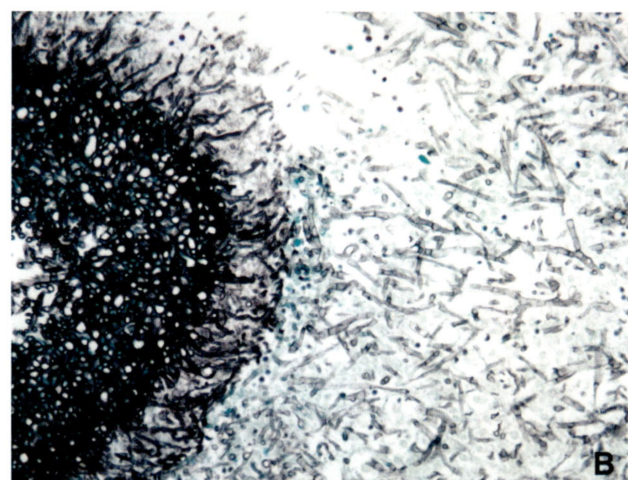

Color Plate 17 (chapter 27). CNPA. (A) Cavity wall is eroded, with penetration of elongated hyphae. (Hematoxylin-eosin stain; magnification, ×100.) (B) Hyphae that have invaded the cavity wall. (GMS stain; magnification, ×200.)

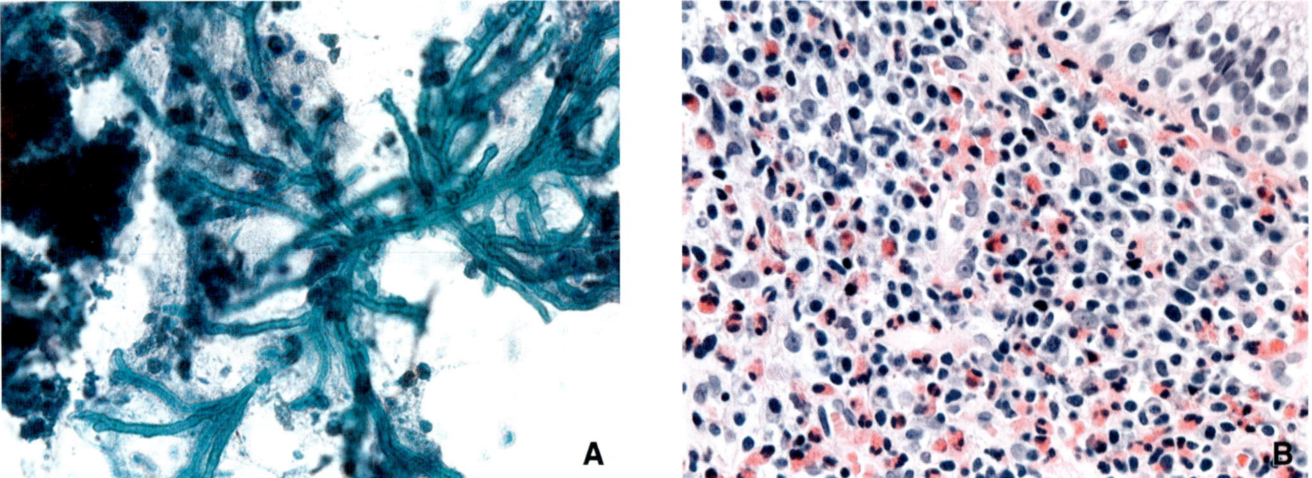

Color Plate 18 (chapter 27). Histopathology of ABPA. (A) Cytological specimen of bronchoalveolar lavage fluid showing a cluster of hyphae with dichotomous bifurcations. (Papanicolaou stain; magnification, ×400.) (B) Histological specimen of a contemporaneous transbronchial biopsy demonstrating a dense inflammatory infiltrate composed of lymphocytes and eosinophils. The basement membrane is mildly thickened and covered with stratified columnar epithelium. (Hematoxylin-eosin stain; magnification, ×400.)

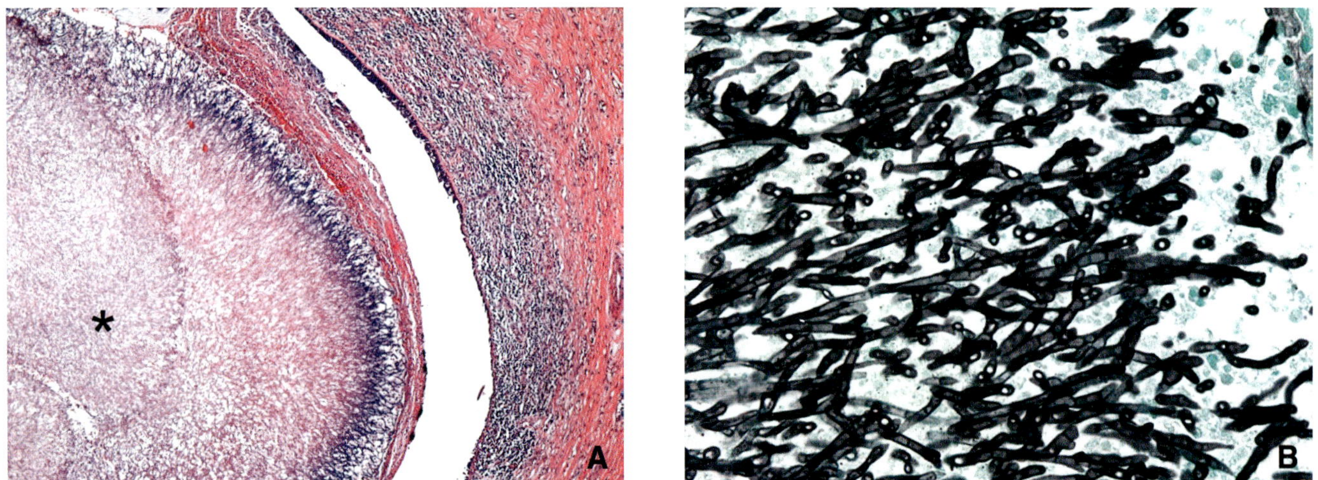

Color Plate 19 (chapter 27). Histopathology of noninvasive pulmonary aspergillosis (aspergilloma). (A) Cavity filled by a fungus ball composed of radially aligned hyphae. The cavity wall is covered with metaplastic epithelium and infiltrated with chronic inflammatory cells. (Hematoxylin-eosin stain; magnification, ×4.) (B) Higher power demonstrates densely intertwined septated hyphae at the periphery of the fungus ball. (GMS stain; magnification, ×400.)

Aspergillus fumigatus and Aspergillosis
Edited by J.-P. Latgé and W. J. Steinbach
©2009 ASM Press, Washington, DC

Chapter 25

Chronic Aspergillosis

DAVID W. DENNING

INTRODUCTION, DEFINITIONS, AND NOMENCLATURE

Unlike immediately life-threatening invasive aspergillosis, several more indolent clinical entities are recognized. An arbitrary duration of disease of at least 3 months distinguishes acute and subacute invasive aspergillosis from chronic aspergillosis. Allergic (or eosinophilic) aspergillosis is also chronic in almost all circumstances, but the term allergic is used to distinguish several eosinophil- and specific immunoglobulin E (IgE)-associated clinical entities from chronic pulmonary and sinus aspergillosis.

Chronic pulmonary aspergillosis was first described in humans (probably an aspergilloma), by Alexander Bennett in 1848 (Bennett, 1848), and numerous descriptive and diagnostic monikers have been used, including pulmonary aspergillosis with cavitation, symptomatic pulmonary aspergilloma, complex aspergilloma, and chronic granulomatous aspergillosis. In the early 1980s, Gefter et al. (1981) and Binder et al. (1982) coined the terms semi-invasive pulmonary aspergillosis and chronic necrotizing pulmonary aspergillosis, respectively. These entities are best regarded as subacute invasive aspergillosis (Hope et al., 2005).

The specific entities embraced within the overall term chronic pulmonary aspergillosis include aspergilloma (depicting a single pulmonary cavity containing a [usually] mobile fungal ball), chronic cavitary pulmonary aspergillosis (CCPA; corresponding to the older surgical term complex aspergilloma, but not necessarily with an aspergilloma present), and a late-stage entity, chronic fibrosing pulmonary aspergillosis.

The term chronic cavitary pulmonary aspergillosis describes patients in whom there is formation and expansion of multiple cavities over time. The term chronic cavitary pulmonary aspergillosis is preferred over the term complex aspergilloma because more than 50% of such patients don't have an aspergilloma visible radiologically, and it is preferred over the term chronic necrotizing pulmonary aspergillosis because pathological appearances are not known in the vast majority of cases and necrotizing features are rarely seen when tissue is obtained. The terminology is an arbitrary classification of a spectrum of disease entities whose precise definition and scope vary considerably.

The first recorded case of a sinus aspergilloma was reported from France in 1791 (Plaignaud, 1791). There are three discrete clinical entities subsumed under the term chronic *Aspergillus* rhinosinusitis, namely, chronic invasive *Aspergillus* rhinosinusitis, fungus ball of the sinus (also termed sinus aspergilloma), and chronic granulomatous *Aspergillus* rhinosinusitis or paranasal *Aspergillus* granuloma. These entities are defined partly clinically, radiologically, and pathologically (Table 1).

It is likely that a form of chronic *Aspergillus* bronchitis exists (i.e., not simply colonization), but precise definition with our current diagnostic tools is problematic. Various cutaneous manifestations are also caused by *Aspergillus* spp., including onychomycosis, external otitis, and primary cutaneous aspergillosis.

With improvements in survival from invasive aspergillosis, many patients receive therapy for months or years with continuing evidence of aspergillosis, usually radiological. In certain respects, such patients can be termed to have chronic aspergillosis and yet be enrolled in studies of new antifungal drugs being evaluated for acute invasive aspergillosis (as salvage or compassionate use). No term has been proposed to cover this situation, although many have attempted to define "salvage" but without a consensus reached. Perhaps the term chronic or acute aspergillosis should be considered.

EPIDEMIOLOGY

Chronic Pulmonary Aspergillosis

A cross-sectional population study to determine the number of cases of chronic pulmonary aspergillosis has

David W. Denning • Medicine and Medical Mycology, University of Manchester, and Education and Research Centre, Wythenshawe Hospital, Manchester M23 9LT, United Kingdom.

Table 1. Chronic aspergillosis syndromes

Category and syndrome
Pulmonary disease
Aspergilloma (single)
CCPA (complex aspergilloma)
Chronic fibrosing pulmonary aspergillosis
Chronic *Aspergillus* bronchitis
Paranasal sinus disease
Fungus ball of paranasal sinuses
Chronic invasive *Aspergillus* sinusitis
Chronic granulomatous *Aspergillus* sinusitis
Miscellaneous diseases
External otitis
Chronic cutaneous aspergillosis
Aspergillus onychomycosis

not been done. It would require a large study using *Aspergillus* antibody testing, which has never been systematically done. Any estimate must rely on the frequency of underlying pulmonary disease, which varies substantially by country, and an approximation would include the following data. Cavities of 2 cm or larger remaining after pulmonary tuberculosis are present in 21 to 35% of patients (Sonnenberg et al., 2001; Lee et al., 2008). Of those with multidrug-resistant tuberculosis (TB) in Africa, 60% have residual cavities (de Valliere et al., 2004). *Aspergillus* precipitins were found in ~25% of patients 1 and 4 years after completion of treatment for pulmonary aspergillosis who have residual pulmonary cavities (Anonymous, 1970). Thus, the rate of chronic pulmonary aspergillosis and/or aspergilloma after pulmonary tuberculosis is approximately 8%, but may be as high as 12%. There were 49 cases of aspergilloma seen over a period of 6 years among 36,340 hospital admissions in India (Singh et al., 1989). The presence of an aspergilloma on chest radiograph underestimates chronic pulmonary aspergillosis by probably 40 to 50%, consistent with continuing symptoms, such as hemoptysis and weight loss (Anonymous, 1970). In addition there may be a stronger association of chronic pulmonary aspergillosis with atypical pulmonary TB than classical TB, although numbers of cases are much lower, even in developed countries where classical tuberculosis is infrequent.

Another important underlying disease of chronic pulmonary aspergillosis is pulmonary sarcoidosis, particularly of the fibrocystic phenotype, and 50% of these patients develop aspergillomas (Wollschlager, 1984). The incidence of sarcoidosis varies from country to country, being most common in northern Europeans and in Blacks. In Japan, for example, the prevalence is 2.4/100,000, but in Sweden it is 64/100,000. The lifetime risk of a U.S. Black person getting sarcoidosis is 2.4%, and in Swedish women it is 1.6%. Approximately 10% develop severe disability as a result of sarcoidosis, usually pulmonary.

Another major category of risk for chronic pulmonary aspergillosis is recurrent pneumothorax and bullous diseases of the lung. The rate of hospital admission in the United Kingdom for pneumothoraxis is 17/100,000 for men and 6/100,000 for women (Gupta et al., 2000), and 50% of patients have recurrences (Sadikot et al., 1997). This estimate suggests a population at risk of aspergillosis of ~5,870 adults annually. It is not known what proportion develop chronic pulmonary aspergillosis, but it is probably <5% and likely <1%. Other lung diseases apparently associated with chronic pulmonary aspergillosis at a low frequency include prior lung surgery, allergic bronchopulmonary aspergillosis, carcinoma of the lung (including postradiotherapy), and emphysema.

Chronic *Aspergillus* Rhinosinusitis

The frequencies of chronic invasive and granulomatous *Aspergillus* rhinosinusitis are not known. They are both exceptionally rare entities in North America and western Europe but are more common in the Middle East (Milosev et al., 1969; Yagi et al., 1999; Dawlatly et al., 1988; Alrajhi et al., 2001) and India (Ramani et al., 1994; Surya Prakash Rao et al., 1984). Both entities are more commonly caused by *Aspergillus flavus* (approximately 90% of cases) than other species (Hedayati et al., 2007). Chronic granulomatous disease usually presents at ages 25 to 38 years (Dawlatly et al., 1988; Yagi et al., 1999).

Fungus ball of the sinus comprised approximately 50% of cases of all forms of fungal rhinosinusitis in a large nasal endoscopy series from Brazil (Dall'Igna et al., 2005), whereas only one fungal ball was found in 349 (0.3%) in a nasal endoscopy series from Australia (Collins et al., 2003). This suggests marked geographical variation in fungus ball of the sinus. Most cases in western Europe are caused by *Aspergillus fumigatus*, whereas elsewhere the fungi implicated vary substantially and include several dematiceous molds.

UNDERLYING DISEASE, PATHOLOGY, AND PATHOGENESIS

Aspergilloma and Fungus Ball of the Sinus

A pulmonary aspergilloma and fungus ball of the sinus are rounded conglomerates of hyphae, mucus, and cellular debris. They may represent a true biofilm. In the case of pulmonary aspergillomas, they are often attached to the wall of a pulmonary cavity, whereas in the sinus they appear to be unattached. In the lung an aspergilloma is contained within a fibrotic and thickened wall and occupies a preexisting pulmonary cavity. The fungal ball is typically mobile, and its size may vary with time, as may the degree of adjacent pleural and cavity wall thickening. Cultures are usually positive for *A. fumigatus* (lung and often sinus) or *A. flavus* (sinus), and isolates with abnormal morphology and/or conidiation and slow growth are relatively common. Some aspergillomas, especially in diabetic patients, are caused by *Aspergillus niger*, in which case oxalic acid crystals may be seen in sputum and the aspergilloma if resected (Severo et al., 1997).

The cause of one or more preexisting pulmonary cavities is most commonly prior tuberculosis, but an aspergilloma may complicate a wide range of other pulmonary diseases which are associated or characterized by cavitation, such as sarcoidosis, histoplasmosis, pulmonary cysts, ankylosing spondylitis, atypical tuberculosis, bronchiectasis, rheumatoid nodules, and adenocarcinoma (Procknow and Loewen, 1960; Stiksa et al., 1976; Wollschlager and Khan, 1984; Butz et al., 1985; McConnochie et al., 1989; McGregor et al., 1989). The distinction between an aspergilloma and CCPA may be somewhat semantic but does reflect the singularity of the aspergilloma as opposed to the multiple cavities typical of CCPA.

No demonstrated risk factors are associated with fungus ball of the sinus. A common association of zinc-based root canal treatments in the upper jaw has been reported (Beck-Mannagetta et al., 1983; Beck-Mannagetta and Necek, 1986), and it may be that the dental materials not only act as a nidus for growth but also directly facilitate fungal growth. It is likely that local anatomical factors, such as temporary or permanent ostial obstruction, are important in facilitating the formation of a fungal ball.

CCPA

Structural lung disease appears to be an initial critical factor in pathogenesis and can be readily documented in the majority of cases of CCPA (Binder et al., 1982; Denning et al., 2003), with mycobacterial infection, emphysema, bullae, asthma, sarcoidosis, pneumoconiosis, thoracic surgery, lung cancer, upper lobe fibrosis complicating ankylosing spondylitis, and *Legionella* spp. infection described (Elliott et al., 1989; Caras and Pluss, 1996; Hafeez et al., 2000; Kato et al., 2002; Denning et al., 2003; Camuset et al., 2007). One or more subtle but important defects in systemic immunity may also be important in the pathogenesis; defects in mannose-binding lectin, surfactant A2, and Toll-like receptor 4 single-nucleotide polymorphisms and alcohol abuse have all been reported in this regard (Crosdale et al., 2001; Vaid et al., 2007; Carvalho et al., 2008). A shift towards a Th2 cytokine production profile was also seen, with single-nucleotide polymorphism analysis showing reduced tumor necrosis factor alpha, low transforming growth factor β, low interleukin-10, and high IL-15 and gamma interferon production (Sambatakou et al., 2006). The precise mechanism of new cavity formation remains unclear. Histologically the cavities show chronic inflammation with localized fibrosis, usually without granulomas, necrosis, or eosinophilic infiltrates. Pleural thickening is common but may not be present, and the stimulus for this is unclear. Localized additional networks of small vessels may be present, and presumably these are responsible for hemoptysis. Why these vessels form is not clear, since they are generally uncommon in pulmonary diseases, with the common exception of bronchiectasis.

Chronic Fibrosing Pulmonary Aspergillosis

The pathogenesis of chronic fibrosing pulmonary aspergillosis is not known (Denning et al., 2003). In cases in which the fibrosis occurs immediately adjacent to cavities caused by *Aspergillus*, the stimulus is probably fungus-mediated with localized pulmonary repair and fibrosis processes operating. In instances where the fibrosis is more distant, i.e., in an uninvolved ipsilateral lobe, the pathogenesis remains obscure.

Aspergillus Bronchitis and Otitis

While acute *Aspergillus* bronchitis and tracheobronchitis have been well described in numerous immunocompromised patients, especially lung transplant recipients, chronic *Aspergillus* bronchitis is underrecognized. It appears be associated with underlying disease, such as cystic fibrosis (Shoseyov et al., 2006) and bronchiectasis. Studies on this topic are lacking. External otitis caused by *Aspergillus* may be related to lack of cerumen production, chronic bacterial otitis, or prior antibiotic therapy, and atopic or other eczematous conditions appear to be predisposing factors (Zaror et al., 1991).

CLINICAL FEATURES AND CRITERIA FOR DIAGNOSIS

Pulmonary

Aspergilloma

There are no symptoms referrable to aspergilloma other than the occurrence of hemoptysis in most (50 to 90%) patients. It is typically infrequent and small in volume but on occasion may be massive and fatal. The appearance of a mobile fungal ball in a pulmonary cavity is highly suggestive of the diagnosis, as is localized pleural thickening overlying a cavity (Greene, 2005). The demonstration of precipitating antibodies to *Aspergillus* spp. is a useful diagnostic test to confirm a fungal etiology, although repetitive *Aspergillus* cultures from sputum may also confirm the fungal etiology. When there is uncertainty about the diagnosis, biopsy of the mass or cavity will yield hyphae if an aspergilloma is present.

CCPA

Patients with CCPA tend to be middle-aged with a symptom complex consisting of weight loss, chronic cough, hemoptysis, fatigue, and shortness of breath (Denning et al., 2003; Camuset et al., 2007). While the clinical signs and symptoms of CCPA are nonspecific, their presence and severity is an important point of distinction from aspergilloma. Although the diagnosis can be inferred from a single chest radiograph, detailed and sequentially acquired radiological data may be required to confirm the typical radiological features, as well as the very slow progression that is characteristic of this entity. Computerized tomography (CT) scans may be useful to further define the precise pattern and extent of disease. Radiological examination usually reveals one or more cavities, typically within the upper lobes, which may or may not contain fungus balls (Fig. 1). New cavity formation or expansion of one or more existing cavities over time is highly characteristic. Pericavitary infiltrates and adjacent pleural thickening are frequently observed

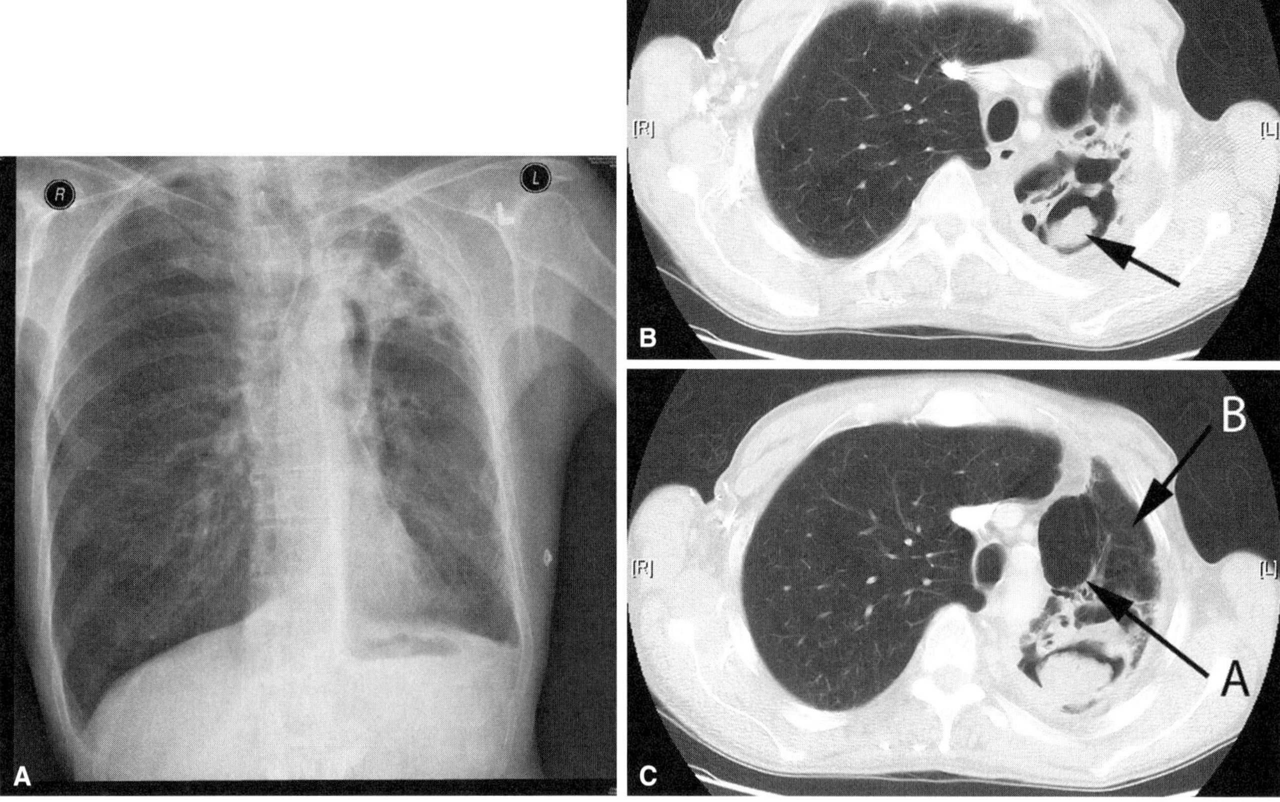

Figure 1. (A) Chest radiograph showing extensive cavitary disease occupying most of the left upper lobe, with some pleural thickening, probably multiple cavities, and no aspergilloma. The trachea is deviated to the left. Some bronchiectasis is also visible in the left lower lobe. (B) CT scan of the thorax in the same patient, showing multicavity disease on the left with a single fungal ball (arrow) in one of the cavities. Moderate pleural thickening is visible around the cavities. (C) CT scan of the thorax in the same patient at a lower level of the thorax, showing a different configuration of cavities (which probably communicate with each other) with the same fungal ball, but larger in this cross-section. Also visible is a large bulla (air space) (arrow A) proximal to the abnormal lung adjacent (arrow B). Much less pleural thickening is visible at this level laterally.

and appear to be indicative of disease activity; these radiological abnormalities may ameliorate with appropriate therapy, leaving residual thin-walled empty cavities.

The demonstration of precipitating antibodies to *Aspergillus* spp. is the cornerstone and prerequisite for the diagnosis (Binder et al., 1982; Denning et al., 2003; Camuset et al., 2007). When precipitins are negative in the context of an otherwise compatible illness, batch-to-batch variability in kit antigen and the prospect of a non-fumigatus *Aspergillus* species as the causative organism should be considered and appropriate testing sought. *Aspergillus* spp. may be grown from respiratory specimens from both the upper and lower respiratory tract (40 to 65% of cases) and, on occasion, the pleural space. Other important diagnostic facets include the demonstration of elevated inflammatory markers, such as erythrocyte sedimentation rate and C-reactive protein, and the exclusion of other infectious, neoplastic, and inflammatory entities which may mimic this syndrome (Denning et al., 2003). There appears to be a particularly strong link with atypical mycobacteria, and these infections may be active simultaneously or sequentially. There is also the possibility of a pyogenic infection in a cavity which may require drainage and appropriate antibacterial therapy.

Chronic fibrosing pulmonary aspergillosis

Chronic fibrosing pulmonary aspergillosis is generally an inferred diagnosis based on a current or prior diagnosis of chronic cavitary disease and/or aspergilloma, extensive upper or lower lobe fibrosis documented on radiology, and poor lung function (Denning et al., 2003). In those cases in which pulmonary biopsies have been done, the histopathology shows extensive fibrosis with a minor chronic inflammatory infiltrate.

Sinusitis or Rhinosinusitis

There are three entities that are subsumed under the broad label of sinusitis or rhinosinusitis: fungus ball of the sinus, chronic invasive rhinosinusitis, and chronic granulomatous rhinosinusitis. Localized fungal colonization of a sinus wall is also clinically recognized.

Fungus ball of the paranasal sinuses

Fungus ball is most commonly caused by *A. fumigatus*, but other *Aspergillus* species occasionally cause this disease, as may *Scedosporium apiospermum* and other rare fungi. It is more common in women (60%) in the middle years of life. It is usually limited to the maxillary sinus but may rarely involve the sphenoid sinus or frontal sinus, although the management of a fungus ball in the sphenoid sinus requires careful consideration. Many patients are asymptomatic or may present with recurrent or chronic sinusitis. The sinus cavity is filled with a fungal ball consisting of multiple layers of fungal hyphae, with very little inflammatory exudate either within the fungal ball or surrounding it (Ferreiro et al., 1997). In the maxillary sinus it is sometimes associated with prior upper jaw root canal work and chronic (bacterial) sinusitis (Stammberger et al., 1984). CT and magnetic resonance (MR) findings are characteristic, as about 90% of CT scans show focal hyperattenuation related to concretions (Zinreich et al., 1988) and the T2-weighted signal on MR scan is decreased, whereas that of bacterial sinusitis is increased. The recommended criteria for diagnosis are shown in Table 2. No tissue invasion is demonstrable histologically or radiologically in cases of a fungal ball of any sinus. Rare instances of isolated neurological deficits of the cranial nerves have been observed with fungus balls, the etiology of which is obscure.

Occasionally a fungal ball is found in the sphenoid sinus (Klossek et al., 1996). The appearance on CT or MR scan may not differ much from chronic invasive *Aspergillus* sinusitis in this location, if no extension beyond the sinus is seen. Removal of a sphenoid fungus ball may precede localized invasion with devastating consequences, such as cavernous sinus thrombosis, carotid artery invasion, or extension into the orbit or brain. In the absence of consensus recommendations, the most conservative approach is to not biopsy the sinus mucosa if it appears normal.

Localized fungal colonization

Sometimes a small crust is found on a sinus wall, especially the maxillary sinus, on which is seen macroscopic fungal growth and from which *Aspergillus* species may be grown. This entity is noninvasive but its management has not been well studied; endoscopic cleaning is appropriate in the first instance (Ferguson, 2000). It may recur, and then saline irrigation is appropriate.

Table 2. Criteria for the diagnosis of fungus ball of the sinus[a]

Criteria
Radiological evidence of sinus opacification with or without calcification
Mucopurulent cheesy or clay-like materials within the sinus
Dense conglomeration of hyphae (fungus ball) separate from the sinus mucosa
Nonspecific chronic inflammation (lymphocytes, plasma cells, eosinophils) of the mucosa
No predominance of eosinophils, no granuloma, no allergic mucin
No histological evidence of fungal invasion of mucosa, blood vessels, or bone visualized microscopically after special stains for fungi

[a] Source: Grosjean and Weber (2007).

Without treatment it may progress to a fungus ball or to chronic invasive rhinosinusitis, although the natural history is not well documented.

Chronic invasive and granulomatous *Aspergillus* rhinosinusitis

Chronic invasive rhinosinusitis is a slowly destructive process that most commonly affects the ethmoid and sphenoid sinuses but may involve any paranasal sinus. Patients are usually somewhat immunocompromised (e.g., diabetes mellitus, human immunodeficiency virus infection) but may be immunocompetent. A common presentation includes chronic nasal discharge and blockage, loss of smell, and persistent headache. Perhaps most commonly, the presentation features local involvement of critical structures. The orbital apex syndrome is characteristic (blindness and proptosis), but facial swelling, cavernous sinus thrombosis, or carotid artery occlusion (Clancy and Nguyen, 1998), pituitary fossa, or skull base invasion have been described (Swift and Denning, 1998). Chronic granulomatous *Aspergillus* sinusitis usually presents with unilateral proptosis (Milosev et al., 1969) caused by a painless, hard, irregular, relatively avascular mass which mimics malignancy. The ethmoid sinuses are the most commonly affected structures, with pan-sinusitis developing in most patients (Yagi et al., 1999; Alrajhi et al., 2001). Symptoms are present for a mean of 16 months before diagnosis. There is frequent local involvement of local structures with intracranial involvement in at least 50% of cases at the time of diagnosis (Alrajhi et al., 2001). Bone erosion is a common finding.

Imaging of the cranial sinuses shows opacification of one or more sinuses, local boney destruction, and invasion of local structures. Both chronic invasive and granulomatous *Aspergillus* sinusitis are characterized by a mass within one or more paranasal sinuses with evidence of invasion of contiguous structures, such as the base of the skull, orbit, and brain. Chronic granulomatous disease may mimic malignancy until the tissue is examined. Some patients with a fungal ball of a sinus or patients with eosinophilic fungal rhinosinusitis have expansion of the boney margin of a sinus, with either boney sclerosis or thinning, but this should not be confused with localized invasion. Some patients develop retro-orbital chronic invasive aspergillosis, which probably originated from the paranasal sinuses, but no sinus disease is visible on imaging.

In chronic invasive disease the sinus mass is composed of friable, necrotic, or purulent material (Milroy et al., 1989; Clancy and Nguyen, 1998; Panda et al., 1998), but in chronic granulomatous disease the mass is firm and difficult to remove surgically (Alrajhi et al., 2001). In chronic invasive disease fungal hyphae are seen invading contiguous structures, whereas in chronic granulomatous disease tissue destruction occurs as a result of expansion of the mass. In chronic invasive disease the histological appearance is that of a sparse low-grade mixed cellular infiltrate without prominent eosinophils, fibrosis, or granulomas, whereas in chronic granulomatous disease there are large numbers of loose, poorly formed granulomas surrounding giant cells containing fungal hyphae (visualized in approximately 50% of cases) with many lymphocytes and plasma cells. Both fibrotic and necrotic histological variants of chronic granulomatous disease have been described (Veress et al., 1973). The vascular appearance, nevertheless, is often abnormal; vessels are characteristically thickened with luminal narrowing due to intimal proliferation with a periarterial, cuff-like chronic inflammatory infiltrate progressing to concentric lamellated dense fibrous tissue without demonstrable hyphal elements (Veress et al., 1973). Hyphae characteristic of *Aspergillus* are seen in both chronic invasive and granulomatous disease and are relatively sparse in both conditions. Vascular invasion is not seen in either condition. The fungal differential diagnosis is wide, as numerous other fungi may cause a similar disease and sphenoid sinusitis is more often caused by bacteria. Cultures of tissue are positive in >50% of cases. Species-specific precipitating antibodies are present in >90% of cases in chronic invasive disease (Chakrabarti and Sharma, 2000). Precipitating antibodies to *Aspergillus* spp. are present in around two-thirds of cases, and a positive IgG radioallergosorbent test to *A. flavus* is also frequently seen in chronic granulomatous disease (Milosev et al., 1969; Yagi et al., 1999). Serological data may be used as indirect evidence to implicate *Aspergillus* spp. when cultures are negative or not obtained.

Aspergillus Bronchitis

Numerous cases of *Aspergillus* tracheobronchitis of differing severity have been described in immunocompromised patients, but few have been noted in nonimmunocompromised patients. One series highlighted this issue in cystic fibrosis patients (Shoseyov et al., 2006). In these patients, deteriorating respiratory function, weight loss, and positive cultures of *A. fumigatus* from their sputum were found, and they improved with antifungal therapy. Some had elevated total IgE and others had positive *Aspergillus* skin tests or specific IgE results. Peribronchial thickening and/or infiltrates were visible on CT scans in some patients. *Aspergillus* precipitins were not measured. The author has seen a few cases of *Aspergillus* bronchitis in the context of other chronic respiratory disease, but little has been written about it in the literature.

Aspergillus Otitis

The symptomatology of all fungal infections of the external ear, of which *Aspergillus* is the most common cause, indicates that pruritus and discharge are most common, with reddened epidermis and lining of the tympanic cavity being common (Kurnatowski and Filipiak, 2001). *A. niger* complex is the most common cause, and *A. fumigatus* isolated in this context should prompt consideration of invasive *Aspergillus* otitis. It may follow mild trauma to the external ear canal from cleaning attempts or other inflammatory processes. Examination usually reveals erythema and discharge in the external canal, sometimes with white plaque-like lesions over the tympanic membrane. Occasionally the infecting *Aspergillus* strain conidiates in situ and the lesion appears black. Rarely, the organism invades beyond the superficial layers of the external ear canal to cause invasive external otitis.

Primary Cutaneous Aspergillosis and Onychomycosis

Most cases of cutaneous aspergillosis are related to disseminated aspergillosis in immunocompromised patients. Less commonly it is seen in burn patients and in premature neonates. Wound infections, particularly associated with *A. flavus*, have also been reported, as are Hickman and other intravenous catheter infections (Pasqualotto and Denning, 2006). This section, however, focuses exclusively on chronic cutaneous aspergillosis in nonimmunocompromised patients. Some cases appear to follow an injury, including burns. The clinical course is chronic and non-life-threatening. Localized swelling or an ulcer appears to be the most common manifestation of infection. *A. fumigatus, A. flavus, A. niger*, and *A. terreus* have all been implicated (Chakrabarti et al., 1998; Ozcan et al., 2003). Biopsy and culture of the affected tissue establish the diagnosis.

Many different species of *Aspergillus* have been reported to cause onychomycosis, including *A. fumigatus, A. versicolor, A. niger, A. terreus*, and some rare species (Rosenthal, 1968; Torres-Rodríguez et al., 1998; Piérard, 2001; Veer et al., 2007). Among nondermatophyte mold onychomycoses, proportional rates of *Aspergillus* onychomycosis vary from 5% to as high as 30% (Hilmioglu-Polat et al., 2005; Romano et al., 2005a, 2005b; Gupta et al., 2007). There are two common patterns of disease, destructive and superficial white onychomycosis (Onsberg et al., 1978; McAleer, 1981; Piraccini and Tosti, 2004), but proximal, lateral, and distal onychomycosis may also be seen (Tosti and Piraccini, 1998; Bonifaz et al., 2007). The particular clinical features suggestive of *Aspergillus* infection of the nail are a chalky, deep white nail with rapid involvement of the lamina and painful perionyxis without suppuration (Gianni and Romano, 2004). The affected nail may have been previously subjected to trauma and is most often a toenail.

ANTIFUNGAL AND SURGICAL TREATMENT AND MANAGEMENT OF COMPLICATIONS

Aspergilloma

Asymptomatic patients should be observed for progression to CCPA (common) or spontaneous resolution (<10% of cases). Patients with single (or simple) aspergillomas generally do well with surgical resection (El-Oakley et al., 1997; Chen et al., 1997; Regnard et al., 2000; Kim et al., 2005; Shiraishi et al., 2006; Demir et al., 2006; Pratap et al., 2007). Patients with life-threatening or significant hemoptysis should undergo resection. All patients should be given pre- and postoperative antifungal agents (e.g., voriconazole) to prevent pleural aspergillosis and bronchopleural fistulae. Prolonged sedation because of an interaction with benzodiazepines may occur with voriconazole. Young women with localized disease may prefer surgery, as azole treatment is contraindicated in pregnancy. More recent surgical series have found substantially lower mortality and morbidity rates, particularly if patients are selected carefully, i.e., rates of <5% (Kim et al., 2005; Shiraishi et al., 2006; Demir et al., 2006; Pratap et al., 2007). Complications include *Aspergillus* empyema, persistent air leak, persistent pleural space, empyema, bronchopleural fistula, respiratory insufficiency, and significant intercurrent infections, although all these are much more common in cases of chronic cavitary disease or complex aspergillomas.

There are some data supporting the use of amphotericin B instillation through flexible plastic catheters (Giron et al., 1995). Repeated instillations are usually necessary (in one study this was done daily for 15 days [Lee et al., 1993]). Communication between the cavity and the airways is usual, so the instilled agent usually leaks into the airways. Repeated instillations are labor-intensive and not very effective for complex and/or bilateral aspergillomas. Recently described is the incorporation of amphotericin B in gelatin or glycerin that solidifies at 37°C (Munk et al., 1993; Giron et al., 1995). In the United Kingdom this approach has become difficult to deliver because of regulations concerning aseptic drug preparation and the unavailability of one key pharmaceutical component.

CCPA

Itraconazole and voriconazole are the preferred oral agents for CCPA (Herbrecht et al., 2002; Denning et al., 2003; Jain and Denning, 2006; Sambatakou et al., 2006; Camuset et al., 2007; Walsh et al., 2008), with posaconazole being substituted for failure, toxicity, or emergence of resistance. Itraconazole capsules are used at a dose of 200 mg twice daily in adults, with confirmation of sufficient drug concentrations in the blood to ensure adequate exposure (e.g., >1 mg/liter by high-performance liquid chromatography or >5 mg/liter by bioassay). Itraconazole suspension is useful for those on treatment with proton pump inhibitors or H2 blockers to optimize bioavailability. Voriconazole is used at the same dose, although many patients require dosage adjustment after plasma drug concentration measurement. About 40% of patients on either itraconazole or voriconazole require a dose or preparation adjustment. Low concentrations of an antifungal agent probably predispose to itraconazole resistance (and azole cross-resistance). Generic preparations of itraconazole may have lower bioavailability (Pasqualotto and Denning, 2007). High concentrations of itraconazole and voriconazole may be associated with adverse events (Boyd et al., 2003; Imhof et al., 2006; Pascual et al., 2008), and drug costs can be saved by lowering the dose. Liver function tests and potassium levels should be monitored frequently in the first 3 months of therapy and rarely become abnormal after 6 months of therapy.

CCPA probably requires life-long therapy; data are sparse, but relapse is common, as expected in patients with immune defects. Resistance to one or more azoles may occur during long-term treatment, and a positive culture while on antifungal therapy is an indication for susceptibility testing, or if such testing is not possible, then a switch of therapy. Itraconazole resistance is considerably more common than voriconazole resistance. Itraconazole-resistant isolates may be cross-resistant to voriconazole, although they usually are not, but may have elevated MICs to posaconazole, the clinical implication of which is not yet clear. Azole-resistant isolates are not cross-resistant to amphotericin B or echinocandins.

Response to therapy can be determined by assessing symptoms of chronic ill health, weight, and pulmonary symptoms, especially cough, sputum volume, and breathlessness (Denning et al., 2003; Jain et al., 2006; Camuset et al., 2007). A falling *Aspergillus* antibody (precipitins) titer is the most useful laboratory parameter to follow. Recurrence of hemoptysis may be a sign of antifungal failure, but lack of resolution may reflect either failure or persistence of a large vascular network which would not be expected to resolve rapidly with control of the fungal infection.

Patients who fail oral therapy or who are very ill with extensive disease may require a 3- to 4-week course of intravenous amphotericin B therapy. Maintenance of improved health on oral azoles is then achievable, even if therapeutic failure with azoles was first documented. Gamma interferon therapy may also be useful (based on anecdotal evidence only) (Denning et al., 2003). Glucocorticoids at modest doses (or anabolic steroids) should be used with caution in CCPA, especially early in the treatment course before the antifungal response can be assessed, but they may have a place in management for some patients who remain chronically ill but have evidence of control of their chronic *Aspergillus* infection.

In patients with moderate or severe hemoptysis, embolization may be appropriate if the patient is not fit to undergo surgery or has extensive disease. In most instances of hemoptysis, abnormal and novel vascular connections to the systemic circulation are implicated. Usually this is the bronchial circulation, but it may be any of the other arteries supplying the chest, e.g., internal, subclavian, or internal mammary arteries (Fig. 2). Aspergillomas also lead to an extensive network of small vessels. Several abnormal connections may exist in a single patient. The objective of embolization is to permanently occlude these vessels. Patients with a communication between an intercostal and the anterior spinal artery can only be embolized safely if the catheter is introduced well past the anterior spinal artery. These patients require a skilled interventional radiologist for these difficult procedures, and the patient has to lie still and flat for 2 to 4 h during the procedure, which can be difficult. Depending on the radiologist, approximately 50 to 90% of embolization procedures are successful (Jardin and Remy, 1988; Corr, 2006). However, a relapse rate of 50% is typical, which can be minimized with long-term antifungal therapy. Complications of local pain, stroke, or chest wall or spinal cord infarction may occur, as well as reactions to the radiological dye.

Temporary relief of hemoptysis may be gained with tranexamic acid, although it is not licensed for this indication. It is widely used throughout the United Kingdom for hemoptysis. Tranexamic acid increases infarction potential, and a low rate of strokes, etc., is reported after its use and it may not be well tolerated.

Surgery is problematic for CCPA, as it usually results in serious complications, and embolization is preferred for serious hemoptysis. Surgical removal of aspergillomas is fraught with difficulty because of the very vascular, adherent pleura and because the remaining chest cavity may become infected with *Aspergillus* (Daly et al., 1986). Surgical removal of pleural aspergillomas and thoracoplasty is also prone to many complications and should be avoided if possible (Massard et al., 1983). In addition, many patients have underlying respiratory

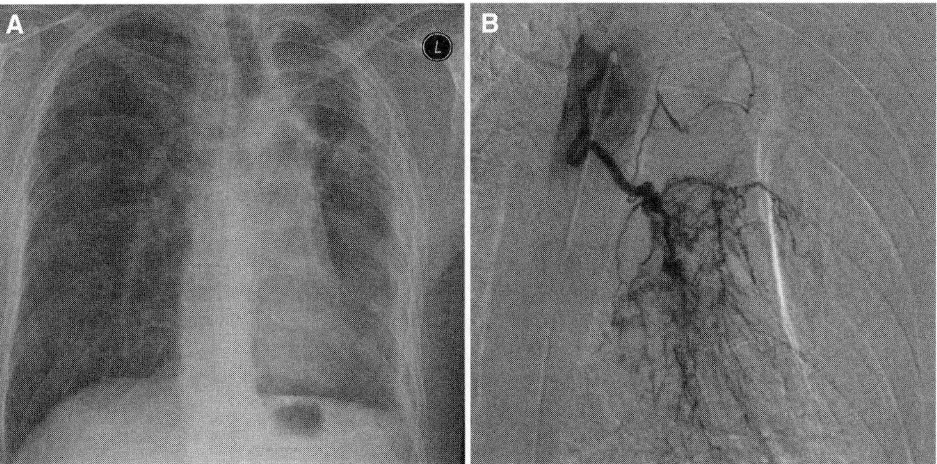

Figure 2. (A) Chest radiograph from a patient with CCPA, showing a single large cavity occupying most of the left upper lobe, with minor overlying pleural thickening and no aspergilloma. There are some pericavitary infiltrates and possibly one or two additional cavities inferior to the main large cavity. (B) Arteriogram of a left bronchial artery approached via a right common femoral approach. An arch aortogram showed some hypervascularity in the left upper lobe. The aorta was explored with a variety of catheters. There were four vessels with abnormal vascularity distally: two bronchial arteries, an intercostal artery, and a large medial branch of the internal mammary artery. The bronchial artery is shown. There is a large hypervascular supply to the mid-portion of the left lung. It was not possible to obtain a stable catheter position, and with the catheter at the orifice there was reflux of contrast into the aorta (just visible). Neither the conventional catheter nor a microcatheter could be advanced deeper into the vessel in order to obtain a satisfactory position for embolization, so this vessel was not embolized, although two others were.

insufficiency, and removal of a lobe of the lung would leave them unacceptably breathless. However, several recent series have described better results, with 2 to 5% mortality and an ~25% complication rate overall (Kim et al., 2005; Shiraishi et al., 2006; Demir et al., 2006; Pratap et al., 2007). This may be preferable to long-term antifungal therapy for some patients.

Fungus Ball of the Sinus

Fungus balls of the maxillary sinus can usually be removed solely with endoscopic surgery. If some involvement of the ipsilateral ethmoid paranasal sinuses is seen, these should also be treated endoscopically at the same time. If calcified or otherwise impossible to remove, a Caldwell Luc procedure is occasionally required. No local or systemic antifungal therapy is required. Recurrence is uncommon. In cases of sphenoid fungal balls, treatment with 4 weeks of oral antifungal therapy at treatment doses (e.g., 400 mg daily of voriconazole or itraconazole) may be the most cautious approach to avoid any serious consequences related to local invasion.

Chronic Invasive and Granulomatous *Aspergillus* Rhinosinusitis

For chronic invasive and granulomatous *Aspergillus* rhinosinusitis, surgical debridement is required, along with aeration of the sinuses. Sometimes this can be done endoscopically, but much more commonly it requires major open surgery. The surgical approach is determined by the CT findings, and extensive involvement of cranial structures, including the brain, may require involvement of neurosurgeons and/or plastic surgeons. In all cases removal of as much of the affected bone and mucosa as possible, without infringing on other major structures, is the objective.

Amphotericin B is still the preferred primary therapy (Chakrabarti and Sharma, 2000), although the duration of therapy may be significantly shortened, with a switch to oral therapy to prevent relapse. Itraconazole has often been used as primary therapy, but it is often not successful (Khoo and Denning, 1995). Sometimes this is because therapeutic serum concentrations are not achieved and use of itraconazole solution may improve matters in some patients. In those who fail itraconazole, a 3- to 6-week course of amphotericin B (>0.8 mg/kg /day) or lipid-associated amphotericin B (3 to 5 mg/ kg/day) usually secures a remission. Voriconazole and posaconazole are alternative oral treatments.

Chronicity and relapse characterize this disease. A rapid reduction in *Aspergillus* antibody (precipitin) titer usually follows surgery. Further reduction in titer occurs with successful medical therapy, although this may take months. Follow-up should continue for about 5 years. Often it is not clear on presentation whether the patient has saprophytic sinusitis or chronic invasive disease, un-

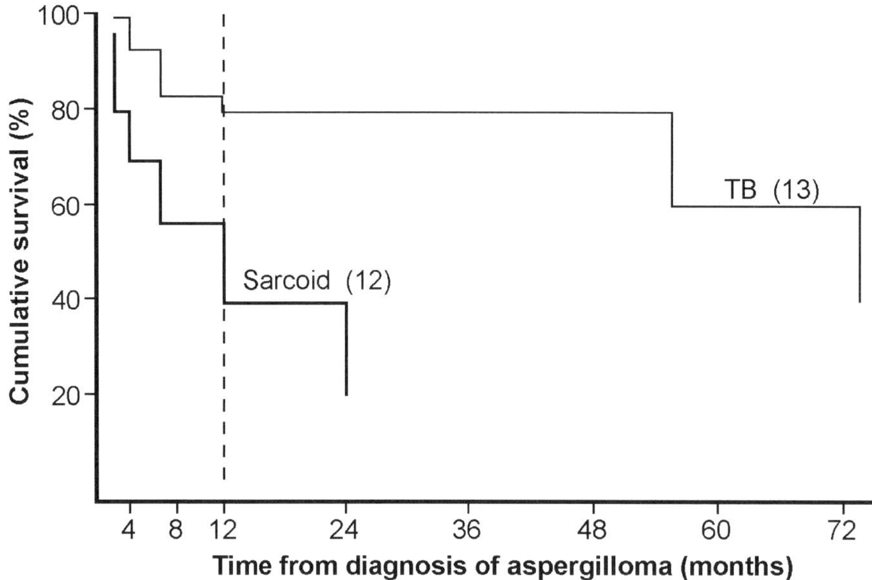

Figure 3. Survival (censored) over 6 years for patients with an "aspergilloma" in sarcoidosis compared to survival after TB. Taken from Tomlinson and Sahn (1987) with permission.

less there is clear-cut bone destruction on CT scanning. In these cases, histology of bone and mucosa should be diagnostic. In both chronic invasive and granulomatous disease, medical therapy should be started early and continued, with scans repeated 4 to 8 weeks after surgery. If there is no improvement or deterioration, then more extensive debridement will be necessary, possibly combined with alternative medical therapy.

Cutaneous Aspergillosis, Otitis, and Onychomycosis

Cutaneous aspergillosis has been treated with oral itraconazole for varying periods of time, but there are few data. External otitis is generally treated with local toilet, local antifungal ointment, such as econazole 1% cream, or other topical antifungal preparations. It may recur, and then oral itraconazole (200 mg twice daily) for 2 to 4 weeks usually suffices.

Several reports have described the efficacy of itraconazole (200 mg daily) for *Aspergillus* onychomycosis (Scher and Barnett, 1990; Tosti et al., 2000) and pulsed terbinafine (Gianni and Romano, 2004). The duration of therapy depends on which nails are affected and the extent of infection. Affected fingernails typically require 3 months of therapy and toenails at least 6 months. Topical amorolfine hydrochloride (0.25%) is not always active against *Aspergillus* species. If only one nail is affected, alternatives include avulsion of the nail or dissolution of the nail with urea paste (Denning et al., 1995).

PROGNOSIS

Historically, patients with chronic pulmonary aspergillosis have high morbidity and usually die early. However, in those patients with major hemoptysis and simple aspergillomas, surgery offers an 84% 5-year survival compared with a 41% survival with conservative therapy (Jewkes et al., 1983). Follow-up of a London series of 85 patients over 5 years in the 1970s and 1980s showed 41 deaths (48%), with diverse causes of death (Jewkes et al., 1983). An older review of patients managed in another institution who had either TB or sarcoidosis as the underlying cause of chronic pulmonary aspergillosis showed much higher and earlier mortality in the sarcoidosis patients (Tomlinson and Sahn, 1987), which was possibly attributable to the corticosteroid therapy presumably given to those with sarcoidosis (Fig. 3). This was in the era before oral antifungal therapy was available. Much-improved surgical and medical results are now reported, with long periods of remission on azole therapy or after surgical resection. The prognosis of a fungal ball of the maxillary sinus is also excellent, with a very low recurrence rate (<5%). Few data are available on the long-term outcomes of chronic invasive and chronic granulomatous sinus aspergillosis.

REFERENCES

Alrajhi, A. A., M. Enani, Z. Mahasin, and Al-Omran. 2001. Chronic invasive aspergillosis of the paranasal sinuses in immunocompetent hosts from Saudi Arabia. *Am. J. Trop. Med. Hyg.* **65:**83–86.

Anonymous. 1970. Aspergilloma and residual tuberculous cavities: the results of a resurvey. *Tubercle* **51:**227–245.

Beck-Mannagetta, J., and D. Necek. 1986. Radiologic findings in aspergillosis of the maxillary sinus. *Oral Surg. Oral Med. Oral Pathol.* **62:**345–349.

Beck-Mannagetta, J., D. Necek, and M. Grasserbauer. 1983. Solitary aspergillosis of maxillary sinus, a complication of dental treatment. *Lancet* **ii:**1260.

Bennett, J. H. 1842. On the parasitic vegetable structures found growing in live animals. *Trans. R. Soc. Edinburgh* **15:**277–279.

Binder, R. E., L. J. Faling, R. D. Pugatch, C. Mahasaen, and G. L. Snider. 1982. Chronic necrotizing pulmonary aspergillosis: a discrete clinical entity. *Medicine* (Baltimore) **61:**109–124.

Bonifaz, A., P. Cruz-Aguilar, and R. M. Ponce. 2007. Onychomycosis by molds. Report of 78 cases. *Eur. J. Dermatol.* **17:**70–72.

Boyd, A. E., S. Modi, S. J. Howard, C. B. Moore, B. G. Keevil, and D. W. Denning. 2004. Adverse reactions to voriconazole. *Clin. Infect. Dis.* **39:**1241–1244.

Butz, R. O., J. R. Zvetina, and B. J. Leininger. 1985. Ten-year experience with mycetomas in patients with pulmonary tuberculosis. *Chest* **87:**356–358.

Camuset, J., H. Nunes, M. C. Dombret, A. Bergeron, P. Henno, B. Philippe, G. Dauriat, G. Mangiapan, A. Rabbat, and J. Cadranel. 2007. Treatment of chronic pulmonary aspergillosis by voriconazole in nonimmunocompromised patients. *Chest* **131:**1435–1441.

Caras, W. E., and J. L. Pluss. 1996. Chronic necrotizing pulmonary aspergillosis: pathologic outcome after itraconazole therapy. *Mayo Clin. Proc.* **71:**25–30.

Carvalho, A., A. C. Pasqualotto, L. Romani, D. W. Denning, and F. Rodrigues. 2008. Polymorphisms in Toll-like receptors (TLR) genes and susceptibility to pulmonary aspergillosis. *J. Infect. Dis.* **197:** 618–621.

Chakrabarti, A., V. Gupta, G. Biswas, B. Kumar, and V. K. Sakhuja. 1998. Primary cutaneous aspergillosis: our experience in 10 years. *J. Infect.* **37:**24–27.

Chakrabarti, A., and S. C. Sharma. 2000. Paranasal sinus mycoses. *Indian J. Chest Dis. Allied Sci.* **42:**293–304.

Chen, J. C., Y. L. Chang, S. P. Luh, J. M. Lee, and Y. C. Lee. 1997. Surgical treatment for pulmonary aspergilloma: a 28 year experience. *Thorax* **52:**810–813.

Clancy, C. J., and M. H. Nguyen. 1998. Invasive sinus aspergillosis in apparently immunocompetent hosts. *J. Infect.* **37:**229–240.

Collins, M. M., S. B. Nair, and J.-P. Womold. 2003. Prevalence of noninvasive fungal sinusitis in South Australia. *Am. J. Rhinol.* **17:** 127–132.

Corr, P. 2006. Management of severe hemoptysis from pulmonary aspergilloma using endovascular embolization. *Cardiovasc. Intervent. Radiol.* **29:**807–810.

Crosdale, D., K. Poulton, W. Ollier, W. Thomson, and D. W. Denning. 2001. Mannose binding lectin gene polymorphisms as a susceptibility factor for chronic necrotising pulmonary aspergillosis. *J. Infect. Dis.* **184:**653–656.

Dall'Igna, C., B. C. Palombini, F. Anselmi, E. Araújo, and D. P. Dall'Igna. 2005. Fungal rhinosinusitis in patients with chronic sinusal disease. *Rev. Bras. Otorhinolaringol.* **71:**712.

Daly, R. C., P. C. Pairolero, J. M. Piehler, V. F. Trastek, W. Spencer Payne, and P. E. Bernatz. 1986. Pulmonary aspergilloma: results of surgical treatment. *J. Thorac. Cardiovasc. Surg.* **92:**981–988.

Dawlatly, E. E., J. T. Anim, S. Sowayan, and A. Y. el-Hassan. 1988. Primary paranasal *Aspergillus* granuloma in Saudi Arabia. *Trop. Geogr. Med.* **40:**247–250.

de Vallière, S., and R. D. Barker. 2004. Residual lung damage after completion of treatment for multidrug-resistant tuberculosis. *Int. J. Tuberc. Lung Dis.* **8:**767–771.

Demir, A., M. Z. Gunluoglu, A. Turna, H. V. Kara, and S. I. Dincer. 2006. Analysis of surgical treatment for pulmonary aspergilloma. *Asian Cardiovasc. Thorac. Ann.* **14:**407–411.

Denning, D. W., E. G. Evans, C. C. Kibbler, M. D. Richardson, M. M. Roberts, T. R. Rogers, D. W. Warnock, and R. E. Warren. 1995. Fungal nail disease: a guide to good practice (report of a Working Group of the British Society for Medical Mycology). *BMJ* **311:** 1277–1281.

Denning, D. W., K. Riniotis, R. Dobrashian, and H. Sambatakou. 2003. Chronic cavitary and fibrosing pulmonary and pleural aspergillosis: case series, proposed nomenclature change, and review. *Clin. Infect. Dis.* **37**(Suppl 3)**:**S265–S280.

Elliott, J. A., L. J. Milne, and D. Cumming. 1989. Chronic necrotising pulmonary aspergillosis treated with itraconazole. *Thorax* **44:**820–821.

El-Oakley, R., P. Petrou, and M. Goldstraw. 1997. Indications and outcome of surgery for pulmonary aspergilloma. *Thorax* **52:**813–815.

Ferguson, B. J. 2000. Definitions of fungal rhinosinusitis. *Otolaryngol. Clin. North Am.* **33:**227–235.

Ferreiro, J. A., B. A. Carlson, and D. T. Cody III. 1997. Paranasal sinus fungus balls. *Head Neck* **19:**481–486.

Gefter, W. B., T. R.Weingrad, D. M. Epstein, R. H. Ochs, and W. T. Miller. 1981. "Semi-invasive" pulmonary aspergillosis: a new look at the spectrum of *Aspergillus* infections of the lung. *Radiology* **140:** 313–321.

Gianni, C., and C. Romano. 2004. Clinical and histological aspects of toenail onychomycosis caused by *Aspergillus* spp.: 34 cases treated with weekly intermittent terbinafine. *Dermatology* **209:**104–110.

Giron, J., C. Poey, P. Fajadet, N. Sans, D. Fourcade, J. P. Senac, J. J. Railhac, M. Dahan, J. Berjaud, M. Krempf, P. Leophonte, A. Didier, R. Escamilla, P. Carles, P. Arlet, and D. Lauque. 1995. Palliative percutaneous treatment of inoperable pulmonary aspergilloma. *Rev. Mal. Respir.* **12:**593–599.

Greene, R. 2005. The radiological spectrum of pulmonary aspergillosis. *Med. Mycol.* **43**(Suppl. 1)**:**S147–S154.

Grosjean, P., and R. Weber. 2007. Fungus balls of the paranasal sinuses: a review. *Eur. Arch. Otorhinolaryngol.* **264:**461–470.

Gupta, D., A. Hansell, T. Nichols, T. Duong, J. G. Ayres, and D. Strachan. 2000. Epidemiology of pneumothorax in England. *Thorax* **55:**666–671.

Gupta, M., N. L. Sharma, A. K. Kanga, V. K. Mahajan, and G. R. Tegta. 2007. Onychomycosis: clinico-mycologic study of 130 patients from Himachal Pradesh, India. *Indian J. Dermatol. Venereol. Leprol.* **73:**389–392.

Gupta, A. K., J. E. Ryder, R. Baran, and R. C. Summerbell. 2003. Non-dermatophyte onychomycosis. *Dermatol. Clin.* **21:**257–268.

Hafeez, I., M. F. Muers, S. A. Murphy, E. G. Evans, R. C. Barton, and P. McWhinney. 2000. Non-tuberculous mycobacterial lung infection complicated by chronic necrotising pulmonary aspergillosis. *Thorax* **55:**717–719.

Hedayati, M. T., A. C. Pasqualotto, P. A. Warn, P. Bowyer, and D. W. Denning. 2007. *Aspergillus flavus*: human pathogen, allergen and mycotoxin producer. *Microbiology* **153:**1677–1692.

Herbrecht, R., D. W. Denning, T. F. Patterson, J. E. Bennett, R. E. Greene, J. W. Oestmann, W. V. Kern, K. A. Marr, P. Ribaud, O. Lortholary, R. Sylvester, R. H. Rubin, J. R. Wingard, P. Stark, C. Durand, D. Caillot, E. Thiel, P. H. Chandrasekar, M. R. Hodges, H. T. Schlamm, P. F. Troke, B. de Pauw, et al. 2000. Randomised comparison of voriconazole and amphotericin B in primary therapy of invasive aspergillosis. *N. Engl. J. Med.* **347:**408–415.

Hilmioglu-Polat, S., D. Y. Metin, R. Inci, T. Dereli, I. Kilinç, and E. Tümbay. 2005. Non-dermatophytic molds as agents of onychomycosis in Izmir, Turkey: a prospective study. *Mycopathologia* **160:** 125–128.

Hope, W. W., T. J. Walsh, and D. W. Denning. 2005. The invasive and saprophytic syndromes due to *Aspergillus* spp. *Med. Mycol.* **43**(Suppl. 1)**:**S207–S238.

Imhof, A., D. J. Schaer, U. Schanz, and U. Schwarz. 2006. Neurological adverse events to voriconazole: evidence for therapeutic drug monitoring. *Swiss Med. Wkly.* **136:**739–742.

Jain, L., and D. W. Denning. 2006. The efficacy and tolerability of voriconazole in the treatment of chronic cavitary pulmonary aspergillosis. *J. Infect.* **52:**e133–e137.

Jardin, M., and J. Remy. 1988. Control of hemoptysis: systemic angiography and anastomoses of the internal mammary artery. *Radiology* **168:**377–383.

Jewkes, J., P. H. Kay, M. Paneth, and K. M. Citron. 1983. Pulmonary aspergilloma: analysis of cavitating invasive pulmonary aspergillosis in immunocompromised patients. *Ann. Thorac. Surg.* **53:**621–624.

Kato, T., I. Usami, H. Morita, M. Goto, M. Hosoda, A. Nakamura, and S. Shima. 2002. Chronic necrotizing pulmonary aspergillosis in pneumoconiosis: clinical and radiologic findings in 10 patients. *Chest* **121:**118–127.

Khoo, S., and D. W. Denning. 1994. *Aspergillus* infection in the acquired immune deficiency syndrome. *Clin. Infect. Dis.* **19**(Suppl. 1): 541–548.

Kim, Y. T., M. C. Kang, S. W. Sung, and J. H. Kim. 2005. Good long-term outcomes after surgical treatment of simple and complex pulmonary aspergilloma. *Ann. Thorac. Surg.* **79:**294–298.

Klossek, J. M., L. Peloquin, P. J. Fourcroy, J. C. Ferrie, and J. P. Fontanel. 1996. Aspergillomas of the sphenoid sinus: a series of 10 cases treated by endoscopic sinus surgery. *Rhinology* **34:**179–183.

Kurnatowski, P., and A. Filipiak. 2001. Otomycosis. Prevalence, clinical symptoms, therapeutic procedure. *Mycoses* **44:**472–479.

Lee, J. J., P. Y. Chong, C. B. Lin, A. H. Hsu, and C. C. Lee. 2008. High resolution chest CT in patients with pulmonary tuberculosis: characteristic findings before and after antituberculous therapy. *Eur. J. Radiol.* **67:**100–104.

Lee, K. S., H. T. Kim, Y. H. Kim, and K. O. Choe. 1993. Treatment of hemoptysis in patients with cavitary aspergilloma of the lung: value of percutaneous instillation of amphotericin B. *Am. J. Roentgenol.* **161:**727–731.

Massard, G., J. M. Roeslin, P. Wihlm, J. P. Dumont, N. Witz, and G. Morand. 1993. Surgical treatment of pulmonary and bronchial aspergilloma. *Ann. Chir.* **47:**141–151.

McAleer, R. 1981. Fungal infections of the nails in Western Australia. *Mycopathologia.* **73:**115–120.

McConnochie, K., M. O'Sullivan, J. F. Khalil, M. H. Pritchard, and A. R. Gibbs. 1989. *Aspergillus* colonization of pulmonary rheumatoid nodule. *Respir. Med.* **83:**157–160.

McGregor, D. H., C. J. Papasian, and P. D. Pierce. 1989. Aspergilloma within cavitating pulmonary adenocarcinoma. *Am. J. Clin. Pathol.* **91:**100–103.

Milosev, B., S. el-Mahgoub, O. A. Aal, and A. M. el-Hassan. 1969. Primary aspergilloma of paranasal sinuses in the Sudan. A review of seventeen cases. *Br. J. Surg.* **56:**132–137.

Milroy, C. M., J. D. Blanshard, S. Lucas, and L. Michaels. 1989. Aspergillosis of the nose and paranasal sinuses. *J. Clin. Pathol.* **42:** 123–127.

Munk, P. L., A. D. Vellet, R. N. Rankin, N. L. Muller, and D. Ahmad. 1993. Intracavitary aspergilloma: transthoracic percutaneous injection of amphotericin gelatin solution. *Radiology* **188:**821–823.

Onsberg, P., D. Stahl, and N. K. Veien. 1978. Onychomycosis caused by *Aspergillus terreus*. *Sabouraudia* **16:**39–46.

Ozcan, M., K. M. Ozcan, A. Karaarslan, and F. Karaarslan. 2003. Concomitant otomycosis and dermatomycoses: a clinical and microbiological study. *Eur. Arch. Otorhinolaryngol.* **260:**24–27.

Panda, N. K., S. C. Sharma, A. Chakrabarti, and S. B. Mann. 1998. Paranasal sinus mycoses in north India. *Mycoses* **41:**281–286.

Pascual, A., T. Calandra, S. Bolay, T. Buclin, J. Bille, and O. Marchetti. 2008. Voriconazole therapeutic drug monitoring in patients with invasive mycoses improves efficacy and safety outcomes. *Clin. Infect. Dis.* **46:**201–211.

Pasqualotto, A. C., and D. W. Denning. 2006. Post-operative aspergillosis. *Clin. Microbiol. Infect.* **12:**1060–1076.

Pasqualotto, A. C., and D. W. Denning. 2007. Generic substitution of itraconazole resulting in clinical failure and resistance. *Int. J. Antimicrob. Agents* **30:**93–94.

Piérard, G. 2001. Onychomycosis and other superficial fungal infections of the foot in the elderly: a pan-European survey. *Dermatology* **202:**220–224.

Piraccini, B. M., and A. Tosti. 2004. White superficial onychomycosis: epidemiological, clinical, and pathological study of 79 patients. *Arch. Dermatol.* **140:**696–701.

Plaignaud, M. 1791. Observation sur fungus de sinus maxillaire. *J. Chir.* **1:**11–116.

Pratap, H., R. K. Dewan, L. Singh, S. Gill, and S. Vaddadi. 2007. Surgical treatment of pulmonary aspergilloma: a series of 72 cases. *Indian J. Chest Dis. Allied Sci.* **49:**23–27.

Procknow, J. J., and D. F. Loewen. 1960. Pulmonary aspergillosis with cavitation secondary to histoplasmosis. *Am. Rev. Respir. Dis.* **82:** 101–111.

Ramani, R., P. Hazarika, R. D. Kapadia, and P. G. Shivananda. 1994. Invasive maxillary aspergillosis in an otherwise healthy individual. *Ear Nose Throat J.* **73:**420–422.

Regnard, J. F., P. Icard, M. Nicolosi, L. Spagiarri, P. Magdeleinat, B. Jauffret, and P. Levasseur. 2000. Aspergilloma: a series of 89 surgical cases. *Ann. Thorac. Surg.* **69:**898–903.

Romano, C., M. Papini, A. Ghilardi, and C. Gianni. 2005. Onychomycosis in children: a survey of 46 cases. *Mycoses* **48:**430–437.

Romano, C., C. Gianni, and E. M. Difonzo. 2005. Retrospective study of onychomycosis in Italy: 1985–2000. *Mycoses* **48:**42–44.

Rosenthal, S. A., R. Stritzler, and J. Vilafane. 1968. Onychomycosis caused by *Aspergillus fumigatus*. Report of a case. *Arch. Dermatol.* **97:**685–687.

Sadikot, R. T., T. Greene, K. Meadows, and A. G. Arnold. 1997. Recurrence of primary spontaneous pneumothorax. *Thorax* **52:** 805–809.

Sambatakou, H., V. Pravica, I. Hutchinson, and D. W. Denning. 2006. Cytokine profiling of pulmonary aspergillosis. *Int. J. Immunogenet.* **33:**297–302.

Scher, R. K., and J. M. Barnett. 1990. Successful treatment of *Aspergillus* flavus onychomycosis with oral itraconazole. *J. Am. Acad. Dermatol.* **23:**749–750.

Severo, L. C., G. R. Geyer, N. S. Porto, M. B. Wagener, and A. T. Londero. 1997. Pulmonary *Aspergillus niger* intracavitary colonization. Report of 23 cases and a review of the literature. *Rev. Iberoam. Micol.* **14:**104–110.

Shiraishi, Y., N. Katsuragi, Y. Nakajima, M. Hashizume, N. Takahashi, and Y. Miyasaka. 2006. Pneumonectomy for complex aspergilloma: is it still dangerous? *Eur. J. Cardiothorac. Surg.* **29:**9–13.

Shoseyov, D., K. G. Brownlee, S. P. Conway, and E. Kerem. 2006. *Aspergillus* bronchitis in cystic fibrosis. *Chest* **130:**222–226.

Singh, P., P. Kumar, R. P. Bhagi, and R. Singla. 1989. Pulmonary aspergilloma: radiological observations. *Indian J. Chest Dis. Allied Sci.* **31:**177–185.

Sonnenberg, P., J. Murray, J. R. Glynn, S. Shearer, B. Kambashi, and P. Godfrey-Faussett. 2001. HIV-1 and recurrence, relapse, and reinfection of tuberculosis after cure: a cohort study in South African mineworkers. *Lancet* **358:**1687–1693.

Stammberger, H., R. Jakse, and F. Beaufort. 1984. Aspergillosis of the paranasal sinuses: X-ray diagnosis, histopathology, and clinical aspects. *Ann. Otol. Rhinol. Laryngol.* **93:**251–256.

Stiksa, G., G. Eklundh, L. Riebe, and B. G. Simonsson. 1976. Bilateral pulmonary aspergilloma in ankylosing spondylitis treated with transthoracic instillations of antifungal agents. *Scand. J. Respir. Dis.* **57:**163–170.

Surjushe, A., R. Kamath, C. Oberai, D. Saple, M. Thakre, S. Dharmshale, and A. Gohil. 2007. A clinical and mycological study of onychomycosis in HIV infection. *Indian J. Dermatol. Venereol. Leprol.* **73**:397–401.

Surya Prakash Rao, G., S. B. Mann, P. Talwar, and M. M. Arora. 1984. Primary mycotic infection of paranasal sinuses. *Mycopathologia* **84**:73–76.

Swift, A. C., and D. W. Denning. 1998. Skull base osteitis following fungal sinusitis. *J. Laryngol. Otol.* **112**:92–97.

Tomlinson, J. R., and S. A. Sahn. 1987. Aspergilloma in sarcoid and tuberculosis. *Chest* **92**:505–508.

Torres-Rodríguez, J. M., N. Madrenys-Brunet, M. Siddat, O. López-Jodra, and T. Jimenez. 1998. *Aspergillus versicolor* as cause of onychomycosis: report of 12 cases and susceptibility testing to antifungal drugs. *J. Eur. Acad. Dermatol. Venereol.* **11**:25–31.

Tosti, A., B. M. Piraccini, and S. Lorenzi. 2000. Onychomycosis caused by nondermatophytic molds: clinical features and response to treatment of 59 cases. *J. Am. Acad. Dermatol.* **42**:217–224.

Tosti, A., and B. M. Piraccini. 1998. Proximal subungual onychomycosis due to *Aspergillus niger*: report of two cases. *Br. J. Dermatol.* **139**:156–157.

Vaid, M., S. Savneet Kaur, H. Sambatakou, T. Madan, D. W. Denning, and P. U. Sarma. 2007. Association of distinct alleles of MBL and SP-A with chronic cavitary pulmonary aspergillosis. *Clin. Chem. Lab. Med.* **45**:183–186.

Veer, P., N. S. Patwardhan, and A. S. Damle. 2007. Study of onychomycosis: prevailing fungi and pattern of infection. *Indian J. Med. Microbiol.* **25**:53–56.

Veress, B., O. A. Malik, A. A. el-Tayeb, S. el-Daoud, E. S. Mahgoub, and A. M. el-Hassan. 1973. Further observations on the primary paranasal *Aspergillus* granuloma in the Sudan: a morphological study of 46 cases. *Am. J. Trop. Med. Hyg.* **22**:765–772.

Walsh, T. J., E. J. Anaissie, D. W. Denning, R. Herbrecht, D. P. Kontoyiannis, K. A. Marr, V. A. Morrison, B. H. Segal, W. J. Steinbach, D. A. Stevens, J. A. van Burik, J. R. Wingard, T. F. Patterson, et al. 2008. Treatment of aspergillosis: clinical practice guidelines of the Infectious Diseases Society of America. *Clin. Infect. Dis.* **46**: 327–360.

Wollschlager, C., and F. Khan. 1984. Aspergillomas complicating sarcoidosis. *Chest* **86**:565–588.

Yagi, H. I., S. A. Gumaa, A. L. Shumo, N. Abdalla, and A. A. Gadir. 1999. Nasosinus aspergillosis in Sudanese patients: clinical features, pathology, diagnosis, and treatment. *J. Otolaryngol.* **28**:90–94.

Zaror, L., O. Fischman, F. A. Suzuki, and R. G. Felipe. 1991. Otomycosis in São Paulo. *Rev. Inst. Med. Trop. Sao Paulo* **33**:169–173.

Zinreich, S. J., D. W. Kennedy, J. Malat, H. D. Curtin, J. I. Epstein, L. C. Huff, A. J. Kumar, M. E. Johns, and A. E. Rosenbaum. 1988. Fungal sinusitis: diagnosis with CT and MR imaging. *Radiology* **169**:439–444.

Aspergillus fumigatus and Aspergillosis
Edited by J.-P. Latgé and W. J. Steinbach

Chapter 26

Allergic Bronchopulmonary Aspergillosis

RICHARD B. MOSS

ASPERGILLUS AND RESPIRATORY DISEASE

The ubiquitous dimorphic fungal genus *Aspergillus* seems unique in its ability to cause both invasive life-threatening respiratory infection and hypersensitivity respiratory illness in humans (Casadevall and Pirofski, 1999). Most *Aspergillus* disease is due to *Aspergillus fumigatus*, but at least 17 of some 175 other species, such as *A. niger, A. terreus, A. nidulans*, and *A. flavus*, have been reported to cause clinical infection or allergic disease (Moss, 2005; Tillie-Leblond and Tonnel, 2005; Zander, 2005). Airborne sampling data suggest *Aspergillus* is the most prevalent of airborne fungal spores. *A. fumigatus* spores 3 to 5 μm in diameter can penetrate to distal human airways, allowing inhalational sensitization to fungal antigens or germination into mycelial growth if conidia are not cleared by local primarily innate host defense mechanisms.

Although *Aspergillus* species grow well at body temperature, *A. fumigatus* can also grow at higher temperatures and low-oxygen environments, such as those found in compost piles (Hall and Denning, 1994; Pitt, 1994). *Aspergillus* spores germinate and grow in the mycelial phase by extension of 7- to 10-μm septate hyphae that branch at an angle of 45°, with *A. fumigatus* capable of rapid growth rates (mean doubling time of <50 min and hyphal extension rates of up to 2 cm/h). Of possible relevance to human respiratory pathogenesis, *A. fumigatus* growth in vitro accelerates in the presence of the glucocorticosteroid hydrocortisone (Ng et al., 1994). *Aspergillus* spores resist degradation in part due to a hydrophobic outer protein layer, the pigment of which confers antiphagocytic protection (Jahn et al., 2000). Most *A. fumigatus* isolates are sensitive to amphotericin B and triazoles, but resistance to both classes of antifungals has been reported (Moore et al., 2000).

SPECTRUM OF LUNG DISEASE DUE TO *ASPERGILLUS*

An astonishing variety of lung diseases secondary to *A. fumigatus* and occasionally other species of *Aspergillus* has been noted in which the elements of the host response seem primarily to drive clinical manifestations. Three main categories of disease linked to host immunocompetence have been recognized: invasive, saprophytic (mycetomal), and allergic (Buckingham and Hansell, 2003; Al-Alawi et al., 2005) (Table 1). Immunocompromised hosts are susceptible to saprophytic and invasive forms of aspergillosis, while immunocompetent hosts with impaired mucociliary clearance and/or other host risk factors are susceptible to allergic and occasionally saprophytic disease (if there has been previous cavitation).

Host defenses against *A. fumigatus* apparently rely upon multiple innate mechanisms of immunity, including the following: mucociliary clearance; phagocytosis by Toll-like receptor-bearing monocytes and macrophages, neutrophils, dendritic cells, and natural killer cells; and soluble factors, such as collectins, antimicrobial peptides, chemokines, and cytokines (Allen et al., 2001; Hartl et al., 2006; Madan et al., 2001; Neth et al., 2000; Walsh et al., 2005). Fungal persistence or breach of innate immune defense leads to onset of adaptive immunity, i.e., antibody- and cell-mediated immune responses (Knutsen, 2003; Ramadan et al., 2005). The allergic disease spectrum arises from an immune response that is functionally maladaptive and includes atopic sensitization with development of mold-induced asthma (Denning et al., 2006), hypersensitivity pneumonitis, and the complex immunologic hypersensitivity lung disease known as allergic bronchopulmonary aspergillosis (ABPA). Overlapping syndromes suggesting a

Richard B. Moss • Center for Excellence in Pulmonary Biology, Dept. of Pediatrics, Stanford University, Palo Alto, CA 94304.

Table 1. Pulmonary diseases caused by *Aspergillus* species

Saprophytic: structurally damaged host (bronchiectasis, cavities, necrotic tissue)
Aspergilloma (mycetoma)
Chronic necrotizing aspergillosis
Aspergillus bronchitis
Allergic: immunocompetent host
Asthma
Allergic bronchopulmonary aspergillosis
Hypersensitivity pneumonitis (allergic alveolitis)
Bronchocentric granulomatosis
Eosinophilic pneumonia
Invasive: immunosuppressed host
Angioinvasive aspergillosis
Acute bronchopneumonia
Pseudomembranous necrotizing tracheobronchitis
Invasive pleural disease

Table 2. Fungi other than *A. fumigatus* associated with allergic bronchopulmonary disease

A. niger
A. flavus
A. nidulans
A. orizae
A. glaucus
Scedosporium apiospermum (anamorph of *Pseudoallescheria boydii*)
Stemphylium lanuginosum
Helminthosporium species
Candida species
Curvularia species
Schizophyllum commune
Dreschslera hawaiiensis
Fusarium vasinfectum
Trichosporon beigelii

spectrum of possible disease not only between subjects but also within subjects have been well described, such as development of saprophytic disease (aspergilloma or mycetoma) in patients with preexisting ABPA and vice versa, and present an extra challenge for the clinician (Al-Alawi et al., 2005; Maguire et al., 1995; Buckingham and Hansell, 2003; Greene, 2005).

A. fumigatus produces a wide variety of virulence factors (Lacadena et al., 2007; Rementeria et al., 2005). These include a variety of exoenzymes which may protect it from host defenses and enhance pathogenicity, such as superoxide dismutase, catalase, phospholipase, alkaline protease, and elastolytic and collagenolytic metalloproteases (Monod et al., 1995). With regard to allergic lung disease, *A. fumigatus* proteases have received particular attention because they induce both respiratory epithelial cell shedding and activation that result in proinflammatory cytokine secretion, which is a likely mechanism allowing penetration of *A. fumigatus* antigens into the submucosa, where the adaptive immune response may be induced or amplified (Kauffman et al., 2000; Tomee et al., 1997; Tomee and Kauffman, 2000).

ABPA

ABPA is one form of immunologic hypersensitivity lung disease induced by *A. fumigatus* or occasionally other *Aspergillus* species. Rarely, non-*Aspergillus* fungi have been implicated in allergic bronchopulmonary mycoses, but the pathogenetic, pathophysiologic, clinical, and radiological features are generally similar (Gondor et al., 1998; Moss, 2005) (Table 2). ABPA was first described as a distinct lung disease syndrome in asthmatic patients and was characterized by productive cough, fever, infiltrates, eosinophilia, and growth of *A. fumigatus* from sputum (Hinson et al., 1952). It was subsequently recognized as a complication of cystic fibrosis (CF) (Mearns et al., 1965). Diagnostic criteria and therapeutic strategies were first worked out in patients with asthma in the 1970s and 1980s, with more recent modifications in patients with CF.

ABPA has distinct pathologic, immunologic, and radiographic features. Pathologically, ABPA is characterized by one or more of the following: mucoid impaction of bronchi, bronchocentric granulomatosis, eosinophilic pneumonia, and/or exudative or obliterative bronchiolitis (Zander, 2005). Immunologically, ABPA is characterized by endobronchial and systemic *A. fumigatus*-specific antibodies of the immunoglobulin G (IgG), IgA, and IgE isotypes, immediate skin test reactivity to *A. fumigatus*, local and peripheral eosinophilia, increased serum interleukin-2 (IL-2) receptor levels, and strikingly elevated total serum IgE levels (Rosenberg et al., 1977; Greenberger, 2002). Radiographically, ABPA is characterized during acute illness episodes by pulmonary infiltrates on chest radiograph, but more sensitive chest computed tomography (CT) shows segmental and subsegmental varicose or cystic bronchiectasis and mucoid impaction and small airway centrilobular nodules (Buckingham and Hansell, 2003; Cortese et al., 2007; Greene, 2005). Central bronchiectasis (i.e., the inner third of the lung) and high-attenuation mucus plugs (i.e., visually denser than normal skeletal tissue) on chest CT scans are often present but not completely specific for ABPA, especially in patients with CF (Agarwal et al., 2007; Goyal et al., 1992; Logan and Muller, 1996; Morozov et al., 2007).

Pathogenesis of ABPA

The earliest phases of ABPA are not well understood. As ABPA is almost exclusive to two diseases, asthma and CF, that share a characteristic of impaired mucociliary clearance (at least with regard to severe, poorly controlled asthma), it is reasonable to infer that inhaled *A. fumigatus* spores may germinate in static mucus plugs or, perhaps, adhere to altered respiratory epithelium. In either case, if normal clearance mechanisms fail, epithelial cell, dendritic cell, and possibly other resident cell cytokine responses are initiated (Alam, 1977; Romagnani, 2001).

A. fumigatus antigens are processed and presented to the adaptive immune system in the lung primarily by dendritic cells, and this process leads to induction of a Th2-dominated immune response as a key feature leading to clinical ABPA (Schuh et al., 2003). In addition to dendritic cells other resident cell types, such as epithelial cells, alveolar macrophages, vascular smooth muscle cells, and fibroblasts, are potential sources of cytokines that drive recruitment of $CD4^+$ T cells and eosinophils into the airway mucosa and submucosa (Kauffman, 2003).

The role of chemotactic cytokines or chemokines in ABPA has received much recent attention, largely resulting from a very useful murine model of chronic ABPA developed by Hogaboam and colleagues in which mice are sensitized systemically and intranasally with soluble *A. fumigatus* antigens followed by intratracheal administration of *A. fumigatus* conidia (Hogaboam et al., 1999, 2000). These mice develop pathologic, physiologic, and immunologic features closely resembling human ABPA. Chemokines have been identified as key orchestrators of T-cell-mediated immune responses, since various chemokine receptors are differentially expressed on Th1, Th2, and regulatory T cells, and several chemokines have been implicated in a variety of allergic diseases. Th2 cells express CCR4, CCR8, and possibly CCR3; Th1 cells express CCR5 and CXCR3. Using a variety of knockout mouse models to examine the effects of particular chemokines and their receptors, it has been shown in this murine ABPA model that CXCR2 and CCR5 receptors are involved in fungal persistence, CCR1, CCR2, CCR4, and CCR5 receptors are involved in initiation or persistence of airway hyperreactivity, CCR1 is involved in airway remodeling, and CCR2 and CCR4 are involved in fungal killing (reviewed by Hartl et al., 2006).

Several chemokine ligand-receptor signaling systems have been critical in inducing or modifying experimental murine ABPA (Blease et al., 2000a, 2000b; Hogaboam et al., 1999). Th2 immune deviation or skewing appears due at least in part to production of chemokines preferentially evoking a Th2 response, with a prominent role for thymus-activated and -regulated chemokine (TARC/CCL17) and macrophage-derived chemokine (MDC/CCL22) effects mediated through their common receptor, CCR4 (Hogaboam et al., 2005; Schuh et al., 2002).

When the roles of chemokines and other cytokines were investigated clinically, it was found that CF and asthma patients with ABPA had significantly higher levels of TARC than CF patients simply colonized with or sensitized to *A. fumigatus*, atopic CF patients, or nonatopic controls. TARC levels increased during ABPA exacerbations, rose earlier and more sharply than IgE levels, and were reduced by corticosteroid therapy (Hartl et al., 2006). A pathogenetic role for chemokines in ABPA is further suggested by findings of differential expression of chemokine receptors in atopic asthmatics with ABPA compared to atopic asthmatics without ABPA: upon in vitro stimulation with *A. fumigatus* extract, CCR4 and CXCR3 are downregulated in the former and upregulated in the latter (Garcia et al., 2007).

$CD4^+$ Th2 cell activation and elaboration of cytokines such as IL-4, IL-5, and IL-13 play crucial roles in mediating the subsequent inflammatory tissue pathology (Knutsen et al., 2004; Kurup et al., 1997; Schuh et al., 2003). A model of how chemokines might induce ABPA, with a focus on the proposed role of TARC/CCL17, is shown in Fig. 1.

The adaptive immune response in ABPA displays characteristic local and systemic biomarkers. There are marked local and systemic humoral immune responses, with highly elevated levels of *Aspergillus*-specific IgE antibodies and potent induction of a polyclonal (i.e., IL-4-driven antigen-nonspecific) IgE response, leading to usually extremely high serum total IgE levels and augmented IgG and IgA *Aspergillus*-specific antibodies. Predominantly local production of *A. fumigatus*-specific IgE and IgA antibodies in bronchoalveolar lavage fluids and bronchial lymphoid follicles has been demonstrated (Greenberger et al., 1988; Slavin et al., 1992).

$CD4^+$ T-cell adaptive immune responses have been studied primarily in peripheral blood lymphocytes from ABPA patients and controls by in vitro evaluation of their surface immunophenotypes and cytokine induction profiles. These have shown a marked *A. fumigatus*-specific $CD4^+$ Th2-skewed response (Knutsen et al., 1998; Skov et al., 1999b). *A. fumigatus*-reactive $CD4^+$ T-cell lines from ABPA patients have an activated immunophenotype ($CD25^+$ $HLA\text{-}DR^+$) and secrete IL-4 but not gamma interferon (i.e., characteristic of a Th2 immune response), while cell lines reactive to control antigens such as tetanus toxoid include Th1 responses (Knutsen et al., 1994a, 1994b). ABPA patients have an increased frequency of circulating *A. fumigatus*-reactive Th2 $CD4^+$ cells. Increased B-cell sensitivity to IL-4 may represent an autocrine mechanism contributing to the IgE production in ABPA (Knutsen et al., 2004a). Studies of murine ABPA models using a variety of approaches, including that cited above and others, with resistant and susceptible murine strains, targeted gene knockouts, and treatments with cytokine-specific monoclonal antibodies, in aggregate suggest that although IL-4 production is necessary for the elevated IgE levels it is not necessary

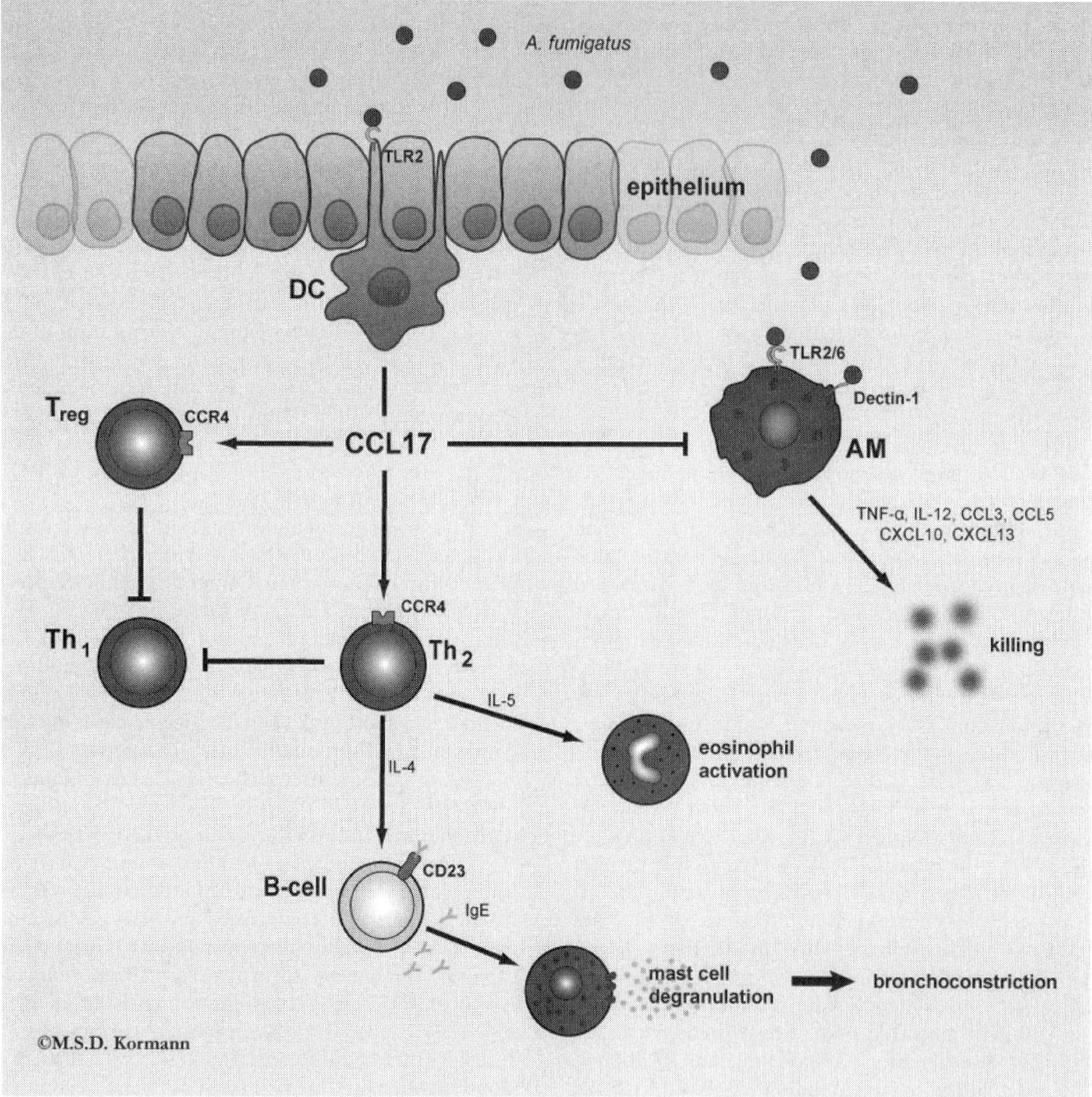

Figure 1. Immunopathogenesis of ABPA. In this schematized hypothesis, hyphal growth of *A. fumigatus* in the airways triggers innate defenses via pattern recognition receptors, such as Toll-like receptors 2 and 4 (TLR2/4), on dendritic cells (DC) and/or pulmonary macrophages (AM). Penetration of *A. fumigatus* antigens is enhanced by epithelial cell activation and shedding induced by *A. fumigatus* proteases. Secretion of chemokines, particularly CCL17 (TARC), from DC stimulated by hyphal forms of *A. fumigatus* results in attraction and activation of T-cell and macrophage subsets via their surface CCR4 receptors. Activated regulatory T cells and Th2 cells dampen Th1 responses, while activated Th2 cells secrete sentinel cytokines IL-4, IL-13, and IL-5, stimulating B cells, airway hyperreactivity and remodeling, and eosinophils, respectively. IL-4 receptors on B cells drive differentiation to IgE-secreting plasmacytes. *A. fumigatus*-specific IgE antibodies arm effector cells, such as mast cells and basophils, to activate and degranulate in the presence of *A. fumigatus* allergens that cross-link their surface-bound IgE, releasing mediators such as histamine and sulfidoleukotrienes. Eosinophils release toxic exoproducts, such as major basic protein, that result in inflammatory tissue damage. Macrophage-mediated fungal killing may also be inhibited by TARC-mediated suppression of Th1 cytokines, such as tumor necrosis factor alpha (TNF-α) and IL-12. Lines with blunted ends indicate inhibition, and lines with arrowed ends indicate activation. From Hartl et al. (2006) with permission of the author and publisher.

for other crucial features, such as airway hyperreactivity, lymphocytic-eosinophilic inflammation, and airway remodeling (Grunig et al., 1998; Kurup et al., 1997). In contrast, CD4$^+$ T cells seem both necessary and sufficient to cause airway hyperreactivity and inflammation in these models (Corry et al., 1998; Kurup and Grunig, 2002).

Eosinophilia is a prominent histopathologic feature of ABPA, although it has been shown that neutrophils are also involved (Gibson et al., 2003; Zander, 2005). Local eosinophilia as well as luminal and interstitial presence of proinflammatory eosinophil exoproducts, such as major basic protein, are present in ABPA lesions (Slavin et al., 1988).

A role for mast cell and/or basophil activation by *A. fumigatus* allergens cross-linking IgE antibodies fixed by high-affinity receptors on the cell surface is suspected from the universal presence of immediate hypersensitivity skin test reactions and in vitro basophil histamine and sulfidoleukotriene release in ABPA patients. However, as these phenomena are also seen in many asthmatic and atopic individuals, their pathogenic role is unclear (Ringer et al., 2007).

Table 3. Allergens of *A. fumigatus*

Allergen[a]	Mass (kDa)	Nature of allergen	% of patients positive for IgE binding[b]
Asp f 1	18	Ribotoxin	83
Asp f 2	37	Fibrinogen binding (?)	90
Asp f 3	19	Peroxisomal protein	94
Asp f 4	30		78
Asp f 5	40	Metalloproteinase	93
Asp f 6	26.5	Mn superoxide dismutase	56
Asp f 7	12		46
Asp f 8	11	Ribosomal protein P2	—[c]
Asp f 9	34		89
Asp f 10	34	Aspartic protease	28
Asp f 11	24	Peptidyl-prolyl isomerase	—
Asp f 12	47	Heat shock protein P90	—
Asp f 13	34	Alkaline serine proteinase	—
Asp f 15	16		—
Asp f 16	43		70
Asp f 17			—
Asp f 18	34	Vacuolar serine proteinase	—
Asp f 22	46	Enolase	—
Asp f 23	44	L3 ribosomal protein	—

[a] Nomenclature for allergens of *A. fumigatus* approved by the International Union of Immunological Societies, Allergen Nomenclature Committee (Kurup, 2005).
[b] Patients allergic to *A. fumigatus* with binding to *A. fumigatus* allergens, reported as the percentage with positive sera (Crameri et al., 1998; Kurup et al., 2000, 2005).
[c] —, not done/not available.

ASPERGILLUS ANTIGENS AND SERODIAGNOSIS OF ABPA

A. fumigatus extracts have long been used in the diagnosis of ABPA, whether in skin testing or in vitro immunoassays. Such extracts are, however, of limited utility due to variable antigenic and allergenic content. They are not standardized and are poorly reproducible. Commercial or other *A. fumigatus* extracts used for skin tests and in vitro antibody assays contain dozens of allergens recognized by IgE antibodies in allergic and ABPA sera (Leser et al., 1992). The development of methods to identify and produce purified, standardized allergens by recombinant DNA technology is important for both diagnosis and understanding the pathogenesis of ABPA. Recombinant DNA technology has allowed for identification and production of many *A. fumigatus* allergens (Table 3) that may play a role in pathogenesis and find use in diagnostic testing (Crameri, 1998; Crameri et al., 2001; Kurup et al., 2001).

Several recombinant *A. fumigatus* allergens show promise for serodiagnosis of ABPA in patients with CF as well as those with asthma, but there is lack of uniformity of results from different studies coming from different regions (Almeida et al., 2006; Casaulta et al., 2005; Crameri et al., 1998; de Oliveria et al., 2007; Hemman et al., 1998b, 1999; Knutsen et al., 2004a; Kurup et al., 2000, 2006; Moser et al., 1994; Nikolaizik et al., 1995, 1996; Sarfati et al., 2006). In general, a distinction between secretory proteins (such as Asp f1, Asp 3, and Asp f5) and cytoplasmic proteins (such as Asp f6, Asp f4, and perhaps Asp f2) is useful in that the secretory proteins may provoke IgE antibodies in allergic asthmatics as well as ABPA patients, while cytoplasmic proteins are likelier to provoke IgE antibodies primarily in ABPA patients, probably expressed and immunogenic as a result of hyphal growth in the lungs. A standardized immunoassay against major *A. fumigatus* recombinant allergens (rAsp f) is now widely available (Crameri et al., 1996). However, several investigators have found that rAsp f2, rAsp f4, and rAsp f6 do not reliably differentiate allergic asthmatics or sensitized CF patients from those with ABPA, and various algorithmic approaches employing combinations of these *A. fumigatus* allergens, sometimes with total IgE included, have been proposed to help differentiate ABPA patients from sensitized asthmatics or CF patients without ABPA (Almeida et al., 2006; Casaulta et al., 2005; de Oliveria et al., 2007; Kurup et al., 2006). In addition, *A. fumigatus* antigen selection based on phage display methods used to generate the widely available rAsp f allergens may underrepresent important *A. fumigatus* allergens of high molecular mass. The latter include 360-kDa catalase and 88-kDa dipeptidyl-peptidase as well as 18-kDa RNase. These latter antigens, recognized in crude form decades earlier as *A. fumigatus* allergens, have now been produced in expression plasmids from products of PCR amplification of *A. fumigatus* cDNA and have been found to differentiate ABPA in asthma and CF patients from nonallergic controls based on a canonical plot statistical approach. However, data on *A. fumigatus*-sensitized non-ABPA atopic controls as well as confirmatory studies are lacking (Sarfati et al., 2006).

Recently, the use of serum TARC/CCL17 levels has been proposed as an alternative diagnostic test for ABPA (Hartl et al., 2006; Latzin et al., 2008). Analysis in two separate cohorts suggests that TARC may be superior to individual rAsp f allergens and total IgE in serodiagnosis of ABPA (Latzin et al., 2008). Although promising, further testing in geographically disparate regions and populations is necessary to validate this proposal. In addition, TARC elevations may precede clinical onset of ABPA by months to years, raising questions about its specificity and diagnostic utility (Latzin et al., 2008). The in vitro release of sulfidoleukotrienes by basophils stimulated with *A. fumigatus* allergens represents another potential approach but lacks sufficient specificity for ABPA (Ringer et al., 2007).

Thus, despite their usefulness and the improvement in diagnostic capabilities they afford, specific IgE antibody immunoassays using rAf Asp allergens lack sufficient sensitivity, specificity, and positive and negative predictive values in multiple regions and populations to replace the complicated current diagnostic criteria combining clinical, imaging, and serologic components discussed below.

Another area of diagnostic difficulty lies in measuring the IgG *A. fumigatus* antibody response. The traditional method, detection of precipitins in double immunodiffusion gel assays using one or more *A. fumigatus* extracts and patient serum, also suffers from lack of standardized antigens, limited sensitivity and specificity, and inability to quantify antibody levels. Solid-phase immunoassays can quantify levels relative to a control pool, but in the past these assays also lacked standardization and absolute quantification (Patterson et al., 1983; Schonheyder and Andersen, 1984). Recently, the same immunoassay system standardized for detection of IgE antibodies to *A. fumigatus* allergens has been adapted to measure IgG antibodies, with promising early results (Van Hoeyvold et al., 2006). However, as with the IgE assays, specificity and sensitivity vary depending on the cutoff value chosen and both the patient and control populations, and the isotype-specific immunoassay results are not always concordant with results from precipitin testing (Van Hoeyveld et al., 2006).

RISK FACTORS FOR ABPA

Since ABPA only occurs in a minority of people with asthma or CF, risk factors for the disease have been sought. Although some environmental aspects have been considered, most research has focused on host factors (Table 4).

Collectin Polymorphisms

A role for other components of the innate immune response in susceptibility to ABPA has been shown by studies of collectins, which are innate opsonins important in fungal clearance. This family includes surfactant proteins A and D (SP-A and SP-D) and mannose-binding lectin (MBL) (Madan et al., 2001, 2005; Neth et al., 2000; Saxena et al., 2003). Studies of genetic polymorphisms of SP-A and MBL genes have yielded results suggesting an important role for collectins in host defense against *A. fumigatus* that are affected by single-nucleotide polymorphisms (SNPs) and/or haplotypes. The T allele and the TT genotype at T1492C and the G allele at G1649C of SP-A2 are found at higher frequencies in persons with ABPA than in controls (Madan et al., 2005; Saxena et al., 2003; Vaid et al., 2007). With regard to MBL, although no differences in the allelic frequencies or genotype distributions of the three reported structural polymorphisms (C868T, G875A, and G884A) have been found, the A allele and AA genotype of the MBL intron 1 SNP G1011A is found more frequently in subjects with ABPA, as well as in subjects with allergic asthma and rhinitis, than in controls (Kaur et al., 2006). Since this allele is associated with higher MBL levels, it may indicate a proinflammatory consequence of augmented MBL activity upon pulmonary challenge with *A. fumigatus.*

IL-10 Polymorphisms

A role for the anti-inflammatory cytokine IL-10 in protection against *A. fumigatus* has been shown in experimental murine ABPA, and a broader role for IL-10 as an anti-inflammatory regulatory cytokine in lung inflammation in both asthma and CF is likely (Bonfield et

Table 4. Risk factors for ABPA

Environmental spore exposure (indoor, compost, ambient seasonal, etc.)
Atopy
CFTR mutations
Surfactant protein A2 polymorphisms (T1492C allele and TT genotype; A1660G and G1649C alleles)
Mannose-binding lectin polymorphisms (G1011A allele)
IL-10 promoter polymorphism (−1082GG genotype)
HLA-DR2 restriction (DRB1*1501 and *1503 alleles) (DRB1*1502 allele is protective)
T-cell receptor Vβ chain restriction (Vβ13) (Vβ1 is protective)
Increased B-cell sensitivity to IL-4
IL-4Rα Ile75Val allele
Colonization with *S. maltophilia* (in CF)

al., 1999; Borish et al., 1996; Grunig et al., 1997). It is therefore notable that the presence of the −1082 GG promoter region IL-10 genotype is associated with susceptibility to both colonization with *A. fumigatus* and to ABPA (Brouard et al., 2005). The −1082G allele has been reported to produce higher levels of IL-10. Peripheral blood T-cell responses to *A. fumigatus* in patients with CF are regulated by IL-10 (Casaulta et al., 2003).

IL-4Rα Polymorphisms

As discussed in the section on pathogenesis, IL-4 plays a major role in induction of allergic immune responses to *A. fumigatus* that characterize ABPA. ABPA patients show increased sensitivity to IL-4 stimulation compared to controls as assayed by upregulation of the B-cell IgE receptor CD23 (Khan et al., 2000; Knutsen et al., 2004b). Since gain-of-function SNPs have been reported in the IL-4 receptor alpha chain (IL-4Rα), these were examined in patients with ABPA. ABPA is associated with an increased frequency of IL-4Rα SNPs; in particular, the extracellular IL-4Rα-binding SNP Ile75Val was found in 80% of ABPA versus 54% of control subjects and was homozygous in 43% of ABPA patients versus 11% of controls (Knutsen et al., 2006).

MHC Alleles

A critical immunogenetic feature of ABPA is inheritance and expression of certain major histocompatibility complex (MHC) alleles that act to regulate $CD4^+$ T-cell responses to *A. fumigatus*. ABPA is much more likely in people with MHC alleles DRB1*1503 and DRB1*1501, while DRB1*1502 is less frequent in ABPA patients than controls (Chauhan et al., 1997, 2000). The roles of HLA-DRB1*1501 and DRB1*1503 as susceptibility alleles for ABPA and HLA DRB1*1502 as a resistance allele have also been convincingly demonstrated in humanized transgenic mice expressing these alleles and challenged with *A. fumigatus* (Koehm et al., 2007).

Thus, both susceptibility and protective MHC alleles for genetic risk of ABPA exist, implying pathogenetic roles for certain *A. fumigatus* antigenic peptide-MHC interactions involving antigen-presenting cells in the lung. The identity of these antigen-presenting cells has not been established but is likely to include dendritic cell populations that sample and present airway antigens. Particular *A. fumigatus* allergens associated with sensitization to *A. fumigatus* and ABPA have been identified (see "Further Diagnostic Considerations for ABPA in CF," below), suggesting certain *A. fumigatus* allergens are candidates for MHC-restricted disease induction (Casaualta et al., 2005; Hemman et al., 1998b). Supporting this, in one murine model some *A. fumigatus* allergens were shown to cause inflammation, airway hyperreactivity, high IgE, eosinophilia, and a Th2 immune response, while others did not (Kurup et al., 2001).

CFTR Mutations

Susceptibility to ABPA appears to be independently increased by mutations in the CF transmembrane conductance regulator (CFTR) gene. Inheritance of CFTR mutations on both alleles causes CF. This may in part explain the higher prevalence of ABPA in CF patients than in asthma patients, although many other phenotypic features may also be responsible. An increased frequency of CFTR mutations has been found in asthmatic non-CF ABPA sample populations in the United States, France, and New Zealand, suggesting a gene dose effect in ABPA susceptibility (Eaton et al., 2002; Marchand et al., 2001; Miller et al., 1996).

The mechanism of this effect has been investigated in animal models. CFTR knockout mice in two independent models have an altered response to *A. fumigatus* challenge that strongly resembles human ABPA. In these murine models CFTR plays a critical role in skewing $CD4^+$ T-cell immune responses to a Th2 phenotype in response to inhaled *A. fumigatus* extract, with higher levels of IL-4 and IgE than age- and genetic background-matched wild-type littermates (Allard et al., 2006; Muller et al., 2006). Recently, correction of the CF defect in CFTR $489X^{-/-}$, gut-corrected FABP-hCFTR$^{+/+}$ mice by recombinant adeno-associated virus vector-mediated gene transfer in a murine model was shown to reduce both IgE and IL-4/IL-13-dependent IgG1 serum levels as well as splenocyte production of several proinflammatory cytokines (e.g., IL-13 and IL-5) upon mitogen stimulation in gene-corrected mice (Mueller et al., 2008). These experiments leave little doubt that CFTR plays a role in allergic responses to *A. fumigatus* that are likely to be clinically relevant in explaining susceptibility to ABPA in response to *A. fumigatus* exposure.

It is also possible that *Aspergillus* may be preferentially selected for colonization of lungs in patients with asthma who are heterozygous for CFTR mutations as well as in CF patients (who are by definition homozygous for CFTR mutations on both alleles). This potential gene dose effect may help explain the higher prevalence of *A. fumigatus* colonization as well as ABPA in CF patients than asthmatics (see below).

Other Risk Factors

A variety of additional risk factors for ABPA have been described.

(i) Since the disease process is driven by respiratory antigenic stimulation, it is likely that environmental factors, e.g., increased inhalational dose exposure to *A. fumigatus* spores, may precipitate disease in susceptible individuals. There is some epidemiologic evidence for this supposition (Beaumont et al., 1984; Krasnick et al., 1995; Radin et al., 1983). Differences in atmospheric humidity do not appear to play a major role, at least with regard to ABPA risk in CF (Skov et al., 2005).

(ii) Analogous to environmental exposure, since ABPA is characterized by the florid allergic response, it is also intuitively likely that atopic individuals are more susceptible; this has been demonstrated for ABPA in patients with CF where atopy (defined as allergic sensitization to common aeroallergens) conveys a relative risk of ~20 for occurrence of ABPA (Nepomuceno et al., 1999). However, contrary findings have also been reported (Skov et al., 2005).

(iii) In addition to MHC susceptibility alleles, one might suspect that phenotypic differences in the T-cell receptor (TCR), part of the molecular complex that transduces the signal delivered by the allergen peptide-MHC complex on the antigen-presenting cell surface, might influence resultant T-cell responses, and this too has been experimentally demonstrated with regard to TCR Vβ chain restriction (Chauhan et al., 2002). It should be added that protective as well as susceptibility MHC alleles and TCR Vβ restrictions for ABPA have been identified (Chauhan et al., 2000, 2002).

(iv) Finally, in patients with CF, who are susceptible to chronic respiratory colonization and infection with a variety of bacterial pathogens, it has been shown that risk for ABPA is strongly associated (odds ratio, 20) with respiratory colonization by *Stenotrophomonas maltophilia* (Ritz et al., 2005) and nontuberculous mycobacterial disease (Kunst et al., 2006).

ABPA IN ASTHMA

ABPA was first recognized in patients with asthma, and diagnostic criteria were developed for this setting (Greenberger, 2002; Rosenberg et al., 1977) (Table 5). Immediate cutaneous reactivity to *Aspergillus* is detectable in 15 to 40% of patients with asthma (reviewed by Maurya et al., 2005). Among asthmatics sensitized to *A. fumigatus*, 25 to 30% will develop ABPA (Agarwal et al., 2006; Eaton et al., 2000; Maurya et al., 2005; Schwartz and Greenberger, 1991). However, APBA is widely underdiagnosed in asthma due to the variability in diagnostic criteria, allergen extracts, imaging modalities, and a low clinical index of suspicion or use of *A. fumigatus* skin test screening. Studies incorporating high-resolution chest CT to detect central bronchiectasis in adult asthma patients screened by *A. fumigatus* skin testing probably have the most validity, and these studies have reported ABPA prevalence rates of 6 to 8% (Agarwal et al., 2006; Eaton et al., 2000; Greenberger and Patterson, 1988; Maurya et al., 2005; Schwartz et al., 1991).

Table 5. Diagnosis of ABPA

Full criteria
1. Asthma
2. Chest roentgenographic infiltrates, current or in past, may be detectable on CT examination when chest film is normal
3. Immediate cutaneous reactivity to *A. fumigatus*
4. Elevated total serum IgE (>417 kU/liter or 1,000 ng/ml)
5. Serum precipitating antibodies to *A. fumigatus*
6. Central (proximal, inner two-thirds) bronchiectasis on high-resolution CT of chest
7. Peripheral blood eosinophilia
8. Elevated serum IgE and IgG antibodies to *A. fumigatus*

Minimal essential criteria
1. Asthma
2. Immediate cutaneous reactivity to *A. fumigatus*
3. Elevated total serum IgE concentration (>417 kU/liter or 1,000 ng/ml)
4. Elevated serum IgE and IgG antibodies to *A. fumigatus*
5. Central bronchiectasis

Use of oral corticosteroids can reduce the total serum IgE concentration to <1,000 ng/ml in some patients with ABPA

Confirmatory findings: *A. fumigatus* in sputum or bronchoalveaolar lavage fluid, expectoration of brown plugs, dual or late skin test reactivity to *A. fumigatus*

High-resolution chest CT is very useful in evaluating for ABPA, as multilobar bronchiectasis, centrilobular nodules, and mucoid impaction are findings highly suggestive of ABPA in asthma (Ward et al., 1999). High-attenuation mucus plugs are relatively specific for ABPA (Agarwal et al., 2007; Goyal et al., 1992; Morozov et al., 2007) but insensitive, occurring in only about 25% of ABPA patients (Logan and Muller, 1996).

Because it is possible to demonstrate serological changes consistent with ABPA in patients who do not have central bronchiectasis, a partial or prodromal ABPA state termed ABPA-serologic (ABPA-S) has been suggested as a nosologic entity along with the conventional or classic ABPA syndrome, which produces pulmonary structural damage (i.e., central bronchiectasis with or without further CT pathology) as well as more overt clinical disease (Greenberger et al., 1993). Based on clinical, spirometric, serologic, and radiologic findings, ABPA-S patients have a milder type of ABPA (Greenberger et al., 1993; Kumar, 2003). In one recent study from northern India, 27% of asthmatics diagnosed with ABPA fell into the ABPA-S category (Agarwal et al., 2006), and in a study of CF patients in Switzerland essentially equal numbers of suspected cases had ABPA-S as had full-blown ABPA (Casaulta et al., 2005). It is unclear what proportion of ABPA-S cases evolves into full-blown ABPA if untreated, or over what period of time, or indeed whether treatment at all is indicated. Many of the diagnostic and treatment dilemmas in

ABPA involve ABPA-S patients who typically show worrisome evidence of pronounced immunologic responses to *A. fumigatus* but have little overt clinical or radiographic disease.

ABPA may have onset in early childhood and can be of acute life-threatening severity, but most cases occur in young adults (Greenberger, 2002; Skowronski and Fitzgerald, 2005; Slavin et al., 1970).

ABPA can be further categorized into five discrete stages (Patterson et al., 1982) (Table 6). These are not necessarily temporal phases. Treatment varies by stage, in that oral glucocorticosteroids are indicated in acute (stage 1) cases and may be tapered and withdrawn when the patient is in remission (stage 2), and addition of an oral triazole antifungal, usually itraconazole, is appropriate for patients with relapse (stage 3) or corticosteroid-dependent asthma (stage 4). The effectiveness of itraconazole as a steroid-sparing and anti-inflammatory agent, presumably via reduction in fungal burden, has been demonstrated in double-blind placebo-controlled randomized clinical trials and a meta-analysis (Wark et al., 2004). Similar data are not available for other antifungal agents or for inhaled corticosteroids. It is not clear if any therapy is effective in patients who have end-stage pulmonary fibrosis (stage 5), but this is rarely if ever seen in pediatric cases (Lee et al., 1987). Further details of treatment are discussed below in the in the section on treatment of ABPA (see also Table 8, below).

ABPA IN CF

A. fumigatus colonization of the airways has been reported in 6 to 57% of patients with CF, a wide range that likely reflects variations in clinical sample acquisition, fungal culturing methods, and patient population differences (Bakare et al., 2003). In what may be the benchmark study, a large multicenter U.S. prospective evaluation in a defined CF patient population (age ≥6 years, *Pseudomonas*-positive, moderate lung disease) using a uniform sample acquisition protocol and methodology with centralized microbiology yielded an *A. fumigatus* sputum colonization prevalence of 25%; the incidence was increased by use of inhaled antibiotics (Burns et al., 1999). *A. fumigatus* strains usually differ between CF patients and can persist long term in colonized patients (Neuveglise et al., 1997). Asthmatic or CF patients can develop ABPA with negative sputum cultures, so the presence of *A. fumigatus* in sputum is not a major criterion for diagnosis, although its presence is supportive (Skov et al., 2000; Stevens et al., 2003) (Table 5).

Among CF patients, reported ABPA prevalence rates are higher than in asthma patients, with up to 16% reported in unselected CF clinic populations (Becker et al., 1996; Brueton et al., 1980; Casaulta et al., 2005; Feanny et al., 1988; Geller et al., 1999; Hutcheson et al., 1996; Laufer et al., 1984; Marchant et al., 1994; Mastella et al., 2000; Mitchell-Heggs et al., 1976; Mroueh and Spock, 1994; Nelson et al., 1979; Nepomuceno et al., 1999; Schonheyder et al., 1988; Simmonds et al., 1990; Skov et al. 1999a; Taccetti et al., 2001; Valletta et al., 1993; Zeaske et al., 1988). The annual USA Cystic Fibrosis Foundation Registry reports a prevalence of 2 to 3% for patients >5 years old, similar to findings in an independent registry (Epidemiologic Study of Cystic Fibrosis), from which a rate of 2% was reported in 14,210 patients (Geller et al., 1999). However, a European Registry of CF study reported a prevalence of 7.8% in 12,447 CF patients from 224 CF centers in nine European countries (range of 2.1% in Sweden to 13.6% in Belgium) (Mastella et al., 2000). A national study in 3,089 Italian CF patients reported ABPA in 6.2% (Taccetti et al., 2001).

A likely source of the variability in reported ABPA prevalence rates among CF patients is the difficulty of case ascertainment. The Cystic Fibrosis Foundation Registry in the United States relies upon an ABPA diagnosis by the reporting center without stipulating any diagnostic criteria. In the U.S. Epidemiologic Study of Cystic Fibrosis registry and European Registry of CF, diagnostic criteria were employed but they varied, and it is not clear whether or how stringently they were applied by

Table 6. Stages of ABPA

Stage	Clinical manifestation(s)	Serology	Chest imaging
1. Acute	Fever, cough, chest pain, hemoptysis, sputum	Very high IgE, eosinophilia usual	Infiltrates (usually upper and middle lobes)
2. Remission	Asymptomatic or stable asthma	Normal or mildly elevated IgE	No infiltrates (off steroids)
3. Exacerbation	Asymptomatic or as in stage 1	As in stage 1	As in stage 1
4. Corticosteroid dependent	Moderate to severe persistent asthma	As in stage 2	As in stage 1 or 2
5. Fibrotic	Hypoxemia, dyspnea	As in stage 2	Fibrosis, cavitation, extensive bronchiectasis

the reporting centers (Geller et al., 1999; Mastella et al., 2000). In addition, ABPA studies in single CF centers have used a wide variety of diagnostic criteria. A survey in 45 United Kingdom CF centers found that *A. fumigatus*-specific IgE was used only 54% of the time and total serum IgE of >1,000 ng/ml was used only 45% of the time (Cunningham et al., 2001). In a large single-center study in Italy, patients were considered to have probable or possible ABPA depending on the number of diagnostic criteria present; by these definitions, 12% of patients had probable ABPA and 11% had possible ABPA (Valletta et al., 1993). Finally, patient status plays a role in case ascertainment. In one study 10% of pulmonary exacerbations in CF patients requiring hospitalization were associated with stage 1 or 2 (acute or relapsed) ABPA (Nepomuceno et al., 1999).

Although there has been uncertainty about the impact of ABPA upon the course of CF, recent longitudinal data incorporating both immunological and extensive pulmonary physiologic measures leave little doubt that ABPA worsens the course of CF (Kraemer et al., 2006). This comports with earlier studies suggesting that an allergic response to *A. fumigatus* in CF (as measured by serum IgE antibodies to *A. fumigatus*) is associated with poorer lung function (Hartl et al., 2006; Kanthan et al., 2007; Nicolai et al., 1990; Wojnarowski et al., 1997). Allergic sensitization, rather than colonization per se, appears to be the marker of worsened lung function (Milla et al., 1996). More widespread use of antifungal drugs may ameliorate this impact, but further study of this interesting approach seems needed (Kanthan et al., 2007). Antifungals also may be beneficial in treating CF patients with worsening lung function and positive *A. fumigatus* sputum cultures but without ABPA, a condition provisionally named *Aspergillus* bronchitis (Shoseyov et al., 2006).

Further Diagnostic Considerations for ABPA in CF

In CF, diagnosis of ABPA is complicated by features of CF that overlap with classic diagnostic criteria of ABPA. These include bronchiectasis, pulmonary infiltrates, obstructive pulmonary physiology and symptoms, and a variable "asthmatic" component of lung disease that can include bronchial hyperreactivity, response to acute bronchodilator, wheezing, and response to steroids (Moss, 2002). Some immunologic features of ABPA are also commonly seen in CF patients, such as *A. fumigatus*-specific immediate skin tests or IgE antibodies, *A. fumigatus*-specific IgG or precipitating antibodies, and elevations in total IgE. Up to 60% of CF patients may have positive *A. fumigatus* skin tests (Casaulta et al., 2005; Hemmann et al., 1998a, 1998b; Valletta et al., 1993; Wojnarowski et al., 1997; Zeaske et al., 1988). Precipitins may be found in up to 25% of CF patients (Hutcheson et al., 1996). Total IgE levels may be elevated in up to 25% of CF patients (Hutcheson et al., 1996; Laufer et al., 1984; Nelson et al., 1979; Nepumuceno et al., 1999). Most children with CF have serum IgG and IgE antibodies to *A. fumigatus* by school age (El-Dahr et al., 1994; Murali et al., 1994).

In addition, these immunologic features can wax and wane over time. Half of CF patients followed up to 12 years showed loss of previously positive AF-specific IgG and/or IgE antibodies (Hutcheson et al., 1996).

On the other hand, in CF large increases in total serum IgE level may provide an additional diagnostic variable, a finding facilitated by the widespread use of longitudinal IgE testing in CF patients. A fourfold rise from stable baseline total IgE to >500 IU/ml, or similar large changes in patients with lower baseline IgE, is strongly suggestive of ABPA in CF (Dorsaneo et al., 2004; Knutsen et al., 2005; Marchant et al., 1994).

Chest CT imaging cannot reliably distinguish ABPA from underlying CF. Thin-section CT may show infiltrates and/or bronchiectasis not seen on plain radiography (Eaton et al., 2000; Lynch, 1998; Neeld et al. 1990; Panchal et al., 1994; Ward et al., 1999). Although bronchiectasis tends to be central in ABPA and more peripheral in CF, some degree of central bronchiectasis is also seen in CF (Reiff et al., 1995; Santis et al., 1991). Varicose and cystic bronchiectasis is more typical of ABPA than CF (in which cylindrical bronchiectasis is more common), but these forms of bronchiectasis are also seen in up to one-third of CF patients (Hansell and Strickland, 1989). High-attenuation mucus plugs are relatively specific for ABPA but are seen in only a minority of patients with ABPA, and mucoid impaction in CF more commonly than not lacks this distinctive feature (Goyal et al., 1992; Logan and Muller, 1996).

Due to these difficulties the classic diagnostic criteria for ABPA have been modified for patients with CF (Table 7). In addition, because of the high incidence of ABPA and difficulty in diagnosis, screening for ABPA in

Table 7. Minimal diagnostic criteria for ABPA in CF[a]

1. Acute or subacute clinical deterioration (cough, wheeze, exercise intolerance, exercise-induced asthma, change in pulmonary function, increased sputum) not attributable to another etiology
2. Total serum IgE >500 IU/ml
3. Immediate skin test reactivity or in vitro demonstration of IgE antibody to *A. fumigatus*
4. One or both of the following:
   a. Serum precipitins or IgG antibody to *A. fumigatus*
   b. New or recent abnormalities on chest radiograph (infiltrates, mucus plugging) or chest CT (bronchiectasis) that do not clear with antibiotics and standard physiotherapy

[a]Source: adapted from Stevens et al. (2003).

CF patients is recommended via annual total serum IgE testing beginning at school age, with further workup of patients showing total IgE levels over 500 IU/ml or levels sharply increased over prior baseline levels (Stevens et al., 2003).

TREATMENT

Treatment of ABPA is essentially the same for asthma and CF (Greenberger, 2002; Stevens et al., 2003) (Table 8). However, in CF treatment is complicated by several additional factors peculiar to CF. First, recognition and therefore prompt treatment are more difficult due to the diagnostic difficulties and clinical similarities discussed above. Second, CF patients are more vulnerable to glucocorticosteroid toxicity, including growth suppression, osteoporosis, and diabetes, which are, unlike in asthma, common independent complications of the underlying CF (Bhudhikanok et al., 1998; Hardin and Moran, 1999; Lai et al., 2000). Third, alterations in drug absorption, distribution, and metabolism and drug interactions are common in CF patients. Absorption of enteric-coated prednisolone may be compromised in CF where the jejunal pH may be lower than desired (Gilbert and Littlewood, 1986). Once absorbed, prednisolone is cleared more rapidly in CF patients, requiring increased dosing or shorter dose intervals (Dove et al., 1992).

Itraconazole is an inhibitor of hepatic cytochrome P450 3A4. Through this mechanism it increases exposure to methylprednisolone but not prednisolone or endogenous cortisol (Lebrun-Vignes et al., 2001; Phillips et al., 1987; Quiroz-Telles et al., 1997; Varis et al., 1998, 1999, 2000a, 2000b). Concomitant use of itraconazole and inhaled budesonide can result in adrenal suppression and even Cushing's syndrome (Bolland et al., 2004; De Wachter et al., 2003a, 2003b; Main et al., 2002; Parmar et al., 2002; Raaska et al., 2002; Skov et al., 2002a).

Oral glucocorticoids are an effective first-line treatment for ABPA and appear as effective in CF as they are in asthmatic APBA (Birx et al., 1984; Fitzsimons et al., 1997; Hutcheson et al., 1996; Knutsen et al., 1994b; Maguire et al., 1988; Marchant et al., 1994; Moss, 2006; Mroueh and Spock, 1994; Nepomuceno et al., 1999; Simmonds et al., 1990a, 1990b; Vilar et al., 2000). An attempt should be made to taper off corticosteroids in 2 to 3 months. But in the European Registry of Cystic Fibrosis survey only 56% of CF patients with ABPA received oral corticosteroids (compared to 15% of CF patients without ABPA); inhaled corticosteroids were given to 75% of the CF patients with ABPA compared to 39% of CF patients without ABPA, despite a lack of published information on efficacy of inhaled corticosteroids for ABPA (Mastella et al., 2000). These data suggest that a well-justified fear of glucocorticosteroid toxicity plays a significant role in clinician treat-

Table 8. Pharmacotherapy of ABPA[a]

Therapeutic category	Indication(s)	Dosing regimen and/or comments
Oral corticosteroids	All patients except those with steroid toxicity	0.5–2.0 mg/kg/day (prednisolone equivalent), orally, 1–2 weeks; max. 60 mg/day. Then begin taper with 0.5–2.0 mg/kg every other day for 1–2 weeks. Attempt to taper off in 2–3 mos. With relapse, increase corticosteroids, add itraconazole, taper corticosteroids when clinical parameters improve.
Oral itraconazole	Patients with slow or poor response to corticosteroids, relapse, corticosteroid dependence, or corticosteroid toxicity	5 mg/kg/day orally, max. dose 400 mg/day, unless itraconazole levels obtained. Twice-a-day dosing when daily dose exceeds 200 mg. Administer 3–6 mos. Monitor liver function in all cases; serum itraconazole levels (steady-state target $\geq$1 μg/ml) if concern for adequate absorption (esp. if concomitant acid suppression therapy), lack of response, or other possible drug interactions. Monitor use or levels of concomitant drugs with potential for drug-drug interactions, esp. adrenal suppression with concomitant budesonide. Oral solution has 50% higher bioavailability than capsule.
Adjunctive therapy	Inhaled corticosteroids, bronchodilators, leukotriene receptor antagonists; no evidence for use in ABPA but may be used for asthma component of APBA; inhibition of cytochrome P450 budesonide catabolism by itraconazole can lead to adrenal insufficiency	Uncontrolled reports of agents with possible efficacy: inhaled amphotericin, oral voriconazole, pulse intravenous methylprednisolone; omalizumab. Environmental manipulation: attempt to find and reduce mold spore exposure in refractory cases.

[a] Source: Stevens et al. (2003).

ment of ABPA in CF patients that prominently includes use of uncertain modalities. Indeed, a worrisome additional example is the suggestion that systemic steroid therapy may place CF patients at higher risk of nontuberculous mycobacterial pulmonary infection (Mussafi et al., 2005).

Serum IgE levels are a useful way to follow a patient's response to pharmacotherapy and guide steroid dosing (Marchant et al., 1994; Nepomuceno et al., 1999; Patterson et al., 1986; Stevens et al., 2003). Consideration should be given to initiating oral corticosteroid therapy if the serum IgE level rises sharply (at least doubles) from a known stable baseline value, even if the absolute level does not reach traditional diagnostic criteria levels of >1,000 ng/ml (417 IU/ml) (Dorsaneo et al., 2004; Knutsen et al., 2005; Marchant et al., 1994). IgE levels alone should not be used to make treatment decisions, however, since the matrix of clinical, imaging, and serologic results creates the therapeutic context. Many clinicians choose to closely follow patients with ABPA-S, for example, rather than treat them with potentially toxic drugs.

Reduction of fungal antigenic burden with antifungal agents is a logical addition to the ABPA treatment program (Leon and Craig, 1999). This approach has been validated by administering itraconazole to asthmatic patients with ABPA in double-blind, placebo-controlled, randomized multicenter studies (Moss, 2006; Stevens et al., 2000; Wark, 2004; Wark et al., 2003, 2004). For CF, uncontrolled studies have suggested similar benefits with good tolerance (Denning et al., 1991; Nepomuceno et al., 1999; Skov et al., 2002b). Itraconazole should be used if a patient has a slow or poor response to glucocorticoids, experiences a relapse, is steroid dependent, or has steroid toxicity (Stevens et al., 2003).

Itraconazole can be given at an adult dose of 200 mg twice a day or a pediatric dose of 5 mg/kg once daily up to 200 mg (Stevens et al., 2003). Itraconazole blood levels (target, $\geq$1 μg/ml) are helpful in patients with a poor response to document adequate absorption and dosing. Administration of itraconazole with cola 1 h before a meal or a gastric acid-suppressing medication improves absorption. Some patients need higher dosing or a switch from capsules to a 2.5-mg/kg twice-a-day liquid itraconazole cyclodextrin suspension (100 mg/5 ml), which has better bioavailability (Conway et al., 2004; Stevens, 1999). Although *A. fumigatus* is typically sensitive to azoles, it may be useful to check sensitivity of sputum isolates if the response is poor, since resistant strains have been reported (Mosquera and Denning, 2002). Itraconazole courses of 3 to 6 months have been recommended, with baseline and periodic monitoring of liver function tests (Stevens et al., 2003).

Therapies for ABPA that have not been adequately evaluated and therefore are considered unproven include inhaled corticosteroids, inhaled amphotericin, newer oral azoles such as voriconazole, intravenous monthly pulse methylprednisolone infusions, immunosuppressive agents such as low-dose methotrexate, and the humanized anti-IgE monoclonal antibody omalizumab (Casey et al., 2002; Hilliard et al., 2005; Thomson et al., 2006; van der Ent et al., 2007). Inhaled corticosteroids and other agents useful in asthma for the asthmatic component of ABPA, as recommended by current asthma guidelines, may be beneficial components of treatment but cannot substitute for validated ABPA-directed drugs.

Finally, the role of environmental remediation and amelioration of fungal exposure in prophylaxis or treatment of ABPA has not been well studied. Measures which may contribute to remission and/or prevent exacerbation include avoidance of outdoor exposures, such as turned compost heaps or moldy hay piles, examination and cleaning of humid moldy indoor areas, and perhaps use of HEPA filters. Measurement of indoor fungal spore counts may be helpful in identifying occult exposure (Beffa et al., 1998; Garrett et al., 1998). In the future, an increasing role for antifungal agents and immunomodulation in treating ABPA seems likely, given the serious problems with chronic systemic glucocorticosteroid therapy and the stubborn diagnostic obstacles that still accompany this disease.

REFERENCES

Agarwal, R., D. Gupta, A. N. Aggarwal, A. K. Saxena, A. Chakrabarti, and S. K. Jindal. 2007. Clinical significance of hyperattenuating mucoid impaction in allergic bronchopulmonary aspergillosis: an analysis of 155 patients. *Chest* **132:**1183–1190.

Agarwal, R., D. Gupta, A. N. Aggarwal, D. Behera, and S. K. Jindal. 2006. Allergic bronchopulmonary aspergillosis: lessons from 126 patients attending a chest clinic in north India. *Chest* **130:**442–448.

Al-Alawi, A., C. F. Ryan, J. D. Flint, and N. L. Muller. 2005. *Aspergillus*-related lung disease. *Can. Respir. J.* **12:**377–387.

Alam, R. 1977. Chemokines in allergic inflammation. *J. Allergy Clin. Immunol.* **99:**273–277.

Allard, J. B., M. E. Poynter, K. A. Marr, L. Cohn, M. Rincon, and L. A. Whittaker. 2006. *Aspergillus fumigatus* generates an enhanced Th2-biased immune response in mice with defective cystic fibrosis transmembrane conductance regulator. *J. Immunol.* **177:**5186–5194.

Allen, M. J., D. R. Voelker, and R. J. Mason. 2001. Interactions of surfactant proteins A and D with *Saccharomyces cerevisiae* and *Aspergillus fumigatus*. *Infect. Immun.* **69:**2037–2044.

Almeida, M. B., M. H. Bussamra, and J. C. Rodrigues. 2006. ABPA diagnosis in cystic fibrosis patients: the clinical utility of IgE specific to recombinant *Aspergillus fumigatus* allergens. *J. Pediatr.* (Rio J) **82:**215–220.

Bakare, N., V. Rickerts, J. Bargon, and G. Just-Nübling. 2003. Prevalence of *Aspergillus fumigatus* and other fungal species in the sputum of adult patients with cystic fibrosis. *Mycoses* **46:**19–23.

Beaumont, F., H. F. Kauffman, H. J. Sluiter, and K. de Vries. 1984. Environmental aerobiological studies in allergic bronchopulmonary aspergillosis. *Allergy* **39:**183–193.

Becker, J. W., W. Burke, G. McDonald, P. A. Greenberger, W. R. Henderson, and M. L. Aitken. 1996. Prevalence of allergic bronchopulmonary aspergillosis and atopy in adult patients with cystic fibrosis. *Chest* **109:**1536–1540.

Beffa, T., F. Staib, J. Lott Fischer, P. F. Lyon, P. Gumowski, O. E. Marfenina, S. Dunoyer-Geindre, F. Georgen, R. Roch-Susuki, L. Gallaz, and J. P. Latgé. 1998. Mycological control and surveillance of biological waste and compost. *Med. Mycol.* 36(Suppl. 1):137–145.

Bhudhikanok, G. S., M. C. Wang, R. Marcus, A. Harkins, R. B. Moss, and L. K. Bachrach. 1998. Bone acquisition and loss in children and adults with cystic fibrosis: a longitudinal study. *J. Pediatr.* **133:**18–27.

Birx, D. L., R. Summers, and M. Berger. 1984. Acute deterioration of pulmonary function in cystic fibrosis illustrating the association of atopy and allergic bronchopulmonary aspergillosis with the underlying disease. *Ann. Allergy* **53:**124–130.

Blease, K., B. Mehrad, T. J. Standiford, N. W. Lukacs, J. Gosling, L. Boring, I. F. Charo, S. L. Kunkel, and C. M. Hogaboam. 2000a. Enhanced pulmonary allergic responses to *Aspergillus* in CCR2 −/− mice. *J. Immunol.* **165:**2603–2611.

Blease, K., B. Mehrad, T. J. Standiford, N. W. Lukacs, S. L. Kunkel, S. W. Chensue, B. Lu, C. J. Gerard, and C. M. Hogaboam. 2000b. Airway remodeling is absent in CCR1 −/− mice during chronic fungal allergic airway disease. *J. Immunol.* **165:**1564–1572.

Bolland, M. J., W. Bagg, M. G. Thomas, J. A. Lucas, R. Ticehurst, and P. N. Black. 2004 Cushing's syndrome due to interaction between inhaled corticosteroids and itraconazole. *Ann. Pharmacother.* **38:**46–49.

Bonfield, T. L., M. W. Konstan, and M. Berger. 1999. Altered respiratory epithelial cell cytokine production in cystic fibrosis. *J. Allergy Clin. Immunol.* **104:**72–78.

Borish, L., A. Aarons, J. Rumbyrt, P. Cvietusa, J. Negri, and S. Wenzel. 1996. Interleukin-10 regulation in normal subjects and patients with asthma. *J. Allergy Clin. Immunol.* **97:**1288–1296.

Brouard, J., N. Knauer, P. Y. Boelle, H. Corvol, A. Henrion-Caude, C. Flamant, F. Bremont, B. Delaisi, J. F. Duhamel, C. Marguet, M. Roussey, M. C. Miesch, K. Chadelat, M. Boule, B. Fauroux, F. Ratjen, H. Grasemann, and A. Clement. 2005. Influence of interleukin-10 on *Aspergillus fumigatus* infection in patients with cystic fibrosis. *J. Infect. Dis.* **191:**1988–1991.

Brueton, M. J., L. P. Ormerod, K. J. Shah, and C. M. Anderson. 1980. Allergic bronchopulmonary aspergillosis complicating cystic fibrosis. *Arch. Dis. Child.* **55:**348–353.

Buckingham, S. J., and D. M. Hansell. 2003. *Aspergillus* in the lung: diverse and coincident forms. *Eur. Radiol.* **13:**1786–1800.

Burns, J. L., J. M. Van Dalfsen, R. M. Shawar, K. L. Otto, R. L. Garber, J. M. Quan, A. B. Montgomery, G. M. Albers, B. W. Ramsey, and A. L. Smith. 1999. Effect of chronic intermittent administration of inhaled tobramycin on respiratory microbial flora in patients with cystic fibrosis. *J. Infect. Dis.* **179:**1190–1196.

Casadevall, A., and L.-A. Pirofski. 1999. Host-pathogen interactions: redefining the basic concepts of virulence and pathogenicity. *Infect. Immun.* **67:**3703–3713.

Casaulta, C., M. H. Schöni, M. Weichel, R. Crameri, M. Jutel, I. Daigle, M. Akdis, K. Blaser, and C. A. Akdis. 2003. IL-10 controls *Aspergillus fumigatus*- and *Pseudomonas aeruginosa*-specific T-cell response in cystic fibrosis. *Pediatr. Res.* **53:**313–319.

Casaulta, C., S. Fluckiger, R. Crameri, K. Blaser, and M. H. Schoeni. 2005. Time course of antibody response to recombinant *Aspergillus fumigatus* antigens in cystic fibrosis with and without ABPA. *Pediatr. Allergy Immunol.* **16:**217–225.

Casey, P., J. Garrett, and T. Eaton. 2002. Allergic bronchopulmonary aspergillosis in a lung transplant patient successfully treated with nebulized amphotericin. *J. Heart Lung Transplant.* **21:**1237–1241.

Chauhan, B., L. Santiago, D. A. Kirschmann, V. Hauptfield, A. P. Knutsen, P. S. Hutcheson, S. L. Woulfe, R. G. Slavin, H. J. Schwartz, and C. J. Bellone. 1997. The association of HLA-DR alleles and T-cell activation with allergic bronchopulmonary aspergillosis. *J. Immunol.* **159:**4072–4076.

Chauhan, B., L. Santiago, P. S. Hutcheson, H. J. Schwartz, E. Spitznagel, M. Castro, R. G. Slavin, and C. J. Bellone. 2000. Evidence for the involvement of two different MHC class II regions in susceptibility or protection in allergic bronchopulmonary aspergillosis. *J. Allergy Clin. Immunol.* **106:**723–729.

Chauhan, B., P. S. Hutcheson, R. G. Slavin, and C. J. Bellone. 2002. T-cell receptor bias in patients with allergic bronchopulmonary aspergillosis. *Hum. Immunol.* **63:**286–294.

Conway, S. P., C. Etherington, D. G. Peckham, K. G. Brownlee, A. Whitehead, and H. Cunliffe. 2004. Pharmacokinetics and safety of itraconazole in patients with cystic fibrosis. *J. Antimicrob. Chemother.* **53:**841–847.

Corry, D. B., G. Grunig, H. Hadeiba, V. P. Kurup, M. L. Warnock, D. Sheppard, D. M. Rennick, and R. M. Locksley. 1998. Requirements for allergen-induced airway hyperreactivity in T- and B-cell-deficient mice. *Mol. Med.* **4:**344–355.

Cortese, G., V. Malfitana, R. Placido, A. Ferrari, B. Grosso, V. De Rose, P. Nespoli, and C. Fava. 2007. Role of chest radiography in the diagnosis of allergic bronchopulmonary aspergillosis in adult patients with cystic fibrosis. *Radiol. Med.* (Torino) **112:**626–636.

Crameri, R., J. Lidholm, H. Grönlund, D. Stüber, K. Blaser, and G. Menz. 1996. Automated specific IgE assay with recombinant allergens: evaluation of the recombinant *Aspergillus fumigatus* allergen I in the Pharmacia Cap System. *Clin. Exp. Allergy* **26:**1411–1419.

Crameri, R. 1998. Recombinant *Aspergillus fumigatus* allergens: from the nucleotide sequences to clinical applications. *Int. Arch. Allergy Immunol.* **115:**99–114.

Crameri, R., S. Hemmann, C. Ismail, G. Menz, and K. Blaser. 1998. Disease-specific recombinant allergens for the diagnosis of allergic bronchopulmonary aspergillosis. *Int. Immunol.* **10:**1211–1216.

Crameri, R., R. Kodzius, Z. Konthur, H. Lehrach, K. Blaser, and G. Walter. 2001. Tapping allergen repertoires by advanced cloning technologies. *Int. Arch. Allergy Immunol.* **124:**43–47.

Cunningham, S., S. L. Madge, and R. Dinwiddie. 2001. Survey of criteria used to diagnose allergic bronchopulmonary aspergillosis in cystic fibrosis. *Arch. Dis. Child.* **84:**89.

Denning, D. W., J. E. Van Wye, N. J. Lewiston, and D. A. Stevens. 1991. Adjunctive therapy of allergic bronchopulmonary aspergillosis with itraconazole. *Chest* **100:**813–819.

Denning, D. W., B. R. O'Driscoll, C. M. Hogaboam, P. Bowyer, and R. M. Niven. 2006. The link between fungi and severe asthma: a summary of the evidence. *Eur. Respir. J.* **27:**615–626.

de Oliveira, E., P. Giavina-Bianchi, L. A. Fonseca, A. T. França, and J. Kalil. 2007. Allergic bronchopulmonary aspergillosis' diagnosis remains a challenge. *Respir. Med.* **101:**2352–2357.

De Wachter, E., J. Vanbesien, I. De Schutter, A. Malfroot, and J. De Schepper. 2003a. Rapidly developing Cushing syndrome in a 4-year-old patient during combined treatment with itraconazole and inhaled budesonide. *Eur. J. Pediatr.* **162:**488–489.

De Wachter, E., A. Malfroot, I. De Schutter, J. Vanbesien, and J. De Schepper. 2003b. Inhaled budesonide induced Cushing's syndrome in cystic fibrosis patients, due to drug inhibition of cytochrome P450. *J. Cyst. Fibros.* **2:**72–75.

Dorsaneo, D., D. Borowitz, J. Sharp, and R. Moss. 2004. Allergic bronchopulmonary aspergillosis with normal serum IgE in a child with cystic fibrosis. *Pediatr. Asthma Allergy Immunol.* **17:**146–150.

Dove, A. M., S. J. Szefler, M. R. Hill, W. J. Jusko, G. L. Larsen, and F. J. Accurso. 1992. Altered prednisolone pharmacokinetics in patients with cystic fibrosis. *J. Pediatr.* **120:**789–794.

Eaton, T., J. Garrett, D. Milne, A. Frankel, and A. U. Wells. 2000. Allergic bronchopulmonary aspergillosis in the asthma clinic. A prospective evaluation of CT in the diagnostic algorithm. *Chest* **118:** 66–72.

Eaton, T. E., P. Weiner Miller, J. E. Garrett, and G. R. Cutting. 2002. Cystic fibrosis transmembrane conductance regulator gene mutations: do they play a role in the aetiology of allergic bronchopulmonary aspergillosis? *Clin. Exp. Allergy* **32:**756–761.

El-Dahr, J. M., R. Fink, R. Selden, L. K. Arruda, T. A. E. Platts-Mills, and P. W. Heymann. 1994. Development of immune responses to *Aspergillus* at an early age in children with cystic fibrosis. *Am. J. Respir. Crit. Care Med.* **150:**1513–1518.

Feanny, S., S. Forsyth, M. Corey, H. Levison, and B. Zimmerman. 1988. Allergic bronchopulmonary aspergillosis in cystic fibrosis: a secretory immune response to a colonizing organism. *Ann. Allergy* **60:**64–68.

Fitzsimons, E. J., R. Aris, and R. Patterson. 1997. Recurrence of allergic bronchopulmonary aspergillosis in the posttransplant lungs of a cystic fibrosis patient. *Chest* **112:**281–282.

Garcia, G., M. Humbert, F. Capel, A. C. Rimaniol, P. Escourrou, D. Emilie, and V. Godot. 2007. Chemokine receptor expression on allergen-specific T cells in asthma and allergic bronchopulmonary aspergillosis. *Allergy* **62:**170–177.

Garrett, M. H., P. R. Rayment, M. A. Hooper, M. J. Abramson, and B. M. Hooper. 1998. Indoor airborne fungal spores, house dampness and associations with environmental factors and respiratory health in children. *Clin. Exp. Allergy* **28:**459–467.

Geller, D. E., H. Kaplowitz, M. J. Light, and A. A. Colin. 1999. Allergic bronchopulmonary aspergillosis in cystic fibrosis: reported prevalence, regional distribution, and patient characteristics. *Chest* **116:**639–646.

Gibson, P. G., P. A. Wark, J. L. Simpson, C. Meldrum, S. Meldrum, N. Saltos, and M. Boyle. 2003. Induced sputum IL-8 gene expression, neutrophil influx and MMP-9 in allergic bronchopulmonary aspergillosis. *Eur. Respir. J.* **21:**582–588.

Gilbert, J., and J. M. Littlewood. 1986. Enteric-coated prednisolone in cystic fibrosis. *Lancet* **ii:**1167–1168.

Gondor, M., M. G. Michaels, and J. D. Finder. 1998. Non-*Aspergillus* allergic bronchopulmonary mycosis in a pediatric patient with cystic fibrosis. *Pediatrics* **102:**1480–1482.

Goyal, R., C. S. White, P A. Templeton, E. J. Britt, and L. J. Rubin. 1992. High attenuation mucous plugs in allergic bronchopulmonary aspergillosis: CT appearance. *J. Comput. Assist. Tomogr.* **16:**649–650.

Greenberger, P. A., and R. Patterson. 1988. Allergic bronchopulmonary aspergillosis and the evaluation of the patient with asthma. *J. Allergy Clin. Immunol.* **81:**646–650.

Greenberger, P. A., L. J. Smith, C. C. Hsu, M. Roberts, and J. L. Liotta. 1988. Analysis of bronchoalveolar lavage in allergic bronchopulmonary aspergillosis: divergent responses of antigen-specific antibodies and total IgE. *J. Allergy Clin. Immunol.* **82:**164–170.

Greenberger, P. A., T. P. Miller, M. Roberts, and L. L. Smith. 1993. Allergic bronchopulmonary aspergillosis in patients with and without evidence of bronchiectasis. *Ann. Allergy* **70:**333–338.

Greenberger, P. A. 2002. Allergic bronchopulmonary aspergillosis. *J. Allergy Clin. Immunol.* **110:**685–692.

Greene, R. 2005. The radiological spectrum of pulmonary aspergillosis. *Med. Mycol.* **43:**S147–S154.

Grünig, G., D. B. Corry, M. W. Leach, B. W. Seymour, V. P. Kurup, and D. M. Rennick. 1997. Interleukin-10 is a natural suppressor of cytokine production and inflammation in a murine model of allergic bronchopulmonary aspergillosis. *J. Exp. Med.* **185:**1089–1099.

Grunig, G., D. B. Corry, R. L. Coffman, D. M. Rennick, and V. P. Kurup. 1998. Animal models of allergic bronchopulmonary aspergillosis. *Immunol. Allergy Clin. North Am.* **18:**661–679.

Hall, L. A., and D. W. Denning. 1994. Oxygen requirements of *Aspergillus* species. *J. Med. Microbiol.* **41:**311–315.

Hansell, D. M., and B. Strickland. 1989. High-resolution computed tomography in pulmonary cystic fibrosis. *Br. J. Radiol.* **62:**1–5.

Hardin, D. S., and A. Moran. 1999. Diabetes mellitus in cystic fibrosis. *Endocrinol. Metab. Clin. North Am.* **28:**787–800.

Hartl, D., K. F. Buckland, and C. M. Hogaboam. 2006. Chemokines in allergic aspergillosis—from animal models to human lung diseases. *Inflamm. Allergy Drug Targets* **5:**219–228.

Hartl, D., P. Latzin, G. Zissel, M. Krane, S. Krauss-Etschmann, and M. Griese. 2006. Chemokines indicate allergic bronchopulmonary aspergillosis in patients with cystic fibrosis. *Am. J. Respir. Crit. Care Med.* **173:**1370–1376.

Hemmann, S., W. H. Nikolaizik, M. H. Schoni, K. Blaser, and R. Crameri. 1998a. Differential IgE recognition of recombinant *Aspergillus fumigatus* allergens by cystic fibrosis patients with allergic bronchopulmonary aspergillosis. *Eur. J. Immunol.* **28:**1155–1160.

Hemmann, S., C. Ismail, K. Blaser, G. Menz, and R. Crameri. 1998b. Skin-test reactivity and isotype-specific immune responses to recombinant Asp f 3, a major allergen of *Aspergillus fumigatus*. *Clin. Exp. Allergy* **28:**860–867.

Hemmann, S., G. Menz, C. Ismail, K. Blaser, and R. Crameri. 1999. Skin test reactivity to 2 recombinant *Aspergillus fumigatus* allergens in *A. fumigatus*-sensitized asthmatic subjects allows diagnostic separation of allergic bronchopulmonary aspergillosis from fungal sensitisation. *J. Allergy Clin. Immunol.* **104:**601–607.

Hilliard, T., S. Edwards, R. Buchdahl, J. Francis, M. Rosenthal, I. Balfour-Lynn, A. Bush, and J. Davies. 2005. Voriconazole therapy in children with cystic fibrosis. *J. Cyst. Fibros.* **4:**215–220.

Hinson, K. F. W., A. J. Moon, and N. S. Plummer. 1952. Bronchopulmonary aspergillosis. A review and a report of eight new cases. *Thorax* **7:**317–333.

Hogaboam, C. M., C. S. Gallinat, D. D. Taub, R. M. Striter, S. L. Kunkel, and N. W. Lukacs. 1999. Immunomodulatory role of C10 chemokine in a murine model of allergic bronchopulmonary aspergillosis. *J. Immunol.* **162:**6071–6079.

Hogaboam, C. M., K. Blease, B. Mehrad, M. L. Steinhauser, T. J. Standiford, S. L. Kunkel, and N. W. Lukacs. 2000. Chronic airway hyperreactivity, goblet cell hyperplasia, and peribronchial fibrosis during allergic airway disease induced by *Aspergillus fumigatus*. *Am. J. Pathol.* **156:**723–732.

Hogaboam, C. M., K. J. Carpenter, J. M. Schuh, and K. F. Buckland. 2005. *Aspergillus* and asthma: any link? *Med. Mycol.* **43:**S197–S202.

Hutcheson, P. S., A. P. Knutsen, A. J. Rejent, and R. G. Slavin. 1996. A 12-year longitudinal study of *Aspergillus* sensitivity in patients with cystic fibrosis. *Chest* **110:**363–366.

Jahn, B., F. Boukhallouk, J. Lotz, K. Langfelder, G. Wanner, and A. A. Brakhage. 2000. Interaction of human phagocytes with pigmentless *Aspergillus* conidia. *Infect. Immun.* **68:**3736–3739.

Kanthan, S. K., A. Bush, M. Kemp, and R. Buchdahl. 2007. Factors effecting impact of *Aspergillus fumigatus* sensitization in cystic fibrosis. *Pediatr. Pulmonol.* **42:**785–793.

Kauffman, H. F., J. F. C. Tomee, M. A. van de Riet, A. J. Timmerman, and P. Borger. 2000. Protease-dependent activation of epithelial cells by fungal allergens leads to morphologic changes and cytokine production. *J. Allergy Clin. Immunol.* **105:**1185–1193.

Kauffman, H. F. 2003. Immunopathogenesis of allergic bronchopulmonary aspergillosis and airway remodeling. *Front. Biosci.* **8:**e190–e196.

Kaur, S., V. K. Gupta, A. Shah, S. Thiel, P. U. Sarma, and T. Madan. 2006. Elevated levels of mannan-binding lectin [corrected] (MBL) and eosinophilia in patients of bronchial asthma with allergic rhinitis and allergic bronchopulmonary aspergillosis associate with a novel intronic polymorphism in MBL. *Clin. Exp. Immunol.* **143:**41441–41449. (Erratum, **144:**552.)

Khan, S., J. S. McClellan, and A. P. Knutsen. 2000. Increased sensitivity to IL-4 in patients with allergic bronchopulmonary aspergillosis. *Int. Arch. Allergy Immunol.* **123:**319–326.

Knutsen, A. P., K. R. Mueller, A. D. Levine, B. Chauhan, P. Hutcheson, and R. G. Slavin. 1994a. Characterization of Asp f1 CD4+ T cell lines in allergic bronchopulmonary aspergillosis. *J. Allergy Clin. Immunol.* **94:**215–221.

Knutsen, A. P., K. R. Mueller, and P. S. Hutcheson. 1994b. Serum anti-*Aspergillus fumigatus* antibodies by immunoblot and ELISA in cystic fibrosis with allergic bronchopulmonary aspergillosis. *J. Allergy Clin. Immunol.* **93:**926–931.

Knutsen, A. P., B. Chauhan, and R. G. Slavin. 1998. Cell-mediated immunity in allergic bronchopulmonary aspergillosis. *Immunol. Allergy Clin. North Am.* **18:**575–599.

Knutsen, A. P. 2003. Lymphocytes in allergic bronchopulmonary aspergillosis. *Front. Biosci.* **8:**589–602.

Knutsen, A. P., P. S. Hutcheson, R. G. Slavin, and V. P. Kurup. 2004a. IgE antibody to *Aspergillus fumigatus* recombinant allergens in cystic fibrosis patients with allergic bronchopulmonary aspergillosis. *Allergy* **59:**198–203.

Knutsen, A. P., P. S. Hutchinson, G. M. Albers, J. Consolino, J. Smick, and V. P. Karup. 2004b. Increased sensitivity to IL-4 in cystic fibrosis patients with allergic bronchopulmonary aspergillosis. *Allergy* **59:**81–87.

Knutsen, A. P., B. Noyes, M. R. Warrier, and J. Consolino. 2005. Allergic bronchopulmonary aspergillosis in a patient with cystic fibrosis: diagnostic criteria when the IgE level is less than 500 IU/mL. *Ann. Allergy Asthma Immunol.* **95:**488–489.

Knutsen, A. P., B. Kariuki, J. D. Consolino, and M. R. Warrier. 2006. IL-4 alpha chain receptor polymorphisms in allergic bronchopulmonary aspergillosis. *Clin. Mol. Allergy* **4:**3.

Koehm, S., R. G. Slavin, P. S. Hutcheson, T. Trejo, C. S. David, and C. J. Bellone. 2007. HLA-DRB1 alleles control allergic bronchopulmonary aspergillosis-like pulmonary responses in humanized transgenic mice. *J. Allergy Clin. Immunol.* **120:**570–577.

Kraemer, R., N. Delosea, P. Ballinari, S. Gallati, and R. Crameri. 2006. Effect of allergic bronchopulmonary aspergillosis on lung function in children with cystic fibrosis. *Am. J. Respir. Crit. Care Med.* **174:**1211–1220.

Krasnick, J., R. Patterson, and M. Roberts. 1995. Allergic bronchopulmonary aspergillosis presenting with cough variant asthma and identifiable source of *Aspergillus fumigatus*. *Ann. Allergy Asthma Immunol.* **75:**344–346.

Kumar, R. 2003. Mild, moderate, and severe forms of allergic bronchopulmonary aspergillosis: a clinical and serologic evaluation. *Chest* **124:**890–892.

Kunst, H., M. Wickremasinghe, A. Wells, and R. Wilson. 2006. Nontuberculous mycobacterial disease and *Aspergillus*-related lung disease in bronchiectasis. *Eur. Respir. J.* **28:**352–357.

Kurup, V. P., J. Guo, P. S. Murali, H. Choi, and J. N. Fink. 1997. Immunopathological responses to *Aspergillus* antigen in interleukin-4 knockout mice. *J. Lab. Clin. Med.* **130:**567–575.

Kurup, V. P., B. Banerjee, S. Hemmann, P. A. Greenberger, K. Blaser, and R. Crameri. 2000. Selected recombinant *Aspergillus fumigatus* allergens bind specifically to IgE in ABPA. *Clin. Exp. Allergy* **30:**988–993.

Kurup, V. P., J.-Q. Xia, R. Crameri, D. A. Rickaby, H. Y. Choi, S. Fluckiger, K. Blaser, C. A. Dawson, and K. J. Kelly. 2001. Purified recombinant *A. fumigatus* allergens induce different responses in mice. *Clin. Immunol.* **98:**327–336.

Kurup, V. P., and G. Grunig. 2002. Animal models of allergic bronchopulmonary aspergillosis. *Mycopathologia* **153:**165–177.

Kurup, V. P. 2005. Aspergillus antigens: which are important? *Med. Mycol.* **43:**S189–S196.

Kurup, V. P., A. P. Knutsen, and R. B. Moss. 2005. *Aspergillus* antigens and immunodiagnosis of allergic bronchopulmonary aspergillosis, p. 137–146. *In* V. S. Kurup (ed.), *Mold Allergy, Biology and Pathogenesis*. Research Signpost, Trivandrum, India.

Kurup, V. P., A. P. Knutsen, R. B. Moss, and N. K. Bansal. 2006. Specific antibodies to recombinant allergens of *Aspergillus fumigatus* in cystic fibrosis patients with ABPA. *Clin. Mol. Allergy* **4:**11.

Lacadena, J., E. Alvarez-García, N. Carreras-Sangrà, E. Herrero-Galán, J. Alegre-Cebollada, L. García-Ortega, M. Oñaderra, J. G. Gavilanes, and A. Martínez del Pozo. 2007. Fungal ribotoxins: molecular dissection of a family of natural killers. *FEMS Microbiol. Rev.* **31:**212–237.

Lai, H. C., S. C. Fitzsimmons, D. B. Allen, M. R. Kosorok, B. J. Rosenstein, P. W. Campbell, and P. M. Farrell. 2000. Risk of persistent growth impairment after alternate-day prednisone treatment in children with cystic fibrosis. *N. Engl. J. Med.* **342:**851–859.

Latzin, P., D. Hartl, N. Regamey, U. Frey, M. H. Schoeni, and C. Casaulta. 2008. Comparison of serum markers for allergic bronchopulmonary aspergillosis in cystic fibrosis. *Eur. Respir. J.* **31:**36–42.

Laufer, P., J. N. Fink, W. T. Bruns, G. F. Unger, J. H. Kalbfleisch, P. A. Greenberger, and R. Patterson. 1984. Allergic bronchopulmonary aspergillosis in cystic fibrosis. *J. Allergy Clin. Immunol.* **73:**44–48.

Lebrun-Vignes, B., V. C. Archer, B. Diquet, J. C. Levron, O. Chosidow, A. J. Puech, and D. Warot. 2001. Effect of itraconazole on the pharmacokinetics of prednisolone and methylprednisolone and cortisol secretion in healthy subjects. *Br. J. Clin. Pharmacol.* **51:**443–450.

Lee, T. M., P. A. Greenberger, R. Patterson, M. Roberts, and J. L. Liotta. 1987. Stage V (fibrotic) allergic bronchopulmonary aspergillosis: a review of 17 cases followed from diagnosis. *Arch. Intern. Med.* **147:**319–323.

Leon, E. E., and T. J. Craig. 1999. Antifungals in the treatment of allergic bronchopulmonary aspergillosis. *Ann. Allergy Asthma Immunol.* **82:**511–517.

Leser, C., H. F. Kauffman, C. Virchow, and G. Menz. 1992. Specific serum immunopatterns in clinical phases of allergic bronchopulmonary aspergillosis. *J. Allergy Clin. Immunol.* **90:**589–599.

Logan, P. M., and N. L. Muller. 1996. High attentuation mucus plugging in allergic bronchopulmonary aspergillosis. *Can. Assoc. Radiol. J.* **47:**374–377.

Lynch, D. A. 1998. Imaging of asthma and allergic bronchopulmonary mycosis. *Radiol. Clin. North Am.* **36:**129–142.

Madan, T., U. Kishore, M. Singh, P. Strong, H. Clark, E. M. Hussain, K. B. Reid, and P. U. Sarma. 2001. Surfactant proteins A and D protect mice against pulmonary hypersensitivity induced by *Aspergillus fumigatus* antigens and allergens. *J. Clin. Investig.* **107:**467–475.

Madan, T., S. Kaur, S. Saxena, M. Singh, U. Kishore, S. Thiel, K. B. Reid, and P. U. Sarma. 2005. Role of collectins in innate immunity against aspergillosis. *Med. Mycol.* **43:**S155–S163.

Maguire, S., P. Moriarty, E. Tempany, and M. FitzGerald. 1988. Unusual clustering of allergic bronchopulmonary aspergillosis in children with cystic fibrosis. *Pediatrics* **82:**835–839.

Maguire, C. P., J. P. Hayes, M. Hayes, J. Masterson, and M. X. FitzGerald. 1995. Three cases of pulmonary aspergilloma in adult patients with cystic fibrosis. *Thorax* **50:**805–806.

Main, K. M., M. Skov, I. B. Sillesen, H. Dige-Petersen, J. Muller, C. Koch, and S. Lanng. 2002. Cushing's syndrome due to pharmacological interaction in a cystic fibrosis patient. *Acta Paediatr.* **91:**1008–1011.

Marchand, E., C. Verellen-Dumoulin, M. Mairesse, L. Delaunois, P. Brancaleone, J. F. Rahier, and O. Vandenplas. 2001. Frequency of cystic fibrosis transmembrane conductance regulator gene mutations and 5T allele in patients with allergic bronchopulmonary aspergillosis. *Chest* **119:**762–767.

Marchant, J. R., J. O. Warner, and A. Bush. 1994. Rise in total IgE as an indicator of allergic bronchopulmonary aspergillosis in cystic fibrosis. *Thorax* **49:**1002–1005.

Mastella, G., M. Rainisio, H. K. Harms, C. Koch, J. Navarro, B. Strandvik, and S. G. McKenzie. 2000. Allergic bronchopulmonary aspergillosis in cystic fibrosis. A European epidemiological study. Epidemiologic Registry of Cystic Fibrosis. *Eur. Respir. J.* **16:**464–471.

Maurya, V., H. C. Gugnani, P. U. Sarma, T. Madan, and A. Shah. 2005. Sensitization to *Aspergillus* antigens and occurrence of allergic bronchopulmonary aspergillosis in patients with asthma. *Chest* **127:** 1252–1259.

Mearns, M., W. Young, and J. Batten. 1965. Transient pulmonary infiltrations in cystic fibrosis due to allergic aspergillosis. *Thorax* **20:** 385–392.

Milla, C. E., C. L. Wielinski, and W. E. Regelmann. 1996. Clinical significance of the recovery of *Aspergillus* species from the respiratory secretions of cystic fibrosis patients. *Pediatr. Pulmonol.* **21:**6–10.

Miller, P. W., A. Hamosh, M. Macej, Jr., P. A. Greenberger, J. MacLean, S. M. Walden, R. G. Slavin, and G. R. Cutting. 1996. Cystic fibrosis transmembrane conductance regulator (CFTR) gene mutations in allergic bronchopulmonary aspergillosis. *Am. J. Hum. Genet.* **59:**45–51.

Mitchell-Heggs, P., M. Mearns, and J. C. Batten. 1976. Cystic fibrosis in adolescents and asthma. *Q. J. Med.* **45:**479–504.

Monod, M., A. Fatih, K. Jaton-Ogay, S. Paris, and J. P. Latgé. 1995. The secreted proteases of pathogenic species of *Aspergillus* and their possible role in virulence. *Can. J. Bot.* 73(Suppl. 1):S1081–S1086.

Moore, C. B., N. Sayers, J. Mosquero, J. Slaven, and D. W. Denning. 2000. Antifungal drug resistance in *Aspergillus. J. Infect.* **41:**203–220.

Morozov, A., K. E. Applegate, S. Brown, and M. Howenstine. 2007. High-attenuation mucus plugs on MDCT in a child with cystic fibrosis: potential cause and differential diagnosis. *Pediatr. Radiol.* **37:** 592–595.

Moser, M., R. Crameri, E. Brust, M. Suter, and G. Menz. 1994. Diagnostic value of recombinant *Aspergillus fumigatus* allergen I/a for skin testing and serology. *J. Allergy Clin. Immunol.* **93:**1–11.

Mosquera, J., and D. W. Denning. 2002. Azole cross-resistance in *Aspergillus fumigatus. Antimicrob. Agents Chemother.* **46:**556–557.

Moss, R. B. 2002. Allergic bronchopulmonary aspergillosis. *Clin. Rev. Allergy* **23:**87–104.

Moss, R. B. 2005. Fungal allergy in cystic fibrosis, p. 93–104. *In* V. S. Kurup (ed.), *Mold Allergy, Biology and Pathogenesis*. Research Signpost, Trivandrum, India.

Moss, R. B. 2006. Critique of trials in allergic bronchopulmonary aspergilllosis and fungal allergy. *Med. Mycol.* **44:**S269–S272.

Mroueh, S., and A. Spock. 1994. Allergic bronchopulmonary aspergillosis in patients with cystic fibrosis. *Chest* **105:**32–36.

Mueller, C., D. Torrez, S. Braag, A. Martino, T. Clarke, M. Campbell-Thompson, and T. R. Flotte. 2008. Partial correction of the CFTR-dependent ABPA mouse model with recombinant adeno-associated virus gene transfer of truncated CFTR gene. *J. Gene Med.* **10:**51–60.

Müller, C., S. A. Braag, J. D. Herlihy, C. H. Wasserfall, S. E. Chesrown, H. S. Nick, M. A. Atkinson, and T. R. Flotte. 2006. Enhanced IgE allergic response to *Aspergillus fumigatus* in CFTR$^{-/-}$ mice. *Lab. Investig.* **86:**130–140.

Murali, P. S., K. Pathial, R. H. Saff, M. L. Splaingard, D. Atluru, V. P. Kurup, and J. N. Fink. 1994. Immune responses to *Aspergillus fumigatus* and *Pseudomonas aeruginosa* antigens in cystic fibrosis and allergic bronchopulmonary aspergillosis. *Chest* **106:**513–519.

Mussaffi, H., J. Rivlin, I. Shalit, M. Ephros, and H. Blau. 2005. Nontuberculous mycobacteria in cystic fibrosis associated with allergic bronchopulmonary aspergillosis and steroid therapy. *Eur. Respir. J.* **25:**324–328.

Neeld, D. A., L. R. Goodman, J. W. Gurney, P. A. Greenberger, and J. N. Fink. 1990. Computerized tomography in the evaluation of allergic bronchopulmonary aspergillosis. *Am. Rev. Respir. Dis.* **142:** 1200–1205.

Nelson, L. A., M. L. Callerame, and R. H. Schwartz. 1979. Aspergillosis and atopy in cystic fibrosis. *Am. Rev. Respir. Dis.* **120:**863–873.

Nepomuceno, I. B., S. Esrig, and R. B. Moss. 1999. Allergic bronchopulmonary aspergillosis in cystic fibrosis: role of atopy and response to itraconazole. *Chest* **115:**364–370.

Neth, O., D. J. Jack, A. W. Dodds, H. Holzel, N. J. Klein, and M. W. Turner. 2000. Mannose binding lectin binds to a range of clinically relevant microorganisms and promotes complement deposition. *Infect. Immun.* **68:**688–693.

Neuvéglise, C., J. Sarfati, J. P. Debeaupuis, H. Vu-Thien, J. Just, J. G. Tournier, and J. P. Latgé. 1997. Longitudinal study of *Aspergillus fumigatus* strains isolated from cystic fibrosis patients. *Eur. J. Clin. Microbiol. Infect. Dis.* **16:**747–750.

Ng, T. T. C., G. D. Robson, and D. W. Denning. 1994. Hydrocortisone-enhanced growth of *Aspergillus* spp.: implications for pathogenesis. *Microbiology* **140:**2475–2480.

Nicolai, T., S. Arleth, A. Spaeth, R. M. Bertele-Harms, and H. K. Harms. 1990. Correlation of IgE antibody titer to *Aspergillus fumigatus* with decreased lung function in cystic fibrosis. *Pediatr. Pulmonol.* **8:**12–15.

Nikolanizik, W. H., M. Moser, R. Crameri, S. Little, J. O. Warner, K. Blaser, and M. H. Schöni. 1995. Identification of allergic bronchopulmonary aspergillosis in cystic fibrosis patients by recombinant *Aspergillus fumigatus* I/a-specific serology. *Am. J. Respir. Crit. Care Med.* **162:**634–639.

Nikolaizik, W. H., R. Crameri, K. Blaser, and R. Crameri. 1996. Skin test reactivity to recombinant *Aspergillus fumigatus* allergen I/a in patients with cystic fibrosis. *Int. Arch. Allergy Immunol.* **111:**403–408.

Panchal, N., C. Pant, R. Bhagat, and A. Shah. 1994. Central bronchiectasis in allergic bronchopulmonary aspergillosis: comparative evaluation of computed tomography of the thorax with bronchography. *Eur. Respir. J.* **7:**1290–1293.

Parmar, J. S., T. Howell, J. Kelly, and D. Bilton. 2002. Profound adrenal suppression secondary to treatment with low dose inhaled steroids and itraconazole in allergic bronchopulmonary aspergillosis in cystic fibrosis. *Thorax* **57:**749–750.

Patterson, R., P. A. Greenberger, R. D. Radin, and M. Roberts. 1982. Allergic bronchopulmonary aspergillosis: staging as an aid to management. *Ann. Intern. Med.* **96:**286–291.

Patterson, R., P. A. Greenberger, A. J. Ricketti, and M. Roberts. 1983. A radioimmunoassay index for allergic bronchopulmonary aspergillosis. *Ann. Intern. Med.* **99:**18–22.

Patterson, R., P. A. Greenberger, J. M. Halwig, J. L. Liotta, and M. Roberts. 1986. Allergic bronchopulmonary aspergillosis: natural history and classification of disease by serologic and roentgenographic studies. *Arch. Intern. Med.* **146:**916–918.

Phillips, P., J. R. Graybill, R. Fetchick, and J. F. Dunn. 1987. Adrenal response to corticotropin during therapy with itraconazole. *Antimicrob. Agents Chemother.* **31:**647–649.

Pitt, J. I. 1994. The current role of *Aspergillus* and *Penicillium* in human and animal health. *J. Med. Vet. Mycol. Suppl.* 32(Suppl. 1): 17–32.

Queiroz-Telles, F., K. S. Purim, C. L. Boguszewski, F. C. Afonso, and H. Graf. 1997. Adrenal response to corticotrophin and testosterone during long-term therapy with itraconazole in patients with chromoblastomycosis. *J. Antimicrob. Chemother.* **40:**899–902.

Raaska, K., M. Niemi, M. Neuvonen, P. J. Neuvonen, and K. T. Kivisto. 2002. Plasma concentrations of inhaled budesonide and its effects on plasma cortisol are increased by the cytochrome P4503A4 inhibitor itraconazole. *Clin. Pharmacol. Ther.* 72:362–369.

Radin, R. C., P. A. Greenberger, R. Patterson, and A. Ghory. 1983. Mould counts and exacerbations of allergic bronchopulmonary aspergillosis. *Clin. Allergy* 13:271–275.

Ramadan, G., B. Davies, V. P. Kurup, and C. A. Keever-Taylor. 2005. Generation of Th1 T cell responses directed to a HLA class II restricted epitope from the *Aspergillus* f16 allergen. *Clin. Exp. Immunol.* 139:257–267.

Reiff, D. B., A. U. Wells, D. H. Carr, P. J. Cole, and D. M. Hansell. 1995. CT findings in bronchiectasis: limited value in distinguishing between idiopathic and specific types. *Am. J. Roentgenol.* 165:261–267.

Rementeria, A., N. López-Molina, A. Ludwig, A. B. Vivanco, B. Joseba, J. Pontón, and J. Garaizar. 2005. Genes and molecules involved in *Aspergillus fumigatus* virulence. *Rev. Iberoam. Micol.* 22: 1–23.

Ringer, S., U. C. Hipler, P. Elsner, F. Zintl, and J. Mainz. 2007. Potential role of the cellular allergen stimulation test (CAST) in diagnosis of allergic bronchopulmonary aspergillosis (ABPA) in cystic fibrosis. *Pediatr. Pulmonol.* 42:314–318.

Ritz, N., R. A. Ammann, C. Casaulta Aebischer, F. Schoeni-Affolter, and M. H. Schoeni. 2005. Risk factors for allergic bronchopulmonary aspergillosis and sensitisation to *Aspergillus fumigatus* in patients with cystic fibrosis. *Eur. J. Pediatr.* 164:577–582.

Romagnani, S. 2001. Cytokines and chemoattractants in allergic inflammation. *Mol. Immunol.* 38:881–885.

Rosenberg, M., M. Rosenberg, R. Mintzer, B. J. Cooper, M. Roberts, and K. E. Harris. 1977. Clinical and immunologic criteria for the diagnosis of allergic bronchopulmonary aspergillosis. *Ann. Intern. Med.* 86:405–414.

Santis, G., M. E. Hodson, and B. Strickland. 1991. High resolution computed tomography in adult cystic fibrosis patients with mild lung disease. *Clin. Radiol.* 44:20–22.

Sarfati, J., M. Monod, P. Recco, A. Sulahian, C. Pinel, E. Candolfi, T. Fontaine, J. P. Debeaupuis, M. Tabouret, and J. P. Latgé. 2006. Recombinant antigens as diagnostic markers for aspergillosis. *Diagn. Microbiol. Infect. Dis.* 55:279–291.

Saxena, S., T. Madan, A. Shah, K. Muralidhar, and P. U. Sarma. 2003. Association of polymorphisms in the collagen region of SP-A2 with increased levels of total IgE antibodies and eosinophilia in patients with allergic bronchopulmonary aspergillosis. *J. Allergy Clin. Immunol.* 111:1001–1007.

Schønheyder, H., and P. Andersen. 1984. IgG antibodies to purified *Aspergillus fumigatus* antigens determined by enzyme-linked immunosorbent assay. *Int. Arch. Allergy Appl. Immunol.* 74:262–269.

Schonheyder, H., T. Jensen, N. Hoiby, and C. Koch. 1988. Clinical and serological survey of pulmonary aspergillosis in patients with cystic fibrosis. *Int. Arch. Allergy Appl. Immunol.* 85:472–477.

Schuh, J. M., C. A. Power, A. E. Proudfoot, S. L. Kunkel, N. W. Lukacs, and C. M. Hogaboam. 2002. Airway hyperresponsiveness, but not airway remodeling, is attenuated during chronic pulmonary allergic responses to *Aspergillus* in CCR4$^{-/-}$ mice. *FASEB J.* 16: 1313–1315.

Schuh, J. M., K. Blease, S. L. Kunkel, and C. M. Hogaboam. 2003. Chemokines and cytokines: axis and allies in asthma and allergy. *Cytokine Growth Factor Rev.* 14:503–510.

Schwartz, H. J., and P. A. Greenberger. 1991. The prevalence of allergic bronchopulmonary aspergillosis in patients with asthma, determined by serologic and radiologic criteria in patients at risk. *J. Lab. Clin. Med.* 117:138–142.

Shoseyov, D., K. G. Brownlee, S. P. Conway, and E. Kerem. 2006. *Aspergillus* bronchitis in cystic fibrosis. *Chest* 130:222–226.

Simmonds, E. J., J. M. Littlewood, and E. G. V. Evans. 1990a. Allergic bronchopulmonary aspergillosis. *Lancet* 335:1229.

Simmonds, E. J., J. M. Littlewood, and E. G. Evans. 1990b. Cystic fibrosis and allergic bronchopulmonary aspergillosis. *Arch. Dis. Child.* 65:507–511.

Skov, M., T. Pressler, H. E. Jensen, N. Hoiby, and C. Koch. 1999s. Specific IgG subclass antibody pattern to *Aspergillus fumigatus* in patients with cystic fibrosis with allergic bronchopulmonary aspergillosis. *Thorax* 54:44–50.

Skov, M., L. K. Poulsen, and C. Koch. 1999b. Increased antigen-specific Th-2 response in allergic bronchopulmonary aspergillosis (ABPA) in patients with cystic fibrosis. *Pediatr. Pulmonol.* 27:74–79.

Skov, M., C. Koch, C. M. Reimert, and L. K. Poulsen. 2000. Diagnosis of allergic bronchopulmonary aspergillosis (ABPA) in cystic fibrosis. *Allergy* 55:50–58.

Skov, M., K. M. Main, I. B. Sillesen, J. Muller, C. Koch, and S. Lanng. 2002a. Iatrogenic adrenal insufficiency as a side-effect of combined treatment of itraconazole and budesonide. *Eur. Respir. J.* 20:127–133.

Skov, M., N. Høiby, and C. Koch. 2002b. Itraconazole treatment of allergic bronchopulmonary aspergillosis in patients with cystic fibrosis. *Allergy* 57:723–728.

Skov, M., K. McKay, C. Koch, and P. J. Cooper. 2005. Prevalence of allergic bronchopulmonary aspergillosis in cystic fibrosis in an area with a high frequency of atopy. *Respir. Med.* 99:887–893.

Skowronski, E., and D. A. Fitzgerald. 2005. Life-threatening allergic bronchopulmonary aspergillosis in a well child with cystic fibrosis. *Med. J. Aust.* 182:482–483.

Slavin, R. G., T. S. Laird, and J. D. Cherry. 1970. Allergic bronchopulmonary aspergillosis in a child. *J. Pediatr.* 76:416–421.

Slavin, R. G., C. W. Bedrossian, P. S. Hutcheson, S. Pittman, L. Salinas-Madrigal, C. C. Tsai, and G. J. Gleich. 1988. A pathologic study of allergic bronchopulmonary aspergillosis. *J. Allergy Clin. Immunol.* 81:718–725.

Slavin, R. G., G. J. Gleich, P. S. Hutcheson, G. M. Kephardt, A. P. Knutsen, and C. C. Tsai. 1992. Localization of IgE to lung germinal lymphoid follicles in a patient with allergic bronchopulmonary aspergillosis. *J. Allergy Clin. Immunol.* 90:1006–1008.

Stevens, D. A. 1999. Itraconazole in cyclodextrin solution. *Pharmacotherapy* 19:603–611.

Stevens, D. A., H. J. Schwartz, J. Y. Lee, B. L. Moskovitz, D. C. Jerome, A. Catanzaro, D. M. Bamberger, A. J. Weinmann, C. U. Tuazon, M. A. Judson, T. A. Platts-Mills, and A. C. DeGraff, Jr. 2000. A randomized trial of itraconazole in allergic bronchopulmonary aspergillosis. *N. Engl. J. Med.* 342:756–762.

Stevens, D. S., R. B. Moss, V. P. Kurup, A. P. Knutsen, P. Greenberger, M. A. Judson, D. W. Denning, R. Crameri, A. Brody, M. Light, M. Skov, G. Maish, and G. Mastella. 2003. Allergic bronchopulmonary aspergillosis in cystic fibrosis. State of the Art: Cystic Fibrosis Foundation Consensus Conference. *Clin. Infect. Dis.* 37: S225–S264.

Taccetti, G., E. Procopio, L. Marianelli, and S. Campana. 2001. Allergic bronchopulmonary aspergillosis in Italian cystic fibrosis patients: prevalence and percentage of positive tests in the employed diagnostic criteria. *Eur. J. Epidemiol.* 16:837–842.

Thomson, J. M., A. Wesley, C. A. Byrnes, and G. M. Nixon. 2006. Pulse intravenous methylprednisolone for resistant allergic bronchopulmonary aspergillosis in cystic fibrosis. *Pediatr. Pulmonol.* 41: 164–170.

Tillie-Leblond, I., and A. B. Tonnel. 2005. Allergic bronchopulmonary aspergillosis. *Allergy* 60:1004–1013.

Tomee, J. F., A. T. J. Wierenhga, P. S. Hiemstra, and H. F. Kauffman. 1997. Proteases from *Aspergillus fumigatus* induce release of proinflammatroy cytokines and cell detachment in airway epithelial cell lines. *J. Infect. Dis.* 176:300–303.

Tomee, J. F. C., and H. F. Kauffman. 2000. Putative virulence factors of *Aspergillus fumigatus*. *Clin. Exp. Allergy* 30:476–484.

Vaid, M., S. Kaur, H. Sambatakou, T. Madan, D. W. Denning, and P. U. Sarma. 2007. Distinct alleles of mannose-binding lectin (MBL) and surfactant proteins A (SP-A) in patients with chronic cavitary pulmonary aspergillosis and allergic bronchopulmonary aspergillosis. *Clin. Chem. Lab. Med.* **45**:18318–18326.

Valletta, E. A., C. Braggion, and G. Mastella. 1993. Sensitization to *Aspergillus* and allergic bronchopulmonary aspergillosis in a cystic fibrosis population. *Pediatr. Asthma Allergy Immunol.* **7**:43–49.

van der Ent, C. K., H. Hoekstra, and G. T. Rijkers. 2007. Successful treatment of allergic bronchopulmonary aspergillosis with recombinant anti-IgE antibody. *Thorax* **62**:276–277.

Van Hoeyveld, E., L. Dupont, and X. Bossuyt. 2006. Quantification of IgG antibodies to *Aspergillus fumigatus* and pigeon antigens by ImmunoCAP technology: an alternative to the precipitation technique? *Clin. Chem.* **52**:1785–1793.

Varis, T., K. M. Kaukonen, K. T. Kivisto, and P. J. Neuvonen. 1998. Plasma concentrations and effects of oral methylprednisolone are considerably increased by itraconazole. *Clin. Pharmacol. Ther.* **64**: 363–368.

Varis, T., K. T. Kivisto, J. T. Backman, and P. J. Neuvonen. 1999. Itraconazole decreases the clearance and enhances the effects of intravenously administered methylprednisolone in healthy volunteers. *Pharmacol. Toxicol.* 85:29–32.

Varis, T., K. T. Kivisto, J. T. Backman, and P. J. Neuvonen. 2000a. The cytochrome P450 3A4 inhibitor itraconazole markedly increases the plasma concentrations of dexamethasone and enhances its adrenal-suppressant effect. *Clin. Pharmacol. Ther.* **68**:487–494.

Varis, T., K. T. Kivisto, and P. J. Neuvonen. 2000b. The effect of itraconazole on the pharmacokinetics and pharmacodynamics of oral prednisolone. *Eur. J. Clin. Pharmacol.* **56**:57–60.

Vilar, M. E. B., N. M. Najib, and I. Chowdhry. 2000. Allergic bronchopulmonary aspergillosis as presenting sign of cystic fibrosis in an elderly man. *Ann. Allergy Asthma Immunol.* 85:70–73.

Walsh, T. J., E. Roilides, K. Cortez, S. Kottilil, J. Bailey, and C. A. Lyman. 2005. Control, immunoregulation, and expression of innate pulmonary host defenses against *Aspergillus fumigatus*. *Med. Mycol.* 43:S165–S172.

Ward, S., L. Heyneman, M. J. Lee, A. N. Leung, D. M. Hansell, and N. L. Muller. 1999. Accuracy of CT in the diagnosis of allergic bronchopulmonary aspergillosis in asthmatic patients. *Am. J. Roentgenol.* **173**:937–942.

Wark, P. A., M. J. Hensley, N. Saltos, M. J. Boyle, R. C. Toneguzzi, G. D. Epid, J. L. Simpson, P. McElduff, and P. G. Gibson. 2003. Anti-inflammatory effect of itraconazole in stable allergic bronchopulmonary aspergillosis: a randomized controlled trial. *J. Allergy Clin. Immunol.* **111**:952–957.

Wark, P. 2004. Pathogenesis of allergic bronchopulmonary aspergillosis and an evidence-based review of azoles in treatment. *Respir. Med.* **98**:915–923.

Wark, P. A., P. G. Gibson, and A. J. Wilson. 2004. Azoles for allergic bronchopulmonary aspergillosis associated with asthma. *Cochrane Database Syst. Rev.* **3**:CD001108.

Wojnarowski, C., I. Eichler, C. Gartner, C. Grabner, W. Mosgoeller, I. Eichler, and R. Ziesche. 1997. Sensitization to *Aspergillus fumigatus* and lung function in children with cystic fibrosis. *Am. J. Respir. Crit. Care Med.* **155**:1902–1907.

Zander, D. S. 2005. Allergic bronchopulmonary aspergillosis: an overview. *Arch. Pathol. Lab. Med.* **129**:924–928.

Zeaske, R., W. T. Bruns, J. N. Fink, P. A. Greenberger, H. Colby, J. L. Liotta, and M. Roberts. 1988. Immune responses to *Aspergillus* in cystic fibrosis. *J. Allergy Clin. Immunol.* **82**:73–77.

V. DIAGNOSIS

Aspergillus fumigatus and Aspergillosis
Edited by J.-P. Latgé and W. J. Steinbach

Chapter 27

Histology and Radiology

REGINALD GREENE, KAZUTOSHI SHIBUYA, AND TSUNIHIRO ANDO

Aspergillus fumigatus is a ubiquitous mold that causes a wide range of diseases in humans, each having characteristic imaging findings that reflect their defining histopathology and pathogenetic sequences. These illnesses can cause innocent saprophyte growth in preexisting chronic lung cysts, lead to allergic lung damage, or create life-threatening invasive infection. In some cases, the imaging findings simulate other serious infectious and noninfectious conditions, and in other cases the findings can even be confused with other categories of aspergillosis. The risk of developing one or another form of aspergillosis is strongly influenced by host factors, including the clinical background, the presence of preexisting underlying lung disease, and the patient's immune status. Other factors include the route of entry of the organism and the size of the inoculum. In this chapter the pathology of the three major categories of pulmonary aspergillosis (invasive, allergic, and saprophytic) and the diagnostic imaging challenges they pose are discussed (Greene, 1981; Soubani and Chandrasekar, 2002).

INVASIVE PULMONARY ASPERGILLOSIS

A. fumigatus rarely causes invasive infection in the immunocompetent host. In the severely immunocompromised host who is at high risk of invasive mold infection, however, it often causes life-threatening primary opportunistic fungal pneumonia. Because the common route of entry is characteristically via pulmonary ventilation, the lung is the most common initial site of infection (Herbrecht et al., 2002). Two unique pathogenetic sequences are responsible for acute invasive pulmonary aspergillosis (IPA). The first, and most important in clinical practice, is angio-invasion, which is responsible for >90% of such infections (Greene et al., 2007; Herbrecht et al., 2002). In this sequence, the infection initiates in the distal airspaces of what is often an otherwise-normal lung. In the second, much less common sequence, airway invasion, infection initiates in the more proximal bronchial airways in what is often previously injured or deformed cystic lung (Franquet et al., 2001). Chronic invasive aspergillosis is an uncommon but potentially life-threatening crossover entity that shares some aspects of both the saprophytic and acute invasive categories of pulmonary aspergillosis and needs to be carefully distinguished (Denning et al., 2003).

Angio-Invasive Pulmonary Aspergillosis

Acute angio-invasive aspergillosis primarily affects patients at high risk of invasive mold infection, in particular those severely immunocompromised by hematopoietic stem cell transplantation (HSCT) or hematologic conditions, particularly when associated with prolonged neutropenia or neutrophil dysfunction (Shibuya et al., 1997; Wakayama et al., 2002). The risk of dissemination and ultimately death are very high (Herbrecht et al., 2002). Among patients with T-cell-dominated immunodeficiency, e.g., solid organ transplant recipients and those with acquired immunodeficiency, the risk of angio-invasive pulmonary aspergillosis is much more sporadic and less common than in the two aforementioned high-risk groups.

Histopathology of early IPA

The histopathology of acute angio-invasive aspergillosis represents the net result of the interaction between the aggression of the invading mold and the host defense reaction to it. Initially, the characteristic lesion is a discrete nodule. It consists of central and peripheral parts: a central zone of coagulation necrosis, invading hyphal forms of *A. fumigatus*, and vascular thrombosis,

Reginald Greene • Dept. of Radiology, Massachusetts General Hospital, Harvard Medical School, Boston, MA 02114. **Kazutoshi Shibuya and Tsunihiro Ando** • Dept. of Surgical Pathology, Toho University School of Medicine, 6-11-1 Omori-Nishi, Ota-Ku, Tokyo 143-8541, Japan.

and a peripheral zone of hemorrhage (Fig. 1 and Color Plate 15). Classically, there is an absence of any associated inflammatory cellular infiltrate in this initial lesion, a characteristic that is attributable to severe neutropenia and/or neutrophil dysfunction (Shibuya et al., 1997, 1999b). The discrete nodule forms the basis for the analogous radiological halo sign, the earliest imaging indicator of angio-invasive aspergillosis.

Radiology of early IPA

By far the most common initial computed tomography (CT) finding of acute angio-invasive aspergillosis is the pulmonary macronodule or mass (≥1 cm diameter); at least one is present in about 90% of patients with IPA (Greene et al., 2007). The macronodule is such a common feature on initial CT of early angio-invasive aspergillosis that its absence argues against such a diagnosis (Greene et al., 2007). At times, nodules less than 1 cm in diameter may be encountered in acute angio-invasive IPA, but generally by the time clinical suspicion warrants a CT scan the nodule is ≥1 cm. The differential diagnosis of the macronodule is wide and includes infections and noninfectious processes by fungi, nocardiosis, tuberculosis, bacterial lung abscess, septic emboli, bland pulmonary infarcts, lung cancer, metastases, lymphoproliferative disorders, and vasculitis.

The halo sign is a special CT finding consisting of the constellation of a macronodule and a perimeter of ground-glass opacity (Fig. 2). In immunocompromised patients at very high risk of invasive mold infection, the halo sign has long been considered an early sign of IPA. On initial CT imaging this finding can be identified in a significant fraction of patients with IPA. In a large series of patients with IPA in whom the halo sign was a criterion for a probable diagnosis of IPA, it was identified in about 60% of all patients with IPA (Greene et al., 2007). The ground-glass component of the halo sign corresponds to hemorrhage surrounding the edge of the discrete nodule identified on histopathology (Shibuya et al., 2006). This sign is highly transitory. For instance, in one longitudinal CT study of patients with IPA in whom 72% of patients had a halo sign on initial CT, only 22% still had a detectable halo sign 10 days later (Caillot et al., 2001).

Differential diagnosis

False-positive halo signs may be produced by edge irregularities of nodules, by technical partial volume ef-

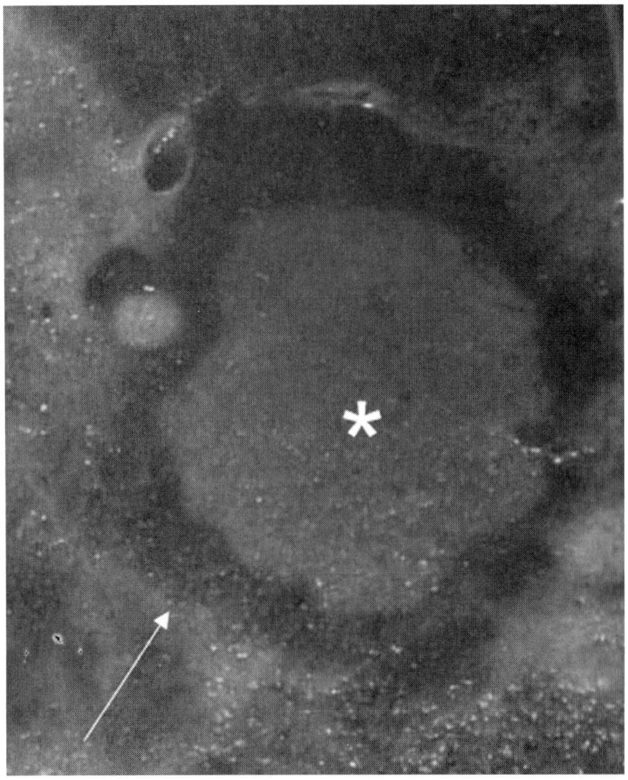

Figure 1. Histopathology of a discrete nodule in IPA. A sharply demarcated nodule (*) surrounded by hemorrhage (arrow) in a whole-lung section of a patient with angio-invasive aspergillosis.

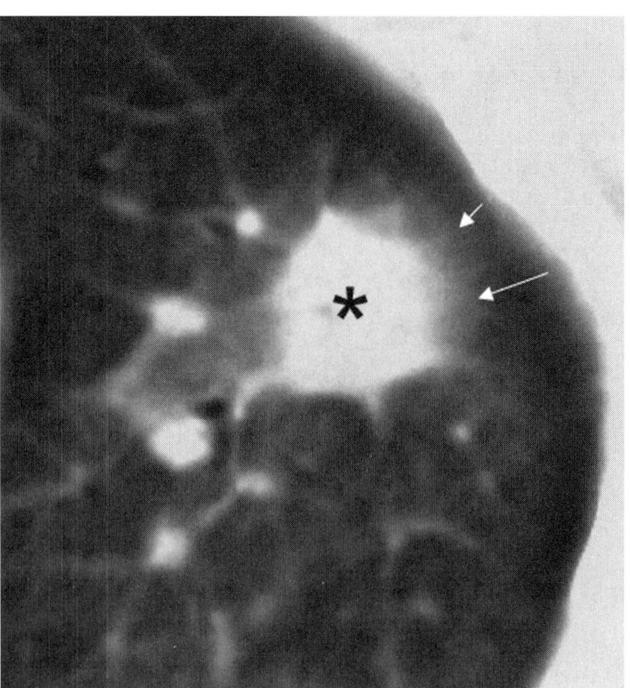

Figure 2. CT halo sign in IPA. The CT image of the lung demonstrates a halo sign in a patient with IPA and underlying hematological malignancy. The sign consists of two parts: first, a solid soft tissue macronodular core (≥1 cm) through which no pulmonary parenchyma is visible (*), and second, a ground-glass perimeter of intermediate density (arrows) through which the pulmonary parenchyma is still visible. Image obtained from Greene et al. (2007) with permission of the publisher.

fects that occur when CT sections are too thick for the size of the pulmonary nodules studied, and when respiratory motion degrades the images.

The CT halo sign is not unique to IPA. It has been reported in other less common angio-invasive mold infections, such as with *Zygomycetes* and even more rare angio-invasive fungi, e.g., *Trichosporon* spp., *Penicillium* spp., and *Fusarium* spp. (Huang and Harris, 1963; Saul et al., 1981; Young et al., 1978). Like *A. fumigatus*, these fungi can also produce nodular metastatic infectious nodules in solid abdominal organs like the liver and spleen, similar to those produced in chronic candidiasis. The halo sign has also been reported in a wide variety of other infections, such as those due to *Coccidioides immitis* and *Candida* spp., *Nocardia* spp., *Mycobacterium tuberculosis*, cytomegalovirus, herpes simplex virus, and angio-invasive bacteria, particularly *Pseudomonas aeruginosa* (Armstrong et al., 1971). Noninfectious conditions also associated with the halo sign include bronchio-alveolar cell carcinoma, lymphoproliferative disorders, metastatic angiosarcoma, Kaposi sarcoma, Wegener granulomatosis, eosinophilic lung disease, and organizing pneumonia (Gaeta et al., 1999; Kim et al., 1999; Primack et al., 1994). While there is a wide variety of conditions other than IPA that are associated with the CT halo sign, it is important to recognize that in the HSCT recipient, or the patient with a hematologic malignancy and neutropenia with suspected mold infection, angio-invasive aspergillosis is by far the most common cause of the CT halo sign.

Some studies indicate that preemptive anti-*Aspergillus* therapy based on the finding of a halo sign can improve outcome in patients with a compatible illness who are at high risk of invasive mold infection (Blum et al., 1994; Caillot et al., 1997). In an analysis of 235 patients with IPA who had an initial chest CT scan, 222 patients presented with at least one macronodule (Greene et al., 2007). Of the 143 of these patients who had an identifiable halo sign and the 79 who had no identifiable halo sign, a significantly higher satisfactory response rate at 12 weeks, irrespective of the treatment arm, was achieved in those with a halo sign (52% versus 29%). Those with halo signs also had better survival rates than those in the no-halo sign group (71% versus 53%) (Greene et al., 2007) (Fig. 3). Favorable treatment response in the halo sign group held true irrespective of the underlying condition category (hematological conditions or nonhematological conditions, baseline neutropenia or no baseline neutropenia, and allogeneic HSCT or no HSCT) (Greene et al., 2007). Since the improved outcome in the halo sign group is arguably related to the earlier stage of the infection as represented by the halo sign, the need for prompt performance of CT scanning and initiation of early preemptive treatment warrants a high priority when the sign is identified in such patients. The evanescent nature of the halo sign further emphasizes the urgency of obtaining a CT scan at the earliest opportunity when mold infection is first suspected.

Histopathology of Late IPA

The characteristic delayed (or transition) lesion of angio-invasive pulmonary aspergillosis occurs after partial recovery of neutrophil function, by which time the discrete nodule has begun to develop liquefaction necrosis limited to its periphery. This process is made possible by blood vessels around the margin of the discrete

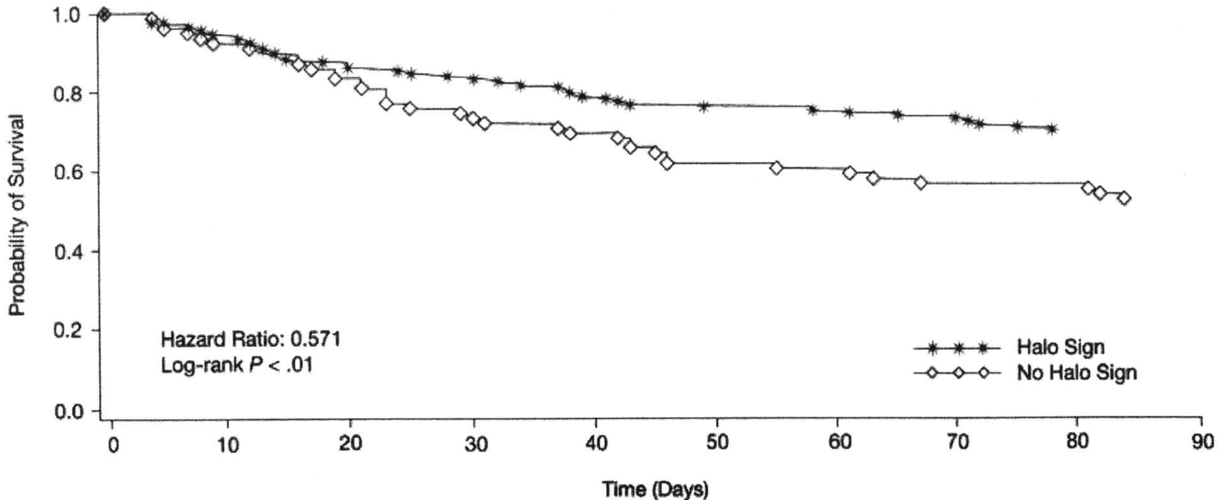

Figure 3. Time to death for patients under treatment for IPA who had a halo sign at presentation (n = 143) or without a halo sign at presentation (n = 79). Reprinted from Greene et al. (2007) with permission of the publisher.

nodule that remain patent. Thus, new neutrophils begin to be delivered to the nodule periphery, thereby facilitating liquefaction necrosis (Color Plate 16). The resulting cavitation separates the lung at the periphery of the persisting central zone of coagulation necrosis. The histopathology of this delayed transition lesion forms the basis for the analogous radiological "air crescent sign," an imaging indicator of late angio-invasive aspergillosis.

Radiology of late IPA

The air crescent sign is a specific type of cavitary lesion in which a semilunar pocket of gas surmounts a macronodule (Curtis et al., 1979) (Fig. 4). Since the appearance of the sign is generally known to coincide with recovery of neutrophil function, it is not surprising that it is found as an initial CT finding in only a small fraction of patients with IPA (~10%) (Greene et al., 2007). Like the halo sign, the air crescent sign is considered a specific indicator of IPA in patients at high risk of invasive mold infection (Aquino et al., 1994; Gefter et al., 1985). One investigation of patients with IPA reported that in the 4- to 10-day period following initial chest CT scan, while the frequency of the halo sign was rapidly diminishing, the frequency of the air crescent sign slowly increased (Caillot et al., 2001). Thick-walled and thin-walled cavities without air crescents can be found in a minority of initial CT studies of patients with IPA, but these findings do not seem to have the same diagnostic predictive value as the air crescent sign, from which they should be differentiated (Godwin et al., 1980).

Differential diagnosis

The differential diagnosis of the air crescent sign is broad, but like the halo sign, angio-invasive aspergillosis is the most common cause when it is found in a patient at high risk of invasive mold infection. The sign can also be found in a variety of other conditions, e.g., nocardiosis, tuberculosis, bacterial lung abscess, cavitary hematoma, and cavitary lung cancer (Ryu and Swensen, 2003; Tuncel, 1984). The nodule of coagulation necrosis in an air crescent sign cavity can closely resemble the fungus ball of saprophytic aspergillosis in a preexisting cyst. The unique clinical circumstances of each of these two patient groups help to differentiate the air crescent sign of IPA from the similar sign of aspergilloma. The two have very different histopathologies and pathogenesis: a patient with one is likely to be clinically well, while a patient with the other condition is likely to be very sick and at high risk of invasive mold infection.

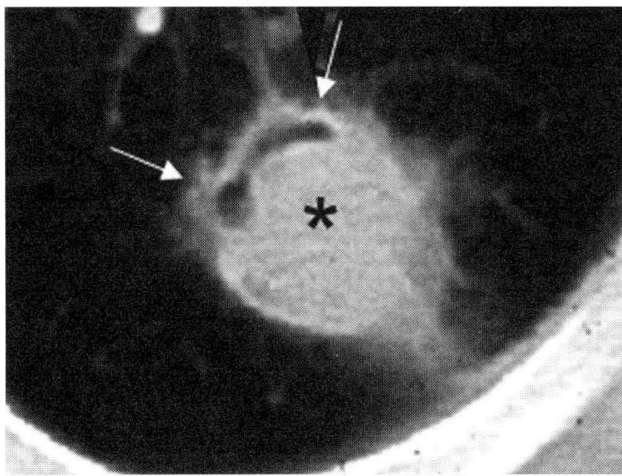

Figure 4. CT air crescent sign in IPA. The CT image demonstrates an air crescent sign in a patient with angio-invasive pulmonary aspergillosis and underlying hematological malignancy. A crescent of air (arrows) surmounts a soft tissue mound of a macronodule (*).

Other less common initial CT imaging findings in IPA include consolidations, consolidative infarcts with or without air bronchograms (about one-third of patients), and small airway opacities (about 1/10 of patients) (Greene et al., 2007). Ancillary CT findings, such as pleural effusion, pericardial effusion, and hilar/mediastinal lymphadenopathy, have not proved to be discriminatory.

The search for a CT halo sign-equivalent finding with magnetic resonance (MR) imaging has been disappointing. Early in the course of IPA, MR performs well with regard to sensitivity. However, specificity is poor and not improved by imaging after intravenous gadolinium (Gd) administration (Blum et al., 1994). On the other hand, with MR imaging later in the course of IPA, i.e., 10 days after onset, there is Gd-diethylenetriamine penta-acetic acid enhancement that often demonstrates a rim area with a characteristic nodular target-like lesion and a "reverse target" on T2-weighted images. These findings are highly suggestive of the later stage of IPA and are not found earlier in the course of disease or in patients with *Pseudomonas* sp. or staphylococcal infection.

Invasion into the great vessels pleura and pericardium, hila, and mediastinum occur relatively infrequently in IPA, but the consequences can be disastrous. If IPA lung lesions abut a large vessel, the possibility of such vessel invasion must be considered. Occasionally, hyphal elements invade adjacent large vessels to cause thrombosis or pseudoaneurysm; these developments increase the risk of blood-borne dissemination, vessel rupture, and exsanguination.

Cerebral aspergillosis generally occurs by direct extension from sino-nasal aspergillosis or systemic hematogenous dissemination from a primary lung infection. Initial brain lesions are analogous in appearance to those in the lung; they are usually well-defined macronodular lesions of low CT attenuation or large-vessel infarcts surrounded by edema. On MR imaging, cerebral lesions usually appear as T2-weighted hyperintense lesions surrounded by edema and exhibit T1-weighted enhancement after intravenous Gd (Cox et al., 1992). In later stages, central necrosis and rim enhancement are the primary findings (Osborn, 1994). Sino-nasal aspergillosis may demonstrate soft tissue masses or abscesses in the nasal cavities or paranasal sinuses, from where there may be extension into the brain or orbits via the cribriform plate, sometimes associated with bone destruction or cavernous sinus thrombosis (Bowen and Post, 1991). Visceral dissemination to the liver, spleen, and kidneys is associated with solid nodular masses, abscesses, or infarcts.

Airway IPA

The less common form of invasive aspergillosis, airway IPA, most often occurs in patients with lesser degrees of neutrophil dysfunction than those in the aforementioned two groups. Many of these patients have evidence of prior airway injury, such as occurs at bronchial anastomoses in lung transplant recipients and along inflamed or infected bronchi after other infections.

Histopathology

Unlike angio-invasive aspergillosis, airway IPA develops axially in the small bronchi and bronchioles, where there is histopathological evidence of deep mycelial invasion to basement membranes (Logan et al., 1994). Among lung transplant recipients, *Aspergillus* spp. can cause airway colonization, tracheobronchitis, and invasive pneumonia. The tracheobronchitic form is common in lung transplant recipients and in patients with AIDS; the spectrum ranges from isolated simple bronchitis to pseudomembranous tracheobronchial aspergillosis (Kramer et al., 1991). In lung transplant recipients, isolated tracheobronchitis often involves the anastomotic site and tends to respond well to antifungal therapy and/or surgical debridement. In a minority of patients this can lead to dissemination. Among lung transplant recipients, only a very small fraction of patients progress from airway colonization to invasive disease, but among those who do so, fewer than half survive the infection (Mehrad et al., 2001).

Imaging

Spread of infection along the airways has the CT appearance of bronchial wall thickening, centrilobular opacities, and thickened and fluid-distended segments of small branching bronchi and bronchioli, i.e., tree-in-bud opacities and/or patches of peribronchial consolidation (Logan et al., 1994) (Fig. 5). While small-airways findings are uncommon in angio-invasive aspergillosis, they are the rule in airway-invasive aspergillosis, where the invasion is dominated by airway imaging findings rather than by macronodules. These findings are nonspecific because they are found in a wide variety of common community-acquired infections and in mycobacterial, cryptococcal, and aspiration pneumonias, all of which are not uncommon in lung transplant recipients and/or AIDS patients.

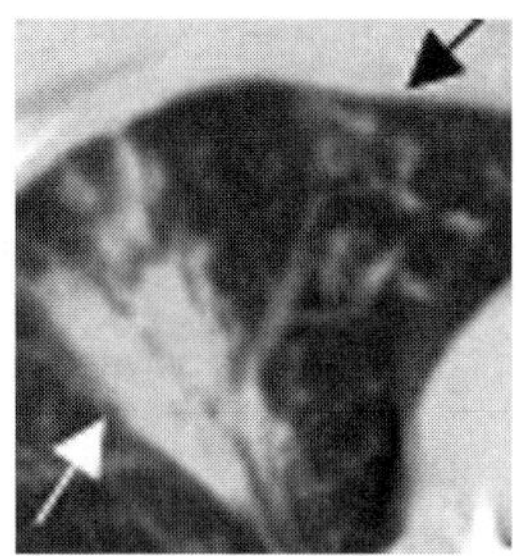

Figure 5. Peribronchovascular CT opacities in airway IPA. The CT image demonstrates a peribronchovascular opacity highlighting branching air bronchograms. There are also small branching bronchiolar opacities in the unconsolidated lung (arrows).

In the lung transplant recipient, isolated tracheobronchitis may be seen as a localized nodule or deformity at a bronchial anastomosis. CT findings of tracheobronchitis may also include bronchial wall thickening, bronchial nodularity, bronchostenosis, bronchiectasis, peribronchial consolidation, and atelectasis.

CHRONIC PULMONARY ASPERGILLOSIS

Chronic IPA (CPA) is an incompletely understood, indolent but progressive *Aspergillus* infection that occurs in preformed or new cysts (or cavities) (Denning et al., 2003; Gefter, 1992). It tends to occur in elderly and/or debilitated patients who might not be otherwise immunodeficient. Progressive constitutional symptoms are characteristic. The underlying chronic cavitary lung disease may be due to prior tuberculosis, bullous lung disease, chronic interstitial lung disease, lung irradiation, surgical lung resection, lung infarction, or cystic fibrosis (Graham-Clarke et al., 1994). End-stage sarcoidosis is a common cause of the cysts associated with CPA. CPA has overlapping characteristics with one or more of the other categories of aspergillosis from which it should be distinguished (Greene, 1981; Soubani and Chandrasekar, 2002).

Histopathology

There is pathologic evidence of local lung tissue invasion, but dissemination does not usually occur (Color

Plates 17A and B). Although the infection tends to remain local, it is resistant to eradication and usually requires long-term anti-*Aspergillus* therapy. Some view CPA as a progression from noninvasive aspergillosis to a low-grade invasive form of aspergillosis when depressed host defense mechanisms facilitate the infection. The cyst wall is usually eroded and invaded by the hyphae. There is often evidence of both acute and chronic inflammation in association with fibrosis and necrosis to various degrees.

Radiology

The characteristic imaging findings of CPA include further enlargement of preexisting cysts, the development of new lung cysts, the appearance of pericystic lung opacities, and the development of new or progressive local pleural thickening. About half of these patients have evidence of saprophytic aspergillomas within the cysts. The findings may simulate activation of postprimary tuberculosis and lung cancer (Fig. 6).

Chronic necrotizing pulmonary aspergillosis (CNPA) is an aggressive subcategory of CPA that has a more rapid course that is sometimes referred to as semi-invasive or subacute pulmonary aspergillosis (Binder et al., 1982; Denning et al., 2003; Yousem, 1997). CNPA can be recognized by the development of new progressive consolidative lung opacities that undergo de novo cyst and cavity formation. The new cysts can later become the host site for an aspergilloma. Alternatively, the cysts may develop thin walls that rapidly expand (Denning et al., 2003). CNPA may demonstrate locally invasive disease wherein mycelia may be identified invading the pleural space. Local blood vessel invasion can lead to hemoptysis. *A. fumigatus* may be recovered from the pleural space when there are bronchopleural fistula and empyema (Denning et al., 2003).

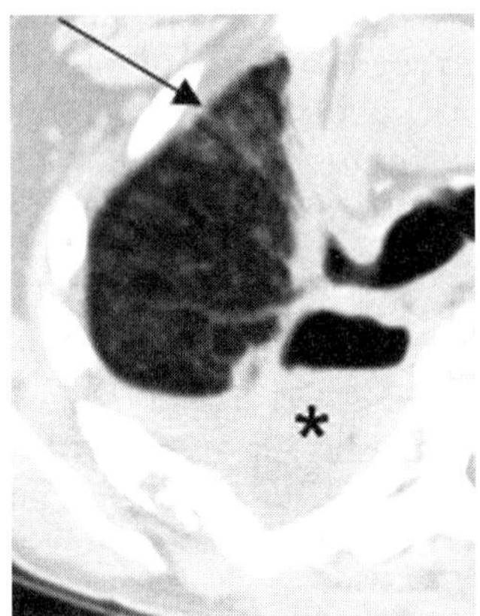

Figure 6. Illustration of CPA. A liquid-filled preexisting cavity (*) from which *Aspergillus* sp. was consistently recovered is shown. Pericystic lung opacities and new pleural thickening developed in the right upper lobe (arrows).

The main challenge in differential diagnosis is to distinguish CPA from serious but unsuspected chronic conditions that require other specific therapy. These include conditions such as upper lobe lung cancers and chronic cavitary lung infections caused by *M. tuberculosis*, nontuberculous mycobacteria, or endemic fungi. A definite diagnosis of CPA often requires biopsy proof to establish that there is local invasion by *Aspergillus* and to exclude tumor or other pulmonary infection. CPA differs from a simple aspergilloma, not only by the presence of associated constitutional symptoms but also by the development of persistent pericystic lung nodules, consolidations, or ground-glass opacities and by the development and/or progression of cavitary disease or pericystic pleural thickening. In simple aspergilloma, the development of cyst enlargement, pericystic lung opacities, and/or pericystic pleural thickening should be regarded as evidence of complicated aspergillosis consistent with CPA. Increased cavity wall thickness alone does not correlate with disease activity (Denning et al., 2003).

ABPA

Allergic bronchopulmonary aspergillosis (ABPA) is the archetype of allergic aspergillosis. The clinical constellation of ABPA consists of chronic asthma, mucus production, and elevated serum antigen levels of *A. fumigatus*. Importantly, the asthmatic component and continuing lung damage are usually responsive to corticosteroid therapy.

Histopathology

ABPA is not an infection but a damaging immune reaction caused by *A. fumigatus* colonization in large and small airways of atopic, immunocompetent patients, including some who also have cystic fibrosis (Shibuya et al., 1999a). Cytological examination of bronchoalveolar lavage fluids can demonstrate hyphae with dichotomous bifurcations in combination with numerous mucin-containing airway eosinophils. Histological studies of mucosal biopsies show evidence of allergic bronchitis without signs of fungal invasion (Color Plates 18A and B).

Radiology

Local hypersensitivity damage tends to affect the large central airways (segmental and subsegmental bronchi), small airways, and the adjacent lung, principally in the upper lobes.

Large airway damage consists mainly of severe bronchiectasis of the segmental and subsegmental bronchi. On chest radiographs large airway bronchiectasis and impaction can take on a "finger-in-glove" appearance due to central bronchiectasis emanating from the pulmonary hila (Color Plates 19A and B). On CT, the damage is seen directly as bronchiectasis filled with mucoid material. The CT attenuation of the mucoid impaction is normally low relative to soft tissue attenuation, but occasionally it is of high attenuation (Agarwal et al., 2007). High attenuation mucus does not seem to correlate with a decreased chance for complete remission. Small-airway damage causes bronchiolar dilatation and impaction that on CT is seen as centrilobular nodules and/or branching tree-in-bud opacities.

Associated damage to the adjacent peripheral lung is evident on CT as transient, migratory patches of consolidation and/or ground-glass opacities that regress spontaneously or after corticosteroid therapy. Other imaging findings include regional and overall lung hyperinflation or atelectasis distal to airway obstructions.

Differential Diagnosis

Corticosteroid-responsive ABPA-related asthma needs to be differentiated from simple chronic asthma. Three main imaging features help to distinguish these two groups of patients (Ward et al., 1999). First, varicose or cystic bronchiectasis of segmental and subsegmental bronchi is present in >90% of patients with ABPA but in only <30% of patients with simple asthma. Second, mucoid impaction of segmental and subsegmental airways occurs in 67% of patients with ABPA but only 4% of those with simple asthma. Third, small-airways abnormalities in the form of centrilobular nodules occur in 93% of patients with ABPA but in only 28% of patients with simple asthma (Ward et al., 1999). When patients with simple chronic asthma have bronchiectasis, it is usually of a mild variety and limited to the cylindrical type, i.e., bronchiectasis with parallel nontapering walls rather than with frank dilatation. Although the imaging features of ABPA described above are highly discriminatory, they do not identify ABPA before airway damage has occurred. In an effort to identify ABPA at an earlier stage, screening CT has been advocated for asthmatics with skin prick hypersensitivity to *A. fumigatus* (Eaton et al., 2000).

ASPERGILLOMA

Simple aspergilloma, or fungus ball, is the most common form of aspergillosis that is detected with imaging studies. It is the archetype of saprophytic (noninvasive) aspergillosis. It results from saprophytic proliferation of *Aspergillus* mycelia within a preformed host lung cyst (or cavity). The patient is generally asymptomatic and immunocompetent (Soubani and Chandrasekar, 2002). As mentioned above, the preformed cysts in which aspergillomas form can be caused by a wide variety of lung diseases. The four common etiologies of the chronic cysts of aspergillosis include post-primary tuberculosis (Glimp and Bayer, 1983; British Thoracic and Tuberculosis Association, 1970), end-stage sarcoidosis (Israel and Ostrow, 1969), bullous emphysema, and bronchiectatic cavities (Solit et al., 1971), such as those found in cystic fibrosis (Ryan et al., 1995) and ABPA. Aspergillomas have also been reported in immunocompetent patients who did not appear to have had a preexisting dilated lung space (Kang et al., 2002).

Histopathology

Mycelial invasion of the lung or vasculature is not a feature of aspergillomas (Rafferty et al., 1983; Tomee et al., 1995). The *Aspergillus* hyphae in the fungus ball are compactly aligned in a radial pattern, and the wall of the cyst is usually eroded or covered with metaplastic respiratory epithelium. A chronic inflammatory infiltrate is present in the wall, but no hyphal invasion occurs when the patient is immunocompetent (Fig. 7A and B).

Radiology

Initial detection is usually made during an incidental chest radiograph, often for evaluation of the underlying disease responsible for the cyst or for some unrelated reason (Glimp and Bayer, 1983). The imaging diagnosis is based on finding a well-defined mycelial

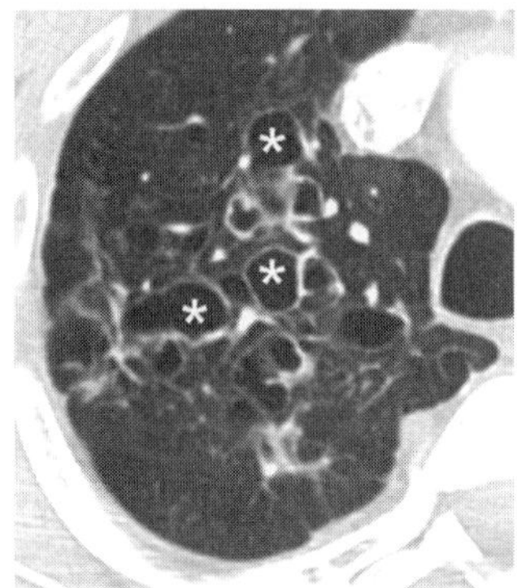

Figure 7. Imaging of ABPA. CT scan in an asthmatic patient with ABPA. There are multiple branching cystic spaces (*) characteristic of central bronchiectasis.

mass in a preformed cyst, most often solitary and in an upper lobe (Glimp and Bayer, 1983; Soubani and Chandrasekar, 2002) (Fig. 8). On CT, air may be visible within the interstices of the hyphal mass. A crescent-shaped cap of air separating the aspergilloma from the cyst wall can usually be identified (Tuncel, 1984). The fungus ball can often, but not always, be shown to move within the cyst on shifting body position from left to right or from prone to supine (Roberts et al., 1987; Soubani and Chandrasekar, 2002). Spontaneous shrinkage or even disappearance can occur in a small fraction of aspergillomas (Gefter, 1992). Enlargement of fungus balls is considered to be rare (British Tuberculosis Association, 1968). The host cysts are generally smooth and thin but may be irregular and thick. The pleural surface overlying the host cyst may be thin or thick, but it does not usually thicken over time unless or until it becomes complicated. Dystrophic, cage-like calcification over the surface of an aspergilloma can occasionally be detected. In a minority of patients, such as those with end-stage sarcoidosis, the aspergilloma may be brought to attention by potentially life-threatening hemoptysis (Glimp and Bayer, 1983; Greenberg et al., 2002). In such patients CT angiography is usually used to search for the source of the hemoptysis, i.e., hypertrophic bronchial arteries supplying the cyst wall. Occlusive metallic coils can then be placed in the dilated bronchial arteries to help control the bleeding. Bleeding may cause CT hyperdense fluid within the cyst or spread into the lung, causing ground-glass or consolidative opacities with air bronchograms.

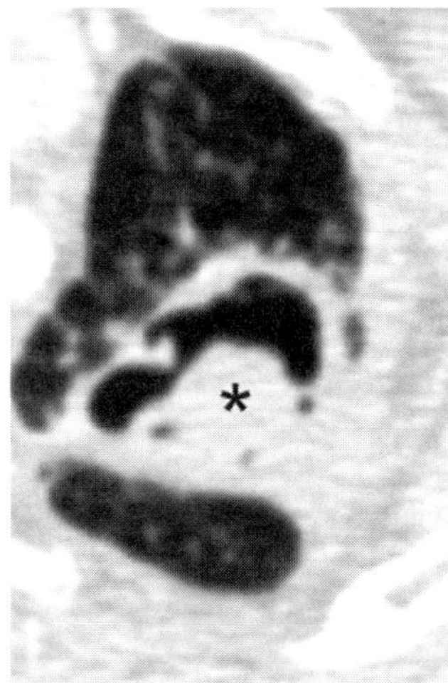

Figure 8. Image of an aspergilloma. A CT scan section demonstrates a fungus ball (*) residing within a preexisting cavity of an asymptomatic patient. The configuration simulates the air crescent sign of angio-invasive aspergillosis.

Differential Diagnosis

Definitive diagnosis requires clinical corroboration based on positive serology precipitins to *A. fumigatus* and/or recovery of *A. fumigatus* in sputum, bronchoalveolar lavage washings, or percutaneous needle aspiration biopsy material. Precipitins to *A. fumigatus* may be negative when the fungus ball is caused by *Aspergillus* species other than *A. fumigatus* or other molds, such as *Zygomycetes* or *Fusaria* (Tomee et al., 1995). The main clinical imaging challenge is to differentiate an aspergilloma (which would not usually require treatment) from other more serious lung conditions (which usually do need specific treatment). The latter includes the late phase of acute angio-invasive aspergillosis (which occurs in a totally different clinical setting) and chronic semi-invasive aspergillosis (patients with chronic, indolent symptomatology). The aspergilloma needs also to be differentiated from other cavitary lung diseases, such as those caused by lung cancer, lung abscess, and cavitary granulomatous vasculitis, such as Wegener's granulomatosis. Lung biopsy and other clinical information may be needed for a definite diagnosis. Predictors of poor prognosis for the saprophytic aspergillomas include an increase in the size or number of cysts (a finding not compatible with the diagnosis of simple aspergilloma), severe underlying lung disease, immunosuppressive therapy, AIDS, sarcoidosis, rising *Aspergillus*-specific immunoglobulin G titer, and repetitive severe hemoptysis (Stevens et al., 2000).

AIDS patients require special attention because they are at high risk to develop aspergillomas and because aspergillomas in these patients tend to develop into CPA. AIDS patients have a high prevalence of abnormally dilated host cysts because of frequent infections by *Pneumocystis jirovicii, M. tuberculosis, Mycobacterium avium intracellulare*, other bacteria, and other fungi. Aspergillomas in AIDS patients, especially those with CD4 counts of <100 cells/liter, are often discovered because of symptoms related to the aspergilloma, such as hemoptysis (Addrizzo-Harris et al., 1997), cough, or fever. Obstructing bronchopulmonary aspergillosis, a unique noninvasive, saprophytic variety of aspergillosis characteristic of AIDS, is diagnosed by massive central bronchial dilatation and mucoid impaction due to noninvasive endo-bronchial fungal overgrowth (Franquet et al., 2001).

CONCLUSION

A. fumigatus causes a wide variety of pulmonary diseases that result in significant morbidity and mortality. The diseases are of particular importance to the clinician because their unique characteristics make them detectable through imaging even when microbiological studies are not definitive. The clinical milieu helps identify the prior probability of such disease while the histopathology is predictive of the imaging findings. When evaluated in an appropriate clinical context, imaging is likely to make the disease specifically recognizable at a time when successful treatment is possible.

REFERENCES

Addrizzo-Harris, D., T. Harkin, G. McGuinness, D. Naidich, and W. Rom. 1997. Pulmonary aspergilloma and AIDS. A comparison of HIV-infected and HIV-negative individuals. *Chest* **111:**612–618.

Agarwal, R., D. Gupta, A. N. Aggarwal, A. K. Saxena, A. Chakrabarti, and S. K. Jindal. 2007. Clinical significance of hyperattenuating mucoid impaction in allergic bronchopulmonary aspergillosis: an analysis of 155 patients. *Chest* **132:**1183–1190.

Aquino, S. L., S. T. Kee, M. L. Warnock, and G. Gamsu. 1994. Pulmonary aspergillosis: imaging findings with pathologic correlation. *Am. J. Roentgenol.* **163:**811–815.

Armstrong, D., L. S. Young, R. D. Meyer, and A. H. Blevins. 1971. Infectious complications of neoplastic disease. *Med. Clin. North Am.* **55:**729–745.

Binder, R. E., L. J. Faling, R. D. Pugatch, C. Mahasaen, and G. L. Snider. 1982. Chronic necrotizing pulmonary aspergillosis: a discrete clinical entity. *Medicine* **61:**109–124.

Blum, U., M. Windfuhr, C. Buitrago-Tellez, G. Sigmund, E. W. Herbst, and M. Langer. 1994. Invasive pulmonary aspergillosis. MRI, CT, and plain radiographic findings and their contribution for early diagnosis. *Chest* **106:**1156–1161.

Bowen, B. C., and M. J. D. Post. 1991. Intracranial infection, p. 501–538. *In* S. W. Atlas (ed.), *Magnetic Resonance Imaging of the Brain and Spine*. Raven Press, New York, NY.

British Thoracic and Tuberculosis Association, Research Committee. 1970. Aspergillosis and residual tuberculous cavities: the results of a resurvey. *Tubercle* **51:**227–245.

British Tuberculosis Association. 1968. *Aspergillus* in persistent lung cavities after tuberculosis: a report from the Research Committee of the British Tuberculosis Association. *Tubercle* **49:**1–11.

Caillot, D., O. Casasnovas, A. Bernard, J. F. Couaillier, C. Durand, B. Cuisenier, E. Solary, F. Piard, T. Petrella, A. Bonnin, G. Couillault, M. Dumas, and H. Guy. 1997. Improved management of invasive pulmonary aspergillosis in neutropenic patients using early thoracic computed tomographic scan and surgery. *J. Clin. Oncol.* **15:**139–147.

Caillot, D., J. F. Couaillier, A. Bernard, O. Casasnovas, D. W. Denning, L. Mannone, J. Lopez, G. Couillault, F. Piard, O. Vagner, and H. Guy. 2001. Increasing volume and changing characteristics of invasive pulmonary aspergillosis on sequential thoracic computed tomography scans in patients with neutropenia. *J. Clin. Oncol.* **19:**253–259.

Cox, J., F. R. Murtagh, A. Wilfong, and J. Brenner. 1992. Cerebral aspergillosis: MR imaging and histopathologic correlation. *Am. J. Neuroradiol.* **13:**1489–1492.

Curtis, A. M., G. F. Smith, and C. E. Ravin. 1979. Air crescent sign of invasive aspergillosis. *Radiology* **133:**17–21.

Denning, D. W., K. Riniotis, R. Dobrashian, and H. Sambatakou. 2003. Chronic cavitary and fibrosing pulmonary and pleural aspergillosis: case series, proposed nomenclature and review. *Clin. Infect. Dis.* **37:**S265–S280.

Eaton, T., J. Garrett, D. Milne, A. Frankel, and A. U. Wells. 2000. Allergic bronchopulmonary aspergillosis in the asthma clinic: a prospective evaluation of CT in the diagnostic algorithm. *Chest* **118:**66–72.

Franquet, T., N. L. Müller, A. Giménez, P. Guembe, J. de la Torre, and S. Bagué. 2001. Spectrum of pulmonary aspergillosis: histologic, clinical, and radiologic findings. *Radiographics* **21:**825–837.

Gaeta, M., A. Blandino, E. Scribano, F. Minutoli, S. Volta, and J. Pandolfo. 1999. Computed tomography halo sign in pulmonary nodules: frequency and diagnostic value. *J. Thorac. Imaging* **14:**109–113.

Gefter, W. B. 1992. The spectrum of pulmonary aspergillosis. *J. Thorac. Imaging* **7:**56–74.

Gefter, W. B., S. M. Albelda, G. H. Talbot, S. L. Gerson, P. A. Cassileth, and W. T. Miller. 1985. Invasive pulmonary aspergillosis and acute leukemia. Limitations in the diagnostic utility of the air crescent sign. *Radiology* **157:**605–610.

Glimp, R., and A. Bayer. 1983. Pulmonary aspergilloma: diagnostic and therapeutic considerations. *Arch. Intern. Med.* **143:**303–308.

Godwin, J. D., W. R. Webb, C. J. Savoca, G. Gamsu, and P. C. Goodman. 1980. Multiple, thin-walled cystic lesions of the lung. *Am. J. Roentgenol.* **135:**593–604.

Grahame-Clarke, C. N., C. M. Roberts, and D. W. Empey. 1994. Chronic necrotizing pulmonary aspergillosis and pulmonary phycomycosis in cystic fibrosis. *Respir. Med.* **88:**465–468.

Greenberg, A. K., J. Knapp, W. N. Rom, and D. J. Addrizzo-Harris. 2002. Clinical presentation of pulmonary mycetoma in HIV-infected patients. *Chest* **122:**886–892.

Greene, R. 1981. The pulmonary aspergilloses: three distinct entities or a spectrum of disease? *Radiology* **140:**527–530.

Greene, R. E., H. T. Schlamm, J. W. Oestmann, P. Stark, C. Durand, O. Lartholary, J. R. Wingard, R. Herbrecht, P. Ribaud, T. F. Patterson, P. F. Troke, D. W. Denning, J. E. Bennett, B. E. de Pauw, and R. H. Rubin. 2007. Imaging findings in acute invasive pulmonary aspergillosis: clinical significance of the halo sign. *Clin. Infect. Dis.* **44:**373–379.

Herbrecht, R., D. W. Denning, T. F. Patterson, J. E. Bennett, R. E. Greene, J. W. Oestmann, W. V. Kern, K. A. Marr, P. Ribaud, O. Lortholary, R. Sylvester, R. H. Rubin, J. R. Wingard, P. Stark, C. Durand, D. Caillot, E. Thiel, P. H. Chandrasekar, M. R. Hodges, H. T. Schlamm, P. F. Troke, B. de Pauw, et al. 2002. Voriconazole versus amphotericin B for primary therapy of invasive aspergillosis. *N. Engl. J. Med.* **347:**408–415.

Huang, S. N., and L. S. Harris. 1963. Acute disseminated penicilliosis: report of a case and review of pertinent literature. *Am. J. Clin. Pathol.* **39:**167–174.

Israel, H. L., and A. Ostrow. 1969. Sarcoidosis and aspergilloma. *Am. J. Med.* **47:**243–250.

Kang, E.-Y., D. H. Kim, O. H. Woo, J.-A. Choi, Y.-W. Oh, and K. H. Kim. 2002. Pulmonary aspergillosis in immunocompetent hosts without underlying lesions of the lung: radiologic and pathologic findings. *Am. J. Roentgenol.* **178:**1395–1399.

Kim, Y., K. S. Lee, K. J. Jung, J. Han, J. S. Kim, and J. S. Suh. 1999. Halo sign on high resolution CT: findings in spectrum of pulmonary diseases with pathologic correlation. *J. Comput. Assist. Tomogr.* **23:**622–626.

Kramer, M. R., D. W. Denning, S. E. Marshall, D. J. Ross, G. Berry, N. J. Lewiston, D. A. Stevens, and J. Theodore. 1991. Ulcerative tracheobronchitis after lung transplantation. A new form of invasive aspergillosis. *Am. Rev. Respir. Dis.* **144:**552–556.

Logan, P. M., S. L. Primack, R. R. Miller, and N. L. Muller. 1994. Invasive aspergillosis of the airways: radiographic, CT, and pathologic findings. *Radiology* **193:**383–388.

Mehrad, B., G. Paciocco, F. J. Martinez, T. C. Ojo, M. D. Iannettoni, and J. P. Lynch III. 2001. Spectrum of *Aspergillus* infection in lung transplant recipients: case series and review of the literature. *Chest* **119:**169–175.

Osborn, A. G. 1994. *Diagnostic Neuroradiology*. Mosby, St. Louis, MO, p. 706–709.

Primack, S. L., T. E. Hartman, K. S. Lee, and N. L. Muller. 1994. Pulmonary nodules and the CT halo sign. *Radiology* **190:**513–515.

Rafferty, P., B. A. Biggs, G. K. Crompton, and I. W. Grant. 1983. What happens to patients with pulmonary aspergilloma? Analysis of 23 cases. *Thorax* **38:**579–583.

Roberts, C. M., K. M. Citron, and B. Strickland. 1987. Intrathoracic aspergilloma: role of CT in diagnosis and treatment. *Radiology* **165:** 123–128.

Ryan, P., D. Stableforth, J. Reynolds, and K. Muhdi. 1995. Treatment of pulmonary aspergilloma in cystic fibrosis by percutaneous instillation of amphotericin B via indwelling catheter. *Thorax* **50:**809–810.

Ryu, J. H., and S. J. Swensen. 2003. Cystic and cavitary lung diseases: focal and diffuse. *Mayo Clin. Proc.* **78:**744–752.

Saul, S. H., T. Khachatoorian, A. Poorsattar, R. L. Myerowitz, S. J. Geyer, A. W. Pasculle, and M. Ho. 1981. Opportunistic *Trichosporon* pneumonia. Association with invasive aspergillosis. *Arch. Pathol. Lab. Med.* **105:**456–459.

Shibuya, K. 1999a. Animal models of *A. fumigatus* infections, p. 130–138. *In* A. A. Brakhage, B. Jahn, and A. Schmidt (ed.), *Aspergillus fumigatus. Contribution to Microbiology*, vol. 2. Karger, Basel, Switzerland.

Shibuya, K. 1999b. Histopathology of experimental invasive pulmonary aspergillosis in rats: pathological comparison of pulmonary lesions induced by specific virulent factor deficient mutants. *Microb. Pathog.* **27:**123–131.

Shibuya, K., T. Ando, M. Wakayama, M. Takaoka, K. Uchida, and S. Naoe. 1997. Pathological spectrum of invasive pulmonary aspergillosis: study of pulmonary lesions of 54 autopsies and the relationship between neutrophilic response and histologic features of lesions in experimental aspergillosis. *Jpn. J. Med. Mycol.* **38:**175–181.

Shibuya, K., S. Paris, T. Ando, H. Nakayama, T. Hatori, and J.-P. Latgé. 2006. Catalases of *Aspergillus fumigatus* and inflammation in aspergillosis. *Jpn. J. Med. Mycol.* **47:**249–255.

Solit, R. W., J. J. McKeown, Jr., S. Smullens, and W. Fraimow. 1971. The surgical implications of intracavitary mycetomas (fungus balls). *J. Thorac. Cardiovasc. Surg.* **62:**411–422.

Soubani, A. O., and P. H. Chandrasekar. 2002. The clinical spectrum of pulmonary aspergillosis. *Chest* **121:**1988–1999.

Stevens, D. A., V. L. Kan, M. A. Judson, V. A. Morrison, S. Dummer, D. W. Denning, J. E. Bennett, T. J. Walsh, T. F. Patterson, and G. A. Pankey. 2000. Practice guidelines for diseases caused by *Aspergillus*, Infectious Diseases Society of America. *Clin. Infect. Dis.* **30:**696–709.

Tomee, J. F., T. S. van der Werf, J. P. Latgé, G. H. Koeter, A. E. Dubois, and H. F. Kauffman. 1995. Serologic monitoring of disease and treatment in a patient with pulmonary aspergilloma. *Am. J. Respir. Crit. Care Med.* **151:**199–204.

Tuncel, E. 1984. Pulmonary air meniscus sign. *Respiration* **46:**139–144.

Wakayama, M., K. Shibuya, T. Ando, T. Oharaseki, K. Takahashi, S. Naoe, and W. F. Coulson. 2002. Deep-seated mycosis as a complication in bone marrow transplantation patients. *Mycoses* **45:**146–151.

Ward, S., L. Heyneman, M. J. Lee, A. N. Leung, D. M. Hansell, and N. L. Muller. 1999. Accuracy of CT in the diagnosis of allergic bronchopulmonary aspergillosis in asthmatic patients. *Am. J. Roentgenol.* **173:**937–942.

Young, N. A., K. H. Kwon-Chung, T. T. Kubota, A. E. Jennings, and R. I. Fisher. 1978. Disseminated infection by *Fusarium moniliforme* during treatment for malignant lymphoma. *J. Clin. Microbiol.* **7:** 589–594.

Yousem, S. 1997. The histological spectrum of chronic necrotizing forms of pulmonary aspergillosis. *Hum. Pathol.* **28:**650–656.

Aspergillus fumigatus and Aspergillosis
Edited by J.-P. Latgé and W. J. Steinbach
©2009 ASM Press, Washington, DC

Chapter 28

Galactomannan and Anti-*Aspergillus* Antibody Detection for the Diagnosis of Invasive Aspergillosis

PAUL E. VERWEIJ

Invasive aspergillosis remains an important cause of morbidity and mortality in patients with compromised immune defenses. However, there has been a significant decrease in mortality in patients with a diagnosis of invasive aspergillosis following hematopoietic stem cell transplantation in recent years, coinciding with multiple changes in transplantation practices, including use of nonmyeloablative conditioning regimens, receipt of peripheral blood stem cells, more prompt diagnosis of invasive aspergillosis, and use of voriconazole (Upton et al., 2007). Another important factor in relation to treatment outcome appears to be the timing of initiation of antifungal therapy. Analysis of the high-resolution computed tomography (CT) images of patients with probable and proven invasive aspergillosis showed that those with a halo sign at initiation of antifungal therapy had a significantly better response and greater survival than those presenting with other CT images (Greene et al., 2007). Since the halo sign is the earliest image in patients presenting with invasive aspergillosis, this finding suggests that patients with a halo sign were treated earlier in the course of their infection, compared to those with other CT results (Verweij et al., 2007).

Early identification of patients who require antifungal therapy is therefore an important goal and requires diagnostic tools that have good performance characteristics and become positive in a very early phase of the infection. Conventional diagnostic techniques, such as microscopy and culture, generally lack sufficient sensitivity and are typically positive when the infection is advanced and the fungal burden high (Hope et al., 2005). The detection of circulating fungal antigens, such as galactomannan, might help in early diagnosis of invasive aspergillosis. A significant number of studies have been published over the past 10 years that evaluated the performance of galactomannan detection using a commercial system, the Platelia *Aspergillus* test (Bio-Rad). Issues related to the detection of galactomannan in the management of patients at high risk of invasive aspergillosis will be discussed in this chapter.

GALACTOMANNAN

Aspergillus contains mannoproteins in the outer cell wall layer, and one of the structures present is the carbohydrate galactomannan. Galactomannan is a family of molecules which are referred to as galactofuranose (Gal-f) antigens (Morelle et al., 2005). The Gal-f antigens contain galactofuranose residues that react with the rat immunoglobulin M (IgM) monoclonal antibody (EB-A2), which is used in the galactomannan enzyme immunoassay (EIA). It was recently shown that a galactofuranose residue is also present in fungal glycoproteins, including phospholipase C and phytase, which consequently also react with the EB-A2 antibody (Mennink-Kersten and Verweij, 2006; Morelle et al., 2005). The release of multiple substances that contain galactofuranose epitopes might increase the sensitivity of the galactomannan EIA in patient samples, but it still remains unclear how *Aspergillus* releases the antigen(s) during infection, which factors have impact on this release, and in which form the antigen(s) circulates in the blood.

In the commercial galactomannan EIA, an EDTA acid pretreatment of serum samples is used to dissociate immune complexes and precipitate proteins. Galactomannan is a polysaccharide that is heat stable and retains its antigenicity through this treatment. A volume of 300 μl of serum is used to retain 50 μl which is actually pipetted into the well of the microtiter plate. The optical density is read by a spectrophotometer, and the galactomannan index is calculated by dividing the

Paul E. Verweij • Dept. of Medical Microbiology, University Medical Center St. Radboud, and Nijmegen University Centre for Infectious Diseases, Nijmegen, The Netherlands.

optical density of the patient's sample by that of a threshold control that contains 1 ng/ml of galactomannan. The cutoff is now 0.5 worldwide (Maertens et al., 2007), although previously different cutoff levels were used in the United States (0.5) and the rest of the world (1.5), despite the fact that identical assays were used. This difference was historical, since the kit first became available in Europe in 1995. At that time a threshold of 1.5 was chosen by the manufacturer and was set relatively high in order to increase the specificity of the assay. Since then, several publications have indicated that a lower cutoff could be used (Herbrecht et al. 2002; Maertens et al., 2004). Although a cutoff of 0.5 is relatively low and close to the galactomannan index found in negative samples, evaluation of the distribution of galactomannan indexes in serum samples from hematology patients without invasive aspergillosis showed that only 22 of 3,691 (0.06%) were above 0.5 (Fig. 1) (Maertens et al., 2007). Depending on the experience in individual centers, higher cutoff values, such as 0.7, or a dynamic cutoff of two consecutive samples with an index of 0.5, have been used as a trigger to perform a diagnostic workup or initiate antifungal therapy (Maertens et al., 2004).

CLINICAL STUDIES

Early studies showed that in approximately two-thirds of patients with hematological malignancy, circulating antigen could be detected before diagnosis was made by other means. However, considerable variability of sensitivity has been reported even among patients with hematological malignancy (Mennink-Kersten et al., 2004). The sensitivity varied between 27.5 and 100% in individual prospective trials, with a specificity generally greater than 85%. The variability may be due to many factors, including the methodology and cutoff value used, study design, and factors relating to the host. Significant heterogeneity within the same patient populations was also noted in a meta-analysis (Pfeiffer et al., 2006). In that study the assay was found to be moderately accurate with an overall sensitivity of 0.71 (95% confidence interval, 0.68 to 0.74) and a specificity of 0.89 (95% confidence interval, 0.88 to 0.90) for proven cases (Pfeiffer et al., 2006). There are many reasons for variability in performance, which were recently reviewed (Mennink-Kersten et al., 2004). Factors that have emerged that cause false-negative reactivity include exposure to mold-active antifungal agents (Marr et al., 2004, 2005). The sensitivity of the galactomannan EIA was reduced to only 20% in patients receiving mold-active antifungal drugs, compared to 87.5% in nonexposed controls (Marr et al., 2005). False-positive reactivity has been reported, especially in patients receiving treatment with the antibacterial agent piperacillin-tazobactam, possibly due to the presence of *Penicillium*-derived galactomannan present in the antibiotic (Mennink-Kersten et al., 2004). Another important factor for the sensitivity of the galactomannan EIA was shown to be the underlying condition of the patient. The highest sensitivity is found in patients with hematological malignancy, especially those who are neutropenic. In liver transplant recipients the sensitivity was only 56% and in lung transplant patients 30% (Fortun et al., 2001; Husain et al., 2004; Kwak et al., 2004). In critically ill patients, of which the minority (22%) were neutropenic, the sensitivity of the galactomannan EIA was 42% (Meersseman et al., 2008). The performance of detection of galactomannan in serum was significantly better in patients with neutropenia compared to non-neutropenic patients (galactomannan index of >0.5 in 70% versus 25%; $P < 0.05$) (Meersseman et al., 2008). Finally, in patients with chronic granulomatous disease, circulating antigen is often not detected, although the number of cases reported is limited (Verweij et al.,

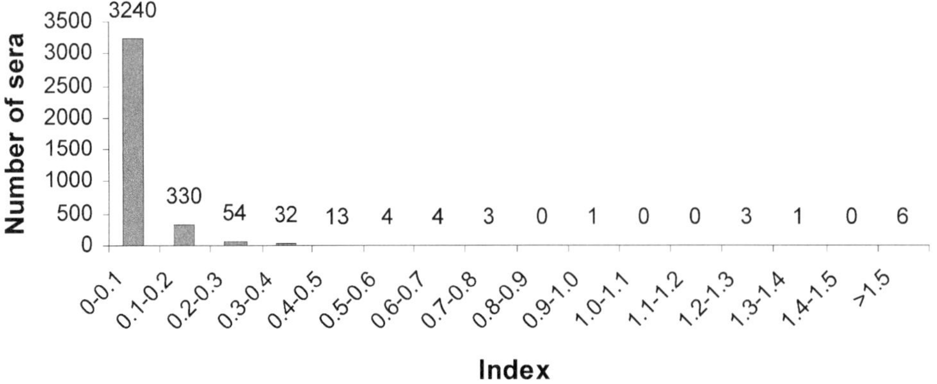

Figure 1. Distribution of galactomannan index in 3,691 serum samples from control samples from patients with hematological malignancy (Maertens et al., 2007).

2000). The pathogenesis of invasive aspergillosis is very different between these patient groups, with rapid angioinvasion in the neutropenic host, as opposed to abscess formation in the patient with chronic granulomatous disease. This might affect the release and leakage of galactomannan from the site of infection to the circulation.

Although the Platelia *Aspergillus* test is standardized for serum, the assay has been increasingly used in nonserum samples, such as bronchoalveolar lavage (BAL) fluid and cerebrospinal fluid (Klont et al., 2004). Several cases have been reported in which a presumptive diagnosis of central nervous system aspergillosis was made by detection of galactomannan in cerebrospinal fluid (Machetti et al., 2000; Verweij et al., 1999). Given the risks of invasive procedures for obtaining tissue from cerebral lesions, detection of galactomannan appears to be a useful aid for diagnosing intracerebral *Aspergillus* infections.

Detection of galactomannan in BAL fluid has been investigated in various patient populations for the diagnosis of invasive pulmonary aspergillosis and was first reported in 1995 with the Platelia *Aspergillus* system (Verweij et al., 1995). The sensitivity was 85% and specificity 100% in patients with hematological malignancy (Becker et al., 2003), 76% and 94%, respectively, in hematopoietic stem cell recipients (Musher et al., 2004), 100% and 98%, respectively, in solid organ transplant recipients (Clancy et al., 2007), and 60% and 95%, respectively, in lung transplant recipients (Husain et al., 2007). In all studies, the sensitivity was better than that of conventional methods, such as microscopy and culture. Despite these encouraging results, the temptation to use the galactomannan EIA by itself to diagnose fungal infection must be tempered by the potential for misdiagnosis (false positives) and by the assay's inability to detect other fungal pathogens that might be present either alone or together with *Aspergillus* species. A diagnostic approach should combine multiple laboratory methods, such as PCR, galactomannan EIA, and culture, to establish a diagnosis of disease (Musher et al., 2004; Sanguinetti et al., 2003).

As with galactomannan detection in serum, treatment of patients with mold-active antifungal agents decreases the performance of galactomannan detection in BAL. In one study galactomannan could not be detected in patients with invasive aspergillosis after 3 days of therapy (Becker et al., 2003). This finding was, however, not confirmed in another study, where the performance of galactomannan detection was not reduced in patients receiving antifungal therapy (Musher et al., 2004).

Recently, galactomannan detection with BAL was found to be useful for the diagnosis of invasive pulmonary aspergillosis in critically ill patients, an increasing entity in this patient population (Meersseman and Van Wijngaerden, 2007). The sensitivity was found to be 88% and specificity was 87% (Meerseman et al., 2008). The sensitivity of galactomannan detection in serum was much lower in this patient group, at 42%. In 11 of 26 proven cases, BAL culture and serum galactomannan remained negative, whereas galactomannan in BAL fluid was positive (Meerseman et al., 2008). However, in a low-risk group of nonneutropenic patients with various underlying diseases, galactomannan testing in BAL fluid was found to be no more sensitive than the combination of BAL microscopy and culture. Although the sensitivity was 100%, the positive predictive value was lower than for these conventional tests (Nguyen et al., 2007). Moreover, galactomannan detection in BAL fluid did not identify any cases of invasive pulmonary aspergillosis or chronic necrotizing pulmonary aspergillosis that were not diagnosed by conventional methods like microscopy and culture and only increased the likelihood of obtaining false-positive results (Nguyen et al., 2007).

The sampling strategies for galactomannan detection differ between different patient groups. In patients with hematological malignancy and in hematopoietic stem cell recipients, repeated sampling of serum has been reported as a feasible approach. These patients are admitted to the hospital during their period of high risk, and the prevalence of invasive aspergillosis is relatively high. Galactomannan detection in this setting is used as a tool to monitor patients, with those patients with a positive result being selected to undergo a diagnostic workup that usually involves high-resolution CT. In solid organ transplant recipients and critically ill patients, galactomannan detection in BAL fluid appears to be more informative than detection in serum. In this setting the assay is used as a diagnostic tool.

DETECTION OF ANTI-*ASPERGILLUS* ANTIBODIES

In the immunocompetent host, the diagnosis of certain manifestations of aspergillosis disease, such as aspergilloma or allergic bronchopulmonary aspergillosis, is based on the presence of specific anti-*Aspergillus* antibodies. It is generally agreed that a rise in the level of circulating antigen or specific antibodies indicates mycelial extension and/or progression of *Aspergillus* infection (Boutboul et al., 2002; Mennink-Kersten et al., 2004; Pazos et al., 2005).

The use of antibody detection in the diagnosis and management of invasive aspergillosis in patients with hematological malignancy has been very limited, following the observation that an antibody response could not be detected (Young and Bennett, 1971) and given the rapidity with which the infection develops. However, changes in the epidemiology of invasive aspergillosis might increase the usefulness of antibody detection and quantification in patients at risk for invasive aspergillosis as well as improvements in the assays used to detect specific antibodies. Besides episodes of neutropenia, the peak occurrence of invasive aspergillosis shifts towards the postengraftment period in patients who have received an allogeneic stem cell transplant, at a time when immune memory is reinitiated (Morgan et al., 2005). This is thought to be associated with changes in transplant practices, such as the use of peripheral blood rather than bone marrow as the stem cell source (Blijlevens et al., 2005). In particular, patients with acute or chronic graft-versus-host disease who require treatment with immunosuppressive agents are at risk for invasive fungal disease (van Burik et al., 2007). In addition, recent clinical surveys showed an increase in the incidence of invasive aspergillosis in critically ill patients without hematological malignancy (Meersseman and Van Wijngaerden, 2007). Since these patients are relatively immunocompetent, anti-*Aspergillus* antibodies may be produced and could be used to diagnose invasive infection. In one study the performance of two antibody tests was compared: immunoelectrophoresis (*Aspergillus* antigens and control *Aspergillus*; Bio-Rad) and either electrosyneresis and the Ouchterlony method (*Aspergillus* antigens and control *Aspergillus*; Bio-Rad) or an enzyme-linked immunosorbent assay (ELISA) with *Aspergillus fumigatus* IgG (Virion-AES Laboratoire) in a group of 88 patients with invasive aspergillosis which included both neutropenic and nonneutropenic patients (Cornillet et al., 2006). Thirty-seven patients (42%) had at least one antibody test, the result of which was positive in 30% of cases. This test was used more often in nonneutropenic patients (58% versus 31% in severe neutropenic patients). In these nonneutropenic patients the sensitivity of antibody testing was 48%, compared with only 6% in patients with severe neutropenia (Cornillet et al., 2006).

In other patient groups, such as solid organ transplant recipients, the detection of anti-*Aspergillus* antibodies might be useful for the diagnosis and management of invasive aspergillosis. In four lung transplant recipients, a close correlation was demonstrated between *A. fumigatus* IgG levels measured in serum and radiographic features, cytologic and microbiological findings, and the clinical diagnosis of fungal disease (Tomee et al., 1996).

A significant problem with the detection of antibody has been the lack of well-validated assays. The presently available assays make use of crude antigens with semiquantitative methods, such as immunoelectrophoresis, counterimmunoelectrophoresis, or hemagglutination (Latgé et al., 1991). ELISA methods using in-house-produced crude antigenic batches also have been introduced, but the lack of standardization between different laboratories makes comparisons of performance in aspergillosis diagnosis difficult to assess because of batch-to-batch variability (Sarfati et al., 2006). The large number of epitopes in crude extracts may compromise specificity. However, recently the use of recombinant *Aspergillus* cell wall antigens has been shown to reliably and reproducibly quantify anti-*Aspergillus* antibodies in immunochemical tests (Sarfati et al., 2006; Weig et al., 2001). In an evaluation of eight recombinant proteins and purified galactomannan, it was shown that specific antibodies against certain proteins (18-kDa RNase, 360-kDa catalase, and 88-kDa dipeptidyl-peptidase V) were present in patients with aspergilloma and allergic bronchopulmonary aspergillosis (Sarfati et al., 2006). The presence of antibodies against these antigens was also determined in patients with documented invasive aspergillosis. An antibody response following invasive infection was not apparent for these antigens, but about half of the patients with proven invasive aspergillosis were found to have high titers of anti-*Aspergillus* antibodies on admission to the hospital, suggesting that these patients had developed an *Aspergillus* infection before treatment with cytotoxic agents or before transplant engraftment but which was not yet clinically manifest. The most discriminative antigen in these patients was catalase. This observation suggests that high-risk patients might develop subclinical invasive aspergillosis in the community which presents as a clinical disease following immunosuppressive treatment in the hospital. This is in accordance with other observations that approximately 50 to 70% of patients have evidence of *Aspergillus* infection prior to diagnosis of the disease (Einsele et al., 1998; Patterson et al., 1997). This observation indicates that different markers can be used in various phases of invasive *Aspergillus* infection (Table 1). These insights will help to design strategies that optimally use the diagnostic potential of different diagnostic tools.

The presence of serum antibodies before clinically manifest invasive fungal disease was also observed for hematology patients with invasive candidiasis. Antiman-

Table 1. Phases in the development of invasive aspergillosis and diagnostic tools that might be used in each phase

Disease phase	Diagnostic tool/intervention
Invasive fungal infection	Antibodies
Invasive fungal disease	Clinical signs and symptoms
	Microscopy and culture
	Antigen
	High-resolution CT
Therapeutic intervention	Antifungal therapy
	Surgery
	Reduction of immune suppression
Follow-up	Antigen
	(Culture)
	High-resolution CT

nan antibodies were found in those patients who developed candidemia after several episodes of neutropenia, compared to control patients and those who presented with candidemia during their first episode of neutropenia. The antimannan antibodies were present in serum before the candidemia was clinically documented, which also suggests that in these patients a subclinical infection had developed during earlier episodes of neutropenia (Verduyn Lunel et al., 2007). These new insights may define a new role for antibody detection in the management of invasive fungal infections.

An alternative approach could be to combine the detection of antigen and antibody for the diagnosis of invasive aspergillosis. Such an approach has been developed for *Candida* spp., for which combined detection of mannan and antimannan antibody has proved useful for the diagnosis of candidemia, and commercial ELISA formats are available. In invasive aspergillosis this approach has been investigated using a purified recombinant cell wall galactomannoprotein, Afmp1p (Yuen et al., 2001). Circulating antigen was detected using an ELISA-based antigen test with rabbit- and guinea pig-derived polyclonal antibodies. The specificity of combined detection was 100% in sera from 138 negative controls, which included blood donors and patients with penicilliosis and candidiasis. The sensitivity of antigen detection and antibody detection in patients with invasive aspergillosis were 53.3% and 33.3%, respectively. Either assay was positive in 86.7% of patients, which is a promising result and justifies further evaluation of this approach (Chan et al., 2002; Woo et al., 2002).

STRATEGIES THAT INCORPORATE SEROLOGIC MARKERS

Empirical administration of antifungals for fever refractory to broad-spectrum antibiotic treatment still remains the standard of care, although the evidence supporting such an approach is very limited (de Pauw, 2005). One could also question if these studies that support an empiric strategy remain applicable given the changes in the epidemiology of invasive fungal diseases and in immunosuppressive treatment regimens for underlying diseases in high-risk patient populations and the increased armamentarium of antifungal drugs with varying spectra of activity.

Alternative approaches have received increased interest, as evidence has become available that prophylactic administration of antifungal agents may prevent invasive fungal disease and improve outcome. Prophylaxis with posaconazole was reported to reduce the frequency of invasive fungal infection in neutropenic patients with acute myeloid leukemia and myelodysplastic syndrome and those receiving treatment for severe graft-versus-host disease (Cornely et al., 2007; Ullmann et al., 2007). In addition, the improvement of diagnostic tests and procedures has prompted a more diagnosis-oriented approach, commonly referred to as a preemptive strategy. These new diagnostic tools include primarily the use of high-resolution CT, since high-resolution CT of the chest proved to be superior to the traditional chest X-ray for detection of pulmonary fungal disease. More importantly, the systematic use of high-resolution CT led to earlier diagnosis of invasive aspergillosis and to improved survival when early diagnosis was combined with early (surgical) interventions (Caillot et al., 1997).

Biological markers can be used to select patients that require a diagnostic workup, i.e., a high-resolution CT scan of the chest. Generally, this approach involves intensive and repeated monitoring of patients during the period of high risk, usually during neutropenia. Such an approach might also be useful in improving the design of clinical trials for evaluation of efficacy of antifungal agents (Rex et al., 2001). It is important to note that when the galactomannan assay is used as a tool to monitor patients, the prevalence of the disease becomes a significant factor in relation to the positive predictive value of the assay. When the prevalence of invasive aspergillosis is low, i.e., lower than 5%, the number of false-positive results will be relatively high. As a consequence, an approach that involves monitoring of patients can best be implemented only in very-high-risk patient populations, such as those with acute myeloid leukemia or myelodysplastic syndrome or those receiving an allogeneic hematopoietic stem cell transplant. The use of biological markers has, however, not been shown to improve outcome in high-risk patients or to improve their prognosis.

There has been increasing interest in strategic studies that compare different approaches to the management of invasive fungal infection, mainly in patients with hematological malignancy. It is unclear if different strategic approaches to the management of high-risk patients result in significant benefits with respect to the number of invasive fungal infections, mortality, or costs. Moreover, the different approaches might have impacts on specific components of the management strategies. For instance, exposure of patients to mold-active azoles was shown to significantly reduce the performance characteristics of galactomannan detection (Marr et al., 2005). The administration of azole prophylaxis therefore significantly reduces the diagnostic benefit of this biomarker.

An autopsy-controlled, nonrandomized pilot study evaluated the feasibility of combining the detection of circulating galactomannan and high-resolution CT for the management of invasive aspergillosis (Maertens et al., 2005). The preemptive strategy was evaluated in patients with acute myeloid leukemia or myelodysplastic syndrome and patients who underwent myeloablative allogeneic hematopoietic stem cell transplantation. Five factors were defined that triggered a diagnostic workup for invasive fungal disease: (i) neutropenic fever refractory to 5 days of broad-spectrum antibiotic treatment or unexplained fever relapsing after at least 48 h of defervescence while the patient was still neutropenic and still receiving antibiotics, (ii) clinical signs and/or symptoms suggestive of invasive fungal infection, (iii) appearance of a new pulmonary infiltrate while the patient was receiving treatment with broad-spectrum antibiotics or steroids, (iv) isolation of molds or demonstration of hyphae in respiratory specimens, and (v) two consecutive galactomannan EIAs with an optical density index of 0.5. Patients presenting with these factors underwent a high-resolution CT of the chest within 24 h of the request and a bronchoscopy with BAL in those patients who were not severely hypoxic (Maertens et al., 2005).

This approach was compared to a fever-driven empiric strategy, based on the calculation of the number of patients that would be classified to receive antifungals due to persistent fever despite treatment with broad-spectrum antibacterial agents. Of 117 febrile neutropenic episodes, 41 patients qualified for classical empirical antifungal therapy (30 episodes of persistent fever and 11 episodes of unexplained relapsing fever), while in only 9 of these study episodes did patients actually receive antifungals. This indicates that the number of patients that receive antifungal therapy can significantly be reduced when a preemptive strategy is used. In addition, patients with invasive fungal disease that did not present with fever and would not have received antifungal therapy in the empiric strategy were detected using the preemptive approach (Maertens et al., 2005). Although the authors concluded that a preemptive approach was feasible, there were several drawbacks. The preemptive strategy failed to detect patients with invasive fungal diseases due to fungi other than *Aspergillus* species (Maertens et al., 2005). With the changing epidemiology of fungi that cause opportunistic invasive mycoses, most notably the reported increase of invasive zygomycosis (Chayakulkeeree et al., 2006), this might restrict the use of this approach.

In the setting of withholding antifungal therapy in patients with febrile neutropenia without a positive galactomannan test or with a negative high-resolution CT, treatment might be delayed compared to the empiric treatment approach. Furthermore, the preemptive approach requires a multidisciplinary approach, and the logistics are critical, as are the commitments of consultant microbiologists, pulmonologists, and radiologists to allow prompt availability of the diagnostic tests and procedures. Although unnecessary treatment with expensive antifungal agents is avoided in a significant proportion of patients, it is unclear if this outweighs the increased costs of diagnostic tests and procedures. Alternatively, the empiric approach has important drawbacks, with fever being a very nonspecific marker for invasive fungal disease. There are many reasons for (persistent) fever, including drug-related causes, mucositis, and slow defervescence in neutropenic patients. Also, patients with invasive fungal disease might not present with fever, as was shown in the study of Maertens et al. (2005). The failure to achieve an etiological diagnosis might result in initiation of treatment with an inappropriate drug. It has been shown in patients with candidemia that inappropriate therapy is associated with increased hospital mortality (Parkins et al., 2007).

A randomized study was recently reported that compared empiric and preemptive strategies in a noninferiority study with a mixed study population of hematology patients with prolonged neutropenia (Cordonnier et al., 2006). Patients were randomized to an empiric strategy where antifungal therapy was initiated in patients with persistent fever irrespective of the symptoms. The group undergoing the preemptive approach received antifungal therapy on the basis of certain symptoms, including pneumonia, shock, skin lesions, sinusitis, unexplained central nervous system symptoms, hepatosplenic microabscesses, orbital inflammation, grade 4 mucositis, *Aspergillus* colonization, or positive serum galactomannan index, using a cutoff of $\geq$1.5. The main results are presented in Table 2.

Table 2. Comparison of empiric and preemptive treatment strategies[a]

Parameter	Empiric strategy	Preemptive strategy	*P* value
Survival at end of study	146 (97.3%)	136 (95.1%)	NS[b]
No. of deaths related to fungal infection	0	3 (43%)	NS
No. of invasive fungal infections (including breakthrough)	4	13	<0.02
Aspergillus	4	8	
Candida	0	5	
No. of patients on antifungal treatment	94 (62.7%)	55 (38.5%)	<0.001
Mean total hospital cost/patient	€25.306	€24.465	NS

[a]Source: Cordonnier et al. (2006).
[b]NS, not significant.

There were no differences in survival at the end of the study, and the number of deaths related to fungal infection was not different between the study groups. However, three patients in the preemptive group died of invasive fungal disease compared to no patients in the empiric group. Significant differences were observed in the number of invasive fungal infections diagnosed in the two groups, with a lower number diagnosed among the patients that were treated empirically. This is, however, a consequence of the nature of the strategies, with the preemptive strategy aimed to diagnose invasive fungal infection. The number of patients receiving antifungal therapy was significantly lower in the preemptive study group, which confirms the observation previously reported by Maertens et al. (2005). Interestingly, the costs per patient in both groups were not significantly different, which suggests that the costs saved in withholding antifungal therapy are similar to the increased costs of diagnostic tests and procedures in the preemptive group (Cordonnier et al., 2006). Based on this study there appears to be no benefit of either strategy, although further studies are required to confirm this finding.

FUTURE DIRECTIONS

All assays for the detection of biomarkers that are currently available clearly have drawbacks, and technical improvements could increase the sensitivity or specificity of the assay. In addition, other targets might be useful, especially if these include antigens that are released very early during growth of *Aspergillus*. The use of recombinant *Aspergillus* cell wall antigens could improve the currently available assays, which are based on crude *Aspergillus* antigens. Changes in the sample volume or pretreatment might also improve the performance of the currently available assays. This was recently shown for the galactomannan EIA, when filtration of a larger volume of plasma (750 μl as opposed to the recommended 300 μl) resulted in improved detection of circulating galactomannan (Mennink-Kersten et al., 2008). Using this new pretreatment method, circulating galactomannan was detected in the plasma of patients with invasive aspergillosis who were negative using the conventional pretreatment method. Also with this new pretreatment method, circulating galactomannan was detected 2 to 17 days earlier than with the conventional method (Mennink-Kersten et al., 2008). This shows that galactomannan might be present in the blood of patients with false-negative galactomannan test results but that the antigen is not detected due to insufficient sensitivity of the Platelia kit. It was previously shown that the various *Aspergillus* species release different levels of galactomannan when grown in vitro, and *A. fumigatus* releases relatively low levels of galactomannan (Mennink-Kersten and Verweij, 2006), which might contribute to low levels in patients with invasive aspergillosis due to this species.

Combining different markers might be useful for improving the performance of individual markers. Several studies have investigated multiple markers, combining antigen with PCR by combining two antigen tests. In an experimental model of invasive aspergillosis due to *Aspergillus terreus*, detection of galactomannan, β-glucan, or using PCR was compared individually and combined. The sensitivity of β-glucan, galactomannan, and PCR was 43%, 78%, and 73%, respectively. When the results of galactomannan and PCR were combined the sensitivity increased to 95%, to 83% for galactomannan combined with β-glucan, and to 95% when all markers were combined (Ahmad et al., 2007). Although this suggests that combining biological markers improves the performance characteristics, this remains to be confirmed in human studies. One such study showed limited accordance between detection of galactomannan and β-glucan in a heterogeneous group of patients with high suspicion of invasive fungal infection (Persat et al., 2008). Another point of concern is the timing of positivity of the biological markers when detection is combined. As an example, the levels of β-glucan and galactomannan are shown for a patient with proven invasive aspergillosis due to *A. fumigatus* in consecutive plasma samples (Fig. 2). β-Glucan was detected in plasma 9 days before galactomannan was detected. The interpretation in the first 9 days is very difficult, as positivity of β-glucan suggests the presence of an invasive fungal infection for which the etiology is unknown. The negative galactomannan results in the first 9 days could be interpreted as the infection likely not being caused by *Aspergillus* species, and this might result in inappropriate treatment. Clearly, more experience is required with combinations of biomarkers, and possibly combined assays would be better for excluding the presence of invasive fungal disease rather than diagnosing one.

Management strategies are commonly based on the perceived risk of invasive fungal infection and the risk of patients in relation to specific host factors, such as underlying disease. The risk of invasive fungal infection in, for instance, patients with acute myeloid leukemia is not the same for each episode of neutropenia (Pagano et al., 2006, 2007). The prevalence of invasive fungal infection was found to be the highest during induction chemotherapy in this patient group, while the risk was lower during consolidation therapy (Pagano et al., 2006). Also, the risk of invasive fungal infection was higher in patients with refractory disease compared to those who were in complete remission (Pagano et al., 2006). The performance of biological markers might be influenced by these factors. Therefore, the above-

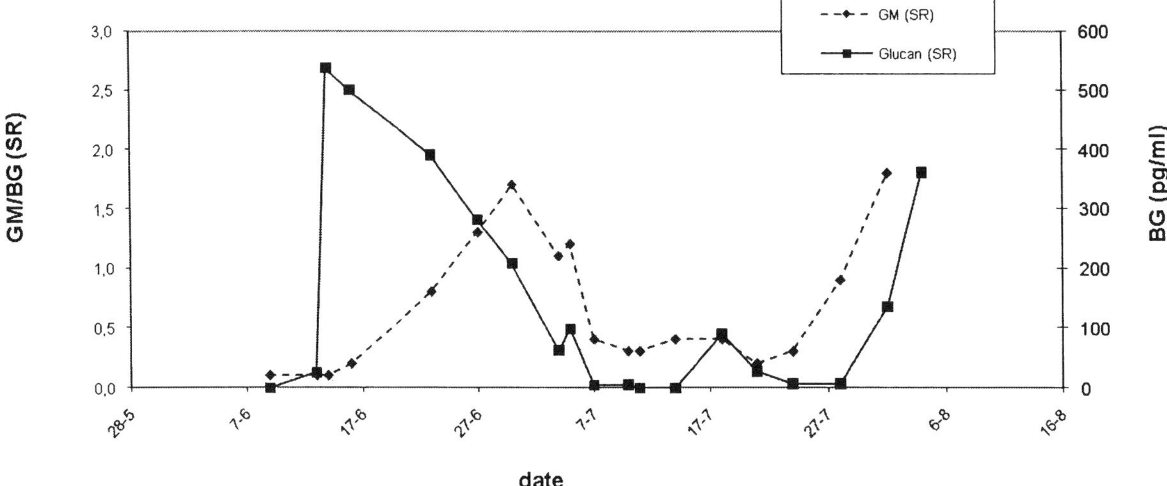

Figure 2. Levels of β-glucan and galactomannan in consecutive plasma samples obtained from a patient with proven invasive aspergillosis due to *A. fumigatus*. β-Glucan was detected first, 9 days before the galactomannan test became positive.

mentioned factors may need to be taken into account in choosing the optimum management strategy and could result in targeting of management strategies. Patients might benefit from prophylaxis during induction chemotherapy, while an empiric or preemptive strategy might be more appropriate during consolidation therapy. However, this fine-tuning of the appropriate management strategies still requires further research.

CONCLUSIONS

In the past decades more insight has been obtained with respect to the performance of the galactomannan EIA in different clinical specimens and in expanding patient populations. New assays have become available that could be combined with galactomannan detection, and the detection of anti-*Aspergillus* antibodies might become more important in the management of high-risk patients. Strategic studies will help to explore the benefits of different management options in order to fully benefit from the diagnostic tests and ultimately to improve the survival and prognosis of patients with invasive aspergillosis.

REFERENCES

Ahmad, S., Z. U. Khan, and A. M. Theyyathel. 2007. Diagnostic value of DNA, (1-3)-beta-D-glucan, and galactomannan detection in serum and bronchoalveolar lavage of mice experimentally infected with *Aspergillus terreus*. *Diagn. Microbiol. Infect. Dis.* **59:**165–171.

Becker, M. J., E. J. Lugtenburg, J. J. Cornelissen, C. Van Der Schee, H. C. Hoogsteden, and S. De Marie. 2003. Galactomannan detection in computerized tomography-based broncho-alveolar lavage fluid and serum in haematological patients at risk for invasive pulmonary aspergillosis. *Br. J. Haematol.* **121:**448–457.

Blijlevens, N. M., J. P. Donnelly, and B. E. de Pauw. 2005. Microbiologic consequences of new approaches to managing hematologic malignancies. *Rev. Clin. Exp. Hematol.* **9:**E2.

Boutboul, F., C. Alberti, T. Leblanc, A. Sulahian, E. Gluckman, F. Derouin, and P. Ribaud. 2002. Invasive aspergillosis in allogeneic stem cell transplant recipients: increasing antigenemia is associated with progressive disease. *Clin. Infect. Dis.* **34:**939–943.

Caillot, D., O. Casasnovas, A. Bernard, J. F. Couaillier, C. Durand, B. Cuisenier, E. Solary, F. Piard, T. Petrella, A. Bonnin, G. Couillaut, M. Dumas, and H. Guy. 1997. Improved management of invasive pulmonary aspergillosis in neutropenic patients using early thoracic computed tomographic scan and surgery. *J. Clin. Oncol.* **15:**139–147.

Chan, C. M., P. C. Woo, A. S. Leung, S. K. Lau, X. Y. Che, L. Cao, and K. Y. Yuen. 2002. Detection of antibodies specific to an antigenic cell wall galactomannoprotein for serodiagnosis of *Aspergillus fumigatus* aspergillosis. *J. Clin. Microbiol.* **40:**2041–2045.

Chayakulkeeree, M., M. A. Ghannoum, and J. R. Perfect. 2006. Zygomycosis: the re-emerging fungal infection. *Eur. J. Clin. Microbiol. Infect. Dis.* **25:**215–229.

Clancy, C. J., R. A. Jaber, H. L. Leather, J. R. Wingard, B. Staley, L. J. Wheat, C. L. Cline, K. H. Rand, D. Schain, M. Baz, and M. H. Nguyen. 2007. Bronchoalveolar lavage galactomannan in diagnosis of invasive pulmonary aspergillosis among solid-organ transplant recipients. *J. Clin. Microbiol.* **45:**1759–1765.

Cordonnier, C., C. Pautas, S. Maury, A. Vekhoff, H. Farhat, F. Suarez, M. Basile, F. Isnard, L. Ades, F. Kuhnoski, O. Reman, S. Chehata, T. De Revel, S. Lepretre, E. Raffoux, S. Bretagne, and M. Schwarzinger. 2006. Empirical versus pre-emptive antifungal therapy in high-risk febrile neutropenic patients: a prospective randomized study, abstr. 2019. *49th Annu. Meet. Am. Soc. Hematol.*, Orlando, FL.

Cornely, O. A., J. Maertens, D. J. Winston, J. Perfect, A. J. Ullmann, T. J. Walsh, D. Helfgott, J. Holowiecki, D. Stockelberg, Y. T. Goh, M. Petrini, C. Hardalo, R. Suresh, and D. Angulo-Gonzalez. 2007. Posaconazole vs. fluconazole or itraconazole prophylaxis in patients with neutropenia. *N. Engl. J. Med.* **356:**348–359.

**Cornillet, A., C. Camus, S. Nimubona, V. Gandemer, P. Tattevin, C. Belleguic, S. Chevrier, C. Meunier, C. Lebert, M. Aupée, S. Caulet-

Maugendre, M. Faucheux, B. Lelong, E. Leray, C. Guiguen, and J. P. Gangneux. 2006. Comparison of epidemiological, clinical, and biological features of invasive aspergillosis in neutropenic and non-neutropenic patients: a 6-year survey. *Clin. Infect. Dis.* **43**:577–584.

de Pauw, B. E. 2005. Between over- and undertreatment of invasive fungal disease. *Clin. Infect. Dis.* **41**:1251–1253.

Einsele, H., K. Quabeck, K. D. Müller, H. Hebart, I. Rothenhöfer, J. Löffler, and U. W. Schaefer. 1998. Prediction of invasive pulmonary aspergillosis from colonisation of lower respiratory tract before marrow transplantation. *Lancet* **352**:1443.

Fortun, J., P. Martin-Davila, M. E. Alvarez, A. Sanchez-Sousa, C. Quereda, E. Navas, R. Barcena, E. Vicente, A. Candelas, A. Honrubia, J. Nuno, V. Pintado, S. Moreno, et al. 2001. *Aspergillus* antigenemia sandwich-enzyme immunoassay test as a serodiagnostic method for invasive aspergillosis in liver transplant recipients. *Transplantation* **71**:145–149.

Greene, R. E., H. T. Schlamm, J. W. Oestmann, P. Stark, C. Durand, O. Lortholary, J. R. Wingard, R. Herbrecht, P. Ribaud, T. F. Patterson, P. F. Troke, D. W. Denning, J. E. Bennett, B. E. de Pauw, and R. H. Rubin. 2007. Imaging findings in acute invasive pulmonary aspergillosis: clinical significance of the halo sign. *Clin. Infect. Dis.* **44**:373–379.

Herbrecht, R., V. Letscher-Bru, C. Oprea, B. Lioure, J. Waller, F. Campos, O. Villard, K. L. Liu, S. Natarajan-Amé, P. Lutz, P. Dufour, J. P. Bergerat, and E. Candolfi. 2002. *Aspergillus* galactomannan detection in the diagnosis of invasive aspergillosis in cancer patients. *J. Clin. Oncol.* **20**:1898–1906.

Hope, W. W., T. J. Walsh, and D. W. Denning. 2005. Laboratory diagnosis of invasive aspergillosis. *Lancet Infect. Dis.* **5**:609–622.

Husain, S., E. J. Kwak, A. Obman, M. M. Wagener, S. Kusne, J. E. Stout, K. R. McCurry, and N. Singh. 2004. Prospective assessment of Platelia *Aspergillus* galactomannan antigen for the diagnosis of invasive aspergillosis in lung transplant recipients. *Am. J. Transplant.* **4**:796–802.

Husain, S., D. L. Paterson, S. M. Studer, M. Crespo, J. Pilewski, M. Durkin, J. L. Wheat, B. Johnson, L. McLaughlin, C. Bentsen, K. R. McCurry, and N. Singh. 2007. *Aspergillus* galactomannan antigen in the bronchoalveolar lavage fluid for the diagnosis of invasive aspergillosis in lung transplant recipients. *Transplantation* **83**:1330–1336.

Klont, R. R., M. A. Mennink-Kersten, and P. E. Verweij. 2004. Utility of *Aspergillus* antigen detection in specimens other than serum specimens. *Clin. Infect. Dis.* **39**:1467–1474.

Kwak, E. J., S. Husain, A. Obman, L. Meinke, J. Stout, S. Kusne, M. M. Wagener, and N. Singh. 2004. Efficacy of galactomannan antigen in the Platelia *Aspergillus* enzyme immunoassay for diagnosis of invasive aspergillosis in liver transplant recipients. *J. Clin. Microbiol.* **42**:435–438.

Latgé, J. P., M. Moutaouakil, J. P. Debeaupuis, J. P. Bouchara, K. Haynes, and M. C. Prévost. 1991. The 18-kilodalton antigen secreted by *Aspergillus fumigatus*. *Infect. Immun.* **59**:2586–2594.

Machetti, M., M. Zotti, L. Veroni, N. Mordini, M. T. Van Lint, A. Bacigalupo, D. Paola, and C. Viscoli. 2000. Antigen detection in the diagnosis and management of a patient with probable cerebral aspergillosis treated with voriconazole. *Transplant. Infect. Dis.* **2**:140–144.

Maertens, J., K. Theunissen, E. Verbeken, K. Lagrou, J. Verhaegen, M. Boogaerts, and J. V. Eldere. 2004. Prospective clinical evaluation of lower cut-offs for galactomannan detection in adult neutropenic cancer patients and haematological stem cell transplant recipients. *Br. J. Haematol.* **126**:852–860.

Maertens, J., K. Theunissen, G. Verhoef, J. Verschakelen, K. Lagrou, E. Verbeken, A. Wilmer, J. Verhaegen, M. Boogaerts, and J. Van Eldere. 2005. Galactomannan and computed tomography-based preemptive antifungal therapy in neutropenic patients at high risk for invasive fungal infection: a prospective feasibility study. *Clin. Infect. Dis.* **41**:1242–1250.

Maertens, J. A., R. Klont, C. Masson, K. Theunissen, W. Meersseman, K. Lagrou, C. Heinen, B. Crépin, J. Van Eldere, M. Tabouret, J. P. Donnelly, and P. E. Verweij. 2007. Optimization of the cutoff value for the *Aspergillus* double-sandwich enzyme immunoassay. *Clin. Infect. Dis.* **44**:1329–1336.

Marr, K. A., S. A. Balajee, L. McLaughlin, M. Tabouret, C. Bentsen, and T. J. Walsh. 2004. Detection of galactomannan antigenemia by enzyme immunoassay for the diagnosis of invasive aspergillosis: variables that affect performance. *J. Infect. Dis.* **190**:641–649.

Marr, K. A., M. Laverdiere, A. Gugel, and W. Leisenring. 2005. Antifungal therapy decreases sensitivity of the *Aspergillus* galactomannan enzyme immunoassay. *Clin. Infect. Dis.* **40**:1762–1769.

Meersseman, W., and E. Van Wijngaerden. 2007. Invasive aspergillosis in the ICU: an emerging disease. *Intensive Care Med.* **33**:1679–1681.

Meersseman, W., K. Lagrou, J. Maertens, A. Wilmer, G. Hermans, S. Vanderschueren, I. Spriet, E. Verbeken, and E. Van Wijngaerden. 2008. Galactomannan in bronchoalveolar lavage fluid: a tool for diagnosing aspergillosis in intensive care unit patients. *Am. J. Respir. Crit. Care Med.* **177**:27–34.

Mennink-Kersten, M. A., J. P. Donnelly, and P. E. Verweij. 2004. Detection of circulating galactomannan for the diagnosis and management of invasive aspergillosis. *Lancet Infect. Dis.* **4**:349–357.

Mennink-Kersten, M. A., and P. E. Verweij. 2006. Non-culture-based diagnostics for opportunistic fungi. *Infect. Dis. Clin. North Am.* **20**:711–727.

Mennink-Kersten, M. A., D. Ruegebrink, R. R. Klont, A. Warris, N. M. Blijlevens, J. P. Donnelly, and P. E. Verweij. 2008. Improved detection of circulating *Aspergillus* antigen using a modified pretreatment procedure. *J. Clin. Microbiol.* **46**:1391–1397.

Morelle, W., M. Bernard, J. P. Debeaupuis, M. Buitrago, M. Tabouret, and J. P. Latgé. 2005. Galactomannoproteins of *Aspergillus fumigatus*. *Eukaryot. Cell* **4**:1308–1316.

Morgan, J., K. A. Wannemuehler, K. A. Marr, S. Hadley, D. P. Kontoyiannis, T. J. Walsh, S. K. Fridkin, P. G. Pappas, and D. W. Warnock. 2005. Incidence of invasive aspergillosis following hematopoietic stem cell and solid organ transplantation: interim results of a prospective multicenter surveillance program. *Med. Mycol.* **43**(Suppl. 1):S49–S58.

Musher, B., D. Fredricks, W. Leisenring, S. A. Balajee, C. Smith, and K. A. Marr. 2004. *Aspergillus* galactomannan enzyme immunoassay and quantitative PCR for diagnosis of invasive aspergillosis with bronchoalveolar lavage fluid. *J. Clin. Microbiol.* **42**:5517–5522.

Nguyen, M. H., R. Jaber, H. L. Leather, J. R. Wingard, B. Staley, L. J. Wheat, C. L. Cline, M. Baz, K. H. Rand, and C. J. Clancy. 2007. Use of bronchoalveolar lavage to detect galactomannan for diagnosis of pulmonary aspergillosis among nonimmunocompromised hosts. *J. Clin. Microbiol.* **45**:2787–2792.

Pagano, L., M. Caira, A. Candoni, M. Offidani, L. Fianchi, B. Martino, D. Pastore, M. Picardi, A. Bonini, A. Chierichini, R. Fanci, C. Caramatti, R. Invernizzi, D. Mattei, M. E. Mitra, L. Melillo, F. Aversa, M. T. Van Lint, P. Falcucci, C. G. Valentini, C. Girmenia, and A. Nosari. 2006. The epidemiology of fungal infections in patients with hematologic malignancies: the SEIFEM-2004 study. *Haematologica* **91**:1068–1075.

Pagano, L., M. Caira, A. Nosari, M. T. Van Lint, A. Candoni, M. Offidani, T. Aloisi, G. Irrera, A. Bonini, M. Picardi, C. Caramatti, R. Invernizzi, D. Mattei, L. Melillo, C. de Waure, G. Reddiconto, L. Fianchi, C. G. Valentini, C. Girmenia, G. Leone, and F. Aversa. 2007. Fungal infections in recipients of hematopoietic stem cell transplants: results of the SEIFEM B-2004 study, Sorveglianza Epidemiologica Infezioni Fungine Nelle Emopatie Maligne. *Clin. Infect. Dis.* **45**:1161–1170.

Parkins, M. D., D. M. Sabuda, S. Elsayed, and K. B. Laupland. 2007. Adequacy of empirical antifungal therapy and effect on outcome among patients with invasive *Candida* species infections. *J. Antimicrob. Chemother.* **60:**613–618.

Patterson, J. E., A. Zidouh, P. Miniter, V. T. Andriole, and T. F. Patterson. 1997. Hospital epidemiologic surveillance for invasive aspergillosis: patient demographics and the utility of antigen detection. *Infect. Control Hosp. Epidemiol.* **18:**104–108.

Pazos, C., J. Pontón, and A. Del Palacio. 2005. Contribution of (1→3)-beta-D-glucan chromogenic assay to diagnosis and therapeutic monitoring of invasive aspergillosis in neutropenic adult patients: a comparison with serial screening for circulating galactomannan. *J. Clin. Microbiol.* **43:**299–305.

Persat, F., S. Ranque, F. Derouin, A. Michel-Nguyen, S. Picot, and A. Sulahian. 2008. Contribution of the (1→3)-beta-D-glucan assay for diagnosis of invasive fungal infections. *J. Clin. Microbiol.* **46:**1009–1013.

Pfeiffer, C. D., J. P. Fine, and N. Safdar. 2006. Diagnosis of invasive aspergillosis using a galactomannan assay: a meta-analysis. *Clin. Infect. Dis.* **42:**1417–1427.

Rex, J. H., T. J. Walsh, M. Nettleman, E. J. Anaissie, J. E. Bennett, E. J. Bow, A. J. Carillo-Munoz, P. Chavanet, G. A. Cloud, D. W. Denning, B. E. de Pauw, J. E. Edwards, Jr., J. W. Hiemenz, C. A. Kauffman, G. Lopez-Berestein, P. Martino, J. D. Sobel, D. A. Stevens, R. Sylvester, J. Tollemar, C. Viscoli, M. A. Viviani, and T. Wu. 2001. Need for alternative trial designs and evaluation strategies for therapeutic studies of invasive mycoses. *Clin. Infect. Dis.* **33:**95–106.

Sanguinetti, M., B. Posteraro, L. Pagano, G. Pagliari, L. Fianchi, L. Mele, M. La Sorda, A. Franco, and G. Fadda. 2003. Comparison of real-time PCR, conventional PCR, and galactomannan antigen detection by enzyme-linked immunosorbent assay using bronchoalveolar lavage fluid samples from hematology patients for diagnosis of invasive pulmonary aspergillosis. *J. Clin. Microbiol.* **41:**3922–3925.

Sarfati, J., M. Monod, P. Recco, A. Sulahian, C. Pinel, E. Candolfi, T. Fontaine, J. P. Debeaupuis, M. Tabouret, and J. P. Latgé. 2006. Recombinant antigens as diagnostic markers for aspergillosis. *Diagn. Microbiol. Infect. Dis.* **55:**279–291.

Tomee, J. F., G. P. Mannes, W. van der Bij, T. S. van der Werf, W. J. de Boer, G. H. Koëter, and H. F. Kauffman. 1996. Serodiagnosis and monitoring of *Aspergillus* infections after lung transplantation. *Ann. Intern. Med.* **125:**197–201.

Ullmann, A. J., J. H. Lipton, D. H. Vesole, P. Chandrasekar, A. Langston, S. R. Tarantolo, H. Greinix, W. Morais de Azevedo, V. Reddy, N. Boparai, L. Pedicone, H. Patino, and S. Durrant. 2007. Posaconazole or fluconazole for prophylaxis in severe graft-versus-host disease. *N. Engl. J. Med.* **356:**335–347.

Upton, A., K. A. Kirby, P. Carpenter, M. Boeckh, and K. A. Marr. 2007. Invasive aspergillosis following hematopoietic cell transplantation: outcomes and prognostic factors associated with mortality. *Clin. Infect. Dis.* **44:**531–540.

van Burik, J. A., S. L. Carter, A. G. Freifeld, K. P. High, K. T. Godder, G. A. Papanicolaou, A. M. Mendizabal, J. E. Wagner, S. Yanovich, and K. A. Kernan. 2007. Higher risk of cytomegalovirus and *Aspergillus* infections in recipients of T cell-depleted unrelated bone marrow: analysis of infectious complications in patients treated with T cell depletion versus immunosuppressive therapy to prevent graft-versus-host disease. *Biol. Blood Marrow Transplant.* **13:**1487–1498.

Verduyn Lunel, F. M., J. P. Donnelly, H. L. van der Lee, N. M. A. Blijlevens, and P. E. Verweij. 2008. Circulating *Candida*-specific antimannan antibodies precede invasive candidiasis in patients undergoing myeloablative chemotherapy. *Clin. Microbiol. Infect.* **14,** in press.

Verweij, P. E., J. P. Latgé, A. J. Rijs, W. J. Melchers, B. E. De Pauw, J. A. Hoogkamp-Korstanje, and J. F. Meis. 1995. Comparison of antigen detection and PCR assay using bronchoalveolar lavage fluid for diagnosing invasive pulmonary aspergillosis in patients receiving treatment for hematological malignancies. *J. Clin. Microbiol.* **33:** 3150–3153.

Verweij, P. E., K. Brinkman, H. P. Kremer, B. J. Kullberg, and J. F. Meis. 1999. *Aspergillus* meningitis: diagnosis by non-culture-based microbiological methods and management. *J. Clin. Microbiol.* **37:** 1186–1189.

Verweij, P. E., C. M. Weemaes, J. H. Curfs, S. Bretagne, and J. F. Meis. 2000. Failure to detect circulating *Aspergillus* markers in a patient with chronic granulomatous disease and invasive aspergillosis. *J. Clin. Microbiol.* **38:**3900–3901.

Verweij, P. E., L. van Die, and J. P. Donnelly. 2007. Halo sign and improved outcome. *Clin. Infect. Dis.* **44:**1666–1667.

Weig, M., M. Frosch, K. Tintelnot, A. Haas, U. Gross, B. Linsmeier, and J. Heesemann. 2001. Use of recombinant mitogillin for improved serodiagnosis of *Aspergillus fumigatus*-associated diseases. *J. Clin. Microbiol.* **39:**1721–1730.

Woo, P. C., C. M. Chan, A. S. Leung, S. K. Lau, X. Y. Che, S. S. Wong, L. Cao, and K. Y. Yuen. 2002. Detection of cell wall galactomannoprotein Afmp1p in culture supernatants of *Aspergillus fumigatus* and in sera of aspergillosis patients. *J. Clin. Microbiol.* **40:** 4382–4387.

Young, R. C., and J. E. Bennett. 1971. Invasive aspergillosis. Absence of detectable antibody response. *Am. Rev. Respir. Dis.* **104:**710–716.

Yuen, K. Y., C. M. Chan, K. M. Chan, P. C. Woo, X. Y. Che, A. S. Leung, and L. Cao. 2001. Characterization of AFMP1: a novel target for serodiagnosis of aspergillosis. *J. Clin. Microbiol.* **39:**3830–3837.

Aspergillus fumigatus and Aspergillosis
Edited by J.-P. Latgé and W. J. Steinbach
©2009 ASM Press, Washington, DC

Chapter 29

Aspergillus PCR

P. Lewis White and Rosemary A. Barnes

The use of molecular techniques to aid in the early diagnosis of invasive aspergillosis (IA) has been extensively reported. However, until recently, little has been done to achieve consensus on an accepted protocol limiting approval of such techniques in a clinical setting. Substantial evidence is required to decide whether molecular techniques are best suited to ruling in or excluding IA and to where molecular testing would be best placed in clinical care pathways with respect to the patient's underlying condition, predisposing factors, and alternative diagnostic tests. Combining this with variations in specimen type, DNA extraction technique, PCR amplification, and minimal evidence for real-time result interpretation highlights the amount of research that still needs to be performed.

In vitro comparison of laboratory-based methods will provide quick results as to optimal methods. However, determination of an ideal specimen type, result interpretation, and the clinical impact may vary with different invasive fungal infections and will require long-term large-scale clinical evaluations to overcome the current low (possibly underdiagnosed) prevalence of disease.

HISTORY OF *ASPERGILLUS* DIAGNOSTIC PCR

The successful application of diagnostic *Aspergillus* PCR has been reported since the early 1990s. Almost 200 articles with a clinical link were published between 1993 and June 2007. Since 1998 more than 20 reviews and an average of more than one article per month have been published, and the interest in diagnostic *Aspergillus* PCR continues to rise (Fig. 1).

The earliest articles were limited to conventional block-based PCR methods and clinical applications focused on respiratory specimens. High sensitivity and specificity were achieved through nested PCR, Southern hybridization, and random fragment length polymorphism analysis (Makimura et al., 1994; Melchers et al., 1994; Nakamura et al., 1994), but false positives were noted with respiratory (bronchoalveolar lavage [BAL]) specimens (Bretagne et al., 1995; Spreadbury et al., 1993; Tang et al., 1993). In an early comparative study of PCR and galactomannan (GM) antigen testing for diagnosis of invasive pulmonary aspergillosis (IPA), Verweij and colleagues investigated the performance of PCR (hybridization and RFLP) and GM enzyme immunoassay (EIA) on BAL specimens and found PCR to be beneficial in the early diagnosis of IPA (Verweij et al., 1995). Interestingly, they proposed a testing algorithm for PCR and clinical tests based on the first GM-positive result. A similar proposal for PCR testing has been suggested recently by another group (Millon et al., 2005).

Yamakami et al. (1996) were the first to use serum for PCR testing. Using a nested PCR they reported a sensitivity of 70% when testing 20 patients with IA, which was slightly superior to a GM enzyme-linked immunosorbent assay (ELISA; 60%). The use of whole blood for fungal PCR was described a year later. Einsele et al. (1997) reported excellent sensitivity (100%) and specificity (98%) when they tested 601 blood specimens from 121 patients (21 with documented fungal disease). The report was also one of the first to highlight the importance of high-frequency PCR screening of patients, with samples being tested two or three times a week. In 1998 a comparison of PCR, β-D-glucan, and GM assay performance in an animal model was described (Hashimoto et al., 1998). PCR provided the highest sensitivity and higher levels of β-D-glucan related to development and progression of IPA. The performance of the GM assay was limited, because latex agglutination was used for detection.

In the late 1990s PCR-ELISA emerged as a highly sensitive and specific technique for testing BAL, blood, and serum specimens (Golbang et al., 1999; Jones et al., 1998; Löffler et al., 1998). Also during this period the

P. Lewis White • NPHS Microbiology Cardiff, University Hospital of Wales, Cardiff CF14 4XN, United Kingdom. **Rosemary A. Barnes** • Dept. of Medical Microbiology, Cardiff University, University Hospital of Wales, Cardiff CF14 4XN, United Kingdom.

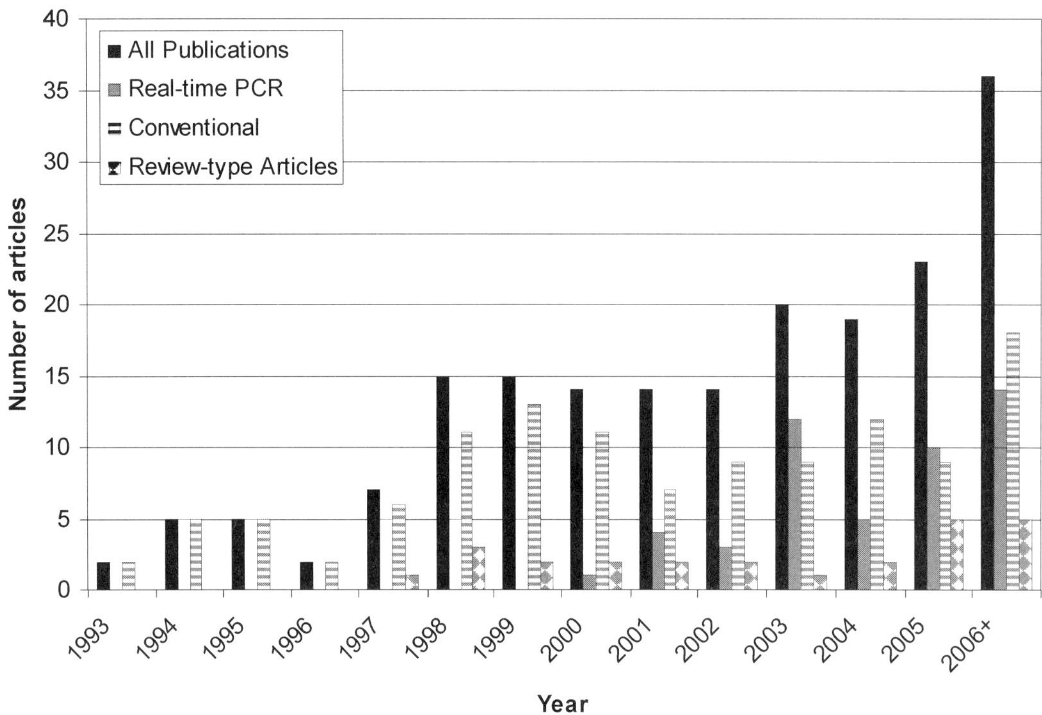

Figure 1. The number of articles describing *Aspergillus* PCR per annum (1993 to 2006).

first papers describing the use of PCR for the diagnosis of cerebral aspergillosis were published. Kami et al. (1999) described the combined use of cerebral diffusion-weighted echo-planar magnetic resonance imaging and cerebrospinal fluid (CSF) PCR to diagnose central nervous system aspergillosis, when conventional computed tomography, magnetic resonance imaging, and culture had failed. Others have recommended combined CSF PCR and CSF GM ELISA (Verweij et al., 1999). Contamination risks associated with fungal PCR were found to be no greater than with any other diagnostic PCR assay, although fungal DNA was found to be contaminating many important components (Zymolase; 10× PCR buffer) involved in molecular mycology (Loeffler et al., 1999; Rimek et al., 1999).

The first real-time PCR method was published in 2000. Using pan-fungal primers and hybridization probes specific for *Candida albicans* and *Aspergillus fumigatus*, Loeffler et al. (2000a) demonstrated the benefits of the LightCycler system for detecting invasive fungal disease and noted most invasive fungal disease patients have a low fungal burden in the blood (<10 CFU/ml). Also, the same year a nested PCR was used to investigate fungal enophthalmitis (Jaeger et al., 2000).

By 2001 the first hydrolysis (TaqMan) probe and nucleic acid sequence-based amplification (NASBA) methods were published. Costa et al. (2001) described the use of two TaqMan systems to detect *A. fumigatus* DNA. They compared the use of serum, white cell pellet, and plasma. Plasma provided a 10-fold-lower DNA yield than either the white cell or serum components (Costa et al., 2001). In 2001, fungal RNA detection was achieved by NASBA. Having a more efficient amplification than PCR, NASBA was able to achieve the same sensitivity as PCR using lower specimen volumes (Loeffler et al., 2001), although real-time technology has largely superceded this methodology. Also in 2001, the combined use of PCR and GM detection was described as an aid in the diagnosis of digestive aspergillosis in an acute myelogenous leukemia patient (Chambon-Pautas et al., 2001).

By 2002 automated nucleic acid extraction systems were introduced. The Roche MagNA Pure LC extractor was evaluated by two groups using EDTA whole blood and serum specimens, respectively (Costa et al., 2002; Loeffler et al., 2002a). The performance of the MagNA Pure system varied with specimen type, functioning optimally when EDTA-whole blood was used. In 2003 and 2004 multiple case reports highlighting the clinical utility of PCR in diagnosis began to appear (McCracken et al., 2003; Polzehl et al., 2005; Sambatakou et al., 2003; Scotter et al., 2004; Willinger et al., 2003; Wu et al., 2003).

By the end of 2005 over 150 papers describing the use of clinical *Aspergillus* PCR methods had been published, but despite this no consensus PCR method had been agreed upon. In 2006 the UK Fungal PCR Consensus Group compared PCR methods in 10 centers within the United Kingdom and Ireland and found two optimal methods (Kami et al., 2001; White et al., 2006a, 2006b). This led to the formation of an International Society for Human and Animal Mycology-associated international group (European *Aspergillus* PCR Initiative), whose aim is to find an optimal, standardized PCR method(s) for large-scale clinical evaluation to allow the inclusion of PCR in future European Organization for Research and Treatment of Cancer, Mycoses Study Group (EORTC/MSG) consensus diagnosis criteria.

BENEFTIS AND LIMITATIONS OF *ASPERGILLUS* PCR

Definitive diagnosis of IA relies on isolation from a sterile site plus histological demonstration of invasion in tissue. Positive cultures are rare, and obtainment of histopathological specimens is often too intrusive to be performed on critically ill patients. PCR can overcome many of the limitations of conventional methods but brings its own limitations. Many of the methods used to extract nucleic acids from bacteria and viruses are not harsh enough to lyse fungal targets, and the fungal burden circulating in fungemia may be much lower than that seen in bacteremia and viremia. Consequently, the PCR must have an enhanced limit of detection, which increases the possibility of false positives, particularly if nested PCR methods are used (Bretagne and Costa, 2006; Verweij, 2005).

The real-time PCR era has improved turnaround times and limited contamination risks by removing the need for postamplification handling and the use of uracil DNA glycosylase (Bretagne and Costa, 2006). Real-time PCR can be used to quantify disease burden, detect point mutations resulting in antifungal resistance, and through multiplexing or melt curve analysis differentiate between genera or species in a single reaction. Combined automated nucleic acid extractors and PCR platforms remove the human variable, allowing better comparison between centers and further minimizing contamination. PCR has also been used in histological studies (see below).

DNA EXTRACTION TECHNIQUES: GENERAL CONSIDERATIONS

An efficient DNA extraction procedure is critical for any molecular diagnostic test. The methodology used in an extraction procedure will vary with volume and specimen type. Larger volumes improve sensitivity; some protocols require up to 10 ml of sample (Lass-Florl et al., 2001; Suarez et al., 2007) but may provide access to more inhibitory substances which may enter at specimen collection (e.g., heparin vacutainers) (Garcia et al., 2002) or may be present in the host as a result of therapy (e.g., heparin used in cardiac bypass surgery) (White and Barnes, 2006). Efficient extraction will remove inhibitors, but it is essential to include an internal control PCR to exclude false-negative results. If the use of larger blood volumes is critical for the successful application of PCR methodology, then its use in neonatal or pediatric populations may be limited.

Antifungal therapy will not alter the extraction efficiency but may reduce the fungal burden to below detectable levels (Lass-Florl et al., 2004; White and Barnes, 2006). The effect of some antifungal agents on fungal cell walls and membranes may actually increase DNA release from fungal cells such that extraction methods not targeting free DNA may give negative results during treatment (Mennink-Kersten et al., 2006).

The intricacy of the method will depend on the specimen, and the experience of the worker has a direct effect on the efficiency of the extraction; automated systems have been introduced to alleviate this problem (Costa et al., 2002; Loeffler et al., 2002a). Fully automated nucleic acid extraction systems are now commonplace in large molecular clinical laboratories and are used routinely for bacterial and viral targets. However, fungal target cell lysis (when whole blood is used) and fungal disruption must be performed prior to automated steps.

Publications comparing *Aspergillus* DNA extraction techniques have shown wide variations in performance, with almost a 10^6-fold difference in DNA recovery (Table 1) (Fredricks et al., 2005). A range of *Aspergillus* DNA extraction procedures targeting both free DNA and fungal cell fragments have been described (see Tables 1 to 5). The fungal DNA extraction yield may be as low as 0.1%, emphasizing the need for optimal extraction methods (O'Sullivan et al., 2003). It is likely that the rate-limiting step of any molecular assay is the efficiency of extraction. For validation of fungal PCR it is important to determine both the optimal specimen and the corresponding DNA extraction technique.

MOLECULAR TESTING OF RESPIRATORY SPECIMENS

The value of respiratory specimens for diagnosis of IA is balanced against the efforts involved in obtaining the specimen and also the possibility of contaminating or colonizing *Aspergillus* spores within the respiratory tract. The clinical relevance of *Aspergillus* isolates cul-

Table 1. Summary of publications comparing *Aspergillus* nucleic acid extraction methods

Reference	Method	Specimen	Detection limit or yield	Assay time (h)
Griffiths et al. (2006)	Freeze-thaw[a]	Serially diluted conidia[b]	10^3	4
	Freeze-boil[a]			4
	Lyticase[a]		10^3	3
	Bead beating[a] (mini-BeadBeater 8)		10^2	2.5
	Lyticase and bead beating[a]		10^3	3.5
	Bead beating[a]	Serially diluted conidia[c,d]	10^1	2.5
	Roche MagNA Pure	Serially diluted conidia[b]	Not specified	3
	Qiagen EZ1		Not specified	0.5
Lugert et al. (2006)	Bead beating[e] (FastPrep Cell Disrupter 120)	Serially diluted conidia[f]	$>9 \times 10^7$	4
	Freeze-boil[g]		9×10^5	4
	Freeze-boil[a]		9×10^5	5
	Zymolase[a]		9×10^4	6–7
	Zymolase and freeze-boil[a]		9×10^2	8
Fredricks et al. (2005)	Nonenzymatic lysis[h]	2.8×10^4 conidia[i]	610 pg/ml	
	Bead beating[j]		2,663 pg/ml	
	Bead beating[e]		2,363 pg/ml	
	Nonenzymatic lysis[k]		137 pg/ml	
	Spheroplast formation and enzymatic lysis[l]		0 pg/ml	
	Hot detergent[m]		29 pg/ml	
Muller et al. (1998)	Lyticase and Novozyme[g]	10^7–10^8 conidia	1.22–10.2 μg/ml	6
	Bead beating (FastPrep Cell Disrupter 120)[n]		2.6–29.9 μg/ml	1
Van Burik et al. (1998a)	Bead beating[g]	30–100 mg mycelial mat	21.6–81.8 μg	2–3
	Liquid nitrogen, bead beating[g]		10.6–45.3 μg	3–4
	Glass beads, CTAB sonication[g]		4.2–6.2 μg	2–3
	CTAB and sonication[g]		3.2–11.2 μg	2–3
	Grinding and CTAB[g]		15.5–15.0 μg	2–3
	Lyticase[g]		8.3–10.4 μg	5–8
Löffler et al. (1997)	Zymolase[o]	Serially diluted conidia[d]	10^1	8
	Zymolase/Qiamp tissue kit (Qiagen)		10^1	4
	Zymolase/Genereleaser (BioVentures)		10^1	3–4
	Zymolase/Puregene D6000 (Gentra)		10^2	5
	Zymolase/Dynabeads DNA direct (Dynal)		10^2	4
	Zymolase/DNAzol (Molecular Research Centre)		10^3	4

[a] Followed by proteinase K digestion and use of a Qiagen DNA mini kit.
[b] Spores suspended in sorbitol buffer.
[c] Spores suspended in Qiagen AL.
[d] Spores suspended in 4 ml EDTA-blood.
[e] Used in combination with a Qbiogene FAST DNA kit.
[f] Spores suspended in phosphate-buffered saline.
[g] Used in combination with phenol-chloroform-isoamyl alcohol purification and alcohol precipitation. CTAB, cetyltrimethylammonium bromide.
[h] MasterPure yeast DNA purification kit (Epicenter, Madison, WI).
[i] Spores suspended in BAL fluid.
[j] Ultraclean soil DNA isolation kit (MoBio Inc., Solano Beach, CA).
[k] MasterPure plant leaf DNA purification kit (Epicenter, Madison, WI).
[l] Yeast cell lysis and GNOME kit (Qbiogene, Irvine, CA).
[m] SoilMaster DNA extraction kit (Epicenter, Madison, WI).
[n] Used in combination with glass milk binding matrix (Bio 101 Inc.).
[o] Used in combination with potassium acetate purification and alcohol precipitation.

tured from the respiratory tract has been determined by various groups; 48 to 56% of *Aspergillus* culture-positive respiratory tract specimens were determined to be proven or probable cases of IA (Garnacho-Montero et al., 2005; Vandewoude et al., 2006). Conversely, less than 50% of BAL specimens from proven cases of IA are culture positive (Levy et al., 1992). Detection of *Aspergillus* spp. by PCR within the respiratory tract should carry the same weight as culture methods, but the improved sensitivity and quick turnaround time make PCR detection favorable.

In respiratory samples the presence of inhaled *Aspergillus* conidia will precede invasive hyphal development during infection. In comparing six extraction methods, Fredricks et al. (2005) showed that bead-beating methods provided the greatest DNA yields from both BAL specimens containing *Aspergillus* conidia and tissue culture media with mycelial growth. Interestingly, these methods provided inferior results when testing for *Candida* spp. Conversely, a spheroplast method that performed particularly poorly when testing for *Aspergillus* conidia in BAL provided optimal results for *Can-*

dida. This may have implications if a pan-fungal approach is required. However, the overall sensitivity when testing BAL specimens from proven and probable cases for lyticase (spheroplast) methods is similar to that for bead-beating methods (78.6 and 79.2%, respectively), indicating that in a clinical setting the use of lyticase is suitable (Table 2).

A recent review of *Aspergillus* PCR on BAL specimens utilized a meta-analysis to review 15 publications (Tuon, 2007). The overall sensitivity (79%) and specificity (94%) of PCR were greater than culture methods. The range in sensitivity (36 to 100%) was broader than that of specificity (75 to 100%), although nine methods generated 100% sensitivity. The heterogeneity results from variable methodologies, populations, and definitions of disease. The overall positive likelihood ratio was 10.41 and the negative likelihood ratio was 0.22, indicating the importance of a positive PCR result in BAL specimens (Tuon, 2007). This needs to be balanced against the fact that up to 25% of BAL specimens from healthy volunteers can be PCR positive (Bart-Delabesse et al., 1997). The significance of a positive *Aspergillus* PCR result for a respiratory specimen should be interpreted with the risk factors and clinical signs for IA and may represent contamination.

The use of PCR-negative BAL specimens to exclude IPA is an important discussion point. It is fairly safe to say that a single one-off nondirected PCR-negative BAL result provides little evidence to withhold antifungal therapy. Conversely, multiple PCR-negative BAL specimens taken from known areas of infection should have a high negative predictive value. The meta-analysis presented an overall sensitivity of 79%, and justifiably the author commented that a negative result therefore should not delay therapy (Tuon, 2007). Within the analysis, four methods presented sensitivities below 70% and nine methods presented 100% sensitivities when testing 44% and 33% of the proven and probable cases of IA, respectively. This highlights two major problems in PCR testing for IA, the lack of a consensus method and the limited number of documented cases available for individual centers, and the author justifiably requested large-scale studies of a standard optimal method to allow a representative evaluation.

The upper airways will have higher levels of exposure to airborne spores and possible contamination,

Table 2. Published *Aspergillus* PCR methods principally testing BAL specimens[i]

Reference	Extraction method	PCR type	PCR target	No. of cases		Performance	
				Proven/ probable IA	Without IA	% Sensitivity	% Specificity
Schaberiter-Gurtner et al. (2007)	Lyticase[a]	RT	ITS2	14	17	100	100
Francesconi et al. (2006)	Lyticase, bead beating[b]	RT	ITS1-2	(Animal model)		80	100
Musher et al. (2004)	Master Pure yeast kit	RT	18S	49	50	67	100
Spiess et al. (2003)	Lyticase[c]	RT	MtCb	11	20	100	100
		Nested	18S	11	20	100	100
Sanguinetti et al. (2003)	DNeasy plant mini kit	RT	18S	20	24	90	100
		Nested	18S	20	24	90	100
Rantakokko-Jalava et al. (2003)	Lyticase, bead beating[d]	RT	MtRNA	11	83	73	93
Kawazu et al. (2003)		RT	18S	(Case study)			
Raad et al. (2002)	SDS-proteinase K[e]	PCR	MtDNA/AlkP	32	199	69	93
Buchheidt et al. (2001)	Lyticase[f]	Nested	18S	13	87	100	90
Hayette et al. (2001)	Proteinase K	Nested	AlkP	10	167	100	96
Skladny et al. (1999)	Lyticase[f]	Nested	18S	21	118	43	97
Jones et al. (1998)	Proteinase K[e]	PCR-E	MtDNA	12	57	100	100
Bretagne et al. (1995)	Hot NaOH	CPCR-SB	MtDNA	3	49	100	76
Verweij et al. (1995)	Bead beating[g]	PCR-SB	18S	7	45	71	84
Melchers et al. (1994)	Bead beating[h]	RT-PCR-SB	18S	6	20[i]	100	95
Tang et al. (1993)		PCR-SB	AlkP	4	18	100	94

[a] Used in combination with the Roche High Pure template prep kit.
[b] Used in combination with the Qiagen DNeasy Plant mini kit.
[c] Used in combination with proteinase K-SDS digestion, phenol-chloroform purification, and isopropanol precipitation.
[d] Used in combination with proteinase K digestion and phenol-ether purification.
[e] Used in combination with phenol-chloroform purification and ethanol precipitation.
[f] Used in combination with phenol-chloroform purification and isopropanol precipitation.
[g] Used in combination with guanidine thiocyanate and silica particles.
[h] Used in combination with phenol-chloroforom-isoamyl alcohol purification and ethanol precipitation.
[i] Abbreviations: RT, real-time; ITS 1-2, internal transcribed spacer region 1 (5.8S rRNA gene) and internal transcribed spacer region 2; MtCb, mitochondrial cytochrome *b*; MtRNA, mitochondrial tRNA; MtDNA, mitochondrial DNA; AlkP, alkaline protease; PCR-E, PCR-ELISA; CPCR-SB, competitive PCR-Southern blotting; PCR-SB, PCR-Southern blotting; RT-PCR-SB, reverse transcription-PCR-Southern blotting.

but testing nasal swabs or washings may be beneficial in patients with sinusitis. However, the performance of PCR to aid in the diagnosis of *Aspergillus* sinusitis is variable. A recent publication showed the negative impact of both GM and PCR testing in diagnosing fungal sinusitis (Kostamo et al., 2007). From seven patients diagnosed with allergic fungal rhinosinusitis-like syndrome, six patients had fungal hyphae and/or *Aspergillus* culture in nasal lavage, mucus, or from tissue. Only four were tested by PCR, of which only two were positive. A similar sensitivity (57%) has been achieved by others when PCR testing seven *Aspergillus* culture-positive nasal lavage specimens (Polzehl et al., 2005). However, in a larger study of 34 patients with histologically proven fungal sinusitis, molecular investigation of maxillary sinus tissue yielded favorable results (Willinger et al., 2003).

On average several hundred *Aspergillus* conidia are inhaled daily by the average human host (Latgé, 1999). A possible method to distinguish between contamination and infection is by obtaining specimens from both infected and uninfected areas. In a case of IPA, Kawazu et al. (2003) reported the use of a real-time PCR to test bronchial specimens. When PCR testing the infected bronchus, the fungal load was 60-fold higher than that in an uninfected bronchus, and the authors proposed that comparisons between areas would allow true positives to be distinguished.

One of the major benefits in testing respiratory specimens is that it allows early detection from the most common origin of infection and preempts invasive processes, such as angio-invasion. In the previous case of IPA, sequential blood samples only became PCR positive at an advanced stage compared to BAL specimens (Kawazu et al., 2003).

A comparison of methods used for *Aspergillus* PCR testing of BAL specimens is shown in Table 2.

MOLECULAR TESTING OF BLOOD SPECIMENS

Since the 1990s the PCR testing of blood specimens has received extensive interest, yet no agreement as to the optimal blood fraction and molecular methodology has been achieved. The benefit of using blood is the ease in obtaining high-frequency large-volume specimens for screening regimens that can combine the use of galactomannan EIA and/or β-D-glucan testing. Intensive sampling is required, particularly as some studies have shown as little as 11% of blood samples from patients with proven or probable IA will be PCR positive (Verweij, 2005). This low positivity rate may be a consequence of the varying degrees of invasion and our limited knowledge of *Aspergillus* pathogenesis, and lack of overt fungemia confounds the issue.

It is clear that the fungal component in blood is generally nonviable and/or very limited, as blood cultures are rarely positive. In an in vitro study Mennink-Kersten and colleagues (2006) showed that DNA was not detected during exponential growth and was only released after hyphal autolysis. They proposed that during infection hyphal damage by the host defenses, antifungal treatment, or autolysis due to nutrient limitation was required for DNA release and that the most likely source of DNA in blood would be free circulating DNA; this has been supported by others (Costa et al., 2001). During antifungal therapy DNA extraction methods not targeting free circulating DNA result in negative results (Mennink-Kersten et al., 2006). Methods targeting fungal fragments circulating in blood (destroying or decanting free DNA) have provided good correlation with proven and probable cases of IA, suggesting a cell-associated fungal DNA source (Loeffler et al., 2000a; White et al., 2006a).

Targeting free circulating DNA, released from lysed fungal cells, allows the use of serum and plasma specimens and basic DNA extraction techniques, excluding bead-beating or lyticase treatment, which will be beneficial if fungal PCR is to attain widespread use. However, the stability of free circulating DNA in blood must be considered. The presence of DNase in blood will impair the ability to detect DNA in a dose-dependent manner (Loeffler et al., 2000b).

Whole blood allows both free DNA and cell-associated DNA to be targeted. Mechanical disruption (bead beating) targets cellular fragments only, with free DNA lost during the aggressive process (White and Barnes, 2006). Utilizing recombinant lyticase with high-speed centrifugation will allow both DNA and fungal fragments to be targeted but adds considerable time and expense.

Whole-blood DNA extraction methods are tedious and complex, involving lysis of the blood fractions, fungal lysis, and DNA purification, but compared to serum and plasma samples they may provide a greater sensitivity. The presence of EDTA may also diminish the activity of DNase, allowing longer-term storage with less degradation (Loeffler et al., 2000b). The debate over whole blood or serum and plasma continues. Little clinical comparison has been performed to provide conclusive evidence. In a small study of three patients with proven IA, 19 specimens were PCR positive in both whole blood and plasma, with a further 22 specimens positive in whole blood only (Loeffler et al., 2000b). This enhancement may be attributed to detection of free DNA plus cell-associated DNA in whole blood, or it may be due to the presence of more circulating free DNA in whole-blood specimens. It has been shown that the quantity of fungal DNA in serum and plasma and the cellular component is equal (Costa et al., 2002).

Combining all blood fractions will increase the overall DNA amount.

The enhanced sensitivity when using whole blood for *Aspergillus* PCR is supported by the comparison of publications that have evaluated either whole blood or serum. Sensitivity is better with whole blood, but specificity is better with serum (Fig. 2). The mean sensitivity and specificity for whole blood are 85% (95% confidence interval [CI], 79 to 89%) and 86% (95% CI, 84 to 88%), compared to 72% (95% CI, 66 to 77%) and 96.5% (95% CI, 94 to 98%) for serum, respectively. Analysis shows that serum methods achieve consistent specificity but show wide variation in sensitivity, whereas whole-blood methods can show variation in both as a result of differences in sample volume, extraction, and amplification procedures. Nested PCR has been used to successfully increase the sensitivities of assays testing serum specimens. In Fig. 2 the two assays with the greatest sensitivities when testing serum are nested methods, and this has been achieved without compromising specificity.

Two assays utilizing whole blood showed poor sensitivities (<50%). The first compared a real-time PCR system with a previous evaluated nested PCR on blood taken from nine proven or probable cases of IA (Speiss et al., 2003). A DNA source was present in all blood specimens as detected by the nested PCR. However, the real-time assay had a much-reduced sensitivity. The reproducible detection limit for the real-time PCR was 15 CFU, compared to 1 to 5 CFU for the nested PCR. This reduction would affect the detection of an infection.

Another paper analyzed the use of PCR during antifungal therapy and showed a 36% reduction in sensitivity when comparing performance pre- and post-initiation of antifungal treatment for 11 proven or probable cases of IA (Lass-Florl et al., 2004). The sensitivity of the assay using BAL or lung tissue remained high (87.5%) despite antifungal therapy, highlighting sample selection issues during therapy. It has been proposed that a high fungal load in tissue correlates with a positive whole-blood PCR result (Loeffler et al., 2002b). In addition to the fungal load within the tissue, the degree of angio-invasion is paramount to detection in blood. Evaluation of *Aspergillus* PCR with regard to varying types of pulmonary aspergillosis revealed that PCR of serum had a much higher sensitivity when detecting IPA than for aspergilloma (Yamakami et al., 1998). In blood the situation remains to be clarified, and long-term studies comparing methods targeting DNA and fungal fragments from the same population are required.

Comparisons of methods used for *Aspergillus* PCR testing of serum and whole-blood specimens are shown in Tables 3 and 4, respectively.

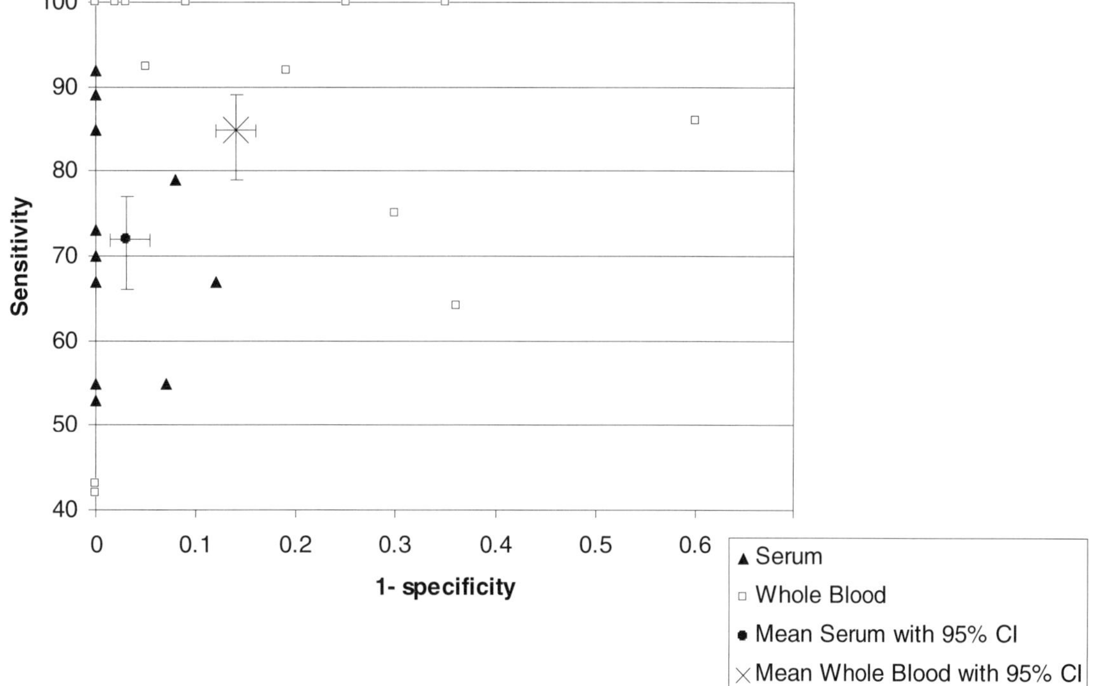

Figure 2. Performance of *Aspergillus* PCR using whole blood versus serum.

Table 3. *Aspergillus* PCR performance when testing serum specimens

Reference	Extraction method	PCR type	PCR target	No. of cases		Performance	
				Proven/ probable IA	No IA	% Sensitivity	% Specificity
Millon et al. (2005)	Roche High Pure template kit	RT	MtDNA	7	11	57	64
Kawazu et al. (2004)	Qiamp blood mini kit	RT	18S	11	125	55	93
Challier et al. (2004)	Qiamp DNA mini kit	RT	28S	26	29 85	100	
Pham et al. (2003)	Guanidine thiocyanate and silica beads	RT	5.8S	41	35	73	100
Costa et al. (2002)	Roche MagNA Pure LC	RT	MtDNA	20	30	70	100
Kami et al. (2001)	Qiamp blood mini kit	RT	18S	33	89	79	92
Williamson et al. (2000)	Lyticase[a]	Nested	28S	13	19	100	79
Kawamura et al. (1999)	Proteinase K	Nested	18S	44	39	89	100
Yamakami et al. (1998)	Proteinase K	Nested	18S	30		53	
Yamakami et al. (1996)	Proteinase K	Nested-SB	18S	20	20	70	100
Bretagne et al. (1998)	Qiamp blood kit	PCR-BIOT	MtDNA	22	19	55	100

[a] Used in combination with the Qiagen Qiamp tissue kit.
[b] Abbreviations: RT, real time; Nested-SB, nested PCR-Southern blotting; PCR-BIOT, PCR amplification combined with detection using biotinylated probes.

TESTING OTHER SPECIMENS (BIOPSY, EYE, AND CSF)

Literature on the use of molecular methods in the diagnosis of other invasive *Aspergillus* infections is limited, as no DNA extraction or PCR amplification methods have been extensively validated and determining performance parameters would be unwise. Conventional and bead-beating methods have been used for CSF specimens and indicated the presence of both free DNA and fungal fragments. Hot detergent extraction methods have proven successful in testing intraocular specimens (Table 5).

In testing tissue specimens it is essential that the tissue be sliced or homogenized and digested to allow the fungal hyphae to be efficiently targeted, and recent research has shown both mechanical disruption and lyticase digestion to be equally effective (Loeffler et al., 2007). The use of fresh tissue is also beneficial with paraffin-embedded specimens, reducing the sensitivity of molecular assays (Lau et al., 2007).

PERFORMANCE IN ICU PATIENTS, NEONATES, AND SOLID ORGAN TRANSPLANT RECIPIENTS

Most *Aspergillus* PCR evaluations have been performed in adult hematological populations. Its use in pediatric hematology, neonatal, intensive care unit (ICU), and solid organ transplant (SOT) recipient populations is limited. Possibly, the requirements of large specimen volumes and the low prevalence of disease have reduced its potential in neonates and pediatric patients and in nonhematological populations, respectively.

PCR has been successfully applied to pediatric hematology patients. It achieved a 75% sensitivity in two studies testing serum from 12 children with proven or probable IA and 28 children with proven or probable invasive funal infection, respectively (Challier et al., 2004; El-Mahallawy et al., 2006). In both studies specificity was high (92%). In a recent study comparing the performance of PCR, GM ELISA, and β-D-glucan in 27 pediatric or neonate patients with probable or possible IA, agreement between PCR and both GM ELISA and β-D-glucan assays was excellent and PCR generated a high negative predictive value (97%) (Velegraki et al., 2007). A pan-fungal PCR ELISA has been used to screen both liver and renal transplant recipients, and in both studies the method achieved good sensitivity (80%). However, in both studies only one case of IA was present (Badiee et al., 2007a, 2007b).

Extensive evaluation is required for PCR to be accepted as a diagnostic tool in pediatric, SOT, or ICU patients, but it is yet to be fully evaluated in the primary

Table 4. *Aspergillus* PCR performance when testing whole blood specimens[i]

Reference	Extraction	PCR type	PCR target	No. of cases		Performance	
				Proven/ probable IA	No IA	% Sensitivity	% Specificity
Einsele et al. (1997)	RCL, WCL, Zymolase[a]	PCR-SB	18S	13	64	100	98
Halliday et al. (2006)	RCL, lyticase[b]	Nested	18S	13	61	100	75
Loeffler et al. (2000a)	RCL, WCL, lyticase[c]	RT	18S	7	50	100	100
Buchheidt et al. (2001)	RCL, lyticase[d]	Nested	18S	36	182	92	81
Spiess et al. (2003)	RCL, WCL, lyticase[d]	RT	MtCb	9	50	56	100
		Nested	18S	9	50	100	100
Jordanides et al. (2005)	RCL, WCL, Zymolase[c]	RT	18S	—	—	—	—
Loeffler et al. (1998)	RCL, WCL, Zymolase[a,c]	PCR-ELISA	18S	14	41	100	88
Van Burik et al. (1998b)	RCL, WCL, lyticase[e]	PCR-SB	18S	5	5	100	100
Hebart et al. (2000a)	RCL, WCL, Zymolase[a]	PCR-SB	18S	10	69	100	65
Lass-Florl et al. (2004)	RCL, WCL, lyticase[a]	PCR-ELISA	18S	24	7	42	100
Loeffler et al. (2001)	RLT lysis buffer, liquid nitrogen, RNA secure[f]	NASBA	18S	—	—	—	—
Klinspor and Jalal (2006)	RCL, WCL, BB[g]	RT	18S	—	—	—	—
White et al. (2006a)	RCL, WCL, BB[h]	Nested RT	28S	14	149	92	95
Lass-Florl et al. (2001)	RCL, WCL, lyticase[c]	PCR ELISA	18S	3	118	75	96
Skladny et al. (1999)	RCL, lyticase[d]	Nested	18S	20	118	100	98

[a] Used in combination with isopropanol precipitation.
[b] Used in combination with a GenElute mammalian genomic DNA kit.
[c] Used in combination with a Qiagen Qiamp tissue kit.
[d] Used in combination with phenol-chloroform and isopropanol.
[e] Used in combination with potassium acetate and isopropanol.
[f] Used in combination with a Qiagen Rneasy mini kit and Qiashredder spin columns.
[g] Used in combination with Roche MagNA Pure DNA isolation.
[h] Used in combination with Roche MagNA Pure total NA isolation.
[i] Abbreviations: BB, bead beating; RT, real time; MtCb, mitochondrial cytochrome *b*; PCR-SB, PCR-Southern blotting; RCL, red blood cell lysis; WCL, white blood cell lysis.

target adult hematology population. The use of PCR in older children should imitate that of the adult population, whereas in neonates independent studies must be performed (Roilides, 2006). Its performance in ICU and SOT populations will vary with the underlying condition of the patient, which will also determine the correct specimen type, and multiple extensive investigations will be required.

ANIMAL MODELS

Several studies have used animal models to investigate the applicability of PCR in diagnosing IA. PCR out-performed GM ELISA when testing serum from eight mice with histologically proven IA, generating sensitivities of 71 and 43%, respectively (Yamakami et al., 1996). Both PCR and ELISA were positive on day 1 after inoculation with 10^6 *Aspergillus* conidia. In a more extensive study using 24 rats and the same methodology, PCR achieved a sensitivity of 20% on day 1 after inoculation and quickly rose to ≥80% on day 2 and thereafter (Hashimoto et al., 1998). Both studies indicated that PCR and ELISA could be used for early diagnosis of disease, but this must be interpreted with caution, as very rarely would patients be inoculated with such a high single dose of conidia.

Table 5. Molecular detection of cerebral aspergillosis, endophthalmic aspergillosis, *Aspergillus* sinusitis, and *Aspergillus*-infected tissue specimens

Reference	Extraction	PCR type	PCR target	Infection	Sample(s)	No. of cases
Khan et al. (2007)	Homogenization and boiling[a]	Nested	18S	CA	Brain tissue	1 proven IFD
Hummel et al. (2006)	Phenol-chloroform	Nested	18S	CA	CSF	2 proven, 2 probable, and 2 possible IA
Komatsu et al. (2004)		PCR-SB, nested-sequencing	18S	CA	CSF	1[b]
Verweij et al. (1999)	Bead beating[c]	PCR-SB	18S	CA	CSF	1 proven IA
Kami et al. (1999)	Proteinase K	Nested-SB	18S	CA	CSF	1 probable IPA, possible CA
Kostamo et al. (2007)	Lyticase, bead beating[d]	RT	MtDNA	CRS	NL, sinus mucus	6 fungal rhinosinusitis
Polzehl et al. (2005)	Hot SDS-NaOH[e]	PCR-SB, nested	18S	CRS	NL	77 CRS
Zeng et al. (2007)	Proteinase K[k]	PCR-RLB	ITS[f]	IA	Sinus, nasal and lung tissue	7 proven IA
Willinger et al. (2003)	Homogenization and boiling[a]	PCR-SB, sequencing	28S	FBS	Maxillary sinus	112 samples
Bagyalakshmi et al. (2007)	Biogene DNA kit	Seminested	ITS[f]	END	Intraocular, corneal scrape	133 classified END
Tarai et al. (2006)	Hot SDS[c]	PCR	ITS[f]	END	Intraocular	50
Ferrer et al. (2001)	Hot SDS[c]	Seminested	ITS[f]	END, KER	Intraocular, corneal scrape	6 END, 3 KER
Jaeger et al. (2000)	QIAmp DNA mini kit	Nested	18S	END	Intraocular	2 fungal END
Anand et al. (2001)	Hot SDS[c]	PCR	28S	END	Intraocular	30
Paterson et al. (2006)	Lyticase[g]	PCR-SB	18S	IFD	Various tissues	56
Scotter et al. (2004)	Lyticase[h]	PCR-ELISA	18S	PER	CAPD	1
Hendolin et al. (2000)	Proteinse K and hot SDS-NaOH[h]	PCR-MLH	ITS[f]	IFD	Various tissues	20
McCracken et al. (2003)	Lyticase and proteinase K[h]	PCR-RFLP, sequencing	18S	CARD	Various tissues	1
Rantakokko-Jalava et al. (2003)	Proteinase K, lyticase, bead beating[i]	RT	MtDNA	IFD	Various tissues	6 cases of proven/ probable IA
Rickerts et al. (2007)	Proteinase K, freeze-boil[j]	Seminested, sequencing	MtDNA	IFD	Mainly lung and sinus tissues	27 cases of proven IFD
Lau et al. (2007)	Proteinase K[k]	PCR, sequencing	ITS1	IFD	Various tissues	62 cases of IFD, 26 cases of IA

[a]Boiled in guanidine thiocyanate-phenol, purified with chloroform-isoamyl alcohol and alcohol precipitation.
[b]Only evidence of infection was from lesions detected by MRI scan.
[c]Used in combination with phenol-chloroform-isoamyl alcohol purification and alcohol precipitation.
[d]Used in combination with the Roche High Pure PCR template prep kit.
[e]Used in combination with the GeneClean II kit (Bio 101).
[f]Refers to ITS1, 5.8S rRNA gene, ITS2.
[g]Used in combination with the TaKaRa DEXPAT kit (TaKaRa Biomedicals, Shiga, Japan) and ethanol precipitation.
[h]Used in combination with Qiamp DNA mini kit (Qiagen).
[i]Used in combination with phenol-ether purification.
[j]Used in combination with the Qiamp tissue kit (Qiagen).
[k]Used in combination with the MagNA Pure LC DNA isolation kit II.
[l]Abbreviations: CA, cerebral aspergillosis; CAPD, continuous ambulatory peritoneal dialysis; CARD, *Aspergillus* endocarditis; CRS, chronic rhinosinusitis; END, endophthalmitis; FBS, fungal ball of the sinuses; KER, keratitis; IFD, invasive fungal disease; ITS, internal transcribed spacer region; MtDNA, mitochondrial DNA; NL, nasal lavage; PCR-MLH, conventional PCR-multiplex liquid hybridization; PCR-SB, conventional PCR-Southern blotting; PCR-RLB, conventional PCR-reverse line blotting; PER, *Aspergillus* peritonitis; RT, real-time PCR.

Initiating infection with a lower inoculum (10^4 conidia) resulted in poor PCR sensitivities (18%) when testing small volumes of whole blood, and PCR positivity rates only improved 7 days after inoculation in association with dissemination of disease (Becker et al., 2000). This correlates to the increasing fungal burden, as PCR-positive blood specimens have been associated with greater fungal loads in tissue, particularly the liver (Loeffler et al., 2002b).

Most experimental models induce IA via the conventional airborne route. Intravenous studies have been performed to allow reproducible induction of fungemia confirmed by *Aspergillus* growth from blood (Hummel et al., 2004). Although not clinically representative, they have been useful in showing the transient nature and rapid clearance of viable organisms from the blood by invasion of parenchymal organs.

Unfortunately none of the papers using animal models compared PCR performance when testing both serum or plasma and whole blood. Discovering the optimal fraction is important for standardization of the *Aspergillus* PCR, and the use of an animal model would be an ideal way to investigate this issue.

CONTAMINATION CONSIDERATIONS

As *Aspergillus* species are ubiquitous, controlling the contamination factor becomes a major issue in developing a suitable area to perform *Aspergillus* PCR. As with all diagnostic PCR systems, universal precautions to prevent contamination should be in place (Kwok and Higuchi, 1989). Potential contamination can be further minimized by the use of real-time systems and uracyl-DNA glycosylase (Bretagne and Costa, 2006).

To minimize contamination with an airborne organism, initial sample processing and manual stages of nucleic acid extraction procedures should be performed in category 2 laminar flow cabinets. If automated nucleic acid extraction systems are used it is important to validate the conidial load required to enter and contaminate the extraction procedure once on the machine. The sealing of windows will minimize exposure to external airflow, which would be particularly important if building work were proceeding in the vicinity. Careful positioning and regular cleaning of air conditioning units to minimize disruption to laminar flow cabinets and to remove any potential contamination reservoirs are also paramount.

In addition to organism-associated contamination, molecular biology reagents have been shown to be contaminated with fungal DNA. Zymolase used to form spheroplasts was found to be contaminated with DNA from *Saccharomyces cerevisiae* (Loeffler et al., 1999; Rimek et al., 1999). Batches of 10× PCR buffer contained DNA from *Acremonium* spp., and batches of conventional partially purified lyticase were also shown to be contaminated with fungal DNA. Proteinase K, widely used in molecular biology, is purified from the environmental fungus *Tritirachium album*; however, contaminating DNA was not observed in this component over a 3-year period (Loeffler et al., 1999). The use of recombinant enzymes has gone some way to resolve the problem of fungal DNA contamination, albeit at great expense. However, certain manufacturers have changed their vector for producing recombinant enzymes from the bacterium *Escherichia coli* to the yeast *Pichia anomalis*, and therefore the fungal DNA contamination issue persists.

If the *Aspergillus* PCR assay used is genus specific and optimized, the reagent contamination mentioned above would not lead to false-positive results. However, if the assay were to utilize pan-fungal primers and an *Aspergillus*-specific probe, then false-negative results might arise if the contamination were great enough to out-compete a low-level true positive.

To our knowledge the only molecular components known to have been contaminated with *Aspergillus* DNA are a lysis buffer and silica spin columns contained in batches of a Qiagen DNA extraction kit. Several members of the UK Fungal PCR Consensus Group experienced *Aspergillus* DNA spin column contamination on a batch basis, and to avoid false-positive clinical results they have recommended the use of alternative automated methods (White et al., 2006b). Others have found the ATL buffer included in QIAmp DNA mini kits to be contaminated with *Aspergillus* material (Jaeger et al., 2000). In addition, *Aspergillus* has been shown to contaminate vacutainers, in particular the sodium citrate, used to draw blood specimens (Williamson, 2001). This may be related to use of *Aspergillus niger* in the production of citric acid. Source-based contamination of this type is almost impossible to control and represents a clinical but not molecular false-positive result (White and Barnes, 2006).

INTERPRETATION OF RESULTS: GENERAL CONSIDERATIONS

The clinical impact of a PCR result will depend on a number of factors and should be interpreted within the context of the clinical findings as defined by the EORTC/MSG consensus criteria (Ascioglu et al., 2002). However, multiple positive PCR results in a high-risk patient without any supporting clinical or microbiological evidence could represent early signs of infection. In screening regimens, positive PCR results preceded clinical diagnosis in hematology patients by up to 30 days (mean, 14 days) (Halliday et al., 2006).

Furthermore, positive PCR results in respiratory specimens may represent colonization or contamination and not current infection, but they should warn the clinician that an etiological agent for IPA is present within the patient. Immunocompromised patients with *Aspergillus* PCR-positive BAL specimens should be considered at risk for IA, and the result may be an indication for preemptive antifungal therapy (Melchers et al., 1994). It is essential that the correct specimen be tested to facilitate early diagnosis. Serial testing of blood specimens for a patient with suspected sinusitis may provide negative results until angio-invasion has occurred. In this case the benefit of early detection may well be lost unless nasal swabs, aspirates, or washings are tested.

The use of antifungal therapy should also be noted, as the sensitivity and negative predictive value of an assay may fall during treatment. Many groups have reported a loss in PCR positivity post-antifungal therapy (Buchheidt et al., 2004; Lass-Florl et al., 2004; Yamakami et al., 1998), and some have also linked this to resolving clinical symptoms (Hebart et al., 2000a, 2000b). This is not always the case, and clinical symptoms have persisted in the presence of negative PCR results (Ferns et al., 2002). Linking PCR negativity with clearance of disease is currently difficult, and negativity may represent the testing of an incorrect specimen under antifungal conditions. Research into the effects of antifungal treatment on PCR performance have shown that testing blood samples during antifungal therapy reduces the sensitivity and negative predictive value of the assay compared to testing lung specimens (Lass-Florl et al., 2004). This suggests that initiation of antifungal therapy enhances the clearance of fungal components from the blood but not from tissue and that the benefit of testing of blood specimens during antifungal therapy is limited (Lass-Florl et al., 2004). Little has been reported on the effects of antifungal prophylaxis on *Aspergillus* PCR. It is possible that if the antifungal levels are satisfactory within a patient, then blood specimens may provide false-negative results or delay PCR diagnosis. However, in our experience, when itraconazole prophylaxis is initiated and levels monitored regularly, positive PCR results generally present earlier than clinical or microbiological criteria.

PCR may be integrated in different clinical care pathways (Boudewijns et al., 2006). It can be used to confirm the diagnosis of IA in a symptomatic patient or, alternatively, it can be used to screen high-risk populations. For the former of these groups a high specificity and positive predictive value are required, whereas for a screening test a high sensitivity and negative predictive value are required. This allows the clinician to confidently withhold antifungal therapy in cases that are consistently negative and for the cases that become positive to request further investigations (additional PCR, chest X-ray, high-resolution computed tomography, or GM ELISA) to confirm the findings (Boudewijns et al., 2006; Buchheidt et al., 2004; Buchheidt and Hummel, 2005).

The criteria to define a patient as PCR positive are yet to be established. Generally the literature supports the presence of two consecutive PCR positive results (Boudewijns et al., 2006; Buchheidt et al., 2004; Florent et al., 2006; Kawazu et al., 2004). However, the use of this parameter has been shown to reduce assay sensitivity (Williamson et al., 2000); consequently, two positive PCR results within a neutropenic episode are also accepted.

Halliday and colleagues (2006) have proposed guidelines for interpreting PCR results in high-risk hematology patients with fever not responsive to broad-spectrum antibiotics based on twice-weekly PCR screening. They proposed that patients who are consistently negative by PCR should be investigated or treated for other infections, and single positive results require repeat testing. Consecutive positives and intermittent positives (within 2 weeks) require antifungal therapy and immediate alternative tests for IA. For intermittent positives (>2 weeks), additional testing is required and antifungal therapy should be withheld pending results. Any patients with proven or probable IA should be treated independent of the PCR results (Halliday et al., 2006). The basis for their guidelines was that "no patient with proven or probable IA had only a single positive PCR test." However, this may not always be the case if the amount of *Aspergillus* DNA in samples is below the reproducibility limits of the PCR, and Millon et al. (2005) proposed that even individual specimens for which a positive result is not replicated may have significance.

It is highly unlikely that an *Aspergillus* PCR alone will provide a definitive answer to the problem of diagnosing IA. It is more feasible and sensible that sensitive and specific PCR testing has its role within care pathways. The combination of host factors and radiological, microbiological, and serological information is essential to provide the highest level of care and as such it is essential that PCR tests be validated for inclusion in future EORTC/MSG criteria (Buchheidt and Hummel, 2005; White and Barnes, 2006).

CONCLUDING REMARKS

PCR has been utilized as an adjunct tool to aid in the diagnosis of IA for almost 2 decades, and the time has arrived for a definitive role(s) for PCR, with accepted standardized methods. The use of external validation, such as quality control for molecular diagnostics panels, is essential to validate the performance of in-house protocols, and despite commercial kits being FDA approved or CE marked, it is sensible that these be validated by a quality control panel on an annual basis. The benefit of commercial kits is their standardized quality-controlled production, but as with all in-house protocols, only limited multicenter evaluations with regard to IA have been performed. For PCR to be accepted by clinicians and included in future consensus criteria, as GM ELISA is now, a large-scale evaluation is critical. Only by combining standardized PCR technology with antigen detection and HRCT in neutropenic care pathways will we enhance the diagnosis of IA in an ever-increasing susceptible population.

REFERENCES

Anand, A. R., H. N. Madhavan, V. Neelam, and T. K. Lily. 2001. Use of polymerase chain reaction in the diagnosis of fungal endophthalmitis. *Opthalmology* **108:**326–330.

Ascioglu, S., J. H. Rex, B. de Pauw, J. E. Bennett, J. Bille, F. Crokaert, D. W. Denning, J. P. Donnelly, J. E. Edwards, Z. Erjavec, D. Fiere, O. Lortholary, J. Maertens, J. F. Meis, T. F. Patterson, J. Ritter, D. Selleslag, P. M. Shah, D. A. Stevens, T. J. Walsh, et al. 2002. Defining opportunistic invasive fungal infections in immunocompromised patients with cancer and hematopoietic stem cell transplants: an international consensus. *Clin. Infect. Dis.* **34:**7–14.

Badiee, P., P. Kordbacheh, A. Alborzi, S. A. Malekhoseini, F. Zeini, H. Mirhendi, and M. Mahmoodi. 2007a. Prospective screening in liver transplant recipients by panfungal PCR-ELISA for early diagnosis of invasive fungal infections. *Liver Transplant.* **13:**1011–1016.

Badiee, P., P. Kordbacheh, A. Alborzi, and S. A. Malekhoseini. 2007b. Invasive fungal infection in renal transplant recipients demonstrated by panfungal polymerase chain reaction. *Exp. Clin. Transplant.* **5:** 624–629.

Bagyalakshmi, R., K. L. Therese, and H. N. Madhavan. 2007. Application of semi-nested polymerase chain reaction targeting internal transcribed spacer region for rapid detection of panfungal genome directly from ocular specimens. *Indian J. Opthalmol.* **55:**261–266.

Bart-Delabesse, E., A. Marmorat-Khuong, J. M. Costa, M. L. Dubreuil-Lemaire, and S. Bretagne. 1997. Detection of *Aspergillus* DNA in bronchoalveolar lavage fluid of AIDS patients by the polymerase chain reaction. *Eur. J. Clin. Microbiol. Infect. Dis.* **16:**24–25.

Becker, M. J., S. de Marie, D. Willemse, H. A. Verbrugh, and I. A. Bakker-Woudenberg. 2000. Quantitative galactomannan detection is superior to PCR in diagnosing and monitoring invasive pulmonary aspergillosis in an experimental rat model. *J. Clin. Microbiol.* **38:** 1434–1438.

Boudewijns, M., P. E. Verweij, and W. J. G. Melchers. 2006. Molecular diagnosis of invasive aspergillosis: the long and winding road. *Future Microbiol.* **1:**283–293.

Bretagne, S., and J. M. Costa. 2006. Towards a nucleic acid based diagnosis in clinical parasitology and mycology. *Clin. Chim. Acta* **363:**221–228.

Bretagne, S., J. M. Costa, E. Bart-Delabesse, N. Dhedin, C. Rieux, and C. Cordonnier. 1998. Comparison of serum galactomannan antigen detection and competitive polymerase chain reaction for diagnosing invasive aspergillosis. *Clin. Infect. Dis.* **26:**1407–1412.

Bretagne, S., J. M. Costa, A. Marmorat-Khuong, F. Poron, C. Cordonnier, M. Vidaud, and J. Fleury-Feith. 1995. Detection of *Aspergillus* species DNA in bronchoalveolar lavage samples by competitive PCR. *J. Clin. Microbiol.* **33:**1164–1168.

Buchheidt, D., and M. Hummel. 2005. *Aspergillus* polymerase chain reaction (PCR) diagnosis. *Med. Mycol.* **43:**S139–S145.

Buchheidt, D., M. Hummel, D. Schleiermacher, B. Spiess, R. Schwerdtfeger, O. A. Cornely, S. Wilhelm, S. Reuter, W. Kern, T. Sudhoff, H. Morz, and R. Hehlmann. 2004. Prospective clinical evaluation of a LightCycler-mediated polymerase chain reaction assay, a nested-PCR assay and a galactomannan enzyme-linked immunosorbent assay for detection of invasive aspergillosis in neutropenic cancer patients and haematological stem cell transplant recipients. *Br. J. Haematol.* **125:**196–202.

Buchheidt, D., C. Baust, H. Skladny, J. Ritter, T. Suedhoff, M. Baldus, W. Seifarth, C. Leib-Moesch, and R. Hehlmann. 2001. Detection of *Aspergillus* species in blood and bronchoalveolar lavage samples from immunocompromised patients by means of 2-step polymerase chain reaction: clinical results. *Clin. Infect. Dis.* **33:**428–435.

Challier, S., S. Boyer, E. Abachin, and P. Berche. 2004. Development of a serum-based Taqman real-time PCR assay for diagnosis of invasive aspergillosis. *J. Clin. Microbiol.* **42:**844–846.

Chambon-Pautas, C., J. M. Costa, M. T. Chaumette, C. Cordonnier, and S. Bretagne. 2001. Galactomannan and polymerase chain reaction for the diagnosis of primary digestive aspergillosis in a patient with acute myeloid leukaemia. *J. Infect.* **43:**213–214.

Costa, C., J. M. Costa, C. Desterke, F. Botterel, C. Cordonnier, and S. Bretagne. 2002. Real-time PCR coupled with automated DNA extraction and detection of galactomannan antigen in serum by enzyme-linked immunosorbent assay for diagnosis of invasive aspergillosis. *J. Clin. Microbiol.* **40:**2224–2227.

Costa, C., D. Vidaud, M. Olivi, E. Bart-Delabesse, M. Vidaud, and S. Bretagne. 2001. Development of two real-time quantitative TaqMan PCR assays to detect circulating *Aspergillus fumigatus* DNA in serum. *J. Microbiol. Methods* **44:**263–269.

Einsele, H., H. Hebart, G. Roller, J. Loffler, I. Rothenhofer, C. A. Muller, R. A. Bowden, J. van Burik, D. Engelhard, L. Kanz, and U. Schumacher. 1997. Detection and identification of fungal pathogens in blood by using molecular probes. *J. Clin. Microbiol.* **35:** 1353–1360.

El-Mahallawy, H. A., H. H. Shaker, H. Ali Helmy, T. Mostafa, and A. Razak Abo-Sedah. 2006. Evaluation of pan-fungal PCR assay and *Aspergillus* antigen detection in the diagnosis of invasive fungal infections in high risk paediatric cancer patients. *Med. Mycol.* **44:**733–739.

Ferns, R. B., H. Fletcher, S. Bradley, S. Mackinnon, C. Hunt, and R. S. Tedder. 2002. The prospective evaluation of a nested polymerase chain reaction assay for the early detection of *Aspergillus* infection in patients with leukaemia or undergoing allograft treatment. *Br. J. Haematol.* **119:**720–725.

Ferrer, C., F. Colom, S. Frases, E. Mulet, J. L. Abad, and J. L. Alio. 2001. Detection and identification of fungal pathogens by PCR and by ITS2 and 5.8S ribosomal DNA typing in ocular infections. *J. Clin. Microbiol.* **39:**2873–2879.

Florent, M., S. Katsahian, A. Vekhoff, V. Levy, B. Rio, J. P. Marie, A. Bouvet, and M. Cornet. 2006. Prospective evaluation of a polymerase chain reaction-ELISA targeted to *Aspergillus fumigatus* and *Aspergillus flavus* for the early diagnosis of invasive aspergillosis in patients with hematological malignancies. *J. Infect. Dis.* **193:**741–747.

Francesconi, A., M. Kasai, R. Petraitiene, V. Petraitis, A. M. Kelaher, R. Schaufele, W. W. Hope, Y. R. Shea, J. Bacher, and T. J. Walsh. 2006. Characterization and comparison of galactomannan enzyme immunoassay and quantitative real-time PCR assay for detection of *Aspergillus fumigatus* in bronchoalveolar lavage fluid from experimental invasive pulmonary aspergillosis. *J. Clin. Microbiol.* **44:** 2475–2480.

Fredricks, D. N., C. Smith, and A. Meier. 2005. Comparison of six DNA extraction methods for recovery of fungal DNA as assessed by quantitative PCR. *J. Clin. Microbiol.* **43:**5122–5128.

Garcia, M. E., J. L. Blanco, J. Caballero, and D. Gargallo-Viola. 2002. Anticoagulants interfere with PCR used to diagnose invasive aspergillosis. *J. Clin. Microbiol.* **40:**1567–1568.

Garnacho-Montero, J., R. Amaya-Villar, C. Ortiz-Leyba, C. Leon, F. Alvarez-Lerma, J. Nolla-Salas, J. R. Iruretagoyena, and F. Barcenilla. 2005. Isolation of *Aspergillus* spp. from the respiratory tract in critically ill patients: risk factors, clinical presentation and outcome. *Crit. Care* **9:**R191–R199.

Golbang, N., J. P. Burnie, and R. C. Matthews. 1999. A polymerase chain reaction enzyme immunoassay for diagnosing infection caused by *Aspergillus fumigatus*. *J. Clin. Pathol.* **52:**419–423.

Griffiths, L. J., M. Anyim, S. R. Doffman, M. Wilks, M. R. Millar, and S. G. Agrawal. 2006. Comparison of DNA extraction methods for *Aspergillus fumigatus* using real-time PCR. *J. Med. Microbiol.* **55:**1187–1191.

Halliday, C., R. Hoile, T. Sorrell, G. James, S. Yadav, P. Shaw, M. Bleakley, K. Bradstock, and S. Chen. 2006. Role of prospective screening of blood for invasive aspergillosis by polymerase chain reaction in febrile neutropenic recipients of haematopoietic stem cell transplants and patients with acute leukaemia. *Br. J. Haematol.* 132:478–486.

Hashimoto, A., Y. Yamakami, P. Kamberi, E. Yamagata, R. Karashima, H. Nagaoka, and M. Nasu. 1998. Comparison of PCR, (1→3)-beta-D-glucan and galactomannan assays in sera of rats with experimental invasive aspergillosis. *J. Clin. Lab. Anal.* 12:257–262.

Hayette, M. P., D. Vaira, F. Susin, P. Boland, G. Christiaens, P. Melin, and P. De Mol. 2001. Detection of *Aspergillus* species DNA by PCR in bronchoalveolar lavage fluid. *J. Clin. Microbiol.* 39:2338–2340.

Hebart, H., J. Löffler, C. Meisner, F. Serey, D. Schmidt, A. Bohme, H. Martin, A. Engel, D. Bunje, W. V. Kern, U. Schumacher, L. Kanz, and H. Einsele. 2000a. Early detection of *Aspergillus* infection after allogeneic stem cell transplantation by polymerase chain reaction screening. *J. Infect. Dis.* 181:1713–1719.

Hebart, H., J. Löffler, H. Reitze, A. Engel, U. Schumacher, T. Klingebiel, P. Bader, A. Böhme, H. Martin, D. Bunjes, W. V. Kern, L. Kanz, and H. Einsele. 2000b. Prospective screening by a pan-fungal polymerase chain reaction assay in patients at risk for fungal infections: implications for the management of febrile neutropenia. *Br. J. Haematol.* 111:635–640.

Hendolin, P. H., L. Paulin, P. Koukila-Kahkola, V. J. Anttila, H. Malmberg, M. Richardson, and J. Ylikoski. 2000. Panfungal PCR and multiplex liquid hybridization for detection of fungi in tissue specimens. *J. Clin. Microbiol.* 38:4186–4192.

Hummel, M., B. Spiess, K. Kentouche, S. Niggemann, C. Bohm, S. Reuter, M. Kiehl, H. Morz, R. Hehlmann, and D. Buchheidt. 2006. Detection of *Aspergillus* DNA in cerebrospinal fluid from patients with cerebral aspergillosis by a nested PCR assay. *J. Clin. Microbiol.* 44:3989–3993.

Hummel, M., C. Baust, M. Kretschmar, T. Nichterlein, D. Schleiermacher, B. Spiess, H. Skladny, H. Morz, R. Hehlmann, and D. Buchheidt. 2004. Detection of *Aspergillus* DNA by a nested PCR assay is superior to blood culture in an experimental murine model of invasive aspergillosis. *J. Med. Microbiol.* 53:803–806.

Jaeger, E. E., N. M. Carroll, S. Choudhury, A. A. Dunlop, H. M. Towler, M. M. Matheson, P. Adamson, N. Okhravi, and S. Lightman. 2000. Rapid detection and identification of *Candida, Aspergillus,* and *Fusarium* species in ocular samples using nested PCR. *J. Clin. Microbiol.* 38:2902–2908.

Jones, M. E., A. J. Fox, A. J. Barnes, B. A. Oppenheim, P. Balagopal, G. R. Morgenstern, and J. H. Scarffe. 1998. PCR-ELISA for the early diagnosis of invasive pulmonary *Aspergillus* infection in neutropenic patients. *J. Clin. Pathol.* 51:652–656.

Jordanides, N. E., E. K. Allan, L. A. McLintock, M. Copland, M. Devaney, K. Stewart, A. N. Parker, P. R. E. Johnson, T. L. Holyoake, and B. L. Jones. 2005. A prospective study of real-time panfungal PCR for the early diagnosis of invasive fungal infection in haemato-oncology patients. *Bone Marrow Transplant.* 35:389–395.

Kami, M., T. Fukui, S. Ogawa, Y. Kazuyama, U. Machida, Y. Tanaka, Y. Kanda, T. Kashima, Y. Yamazaki, T. Hamaki, S. Mori, H. Akiyama, Y. Mutou, H. Sakamaki, K. Osumi, S. Kimura, and H. Hirai. 2001. Use of real-time PCR on blood samples for diagnosis of invasive aspergillosis. *Clin. Infect. Dis.* 33:1504–1512.

Kami, M., I. Shirouzu, K. Mitani, S. Ogawa, T. Matsumura, Y. Kanda, T. Masumoto, T. Saito, Y. Tanaka, K. Maki, H. Honda, S. Chiba, K. Ohtomo, H. Hirai, and Y. Yazaki. 1999. Early diagnosis of central nervous system aspergillosis with combination use of cerebral diffusion-weighted echo-planar magnetic resonance image and polymerase chain reaction of cerebrospinal fluid. *Intern. Med.* 38:45–48.

Kawamura, S., S. Maesaki, T. Noda, Y. Hirakata, K. Tomono, T. Tashiro, and S. Kohno. 1999. Comparison between PCR and detection of antigen in sera for diagnosis of pulmonary aspergillosis. *J. Clin. Microbiol.* 37:218–220.

Kawazu, M., Y. Kanda, Y. Nannya, K. Aoki, M. Kurokawa, S. Chiba, T. Motokura, H. Hirai, and S. Ogawa. 2004. Prospective comparison of the diagnostic potential of real-time PCR, double-sandwich enzyme-linked immunosorbent assay for galactomannan, and a (1→3)-beta-D-glucan test in weekly screening for invasive aspergillosis in patients with hematological disorders. *J. Clin. Microbiol.* 42:2733–2741.

Kawazu, M., Y. Kanda, S. Goyama, M. Takeshita, Y. Nannya, M. Niino, Y. Komeno, T. Nakamoto, M. Kurokawa, S. Tsujino, S. Ogawa, K. Aoki, S. Chiba, T. Motokura, N. Ohishi, and H. Hirai. 2003. Rapid diagnosis of invasive pulmonary aspergillosis by quantitative polymerase chain reaction using bronchial lavage fluid. *Am. J. Hematol.* 72:27–30.

Khan, Z. U., S. Ahmad, E. Mokaddas, T. Said, M. P. Nair, M. A. Halim, M. R. Nampoory, and M. R. McGinnis. 2007. Cerebral aspergillosis diagnosed by detection of *Aspergillus flavus*-specific DNA, galactomannan and (1→3)-beta-D-glucan in clinical specimens. *J. Med. Microbiol.* 56:129–132.

Klinspor, L., and S. Jalal. 2006. Molecular detection and identification of *Candida* and *Aspergillus* spp. from clinical samples using real-time PCR. *Clin. Microbiol. Infect.* 12:745–753.

Komatsu, H., T. Fujisawa, A. Inui, K. Horiuchi, H. Hashizume, T. Sogo, and I. Sekine. 2004. Molecular diagnosis of cerebral aspergillosis by sequence analysis with panfungal polymerase chain reaction. *J. Pediatr. Hematol. Oncol.* 26:40–44.

Kostamo, K., M. Richardson, E. Eerola, K. Rantakokko-Jalava, T. Meir, H. Malmberg, and E. Toskala. 2007. Negative impact of *Aspergillus* galactomannan and DNA detection in the diagnosis of fungal rhinosinusitis. *J. Med. Microbiol.* 56:1322–1327.

Kwok, S., and R. Higuchi. 1989. Avoiding false positives with PCR. *Nature* 339:237–238.

Lass-Flörl, C., J. Aigner, E. Gunsilius, A. Petzer, D. Nachbaur, G. Gastl, H. Einsele, J. Löffler, M. P. Dietrich, and R. Würzner. 2001. Screening for *Aspergillus* spp. using polymerase chain reaction of whole blood samples from patients with haematological malignancies. *Br. J. Haematol.* 113:180–184.

Lass-Florl, C., E. Gunsilius, G. Gastl, H. Bonatti, M. C. Freund, A. Gschwendtner, G. Kropshofer, M. P. Dierich, and A. Petzer. 2004. Diagnosing invasive aspergillosis during antifungal therapy by PCR analysis of blood samples. *J. Clin. Microbiol.* 42:4154–4157.

Latgé, J.-P. 1999. *Aspergillus fumigatus* and aspergillosis. *Clin. Microbiol. Rev.* 12:310–350.

Lau, A., S. Chen, T. Sorrell, D. Carter, R. Malik, P. Martin, and C. Halliday. 2007. Development and clinical application of a panfungal PCR assay to detect and identify fungal DNA in tissue specimens. *J. Clin. Microbiol.* 45:380–385.

Levy, H., D. A. Horak, B. R. Tegtmeier, S. B. Yokota, and S. J. Forman. 1992. The value of bronchoalveolar lavage and bronchial washings in the diagnosis of invasive pulmonary aspergillosis. *Respir. Med.* 86:243–248.

Loeffler, J., A. C. Vallor, S. Reichwen, W. Heinz, W. R. Kirkpatrick, L. K. Najvar, J. R. Graybill, H. Einsele, and T. F. Patterson. 2007. Comparison of DNA extraction methods utilizing real-time PCR-based amplification for the detection of *Aspergillus fumigatus* in murine lung and serum specimens, abstr. M576. *Abstr. 47th Intersci. Conf. Antimicrob. Agents Chemother.*, 17 to 20 September 2007.

Loeffler, J., K. Schmidt, H. Hebart, U. Schumacher, and H. Einsele. 2002a. Automated extraction of genomic DNA from medically important yeast species and filamentous fungi by using the MagNA Pure LC system. *J. Clin. Microbiol.* 40:2240–2243.

Loeffler, J., K. Kloepfer, H. Hebart, L. Najvar, J. R. Graybill, W. R. Kirkpatrick, T. F. Patterson, K. Dietz, R. Bialek, and H. Einsele. 2002b. Polymerase chain reaction detection of *Aspergillus* DNA in experimental models of invasive aspergillosis. *J. Infect. Dis.* **185**: 1203–1206.

Loeffler, J., H. Hebart, P. Cox, N. Flues, U. Schumacher, and H. Einsele. 2001. Nucleic acid sequence-based amplification of *Aspergillus* RNA in blood samples. *J. Clin. Microbiol.* **39**:1626–1629.

Loeffler, J., N. Henke, H. Hebart, D. Schmidt, L. Hagmeyer, U. Schumacher, and H. Einsele. 2000a. Quantification of fungal DNA by using fluorescence resonance energy transfer and the LightCycler system. *J. Clin. Microbiol.* **38**:586–590.

Loeffler, J., H. Hebart, U. Brauchle, U. Schumacher, and H. Einsele. 2000b. Comparison between plasma and whole blood specimens for detection of *Aspergillus* DNA by PCR. *J. Clin. Microbiol.* **38**:3830–3833.

Loeffler, J., H. Hebart, R. Bialek, L. Hagmeyer, D. Schmidt, F. P. Serey, M. Hartmann, J. Eucker, and H. Einsele. 1999. Contaminations occurring in fungal PCR assays. *J. Clin. Microbiol.* **37**:1200–1202.

Löffler, J., H. Hebart, S. Sepe, U. Schumacher, T. Klingebiel, and H. Einsele. 1998. Detection of PCR-amplified fungal DNA by using a PCR-ELISA system. *Med. Mycol.* **36**:275–279.

Löffler, J., H. Hebart, U. Schumacher, H. Reitze, and H. Einsele. 1997. Comparison of different methods for extraction of DNA of fungal pathogens from cultures and blood. *J. Clin. Microbiol.* **35**: 3311–3312.

Lugert, R., C. Schettler, and U. Gross. 2006. Comparison of different protocols for DNA preparation and PCR for the detection of fungal pathogens in vitro. *Mycoses* **49**:298–304.

Makimura, K., S. Y. Murayama, and H. Yamaguchi. 1994. Specific detection of *Aspergillus* and *Penicillium* species from respiratory specimens by polymerase chain reaction (PCR). *Jpn. J. Med. Sci. Biol.* **47**:141–156.

McCracken, D., R. Barnes, C. Poynton, P. L. White, N. Isik, and D. Cook. 2003. Polymerase chain reaction aids in the diagnosis of an unusual case of *Aspergillus niger* endocarditis in a patient with acute myeloid leukaemia. *J. Infect.* **47**:344–347.

Melchers, W. J., P. E. Verweij, P. van den Hurk, A. van Belkum, B. E. De Pauw, J. A. Hoogkamp-Korstanje, and J. F. Meis. 1994. General primer-mediated PCR for detection of *Aspergillus* species. *J. Clin. Microbiol.* **32**:1710–1717.

Mennink-Kersten, M. A., D. Ruegebrink, N. Wasei, W. J. Melchers, and P. E. Verweij. 2006. In vitro release by *Aspergillus fumigatus* of galactofuranose antigens, 1,3-beta-D-glucan, and DNA, surrogate markers used for diagnosis of invasive aspergillosis. *J. Clin. Microbiol.* **44**:1711–1718.

Millon, L., R. Piarroux, E. Deconinck, C. E. Bulabois, F. Grenouillet, P. Rohrlich, J. M. Costa, and S. Bretagne. 2005. Use of real-time PCR to process the first galactomannan-positive serum sample in diagnosing invasive aspergillosis. *J. Clin. Microbiol.* **43**:5097–5101.

Muller, F. M., K. E. Werner, M. Kasai, A. Francesconi, S. J. Chanock, and T. J. Walsh. 1998. Rapid extraction of genomic DNA from medically important yeasts and filamentous fungi by high-speed cell disruption. *J. Clin. Microbiol.* **36**:1625–1629.

Musher, B., D. Fredricks, W. Leisenring, S. A. Balajee, C. Smith, and K. A. Marr. 2004. *Aspergillus* galactomannan enzyme immunoassay and quantitative PCR for diagnosis of invasive aspergillosis with bronchoalveolar lavage fluid. *J. Clin. Microbiol.* **42**:5517–5522.

Nakamura, H., Y. Shibata, Y. Kudo, S. Saito, H. Kimura, and H. Tomoike. 1994. Detection of *Aspergillus fumigatus* DNA by polymerase chain reaction in the clinical samples from individuals with pulmonary aspergillosis. *Rinsho Byori* **42**:676–681. (In Japanese.)

O'Sullivan, C. E., M. Kasai, A. Francesconi, V. Petraitis, R. Petraitiene, A. M. Kelaher, A. A. Sarafandi, and T. J. Walsh. 2003. Development and validation of a quantitative real-time PCR assay using fluorescence resonance energy transfer technology for detection of *Aspergillus fumigatus* in experimental invasive aspergillosis. *J. Clin. Microbiol.* **41**:5676–5682.

Paterson, P. J., S. Seaton, T. D. McHugh, J. McLaughlin, M. Potter, H. G. Prentice, and C. C. Kibbler. 2006. Validation and clinical application of molecular methods for the identification of molds in tissue. *Clin. Infect. Dis.* **42**:51–56.

Pham, A. S., J. J. Tarrand, G. S. May, M. S. Lee, D. P. Kontoyiannis, and X. Y. Han. 2003. Diagnosis of invasive mold infection by real-time quantitative PCR. *Am. J. Clin. Pathol.* **119**:38–44.

Polzehl, D., M. Weschta, A. Podbielski, H. Riechelmann, and D. Rimek. 2005. Fungus culture and PCR in nasal lavage samples of patients with chronic rhinosinusitis. *J. Med. Microbiol.* **54**:31–37.

Raad, I., H. Hanna, A. Huaringa, D. Sumoza, R. Hachem, and M. Albitar. 2002. Diagnosis of invasive pulmonary aspergillosis using polymerase chain reaction-based detection of aspergillus in BAL. *Chest* **121**:1171–1176.

Rantakokko-Jalava, K., S. Laaksonen, J. Issakainen, J. Vauras, J. Nikoskelainen, M. K. Viljanen, and J. Salonen. 2003. Semiquantitative detection by real-time PCR of *Aspergillus fumigatus* in bronchoalveolar lavage fluids and tissue biopsy specimens from patients with invasive aspergillosis. *J. Clin. Microbiol.* **41**:4304–4311.

Rickerts, V., S. Mousset, E. Lambrecht, K. Tintelnot, R. Schwerdtfeger, E. Presterl, V. Jacobi, G. Just-Nubling, and R. Bialek. 2007. Comparison of histopathological analysis, culture, and polymerase chain reaction assays to detect invasive mold infections from biopsy specimens. *Clin. Infect. Dis.* **44**:1078–1083.

Rimek, D., A. P. Garg, W. H. Haas, and R. Kappe. 1999. Identification of contaminating fungal DNA sequences in zymolyase. *J. Clin. Microbiol.* **37**:830–831.

Roilides, E. 2006. Early diagnosis of invasive aspergillosis in infants and children. *Med. Mycol.* **44**:S199–S205.

Sambatakou, H., M. Guiver, and D. Denning. 2003. Pulmonary aspergillosis in a patient with chronic granulomatous disease: confirmation by polymerase chain reaction and serological tests, and successful treatment with voriconazole. *Eur. J. Clin. Microbiol. Infect. Dis.* **22**:681–685.

Sanguinetti, M., B. Posteraro, L. Pagano, G. Pagliari, L. Fianchi, L. Mele, M. La Sorda, A. Franco, and G. Fadda. 2003. Comparison of real-time PCR, conventional PCR, and galactomannan antigen detection by enzyme-linked immunosorbent assay using bronchoalveolar lavage fluid samples from hematology patients for diagnosis of invasive pulmonary aspergillosis. *J. Clin. Microbiol.* **41**:3922–3925. (Erratum, **43**:3588, 2005.)

Schabereiter-Gurtner, C., B. Selitsch, M. L. Rotter, A. M. Hirschl, and B. Willinger. 2007. Development of novel real-time PCR assays for detection and differentiation of eleven medically important *Aspergillus* and *Candida* species in clinical specimens. *J. Clin. Microbiol.* **45**:906–914.

Scotter, J. M., J. M. Stevens, S. T. Chambers, K. L. Lynn, and W. N. Patton. 2004. Diagnosis of *Aspergillus* peritonitis in a renal dialysis patient by PCR and galactomannan detection. *J. Clin. Pathol.* **57**: 662–664.

Skladny, H., D. Buchheidt, C. Baust, F. Krieg-Schneider, W. Seifarth, C. Leib-Mosch, and R. Hehlmann. 1999. Specific detection of *Aspergillus* species in blood and bronchoalveolar lavage samples of immunocompromised patients by two-step PCR. *J. Clin. Microbiol.* **37**:3865–3871.

Spiess, B., D. Buchheidt, C. Baust, H. Skladny, W. Seifarth, U. Zeilfelder, C. Leib-Mosch, H. Morz, and R. Hehlmann. 2003. Development of a LightCycler PCR assay for detection and quantification of *Aspergillus fumigatus* DNA in clinical samples from neutropenic patients. *J. Clin. Microbiol.* **41**:1811–1818.

Spreadbury, C., D. Holden, A. Aufauvre-Brown, B. Bainbridge, and J. Cohen. 1993. Detection of *Aspergillus fumigatus* by polymerase chain reaction. *J. Clin. Microbiol.* **31**:615–621. (Erratum, **32**:2039, 1994.)

Suarez, F., O. Lortholary, S. Buland, D. Ghez, P. Berche, and M. E. Bougnoux. 2007. Use of large serum volume for RT PCR assay improves early diagnosis of invasive aspergillosis in high-risk adult haematology patients, abstr. M-562. *Abstr. 47th Intersci. Conf. Antimicrob. Agents Chemother.*, 17 to 20 September 2007.

Tang, C. M., D. W. Holden, A. Aufauvre-Brown, and J. Cohen. 1993. The detection of *Aspergillus* spp. by the polymerase chain reaction and its evaluation in bronchoalveolar lavage fluid. *Am. Rev. Respir. Dis.* **148:**1313–1317.

Tarai, B., A. Gupta, P. Ray, M. R. Shivaprakash, and A. Chakrabarti. 2006. Polymerase chain reaction for early diagnosis of post-operative fungal endophthalmitis. *Indian J. Med. Res.* **123:**671–678.

Tuon, F. F. 2007. A systematic literature review on the diagnosis of invasive aspergillosis using the polymerase chain reaction (PCR) from broncholaveolar lavage clinical samples. *Rev. Iberoam. Micol.* **24:**89–94.

Van Burik, J. A. H., R. W. Schreckhise, T. C. White, R. A. Bowden, and D. Myerson. 1998a. Comparison of six extraction techniques for isolation of DNA from filamentous fungi. *Med. Mycol.* **36:**299–303.

Van Burik, J. A., D. Myerson, R. W. Schreckhise, and R. A. Bowden. 1998b. Panfungal PCR assay for detection of fungal infection in human blood specimens. *J. Clin. Microbiol.* **36:**1169–1175.

Vandewoude, K. H., S. I. Blot, P. Depuydt, D. Benoit, W. Temmerman, F. Colardyn, and D. Vogelaers. 2006. Clinical relevance of *Aspergillus* isolation from respiratory tract samples in critically ill patients. *Crit. Care* **10:**R31.

Velegraki, A., S. Polychronopoulou, A. Velegraki, M. Anagostakou, E. C. Alexopoulos, E. Liatsis, B. Kitra, A. Parcharidou, J. Persiteri, V. Papadakis, P. Kalambalikis, R. Sklavou, C. Hazicostantinou, G. Paphitou, C. Petropoulou, G. Arsenis, S. Kritikou, A. Milioni, A. Hatzis, and S. Graphakos. 2007. PCR, galactomannan and beta glucan assays (Fungitell) for screening neonatal and paediatric ICU patients, paediatric BMT and patients with cancer and haematological malignancies for aspergillosis, abstr. 0.06. *Med. Mycol. Suppl.* **19:** S34.

Verweij, P. E. 2005. Advances in diagnostic testing. *Med. Mycol.* **43:** S121–S124.

Verweij, P. E., K. Brinkman, H. P. Kremer, B. J. Kullberg, and J. F. Meis. 1999. *Aspergillus* meningitis: diagnosis by non-culture-based microbiological methods and management. *J. Clin. Microbiol.* **37:** 1186–1189.

Verweij, P. E., J. P. Latgé, A. J. Rijs, W. J. Melchers, B. E. De Pauw, J. A. Hoogkamp-Korstanje, and J. F. Meis. 1995. Comparison of antigen detection and PCR assay using bronchoalveolar lavage fluid for diagnosing invasive pulmonary aspergillosis in patients receiving treatment for hematological malignancies. *J. Clin. Microbiol.* **33:** 3150–3153.

White, L., and R. A. Barnes. 2006. *Aspergillus* PCR: platforms, strengths and weaknesses. *Med. Mycol.* **44:**S191–S198.

White, P. L., C. J. Linton, M. D. Perry, E. M. Johnson, and R. A. Barnes. 2006a. The evolution and evaluation of a whole blood polymerase chain reaction assay for the detection of invasive aspergillosis in hematology patients in a routine clinical setting. *Clin. Infect. Dis.* **42:**479–486.

White, P. L., R. Barton, M. Guiver, C. J. Linton, S. Wilson, M. Smith, B. L. Gomez, M. J. Carr, P. T. Kimmitt, S. Seaton, K. Rajakumar, T. Holyoake, C. C. Kibbler, E. Johnson, R. P. Hobson, B. Jones, and R. A. Barnes. 2006b. A consensus on fungal polymerase chain reaction diagnosis: a United Kingdom-Ireland evaluation of polymerase chain reaction methods for detection of systemic fungal infections. *J. Mol. Diagn.* **8:**376–384.

Williamson, E. C. M. 2001. Molecular approaches to fungal infections in immunocompromised patients. M.D. thesis, University of Bristol, Bristol, United Kingdom.

Williamson, E. C., J. P. Leeming, H. M. Palmer, C. G. Steward, D. Warnock, D. I. Marks, and M. R. Millar. 2000. Diagnosis of invasive aspergillosis in bone marrow transplant recipients by polymerase chain reaction. *Br. J. Haematol.* **108:**132–139.

Willinger, B., A. Obradovic, B. Selitsch, J. Beck-Mannagetta, W. Buzina, H. Braun, P. Apfalter, A. M. Hirschl, A. Makristathis, and M. Rotter. 2003. Detection and identification of fungi from fungus balls of the maxillary sinus by molecular techniques. *J. Clin. Microbiol.* **41:**581–585.

Wu, Y. P., R. Wei, and J. Verhoef. 2003. Real time assay of *Aspergillus* should be used in SARS patients receiving corticosteroids. *BMJ* **327:** 1405.

Yamakami, Y., A. Hashimoto, E. Yamagata, P. Kamberi, R. Karashima, H. Nagai, and M. Nasu. 1998. Evaluation of PCR for detection of DNA specific for *Aspergillus* species in sera of patients with various forms of pulmonary aspergillosis. *J. Clin. Microbiol.* **36:** 3619–3623.

Yamakami, Y., A. Hashimoto, I. Tokimatsu, and M. Nasu. 1996. PCR detection of DNA specific for *Aspergillus* species in serum of patients with invasive aspergillosis. *J. Clin. Microbiol.* **34:**2464–2468.

Zeng, X., F. Kong, C. Halliday, S. Chen, A. Lau, G. Playford, and T. C. Sorrell. 2007. Reverse line blot hybridization assay for the identification of medically important fungi from culture and clinical specimens. *J. Clin. Microbiol.* **45:**2872–2880.

VI. THERAPY

Aspergillus fumigatus and Aspergillosis
Edited by J.-P. Latgé and W. J. Steinbach

Chapter 30

Antifungal Polyenes

ANDREAS H. GROLL AND THOMAS J. WALSH

Based on their broad-spectrum, concentration-dependent fungicidal action in vitro, potent dose-dependent activity in a large number of animal models, and well-documented clinical efficacy, the class of antifungal polyenes remains an important option for management of invasive *Aspergillus* spp. infections. For decades, amphotericin B deoxycholate (DAMB) has been the cornerstone for treatment and prevention of life-threatening fungal infections. Its clinical utility, however, is hampered by dose-dependent renal toxicity, infusion-associated reactions, and thereby, limited therapeutic efficacy. The development of novel, less-toxic lipid-based polyene formulations in the late 1980s and early 1990s was a major advance in the field of antifungal chemotherapy, particularly for patients with invasive aspergillosis. This chapter is devoted to the pharmacology of the antifungal polyenes, with an emphasis on their role in the management of invasive *Aspergillus* infections.

DAMB

Amphotericin B is a naturally occurring antifungal antibiotic which was first isolated in the mid-1950s from a strain of *Streptomyces nodosus* obtained from soil of the Orinoco River valley in Venezuela (Gold et al., 1955). Its chemical structure is that of a polyene macrolide and consists of seven conjugated double bounds, an internal ester, a free carboxyl group, and a glycoside side chain with a primary amino group (Fig. 1). Amphotericin B is amphoteric, possessing one basic and one acidic group, and like all polyenes, has clearly demarcated hydrophilic (polyol) and hydrophobic (polyene) regions (Hamilton-Miller, 1973). It is not orally or intramuscularly absorbed and is virtually insoluble in water. For parenteral use, amphotericin B has been complexed with sodium deoxycholate, forming a micellar solution in aqueous environments (DAMB) (Barton et al., 1958). Methyl esters of amphotericin B, developed and evaluated in the early 1980s, were found to have serious cerebral toxicity and were therefore not further pursued (Groll et al., 1998a).

Mechanism of Action

In common with other polyenes, such as nystatin, amphotericin B primarily acts by binding to ergosterol, the principal sterol in the cell membrane of most fungi. The interaction with ergosterol results in the formation of transmembrane channels, which leads to an efflux of protons and monovalent cations, depolarization of the membrane, and ultimately, cell death (Gale, 1974; Pallacios and Serrano, 1978) (Fig. 2). Although with less avidity, the compound also binds to cholesterol of mammalian membranes, which accounts for most of its toxicity (Medoff and Kobayashi, 1980; Vertut-Croquin et al., 1983). To gain access to ergosterol, amphotericin B must first pass the fungal cell wall, which primarily consists of a rigid mesh of glucans and chitin. It is yet unclear how this is accomplished and whether this transfer may play a role in resistance (Gale, 1986).

A second mechanism of action of amphotericin B may involve oxidative damage of the cell through a cascade of oxidative reactions linked to its own oxidation, with formation of free radicals or an increase in membrane permeability. In addition to its antifungal activity, amphotericin B has immunomodulatory effects on lymphocytes and phagocytic cells that are also related to oxidation-dependent events (Brajtburg et al., 1990; Mozzafarian et al., 1997; Sokol-Anderson et al., 1986; Tohyama et al., 1996; Wilson et al., 1991). To a small extent, amphotericin B also inhibits membrane-associated enzymes, such as proton ATPase in fungal cells and Na^+/K^+-ATPase in mammalian cells (Brajtburg and Bolard, 1996). The polyenes do not penetrate past

Andreas H. Groll • Immunocompromised Host Section, Pediatric Oncology Branch, National Cancer Institute, National Institutes of Health, Bethesda, MD 20892. **Thomas J. Walsh** • Infectious Disease Research Program, Center for Bone Marrow Transplantation and Dept. of Pediatric Hematology/Oncology, University Children's Hospital, 48149 Muenster, Germany.

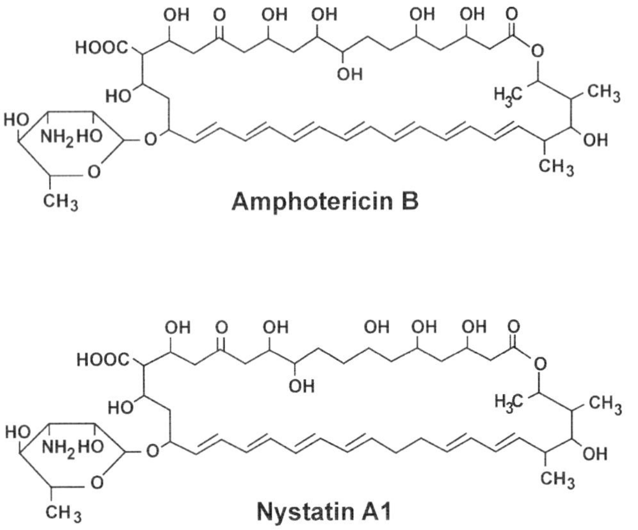

Amphotericin B

Nystatin A1

Figure 1. Structural formulas for amphotericin B (top) and nystatin A1 (bottom).

the fungal cell membrane and do not appear to have direct effects on intermediary metabolism or nucleic acid synthesis (Bates, 1993).

Spectrum of Activity

Amphotericin B possesses a broad spectrum of antifungal activity which includes most fungi pathogenic to humans.

True microbiological resistance to antifungal polyenes has been associated with quantitative or qualitative alterations in the sterol composition of the fungal cell membrane, but it may also be related to phenotype switching, such as increased catalase activity with decreased susceptibility to oxidative damage (Chamilos et al., 2005; Groll et al., 2003; Hamilton-Miller, 1972; Kelly et al., 1984; Pierce et al., 1971; Subden et al., 1978; Woods, 1971). Resistance to amphotericin B remains rare in *Candida* spp. other than *C. lusitaniae*, although the compound appears somewhat less active against *C. guilliermondii, C. tropicalis*, and *C. parapsilosis* (Brajtburg et al., 1990; Pfaller et al., 2007; Walsh et al., 1996). Resistant clinical yeast isolates have been isolated from immunocompromised patients who had received amphotericin B for prolonged time periods (Wingard, 1994).

Aspergillus spp. and other opportunistic molds tend to have more variable susceptibility to amphotericin B (Chamilos et al., 2005). *Aspergillus terreus* (Iwen et al., 1998; Sutton et al., 1999; Walsh et al., 2003), *Aspergillus nidulans* (Kontoyiannis et al., 2002), and some of the emerging pathogens, such as *Fusarium* spp. (Boutati et al., 1997; Reuben et al., 1989), *Scedosporium apiospermum* (Travis et al., 1985; Walsh et al., 1995), *Scedosporium prolificans* (Berenguer et al., 1997; Maertens et al., 2000), and dematiaceous fungi (Groll and Walsh, 2001), may be completely resistant to amphotericin B at concentrations achievable in patients by maximum tol-

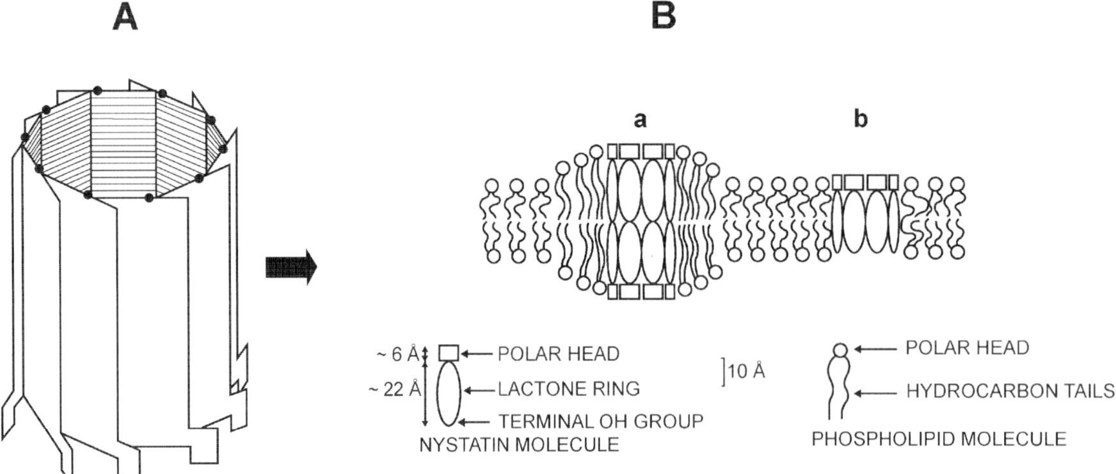

Figure 2. (A) Schematic of a (hypothetical) single-length amphotericin B or nystatin channel. Each polyene molecule is depicted as a plane. The hydrophilic polyhydroxyl polar regions of the molecules face the center of the channel (shaded surface); the exterior is completely nonpolar. The cleft between each of two polyene molecules can accommodate a sterol molecule. The black dots represent the terminal OH group, and the protuberances represent the amino sugar. (B) Diagram of a (hypothetical) double-length (a) and a single-length (b) channel. A normal bilayer is depicted between both channels. Double-length channels are formed from two single-length channels hydrogen bonded in the middle of the membrane through the ring of hydroxyl groups (black dots). Double-length channels only occur when a polyene has access to both sides of the bilayer, i.e., under certain experimental conditions in vitro. (Adapted from Kleinberg et al., 1984.)

erated dosages. Acquisition of secondary resistance is uncommon and has not been a clinical problem (Groll et al., 1998a).

Pharmacodynamics

In time-kill studies, amphotericin B displays concentration-dependent fungicidal activity against susceptible *Candida albicans, Cryptococcus neoformans*, and *Aspergillus fumigatus* (Klepser et al., 1997, 1998; Ralph et al., 1991; Krishnan et al., 2005) (Fig. 3). Prolonged postantifungal effects of amphotericin B of up to 12-h duration have been demonstrated in *C. albicans, C. neoformans*, and *A. fumigatus* (Ernst et al., 2000; Turnidge et al., 1994; Manavathu et al., 2004). Studies in laboratory animals support the concentration-dependent kill kinetics of amphotericin B in vitro (Francis et al., 1994). In neutropenic pharmacokinetic and pharmacodynamic mouse models of disseminated candidiasis and pulmonary aspergillosis, the C_{max}/MIC ratio was the parameter that provided the best correlation with outcome as measured by the residual organismal burden in tissue (Andes et al., 2001; Wiederholt et al., 2006). These laboratory findings indicate that large doses will be most effective and that achievement of optimal peak concentrations is important.

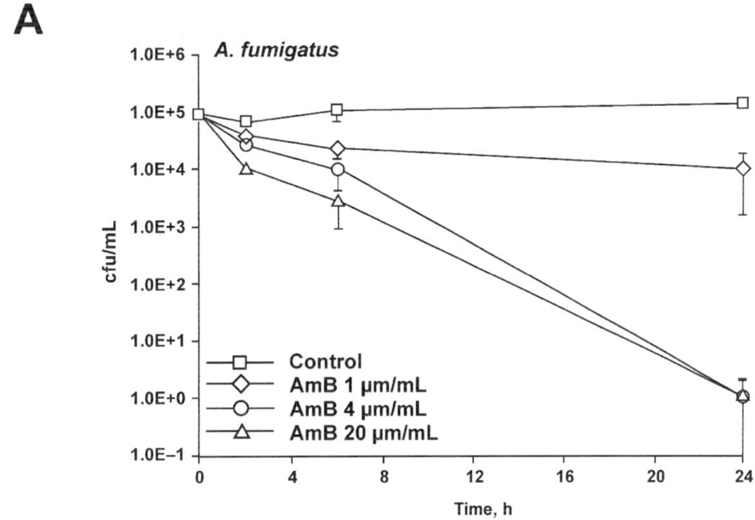

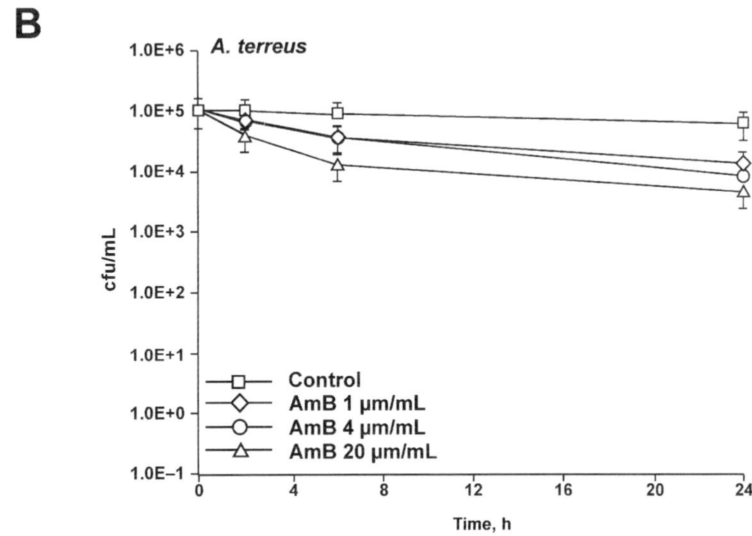

Figure 3. Time kill assay of *A. fumigatus* versus *A. terreus* in antibiotic medium 3. Whereas *A. fumigatus* demonstrated concentration-dependent fungicidal activity of amphotericin B with an approximately 10^5-fold reduction in viable CFU by 24 h at drug concentrations of ≥4 μg/ml, there was less than 1 log kill of *A. terreus* at all concentrations of amphotericin B tested. (Adapted from Walsh et al., 2003.)

Pharmacokinetics

After intravenous administration, amphotericin B rapidly dissociates from its vehicle and becomes highly protein bound before it is distributed into tissues (Bekersky et al., 2002a; Christiansen et al., 1985). The disposition of the compound follows a three-compartment model with rapid initial clearance from plasma followed by a biphasic pattern of elimination with a beta half-life of 24 to 48 h and a prolonged terminal (gamma) half-life of up to 15 days (Atkinson and Bennet, 1978; Bekersky et al., 2002b) (Fig. 4). Tissue levels of amphotericin B in laboratory animals are highest in liver, spleen, bone marrow, kidney, and lung; concentrations in body fluids other than plasma are generally low (Groll et al., 2003; Jagdis et al., 1972; Lawrence et al., 1980). Despite mostly undetectable concentrations in the cerebrospinal fluid and comparatively low concentrations in brain tissue across all species, amphotericin B is effective in the treatment of fungal infections of the central nervous system (CNS) (Bennet et al., 1979). Amphotericin B is not metabolized by the liver but is slowly excreted in unchanged form into urine and bile (Bekersky et al., 2002a, 2002b; Craven et al., 1979; Reynolds et al., 1993); following administration of a single dose of 0.6 mg/kg of body weight to human volunteers, two-thirds of amphotericin B was excreted into urine (21%) and feces (43%), with greater than 90% accounted for in mass balance calculations at 1 week (Bekersky et al., 2002b). Dose adjustment is not necessary in patients with unrelated renal or hepatic dysfunction. Because of its high protein binding, hemodialysis usually does not affect plasma concentrations of amphotericin B (Daneshmend and Warnock, 1983).

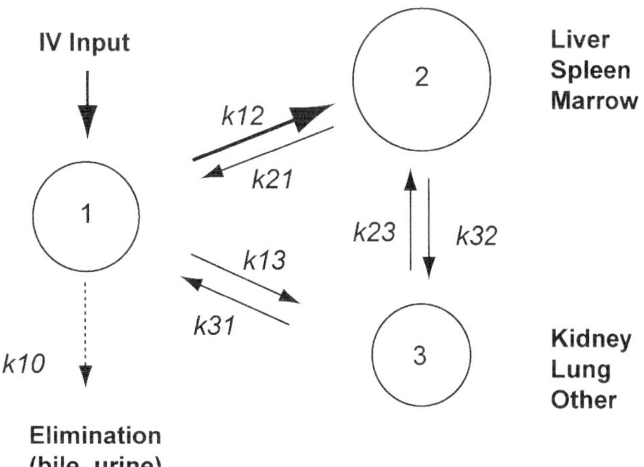

Figure 4. Simplified model of the distribution of amphotericin B after intravenous administration. From the central plasma compartment, independent of the formulation, the drug distributes into three hypothetical peripheral tissue compartments which display quantitatively different uptake. There is a slow redistribution from those peripheral compartments back into plasma. Elimination occurs in unchanged form from the plasma pool via the kidney and bile fluid. HC, high-concentration compartment; IC, intermediate-concentration compartment; LC, low-concentration compartment.

Pharmacokinetic data in pediatric age groups are characterized by high interindividual variability, which may be related to differences in underlying diseases and modes of administration (Chiou et al., 2007). However, infants and children appear to clear the drug from plasma more rapidly than adults, as indicated by a significant negative correlation between age and clearance in two studies (Benson and Nahata, 1989; Koren et al., 1988). Whether the enhanced clearance from the bloodstream has implications for dosing remains unknown. Currently, dosage recommendations for all pediatric age groups do not differ from those for adult patients (Chiou et al., 2007). Amphotericin B crosses the placenta readily and has been the preferred drug in life-threatening fungal infections in pregnant women (Dean et al., 1994).

Adverse Effects

Infusion-related adverse events and nephrotoxicity are major problems of treatment with DAMB and may curtail successful therapy. In a large multicenter study in neutropenic patients receiving empirical antifungal therapy at doses ranging from 0.3 to 1.2 mg/kg/day, 73% of 343 prospectively monitored patients exhibited infusion-related reactions and 34% had increases in serum creatinine during treatment (Walsh et al., 1999a).

Infusion-related reactions (fever, rigor, chills, myalgia, arthralgia, nausea, vomiting, and headache) are not histamine-related events (Cleary et al., 2003) but are thought to be mediated by the release of cytokines from monocytes in response to the drug (Arning et al., 1995). They have a tendency to subside over time despite continuous use of the drug (Walsh et al., 1996) and may be blunted by decreasing the infusion rate (Ellis et al., 1992) or by premedication with corticosteroids, meperidine, or pethidine, and acetaminophen (Walsh et al., 1996). Less common acute adverse effects are hypotension, hypertension, flushing, and vestibular disturbances; bronchospasm, true anaphylaxis, apnea, and convulsions are rare (Table 1) (Sawaya et al., 1995). Cardiac arrhythmias and cardiac arrest may occur due to acute potassium release during rapid infusion (<60 min), particularly in patients with hyperkalemia and/or renal impairment (Butler et al., 1966; Groll et al., 2003).

The hallmarks of amphotericin B-associated nephrotoxicity are azotemia and wasting of potassium and magnesium; tubular acidosis and impaired urinary concentration ability are rarely of clinical significance (Butler et al., 1964; Goldman et al., 2004; Sawaya et al.,

Table 1. Adverse effects of amphotericin B

Organ system	Effects[a]
Skin and appendages	Flushing; skin rash
Gastrointestinal tract	**Nausea, vomiting, anorexia**
Hepatobiliary system	Bilirubinemia; increases in serum levels of alkaline phosphatase and hepatic transaminases
Kidney	**Azotemia**; tubular acidosis, hypostenuria, **hypokalemia, hypomagnesemia**
Bone marrow	**Normochromic, normocytic anemia;** thrombocytopenia
Nervous system	Headache; convulsions; vestibular disturbances; leukencephalopathy
Cardiovascular	Thrombophlebitis; hypotension, hypertension, cardiac arrhythmia, cardiac arrest; Raynaud's phenomenon
Respiratory	Bronchospasm; dyspnea, chest tightness, hypoxemia; respiratory arrest
Mucsuloskeletal	Arthralgia, myalgia, back pain
Immunologic	Anaphylaxis
Other	**Fever, shaking chills, rigor**

[a] Common adverse events are shown in bold.

1995). Relevant electrolyte wasting occurs in approximately 12% of prospectively monitored patients (Walsh et al., 1999a); of note, hypokalemia can be quite refractory to replacement until hypomagnesemia is corrected (Sawaya et al., 1995). Azotemia is common and of considerable clinical impact: of 239 immunosuppressed patients receiving DAMB for suspected or proven aspergillosis for a median duration of treatment of 15 days, the creatinine level doubled in 53% of patients and exceeded 2.5 mg/dl in 29%; 14.5% underwent dialysis, and 60% died. A multivariate Cox proportional hazards analysis showed that patients whose creatinine level exceeded 2.5 mg/dl and allogeneic bone marrow transplant patients were at greatest risk for requiring hemodialysis. Use of hemodialysis, duration of amphotericin B use, and concomitant use of nephrotoxic agents were associated with greater risk of death (Wingard et al., 1999). Of note, nephrotoxicity is somewhat less frequent (28 to 34%) in the setting of empirical antifungal therapy with shorter treatment periods, as evidenced in large prospective (Walsh et al., 1999a) and retrospective (Harbarth et al., 2001) studies. It is unclear whether the rate of azotemia is lower in children than in adults; in contemporary series in premature neonates, the incidence of azotemia ranged from 0 to 15%, indicating that the compound is much better tolerated in this setting than reported earlier during its use (Chiou et al., 2007).

While treatment with DAMB has the potential to lead to renal failure and dialysis (Wingard et al., 1999), azotemia often stabilizes on therapy and is usually reversible after discontinuation of the drug (Walsh et al., 1996). Avoiding concomitant nephrotoxic agents, appropriate hydration, and normal saline loading (10 to 15 ml NaCl/kg/day) may greatly lessen the likelihood and severity of azotemia associated with amphotericin B therapy (Arning and Scharf, 1989; Heidermann et al., 1983; Girmenia et al., 2005).

Other potentially relevant side effects associated with the use of DAMB include a demyelinating encephalopathy in bone marrow transplant patients conditioned with total body irradiation and/or receiving concurrent cyclosporine therapy (Mott et al., 1995) and a normocytic, normochromic anemia after chronic administration (Daneshmend and Warnock, 1983). DAMB is locally irritating; hence, a central line should be used for infusion, and local instillations should only be considered in conjunction with expert consultation.

Drug Interactions

Drug interactions due to shared metabolic pathways are unknown for amphotericin B. Hypokalemia may be aggravated by corticosteroids and, in turn, can potentiate digoxin toxicity, cause rhabdomyolysis, and enhance the effects of nondepolarizing muscle relaxants (Daneshmand and Warnock, 1983). Similarly, hypomagnesemia may become especially profound in cancer patients with platinum-associated nephropathy. Impairment of glomerular filtration by amphotericin B may enhance plasma levels and thereby toxicity of many renally cleared drugs, such as aminoglycosides, glycopeptides, calcineurin inhibitors, and fluorocytosine (Groll et al., 1998a). Finally, the simultaneous infusion of granulocytes has been associated with acute pulmonary reactions (Wright et al., 1981) and should therefore be avoided.

Clinical Indications

With the advent of new antifungal agents and following the completion of pivotal phase III clinical trials, few indications are left in developed countries for antifungal treatment of opportunistic mycoses with DAMB. These may include induction therapy for cryptococcal meningitis, severe forms of the endemic mycoses, and perhaps uncomplicated candidemia in patients without renal impairment. As a principle, treatment should be started at the full target dosage of 0.7 to 1.0 mg/kg/day with careful bedside monitoring during the first hour of infusion to allow for prompt intervention for infusion-related reactions (Groll et al., 2003).

Despite its decade-long use, responses to treatment with DAMB for invasive aspergillosis have been difficult to assess due to the lack of clinical trials and uniform definitions. In a retrospective cohort analysis in 261 immunocompromised patients with proven or probable invasive aspergillosis who received DAMB at the recommended dosage range of 1.0 to 1.5 mg/kg/day, the response rate was 23% and the survival rate was 28% (White et al., 1997). Likewise, in a large literature analysis, patients who received treatment with DAMB had a case fatality rate of 65% (Lin et al., 2001). As expected, outcomes in the two conducted randomized, prospective clinical trials that compared DAMB with amphotericin B colloidal dispersion (ABCD) (Bowden et al., 2002) and voriconazole (Herbrecht et al., 2002) were slightly better: response rates ranged between 30 and 50% and overall mortality rates were between 42 and 45% at 3 months. While ABCD had similar efficacy to DAMB, it did not receive approval as a first-line agent based on a higher rate of infusion-associated reactions, and voriconazole has become the new standard of care in first-line treatment of invasive aspergillosis: initial therapy with voriconazole led to better responses (52.8 versus 31.6%) and improved survival (70.8% versus 57.9%) and resulted in fewer severe side effects than the standard approach of initial therapy with amphotericin B (Herbrecht et al., 2002). Based on this pivotal trial, the overall improved safety profile of the lipid formulations of amphotericin B, and further alternative drugs, there is virtually no situation left in which DAMB could be indicated for treatment of invasive aspergillosis.

DAMB IN LIPID EMULSIONS

Several small studies on the use of DAMB mixed with fat emulsions for parenteral nutrition have been published. This approach is founded on experimental work performed by Kirsh et al. (1988) and corroborated by Chavanet et al. (1992) which demonstrated reduced in vitro and in vivo toxicity of this formulation without loss of antifungal efficacy in models of systemic candidiasis.

In patients, doses of 0.7 to 2.2 mg of DAMB/kg/day, diluted in 20% parenteral lipid emulsion to final concentrations of 0.16 to 2.0 mg amphotericin B/ml, have been administered at different infusion rates. In comparison to DAMB, lipid emulsion-based DAMB exhibited lower C_{max} and area under the time-concentration curve (AUC) values with enhanced clearance and a larger V_d, suggesting more rapid uptake by the mononuclear phagocytic system (MPS) (Ayjestaran et al., 1996; Chavanet et al., 1992; Heinemann et al., 1997c). In comparison to DAMB, the lipid emulsion preparation was associated with reduced infusion-related and renal toxicity in the majority of reports (Caillot et al., 1994; Chavanet et al., 1992; Moreau et al., 1992; Pascual et al., 1995; Sorkine et al., 1996). Other investigators, however, found no evidence for an improved toxicity profile but higher renal (Joly et al., 1996a) and pulmonary (Schoeffski et al., 1996) toxicity, and the physicochemical behavior of DAMB in the fat emulsion is highly controversial (Lopez et al., 1996; Ranchere et al., 1996; Shadkhan et al., 1996; Sievers et al., 1996; Trissel, 1995).

The majority of comparative studies were conducted in neutropenic patients requiring empirical antifungal treatment; their sample size, however, was too small to allow for any comparison of efficacy (Cailllot et al., 1994; Moreau et al., 1992; Pascual et al., 1995; Schoeffski et al., 1996). In 30 critically ill patients with suspected or proven *Candida* infections (Sorkine et al., 1996) and in 10 human immunodeficiency virus (HIV)-infected individuals with oral candidiasis (Chavanet et al., 1992), DAMB in lipid emulsion appeared as effective as conventionally prepared DAMB. A noncomparative trial in 36 neutropenic cancer patients with documented candidemia reported a response rate of 75% and attributable mortality of 25% (Caillot et al., *Abstr. 36th Int. Conf. Antimicrob. Agents Chemother.*, abstr. J48, 1996). Treatment results in HIV-associated cryptococcal meningitis did not show any apparent advantage over conventional DAMB (Joly et al., 1996a, 1996b).

It is not clear whether DAMB mixed into lipid emulsion has an increased therapeutic index. Methods of drug preparation have not been standardized, and the formulation may be unstable. In essence, administration of DAMB mixed into lipid emulsions is an unapproved drug treatment and should not be used.

LIPID FORMULATIONS OF AMPHOTERICIN B

While the large multilamellar liposomal formulation of amphotericin B developed by Lopez-Berestein and coworkers and a small unilamellar liposomal formulation studied at the Institute Jules Bordet in Brussels, Belgium, were eventually not commercially developed (Groll et al., 1998a), three novel and distinct lipid formulations were approved throughout the world during the 1990s: ABCD (Amphocil or Amphotec), amphotericin B lipid complex (ABLC; Abelcet), and a small unilamellar vesicle liposomal formulation (LAMB; AmBisome). Because of their reduced nephrotoxicity in comparison to DAMB, these compounds allow for the safer delivery of the parent with similar or greater antifungal efficacy (Hiemenz and Walsh, 1996; Wong-Beringer et al., 1997; Groll et al., 1998b).

Principles of Drug Distribution

The vehicles of the lipid formulations of amphotericin B are composed of biodegradable lipid molecules that are composed of hydrophilic heads and hydrophobic tails. These amphiphilic lipid molecules can either directly complex with amphotericin B in a 1:1 molar ratio to form a colloidal dispersion of microscopic disc-shaped particles, or they can arrange themselves in water into bilayered membranes to form uni- or multilamellar spherical lipid vesicles (liposomes) or membrane-like lipid complexes in which amphotericin B can be integrated at varying molar ratios. In either circumstance, when incorporated into these water-soluble carriers, amphotericin B becomes soluble in aqueous environments and available for distribution in the body (Braijtburg and Bolard, 1996; Janknegt et al., 1992; Hiemenz and Walsh 1996).

Physicochemical characteristics such as size, lipid composition, molar ratio of lipids, bilayer rigidity, and electrical charge all play an important role (Janknegt et al., 1992). After intravenous injection, the lipid formulations may either undergo phase transition in the blood or be taken up after variable periods of time by cells of the MPS that lines the bloodstream (i.e., macrophages in liver, spleen, and lungs) or by monocytes targeted for sites of inflammation (Ostro and Cullis, 1989; Jangknegt et al., 1992). Particle size is a pivotal determinant in the disposition process. For example, small unilamellar vesicles are slowly taken up by the MPS, have a long circulation half-life, and may be taken up by non-reticuloendothelial tissues. In contrast, large vesicles and membrane-like complexes are rapidly and efficiently taken up by the MPS, and they tend to have a short half-life in the blood (Hwang et al., 1987; Senior, 1987). Independent of the formulation, distribution into the MPS is followed by a slow redistribution of the parent compound into the plasma pool. Of note, the uptake by the MPS is a saturable process, and the half-lives of some lipid formulations may depend on the total dose of the compound administered (Ostro and Cullis, 1989) (Fig. 4).

Pharmacokinetics and Pharmacodynamics

Each of the lipid formulations of amphotericin B possesses distinct physicochemical and pharmacokinetic properties (Table 2; Fig. 5). All three, however, preferentially distribute to organs of the MPS. While the micellar dispersion of ABCD behaves kinetically very similar to DAMB, the small unilamellar liposomal preparation has a prolonged circulation time in plasma, achieves strikingly high peak plasma drug concentrations and AUC values, and is only slowly taken up by the MPS. In contrast, the large ribbon-like aggregates of ABLC are efficiently opsonized by plasma proteins and rapidly taken up by the MPS, resulting in lower peak plasma and AUC values. The multilamellar liposomal formulation of nystatin, for comparison, displays another disposition pattern, with high peak plasma drug concentrations but rapid elimination from the bloodstream (Groll et al., 1998b, 2000a; Hiemenz and Walsh, 1997).

At their target, the lipid formulations appear to act in a more selective fashion than DAMB: in vitro studies have shown reduced lysis of human erythrocytes or damage to tubular renal cells in comparison to DAMB but retained activity against fungal cells (Braijtburg and Bolard, 1996). These differences may be due to a selective drug transfer by an enhanced physicochemical interaction of the carrier with fungal cell membranes (Adler-Moore et al., 1993) or by the action of fungal or inflammatory cell-derived lipases liberating amphotericin B from its carrier (Perkins et al., 1992; Swenson et al., 1998). Additionally, differences in both the dissociation of free amphotericin B from its carrier and the degree of aggregation of free amphotericin B molecules have been proposed to account for the observed selectivity of LAMB (Braijtburg and Bolard, 1996).

Whether and how the distinct physicochemical and pharmacokinetic features of each formulation translate

Table 2. Physicochemical properties of the four available amphotericin B formulations and of liposomal nystatin and multidose pharmacokinetics after administration of doses considered equivalent[a]

Parameter	DAMB	ABCD	ABLC	LAMB	LNYS
Lipids (ratio)	Deoxycholate	Chol-sulfate	DMPC-DMPG (7:3)	HPC-Chol-DSPG (2:1:0.8)	DMPC-DMPG (7:3)
Mol% polyene	34	50	50	10	10
Configuration	Micellar	Disc-like	Ribbon-like	SUV	MLV
Particle diameter (μm)	<0.4	0.11–0.14	1.6–11	0.08	0.3
Std. dose (mg/kg)	1	5	5	5	4
Mean C_{max} (μg/ml)	2.9	3.1	1.7	58	24
Mean AUC_{0-24} (μg/ml·h)	36	43	14	713	80
Mean V_d (liters/kg)	1.1	4.3	131	0.22	0.17
Mean Clt (liters/h/kg)	0.028	0.117	0.476	0.017	0.051

[a]Abbreviations: Chol, cholesterol; HPC, hydrogenated phosphatidylcholine; SUV, small unilamellar vesicles; MLV, multilamellar vesicles (true liposomes); Clt, total clearance. Data were compiled from Cossum et al. (1996) and Groll et al. (2003). Note that data were obtained in different (adult) patient populations and after different rates of infusion.

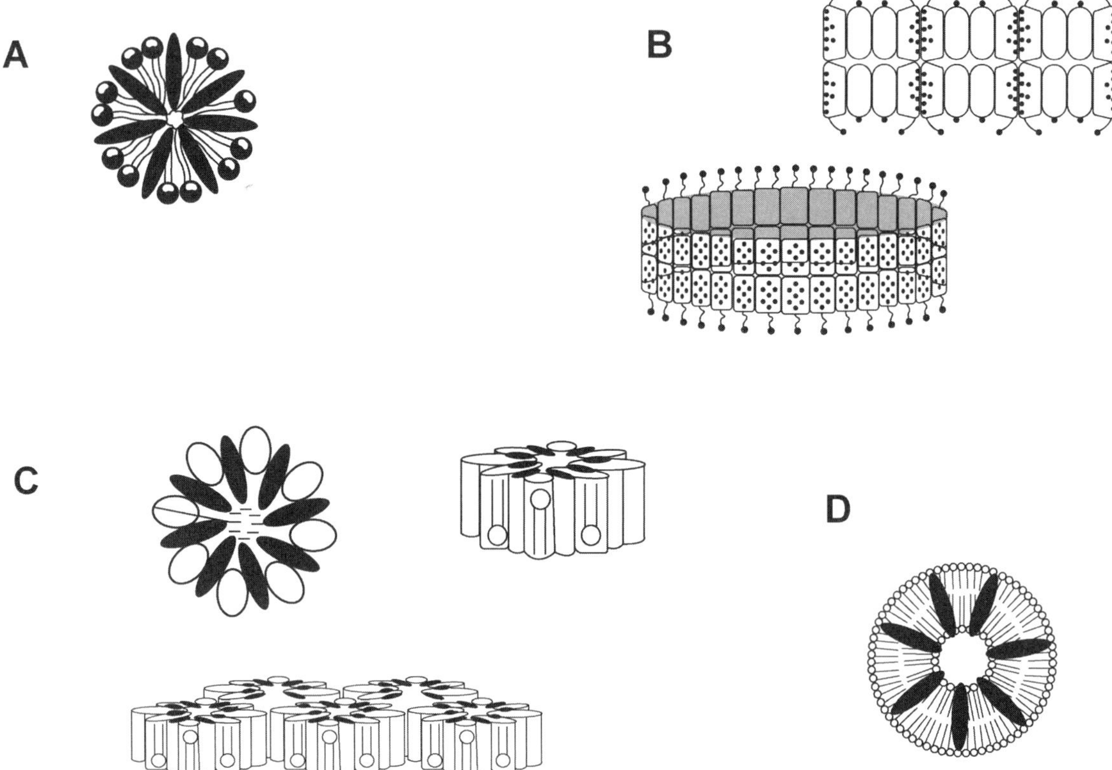

Figure 5. Schematic of the molecular composition of amphotericin B formulations. (A) DAMB forms aggregates of mixed micelles containing amphotericin B and deoxycholate. (B) ABCD forms disk-like colloidal structures composed of amphotericin B and cholesterylsulfate. (C) ABLC is composed of DMPC and DMPG in a 7:3 molar ratio complexed with amphotericin B and forms large ribbon-like structures. (D) LAMB consists of small, unilamellar vesicles made up of hydrogenated soy phosphatidyl choline and disteaoryl phosphatidyl glycerol stabilized by cholesterol and combined with amphotericin B. Liposomal nystatin (not shown) consists of multilamellar liposomes composed of DMPC and DMPG combined with nystatin.

into different pharmacodynamic properties in target sites of invasive *Aspergillus* infections in vivo are mostly unknown. In preclinical single- and multidose distribution studies in rodents spanning the entire dosing interval and using equimolar doses of 1 mg/kg of amphotericin B, lung levels achieved by ABCD and LAMB were lower and those achieved by ABLC were similar or slightly higher than those obtained with DAMB. However, after multiple dosing at safely tolerated doses that were 5-fold higher (ABCD and LAMB) to 10-fold higher (ABLC), drug accumulation in the lung clearly exceeded that achieved by 1 mg/kg DAMB (Clark et al., 1991; Fielding et al., 1991; Olsen et al., 1991; Proffitt et al., 1991; Wang et al., 1995). Notably, these differences in lung distribution are consistent with the results of pulmonary infection models, in which the lipid formulations required at least fivefold-higher doses to produce equivalent or superior reductions of the pulmonary tissue burden than standard doses of DAMB (Fig. 6) (Allende et al., 1994; Clemons et al., 1991, 1993; Francis et al., 1994) and with pharmacokinetic-pharmacodynamic assessments in a murine *Candida* kidney target model (Andes et al., 2006). Experimental investigation of the compartmentalized intrapulmonary pharmacokinetics revealed strikingly different patterns among the four formulations at therapeutic dosages (DAMB at 1 mg/kg, all others at 5 mg/kg). Whereas the disposition of ABCD was similar to that of DAMB, ABLC showed prominent accumulation in lung tissue and pulmonary alveolar macrophages, while LAMB achieved the highest concentrations in epithelial lining fluid (Groll et al., 2006). In a murine model of invasive pulmonary aspergillosis, ABLC appeared to deliver amphotericin B to the lung more rapidly than LAMB at 5 mg/kg, resulting in faster *Aspergillus* clearance; however, this difference was equalized following further dose escalation to 10 mg/kg (Lewis et al., 2007). The clinical relevance of these differences, however, remains to be explored. Finally, considering the CNS as a target site of invasive aspergillosis, in a pharmacokinetic/pharmacodynamic rabbit model of hematogenous *C. albicans* meningoencephalitis, there was a strong, concentration- and time-dependent correlation of plasma drug exposure with antifungal efficacy, indicat-

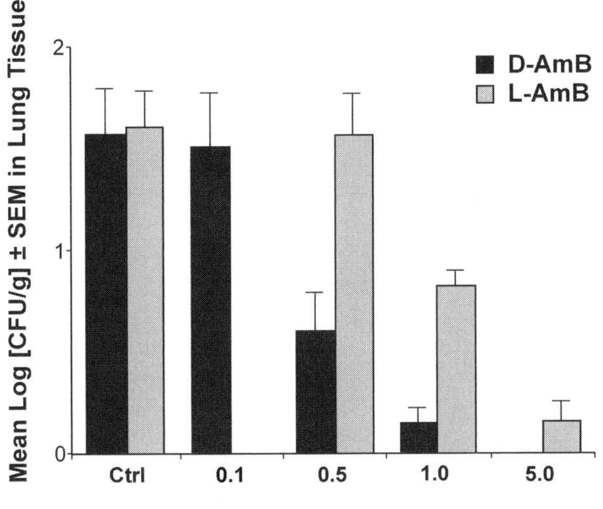

Figure 6. Response of primary pulmonary aspergillosis in rabbits to antifungal therapy with DAMB or LAMB as measured by the residual burden of *A. fumigatus* in lung tissue. Note that at similar doses LAMB was not as effective as DAMB and that similar efficacy was achieved only at higher dosages. (Modified from Francis et al., 1994.)

ing a potential advantage of LAMB for treatment of CNS infections (Groll et al., 2000b).

Based on the predominant importance of the drug carrier for antifungal efficacy, there is no role of in vitro testing of the lipid formulations; instead, the in vitro activity of the parent is used according to current CLSI recommendations. This is well exemplified by pharmacodynamic correlations in experimental *Candida* meningoencephalitis: while pharmacodynamic parameters derived from the MIC of free amphotericin B are highly predictive of antifungal efficacy, parameters derived from MICs of individual formulations are not predictive (Groll et al., 2000b). However, the principal antifungal efficacy of the lipid formulations has been demonstrated in several models of invasive fungal infections in both normal and immunocompromised animals, and these have been reviewed in detail elsewhere (Adler-Moore and Proffitt, 2002; Braijtburg and Bolard, 1996; Clemons and Stevens, 2005; Groll et al., 1998a; Hiemenz and Walsh, 1996; Janknegt et al, 1992; Patterson 2005).

Safety and Antifungal Efficacy

Safety and antifungal efficacy of ABCD, ABLC, and LAMB have been demonstrated in an array of phase II and III clinical trials in immunocompromised, mostly adult, patients with a wide spectrum of underlying disorders (Cornely et al., 2007; Kuse et al., 2007; Ringden et al., 1993; Walsh et al., 1998b; White et al., 1998). The overall response rates in these trials ranged from 53 to 84% in patients with invasive candidiasis and 34 to 59% in patients with presumed or documented invasive aspergillosis. A few randomized, controlled trials have been completed in which one of the new formulations has been compared with DAMB. These studies have consistently shown at least equivalent therapeutic efficacy and reduced nephrotoxicity of the investigated lipid formulation. Infusion-related side effects of fever, chills, and rigor appear to be less frequent only with LAMB (Bowden et al., 2002; Fleming et al., 2001; Walsh et al., 1999a; Anaissie et al., *Abstr. 35th Int. Conf. Antimicrob. Agents Chemother.*, abstr. LM21, 1995). Individual cases of substernal chest discomfort, respiratory distress, and of sharp flank pain have been noted during or following infusion of ABLC and LAMB (Garmacho-Montero et al., 1998; Johnson et al., 1998b; Roden et al., 2002), and in comparative studies, hypoxic episodes associated with fever and chills were more frequent in ABCD recipients than in recipients of the deoxycholate formulation (Bowden et al., 2002; White et al., 1998). Increases in serum bilirubin and alkaline phosphatase have been observed with all three formulations and also increases in serum transaminases with LAMB. However, no case of fatal liver disease has occurred (Fisher et al., 2005; Groll et al., 2003). There have been observations of clinical pancreatitis in association with LAMB (Walsh et al., 2001; Stuecklin-Utsch et al., 2002) and cases of pseudo-hyperphosphatemia (Lane et al., 2008). Of note, strict adherence to the recommended infusion time and careful handling of amphotericin B remain critical: a cardiopulmonary arrest after too rapid an infusion of ABLC (Barcia, 1998) and a dispensing and administration error that caused DAMB to be given instead of LAMB (Mohr et al., 2005) have been reported; both events were fatal.

ABCD

ABCD is a complex of amphotericin B and sodium cholesteryl sulfate in an approximate 1:1 molar ratio and forms disk-like colloidal structures approximately 122 nm in diameter and a thickness of 4 nm upon dissolution (Table 2; Fig. 5) (Guo et al., 1991). The compound's in vitro activity appears essentially similar to that of DAMB (Hanson et al., 1992).

Pharmacokinetics

Population-based multidose pharmacokinetic data in 51 adult and pediatric bone marrow transplant patients with systemic fungal infections and treated with ABCD best fit into a two-compartment open model of distribution. Over the investigated dose range of 0.5 to 8.0 mg/kg/day, ABCD had nonlinear pharmacokinetics with less than proportional increases in C_{max} and AUC_{0-24} values. Both plasma drug clearance and the volume of distribution increased with escalating doses; the overall average terminal elimination half-life was 29 h (Table 2). Estimated parameters in five children <13 years of age enrolled at the 7.0- and 7.5-mg/kg/day dose levels were not significantly different from those obtained in a dose-matched cohort of adult patients; under steady-state conditions, the mean AUC was 7.10 μg/ml·h (normalized to a 1-mg/kg/day dose), the mean V_d was 4.57 liters/kg, and the mean total clearance was 0.144 liters/h/kg (Amantea et al., 1995).

Safety and Antifungal Efficacy

A phase I study in 75 bone marrow transplant patients with documented invasive fungal infection established a maximally tolerated dose (MTD) of 7.5 mg/kg/day; dose-limiting events were rigors, chills, and hypotension in three of five patients treated at 8 mg/kg/day. Infusion-related toxicities occurred in 68% of all patients, but there was no appreciable renal toxicity (Bowden et al., 1996). The combined safety analysis from five open-label phase I/II studies that included 572, mostly pretreated patients receiving ABCD at a median daily dose of 3.8 mg/kg (range, 0.1 to 9.1 mg/kg) for a median of 16 days (range, 1 to 409 days) reported infusion-related adverse effects with at least one infusion in 62% of patients. Mean serum creatinine levels at the end of therapy were not different from those at baseline. Adverse effects attributable to ABCD requiring the discontinuation of therapy occurred in 70 patients (12.2%). The most frequent of these were infusion-related adverse events (5.4%), followed by elevated creatinine levels (3.3%) and abnormal liver function tests (1.4%) (Herbrecht, 1997). A subanalysis has been reported for 133 renally compromised patients who had experienced nephrotoxicity while being treated with DAMB or who had preexisting renal disease. Treatment with ABCD was administered at a median daily dose of 3.8 mg/kg (range, 0.1 to 5.5 mg/kg) for a median of 14 days (range, 1 to 204 days). Infusion-related events occurred at least once in 74 patients (54%), but the proportion of patients experiencing such events dropped from 43% on day 1 to 18% on day 7 of treatment. Six patients discontinued treatment prematurely because of renal toxicity (4.5%). However, although increases in serum creatinine levels occurred, there was no overall trend for increased creatinine levels during treatment, even in patients receiving concurrent cyclosporine, and there was no difference in serum potassium and serum magnesium between baseline and end-of-treatment values (Anaissie et al., 1998).

Across dose levels (0.5 to 8 mg/kg/day) and infection types, complete or partial responses of suspected or documented invasive fungal infections to ABCD therapy have been reported in approximately 50% of patients (Anaissie et al., 1998; Bowden et al., 1996; Oppenheim et al., 1995). The majority of patients enrolled in those trials had a malignancy as an underlying disorder or had undergone bone marrow transplantation; most had either been refractory or intolerant to DAMB or had preexisting renal impairment. Retrospective, pooled subgroup analyses of patients treated with ABCD on open-label phase I/II protocols have documented response rates of 26% (5 of 19) in disseminated candidiasis, 70% (37 of 53) in candidemia (Noskin et al., 1998), and 60% (12 of 20) in zygomycosis (Herbrecht et al., 2001). ABCD has been used with success in a few patients with cryptococcal meningitis and coccidioidomycosis (Anaissie et al., 1998; Hostetter et al., 1992; Oppenheim et al., 1995; Valero et al., 1995) and was highly successful as a short-term treatment of Mediterranean and Indian visceral leishmaniasis (Gaeta et al., 2000; Sundar et al., 2006).

Treatment of Invasive Aspergillosis

In order to assess the efficacy of ABCD in invasive *Aspergillus* infections, 82 patients with proven or probable aspergillosis who were treated in clinical trials with ABCD were compared retrospectively with 261 patients treated with amphotericin B at six cancer or transplant centers (Table 3). ABCD was given in a daily dose of 0.5 to 8 mg/kg with a median duration of treatment of 24 days and a median cumulative dose of 5.9 mg/kg. Response rates (48.8%) and survival rates (50%) among ABCD-treated patients were higher than those among amphotericin B-treated patients (23.4% and 28.4%, respectively; $P < .001$). Renal dysfunction developed less frequently in ABCD recipients than in amphotericin B recipients (8.2% versus 43.1%, respectively; $P < .001$) (White et al., 1997). Based on these data, a randomized, double-blind, multicenter trial was initiated that compared ABCD (6 mg/kg/day) with DAMB (1.0 to 1.5 mg/kg/day) for first-line treatment of invasive aspergillosis in 174 patients. The median duration of therapy was 13.0 days (range, 1 to 357 days) in the ABCD cohort and 14.5 days (range, 1 to 87 days) in the DAMB group. For evaluable patients ($n = 103$) in the ABCD and DAMB treatment groups, respective rates of therapeutic response (52% versus 51%; $P = 1.0$), mortality (36% versus 45%; $P = 0.4$), and death due to fungal infection (32% versus 26%; $P = 0.7$) were similar. Renal toxicity was lower (25% versus 49%; $P = 0.002$) and the median time to onset of nephrotoxicity was longer (301 versus 22 days; $P < 0.001$) in patients treated with ABCD. Rates of drug-related toxicity in patients receiving ABCD and DAMB were, respectively, 53% versus 30% (chills), 27% versus 16% (fever), 1% versus 4%

Table 3. Summary of clinical efficacy of amphotericin B lipid formulations against invasive *Aspergillus* infections

Formulation and study design (reference)	End points of efficacy	Main results
ABCD		
Retrospective comparison of 82 patients with proven/probable invasive aspergillosis treated with ABCD in open-label salvage trials and 261 patients treated with DAMB at six cancer or transplant centers (White et al., 1997)	Complete/partial responses and stable disease in patients receiving treatment for ≥7 days; overall mortality at day +120	Response rates (48.8 vs 23.4%) and survival rates (50 vs 28.4%) among ABCD-treated patients were higher than among DAMB-treated patients, while renal dysfunction was less common (8.2 vs 43.1%, respectively).
Randomized, double-blind multicenter study comparing ABCD at 6 mg/kg/day vs DAMB at 1.0–1.5 mg/kg/day for first-line treatment of invasive aspergillosis in 174 patients (Bowden et al., 2002)	Complete/partial responses and stable disease in patients receiving treatment for ≥7 days; overall mortality at day +84	Response rates in evaluable patients (n = 103) and overall mortality were similar in ABCD- and DAMB-treated patients (52 vs. 51% and 36 vs. 45%, respectively), while renal toxicity was less frequent; infusion-related reactions were more common in ABCD-treated patients.
ABLC		
Open-label, noncomparative salvage study of ABLC in adult and pediatric patients with invasive mycoses, including 130 patients with probable/proven invasive aspergillosis (Walsh et al., 1998b; Hiemenz et al., 1995)	Complete/partial responses at end of therapy in patients receiving treatment for ≥5 days	Response rates were 42% in entire population of 130 evaluable patients and 56% in subset of 25 evaluable pediatric patients; these response rates were higher than those in a historical control group of 60 patients treated with DAMB.
Phase IV, prospective postapproval registry of ABLC treatment in a total of 3,514 patients, including 398 patients with proven or suspected invasive aspergillosis (Chandresakar and Ito, 2005)	Complete/partial responses at end of therapy in patients receiving treatment for ≥4 doses	Response rates were 44% in entire population of patients; response rates were similar for patients who received ABLC as first-line (n = 139) or as second-line therapy (n = 216); response rates in *A. terreus* infections were 37%.
LAMB		
Open, randomized multicenter trial comparing efficacy of two doses of LAMB (1 and 4 mg/kg/day) for treatment of proven or probable invasive aspergillosis in 120 neutropenic patients (Ellis et al., 1998)	Complete/partial responses and stable disease at end of therapy in patients receiving ≥1 dose; overall mortality at 2 mos post-start of therapy	Response rates were 64% in 41 eligible LAMB 1 recipients and 48% in 46 eligible LAMB 4 recipients; overall mortality was 42 and 49% in the two cohorts; rate of patients with adverse events was slightly higher in the LAMB 4 cohort.
Double-blind, randomized trial in 201 patients with proven or probable invasive mold infection comparing LAMB as first-line therapy at either 3 or 10 mg/kg/day for 14 days, followed by 3 mg/kg/day (Cornely et al., 2007)	Complete/partial responses at end of study drug treatment in patients receiving ≥1 dose; overall mortality at 12 wks post-start of therapy	Complete or partial response was achieved in 50 and 46% of patients in the 3- and 10-mg/kg groups, respectively; mortality at 12 wks was 28 and 41% in the two treatment arms; rates of nephrotoxicity and hypokalemia were significantly higher in the high-dose group.

[a] Note that there were differences in disease definitions and outcome assessment methods across the clinical trials.

(hypoxia), and 22% versus 24% (toxicity requiring study drug discontinuation). These data demonstrate equivalent efficacy and superior renal safety but more frequent infusion-related chills and fever following treatment with ABCD (Bowden et al., 2002).

Of note, an identical pattern of equivalent efficacy and reduced nephrotoxicity but increased infusion-related reactions was found in a large double-blind, randomized trial comparing ABCD with DAMB for empirical antifungal treatment in febrile neutropenic patients (White et al., 1998). Similarly, infusion-related reactions led to the early termination of a randomized clinical trial that compared ABCD with fluconazole for the prevention of fungal infections in neutropenic patients with hematological malignancies (Timmers et al., 2000).

Clinical Indications

The available clinical data indicate that ABCD is less nephrotoxic than DAMB and effective against invasive aspergillosis. However, the use of this agent is compromised by occurrence of infusion-related reactions that were more frequent and severe than with DAMB in randomized trials. The FDA-approved indication is for treatment of probable or proven invasive aspergillosis refractory to or intolerant of DAMB, and the approved dosage is 3 to 4 mg/kg/day, administered over 2 h. Similar to DAMB, treatment should be started with the full target dose under careful clinical monitoring, and premedication is advised.

In patients with renal dysfunction, dose reduction is generally not required, unless there is evidence of relevant drug-induced nephrotoxicity during treatment. Experience with ABCD in patients undergoing dialysis and in patients with significant hepatic dysfunction has not been reported. A greater number of children ≥3 months of age and a limited number of very low birth weight infants have been treated with ABCD, but no formal phase I/II pediatric trial has been conducted. Overall, the data indicate no fundamental differences in disposition, safety, and antifungal efficacy of ABCD in comparison with adult populations (Chiou et al., 2007; Sandler et al., 2000). There is no experience with the use of ABCD in pregnant women.

ABLC

ABLC is composed of dimyristoyl phosphatidylcholine and dimyristoyl phospatidylglycerol (DMPC and DMPG) in a 7:3 molar ratio complexed with amphotericin B in a 1:1 molar ratio of lipid to amphotericin B and forms large ribbon-like structures of 1.6 to 11 μm in the greatest diameter (Table 2; Fig. 5). ABLC has fungicidal activity in vitro against a variety of clinically relevant molds and yeasts (Perkins et al., 1992; Johnson et al., 1998a).

Pharmacokinetics

The plasma pharmacokinetics of ABLC in adults are nonlinear: V_d and clearance (Cl) increase with increasing doses, resulting in less than proportional increases in plasma drug concentrations. In both adults and children, when compared to equivalent doses of DAMB, peak plasma drug levels and AUC values after administration of ABLC are lower, clearance from blood is more rapid, and the V_d is larger (Table 2) (Adedoyin et al., 1997; Kan et al., 1991; Walsh et al., 1997; Wuerthwein et al., 2005). The rapid clearance from blood and the large V_d both reflect rapid and extensive uptake by the MPS. In a cohort of six children treated with ABLC for hepatosplenic candidiasis at 2.5 mg/kg/day, steady state was achieved by day 7; following the final dose, the mean AUC_{0-24} was 11.9 μg·h/ml, mean C_{max} in whole blood was 1.69 μg/ml, and clearance was 0.218 liters/kg/h (Walsh et al., 1997). A population-based pharmacokinetic study in premature neonates with invasive candidiasis showed that the disposition of ABLC in neonates was similar to that observed in other age groups; weight was the only factor that influenced clearance (Wuerthwein et al., 2005).

Safety and Antifungal Efficacy

A total of 556 patients who had a proven or presumptive invasive fungal infection refractory or intolerant to conventional amphotericin B were treated with ABLC on an emergency use protocol at a median daily dose of 4.9 mg/kg for a median of 22 days (range, 1 to 510 days). Clinical response rates were 67% (28 of 42) for disseminated candidiasis and 42% (55 of 130) for aspergillosis; of note, 17 of 24 (71%) patients with zygomycosis and 9 of 11 (82%) patients with fusariosis responded to treatment. Response rates varied according to the pattern of invasive fungal infection, underlying condition, and reason for enrollment (intolerance versus refractory infection). Although the serum creatinine increased in a number of individual patients, the mean serum creatinine decreased significantly from baseline in the entire population and also in the subgroup with initial renal dysfunction (Walsh et al., 1998b). In the subgroup analysis of bone marrow transplant recipients, response rates were 70% (14 of 20) in patients with candidiasis and 38% (11 of 29) in patients with aspergillosis (Wingard et al., 1997); in the subgroup analysis of 111 pediatric patients, the respective response rates were 81% (22 of 27) and 56% (14 of 25) in the 54 patients eligible for evaluation of efficacy (Walsh et al., 1999b).

A randomized, controlled multicenter trial, which included neutropenic cancer patients, enrolled a total of 213 patients with invasive candidiasis. Patients were randomized 2:1 to receive either ABLC at 5 mg/kg/day or DAMB (0.6 to 1.0 mg/kg/day). Response rates were similar (63 versus 68%) in both arms. Nephrotoxicity was significantly lower in the ABLC group, whereas other adverse effects occurred at similar frequencies; unfortunately, however, the complete results of this clinical trial have not been fully published (Anaissie et al., 35th ICAAC). The efficacy of ABLC in the treatment of invasive candidiasis was corroborated by data from >900 patients with invasive candidiasis accrued in the phase IV program of the manufacturer, with clinical responses (cured or improved) of 63% and 62%, respectively, in invasive infection with *C. albicans* and non-albicans *Candida* species (Ito and Hooshmand-Rad, 2005). ABLC was well tolerated and demonstrated efficacy in a cohort of pediatric cancer patients with hepatosplenic candidiasis (Walsh et al., 1997) and in premature neonates with invasive *Candida* infections (Wuerthwein et al., 2005).

The efficacy of ABLC in the treatment of non-*Aspergillus* molds was assessed on the basis of a retrospective analysis of data from the manufacturer's phase

IV program. Of 64 patients with zygomycosis, 33 (52%) were cured or improved; the response rate in patients with invasive fusariosis was 46% (12 of 26), and that in patients with other non-*Aspergillus* infections, mostly of dimorphic molds, was 61% (48 of 79) (Perfect, 2005). In an early comparative trial in 55 patients with AIDS-associated cryptococcal meningitis, a complete clinical response was noted in 18 of 21 patients (86%) treated with ABLC, and conversion to negative cerebrospinal fluid (CSF) cultures was documented in 8 (42%). At the end of treatment, however, 42% of patients who received ABLC at 5 mg/kg had persistent positivity of CSF cultures (Sharkey et al., 1996). Among 106 patients with mostly cerebral cryptococcosis accrued in the manufacturer's phase IV program, clinical responses (cured or improved) were achieved in 67 of 101 (66%) evaluable patients, with only minor differences between patients with CNS and extracerebral infection, HIV infection or no HIV infection, and first- or second-line treatment (Badour et al., 2005). ABLC has also been used with success for antimony-refractory Indian visceral leishmanisis (Sundar et al., 1998).

Despite reduced nephrotoxicity when directly compared to DAMB (0.6 to 1.0 mg/kg) (Anaissie et al., 35th ICAAC) and the tolerance of cumulative doses of up to 70 g (Kline et al., 1995), renal toxicity may still occur (Walsh et al., 1998a; Anaissie et al., 35th ICAAC). Increases in serum creatinine levels from normal at baseline to ≥2.0 mg/dl were observed in approximately 25% of patients treated with ABLC at 5 mg/kg/day, and infusion-related events were noted in approximately 60% (Anaissie et al., 35th ICAAC). Adverse events leading to the discontinuation of ABCD occurred at a frequency of 6 to 9% (Walsh et al., 1998b, 1999a; Anaissie et al., 35th ICAAC). Registry data on renal safety for 3,514 patients with invasive fungal infections treated with ABLC from 1996 to 2000 at >160 institutions post-regulatory approval showed a median change in predicted creatinine clearance from baseline to the end of therapy of −3 ml/min (range, −119 to 118 ml/min), a doubling of serum creatinine level in 13%, and new dialysis in 3% of the 3,514 patients. Concomitant treatment with potentially nephrotoxic agents and a baseline serum creatinine level of <2 mg/dl were factors predisposing for the development of nephrotoxicity (Alexander and Wingard, 2005). Among 548 children and adolescents 0 to 20 years of age who were enrolled into the registry, no significant difference between the rates of new hemodialysis versus baseline hemodialysis was observed. Elevations in serum creatinine of >1.5× baseline and >2.5× baseline values were seen in 24.8 and 8.8% of all patients, respectively (Wiley et al., 2005). In a comparison of 572 elderly patients >65 years of age versus 2,930 patients ≤65 years, both groups showed a 0.1-mg/dl median serum creatinine change from baseline to the end of therapy (Hooshmand-Rad et al., 2005). Finally, no increased risk for renal dysfunction was observed in a comparison of the renal effects of high-dosage/long-duration ABLC therapy (>5 mg/kg/day for >12 days; 309 patients) with those of low-dosage/short-duration ABLC therapy (≤5 mg/kg/day for ≤12 days; 1,417 patients), suggesting that higher ABLC dosages are similarly tolerated by the kidney as lower dosages (Hooshmand-Rad et al., 2004).

A double-blind, randomized study compared the safety of ABLC at a dose of 5 mg/kg/day (n = 78), LAMB at a dose of 3 mg/kg/day (n = 85), or LAMB at a dose of 5 mg/kg/day (n = 81). The median duration of therapy ranged from 7.5 to 8.6 days. In this trial, LAMB (3 and 5 mg/kg/day) had lower rates of fever (23.5 and 19.8% versus 57.7% on day 1; $P < 0.001$), chills or rigors (18.8 and 23.5% versus 79.5% on day 1; $P < 0.001$), nephrotoxicity (defined as a serum creatinine of two times the baseline value; 14.1 and 14.8% versus 42.3%; $P < 0.01$), and toxicity-related discontinuations of therapy (12.9 and 12.3% versus 32.1%; P = 0.004). After day 1, infusion-related reactions were less frequent with ABLC, but chills or rigors were still higher (21.0 and 24.3% versus 50.7%; $P < 0.001$). Hepatotoxicity occurred in 11.5% of patients, with no differences between the three cohorts. Therapeutic success was similar in all three groups. While the higher rate of infusion-related reactions of ABLC is overall supported by the cumulative data of noncomparative clinical studies, the unusually high rate of nephrotoxicity and discontinuations is not and is more difficult to interpret (Wingard et al., 2000).

Treatment of Invasive Aspergillosis

Response rates after second-line treatment with ABLC for proven or presumptive invasive *Aspergillus* infection refractory or intolerant to conventional amphotericin B in the compassionate use program were 42% in the entire cohort of 130 evaluable patients (Walsh et al., 1998b) and 56% in the subset of the 25 evaluable pediatric patients (Walsh et al., 1999b) (Table 3). In a different analysis comprising 151 patients with probable or definite invasive aspergillosis, the overall response rate to ABLC was superior to that of a historical control group treated with conventional amphotericin B (40 versus 23%; $P < 0.05$) (Hiemenz et al., 1995). In 29 hematopoietic stem cell transplant patients, response rates in presumed or proven invasive aspergillosis were 38% (Wingard, 1997), and the response rate was 47% among 39 solid organ transplant recipients (Linden et al., 2000).

Within the phase IV registry, the efficacy and renal safety of ABLC were also assessed in 398 patients with invasive aspergillosis. The most common underlying conditions were hematopoietic stem cell transplantation (25%), hematologic malignancy (25%), and solid organ transplantation (27%). The most common reason for administration of ABLC was lack of response to prior antifungal therapy. Overall, 65% of patients had a favorable clinical response: 44% were cured or improved, and 21% were stabilized. Clinical responses were similar for patients who received ABLC as either first-line (n = 139) or second-line (n = 216) therapy. Patients infected with *Aspergillus terreus*, an innately polyene-resistant species, experienced a 37% response rate. Changes in serum creatinine levels were not clinically significant in most patients; however, dialysis was initiated in seven patients, of whom six had prior antifungal therapy or preexisting renal disease (Chandresakar and Ito, 2005). Among 85 allogeneic hematopoietic stem cell transplant recipients with invasive aspergillosis treated with ABLC identified from the registry, the response rate to ABLC was 31% (26 of 85) overall and 21% (5 of 24) in patients with graft-versus-host disease. The overall response rate to first-line ABLC treatment was 41% (11 of 27) (Ito et al., 2005). Similarly, in the subset of children and adolescents 0 to 20 years of age, the response rates (cured plus improved) in patients with proven invasive aspergillosis were 40.5 and 37.5% in transplant and nontransplant patients, respectively. When stable responses were added, the response rates were 48.6 and 71.9%, respectively (Wiley et al., 2005).

Clinical Indications

The published data indicate that ABLC is less nephrotoxic than DAMB and effective against invasive opportunistic mycoses. The experience with ABLC in the treatment of fungal diseases caused by endemic fungi, however, is limited. Compared to DAMB, infusion-related reactions appear to be similar in frequency and extent. The FDA-approved indication is for treatment of invasive fungal infections refractory to or intolerant of DAMB, and the approved dosage is 5 mg/kg/day administered over 2 h. Similar to DAMB, treatment should be started with the full target dose under careful clinical monitoring, and premedication may be advisable.

In patients with renal dysfunction, dose reduction is generally not required, unless there is evidence of relevant drug-induced nephrotoxicity during treatment. Experience with ABLC in patients undergoing dialysis and in patients with significant hepatic dysfunction has not been reported. The pharmacokinetics of ABLC have been investigated in small phase I/II studies in children and neonates (Walsh et al., 1997; Wuerthwein et al., 2005), and a large number of children including at least 30 very low birth weight infants have been treated with ABLC (Walsh et al., 1999b; Wiley et al., 2005; Wuerthwein et al., 2005). Overall, the data indicate no fundamental differences in disposition, safety, and antifungal efficacy of ABLC in pediatric patients in comparison with adult populations (Chiou et al., 2007). There is no experience with the use of ABCD in pregnant women.

LAMB

LAMB consists of small, unilamellar vesicles (liposomes) of 60 to 80 nm which are composed of hydrogenated soy phosphatidyl choline and disteaoryl phosphatidyl glycerol stabilized by cholesterol and combined with amphotericin B in a 2:0.8:1:0.4 molar ratio (Table 2; Fig. 5). When tested in vitro against a number of clinical isolates, its antifungal spectrum appears comparable to that of DAMB (Anaissie et al., 1991).

Pharmacokinetics

Early pharmacokinetic data from 10 adult patients treated at a dose of 2.8 to 3.0 mg/kg/day demonstrated an 8.4-fold-higher median C_{max} (14.4 versus 1.7 μg/ml) and a 9-fold-higher median AUC (171 versus 18.5 μg·h/ml) compared to data from six patients treated with an approximately 3-fold-lower dose of 1 mg/kg/day of conventional DAMB (Heinemann et al., 1997b). In 24 persistently febrile neutropenic adult patients who received the compound as empirical antifungal therapy in a phase I/II sequential dose escalation trial at 1.0, 2.5, 5.0, and 7.5 mg/kg, the mean AUC on the first day of treatment increased disproportionately from 32, 71, and 294 to 534 μg·h/ml and the mean plasma clearance tended to decrease from 39, 51, and 21 to 25 ml/h/kg, respectively (Walsh et al., 1998a). Further dose escalation to 10, 12.5, and 15 mg/kg/day in a subsequent phase I/II trial in patients with invasive mold infection, however, revealed dose-related nonlinear saturation-like pharmacokinetics. Mean AUC and C_{max} values reached maximum values following administration of 10 mg/kg/day and declined at 12.5 and 15 mg/kg/day (Walsh et al., 2001).

In order to better understand the disposition of LAMB, the pharmacokinetics, excretion, and mass balance of LAMB (2mg/kg) and conventional DAMB (0.6 mg/kg) were investigated in healthy volunteers. Both formulations had triphasic plasma profiles with long terminal half-lives (LAMB, 152 ± 116 h; DAMB, 127 ± 30 h), but plasma concentrations were higher after administration of LAMB (C_{max}, 22.9 ± 10 versus 1.4 ± 0.2 μg/ml). LAMB had a central compartment volume close to that of plasma and a volume of distribution at steady state smaller than that of DAMB. Total clearances were similar, but renal and fecal clearances of

LAMB were 10-fold lower than those of DAMB. Two-thirds of the DAMB was excreted unchanged in the urine (20.6%) and feces (42.5%), with >90% accounted for in mass balance calculations at 1 week, suggesting that metabolism plays no major role in elimination. In contrast, <10% of LAMB was excreted unchanged. No metabolites were observed by high-performance liquid chromatography or mass spectrometry (Bekersky et al., 2002a, 2002b). Protein-binding studies of both formulations revealed lower exposures to both unbound and nonliposomal drug following LAMB, with most of amphotericin B in plasma remaining liposome associated (97% at 4 h and 55% at 168 h). Although LAMB markedly reduced the total urinary and fecal recoveries of amphotericin B, urinary and fecal clearances based on unbound compound were similar for both formulations. Unbound drug urinary clearances were equal to the glomerular filtration rate, and tubular transit rates were <16% of the urinary excretion rate, suggesting that net filtration of unbound drug is the mechanism of renal clearance for both LAMB and DAMB in humans (Bekersky et al., 2002a).

Safety and Antifungal Efficacy

In the safety analysis of the initial multicenter compassionate use trial in Europe that included 133 courses of treatment with LAMB (mean maximal dose, 2.1 mg/kg/day; range, 0.45 to 5.0 mg/kg; mean duration of treatment, 21 days) in patients with invasive fungal infections refractory or intolerant to DAMB, increases in serum creatinine from normal at baseline occurred in 15% of patients; however, 17 of 50 patients with initially elevated creatinine levels had a return to normal at end of treatment. Hypokalemia was noted in 18% and infusion-related toxicity in less than 1% (Meunier et al., 1991). The combined safety analysis of similar trials in bone marrow and solid organ transplant patients (n = 187) revealed a frequency of definitely infusion-related side effects of 4% and increases in serum creatinine in 31% of the patients. Therapy with LAMB had to be discontinued due to side effects in 3% of cases (Ringden et al., 1994).

In a phase I/II, open-label, sequential dose escalation trial conducted at the U.S. National Cancer Institute in 36 persistently febrile neutropenic adults who received 1.0, 2.5, 5.0, or 7.5 mg of LAMB, infusion-related side effects occurred in 15 (5%) of all 331 infusions, and only 2 patients (5%) required premedication. Serum creatinine, potassium, and magnesium levels were not significantly changed from baseline in any of the dosage cohorts, and there was no net increase in serum transaminase levels (Walsh et al., 1998b). A subsequent phase I/II, sequential dose escalation trial explored the compound's MTD. A total of 43 patients with probable or proven invasive mold infection received LAMB at 7.5, 10, 12.5, and 15 mg/kg/day. The number of infusions ranged from 1 to 83, with a median duration of 11 days. The MTD was at least 15 mg/kg. Infusion-related reactions of fever occurred in 8 (19%) and chills and rigors occurred in 5 (12%) of 43 patients. Serum creatinine increased two times above baseline in 32% of patients, but this increase was not dose related. Hepatotoxicity developed in one patient. Altogether, the most common adverse events included fever (48%), increased creatinine level (46%), hypokalemia (39%), chills (32%), and abdominal pain (25%), with no obvious dose dependency. A total of six patients (14%) discontinued the study drug due to a possibly related adverse event. The reasons included elevated serum creatinine, renal failure, pancreatitis, hyperbilirubinemia, and hypotension associated with infusion. Of note, discontinuation was unrelated to the dosage group (Walsh et al., 2001).

A combined analysis of two parallel, prospective, open-label, randomized, multicenter comparisons of LAMB (1 or 3 mg/kg/day) and DAMB (1 mg/kg/day) as empirical antifungal therapy in 338 persistently febrile neutropenic adults and children showed fewer severe drug-related adverse effects with LAMB. There was significantly less hypokalemia in patients treated with LAMB, and nephrotoxicity, defined as a 100% or more increase in serum creatinine from baseline, occurred significantly less often with LAMB (11%) than with DAMB (24%) (Prentice et al., 1997). In a second large-scale, randomized, double-blind, multicenter comparison of LAMB (3.0 mg/kg/day) with DAMB (0.6 mg/kg/day) for empirical antifungal therapy in 787 patients evaluable for safety and tolerance, significantly fewer patients who received LAMB had infusion-related fever (17 versus 44%), chills or rigors (18 versus 54%), or other reactions, including hypotension, hypertension, and hypoxia. Nephrotoxicity (defined as a serum creatinine concentration twice the upper limit of normal) was significantly less frequent among patients treated with LAMB (19%) than among those treated with DAMB (34%) (Walsh et al., 1999a).

The first efficacy data in patients with documented or presumed invasive fungal infections not responding or intolerant to DAMB have been reported from three larger studies performed in Europe; the most common underlying conditions were malignancy and solid organ or bone marrow transplantation (Mills et al., 1994; Ng et al., 1995; Ringden et al., 1991). At doses ranging from 0.5 to 5.0 mg/kg, the reported overall response rates were approximately 60% (Mills et al., 1994; Ringden et al., 1991); response rates in evaluable patients with documented or presumed infections due to *Candida* spp. were 84% (Ringden et al., 1991), and those in patients with documented or presumed *Aspergillus* infections ranged from 58 to 77% (Mills et al., 1994; Ng et al., 1995; Ringden et al., 1991). Treatment with LAMB at 3 mg/kg/day was effective and well-tolerated against AIDS-associated cryptococcosis in an open, noncomparative study (Coker et al., 1993), and at 4 mg/kg/day it was equally effective but less nephrotoxic than DAMB (0.7 mg/kg/day) in a small, randomized comparative trial (Leenders et al., 1997); the latter study also demonstrated more rapid clearing from CSF of LAMB compared to DAMB. Based on cumulative clinical experience, LAMB at doses of at least 5 mg/kg/day is considered a standard option for treatment for zygomycosis (Groll and Walsh, 2001). Clinical trials and experience also demonstrate high efficacy and low toxicity for LAMB in immunocompetent and immunocompromised adult and pediatric patients with leishmaniasis (Bermann et al., 1998; Bern et al., 2006; Cascio et al., 2004; Davidson et al., 1994; Musa et al., 2005).

Few adequately designed, randomized clinical trials have been conducted with LAMB in the therapeutic antifungal setting. In a randomized, double-blind multicenter trial in 81 AIDS patients with disseminated histoplasmosis, induction therapy with LAMB (3 mg/kg/day) achieved better response rates and survival than DAMB (0.7 mg/kg/day) and was better tolerated (Johnson et al., 2002). More recently, a double-blind, randomized, multinational noninferiority study compared micafungin (100 mg/day) with LAMB (3 mg/kg/day) as first-line treatment of candidemia and invasive candidosis in 531 patients. While both treatments were equally effective, there were fewer treatment-related adverse events with micafungin (Kuse et al., 2007).

Treatment of Invasive Aspergillosis

In early studies in patients with documented or presumed invasive aspergillosis not responding or intolerant to DAMB, response rates in the range of 58 to 77% have been reported (Mills et al., 1994; Ng et al., 1995; Ringden et al., 1991).

Randomized, prospective comparative data are limited to two phase III clinical trials that compared different doses of LAMB for primary treatment of invasive aspergillosis (Table 3). The first was an open, randomized multicenter trial orchestrated by the European Organization for Research and Treatment of Cancer that compared the efficacy of two doses of LAMB for treatment of proven or probable invasive aspergillosis in neutropenic patients (Ellis et al., 1998). A total of 120 patients were randomized to receive either 1 or 4 mg/kg/day of LAMB (referred to as groups LAMB 1 and 4, respectively); 87 patients were available for evaluation. The median duration of treatment in the cohorts was 18 and 19 days, respectively. Fifteen of 41 (LAMB 1) and 25 of 46 (LAMB 4) patients experienced at least one adverse event during treatment, but the numbers of events per patient were similar. Renal toxicity that was definitely related to LAMB therapy occurred in 1 of 41 (2%) patients treated with LAMB 1 and 5 of 46 (11%) patients treated with LAMB 4. Thirty-nine of 41 patients (LAMB 1) versus 44 of 46 patients (LAMB 4) completed their regimens without interruptions due to toxicity; only one patient's treatment (LAMB 4) was permanently discontinued because of drug-related toxicity. No patient died as a result of LAMB toxicity. Overall, LAMB was effective in 50 to 60% of patients; however, the number of cases with proven invasive aspergillosis was too low to allow for a meaningful comparison of the two doses.

In the second, double-blind comparative trial, which had a similar study design as the pivotal phase III trial of voriconazole (Herbrecht et al., 2002), patients with proven or probable invasive mold infection were randomized to receive LAMB at either 3 or 10 mg/kg/day for 14 days, followed by 3 mg/kg/day. The primary end point was a favorable (i.e., complete or partial) response at the end of study drug treatment. Of 201 patients with confirmed invasive mold infection, 107 received the 3-mg/kg/day dose and 94 received the 10-mg/kg/day dose. Invasive aspergillosis accounted for 97% of cases. Hematological malignancies were present in 93% of patients, and 73% of patients were neutropenic at baseline. A favorable response was achieved in 50% and 46% of patients in the 3- and 10-mg/kg/day groups, respectively (difference, 4%; 95% confidence interval, −10 to 18%; $P > 0.05$); the respective survival rates at 12 weeks were 72 and 59% (difference, 13%; 95% confidence interval, −0.2 to 26%; $P > 0.05$). Significantly higher rates of nephrotoxicity and hypokalemia were seen in the high-dose group (Cornely et al., 2007). Thus, while a dose of 3 mg/kg/day of LAMB appeared to have similar efficacy in primary treatment of invasive aspergillosis as voriconazole, dose escalation to 10 mg/kg/day for 14 days was not beneficial in a therapeutic sense but more toxic.

More recently, combination therapy with LAMB has been explored as an option to improve the poor outcome with invasive aspergillosis (Aliff et al., 2003; Kontoyiannis et al., 2003). In a prospective, open-label pilot study, 30 patients with proven or probable invasive aspergillosis were randomized to receive either a combination of LAMB at the standard dose (3 mg/kg/day) and caspofungin at the standard dose or monotherapy with a high-dose LAMB regimen (10 mg/kg/day). At the end of treatment, there were significantly more partial or complete responses ($P = 0.028$) in the combination group (10 of 15 patients [67%]) compared with the high-dose monotherapy group (4 of 15 patients [27%]). Survival rates at 12 weeks after inclusion were 100 and

80%, respectively. A twofold increase in serum creatinine occurred in 4 of 17 patients (23%) who received monotherapy and 1 of 15 patients (7%) who received combination therapy (Caillot et al., 2007). Although the number of patients was low, the results of this trial suggest a potential benefit of combination therapy with LAMB and may serve as a platform for a more definitive phase III trial.

Due to consistently noted antagonistic effects in vitro and in vivo (Meletiadis et al., 2006), the combination of amphotericin B with triazoles is not advised for treatment of invasive aspergillosis (Groll and Kolve, 2004). Similarly, while there are anecdotal data for a benefit of combining amphotericin B with flucytosine, this combination has never been systematically investigated (Groll et al., 1998a; Groll and Kolve, 2004).

Clinical Indications

A large body of data indicate that LAMB is less nephrotoxic and better tolerated than DAMB and is effective against most invasive opportunistic fungal infections. The experience with endemic mycoses other than histoplasmosis, however, is limited. Whether LAMB can be considered a first-line choice in treatment of invasive *Candida* and *Aspergillus* infections is a matter of regulatory and academic rather than clinical debate. The FDA-approved doses are 3 mg/kg/day for empirical antifungal therapy in febrile neutropenic patients, 3 to 5 mg/kg/day for therapy of invasive infections (i.e., candidiasis, aspergillosis, and cryptococcosis) intolerant or refractory to DAMB, and 6 mg/kg/day for HIV-associated cryptococcal meningitis, administered over 2 h. Treatment should be started with the full target dose under clinical monitoring; premedication is only needed in patients with prior infusion-related reactions.

In patients with renal dysfunction, dose reduction is generally not required unless there is evidence of relevant drug-induced nephrotoxicity during treatment. Case reports have confirmed that LAMB is not significantly removed from the circulation in patients undergoing dialysis (Humphrey et al., 1994; Tomlin et al., 1995; Heinemann et al., 1997b). Enhanced pulmonary accumulation has been reported in a patient with acute liver failure; however, the clinical significance of such accumulation is unclear (Heinemann et al., 1997a). Formal and population-based pharmacokinetic studies indicate that the disposition of LAMB in pediatric patients beyond the neonatal period is not substantially different from that in adults and that weight is the covariate that determines clearance and volume of distribution (Hong et al., 2006; N. L. Seibel, A. Shad, I. Bekersky, C. Gonzalez, A. H. Groll, L. Wood, P. Jarosinski, D. Buell, and T. J. Walsh, submitted for publication). A large number of children ≥3 months of age have been treated with LAMB with no evidence for any difference in efficacy or safety compared to adults (Chiou et al., 2007; Kolve et al., 2005; Meunier et al., 1991; Prentice et al., 1997; Ringden et al., 1993; Walsh et al., 1999b). In addition, a considerable number of children ≤3 months of age, including very low (<1,500 g) birth weight infants have safely received LAMB at doses of up to 7 mg/kg over prolonged periods of time (Chiou et al., 2007; Juster-Reichert et al., 2001, 2003; Scarcella et al., 1998). LAMB has been used for treatment of visceral leishmaniasis in pregnancy and appears to be safe and effective for pregnant women and their fetuses (Mueller et al., 2006).

LIPOSOMAL NYSTATIN

Nystatin, the first antifungal polyene, was discovered as a natural fermentation product of *Streptomyces noursei* in the early 1950s (Hazen and Brown, 1950). It is a tetraen-diene macrolide, and its carbon skeleton differs from that of amphotericin B only by the lack of one double bond (Fig. 1). Although generally less potent on a molar basis, the compound has broad-spectrum fungicidal activity similar to that of amphotericin B; resistance in clinical isolates is rare (Carillo-Munoz et al., 1999). While nystatin has been available for many years for topical use, problems with solubilization and toxicity precluded its development for systemic treatment (Groll et al., 1998a).

However, laboratory research carried out by Mehta and coworkers in the mid 1980s demonstrated that incorporation of nystatin into multilamellar liposomes consisting of DMPC and DMPG protects human erythrocytes from toxicity while preserving the compound's antifungal activity in vitro (Mehta et al., 1987). The liposomal formulation of nystatin was well-tolerated and effective in murine screening models of disseminated candidiasis (Mehta et al., 1987) and disseminated aspergillosis (Denning et al., 1999; Wallace et al., 1997) and showed promising activity in persistently neutropenic rabbit models of invasive pulmonary aspergillosis (Groll et al., 1999a) and subacute disseminated candidiasis (Groll et al., 1999b). Pharmacokinetic studies in healthy rabbits established nonlinear pharmacokinetics with decreasing clearance at higher dose levels; the compound reached relatively high peak plasma levels and was then rapidly distributed and eliminated from plasma with a half-life of 1 to 2 h (Groll et al., 2000a).

Clinical Trials

The plasma pharmacokinetics of liposomal nystatin (Nyotran) were investigated in HIV-infected patients at doses ranging from 2 to 7 mg/kg. After achievement of comparatively high peak plasma drug concentrations in the range of 4.8 to 24.1 μg/ml, the drug was rapidly distributed and eliminated from plasma with a half-life of 5 to 7 h, exhibiting a pharmacokinetic profile that is different from all four amphotericin B formulations (Cossum et al., *Abstr. 36th Int. Conf. Antimicrob. Agents Chemother.*, abstr. A88, 1996) (Table 2). In a phase I study in 32 patients with hematological malignancies and refractory febrile neutropenia, liposomal nystatin was relatively well-tolerated at multiple doses of up to 8 mg/kg without reaching an MTD. Nephrotoxicity occurred at the higher end of the dose range but did not exceed grade II (Boutati et al., *Abstr. 35th Int. Conf. Antimicrob. Agents Chemother.*, abstr. LM22, 1995). In the combined analysis of randomized trials involving 538 patients with persistent fever and neutropenia, liposomal nystatin (2 mg/kg) was as effective as conventional amphotericin B (0.6 to 0.8 mg/kg) but less nephrotoxic (Powles et al., *Abstr. 39th Int. Conf. Antimicrob. Agents Chemother.*, abstr. LB-4, 1999). In a therapeutic setting, liposomal nystatin at 2 to 4 mg/kg/day showed promising antifungal efficacy in a phase II study in 109 nonneutropenic patients with candidemia, with successful outcomes in 60 of 72 evaluable patients (83%) (Williams et al., *Abstr. 39th Int. Conf. Antimicrob. Agents Chemother.*, abstr.1420, 1999). Finally, liposomal nystatin was also investigated as salvage therapy in 33 patients with probable or definite invasive aspergillosis who were either intolerant or refractory to conventional amphotericin B. The 26 patients eligible for analysis of efficacy received liposomal nystatin at a median dose 4 mg/kg once daily (range, 1.5 to 5 mg) for a median of 22 days (range, 1 to 48 days). Complete or partial responses were noted in 7 of 25 evaluable patients (28%); 8 of 25 evaluable patients were alive on day 30 after the end of treatment (32%) (Offner et al., 2004).

Status of Development

The clinical development of liposomal nystatin (Nyotran) was not further pursued, as the compound proved less effective than conventional amphotericin B in the pivotal phase III trial in patients with cryptococcal meningoencephalitis (unpublished data). Unfortunately, this clinical trial was initiated without consideration and exploration of the particular pharmacokinetics of this liposomal compound in the CNS and without the prior conduct of an explorative animal study for this indication.

REFERENCES

Adedoyin, A., J. F. Bernardo, C. E. Swenson, L. E. Bolsack, G. Horwith, S. DeWit, E. Kelly, J. Klasterksy, J. P. Sculier, D. DeValeriola, E. Anaissie, G. Lopez-Berestein, A. Llanos-Cuentas, A. Boyle, and R. A. Branch. 1997. Pharmacokinetic profile of ABELCET (amphotericin B lipid complex injection): combined experience from phase I and phase II studies. *Antimicrob. Agents Chemother.* **41:**2201–2208.

Adler-Moore, J. P., and R. T. Proffitt. 2002. AmBisome: liposomal formulation, structure, mechanism of action and pre-clinical experience. *J. Antimicrob. Chemother.* **49**(Suppl. 1):21–30.

Adler-Moore, J. P., G. Fujii, M. J. A. Lee, A. Satorius, A. Bailey, and R. Proffitt. 1993. In vitro and in vivo interactions of AmBisome with pathogenic fungi. *J. Liposome Res.* **3:**151–163.

Alexander, B. D., and J. R.Wingard. 2005. Study of renal safety in amphotericin B lipid complex-treated patients. *Clin. Infect. Dis.* **40**(Suppl. 6):S414–421.

Aliff, T. B., P. G. Maslak, J. G. Jurcic, M. L. Heaney, K. N. Cathcart, K. A. Sepkowitz, and M. A. Weiss. 2003. Refractory *Aspergillus* pneumonia in patients with acute leukemia: successful therapy with combination caspofungin and liposomal amphotericin. *Cancer* **97:**1025–1032.

Allende, M. C., J. W. Lee, P. Francis, K. Garrett, H. Dollenberg, J. Berenguer, C. A. Lyman, P. A. Pizzo, and T. J. Walsh. 1994. Dose-dependent antifungal activity and nephrotoxicity of amphotericin B colloidal dispersion in experimental pulmonary aspergillosis. *Antimicrob. Agents Chemother.* **38:**518–522.

Amantea, M. A., R. A. Bowden, A. Forrest, P. K. Working, M. S. Newman, and R. D. Mamelok. 1995. Population pharmacokinetics and renal function-sparing effects of amphotericin B colloidal dispersion in patients receiving bone marrow transplants. *Antimicrob. Agents Chemother.* **39:**2042–2047.

Anaissie, E. J., G. N. Mattiuzzi, C. B. Miller, G. A. Noskin, M. J. Gurwith, R. D. Mamelok, and L. A. Pietrelli. 1998. Treatment of invasive fungal infections in renally impaired patients with amphotericin B colloidal dispersion. *Antimicrob. Agents Chemother.* **42:**606–611.

Anaissie, E., V. Paetznick, R. Proffitt, J. Adler-Moore, and G. P. Bodey. 1991. Comparison of the in vitro antifungal activity of free and liposomal-encapsulated amphotericin B. *Eur. J. Clin. Microbiol. Infect. Dis.* **10:**665–668.

Andes, D., N. Safdar, K. Marchillo, and R. Conklin. 2006. Pharmacokinetic-pharmacodynamic comparison of amphotericin B (AMB) and two lipid-associated AMB preparations, liposomal AMB and AMB lipid complex, in murine candidiasis models. *Antimicrob. Agents Chemother.* **50:**674–684.

Andes, D., T. Stamsted, and R. Conklin. 2001. Pharmacodynamics of amphotericin B in a neutropenic-mouse disseminated-candidiasis model. *Antimicrob. Agents Chemother.* **45:**922–926.

Arning, M., K. O. Kliche, A. H. Heer-Sonderhoff, and A. Wehmeier. 1996. Infusion-related toxicity of three different amphotericin B formulations and its relation to cytokine plasma levels. *Mycoses* **38:**459–465.

Arning, M., and R. E. Scharf. 1989. Prevention of amphotericin B induced nephrotoxicity by loading with sodium-chloride: a report of 1291 days of treatment with amphotericin B without renal failure. *Klin. Wochenschr.* **67:**1020–1028.

Atkinson, A. J., Jr., and J. E. Bennett. 1978. Amphotericin B pharmacokinetics in humans. *Antimicrob. Agents Chemother.* **13:**271–276.

Ayestarán, A., R. M. López, J. B. Montoro, A. Estíbalez, I. Pou, A. Julià, A. López, and B. Pascual. 1996. Pharmacokinetics of conventional formulation versus fat emulsion formulation of amphotericin B in a group of patients with neutropenia. *Antimicrob. Agents Chemother.* **40:**609–612.

Baddour, L. M., J. R. Perfect, and L. Ostrosky-Zeichner. 2005. Successful use of amphotericin B lipid complex in the treatment of cryptococcosis. *Clin. Infect. Dis.* **40**(Suppl. 6):S409–S413.

Barcia, J. P. 1998. Hyperkalemia associated with rapid infusion of conventional and lipid complex formulations of amphotericin B. *Pharmacotherapy* **18**:874–876.

Barton, E., H. Zinnes, and R. A. Moe. 1958. Studies on a new solubilized preparation of amphotericin B, p. 53–57. *In Antibiotics Annual, 1957–1958.* Medical Encyclopedia, Inc., New York, NY.

Bates, J. H. 1993. Amphotericin B, amphotericin B methylester, and other polyenes, p. 295–312. *In* G. A. Sarosi and S. F. Davies (ed.), *Fungal Diseases of the Lung*, 2nd ed. Raven Press Ltd., New York, NY.

Bekersky, I., R. M. Fielding, D. E. Dressler, S. Kline, D. N. Buell, and T. J. Walsh. 2001. Pharmacokinetics, excretion, and mass balance of ^{14}C after administration of ^{14}C-cholesterol-labeled AmBisome to healthy volunteers. *J. Clin. Pharmacol.* **41**:963–971.

Bekersky, I., R. M. Fielding, D. E. Dressler, J. W. Lee, D. N. Buell, and T. J. Walsh. 2002a. Plasma protein binding of amphotericin B and pharmacokinetics of bound versus unbound amphotericin B after administration of intravenous liposomal amphotericin B (AmBisome) and amphotericin B deoxycholate. *Antimicrob. Agents Chemother.* **46**:834–840.

Bekersky, I., R. M. Fielding, D. E. Dressler, J. W. Lee, D. N. Buell, and T. J. Walsh. 2002b. Pharmacokinetics, excretion, and mass balance of liposomal amphotericin B (AmBisome) and amphotericin B deoxycholate in humans. *Antimicrob. Agents Chemother.* **46**:828–833.

Bennett, J. E., W. E. Dismukes, R. J. Duma, G. Medoff, M. A. Sande, H. Gallis, J. Leonard, B. T. Fields, M. Bradshaw, H. Haywood, Z. A. McGee, T. R. Cate, C. G. Cobbs, J. F. Warner, and D. W. Alling. 1979. A comparison of amphotericin B alone and in combination with flucytosine in the treatment of cryptococcal meningitis. *N. Engl. J. Med.* **301**:126–131.

Benson, J. M., and M. C. Nahata. 1989. Pharmacokinetics of amphotericin B in children. *Antimicrob. Agents Chemother.* **33**:1989–1993.

Berenguer, J., J. L. Rodríguez-Tudela, C. Richard, M. Alvarez, M. A. Sanz, L. Gaztelurrutia, J. Ayats, and J. V. Martinez-Suarez. 1997. Deep infections caused by *Scedosporium prolificans*. A report on 16 cases in Spain and a review of the literature. *Medicine* (Baltimore) **76**:256–265.

Berman, J. 1998. Chemotherapy of leishmaniasis: recent advances in the treatment of visceral disease. *Curr. Opin. Infect. Dis.* **11**:707–710.

Bern, C., J. Adler-Moore, J. Berenguer, M. Boelaert, M. den Boer, R. N. Davidson, C. Figueras, L. Gradoni, D. A. Kafetzis, K. Ritmeijer, E. Rosenthal, C. Royce, R. Russo, S. Sundar, and J. Alvar. 2006. Liposomal amphotericin B for the treatment of visceral leishmaniasis. *Clin. Infect. Dis.* **43**:917–924.

Boutati, E. I., and E. J. Anaissie. 1997. *Fusarium*, a significant emerging pathogen in patients with hematologic malignancy: ten years' experience at a cancer center and implications for management. *Blood* **90**:999–1008.

Bowden, R., P. Chandrasekar, M. H. White, X. Li, L. Pietrelli, M. Gurwith, J. A. van Burik, M. Laverdiere, S. Safrin, and R. Wingard. 2002. A double-blind, randomized, controlled trial of amphotericin B colloidal dispersion versus amphotericin B for treatment of invasive aspergillosis in immunocompromised patients. *Clin. Infect. Dis.* **35**:359–366.

Bowden, R. A., M. Cays, T. Gooley, R. D. Mamelok, and J. A. van Burik. 1996. Phase I study of amphotericin B colloidal dispersion for the treatment of invasive fungal infections after marrow transplant. *J. Infect. Dis.* **173**:1208–1215.

Braijtburg, J., and J. Bolard. 1996. Carrier effects on biological activity of amphotericin B. *Clin. Microbiol. Rev.* **9**:512–531.

Brajtburg, J., W. G. Powderly, G. S. Kobayashi, and G. Medoff. 1990. Amphotericin B: current understanding of mechanisms of action. *Antimicrob. Agents Chemother.* **34**:183–188.

Butler, W. T., J. E. Bennett, D. W. Alling, P. T. Wertlake, J. P. Utz, and G. J. Hill. 1964. Nephrotoxicity of amphotericin B: early and late effects in 81 patients. *Ann. Intern. Med.* **62**:175–187.

Butler, W. T. 1966. Pharmacology, toxicology and therapeutic usefulness of amphotericin B. *JAMA* **195**:127–131.

Caillot, D., G. Reny, E. Solary, O. Casasnovas, P. Chavanet, B. Bonnotte, L. Perello, M. Dumas, F. Entezam, and H. Guy. 1994. A controlled trial of the tolerance of amphotericin B infused in dextrose or Intralipid in patients with hematological malignancies. *J. Antimicrob. Chemother.* 33:603–613.

Caillot, D., A. Thiébaut, R. Herbrecht, S. de Botton, A. Pigneux, F. Bernard, J. Larché, F. Monchecourt, S. Alfandari, and L. Mahi. 2007. Liposomal amphotericin B in combination with caspofungin for invasive aspergillosis in patients with hematologic malignancies: a randomized pilot study (Combistrat trial). *Cancer* **110**:2740–2746.

Carrillo-Muñoz, A. J., G. Quindós, C. Tur, M. T. Ruesga, Y. Miranda, O. del Valle, P. A. Cossum, and T. L. Wallace. 1999. *In vitro* antifungal activity of liposomal nystatin in comparison with nystatin, amphotericin B cholesteryl sulphate, liposomal amphotericin B, amphotericin B lipid complex, amphotericin B deoxycholate, fluconazole and itraconazole. *J. Antimicrob. Chemother.* **44**:397–401.

Cascio, A., L. di Martino, P. Occorsio, R. Giacchino, S. Catania, A. R. Gigliotti, C. Aiassa, C. Iaria, S. Giordano, C. Colomba, V. F. Polara, L. Titone, L. Gradoni, M. Gramiccia, and S. Antinori. 2004. A 6 day course of liposomal amphotericin B in the treatment of infantile visceral leishmaniasis: the Italian experience. *J. Antimicrob. Chemother.* **54**:217–220.

Chamilos, G., and D. P. Kontoyiannis. 2005. Update on antifungal drug resistance mechanisms of *Aspergillus fumigatus*. *Drug Resist. Update* **8**:344–358.

Chandrasekar, P. H., and J. L. Ito. 2005. Amphotericin B lipid complex in the management of invasive aspergillosis in immunocompromised patients. *Clin. Infect. Dis.* **40**(Suppl. 6):S392–S400.

Chavanet, P., M. Duong, M. Buisson, H. Hamel, C. Dubois, A. Bonnin, and H. Portier. 1997. In vivo activity and tolerance of conventional formulation versus fat emulsion formulation of amphotericin B in experimental disseminated candidiasis in neutropenic rabbits. *J. Antimicrob. Chemother.* 39:427–430.

Chavanet, P. Y., I. Garry, N. Charlier, D. Caillot, J. P. Kisterman, M. D'Athis, and H. Portier. 1992. Trial of glucose versus fat emulsion in preparation of amphotericin B for use in HIV infected patients with candidiasis. *BMJ* **305**:921–925.

Chiou, C. C., T. J. Walsh, and A. H. Groll. 2007. Clinical pharmacology of antifungal agents in pediatric patients. *Expert Opin. Pharmacother.* 8:2465–2489.

Christiansen, K. J., E. M. Bernard, J. W. M. Gold, and D. Armstrong. 1985. Distribution and activity of amphotericin B in humans. *J. Infect. Dis.* **152**:1037–1043.

Clark, J. M., R. R. Whitney, S. J. Olsen, R. J. George, M. R. Swerdel, L. Kunselman, and D. P. Bonner. 1991. Amphotericin B lipid complex therapy of experimental fungal infections in mice. *Antimicrob. Agents Chemother.* **35**:615–621.

Cleary, J. D., M. Schwartz, P. D. Rogers, J. de Mestral, and S. W. Chapman. 2003. Effects of amphotericin B and caspofungin on histamine expression. *Pharmacotherapy* **23**:966–973.

Clemons, K. V., and D. A. Stevens. 2005. The contribution of animal models of aspergillosis to understanding pathogenesis, therapy and virulence. *Med. Mycol.* 43(Suppl. 1):S101–S110.

Clemons, K. V., and D. A. Stevens. 1991. Comparative efficacies of amphotericin B lipid complex and amphotericin B deoxycholate suspension against murine blastomycosis. *Antimicrob. Agents Chemother.* 35:2144–2146.

Clemons, K. V., and D. A. Stevens. 1993. Therapeutic efficacy of a liposomal formulation of amphotericin B (AmBisome) against murine blastomycosis. *J. Antimicrob. Chemother.* **32**:465–472.

Coker, R. J., M. Viviani, B. G. Gazzard, B. Du Pont, H. D. Pohle, S. M. Murphy, J. Atouguia, J. L. Champalimaud, and J. R. Harris. 1993. Treatment of cryptococcosis with liposomal amphotericin B in 23 patients with AIDS. *AIDS* **7**:829–835.

Cornely, O. A., J. Maertens, M. Bresnik, R. Ebrahimi, A. J. Ullmann, E. Bouza, C. P. Heussel, O. Lortholary, C. Rieger, A. Boehme, M. Aoun, H. A. Horst, A. Thiebaut, M. Ruhnke, D. Reichert, N. Vianelli, S. W. Krause, E. Olavarria, R. Herbrecht, et al. 2007. Liposomal amphotericin B as initial therapy for invasive mold infection: a randomized trial comparing a high-loading dose regimen with standard dosing (AmBiLoad trial). *Clin. Infect. Dis.* **44**:1289–1297.

Craven, P. C., T. M. Ludden, D. J. Drutz, W. Rogers, K. A. Haegele, and H. B. Skrdlant. 1979. Excretion pathways of amphotericin B. *J. Infect. Dis.* **140**:329–341.

Daneshmend, T. K., and D. W. Warnock. 1983. Clinical pharmacokinetics of systemic antifungal drugs. *Clin. Pharmacokin.* **8**:17–42.

Davidson, R. N., L. Di Martino, L. Gradoni, R. Giacchino, R. Russo, G. B. Gaeta, R. Pempinello, S. Scott, F. Raimondi, and A. Cascio. 1994. Liposomal amphotericin B (AmBisome) in Mediterranean visceral leishmaniasis: a multi-centre trial. *Q. J. Med.* **87**:75–81.

Dean, J. L., J. E. Wolf, A. C. Ranzini, and M. A. Laughlin. 1994. Use of amphotericin B during pregnancy: case report and review. *Clin. Infect. Dis.* **18**:364–368.

Denning, D. W., and P. Warn. 1999. Dose range evaluation of liposomal nystatin and comparisons with amphotericin B and amphotericin B lipid complex in temporarily neutropenic mice infected with an isolate of *Aspergillus fumigatus* with reduced susceptibility to amphotericin B. *Antimicrob. Agents Chemother.* **43**:2592–2599.

Ellis, M., D. Spence, B. de Pauw, F. Meunier, A. Marinus, L. Collette, R. Sylvester, J. Meis, M. Boogaerts, D. Selleslag, V. Krcmery, W. von Sinner, P. MacDonald, C. Doyen, and B. Vandercam. 1998. An EORTC international multicenter randomized trial (EORTC number 19923) comparing two dosages of liposomal amphotericin B for treatment of invasive aspergillosis. *Clin. Infect. Dis.* **27**:1406–1412.

Ellis, M. E., A. A. Al-Hokail, H. M. Clink, M. A. Padmos, P. Ernst, D. G. Spence, W. N. Tharpe, and V. F. Hillier. 1992. Double-blind randomized study of the effect of infusion rates on toxicity of amphotericin B. *Antimicrob. Agents Chemother.* **36**:172–179.

Ernst, E. J., M. E. Klepser, and M. A. Pfaller. 2000. Postantifungal effects of echinocandin, azole, and polyene antifungal agents *Candida albicans* and *Cryptococcus neoformans. Antimicrob. Agents Chemother.* **44**:1108–1111.

Fielding, R. M., P. C. Smith, L. H. Wang, J. Porter, and L. S. S. Guo. 1991. Comparative pharmacokinetics of amphotericin B after administration of a novel colloidal delivery system, ABCD, and a conventional formulation to rats. *Antimicrob. Agents Chemother.* **35**:1208–1213.

Fischer, M. A., W. C. Winkelmayer, R. H. Rubin, and J. Avorn. 2005. The hepatotoxicity of antifungal medications in bone marrow transplant recipients. *Clin. Infect. Dis.* **41**:301–307.

Fleming, R. V., H. M. Kantarjian, R. Husni, K. Rolston, J. Lim, I. Raad, S. Pierce, J. Cortes, and E. Estey. 2001. Comparison of amphotericin B lipid complex (ABLC) vs. ambisome in the treatment of suspected or documented fungal infections in patients with leukemia. *Leuk. Lymphoma* **40**:511–520.

Francis, P., J. W. Lee, A. Hoffman, J. Peter, A. Francesconi, J. Bacher, J. Shelhamer, P. A. Pizzo, and T. J. Walsh. 1994. Efficacy of unilamellar liposomal amphotericin B in treatment of pulmonary aspergillosis in persistently granulocytopenic rabbits: the potential role of bronchoalveolar o-mannitol and serum galactomannan as markers of infection. *J. Infect. Dis.* **169**:356–368.

Gaeta, G. B., A. Maisto, D. Di Caprio, A. Scalone, G. Pasquale, F. M. Felaco, D. Galante, and L. Gradoni. 2000. Efficacy of amphotericin B colloidal dispersion in the treatment of Mediterranean visceral leishmaniasis in immunocompetent adult patients. *Scand. J. Infect. Dis.* **32**:675–677.

Gale, E. F. 1986. Nature and development of phenotypic resistance to amphotericin B in *Candida albicans. Adv. Microb. Physiol.* **27**:278–320.

Gale, E. F. 1974. The release of potassium ions from *Candida albicans* in the presence of polyene antibiotics. *J. Gen. Microbiol.* **80**:451–465.

Garnacho-Montero, J., C. Ortiz-Leyba, J. L. Garcia Garmendia, and F. Jimenez. 1998. Life-threatening adverse event after amphotericin B lipid complex treatment in a patient treated previously with amphotericin B deoxycholate. *Clin. Infect. Dis.* **26**:1016.

Girmenia, C., G. Cimino, F. Di Cristofano, A. Micozzi, G. Gentile, and P. Martino. 2005. Effects of hydration with salt repletion on renal toxicity of conventional amphotericin B empirical therapy: a prospective study in patients with hematological malignancies. *Support. Care Cancer* **13**:987–992.

Gold, W., H. A. Stout, and I. F. Pagona. 1955. Amphotericins A and B, antifungal antibiotics produced by a streptomycete. I. In vitro studies, p. 579–586. *In Antibiotics Annual, 1955–1956.* Medical Encyclopedia, Inc., New York, NY.

Goldman, R. D., and K. Koren. 2004. Amphotericin B nephrotoxicity in children. *J. Pediatr. Hematol. Oncol.* **26**:421–426.

Groll, A. H., J. C. Gea-Banacloche, A. Glasmacher, G. Just-Nuebling, G. Maschmeyer, and T. J. Walsh. 2003. Clinical pharmacology of antifungal compounds. *Infect. Dis. Clin. North Am.* **17**:159–191.

Groll, A. H., C. E. Gonzalez, N. Giri, K. Kligys, W. Love, J. Peter, E. Feuerstein, J. Bacher, S. C. Piscitelli, and T. J. Walsh. 1999a. Liposomal nystatin against experimental pulmonary aspergillosis in persistently neutropenic rabbits: efficacy, safety and noncompartmental pharmacokinetics. *J. Antimicrob. Chemother.* **43**:95–103.

Groll, A. H., C. A. Lyman, V. Petraitis, R. Petraitiene, D. Armstrong, D. Mickiene, R. M. Alfaro, R. L. Schaufele, T. Sein, J. Bacher, and T. J. Walsh. 2006. Compartmentalized intrapulmonary pharmacokinetics of amphotericin B and its lipid formulations. *Antimicrob. Agents Chemother.* **50**:3418–3423.

Groll, A. H., D. Mickiene, K. Werner, R. Petraitiene, V. Petraitis, M. Calendario, A. Field-Ridley, J. Crisp, S. C. Piscitelli, and T. J. Walsh. 2000a. Compartmental distribution of multilamellar liposomal nystatin in rabbits. *Antimicrob. Agents Chemother.* **44**:950–957.

Groll, A. H., V. Petraitis, R. Petraitiene, A. Field-Ridley, M. Calendario, J. Bacher, S. C. Piscitelli, and T. J. Walsh. 1999b. Safety and efficacy of multilamellar liposomal nystatin against disseminated candidiasis in persistently neutropenic rabbits. *Antimicrob. Agents Chemother.* **43**:2463–2467.

Groll, A. H., S. C. Piscitelli, and T. J. Walsh. 1998a. Clinical pharmacology of systemic antifungal agents: a comprehensive review of agents in clinical use, current investigational compounds, and putative targets for antifungal drug development. *Adv. Pharmacol.* **44**:343–500.

Groll, A. H., and T. J. Walsh. 2001. Uncommon opportunistic fungi: new nosocomial threats. *Clin. Microbiol. Infect.* 7(Suppl. 2):8–24.

Groll, A. H., N. Giri, V. Petraitis, R. Petraitiene, M. Candelario, J. S. Bacher, S. C. Piscitelli, and T. J. Walsh. 2000b. Comparative central nervous system distribution and antifungal activity of lipid formulations of amphotericin B in rabbits. *J. Infect. Dis.* **182**:274–282.

Groll, A. H., and H. Kolve. 2004. Antifungal agents: in vitro susceptibility testing, pharmacodynamics, and prospects for combination therapy. *Eur. J. Clin. Microbiol. Infect. Dis.* **23**:256–270.

Groll, A. H., F. M. Muller, S. C. Piscitelli, and T. J. Walsh. 1998b. Lipid formulations of amphotericin B: clinical perspectives for the management of invasive fungal infections in children with cancer. *Klin. Padiatr.* **210**:264–273.

Groll, A. H., S. C. Piscitelli, and T. J. Walsh. 2001. Antifungal pharmacodynamics. Concentration-effect relationships in vitro and in vivo. *Pharmacotherapy* **21**(Suppl. 8):133–148.

Guo, L. S. S., R. M. Fielding, and D. D. Lasic. 1991. Novel antifungal drug delivery: stable amphotericin B cholesteryl sulfate discs. *Int. J. Pharm.* 75:45–54.

Hamilton-Miller, J. M. T. 1972. Sterols from polyene-resistant mutants of *Candida albicans*. *J. Gen. Microbiol.* **73**:201–203.

Hamilton-Miller, J. M. T. 1973. Chemistry and biology of the polyene macrolide antibiotics. *Bacteriol. Rev.* 37:166–196.

Hanson, L. H., and D. A. Stevens. 1992. Comparison of antifungal activity of amphotericin B deoxycholate with that of amphotericin B cholesteryl sulfate colloidal dispersion. *Antimicrob. Agents Chemother.* **36**:486–488.

Harbarth, S., S. L. Pestotnik, J. F. Lloyd, J. P. Burke, and M. H. Samore. 2001. The epidemiology of nephrotoxicity associated with conventional amphotericin B therapy. *Am. J. Med.* **111**:528–534.

Haze, E. L., and R. Brown. 1950. Two antifungal agents produced by a soil actinomycete. *Science* **112**:423.

Heidemann, H. T., J. F. Gerkens, W. A. Spickard, E. K. Jackson, and R. A. Branch. 1983. Amphotericin B nephrotoxicity in humans decreased by salt repletion. *Am. J. Med.* 75:476–481.

Heinemann, V., D. Bosse, U. Jehn, A. Debus, K. Wachholz, H. Forst, and W. Wilmanns. 1997a. Enhanced pulmonary accumulation of liposomal amphotericin B (AmBisome) in acute liver transplant failure. *J. Antimicrob. Chemother.* **40**:295–298.

Heinemann, V., D. Bosse, U. Jehn, B. Kähny, K. Wachholz, A. Debus, P. Scholz, H. J. Kolb, and W. Wilmanns. 1997b. Pharmacokinetics of liposomal amphotericin B (AmBisome) in critically ill patients. *Antimicrob. Agents Chemother.* **41**:1275–1280.

Heinemann, V., B. Kähny, U. Jehn, D. Mühlbayer, A. Debus, K. Wachholz, D. Bosse, H. J. Kolb, and W. Wilmanns. 1997c. Serum pharmacology of amphotericin B applied in lipid emulsion. *Antimicrob. Agents Chemother.* **41**:728–732.

Herbrecht, R., D. W. Denning, T. F. Patterson, J. E. Bennett, R. E. Greene, J. W. Oestmann, W. V. Kern, K. A. Marr, P. Ribaud, O. Lortholary, R. Sylvester, R. H. Rubin, J. R. Wingard, P. Stark, C. Durand, D. Caillot, E. Thiel, P. H. Chandrasekar, M. R. Hodges, H. T. Schlamm, P. F. Troke, B. de Pauw, et al. 2002. Voriconazole versus amphotericin B for primary therapy of invasive aspergillosis. *N. Engl. J. Med.* 347:408–415.

Herbrecht, R., V. Letscher-Bru, R. A. Bowden, S. Kusne, E. J. Anaissie, J. R. Graybill, G. A. Noskin, F. Oppenheim, E. Andrès, and L. A. Pietrelli. 2001. Treatment of 21 cases of invasive mucormycosis with amphotericin B colloidal dispersion. *Eur. J. Clin. Microbiol. Infect. Dis.* **20**:460–466.

Herbrecht, R. 1997. Safety of amphotericin B colloidal dispersion. *Eur. J. Clin. Microbiol. Infect. Dis.* **16**:74–80.

Hiemenz, J. W., J. Lister, and E. J. Anaissie. 1995. Emergency use of amphotericin B lipid complex (ABLC) in the treatment of patients with aspergillosis: historical control comparison with amphotericin B. *Blood* **86**(Suppl. 1):849a.

Hiemenz, J. W., and T. J. Walsh. 1996. Lipid formulations of amphotericin B: recent progress and future directions. *Clin. Infect. Dis.* **22**(Suppl. 2):S133–S144.

Hong, Y., P. J. Shaw, C. E. Nath, S. P. Yadav, K. R. Stephen, J. W. Earl, and A. J. McLachlan. 2006. Population pharmacokinetics of liposomal amphotericin B in pediatric patients with malignant diseases. *Antimicrob. Agents Chemother.* **50**:935–942.

Hooshmand-Rad, R., A. Chu, V. Gotz, J. Morris, S. Batty, and A. Freifeld. 2005. Use of amphotericin B lipid complex in elderly patients. *J. Infect.* **50**:277–287.

Hooshmand-Rad, R., M. D. Reed, A. Chu, V. Gotz, J. A. Morris, J. Weinberg, and E. A. Dominguez. 2004. Retrospective study of the renal effects of amphotericin B lipid complex when used at higher-than-recommended dosages and longer durations compared with lower dosages and shorter durations in patients with systemic fungal infections. *Clin. Ther.* **26**:1652–1662.

Humphrey, H., D. A. Oliver, R. Winter, and D. W. Warnock. 1994. Liposomal amphotericin B and continuous venous-venous hemofiltration. *J. Antimicrob. Chemother.* 33:1070–1071.

Hwang, K. J., M. M. Padki, D. D. Chow, H. E. Essien, J. Y. Lai, and P. L. Beaumier. 1987. Uptake of small liposomes by non-reticuloendothelial tissues. *Biochim. Biophys. Acta* **901**:88–96.

Ito, J. I., P. H. Chandrasekar, and R. Hooshmand-Rad. 2005. Effectiveness of amphotericin B lipid complex (ABLC) treatment in allogeneic hematopoietic cell transplant (HCT) recipients with invasive aspergillosis (IA). *Bone Marrow Transplant.* 36:873–877.

Ito, J. I., and R. Hooshmand-Rad. 2005. Treatment of *Candida* infections with amphotericin B lipid complex. *Clin. Infect. Dis.* **40**(Suppl. 6):S384–S391.

Iwen, P. C., M. E. Rupp, A. N. Langnas, E. C. Reed, and S. H. Hinrichs. 1998. Invasive pulmonary aspergillosis due to *Aspergillus terreus:* 12-year experience and review of the literature. *Clin. Infect. Dis.* **26**:1092–1097.

Jagdis, F. A., P. D. Hoeprich, R. M. Lawrence, and C. P. Schaffner. 1977. Comparative pharmacology of amphotericin B and amphotericin B methyl ester in the non-human primate, *Macacca mulatta*. *Antimicrob. Agents Chemother.* **12**:582–590.

Janknegt, R., S. deMarie, I. A. Bakker-Woudenberg, and D. J. Crommelin. 1992. Liposomal and lipid formulations of amphotericin B. Clinical pharmacokinetics. *Clin. Pharmacokinet.* **23**:279–291.

Johnson, E. M., J. O. Ojwang, A. Szekely, T. L. Wallace, and D. W. Warnock. 1998a. Comparison of in vitro antifungal activities of free and liposome-encapsulated nystatin with those of four amphotericin B formulations. *Antimicrob. Agents Chemother.* **42**:1412–1416.

Johnson, P. C., L. J. Wheat, G. A. Cloud, M. Goldman, D. Lancaster, D. M. Bamberger, W. G. Powderly, R. Hafner, C. A. Kauffman, and W. E. Dismukes. 2002. Safety and efficacy of liposomal amphotericin B compared with conventional amphotericin B for induction therapy of histoplasmosis in patients with AIDS. *Ann. Intern. Med.* 137:105–109.

Johnson, M. D., R. H. Drew, and J. R. Perfect. 1998b. Chest discomfort associated with liposomal amphotericin B: report of three cases and review of the literature. *Pharmacotherapy* **18**:1053–1061.

Joly, V., P. Aubry, A. Ndayiragide, I. Carrière, E. Kawa, N. Mlika-Cabanne, J. P. Aboulker, J. P. Coulaud, B. Larouze, and P. Yeni. 1996a. Randomized comparison of amphotericin deoxycholate dissolved in dextrose or intralipid for the treatment of AIDS-associated cryptococcal meningitis. *Clin. Infect. Dis.* **23**:556–562.

Joly, V., C. Geoffray, J. Reynes, C. Goujard, D. Méchali, C. Maslo, F. Raffi, and P. Yeni. 1996b. Amphotericin B in lipid emulsion for the treatment of cryptococcal meningitis in AIDS patients. *J. Antimicrob. Chemother.* **38**:117–126.

Juster-Reicher, A., O. Flidel-Rimon, M. Amitay, S. Even-Tov, E. Shinwell, and E. Leibovitz. 2003. High-dose liposomal amphotericin B in the therapy of systemic candidiasis in neonates. *Eur. J. Clin. Microbiol. Infect. Dis.* **22**:603–607.

Juster-Reicher, A., E. Leibovitz, N. Linder, M. Amitay, O. Flidel-Rimon, S. Even-Tov, B. Mogilner, and A. Barzilai. 2001. Liposomal amphotericin B (AmBisome) in the treatment of neonatal candidiasis in very low birth weight infants. *Infection* **28**:223–226.

Kan,V. L., J. E. Bennett, M. A. Amantea, M. C. Smolskis, E. McManus, D. M. Grasela, and J. W. Sherman. 1991. Comparative safety, tolerance, and pharmacokinetics of amphotericin B lipid complex and amphotericin B deoxycholate in healthy male volunteers. *J. Infect. Dis.* **164**:418–421.

Kelly, S. L., D. C. Lamb, M. Taylor, A. J. Corran, B. C. Baldwin, and W. G. Powderly. 1994. Resistance to amphotericin B associated with defective sterol delta 8→7 isomerase in a *Cryptococcus neoformans* strain from an AIDS patient. *FEMS Microbiol. Lett.* **122**: 39–42.

Kirsh, R., R. Goldstein, J. Tarloff, D. Parris, J. Hook, N. Hanna, P. Bugelski, and G. Poste. 1988. An emulsion formulation of amphotericin B improves the therapeutic index when treating systemic murine candidiasis. *J. Infect. Dis.* **158**:1065–1070.

Kleinberg, M. E., and A. Finkelstein. 1984. Single-length and double-length channels formed by nystatin in lipid bilayer membranes. *J. Membr. Biol.* **80**:257–269.

Klepser, M. E., E. J. Wolfe, R. N. Jones, C. H. Nightingale, and M. A. Pfaller. 1997. Antifungal pharmacodynamic characteristics of fluconazole and amphotericin B tested against *Candida albicans*. *Antimicrob. Agents Chemother.* **41**:1392–1395.

Klepser, M. E., E. J. Wolfe, and M. A. Pfaller. 1998. Antifungal pharmacodynamic characteristics of fluconazole and amphotericin B against *Cryptococcus neoformans*. *J. Antimicrob. Chemother.* **41**: 397–401.

Kline, S., T. A. Larsen, L. Fieber, R. Fishbach, M. Greenwood, R. Harris, M. W. Kline, P. O. Tennican, and E. N. Janoff. 1995. Limited toxicity of prolonged therapy with high doses of amphotericin B lipid complex. *Clin. Infect. Dis.* **21**:1154–1158.

Kolve, H., J. Ritter, H. Juergens, and A. H. Groll. 2005. Safety, tolerance and outcome of treatment with liposomal amphotericin B in pediatric cancer/HSCT patients. *Bone Marrow Transplant.* **35**(Suppl. 2):S264–S265.

Kontoyiannis, D. P., R. Hachem, R. E. Lewis, G. A. Rivero, H. A. Torres, J. Thornby, R. Champlin, H. Kantarjian, G. P. Bodey, and I. I. Raad. 2003. Efficacy and toxicity of caspofungin in combination with liposomal amphotericin B as primary or salvage treatment of invasive aspergillosis in patients with hematologic malignancies. *Cancer* **98**:292–299.

Kontoyiannis, D. P., R. E. Lewis, G. S. May, N. Osherov, and M. G. Rinaldi. 2002. *Aspergillus nidulans* is frequently resistant to amphotericin B. *Mycoses* **45**:406–407.

Koren, G., A. Lau, J. Klein, C. Golas, M. Bologa-Campeanu, S. Soldin, S. M. MacLeod, and C. Prober. 1988. Pharmacokinetics and adverse effects of amphotericin B in infants and children. *J. Pediatr.* **113**:559–563.

Krishnan, S., E. K. Manavathu, and P. H. Chandrasekar. 2005. A comparative study of fungicidal activities of voriconazole and amphotericin B against hyphae of *Aspergillus fumigatus*. *J. Antimicrob. Chemother.* **55**:914–920.

Kuse, E. R., P. Chetchotisakd, C. A. da Cunha, M. Ruhnke, C. Barrios, D. Raghunadharao, J. S. Sekhon, A. Freire, V. Ramasubramanian, I. Demeyer, M. Nucci, A. Leelarasamee, F. Jacobs, J. Decruyenaere, D. Pittet, A. J. Ullmann, L. Ostrosky-Zeichner, O. Lortholary, S. Koblinger, H. Diekmann-Berndt, O. A. Cornely, et al. 2007. Micafungin versus liposomal amphotericin B for candidaemia and invasive candidosis: a phase III randomised double-blind trial. *Lancet* **369**:1519–1527.

Lane, J. W., N. N. Rehak, G. L. Hortin, T. Zaoutis, P. R. Krause, and T. J. Walsh. 2008. Pseudohyperphosphatemia associated with high-dose liposomal amphotericin B therapy. *Clin. Chim. Acta* **387**: 145–149.

Lawrence, R. M., P. D. Hoeprich, F. A. Jagdis, N. Monji, A. C. Huston, and C. P. Schaffner. 1980. Distribution of doubly radiolabelled amphotericin B methyl ester and amphotericin B in the non-human primate, *Macaca mulatta*. *J. Antimicrob. Chemother.* **6**:241–249.

Leenders, A. C., P. Reiss, P. Portegies, K. Clezy, W. C. Hop, J. Hoy, J. C. Borleffs, T. Allworth, R. H. Kauffmann, P. Jones, F. P. Kroon, H. A. Verbrugh, and S. de Marie. 1997. Liposomal amphotericin B (AmBisome™) compared with amphotericin B both followed by oral fluconazole in the treatment of AIDS-associated cryptococcal meningitis. *AIDS* **11**:1463–1471.

Lewis, R. E., G. Liao, J. Hou, G. Chamilos, R. A. Prince, and D. P. Kontoyiannis. 2007. Comparative analysis of amphotericin B lipid complex and liposomal amphotericin B kinetics of lung accumulation and fungal clearance in a murine model of acute invasive pulmonary aspergillosis. *Antimicrob. Agents Chemother.* **51**:1253–1258.

Lin, S. J., J. Schranz, and S. M. Teutsch. 2001. Aspergillosis case-fatality rate: systematic review of the literature. *Clin. Infect. Dis.* **32**:358–366.

Linden, P., P. Williams, and K. M. Chan. 2000. Efficacy and safety of amphotericin B lipid complex injection (ABLC) in solid-organ transplant recipients with invasive fungal infections. *Clin. Transplant.* **14**:329–339.

Lopez, R. M., A. Ayestaran, L. Pou, J. B. Montoro, M. Hernandez, and I. Caragol. 1996. Stability of amphotericin B in an extemporaneously prepared i.v. fat emulsion. *Am. J. Health Syst. Pharm.* **53**: 2724–2727.

Maertens, J., K. Lagrou, H. Deweerdt, I. Surmont, G. E. Verhoef, J. Verhaegen, and M. A. Boogaerts. 2000. Disseminated infection by *Scedosporium prolificans*: an emerging fatality among haematology patients. Case report and review. *Ann. Hematol.* **79**:340–344.

Manavathu, E. K., M. S. Ramesh, I. Baskaran, L. T. Ganesan, and P. H. Chandrasekar. 2004. A comparative study of the post-antifungal effect (PAFE) of amphotericin B, triazoles and echinocandins on *Aspergillus fumigatus* and *Candida albicans*. *J. Antimicrob. Chemother.* **53**:386–389.

Medoff, G., and G. S. Kobayashi. 1980. Strategies in the treatment of systemic fungal infections. *N. Engl. J. Med.* **302**:1451–1455.

Mehta, R. T., R. L. Hopfer, L. A. Gunner, R. L. Juliano, and G. Lopez-Berestein. 1987. Formulation, toxicity, and antifungal activity in vitro of liposome-encapsulated nystatin as therapeutic agent for systemic candidiasis. *Antimicrob. Agents Chemother.* **31**:1897–1900.

Mehta, R. T., R. L. Hopfer, T. McQueen, R. L. Juliano, and G. Lopez-Berestein. 1997. Toxicity and therapeutic effects in mice of liposome-encapsulated nystatin for systemic fungal infections. *Antimicrob. Agents Chemother.* **31**:1901–1903.

Meletiadis, J., V. Petraitis, R. Petraitiene, P. Lin, T. Stergiopoulou, A. M. Kelaher, T. Sein, R. L. Schaufele, J. Bacher, and T. J. Walsh. 2006. Triazole-polyene antagonism in experimental invasive pulmonary aspergillosis: in vitro and in vivo correlation. *J. Infect. Dis.* **194**:1008–1018.

Meunier, F., H. G. Prentice, and O. Ringden. 1991. Liposomal amphotericin B (AmBisome™): safety data from a phase II/III clinical trial. *J. Antimicrob. Chemother.* **28**(Suppl. B):83–91.

Mills, W., R. Chopra, D. C. Linch, and A. H. Goldstone. 1994. Liposomal amphotericin B in the treatment of fungal infections in neutropenic patients: a single-center experience of 133 episodes in 116 patients. *Br. J. Haematol.* **86**:754–760.

Mohr, J. F., A. C. Hall, C. D. Ericsson, and L. Ostrosky-Zeichner. 2005. Fatal amphotericin B overdose due to administration of nonlipid formulation instead of lipid formulation. *Pharmacotherapy* **25**: 426–428.

Moreau, P., N. Milpied, N. Fayette, J. F. Ramée, and J. L. Harousseau. 1992. Reduced renal toxicity and improved clinical tolerance of amphotericin B mixed with intralipid compared with conventional amphotericin B in neutropenic patients. *J. Antimicrob. Chemother.* **30**:535–541.

Mott, S. H., R. J. Packer, L. G. Vezina, S. Kapur, P. A. Dinndorf, J. A. Conry, M. R. Pranzatelli, and R. R. Quinones. 1995. Encephalopathy with parkinsonian features in children following bone marrow transplantation and high dose amphotericin B. *Ann. Neurol.* **37**: 810–814.

Mozaffarian, N., J. W. Berman, and A. Casadevall. 1997. Enhancement of nitric oxide synthesis by macrophages represents an additional mechanism of action for amphotericin B. *Antimicrob. Agents Chemother.* **41**:1825–1829.

Mueller, M., M. Balasegaram, Z. Koummuki, K. Ritmeijer, M. R. Santana, and R. Davidson. 2006. A comparison of liposomal amphotericin B with sodium stibogluconate for the treatment of visceral leishmaniasis in pregnancy in Sudan. *J. Antimicrob. Chemother.* 58:811–815.

Musa, A. M., E. A. Khalil, F. A. Mahgoub, S. Hamad, A. M. Elkadaru, and A. M. El Hassan. 2005. Efficacy of liposomal amphotericin B (AmBisome) in the treatment of persistent post-kala-azar dermal leishmaniasis (PKDL). *Ann. Trop. Med. Parasitol.* **99:**563–569.

Ng, T. C. C., and D. W. Denning. 1995. Liposomal amphotericin B (AmBisome™) therapy in invasive fungal infections: evaluation of United Kingdom compassionate use data. *Arch. Intern. Med.* **155:** 1093–1098.

Noskin, G. A., L. Pietrelli, G. Coffey, M. Gurwith, and L. J. Liang. 1998. Amphotericin B colloidal dispersion for treatment of candidemia in immunocompromised patients. *Clin. Infect. Dis.* **26:**461–467.

Offner, F., V. Krcmery, M. Boogaerts, C. Doyen, D. Engelhard, P. Ribaud, C. Cordonnier, B. de Pauw, S. Durrant, J. P. Marie, P. Moreau, H. Guiot, G. Samonis, R. Sylvester, R. Herbrecht, et al. 2004. Liposomal nystatin in patients with invasive aspergillosis refractory to or intolerant of amphotericin B. *Antimicrob. Agents Chemother.* **48:**4808–4812.

Olsen, S. J., M. R. Swerdel, B. Blue, J. M. Clark, and D. P. Bonner. 1991. Tissue distribution of amphotericin B lipid complex in laboratory animals. *J. Pharm. Pharmacol.* **43:**831–835.

Oppenheim, B. A., R. Herbrecht, and S. Kusne. 1995. The safety and efficacy of amphotericin B colloidal dispersion in the treatment of invasive mycoses. *Clin. Infect. Dis.* **21:**1145–1153.

Ostro, M. J., and P. R. Cullis. 1989. Use of liposomes as injectable drug delivery systems. *Am. J. Hosp. Pharm.* **46:**1576–1587.

Palacios, J., and R. Serrano. 1978. Proton permeability induced by polyene antibiotics. A plausible mechanism for their inhibition of maltose fermentation in yeast. *FEBS Microbiol. Lett.* **91:**198–201.

Pascual, B., A. Ayestaran, J. B. Montoro, J. Oliveras, A. Estibalez, A. Julia, and A. Lopez. 1995. Administration of lipid emulsion versus conventional amphotericin B in patients with neutropenia. *Ann. Pharmacother.* **29:**1197–1201.

Patterson, T. F. 2005. The future of animal models of invasive aspergillosis. *Med. Mycol.* **43**(Suppl. 1)**:**S115–S119.

Perfect, J. R. 2005. Treatment of non-*Aspergillus* moulds in immunocompromised patients, with amphotericin B lipid complex. *Clin. Infect. Dis.* **40**(Suppl. 6)**:**S401–S408.

Perkins, W. R., S. R. Minchey, L. T. Boni, C. E. Swenson, M. C. Popescu, R. F. Pasternack, and A. S. Janoff. 1992. Amphotericin B-phospholipid interactions responsible for reduced mammalian cell toxicity. *Biochim. Biophys. Acta* **1107:**271–282.

Pfaller, M. A., D. J. Diekema, G. W. Procop, and M. G. Rinaldi. 2007. Multicenter comparison of the VITEK 2 antifungal susceptibility test with the CLSI broth microdilution reference method for testing amphotericin B, flucytosine, and voriconazole against *Candida* spp. *J. Clin. Microbiol.* **45:**3522–3528.

Pierce, A. M., H. D. Pierce, A. M. Unrau, and A. C. Oehlschlaeger. 1978. Lipid composition and polyene antibiotic resistance of *Candida albicans. Can. J. Biochem.* **56:**135–142.

Prentice, H. G., I. M. Hann, R. Herbrecht, M. Aoun, S. Kvaloy, D. Catovsky, C. R. Pinkerton, S. A. Schey, F. Jacobs, A. Oakhill, R. F. Stevens, P. J. Darbyshire, and B. E. Gibson. 1997. A randomized comparison of liposomal versus conventional amphotericin B for the treatment of pyrexia of unknown origin in neutropenic patients. *Br. J. Haematol.* **98:**711–718.

Proffitt, R. T., A. Satorius, S. M. Chiang, L. Sullivan, and J. P. Adler-Moore. 1991. Pharmacology and toxicology of a liposomal formulation of amphotericin B (AmBisome) in rodents. *J. Antimicrob. Chemother.* **28**(Suppl. B)**:**49–61.

Ralph, E. D., A. M. Khazindar, K. R. Barber, and C. W. Grant. 1991. Comparative in vitro effects of liposomal amphotericin B, amphotericin B-deoxycholate, and free amphotericin B against fungal strains determined by using MIC and minimal lethal concentration susceptibility studies and time-kill curves. *Antimicrob. Agents Chemother.* **35:**188–191.

Ranchere, J. Y., J. F. Latour, C. Fuhrmann, C. Lagallarde, and F. Loreuil. 1996. Amphotericin B Intralipid formulation: stability and particle size. *J. Antimicrob. Chemother.* **37:**1165–1169.

Reuben, A., E. Anaissie, P. E. Nelson, R. Hashem, C. Legrand, D. H. Ho, and G. P. Bodey. 1989. Antifungal susceptibility of 44 clinical isolates of *Fusarium* species determined by using a broth microdilution method. *Antimicrob. Agents Chemother.* **33:**1647–1649.

Reynolds, E. S., Z. M. Tomkiewicz, and G. J. Dammin. 1963. The renal lesion related to amphotericin B treatment for coccidioidomycosis. *Med. Clin. North Am.* **47:**1149–1154.

Ringdén, O., E. Andström, M. Remberger, B. M. Svahn, and J. Tollemar. 1994. Safety of liposomal amphotericin B (AmBisome™) in 187 transplant recipients treated with cyclosporin. *Bone Marrow Transplant.* **14**(Suppl. 5)**:**S10–S14.

Ringdén, O., F. Meunier, J. Tollemar, P. Ricci, S. Tura, E. Kuse, M. A. Viviani, N. C. Gorin, J. Klastersky, and P. Fenaux. 1991. Efficacy of amphotericin B encapsulated in liposome (AmBisome™) in the treatment of invasive fungal infections in immunocompromised patients. *J. Antimicrob. Chemother.* **28**(Suppl. B)**:**73–82.

Roden, M. M., L. D. Nelson, T. A. Knudsen, P. F. Jarosinski, J. M. Starling, S. E. Shiflett, K. Calis, R. DeChristoforo, G. R. Donowitz, D. Buell, and T. J. Walsh. 2003. Triad of acute infusion-related reactions associated with liposomal amphotericin B: analysis of clinical and epidemiological characteristics. *Clin. Infect. Dis.* **36:**1213–1220.

Sandler, E. S., M. M. Mustafa, I. Tkaczewski, M. L. Graham, V. A. Morrison, M. Green, M. Trigg, M. Abboud, V. M. Aquino, M. Gurwith, and L. Pietrelli. 2000. Use of amphotericin B colloidal dispersion in children. *J. Pediatr. Hematol. Oncol.* **22:**242–246.

Sawaya, B. P., J. P. Briggs, and J. Schnermann. 1995. Amphotericin B nephrotoxicity: the adverse consequences of altered membrane properties. *J. Am. Soc. Nephrol.* **6:**154–164.

Scarcella, A., M. B. Pasquariello, B. Giugliano, M. Vendemmia, and A. de Lucia. 1998. Liposomal amphotericin B treatment for neonatal fungal infections. *Pediatr. Infect. Dis. J.* **17:**146–148.

Schöffski, P., M. Freund, R. Wunder, D. Petersen, C. H. Köhne, H. Hecker, U. Schubert, and A. Ganser. 1998. Safety and toxicity of amphotericin B in glucose 5% or intralipid 20% in neutropenic patients with pneumonia or fever of unknown origin: randomised study. *BMJ* **317:**379–384.

Senior, J. H. 1987. Fate and behavior of liposomes in vivo: a review of controlling factors. *Crit. Rev. Ther. Drug Carrier Syst.* **3:**123–193.

Shadkhan, Y., E. Segal, A. Bor, Y. Gov, M. Rubin, and D. Lichtenberg. 1996. The use of commercially available lipid emulsions for the preparation of amphotericin B-lipid admixtures. *J. Antimicrob. Chemother.* **39:**655–658.

Sharkey, P. K., J. R. Graybill, E. S. Johnson, S. G. Hausrath, P. B. Pollard, A. Kolokathis, D. Mildvan, P. Fan-Havard, R. H. Eng, T. F. Patterson, J. C. Pottage, Jr., M. S. Simberkoff, J. Wolf, R. D. Meyer, R. Gupta, L. W. Lee, and D. S. Gordon. 1996. Amphotericin B lipid complex compared with amphotericin B in the treatment of cryptococcal meningitis in patients with AIDS. *Clin. Infect. Dis.* **22:** 315–321.

Sievers, T. M., B. M. Kubak, and A. Wong-Beringer. 1996. Safety and efficacy of Intralipid emulsions of amphotericin B. *J. Antimicrob. Chemother.* **38:**333–347.

Sokol-Anderson, M. L., J. Brajtburg, and G. Medoff. 1986. Amphotericin B-induced oxidative damage and killing of *Candida albicans. J. Infect. Dis.* **154:**76–83.

Sorkine, P., H. Nagar, A. Weinbroum, A. Setton, E. Israitel, A. Scarlatt, A. Silbiger, V. Rudick, Y. Kluger, and P. Halpern. 1996. Administration of amphotericin B in lipid emulsion decreases nephrotoxicity: results of a prospective, randomized, controlled study in critically ill patients. *Crit. Care Med.* **24:**1311–1315.

Stuecklin-Utsch, A., C. Hasan, U. Bode, and G. Fleischhack. 2002. Pancreatic toxicity after liposomal amphotericin B. *Mycoses* **45:** 170–173.

Sundar, S., A. K. Goyal, D. K. More, M. K. Singh, and H. W. Murray. 1998. Treatment of antimony-unresponsive Indian visceral leishmaniasis with ultra-short courses of amphotericin-B-lipid complex. *Ann. Trop. Med. Parasitol.* **92:**755–764.

Sundar, S., L. B. Gupta, V. Rastogi, G. Agrawal, and H. W. Murray. 2000. Short-course, cost-effective treatment with amphotericin B-fat emulsion cures visceral leishmaniasis. *Trans. R. Soc. Trop. Med. Hyg.* **94:**200–204.

Sutton, D. A., S. E. Sanche, S. G. Revankar, A. W. Fothergill, and M. G. Rinaldi. 1999. In vitro amphotericin B resistance in clinical isolates of *Aspergillus terreus*, with a head-to-head comparison to voriconazole. *J. Clin. Microbiol.* **37:**2343–2345.

Swenson, C. E., W. R. Perkins, P. Roberts, I. Ahmad, R. Stevens, D. A. Stevens, and A. S. Janoff. 1998. In vitro and in vivo antifungal activity of amphotericin B lipid complex: are phospholipases important? *Antimicrob. Agents Chemother.* **42:**767–771.

Timmers, G. J., S. Zweegman, A. M. Simoons-Smit, A. C. van Loenen, D. Touw, and P. C. Huijgens. 2000. Amphotericin B colloidal dispersion (Amphocil) vs fluconazole for the prevention of fungal infections in neutropenic patients: data of a prematurely stopped clinical trial. *Bone Marrow Transplant.* **25:**879–884.

Tohyama, M., K. Kawakami, and A. Saito. 1996. Anticryptococcal effect of amphotericin B is mediated through macrophage production of nitric oxide. *Antimicrob. Agents Chemother.* **40:**1919–1923.

Tomlin, M., and G. S. Priestley. 1995. Elimination of liposomal amphotericin B by hemodiafiltration. *Intensive Care Med.* **21:**699–700.

Travis, L. B., G. D. Roberts, and W. R. Wilson. 1985. Clinical significance of *Pseudallescheria boydii*: a review of 10 years experience. *Mayo Clin. Proc.* **60:**531–537.

Trissel, L. A. 1995. Amphotericin B does not mix with fat emulsion. *Am. J. Health Syst. Pharm.* **52:**1463–1464.

Turnidge, J. D., S. Gudmundsson, B. Vogelman, and W. A. Craig. 1994. The postantibiotic effect of antifungal agents against common pathogenic yeasts. *J. Antimicrob. Chemother.* **34:**83–92.

Valero, G., and J. R. Graybill. 1995. Successful treatment of cryptococcal meningitis with amphotericin B colloidal dispersion: report of four cases. *Antimicrob. Agents Chemother.* **39:**2588–2590.

Vertut-Croquin, A., J. Bolard, M. Chabert, and C. Gary-Bobo. 1983. Differences in the interaction of the polyene antibiotic amphotericin B with cholesterol- or ergosterol-containing phospholipid vesicles. A circular dichroism and permeability study. *Biochemistry* **22:**2939–2944.

Wallace, T. L., V. Paetznick, P. A. Cossum, G. Lopez-Berestein, J. H. Rex, and E. Anaissie. 1997. Activity of liposomal nystatin against disseminated *Aspergillus fumigatus* infection in neutropenic mice. *Antimicrob. Agents Chemother.* **41:**2238–2243.

Walsh, T. J., V. Yeldandi, M. McEvoy, C. Gonzalez, S. Chanock, A. Freifeld, N. I. Seibel, P. O. Whitcomb, P. Jarosinski, G. Boswell, I. Bekersky, A. Alak, D. Buell, J. Barret, and W. Wilson. 1998a. Safety, tolerance, and pharmacokinetics of a small unilamellar liposomal formulation of amphotericin B (AmBisome) in neutropenic patients. *Antimicrob. Agents Chemother.* **42:**2391–2398.

Walsh, T. J., R. W. Finberg, C. Arndt, J. Hiemenz, C. Schwartz, D. Bodensteiner, P. Pappas, N. Seibel, R. N. Greenberg, S. Dummer, M. Schuster, J. S. Holcenberg, et al. 1999a. Liposomal amphotericin B for empirical therapy in patients with persistent fever and neutropenia. *N. Engl. J. Med.* **340:**764–771.

Walsh, T. J., C. Gonzalez, C. A. Lyman, S. J. Channock, and P. A. Pizzo. 1996. Invasive fungal infections in children: recent advances in diagnosis and treatment. *Adv. Pediatr. Infect. Dis.* **11:**187–290.

Walsh, T. J., J. L. Goodman, P. Pappas, I. Bekersky, D. N. Buell, M. Roden, J. Barrett, and E. J. Anaissie. 2001. Safety, tolerance, and pharmacokinetics of high-dose liposomal amphotericin B (AmBisome) in patients infected with *Aspergillus* species and other filamentous fungi: maximum tolerated dose study. *Antimicrob. Agents Chemother.* **45:**3487–3496.

Walsh, T. J., J. W. Hiemenz, N. I. Seibel, J. R. Perfect, G. Horwith, L. Lee, J. L. Silber, M. J. DiNubile, A. Reboli, E. Bow, J. Lister, and E. J. Anaissie. 1998b. Amphotericin B lipid complex for invasive fungal infections: analysis of safety and efficacy in 556 cases. *Clin. Infect. Dis.* 261383–1396.

Walsh, T. J., N. L. Seibel, C. Arndt, R. E. Harris, M. J. Dinubile, A. Reboli, J. Hiemenz, and S. J. Chanock. 1999b. Amphotericin B lipid complex in pediatric patients with invasive fungal infections. *Pediatr. Infect. Dis. J.* **18:**702–708.

Walsh, T. J., P. Whitcomb, S. Piscitelli, W. D. Figg, S. Hill, S. J. Chanock, P. Jarosinski, R. Gupta, and P. A. Pizzo. 1997. Safety, tolerance, and pharmacokinetics of amphotericin B lipid complex in children with hepatosplenic candidiasis. *Antimicrob. Agents Chemother.* **41:**1944–1948.

Walsh, T. J., J. Peter, D. A. McGough, A. W. Fothergill, M. G. Rinaldi, and P. A. Pizzo. 1995. Activities of amphotericin B and antifungal azoles alone and in combination against *Pseudallescheria boydii. Antimicrob. Agents Chemother.* **39:**1361–1364.

Walsh, T. J., V. Petraitis, R. Petraitiene, A. Field-Ridley, D. Sutton, M. Ghannoum, T. Sein, R. Schaufele, J. Peter, J. Bacher, H. Casler, D. Armstrong, A. Espinel-Ingroff, M. G. Rinaldi, and C. A. Lyman. 2003. Experimental pulmonary aspergillosis due to *Aspergillus terreus*: pathogenesis and treatment of an emerging fungal pathogen resistant to amphotericin B. *J. Infect. Dis.* **188:**305–319.

Wang, L. H., R. M. Fielding, P. C. Smith, and L. S. S. Guo. 1995. Comparative tissue distribution and elimination of amphotericin B colloidal dispersion (Amphocil®) and Fungizone® after repeated dosing in rats. *Pharm. Res.* **12:**275–283.

White, M. H., E. J. Anaissie, S. Kusne, J. R. Wingard, J. W. Hiemenz, A. Cantor, M. Gurwith, C. Du Mond, R. D. Mamelok, and R. A. Bowden. 1997. Amphotericin B colloidal dispersion vs. amphotericin B as therapy for invasive aspergillosis. *Clin. Infect. Dis.* **24:**635–642.

White, M. H., R. A. Bowden, F. Sandler, M. L. Graham, G. A. Noskin, J. R. Wingard, M. Goldman, J. A. van Burik, A. McCabe, J. S. Lin, M. Gurwith, and C. B. Miller. 1998. Randomized, double-blind clinical trial of amphotericin B colloidal dispersion vs. amphotericin B in the empiric treatment of fever and neutropenia. *Clin. Infect. Dis.* **27:**296–302.

Wiederhold, N. P., V. H. Tam, J. Chi, R. A. Prince, D. P. Kontoyiannis, and R. E. Lewis. 2006. Pharmacodynamic activity of amphotericin B deoxycholate is associated with peak plasma concentrations in a neutropenic murine model of invasive pulmonary aspergillosis. *Antimicrob. Agents Chemother.* **50:**469–473.

Wiley, J. M., N. L. Seibel, and T. J. Walsh. 2005. Efficacy and safety of amphotericin B lipid complex in 548 children and adolescents with invasive fungal infections. *Pediatr. Infect. Dis.* **24:**167–174.

Wilson, E., L. Tharson, and D. P. Speert. 1991. Enhancement of macrophage superoxide anion production by amphotericin B. *Antimicrob. Agents Chemother.* **35:**796–800.

Wingard, J. R., P. Kubilis, L. Lee, G. Yee, M. White, L. Walshe, R. Bowden, E. Anaissie, J. Hiemenz, and J. Lister. 1999. Clinical significance of nephrotoxicity in patients treated with amphotericin B for suspected or proven aspergillosis. *Clin. Infect. Dis.* **29:**1402–1407.

Wingard, J. R., M. H. White, E. Anaissie, J. Raffalli, J. Goodman, A. Arrieta, et al. 2000. A randomized, double-blind comparative trial evaluating the safety of liposomal amphotericin B versus amphotericin B lipid complex in the empirical treatment of febrile neutropenia. *Clin. Infect. Dis.* **31:**1155–1163.

Wingard, J. R. 1997. Efficacy of amphotericin B lipid complex injection (ABLC) in bone marrow transplant recipients with life-threatening systemic mycoses. *Bone Marrow Transplant.* **19:**343–347.

Wingard, J. R. 1994. Infections due to resistant *Candida* species in patients with cancer who are receiving chemotherapy. *Clin. Infect. Dis.* **19**(Suppl. 1):S49–S53.

Wong-Beringer, A., R. A. Jacobs, and B. J. Guglielmo. 1997. Lipid formulations of amphotericin B: clinical efficacy and toxicities. *Clin. Infect. Dis.* **27:**603–618.

Woods, R. A. 1971. Nystatin-resistant mutants of yeast: alterations in sterol content. *J. Bacteriol.* **108:**69–73.

Wright, D. G., K. J. Robichaud, P. A. Pizzo, and A. B. Deisseroth. 1981. Lethal pulmonary reactions associated with the combined use of amphotericin B and leukocyte transfusions. *N. Engl. J. Med.* **304:** 1185–1189.

Wurthwein, G., A. H. Groll, G. Hempel, F. C. Adler-Shohet, J. M. Lieberman, and T. J. Walsh. 2005. Population pharmacokinetics of amphotericin B lipid complex in neonates. *Antimicrob. Agents Chemother.* **49:**5092–5098.

Aspergillus fumigatus and Aspergillosis
Edited by J.-P. Latgé and W. J. Steinbach
©2009 ASM Press, Washington, DC

Chapter 31

Azoles

RAOUL HERBRECHT AND YASMINE NIVOIX

Antifungal azoles are the most widely used antifungal agents. They all act by inhibiting the fungal cytochrome P450 (CYP450)-dependent enzyme, lanosterol 14α-demethylase. Despite similar mechanisms of action, the various components of this family differ considerably by their spectrum, water or lipid solubility, pharmacokinetic profile, and tolerability, and consequently their clinical indications.

The azole antifungals are divided into two groups:

(i) The imidazoles are represented by bifonazole, butoconazole, clotrimazole, econazole, fenticonazole, isoconazole, ketoconazole, lanoconazole, miconazole, omoconazole, oxiconazole, sertaconazole, sulconazole, and tioconazole. Most of these agents are only available as topical agents.

(ii) The triazoles include terconazole, fluconazole, saperconazole, itraconazole, voriconazole, posaconazole, ravuconazole, isavuconazole, and albaconazole. The triazoles have a greater affinity for fungal than mammalian CYP450 enzymes and are therefore safer for systemic use compared to the imidazoles.

The imidazoles are not indicated for the treatment of aspergillosis. Of the triazoles, fluconazole has no activity at all against *Aspergillus* spp., saperconazole clinical development has been stopped, and terconazole is only used as a topical formulation for vulvovaginal candidiasis. This chapter will focus on the remaining agents and their role in treating invasive aspergillosis.

MECHANISMS OF ACTION OF ANTIFUNGAL AZOLES

Azole antifungal agents act by inhibiting ergosterol synthesis. Ergosterol is the predominant cell membrane sterol of most fungi, in contrast to mammalian cells, where the membrane is mainly composed of cholesterol. Azole antifungals inhibit the fungal CYP450-dependent enzyme lanosterol 14α-demethylase. This inhibition results in an interruption of ergosterol synthesis and the accumulation of toxic 14-methylsterol intermediates. Both events cause an alteration of the fungal cell membrane. Antifungal azole agents are mostly fungistatic rather than fungicidal. As azole antifungals can also inhibit many mammalian CYP450-dependent isoenzymes involved in hormone synthesis or drug metabolism, there is potential for toxicity or drug-drug interactions.

ITRACONAZOLE

Itraconazole was introduced for use in patients in 1987, at a time when amphotericin B was the only anti-*Aspergillus*-active agent available for systemic use (Glasmacher and Prentice, 2006). Itraconazole was initially available only as capsules containing sugar-coated pellets and later as a solution in hydroxypropyl-β-cyclodextrin for oral and intravenous use (Table 1).

Spectrum

The spectrum of activity of itraconazole includes *Candida* spp. (with high resistance rates in *C. glabrata* and *C. krusei*), *Cryptococcus neoformans, Aspergillus* spp., dimorphic fungi, and agents of dermatomycosis. Itraconazole has limited in vitro activity against zygomycetes (Almyroudis et al., 2007). Itraconazole MIC_{90}s range from 1 to >8 μg/ml for *Aspergillus* spp. and are therefore higher than for the other anti-*Aspergillus* azoles (Table 2).

Animal Studies

The efficacy of itraconazole has been demonstrated for oral as well as for intravenous formulations in many animal models of prophylaxis or treatment of invasive

Raoul Herbrecht • Dept. of Oncology and Hematology, Hôpital de Hautepierre, 67098 Strasbourg, France. **Yasmine Nivoix** • Dept. of Pharmacy and Pharmacology, Hôpital de Hautepierre, 67098 Strasbourg, France.

Table 1. Availability of anti-*Aspergillus* azoles and their approved indications

Drug[a]	Trade name	Water solubility	Formulation	Main solubilizing agent	Indication[b] in aspergillosis	
					FDA	EMEA
Itraconazole	Sporanox	Insoluble	Oral: - capsule, 100 mg - suspension, 10 mg/ml, 1,500 mg/bottle Intravenous: 250 mg/vial	Oral suspension and intravenous form: hydroxypropyl-β-cyclodextrin	Capsules: Aspergillosis, pulmonary and extrapulmonary, in patients who are intolerant of or are refractory to amphotericin B therapy Oral solution: Empiric therapy in febrile neutropenic patients with suspected fungal infections Intravenous: Empiric therapy in febrile neutropenic patients with suspected fungal infections	Treatment of aspergillosis
Voriconazole	Vfend	Low	Oral: - tablet, 50 or 200 mg - suspension, 40 mg/ml, 3,000 mg/bottle Intravenous: 200 mg/vial	Oral suspension: powder dispersed in water (xanthan gum as wetting and suspending agent) Intravenous form: sulfobutyl ether β-cyclodextrin sodium	Treatment of invasive aspergillosis; therapy may be instituted before results of cultures and other laboratory studies are known; however, once these results become available, antifungal therapy should be adjusted accordingly.	Treatment of invasive aspergillosis
Posaconazole	Noxafil	Insoluble	Oral: suspension, 40 mg/ml, 4,200 mg/bottle Intravenous: under development	Oral suspension: micronized powder dispersed in water (xanthan gum as wetting and suspending agent)	Prophylaxis in hematopoietic stem cell transplant recipients with graft-versus-host disease and in patients with hematologic malignancies with prolonged neutropenia from chemotherapy	Invasive aspergillosis in patients with disease that is refractory to amphotericin B or itraconazole or in patients intolerant of these drugs
Isavuconazole	NA[c]	Soluble prodrug	Oral and intravenous	Water-soluble prodrug	Ongoing phase III, double-blind, randomized study to evaluate safety and efficacy of isavuconazole vs voriconazole for primary treatment of invasive fungal disease caused by *Aspergillus* spp. or other filamentous fungi	
Ravuconazole	NA	Poorly soluble	Oral and intravenous	Water-soluble prodrug for intravenous use	Not yet in phase III clinical trials	

[a] Albaconazole is not included in this table due to insufficient data available in the literature.
[b] Indications other than aspergillosis are not listed. All formulations are not approved for all the indications listed here.
[c] NA, not available.

aspergillosis (Arrese et al., 1994; Dannaoui et al., 1999, 2000; Patterson et al., 1993; Schmitt et al., 1992). Usually, itraconazole has activity similar to amphotericin B, with the exception of infections caused by *Aspergillus terreus*, against which amphotericin B is poorly active. A correlation between in vitro resistance to itraconazole and failure of therapy has been shown (Dannaoui et al., 1999).

Table 2. In vitro activities of azoles and amphotericin B against *Aspergillus* spp. from clinical or environmental sources

Azole	MIC$_{90}$ (μg/ml) of azole against:					
	A. fumigatus	*A. flavus*	*A. niger*	*A. terreus*	*A. versicolor*	All *Aspergillus* spp.
Itraconazole	1–>8	0.5–2	1–2	0.25–0.5	2	1–>8
Voriconazole	0.5–1	0.5–1	1–2	0.25–1	0.5–1	0.5–1
Posaconazole	0.25–0.5	0.25–0.75	0.25–1	0.12–0.25	0.5–1	0.5
Isavuconazole	2	1	2	0.5	NA	2
Ravuconazole	0.5–1	1	2	0.25	1	0.5–1
Albaconazole	0.12	0.25	0.5	NA	NA	NA
Amphotericin B	0.5–4	0.5–>2	0.12–1	1–>2	1–2	1–2

[a]Data were obtained from Messer et al. (2006), Warn et al. (2006a), Diekema et al. (2003), Pfaller et al. (2002), Capilla et al. (2001), Panagopoulou et al. (2007), Guinea et al. (2005), and Manavathu et al. (2000). NA, not available.

Pharmacokinetics and Metabolism

Itraconazole is highly lipophilic. Absorption from capsules is limited and has been reported to be as low as 22% (Table 3) (Glasmacher and Prentice, 2006). Absorption of itraconazole capsules is enhanced by food and acidic beverages (e.g., a cola beverage) and is markedly decreased by concomitant antacid drugs. In contrast to capsules, itraconazole in oral solution requires no dissolution, thus its absorption is not influenced by gastric pH, and the bioavailability of the oral solution is 30 to 40% higher than that of the capsule formulation (Barone et al., 1998a; Van de Velde et al., 1996; Willems et al., 2001). Unlike the capsule formulation, absorption of itraconazole oral solution is highest in the fasted state (Barone et al., 1998b; Van de Velde et al., 1996). Steady state is achieved within 14 days of oral therapy (Barone et al., 1998b).

Following oral administration, itraconazole is highly protein bound (99%) and widely distributed in the body. It is extensively metabolized in the liver into many metabolites by CYP3A4. The main metabolite, hydroxyitraconazole, has considerable antifungal activity (Heykants et al., 1989). Hydroxyitraconazole accumulates at approximately twice the rate of itraconazole. Fecal excretion of the parent drug varies between 3 and 18% of the dose. Renal excretion of the parent drug is less than 0.03% of the dose. About 40% of the dose is excreted as inactive metabolites in the urine. Itraconazole and hydroxyitraconazole are both inhibitors of CYP3A4.

Table 3. Main pharmacokinetics characteristics and recommended dosages for itraconazole, voriconazole, and posaconazole[a]

Azole	Water solubility	Oral bioavailability	Half-life (h)	Time to steady state	Recommended dosage in aspergillosis	Specific recommendation(s)
Itraconazole	Insoluble	22% (capsules), 55% (oral solution)	20	14 days (oral therapy), 2 days (intravenous form)	Capsules: 200–400 mg/day Oral solution and intravenous: 2 × 200 mg on day 1, 2 then 200 mg/d	Capsules: take with full meal; avoid antacid therapy Suspension: take in fasted condition
Voriconazole	Low	90–96%	6–12	5 days in absence of loading dose; 3 days after loading dose	Intravenous: 2 × 6 mg/kg on day 1 then 2 × 4 mg/kg/day Oral: 2 × 200 mg/day[b] (2 × 100 mg/day if wt <40 kg)	Tablets and suspension: take 1 h before or 1 h after meal Intravenous: contraindicated if renal impairment unless benefit/risk justifies use of intravenous form
Posaconazole	Insoluble	Unknown	25–35	7–10 days	800 mg/day (4 × 200 mg or 2 × 400 mg) for curative therapy 3 × 200 mg for prophylaxis	Take oral suspension with food or a nutritional supplement

[a]Data were obtained from Glasmacher and Prentice (2006) and FDA (2003, 2004a, 2004b, 2008).
[b]It is not recommended that treatment start with the oral form. If the intravenous form is contraindicated, a doubling of the oral dose on day 1 should be considered.

Itraconazole pharmacokinetics are not significantly affected by renal function (Boelaert et al., 1988). There is no need for a dose reduction in patients with impaired renal function. However, the intravenous formulation is contraindicated in patients with a creatinine clearance below 30 ml/min, due to the accumulation of the solubilizing agent, hydroxypropyl-β-cyclodextrin (FDA, 2004b). Itraconazole cannot be removed by hemodialysis or continuous ambulatory peritoneal dialysis. In patients with a minor or moderate degree of hepatic insufficiency, initial dosing of itraconazole does not need to be changed. Monitoring of serum drug levels is recommended to adjust further dosing, targeting a trough concentration of 500 to 2,000 ng/ml.

Clinical Efficacy

Early data were obtained from uncontrolled trials or compassionate use programs conducted at a time when there were no consensus definitions for invasive fungal infection. Numbers of probable and proven infections were low in these studies, subjects had various degrees of immunosuppression, and dosages varied over a wide range (Maertens and Boogaerts, 2005). A small randomized study (32 evaluable patients) showed no difference in efficacy in a comparison of itraconazole (capsules at a dose of 400 mg/day) versus a low dose of amphotericin B plus flucytosine for the treatment of suspected fungal infections in neutropenic patients (van't Wout et al., 1991).

An open trial was conducted by the Mycosis Study Group in 76 evaluable patients with invasive aspergillosis and various underlying conditions that involved immunosuppression. Itraconazole capsules were given at a dose of 600 mg/day for 4 days followed by 400 mg/day (Denning et al., 1994). The overall favorable response rate was 39%, suggesting at the time that itraconazole could be an alternative to amphotericin B deoxycholate for invasive aspergillosis and that a controlled trial should be performed. This trial has never been performed because of the development of more promising agents, such as the lipid formulations of amphotericin B and voriconazole, and also because of the erratic absorption of itraconazole capsules.

In the 1990s, itraconazole capsules were mainly used, with some satisfactory results as oral follow-on therapy in patients with invasive aspergillosis who responded to amphotericin B and also in patients with chronic necrotizing aspergillosis, aspergilloma, and allergic bronchopulmonary aspergillosis who required a long duration of therapy (de Almeida et al., 2006; Denning, 1996; Denning et al., 1991, 2003; Wark et al., 2004). The further development of an oral solution and an intravenous solution led to the assessment of itraconazole in empiric and in prophylactic therapy of invasive fungal infections in high-risk hematological patients and solid organ transplant recipients.

Intravenous itraconazole followed by oral solution was compared to amphotericin B deoxycholate for the treatment of persistent febrile neutropenia in an open randomized trial (Boogaerts et al., 2001). Forty-seven percent of the patients responded to therapy in the itraconazole arm compared to 38% in the amphotericin B arm, and this difference was close to statistical significance. Breakthrough infection and death rates were similar in both arms. Itraconazole was better tolerated with fewer infusion-related events, less nephrotoxicity, and fewer discontinuations due to an adverse event. Patients receiving itraconazole more often had an increase in bilirubin, and this study led to the approval of itraconazole for empiric therapy.

Several studies have now been conducted to assess the prophylactic efficacy of itraconazole in neutropenic patients. Only one study convincingly demonstrated a benefit of itraconazole (given first intravenously and then as an oral solution) compared to fluconazole (given intravenously or orally) (Winston et al., 2003). The incidence of all invasive fungal infections was significantly reduced, mostly by reduction of yeast infections. The incidence of aspergillosis was numerically lower but not statistically significantly different. A meta-analysis of 13 prophylaxis studies confined to itraconazole compared to fluconazole, oral amphotericin B, oral nystatin, or a placebo confirmed the efficacy of itraconazole in reducing the incidence of all invasive fungal infections, invasive yeast infections, and invasive aspergillosis (Glasmacher et al., 2003). Those benefits were derived mainly from trials using the oral or intravenous hydroxypropyl-β-cyclodextrine solution of the drug. Fungus-related mortality was also significantly reduced, but all-cause mortality was similar in the itraconazole and control arms. A bioavailable daily dose-response analysis showed that the beneficial effect was only seen in patients receiving at least 400 mg/day of oral solution or 200 mg/day of intravenous formulation.

Itraconazole oral solution has been compared to fluconazole for prophylaxis of fungal infections in liver transplant recipients (Winston and Busuttil, 2002). Efficacy in terms of decrease in fungal colonization, occurrence of proven fungal infections, and mortality from fungal infection was similar in both arms. Itraconazole therapy was associated with more frequent gastrointestinal adverse events, and hepatic tolerance was satisfactory in both arms.

Patients with chronic granulomatous disease benefit from prophylaxis with itraconazole, as shown in a randomized placebo-controlled trial in 39 patients (Gallin et al., 2003). Patients younger than 13 years and weighing less than 50 kg received 100 mg/day itraconazole; the others received 200 mg/day. Only one patient (noncompliant with the treatment) developed an infection while receiving itraconazole, compared to seven who received the placebo. Importantly, six of the eight fungal infections were aspergillosis.

Safety and Tolerability

Itraconazole capsules are generally well tolerated. Itraconazole oral solution is associated with digestive tract intolerance, mainly nausea, vomiting, diarrhea, and abdominal pain, which can lead to a very high rate of discontinuation (up to 36%) (Glasmacher et al., 2006). These events are more likely to be due to the vehicle, hydroxypropyl-β-cyclodextrin, than to the antifungal agent itself. Hypokalemia has been reported in several trials (Glasmacher et al., 2006; Sharkey et al., 1991; Wheat et al., 1993), and it can be severe, requiring treatment discontinuation or leading to cardiac complications. Itraconazole is generally associated with only mild hepatic toxicity, mainly hyperbilirubinemia (Boogaerts et al., 2001; Glasmacher et al., 2003). However, severe hepatic events have been reported, especially in allogeneic hematopoietic stem cell transplant recipients receiving itraconazole prophylaxis concomitantly with cyclophosphamide (FDA, 2001; Marr et al., 2004a). Analysis in a subset of the allogeneic stem cell transplant recipients demonstrated higher exposure to toxic cyclophosphamide metabolites among recipients of itraconazole compared with the fluconazole group, explaining the conditioning-related liver toxicities (Marr et al., 2004b).

Negative inotropic effects were seen when itraconazole was administered intravenously to dogs and healthy human volunteers (FDA, 2001). Data from the FDA's Adverse Event Reporting System also suggest that use of itraconazole is associated with congestive heart failure in patients. Fifty-eight cases suggestive of congestive heart failure in association with the use of itraconazole were summarized in a report by Ahmad et al. (2001). The labeling of itraconazole has been changed to alert physicians to this finding, and itraconazole is now contraindicated for the treatment of superficial fungal infections in patients with signs of congestive heart failure or other ventricular dysfunction (FDA, 2004a).

As with other azoles, neurotoxicity, sometimes very severe, has been observed when itraconazole is concomitantly administered with vincristine (Bermudez et al., 2005; Bohme et al., 1995; Singh and Cundy, 2005; Tsiodras et al., 2005). The proposed mechanism is most likely attributable to either inhibition by the azole of CYP3A4 enzymes which are involved in vincristine metabolism or the blockade of P-glycoprotein pumps which participate in the cellular efflux of vinca alkaloids (Chan, 1998; Miyama et al., 1998). This interaction with vincristine is likely to occur with other vinca alkaloids, as they share similar metabolism pathways. Rare cases of neuropathy have also been reported in the absence of concomitant vincristine administration.

Breakthrough Fungal Infections

Multiple cases of yeast breakthrough infections, including several cases of *Trichosporon* sp. fungemia, and filamentous fungi breakthrough infections have been observed during itraconazole therapy (Glasmacher et al., 1999; Krcmery et al., 1998). They have mainly been associated with low serum drug levels. More recently it has been suggested that, similarly to voriconazole, itraconazole therapy might also be a risk for developing zygomycosis (Pagano et al., 2005).

Use and Administration in Aspergillosis

Itraconazole oral and intravenous solutions are approved for empiric therapy of persistent febrile neutropenia. The recommended adult dose is 200 mg twice a day for 2 days, followed by 200 mg once daily for up to 14 days for the intravenous solution or until resolution of the neutropenia for the oral solution. Duration of intravenous therapy is restricted, as the safety of intravenous formulation use exceeding 28 days is not known.

Capsule and intravenous formulations are approved for the treatment of aspergillosis with, according to the FDA label information, a restriction to patients who are refractory or intolerant to amphotericin B. The intravenous form is more appropriate for acute and life-threatening infections. Recommended intravenous invasive aspergillosis dosing is the same as for empiric therapy. When capsules are given, the daily dose should be 200 to 400 mg with a loading dose of 200 mg three times a day for the first 3 days. Itraconazole is not approved for prophylaxis of aspergillosis, although it has been used by many centers for this indication.

Guidelines

Itraconazole is considered an option for primary and salvage therapy of invasive aspergillosis (grade C III) and empiric treatment of persistent febrile neutropenia (grade C I) in the guidelines of the European Conference on Infections in Leukemia (Herbrecht et al., 2007; Marchetti et al., 2007). The same group assigned a rating of grade B I (generally recommended) for prophylaxis in hematopoietic stem cell transplant patients and grade C I (optional) for prophylaxis during induction chemotherapy for acute leukemia (Maertens et al., 2007). The Infectious Diseases Society of America recommends itraconazole for primary therapy in chronic cavitary pulmonary aspergillosis with a similar strength as for voriconazole and as first choice for allergic bronchopulmonary aspergillosis and for allergic sinusitis when medical therapy is needed (Walsh et al., 2008). Itraconazole is also proposed as an alternative for invasive aspergillosis, prophylaxis against aspergillosis, and aspergilloma (when medical therapy is needed).

VORICONAZOLE

Voriconazole is available for oral administration as tablets, as an oral suspension, and as an intravenous formulation using sulfobutyl ether β-cyclodextrin sodium as the solubilizing agent (Table 1).

Spectrum

Voriconazole has broad-spectrum activity against most pathogenic yeasts, dimorphic fungi, and *Aspergillus* spp. and opportunistic molds, excluding *Zygomycetes* (Espinel-Ingroff, 1998; Guinea et al., 2008; Verweij et al., 2002). Voriconazole MIC_{90}s range from 0.5 to 2 μg/ml for *A. fumigatus, A. flavus, A. niger*, and *A. nidulans*, and minimum fungicidal concentrations for *A. fumigatus* and *A. flavus* range from 0.5 to 8.0 μg/ml (Table 2). Voriconazole also shows activity against itraconazole- and amphotericin B-resistant isolates of *A. fumigatus* and amphotericin B-resistant *A. terreus*, with no apparent cross-resistance to these antifungal agents.

Animal Studies

The efficacy of voriconazole has mostly been assessed in guinea pigs. Administration of voriconazole to mice results in very low, often undetectable, serum drug levels due to a combination of high clearance and extensive metabolism by CYP450 isoenzymes (Sugar and Liu, 2000). Administration of grapefruit juice by gavage or in drinking water to mice is effective in producing measurable serum levels of voriconazole, which have been correlated with treatment efficacy (Graybill et al., 2003; Sugar and Liu, 2001). The impact of grapefruit juice is attributed to its inhibitory effect of gut mucosal CYP3A4.

In neutropenic animals with invasive aspergillosis, voriconazole treatment led to a significant decrease in tissue fungal burden compared with untreated control animals (Kirkpatrick et al., 2000b). Furthermore, the reduction in fungal burden was shown to be dose dependent, and clearance of *Aspergillus* from tissue was greater with voriconazole than with itraconazole or amphotericin B. Voriconazole administered intraperitoneally also demonstrated efficacy in prophylaxis and treatment of experimental *A. fumigatus* endocarditis in guinea pigs, while itraconazole given at the same dosage failed to prevent or treat the infection (Martin et al., 1997).

Pharmacokinetics and Metabolism

The pharmacokinetics of voriconazole have been investigated in healthy volunteers and patients given single and multiple doses of the drug orally and intravenously (Leveque et al., 2006). Voriconazole is absorbed rapidly after oral administration, reaching maximum serum concentrations within 1 to 2 h. Administering voriconazole with food delays the time to reach maximum plasma levels to approx 2.5 h, and fat decreases the bioavailability of voriconazole, with reductions of approximately 30% in the area under the time-concentration curve (AUC) and peak plasma concentration. Bioavailability of 90 to 96% has been reported in healthy volunteers (Table 3). Absorption is not affected by antacid agents (FDA, 2008).

Steady-state plasma drug concentrations of 2.2 to 3.4 μg/ml are achieved after oral administration of 200 mg of voriconazole twice daily (Herbrecht, 2004b). Interpatient variability in peak serum drug concentration and AUC values has been observed after multiple doses; therefore, a wide range of plasma drug concentrations can be expected in patients treated with the same dose of voriconazole. Steady-state trough plasma drug concentrations are reached after 5 days of oral or intravenous administration but can be achieved after 3 days if a loading dose is given.

Voriconazole exhibits nonlinear pharmacokinetics in adult patients due to saturable first-pass metabolism and systemic clearance; thus, peak plasma drug concentrations and the AUC increase disproportionately with increasing dose. It is estimated that, on average, increasing the oral dose in healthy subjects from 200 to 300 mg leads to a 2.5-fold increase in exposure (AUC), while increasing the intravenous dose from 3 to 4 mg/kg produces a 2.3-fold increase in exposure (FDA, 2008). The mean elimination half-life of orally administered voriconazole is 6 to 12 h, although this is also dose dependent.

Voriconazole binds moderately to plasma proteins (58 to 65%) (Leveque et al., 2006). The drug penetrates well into body fluids, including cerebrospinal fluid, and has a high volume of distribution of 4.6 liters/kg, implying wide tissue distribution. Concentrations in cerebrospinal fluid are reported to be between 29% and 68% of concurrent plasma drug levels. High concentrations of voriconazole have been reported in brain, ocular tissue, and hepatic tissue.

Patients with mild to moderate hepatic impairment have an AUC threefold higher than that of healthy patients. High accumulation of voriconazole has been reported in a patient with liver cirrhosis (Weiler et al.,

2007). It is recommended that the standard loading dose regimens be used but that the maintenance dose be halved in patients with mild to moderate hepatic cirrhosis (Child-Pugh class A and B) receiving voriconazole. The drug is contraindicated in patients with severe hepatic dysfunction (Child-Pugh class C).

Patients with renal disease had voriconazole pharmacokinetics and plasma protein binding levels similar to those of healthy volunteers. Renal impairment was found to decrease the clearance of sulfobutyl ether β-cyclodextrin sodium, which is used as the solubilizing agent in the intravenous preparation. Accumulation of sulfobutyl ether β-cyclodextrin sodium has been associated with histological changes in renal tissue in animals, although studies in human volunteers and a limited number of patients undergoing hemodialysis suggest that sulfobutyl ether β-cyclodextrin sodium has no toxic effects (von Mach et al., 2006). Data from five patients with end-stage renal disease on peritoneal dialysis receiving oral voriconazole indicated that the passage of voriconazole in the peritoneal dialysate is minimal, less than 1% of the dose in 24 h (Peng and Lien, 2005). No dose adjustment of the oral formulation is needed for patients on peritoneal dialysis. Similarly, it has been shown that there is no need for a dose adjustment in patients undergoing hemodialysis or continuous venovenous hemodiafiltration (FDA, 2008; Fuhrmann et al., 2007; Robatel et al., 2004).

Pediatric patients have a higher capacity for elimination of voriconazole per kilogram of body weight than do adult healthy volunteers, and doses greater than 4 mg/kg of body weight are required in children to achieve exposures consistent with those in adults following doses of 3 mg/kg (Walsh et al., 2004). For further details, see the discussion of pediatric aspergillosis in chapter 40.

Metabolism of voriconazole takes place in the liver via the hepatic CYP isoenzymes CYP3A4, CYP2C19, and CYP2C9. Three major and five minor metabolites have been identified which are eliminated within 48 h in the urine (~80% of dose) and feces (~20% of dose) (Roffey et al., 2003). Less than 2% of unchanged drug is detected in urine or feces. The major metabolite is voriconazole *N*-oxide, which represents approximately 72% of the metabolites in plasma. Voriconazole *N*-oxide has minimal antifungal activity but can inhibit the metabolism of other CYP2C9 and CYP3A4 substrates. The major enzyme involved in the metabolism of voriconazole is CYP2C19, and this enzyme exhibits genetic polymorphism. Between 15 and 20% of Asians are expected to be poor metabolizers of voriconazole and have a consequent fourfold-higher exposure to voriconazole. Conversely, only 3 to 5% of whites and blacks are expected to show genetic polymorphism and to be poor metabolizers of voriconazole.

Clinical Efficacy

The clinical efficacy of voriconazole has been investigated in phase II/III trials and a compassionate use program in both adult and pediatric patients with invasive fungal infections and in patients with febrile neutropenia. Clinical data collected to date suggest that voriconazole has a major role in the treatment of invasive aspergillosis, in addition to its activity in oropharyngeal and esophageal candidiasis, candidemia, and in some of the more unusual fungal infections, such as fusariosis and scedosporiosis, where treatment options are limited (Herbrecht, 2004b; Johnson and Kauffman, 2003).

Voriconazole has been used as primary therapy and as salvage therapy in immunosuppressed patients with invasive aspergillosis with encouraging results. An open, noncomparative study of voriconazole was conducted in 116 evaluable patients with proven or probable invasive aspergillosis (Denning et al., 2002). In this study, voriconazole was given as either primary therapy or as salvage therapy when treatment with another antifungal agent was considered ineffective or toxic. Complete or partial responses were seen in 58% of hematology patients, 26% of patients who had undergone hematopoietic stem cell transplantation, 16% of patients with cerebral aspergillosis, 50% of patients with disseminated aspergillosis, and 60% of patients with pulmonary or tracheal aspergillosis.

The largest study ever performed in invasive aspergillosis patients compared voriconazole with amphotericin B for primary therapy in 277 patients with invasive aspergillosis (Herbrecht et al., 2002). Patients received standard therapy with voriconazole (intravenous for at least 7 days and then oral) or amphotericin B deoxycholate. Partial (significant clinical improvement and at least 50% decrease in size of radiological lesions) or complete responses were considered a successful outcome. Patients given voriconazole had a higher successful outcome rate, with a 53% response (complete 21%, partial 32%) compared to a 32% response in those given amphotericin B (complete 17%, partial 15%). Survival at 12 weeks after treatment was significantly higher in the voriconazole group (71%), compared with only 58% survival in the amphotericin B group. Voriconazole was more effective than amphotericin B regardless of the underlying condition (allogeneic stem cell transplantation versus leukemia versus other immunosuppressive conditions), the site of infection (pulmonary versus extrapulmonary), the presence or absence of neutropenia, or the degree of certainty of the diagnosis of aspergillosis (probable versus definite). Among patients treated with voriconazole, the highest favorable response rate was seen in patients with a hematological malignancy (63%) and the lowest response rate was in allogeneic hematopoietic stem cell transplant recipients (32%). The median duration of therapy with voriconazole was 77 days in this trial, with a median of 10 days of intravenous therapy. Similar results have been reported in pediatric patients, including patients with chronic granulomatous disease (van 't Hek et al., 1998; Walsh et al., 2002a).

A retrospective analysis of 82 cases of central nervous system aspergillosis treated with voriconazole demonstrated a 35% complete or partial response rate (Schwartz et al., 2005). Most patients had received previous antifungal therapy with another agent. The complete or partial response rate was highest in patients with a hematological malignancy (54%) and lowest in solid organ (36%) or hematopoietic stem cell (16%) transplant recipients. In contrast, long-term survival was lowest in patients with a hematological malignancy (15%), but most causes of death were related to the progression of the underlying malignancy, while in solid organ or in hematopoietic stem cell transplant recipients the primary cause of death was mainly invasive aspergillosis. Neurosurgical intervention at any time after initial diagnosis of central nervous system aspergillosis was associated with improved survival compared with patients without neurosurgery.

Voriconazole has also proven to be effective in bone aspergillosis (Mouas et al., 2005), where most cases were spondylodiskitis or osteomyelitis. At end of therapy 11 of 20 (55%) patients had a favorable response. This rate, similar to the response rate observed in pulmonary aspergillosis, suggests a satisfactory tissue penetration in bone lesions. The median duration of therapy was 6 months (range, 11 weeks to 14 months) in responding patients.

There has been no large randomized study of prophylaxis in high-risk patients (leukemia patients undergoing induction or consolidation chemotherapy or hematopoietic stem cell transplant recipients) published so far.

One study compared voriconazole with liposomal amphotericin B for the empiric treatment of invasive fungal infection in neutropenic patients with persistent fever (Walsh et al., 2002b). Overall response rates were 26% in those receiving voriconazole and 31% in those given liposomal amphotericin B (95% confidence interval for the difference, −10.6 to 1.6%). This 95% confidence interval falls just outside the predefined lower limit of −10%. As a consequence, voriconazole failed to fulfill the protocol-defined criterion for noninferiority to liposomal amphotericin B with respect to overall response to empirical therapy. However, a statistically significant difference was reported in favor of voriconazole in the number of patients who developed a breakthrough fungal infection. Furthermore, voriconazole was better tolerated than liposomal amphotericin B. Voriconazole has not been approved for this indication but is nevertheless recommended by the Infectious Diseases Society of America guidelines, similarly to liposomal amphotericin B and caspofungin, for empiric therapy in persistent febrile neutropenia (Walsh et al., 2008).

Little data are available for subacute invasive or chronic pulmonary aspergillosis. In three series of 36, 24, and 11 assessable cases, there was evidence of efficacy of voriconazole given orally at a dose of 200 mg twice daily (Camuset et al., 2007; Jain and Denning, 2006; Sambatakou et al., 2006). Most of these patients had previously failed treatment with itraconazole or amphotericin B.

Safety and Tolerability

Voriconazole has a good safety profile. The most frequent adverse reactions reported after administration of oral or intravenous voriconazole are visual disturbances, hepatic abnormalities, and skin reactions. Visual disturbances (enhanced light perception, blurred vision, photophobia, or altered color discrimination) are usually mild to moderate in severity, occur shortly after dosing, and usually last up to 30 min. Visual disturbances have been reported in approximately 30 to 45% of patients receiving voriconazole in clinical studies and are most common during the first days of therapy (Herbrecht, 2004b; Herbrecht et al., 2002). Adverse visual events rarely lead to discontinuation of therapy. The retina has been shown to be the site of these side effects, with decreased amplitude of electroretinogram wave forms in dogs and humans. However, no long-term visual sequelae have been reported.

Elevations in hepatic enzymes occur in 4 to 27% of patients receiving voriconazole and are generally associated with high plasma drug levels and/or doses of the drug (Herbrecht, 2004b; Johnson and Kauffman, 2003). The usual pattern is of increases in aspartate aminotransferase and alanine aminotransferase, but increases in alkaline phosphatase and bilirubin levels have also been reported. Rare cases of severe hepatic reactions have been reported during clinical trials, including clinical hepatitis, cholestasis, and fulminant hepatic failure with resulting fatalities. Pooled safety data showed that 0.9% of patients receiving voriconazole developed hepatic failure leading to death, which is similar to the prevalence reported for comparator drugs. These events are more frequent in patients with other serious underlying medical conditions, particularly hematological malignancies, and solid organ or hematopoietic stem cell transplantation (Husain et al., 2006). High-risk patients, particularly those with preexisting hepatic dysfunction, should be monitored closely during voriconazole therapy, and voriconazole should be discontinued if clinical signs and symptoms consistent with liver failure appear that might be attributable to the drug. No significant relationship between CYP2C9, CYP2C19, or CYP3A5 polymorphisms and serum liver enzyme levels was observed in patients treated with voriconazole (Levin et al., 2007).

Skin reactions, mostly rashes, are the second most common adverse event due to voriconazole (Herbrecht, 2004b; Johnson and Kauffman, 2003). They have been reported in 1 to 19% of patients receiving voriconazole. Photosensitivity can occur, and patients taking voriconazole should be advised to avoid strong, direct sunlight during the course of their treatment. Facial erythema, cheilitis, and discoid lupus erythematous-like lesions have been described in patients receiving prolonged voriconazole therapy; these reactions resolved after discontinuation of the drug. Rare cases of Stevens-Johnson syndrome, toxic epidermal necrolysis, and erythema multiform have also been reported.

Other adverse reactions reported with voriconazole include nausea, vomiting, diarrhea, headaches, and visual hallucinations or confusion. These neurological events occurred in approximately 7% of the patients receiving voriconazole in a large clinical trial and are not related to the visual disturbances (Herbrecht et al., 2002). A case of painful peripheral neuropathy has been reported in the absence of concomitant neurotoxic drug administration (Tsiodras et al., 2005). Similar to itraconazole, concomitant administration of voriconazole and vinca alkaloids should be avoided.

Breakthrough Fungal Infections

Several case reports or short series have identified breakthrough fungal infections in patients receiving voriconazole. These breakthrough infections are usually associated with long-term therapy with voriconazole (Alexander et al., 2005; Imhof et al., 2004; Trifilio et al., 2007b). *Candida glabrata* and *Zygomycetes* accounted for most of the cases.

Voriconazole therapy has been identified as a risk factor for developing zygomycosis (Marty et al., 2004; Trifilio et al., 2007c). In most of these cases, voriconazole was administered for primary or secondary prophylaxis in hematopoietic stem cell transplant recipients, and most patients also had several of the previously established risk factors for zygomycosis. In a multivariate analysis, voriconazole therapy as well as diabetes mellitus and malnutrition were independently associated with the occurrence of zygomycosis (Kontoyiannis et al., 2005). Another group also found a correlation between occurrence of zygomycosis and azole prophylaxis. In this latter work, however, none of the patients had received voriconazole, but several were on therapy with itraconazole or fluconazole (Pagano et al., 2005). The authors suggested that the increased rate of zygomycosis was more likely to be related to recent improved diagnostic procedures of this disease rather than a selection of the agents of zygomycosis by voriconazole.

Interestingly, *Aspergillus ustus* infections occurred in two hematopoietic stem cell transplant recipients treated with voriconazole (Pavie et al., 2005). The two strains had an MIC greater than 4 μg/ml for voriconazole and itraconazole, reflecting their usual resistance to the azole antifungal agents. One of these infections had a fatal outcome; the second case was cured by a combination of liposomal amphotericin B and caspofungin.

Use and Administration in Aspergillosis

Voriconazole is approved for the treatment of invasive aspergillosis but not for prophylaxis or empiric therapy of persistent febrile neutropenia (FDA, 2008). Voriconazole is administered as an initial loading dose followed by maintenance dosing. In patients with normal hepatic and renal function, voriconazole should be administered in adults as an intravenous loading dose of 6 mg/kg once every 12 h for the first 24 h, followed by maintenance therapy of 4 mg/kg intravenously once every 12 h or 200 mg orally once every 12 h in patients weighing ≥40 kg and 100 mg orally once every 12 h in those weighing <40 kg. It is not recommended to start therapy with the oral formulation. If the intravenous formulation is contraindicated, a doubling of the oral dose on day 1 should be considered. The oral and intravenous maintenance doses can both be increased by 50% in patients with refractory disease. Oral doses should be taken at least 1 h before or 1 h after eating. Patients should be warned about the possible visual disturbances associated with voriconazole and should be advised not to drive at night or operate machinery.

Patients with mild to moderate hepatic cirrhosis (Child-Pugh class A and B) should receive the normal loading dose of voriconazole, but the maintenance dose should be reduced by half. There are no pharmacokinetic data available for patients with severe hepatic cirrhosis, and voriconazole is therefore contraindicated in these patients (Child-Pugh class C). Patients with renal insufficiency and reduced creatinine clearance (<50 ml/min) should take oral voriconazole and should not be given the intravenous preparation, to prevent accumulation of the solubilizing agent, sulfobutyl ether β-cyclodextrin sodium, unless the benefit/risk ratio to the patient justifies use of the intravenous form.

Guidelines

According to international guidelines (of the Infectious Diseases Society of America and the European Conference on Infections in Leukemia), voriconazole is strongly recommended for primary therapy of invasive pulmonary aspergillosis (Herbrecht et al., 2007; Walsh et al., 2008). Voriconazole is also recommended for extrapulmonary infections, including central nervous system aspergillosis with the exception of endophthalmitis, for which intraocular amphotericin B and partial vitrectomy are indicated (Walsh et al., 2008). Although not approved for these indications, voriconazole is also recommended for empiric therapy of febrile neutropenia and for chronic cavitary aspergillosis—but not for prophylaxis—in the guidelines of the Infectious Diseases Society of America (Walsh et al., 2008). Guidelines from the European Conference on Infection in Leukemia assigned grade B I (generally recommended with evidence from one well-executed randomized trial) to voriconazole for empiric therapy (Marchetti et al., 2007).

POSACONAZOLE

Posaconazole is an extended-spectrum triazole that is currently only available as an oral suspension (Table 1). An intravenous formulation is under development. Like other triazoles, posaconazole inhibits fungal ergosterol synthesis. It is more selective for fungal CYP450 enzyme systems and has demonstrated activity against many of the CYP51A1 mutants described to date.

Spectrum

Posaconazole is active against *Candida* spp., including *C. krusei* and *C. glabrata, Cryptococcus neoformans*, dimorphic fungi such as *Coccidioides immitis* and *Histoplasma* spp., *Aspergillus* spp., *Fusarium* spp., and some other filamentous fungi involved in human disease. Posaconazole has potent activity against *Aspergillus* species, inhibiting >90% of isolates at concentrations of ≤1 μg/ml (Table 2). Importantly, the spectrum of activity is extended to *Zygomycetes*, and clinical trials have shown efficacy of posaconazole in invasive zygomycosis (Almyroudis et al., 2007; Greenberg et al., 2006; van Burik et al., 2006).

Animal Studies

Posaconazole has shown activity in several models of experimental pulmonary, cerebral, and disseminated aspergillosis. In a rabbit model of *A. fumigatus* infection, posaconazole administered at 10 mg/kg/day significantly improved survival compared with the same dosage of itraconazole or in untreated controls (Kirkpatrick et al., 2000a). In this model, tissue burden reduction with posaconazole was similar to that observed with amphotericin B. In a very sophisticated model of pulmonary aspergillosis in rabbits, animals treated with posaconazole showed a significant improvement in survival and significant reductions in pulmonary infarct scores, total lung weights, numbers of pulmonary CFU per gram, numbers of computed tomography-monitored pulmonary lesions, and levels of galactomannan antigenemia (Petraitiene et al., 2001).

Efficacy of posaconazole was also demonstrated in steroid-pretreated mice infected with *A. flavus* and in persistently neutropenic rabbits infected with *A. terreus* (Najvar et al., 2004; Walsh et al., 2003). In this latter model, posaconazole was as effective as itraconazole and superior to liposomal amphotericin B. Posaconazole had demonstrated activity against an itraconazole-resistant strain of *A. fumigatus* in a neutropenic mouse model (Oakley et al., 1997). However, higher doses of posaconazole were required to improve survival when the strain was resistant to itraconazole compared to an itraconazole-sensitive strain, suggesting a degree of cross-resistance. Similar to amphotericin B, posaconazole also prolonged survival and reduced *A. fumigatus* burdens in the brain and kidneys in a murine model of central nervous system aspergillosis compared with caspofungin, itraconazole, or untreated controls (Imai et al., 2004).

Interestingly, experimental models have been used to investigate the prophylactic activity of posaconazole in a model of *A. fumigatus* pulmonary infection in rabbits and mice (Cacciapuoti et al., 2000; Petraitiene et al., 2001). Animals receiving prophylactic posaconazole showed a reduction in infarct scores, total lung weights, and organism clearance from lung tissue in comparison to untreated controls. The prophylactic effective dose is three to four times lower than the therapeutically active dose.

Pharmacokinetics and Metabolism

Following administration of rising single or multiple doses to healthy subjects, posaconazole exhibits dose-proportional pharmacokinetics up to 800 mg/day (Courtney et al., 2003; Ezzet et al., 2005). No increase in the AUC is observed when the posaconazole dose is increased further. Posaconazole exposure is enhanced two to four times when administered with food, with peak plasma drug concentrations attained at ~5 to 6 h postdose (Courtney et al., 2004). Similarly, administration of posaconazole with a nutritional supplement increases approximately threefold the maximum plasma concentration and the AUC.

The recommended dose for treatment of an invasive fungal infection is 800 mg/day, usually given in two doses of 400 mg. When food intake is limited because of mucositis or other gastrointestinal dysfunction, dividing the daily dose in four enhances slightly the bioavailability. Posaconazole has extensive tissue distribution and a long half-life (Table 3). Steady-state concentrations are achieved within 7 to 10 days. Posaconazole is primarily metabolized in the liver, where it undergoes glucuronidation and transformation into other biologically inactive metabolites (Krieter et al., 2004). Approximately 14% of an administered dose is excreted as multiple glucuronidated derivatives in the urine; an additional 77% is eliminated as unchanged drug in the feces. Minor amounts are excreted as unchanged drug in the urine.

Posaconazole pharmacokinetics are not significantly influenced by age, ethnicity, or renal or hepatic function (Herbrecht, 2004a). No dose adjustment is necessary to accommodate for differences in these patient factors. Furthermore, renal or hepatic impairment has no significant influence on the single-dose pharmacokinetics of posaconazole.

Posaconazole is not primarily metabolized by CYP enzymes (Herbrecht, 2004a). Thus, coadministering drugs that interact with the CYP enzyme system is unlikely to alter posaconazole plasma concentrations. An evaluation of the effects of posaconazole on the various CYP enzymes demonstrated an inhibitory effect on CYP3A4 activity but no influence on the activity of the other isoenzymes (CYP1A2, CYP2C8/9, CYP2D6, or CYP2E1). In contrast to other azoles that are known to inhibit a variety of CYP drug-metabolizing isoenzymes, posaconazole is unique in that its effects on CYP enzyme activity are relatively limited. Therefore, posaconazole may have the potential for fewer drug interactions compared with other azole antifungal agents.

Clinical Efficacy

Two large randomized studies demonstrated efficacy of posaconazole given for prophylaxis to high-risk patients (Cornely et al., 2007; Ullmann et al., 2007). The first study compared oral posaconazole (200 mg, three times a day) to oral fluconazole in 600 allogeneic stem cell transplant recipients with graft-versus-host disease who received immunosuppressive therapy (Ullmann et al., 2007). Posaconazole was as effective as fluconazole in preventing all invasive fungal infections and was superior to fluconazole in preventing proven or probable invasive aspergillosis. Posaconazole significantly reduced the fungus-related mortality, but overall mortality was similar in both arms. These results allow the conclusion of superiority of posaconazole over fluconazole.

The second study compared posaconazole (200 mg, three times a day) to either fluconazole or itraconazole in 602 patients with prolonged neutropenia following intensive chemotherapy for acute myeloblastic leukemia or myelodysplastic syndrome (Cornely et al., 2007). Patients received prophylaxis with each cycle of chemotherapy until recovery from neutropenia and complete remission, until occurrence of an invasive fungal infection, or for up to 12 weeks. Fewer patients who received posaconazole had a probable or proven invasive fungal infection (2% versus 8% in the control group) and fewer patients had aspergillosis (1% versus 7% in the control group), thus fulfilling statistical criteria for superiority. Importantly, the overall survival rate was significantly higher in patients treated with posaconazole. The estimated number needed to treat with posaconazole to prevent 1 invasive fungal infection was 16, and the number needed to treat to prevent 1 death was 14.

These are the first studies showing convincingly that aspergillosis can be prevented by a medication in patients with severe immune deficiency. Posaconazole prophylaxis is therefore a new standard of care in leukemic patients or hematopoietic stem cell transplant recipients wherever the incidence of aspergillosis is sufficiently high to warrant prophylaxis. Due to the high cost of systematic prophylaxis, there still remains controversy about the generalization of such strategies (De Pauw and Donnelly, 2007). Centers with a low incidence of invasive fungal infections and an easy access to chest computed tomography scans and biomarker tests such as *Aspergillus* galactomannan or β-D-glucan detection may prefer a preemptive strategy based on an early diagnosis of an invasive fungal infection and early initiation of the antifungal therapy rather than giving the drug prophylactically to many patients who do not need it.

The efficacy and safety of a posaconazole oral suspension (200 mg/day 4 times a day or 400 mg twice a day) have been investigated in an open-label study in patients with invasive aspergillosis who were refractory to or intolerant of conventional antifungal therapy (amphotericin B or itraconazole) (Walsh et al., 2007). Data from external control cases were collected retrospectively to provide a comparative reference group. The median duration of posaconazole therapy was 56 days (range, 3 to 360 days). Forty-two percent of the 107 patients treated with posaconazole responded favorably to the therapy, compared to only 26% in the external control group. Posaconazole conferred a survival benefit compared to the control group. Interestingly, a dose-response analysis showed a 75% response rate in the quartile of patients with the highest serum drug levels and only a 24% response rate in the quartile of patients with the lowest serum drug levels. A major variability in serum drug levels (ratio of 1:10) has been shown in patients taking the same dosage of oral posaconazole, suggesting differences in absorption and or metabolism of the antifungal drug. Following this study, posaconazole was approved in Europe for salvage therapy in invasive aspergillosis.

Safety and Tolerability

Posaconazole oral solution is usually well accepted by patients. The two large randomized prophylaxis trials provided a substantial set of safety information (Cornely et al., 2007; Ullmann et al., 2007). Posaconazole has a safety profile similar to fluconazole. The most frequent adverse events were gastrointestinal events, such as nausea, vomiting, diarrhea, anorexia, and abdominal pain. Of note, one case of torsades de pointe and one case of QTc prolongation were attributed to posaconazole by investigators in one of the prophylactic studies (Cornely et al., 2007). Hepatic toxicity is low and rarely requires discontinuation of the treatment (Walsh et al., 2007). Skin rashes have been reported in 4% of patients. As for other azoles, posaconazole may increase the neurotoxicity of vinca alkaloids (Mantadakis et al., 2007).

Breakthrough Fungal Infections

So far, few cases of breakthrough infection during posaconazole therapy have been reported. The activity of posaconazole against *Zygomycetes* should considerably reduce the risk of breakthrough zygomycosis.

Use and Administration in Aspergillosis

Posaconazole has been approved by the FDA for prophylaxis of invasive *Aspergillus* and *Candida* infections in patients 13 years of age and older who are at high risk of developing these infections due to being severely immunocompromised, such as hematopoietic stem cell transplant recipients with graft-versus-host disease or those with hematologic malignancies with prolonged neutropenia from chemotherapy (FDA, 2006). In Europe, posaconazole is also approved for invasive aspergillosis in patients with disease that is refractory to amphotericin B or itraconazole or in patients who are intolerant of these medicinal products. The recommended dose is 200 mg three times a day for prophylaxis and 800 mg/day (divided in two or four doses) for aspergillosis. To optimize absorption, each dose of posaconazole should be administered with a full meal or a liquid nutritional supplement (FDA, 2006). Alternative antifungal therapy must be considered in patients not able to ingest food or an oral nutritional supplement.

Guidelines

Posaconazole is strongly recommended as the first choice for prophylaxis against invasive aspergillosis in the guidelines of both the Infectious Diseases Society of America and the European Conference on Infections in Leukemia (Maertens et al., 2007; Walsh et al., 2008). Posaconazole is not recommended for primary therapy of invasive aspergillosis but is generally recommended for salvage therapy, at the same level of strength as lipid formulations of amphotericin B, caspofungin (Herbrecht et al., 2007), and itraconazole (Walsh et al., 2008).

RAVUCONAZOLE

Ravuconazole is being developed as an oral formulation and as a water-soluble prodrug (ravuconazole di-lysine phosphoester) for intravenous administration (Table 1). Ravuconazole exhibits good activity in vitro against *Candida* spp. (including *C. glabrata* and *C. krusei*), *C. neoformans, Trichosporon beigelii, Aspergillus* spp., and dermatophytes. The MIC_{90}s were lower for ravuconazole (0.39 μg/ml) than for itraconazole (0.78 μg/ml) in 17 clinical isolates of *A. fumigatus* (Table 2) (Hata et al., 1996b). However, approximately 2% of the isolates had MICs of >4 μg/ml in an analysis of 575 *A. fumigatus* isolates (Cuenca-Estrella et al., 2005). Ravuconazole is active against other clinically relevant filamentous fungi, including some *Zygomycetes* (Cuenca-Estrella et al., 2005; Minassian et al., 2003), yet *Scedosporium* spp. and *Fusarium* spp. are resistant. Efficacy has been shown in experimental aspergillosis in mice, rabbits, and guinea pigs (Hata et al., 1996a; Kirkpatrick et al., 2002; Petraitiene et al., 2004).

The pharmacokinetics has mainly been studied in animals, and few data are available for humans (Andes et al., 2003; Groll et al., 2005; Mikamo et al., 2002; Petraitiene et al., 2004). Most data have been presented at meetings and are not yet published. In humans, absorption of oral ravuconazole ranges from 47 to 74%. A two- to fourfold increase has been observed when ravuconazole is administered with a high-fat meal (Pasqualotto and Denning, 2008). Ravuconazole is highly protein bound (95 to 98%), has linear pharmacokinetics, and is remarkable by its very long elimination half-life (76 to 202 h) (Aperis and Mylonakis, 2006; Gupta et al., 2005; Pasqualotto and Denning, 2008). Due to this long half-life, there is a 10-fold accumulation after daily dosing for 14 days.

Clinical efficacy data are restricted to a phase I/II dose range-finding trial in onychomycosis and a phase II trial in oropharyngeal candidiasis (Gupta et al., 2005; Pasqualotto and Denning, 2008). Partial results of a phase I/II prophylactic trial in allogeneic hematopoietic stem cell transplant recipients suggested a linear dose response for plasma drug concentration (Pasqualotto and Denning, 2008). Headache, abdominal pain, diarrhea, pruritis, and rash are the most common adverse events reported so far.

ISAVUCONAZOLE

Isavuconazole is a new broad-spectrum oral and intravenous azole currently in phase III clinical trials (Table 1). In vitro, isavuconazole is active against *Candida* spp., including fluconazole-resistant strains, and against the *Aspergillus* species most frequently involved in human infections, including *A. terreus* (Table 2) (Guinea et al., 2008). MICs are similar for isavuconazole and voriconazole. Isavuconazole also has activity against dermatophytes and some *Zygomycetes*. Isavuconazole demonstrated impressive antifungal activity against *A. flavus* in a model of disseminated infection in mice (Warn et al., 2006b).

The drug is being developed as a water-soluble prodrug (BAL8557) suitable for oral and intravenous administration. The prodrug is rapidly and almost totally (>99%) converted by esterases into BAL8728 (the prodrug cleavage product) and to the active drug (isavuconazole; BAL4815). The drug is characterized by a long elimination half-life (56 to 77 h after oral administration and 76 to 104 h after intravenous administration) (Pasqualotto and Denning, 2008). The most frequent adverse events reported in phase II studies were headache, nasopharyngitis, and rhinitis. Phase III studies are ongoing for invasive aspergillosis (compared to voriconazole) and for invasive candidiasis (compared to caspofungin followed by oral voriconazole).

ALBACONAZOLE

Albaconazole was developed as an oral solution and as a topical agent (Table 1). It was highly active in vitro against 77 strains of various opportunistic filamentous fungi (Capilla et al., 2001). MIC_{90}s against *Aspergillus* isolates were very low (Table 2), and albaconazole was more active than amphotericin B against all fungi tested except for *Fusarium solani* and *Scytalidium* spp. Potent in vitro activity has also been shown against *Candida* spp., including strains with decreased susceptibility to fluconazole, and against *C. neoformans* (Miller et al., 2004; Ramos et al., 1999).

Good antifungal activity was confirmed in preliminary experimental models of aspergillosis and candidiasis in rats (Bartroli et al., 1998). Albaconazole improved in a dose-dependent manner the survival of rabbits infected with *Scedosporium prolificans* (Capilla et al., 2003). Data on the pharmacokinetic profile of albaconazole are available mostly from studies in animals (Bartroli et al., 1998). Oral bioavailability is excellent in rats and in dogs (Pasqualotto and Denning, 2008). In humans, albaconazole is rapidly absorbed (less than 1 h). The pharmacokinetics is nonlinear, and the plasma half-life is 30 to 56 h (Pasqualotto and Denning, 2008).

Clinical efficacy data are restricted to vulvovaginal candidiasis, in which a low dose of 40 mg albaconazole oral solution showed better efficacy than 150 mg of fluconazole. No serious adverse event has been reported (Pasqualotto and Denning, 2008), and no data are available for aspergillosis.

AZOLES AND DRUG MONITORING

Various factors interfere with the pharmacokinetics and metabolism of antifungal azoles. These conditions include the formulation of the antifungal agent, variation in gastric pH, digestive tract dysfunction, genetic polymorphism of CYP450 isoenzymes, saturation of intestinal absorption or of metabolism, potential liver disease, and potential interactions with food or with other drugs. The large variations in itraconazole, voriconazole, or posaconazole serum levels may be associated with decreased efficacy, increased toxicity, or occurrence of breakthrough infections (Goodwin and Drew, 2008; Pascual et al., 2008; Pasqualotto et al., 2008; Trifilio et al., 2007a, 2007b, 2007c).

Itraconazole serum levels should be monitored in all patients treated with itraconazole capsules for a prolonged period for invasive fungal infections such as aspergillosis or histoplasmosis (Goodwin and Drew, 2008). A trough serum level of 500 ng/ml of itraconazole, measured by high-performance liquid chromatography, has been considered adequate for prophylaxis (Glasmacher et al., 2003). A trough concentration of itraconazole around 1,000 ng/ml should be considered a target concentration for activity against invasive aspergillosis (Glasmacher and Prentice, 2006). The additional concentration of the active first-pass metabolite of itraconazole, hydroxyitraconazole, more than doubles the total serum drug concentration.

Large variabilities in voriconazole serum levels have been documented in patients (Pascual et al., 2007; Trifilio et al., 2007a). Underexposure to voriconazole can result in failure to cure the infection and increased risk of breakthrough infection. A failure rate of 46% has been observed in patients with trough levels of ≤1,000 ng/ml, while only 12% of the patients with a serum drug level of >1,000 ng/ml failed to respond (Pascual et al., 2008). In another study, it was shown that breakthrough *Candida* spp. infections were more likely to occur in patients with serum drug levels of <2,000 ng/ml (Trifilio et al., 2007b). Overexposure also needs to be documented in patients with unexpected toxicity: a correlation has been found between high serum drug levels (>5,500 ng/ml) and occurrence of encephalopathy (Pascual et al., 2008). As under- and overexposure lead to failures to respond or to adverse events, monitoring of voriconazole through serum drug levels is recommended to optimize its safety and efficacy (Walsh et al., 2008).

Little information is available regarding the use of posaconazole drug monitoring. Monitoring serum levels in a clinical trial showed variability in trough serum drug levels of more than a 1:10 ratio (Walsh et al., 2007). The highest serum drug levels were associated with the highest response rates. Conversely, 25% of the patients had very low serum drug levels, and only 24% of them responded to therapy. Early identification of these patients allows the switch to another therapy to increase the probability of efficacy of the antifungal treatment.

REFERENCES

Ahmad, S. R., S. J. Singer, and B. G. Leissa. 2001. Congestive heart failure associated with itraconazole. *Lancet* **357:**1766–1767.

Alexander, B. D., W. A. Schell, J. L. Miller, G. D. Long, and J. R. Perfect. 2005. *Candida glabrata* fungemia in transplant patients receiving voriconazole after fluconazole. *Transplantation* **80:**868–871.

Almyroudis, N. G., D. A. Sutton, A. W. Fothergill, M. G. Rinaldi, and S. Kusne. 2007. In vitro susceptibilities of 217 clinical isolates of zygomycetes to conventional and new antifungal agents. *Antimicrob. Agents Chemother.* **51:**2587–2590.

Andes, D., K. Marchillo, T. Stamstad, and R. Conklin. 2003. In vivo pharmacodynamics of a new triazole, ravuconazole, in a murine candidiasis model. *Antimicrob. Agents Chemother.* **47:**1193–1199.

Aperis, G., and E. Mylonakis. 2006. Newer triazole antifungal agents: pharmacology, spectrum, clinical efficacy and limitations. *Expert Opin. Investig. Drugs* **15:**579–602.

Arrese, J. E., P. Delvenne, J. Van Cutsem, C. Pierard-Franchimont, and G. E. Pierard. 1994. Experimental aspergillosis in guinea pigs: influence of itraconazole on fungaemia and invasive fungal growth. *Mycoses* **37:**117–122.

Barone, J. A., B. L. Moskovitz, J. Guarnieri, A. E. Hassell, J. L. Colaizzi, R. H. Bierman, and L. Jessen. 1998a. Enhanced bioavailability of itraconazole in hydroxypropyl-β-cyclodextrin solution versus capsules in healthy volunteers. *Antimicrob. Agents Chemother.* **42:**1862–1865.

Barone, J. A., B. L. Moskovitz, J. Guarnieri, A. E. Hassell, J. L. Colaizzi, R. H. Bierman, and L. Jessen. 1998b. Food interaction and steady-state pharmacokinetics of itraconazole oral solution in healthy volunteers. *Pharmacotherapy* **18:**295–301.

Bartroli, J., E. Turmo, M. Alguero, E. Boncompte, M. L. Vericat, L. Conte, J. Ramis, M. Merlos, J. Garcia-Rafanell, and J. Forn. 1998. New azole antifungals. 3. Synthesis and antifungal activity of 3-substituted-4(3*H*)-quinazolinones. *J. Med. Chem.* **41:**1869–1882.

Bermudez, M., J. L. Fuster, E. Llinares, A. Galera, and C. Gonzalez. 2005. Itraconazole-related increased vincristine neurotoxicity: case report and review of literature. *J. Pediatr. Hematol. Oncol.* **27:**389–392.

Boelaert, J., M. Schurgers, E. Matthys, R. Daneels, A. van Peer, K. de Beule, R. Woestenborghs, and J. Heykants. 1988. Itraconazole pharmacokinetics in patients with renal dysfunction. *Antimicrob. Agents Chemother.* **32:**1595–1597.

Bohme, A., A. Ganser, and D. Hoelzer. 1995. Aggravation of vincristine-induced neurotoxicity by itraconazole in the treatment of adult ALL. *Ann. Hematol.* **71:**311–312.

Boogaerts, M., D. J. Winston, E. J. Bow, G. Garber, A. C. Reboli, A. P. Schwarer, N. Novitzky, A. Boehme, E. Chwetzoff, and K. de Beule. 2001. Intravenous and oral itraconazole versus intravenous amphotericin B deoxycholate as empirical antifungal therapy for persistent fever in neutropenic patients with cancer who are receiving broad-spectrum antibacterial therapy. A randomized, controlled trial. *Ann. Intern. Med.* **135:**412–422.

Cacciapuoti, A., D. Loebenberg, E. Corcoran, F. Menzel, Jr., E. L. Moss, Jr., C. Norris, M. Michalski, K. Raynor, J. Halpern, C. Mendrick, B. Arnold, B. Antonacci, R. Parmegiani, T. Yarosh-Tomaine, G. H. Miller, and R. S. Hare. 2000. In vitro and in vivo activities of SCH 56592 (posaconazole), a new triazole antifungal agent, against *Aspergillus* and *Candida*. *Antimicrob. Agents Chemother.* **44:**2017–2022.

Camuset, J., H. Nunes, M. C. Dombret, A. Bergeron, P. Henno, B. Philippe, G. Dauriat, G. Mangiapan, A. Rabbat, and J. Cadranel. 2007. Treatment of chronic pulmonary aspergillosis by voriconazole in nonimmunocompromised patients. *Chest* **131:**1435–1441.

Capilla, J., M. Ortoneda, F. J. Pastor, and J. Guarro. 2001. In vitro antifungal activities of the new triazole UR-9825 against clinically important filamentous fungi. *Antimicrob. Agents Chemother.* **45:**2635–2637.

Capilla, J., C. Yustes, E. Mayayo, B. Fernandez, M. Ortoneda, F. J. Pastor, and J. Guarro. 2003. Efficacy of albaconazole (UR-9825) in treatment of disseminated *Scedosporium prolificans* infection in rabbits. *Antimicrob. Agents Chemother.* **47:**1948–1951.

Chan, J. D. 1998. Pharmacokinetic drug interactions of vinca alkaloids: summary of case reports. *Pharmacotherapy* **18:**1304–1307.

Cornely, O. A., J. Maertens, D. J. Winston, J. Perfect, A. J. Ullmann, T. J. Walsh, D. Helfgott, J. Holowiecki, D. Stockelberg, Y. T. Goh, M. Petrini, C. Hardalo, R. Suresh, and D. Angulo-Gonzalez. 2007. Posaconazole vs. fluconazole or itraconazole prophylaxis in patients with neutropenia. *N. Engl. J. Med.* **356:**348–359.

Courtney, R., S. Pai, M. Laughlin, J. Lim, and V. Batra. 2003. Pharmacokinetics, safety, and tolerability of oral posaconazole administered in single and multiple doses in healthy adults. *Antimicrob. Agents Chemother.* **47:**2788–2795.

Courtney, R., D. Wexler, E. Radwanski, J. Lim, and M. Laughlin. 2004. Effect of food on the relative bioavailability of two oral formulations of posaconazole in healthy adults. *Br. J. Clin. Pharmacol.* **57:**218–222.

Cuenca-Estrella, M., A. Gomez-Lopez, E. Mellado, G. Garcia-Effron, A. Monzon, and J. L. Rodriguez-Tudela. 2005. In vitro activity of ravuconazole against 923 clinical isolates of nondermatophyte filamentous fungi. *Antimicrob. Agents Chemother.* **49:**5136–5138.

Dannaoui, E., E. Borel, F. Persat, M. F. Monier, M. A. Piens, et al. 1999. In-vivo itraconazole resistance of *Aspergillus fumigatus* in systemic murine aspergillosis. *J. Med. Microbiol.* **48:**1087–1093.

Dannaoui, E., E. Borel, F. Persat, M. A. Piens, and S. Picot. 2000. Amphotericin B resistance of *Aspergillus terreus* in a murine model of disseminated aspergillosis. *J. Med. Microbiol.* **49:**601–606.

de Almeida, M. B., M. H. Bussamra, and J. C. Rodrigues. 2006. Allergic bronchopulmonary aspergillosis in paediatric cystic fibrosis patients. *Paediatr. Respir. Rev.* **7:**67–72.

Denning, D. W. 1996. Therapeutic outcome in invasive aspergillosis. *Clin. Infect. Dis.* **23:**608–615.

Denning, D. W., J. Y. Lee, J. S. Hostetler, P. Pappas, C. A. Kauffman, D. H. Dewsnup, J. N. Galgiani, J. R. Graybill, A. M. Sugar, A. Catanzaro, et al. 1994. Multicenter trial of oral itraconazole therapy for invasive aspergillosis. *Am. J. Med.* **97:**135–144.

Denning, D. W., P. Ribaud, N. Milpied, D. Caillot, R. Herbrecht, E. Thiel, A. Haas, M. Ruhnke, and H. Lode. 2002. Efficacy and safety of voriconazole in the treatment of acute invasive aspergillosis. *Clin. Infect. Dis.* **34:**563–571.

Denning, D. W., K. Riniotis, R. Dobrashian, and H. Sambatakou. 2003. Chronic cavitary and fibrosing pulmonary and pleural aspergillosis: case series, proposed nomenclature change, and review. *Clin. Infect. Dis.* 37(Suppl. 3):S265–S280.

Denning, D. W., J. E. Van Wye, N. J. Lewiston, and D. A. Stevens. 1991. Adjunctive therapy of allergic bronchopulmonary aspergillosis with itraconazole. *Chest* **100:**813–819.

De Pauw, B. E., and J. P. Donnelly. 2007. Prophylaxis and aspergillosis: has the principle been proven? *N. Engl. J. Med.* **356:**409–411.

Diekema, D. J., S. A. Messer, R. J. Hollis, R. N. Jones, and M. A. Pfaller. 2003. Activities of caspofungin, itraconazole, posaconazole, ravuconazole, voriconazole, and amphotericin B against 448 recent clinical isolates of filamentous fungi. *J. Clin. Microbiol.* **41:**3623–3626.

Espinel-Ingroff, A. 1998. In vitro activity of the new triazole voriconazole (UK-109,496) against opportunistic filamentous and dimorphic fungi and common and emerging yeast pathogens. *J. Clin. Microbiol.* 36:198–202.

Ezzet, F., D. Wexler, R. Courtney, G. Krishna, J. Lim, and M. Laughlin. 2005. Oral bioavailability of posaconazole in fasted healthy subjects: comparison between three regimens and basis for clinical dosage recommendations. *Clin. Pharmacokinet.* **44:**211–220.

Food and Drug Administration. 2001. Advisory for fungal drugs. *FDA Consum.* 35:4.

Food and Drug Administration. 2003. Sporanox (itraconazole) oral solution label information. http://www.fda.gov/cder/foi/label/2003/20657slr010_sporanox_lbl.pdf. Accessed 15 March 2008.

Food and Drug Administration. 2004a. Sporanox (itraconazole) capsules label information. http://www.fda.gov/cder/foi/label/2004/20083s034,035lbl.pdf. Accessed 15 March 2008.

Food and Drug Administration. 2004b. Sporanox (itraconazole) injection label information. http://www.fda.gov/cder/foi/label/2004/20966s011lbl.pdf. Accessed 15 March 2008.

Food and Drug Administration. 2006. Noxafil oral suspension product information. http://www.fda.gov/cder/foi/label/2006/022027lbl.pdf. Accessed 15 March 2008.

Food and Drug Administration. 2008. Vfend (voriconazole) iv, tablets and oral supension label information. http://www.fda.gov/cder/foi/label/2008/021266s023,021267s024,021630s013lbl.pdf. Accessed 15 March 2008.

Fuhrmann, V., P. Schenk, W. Jaeger, M. Miksits, N. Kneidinger, J. Warszawska, U. Holzinger, R. Kitzberger, and F. Thalhammer. 2007. Pharmacokinetics of voriconazole during continuous venovenous haemodiafiltration. *J. Antimicrob. Chemother.* **60:**1085–1090.

Gallin, J. I., D. W. Alling, H. L. Malech, R. Wesley, D. Koziol, B. Marciano, E. M. Eisenstein, M. L. Turner, E. S. DeCarlo, J. M. Starling, and S. M. Holland. 2003. Itraconazole to prevent fungal infections in chronic granulomatous disease. *N. Engl. J. Med.* **348:** 2416–2422.

Glasmacher, A., O. Cornely, A. J. Ullmann, U. Wedding, H. Bodenstein, H. Wandt, C. Boewer, R. Pasold, H. H. Wolf, M. Hanel, G. Dolken, C. Junghanss, R. Andreesen, and H. Bertz. 2006. An open-label randomized trial comparing itraconazole oral solution with fluconazole oral solution for primary prophylaxis of fungal infections in patients with haematological malignancy and profound neutropenia. *J. Antimicrob. Chemother.* 57:317–325.

Glasmacher, A., C. Hahn, C. Leutner, E. Molitor, E. Wardelmann, C. Losem, T. Sauerbruch, G. Marklein, and I. G. Schmidt-Wolf. 1999. Breakthrough invasive fungal infections in neutropenic patients after prophylaxis with itraconazole. *Mycoses* **42:**443–451.

Glasmacher, A., and A. Prentice. 2006. Current experience with itraconazole in neutropenic patients: a concise overview of pharmacological properties and use in prophylactic and empirical antifungal therapy. *Clin. Microbiol. Infect.* 12(Suppl. 7):84–90.

Glasmacher, A., A. Prentice, M. Gorschluter, S. Engelhart, C. Hahn, B. Djulbegovic, and I. G. Schmidt-Wolf. 2003. Itraconazole prevents invasive fungal infections in neutropenic patients treated for hematologic malignancies: evidence from a meta-analysis of 3,597 patients. *J. Clin. Oncol.* **21:**4615–4626.

Goodwin, M. L., and R. H. Drew. 2008. Antifungal serum concentration monitoring: an update. *J. Antimicrob. Chemother.* **61:**17–25.

Graybill, J. R., L. K. Najvar, G. M. Gonzalez, S. Hernandez, and R. Bocanegra. 2003. Improving the mouse model for studying the efficacy of voriconazole. *J. Antimicrob. Chemother.* **51:**1373–1376.

Greenberg, R. N., K. Mullane, J. A. van Burik, I. Raad, M. J. Abzug, G. Anstead, R. Herbrecht, A. Langston, K. A. Marr, G. Schiller, M. Schuster, J. R. Wingard, C. E. Gonzalez, S. G. Revankar, G. Corcoran, R. J. Kryscio, and R. Hare. 2006. Posaconazole as salvage therapy for zygomycosis. *Antimicrob. Agents Chemother.* **50:**126–133.

Groll, A. H., D. Mickiene, V. Petraitis, R. Petraitiene, A. Kelaher, A. Sarafandi, G. Wuerthwein, J. Bacher, and T. J. Walsh. 2005. Compartmental pharmacokinetics and tissue distribution of the antifungal triazole ravuconazole following intravenous administration of its di-lysine phosphoester prodrug (BMS-379224) in rabbits. *J. Antimicrob. Chemother.* **56:**899–907.

Guinea, J., T. Pelaez, L. Alcala, M. J. Ruiz-Serrano, and E. Bouza. 2005. Antifungal susceptibility of 596 *Aspergillus fumigatus* strains isolated from outdoor air, hospital air, and clinical samples: analysis by site of isolation. *Antimicrob. Agents Chemother.* **49:**3495–3497.

Guinea, J., T. Pelaez, S. Recio, M. Torres-Narbona, and E. Bouza. 2008. In vitro antifungal activities of isavuconazole (BAL4815), voriconazole, and fluconazole against 1,007 isolates of zygomycete, *Candida, Aspergillus, Fusarium*, and *Scedosporium* species. *Antimicrob. Agents Chemother.* **52:**1396–1400.

Gupta, A. K., C. Leonardi, R. R. Stoltz, P. F. Pierce, and B. Conetta. 2005. A phase I/II randomized, double-blind, placebo-controlled, dose-ranging study evaluating the efficacy, safety and pharmacokinetics of ravuconazole in the treatment of onychomycosis. *J. Eur. Acad. Dermatol. Venereol.* **19:**437–443.

Hata, K., J. Kimura, H. Miki, T. Toyosawa, M. Moriyama, and K. Katsu. 1996a. Efficacy of ER-30346, a novel oral triazole antifungal agent, in experimental models of aspergillosis, candidiasis, and cryptococcosis. *Antimicrob. Agents Chemother.* **40:**2243–2247.

Hata, K., J. Kimura, H. Miki, T. Toyosawa, T. Nakamura, and K. Katsu. 1996b. In vitro and in vivo antifungal activities of ER-30346, a novel oral triazole with a broad antifungal spectrum. *Antimicrob. Agents Chemother.* **40:**2237–2242.

Herbrecht, R. 2004a. Posaconazole: a potent, extended-spectrum triazole anti-fungal for the treatment of serious fungal infections. *Int. J. Clin. Pract.* 58:612–624.

Herbrecht, R. 2004b. Voriconazole: therapeutic review of a new azole antifungal. *Expert Rev. Anti Infect. Ther.* 2:485–497.

Herbrecht, R., D. W. Denning, T. F. Patterson, J. E. Bennett, R. E. Greene, J. W. Oestmann, W. V. Kern, K. A. Marr, P. Ribaud, O. Lortholary, R. Sylvester, R. H. Rubin, J. R. Wingard, P. Stark, C. Durand, D. Caillot, E. Thiel, P. H. Chandrasekar, M. R. Hodges, H. T. Schlamm, P. F. Troke, and B. de Pauw. 2002. Voriconazole versus amphotericin B for primary therapy of invasive aspergillosis. *N. Engl. J. Med.* **347:**408–415.

Herbrecht, R., U. Fluckiger, B. Gachot, P. Ribaud, A. Thiebaut, and C. Cordonnier. 2007. Treatment of invasive *Candida* and invasive *Aspergillus* infections in adult haematological patients. *Eur. J. Cancer Suppl.* 5:49–52.

Heykants, J., A. van Peer, V. Van de Velde, P. Van Rooy, W. Meuldermans, K. Lavrijsen, R. Woestenborghs, J. Van Cutsem, and G. Cauwenbergh. 1989. The clinical pharmacokinetics of itraconazole: an overview. *Mycoses* 32(Suppl. 1):67–87.

Husain, S., D. L. Paterson, S. Studer, J. Pilewski, M. Crespo, D. Zaldonis, K. Shutt, D. L. Pakstis, A. Zeevi, B. Johnson, E. J. Kwak, and K. R. McCurry. 2006. Voriconazole prophylaxis in lung transplant recipients. *Am. J. Transplant.* 6:3008–3016.

Imai, J. K., G. Singh, K. V. Clemons, and D. A. Stevens. 2004. Efficacy of posaconazole in a murine model of central nervous system aspergillosis. *Antimicrob. Agents Chemother.* 48:4063–4066.

Imhof, A., S. A. Balajee, D. N. Fredricks, J. A. Englund, and K. A. Marr. 2004. Breakthrough fungal infections in stem cell transplant recipients receiving voriconazole. *Clin. Infect. Dis.* 39:743–746.

Jain, L. R., and D. W. Denning. 2006. The efficacy and tolerability of voriconazole in the treatment of chronic cavitary pulmonary aspergillosis. *J. Infect.* 52:e133–e137.

Johnson, L. B., and C. A. Kauffman. 2003. Voriconazole: a new triazole antifungal agent. *Clin. Infect. Dis.* 36:630–637.

Kirkpatrick, W. R., R. K. McAtee, A. W. Fothergill, D. Loebenberg, M. G. Rinaldi, and T. F. Patterson. 2000a. Efficacy of SCH56592 in a rabbit model of invasive aspergillosis. *Antimicrob. Agents Chemother.* 44:780–782.

Kirkpatrick, W. R., R. K. McAtee, A. W. Fothergill, M. G. Rinaldi, and T. F. Patterson. 2000b. Efficacy of voriconazole in a guinea pig model of disseminated invasive aspergillosis. *Antimicrob. Agents Chemother.* 44:2865–2868.

Kirkpatrick, W. R., S. Perea, B. J. Coco, and T. F. Patterson. 2002. Efficacy of ravuconazole (BMS-207147) in a guinea pig model of disseminated aspergillosis. *J. Antimicrob. Chemother.* 49:353–357.

Kontoyiannis, D. P., M. S. Lionakis, R. E. Lewis, G. Chamilos, M. Healy, C. Perego, A. Safdar, H. Kantarjian, R. Champlin, T. J. Walsh, and I. I. Raad. 2005. Zygomycosis in a tertiary-care cancer center in the era of *Aspergillus*-active antifungal therapy: a case-control observational study of 27 recent cases. *J. Infect. Dis.* 191: 1350–1360.

Krcmery, V., Jr., E. Oravcova, S. Spanik, M. Mrazova-Studena, J. Trupl, A. Kunova, K. Stopkova-Grey, E. Kukuckova, I. Krupova, A. Demitrovicova, and K. Kralovicova. 1998. Nosocomial breakthrough fungaemia during antifungal prophylaxis or empirical antifungal therapy in 41 cancer patients receiving antineoplastic chemotherapy: analysis of aetiology risk factors and outcome. *J. Antimicrob. Chemother.* 41:373–380.

Krieter, P., B. Flannery, T. Musick, M. Gohdes, M. Martinho, and R. Courtney. 2004. Disposition of posaconazole following single-dose oral administration in healthy subjects. *Antimicrob. Agents Chemother.* 48:3543–3551.

Leveque, D., Y. Nivoix, F. Jehl, and R. Herbrecht. 2006. Clinical pharmacokinetics of voriconazole. *Int. J. Antimicrob. Agents* 27: 274–284.

Levin, M. D., J. G. den Hollander, B. van der Holt, B. J. Rijnders, M. van Vliet, P. Sonneveld, and R. H. van Schaik. 2007. Hepatotoxicity of oral and intravenous voriconazole in relation to cytochrome P450 polymorphisms. *J. Antimicrob. Chemother.* 60:1104–1107.

Maertens, J., and M. Boogaerts. 2005. The place for itraconazole in treatment. *J. Antimicrob. Chemother.* 56(Suppl. 1):i33–i38.

Maertens, J. A., P. Frère, C. Lass-Floerl, W. Heinz, and O. A. Cornely. 2007. Primary antifungal prophylaxis in leukemia patients. *Eur. J. Cancer Suppl.* 5:43–48.

Manavathu, E. K., J. L. Cutright, D. Loebenberg, and P. H. Chandrasekar. 2000. A comparative study of the in vitro susceptibilities of clinical and laboratory-selected resistant isolates of *Aspergillus* spp. to amphotericin B, itraconazole, voriconazole and posaconazole (SCH 56592). *J. Antimicrob. Chemother.* 46:229–234.

Mantadakis, E., G. Amoiridis, A. Kondi, and M. Kalmanti. 2007. Possible increase of the neurotoxicity of vincristine by the concurrent use of posaconazole in a young adult with leukemia. *J. Pediatr. Hematol. Oncol.* 29:130.

Marchetti, O., C. Cordonnier, and T. Calandra. 2007. Empirical antifungal therapy in neutropenic cancer patients with persistent fever. *Eur. J. Cancer Suppl.* 5:32–42.

Marr, K. A., F. Crippa, W. Leisenring, M. Hoyle, M. Boeckh, S. A. Balajee, W. G. Nichols, B. Musher, and L. Corey. 2004a. Itraconazole versus fluconazole for prevention of fungal infections in patients receiving allogeneic stem cell transplants. *Blood* 103:1527–1533.

Marr, K. A., W. Leisenring, F. Crippa, J. T. Slattery, L. Corey, M. Boeckh, and G. B. McDonald. 2004b. Cyclophosphamide metabolism is affected by azole antifungals. *Blood* 103:1557–1559.

Martin, M. V., J. Yates, and C. A. Hitchcock. 1997. Comparison of voriconazole (UK-109,496) and itraconazole in prevention and treatment of *Aspergillus fumigatus* endocarditis in guinea pigs. *Antimicrob. Agents Chemother.* 41:13–16.

Marty, F. M., L. A. Cosimi, and L. R. Baden. 2004. Breakthrough zygomycosis after voriconazole treatment in recipients of hematopoietic stem-cell transplants. *N. Engl. J. Med.* 350:950–952.

Messer, S. A., R. N. Jones, and T. R. Fritsche. 2006. International surveillance of *Candida* spp. and *Aspergillus* spp.: report from the SENTRY Antimicrobial Surveillance Program (2003). *J. Clin. Microbiol.* 44:1782–1787.

Mikamo, H., X. H. Yin, Y. Hayasaki, Y. Shimamura, K. Uesugi, N. Fukayama, M. Satoh, and T. Tamaya. 2002. Penetration of ravuconazole, a new triazole antifungal, into rat tissues. *Chemotherapy* 48:7–9.

Miller, J. L., W. A. Schell, E. A. Wills, D. L. Toffaletti, M. Boyce, D. K. Benjamin, Jr., J. Bartroli, and J. R. Perfect. 2004. In vitro and in vivo efficacies of the new triazole albaconazole against *Cryptococcus neoformans*. *Antimicrob. Agents Chemother.* 48:384–387.

Minassian, B., E. Huczko, T. Washo, D. Bonner, and J. Fung-Tomc. 2003. In vitro activity of ravuconazole against *Zygomycetes, Scedosporium* and *Fusarium* isolates. *Clin. Microbiol. Infect.* 9:1250–1252.

Miyama, T., H. Takanaga, H. Matsuo, K. Yamano, K. Yamamoto, T. Iga, M. Naito, T. Tsuruo, H. Ishizuka, Y. Kawahara, and Y. Sawada. 1998. P-glycoprotein-mediated transport of itraconazole across the blood-brain barrier. *Antimicrob. Agents Chemother.* 42: 1738–1744.

Mouas, H., I. Lutsar, B. Dupont, O. Fain, R. Herbrecht, F. X. Lescure, and O. Lortholary. 2005. Voriconazole for invasive bone aspergillosis: a worldwide experience of 20 cases. *Clin. Infect. Dis.* 40:1141–1147.

Najvar, L. K., A. Cacciapuoti, S. Hernandez, J. Halpern, R. Bocanegra, M. Gurnani, F. Menzel, D. Loebenberg, and J. R. Graybill. 2004. Activity of posaconazole combined with amphotericin B against *Aspergillus flavus* infection in mice: comparative studies in two laboratories. *Antimicrob. Agents Chemother.* 48:758–764.

Oakley, K. L., G. Morrissey, and D. W. Denning. 1997. Efficacy of SCH-56592 in a temporarily neutropenic murine model of invasive aspergillosis with an itraconazole-susceptible and an itraconazole-resistant isolate of *Aspergillus fumigatus*. *Antimicrob. Agents Chemother.* 41:1504–1507.

Pagano, L., B. Gleissner, and L. Fianchi. 2005. Breakthrough zygomycosis and voriconazole. *J. Infect. Dis.* 192:1496–1497.

Panagopoulou, P., J. Filioti, E. Farmaki, A. Maloukou, and E. Roilides. 2007. Filamentous fungi in a tertiary care hospital: environmental surveillance and susceptibility to antifungal drugs. *Infect. Control Hosp. Epidemiol.* 28:60–67.

Pascual, A., T. Calandra, S. Bolay, T. Buclin, J. Bille, and O. Marchetti. 2008. Voriconazole therapeutic drug monitoring in patients with invasive mycoses improves efficacy and safety outcomes. *Clin. Infect. Dis.* 46:201–211.

Pascual, A., V. Nieth, T. Calandra, J. Bille, S. Bolay, L. A. Decosterd, T. Buclin, P. A. Majcherczyk, D. Sanglard, and O. Marchetti. 2007. Variability of voriconazole plasma levels measured by new high-performance liquid chromatography and bioassay methods. *Antimicrob. Agents Chemother.* 51:137–143.

Pasqualotto, A. C., and D. W. Denning. 2008. New and emerging treatments for fungal infections. *J. Antimicrob. Chemother.* 61(Suppl. 1):i19–i30.

Pasqualotto, A. C., M. Shah, R. Wynn, and D. W. Denning. 2008. Voriconazole plasma monitoring. *Arch. Dis. Child.* **93:**578–581.

Patterson, T. F., A. W. Fothergill, and M. G. Rinaldi. 1993. Efficacy of itraconazole solution in a rabbit model of invasive aspergillosis. *Antimicrob. Agents Chemother.* **37:**2307–2310.

Pavie, J., C. Lacroix, D. G. Hermoso, M. Robin, C. Ferry, A. Bergeron, M. Feuilhade, F. Dromer, E. Gluckman, J. M. Molina, and P. Ribaud. 2005. Breakthrough disseminated *Aspergillus ustus* infection in allogeneic hematopoietic stem cell transplant recipients receiving voriconazole or caspofungin prophylaxis. *J. Clin. Microbiol.* **43:**4902–4904.

Peng, L. W., and Y. H. Lien. 2005. Pharmacokinetics of single, oral-dose voriconazole in peritoneal dialysis patients. *Am. J. Kidney Dis.* **45:**162–166.

Petraitiene, R., V. Petraitis, A. H. Groll, T. Sein, S. Piscitelli, M. Candelario, A. Field-Ridley, N. Avila, J. Bacher, and T. J. Walsh. 2001. Antifungal activity and pharmacokinetics of posaconazole (SCH 56592) in treatment and prevention of experimental invasive pulmonary aspergillosis: correlation with galactomannan antigenemia. *Antimicrob. Agents Chemother.* **45:**857–869.

Petraitiene, R., V. Petraitis, C. A. Lyman, A. H. Groll, D. Mickiene, J. Peter, J. Bacher, K. Roussillon, M. Hemmings, D. Armstrong, N. A. Avila, and T. J. Walsh. 2004. Efficacy, safety, and plasma pharmacokinetics of escalating dosages of intravenously administered ravuconazole lysine phosphoester for treatment of experimental pulmonary aspergillosis in persistently neutropenic rabbits. *Antimicrob. Agents Chemother.* **48:**1188–1196.

Pfaller, M. A., S. A. Messer, R. J. Hollis, and R. N. Jones. 2002. Antifungal activities of posaconazole, ravuconazole, and voriconazole compared to those of itraconazole and amphotericin B against 239 clinical isolates of *Aspergillus* spp. and other filamentous fungi: report from SENTRY Antimicrobial Surveillance Program, 2000. *Antimicrob. Agents Chemother.* **46:**1032–1037.

Ramos, G., M. Cuenca-Estrella, A. Monzon, and J. L. Rodriguez-Tudela. 1999. In-vitro comparative activity of UR-9825, itraconazole and fluconazole against clinical isolates of *Candida* spp. *J. Antimicrob. Chemother.* **44:**283–286.

Robatel, C., M. Rusca, C. Padoin, O. Marchetti, L. Liaudet, and T. Buclin. 2004. Disposition of voriconazole during continuous veno-venous haemodiafiltration (CVVHDF) in a single patient. *J. Antimicrob. Chemother.* **54:**269–270.

Roffey, S. J., S. Cole, P. Comby, D. Gibson, S. G. Jezequel, A. N. Nedderman, D. A. Smith, D. K. Walker, and N. Wood. 2003. The disposition of voriconazole in mouse, rat, rabbit, guinea pig, dog, and human. *Drug Metab. Dispos.* **31:**731–741.

Sambatakou, H., B. Dupont, H. Lode, and D. W. Denning. 2006. Voriconazole treatment for subacute invasive and chronic pulmonary aspergillosis. *Am. J. Med.* **119:**527–524.

Schmitt, H. J., F. Edwards, J. Andrade, Y. Niki, and D. Armstrong. 1992. Comparison of azoles against aspergilli in vitro and in an experimental model of pulmonary aspergillosis. *Chemotherapy* **38:** 118–126.

Schwartz, S., M. Ruhnke, P. Ribaud, L. Corey, T. Driscoll, O. A. Cornely, U. Schuler, I. Lutsar, P. Troke, and E. Thiel. 2005. Improved outcome in central nervous system aspergillosis, using voriconazole treatment. *Blood* **106:**2641–2645.

Sharkey, P. K., M. G. Rinaldi, J. F. Dunn, T. C. Hardin, R. J. Fetchick, and J. R. Graybill. 1991. High-dose itraconazole in the treatment of severe mycoses. *Antimicrob. Agents Chemother.* **35:**707–713.

Singh, R., and T. Cundy. 2005. Itraconazole-induced painful neuropathy in a man with type 1 diabetes. *Diabetes Care* **28:**225.

Sugar, A. M., and X. P. Liu. 2000. Effect of grapefruit juice on serum voriconazole concentrations in the mouse. *Med. Mycol.* **38:**209–212.

Sugar, A. M., and X. P. Liu. 2001. Efficacy of voriconazole in treatment of murine pulmonary blastomycosis. *Antimicrob. Agents Chemother.* **45:**601–604.

Trifilio, S., G. Pennick, J. Pi, J. Zook, M. Golf, K. Kaniecki, S. Singhal, S. Williams, J. Winter, M. Tallman, L. Gordon, O. Frankfurt, A. Evens, and J. Mehta. 2007a. Monitoring plasma voriconazole levels may be necessary to avoid subtherapeutic levels in hematopoietic stem cell transplant recipients. *Cancer* **109:**1532–1535.

Trifilio, S., S. Singhal, S. Williams, O. Frankfurt, L. Gordon, A. Evens, J. Winter, M. Tallman, J. Pi, and J. Mehta. 2007b. Breakthrough fungal infections after allogeneic hematopoietic stem cell transplantation in patients on prophylactic voriconazole. *Bone Marrow Transplant.* **40:**451–456.

Trifilio, S. M., C. L. Bennett, P. R. Yarnold, J. M. McKoy, J. Parada, J. Mehta, G. Chamilos, F. Palella, L. Kennedy, K. Mullane, M. S. Tallman, A. Evens, M. H. Scheetz, W. Blum, and D. P. Kontoyiannis. 2007c. Breakthrough zygomycosis after voriconazole administration among patients with hematologic malignancies who receive hematopoietic stem-cell transplants or intensive chemotherapy. *Bone Marrow Transplant.* **39:**425–429.

Tsiodras, S., R. Zafiropoulou, E. Kanta, C. Demponeras, N. Karandreas, and E. K. Manesis. 2005. Painful peripheral neuropathy associated with voriconazole use. *Arch. Neurol.* **62:**144–146.

Ullmann, A. J., J. H. Lipton, D. H. Vesole, P. Chandrasekar, A. Langston, S. R. Tarantolo, H. Greinix, D. A. Morais, V. Reddy, N. Boparai, L. Pedicone, H. Patino, and S. Durrant. 2007. Posaconazole or fluconazole for prophylaxis in severe graft-versus-host disease. *N. Engl. J. Med.* **356:**335–347.

van Burik, J. A., R. S. Hare, H. F. Solomon, M. L. Corrado, and D. P. Kontoyiannis. 2006. Posaconazole is effective as salvage therapy in zygomycosis: a retrospective summary of 91 cases. *Clin. Infect. Dis.* **42:**e61–e65.

Van de Velde, V. J., A. P. Van Peer, J. J. Heykants, R. J. Woestenborghs, P. Van Rooy, K. L. De Beule, and G. F. Cauwenbergh. 1996. Effect of food on the pharmacokinetics of a new hydroxypropyl-beta-cyclodextrin formulation of itraconazole. *Pharmacotherapy* **16:**424–428.

van 't Hek, L. G., P. E. Verweij, C. M. Weemaes, R. van Dalen, J. B. Yntema, and J. F. Meis. 1998. Successful treatment with voriconazole of invasive aspergillosis in chronic granulomatous disease. *Am. J. Respir. Crit. Care Med.* **157:**1694–1696.

van't Wout, J. W., I. Novakova, C. A. Verhagen, W. E. Fibbe, B. E. De Pauw, and J. W. van der Meer. 1991. The efficacy of itraconazole against systemic fungal infections in neutropenic patients: a randomised comparative study with amphotericin B. *J. Infect.* **22:** 45–52.

Verweij, P. E., D. T. te Dorsthorst, A. J. Rijs, H. G. Vries-Hospers, and J. F. Meis. 2002. Nationwide survey of in vitro activities of itraconazole and voriconazole against clinical *Aspergillus fumigatus* isolates cultured between 1945 and 1998. *J. Clin. Microbiol.* **40:** 2648–2650.

von Mach, M. A., J. Burhenne, and L. S. Weilemann. 2006. Accumulation of the solvent vehicle sulphobutylether beta cyclodextrin sodium in critically ill patients treated with intravenous voriconazole under renal replacement therapy. *BMC Clin. Pharmacol.* **6:**6.

Walsh, T. J., E. J. Anaissie, D. W. Denning, R. Herbrecht, D. P. Kontoyiannis, K. A. Marr, V. A. Morrison, B. H. Segal, W. J. Steinbach, D. A. Stevens, J. A. van Burik, J. R. Wingard, and T. F. Patterson. 2008. Treatment of aspergillosis: clinical practice guidelines of the Infectious Diseases Society of America. *Clin. Infect. Dis.* **46:**327–360.

Walsh, T. J., M. O. Karlsson, T. Driscoll, A. G. Arguedas, P. Adamson, X. Saez-Llorens, A. J. Vora, A. C. Arrieta, J. Blumer, I. Lutsar, P. Milligan, and N. Wood. 2004. Pharmacokinetics and safety of intravenous voriconazole in children after single- or multiple-dose administration. *Antimicrob. Agents Chemother.* **48:** 2166–2172.

Walsh, T. J., I. Lutsar, T. Driscoll, B. Dupont, M. Roden, P. Ghahramani, M. Hodges, A. H. Groll, and J. R. Perfect. 2002a. Voriconazole in the treatment of aspergillosis, scedosporiosis and other invasive fungal infections in children. *Pediatr. Infect. Dis. J.* **21:**240–248.

Walsh, T. J., P. Pappas, D. J. Winston, H. M. Lazarus, F. Petersen, J. Raffalli, S. Yanovich, P. Stiff, R. Greenberg, G. Donowitz, M. Schuster, A. Reboli, J. Wingard, C. Arndt, J. Reinhardt, S. Hadley, R. Finberg, M. Laverdiere, J. Perfect, G. Garber, G. Fioritoni, E. Anaissie, and J. Lee. 2002b. Voriconazole compared with liposomal amphotericin B for empirical antifungal therapy in patients with neutropenia and persistent fever. *N. Engl. J. Med.* **346:**225–234.

Walsh, T. J., V. Petraitis, R. Petraitiene, A. Field-Ridley, D. Sutton, M. Ghannoum, T. Sein, R. Schaufele, J. Peter, J. Bacher, H. Casler, D. Armstrong, A. Espinel-Ingroff, M. G. Rinaldi, and C. A. Lyman. 2003. Experimental pulmonary aspergillosis due to *Aspergillus terreus*: pathogenesis and treatment of an emerging fungal pathogen resistant to amphotericin B. *J. Infect. Dis.* **188:**305–319.

Walsh, T. J., I. Raad, T. F. Patterson, P. Chandrasekar, G. R. Donowitz, R. Graybill, R. E. Greene, R. Hachem, S. Hadley, R. Herbrecht, A. Langston, A. Louie, P. Ribaud, B. H. Segal, D. A. Stevens, J. A. van Burik, C. S. White, G. Corcoran, J. Gogate, G. Krishna, L. Pedicone, C. Hardalo, and J. R. Perfect. 2007. Treatment of invasive aspergillosis with posaconazole in patients who are refractory to or intolerant of conventional therapy: an externally controlled trial. *Clin. Infect. Dis.* **44:**2–12.

Wark, P. A., P. G. Gibson, and A. J. Wilson. 2004. Azoles for allergic bronchopulmonary aspergillosis associated with asthma. *Cochrane Database Syst. Rev.* **3:**CD001108.

Warn, P. A., A. Sharp, and D. W. Denning. 2006a. In vitro activity of a new triazole BAL4815, the active component of BAL8557 (the water-soluble prodrug), against *Aspergillus* spp. *J. Antimicrob. Chemother.* **57:**135–138.

Warn, P. A., A. Sharp, J. Mosquera, J. Spickermann, A. Schmitt-Hoffmann, M. Heep, and D. W. Denning. 2006b. Comparative in vivo activity of BAL4815, the active component of the prodrug BAL8557, in a neutropenic murine model of disseminated *Aspergillus flavus*. *J. Antimicrob. Chemother.* **58:**1198–1207.

Weiler, S., H. Zoller, I. Graziadei, W. Vogel, R. Bellmann-Weiler, M. Joannidis, and R. Bellmann. 2007. Altered pharmacokinetics of voriconazole in a patient with liver cirrhosis. *Antimicrob. Agents Chemother.* **51:**3459–3460.

Wheat, J., R. Hafner, M. Wulfsohn, P. Spencer, K. Squires, W. Powderly, B. Wong, M. Rinaldi, M. Saag, R. Hamill, R. Murphy, P. Connolly-Stringfield, N. Briggs, and S. Owens. 1993. Prevention of relapse of histoplasmosis with itraconazole in patients with the acquired immunodeficiency syndrome. *Ann. Intern. Med.* **118:**610–616.

Willems, L., R. van der Geest, and K. de Beule. 2001. Itraconazole oral solution and intravenous formulations: a review of pharmacokinetics and pharmacodynamics. *J. Clin. Pharm. Ther.* **26:**159–169.

Winston, D. J., and R. W. Busuttil. 2002. Randomized controlled trial of oral itraconazole solution versus intravenous/oral fluconazole for prevention of fungal infections in liver transplant recipients. *Transplantation* **74:**688–695.

Winston, D. J., R. T. Maziarz, P. H. Chandrasekar, H. M. Lazarus, M. Goldman, J. L. Blumer, G. J. Leitz, and M. C. Territo. 2003. Intravenous and oral itraconazole versus intravenous and oral fluconazole for long-term antifungal prophylaxis in allogeneic hematopoietic stem-cell transplant recipients. A multicenter, randomized trial. *Ann. Intern. Med.* **138:**705–713.

Aspergillus fumigatus and Aspergillosis
Edited by J.-P. Latgé and W. J. Steinbach

Chapter 32

Echinocandins in the Treatment of Aspergillosis

JOHAN MAERTENS AND VINCENT MAERTENS

The incidence of invasive *Aspergillus* infections has increased dramatically over the last 2 decades, not only in patients with an underlying hematologic disorder and in hematopoietic stem cell transplant recipients (Marr et al., 2002; Pagano et al., 2006, 2007), but also in solid organ transplant recipients (Silveira and Husain, 2007) and in patients hospitalized in medical intensive care facilities (Meersseman et al., 2007). Many of these latter patients, including those with autoimmune disorders, chronic obstructive lung disease, and liver cirrhosis, have compromised organ function, including renal impairment. The high crude mortality rate of these infections results in part from difficulties in securing an early diagnosis (Hope et al., 2005). However, high mortality rates result also from shortcomings in the available therapeutic arsenal (Maertens et al., 2002). Amphotericin B deoxycholate, flucytosine, and itraconazole are associated with low success rates and are hampered by serious infusion- or drug-related toxicities, difficult-to-manage drug interactions, and pharmacokinetic problems (Maertens et al., 2002). Lipid formulations of amphotericin B display a better toxicity profile and, therefore, in spite of their high acquisition cost, may even be cost-effective when compared with conventional amphotericin B (Bates et al., 2001; Cagnoni et al., 2000). It must, however, be mentioned that each of the lipid formulations (liposomal amphotericin B, amphotericin B lipid complex, and amphotericin B colloidal dispersion) confers markedly distinct biochemical, pharmacokinetic, and pharmacodynamic properties. Also, neither the amphotericin B colloidal dispersion nor amphotericin B lipid complex shows a reduced incidence of acute infusion-related side effects compared to the parent compound (Frothingham, 2002; Wingard, 2002). Voriconazole, the current drug of choice for treating invasive *Aspergillus* infections (Herbrecht et al., 2002), displays nonlinear pharmacokinetics, underscoring the need for therapeutic drug monitoring (Pascual et al., 2008; Imhof et al., 2006), and displays a number of azole class-related side effects, including hepatic side effects (Eiden et al., 2007). In addition, the use of the drug is hampered by reversible visual disturbances and by the recommendation not to use the intravenous formulation in patients with renal impairment (von Mach et al., 2006).

THE FUNGAL CELL WALL: AN ATTRACTIVE TARGET FOR ANTIFUNGAL THERAPY

Given these drawbacks, researchers were forced to look for agents with a new mode of action. Ideally, a highly selective therapeutic target needed to be identified that is essential to the pathogen but absent or at least sufficiently different in nonfungal eukaryotic cells in order to avoid cross-inhibition. This new class of antifungals needed to cover a wide range of fungal organisms, since often therapy is started empirically, before the identification of the causative pathogen. In addition, given the long-lasting immune deficiency in many at-risk patients, the new drug needed to be fungicidal rather than fungistatic.

Although fungi share many biochemical targets with other eukaryotic cells, including mammalian cells, the fungal cell wall appears to be a unique structure that is essential for the survival of the fungus and that fulfills the criterion of selectivity. Human cells have no comparable wall; it is therefore unlikely that human cells carry analogous enzymatic machinery as that needed to synthesize fungal cell walls (Debono and Gordee, 1994; Maertens and Boogaerts, 2000; Denning, 2003; Kurtz and Rex, 2001). Since the discovery that penicillin inhibits bacterial cell wall biosynthesis, researchers have made tremendous efforts to discover equivalent agents to interfere with fungal cell wall synthesis. Clearly, this is an area of research where considerable progress has been achieved over the past decade.

Johan Maertens and Vincent Maertens • Dept. of Haematology, Acute Leukaemia and Hematopoietic Stem Cell Transplantation Unit, University Hospitals Leuven, Campus Gasthuisberg, Herestraat 49, B-3000 Leuven, Belgium.

For a detailed description of the cell wall ultrastructure and function, the reader is referred to chapter 2 of this volume. In brief, the fungal cell wall is a rigid but dynamic structure consisting basically of three macromolecular components: a complex network of chitinous and glucan microfibrils to prevent osmotic distension of the fungal cell and a high proportion of embedded or interstitial mannoproteins that play a key role in the wall's porosity, antigenicity, and fungal virulence. Glucans, the most abundant elements in the cell wall of many fungi, are mixtures of β-1,3- and β-1,6-linked glucose polymers. β-1,3-D-Glucan is a glucose homopolymer that forms a rope-like fibril of three helically entwined coiling chains of β-1,3-linked residues, with occasional β-1,6-linked side chains. Further aggregation of these chains provides rigidity and appears necessary to conserve the integrity of the cell wall. β-1,3-D-Glucan synthase, the membrane-bound enzyme complex that catalyzes this polymerization, is composed of at least two functional subunits: a large integral membrane catalytic subunit, Fks1p, that incorporates glucose from UDP-glucose into glucan fibrils before pushing them through a pore in the plasma membrane, and a small soluble regulatory unit encoded by RHO1, which has intrinsic GTPase activity and associates with and activates the catalytic unit (Debono and Gordee, 1994; Maertens and Boogaerts, 2000; Denning, 2003; Kurtz and Rex, 2001).

During screening programs for new antibiotics in the 1970s, researchers identified several classes of naturally occurring compounds that could inhibit fungal glucan synthesis, including (i) the papulacandins, liposaccharide inhibitors isolated from *Papularia sphaerosperma*, and (ii) the echinocandin family (echinocandins, cilofungin, aculeacins, pneumocandins, mulundocandins, and WF-11899) (Hector, 1993). The clinical development of the papulacandins has been suspended because of the limited in vitro spectrum of activity and the lack of activity in animal models (Traxler et al., 1987). In contrast, the echinocandins and pneumocandins, structurally related lipopeptides with a cyclic peptide core and a number of attached lipid chains, proved to be fungicidal in vitro and in animal models. Although the natural product, echinocandin B, the first antifungal representative of this class, induced hemolysis, simple substitution of the linoleoyl group with synthetic side chains resulted in nonlytic analogs. In particular, cilofungin (LY121019), the 4-*n*-octyloxybenzoyl echinocandin B derivative, appeared very promising; the drug was more than 10-fold less hemolytic than the parent compound and retained excellent fungicidal activity against *Candida* spp. and *Aspergillus fumigatus*. However, cilofungin was abandoned in phase II clinical trials because of renal acidosis associated with the polyethylene glycol vehicle in which the drug was given (Debono et al., 1988). Despite this disappointment, several companies have continued their search for semisynthetic analogs with improved pharmacological properties and identified three water-soluble bioactive derivatives: caspofungin (caspofungin acetate; Cancidas; derived from pneumocandin B; Merck & Co., Inc.), anidulafungin (Eraxis; derived from echinocandin B_0; Pfizer, Inc. [formerly Versicor and Vicuron]), and micafungin (Mycamine; Astellas Pharma [formerly Fujisawa Healthcare, Inc.]). These analogs are more potent and display a broader antifungal spectrum than their respective natural counterpart. These agents primarily act as specific and noncompetitive inhibitors of β-1,3-D-glucan synthase. The resulting depletion of cell wall glucan leads to osmotic shock, "ballooning," and ultimately lysis of the cell. Accordingly, echinocandins have been shown to act fungicidally against *Candida* spp. in vitro and in vivo. However, echinocandins do not fully inhibit growth of *Aspergillus* species and yet induce prominent morphologic changes at the extremities of the hyphae at or above the minimal effective concentration; as such, echinocandins act fungistatically against *Aspergillus* spp. (Bowman et al., 2002). In addition, recent data have demonstrated that echinocandins have immunomodulating modes of action (Kinoshita et al., 2006). The main mechanism of action is distinct from the azoles and polyenes, which target cell membrane ergosterol biosynthesis.

IN VITRO SPECTRUM OF ACTIVITY

Caspofungin demonstrates enhanced in vitro activity (MIC_{90} of 0.12 μg/ml) against clinical isolates of *Aspergillus*, including *A. fumigatus* and *A. flavus*, compared with itraconazole, amphotericin B, and flucytosine (Pfaller et al., 1998). These results have been confirmed with testing against *Aspergillus* isolates obtained from patients included in clinical trials; for all species, the MIC_{80} geometric means are below 1 μg/ml at 24 h. Compared to controls, caspofungin also significantly prolongs the survival of immunosuppressed and neutropenic mice with disseminated or pulmonary aspergillosis (Abruzzo et al., 1997, 2000). Anidulafungin has also demonstrated excellent in vitro activity against several species of *Aspergillus* (e.g., MIC_{90} of ≤0.03 μg/ml for *A. fumigatus*) (Serrano et al., 2003). Anidulafungin prolongs survival and reduces antigenemia in rabbits with disseminated aspergillosis, including persistently neutropenic rabbits (Petraitis et al., 1998). Micafungin has potent in vitro inhibitory activity against *Aspergillus* species at lower concentrations than amphotericin B and itraconazole (Tawara et al., 2000).

In most studies, susceptibility testing was performed according to a modification of the National

Committee for Clinical Laboratory Standards method M38-P. It is important to remember that in vitro antifungal susceptibility testing for echinocandins has not yet been standardized. Therefore, results of in vitro susceptibility tests do not necessarily correlate with in vivo or clinical outcomes.

As already mentioned, it is difficult to characterize the nature of the activity of echinocandins against *Aspergillus* spp. In the presence of caspofungin, blunting and abnormal branching of the hyphae in actively growing areas of the cell are observed, primarily at the tips and branching points. These observations are consistent with the mechanism of action but do not fit the classical definitions of a fungicidal or fungistatic agent. However, data from humans and animal models show that the drug is efficacious, even in animals with prolonged immunosuppression. Contrary to *Candida* animal studies, reduction of CFU in tissue correlates poorly with fungal burden in *Aspergillus* models. Using a quantitative PCR-based assay to monitor disease progression and measure drug efficacy, Bowman et al. demonstrated that both caspofungin and amphotericin B reduced the *A. fumigatus* burden in infected kidneys to the limit of detection for the quantitative PCR assay (Bowman et al., 2001).

To date, there has been no antagonism noted in preclinical studies between echinocandins and other antifungal therapies. However, conflicting results have been seen in studies evaluating synergy or additivity with echinocandins in vitro and in animal models. The frequency of echinocandin resistance in clinical isolates of *Aspergillus* species remains unknown; so far, there are no documented reports of clinical failure due to echinocandin resistance in aspergillosis. However, as recently demonstrated, an S678P substitution in Fks1p appears to be sufficient to confer echinocandin resistance in *A. fumigatus* (Rocha et al., 2007).

COMPARATIVE HUMAN PHARMACOKINETICS

Echinocandins are large molecules with a relative molecular weight of approximately 1,200. Due to their large molecule size and poor oral bioavailability (less than 10%), echinocandins are available for parenteral administration only (Denning, 2003). All commercially available candins display linear pharmacokinetic profiles. In humans, echinocandins are extensively bound to plasma proteins (Denning, 2003). Table 1 compares the pharmacokinetic data of caspofungin, micafungin, and anidulafungin.

There are no reported clinically meaningful alterations of pharmacokinetics in patients with mild, mod-

Table 1. Comparison of echinocandins in adults

Characteristic	Caspofungin	Micafungin	Anidulafungin
Origin	*Glarea lozoyensis*	*Coleophoma empetri*	*Aspergillus nidulans*
Precursor	Pneumocandin B_0	FR901379	Echinocandin B_0
Brand name	Cancidas	Mycamine	Eraxis
Chemical synonyms	MK-0991 MK 991/M991 L-743,872 L-743872	FK463 FK463 FK463	VER-002 LY303366 LY-303366
Chemical formula	$C_{52}H_{88}N_{10}O_{15}$	$C_{56}H_{71}N_9O_{23}S$	$C_{58}H_{73}N_7O_{17}$
Relative molecular weight	1,213.4	1,292.26	1,140.3
Required daily dose	70 mg, loading dose 50 mg, daily dose	50–150 mg, daily dose	200 mg, loading dose 100 mg, daily dose
Available as:	Lyophilized powder: no	Lyophilized powder: yes	Reconstituted: yes
Stability at room temp	24 h	24 h	24 h
Administration time	~1 h	~1 h	At least 1 h
Plasma protein binding	96.5%	99.5%	80%
C_{max}, μg/ml (drug dose)	12 (70 mg)	7.1 (75 mg)	7.5 (200 mg)
AUC_{0-24} (mg·h/liter)	93.5	59.9	104.5
$t_{1/2}$ (h)	10	13	25.6
CL (ml/min/kg)	0.15	0.16	0.16
Vd_{ss} (liters)	9.5	14	33.4
Metabolism	Hepatic	Hepatic	Nonenzymatic degradation
Dose adaptation in renal failure	No	No	No
Dose adjustment during dialysis	No	No	No
Dose adaptation in:			
Moderate hepatic dysfunction	35 mg/day	No	No
Severe hepatic dysfunction	No data	No data	No

erate, advanced, or end-stage renal insufficiency. Therefore, no dose adjustment is necessary for patients with any degree of renal impairment. Echinocandins are not removed from the blood by hemodialysis or ultrafiltration, and supplementary dosing is not required following hemodialysis (Denning, 2003).

In adults, there are no clinically meaningful alterations in pharmacokinetics with age (for pediatric use, see chapter 40), gender, or race. There have been no well-controlled studies in pregnant women, and it is not known whether echinocandins are excreted in human milk.

DRUG INTERACTIONS

Few drug interactions have been described for echinocandins, mainly because echinocandins are neither inhibitors of nor substrates for enzymes of the cytochrome P450 system. Coadministration of caspofungin and itraconazole did not alter the pharmacokinetics of either drug, suggesting that caspofungin is not subject to drug interactions based on CYP3A4 inhibition (Kulemann et al., 2005). Caspofungin also did not influence the pharmacokinetics of amphotericin B or mycophenolate mofetil.

In clinical studies performed in healthy volunteers, cyclosporine A (CyA) increased the area under the time-concentration curve (AUC) of caspofungin by ~35%, due to a decreased hepatic uptake of caspofungin, whereas the plasma levels of CyA remained unchanged. In 5 of 12 subjects, coadministration of both drugs resulted in an increase in liver enzymes (alanine aminotransferase) of less than or equal to three times the upper limit of normal, which resolved with discontinuation of the product. Hence, coadministration of caspofungin and CyA was not recommended by the manufacturer, unless the benefit outweighed the risks. However, data from allogeneic stem cell transplant recipients failed to demonstrate a clinically meaningful interaction between caspofungin and CyA (Marr et al., 2004a; Sanz-Rodriguez et al., 2004). Caspofungin decreases the AUC of tacrolimus by ~25%. Coadministration of CyA or tacrolimus with micafungin or anidulafungin does not result in clinically significant interactions.

Rifampin has both an inhibitory and an induction effect on caspofungin disposition. Inhibition occurs probably through blockage of caspofungin uptake. However, this same uptake mechanism can be induced by rifampin with continued dosing. Caspofungin should therefore be given at an increased daily dose of 70 mg when coadministered with rifampin (Stone et al., 2004).

CLINICAL DATA IN THE ADULT POPULATION: CASPOFUNGIN

Salvage Data

The response rate to caspofungin therapy in invasive aspergillosis was illustrated in an open-label, noncomparative study in 83 evaluable patients with definite or probable aspergillosis who were immunocompromised primarily due to hematological malignancy (Maertens et al., 2004). A majority of patients (85.5%) were refractory to standard therapies, which included amphotericin B (conventional or lipid-based formulations) or triazoles; the remainder of the patients (14.5%) were unable to tolerate these agents due to development of adverse events, especially nephrotoxicity. According to an expert panel assessment, a favorable response to caspofungin occurred in 37 of 83 (45%) patients. A favorable response was defined as a complete resolution or clinically meaningful improvement of all signs and symptoms and associated radiographic or bronchoscopic findings. As expected, the response rate was higher in patients who received caspofungin for at least 7 days (37 of 66 patients [56% response rate]), in patients intolerant of other antifungals (9 of 12 patients [75% response rate]), and in patients with pulmonary aspergillosis (32 of 64 [50%] patients, versus 5 of 19 [26%] patients with extrapulmonary aspergillosis). Caspofungin also produced a favorable response in patients considered difficult to treat, such as those with hematological malignancy or those undergoing solid organ transplantation (25 of 60 patients [41.7%] and 4 of 9 patients [44%], respectively). In contrast, fewer patients who had undergone allogeneic hematopoietic stem cell transplantation responded favorably to caspofungin compared with the overall response rate (3 of 21 patients [14.3%] versus 37 of 83 patients [45%], respectively). Nevertheless, the findings of this study are particularly remarkable given the unusually high number of refractory cases compared to previously published salvage studies. These data led to the first approval of an echinocandin in North America and the European Union for the salvage treatment of invasive aspergillosis.

Compassionate Use Studies

Four compassionate use studies conducted prior to licensure in several different countries showed that caspofungin produced favorable responses in a variety of patients with invasive aspergillosis, most of whom demonstrated refractoriness to or intolerance of prior systemic antifungal therapy (Kartsonis et al., 2005; Glasmacher et al., 2006; Sanz-Rodriguez et al., 2001; Morrissey et al., 2004). In a worldwide compassionate use study (Kartsonis et al., 2005), 20 of 45 (44%) patients with probable or definite invasive aspergillosis refractory or intolerant to at least one amphotericin B formulation responded favorably to caspofungin treatment. This included a complete response in 9 of 45 (20%) and a partial response in 11 of 45 (24%) patients. The response to caspofungin was consistent in both neutropenic and nonneutropenic patients as well as across diverse sites of infection and patients with various underlying diseases. Other compassionate use studies conducted in Germany (Glasmacher et al., 2006), Spain (Sanz-Rodriguez et al., 2001), and Australia (Morrissey et al., 2004) revealed similar consistent benefits of caspofungin therapy in proven or probable invasive aspergillosis. Caspofungin therapy produced a favorable response in 26 of 65 (40%) patients with invasive aspergillosis, 31 of 51 (61%) patients with invasive fungal infection with or without neutropenia, and 12 of 18 (67%) patients with hematological disorders and proven or probable invasive mycoses.

Empirical Therapy

In patients with persistent febrile neutropenia, a population that often receives empiric therapy for the early treatment of a possible invasive fungal infection, caspofungin demonstrated efficacy comparable to liposomal amphotericin B, according to the results of a double-blind, randomized comparative trial (Walsh et al., 2004). This study involved only patients with an underlying hematological malignancy that had undergone chemotherapy or hematopoietic stem cell transplantation and had persistent neutropenia and fever despite at least 4 days of broad-spectrum parenteral antibiotic therapy. Based on a composite end point assessment, caspofungin proved to be noninferior to liposomal amphotericin B. However, caspofungin-treated patients had significantly better outcomes than liposomal amphotericin B-treated patients with respect to survival for >7 days after treatment, successful treatment of baseline fungal infection, and fewer discontinuations due to toxicity or lack of efficacy. Although the numbers of patients with documented specific fungal infections were small compared to the overall study cohort, an analysis of patients with baseline *Aspergillus* infections revealed that caspofungin therapy produced a successful outcome in 5 of 12 (41.7%) patients. As illustrated in a Kaplan-Meier analysis of survival data, caspofungin provided a survival benefit compared with liposomal amphotericin B over the course of the study. Overall, the results of this study indicate that caspofungin represents an effective and better-tolerated alternative to liposomal amphotericin B in patients with fever and neutropenia.

The results of another smaller study confirmed the utility of caspofungin in treating patients with fever of unknown origin or invasive fungal infection following stem cell transplantation (Trenschel et al., 2005). This study was conducted in a total of 31 patients who were recalcitrant to or intolerant of prior antifungal therapy, including amphotericin B, liposomal amphotericin B, fluconazole, itraconazole, or voriconazole. Caspofungin therapy for a median of 13 days resulted in complete resolution or clinical stabilization in 74% of patients with fever of unknown origin or invasive fungal infection following stem cell transplantation. A favorable outcome to caspofungin occurred in 15 of 16 (94%) patients with fever of unknown origin and in 8 of 15 (53%) patients with invasive fungal infection (5 with proven, 2 with probable, and 1 with possible fungal infection). However, only two of these five patients were actually shown to have *A. fumigatus* as the invading pathogen; the outcomes for these specific patients were not reported.

Primary Therapy of Invasive Aspergillosis

An Italian single-center study evaluated the antifungal efficacy of caspofungin as first-line therapy for invasive pulmonary fungal disease in 32 immunocompromised patients with hematologic malignancies (Candoni et al., 2005). The study cohort comprised 7 patients (22%) with proven pulmonary invasive aspergillosis and 25 patients (78%) with a diagnosis of probable invasive fungal infection with pulmonary localization (according to European Organization for Research and Treatment of Cancer criteria) (Sanz-Rodriguez et al., 2004). Caspofungin (given for a median duration of 20 days; range, 8 to 64 days) produced an overall response rate of 56% (18 of 32 patients), with complete responses in 12 of 18 and partial responses in 6 of 18; 2 patients (6%) had stable disease. Approximately 38% (12 of 32) did not respond, and of these, 7 died of mycotic infection.

More recently, the European Organization for Research and Treatment of Cancer reported on their multicenter clinical trial that evaluated caspofungin as first-line treatment for proven and probable invasive aspergillosis in 61 evaluable patients with hematological cancer (Viscoli et al., 2007). In this population of predominantly neutropenic patients (mostly with acute leukemia) and often with uncontrolled cancer (75%) and microbiologically documented invasive aspergillosis, caspofungin achieved an end-of-treatment response rate of 33% and a 12-week survival rate of 54%.

Combination Therapy

As new antifungal agents with unique mechanisms of action have been developed, scientific enthusiasm for combination therapy has resurfaced (Steinbach et al., 2003). Indeed, combination treatment is attractive from the perspectives of synergistic potential, relative safety, and lack of overlapping toxicities. However, to date, there are few or no convincing clinical data to support use of combination antifungal therapy for mold infections.

Initial reports of combination therapy in invasive aspergillosis have focused on refractory disease (Aliff et al., 2003; Kontoyiannis et al., 2003). In a retrospective study, 16 hematopoietic stem cell transplant or cytotoxic chemotherapy recipients with proven or probable invasive aspergillosis refractory to or intolerant of at least 7 days of amphotericin B were managed with a combination of voriconazole and caspofungin (Marr et al., 2004b). This cohort was compared to a historical group of patients (n = 31) who had received voriconazole monotherapy as salvage therapy. Use of combination therapy was associated with improved 3-month survival compared with voriconazole alone (hazard ratio [HR], 0.42; P = 0.048). Combination therapy was also associated with a reduced overall mortality compared with voriconazole alone (HR, 0.27; P = 0.008). However, at 1 year, the overall survival rate between the two groups was equivalent (P = 0.26), although fewer patients died of aspergillosis in the combination group than in the voriconazole group (Marr et al., 2005). More recently, data became available from a multicenter, noncomparative study of caspofungin (70 mg daily) combined with other antifungal agents in 53 adult patients with invasive aspergillosis that was refractory (n = 46) or intolerant (n = 7) to first-line therapy (Maertens et al., 2006). Sixteen patients (30%) received caspofungin with a polyene, 7 (13%) patients received caspofungin with itraconazole, and the majority of patients (n = 30; 57%) received caspofungin with voriconazole. The average duration of combination therapy was 31.3 days (range, 1 to 196 days). Overall success at the end of combination therapy was 55% for patients receiving >1 day of combination therapy and 66% in patients receiving >7 days of combination therapy. Fifty-seven percent of patients with baseline neutropenia and 54% of allogeneic transplant recipients responded favorably. A sustained response in 49% (25 of 51) was maintained at day 84 from study therapy. Combination therapy was generally well-tolerated, with only two serious voriconazole-related adverse events.

Clearly, these studies help reassure us that we are not doing any harm to the patient, but they do not allow us to determine which single or combination regimens are most successful. These findings indicate that salvage therapy with caspofungin plus voriconazole may result in higher survival rates than salvage therapy with either drug alone and that randomized trials are warranted to investigate further the clinical utility of this combination for primary therapy of aspergillosis. In a small clinical trial, the combination of caspofungin plus voriconazole as primary therapy for invasive aspergillosis in solid organ transplant recipients was compared to historical controls managed with a lipid formulation of amphotericin B (Singh et al., 2006). Ninety-day survival was 67.5% among those who received combination therapy, versus 51% in controls (HR, 0.57; P = 0.117). However, in transplant recipients with renal failure and in those with *A. fumigatus* infection, combination therapy was independently associated with an improved 90-day survival. More recently, a small comparative study of liposomal amphotericin B (3 mg/kg) plus caspofungin (n = 15) versus liposomal amphotericin B alone (10 mg/kg; n = 15) showed a higher favorable response rate at the end of therapy in the combination arm (67%) than in the monotherapy arm (27%); unfortunately, this study was severely underpowered for any definitive conclusions (Caillot et al., 2007). A randomized, comparative, clinical trial is urgently needed to make evidence-based conclusions regarding the benefits and risks of combination therapy versus voriconazole monotherapy.

CLINICAL DATA IN THE ADULT POPULATION: ANIDULAFUNGIN

The clinical experience with anidulafungin in aspergillosis is limited to a single study that evaluated the safety and tolerability of combination therapy with liposomal amphotericin B in 30 severely immunocompromised patients (Herbrecht et al., 2004). There were no unanticipated adverse events in the first 17 patients. Only two severe adverse events were believed to be drug related: one case of renal failure attributed to liposomal amphotericin B and one case of abnormal liver function values. The overall death rate was high (9 of 17), but no death was associated with a drug-related adverse event. To date, overall efficacy results from this study have not been reported.

CLINICAL DATA IN THE ADULT POPULATION: MICAFUNGIN

A phase II noncomparative study evaluated the efficacy of micafungin as primary therapy (n = 29; 17 patients received micafungin in combination) or salvage therapy (n = 196; 174 patients received micafungin in combination) for invasive aspergillosis, with outcome data reported for both monotherapy and combination therapy (Denning et al., 2006). For adult patients, the mean daily dose of study drug was 111.4 mg/day. Overall, the favorable response rate at the end of therapy was 35.6%: 37.9% in the primary therapy group responded and approximately 35% in the salvage group responded.

TOLERABILITY AND SAFETY

Echinocandins are generally very well tolerated both in the healthy subjects of the clinical pharmacology studies and in patients with a wide spectrum of diseases and receiving many concomitant medications who were included in the clinical trials. In clinical studies of patients with invasive aspergillosis, caspofungin and micafungin are well-tolerated for prolonged treatment durations (up to 162 days for caspofungin), and the favorable safety profile was maintained with extended therapy. Drug-related clinical adverse experiences are typically mild and self-limiting, including fever, headache, nausea and vomiting, (thrombo-)phlebitis, and rash (Maertens et al., 2006). So far, only one case of anaphylaxis has been described. Overall, infusion-related (fever, chills, rigors) or histamine-like reactions are extremely rare. To date, the maximum tolerated dose has not been defined for any of the echinocandins.

The most frequent drug-related laboratory adverse effects are liver function test abnormalities, which occur at rates similar to those seen with fluconazole, and decreases in hemoglobin and hematocrit. Most of the elevations in hepatic enzyme levels are transient and do not limit therapy. In an in vitro hemolysis assay, caspofungin was found to be relatively nonhemolytic against human red blood cells (similar to distilled water and less toxic than low doses of amphotericin B). Also, in contrast to amphotericin B, drug-related elevations in serum creatinine are extremely uncommon in echinocandin-treated subjects and are similar to those seen with fluconazole. In all clinical trials, very few adverse experiences led to discontinuation of therapy for the echinocandin recipients (Maertens et al., 2006; Vazquez and Sobel, 2006; Chandrasekar and Sobel, 2006).

CONCLUSION

Antifungal agents can be classified by their site of action in fungal cells, which can have important implications for efficacy, safety, and tolerability. Until recently, available agents included the polyenes, nucleoside analogs, and the azoles. With the exception of 5-fluorocytosine, all of these agents act by interfering with the structural or functional integrity of the fungal plasma membrane. However, the nonselective nature of this therapeutic target results in concomitant cross-inhibition (or toxicity) in mammalian cells. The echinocandins interfere with the synthesis of the fungal cell wall, a target not present in mammalian cells. To date, three echinocandins are commercially available; they all inhibit β-1,3-D-glucan synthesis, exert fungistatic activity against *Aspergillus* species, and are generally well-tolerated and safe. Clinical data support the use of caspofungin, the first representative of this new class, in the salvage treatment of invasive aspergillosis and in the empirical treatment of refractory neutropenic fever. The novel mechanism of action has boosted the enthusiasm for combination antifungal therapy in high-risk patient groups.

REFERENCES

Abruzzo, G. K., A. M. Flattery, C. J. Gill, L. Kong, J. G. Smith, V. B. Pikounis, J. M. Balkovec, A. F. Bouffard, J. F. Dropinski, H. Rosen, H. Kropp, and K. Bartizal. 1997. Evaluation of the echinocandin antifungal MK-0991 (L-743,872): efficacies in mouse models of disseminated aspergillosis, candidiasis, and cryptococcosis. *Antimicrob. Agents Chemother.* **41**:2333–2338.

Abruzzo, G. K., C. J. Gill, A. M. Flattery, L. Kong, C. Leighton, J. G. Smith, V. B. Pikounis, K. Bartizal, and H. Rosen. 2000. Efficacy of echinocandin caspofungin against disseminated aspergillosis and candidiasis in cyclophosphamide-induced immunosuppressed mice. *Antimicrob. Agents Chemother.* **44**:2310–2318.

Aliff, T. B., P. G. Maslak, J. G. Jurcic, M. L. Heaney, K. N. Cathcart, K. A. Sepkowitz, and M. A. Weiss. 2003. Refractory *Aspergillus* pneumonia in patients with acute leukemia: successful therapy with combination caspofungin and liposomal amphotericin. *Cancer* **97**: 1025–1032.

Bates, D. W., L. Su, D. T. Yu, G. M. Chertow, D. L. Seger, D. R. Gomes, E. J. Dasbach, and R. Platt. 2001. Mortality and costs of acute renal failure associated with amphotericin B therapy. *Clin. Infect. Dis.* **32**:686–693.

Bowman, J. C., G. K. Abruzzo, J. W. Anderson, A. M. Flattery, C. J. Gill, V. B. Pikounis, D. M. Schmatz, P. A. Liberator, and C. M. Douglas. 2001. Quantitative PCR assay to measure *Aspergillus fumigatus* burden in a murine model of disseminated aspergillosis: demonstration of efficacy of caspofungin acetate. *Antimicrob. Agents Chemother.* **45**:3474–3481.

Bowman, J. C., P. S. Hicks, M. B. Kurtz, H. Rosen, D. M. Schmatz, P. A. Liberator, and C. M. Douglas. 2002. The antifungal echinocandin caspofungin acetate kills growing cells of *Aspergillus fumigatus* in vitro. *Antimicrob. Agents Chemother.* **46**:3001–3012.

Cagnoni, P. J., T. J. Walsh, M. M. Prendergast, D. Bodensteiner, S. Hiemenz, R. N. Greenberg, C. A. Arndt, M. Schuster, N. Seibel, V. Yeldandi, and T. B. Tong. 2000. Pharmacoeconomic analysis of liposomal amphotericin B versus conventional amphotericin B in the empirical treatment of persistently neutropenic patients. *J. Clin. Oncol.* **18**:2476–2483.

Caillot, D., A. Thiébaut, R. Herbrecht, S. de Botton, A. Pigneux, F. Bernard, J. Larché, F. Monchecourt, S. Alfandari, and L. Mahi. 2007. Liposomal amphotericin B in combination with caspofungin for invasive aspergillosis in patients with hematologic malignancies: a randomized pilot study (Combistrat trial). *Cancer* **110**:2740–2746.

Candoni, A., R. Mestroni, D. Damiani, M. Tiribelli, A. Michelutti, F. Silvestri, M. Castelli, P. Viale, and R. Fanin. 2005. Caspofungin as first line therapy of pulmonary invasive fungal infections in 32 immunocompromised patients with hematologic malignancies. *Eur. J. Haematol.* **75**:227–233.

Chandrasekar, P. H., and J. D. Sobel. 2006. Micafungin: a new echinocandin. *Clin. Infect. Dis.* **42**:1171–1178.

Debono, M., B. J. Abbott, J. R. Turner, L. C. Howard, R. S. Gordee, A. S. Hunt, M. Barnhart, R. M. Molloy, K. E. Willard, and D. Fukuda. 1988. Synthesis and evaluation of LY121019, a member of a series of semisynthetic analogues of the antifungal lipopeptide echinocandin B. *Ann. N. Y. Acad. Sci.* **544**:152–167.

Debono, M., and R. S. Gordee. 1994. Antibiotics that inhibit fungal cell wall development. *Ann. Rev. Microbiol.* **48:**471–497.

Denning, D. W. 2003. Echinocandin antifungal drugs. *Lancet* **362:** 1142–1151.

Denning, D. W., K. A. Marr, W. M. Lau, D. P. Facklam, V. Ratanatharathorn, C. Becker, A. J. Ullmann, N. L. Seibel, P. M. Flynn, J. A. van Burik, D. N. Buell, and T. F. Patterson. 2006. Micafungin (FK463), alone or in combination with other systemic antifungal agents, for the treatment of acute invasive aspergillosis. *J. Infect.* **53:** 337–349.

Eiden, C., H. Peyrière, M. Cociglio, S. Djezzar, S. Hansel, J. P. Blayac, D. Hillaire-Buys, and Network of the French Pharmacovigilance Database. 2007. Adverse effects of voriconazole: analysis of the French Pharmacovigilance Database. *Ann. Pharmacother.* **41:**755–763.

Frothingham, R. 2002. Lipid formulations of amphotericin B for empirical treatment of fever and neutropenia. *Clin. Infect. Dis.* **35:** 896–897.

Glasmacher, A., O. A. Cornely, K. Orlopp, S. Reuter, S. Blaschke, M. Eichel, G. Silling, B. Simons, G. Egerer, M. Siemann, M. Florek, R. Schnitzler, P. Ebeling, J. Ritter, H. Reinel, P. Schütt, H. Fischer, C. Hahn, and G. Just-Nuebling. 2006. Caspofungin treatment in severely ill, immunocompromised patients: a case-documentation study of 118 patients. *J. Antimicrob. Chemother.* **57:**127–134.

Hector, R. F. 1993. Compounds active against cell walls of medically important fungi. *Clin. Microbiol. Rev.* **6:**1–21.

Herbrecht, R., D. W. Denning, T. F. Patterson, J. E. Bennett, R. E. Greene, J. W. Oestmann, W. V. Kern, K. A. Marr, P. Ribaud, O. Lortholary, R. Sylvester, R. H. Rubin, J. R. Wingard, P. Stark, C. Durand, D. Caillot, E. Thiel, P. H. Chandrasekar, M. R. Hodges, H. T. Schlamm, P. F. Troke, B. de Pauw, et al. 2002. Voriconazole versus amphotericin B for primary therapy of invasive aspergillosis. *N. Engl. J. Med.* **347:**408–415.

Herbrecht, R., D. Graham, M. Schuster, T. Henkel, D. Krause, J. Schranz, J. Garbino, D. Caillot, J. Reinhardt, and J. Maertens. 2004. Safety and tolerability of combination anidulafungin and liposomal amphotericin B for the treatment of invasive aspergillosis. *Biol. Blood Marrow Transplant.* **10:**91.

Hope, W. W., T. J. Walsh, and D. W. Denning. 2005. Laboratory diagnosis of invasive aspergillosis. *Lancet Infect. Dis.* **5:**609–622.

Imhof, A., D. J. Schaer, U. Schanz, and U. Schwarz. 2006. Neurological adverse events to voriconazole: evidence for therapeutic drug monitoring. *Swiss Med. Wkly.* **136:**739–742.

Kartsonis, N. A., A. J. Saah, L. C. Joy, A. F. Taylor, and C. A. Sable. 2005. Salvage therapy with caspofungin for invasive aspergillosis: results from the caspofungin compassionate use study. *J. Infect.* **50:** 196–205.

Kinoshita, K., H. Iwasaki, H. Uzui, and T. Ueda. 2006. Candin family antifungal agent micafungin (FK463) modulates the inflammatory cytokine production stimulated by lipopolysaccharide in THP-1 cells. *Transl. Res.* **148:**207–213.

Kontoyiannis, D. P., R. Hachem, R. E. Lewis, G. A. Rivero, H. A. Torres, J. Thornby, R. Champlin, H. Kantarjian, G. P. Bodey, and I. Raad. 2003. Efficacy and toxicity of caspofungin in combination with liposomal amphotericin B as primary or salvage treatment of invasive aspergillosis in patients with hematologic malignancies. *Cancer* **98:**292–299.

Kulemann, V., M. Bauer, W. Graninger, and C. Joukhadar. 2005. Safety and potential of drug interactions of caspofungin and voriconazole in multimorbid patients. *Pharmacology* **75:**165–178.

Kurtz, M. B., and J. H. Rex. 2001. Glucan synthase inhibitors as antifungal agents. *Adv. Protein Chem.* **56:**423–475.

Maertens, J., and M. Boogaerts. 2000. Fungal cell wall inhibitors. *Curr. Pharm. Des.* **6:**225–239.

Maertens, J., K. Theunissen, and M. Boogaerts. 2002. Invasive aspergillosis: focus on new approaches and new therapeutic agents. *Curr. Med. Chem. Anti Infect. Agents* **1:**65–81.

Maertens, J., I. Raad, G. Petrikkos, M. Boogaerts, D. Selleslag, F. B. Petersen, C. A. Sable, N. A. Kartsonis, A. Ngai, A. Taylor, T. F. Patterson, D. W. Denning, T. J. Walsh, et al. 2004. Efficacy and safety of caspofungin for treatment of invasive aspergillosis in patients refractory to or intolerant of conventional antifungal therapy. *Clin. Infect. Dis.* **39:**1563–1571.

Maertens, J., A. Glasmacher, R. Herbrecht, A. Thiebaut, C. Cordonnier, B. H. Segal, J. Killar, A. Taylor, N. Kartsonis, T. F. Patterson, M. Aoun, D. Caillot, C. Sable, et al. 2006. Multicenter, noncomparative study of caspofungin in combination with other antifungals as salvage therapy in adults with invasive aspergillosis. *Cancer* **107:** 2888–2897.

Marr, K. A., R. A. Carter, F. Crippa, A. Wald, and L. Corey. 2002. Epidemiology and outcome of mould infections in hematopoietic stem cell transplant recipients. *Clin. Infect. Dis.* **34:**909–917.

Marr, K. A., R. Hachem, G. Papanicolaou, J. Somani, J. M. Arduino, C. J. Lipka, A. L. Ngai, N. Kartsonis, J. Chodakewitz, and C. Sable. 2004a. Retrospective study of the hepatic safety profile of patients concomitantly treated with caspofungin and cyclosporin A. *Transplant. Infect. Dis.* **6:**110–116.

Marr, K. A., M. Boeckh, R. A. Carter, H. W. Kim, and L. Corey. 2004b. Combination antifungal therapy for invasive aspergillosis. *Clin. Infect. Dis.* **39:**797–802.

Marr, K. A., M. Boeckh, and H. W. Kim. 2005. Combination antifungal therapy for invasive aspergillosis. *Clin. Infect. Dis.* **40:**1075–1076. (Author reply.)

Meersseman, W., K. Lagrou, J. Maertens, and E. Van Wijngaerden. 2007. Invasive aspergillosis in the intensive care unit. *Clin. Infect. Dis.* **45:**205–216.

Morrissey, C. O., M. A. Slavin, M. A. O'Reilly, J. Daffy, and L. Coyle. 2004. Caspofungin (CAS) as salvage therapy (Rx) for invasive aspergillosis (IA): results of the Australian Compassionate Access Program (CAP), abstr. M-670. *44th Intersci. Conf. Antimicrob. Agents Chemother.*, 30 October to 2 November 2004. ASM, Washington, DC.

Pagano, L., M. Caira, A. Nosari, M. T. Van Lint, A. Candoni, M. Offidani, T. Aloisi, G. Irrera, A. Bonini, M. Picardi, C. Caramatti, R. Invernizzi, D. Mattei, L. Melillo, C. de Waure, G. Reddiconto, L. Fianchi, C. G. Valentini, C. Girmenia, G. Leone, and F. Aversa. 2007. Fungal infections in recipients of hematopoietic stem cell transplants: results of the SEIFEM B-2004 study. *Clin. Infect. Dis.* **45:**1161–1170.

Pagano, L., M. Caira, A. Candoni, M. Offidani, L. Fianchi, B. Martino, D. Pastore, M. Picardi, A. Bonini, A. Chierichini, R. Fanci, C. Caramatti, R. Invernizzi, D. Mattei, M. E. Mitra, L. Melillo, F. Aversa, M. T. Van Lint, P. Falcucci, C. G. Valentini, C. Girmenia, and A. Nosari. 2006. The epidemiology of fungal infections in patients with hematologic malignancies: the SEIFEM-2004 study. *Haematologica* **91:**1068–1075.

Pascual, A., T. Calandra, S. Bolay, T. Buclin, J. Bille, and O. Marchetti. 2008. Voriconazole therapeutic drug monitoring in patients with invasive mycoses improves efficacy and safety outcomes. *Clin. Infect. Dis.* **46:**201–211.

Petraitis, V., R. Petraitiene, A. H. Groll, A. Bell, D. P. Callender, T. Sein, R. L. Schaufele, C. L. McMillian, J. Bacher, and T. J. Walsh. 1998. Antifungal efficacy, safety, and single-dose pharmacokinetics of LY303366, a novel echinocandin B, in experimental pulmonary aspergillosis in persistently neutropenic rabbits. *Antimicrob. Agents Chemother.* **42:**2898–2905.

Pfaller, M. A., F. Marco, S. A. Messer, and R. N. Jones. 1998. In vitro activity of two echinocandin derivatives, LY303366 and MK-0991 (L-743,792), against clinical isolates of *Aspergillus, Fusarium, Rhizopus* and other filamentous fungi. *Diagn. Microbiol. Infect. Dis.* **30:**251–255.

Rocha, E. M., G. Garcia-Effron, S. Park, and D. S. Perlin. 2007. A Ser678Pro substitution in Fks1p confers resistance to echinocandin drugs in *Aspergillus fumigatus*. *Antimicrob. Agents Chemother.* 51: 4174–4176.

Sanz-Rodriguez, C., R. Martino, and M. Canales. 2002. Caspofungin therapy in documented fungal infections in patients with hematological disorders: Spanish experience before licensure of the drug. Abstr. P-2483. *Annu. Meet. Am. Soc. Hematol.*, 7 to 11 December 2002, Philadelphia, PA.

Sanz-Rodriguez, C., M. Lopez-Duarte, M. Jurado, J. Lopez, R. Arranz, J. M. Cisneros, M. L. Martino, P. J. Garcia-Sanchez, P. Morales, T. Olivé, M. Rovira, and C. Solano. 2004. Safety of the concomitant use of caspofungin and cyclosporin A in patients with invasive fungal infections. *Bone Marrow Transplant.* 34:13–20.

Serrano, M., A. Valverde-Conde, M. Chávez, S. Bernal, R. M. Clara, J. Pemán, M. Ramirez, and E. Martín-Mazuelos. 2003. In vitro activity of voriconazole, itraconazole, caspofungin, anidulafungin (VER002, LY303366) and amphotericin B against *Aspergillus* spp. *Diagn. Microbiol. Infect. Dis.* 45:131–135.

Silveira, F. P., and S. Husain. 2007. Fungal infections in solid organ transplantation. *Med. Mycol.* 45:305–320.

Singh, N., A. P. Limaye, G. Forrest, N. Safdar, P. Muñoz, K. Pursell, S. Houston, F. Rosso, J. G. Montoya, P. Patton, R. Del Busto, J. M. Aguado, R. A. Fisher, G. B. Klintmalm, R. Miller, M. M. Wagener, R. E. Lewis, D. P. Kontoyiannis, and S. Husain. 2006. Combination of voriconazole and caspofungin as primary therapy for invasive aspergillosis in solid organ transplant recipients: a prospective, multicenter, observational study. *Transplantation* 81:320–326.

Steinbach, W. J., D. A. Stevens, and D. W. Denning. 2003. Combination and sequential antifungal therapy for invasive aspergillosis: review of published in vitro and in vivo interactions and 6281 clinical cases from 1966 to 2001. *Clin. Infect. Dis.* 37(Suppl. 3):S188–S224.

Stone, J. A., E. M. Migoya, L. Hickey, G. A. Winchell, P. J. Deutsch, K. Ghosh, A. Freeman, S. Bi, R. Desai, S. C. Dilzer, K. C. Lasseter, W. K. Kraft, H. Greenberg, and S. A. Waldman. 2004. Potential for interactions between caspofungin and nelfinavir or rifampin. *Antimicrob. Agents Chemother.* 48:4306–4314.

Tawara, S., F. Ikeda, K. Maki, Y. Morishita, K. Otomo, N. Teratani, T. Goto, M. Tomishima, H. Ohki, A. Yamada, K. Kawabata, H. Takasugi, K. Sakane, H. Tanaka, F. Matsumoto, and S. Kuwahara. 2000. In vitro activities of a new lipopeptide antifungal agent, FK463, against a variety of clinically important fungi. *Antimicrob. Agents Chemother.* 44:57–62.

Traxler, P., W. Tosch, and O. Zak. 1987. Papulacandins: synthesis and biological activity of papulacandin B derivatives. *J. Antibiot.* 40: 1146–1164.

Trenschel, R., M. Ditschkowski, A. H. Elmaagacli, M. Koldehoff, H. Ottinger, N. Steckel, M. Hlinka, R. Peceny, P. M. Rath, H. Dermoumi, and D. W. Beelen. 2005. Caspofungin as second-line therapy for fever of unknown origin or invasive fungal infection following allogeneic stem cell transplantation. *Bone Marrow Transplant.* 35:583–586.

Vazquez, J. A., and J. D. Sobel. 2006. Anidulafungin: a novel echinocandin. *Clin. Infect. Dis.* 43:215–222.

Viscoli, C., R. Herbrecht, H. Akan, L. Baila, C. Doyen, A. Gallamini, A. Giagounidis, O. Marchetti, R. Martino, L. Meert, M. Paesmans, M. Shivaprakash, A. J. Ullmann, J. A. Maertens, et al. 2007. Caspofungin as first-line therapy of invasive aspergillosis in haematological patients: a study of the EORTC infectious diseases group. Abstr. O-43. *3rd Trends Med. Mycol.*, 28 to 31 October 2007, Turin, Italy.

von Mach, M. A., J. Burhenne, and L. S. Weilemann. 2006. Accumulation of the solvent vehicle sulphobutylether beta cyclodextrin sodium in critically ill patients treated with intravenous voriconazole under renal replacement therapy. *BMC Clin. Pharmacol.* 6:6.

Walsh, T. J., H. Teppler, G. R. Donowitz, J. A. Maertens, L. R. Baden, A. Dmoszynska, O. A. Cornely, M. R. Bourque, R. J. Lupinacci, C. A. Sable, and B. E. dePauw. 2004. Caspofungin versus liposomal amphotericin B for empirical antifungal therapy in patients with persistent fever and neutropenia. *N. Engl. J. Med.* 351: 1391–1402.

Wingard, J. R. 2002. Lipid formulations of amphotericins: are you a lumper or a splitter? *Clin. Infect. Dis.* 35:891–895.

Aspergillus fumigatus and Aspergillosis
Edited by J.-P. Latgé and W. J. Steinbach

Chapter 33

Antifungal Drug Interactions

RUSSELL E. LEWIS

"All things are poison and nothing is without poison, only the dose permits something not to be poisonous."

Paracelsus (1490–1541)

Drug interactions are a recurring problem in the treatment of invasive aspergillosis and they can severely limit treatment options for this life-threatening infection. While the incidence of drug interactions in patients with aspergillosis is unknown, recent surveys from large cancer treatment centers have suggested that nearly one-half of hospitalized patients have potentially serious interactions involving drugs used for the treatment of comorbid conditions or infection (Riechelmann, 2007; Riechelmann and Saad, 2006). Retrospective studies from the United States (Yu et al., 2005) and France (Depont et al., 2007) have reported the frequency of drug interactions with antifungals to be as high as 70 to 77%, although serious adverse effects resulting from these interactions were uncommon (1 to 14%). The risk of an adverse effect with a drug interaction is further increased in patients with advanced age, malnutrition, malabsorption, chronic illness, hepatic or renal dysfunction, polypharmacy, or concomitant use of drugs with a narrow therapeutic index or in patients whose care is provided by multiple specialists or prescribers (Lewis, 2006; Riechelmann and Saad, 2006). Because many or all of these risk factors are present in the patient populations predisposed to invasive fungal infections, drug interactions should be anticipated in any patient receiving systemic antifungal therapy for invasive aspergillosis.

Drug interactions can arise through a variety of mechanisms that are classified into three groups: pharmaceutical (drug acting on drug), pharmacodynamic (drug acting on the host and fungal pathogen), or pharmacokinetic (host acting on the drug) (Table 1) (Gallicano and Drusano, 2005). Pharmaceutical interactions are concerned primarily with drug stability and compatibility in intravenous admixtures, i.e., the potential for one drug to cause precipitation and/or inactivation of a second drug when the two are mixed together. Amphotericin B formulations are prone to pharmaceutical interactions due to the limited solubility of the polyene molecule and its inherent instability when complexed with either deoxycholate or phospholipid carriers. Consequently, amphotericin B formulations generally should not be mixed with other drugs and should be diluted only in 5% dextrose water, since any precipitation increases the risk of toxic reactions (Grillot and Lebeau, 2005). Echinocandins are also incompatible with some solutions (e.g., caspofungin with 5% dextrose) or may precipitate when mixed with other medications. As such, precautions need to be taken to ensure other drugs are not added to intravenous admixtures unless their compatibility has been previously documented.

Pharmacodynamic interactions encompass the combined biological effects of drugs and, in the case of antifungals, how these combinations affect the fungal pathogen (Gallicano and Drusano, 2005). From a therapeutic standpoint, some interactions are desirable, as is the case when the combined use of two or more drugs with different mechanisms of action produces synergistic lethality in fungi. However, it is also possible for one drug to antagonize the lethal effects of the second agent, resulting in activity that is no better than the least active antifungal agent used alone. Drug interactions can also have unwanted effects on the biology of the host (toxicodynamics), producing additive damage to organs such as the liver and kidney. Amphotericin B-associated nephrotoxicity and electrolyte disturbances are increased in patients who have already received drugs that damage renal proximal and distal tubular membranes (e.g., aminoglycosides, cyclosporine, tacrolimus, and foscarnet) (Wingard et al., 1999). Similarly, hepatic toxicity with *Aspergillus*-active triazoles is more common in transplant patients who have preexisting liver dysfunc-

Russell E. Lewis • Dept. of Clinical Sciences and Administration, The University of Houston College of Pharmacy, and The University of Texas M.D. Anderson Cancer Center, 1441 Moursund St. #424, Houston, TX 77030.

Table 1. Antifungal characteristics that predispose to potential adverse drug interactions

Pharmaceutical
Limited aqueous stability or solubility in intravenous admixtures (amphotericin B, echinocandins)
Chelation interaction with polyanions? (itraconazole-sucralfate, antacids)
High degree of protein binding (amphotericin B, itraconazole, posaconazole echinocandins)
Pharmacodynamic
Antagonistic mechanisms of action (azoles, amphotericin B)
Relatively narrow therapeutic index (amphotericin B, voriconazole)
Additive renal toxicity (amphotericin B)
Additive hepatic toxicity (all antifungals)
Pharmacokinetic
Weak bases requiring acidic pH for gastric dissolution (itraconazole, posaconazole)
Saturable bioavailability of oral formulations (itraconazole, posaconazole)
Wide inter- and intrapatient pharmacokinetic variability (itraconazole, voriconazole, posaconazole)
Renal elimination (intravenous voriconazole vehicle [sulfobutyl ether β-cyclodextrin])
Extensive hepatic biotransformation (itraconazole, voriconazole)
Substrates or inhibitors of mammalian cytochrome P450 (itraconazole, voriconazole, posaconazole)
Nonlinear metabolism and elimination profile (voriconazole)

tion following high-dose chemotherapy or with the onset of graft-versus-host disease (Tan et al., 2006). While fulminant hepatic failure with antifungals is uncommon, many patients will require treatment interruptions or a switch to other antifungals until liver function stabilizes.

Pharmacokinetic interactions are by far the most common mechanism of drug interactions in patients with invasive aspergillosis and occur when one drug alters the absorption, distribution, metabolism, and/or excretion of a second drug (Gallicano and Drusano, 2005). While many of these interactions are not clinically significant, some interactions affecting the pharmacokinetics of antifungals used to treat *Aspergillus* infection, or interactions that decrease the metabolism or elimination of a drug with a narrow therapeutic index for efficacy (e.g., antiretrovirals) or safety (immunosuppressants, chemotherapy) have the potential for serious harm to the patient (Lewis, 2006). Early indications of drug toxicity arising from pharmacokinetic drug interactions in a patient (e.g., electrolyte disturbances, increases in markers of hepatic toxicity, mental status changes, etc.) often present with subtle clinical signs that are difficult to distinguish from underlying diseases or infection. Therefore, knowledge of these potential interactions and a proactive approach towards their management is essential for the safe and effective use of systemic antifungal therapy in patients with invasive aspergillosis. The remainder of this chapter will focus in detail on mechanisms and clinical implications of pharmacokinetic drug-drug interactions in the patient with aspergillosis.

ANTIFUNGAL INTERACTIONS IN THE GASTROINTESTINAL TRACT

Gastric pH and Absorption

Absorption of orally administered antifungals occurs primarily in the upper small intestine (duodenum and jejunum) through passive diffusion, although other processes, including active, facilitated, or ion-pair transport and endocytosis, may also play a role (Kashuba and Bertino, 2005). The rate of drug absorption by passive diffusion is heavily influenced by drug solubility, particle size, contact time with the absorption surface, the integrity of the intestinal membranes, and hydrogen ion concentration (pH) (Kashuba and Bertino, 2005). As a result, disease states or drug therapies influencing these variables can have a significant impact on the absorptive process of oral antifungal agents.

Gastric pH is a critical factor in the absorption of weakly basic drugs with limited aqueous solubility, such as itraconazole (pK_a 3.7) and posaconazole (pK_a 3.6), which are maximally absorbed when gastric pH is 1 to 4 (Lange et al., 1997). Absorption of other triazoles, such as fluconazole (pK_a 2) and voriconazole (pK_a 1.8), are less affected by gastric pH because of their improved aqueous solubility and lower pK_a values (Gubbins et al., 2005). Hence, a number of acid suppression therapies have the potential to interfere with itraconazole and posaconazole absorption, but not with fluconazole or voriconazole. While antacids may have only modest effects on drug dissolution rates due to their transient (0.5 to 2 h) and modest alkalinization (1 to 2 pH units) effects, H2 antagonists and proton pump inhibitors (PPIs) can dose dependently maintain gastric pH at >5 for nearly 20 h (Burget et al., 1990). Coadministration of itraconazole capsules with famotidine or omeprazole reduces the plasma itraconazole area under the concentration-time curve from 0 to 24 h (AUC_{0-24}) by 30 to 60% (Kanda et al., 1998; Lim et al., 1993). Similarly, the coadministration of an early investigational posaconazole tablet formulation with cimetidine reduced the posaconazole plasma AUC by approximately 40% (Schering-Plough, Inc., 2006). Furthermore, in a population pharmacokinetic analysis of patients receiving a posaconazole suspension formulation, average posaconazole plasma concentrations were ~30% lower in patients receiving PPIs compared to matched patients who were not receiving PPI therapy (range, 20 to 40%) (Schering-Plough, Inc., 2006). Generally a change in the extent of a medication's absorption by more than 20% is considered clinically significant (Kashuba and Bertino, 2005).

Formulation approaches for itraconazole and posaconazole can overcome, to varying degrees, the poor solubility characteristics of these agents. Solid dosage forms for itraconazole are administered as antifungal-coated sugar spheres in a capsule formulation to maximize the available surface area for drug dissolution. Similarly, posaconazole suspension consists of micronized drug particles dispersed with nonionic surfactant and thickening agents to improve the dispersion characteristics of the drug (Schering-Plough, Inc., 2002). Perhaps the most successful approach for overcoming incomplete drug dissolution at high gastric pH is to presolubilize the drugs into solution using supramolecular carriers, such as cyclodextrins. Itraconazole is available as a solution formulation dissolved in a 40% hydroxypropyl-β-cyclodextrin vehicle, which achieves a 30% higher AUC_{0-24} under fed conditions compared to the capsules and 60% higher AUC_{0-24} compared to capsules when administered under fasting conditions (Van de Velde et al., 1996). The principal drawback of the solution formulation is that >50% of the cyclodextrin passes through the intestinal tract unabsorbed (Stevens, 1999). Like any polysaccharide, unchanged cyclodextrin passing through the gut stimulates intestinal secretion and gastrointestinal propulsion, resulting in nausea and/or osmotic diarrhea (Stevens, 1999). Cyclodextrin-related gastrointestinal effects further increase with escalating itraconazole doses, with few patients tolerating doses of the oral solution exceeding 400 mg/day (Glasmacher et al., 1999). Therefore, itraconazole solution may not be an acceptable option for many patients who are already experiencing nausea or gastrointestinal distress.

Another approach that has been used with some success to overcome itraconazole and posaconazole's pH-impaired absorption is the coadministration of these triazoles with an acidic beverage (i.e., cola or orange juice) in divided doses (Poirier and Cheymol, 1998). This strategy is typically more effective than simple dose escalation (Poirier and Cheymol, 1998). Absorption of posaconazole is saturated at 800 mg/day, but the bioavailability of an 800-mg/day regimen can be improved by 98% with twice-daily dosing with 400 mg and 220% with four-times-daily dosing with 200 mg, relative to once-daily dosing with 800 mg (Courtney et al., 2003). This bioavailability of solid itraconazole and posaconazole dosage forms is further improved if the drugs are administered with a high-fat meal, which increases gastric retention time and therefore allows more time for dissolution (Courtney et al., 2003). For posaconazole, coadministration of the suspension and a high-fat meal (>50% calories from fat) increased the mean AUC and maximum observed plasma concentration (C_{max}) values fourfold compared with fasted conditions ($P < 0.001$) (Courtney et al., 2004). Even a nonfat meal enhances posaconazole oral bioavailability, as mean AUC and C_{max} values increased 2.6- and 3-fold, respectively, compared with fasted conditions ($P < 0.001$) (Courtney et al., 2004).

Ultimately, it is important for clinicians to recognize that many patients who develop invasive aspergillosis already have factors impairing drug absorption that are multiplicative with drug interactions. For example, an acid suppression agent alone may not cause substantive reductions (<20%) in posaconazole absorption in a healthy volunteer, but in a patient already receiving mucotoxic chemotherapy who has diarrhea and a poor appetite the effects of acid suppression may be much more prominent (~50%) (Walsh et al., 2007). Therefore, data included in drug labeling or published pharmacokinetic studies in healthy volunteers should never be used in lieu of careful monitoring and a high index of suspicion for drug malabsorption with either itraconazole or posaconazole, particularly if the patient has underlying gastrointestinal conditions.

Presystemic Biotransformation and Clearance

Although absorption is the principal function of the intestinal mucosa, the capacity of intestinal enterocytes to biotransform (metabolize) and efflux xenobiotics before reaching the systemic circulation can be an important contributor to pharmacokinetic variability and drug interactions with oral antifungal therapy (Kashuba and Bertino, 2005; Gubbins et al., 2005). Two mechanisms in intestinal enterocytes have specifically been identified as important modulators of presystemic clearance and metabolism for itraconazole and possibly other mold-active azoles: P-glycoprotein (P-gp) and cytochrome P450 3A4/3A5 (CYP3A4/3A5) (Fig. 1).

P-gp, the product of the multidrug resistance gene (*MDR1*), is a member of the ATP-binding cassette superfamily of transport proteins that utilize ATP to translocate a wide range of substrates across biological membranes, pumping drugs out of the cell into apical or basolateral extracellular fluids (Meletiadis et al., 2006; Penzak, 2005). P-gp is believed to function primarily as a detoxification pump; however, some in vitro and in vivo evidence suggests that P-gp can also serve as a drug transporter. Itraconazole and posaconazole, but not voriconazole, are both inhibitors and substrates of P-gp (Wang et al., 2002). In vitro studies have suggested, however, that posaconazole is a weaker inhibitor of P-gp than itraconazole (Sansone-Parsons et al., 2007a). The density of P-gp expression in the gastrointestinal tract exhibits wide intra- and interpatient variability that

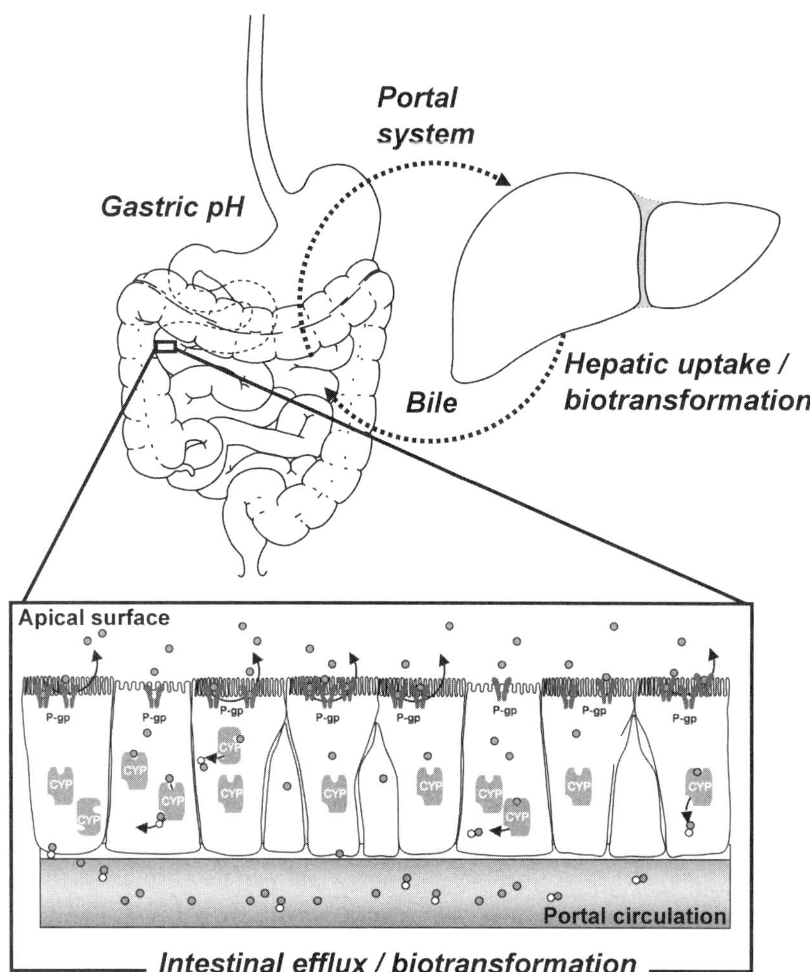

Figure 1. Mechanisms of pharmacokinetic variability and drug-drug interactions in the gastrointestinal and hepatobiliary tracts. Most antifungal agents are absorbed by passive diffusion. Intestinal enterocytes can efflux drug out of the cell by action of the P-gp pump. Mixed function oxidases (primarily CYP3A4/3A5) oxidize the drug before it reaches the portal circulation. The overlapping substrate specificity of P-gp and CYP3A4/5 represents a complementary barrier against drug absorption that serves as a possible source for drug interactions with orally absorbed antifungal agents.

is affected by diet, underlying disease, drug therapy, and genetics (Hall et al., 1999). Furthermore, several genetic polymorphisms of *MDR1* have been associated with variability in P-gp activity among various racial groups (Schaeffeler et al., 2001). However, the pharmacokinetic impact of this genotypic variability is more evident with other P-gp substrates administered with itraconazole or posaconazole (e.g., cyclosporine or tacrolimus) than with the pharmacokinetics of the antifungals alone (Sansone-Parsons et al., 2007b; Shon et al., 2005).

The other major mechanism of presystemic clearance in the intestinal lumen is oxidation catalyzed by CYP3A4/3A5, which in the case of itraconazole results in the production of both inactive and active metabolites. Hydroxy-itraconazole is a by-product of CYP3A4 biotransformation that is found at twofold higher concentrations in the bloodstream compared to the parent compound after oral dosing, suggesting a major contribution to the overall mycological activity of itraconazole (Poirier and Cheymol, 1998). Like P-gp, itraconazole is both a substrate and inhibitor of CYP3A4/3A5 (Gubbins et al., 2005). The cellular colocation of P-gp and CYP3A4/3A5 in enterocytes and overlapping substrate specificity suggest that the two mechanisms work synergistically to form a coordinated intestinal barrier to a variety of xenobiotics, including itraconazole (Hall et al., 1999). Additionally, expression of CYP3A4/3A5 in enterocytes exhibits similar intra- and interpatient variability as seen with P-gp and is not under coordinate regulation of CYP3A4 expression in the liver (Lown et al., 1994). Therefore, it is difficult to predict what effects increased drug doses, concomitant medications, or chemotherapy-induced changes in the intestinal lumen may have on the absorption of triazoles.

ANTIFUNGAL INTERACTIONS INVOLVING HEPATIC BIOTRANSFORMATION

The principal site for drug biostransformation is the liver, where lipophilic molecules are transformed into ionized metabolites for renal elimination. This transformation generally occurs via two different types of reactions: (i) phase I (nonsynthetic) reactions, which include oxidation, reduction, and hydrolysis, and (ii) phase II synthetic reactions, which involve conjugation with other molecules (i.e., glucoronidation, sulfation) to improve aqueous solubility (Kashuba and Bertino, 2005). Induction of these pathways by other drugs can increase the clearance of antifungals, potentially resulting in subtherapeutic bloodstream drug concentrations or, alternatively, inhibition of these reactions may cause drug accumulation, predisposing patients to excessive toxicity (Gubbins et al., 2005).

Most oxidative reactions are catalyzed by a superfamily of mixed function monooxygenases called the cytochrome P450 system. Although 14 human families of CYP enzymes have been identified, 95% of all drug oxidation occurs through the action of six CYP enzymes: CYP1A2, CYP2C8/9, CYP2C19, CYP2D6, CYP2E1, and CYP3A4/5 (Kashuba and Bertino, 2005). The CYP3A4 pathway is most frequently associated with severe drug interactions by virtue of the fact that it is responsible for metabolism of 50% of clinically utilized drugs (Smith et al., 1998). In terms of drugs used for the treatment of invasive aspergillosis, itraconazole and voriconazole undergo extensive phase I metabolism by CYP3A4 and 2C19, respectively (Table 2) (Gubbins et al., 2005). On the other hand, posaconazole is eliminated by UDP glucoronidation, a phase II mechanism that allows biliary and fecal elimination of the lipophilic triazole (Courtney et al., 2005). Neither amphotericin B nor the echinocandins are major substrates or inhibitors of phase I/II biotransformation in the liver (Groll and Walsh, 1998; Wiederhold and Lewis, 2007).

Similar to P-gp, considerable intra- and interpatient variability in CYP enzyme activity is due to medical, environmental, and dietary factors (Smith et al., 1998). CYP enzymes also show large interindividual differences in activities due to genetic variants constituting multiallelic systems that express a variety of phenotypes. Specifically, clinically significant genetic polymorphisms have been described for CYD2D6, CYP2C9, and CYP2C19 (Kashuba and Bertino, 2005; Smith et al., 1998). The frequency of polymorphisms also differs among different racial populations. Unlike the unimodal population distribution of CYP3A4, 15 to 30% of the Asian population and 3 to 5% of Caucasians carry a homozygous allele for limited CYP2C19 biotransformation (Kashuba and Bertino, 2005; Meletiadis et al., 2006; Smith et al., 1998). Voriconazole pharmacokinetics are clearly affected by these polymorphisms in CYP2C19 (Eiden et al., 2007). Subpopulations of patients who are homozygous poor metabolizers through the CYP2C19 pathway may experience, on average, fourfold higher serum concentrations of voriconazole compared to other patients who are heterozygous extensive or homozygous extensive metabolizers (Ikeda et al., 2004). Clearance of triazole antifungals is markedly increased by drugs that accelerate CYP450 metabolism (Table 3). Concomitant use of drugs such as rifampin can result in low (Nicolau et al., 1995) or undetectable levels of the triazole in the bloodstream (Bonay et al., 1993; Ducharme et al., 1995; Krishna et al., 2007) that often cannot be overcome with higher drug doses.

All azoles are also reversible inhibitors of CYP enzymes in humans (Gubbins et al., 2005). This inhibition is a collateral effect of their antifungal mechanism of inhibiting the homologous fungal CYP P450 enzyme involved in ergosterol biosynthesis, 14α-demethylase. The most important drug interactions seen with azole antifungals typically arise from inhibition of CYP3A4, which plays a critical role in the metabolism of a broad array of drug therapies used for cardiovascular disease, endocrine disorders including hyperglycemia, anesthesia, psychiatric disorders, epilepsy, cancer chemotherapy, immunosuppression used in transplantation, and treatment of infectious diseases (Gubbins et al., 2005). A

Table 2. Comparative CYP P450 interaction profiles of the triazole antifungals

Triazole	Level of inhibition[a]					
	CYP3A4		CYP2C8/9		CYP2C19	
	Substrate	Inhibitor	Substrate	Inhibitor	Substrate	Inhibitor
Fluconazole	+	+	++	++		
Itraconazole	+++	+++	+	+		
Voriconazole	+	++	+		+++	++
Posaconazole		++				

[a] +, weak inhibitor; +++, strong inhibitor. Data were extracted from Gubbins et al. (2005).

Table 3. Summary of common drug interactions affecting hepatic biotransformation and clearance in patients with aspergillosis

Interaction	Example(s)	Effect(s)
Voriconazole, itraconazole, or posaconazole *plus* hepatic biotransformation (phase I or II) inducers	Carbamazepine, phenobarbital, phenytoin, rifabutin, rifampin, nevirapine	Increased clearance rate of triazole antifungal, low or undetectable antifungal plasma concentrations
Voriconazole, itraconazole, or posaconazole *plus* hepatic biotransformation (CYP3A4) inhibitors	Statins (hepatically metabolized), cyclosporine, tacrolimus, sirolimus, protease inhibitors (saquinavir, ritonavir), wafarin, Ca^{2+} channel blockers (diltiazem, verapamil, nifedipine, nisoldipine)	Decreased clearance rate of CYP substrate leading to potentially toxic drug exposures
Caspofungin *plus* hepatic biotransformation inducers	Rifampin	Caspofungin AUC decreased by 30%
Caspofungin *plus* OATP (?) substrates	Cyclosporine (?)	Caspofungin AUC increased by 30%
Micafungin *plus* OATP (?) substrates	Cyclosporine (?), nifedipine, sirolimus	Decreased clearance rate of substrates

general overview of interactions associated with antifungal therapy is presented in Table 3.

Antifungal CYP3A4/5 Interactions with Chemotherapy Agents or Immunosuppressants

Some chemotherapy agents and immunosuppressants used in solid organ or hematopoetic stem cell transplantation are extensively cleared through CYP3A4 biotransformation. Inhibition of CYP3A4 by triazole antifungals has the potential for marked increases in serum drug concentrations and toxicity with the relatively narrow therapeutic indices of these drugs (Saad et al., 2006). Therefore, concomitant use of these medications with antifungal therapies, particularly triazole antifungals used in the treatment of invasive aspergillosis, requires careful dosage adjustment and close monitoring both before and after discontinuation of the triazole. Trough concentrations of cyclosporine are increased by approximately twofold in patients who are administered either itraconazole or voriconazole (Romero et al., 2002; Saad et al., 2006). However, in some patients trough concentrations may increase by as much as threefold, highlighting the need for individualized dosing and serum drug concentration monitoring. Therefore, it is generally recommended that doses of cyclosporine be reduced by 50 to 60% at the initiation of either itraconazole or voriconazole administration (Romero et al., 2002; Saad et al., 2006). For posaconazole, a small case series suggested that cyclosporine doses should be reduced by one-third in heart transplant patients (Sansone-Parsons et al., 2007b).

While tacrolimus pharmacokinetics in the setting of triazole therapy are less well characterized, a two-thirds reduction in the tacrolimus dose is recommended in patients who are started on itraconazole, voriconazole, or posaconazole therapy (Billaud et al., 1998; Capone et al., 1998, 1999; Pfizer, Inc., 2001; Sansone-Parsons et al., 2007b; Wood et al., 2001). The inhibitory effects of triazoles appear to be even more pronounced with sirolimus. In a single-blinded, randomized placebo-controlled crossover study of healthy patients, voriconazole at 400 mg twice daily on day 1 followed by 200 mg twice daily for 8 days increased the AUC of sirolimus by 11-fold (Pfizer, Inc., 2001). Based on these findings, the coadministration of sirolimus and voriconazole is contraindicated. However, a recent case series suggested that a 90% reduction in the sirolimus dosage might allow both drugs to be administered safely with careful monitoring (Marty et al., 2006).

Some chemotherapy agents are biotransformed by CYP enzymes into both active and toxic metabolites or rely on CYP biotransformation for clearance. Concomitant use of a triazole antifungal, particularly in patients receiving high-dose cyclophosphamide, busulfan, or vinca alkaloids, can produce an overdose of chemotherapy, causing excessive organ and hematological toxicities and a possibly lower therapeutic response, leading to malignancy relapse. Results from a multicenter trial comparing the prophylactic efficacy of fluconazole and itraconazole in allogeneic stem cell transplant recipients were complicated by higher serum bilirubin and creatinine values in patients who received itraconazole concurrently with cyclophosphamide conditioning (Marr et al., 2004). Analysis of cyclophosphamide metabolism in a subset of patients revealed a higher exposure of toxic metabolites among itraconazole recipients compared with fluconazole (Marr et al., 2004). As a result, many transplant centers withhold the use of triazole antifungals until 48 to 72 h after the completion of conditioning chemotherapy to reduce the risk of pharmacokinetic and pharmacodynamic interactions (Lewis, 2006).

ANTIFUNGAL INTERACTIONS INVOLVING DRUG TRANSPORTERS

Organic anion transporting polypeptides (OATPs) represent a group of membrane carriers that transport a wide spectrum of amphipathic substrates. Currently, nine OATPs have been identified in humans and are expressed in various tissues including the liver, blood-brain barrier, choroids plexus, lung, heart, intestine, kidney, placenta, and testes (Kim, 2003; Meletiadis, 2006). OATPs transport a variety of diverse compounds and function as uptake transporters that facilitate the influx of compounds (Kim, 2003). The OATP family appears to play a major role in hepatobiliary drug extraction as they are prominently located on the basolateral (sinusoidal) membranes of hepatocytes (Kim, 2003). The echinocandin caspofungin has been reported to be a substrate of the OATP-1B1 (OATP-C) transporter in rats (Sandhu et al., 2005). This may explain in part why caspofungin exhibits modest interactions with cyclosporine (increases the caspofungin AUC by 35%) and rifampin (decreases the caspofungin AUC by 30%), despite a lack of substrate or inhibitory activity in human CYP3A4 enzymes (Saad et al., 2006; Sanz-Rodriguez et al., 2004). It is not known whether micafungin and anidulafungin have similar affinities for this transporter, given their lack of clinically significant interactions with rifampin (Lewis, 2006). Caspofungin can decrease tacrolimus concentrations by 20%, although OATP mechanisms have not been clearly implicated with this interaction. Currently, additional work is under way to identify the precise roles and substrate specificities of various OATPs in the liver and other tissues in humans.

ANTIFUNGAL INTERACTIONS INVOLVING RENAL ELIMINATION

Amphotericin B-induced nephrotoxicity can be additive to that caused by other agents that cause damage to the distal tubules of the kidney (i.e., cyclosporine, tacrolimus, aminoglycosides, cisplatin) or can lead to the accumulation of toxic serum concentrations of renally eliminated drugs, such as flucytosine (Groll and Walsh, 1998; Gubbins et al., 2005). Electrolyte disturbances that arise and that accompany distal tubular damage (hypokalemia, hypomagnesemia) can also be additive to other medications (i.e., loop and thiazide diuretics, corticosteroids) or augment the pharmacological effect of drugs such as antiarrhythmics, nondepolarizing skeletal muscle relaxants, and digoxin with harmful results. Corticosteroids enhance amphotericin B-induced hypokalemia and have contributed to reversible cardiomyopathy in part due to the augmentation of corticosteroid hypokalemic and salt and water retention effects (Chung and Koenig, 1971). Therefore, this combination should be avoided whenever possible in patients with a history of congestive heart failure.

Frequently, drug interactions that arise from amphotericin B-associated nephrotoxicity are unavoidable in patients with invasive aspergillosis. Management of these interactions focuses on preventative measures to limit the severity of the reaction and careful monitoring and supplementation of electrolyte deficiencies. Lipid formulations of amphotericin B should be substituted for amphotericin B deoxycholate whenever possible in patients with underlying renal dysfunction or receiving concomitant nephrotoxic agents. Hydration with normal saline before and after amphotericin B infusions, a practice known as sodium loading, may also be helpful for blunting tubular-glomerular feedback mechanisms that accelerate decreases in glomerular filtration (Branch, 1988; Heidemann et al., 1983). Careful monitoring of renal function (serum creatinine, blood urea nitrogen, and electrolytes, i.e., K^+, Mg^{2+}, and PO_4) is essential in invasive aspergillosis patients receiving amphotericin B-based therapy. Blood or serum drug concentration monitoring of renally eliminated drugs with a narrow therapeutic index (i.e., aminoglycosides, flucytosine) requires intensive monitoring even in the absence of glomerular filtration changes, as shifts in serum drug concentrations of these drugs generally precede changes in serum creatinine levels.

PHARMACOKINETIC INTERACTIONS AFFECTING CARDIAC CONDUCTION

Although cardiac toxicity is a relatively uncommon adverse effect, it is the most severe consequence of antifungal drug interactions that can be encountered in patients with invasive aspergillosis. Antifungals have been associated with sudden cardiac death resulting from polymorphic ventricular tachyarrhythmias called torsades de pointes (TdP). Prolongation of the QT interval, an indication of delayed cardiac repolarization, has been demonstrated to play a key role in drug-induced TdP. Many drugs, including some antifungals, have the potential to block the product of the human ether-a-go-go (HERG) gene, which encodes a rapid component of the delayed rectified potassium channel (I_{kr}). The blockade of HERG-encoded I_{kr} results in accumulation of potassium within the myocyte that delays cardiac repolarization (Owens, 2004). Electrical instability that occurs during this delayed repolarization phase may abruptly reverse its course, resulting in premature ventricular complex on electrocardiogram. If the pathological process is repetitive or self-sustaining, a persistent polymorphic ventricular arrhythmia is produced, leading to syncope, dizziness, palpitations, or sudden death (Owens, 2004).

Although patients often receive drugs that directly cause QT prolongation, it is uncommon for QT prolongation itself to result in TdP (Owens, 2004). The extent of QT interval prolongation is governed by drug-related and host-specific risk factors. Drug-related factors that can influence the risk of TdP include the intrinsic QT prolongation potential of the drug itself (ability to inhibit I_{kr} at physiologically relevant concentrations), comedications associated with QT prolongation, metabolic drug interactions, particularly those involving CYP2C19 or CYP3A4 that result in supratherapeutic concentrations of a second coadministered drug known to inhibit I_{kr}, drug dose, and route of administration (intravenous > oral) (Owens, 2004). Host-specific factors include electrolyte derangements (hypokalemia, hypomagnesemia, hypocalcemia), increased age, female gender, underlying structural heart disease (myocardial infarction, congestive heart failure), bradycardia, underlying organ dysfunction with failure to adjust dosage accordingly for drugs known to prolong the QT interval, hypothyroidism, central nervous system tumor or infection, and obesity (Owens, 2004). More recently, a genetic predisposition has been identified for inherited long QT syndrome that results in symptomatic or asymptomatic decreases in cardiac repolarization reserve. This electrophysiological abnormality may remain clinically silent until a pharmacological or other stimulus surfaces, thus unmasking the cardiac ion mutation and resulting in symptomatic arrhythmias (Chiang, 2004; Chiang and Roden, 2000). Indeed, ketoconazole, fluconazole, and voriconazole have been implicated in cases of drug-induced TdP in patients with clinically asymptomatic long QT syndrome (Eiden et al., 2007; Owens, 2004).

Although current antifungals used in the treatment of invasive aspergillosis have shown minimal HERG inhibition at clinically achievable levels, direct I_{kr} inhibition cannot be excluded at high drug exposures (Fig. 2) (Owens, 2004). Furthermore, in preclinical testing of voriconazole, high dosages administered to dogs resulted in arrhythmias, premature ventricular complex, and prolonged QT intervals (Pfizer, Inc., 2001). Phase I studies of voriconazole in healthy humans revealed 28% of patients with QTc intervals elevated ≥30 ms above baseline but <60 ms, compared to 19% in the placebo group (Pfizer, Inc., 2001). One sudden death was also reported in phase III studies as a result of cardiac arrest following voriconazole administration, although the patient had multiple risk factors, including hypokalemia and prior history of benign ventricular arrhythmias (Baildon et al., 2001). QTc prolongation with posaconazole has been reported in similar percentages as in fluconazole- and itraconazole-treated patients, and one patient receiving posaconazole prophylaxis during a clinical trial who had severe electrolyte abnormalities developed TdP that was not fatal (Schering-Plough, Inc., 2006).

A more likely scenario for antifungal therapy-associated QT prolongation is through the inhibition of

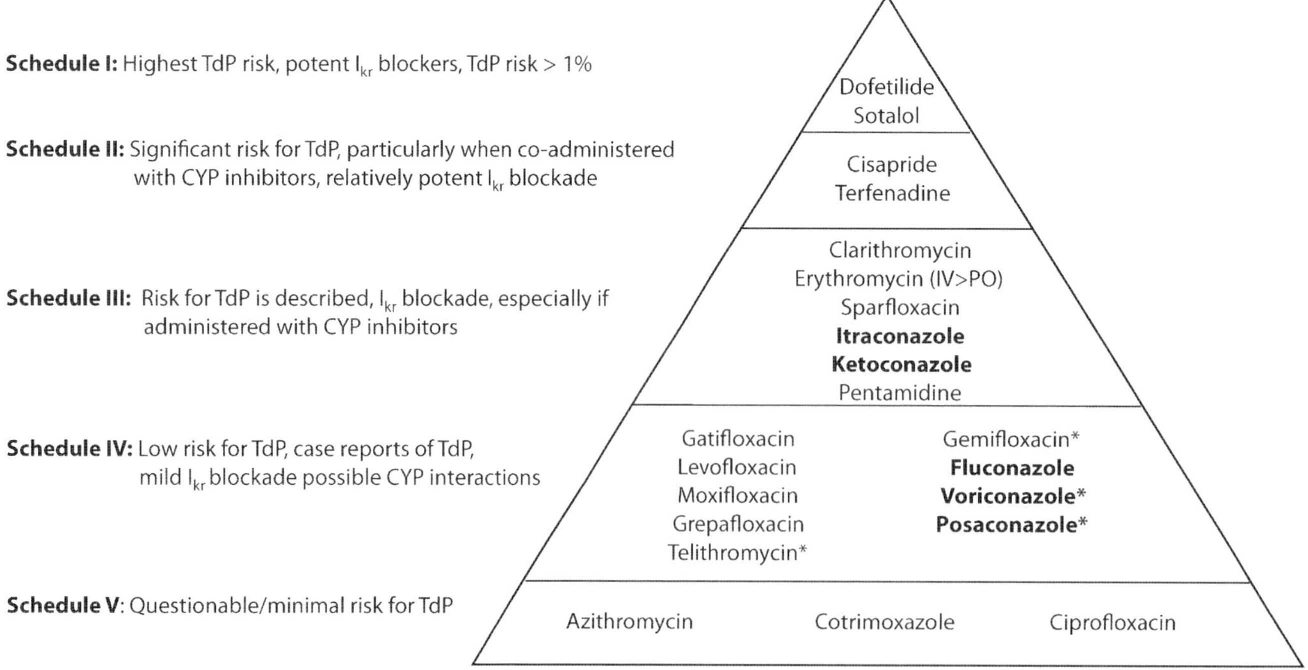

Figure 2. TdP risk stratification schedules for antimicrobial agents. *, antimicrobial agent that is new to the market or still investigational and has minimal to no postmarketing data; based on additional data, the drug may be recategorized in a higher or lower schedule. I_{kr}, rapid component of the delayed rectifier potassium current. Adapted from Owens (2004) with the permission of the publisher.

P450 metabolism of drugs with more potent effects on QT prolongation (i.e., class I and II agents) (Fig. 2) (Owens, 2004). Case reports of antifungal agent-associated TdP most commonly result from combinations of either ketoconazole or itraconazole with terbinafine, astemizole, cisapride, or quinidine derivatives. These drugs should be considered contraindicated in any patient receiving itraconazole, voriconazole, or posaconazole. Any patient receiving triazoles should be monitored for dosage adjustments, potential drug interactions, and electrolyte disturbances to minimize the possibility of a proarrhythmia (Owens, 2004). Avoidance of azoles would be prudent in any patient with inherited or drug- or disease-induced poor repolarization reserve (Eiden et al., 2007).

MANAGEMENT OF CLINICALLY SIGNIFICANT DRUG INTERACTIONS

Adverse drug reactions should be considered in any patient with unexplained changes in renal or hepatic function or cognitive disturbances. Central nervous system toxicities, in particular, can arise with excessive cyclosporine or tacrolimus exposure following the initiation of azole therapy and cannot necessarily be ruled out by "therapeutic" serum drug concentrations (Wong et al., 2003). In some patients, a trial switch to alternative antifungal agents with a noninteracting drug or holding doses of the interacting medication may be the only available options for managing a suspected interaction. Other steps for preventing or minimizing the effect of an interaction include (i) avoiding one or more of the interacting drugs, (ii) substitution of one of the drugs with a noninteracting drug, (iii) staggering the time or modifying the dose strength or interval of drug administration, (iv) changing the route of administration, and/or (v) more intensive monitoring for drug serum levels or toxicity.

Laboratory monitoring plays an essential role in the management of drug interactions in patients with invasive aspergillosis. Beyond routine electrolyte and serum drug level monitoring of agents with a narrow therapeutic index, measurement of antifungal agent serum drug concentrations, in select situations, can provide useful information concerning inadequate drug exposures (Summers et al., 1997). Typically, four criteria must be fulfilled to justify the use of serum drug concentrations to guide drug dosing (Summers et al., 1997). First, an assay with appropriate sensitivity, specificity, and "turnaround time" from the clinical laboratory must be available for analysis of the drug in question. Second, it must be more practical to make dosing adjustments on the basis of a serum drug concentration rather than other more immediately available markers (e.g., estimated creatinine clearance [serum creatinine] for fluconazole). Third, the drug must exhibit significant interpatient variability in pharmacokinetics to an extent that drug concentrations cannot be assumed from empirical dosing strategies. The fourth requirement, and often most problematic, is that a correlation must be established between drug concentrations and clinical efficacy and toxicity.

Although antifungal therapy does not typically fulfill all four criteria, monitoring in select situations may be helpful for certain agents where noncompliance, drug interactions, or unexpected toxicity is encountered (Goodwin and Drew, 2008; Summers et al., 1997). For example, measurement of a single trough concentration once the patient reaches steady state (i.e., after 7 days of therapy) with voriconazole could potentially identify subtherapeutic antifungal agent concentrations (dosing) in a patient who is also receiving phenytoin (a CYP3A4 inducer). Intensive monitoring is also paramount when antifungal therapy or a potentially interacting medication is stopped. Serum drug monitoring should be considered for any patient with documented invasive aspergillosis and suspected malabsorption or with potentially subtherapeutic concentrations due to excessive clearance. Cryptic hepatic, central nervous system, or cardiac toxicities could also prompt determination of voriconazole serum drug concentrations if therapy cannot be discontinued (Pascual et al., 2007). A summary of recommendations for performing and interpreting antifungal serum drug concentrations is outlined in Table 4.

Drug Interaction Databases

Not all drug interactions are known or reported in the literature and new drug interactions are continually being reported, making it impossible for clinicians to keep abreast of all the possible drug interactions. Consequently, many clinicians utilize computer resources to capture and interpret relevant interactions with antifungal agents. These resources are only as reliable as the databases are accurate, current, and comprehensive. In fact, studies have suggested 33% of relevant drug interactions are not identified by computer software (Cavuto et al., 1996; Hazlet et al., 2001). This problem is compounded by the fact that many software programs used at the point of prescription order entry provide frequent alerts of unimportant drug interactions. Pharmacists and other practitioners are likely to ignore excessive warnings of irrelevant drug interactions, which may also lead to potential adverse outcomes (Gaddis et al., 2002). Therefore, computer programs are only a complementary screening tool but cannot replace sound clinical judgment and knowledge of interaction mechanisms. The best computer databases and programs provide supporting information and key references that allow cli-

Table 4. Recommendations for therapeutic drug monitoring of antifungal therapy in patients with aspergillosis

Agent	Justified in select situations?	Target range (μg/ml)	Timing of sample
Amphotericin B	No	NA[a]	NA
Echinocandins	No	NA	NA
Itraconazole	Yes, ensure absorption	>0.5[b]	Trough after 7 days of therapy
Voriconazole	May be helpful for dosing guidance in patients	1–5[c]	Trough after 7 days of therapy
Posaconazole	Yes, ensure absorption, efficacy	>0.25–0.7[d]	3–5 h after oral dose

[a] NA, not applicable.
[b] The itraconazole target range was extrapolated from Glasmacher et al. (1999) and Poirier and Cheymol (1998).
[c] The voriconazole target range was extrapolated from Pascual et al. (2007), Smith et al. (2006), and Tan et al. (2006).
[d] The posaconazole target range was extrapolated from Schering-Plough (2006) and Walsh et al. (2007).

nicians to make an informed decision of the risks to the patient when two potentially interacting medications are required. Several recent reviews of popular drug interaction programs offer a more detailed discussion (Barrons, 2004; Clauson et al., 2007; Perkins et al., 2006).

SUMMARY

Patients with invasive aspergillosis have multiple risk factors for potentially harmful drug interactions. Although some drug interactions may be trivial or have minor effects, interactions that enhance the metabolism of antifungals used to treat *Aspergillus* infection, interactions that affect drugs with a narrow therapeutic index (i.e., immunosuppressants, chemotherapy, antiretrovirals), and interactions that increase cardiac QT prolongation should always be considered important and managed in a proactive fashion. New treatment options and expanded opportunities for laboratory support, including therapeutic drug monitoring and pharmacogenomic assessment that will allow clinicians to stratify a patient's risk for drug adverse effects as well as drug interactions, will become increasingly important tools in the prevention of drug interactions and adverse effects in patients with invasive aspergillosis.

REFERENCES

Baildon, R., T. Patterson, H. W. Boucher, R. Tiernan, and J. H. Powers. 2001. NDA 21-266, Vfend™ (voriconazole) tablets, and NDA 21-267, Vfend™ I.V. (voriconazole) for infusion. [Slide, October 4, 2001.] (Online.) http://www.fda.gov/ohrms/dockets/ac/01/slides/3792s2.htm

Barrons, R. 2004. Evaluation of personal digital assistant software for drug interactions. *Am. J. Health Syst. Pharm.* **61**:380–385.

Billaud, E. M., R. Guillemain, F. Tacco, and P. Chevalier. 1998. Evidence for a pharmacokinetic interaction between itraconazole and tacrolimus in organ transplant patients. *Br. J. Clin. Pharmacol.* **46**:271–272.

Bonay, M., A. P. Jonville-Bera, P. Diot, E. Lemarie, M. Lavandier, and E. Autret. 1993. Possible interaction between phenobarbital, carbamazepine and itraconazole. *Drug Saf.* **9**:309–311.

Branch, R. A. 1988. Prevention of amphotericin B-induced renal impairment: a review on the use of sodium supplementation. *Arch. Intern. Med.* **148**:2389–2394.

Burget, D. W., S. G. Chiverton, and R. H. Hunt. 1990. Is there an optimal degree of acid suppression for healing of duodenal ulcers? A model of the relationship between ulcer healing and acid suppression. *Gastroenterology* **99**:345–351.

Capone, D., A. Gentile, P. Imperatore, G. Palmiero, and V. Basile. 1999. Effects of itraconazole on tacrolimus blood concentrations in a renal transplant recipient. *Ann. Pharmacother.* **33**:1124–1125.

Capone, D., A. Gentile, P. Vajro, P. Stanziale, P. Imperatore, C. De Silva, T. Pellegrino, and V. Basile. 1998. Therapeutic monitoring of tacrolimus in pediatric and adult transplanted patients. *J. Chemother.* **10**:176–178.

Cavuto, N. J., R. L. Woosley, and M. Sale. 1996. Pharmacies and prevention of potentially fatal drug interactions. *JAMA* **275**:1086–1087.

Chiang, C. E. 2004. Congenital and acquired long QT syndrome. Current concepts and management. *Cardiol. Rev.* **12**:222–234.

Chiang, C. E., and D. M. Roden. 2000. The long QT syndromes: genetic basis and clinical implications. *J. Am. Coll. Cardiol.* **36**:1–12.

Chung, D. K., and M. G. Koenig. 1971. Reversible cardiac enlargement during treatment with amphotericin B and hydrocortisone. Report of three cases. *Am. Rev. Respir. Dis.* **103**:831–841.

Clauson, K. A., H. H. Polen, and W. A. Marsh. 2007. Clinical decision support tools: performance of personal digital assistant versus online drug information databases. *Pharmacotherapy* **27**:1651–1658.

Courtney, R., S. Pai, M. Laughlin, J. Lim, and V. Batra. 2003. Pharmacokinetics, safety, and tolerability of oral posaconazole administered in single and multiple doses in healthy adults. *Antimicrob. Agents Chemother.* **47**:2788–2795.

Courtney, R., A. Sansone, W. Smith, T. Marbury, P. Statkevich, M. Martinho, M. Laughlin, and S. Swan. 2005. Posaconazole pharmacokinetics, safety, and tolerability in subjects with varying degrees of chronic renal disease. *J. Clin. Pharmacol.* **45**:185–192.

Courtney, R., D. Wexler, E. Radwanski, J. Lim, and M. Laughlin. 2004. Effect of food on the relative bioavailability of two oral formulations of posaconazole in healthy adults. *Br. J. Clin. Pharmacol.* **57**:218–222.

Depont, F., F. Vargas, H. Dutronc, E. Giauque, J. M. Ragnaud, T. Galperine, A. Abouelfath, R. Valentino, M. Dupon, G. Hebert, and N. Moore. 2007. Drug-drug interactions with systemic antifungals in clinical practice. *Pharmacoepidemiol. Drug Saf.* **16**:1227–1233.

Ducharme, M. P., R. L. Slaughter, L. H. Warbasse, P. H. Chandrasekar, V. Van de Velde, G. Mannens, and D. J. Edwards. 1995. Itraconazole and hydroxyitraconazole serum concentrations are reduced more than tenfold by phenytoin. *Clin. Pharmacol. Ther.* **58**:617–624.

Eiden, C., H. Peyriere, R. Tichit, M. Cociglio, P. Amedro, J. P. Blayac, G. Margueritte, and D. Hillaire-Buys. 2007. Inherited long QT syndrome revealed by antifungals drug-drug interaction. *J. Clin. Pharm. Ther.* **32:**321–324.

Gaddis, G. M., T. R. Holt, and M. Woods. 2002. Drug interactions in at-risk emergency department patients. *Acad. Emerg. Med.* **9:** 1162–1167.

Gallicano, K., and G. L. Drusano. 2005. Introduction to drug interactions, p. 1–12. *In* S. Piscitelli and K. Rodvold (ed.), *Drug Interactions in Infectious Diseases*, 2nd ed. Humana Press, Totowa, NJ.

Glasmacher, A., C. Hahn, E. Molitor, G. Marklein, T. Sauerbruch, and I. G. H. Schmidt-Wolf. 1999. Itraconazole trough concentrations in antifungal prophylaxis with six different dosing regimens using hydroxypropyl-beta-cyclodextrin oral solution or coated-pellet capsules. *Mycoses* **42:**591–600.

Goodwin, M. L., and R. H. Drew. 2008. Antifungal serum concentration monitoring: an update. *J. Antimicrob. Chemother.* **61:**17–25.

Grillot, R., and B. Lebeau. 2005 Systemic antifungal agents, p. 1260–1287. *In* A. Bryskier (ed.), *Antimicrobial Agents: Antibacterials and Antifungals*. ASM Press, Washington, DC.

Groll, A., and T. J. Walsh. 1998. Pharmacology of antifungal compounds. *Adv. Pharmacol.* **44:**343–500.

Gubbins, P. O., S. A. McConnell, and J. R. Amsden. 2005 Antifungal agents, p. 289–338. *In* S. Piscitelli and K. Rodvold (ed.), *Drug Interactions in Infectious Diseases*, 2nd ed. Humana Press, Totowa, NJ.

Hall, S. D., K. E. Thummel, P. B. Watkins, K. S. Lown, L. Z. Benet, M. F. Paine, R. R. Mayo, D. K. Turgeon, D. G. Bailey, R. J. Fontana, and S. A. Wrighton. 1999. Molecular and physical mechanisms of first-pass extraction. *Drug Metab. Dispos.* **27:**161–166.

Hazlet, T. K., T. A. Lee, P. D. Hansten, and J. R. Horn. 2001. Performance of community pharmacy drug interaction software. *J. Am. Pharm. Assoc.* **41:**200–204.

Heidemann, H. T., J. F. Gerkens, W. A. Spickard, E. K. Jackson, and R. A. Branch. 1983. Amphotericin B nephrotoxicity in humans decreased by salt repletion. *Am. J. Med.* **75:**476–481.

Ikeda, Y., K. Umemura, K. Kondo, K. Sekiguchi, S. Miyoshi, and M. Nakashima. 2004. Pharmacokinetics of voriconazole and cytochrome P450 2C19 genetic status. *Clin. Pharmacol. Ther.* **75:**587–588.

Kanda, Y., M. Kami, T. Matsuyama, K. Mitani, S. Chiba, Y. Yazaki, and H. Hirai. 1998. Plasma concentration of itraconazole in patients receiving chemotherapy for hematological malignancies: the effect of famotidine on the absorption of itraconazole. *Hematol. Oncol.* **16:**33–37.

Kashuba, D. M., and J. S. Bertino. 2005. Mechanisms of drug interactions I, p. 13–39. *In* S. Piscitelli and K. Rodvold (ed.), *Drug Interactions in Infectious Diseases*, 2nd ed. Humana Press, Totowa, NJ.

Kim, R. B. 2003. Organic anion-transporting polypeptide (OATP) transporter family and drug disposition. *Eur. J. Clin. Investig.* **33**(Suppl. 2)**:**1–5.

Krishna, G., A. Parsons, B. Kantesaria, and T. Mant. 2007. Evaluation of the pharmacokinetics of posaconazole and rifabutin following co-administration to healthy men. *Curr. Med. Res. Opin.* **23:** 545–552.

Lange, D., J. H. Pavao, J. Wu, and M. Klausner. 1997. Effect of a cola beverage on the bioavailability of itraconazole in the presence of H2 blockers. *J. Clin. Pharmacol.* **37:**535–540.

Lewis, R. E. 2006. Managing drug interactions in the patient with aspergillosis. *Med. Mycol.* **44**(Suppl.)**:**349–356.

Lim, S. G., A. M. Sawyerr, M. Hudson, J. Sercombe, and P. E. Pounder. 1993. Short report: the absorption of fluconazole and itraconazole under conditions of low intragastric acidity. *Aliment. Pharmacol. Ther.* **7:**317–321.

Lown, K. S., J. C. Kolars, K. E. Thummel, J. L. Barnett, K. L. Kunze, S. Wrighton, and P. B. Watkins. 1994. Interpatient heterogeneity in expression of CYP3A4 and CYP3A5 in small bowel. Lack of prediction by the erythromycin breath test. *Drug Metab. Dispos.* **22:** 947–955.

Marr, K. A., W. Leisenring, F. Crippa, J. T. Slattery, L. Corey, M. Boeckh, and G. B. McDonald. 2004. Cyclophosphamide metabolism is affected by azole antifungals. *Blood* **103:**1557–1559.

Marty, F. M., C. M. Lowry, C. S. Cutler, B. J. Campbell, K. Fiumara, L. R. Baden, and J. H. Antin. 2006. Voriconazole and sirolimus co-administration after allogeneic hematopoietic stem cell transplantation. *Biol. Blood Marrow Transplant.* **12:**552–559.

Meletiadis, J., S. Chanock, and T. J. Walsh. 2006. Human pharmacogenomic variations and their implications for antifungal efficacy. *Clin. Microbiol. Rev.* **19:**763–787.

Nicolau, D. P., H. M. Crowe, C. H. Nightingale, and R. Quintiliani. 1995. Rifampin-fluconazole interaction in critically ill patients. *Ann. Pharmacother.* **29:**994–996.

Owens, R. C., Jr. 2004. QT prolongation with antimicrobial agents: understanding the significance. *Drugs* **64:**1091–1124.

Pascual, A., T. Calandra, S. Bolay, T. Buclin, J. Bille, and O. Marchetti. 2007. Voriconazole therapeutic drug monitoring in patients with invasive mycoses improves safety and efficacy outcomes. *Clin. Infect. Dis.* **46:**201–211.

Penzak, S. R. 2005 Mechanisms of drug interactions II, p. 41–82. *In* S. Piscitelli and K. Rodvold (ed.), *Drug Interactions in Infectious Diseases*, 2nd ed. Humana Press, Totowa, NJ.

Perkins, N. A., J. E. Murphy, D. C. Malone, and E. P. Armstrong. 2006. Performance of drug-drug interaction software for personal digital assistants. *Ann. Pharmacother.* **40:**850–855.

Pfizer, Inc. 2001. New drug application 21-266, Vfend (voriconazole) tablets, and new drug application 21-267, Vfend I.V. (voriconazole) for infusion. Accessed 30 December 2007. http://www.fda.gov/cder/foi/nda/2002/21-266_21-267_Vfend.htm.

Poirier, J. M., and G. Cheymol. 1998. Optimisation of itraconazole therapy using target drug concentrations. *Clin. Pharmacokinet.* **35:** 461–473.

Riechelmann, R. P. 2007. Drug combinations with the potential to interact among cancer patients. *Support. Care Cancer* **15:**1113–1114.

Riechelmann, R. P., and E. D. Saad. 2006. A systematic review on drug interactions in oncology. *Cancer Investig.* **24:**704–712.

Romero, A. J., P. Le Pogamp, L. G. Nilsson, and N. Wood. 2002. Effect of voriconazole on the pharmacokinetics of cyclosporine in renal transplant patients. *Clin. Pharmacol. Ther.* **71:**226–234.

Saad, A. H., D. D. DePestel, and P. L. Carver. 2006. Factors influencing the magnitude and clinical significance of drug interactions between azole antifungals and select immunosuppressants. *Pharmacotherapy* **26:**1730–1744.

Sandhu, P., W. Lee, X. Xu, B. F. Leake, M. Yamazaki, J. A. Stone, J. H. Lin, P. G. Pearson, and R. B. Kim. 2005. Hepatic uptake of the novel antifungal agent caspofungin. *Drug Metab. Dispos.* **33:** 676–682.

Sansone-Parsons, A., G. Krishna, M. Martinho, B. Kantesaria, S. Gelone, and T. G. Mant. 2007a. Effect of oral posaconazole on the pharmacokinctics of cyclosporine and tacrolimus. *Pharmacotherapy* **27:**825–834.

Sansone-Parsons, A., G. Krishna, J. Simon, P. Soni, B. Kantesaria, J. Herron, and R. Stoltz. 2007b. Effects of age, gender, and race/ethnicity on the pharmacokinetics of posaconazole in healthy volunteers. *Antimicrob. Agents Chemother.* **51:**495–502.

Sanz-Rodriguez, C., M. Lopez-Duarte, M. Jurado, J. Lopez, R. Arranz, J. M. Cisneros, M. L. Martino, P. J. Garcia-Sanchez, P. Morales, T. Olive, M. Rovira, and C. Solano. 2004. Safety of the concomitant use of caspofungin and cyclosporin A in patients with invasive fungal infections. *Bone Marrow Transplant.* **34:**13–20.

Schaeffeler, E., M. Eichelbaum, U. Brinkmann, A. Penger, S. Asante-Poku, U. M. Zanger, and M. Schwab. 2001. Frequency of C3435T polymorphism of MDR1 gene in African people. *Lancet* **358**:383–384.

Schering-Plough, Inc. 2006. New drug application 022003, Noxafil (posaconazole) oral suspension. Accessed 30 December 2007. http://www.accessdata.fda.gov/scripts/cder/drugsatfda/index.cfm?fuseaction=Search.Label_ApprovalHistory#apphist.

Schering-Plough, Inc. 26 February 1998. Antifungal composition with enhanced bioavailability. Patent WO/1998/007429. http://wipo.int/pctdb/en.

Shon, J. H., Y. R. Yoon, W. S. Hong, P. M. Nguyen, S. S. Lee, Y. G. Choi, I. J. Cha, and J. G. Shin. 2005. Effect of itraconazole on the pharmacokinetics and pharmacodynamics of fexofenadine in relation to the MDR1 genetic polymorphism. *Clin. Pharmacol. Ther.* **78**:191–201.

Smith, G., M. J. Stubbins, L. W. Harries, and C. R. Wolf. 1998. Molecular genetics of the human cytochrome P450 monooxygenase superfamily. *Xenobiotica* **28**:1129–1165.

Smith, J., N. Safdar, V. Knasinski, W. Simmons, S. M. Bhavnani, P. G. Ambrose, and D. Andes. 2006. Voriconazole therapeutic drug monitoring. *Antimicrob. Agents Chemother.* **50**:1570–1572.

Stevens, D. A. 1999. Itraconazole in cyclodextrin solution. *Pharmacotherapy* **19**:603–611.

Summers, K. K., T. C. Hardin, S. J. Gore, and J. R. Graybill. 1997. Therapeutic drug monitoring of systemic antifungal therapy. *J. Antimicrob. Chemother.* **40**:753–764.

Tan, K., N. Brayshaw, K. Tomaszewski, P. Troke, and N. Wood. 2006. Investigation of the potential relationships between plasma voriconazole concentrations and visual adverse events or liver function test abnormalities. *J. Clin. Pharmacol.* **46**:235–243.

Van de Velde, V. J., A. P. Van Peer, J. J. Heykants, R. J. Woestenborghs, P. Van Rooy, K. L. De Beule, and G. F. Cauwenbergh. 1996. Effect of food on the pharmacokinetics of a new hydroxypropyl-beta-cyclodextrin formulation of itraconazole. *Pharmacotherapy* **16**:424–428.

Walsh, T. J., I. I. Raad, T. F. Patterson, P. Chandrasekar, G. R. Donowitz, R. Graybill, R. E. Greene, R. Hachem, S. Hadley, R. Herbrecht, A. Langston, A. Louie, P. Ribaud, B. H. Segal, D. A. Stevens, J. A. van Burik, C. S. White, G. Corcoran, J. Gogate, G. Krishna, L. Pedicone, C. Hardalo, and J. R. Perfect. 2007. Treatment of invasive aspergillosis with posaconazole in patients who are refractory to or intolerant of conventional therapy: an externally controlled trial. *Clin. Infect. Dis.* **44**:2–12.

Wang, E. J., K. Lew, C. N. Casciano, R. P. Clement, and W. W. Johnson. 2002. Interaction of common azole antifungals with P glycoprotein. *Antimicrob. Agents Chemother.* **46**:160–165.

Wiederhold, N. P., and J. S. Lewis II. 2007. The echinocandin micafungin: a review of the pharmacology, spectrum of activity, clinical efficacy and safety. *Expert Opin. Pharmacother.* **8**:1155–1166.

Wingard, J. R., P. Kubilis, L. Lee, G. Yee, M. White, L. Walshe, R. Bowden, E. Anaissie, J. Hiemenz, and J. Lister. 1999. Clinical significance of nephrotoxicity in patients treated with amphotericin B for suspected or proven aspergillosis. *Clin. Infect. Dis.* **29**:1402–1407.

Wong, R., G. Z. Beguelin, M. de Lima, S. A. Giralt, C. Hosing, C. Ippoliti, A. D. Forman, A. J. Kumar, R. D. Champlin, and D. Couriel. 2003. Tacrolimus-associated posterior reversible encephalopathy syndrome after allogeneic haematopoietic stem cell transplantation. *Br. J. Haematol.* **122**:128–134.

Wood, N., K. Tan, R. Allan, A. Fielding, and D. Nichols. 2001. Effect of voriconazole on the pharmacokinetics of tacrolimus. *Abstr. 41st Intersci. Conf. Antimicrob. Agents Chemother.*, Chicago, IL. American Society for Microbiology, Washington, DC.

Yu, D. T., J. F. Peterson, D. L. Seger, W. C. Gerth, and D. W. Bates. 2005. Frequency of potential azole drug-drug interactions and consequences of potential fluconazole drug interactions. *Pharmacoepidemiol. Drug Saf.* **14**:755–767.

Aspergillus fumigatus and Aspergillosis
Edited by J.-P. Latgé and W. J. Steinbach
©2009 ASM Press, Washington, DC

Chapter 34

Antifungal Mechanisms of Action and Resistance

DAVID S. PERLIN AND EMILIA MELLADO

The incidence of invasive fungal infections (IFIs) has increased in the past decade, especially those caused by *Aspergillus* species (Marr et al., 2002; Bhatti et al., 2006; Maschmeyer et al., 2007), and *Aspergillus fumigatus* accounts for most of these infections (Marr et al., 2002; Maschmeyer et al., 2007). Invasive aspergillosis (IA) commonly develops in recipients of allogeneic hematopoetic stem cell transplantation and in patients with persistent neutropenia. Unlike other IFIs, *Aspergillus* isolates causing infection are infrequently cultured from blood but are most often recovered from respiratory specimens, either before antifungal therapy or at autopsy (Denning, 2000).

The development of antifungal resistance is complex and relies on multiple host and microbial risk factors (White et al., 1998). The widespread use of triazole antifungal drugs for therapy and prophylaxis contributes to antifungal resistance, as there is strong selection pressure for the development of resistant organisms. In *A. fumigatus*, specific resistance to triazole drugs results from selection pressure caused by drug exposure. This paradigm is supported by the observation that isolates with high MICs are not observed in historic collections of *A. fumigatus* prior to the introduction of itraconazole (Verweij et al., 2002). It is also well-documented in patients undergoing triazole therapy for whom the primary infecting strain developed resistance (Dannaoui et al., 2001, 2004b; Chen et al., 2005; Howard et al., 2006). There remains a strong relationship between drug exposure and the emergence of resistance. For this reason, it has been suggested that antifungal prophylaxis with triazole antifungals should be used with caution and only in patients at high risk for IFI. This concern arises from the enhanced use of oral itraconazole, and more recently voriconazole and posaconazole, for prophylaxis in high-risk neutropenic patients or for prolonged prophylaxis after allogeneic stem cell transplantation (Glasmacher and Prentice, 2005; Prentice et al., 2006) and long-term therapy of patients with allergic bronchopulmonary aspergillosis (Judson and Stevens, 2000; Stevens et al., 2000; Denning, 2001).

When antifungal resistance emerges, it can be drug specific or, more usually, class specific, displaying triazole cross-resistance among structurally diverse drugs (Mosquera and Denning, 2002; Gomez-Lopez et al., 2003; Howard et al., 2006; Verweij et al., 2007a). This latter multitriazole-resistant property is now being observed more commonly (Howard et al., 2006; Verweij et al. 2007b).

ANTIFUNGAL MODE OF ACTION

The treatment options for *Aspergillus* infections are effectively limited to three classes of antifungal drugs: triazoles (itraconazole, voriconazole, and posaconazole), polyenes (amphotericin B [AMB]), and echinocandins (caspofungin, micafungin, and anidulafungin). In general, the new triazole antifungal drugs (voriconazole and posaconazole) are the most promising therapeutic options (Torres et al., 2005; Maschmeyer et al., 2007) (Table 1). The most widely used drugs, triazoles and polyenes, target ergosterol, a predominant sterol within the fungal cell plasma membrane. Each class has its drawbacks, as the polyenes are also toxic to the host while the triazoles are more vulnerable to resistance. The polyene antibiotic AMB, which has been in use for more than 50 years, is active against a wide spectrum of molds. It binds preferentially to ergosterol over other types of membrane sterols and forms an aqueous pore resulting in potassium leakage, disruption of the proton gradient, and oxidative damage (Gallis et al., 1990). For decades, AMB-deoxycholate was the standard therapy for IA. However, AMB therapy is severely limited by adverse side effects, notably, renal and infusion-related toxicities (Sabra and Branch, 1990). These adverse ef-

David S. Perlin • Public Health Research Institute, New Jersey Medical School-UMDNJ, International Center for Public Health, 225 Warren Street, Newark, NJ 07103. **Emilia Mellado** • Servicio de Micologia, Centro Nacional de Microbiologia, Instituto de Salud Carlos III, Carretera Majadahonda-Pozuleo Km2, 28.220 Majadahonda, Madrid, Spain.

Table 1. Principal antifungal drugs, modes of action, and specific targets

Group	Mode of action	Compound(s) approved or in clinical trials[a]	Specific target (enzyme)
Polyenes	Bind ergosterol and form pores	Amphotericin B lipid formulations	Membrane ergosterol
Triazoles	Inhibit ergosterol biosynthesis	Fluconazole Ketoconazole Itraconazole Voriconazole Posaconazole Ravuconazole* Isavuconazole* Albaconazole*	14α–sterol demethylase (Erg11/Cyp51)
Echinocandins	Inhibit cell wall biosynthesis	Caspofungin Micafungin Anidulafungin Aminocandin*	1,3–β-D-glucan synthase (Fsk1/Fsk2)
Allylamines	Inhibit ergosterol biosynthesis	Terbinafine	Squalene epoxidase (Erg1)

[a] Compounds currently in clinical trials are indicated by an asterisk.

fects have been ameliorated with less toxic lipid complex or liposomal formulations to deliver the same parent AMB antifungal drug (Walsh et al., 1999). Three lipid formulations are approved for the treatment of IA: AMB lipid complex, AMB colloidal dispersion, and liposomal AMB (Rapp, 2004). Liposomal AMB can be used as an alternative for primary treatment (Cornely et al., 2007), and AMB remains an important treatment option for IA (Stevens et al., 2000).

The chemical families that inhibit C-14 demethylation are the imidazoles and triazoles. Collectively, these compounds are referred to as sterol demethylation inhibitors, and they are widely used both clinically and agriculturally. The triazole antifungal drugs comprise the largest class of compounds in clinical practice. This class of antifungals significantly advanced medical mycology, owing to their broad-spectrum action with reduced toxicity compared to AmB. They specifically target and inhibit the cytochrome P450-dependent enzyme lanosterol 14α-demethylase, which is involved in the biosynthesis of ergosterol, a fungus-specific sterol that maintains membrane function. Since the synthesis of the first azole compounds (imidazoles, e.g., miconazole, clotrimazole, and ketoconazole), chemical modifications have been made in order to increase their activity and reduce cell toxicity, giving rise to the triazoles: fluconazole, itraconazole, voriconazole, and posaconazole.

Fluconazole, a first-generation triazole drug, is weakly active against *Aspergillus*. The second-generation triazole drugs, itraconazole, voriconazole, ravuconazole, and posaconazole, have good in vitro and in vivo activities against *A. fumigatus* and other *Aspergillus* spp. (Denning et al., 1990; Boucher et al., 2004; Torres et al., 2005; Scott and Simpson, 2007). Voriconazole was shown to be more statistically superior to AMB in the treatment of IA (Herbrecht et al., 2002), which has led to improved survival rates in patients with central nervous system aspergillosis (Ruhnke et al., 2007). In the United States and Europe, voriconazole has been approved by the Food and Drug Administration and the European Medicines Agency, and it is has now replaced AMB as the primary therapeutic option and "gold standard" for therapy of IA (Maschmeyer and Ruhnke, 2004; Maschmeyer et al., 2007; Scott and Simpson, 2007). Itraconazole, voriconazole, and posaconazole can be used for the prophylaxis of IFI in high-risk individuals (Maschmeyer et al., 2007; Metcalf and Dockrell, 2007). In addition, new triazole drugs are undergoing early clinical evaluation (e.g., ravuconazole and albaconazole) (Aperis and Mylonakis, 2006). Also, Basilea Pharmaceutica is developing BAL-8557 (isavuconazole), a water-soluble pro-drug of the triazole BAL-4815, for the potential treatment of fungal infections (Odds, 2006; Seifert et al., 2007).

The echinocandin lipopeptides (caspofungin, micafungin, and anidulafungin) are the newest drug class to enter the market. They inhibit β-1,3-D-glucan synthase, a key enzyme responsible for biosynthesis of the major cell wall glucan biopolymer. The echinocandins are broadly active against a wide range of fungi without cross-resistance to existing antifungal agents and therefore are effective against azole-resistant yeasts and molds (Morrison, 2006; Ghannoum et al., 2007). Importantly, due to their critical effect on the cell wall, echinocandins are fungicidal with yeasts. They are fungistatic against *Aspergillus* and only appear to block the growing tips of hyphae (Kurtz et al., 1994; Bowman et al., 2002). The U.S. Food and Drug Administration approved caspofungin in 2002, and later micafungin (2005) and anidulafungin (2006) were licensed for clinical use. Caspofungin was approved for the treatment of invasive aspergillosis in patients refractory to or intolerant of other therapies (Kartsonis et al., 2003; Maertens et al., 2004), as well as for empirical therapy of febrile neutropenia (Walsh et al., 2004; Hope et al., 2007). Micafungin and anidulafungin have shown efficacy in animal models and limited patient studies against IA (Denning et al., 2006; Vehreschild and Cornely, 2006). A new echinocandin, aminocandin, has demonstrated potent antifungal activities against *Aspergillus* isolates, suggesting that aminocandin could be an important addition to this class of antifungal drugs for the treatment of invasive fungal disease (Isham and Ghannoum, 2006).

Terbinafine, an allylamine class antifungal drug, competitively inhibits squalene epoxidase (Erg1), which blocks the first step of the ergosterol biosynthesis pathway. It is typically used to treat infections of the nail (Gupta and Tu, 2006). In vitro susceptibility testing shows that MIC values for terbinafine vary according to *Aspergillus* species (Garcia-Effron et al., 2004). Its efficacy in the treatment of refractory pulmonary aspergillosis has been rarely reported (Schiraldi et al., 1996), and it may represent a possible second-line therapeutic option for some selected cases.

Finally, the availability of new antifungal agents with reduced toxicities is opening new opportunities for combination therapy of IA. Since azoles and echinocandins target different cellular sites, combination therapies are being investigated as possible alternative therapeutic options (Dannaoui et al., 2004a; Marr et al., 2004). Randomized prospective studies of combination antifungal therapy in mold infections are lacking, but some preliminary trials have provided supportive evidence for this approach (Johnson and Perfect, 2007).

DEFINING AND MEASURING DRUG RESISTANCE

Resistance in molds has been identified as either clinical resistance or microbiological resistance, and both terms need to be defined. Clinical resistance or treatment failure is considered when signs and symptoms of a fungal infection persist despite adequate delivery of a tolerable concentration of drug. It depends on both the drug that is used to treat the infection and microbiologic and host factors. On the other hand, microbiological resistance is solely dependent on the microorganism, and it can be present without prior antifungal therapy. Intrinsic or innate resistance occurs when all isolates from a fungal species are inherently resistant to a drug (e.g., *A. fumigatus* resistance to fluconazole), while primary resistance refers to fungi that are normally susceptible to a drug but a subset of isolates show antifungal resistance without previous drug exposure (i.e., *Candida glabrata* resistance to fluconazole). Alternatively, secondary or acquired resistance emerges in previously susceptible isolates during or after antifungal drug exposure. In *A. fumigatus*, this kind of resistance is mainly associated with exposure to triazole drugs.

The problem with relying on phenotypic end points (in vitro growth inhibition) to determine clinical resistance is that in vitro susceptibility does not always predict successful outcome (Sanglard and Odds, 2002). In vitro susceptibility testing is able to categorize an organism as susceptible or resistant, but its clinical value is limited.

Currently, there are two reference methods for determining mold antifungal susceptibility testing. The first, published by the Clinical Laboratory Standards Institute (formerly the National Committee for Clinical Laboratory Standards), is covered by standard document M38-A (approved standard M38-A) (NCCLS, 2002). The second, developed by the Antifungal Susceptibility Testing Subcommittee of the European Committee on Antibiotic Susceptibility Testing, is in its final approval phase (http://www.srga.org/Eucastwt/eucastdefinitions.htm). Both protocols utilize broth microdilution to assess phenotypic end points as minimal drug concentrations that prevent growth. The echinocandin drugs fail to produce complete growth arrest in molds. Therefore, an alternative in vitro susceptibility end point called the minimal effective concentration has been described as the minimum echinocandin concentration resulting in morphological alterations of hyphal growth (Kurtz et al., 1998).

The ultimate role of susceptibility testing is to detect potential clinically resistant isolates. In that sense, both procedures have been extensively employed, and it has been demonstrated that they identify resistant samples and/or molds with high MICs of antifungal drugs (Cuenca-Estrella et al., 2006). Although susceptibility testing methods have been standardized and have been applied to large collections of clinical isolates, it has been difficult to assign in vitro breakpoints of clinical resistance to molds. To date, an MIC of >8 mg/liter for in vitro susceptibility of *A. fumigatus* strains to itraconazole connotes resistance (Denning et al., 1997b).

ANTIFUNGAL DRUG RESISTANCE MECHANISMS IN *A. FUMIGATUS*

The characterization of phenotypic resistance in clinical isolates can help in evaluating their frequency and suggest possible therapeutic options. However, as growth inhibition assays can be misleading, in vitro resistance identification should be combined with the study of molecular mechanisms of resistance. Advances in molecular genetics and genomics have facilitated the analysis and characterization of a number of genes which are implicated in *A. fumigatus* antifungal drug resistance (*erg1*, *cyp51s*, and *fks1*).

Polyenes

In *A. fumigatus*, the association between in vitro susceptibility to AMB and clinical outcome is unclear, despite the presence of clinical isolates with high MICs (Verweij et al., 1998; Johnson et al., 2000). Secondary resistance to AMB is generally not observed, even in patients failing therapy. Yet, resistant mutants can be spontaneously induced in the laboratory (Manavathu et al., 2001), and there was a report of AMB resistance following prior exposure to itraconazole (Schaffner and Bohler, 1993). Acquired resistance to AMB has been most extensively evaluated in yeasts and is associated with mutations in *erg3* (C-5 sterol desaturase), which is linked to qualitative and quantitative alterations of membrane lipids and an absence of ergosterol (Kelly et al., 1997). However, the deletion of three *erg3* genes in *A. fumigatus* did not change the susceptibility phenotype to AMB, even when the mutants showed a marked alteration of sterol composition and decreased total ergosterol (Alcazar-Fuoli et al., 2006).

Azoles

The majority of *A. fumigatus* isolates are susceptible in vitro to triazole drugs. However, resistant strains have been well-documented, and recently an increasing number of strains with cross-resistance to azoles have been reported (Verweij et al., 2007a). A variety of surveillance studies have demonstrated that triazole-resistant isolates occur at a low frequency, which is generally less than 2% for itraconazole (MIC, >16 mg/liter) and lower for voriconazole and posaconazole (Verweij et al., 1998; Dannaoui et al., 1999a, 1999b, 2004a; Moore et al., 2000; Gomez-Lopez et al., 2003; Kaya and Kiraz, 2007; Quindos et al., 2007).

In yeast, a majority of reports on azole drug resistance are related to increased efflux of drugs due to overexpression of efflux pumps (Sanglard and Odds, 2002; White et al., 2002; Cannon et al., 2007). In addition, a number of reports have identified polymorphisms in the *erg11* gene (homologous to *cyp51s*) from clinical *Candida albicans* isolates that are responsible for and/or associated with fluconazole resistance (Marichal et al., 1999; Sanglard et al., 2003). Both mechanisms can contribute to resistance in the same strain in yeasts (Perea et al., 2001). Resistance of non-*Aspergillus* filamentous fungi to demethylase inhibitors used for agricultural purposes has been mainly associated with amino acid modifications in the azole target Cyp51 (Delye et al., 1997a, 1997b; Wyand and Brown, 2005; Cools et al., 2006) or with its overexpression (Hamamoto et al., 2000; Ma et al., 2006; Luo and Schnabel, 2008).

In *A. fumigatus*, azole drug resistance has been described for both clinical strains and laboratory mutants (Denning et al., 1997a), and complex cross-resistance patterns reflect different underlying resistance mechanisms (Mosquera and Denning, 2002). Like yeasts, resistance can arise from either modification of the target site, encoded by Cyp51A (Edlind et al., 2001; Mellado et al., 2001, 2004, 2005, 2007; Diaz-Guerra et al., 2003; Mann et al., 2003; Nascimento et al., 2003; Garcia-Effron et al., 2005; Howard et al., 2006) or overexpression of drug transporters of the ABC and/or MFS type (Tobin et al., 1997; Slaven et al., 2002; Nascimento et al., 2003). However, only the former mechanism (modification of the target site) appears to contribute to clinical resistance. Like other fungi, *A. fumigatus* contains two different but related Cyp51 proteins encoded by *cyp51A* and *cyp51B* (Mellado et al., 2001). The importance of both proteins in sterol biosynthesis and their significance in the susceptibility to azole drugs have been studied extensively. Targeted disruption of the *cyp51A* gene in itraconazole-resistant strains of *A. fumigatus* restored drug susceptibility, confirming that Cyp51A is the target of these antifungal agents (Mellado et al., 2005). On the other hand, Cyp51B is important for growth rate and shape maintenance in *A. fumigatus*, but its role in azole susceptibility remains unclear (E. Mellado, unpublished data).

Clinical isolates resistant to azoles are associated with specific mutations in *cyp51A*, which in turn correlate with different susceptibility profiles (Fig. 1). Triazole drugs show complex cross-resistance, reflecting underlying resistance mechanisms (Mosquera and Denning, 2002). One azole-resistant profile is due to different amino acid substitutions at glycine 54 (G54E, G54V, G54R, and G54W). These amino acid changes confer resistance to itraconazole and yield high MICs of posaconazole but not of voriconazole or ravuconazole (Diaz-Guerra et al., 2003; Mann et al., 2003; Nascimento et al., 2003). However, mutations at methionine 220 (M220V, M220K, M220T, and M220I) result in resistance to itraconazole and reduced susceptibility to posaconazole, voriconazole, and ravuconazole (Mellado et al., 2004). Also, a G138C mutation was described in itraconazole- and voriconazole-resistant strains isolated from a patient who received long-term treatment with both drugs (Howard et al., 2006). The precise manner in which Cyp51A amino acid substitutions affect triazole binding is not known. However, the fact that differences in amino acid substitutions can, sometimes, completely change the MIC for a specific azole drug seems to be related to both the drug's chemical composition and conformation and its interaction with the enzyme.

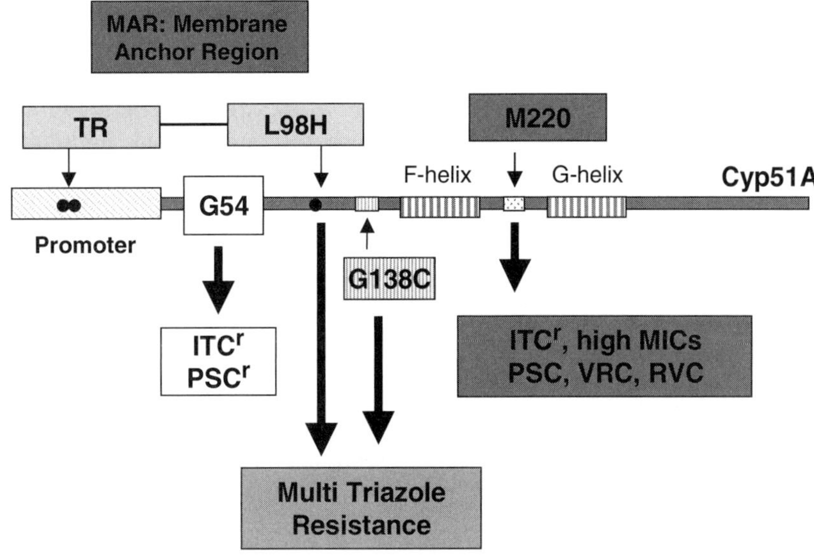

Figure 1. *A. fumigatus* Cyp51A-related resistance mechanisms to azoles. The antifungal drugs noted are as follows: ITC, itraconazole; VRC, voriconazole; PSC, posaconazole; RVC, ravuconazole. The "r" superscript connotes resistance.

Some of these possibilities have been explored in a structural model of the 14α sterol demethylase (Cyp51A) from *A. fumigatus* based on homology modeling with the X-ray structure of the Cyp51 ortholog from *Mycobacterium tuberculosis* (Xiao et al., 2004). In this model, direct drug interactions with amino acid side groups conferring resistance vary according to the chemical structures of the drugs, which helps account for the differences observed between amino acid substitutions and azole MICs. However, other studies suggested that amino acid substitutions involved in azole resistance are located in regions of the protein involved in conformational changes rather than in direct drug binding (Podust et al., 2001). Further evidence will have to await an X-ray structure of the *A. fumigatus* Cyp51A complexed to drug.

Recently, a new mechanism of resistance was found in multiple-azole-resistant strains isolated from nine patients with different underlying diseases (Verweij et al., 2007b). This novel mechanism involves a substitution of leucine 98 for histidine (L98H) linked to a duplication in tandem of a 34-bp repeat in the *cyp51A* promoter region that causes a six- to eightfold increase in the expression of the *cyp51A* gene (Mellado et al., 2007). However, it is important to point out that half of the resistant strains were isolated from patients who had no known azole exposure. The resistance mechanisms shown by these strains are similar to the mechanisms operating in plant pathogens. As molds causing human infections are ubiquitous, the development of azole resistance could be linked to the use of demethylase inhibitors in other settings, such as in agricultural environments.

Echinocandins

Echinocandin resistance has been well-documented with *Candida* spp. (Hakki et al., 2006; Laverdiere et al., 2006; Miller et al., 2006), but it has not been observed for *A. fumigatus* infections. However, *Aspergillus* infections refractory to caposofungin therapy have been noted for *A. lentulus* (Balajee et al., 2005) and *A. flavus* (Eschertzhuber et al., 2008). Resistance to echinocandin drugs among clinical isolates is associated with amino acid substitutions in two "hot spot" regions of Fks1, the major subunit of glucan synthase (Perlin, 2007). The mutations, yielding highly elevated MIC values, are genetically dominant and confer cross-resistance to all echinocandin drugs. Prominent Fks1 mutations decrease the drug sensitivity of glucan synthase by several log orders of magnitude, and strains harboring such mutations require a comparable level of drug to reduce fungal organ burdens in animal infection models (Balashov et al., 2006). The Fks1-mediated resistance mechanism is conserved in a wide variety of *Candida* spp. and may account for intrinsic reduced susceptibility of certain species (*C. parapsilosis*). Fks1 mutations confer resistance in both yeasts and molds, suggesting that this mechanism is conserved among most fungi (Rocha et al., 2007).

Allylamines

Terbinafine belongs to the allylamine class of antifungals, which inhibit squalene epoxidase (Erg1). This enzyme acts in the first step of the ergosterol biosynthesis pathway, and it is specific for the fungi kingdom. The first resistance mechanism associated with terbinafine was described in *A. nidulans* (Graminha et al., 2004). A point mutation, F391L, in the squalene epoxidase enzyme conferred a strong terbinafine resistance phenotype. This equivalent mutation was introduced in the homologous gene of *A. fumigatus* and resulted in resistance to terbinafine (Rocha et al., 2006).

PRIMARY RESISTANCE IN NON-FUMIGATUS *ASPERGILLUS* SPECIES

Although the most common cause of IA is *A. fumigatus*, there have been increasing reports of infection due to non-fumigatus *Aspergillus* species. This shift in infecting species is worrisome, since many of these organisms are resistant to one or more antifungals. Primary resistance to AMB has been reported for strains of *Aspergillus terreus* (Dannaoui et al., 2000; Gomez-Lopez et al., 2003; Steinbach et al., 2004a, 2004b), *A. flavus* (Koss et al., 2002), and *Aspergillus ustus* (Azzola et al., 2004). It should be noted that in some clinical cases, treatment failure with amphotericin B may result from insufficient drug delivery rather than true microbial resistance (Paterson et al., 2003). Breakthrough *A. ustus* infections have been reported in allogeneic stem cell transplant recipients while they were receiving either voriconazole or caspofungin (Pavie et al., 2005). Resistance to echinocandin drugs is limited to primary resistance described for *A. lentulus* (Balajee et al., 2005b) and a single case report of a caspofungin-resistant *A. flavus* strain recovered from a patient failing echinocandin therapy (Eschertzhuber et al., 2008). Recently, an *Aspergillus allilaceus* isolate presenting combined reduced in vitro susceptibilities to AMB and caspofungin was isolated from a patient with IA who failed antifungal therapy (Balajee et al., 2007).

A matter of concern is the increasing reports of isolation of genetic variants of *A. fumigatus*, like *A. lentulus*, which were originally identified as poorly sporulating strains of *A. fumigatus* (Balajee et al., 2005a, 2006). These strains show pleiotropic reduced in vitro susceptibilities to multiple antifungal drugs, but mainly to voriconazole (voriconazole MICs, >4.0 mg/liter) (Balajee et al., 2006; Mellado et al., 2006). All of these atypical *Aspergillus* strains belong to the *Aspergillus* section *Fumigati*, which includes at least 20 species of *Neosartorya* and five asexual aspergilli (Varga et al., 2000). However, only a few of them have been implicated in human deep tissue infections, and in most cases antifungal susceptibility profiles and clinical outcome data are lacking (Padhye et al., 1994; Lonial et al., 1997; Guarro et al., 2002; Jarv et al., 2004). These strains are difficult to distinguish, although Balajee and colleagues correctly identified *Neosartorya pseudofischeri* by sequencing portions of the β-tubulin and the rodlet A genes (Balajee et al., 2005a). This same approach was used to identify a new *Aspergillus* species belonging to this group (*A. lentulus*), based on multilocus sequence typing of five genes (the β-tubulin gene, the rodlet A gene, the salt-responsive gene, the mitochondrial cytochrome *b* gene, and the internal transcribed spacer regions) (Balajee et al., 2005b). Another laboratory reported the isolation of *Aspergillus* section *Fumigati* from a variety of clinical specimens. Several isolates had high MICs of AMB, and a variety of susceptibility patterns was found for azole drugs, including some with MICs of ≥4 mg/liter, suggesting resistance. However, correlations with in vivo data are lacking (Mellado et al., 2006).

FUTURE PERSPECTIVES

As triazole resistance in *A. fumigatus* has become a recognized problem that limits effective therapy, susceptibility testing in clinical labs is still performed sporadically. Similarly, local or regional surveillance of resistance is even less common. A more efficient approach would be to take advantage of advances in molecular diagnostic technology to detect well-defined resistance mechanisms which can serve as surrogate markers for phenotypic resistance. *A. fumigatus* is particularly amenable to this approach, because clinical triazole resistance is limited to amino acid substitutions in the drug target Cyp51A. Furthermore, individual mutations confer a particular triazole resistance pattern with different triazole drugs. Therefore, specific mutations are surrogate markers for phenotypic triazole resistance. Presently, a wide range of amplification and detection platforms exist which enable high-throughput sampling with rapid detection and are well-suited for routine clinical microbiological laboratories (Balashov et al., 2006; Garcia-Effron et al., 2007).

CONCLUSION

A. fumigatus resistance to antifungal drugs is a recognized problem that occurs at a low frequency, but at an increasing rate in some countries. Resistance is largely confined to azole class drugs, as resistance to polyene or echinocandin drugs is uncommon. However, as the number of immunocompromised patients increases, resistance among clinical strains may become more abundant, especially with the expanded use of second-generation triazole drugs for prophylaxis and empirical therapy. Therefore, epidemiological surveillance is necessary to detect trends in *A. fumigatus* azole resistance and to anticipate resistance emergence against new antifungal drugs in clinical use. Finally, the study of antifungal resistance mechanisms at the molecular level is important to understand better the nature of resistance and to develop clinical and pharmacological strategies to avoid resistance in the future.

Acknowledgments. E. Mellado is supported by national project grants from the Instituto de Salud Carlos III (MPY1175/06) and the Ministerio de Educacion y Ciencia (SAF2005-06541). E.M. also has funding from the European Union (EU-STREP project LSHM-CT-2005-518199). D. Perlin is supported by NIH grants AI066561 and AI069397.

REFERENCES

Alcazar-Fuoli, L., E. Mellado, G. Garcia-Effron, M. J. Buitrago, J. F. Lopez, J. O. Grimalt, J. M. Cuenca-Estrella, and J. L. Rodriguez-Tudela. 2006. *Aspergillus fumigatus* C-5 sterol desaturases Erg3A and Erg3B: role in sterol biosynthesis and antifungal drug susceptibility. *Antimicrob. Agents Chemother.* **50:**453–460.

Aperis, G., and E. Mylonakis. 2006. Newer triazole antifungal agents: pharmacology, spectrum, clinical efficacy and limitations. *Expert Opin. Investig. Drugs* **15:**579–602.

Azzola, A., J. R. Passweg, J. M. Habicht, L. Bubendorf, M. Tamm, A. Gratwohl, and G. Eich. 2004. Use of lung resection and voriconazole for successful treatment of invasive pulmonary *Aspergillus ustus* infection. *J. Clin. Microbiol.* **42:**4805–4808.

Balajee, S. A., J. Gribskov, M. Brandt, J. Ito, A. Fothergill, and K. A. Marr. 2005a. Mistaken identity: *Neosartorya pseudofischeri* and its anamorph masquerading as *Aspergillus fumigatus*. *J. Clin. Microbiol.* **43:**5996–5999.

Balajee, S. A., J. L. Gribskov, E. Hanley, D. Nickle, and K. A. Marr. 2005b. *Aspergillus lentulus* sp. nov., a new sibling species of *A. fumigatus*. *Eukaryot. Cell* **4:**625–632.

Balajee, S. A., M. D. Lindsley, N. Iqbal, J. Ito, P. G. Pappas, and M. E. Brandt. 2007. Nonsporulating clinical isolate identified as *Petromyces alliaceus* (anamorph *Aspergillus alliaceus*) by morphological and sequence-based methods. *J. Clin. Microbiol.* **45:**2701–2703.

Balajee, S. A., D. Nickle, J. Varga, and K. A. Marr. 2006. Molecular studies reveal frequent misidentification of *Aspergillus fumigatus* by morphotyping. *Eukaryot. Cell* **5:**1705–1712.

Balashov, S. V., S. Park, and D. S. Perlin. 2006. Assessing resistance to the echinocandin antifungal drug caspofungin in *Candida albicans* by profiling mutations in FKS1. *Antimicrob. Agents Chemother.* **50:**2058–2063.

Bhatti, Z., A. Shaukat, N. G. Almyroudis, and B. H. Segal. 2006. Review of epidemiology, diagnosis, and treatment of invasive mould infections in allogeneic hematopoietic stem cell transplant recipients. *Mycopathologia* **162:**1–15.

Boucher, H. W., A. H. Groll, C. C. Chiou, and T. J. Walsh. 2004. Newer systemic antifungal agents: pharmacokinetics, safety and efficacy. *Drugs* **64:**1997–2020.

Bowman, J. C., P. S. Hicks, M. B. Kurtz, H. Rosen, D. M. Schmatz, P. A. Liberator, and C. M. Douglas. 2002. The antifungal echinocandin caspofungin acetate kills growing cells of *Aspergillus fumigatus* in vitro. *Antimicrob. Agents Chemother.* **46:**3001–3012.

Cannon, R. D., E. Lamping, A. R. Holmes, K. Niimi, K. Tanabe, M. Niimi, and B. C. Monk. 2007. *Candida albicans* drug resistance another way to cope with stress. *Microbiology* **153:**3211–3217.

Chen, J., H. Li, R. Li, D. Bu, and Z. Wan. 2005. Mutations in the *cyp51A* gene and susceptibility to itraconazole in *Aspergillus fumigatus* serially isolated from a patient with lung aspergilloma. *J. Antimicrob. Chemother.* **55:**31–37.

Cools, H. J., B. A. Fraaije, S. H. Kim, and J. A. Lucas. 2006. Impact of changes in the target P450 CYP51 enzyme associated with altered triazole-sensitivity in fungal pathogens of cereal crops. *Biochem. Soc. Trans.* **34:**1219–1222.

Cornely, O. A., J. Maertens, M. Bresnik, R. Ebrahimi, A. J. Ullmann, E. Bouza, C. P. Heussel, O. Lortholary, C. Rieger, A. Boehme, M. Aoun, H. A. Horst, A. Thiebaut, M. Ruhnke, D. Reichert, N. Vianelli, S. W. Krause, E. Olavarria, R. Herbrecht, et al. 2007. Liposomal amphotericin B as initial therapy for invasive mold infection: a randomized trial comparing a high-loading dose regimen with standard dosing (AMBiLoad trial). *Clin. Infect. Dis.* **44:**1289–1297.

Cuenca-Estrella, M., A. Gomez-Lopez, E. Mellado, M. J. Buitrago, A. Monzon, and J. L. Rodriguez-Tudela. 2006. Head-to-head comparison of the activities of currently available antifungal agents against 3,378 Spanish clinical isolates of yeasts and filamentous fungi. *Antimicrob. Agents Chemother.* **50:**917–921.

Dannaoui, E., E. Borel, M. F. Monier, M. A. Piens, S. Picot, and F. Persat. 2001. Acquired itraconazole resistance in *Aspergillus fumigatus*. *J. Antimicrob. Chemother.* **47:**333–340.

Dannaoui, E., E. Borel, F. Persat, M. F. Monier, and M. A. Piens. 1999a. In-vivo itraconazole resistance of *Aspergillus fumigatus* in systemic murine aspergillosis. *J. Med. Microbiol.* **48:**1087–1093.

Dannaoui, E., E. Borel, F. Persat, M. A. Piens, and S. Picot. 2000. Amphotericin B resistance of *Aspergillus terreus* in a murine model of disseminated aspergillosis. *J. Med. Microbiol.* **49:**601–606.

Dannaoui, E., O. Lortholary, and F. Dromer. 2004a. In vitro evaluation of double and triple combinations of antifungal drugs against *Aspergillus fumigatus* and *Aspergillus terreus*. *Antimicrob. Agents Chemother.* **48:**970–978.

Dannaoui, E., J. Meletiadis, A. M. Tortorano, F. Symoens, N. Nolard, M. A. Viviani, M. A. Piens, B. Lebeau, P. E. Verweij, and R. Grillot. 2004b. Susceptibility testing of sequential isolates of *Aspergillus fumigatus* recovered from treated patients. *J. Med. Microbiol.* **53:**129–134.

Dannaoui, E., F. Persat, M. F. Monier, E. Borel, M. A. Piens, and S. Picot. 1999b. In-vitro susceptibility of *Aspergillus* spp. isolates to amphotericin B and itraconazole. *J. Antimicrob. Chemother.* **44:**553–555.

Delye, C., F. Laigret, and M. F. Corio-Costet. 1997a. Cloning and sequence analysis of the eburicol 14α-demethylase gene of the obligate biotrophic grape powdery mildew fungus. *Gene* **195:**29–33.

Delye, C., F. Laigret, and M. F. Corio-Costet. 1997b. A mutation in the 14 alpha-demethylase gene of *Uncinula necator* that correlates with resistance to a sterol biosynthesis inhibitor. *Appl. Environ. Microbiol.* **63:**2966–2970.

Denning, D. W. 2000. Early diagnosis of invasive aspergillosis. *Lancet* **355:**423–424.

Denning, D. W. 2001. Chronic forms of pulmonary aspergillosis. *Clin. Microbiol. Infect.* **7:**25–31.

Denning, D. W., K. A. Marr, W. M. Lau, D. P. Facklam, V. Ratanatharathorn, C. Becker, A. J. Ullmann, N. L. Seibel, P. M. Flynn, J. A. van Burik, D. N. Buell, and T. F. Patterson. 2006. Micafungin (FK463), alone or in combination with other systemic antifungal agents, for the treatment of acute invasive aspergillosis. *J. Infect.* **53:**337–349.

Denning, D. W., S. A. Radford, K. L. Oakley, L. Hall, E. M. Johnson, and D. W. Warnock. 1997a. Correlation between in-vitro susceptibility testing to itraconazole and in-vivo outcome of *Aspergillus fumigatus* infection. *J. Antimicrob. Chemother.* 40:401–414.

Denning, D. W., R. M. Tucker, L. H. Hanson, and D. A. Stevens. 1990. Itraconazole in opportunistic mycoses: cryptococcosis and aspergillosis. *J. Am. Acad. Dermatol.* 23:602–607.

Denning, D. W., K. Venkateswarlu, K. L. Oakley, M. J. Anderson, N. J. Manning, D. A. Stevens, D. W. Warnock, and S. L. Kelly. 1997b. Itraconazole resistance in *Aspergillus fumigatus. Antimicrob. Agents Chemother.* 41:1364–1368.

Diaz-Guerra, T. M., E. Mellado, M. Cuenca-Estrella, and J. L. Rodriguez-Tudela. 2003. A point mutation in the 14α-sterol demethylase gene *cyp51A* contributes to itraconazole resistance in *Aspergillus fumigatus. Antimicrob. Agents Chemother.* 47:1120–1124.

Edlind, T. D., K. W. Henry, K. A. Metera, and S. K. Katiyar. 2001. *Aspergillus fumigatus* CYP51 sequence: potential basis for fluconazole resistance. *Med. Mycol.* 39:299–302.

Eschertzhuber, S., C. Velik-Salchner, C. Hoermann, D. Hoefer, and C. Lass-Florl. 2008. Caspofungin-resistant *Aspergillus flavus* after heart transplantation and mechanical circulatory support: a case report. *Transplant. Infect. Dis.* 10:190–192.

Gallis, H. A., R. H. Drew, and W. W. Pickard. 1990. Amphotericin B: 30 years of clinical experience. *Rev. Infect. Dis.* 12:308–329.

Garcia-Effron, G., A. Gomez-Lopez, E. Mellado, A. Monzon, J. L. Rodriguez-Tudela, and M. Cuenca-Estrella. 2004. In vitro activity of terbinafine against medically important non-dermatophyte species of filamentous fungi. *J. Antimicrob. Chemother.* 53:1086–1089.

Garcia-Effron, G., E. Mellado, A. Gomez-Lopez, L. Alcazar-Fuoli, M. Cuenca-Estrella, and J. L. Rodriguez-Tudela. 2005. Differences in interactions between azole drugs related to modifications in the 14-α sterol demethylase gene (*cyp51A*) of *Aspergillus fumigatus. Antimicrob. Agents Chemother.* 49:2119–2121.

Garcia-Effron, G., S. Park, E. Mellado, and D. S. Perlin. 2007. A new molecular diagnostic assay for rapid detection of triazole resistance in *Aspergillus fumigatus*, abstr. M-518. *Abstr. 47th Intersci. Conf. Antimicrob. Agents Chemother.*, Chicago, IL, 17 to 20 September 2007.

Ghannoum, M. A., H. G. Kim, and L. Long. 2007. Efficacy of aminocandin in the treatment of immunocompetent mice with haematogenously disseminated fluconazole-resistant candidiasis. *J. Antimicrob. Chemother.* 59:556–559.

Glasmacher, A., and A. G. Prentice. 2005. Evidence-based review of antifungal prophylaxis in neutropenic patients with haematological malignancies. *J. Antimicrob. Chemother.* 56(Suppl. 1):i23–i32.

Gomez-Lopez, A., G. Garcia-Effron, E. Mellado, A. Monzon, J. L. Rodriguez-Tudela, and M. Cuenca-Estrella. 2003. In vitro activities of three licensed antifungal agents against spanish clinical isolates of *Aspergillus* spp. *Antimicrob. Agents Chemother.* 47:3085–3088.

Graminha, M. A., E. M. Rocha, R. A. Prade, and N. M. Martinez-Rossi. 2004. Terbinafine resistance mediated by salicylate 1-monooxygenase in *Aspergillus nidulans. Antimicrob. Agents Chemother.* 48:3530–3535.

Guarro, J., E. G. Kallas, P. Godoy, A. Karenina, J. Gene, A. Stchigel, and A. L. Colombo. 2002. Cerebral aspergillosis caused by *Neosartorya hiratsukae*, Brazil. *Emerg. Infect. Dis.* 8:989–991.

Gupta, A. K., and L. Q. Tu. 2006. Therapies for onychomycosis: a review. *Dermatol. Clin.* 24:375–379.

Hakki, M., J. F. Staab, and K. A. Marr. 2006. Emergence of a *Candida krusei* isolate with reduced susceptibility to caspofungin during therapy. *Antimicrob. Agents Chemother.* 50:2522–2524.

Hamamoto, H., K. Hasegawa, R. Nakaune, Y. J. Lee, Y. Makizumi, K. Akutsu, and T. Hibi. 2000. Tandem repeat of a transcriptional enhancer upstream of the sterol 14α-demethylase gene (CYP51) in *Penicillium digitatum. Appl. Environ. Microbiol.* 66:3421–3426.

Herbrecht, R., D. W. Denning, T. F. Patterson, J. E. Bennett, R. E. Greene, J. W. Oestmann, W. V. Kern, K. A. Marr, P. Ribaud, O. Lortholary, R. Sylvester, R. H. Rubin, J. R. Wingard, P. Stark, C. Durand, D. Caillot, E. Thiel, P. H. Chandrasekar, M. R. Hodges, H. T. Schlamm, P. F. Troke, B. de Pauw, et al. 2002. Voriconazole versus amphotericin B for primary therapy of invasive aspergillosis. *N. Engl. J. Med.* 347:408–415.

Hope, W. W., S. Shoham, and T. J. Walsh. 2007. The pharmacology and clinical use of caspofungin. *Expert Opin. Drug. Metab. Toxicol.* 3:263–274.

Howard, S. J., I. Webster, C. B. Moore, R. E. Gardiner, S. Park, D. S. Perlin, and D. W. Denning. 2006. Multi-azole resistance in *Aspergillus fumigatus. Int. J. Antimicrob. Agents* 28:450–453.

Isham, N., and M. A. Ghannoum. 2006. Determination of MICs of aminocandin for *Candida* spp. and filamentous fungi. *J. Clin. Microbiol.* 44:4342–4344.

Jarv, H., J. Lehtmaa, R. C. Summerbell, E. S. Hoekstra, R. A. Samson, and P. Naaber. 2004. Isolation of *Neosartorya pseudofischeri* from blood: first hint of pulmonary aspergillosis. *J. Clin. Microbiol.* 42: 925–928.

Johnson, E. M., K. L. Oakley, S. A. Radford, C. B. Moore, P. Warn, D. W. Warnock, and D. W. Denning. 2000. Lack of correlation of in vitro amphotericin B susceptibility testing with outcome in a murine model of *Aspergillus* infection. *J. Antimicrob. Chemother.* 45: 85–93.

Johnson, M. D., and J. R. Perfect. 2007. Combination antifungal therapy: what can and should we expect? *Bone Marrow Transplant.* 40: 297–306.

Judson, M. A., and D. A. Stevens. 2001. Current pharmacotherapy of allergic bronchopulmonary aspergillosis. *Expert Opin. Pharmacother.* 2:1065–1071.

Kartsonis, N. A., J. Nielsen, and C. M. Douglas. 2003. Caspofungin: the first in a new class of antifungal agents. *Drug Resist. Update* 6: 197–218.

Kaya, A. D., and N. Kiraz. 2007. In vitro susceptibilities of *Aspergillus* spp. causing otomycosis to amphotericin B, voriconazole and itraconazole. *Mycoses* 50:447–450.

Kelly, S. L., D. C. Lamb, D. E. Kelly, N. J. Manning, J. Loeffler, H. Hebart, U. Schumacher, and H. Einsele. 1997. Resistance to fluconazole and cross-resistance to amphotericin B in *Candida albicans* from AIDS patients caused by defective sterol δ5,6-desaturation. *FEBS Lett.* 400:80–82.

Koss, T., B. Bagheri, C. Zeana, M. F. Romagnoli, and M. E. Grossman. 2002. Amphotericin B-resistant *Aspergillus flavus* infection successfully treated with caspofungin, a novel antifungal agent. *J. Am. Acad. Dermatol.* 46:945–947.

Kurtz, M. B., I. B. Heath, J. Marrinan, S. Dreikorn, J. Onishi, and C. Douglas. 1994. Morphological effects of lipopeptides against *Aspergillus fumigatus* correlate with activities against (1,3)-beta-D-glucan synthase. *Antimicrob. Agents Chemother.* 38:1480–1489.

Laverdiere, M., R. G. Lalonde, J. G. Baril, D. C. Sheppard, S. Park, and D. S. Perlin. 2006. Progressive loss of echinocandin activity following prolonged use for treatment of *Candida albicans* oesophagitis. *J. Antimicrob. Chemother.* 57:705–708.

Lonial, S., L. Williams, G. Carrum, M. Ostrowski, and P. McCarthy, Jr. 1997. *Neosartorya fischeri*: an invasive fungal pathogen in an allogeneic bone marrow transplant patient. *Bone Marrow Transplant.* 19:753–755.

Luo, C. X., and G. Schnabel. 2008. The cytochrome P450 lanosterol 14α-demethylase gene is a demethylation inhibitor fungicide resistance determinant in *Monilinia fructicola* field isolates from Georgia. *Appl. Environ. Microbiol.* 74:359–366.

Ma, Z., T. J. Proffer, J. L. Jacobs, and G. W. Sundin. 2006. Overexpression of the 14α-demethylase target gene (*CYP51*) mediates fungicide resistance in *Blumeriella jaapii. Appl. Environ. Microbiol.* 72:2581–2585.

Maertens, J., I. Raad, G. Petrikkos, M. Boogaerts, D. Selleslag, F. B. Petersen, C. A. Sable, N. A. Kartsonis, A. Ngai, A. Taylor, T. F. Patterson, D. W. Denning, T. J. Walsh, et al. 2004. Efficacy and safety of caspofungin for treatment of invasive aspergillosis in patients refractory to or intolerant of conventional antifungal therapy. *Clin. Infect. Dis.* 39:1563–1571.

Manavathu, E. K., O. C. Abraham, and P. H. Chandrasekar. 2001. Isolation and in vitro susceptibility to amphotericin B, itraconazole and posaconazole of voriconazole-resistant laboratory isolates of *Aspergillus fumigatus*. *Clin. Microbiol. Infect.* 7:130–137.

Mann, P. A., R. M. Parmegiani, S. Q. Wei, C. A. Mendrick, X. Li, D. Loebenberg, B. DiDomenico, R. S. Hare, S. S. Walker, and P. M. McNicholas. 2003. Mutations in *Aspergillus fumigatus* resulting in reduced susceptibility to posaconazole appear to be restricted to a single amino acid in the cytochrome P450 14α-demethylase. *Antimicrob. Agents Chemother.* 47:577–581.

Marichal, P., L. Koymans, S. Willemsens, D. Bellens, P. Verhasselt, W. Luyten, M. Borgers, F. C. Ramaekers, F. C. Odds, and H. V. Bossche. 1999. Contribution of mutations in the cytochrome P450 14α-demethylase (Erg11p, Cyp51p) to azole resistance in *Candida albicans*. *Microbiology* 145:2701–2713.

Marr, K. A., M. Boeckh, R. A. Carter, H. W. Kim, and L. Corey. 2004. Combination antifungal therapy for invasive aspergillosis. *Clin. Infect. Dis.* 39:797–802.

Marr, K. A., T. Patterson, and D. Denning. 2002. Aspergillosis. Pathogenesis, clinical manifestations, and therapy. *Infect. Dis. Clin. North Am.* 16:875–894.

Maschmeyer, G., A. Haas, and O. A. Cornely. 2007. Invasive aspergillosis: epidemiology, diagnosis and management in immunocompromised patients. *Drugs* 67:1567–1601.

Maschmeyer, G., and M. Ruhnke. 2004. Update on antifungal treatment of invasive *Candida* and *Aspergillus* infections. *Mycoses* 47: 263–276.

Mellado, E., L. Alcazar-Fuoli, G. García-Effrón, A. Alastruey-Izquierdo, M. Cuenca-Estrella, and J. L. Rodriguez-Tudela. 2006. New resistance mechanisms to azole drugs in *Aspergillus fumigatus* and emergence of antifungal drug-resistant *A. fumigatus* atypical strains. *Med. Mycol.* 44:S367–S371.

Mellado, E., T. M. Diaz-Guerra, M. Cuenca-Estrella, and J. L. Rodriguez-Tudela. 2001. Identification of two different 14-α sterol demethylase-related genes (*cyp51A* and *cyp51B*) in *Aspergillus fumigatus* and other *Aspergillus* species. *J. Clin. Microbiol.* 39:2431–2438.

Mellado, E., G. Garcia-Effron, L. Alcazar-Fuoli, M. Cuenca-Estrella, and J. L. Rodriguez-Tudela. 2004. Substitutions at methionine 220 in the 14α-sterol demethylase (Cyp51A) of *Aspergillus fumigatus* are responsible for resistance in vitro to azole antifungal drugs. *Antimicrob. Agents Chemother.* 48:2747–2750.

Mellado, E., G. Garcia-Effron, L. Alcazar-Fuoli, W. J. Melchers, P. E. Verweij, M. Cuenca-Estrella, and J. L. Rodriguez-Tudela. 2007. A new *Aspergillus fumigatus* resistance mechanism conferring in vitro cross-resistance to azole antifungals involves a combination of *cyp51A* alterations. *Antimicrob. Agents Chemother.* 51:1897–1904.

Mellado, E., G. Garcia-Effron, M. J. Buitrago, L. Alcazar-Fuoli, M. Cuenca-Estrella, and J. L. Rodriguez-Tudela. 2005. Targeted gene disruption of the 14-α sterol demethylase (*cyp51A*) in *Aspergillus fumigatus* and its role in azole drug susceptibility. *Antimicrob. Agents Chemother.* 49:2536–2538.

Metcalf, S. C., and D. H. Dockrell. 2007. Improved outcomes associated with advances in therapy for invasive fungal infections in immunocompromised hosts. *J. Infect.* 55:287–299.

Miller, C. D., B. W. Lomaestro, S. Park, and D. S. Perlin. 2006. Progressive esophagitis caused by *Candida albicans* with reduced susceptibility to caspofungin. *Pharmacotherapy* 26:877–880.

Moore, C. B., N. Sayers, J. Mosquera, J. Slaven, and D. W. Denning. 2000. Antifungal drug resistance in *Aspergillus*. *J. Infect.* 41:203–220.

Morrison, V. A. 2006. Echinocandin antifungals: review and update. *Expert Rev. Anti Infect.Ther.* 4:325–342.

Mosquera, J., and D. W. Denning. 2002. Azole cross-resistance in *Aspergillus fumigatus*. *Antimicrob. Agents Chemother.* 46:556–557.

Nascimento, A. M., G. H. Goldman, S. Park, S. A. Marras, G. Delmas, U. Oza, K. Lolans, M. N. Dudley, P. A. Mann, and D. S. Perlin. 2003. Multiple resistance mechanisms among *Aspergillus fumigatus* mutants with high-level resistance to itraconazole. *Antimicrob. Agents Chemother.* 47:1719–1726.

NCCLS. 2002. Reference method for broth dilution antifungal susceptibility testing of filamentous fungi, approved standard document M38-A. National Committee for Clinical Laboratory Standards, Wayne, PA.

Odds, F. C. 2006. Drug evaluation. BAL-8557: a novel broad-spectrum triazole antifungal. *Curr. Opin. Investig. Drugs* 7:766–772.

Padhye, A. A., J. H. Godfrey, F. W. Chandler, and S. W. Peterson. 1994. Osteomyelitis caused by *Neosartorya pseudofischeri*. *J. Clin. Microbiol.* 32:2832–2836.

Paterson, P. J., S. Seaton, H. G. Prentice, and C. C. Kibbler. 2003. Treatment failure in invasive aspergillosis: susceptibility of deep tissue isolates following treatment with amphotericin B. *J. Antimicrob. Chemother.* 52:873–876.

Pavie, J., C. Lacroix, D. G. Hermoso, M. Robin, C. Ferry, A. Bergeron, M. Feuilhade, F. Dromer, E. Gluckman, J. M. Molina, and P. Ribaud. 2005. Breakthrough disseminated *Aspergillus ustus* infection in allogeneic hematopoietic stem cell transplant recipients receiving voriconazole or caspofungin prophylaxis. *J. Clin. Microbiol.* 43:4902–4904.

Perea, S., J. L. Lopez-Ribot, W. R. Kirkpatrick, R. K. McAtee, R. A. Santillan, M. Martinez, D. Calabrese, D. Sanglard, and T. F. Patterson. 2001. Prevalence of molecular mechanisms of resistance to azole antifungal agents in *Candida albicans* strains displaying high-level fluconazole resistance isolated from human immunodeficiency virus-infected patients. *Antimicrob. Agents Chemother.* 45:2676–2684.

Perlin, D. S. 2007. Resistance to echinocandin-class antifungal drugs. *Drug Resist. Update* 10:121–130.

Podust, L. M., J. Stojan, T. L. Poulos, and M. R. Waterman. 2001. Substrate recognition sites in 14α-sterol demethylase from comparative analysis of amino acid sequences and X-ray structure of *Mycobacterium tuberculosis* CYP51. *J. Inorg. Biochem.* 87:227–235.

Prentice, A. G., A. Glasmacher, and B. Djulbegovic. 2006. In meta-analysis itraconazole is superior to fluconazole for prophylaxis of systemic fungal infection in the treatment of haematological malignancy. *Br. J. Haematol.* 132:656–658.

Quindos, G., A. J. Carrillo-Munoz, E. Eraso, E. Canton, and J. Peman. 2007. In vitro antifungal activity of voriconazole: new data after the first years of clinical experience. *Rev. Iberoam. Micol.* 24: 198–208.

Rapp, R. P. 2004. Changing strategies for the management of invasive fungal infections. *Pharmacotherapy* 24:4S–28S.

Rocha, E. M., G. Garcia-Effron, S. Park, and D. S. Perlin. 2007. A Ser678Pro substitution in Fks1p confers resistance to echinocandin drugs in *Aspergillus fumigatus*. *Antimicrob. Agents Chemother.* 51: 4174–4176.

Rocha, E. M., R. E. Gardiner, S. Park, N. M. Martinez-Rossi, and D. S. Perlin. 2006. A Phe389Leu substitution in *ergA* confers terbinafine resistance in *Aspergillus fumigatus*. *Antimicrob. Agents Chemother.* 50:2533–2536.

Ruhnke, M., G. Kofla, K. Otto, and S. Schwartz. 2007. CNS aspergillosis: recognition, diagnosis and management. *CNS Drugs* 21: 659–676.

Sabra, R., and R. A. Branch. 1990. Amphotericin B nephrotoxicity. *Drug Saf.* 5:94–108.

Sanglard, D., F. Ischer, T. Parkinson, D. Falconer, and J. Bille. 2003. *Candida albicans* mutations in the ergosterol biosynthetic pathway and resistance to several antifungal agents. *Antimicrob. Agents Chemother.* 47:2404–2412.

Sanglard, D., and F. C. Odds. 2002. Resistance of *Candida* species to antifungal agents: molecular mechanisms and clinical consequences. *Lancet Infect. Dis.* **2**:73–85.

Schaffner, A., and A. Bohler. 1993. Amphotericin B refractory aspergillosis after itraconazole: evidence for significant antagonism. *Mycoses* **36**:421–424.

Schiraldi, G. F., S. L. Cicero, M. D. Colombo, D. Rossato, M. Ferrarese, and E. Soresi 1996. Refractory pulmonary aspergillosis: compassionate trial with terbinafine. *Br. J. Dermatol.* **134**(Suppl. 46):25–29.

Scott, L. J., and D. Simpson. 2007. Voriconazole: a review of its use in the management of invasive fungal infections. *Drugs* **67**:269–298.

Seifert, H., U. Aurbach, D. Stefanik, and O. Cornely. 2007. In vitro activities of isavuconazole and other antifungal agents against *Candida* bloodstream isolates. *Antimicrob. Agents Chemother.* **51**:1818–1821.

Slaven, J. W., M. J. Anderson, D. Sanglard, G. K. Dixon, J. Bille, I. S. Roberts, and D. W. Denning. 2002. Increased expression of a novel *Aspergillus fumigatus* ABC transporter gene, *atrF*, in the presence of itraconazole in an itraconazole resistant clinical isolate. *Fungal Genet. Biol.* **363**:199–206.

Steinbach, W. J., D. K. Benjamin, Jr., D. P. Kontoyiannis, J. R. Perfect, I. Lutsar, K. A. Marr, M. S. Lionakis, H. A. Torres, H. Jafri, and T. J. Walsh. 2004a. Infections due to *Aspergillus terreus*: a multicenter retrospective analysis of 83 cases. *Clin. Infect. Dis.* **39**:192–198.

Steinbach, W. J., J. R. Perfect, W. A. Schell, T. J. Walsh, and D. K. Benjamin, Jr. 2004b. In vitro analyses, animal models, and 60 clinical cases of invasive *Aspergillus terreus* infection. *Antimicrob. Agents Chemother.* **48**:3217–3225.

Stevens, D. A., V. L. Kan, M. A. Judson, V. A. Morrison, S. Dummer, D. W. Denning, J. E. Bennett, T. J. Walsh, T. F. Patterson, and G. A. Pankey. 2000. Practice guidelines for diseases caused by *Aspergillus*, Infectious Diseases Society of America. *Clin. Infect. Dis.* **30**:696–709.

Tobin, M. B., R. B. Peery, and P. L. Skatrud. 1997. Genes encoding multiple drug resistance-like proteins in *Aspergillus fumigatus* and *Aspergillus flavus*. *Gene* **200**:11–23.

Torres, H. A., R. Y. Hachem, R. F. Chemaly, D. P. Kontoyiannis, and I. I. Raad. 2005. Posaconazole: a broad-spectrum triazole antifungal. *Lancet Infect. Dis.* **5**:775–785.

Varga, J., Z. Vida, B. Toth, F. Debets, and Y. Horie. 2000. Phylogenetic analysis of newly described *Neosartorya* species. *Antonie Leeuwenhoek* 77:235–239.

Vehreschild, J. J., and O. A. Cornely. 2006. Micafungin sodium, the second of the echinocandin class of antifungals: theory and practice. *Future Microbiol.* **1**:161–170.

Verweij, P. E., D. T. Te Dorsthorst, A. J. Rijs, H. G. De Vries-Hospers, and J. F. Meis. 2002. Nationwide survey of in vitro activities of itraconazole and voriconazole against clinical *Aspergillus fumigatus* isolates cultured between 1945 and 1998. *J. Clin. Microbiol.* **40**:2648–2650.

Verweij, P. E., E. Mellado, and W. J. Melchers. 2007a. Multiple-triazole-resistant aspergillosis. *N. Engl. J. Med.* **356**:1481–1483.

Verweij, P. E., M. Mensink, A. J. Rijs, J. P. Donnelly, J. F. Meis, and D. W. Denning. 1998. In-vitro activities of amphotericin B, itraconazole and voriconazole against 150 clinical and environmental *Aspergillus fumigatus* isolates. *J. Antimicrob. Chemother.* **42**:389–392.

Verweij, P., H. A. L. Van der Lee, J. Kuijpers, E. Snelders, A. J. M. M. Rijs, and W. J. G. Melchers. 2007b. Epidemiology of multiple-triazole-resistant *Aspergillus fumigatus*, abstr. M-2018. *Abstr. 47th Intersci. Conf. Antimicrob. Agents Chemother.*, Chicago, IL, 17 to 20 September 2007.

Walsh, T. J., R. W. Finberg, C. Arndt, J. Hiemenz, C. Schwartz, D. Bodensteiner, P. Pappas, N. Seibel, R. N. Greenberg, S. Dummer, M. Schuster, J. S. Holcenberg, et al. 1999. Liposomal amphotericin B for empirical therapy in patients with persistent fever and neutropenia. *N. Engl. J. Med.* **340**:764–771.

Walsh, T. J., H. Teppler, G. R. Donowitz, J. A. Maertens, L. R. Baden, A. Dmoszynska, O. A. Cornely, M. R. Bourque, R. J. Lupinacci, C. A. Sable, and B. E. dePauw. 2004. Caspofungin versus liposomal amphotericin B for empirical antifungal therapy in patients with persistent fever and neutropenia. *N. Engl. J. Med.* **351**:1391–1402.

White, T. C., S. Holleman, F. Dy, L. F. Mirels, and D. A. Stevens. 2002. Resistance mechanisms in clinical isolates of *Candida albicans*. *Antimicrob. Agents Chemother.* **46**:1704–1713.

White, T. C., K. A. Marr, and R. A. Bowden. 1998. Clinical, cellular, and molecular factors that contribute to antifungal drug resistance. *Clin. Microbiol. Rev.* **11**:382–402.

Wyand, R. A., and J. K. Brown. 2005. Sequence variation in the *CYP51* gene of *Blumeria graminis* associated with resistance to sterol demethylase inhibiting fungicides. *Fungal Genet. Biol.* **42**:726–735.

Xiao, L., V. Madison, A. S. Chau, D. Loebenberg, R. E. Palermo, and P. M. McNicholas. 2004. Three-dimensional models of wild-type and mutated forms of cytochrome P450 14α-sterol demethylases from *Aspergillus fumigatus* and *Candida albicans* provide insights into posaconazole binding. *Antimicrob. Agents Chemother.* **48**:568–574.

Aspergillus fumigatus and Aspergillosis
Edited by J.-P. Latgé and W. J. Steinbach
©2009 ASM Press, Washington, DC

Chapter 35

Immunotherapy

BRAHM H. SEGAL AND LUIGINA R. ROMANI

Aspergillus species are ubiquitous soil inhabitants whose conidia we inhale on a regular basis, and they are normally harmless to immunocompetent individuals. Acute invasive aspergillosis, as distinguished from allergic diseases and chronic necrotizing aspergillosis, is principally a disease of the highly immunocompromised. Mortality from invasive aspergillosis increased by severalfold in the 1980s and 1990s (Denning, 2003; McNeil et al., 2001), a reflection of more patients undergoing treatment for hematologic malignancies and allogeneic hematopoietic stem cell transplantation (HSCT). Deficits in host defense which render persons susceptible to invasive aspergillosis are complex but can be broadly divided into the following categories: (i) neutropenia, (ii) qualitative deficits in phagocyte function, and (iii) deficits in cell-mediated immunity (Table 1). Patient groups at highest risk for invasive aspergillosis are the following: those with prolonged and severe neutropenia (such as occurs following chemotherapy for acute leukemia), allogeneic HSCT recipients with significant graft-versus-host disease (GVHD), lung transplant recipients, and patients with chronic granulomatous disease (CGD) (Segal and Walsh, 2006).

There has been significant improvement in the antifungal armamentarium that is relevant to both therapy of invasive aspergillosis (Herbrecht et al., 2002) and prevention of fungal disease in high-risk patients (Cornely et al., 2007; Ullmann et al., 2007; van Burik et al., 2004). Strategies to augment immunity against fungal pathogens are complementary to the important advances in drug development. Indeed, part of the antifungal effects of antifungal agents may occur via immunomodulation.

EPIDEMIOLOGY AND IMMUNOPATHOLOGY OF INVASIVE ASPERGILLOSIS

Immunotherapeutic strategies for invasive aspergillosis should be understood in the context of host defense deficits (primary and iatrogenic immunosuppression) in high-risk patients and immunopathology of fungal disease. The risk of invasive aspergillosis is strongly related to the duration and degree of neutropenia. Most cases of invasive aspergillosis complicating neutropenia occur in patients receiving potent cytotoxic regimens for hematologic malignancies and myeloablative HSCT (Denning et al., 1998; Gerson et al., 1984; Wald et al., 1997). Invasive aspergillosis is also a major cause of mortality in patients with aplastic anemia and refractory neutropenia (Weinberger et al., 1992).

Several studies have reported the predominance of invasive aspergillosis cases occurring in the postengraftment rather than the neutropenic period in allogeneic HSCT recipients (Baddley et al., 2001; Grow et al., 2002; Jantunen et al., 1997; Marr et al., 2002; Martino et al., 2002; McWhinney et al., 1993; Shaukat et al., 2005; Wald et al., 1997; Yuen et al., 1997). The regimens used to control GVHD (e.g., high-dose corticosteroids, tumor necrosis factor alpha [TNF-α] inhibition) have broad immunosuppressive properties that affect both innate macrophage and neutrophil functions and antigen-driven immunity.

The key importance of neutrophils in host defense against invasive aspergillosis is demonstrated in CGD, an inherited disorder of the NADPH oxidase complex in which phagocytes are defective in generating the reactive oxidant superoxide anion and its downstream metabolites (Segal et al., 2000). Neutrophil proteases released from secondary granules in neutrophils following NADPH oxidase activation likely play the major role in defending against bacterial and fungal pathogens (Reeves et al., 2002; Tkalcevic et al., 2000). Invasive aspergillosis has been reported to be the most important cause of mortality in CGD (Cohen et al., 1981; Mouy et al., 1994; Segal et al., 1998). Despite the routine use of gamma interferon (IFN-γ) prophylaxis, fungal infections have remained a persistent problem, with an incidence of 0.1 fungal infections per patient year (Wink-

Brahm H. Segal • Dept. of Medicine and Dept. of Immunology, Roswell Park Cancer Institute, Buffalo, NY 14263. **Luigina R. Romani** • Microbiology Section, Dept. of Experimental Medicine, University of Perugia, Perugia, Italy.

Table 1. Patients at risk for invasive aspergillosis

Patients	Immunologic deficits and comments
Acute myelogenous leukemia, myelodysplastic syndrome, myeloablative hematopoietic stem cell transplantation (early period), aplastic anemia, and bone marrow failure from other etiologies	Neutropenia
Acute lymphoblastic leukemia	Neutropenia and corticosteroids
Allogeneic hematopoietic stem cell transplantation with graft-versus-host disease (GVHD)	Global immune impairment of innate and T-cell immunity related to severity of GVHD and intensity of immunosuppressive therapy
Solid organ transplant recipients	Risk related to intensity of suppression of T-cell immunity to treat allograft rejection. Lung transplant recipients are at highest risk.
Alemtuzumab	Most commonly administered for refractory chronic lymphocytic leukemia, where it is associated with neutropenia and prolonged T-cell depletion. Also used to prevent GVHD and rejection of solid organ allografts
Advanced AIDS (generally CD4 count of $<50/\mu l$)	Uncommonly complicated by invasive aspergillosis in absence of coexisting neutropenia or systemic corticosteroids
Chronic granulomatous disease	Inherited disorder of NADPH oxidase; neutrophil numbers are normal but function is impaired

elstein et al., 2000). The frequency of aspergillosis may decrease with mold-active prophylaxis (Gallin et al., 2003).

There has been an increased appreciation that acute invasive aspergillosis encompasses a number of disease entities with distinct clinical and histopathologic characteristics that reflect host defense responses (Segal and Walsh, 2006). Animal models of invasive aspergillosis and experience in patients suggest that *Aspergillus* infection during neutropenia is pathologically and immunologically distinct from infection in the absence of neutropenia in the setting of potent immunosuppressive regimens. Berenguer et al. (1995) reported that in profoundly neutropenic rabbits challenged with *Aspergillus fumigatus*, pulmonary lesions consisted predominantly of coagulative necrosis, intraalveolar hemorrhage, and scant mononuclear inflammatory infiltrate. In contrast, pulmonary foci in rabbits treated with cyclosporine plus corticosteroids (modeling the immunosuppression in patients undergoing hematopoietic and solid organ transplantation) consisted mainly of neutrophilic and monocytic infiltrates, inflammatory necrosis, and scant intraalveolar hemorrhage. Studies of murine aspergillosis have produced similar findings and demonstrated the key role of neutrophils in controlling fungal burden and protecting against vascular invasion (Balloy et al., 2005; Stephens-Romero et al., 2005).

In our single-center review of invasive mold infections in allogeneic HSCT recipients, 21 of 22 cases (95%) were diagnosed after neutrophil recovery (Shaukat et al., 2005). All had received systemic corticosteroids within 1 month prior to diagnosis of mold infection. *Aspergillus* species were isolated in 18 (82%) cases. In contrast to animal studies, the predominant lung histopathology was coagulative necrosis with sparse inflammation, similar to the histology associated with neutropenic hosts. Hyphal angioinvasion was observed in some of these cases. Analysis of a larger pathologic database of invasive aspergillosis at the NIH showed similar findings (Stergiopoulou et al., 2007). Lesions of invasive pulmonary aspergillosis in neutropenic patients and HSCT recipients were similar and consisted predominantly of angioinvasion and intraalveolar hemorrhage. In the nonneutropenic, non-HSCT cohort, lesions of invasive pulmonary aspergillosis consisted mainly of neutrophilic and monocytic infiltrates and inflammatory necrosis.

In contrast to our results, Chamilos et al. (2006) reported in a large autopsy series that a low fungal burden and high inflammatory pattern characterized invasive pulmonary aspergillosis associated with GVHD, whereas a high fungal burden and coagulative necrosis predominated during neutropenia. We speculate that the predominance of coagulative necrosis observed in nonneutropenic allogeneic HSCT recipients in our series may reflect the high doses of corticosteroids used to treat GVHD, which may disable leukocyte trafficking and hyphal killing.

In CGD, invasive aspergillosis is characterized by a robust pyogranulomatous response. Phagocyte trafficking to the site of infection is intact, but the fungicidal activity of phagocytes is disabled. Despite parenchymal invasive disease, hyphal vascular invasion is not a feature of pulmonary aspergillosis in CGD patients (Segal and Holland, 2003) or mouse models (Chang et al., 1998; Dennis et al., 2006). Bignell et al. (2005) reported that even low-virulence mutant strains of *Aspergillus nidulans* caused mortality in experimental pulmonary aspergillosis in mice with CGD as a result of an excessive inflammatory response. This finding emphasizes the unique pathophysiologic features of aspergillosis in mice with CGD that must be considered in studies of pathogen virulence and immune augmentation.

IMMUNOTHERAPY

We will discuss the following strategies for immunotherapy for invasive aspergillosis: (i) augmentation of neutrophil number; (ii) pathogen recognition receptor (PRR) ligands; (iii) cytokine administration and depletion; and (iv) vaccination (Table 2). No immunotherapeutic strategy has been shown to be of benefit in invasive aspergillosis in clinical trials. Therefore, these approaches should be viewed as promising areas for future research.

Augmentation of Neutrophil Numbers

Colony-stimulating factors

Normal myelopoiesis requires myeloid stem cells. Under the influence of stem cell factor, interleukin-3 (IL-3), and granulocyte-macrophage colony-stimulating factor (GM-CSF), these stem cells give rise to the colony-forming granulocyte-macrophage. Granulocyte colony-stimulating factor (G-CSF) acts at a later stage in concert with other growth factors to specifically drive granulopoiesis. Multiple randomized clinical trials of prophylactic recombinant G-CSF and GM-CSF have shown the benefit of CSFs in reducing the time to neutrophil recovery and the duration of fever and hospitalization in patients with acute myelogenous leukemia (Rowe, 1998). In one randomized study in patients receiving chemotherapy for acute myelogenous leukemia, prophylaxis with GM-CSF led to a lower frequency of fatal fungal infections compared to placebo and reduced overall early mortality (Rowe et al., 1995). However, no other randomized study of prophylactic CSFs has demonstrated a survival advantage compared to placebo in patients with hematological malignancies. A meta-analysis of randomized trials of prophylactic G-CSF and GM-CSF in autologous and allogeneic HSCT recipients showed that CSFs were associated with a small reduction in the risk of documented infections but did not affect infection or treatment-related mortality (Dekker et al., 2006). The American Society of Clinical Oncology has established authoritative guidelines related to the use of prophylactic CSFs in standard practice (Smith et al., 2006).

Table 2. Summary of immune augmentation strategies against aspergillosis

Aim	Strategy[a]
Increase neutrophil number	• Colony-stimulating factors (CSFs) • Granulocyte transfusions • Experimental myeloid transfusion studies
Augment neutrophil and macrophage function	• Colony-stimulating factors (CSFs) • Cytokine administration or depletion (e.g., IFN-γ, depletion of IL-17) • Toll-like receptor (TLR) ligands • Pentraxin-3
Augment T-cell immunity	• Vaccination • Adoptive transfer of T-cell populations • Cytokine administration (e.g., IFN-γ) • TLR ligands • Pentraxin-3
Augment humoral immunity	• Vaccination • Adoptive transfer
Complement activation	• Mannose-binding lectin

[a] Note that none of the immune augmentation strategies has been proven to be of benefit in preventing or treating invasive aspergillosis in clinical trials. Most of the above immunotherapies have only been evaluated in animal models of aspergillosis.

The rationale for CSFs for established infections (as opposed to prophylaxis) stems from both the quantitative and the qualitative effects of these agents on phagocytic cells. In neutropenic patients with life-threatening infections, survival is strongly influenced by the rapidity of neutrophil recovery. Thus, CSFs and granulocyte transfusions may be used in these settings to augment the number of circulating neutrophils. Randomized trials have not shown a benefit of CSFs as adjunct therapy for uncomplicated neutropenic fever. Although the benefit of a CSF for established infection is unproven, it may be considered in the setting of profound neutropenia (absolute neutrophil count, $<100/\mu l$), uncontrolled primary disease, and in serious infections, such as pneumonia, hypotension, multiorgan dysfunction, and invasive fungal infection.

Another gap in knowledge is whether CSFs are safe and effective as either prophylaxis or adjunctive therapy in nonleukopenic patients with severe impairment in phagocyte function. Intensive immunosuppressive corticosteroid-based regimens for GVHD cause global impairment of phagocyte effector functions and disable reconstitution of antigen-specific immunity, though circulating neutrophil counts are generally normal. In theory, GM-CSF may augment qualitative macrophage and neutrophil function that may protect against infections. There are no data to support prophylactic CSFs in nonneutropenic patients, and they should not be used as prophylaxis in this setting outside of a clinical trial.

Granulocyte transfusions

The rationale for granulocyte transfusions is to provide supportive therapy for the neutropenic patient with a life-threatening infection by augmenting the number of circulating neutrophils until autologous myeloid regeneration occurs. In the 1970s, apheresis technology for harvesting large numbers of donor granulocytes became available. Controlled trials of granulocyte transfusions as adjuvant therapy in neutropenic patients produced mixed results. In the 1980s, the enthusiasm for granulocyte transfusions waned as more effective antibiotics became available, survival from serious bacterial infections improved, and recombinant growth factors reduced the duration of neutropenia. In addition, concerns about the toxicity of granulocyte transfusions, including acute pulmonary reactions, HLA alloimmunization (which could render patients refractory to platelet transfusions and potentially impair myeloid engraftment following HSCT), and transfusion-associated infections (particularly with cytomegalovirus [CMV]), outweighed the perceived benefits.

Today, the impetus to reexamine the role of granulocyte transfusions stems largely from improvements in donor mobilization methods. Bensinger et al. (1993) showed that G-CSF mobilization significantly increased the granulocyte yield and resulted in improved circulating neutrophil levels in neutropenic recipients. Using a standard continuous flow centrifugation apparatus, the mean absolute neutrophil yield per collection was typically in the range of 8×10^{10} cells when both G-CSF and dexamethasone were used in the donor preparatory regimen. Higher numbers of harvested neutrophils correlated with higher posttransfusion neutrophil counts. Furthermore, the increase in circulating neutrophils tends to be sustained for 24 to 30 h following transfusion, as a consequence of a prolonged circulating half-life of G-CSF-mobilized granulocytes (Dale et al., 1998). The qualitative functions of G-CSF- and steroid-mobilized neutrophils are intact, based on in vitro bactericidal activity, respiratory burst, migration to experimental skin chambers, and localization to sites of inflammation.

Successful outcomes using granulocyte transfusions have been described in patients with life-threatening fungal infections in small series and in case reports. A phase I/II trial using G-CSF-mobilized granulocyte transfusions for refractory fungal infections in neutropenic patients with hematological malignancies reported favorable responses in 11 of 15 patients (Dignani et al., 1997). Peters et al. (1999) evaluated granulocyte transfusions (G-CSF or prednisolone mobilized) in 30 patients with neutropenia and life-threatening, refractory infections. Infections cleared in 20 of 30 patients, including 5 of 9 patients with invasive aspergillosis. No benefit of granulocyte transfusions was noted in neutropenic HSCT recipients with invasive mold infection in a retrospective series in which G-CSF donor granulocyte mobilization was not used (Bhatia et al., 1994).

Price et al. (2000) conducted a phase I/II study of granulocyte transfusions derived from unrelated, non-HLA-matched, community donors following G-CSF and dexamethasone mobilization. Chills, fever, and oxygen desaturation of $\geq$3% occurred in association with 7% of transfusions but did not limit therapy. Eight of 11 patients with bacterial infections or candidemia survived, but all 8 patients with invasive mold infection died. This study showed the safety and feasibility of using community donors for granulocytapheresis donations.

In the absence of modern, prospective, randomized studies, when should granulocyte transfusions be considered? Currently, there is no justification (outside of a clinical trial) to use granulocyte transfusions either as prophylaxis or in cases of documented infections that are likely to respond to conventional therapy. We reserve granulocyte transfusions for patients with prolonged neutropenia and life-threatening infections refractory to conventional therapy. Filamentous fungi are likely to constitute the majority of such refractory infections. Infusions of amphotericin B should be separated by several hours from granulocyte transfusions to avoid pulmonary toxicity. In some highly alloimmunized patients, transfused granulocytes are rapidly consumed and are likely to have more toxicity than benefit. In allogeneic transplants in which the donor and recipient are CMV seronegative, using CMV-seronegative granulocyte donors is advised (Nichols et al., 2002).

The Transfusion Medicine and Hemostasis Network of the National Heart, Lung and Blood Institute is in the planning stages of a randomized study of adjunctive granulocyte transfusions in neutropenic patients with severe bacterial and fungal infections. This study is expected to definitively evaluate the benefits and risks of adjunctive granulocyte transfusions.

Experimental strategies for myeloid transfusions

The myeloid progenitors, common myeloid progenitors and granulocyte-monocyte progenitors, have recently been identified. The addition of these progenitors to hematopoietic grafts in mice that had been rendered neutropenic conferred protection against challenge with *Pseudomonas aeruginosa* and *A. fumigatus* (BitMansour et al., 2002). Novel strategies such as this approach to accelerate neutrophil recovery merit further study.

Spellberg et al. (2005, 2007) developed a myeloid transfusion approach using HL-60 cells, a myeloid cell line. Activated and irradiated HL-60 cells had similar properties to granulocytes, including respiratory burst activity and candidacidal activity in vitro. These cells had virtually no replicative capacity and were safe in mice. Infusion of irradiated, activated HL-60 cells improved survival of neutropenic, candidemic mice. This approach has promise for adaptation to the clinic.

PRR LIGANDS

Alveolar macrophages constitute the first line of host defense against aerosolized conidia. Following germination, neutrophils are the dominant host defense arm against the hyphal stage. There are several classes of innate PRRs that recognize fungal motifs. Examples include Toll-like receptors (TLRs), dectin-1, pentraxins, collectins, SP-A, SP-D, and mannose-binding lectin (Hogaboam et al., 2004), classical C-type lectins, and lactosyceramide (Brown, 2006; Mukhopadhyay et al., 2004).

TLRs are a conserved family of receptors that recognize common protein and DNA pattern motifs present on microbial pathogens and initiate signaling events related to cytokine production and T-cell and dendritic cell (DC) maturation. During the phagocytosis of pathogens, TLRs recognize pathogen-specific motifs within the vacuole, distinguish between pathogens, and trigger an inflammatory response appropriate to defense against the specific organism (Ozinsky et al., 2000; Underhill et al., 1999). TLRs have homology to IL-1R1 and share a similar signaling cascade, leading to activation of NF-κB and mitogen-activated protein kinases, a process that mediates gene expression and regulation of inflammatory responses. TLR-dependent antifungal pathways are highly conserved in nature, as demonstrated by their presence in *Drosophila melanogaster* (Lemaitre et al., 1996; Tauszig-Delamasure et al., 2002). TLRs recognize motifs on *Candida* spp. (Netea et al., 2002) and *Cryptococcus* spp. (Shoham et al., 2001) and regulate inflammatory responses.

Brown et al. identified dectin-1 as a major innate immune recognition receptor and immunomodulator of β-glucans (Brown and Gordon, 2001; Brown et al., 2002, 2003). Dectin-1 is a natural killer (NK) cell receptor-like C-type lectin expressed at high levels within the pulmonary and gastrointestinal tract at portals of pathogen entry, suggesting a potential role for pathogen recognition and host defense (Brown, 2006). After binding to particulate glucans (e.g., zymosan), dectin-1 stimulates the production of IL-12, IL-10, TNF-α, and macrophage inflammatory protein 2. It also induces ligand uptake through phagocytosis and NADPH oxidase activity (Gantner et al., 2003). Dectin-1 contains an immunoreceptor tyrosine-based activation motif in its cytoplasmic tail that is involved in cellular activation (Gantner et al., 2003). Dectin-1 collaborates with TLR2 to induce TNF-α and IL-12 following zymosan stimulation (Gantner et al., 2003). Dectin-1 can recognize β-glucan motifs on several fungal species, including *Candida* spp. (Brown et al., 2003), *Pneumocystis jirovecii* (Steele et al., 2003), *Coccidiodes* spp. (Viriyakosol et al., 2005), and *Aspergillus* spp. (Brown, 2006). Based on the ability of dectin-1 to recognize immunomodulatory fungal cell wall products, facilitate phagocytosis and fungal killing, induce NADPH oxidase activation, and in cooperation with TLRs stimulate and regulate cytokine responses, dectin-1 has been posited to play a role in fungal recognition and antifungal immunity (Brown, 2006). Definitive evidence for a role of dectin-1 in antifungal immunity should be derived from genetically engineered dectin-$1^{-/-}$ mice.

Gantner et al. (2003) demonstrated a cooperative interaction between dectin-1, TLRs, and NADPH oxidase activation in response to zymosan. During macrophage and DC recognition of zymosan, both dectin-1 and TLR2 were recruited to phagosomes, where dectin-1 binds to β-glucans and TLR2/CD14 recognize other components of the fungal cell wall. Dectin-1 enhanced TLR2-mediated activation of NF-κB in response to zymosan. Dectin-1 and TLR2 cooperatively interacted in the activation of macrophages and DCs following zymosan challenge and in mediating production of IL-12 and TNF-α. Additionally, dectin-1 was required for zymosan-mediated activation of NADPH oxidase, a response that was primed by TLR4 activation.

Fungal β-glucans act as a trigger for the induction of inflammatory responses in macrophages through their time-dependent exposure on the surface of germinating conidia (Gersuk et al., 2006; Hohl et al., 2005; Steele et al., 2005). Dectin-1 and TLRs permit macrophages to distinguish between *A. fumigatus* conidia and hyphae. Whereas conidia ingested by macrophages did not stimulate NADPH oxidase or an inflammatory response, early germinated hyphae stimulated NF-κB, secretion of proinflammatory cytokines, and NADPH oxidase activation in human and mouse macrophages (Gersuk et al., 2006). Germination rendered fungal cell wall β-glucans accessible to dectin-1, and dectin-1 binding to germ tubes augmented TLR2-mediated stimulation of cytokines (Gersuk et al., 2006).

Ligation of specific PRRs to modulate the inflammatory response to *Aspergillus* is an intriguing strategy, in both invasive and allergic aspergillosis. PRR ligands can affect innate and antigen-driven immunity at different levels. They may augment phagocytic antifungal effector functions and induce maturation of DCs, leading to increased *Aspergillus* antigen display and modulation of T-cell responses, which may be useful in vaccine-based strategies.

Local delivery of CpG oligodeoxynucleotides (which signal through TLR9) and the Asp f 16 *Aspergillus* allergen resulted in activation of airway DCs capable of inducing Th1 priming and resistance to the fungus (Bozza et al., 2002). Thymosin α1, a naturally occurring thymic peptide, induced maturation and IL-12 production in DCs pulsed with *Aspergillus*, an effect mediated by distinct TLRs (Romani et al., 2004). Thymosin α1 augmented Th1 immunity against *Aspergillus*, accelerated myeloid recovery in neutropenic mice, and was protective against *Aspergillus* challenge in murine bone marrow transplant recipients.

Recognition of *Aspergillus* motifs and activation of neutrophils are coordinated by distinct members of the TLR family, each likely activating specialized antifungal effector functions and inflammatory responses (Bellocchio et al., 2004). Indeed, liposomal amphotericin B, in addition to its intrinsic antifungal activity, may activate antifungal resistance by activating TLR4 in neutrophils (Bellocchio et al., 2005). These studies provide a rationale to stimulate or inhibit specific classes of TLRs as a means of enhancing both innate and antigen-specific immunity to fungi.

PENTRAXIN 3

Pentraxins are a superfamily of conserved proteins characterized by a cyclic multimeric structure. Pentraxin 3 (PTX3) is an innate pathogen recognition protein that binds to specific motifs on *P. aeruginosa*, *Salmonella enterica* serovar Typhimurium, and *A. fumigatus*. PTX3-deficient mice were highly susceptible to *Aspergillus* infection (Garlanda et al., 2002). These mice demonstrated defective recognition of conidia by alveolar macrophages and DCs, as well as inappropriate induction of type 2 cytokine responses. Administration of PTX3 protected against *Aspergillus* challenge in murine T-cell-depleted allogeneic bone marrow transplant recipients (Garlanda et al., 2002) and potentiated the protective effect of subtherapeutic amphotericin B (Gaziano et al., 2004).

CYTOKINE ADMINISTRATION AND DEPLETION: WHAT'S NEW?

Protective immunity against *Aspergillus* is achieved by the integration of two distinct arms of the immune system, the innate and adaptive responses. Most of the innate mechanisms are inducible upon infection, and their activation requires specific recognition of invariant evolutionarily conserved molecular structures shared by large groups of pathogens by PRRs. As the balance between proinflammatory and anti-inflammatory signaling is a prerequisite for successful host-fungus interactions, the activation of PRRs pivotally contributes to inflammation and immunity to the fungus. However, although the decision of how to respond will still be primarily determined by interactions between pathogens and cells of the innate immune system, the actions of T cells will feed back into this dynamic equilibrium to regulate inflammation and subsequent immune responses. Although inflammation is an essential component of the protective response to fungi, its dysregulation may significantly worsen fungal diseases and limit protective antifungal immune responses.

In the last 2 decades, the immunopathogenesis of aspergillosis and associated diseases was explained primarily in terms of the Th1/Th2 balance. However, while the pathogenetic role of either subset may still hold true, the role of reciprocal regulation of both subsets is apparently outdated. A number of observations suggest that the Th1 IL-12–IFN-γ axis may not be as central to fungal immunity as currently believed, i.e., other cytokine pathways may be the main players. The newly described Th17 pathway, which has an inflammatory role previously attributed to uncontrolled Th1 cell reactivity, and regulatory T cells, capable of fine-tuning protective antimicrobial immunity to minimize harmful immune pathology, have become integral components of the immune response to the fungus. Indoleamine dioxygenase (IDO) and tryptophan catabolites contribute to such a homeostatic condition by providing the host with immune defense mechanisms adequate for protection, without necessarily eliminating fungal pathogens—which would impair immune memory—or causing an unacceptable level of tissue damage. These new findings provide a molecular connection between the failure to resolve inflammation and lack of antifungal immune resistance and point to strategies for immune therapy of fungal infections that could attempt to limit inflammation in order to stimulate an effective immune response (Romani and Puccetti, 2007). Figure 1 illustrates the possible contribution of T-cell responses to inflammation and immunity to *Aspergillus*.

Recombinant IFN-γ

There are several potential cytokines and chemokines that can be exploited as immunotherapy for invasive aspergillosis. We will focus on IFN-γ and IL-17. Interferons are immune modulators that regulate the expression of numerous genes that mediate inflammation. Exposure to various pathogens can stimulate at least two patterns of cytokine production by $CD4^+$ T cells. Th1 cells are defined by production of IFN-γ, lymphotoxin, and IL-2, and Th2 cells are defined by production of IL-4, IL-5, and IL-13. Several laboratories have shown that IFN-γ augments the antifungal activity of effector cells

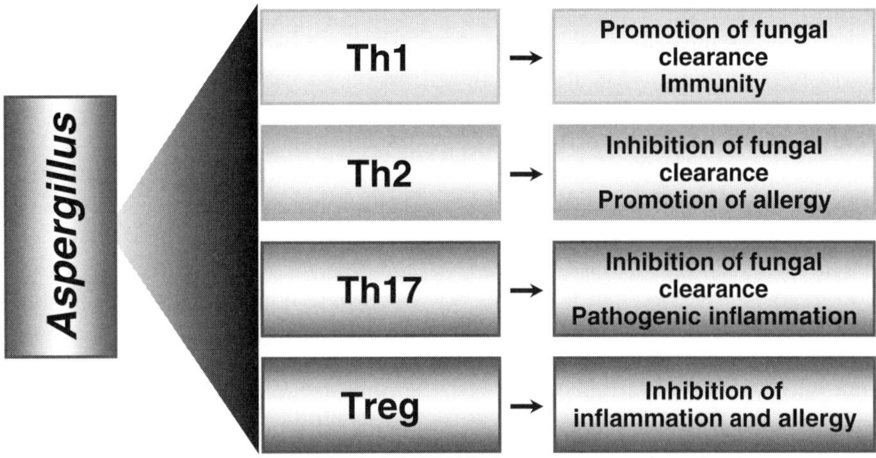

Figure 1. Model of contribution of T cells to host defense and inflammation against *Aspergillus*.

(macrophages and neutrophils). Studies in vitro and from animal models (Nagai et al., 1995) and limited patient data provide a rationale for adjunctive IFN-γ for invasive aspergillosis. Recombinant IFN-γ (rIFN-γ) augmented the human neutrophil oxidative response and killing of *A. fumigatus* hyphae in vitro and acted additively with G-CSF (Roilides et al., 1993a). It prevented corticosteroid-mediated suppression of neutrophil killing of hyphae (Roilides et al., 1993b). rIFN-γ also enhanced killing of *A. fumigatus* hyphae by human monocytes (Roilides et al., 1994). In addition to augmenting the function of phagocytes, IFN-γ is the signature cytokine for type 1 cellular immunity. Augmentation of cellular immunity has been shown to be protective in experimental aspergillosis (Cenci et al., 2000).

IFN-γ is licensed as a prophylactic agent in patients with CGD, an inherited disorder of phagocyte NADPH oxidase characterized by recurrent life-threatening bacterial and fungal infections (Anonymous, 1991) and in osteopetrosis. In addition, IFN-γ in combination with antimycobacterial agents had a positive effect on patients with refractory nontuberculous mycobacterial infection and defective IFN-γ production (Holland et al., 1994). IFN-γ confers protection against a variety of experimental fungal infections in animals (Segal et al., 2006). Adjunctive IFN-γ showed promising results in a pilot study of AIDS-associated cryptococcal meningitis (Pappas et al., 2001). Dignani et al. (2005) reported successful outcomes using rIFN-γ paired with CSFs in four patients with leukemia and refractory fungal disease. One concern about rIFN-γ in allogeneic HSCT recipients is the potential for worsening GVHD. Although preliminary results suggest that rIFN-γ may be safe in allogeneic HSCT recipients (Fleming et al., 1997; Safdar et al., 2005), the safety of rIFN-γ cannot be definitively predicted based on this limited database. It was disappointing that a randomized trial evaluating rIFN-γ as adjunctive therapy for invasive aspergillosis was prematurely terminated before any patient was enrolled and before Institutional Review Board approval at most of the study sites.

We reserve rIFN-γ for patients with life-threatening invasive mold infections refractory to standard antifungal therapy. Such decisions are necessarily based on retrospective analyses and anecdotal data. Pairing rIFN-γ with G-CSF or GM-CSF is another reasonable option in the setting of refractory fungal disease, although we emphasize that the clinical experience is anecdotal and that the efficacy of this approach has not been established.

IL-17 and IDO

IL-17A-producing lymphocytes, known as neutrophil regulatory T cells or Tn cells, play an important role in neutrophilic inflammation and autoimmune diseases. About 60% of Tn cells are γδ T cells, 25% are NK T-like cells, and the remainder are CD4 T cells (Ley et al., 2006). These latter cells are also termed Th-17 cells, a distinct subset of T helper cells that play a role in autoimmune diseases. Th-17 cells are induced and expanded by transforming growth factor β, IL-6, and IL-23 through a STAT-3 and ROR-γT nuclear receptor-dependent pathway. IL-17 stimulates production of G-CSF, GM-CSF, TNF-α, and chemokines that mediate neutrophil recruitment.

Th17 and regulatory T cells (Tregs) mediate opposing pathways (Reiner, 2007). Augmentation of the Th17 pathway is associated with systemic inflammatory diseases, such as rheumatoid arthritis and inflammatory bowel disease, whereas the main function of Tregs is to induce immunologic tolerance to control excessive inflammation. IDO is the rate-limiting enzyme in tryptophan degradation along the kynurenine pathway. The importance of tryptophan metabolism in both antimicrobial resistance and protective tolerance is an area of substantial interest. IDO and the other kynurenine pathway enzymes represent not only effector host defense pathways but also means of generating Tregs with anti-inflammatory, tolerogenic activity (Belladonna et al., 2006). IDO creates a tolerogenic environment both by suppression of T cells and augmentation of local Treg-mediated immunosuppression (Mellor and Munn, 2004).

IDO plays an important role in Treg-mediated suppression of neutrophilic inflammation during the early immune response to experimental *Aspergillus* challenge and in attenuating subsequent allergic responses (Montagnoli et al., 2006). In contrast, the IL-23/IL-17 pathway augments neutrophilic inflammation and, paradoxically, disables antifungal host defense in murine candidiasis and aspergillosis (Zelante et al., 2007). We recently demonstrated that aspergillosis in CGD mice is characterized by augmented IL-17 and diminished IDO-mediated inflammatory responses and that host defense against aspergillosis could be ameliorated by neutralizing IL-17 and by administration of L-kynurenine (Romani et al., 2008). Taken together, manipulation of IL-17 and Treg-driven inflammatory responses are promising immunomodulatory strategies for both invasive and allergic aspergillosis.

VACCINATION

Vaccine development is a priority for opportunistic fungal pathogens that afflict patients with hematologic malignancies and allogeneic HSCT recipients (Stevens, 2004). In this section, we will only briefly discuss challenges related to vaccine development and discuss a few approaches illustrative of the promising research in the field. One impediment to vaccine development is that those patients who are most susceptible to opportunistic infections are least able to mount protective responses. Another impediment relates to the limited number of licensed vaccine adjuvants. Candidate adjuvants that act on multiple innate and antigen-specific host defense pathways are likely to be the most effective in protecting against opportunistic fungal infections. The definition of adjuvants has mostly been restricted to those that stimulate antibody titers (e.g., pneumococcus) or in the case of the bacillus Calmette-Guérin vaccine, delayed-type hypersensitivity responses. More recently, the concept of adjuvants has been expanded to include soluble mediators and antigenic carriers (e.g., endotoxin, Flt3L, heat shock proteins) that activate antigen-presenting cells and stimulate innate and cellular immunity (Gamvrellis et al., 2004).

Vaccine-based strategies have been effective in immunocompromised animal models. In mice, the importance of cell-mediated immunity against *Aspergillus* infection (an extracellular pathogen) has become well-established (Cenci et al., 1997, 1999). Immunization of immunocompetent mice with an *Aspergillus* crude filtrate resulted in memory responses mediated by antigen-specific, Th1-committed $CD4^+$ T cells (Cenci et al., 2000). Adoptive transfer of these cells conferred protection to neutropenic mice, establishing a proof of principle regarding cellular immunity as a target for immune augmentation in invasive aspergillosis (Cenci et al., 2000). This study also showed that the dichotomy in which the host defense against extracellular pathogens (such as *Aspergillus*) is humoral while the defense against intracellular pathogens is cellular is overly simplistic.

Torosantucci et al. (2005) developed a fungal vaccine consisting of laminarin, a poorly immunogenic β-glucan preparation (β-glucan is a cell wall constituent in fungi, plants, and algae), conjugated to diphtheria toxoid. The vaccine was protective in experimental candidiasis and aspergillosis. Protection was, at least in part, mediated by anti-β-glucan antibodies that could be adoptively transferred to naïve mice. Since β-glucan is a ubiquitous cell wall constituent in fungi, this vaccine may be protective against a broad spectrum of fungal pathogens.

Exploitation of *Aspergillus*-pulsed DCs for pathogen immunity in hematopoietic transplantation has recently been reported (Bozza et al., 2003; Romani et al., 2006). Because host DC function is impaired during the immediate period posttransplantation, the administration of donor DCs may be useful for the educational program of recovering T cells. We found specialization and complementarity in priming and tolerization by the different DC subsets, with $CD80^+$ plasmacytoid DCs fulfilling the requirement for (i) Th1/Treg antifungal priming, (ii) tolerization toward alloantigens, and (iii) diversion from alloantigen-specific to antigen-specific T-cell responses in the presence of donor T lymphocytes. Thus, transplantation tolerance and concomitant pathogen clearance could be achieved through the therapeutic induction of antigen-specific Treg cells via instructive immunotherapy with pathogen-conditioned donor DCs (Montagnoli et al., 2008).

IMMUNOLOGIC EFFECTS OF ANTIFUNGALS

Antifungal agents have immunomodulatory effects that may be clinically relevant and exploitable for immunotherapeutic strategies. One example of the immunologic effect of antifungals is the infusional toxicity of amphotericin B that likely results from release of proinflammatory cytokines from monocytes (Arning et al., 1995; Rogers et al., 1999). Other antifungal agents may have more indirect effects on host cell responses. Amphotericin B deoxycholate and liposomal amphotericin B have distinct effects on TLR signaling and antifungal activity of murine neutrophils (Bellocchio et al., 2005). Empty liposomes attenuate the immunopathology in experimental aspergillosis (Lewis et al., 2007). Echinocandins cause structural changes in the fungal cell wall, including blebbing and cell wall rupture (Dennis et al., 2006; Petraitis et al., 1998). Fungal cell wall constituents are recognized by specific host cell PRRs and elicit immunologic responses (Brown, 2006; Gantner et al., 2003; Gersuk et al., 2006; Graham et al., 2006; Hohl et al., 2005; Netea et al., 2006). Both fungal burden and inflammatory responses should be included as pharmacodynamic end points in the preclinical evaluation of antifungal drugs, and immunologic correlates should be included as secondary end points in future clinical trials.

CONCLUSIONS

A few years ago, Stevens (2004) wrote a thoughtful paper entitled, "Vaccinate against aspergillosis! A call to arms of the immune system." As we have discussed, there are many promising strategies for immune augmentation against aspergillosis, vaccination being one approach. The greatest challenge in paving the way from promising results in experimental models to the clinic may be market forces (Segal and Walsh, 2006). Aspergillosis is an uncommon disease that predominantly affects the highly immunocompromised. The heterogeneity of patient populations at risk for invasive aspergillosis is another challenge in developing broadly effective immunotherapeutics. Therefore, the expense involved in developing the clinical potential of an immune-based strategy for invasive aspergillosis may be considered too high by industry when weighed against the modest expected financial gain, even if the approach were highly effective. A cooperative relationship between academia, industry, and government (with financial incentives for antifungal immunotherapeutic development) will likely be required to bring promising immunotherapies for invasive aspergillosis to fruition.

REFERENCES

Anonymous. 1991. A controlled trial of interferon gamma to prevent infection in chronic granulomatous disease. The International Chronic Granulomatous Disease Cooperative Study Group. *N. Engl. J. Med.* **324:**509–516.

Arning, M., K. O. Kliche, A. H. Heer-Sonderhoff, and A. Wehmeier. 1995. Infusion-related toxicity of three different amphotericin B formulations and its relation to cytokine plasma levels. *Mycoses* **38:** 459–465.

Baddley, J. W., T. P. Stroud, D. Salzman, and P. G. Pappas. 2001. Invasive mold infections in allogeneic bone marrow transplant recipients. *Clin. Infect. Dis.* **32:**1319–1324.

Balloy, V., M. Huerre, J. P. Latgé, and M. Chignard. 2005. Differences in patterns of infection and inflammation for corticosteroid treatment and chemotherapy in experimental invasive pulmonary aspergillosis. *Infect. Immun.* **73:**494–503.

Belladonna, M. L., U. Grohmann, P. Guidetti, C. Volpi, R. Bianchi, M. C. Fioretti, R. Schwarcz, F. Fallarino, and P. Puccetti. 2006. Kynurenine pathway enzymes in dendritic cells initiate tolerogenesis in the absence of functional IDO. *J. Immunol.* **177:**130–137.

Bellocchio, S., R. Gaziano, S. Bozza, G. Rossi, C. Montagnoli, K. Perruccio, M. Calvitti, L. Pitzurra, and L. Romani. 2005. Liposomal amphotericin B activates antifungal resistance with reduced toxicity by diverting Toll-like receptor signalling from TLR-2 to TLR-4. *J. Antimicrob. Chemother.* **55:**214–222.

Bellocchio, S., S. Moretti, K. Perruccio, F. Fallarino, S. Bozza, C. Montagnoli, P. Mosci, G. B. Lipford, L. Pitzurra, and L. Romani. 2004. TLRs govern neutrophil activity in aspergillosis. *J. Immunol.* **173:**7406–7415.

Bensinger, W. I., T. H. Price, D. C. Dale, F. R. Appelbaum, R. Clift, K. Lilleby, B. Williams, R. Storb, E. D. Thomas, and C. D. Buckner. 1993. The effects of daily recombinant human granulocyte colony-stimulating factor administration on normal granulocyte donors undergoing leukapheresis. *Blood* **81:**1883–1888.

Berenguer, J., M. C. Allende, J. W. Lee, K. Garrett, C. Lyman, N. M. Ali, J. Bacher, P. A. Pizzo, and T. J. Walsh. 1995. Pathogenesis of pulmonary aspergillosis. Granulocytopenia versus cyclosporine and methylprednisolone-induced immunosuppression. *Am. J. Respir. Crit. Care Med.* **152:**1079–1086.

Bhatia, S., J. McCullough, E. H. Perry, M. Clay, N. K. Ramsay, and J. P. Neglia. 1994. Granulocyte transfusions: efficacy in treating fungal infections in neutropenic patients following bone marrow transplantation. *Transfusion* **34:**226–232.

Bignell, E., S. Negrete-Urtasun, A. M. Calcagno, H. N. Arst, Jr., T. Rogers, and K. Haynes. 2005. Virulence comparisons of *Aspergillus nidulans* mutants are confounded by the inflammatory response of p47$^{phox-/-}$ mice. *Infect. Immun.* **73:**5204–5207.

BitMansour, A., S. M. Burns, D. Traver, K. Akashi, C. H. Contag, I. L. Weissman, and J. M. Brown. 2002. Myeloid progenitors protect against invasive aspergillosis and *Pseudomonas aeruginosa* infection following hematopoietic stem cell transplantation. *Blood* **100:**4660–4667.

Bozza, S., R. Gaziano, G. B. Lipford, C. Montagnoli, A. Bacci, P. Di Francesco, V. P. Kurup, H. Wagner, and L. Romani. 2002. Vaccination of mice against invasive aspergillosis with recombinant *Aspergillus* proteins and CpG oligodeoxynucleotides as adjuvants. *Microbes Infect.* **4:**1281–1290.

Bozza, S., K. Perruccio, C. Montagnoli, R. Gaziano, S. Bellocchio, E. Burchielli, G. Nkwanyuo, L. Pitzurra, A. Velardi, and L. Romani. 2003. A dendritic cell vaccine against invasive aspergillosis in allogeneic hematopoietic transplantation. *Blood* **5:**5.

Brown, G. D. 2006. Dectin-1: a signalling non-TLR pattern-recognition receptor. *Nat. Rev. Immunol.* **6:**33–43.

Brown, G. D., and S. Gordon. 2001. Immune recognition. A new receptor for beta-glucans. *Nature* **413:**36–37.

Brown, G. D., J. Herre, D. L. Williams, J. A. Willment, A. S. Marshall, and S. Gordon. 2003. Dectin-1 mediates the biological effects of beta-glucans. *J. Exp. Med.* **197:**1119–1124.

Brown, G. D., P. R. Taylor, D. M. Reid, J. A. Willment, D. L. Williams, L. Martinez-Pomares, S. Y. Wong, and S. Gordon. 2002. Dectin-1 is a major beta-glucan receptor on macrophages. *J. Exp. Med.* **196:**407–412.

Cenci, E., A. Mencacci, A. Bacci, F. Bistoni, V. P. Kurup, and L. Romani. 2000. T cell vaccination in mice with invasive pulmonary aspergillosis. *J. Immunol.* **165:**381–388.

Cenci, E., A. Mencacci, G. Del Sero, A. Bacci, C. Montagnoli, C. F. d'Ostiani, P. Mosci, M. Bachmann, F. Bistoni, M. Kopf, and L. Romani. 1999. Interleukin-4 causes susceptibility to invasive pulmonary aspergillosis through suppression of protective type I responses. *J. Infect. Dis.* **180:**1957–1968.

Cenci, E., S. Perito, K. H. Enssle, P. Mosci, J. P. Latgé, L. Romani, and F. Bistoni. 1997. Th1 and Th2 cytokines in mice with invasive aspergillosis. *Infect. Immun.* **65:**564–570.

Chamilos, G., M. Luna, R. E. Lewis, G. P. Bodey, R. Chemaly, J. J. Tarrand, A. Safdar, I. I. Raad, and D. P. Kontoyiannis. 2006. Invasive fungal infections in patients with hematologic malignancies in a tertiary care cancer center: an autopsy study over a 15-year period (1989–2003). *Haematologica* **91:**986–989.

Chang, Y. C., B. H. Segal, S. M. Holland, G. F. Miller, and K. J. Kwon-Chung. 1998. Virulence of catalase-deficient aspergillus nidulans in $p47^{phox-/-}$ mice. Implications for fungal pathogenicity and host defense in chronic granulomatous disease. *J. Clin. Investig.* **101:**1843–1850.

Cohen, M. S., R. E. Isturiz, H. L. Malech, R. K. Root, C. M. Wilfert, L. Gutman, and R. H. Buckley. 1981. Fungal infection in chronic granulomatous disease. The importance of the phagocyte in defense against fungi. *Am. J. Med.* **71:**59–66.

Cornely, O. A., J. Maertens, D. J. Winston, J. Perfect, A. J. Ullmann, T. J. Walsh, D. Helfgott, J. Holowiecki, D. Stockelberg, Y. T. Goh, M. Petrini, C. Hardalo, R. Suresh, and D. Angulo-Gonzalez. 2007. Posaconazole vs. fluconazole or itraconazole prophylaxis in patients with neutropenia. *N. Engl. J. Med.* **356:**348–359.

Dale, D. C., W. C. Liles, C. Llewellyn, E. Rodger, and T. H. Price. 1998. Neutrophil transfusions: kinetics and functions of neutrophils mobilized with granulocyte-colony-stimulating factor and dexamethasone. *Transfusion* **38:**713–721.

Dekker, A., S. Bulley, J. Beyene, L. L. Dupuis, J. J. Doyle, and L. Sung. 2006. Meta-analysis of randomized controlled trials of prophylactic granulocyte colony-stimulating factor and granulocyte-macrophage colony-stimulating factor after autologous and allogeneic stem cell transplantation. *J. Clin. Oncol.* **24:**5207–5215.

Denning, D. W. 2003, posting date. Introduction: the *Aspergillus fumigatus* Genome Database. The Institute for Genomic Research, Rockville, MD. www.tigr.org/tdb/e2k1/afu1/intro.shtml.

Denning, D. W., A. Marinus, J. Cohen, D. Spence, R. Herbrecht, L. Pagano, C. Kibbler, V. Kcrmery, F. Offner, C. Cordonnier, U. Jehn, M. Ellis, L. Collette, R. Sylvester, et al. 1998. An EORTC multicentre prospective survey of invasive aspergillosis in haematological patients: diagnosis and therapeutic outcome. *J. Infect.* **37:**173–180.

Dennis, C. G., W. R. Greco, Y. Brun, R. Youn, H. K. Slocum, R. J. Bernacki, R. Lewis, N. Wiederhold, S. M. Holland, R. Petraitiene, T. J. Walsh, and B. H. Segal. 2006. Effect of amphotericin B and micafungin combination on survival, histopathology, and fungal burden in experimental aspergillosis in the $p47^{phox-/-}$ mouse model of chronic granulomatous disease. *Antimicrob. Agents Chemother.* **50:**422–427.

Dignani, M. C., E. J. Anaissie, J. P. Hester, S. O'Brien, S. E. Vartivarian, J. H. Rex, H. Kantarjian, D. B. Jendiroba, B. Lichtiger, B. S. Andersson, and E. J. Freireich. 1997. Treatment of neutropenia-related fungal infections with granulocyte colony-stimulating factor-elicited white blood cell transfusions: a pilot study. *Leukemia* **11:**1621–1630.

Dignani, M. C., J. H. Rex, K. W. Chan, G. Dow, M. deMagalhaes-Silverman, A. Maddox, T. Walsh, and E. Anaissie. 2005. Immunomodulation with interferon-gamma and colony stimulating factors for refractory fungal infections in patients with leukemia. *Cancer* **104:**199–204.

Fleming, R. V., E. J. Anaissie, H. M. Kantarjian, C. Savary, and J. H. Rex. 1997. Interferon-gamma (IFN-γ) plus granulocyte-colony stimulating factor (G-CSF) in the treatment of refractory fungal infections: a pilot study, abstr. G-31. *37th Intersci. Conf. Antimicrob. Agents Chemother.*, Toronto, Canada.

Gallin, J. I., D. W. Alling, H. L. Malech, R. Wesley, D. Koziol, B. Marciano, E. M. Eisenstein, M. L. Turner, E. S. DeCarlo, J. M. Starling, and S. M. Holland. 2003. Itraconazole prophylaxis for fungal infections in chronic granulomatous disease of childhood. *N. Engl. J. Med.* **348:**2416–2422.

Gamvrellis, A., D. Leong, J. C. Hanley, S. D. Xiang, P. Mottram, and M. Plebanski. 2004. Vaccines that facilitate antigen entry into dendritic cells. *Immunol. Cell Biol.* **82:**506–516.

Gantner, B. N., R. M. Simmons, S. J. Canavera, S. Akira, and D. M. Underhill. 2003. Collaborative induction of inflammatory responses by dectin-1 and Toll-like receptor 2. *J. Exp. Med.* **197:**1107–1117.

Garlanda, C., E. Hirsch, S. Bozza, A. Salustri, M. De Acetis, R. Nota, A. Maccagno, F. Riva, B. Bottazzi, G. Peri, A. Doni, L. Vago, M. Botto, R. De Santis, P. Carminati, G. Siracusa, F. Altruda, A. Vecchi, L. Romani, and A. Mantovani. 2002. Non-redundant role of the long pentraxin PTX3 in anti-fungal innate immune response. *Nature* **420:**182–186.

Gaziano, R., S. Bozza, S. Bellocchio, K. Perruccio, C. Montagnoli, L. Pitzurra, G. Salvatori, R. De Santis, P. Carminati, A. Mantovani, and L. Romani. 2004. Anti-*Aspergillus fumigatus* efficacy of pentraxin 3 alone and in combination with antifungals. *Antimicrob. Agents Chemother.* **48:**4414–4421.

Gerson, S. L., G. H. Talbot, S. Hurwitz, B. L. Strom, E. J. Lusk, and P. A. Cassileth. 1984. Prolonged granulocytopenia: the major risk factor for invasive pulmonary aspergillosis in patients with acute leukemia. *Ann. Intern. Med.* **100:**345–351.

Gersuk, G. M., D. M. Underhill, L. Zhu, and K. A. Marr. 2006. Dectin-1 and TLRs permit macrophages to distinguish between different *Aspergillus fumigatus* cellular states. *J. Immunol.* **176:**3717–3724.

Graham, L. M., S. V. Tsoni, J. A. Willment, D. L. Williams, P. R. Taylor, S. Gordon, K. Dennehy, and G. D. Brown. 2006. Soluble dectin-1 as a tool to detect beta-glucans. *J. Immunol. Methods* **314:**164–169.

Grow, W. B., J. S. Moreb, D. Roque, K. Manion, H. Leather, V. Reddy, S. A. Khan, K. J. Finiewicz, H. Nguyen, C. J. Clancy, P. S. Mehta, and J. R. Wingard. 2002. Late onset of invasive aspergillus infection in bone marrow transplant patients at a university hospital. *Bone Marrow Transplant.* **29:**15–19.

Herbrecht, R., D. W. Denning, T. F. Patterson, J. E. Bennett, R. E. Greene, J. W. Oestmann, W. V. Kern, K. A. Marr, P. Ribaud, O. Lortholary, R. Sylvester, R. H. Rubin, J. R. Wingard, P. Stark, C. Durand, D. Caillot, E. Thiel, P. H. Chandrasekar, M. R. Hodges, H. T. Schlamm, P. F. Troke, and B. de Pauw. 2002. Voriconazole versus amphotericin B for primary therapy of invasive aspergillosis. *N. Engl. J. Med.* **347:**408–415.

Hogaboam, C. M., K. Takahashi, R. A. Ezekowitz, S. L. Kunkel, and J. M. Schuh. 2004. Mannose-binding lectin deficiency alters the development of fungal asthma: effects on airway response, inflammation, and cytokine profile. *J. Leukoc. Biol.* **75:**805–814.

Hohl, T. M., H. L. Van Epps, A. Rivera, L. A. Morgan, P. L. Chen, M. Feldmesser, and E. G. Pamer. 2005. *Aspergillus fumigatus* triggers inflammatory responses by stage-specific beta-glucan display. *PLoS Pathog.* **1:**e30.

Holland, S. M., E. M. Eisenstein, D. B. Kuhns, M. L. Turner, T. A. Fleisher, W. Strober, and J. I. Gallin. 1994. Treatment of refractory disseminated nontuberculous mycobacterial infection with interferon gamma. A preliminary report. *N. Engl. J. Med.* **330:**1348–1355.

Jantunen, E., P. Ruutu, L. Niskanen, L. Volin, T. Parkkali, P. Koukila-Kahkola, and T. Ruutu. 1997. Incidence and risk factors for invasive fungal infections in allogeneic BMT recipients. *Bone Marrow Transplant.* **19:**801–808.

Lemaitre, B., E. Nicolas, L. Michaut, J. M. Reichhart, and J. A. Hoffmann. 1996. The dorsoventral regulatory gene cassette spatzle/Toll/cactus controls the potent antifungal response in *Drosophila* adults. *Cell* **86:**973–983.

Lewis, R. E., G. Chamilos, R. A. Prince, and D. P. Kontoyiannis. 2007. Pretreatment with empty liposomes attenuates the immunopathology of invasive pulmonary aspergillosis in corticosteroid-immunosuppressed mice. *Antimicrob. Agents Chemother.* **51:**1078–1081.

Ley, K., E. Smith, and M. A. Stark. 2006. IL-17A-producing neutrophil-regulatory Tn lymphocytes. *Immunol. Res.* **34:**229–242.

Marr, K. A., R. A. Carter, M. Boeckh, P. Martin, and L. Corey. 2002. Invasive aspergillosis in allogeneic stem cell transplant recipients: changes in epidemiology and risk factors. *Blood* **100:**4358–4366.

Martino, R., M. Subira, M. Rovira, C. Solano, L. Vazquez, G. F. Sanz, A. Urbano-Ispizua, S. Brunet, and R. De la Camara. 2002. Invasive fungal infections after allogeneic peripheral blood stem cell transplantation: incidence and risk factors in 395 patients. *Br. J. Haematol.* **116:**475–482.

McNeil, M. M., S. L. Nash, R. A. Hajjeh, M. A. Phelan, L. A. Conn, B. D. Plikaytis, and D. W. Warnock. 2001. Trends in mortality due to invasive mycotic diseases in the United States, 1980–1997. *Clin. Infect. Dis.* **33:**641–647.

McWhinney, P. H., C. C. Kibbler, M. D. Hamon, O. P. Smith, L. Gandhi, L. A. Berger, R. K. Walesby, A. V. Hoffbrand, and H. G. Prentice. 1993. Progress in the diagnosis and management of aspergillosis in bone marrow transplantation: 13 years' experience. *Clin. Infect. Dis.* **17:**397–404.

Mellor, A. L., and D. H. Munn. 2004. IDO expression by dendritic cells: tolerance and tryptophan catabolism. *Nat. Rev. Immunol.* **4:** 762–774.

Montagnoli, C., F. Fallarino, R. Gaziano, S. Bozza, S. Bellocchio, T. Zelante, W. P. Kurup, L. Pitzurra, P. Puccetti, and L. Romani. 2006. Immunity and tolerance to *Aspergillus* involve functionally distinct regulatory T cells and tryptophan catabolism. *J. Immunol.* **176:**1712–1723.

Montagnoli, C., K. Perruccio, S. Bozza, P. Bonifazi, T. Zelante, A. De Luca, S. Moretti, C. D'Angelo, F. Bistoni, M. Martelli, F. Aversa, A. Velardi, and L. Romani. 2008. Provision of antifungal immunity and concomitant alloantigen tolerization by conditioned dendritic cells in experimental hematopoietic transplantation. *Blood Cells Mol. Dis.* **40:**55–62.

Mouy, R., F. Veber, S. Blanche, J. Donadieu, R. Brauner, J. C. Levron, C. Griscelli, and A. Fischer. 1994. Long-term itraconazole prophylaxis against *Aspergillus* infections in thirty-two patients with chronic granulomatous disease. *J. Pediatr.* **125:**998–1003.

Mukhopadhyay, S., J. Herre, G. D. Brown, and S. Gordon. 2004. The potential for Toll-like receptors to collaborate with other innate immune receptors. *Immunology* **112:**521–530.

Nagai, H., J. Guo, H. Choi, and V. Kurup. 1995. Interferon-gamma and tumor necrosis factor-alpha protect mice from invasive aspergillosis. *J. Infect. Dis.* **172:**1554–1560.

Netea, M. G., N. A. Gow, C. A. Munro, S. Bates, C. Collins, G. Ferwerda, R. P. Hobson, G. Bertram, H. B. Hughes, T. Jansen, L. Jacobs, E. T. Buurman, K. Gijzen, D. L. Williams, R. Torensma, A. McKinnon, D. M. Maccallum, F. C. Odds, J. W. Van der Meer, A. J. Brown, and B. J. Kullberg. 2006. Immune sensing of *Candida albicans* requires cooperative recognition of mannans and glucans by lectin and Toll-like receptors. *J. Clin. Investig.* **116:**1642–1650.

Netea, M. G., C. A. Van Der Graaf, A. G. Vonk, I. Verschueren, J. W. Van Der Meer, and B. J. Kullberg. 2002. The role of toll-like receptor (TLR) 2 and TLR4 in the host defense against disseminated candidiasis. *J. Infect. Dis.* **185:**1483–1489.

Nichols, W. G., T. Price, and M. Boeckh. 2002. Cytomegalovirus infections in cancer patients receiving granulocyte transfusions. *Blood* **99:**3483–3484.

Ozinsky, A., D. M. Underhill, J. D. Fontenot, A. M. Hajjar, K. D. Smith, C. B. Wilson, L. Schroeder, and A. Aderem. 2000. The repertoire for pattern recognition of pathogens by the innate immune system is defined by cooperation between toll-like receptors. *Proc. Natl. Acad. Sci. USA* **97:**13766–13771.

Pappas, P. G., B. Bustamante, E. Ticona, R. Hamill, P. Johnson, A. Reboli, J. Aberg, R. Hasbun, and H. H. Hsu. 2004. Recombinant interferon-gamma 1b as adjunctive therapy for AIDS-related acute cryptococcal meningitis. *J. Infect. Dis.* **189:**2185–2191.

Peters, C., M. Minkov, S. Matthes-Martin, U. Potschger, V. Witt, G. Mann, P. Hocker, N. Worel, J. Stary, T. Klingebiel, and H. Gadner. 1999. Leucocyte transfusions from rhG-CSF or prednisolone stimulated donors for treatment of severe infections in immunocompromised neutropenic patients. *Br. J. Haematol.* **106:**689–696.

Petraitis, V., R. Petraitiene, A. H. Groll, A. Bell, D. P. Callender, T. Sein, R. L. Schaufele, C. L. McMillian, J. Bacher, and T. J. Walsh. 1998. Antifungal efficacy, safety, and single-dose pharmacokinetics of LY303366, a novel echinocandin B, in experimental pulmonary aspergillosis in persistently neutropenic rabbits. *Antimicrob. Agents Chemother.* **42:**2898–2905.

Price, T. H., R. A. Bowden, M. Boeckh, J. Bux, K. Nelson, W. C. Liles, and D. C. Dale. 2000. Phase I/II trial of neutrophil transfusions from donors stimulated with G-CSF and dexamethasone for treatment of patients with infections in hematopoeitic stem cell transplantation. *Blood* **95:**3302–3309.

Reeves, E. P., H. Lu, H. L. Jacobs, C. G. Messina, S. Bolsover, G. Gabella, E. O. Potma, A. Warley, J. Roes, and A. W. Segal. 2002. Killing activity of neutrophils is mediated through activation of proteases by K^+ flux. *Nature* **416:**291–297.

Reiner, S. L. 2007. Development in motion: helper T cells at work. *Cell* **129:**33–36.

Rogers, P. D., R. E. Kramer, S. W. Chapman, and J. D. Cleary. 1999. Amphotericin B-induced interleukin-1β expression in human monocytic cells is calcium and calmodulin dependent. *J. Infect. Dis.* **180:** 1259–1266.

Roilides, E., A. Holmes, C. Blake, D. Venzon, P. A. Pizzo, and T. J. Walsh. 1994. Antifungal activity of elutriated human monocytes against *Aspergillus fumigatus* hyphae: enhancement by granulocyte-macrophage colony-stimulating factor and interferon-gamma. *J. Infect. Dis.* **170:**894–899.

Roilides, E., K. Uhlig, D. Venzon, P. A. Pizzo, and T. J. Walsh. 1993a. Enhancement of oxidative response and damage caused by human neutrophils to *Aspergillus fumigatus* hyphae by granulocyte colony-stimulating factor and gamma interferon. *Infect. Immun.* **61:**1185–1193.

Roilides, E., K. Uhlig, D. Venzon, P. A. Pizzo, and T. J. Walsh. 1993b. Prevention of corticosteroid-induced suppression of human polymorphonuclear leukocyte-induced damage of *Aspergillus fumigatus* hyphae by granulocyte colony-stimulating factor and gamma interferon. *Infect. Immun.* **61:**4870–4877.

Romani, L., F. Bistoni, R. Gaziano, S. Bozza, C. Montagnoli, K. Perruccio, L. Pitzurra, S. Bellocchio, A. Velardi, G. Rasi, P. Di Francesco, and E. Garaci. 2004. Thymosin alpha 1 activates dendritic cells for antifungal Th1 resistance through toll-like receptor signaling. *Blood* **103:**4232–4239.

Romani, L., F. Bistoni, K. Perruccio, C. Montagnoli, R. Gaziano, S. Bozza, P. Bonifazi, G. Bistoni, G. Rasi, A. Velardi, F. Fallarino, E. Garaci, and P. Puccetti. 2006. Thymosin α1 activates dendritic cell tryptophan catabolism and establishes a regulatory environment for balance of inflammation and tolerance. *Blood* **108:**2265–2274.

Romani, L., F. Fallarino, A. De Luca, C. Montagnoli, C. D'Angelo, T. Zelante, C. Vacca, F. Bistoni, M. C. Fioretti, U. Grohmann, B. H. Segal, and P. Puccetti. 2008. Defective tryptophan catabolism underlies inflammation in mouse chronic granulomatous disease. *Nature* **451:**211–215.

Romani, L., and P. Puccetti. 2007. Controlling pathogenic inflammation to fungi. *Expert Rev. Anti Infect. Ther.* **5:**1007–1017.

Rowe, J. M. 1998. Treatment of acute myeloid leukemia with cytokines: effect on duration of neutropenia and response to infections. *Clin. Infect. Dis.* **26:**1290–1294.

Rowe, J. M., J. W. Andersen, J. J. Mazza, J. M. Bennett, E. Paietta, F. A. Hayes, D. Oette, P. A. Cassileth, E. A. Stadtmauer, and P. H. Wiernik. 1995. A randomized placebo-controlled phase III study of granulocyte-macrophage colony-stimulating factor in adult patients (>55 to 70 years of age) with acute myelogenous leukemia: a study of the Eastern Cooperative Oncology Group (E1490). *Blood* **86:** 457–462.

Safdar, A., G. Rodriguez, N. Ohmagari, D. P. Kontoyiannis, K. V. Rolston, I. I. Raad, and R. E. Champlin. 2005. The safety of interferon-gamma-1b for invasive fungal infections after hematopoietic stem cell transplantation. *Cancer* **103:**731–739.

Segal, B. H., E. S. DeCarlo, K. J. Kwon-Chung, H. L. Malech, J. I. Gallin, and S. M. Holland. 1998. *Aspergillus nidulans* infection in chronic granulomatous disease. *Medicine* (Baltimore) **77:**345–354.

Segal, B. H., and S. M. Holland. 2003. Invasive aspergillosis in chronic granulomatous disease. The *Aspergillus* website. http://www.aspergillus.man.ac.uk.

Segal, B. H., J. Kwon-Chung, T. J. Walsh, B. S. Klein, M. Battiwalla, N. G. Almyroudis, S. M. Holland, and L. Romani. 2006. Immunotherapy for fungal infections. *Clin. Infect. Dis.* **42:**507–515.

Segal, B. H., T. L. Leto, J. I. Gallin, H. L. Malech, and S. M. Holland. 2000. Genetic, biochemical, and clinical features of chronic granulomatous disease. *Medicine* (Baltimore) **79:**170–200.

Segal, B. H., and T. J. Walsh. 2006. Current approaches to diagnosis and treatment of invasive aspergillosis. *Am. J. Respir. Crit. Care Med.* **173:**707–717.

Shaukat, A., F. Bakri, P. Young, T. Hahn, D. Ball, M. R. Baer, M. Wetzler, J. L. Slack, P. Loud, M. Czuczman, P. L. McCarthy, T. J. Walsh, and B. H. Segal. 2005. Invasive filamentous fungal infections in allogeneic hematopoietic stem cell transplant recipients after recovery from neutropenia: clinical, radiologic, and pathologic characteristics. *Mycopathologia* **159:**181–188.

Shoham, S., C. Huang, J. M. Chen, D. T. Golenbock, and S. M. Levitz. 2001. Toll-like receptor 4 mediates intracellular signaling without TNF-alpha release in response to *Cryptococcus neoformans* polysaccharide capsule. *J. Immunol.* **166:**4620–4626.

Smith, T. J., J. Khatcheressian, G. H. Lyman, H. Ozer, J. O. Armitage, L. Balducci, C. L. Bennett, S. B. Cantor, J. Crawford, S. J. Cross, G. Demetri, C. E. Desch, P. A. Pizzo, C. A. Schiffer, L. Schwartzberg, M. R. Somerfield, G. Somlo, J. C. Wade, J. L. Wade, R. J. Winn, A. J. Wozniak, and A. C. Wolff. 2006. 2006 update of recommendations for the use of white blood cell growth factors: an evidence-based clinical practice guideline. *J. Clin. Oncol.* **24:**3187–3205.

Spellberg, B. J., M. Collins, V. Avanesian, M. Gomez, J. E. Edwards, Jr., C. Cogle, D. Applebaum, Y. Fu, and A. S. Ibrahim. 2007. Optimization of a myeloid cell transfusion strategy for infected neutropenic hosts. *J. Leukoc. Biol.* **81:**632–641.

Spellberg, B. J., M. Collins, S. W. French, J. E. Edwards, Jr., Y. Fu, and A. S. Ibrahim. 2005. A phagocytic cell line markedly improves survival of infected neutropenic mice. *J. Leukoc. Biol.* **78:**338–344.

Steele, C., L. Marrero, S. Swain, A. G. Harmsen, M. Zheng, G. D. Brown, S. Gordon, J. E. Shellito, and J. K. Kolls. 2003. Alveolar macrophage-mediated killing of *Pneumocystis carinii* f. sp. *muris* involves molecular recognition by the dectin-1 beta-glucan receptor. *J. Exp. Med.* **198:**1677–1688.

Steele, C., R. R. Rapaka, A. Metz, S. M. Pop, D. L. Williams, S. Gordon, J. K. Kolls, and G. D. Brown. 2005. The beta-glucan receptor dectin-1 recognizes specific morphologies of *Aspergillus fumigatus. PLoS Pathog.* **1:**e42.

Stephens-Romero, S. D., A. J. Mednick, and M. Feldmesser. 2005. The pathogenesis of fatal outcome in murine pulmonary aspergillosis depends on the neutrophil depletion strategy. *Infect. Immun.* **73:**114–125.

Stergiopoulou, T., E. Roilides, M. Rowden, B. H. Segal, and T. J. Walsh. 2007. Host-dependent patterns of tissue injury in invasive pulmonary aspergillosis. *Am. J. Clin. Pathol.* **127:**349–355.

Stevens, D. A. 2004. Vaccinate against aspergillosis! A call to arms of the immune system. *Clin. Infect. Dis.* **38:**1131–1136.

Tauszig-Delamasure, S., H. Bilak, M. Capovilla, J. A. Hoffmann, and J. L. Imler. 2002. *Drosophila* MyD88 is required for the response to fungal and gram-positive bacterial infections. *Nat. Immunol.* **3:** 91–97.

Tkalcevic, J., M. Novelli, M. Phylactides, J. P. Iredale, A. W. Segal, and J. Roes. 2000. Impaired immunity and enhanced resistance to endotoxin in the absence of neutrophil elastase and cathepsin G. *Immunity* **12:**201–210.

Torosantucci, A., C. Bromuro, P. Chiani, F. De Bernardis, F. Berti, C. Galli, F. Norelli, C. Bellucci, L. Polonelli, P. Costantino, R. Rappuoli, and A. Cassone. 2005. A novel glyco-conjugate vaccine against fungal pathogens. *J. Exp. Med.* **202:**597–606.

Ullmann, A. J., J. H. Lipton, D. H. Vesole, P. Chandrasekar, A. Langston, S. R. Tarantolo, H. Greinix, W. Morais de Azevedo, V. Reddy, N. Boparai, L. Pedicone, H. Patino, and S. Durrant. 2007. Posaconazole or fluconazole for prophylaxis in severe graft-versus-host disease. *N. Engl. J. Med.* **356:**335–347.

Underhill, D. M., A. Ozinsky, A. M. Hajjar, A. Stevens, C. B. Wilson, M. Bassetti, and A. Aderem. 1999. The Toll-like receptor 2 is recruited to macrophage phagosomes and discriminates between pathogens. *Nature* **401:**811–815.

van Burik, J. A., V. Ratanatharathorn, D. E. Stepan, C. B. Miller, J. H. Lipton, D. H. Vesole, N. Bunin, D. A. Wall, J. W. Hiemenz, Y. Satoi, J. M. Lee, and T. J. Walsh. 2004. Micafungin versus fluconazole for prophylaxis against invasive fungal infections during neutropenia in patients undergoing hematopoietic stem cell transplantation. *Clin. Infect. Dis.* **39:**1407–1416.

Viriyakosol, S., J. Fierer, G. D. Brown, and T. N. Kirkland. 2005. Innate immunity to the pathogenic fungus *Coccidioides posadasii* is dependent on Toll-like receptor 2 and dectin-1. *Infect. Immun.* **73:** 1553–1560.

Wald, A., W. Leisenring, J. A. van Burik, and R. A. Bowden. 1997. Epidemiology of *Aspergillus* infections in a large cohort of patients undergoing bone marrow transplantation. *J. Infect. Dis.* **175:**1459–1466.

Weinberger, M., I. Elattar, D. Marshall, S. M. Steinberg, R. L. Redner, N. S. Young, and P. A. Pizzo. 1992. Patterns of infection in patients with aplastic anemia and the emergence of *Aspergillus* as a major cause of death. *Medicine* (Baltimore) **71:**24–43.

Winkelstein, J. A., M. C. Marino, R. B. Johnston, Jr., J. Boyle, J. Curnutte, J. I. Gallin, H. L. Malech, S. M. Holland, H. Ochs, P. Quie, R. H. Buckley, C. B. Foster, S. J. Chanock, and H. Dickler. 2000. Chronic granulomatous disease: report on a national registry of 368 patients. *Medicine* (Baltimore) **79:**155–169.

Yuen, K. Y., P. C. Woo, M. S. Ip, R. H. Liang, E. K. Chiu, H. Siau, P. L. Ho, F. F. Chen, and T. K. Chan. 1997. Stage-specific manifestation of mold infections in bone marrow transplant recipients: risk factors and clinical significance of positive concentrated smears. *Clin. Infect. Dis.* **25:**37–42.

Zelante, T., A. De Luca, P. Bonifazi, C. Montagnoli, S. Bozza, S. Moretti, M. L. Belladonna, C. Vacca, C. Conte, P. Mosci, F. Bistoni, P. Puccetti, R. A. Kastelein, M. Kopf, and L. Romani. 2007. IL-23 and the Th17 pathway promote inflammation and impair antifungal immune resistance. *Eur. J. Immunol.* **37:**2695–2706.

VII. TIMING OF ANTIFUNGAL THERAPY

Aspergillus fumigatus and Aspergillosis
Edited by J.-P. Latgé and W. J. Steinbach
©2009 ASM Press, Washington, DC

Chapter 36

Prophylaxis for Aspergillosis

Jo-Anne H. Young

Prophylaxis of aspergillosis involves manipulation of the environment to create minimal exposure to both airborne and food-borne spores, as well as pharmacologic prophylaxis during periods of greatest risk. Prophylaxis against *Aspergillus* infections is important for several patient cohorts, primarily hematopoietic stem cell transplant (HSCT) recipients and burn victims, and to a lesser extent leukemia patients undergoing chemotherapy, lung transplant recipients, and liver transplant recipients. These patients require prophylaxis against *Aspergillus* infection during specific time periods. An understanding of the natural history of this infection within each of these at-risk cohorts helps define the periods of greatest risk (see chapter 38 for solid organ transplant recipients and chapter 39 for guidelines regarding those entering transplant programs with prolonged neutropenia, e.g., allogeneic recipients or patients with aplastic anemia, Fanconi anemia, or myelodysplastic syndrome, or heavy prior chemotherapy treatments). In nonoutbreak settings, the incidence of invasive aspergillosis should not be higher than 15%, even among HSCT recipients. This chapter emphasizes prophylaxis rather than preemptive algorithms. Treatment is discussed in chapter 37.

MINIMIZING EXPOSURE TO AIRBORNE SPORES

In order to prevent inhaled *Aspergillus* infections, HSCT recipients receive the most profoundly immunosuppressive portion of their treatment in a filtered air environment. In order to prevent primary cutaneous *Aspergillus* infections, liver transplant recipients and burn victims may receive interventions such as wound management and skin grafting in operating theaters with filtered air (Levenson et al., 1991). Importantly for burn victims, prevention of wound colonization is the key to prophylaxis for infection.

There are two means of providing this type of environment: laminar air flow and high-efficiency particulate air (HEPA) filtration. Laminar air flow is a term that encompasses reverse isolation with gut decontamination procedures and oral, nonabsorbable prophylactic antimicrobials, as well as filtered air. It was prospectively evaluated in leukemia patients and patients with aplastic anemia undergoing marrow transplantation in the 1970s and has been shown to be associated with less septicemia and fewer major infections (Buckner et al., 1978; Navari et al., 1984; Schimpff et al., 1975; Petersen et al., 1986, 1987, 1988). Since the 1970s, many changes have occurred in the medical care of cancer patients and HSCT recipients. The important component of laminar air flow is filtered air. Bone marrow transplant recipients were found to have a 10-fold-greater incidence of nosocomial *Aspergillus* infection than other immunocompromised patient populations when housed outside of rooms with HEPA filtration. In one study, the use of whole-wall HEPA filtration units with horizontal laminar flow in patient rooms reduced the number of *Aspergillus* organisms in the air to a very low 0.009 CFU/m^3 (Sherertz et al., 1987). Several of the components of laminar air flow are primarily of historical interest at this point in time, as newer treatment facilities have incorporated HEPA filtration into the building structure. Reverse isolation, gut decontamination procedures, and nonabsorbable antimicrobials such as oral paromomycin and oral gentamicin are now uncommonly used.

Filtration with HEPA provides effective prophylaxis against *Aspergillus* under normal conditions (Berthelot et al., 2006; Wald et al., 1997). However, renovations and construction in the vicinity of the hospital and/or clinic may increase the risk of release of spore-bearing dust (Wald et al., 1997). There is a strong association between building renovation and an increase in environmental *Aspergillus* contamination. The addi-

Jo-Anne H. Young • Division of Infectious Disease and International Medicine, Dept. of Medicine, University of Minnesota, MMC 250, 420 Delaware Street S.E., Minneapolis, MN 55455.

tion of temporary laminar airflow units to HEPA filtration is important in preventing environmental *Aspergillus* contamination during hospital construction or renovation (Cornet et al., 1999). Environmental surveys of airborne contamination related to construction are helpful in facilitating prevention of nosocomial aspergillosis outbreaks.

Severely immunocompromised patients should use high-efficiency respiratory protection devices (e.g., N95 respirators) when leaving HEPA-protected rooms if dust-generating activities are ongoing in the building. Many patients will ask if portable HEPA units are helpful in the home environment. These are generally not helpful because each room in the house that the patient will be using must have an individual unit that is sized properly for the room. In addition, any time spent outdoors will negate the effect of the in-house HEPA filtration.

Potted plants, plant material, and marijuana have been suggested as environmental sources of airborne *Aspergillus* spores. Environmental studies as well as clinical case reports have linked *Aspergillus fumigatus* spores to potted plants (Summerbell et al., 1989; Staib et al., 1978, 1980). Since indoor plant soils constitute a serious mycotic hazard to the profoundly immunosuppressed patient, both plants and cut flowers are not permitted on the wards of or in patient rooms on HSCT floors (Dykewicz, 2001a, 2001b). Marijuana should be avoided (Chusid et al., 1975; Kagen, 1981; Hamadeh et al., 1988). In one case, a 34-year-old man smoked marijuana heavily for several weeks prior to the diagnosis of pulmonary aspergillosis on the 75th day after marrow transplantation. Cultures of his marijuana revealed *A. fumigatus* with morphology and growth characteristics identical to the organism grown from his open lung biopsy specimen. Despite aggressive antifungal therapy, he died with disseminated disease (Hamadeh et al., 1988). In another case, a 46-year-old patient with acute myeloid leukemia presented with fever, chills, dry cough, and hypoxemia. Before becoming acutely ill, the patient smoked daily tobacco mixed with marijuana from a hookah bottle (water pipe device for smoking). Tobacco cultures yielded heavy growth of *Aspergillus* species (Szyper-Kravitz et al., 2001). Immunosuppression from renal transplantation predisposed a patient to the development of invasive aspergillosis when exposure was sufficiently prolonged during habitual marijuana smoking (Marks et al., 1996).

For burn victims, an additional prophylactic measure to avoid airborne spores along with air filtration is topical wound management to prevent colonization (Levenson et al., 1991; Stone et al., 1979). Absorbent dressings containing ionic silver are microbicidal against aerobic and anaerobic bacteria, yeasts, and filamentous fungi in the management of partial thickness burns (Wright et al., 1999; Bowler et al., 2004). In addition, periodic decontamination of the environment of the burn care unit is important (Mousa et al., 1999; Chakrabarti et al., 1992).

MINIMIZING EXPOSURE TO FOOD-BORNE SPORES

Health food supplements which are not regulated by the Food and Drug Administration may contain mold spores (Halt, 1998; Efuntoye, 1996; Hitokoto et al., 1978). Such "medications" should be discontinued prior to initiating intensive immunosuppression in order to prevent mold infections of internal organs that can be associated with consumption of various naturopathic products (Oliver et al., 1996). Foods that may carry *Aspergillus* spores include pepper, tea, freeze-dried soup, Beano, miso (bean) paste, soy sauce, and blue cheese (Bouakline et al., 2000; Christensen et al., 1967; Seenappa and Kempton, 1980; Madhyastha and Bhat, 1984; De Bock et al., 1989; Vargas et al., 1990; Eccles and Scott, 1992; Mandeel, 2005; Kino et al., 1982). Common sense measures should be reviewed with the at-risk patient so that the patient recognizes restricted foods and herbal supplements (Dykewicz, 2001). The "neutropenic diet" is a hospital intervention that excludes certain foods, especially fresh fruits and vegetables, from the diets of immunosuppressed patients to reduce patients' exposures to bacteria during neutropenia. There is inconsistent compliance with this low-microbial, restricted diet. This diet likely has an impact on the prevention of bacterial infections rather than aspergillosis (Moody et al., 2002; Wilson, 2002; DeMille et al., 2006).

TOPICAL PHARMACOLOGIC PROPHYLAXIS

Aerosolized administration of amphotericin B has been used to prevent aspergillosis among transplant recipients (Drew et al., 2004; Conneally et al., 1990; Minari et al., 2002). In particular, lung transplant recipients are at risk of *Aspergillus* infection at the tracheobronchial anastamosis, a condition that is diagnosed at bronchoscopy and which may have no findings on imaging studies (see chapter 38) (Singh and Husain, 2003; Mehrad et al., 2001; Birsan et al., 1998; Kramer et al., 1991). A prospective trial with 100 lung transplant recipients randomized to aerosolized generic amphotericin B administered at 25 mg or amphotericin B lipid complex at 50 mg showed similar results for the two groups (Drew et al., 2004). The planned treatment was for one nebulizer every day for 4 days and then once per week for 7 weeks. Drug was discontinued for intolerance in

12% of the generic amphotericin B group and 6% of the amphotericin B lipid complex-treated group. Primary prophylaxis failure within 2 months of study drug initiation was observed in 14% of generic amphotericin B-treated and 12% of amphotericin B lipid complex-treated patients. Two percent of patients experienced primary prophylaxis failure with documented *Aspergillus* infections within the follow-up period.

In patients with neutropenia of greater than 10 days' duration, inhaled amphotericin B prevented invasive aspergillosis (Conneally et al., 1990). Amphotericin B prophylaxis was administered twice daily to 34 patients during 144 episodes of neutropenia, with no cases of invasive aspergillosis. In 2 years prior to institution of prophylaxis, 11% of historical controls developed invasive aspergillosis, but these patients were in rooms without HEPA filtration. Since HSCT recipients are now often housed in rooms or entire wards with HEPA-filtered air, the usefulness of nebulized amphotericin B cannot be clearly delineated for patients with malignancies.

SYSTEMIC PHARMACOLOGIC PROPHYLAXIS

Invasive aspergillosis has long been recognized as an important contributor to morbidity and mortality among cancer patients undergoing immunosuppressive chemotherapy. The first systemic agent used for treatment of aspergillosis was the polyene drug amphotericin B, discovered in the late 1950s (see chapter 30). Lipid formulations with less dose-related nephrotoxicity but not necessarily less infusion-related toxicity were developed in the late 1980s. Due to these toxicities, the drug was not used for prophylaxis. However, treatment of prolonged fever during cancer-associated neutropenia from chemotherapy with empirical amphotericin B led to improved patient outcomes by the early 1980s (Stein et al., 1982; Pizzo et al., 1982), primarily from a reduction in systemic yeast infections. Prophylaxis, rather than preemptive therapy, with amphotericin B gained enthusiasm in the late 1980s. Administration of low-dose prophylactic amphotericin B reduced but did not eliminate the incidence and mortality associated with invasive aspergillosis in patients undergoing allogeneic marrow transplantation (Rousey et al., 1991; O'Donnell et al., 1994). In a small trial, a low dose of a liposomal formulation of amphotericin B (1 mg/kg of body weight/day of AmBisome) was not different from placebo in the prophylaxis of bone marrow transplant recipients (Tollemar et al., 1993). With low-dose amphotericin B, cyclosporine doses needed to be monitored to maintain target levels (O'Donnell et al., 1994), and breakthrough infections still occurred.

The triazole antifungal agents were the next major improvement in pharmacologic control of *Aspergillus* infections (see chapter 31). They have three nitrogens in the five-membered azole ring and as a class inhibit ergosterol synthesis, targeting the enzyme that is necessary for conversion of lanosterol to ergosterol. This enzyme, C-14α demethylase, is dependent on fungal cytochrome P450; azole interactions with mammalian cell cytochrome P450 enzymes are known to cause side effects (see chapters 31 and 33). In the late 1980s to early 1990s, the introduction of fluconazole was a major advance in azole antifungal therapy because its small molecular size and low lipophilicity allowed for ease of administration in the setting of good efficacy against *Candida albicans* (Grant and Clissold, 1990).

As fluconazole became available in the late 1980s, two controlled trials showed that prophylactic fluconazole decreased the incidence of *C. albicans* disease for bone marrow transplant patients (Goodman et al., 1992; Slavin et al., 1995). Since *C. albicans* was the dominant fungal infection affecting transplant recipients and since fluconazole is less toxic than amphotericin B, amphotericin B products were relegated to empirical treatment of fevers that were present despite fluconazole prophylaxis. At the time these studies were designed, a fluconazole dose of 400 mg daily was considered to have potential *Aspergillus* activity, which we now know to not be the case.

Fluconazole prophylaxis has changed the epidemiology of fungal infection following HSCT and subsequently changed the times when *Aspergillus* prophylaxis is needed for these patients. In an autopsy study, fungal infections were found in 40% of 355 HSCT recipients transplanted between 1990 and 1994 (van Burik et al., 1998). Overall, the proportion of autopsies with an invasive fungal infection of any type was not different for those patients receiving no fluconazole prophylaxis (43%) versus those treated with prophylactic fluconazole (37%). A decrease in *Candida* infections (27 versus 8%; $P < 0.001$) was offset by a larger number of *Aspergillus* ($n = 80$) and *Zygomycetes* ($n = 4$) mold infections (19 versus 28%; $P = 0.03$). Prevention of yeast infection allows the bone marrow transplant patient to live longer to develop the more indolent *Aspergillus* infection: the median time interval between last transplantation and death was 26 days for the 62 patients who had a yeast infection present at autopsy, while the number of days to death was 62 days for the 84 patients with a mold infection found at autopsy. For this reason, and with the expanding armamentarium of antifungal agents that have become available over the last decade, prophylaxis of invasive aspergillosis for patients with hematological malignancies undergoing therapy has been studied with agents less toxic than amphotericin B.

In the 1990s, the main mold-active triazole tested for prophylaxis against *Aspergillus* was itraconazole. Itraconazole is an azole with a lipophilic four-ring tail that allows additional contact within CYP51, which broadens the spectrum of fungal activity to include molds. It is available as an oral capsule that requires an acidic environment for absorption and as an oral solution. The intravenous solution, which may clog administration tubing, is not always available. Metabolites have antifungal activity, so laboratory reports of drug levels should be interpreted as a "combined active azole" level for purposes of demonstrating minimum blood drug levels. When itraconazole was tested versus fluconazole for prophylaxis against invasive aspergillosis among liver transplant recipients, the rates of proven invasive fungal infection of 9% versus 4% were not different between study arms ($P = 0.25$), and more gastrointestinal side effects occurred among itraconazole-treated patients (Winston and Busuttil, 2002).

In one study of *Aspergillus* prophylaxis for neutropenic patients, treatment of 65% of patients treated prophylactically with itraconazole oral solution was considered successful, compared with 53% of patients treated with a combination of oral amphotericin B capsules plus oral nystatin suspension (Boogaerts et al., 2001). Proven deep fungal infections occurred in 5% of patients in each group. In another study, 201 neutropenic patients with hematological malignancies treated prophylactically with itraconazole at 2.5 mg/kg every 12 h had lower rates of proven or suspected cases of deep fungal infection than 204 patients treated prophylactically with placebo (Menichetti et al., 1999). Paradoxically, invasive aspergillosis was documented in more itraconazole recipients ($n = 4$) than placebo recipients ($n = 1$). Side effects causing drug interruption occurred in 18% of itraconazole recipients and 13% of placebo recipients.

Itraconazole is effective as prophylaxis for allogeneic HSCT recipients if the gastrointestinal side effects can be tolerated. Proven invasive fungal infections occurred in 9% of 71 itraconazole prophylaxis patients and in 25% of 67 fluconazole prophylaxis patients during the first 180 days after HSCT (difference, −16%; $P = 0.01$) (Winston et al., 2003). Prophylaxis was given from day 1 until day 100 after HSCT (Winston et al., 2003). This is in contrast to the two fluconazole prophylaxis trials which initiated the antifungal prophylaxis drug at the onset of HSCT conditioning (Goodman et al., 1992; Slavin et al., 1995). In another HSCT study of itraconazole versus fluconazole prophylaxis, when itraconazole prophylaxis was initiated at the onset of conditioning (Marr et al., 2004), a unique problem was identified. Patients who received itraconazole developed higher serum bilirubin and creatinine values in the first 20 days after HSCT, with the highest values in patients who received itraconazole concurrently with cyclophosphamide conditioning (Marr et al., 2004). There were no differences demonstrated in the incidence of invasive fungal infections by the end of follow-up (fluconazole, 16%, versus itraconazole, 13%; $P = 0.46$). In the analysis of a concurrent study at this center that examined cyclophosphamide metabolites, recipients of itraconazole had higher exposure to toxic cyclophosphamide metabolites than recipients of fluconazole (Marr et al., 2004). Future clinical trials, which would involve the advanced-generation azole antifungals voriconazole and posaconazole, would no longer use triazole antifungal agents together with cyclophosphamide.

Voriconazole is a twice-daily agent that is the gold standard for treatment of invasive aspergillosis (Herbrecht et al., 2002). Voriconazole is available both orally and intravenously, with oral therapy preferred when the creatinine clearance is less than 50 ml/min. Dosing of voriconazole in adult patients includes loading with 6 mg/kg/dose for two doses, followed by a drop to 3 to 4 mg/kg/dose. Due to its hydrophilicity, it has the best central nervous system penetration of the antifungal agents. Invasive aspergillosis disease is effectively treated with voriconazole (Herbrecht et al., 2002; Baden et al., 2003), so the use of voriconazole for *Aspergillus* prophylaxis seems logical. Voriconazole may also protect against other molds, such as *Fusarium* (Bigley et al., 2004; Sagnelli et al., 2006; Durand et al., 2005; Vincent et al., 2003), and an evaluation of this agent for prophylaxis against invasive aspergillosis has only recently been completed.

The Blood & Marrow Transplant Clinical Trials Network trial compared prophylaxis with voriconazole versus fluconazole (Wingard et al., 2007). The study combined the randomized prophylaxis agents with monitoring for the development of invasive infection with the *Aspergillus* galactomannan antigen assay, but results of this portion of the analysis are not yet available. Due to the alterations in cyclophosphamide metabolism demonstrated by one of the itraconazole studies (Marr et al., 2004a, 2004b), this ongoing study started the antifungal prophylaxis study drug after conditioning, when treatment with cyclophosphamide was complete (Wingard, 2004). Study drugs were given for 100 days from the time of transplant; in those receiving prednisone at a dose of 1 mg/kg/day at day 100 or for recipients of T-cell-depleted grafts with T4 counts less than 200 cells/mm^3 at day 100, drugs were administered for 180 days (Wingard et al., 2007). Microbiologically documented *Aspergillus* infections decreased from 16 cases among 295 patients treated prophylactically with fluconazole to 7 cases among 305 treated prophylactically with voriconazole ($P = 0.05$) (Wingard et al., 2007). Event-free and overall survival rates were similar in both arms at 6 and 12 months. There were no differences in fungus-free survival rates in patients who received prophylactic fluconazole or voriconazole when intensive monitoring and early empirical therapy were employed in standard-risk allogeneic HSCT recipients.

Outside of the randomized trial setting, voriconazole prophylaxis appears to be successful (Siwek et al., 2006). One study reported no cases of invasive aspergillosis among 92 HSCT recipients who received voriconazole prophylaxis. There was a 10% rate of infection among 223 HSCT recipients who received other systemic antifungal agents for prophylaxis (Siwek et al., 2006).

Posaconazole is another advanced-generation azole (Petraitiene et al., 2001). The compound has been modified in four locations to cause fewer side effects to humans than itraconazole, and the drug that is being manufactured is purified to the one of the four isomers with the most favorable pharmacokinetics and *Aspergillus* activity. Patients with acute myelogenous leukemia or myelodysplastic syndrome were randomized to receive fungal prophylaxis with posaconazole or the choice of fluconazole or itraconazole with each cycle of chemotherapy until recovery from neutropenia (Cornely et al., 2007). Proven or probable invasive fungal infections were reported in 7 of 304 patients (2%) in the posaconazole group and 25 of 298 patients (8%) in the fluconazole-itraconazole group (absolute reduction in the posaconazole group, −6%). Significantly fewer patients in the posaconazole group had invasive aspergillosis (1 versus 7%). Serious adverse events that were possibly or probably related to treatment occurred in 6% of the posaconazole group and 2% of the fluconazole-itraconazole group ($P = 0.01$). Posaconazole also improved overall survival (Cornely et al., 2007).

Posaconazole was also tested versus fluconazole as antifungal prophylaxis during a 16-week period of biopsy-proven graft-versus-host disease (GVHD) of HSCT recipients. These patients recovered from absolute neutropenia but were starting more intensive immune suppression treatments for GVHD (Wald et al., 1997). This international trial compared 301 patients receiving prophylaxis with posaconazole to 299 patients receiving prophylaxis with fluconazole for the first 16 weeks of treatment for GVHD (Cornely et al., 2007). Posaconazole was superior to fluconazole in preventing proven or probable invasive aspergillosis, decreasing the number of infections from 21 to 7 cases ($P = 0.006$) and reducing the rate of deaths related to fungal infections (Ullmann et al., 2007).

Posaconazole is the only oral antifungal agent which shows activity in the treatment of zygomycosis infections (van Burik et al., 2006; Greenberg et al., 2006). The zygomycetes can create a clinical picture that mimics aspergillosis. Zygomycosis infections are of concern because empirical treatment with fluconazole or voriconazole will not cover this class of fungal organisms. Zygomycosis infections did not break through on the posaconazole GVHD prophylaxis trial (Ullmann et al., 2007), although it is important to emphasize that these infections are rare. Since intravenous amphotericin B products are the other group of prophylaxis or treatment agents for zygomycetes, posaconazole is the one oral agent that can serve as a relatively nontoxic form of pharmacologic prophylaxis against both aspergillosis and zygomycosis infections.

The echinocandins are a new class of intravenous antifungal agents (see chapter 32). They work by inhibiting synthesis of an integral component of the fungal cell wall, β-1,3-D-glucan. Three echinocandins are approved by the Food and Drug Administration: caspofungin, micafungin, and anidulafungin. Only micafungin has been studied in the setting of prophylaxis for invasive aspergillosis.

Micafungin does not have any identified differential inhibition of hepatic cytochrome P450 isoenzymes to affect cyclophosphamide metabolism or conditioning-related toxicities. When it was tested as antifungal prophylaxis for HSCT recipients, the study drug was started at the onset of conditioning, i.e., as cyclophosphamide treatment was beginning, and until up to 5 days after engraftment. Micafungin was tested at 50 mg daily against fluconazole at 400 mg daily during the neutropenic phase of HSCT (van Burik et al., 2004). More HSCT recipients on the micafungin prophylaxis arm met the composite end point of treatment success (340 of 425 [80%]) than with fluconazole (336 of 457 [73.5%]; $P = 0.03$). There was a trend toward fewer cases of invasive aspergillosis among patients receiving micafungin (micafungin, 1 case, versus fluconazole, 7 cases; $P = 0.07$). More patients were also discontinued from fluconazole to start empirical amphotericin therapy because of persistent fever during neutropenia than patients discontinued from micafungin (21 versus 15%; $P = 0.018$).

Importantly, it was demonstrated that micafungin could be used concurrently with cyclophosphamide as prophylaxis against fungal infections, including aspergillosis (van Burik et al., 2004). All of the echinocandins and the vast majority of amphotericin B agents are intravenous drugs. When cancer patients and HSCT recipients have too much nausea and mucositis to be able to tolerate an oral agent for prophylaxis against aspergillosis, an intravenous agent may be preferred. Based on clinical trials performed to date, the anti-*Aspergillus* agent of choice during cyclophosphamide therapy in the preengraftment phase of HSCT is micafungin.

For continuous antifungal prophylaxis during a year-long HSCT process, there may need to be an initial regimen that contains some intravenous agents, followed by a switch to oral agents once mucositis has cleared and the patient can swallow again.

The sequential use of different antifungal agents may be needed in order to easily convert from an intravenous to an entirely oral regimen. Later following HSCT, if mold coverage is desired during prolonged immunosuppression, voriconazole, posaconazole, or itraconazole can be used. The triazole agent chosen may have to be individualized based on side effects: voriconazole may have too many vision issues for a particular patient, or itraconazole may have too many absorption issues as well as gut side effects, etc.

Of note, invasive aspergillosis prophylaxis has become increasingly important during GVHD therapy, paradoxically because the success of immunosuppressive practice has led to improved survival. New symptoms should be rapidly evaluated. During immunosuppressive therapy for GVHD, prophylaxis against inhaled molds is continued until 3 months after the cessation of immunosuppressive medications for GVHD, or indefinitely for patients who are anatomically or functionally asplenic (chronic GVHD). These are usually outpatients who are using oral medication regimens, so a triazole is a good prophylaxis choice. In some senses, it really does not matter whether the drug is itraconazole, voriconazole, or posaconazole, as long as the patient is tolerating and compliant with the medication used. In patients with limited chronic GVHD symptoms, i.e., taking prednisone equivalents of ≤0.5 mg/kg every other day, ongoing prophylactic fungal medication is not really so critical.

SECONDARY ANTIFUNGAL PROPHYLAXIS

As management of cases of invasive aspergillosis has improved, an increasing number of patients are recovering from the infection and can subsequently require additional chemotherapy or undergo HSCT (Sipsas and Kontoyiannis, 2006). These patients are at high risk for infection relapse unless secondary prophylaxis is used (Sipsas and Kontoyiannis, 2006; Cordonnier et al., 2004; Uriz et al., 2007). Secondary prophylaxis should be started at the onset of neutropenia in patients with a history of invasive aspergillosis who are entering major periods of immunosuppression, such as HSCT candidates. The agents used in most reports of secondary prophylaxis are voriconazole and/or amphotericin B (Sipsas and Kontoyiannis, 2006; Cordonnier et al., 2004; Uriz et al., 2007), although caspofungin has also been reported to be successful in case reports (Ifran et al., 2005). Heart transplant recipients with an episode of invasive aspergillosis 2 months before or after the transplantation date have received successful secondary prophylaxis using itraconazole (Munoz et al., 2004). Adjunctive measures such as nonmyeloablative conditioning procedures and granulocyte transfusions may be used. Granulocyte transfusions have not been proven effective for secondary prevention of fungal infection during profound neutropenia.

CONCLUSIONS

The HSCT recipient and patients with leukemia undergoing chemotherapy are at risk for exogenously acquired *Aspergillus* infections. Lung and liver transplant recipients, as well as burn victims, are the other patient groups that will require invasive aspergillosis prophylaxis at select times. Prophylaxis against these infections requires minimization of exposure to airborne and foodborne spores as well as pharmacologic agents. HEPA filtration has become the gold standard method for reduction of exposure to airborne spores among hospitalized patients. HEPA filtration of operating rooms may be used to prevent primary cutaneous aspergillosis of wounds for liver transplant recipients and burn victims. HEPA filtration can be augmented with portable laminar air flow units if there is construction or renovation in the patient treatment area. At-risk malignancy patients benefit from replacement of systemic pharmacologic fluconazole antifungal prophylaxis with an agent that possesses antifungal activity against molds as well, such as a triazole (itraconazole, voriconazole, or posaconazole), an echinocandin (micafungin), or a polyene (amphotericin B). Triazoles should not be used during cyclophosphamide therapy. All of the echinocandins (such as micafungin) and most amphotericin B agents are intravenous drugs. When cancer patients and HSCT recipients have too much nausea and mucositis to be able to tolerate an oral agent, an intravenous agent may be preferred. For continuous antifungal prophylaxis of many months' duration, the at-risk patient may require a regimen that contains time periods with intravenous agents and time periods with oral agents. The sequential use of different antifungal agents may be needed to provide prophylaxis during the entire at-risk period.

REFERENCES

Baden, L. R., J. T. Katz, J. A. Fishman, C. Koziol, A. DelVecchio, M. Doran, and R. H. Rubin. 2003. Salvage therapy with voriconazole for invasive fungal infections in patients failing or intolerant to standard antifungal therapy. *Transplantation* **76**:1632–1637.

Berthelot, P., P. Loulergue, H. Raberin, M. Turco, C. Mounier, R. Tran Manh Sung, F. Lucht, B. Pozzetto, and D. Guyotat. 2006. Efficacy of environmental measures to decrease the risk of hospital-acquired aspergillosis in patients hospitalised in haematology wards. *Clin. Microbiol. Infect.* **12**:738–744.

Bigley, V. H., R. F. Duarte, R. D. Gosling, C. C. Kibbler, S. Seaton, and M. Potter. 2004. *Fusarium dimerum* infection in a stem cell transplant recipient treated successfully with voriconazole. *Bone Marrow Transplant.* 34:815–817.

Birsan, T., S. Taghavi, and W. Klepetko. 1998. Treatment of *Aspergillus*-related ulcerative tracheobronchitis in lung transplant recipients. *J. Heart Lung Transplant.* 17:437–438.

Boogaerts, M., J. Maertens, A. van Hoof, R. de Bock, G. Fillet, M. Peetermans, D. Selleslag, B. Vandercam, K. Vandewoude, P. Zachee, and K. De Beule. 2001. Itraconazole versus amphotericin B plus nystatin in the prophylaxis of fungal infections in neutropenic cancer patients. *J. Antimicrob. Chemother.* 48:97–103.

Bouakline, A., C. Lacroix, N. Roux, J. P. Gangneux, and F. Derouin. 2000. Fungal contamination of food in hematology units. *J. Clin. Microbiol.* 38:4272–4273.

Bowler, P. G., S. A. Jones, M. Walker, and D. Parsons. 2004. Microbicidal properties of a silver-containing hydrofiber dressing against a variety of burn wound pathogens. *J. Burn Care Rehabil.* 25:192–196.

Buckner, C. D., R. A. Clift, J. E. Sanders, J. D. Meyers, G. W. Counts, V. T. Farewell, and E. D. Thomas. 1978. Protective environment for marrow transplant recipients: a prospective study. *Ann. Intern. Med.* 89:893–901.

Chakrabarti, A., N. Nayak, P. S. Kumar, P. Talwar, P. S. Chari, and D. Panigrahi. 1992. Surveillance of nosocomial fungal infections in a burn care unit. *Infection* 20:132–135.

Christensen, C. M., H. A. Fanse, G. H. Nelson, F. Bates, and C. J. Mirocha. 1967. Microflora of black and red pepper. *Appl. Microbiol.* 15:622–626.

Chusid, M. J., J. A. Gelfand, C. Nutter, and A. S. Fauci. 1975. Pulmonary aspergillosis, inhalation of contaminated marijuana smoke, chronic granulomatous disease. *Ann. Intern. Med.* 82:682–683.

Conneally, E., M. T. Cafferkey, P. A. Daly, C. T. Keane, and S. R. McCann. 1990. Nebulized amphotericin B as prophylaxis against invasive aspergillosis in granulocytopenic patients. *Bone Marrow Transplant.* 5:403–406.

Cordonnier, C., S. Maury, C. Pautas, J. N. Bastie, S. Chehata, S. Castaigne, M. Kuentz, S. Bretagne, and P. Ribaud. 2004. Secondary antifungal prophylaxis with voriconazole to adhere to scheduled treatment in leukemic patients and stem cell transplant recipients. *Bone Marrow Transplant.* 33:943–948.

Cornely, O. A., J. Maertens, D. J. Winston, J. Perfect, A. J. Ullmann, T. J. Walsh, D. Helfgott, J. Holowiecki, D. Stockelberg, Y. T. Goh, M. Petrini, C. Hardalo, R. Suresh, and D. Angulo-Gonzalez. 2007. Posaconazole vs. fluconazole or itraconazole prophylaxis in patients with neutropenia. *N. Engl. J. Med.* 356:348–359.

Cornet, M., V. Levy, L. Fleury, J. Lortholary, S. Barquins, M. H. Coureul, E. Deliere, R. Zittoun, G. Brucker, and A. Bouvet. 1999. Efficacy of prevention by high-efficiency particulate air filtration or laminar airflow against Aspergillus airborne contamination during hospital renovation. *Infect. Control Hosp. Epidemiol.* 20:508–513.

De Bock, R., I. Gyssens, M. Peetermans, and N. Nolard. 1989. *Aspergillus* in pepper. *Lancet* ii:331–332.

DeMille, D., P. Deming, P. Lupinacci, and L. A. Jacobs. 2006. The effect of the neutropenic diet in the outpatient setting: a pilot study. *Oncol. Nurs. Forum* 33:337–343.

Drew, R. H., E. Dodds Ashley, D. K. Benjamin, Jr., R. Duane Davis, S. M. Palmer, and J. R. Perfect. 2004. Comparative safety of amphotericin B lipid complex and amphotericin B deoxycholate as aerosolized antifungal prophylaxis in lung-transplant recipients. *Transplantation* 77:232–237.

Durand, M. L., I. K. Kim, D. J. D'Amico, J. I. Loewenstein, E. H. Tobin, S. J. Kieval, S. S. Martin, D. T. Azar, F. S. Miller III, B. J. Lujan, and J. W. Miller. 2005. Successful treatment of *Fusarium* endophthalmitis with voriconazole and *Aspergillus* endophthalmitis with voriconazole plus caspofungin. *Am. J. Ophthalmol.* 140:552–554.

Dykewicz, C. A. 2001. Hospital infection control in hematopoietic stem cell transplant recipients. *Emerg. Infect. Dis.* 7:263–267.

Dykewicz, C. A. 2001. Summary of the guidelines for preventing opportunistic infections among hematopoietic stem cell transplant recipients. *Clin. Infect. Dis.* 33:139–144.

Eccles, N. K., and G. M. Scott. 1992. *Aspergillus* in pepper. *Lancet* 339:618.

Efuntoye, M. O. 1996. Fungi associated with herbal drug plants during storage. *Mycopathologia* 136:115–118.

Goodman, J. L., D. J. Winston, R. A. Greenfield, P. H. Chandrasekar, B. Fox, H. Kaizer, R. K. Shadduck, T. C. Shea, P. Stiff, D. J. Friedman, W. G. Powderly, J. L. Silber, H. Horowitz, A. Lichtin, S. N. Wolff, K. F. Mangan, S. M. Silver, D. Weisdorf, W. G. Ho, G. Gilbert, and D. Buell. 1992. A controlled trial of fluconazole to prevent fungal infections in patients undergoing bone marrow transplantation. *N. Engl. J. Med.* 326:845–851.

Grant, S. M., and S. P. Clissold. 1990. Fluconazole. A review of its pharmacodynamic and pharmacokinetic properties, and therapeutic potential in superficial and systemic mycoses. *Drugs* 39:877–916.

Greenberg, R. N., K. Mullane, J. A. van Burik, I. Raad, M. J. Abzug, G. Anstead, R. Herbrecht, A. Langston, K. A. Marr, G. Schiller, M. Schuster, J. R. Wingard, C. E. Gonzalez, S. G. Revankar, G. Corcoran, R. J. Kryscio, and R. Hare. 2006. Posaconazole as salvage therapy for zygomycosis. *Antimicrob. Agents Chemother.* 50:126–133.

Halt, M. 1998. Moulds and mycotoxins in herb tea and medicinal plants. *Eur. J. Epidemiol.* 14:269–274.

Hamadeh, R., A. Ardehali, R. M. Locksley, and M. K. York. 1988. Fatal aspergillosis associated with smoking contaminated marijuana, in a marrow transplant recipient. *Chest* 94:432–433.

Herbrecht, R., D. W. Denning, T. F. Patterson, J. E. Bennett, R. E. Greene, J. W. Oestmann, W. V. Kern, K. A. Marr, P. Ribaud, O. Lortholary, R. Sylvester, R. H. Rubin, J. R. Wingard, P. Stark, C. Durand, D. Caillot, E. Thiel, P. H. Chandrasekar, M. R. Hodges, H. T. Schlamm, P. F. Troke, and B. de Pauw. 2002. Voriconazole versus amphotericin B for primary therapy of invasive aspergillosis. *N. Engl. J. Med.* 347:408–415.

Hitokoto, H., S. Morozumi, T. Wauke, S. Sakai, and H. Kurata. 1978. Fungal contamination and mycotoxin detection of powdered herbal drugs. *Appl. Environ. Microbiol.* 36:252–256.

Ifran, A., K. Kaptan, and C. Beyan. 2005. Efficacy of caspofungin in prophylaxis and treatment of an adult leukemic patient with invasive pulmonary aspergillosis in allogeneic stem cell transplantation. *Mycoses* 48:146–148.

Kagen, S. L. 1981. *Aspergillus*: an inhalable contaminant of marihuana. *N. Engl. J. Med.* 304:483–484.

Kino, T., J. Chihara, K. Mitsuyasu, M. Kitaichi, K. Honda, M. Kado, T. Izumi, S. Oshima, I. Uesaka, K. Maeda, M. Kurozumi, and S. Oota. 1982. A case of allergic bronchopulmonary aspergillosis caused by *Aspergillus oryzae* which is used for brewing bean paste (miso) and soy sauce (shoyu). *Nihon Kyobu Shikkan Gakkai Zasshi* 20:467–475.

Kramer, M. R., D. W. Denning, S. E. Marshall, D. J. Ross, G. Berry, N. J. Lewiston, D. A. Stevens, and J. Theodore. 1991. Ulcerative tracheobronchitis after lung transplantation. A new form of invasive aspergillosis. *Am. Rev. Respir. Dis.* 144:552–556.

Levenson, C., P. Wohlford, J. Djou, S. Evans, and B. Zawacki. 1991. Preventing postoperative burn wound aspergillosis. *J. Burn Care Rehabil.* 12:132–135.

Madhyastha, M. S., and R. V. Bhat. 1984. *Aspergillus parasiticus* growth and aflatoxin production on black and white pepper and the inhibitory action of their chemical constituents. *Appl. Environ. Microbiol.* 48:376–379.

Mandeel, Q. A. 2005. Fungal contamination of some imported spices. *Mycopathologia* 159:291–298.

Marks, W. H., L. Florence, J. Lieberman, P. Chapman, D. Howard, P. Roberts, and D. Perkinson. 1996. Successfully treated invasive pulmonary aspergillosis associated with smoking marijuana in a renal transplant recipient. *Transplantation* **61:**1771–1774.

Marr, K. A., F. Crippa, W. Leisenring, M. Hoyle, M. Boeckh, S. A. Balajee, W. G. Nichols, B. Musher, and L. Corey. 2004. Itraconazole versus fluconazole for prevention of fungal infections in patients receiving allogeneic stem cell transplants. *Blood* **103:**1527–1533.

Marr, K. A., W. Leisenring, F. Crippa, J. T. Slattery, L. Corey, M. Boeckh, and G. B. McDonald. 2004. Cyclophosphamide metabolism is affected by azole antifungals. *Blood* **103:**1557–1559.

Mehrad, B., G. Paciocco, F. J. Martinez, T. C. Ojo, M. D. Iannettoni, and J. P. Lynch III. 2001. Spectrum of *Aspergillus* infection in lung transplant recipients: case series and review of the literature. *Chest* **119:**169–175.

Menichetti, F., A. Del Favero, P. Martino, G. Bucaneve, A. Micozzi, C. Girmenia, G. Barbabietola, L. Pagano, P. Leoni, G. Specchia, A. Caiozzo, R. Raimondi, F. Mandelli, et al. 1999. Itraconazole oral solution as prophylaxis for fungal infections in neutropenic patients with hematologic malignancies: a randomized, placebo-controlled, double-blind, multicenter trial. *Clin. Infect. Dis.* **28:**250–255.

Minari, A., R. Husni, R. K. Avery, D. L. Longworth, M. DeCamp, M. Bertin, R. Schilz, N. Smedira, M. T. Haug, A. Mehta, and S. M. Gordon. 2002. The incidence of invasive aspergillosis among solid organ transplant recipients and implications for prophylaxis in lung transplants. *Transplant. Infect. Dis.* **4:**195–200.

Moody, K., M. E. Charlson, and J. Finlay. 2002. The neutropenic diet: what's the evidence? *J. Pediatr. Hematol. Oncol.* **24:**717–721.

Mousa, H. A., S. M. Al-Bader, and D. A. Hassan. 1999. Correlation between fungi isolated from burn wounds and burn care units. *Burns* **25:**145–147.

Munoz, P., C. Rodriguez, E. Bouza, J. Palomo, J. F. Yanez, M. J. Dominguez, and M. Desco. 2004. Risk factors of invasive aspergillosis after heart transplantation: protective role of oral itraconazole prophylaxis. *Am. J. Transplant.* **4:**636–643.

Navari, R. M., C. D. Buckner, R. A. Clift, R. Storb, J. E. Sanders, P. Stewart, K. M. Sullivan, B. Williams, G. W. Counts, J. D. Meyers, et al. 1984. Prophylaxis of infection in patients with aplastic anemia receiving allogeneic marrow transplants. *Am. J. Med.* **76:**564–572.

O'Donnell, M., G. M. Schmidt, B. R. Tegtmeier, C. Faucett, J. L. Fahey, J. Ito, A. Nademanee, J. Niland, P. Parker, E. P. Smith, D. S. Snyder, A. S. Stein, K. G. Blume, and S. J. Forman. 1994. Prediction of systemic fungal infection in allogeneic marrow recipients: impact of amphotericin prophylaxis in high-risk patients. *J. Clin. Oncol.* **12:**827–834.

Oliver, M. R., W. C. Van Voorhis, M. Boeckh, D. Mattson, and R. A. Bowden. 1996. Hepatic mucormycosis in a bone marrow transplant recipient who ingested naturopathic medicine. *Clin. Infect. Dis.* **22:**521–524.

Petersen, F., M. Thornquist, C. Buckner, G. Counts, N. Nelson, J. Meyers, R. Clift, and E. Thomas. 1988. The effects of infection prevention regimens on early infectious complications in marrow transplant patients: a four arm randomized study. *Infection* **16:**199–208.

Petersen, F. B., C. D. Buckner, R. A. Clift, S. Lee, N. Nelson, G. W. Counts, J. D. Meyers, J. E. Sanders, P. S. Stewart, W. I. Bensinger, et al. 1986. Laminar air flow isolation and decontamination: a prospective randomized study of the effects of prophylactic systemic antibiotics in bone marrow transplant patients. *Infection* **14:**115–121.

Petersen, F. B., C. D. Buckner, R. A. Clift, N. Nelson, G. W. Counts, J. D. Meyers, and E. D. Thomas. 1987. Infectious complications in patients undergoing marrow transplantation: a prospective randomized study of the additional effect of decontamination and laminar air flow isolation among patients receiving prophylactic systemic antibiotics. *Scand. J. Infect. Dis.* **19:**559–567.

Petraitiene, R., V. Petraitis, A. H. Groll, T. Sein, S. Piscitelli, M. Candelario, A. Field-Ridley, N. Avila, J. Bacher, and T. J. Walsh. 2001. Antifungal activity and pharmacokinetics of posaconazole (SCH 56592) in treatment and prevention of experimental invasive pulmonary aspergillosis: correlation with galactomannan antigenemia. *Antimicrob. Agents Chemother.* **45:**857–869.

Pizzo, P. A., K. J. Robichaud, F. A. Gill, and F. G. Witebsky. 1982. Empiric antibiotic and antifungal therapy for cancer patients with prolonged fever and granulocytopenia. *Am. J. Med.* **2:**101–110.

Rousey, S. R., S. Russler, M. Gottlieb, and R. C. Ash. 1991. Low-dose amphotericin B prophylaxis against invasive *Aspergillus* infections in allogeneic marrow transplantation. *Am. J. Med.* **91:**484–492.

Sagnelli, C., L. Fumagalli, A. Prigitano, P. Baccari, P. Magnani, and A. Lazzarin. 2006. Successful voriconazole therapy of disseminated *Fusarium verticillioides* infection in an immunocompromised patient receiving chemotherapy. *J. Antimicrob. Chemother.* 57:796–798.

Schimpff, S. C., W. H. Greene, V. M. Young, C. L. Fortner, N. Cusack, J. B. Block, and P. H. Wiernik. 1975. Infection prevention in acute nonlymphocytic leukemia. Laminar air flow room reverse isolation with oral, nonabsorbable antibiotic prophylaxis. *Ann. Intern. Med.* 82:351–358.

Seenappa, M., and A. G. Kempton. 1980. *Aspergillus* growth and aflatoxin production on black pepper. *Mycopathologia* **70:**135–137.

Sherertz, R. J., A. Belani, B. S. Kramer, G. J. Elfenbein, R. S. Weiner, M. L. Sullivan, R. G. Thomas, and G. P. Samsa. 1987. Impact of air filtration on nosocomial *Aspergillus* infections. Unique risk of bone marrow transplant recipients. *Am. J. Med.* **83:**709–718.

Singh, N., and S. Husain. 2003. *Aspergillus* infections after lung transplantation: clinical differences in type of transplant and implications for management. *J. Heart Lung Transplant.* **22:**258–266.

Sipsas, N. V., and D. P. Kontoyiannis. 2006. Clinical issues regarding relapsing aspergillosis and the efficacy of secondary antifungal prophylaxis in patients with hematological malignancies. *Clin. Infect. Dis.* **42:**1584–1591.

Siwek, G. T., M. A. Pfaller, P. M. Polgreen, S. Cobb, P. Hoth, M. Magalheas-Silverman, and D. J. Diekema. 2006. Incidence of invasive aspergillosis among allogeneic hematopoietic stem cell transplant patients receiving voriconazole prophylaxis. *Diagn. Microbiol. Infect. Dis.* **55:**209–212.

Slavin, M. A., B. Osborne, R. Adams, M. J. Levenstein, H. G. Schoch, A. R. Feldman, J. D. Meyers, and R. A. Bowden. 1995. Efficacy and safety of fluconazole prophylaxis for fungal infections after marrow transplantation—a prospective, randomized, double-blind study. *J. Infect. Dis.* **171:**1545–1552.

Staib, F., S. K. Mishra, and A. Blisse. 1980. Interaction between aspergilli and streptomycetes in the soil of potted indoor plants: a preliminary report (contribution to the epidemiology of human aspergillosis). *Mycopathologia* **70:**9–12.

Staib, F., B. Tompak, D. Thiel, and A. Blisse. 1978. *Aspergillus fumigatus* and *Aspergillus niger* in two potted ornamental plants, cactus (*Epiphyllum truncatum*) and clivia (*Clivia miniata*). Biological and epidemiological aspects. *Mycopathologia* **66:**27–30.

Stein, R. S., J. Kayser, and J. M. Flexner. 1982. Clinical value of empirical amphotericin B in patients with acute myelogenous leukemia. *Cancer* **50:**2247–2251.

Stone, H. H., J. Z. Cuzzell, L. D. Kolb, M. S. Moskowitz, and J. E. McGowan, Jr. 1979. *Aspergillus* infection of the burn wound. *J. Trauma* **19:**765–767.

Summerbell, R. C., S. Krajden, and J. Kane. 1989. Potted plants in hospitals as reservoirs of pathogenic fungi. *Mycopathologia* **106:**13–22.

Szyper-Kravitz, M., R. Lang, Y. Manor, and M. Lahav. 2001. Early invasive pulmonary aspergillosis in a leukemia patient linked to *As-*

pergillus contaminated marijuana smoking. *Leuk. Lymphoma* **42:** 1433–1437.

Tollemar, J., O. Ringden, S. Andersson, B. Sundberg, P. Ljungman, and G. Tyden. 1993. Randomized double-blind study of liposomal amphotericin B (Ambisome) prophylaxis of invasive fungal infections in bone marrow transplant recipients. *Bone Marrow Transplant.* **12:**577–582.

Ullmann, A. J., J. H. Lipton, D. H. Vesole, P. Chandrasekar, A. Langston, S. R. Tarantolo, H. Greinix, W. Morais de Azevedo, V. Reddy, N. Boparai, L. Pedicone, H. Patino, and S. Durrant. 2007. Posaconazole or fluconazole for prophylaxis in severe graft-versus-host disease. *N. Engl. J. Med.* **356:**335–347.

Uriz, J., N. G. de Andoin, C. Calvo, J. Landa, E. Onate, A. Nogues, and R. Guerrero. 2007. Secondary prophylaxis with voriconazole in a leukemic patient with pulmonary aspergillosis. *Pediatr. Infect. Dis. J.* **26:**971–972.

van Burik, J.-A., W. Leisenring, D. Myerson, R. Hackman, H. Shulman, G. Sale, R. Bowden, and G. McDonald. 1998. The effect of prophylactic fluconazole on the clinical spectrum of fungal diseases in bone marrow transplant recipients with special attention to hepatic candidiasis: an autopsy study of 355 patients. *Medicine* (Baltimore) **77:**246–254.

van Burik, J.-A., R. S. Hare, H. F. Solomon, M. L. Corrado, and D. P. Kontoyiannis. 2006. Posaconazole is effective as salvage therapy in zygomycosis: a retrospective summary of 91 cases. *Clin. Infect. Dis.* **42:**e61–e65.

van Burik, J.-A., V. Ratanatharathorn, D. E. Stepan, C. B. Miller, J. H. Lipton, D. H. Vesole, N. Bunin, D. A. Wall, J. W. Hiemenz, Y. Satoi, J. M. Lee, and T. J. Walsh. 2004. Micafungin versus fluconazole for prophylaxis against invasive fungal infections during neutropenia in patients undergoing hematopoietic stem cell transplantation. *Clin. Infect. Dis.* **39:**1407–1416.

Vargas, S., W. T. Hughes, and M. A. Giannini. 1990. *Aspergillus* in pepper. *Lancet* **336:**881.

Vincent, A. L., J. E. Cabrero, J. N. Greene, and R. L. Sandin. 2003. Successful voriconazole therapy of disseminated *Fusarium solani* in the brain of a neutropenic cancer patient. *Cancer Control* **10:**414–419.

Wald, A., W. Leisenring, J. A. van Burik, and R. A. Bowden. 1997. Epidemiology of *Aspergillus* infections in a large cohort of patients undergoing bone marrow transplantation. *J. Infect. Dis.* **175:**1459–1466.

Wilson, B. J. 2002. Dietary recommendations for neutropenic patients. *Semin. Oncol. Nurs.* **18:**44–49.

Wingard, J., S. Carter, T. Walsh, J. Kurtzberg, T. Small, I. Gersten, A. Mendizabal, H. Leather, D. Confer, L. Baden, R. Maziarz, E. Stadtmauer, J. Bolanos-Meade, J. Brown, J. DiPersio, M. Boeckh, K. Marr, et al. 2007. Results of a randomized, double-blind trial of fluconazole vs. voriconazole for the prevention of invasive fungal infections in 600 allogeneic blood and marrow transplant patients. *Blood* **11:**A163.

Wingard, J. R. 2004. Design issues in a prospective randomized double-blinded trial of prophylaxis with fluconazole versus voriconazole after allogeneic hematopoietic cell transplantation. *Clin. Infect. Dis.* **39**(Suppl. 4)**:**S176–S180.

Winston, D. J., and R. W. Busuttil. 2002. Randomized controlled trial of oral itraconazole solution versus intravenous/oral fluconazole for prevention of fungal infections in liver transplant recipients. *Transplantation* **74:**688–695.

Winston, D. J., R. T. Maziarz, P. H. Chandrasekar, H. M. Lazarus, M. Goldman, J. L. Blumer, G. J. Leitz, and M. C. Territo. 2003. Intravenous and oral itraconazole versus intravenous and oral fluconazole for long-term antifungal prophylaxis in allogeneic hematopoietic stem-cell transplant recipients. A multicenter, randomized trial. *Ann. Intern. Med.* **138:**705–713.

Wright, J. B., K. Lam, D. Hansen, and R. E. Burrell. 1999. Efficacy of topical silver against fungal burn wound pathogens. *Am. J. Infect. Control* **27:**344–350.

Aspergillus fumigatus and Aspergillosis
Edited by J.-P. Latgé and W. J. Steinbach
©2009 ASM Press, Washington, DC

Chapter 37

Therapy of Invasive Aspergillosis: Current Consensus and Controversies

DIMITRIOS P. KONTOYIANNIS AND KIEREN A. MARR

Invasive aspergillosis (IA) has emerged as an important infection, accounting for 20 to 40% of systemic fungal infections in patients with acute leukemia, 10 to 20% in blood and marrow transplant (BMT) recipients, and 5 to 15% in solid organ transplant recipients (Kontoyiannis and Bodey, 2002; Latgé, 1999; Patterson et al., 2000; Upton et al., 2007). Important advances have been made in the management of IA, including the introduction of several new agents with activity against this opportunistic mycosis (Table 1) (Groll et al., 1998) and the publication of several studies on primary and salvage therapy that provide a reasonable basis for treatment guidelines (Table 2; Fig. 1 and 2) (Groll et al., 1998). IA remains the formidable frontier in modern mycology. This infection continues to pose challenges in its management, including the following: patients with active underlying disease (e.g., leukemia, graft-versus-host disease, or pleiotropic immune defects) leading to poor host immunity; patients with multiple comorbidities, especially older patients, and increased drug toxicities; widespread use of antifungal prophylaxis, resulting in selection pressure for resistance; diagnostic tests that lack sufficient sensitivity and specificity; lack of autopsy data; a heterogeneous patient population undergoing multiple interventions, either simultaneously or sequentially, making extrapolation of results from published trials difficult. This chapter attempts to provide a conceptual framework for the consensus and controversies in primary and salvage therapies for this infection. We have not focused on individual agents or classes of agents, or diagnostics, as these topics are covered in other chapters.

CURRENT CONSENSUS

Current Consensus 1: Voriconazole and Lipid Formulations of AMB Are Reasonable Primary Choices, Especially for Early Therapy of IA

There is a paucity of well-conducted, controlled clinical studies of the treatment of IA. A large randomized multicenter study comparing the triazole voriconazole (VRC) to amphotericin-deoxycholate (DAMB; given at relatively high doses of 1 to 1.5 mg/kg of body weight/day), followed by other licensed antifungal therapy (mostly lipid formulations of AMB) for patients with lack of response or toxicity, showed a survival advantage in patients randomized to the VRC arm (Herbrecht et al., 2002). The superiority of the VRC-based strategy was consistent in all subgroups studied. That study showed for the first time that a triazole such as VRC is an effective primary therapy for IA and that starting with a toxic agent such as DAMB is detrimental to the patient (Patterson et al., 2005). In addition, the response rates to a lipid formulation of AMB (liposomal AMB) given as primary therapy were comparable to the one of VRC in a recently published study (Cornely et al., 2007). While some in the mycology community would have liked to know how VRC directly compares with less toxic formulations of AMB, the results of this unique trial propelled VRC forward as the preferred agent in the majority of patients with documented IA (Walsh et al., 2008). Unfortunately, a head-to-head comparison of a triazole versus a lipid formulation of AMB as primary therapy is unlikely to occur.

Dimitrios P. Kontoyiannis • Dept. of Infectious Diseases, Infection Control, and Employee Health and Dept. of Bone Marrow Transplantation, The University of Texas M. D. Anderson Cancer Center, Houston, TX 77030. **Kieren A. Marr** • Comprehensive Transplant and Oncology Infectious Diseases Program, Dept. of Medicine, Johns Hopkins University, Baltimore, MD 21205.

Table 1. Systemic antifungal therapies as choices in the treatment of aspergillosis[a]

Antifungals	Usual adult dose	Mechanism of action	Toxicities	Spectrum, comments
Polyenes				
AMB	0.25–1.5 mg/kg i.v. q24h	Binds to ergosterol and intercalates the fungal cell membrane, resulting in increased membrane permeability to univalent and divalent cations	Acute: fever, chills, rigor, arthralgia with infusion; thrombophlebitis, dyspnea (rare), arrhythmias (rare) Delayed: azotemia (26%), tubular acidosis, hypokalemia, hypomagnesemia, anemia	Has activity for most *Aspergillus* species with the exception of *A. terreus*; broad spectrum of activity makes it reasonable choice for preemptive therapy, as it covers other molds (e.g., Zygomycetes). Nephrotoxicity is the dose-limiting side effect, reduced with lipid AMB formulations and saline pre- and post-hydration. Infusion-related reactions: AmBisome < Abelcet < Amphotec AmBisome considered the preferred formulation for central nervous system aspergillosis.
Lipid formulations of AMB				
LAMB	3–10 mg/kg q24h			
ABLC	5 mg/kg q24h			
ABCD	3–4 mg/k q24h			
Triazoles				
Itraconazole	200–400 mg p.o. q24h 200–400 mg i.v. q12h, then q24h Divided doses recommended at 400 mg/day or greater	More selective for fungal P450 demethylase, leading to ergosterol depletion in *Aspergillus* fungal membrane	Gastrointestinal (20%), including nausea, vomiting, and diarrhea, rash (2%), taste disturbance (with oral solution), transient increases in hepatic enzymes, severe hepatotoxicity (rare), alopecia, inhibition of adrenal steroid synthesis (especially at doses >600 mg/day) Accumulation of hydroxy-propyl-β-cyclodextran vehicle in patients with creatine clearance of <30 ml/min (i.v. formulation); consider switch to oral therapy if possible. Congestive heart failure (rare)	Spectrum similar to fluconazole with enhanced activity against *C. krusei* and *Aspergillus*. Not active against *Fusarium* and Zygomycetes. Drug of choice for mild to moderate infections caused by endemic dimorphic fungi. Bioavailability of oral solution is improved over capsules by 30% under fed conditions and 60% in fasting conditions. Potent inhibitor of mammalian cytochrome P450 enzymes. Serum level monitoring is occasionally recommended; trough levels measured by high performance liquid chromatography should exceed 0.5 μg/ml.
Voriconazole	6 mg/kg i.v. q12h for 2 doses, then 4 mg/kg q12h 200 mg p.o. q12h if ≥40 kg, 100 mg p.o. q12h if <40 kg	Similar to fluconazole, but higher affinity for fungal 14α-demethylase	Transient visual disturbances (reported up to 30%), rash, hallucinations (2%), transient increases in hepatic enzymes, severe hepatotoxicity (rare) Accumulation of sulfobutyl ester cyclodextran vehicle may occur in patients with creatinine clearance of <50 ml/min receiving i.v. formulation.	Spectrum similar to itraconazole with enhanced activity against *Aspergillus*, *Fusarium*, and *Scedosporium apiospermum*. Retains activity against some fluconazole-resistant *C. glabrata* but cross-resistance is possible. Active against *C. krusei*. Lacks activity against Zygomycetes. Considered by many experts as the initial drug of choice for invasive aspergillosis Inhibitor of mammalian cytochrome P450 enzymes

Continued on following page

Table 1. *Continued*

Antifungals	Usual adult dose	Mechanism of action	Toxicities	Spectrum, comments
Posaconazole	200 mg p.o. q6h for 7 days, then 400 mg q12h Dose-proportional saturable oral absorption	Similar to voriconazole	Gastrointestinal (5–15%), fever, headache, musculoskeletal pain (5%)	Spectrum similar to voriconazole with enhanced activity against *Fusarium*, Zygomycetes, and black molds (Phaeohyphomycetes) Inhibitor of mammalian cytochrome P450 3A4
Echinocandins				
Caspofungin	70 mg i.v. on day 1, then 50 mg q24h	Inhibition of cell wall glucan synthesis, leading to osmotic instability of fungal cell	Fever, chills, phlebitis, thrombophlebitis (peripheral line), rash; drug concentrations decreased with P450 3A4 inducers; decreases tacrolimus blood levels by 25%. Mostly salvage data	Spectrum includes most *Candida* species including fluconazole-resistant *Candida krusei* and *Candida glabrata*. Higher dosages may be required for *C. parapsilosis*. Active against *Aspergillus* species. Not active against *Cryptococcus neoformans*, *Fusarium*, Zygomycetes, or black molds (Phaeohyphomycetes).
Micafungin	50–100 mg i.v. q24h	Similar to caspofungin	Similar to caspofungin, limited data for primary or salvage therapy	Similar to caspofungin
Anidulafungin	200 mg i.v. on day 1 then 100 mg/day	Similar to caspofungin	Similar to caspofungin, limited human data for activity	Similar to caspofungin

[a] Abbreviations: p.o., orally; i.v., intravenously; q6h, every 6 h; q12h, every 12 h; q24h, every 24 h; P450, cytochrome P450; LAMB, liposomal AMB; ABLC, AMB lipid complex; ABCD, AMB cholesterol dispersion.

Current Consensus 2: Early Use of CT Is Critical, at Least in Neutropenic Patients

One of the most important advances to improved outcomes in IA has been the systematic use of high-resolution chest computed tomography (CT) in high-risk hematology patients. Several lines of evidence support this approach: the results of the recent randomized study of high-dose (10 mg/kg/day) versus lower-dose (3 mg/kg/day) liposomal AMB demonstrated that starting therapy in the phase of "probable" IA based on the halo sign with or without a positive *Aspergillus* galactomannan (GM) results in response rates similar to those observed with VRC primary therapy (Cornely et al., 2007). The importance of early therapy was echoed by the results of a post hoc analysis of the Herbert et al. study, in which responses to either VRC (62%) or DAMB (41%) were clearly better than responses seen

Table 2. Outline for the treatment of IA

Specific treatment	Comment(s)
Primary therapy[a]	Every effort should be made to confirm diagnosis.
VOR, *or*	VOR can be administered as oral therapy in patients taking oral medications.
Liposomal AMB	
Salvage therapy	Optimal strategy not defined by appropriate clinical trials; approach should be individualized based on various criteria, including host immunosuppression, underlying disease, site of infection, antifungal dosing, therapeutic monitoring of drug levels (azoles), switch to intravenous therapy, and/or a switch to another drug class or combination.
Lipid formulations of AMB (e.g., liposomal AMB), *or*	Because of prolonged treatment courses, lipid formulations are preferred for AMB-based therapy.
Caspofungin, *or*	
Triazoles: POSA, VOR, or ITC	Use of triazoles should take into account pharmacokinetic considerations.
Combination therapy	Preclinical studies suggest mold-active azoles and echinocandins have enhanced activity against *Aspergillus*; the role of combination therapy remains uncertain and warrants a prospective controlled clinical trial.

[a] Based on guidelines published by the Infectious Diseases Society of America (Walsh et al., 2008).

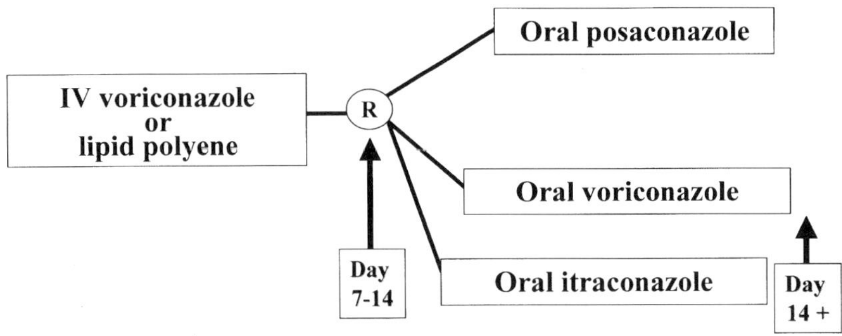

Figure 1. Primary therapy of IA.

when each drug was started later (42% for VRC, 29% for DAMB) (Greene et al., 2007). Uncontrolled studies in leukemic neutropenic patients also have shown that systematic use of high-resolution chest CT (e.g., on day 4 of refractory fever in a high-risk patient despite broad-spectrum antibacterials) coupled with early initiation of antifungal therapy and, in many patients, of adjunct surgery results in a survival advantage (Caillot et al., 1997). The window of opportunity of early diagnosis by CT is rather narrow, as the halo sign is transient (Caillot et al., 1997; Verweij et al., 2007b).

Current Consensus 3: DAMB Is a Drug of the Past for Treatment of IA

The modern way of giving AMB-based therapy, especially if a patient is to be committed to weeks or even months of therapy, is with lipid AMB formulations, as these agents have lower nephrotoxicity and infusion-related toxicity (Ostrosky-Zeichner et al., 2003; Wingard, 2002). Although their daily acquisition prices are much higher than those of DAMB, in high-risk patients with IA, especially in the BMT setting, these agents appear cost-effective (Cagnoni et al., 1997), as the related nephrotoxicity frequently appears to be clinically significant.

CURRENT CONTROVERSIES

Despite the fact that IA is a well-studied infection, controversies abound. A key problem is the heterogeneity of the host and the fact that multiple interventions are performed at the same time. Therefore, it is very difficult to dissect the relative contributions of each factor to the overall outcome. Our inabilities to diagnose the infection early, to quantify the net state of immunosuppression and comorbidities, and to define the effects of prior, concomitant, and subsequent antifungal exposure(s) create complex scenarios and necessitate individualized therapeutic decisions.

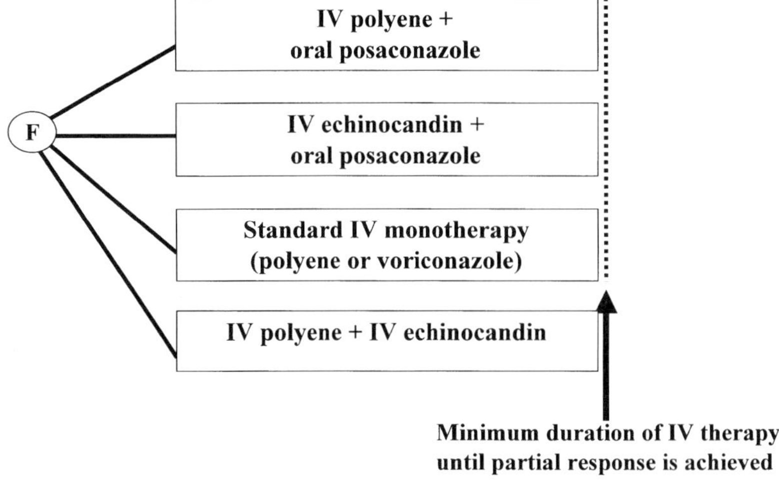

Figure 2. Pharmacologic options for salvage therapy of IA with VRC monotherapy.

Current Controversy 1: How Does One Deal with Voriconazole Failures in Documented IA?

In fact, do we know how to define "failure"? If we do, do we understand its mechanics? Specifically, what are the contributions? Is it "host failure," poor pharmacology associated with suboptimal VRC levels (Pascual et al., 2007; Trifilio et al., 2007a, 2007b), drug-resistant *Aspergillus* spp. (Verweij et al., 2007a), VRC-resistant molds such as the Zygomycetes (Kontoyiannis and Lewis, 2006; Kontoyiannis et al., 2005), or some combination of the above? Today, fewer patients with IA in the setting of BMT or leukemia undergo an autopsy, challenging our confidence in the performance of drugs and diagnostic tests for this infection (Chamilos et al., 2006). Attempts to streamline our understanding of what constitutes response or failure to antifungal therapy have been recently published (Segal et al., 2008).

Current Controversy 2: How Does One Integrate Serum Azole Monitoring into the Therapeutic Algorithms and Therapeutic Decisions for IA?

A major advance in the treatment of IA has been the arrival of new treatment options, including new triazole antifungals with improved activity against *Aspergillus* species, such as VRC (Denning et al., 2002; Herbrecht et al., 2002), posaconazole (POSA) (Walsh et al., 2007), and even itraconazole (ITC) (Caillot, 2003; Stevens and Lee, 1997). The new triazoles show impressive activity in vitro and in animal models of IA (Boucher et al., 2004), and the availability of oral formulations allows their use for long-term therapy. However, there is emerging concern about their bioavailability because of interactions with other concomitantly given medications (VRC, ITC), pharmacogenetic differences in metabolism (VRC), and impaired oral absorption in the setting of mucositis or poor oral intake (POSA) (Lewis, 2008; Trifilio et al., 2007b). In a patient with IA treated with a triazole alone who has evidence of smoldering progression of IA, an attempt to exclude a superinfection and/or low triazole drug exposures should be made. Whether another class of agent(s) should be used is an individualized decision that depends on the stability of the clinical picture, degree of immunosuppression, potential involvement of anatomical sanctuary sites (e.g., the central nervous system), major organ (liver, kidney) comorbidities, and whether azole levels can be routinely monitored in a timely fashion.

Current Controversy 3: How Does One Interpret the Role of POSA, ITC, or the Echinocandins and Other Agents Reported To Have Activity as Salvage Agents?

It is difficult to interpret the results of salvage therapy studies in patients with IA. The status of the patient's host defenses is critical to the therapeutic response. Neutropenic patients typically do not respond unless their neutrophil counts recover during therapy, a factor not always reported in the literature. Solid organ transplant recipients often have better response rates because their immunosuppressive therapy can be modified temporarily. Some patients are left with residual lesions after the infection has resolved, and these lesions may relapse. It is important to emphasize that most of the data about salvage therapy of single (Caillot, 2003; Denning et al., 2002; Maertens et al., 2004; Walsh et al., 1998, 2007) or combination therapy (Aliff et al., 2003; Kontoyiannis et al., 2003a; Marr et al., 2004b; Raad et al., 2008) have been generated in patients who did not respond or were intolerant to AMB-based therapy. These data might not be extrapolated to patients with IA who did not respond or were intolerant to VRC given as primary therapy. Specifically, the issue of cross-resistance or tolerance between the new triazoles can be an important one (Verweij et al., 2007a) and invites the question of whether POSA is effective in VRC failures in view of the reports of multi-azole resistance.

Current Controversy 4: Do We Agree on the Best Agent for "Preemptive" Treatment of IA? How Should Local Epidemiology Influence Such Decisions?

In view of the evolving complex epidemiology of invasive mold infections that involves not only IA but mixed fungal infections (Chamilos et al., 2006) and infections with other genera (e.g., the *Zygomycetes*) (Kontoyiannis et al., 2005a; Trifilio et al., 2007b), should we always start these patients on a VRC-based preemptive regimen, or is it more prudent to include *Zygomycetes* coverage with AMB? Missing the diagnosis of zygomycosis, a less frequent infection that clinically mimics IA (Chamilos et al., 2005; Kontoyiannis and Lewis, 2006; Kontoyiannis et al., 2005a), might have devastating consequences, as delaying effective therapy by 6 days doubles the mortality rate (Chamilos et al., 2008). On the other hand, the emergence of *Aspergillus terreus*, an *Aspergillus* species that tends to respond less to AMB than to VRC (Hachem et al., 2004; Steinbach et al., 2004), emphasizes the importance of close monitoring of evolving local epidemiology. Microbial epidemiology is a moving target, influenced by exposure, profiles of patients at risk, immunosuppression practices, and antifungal selection pressure; as this knowledge also requires reliable fungus identification, diligent diagnostics cannot be overemphasized. Although preemptive approaches instead of empiric antifungal therapy have been increasingly employed in settings where the pretest probability of early IA is high, it is still unclear whether such a strategy is beneficial. However, if one decides to employ a lipid formulation-AMB-based initial therapy, are there toxicity or pharmacokinetic/pharmacodynamic reasons to choose one lipid AMB formulation over another? Most of the available data have been derived from indirect comparisons and suggest that all lipid formulations of AMB, when given at the standard dosage of 5 mg/kg/day, appear to have comparable efficacy in the setting of empiric antifungal therapy (Wingard et al., 1999). However, differences in rapidity of lung concentrations of AMB lipid complex versus AmBisome have been recently shown in experimental models of IA (Becker et al., 2002; Lewis et al., 2007). Whether these differences are translated to clinical benefit, especially in the setting of IA caused by AMB-resistant species such as *A. terreus*, remains to be seen.

Current Controversy 5: Will Antifungal Combinations Live up to Their Promise?

VRC has a 66% failure rate in the BMT patient population (Herbrecht et al. 2002), and the availability of new agents has made the concept of combination therapy theoretically appealing (Kontoyiannis and Lewis, 2004). To date, no clinical studies have convincingly determined whether antifungal combinations are more beneficial than therapy using one agent for IA, and results of published observations are mixed. For instance, one retrospective single-institution experience suggested that the combinations of ITC (Kontoyiannis et al., 2003b) or caspofungin with lipid formulations of AMB were not beneficial for salvage therapy in refractory IA (Kontoyiannis et al., 2003a). In contrast, a small recently published prospective study suggested that the combination of caspofungin with modest doses (3 mg/kg/day) of liposomal AMB compared to high doses (10 mg/kg/day) of liposomal AMB were beneficial as primary therapy of IA (Caillot et al., 2007). In addition, uncontrolled studies in patients with IA in the setting of solid organ transplantation (Singh et al., 2006) and BMT (Marr et al., 2004b) have indicated that the combination of VRC and caspofungin appears beneficial as salvage therapy. Again, it is unclear if these promising results can be extrapolated in VRC-refractory IA. This is one of the most important areas for clinical investigation in IA. The benefits of combination therapies need to be subjected to the rigor of a careful, prospective randomized trial that will incorporate a sizeable number of IA patients at high risk for a poor outcome.

Current Controversy 6: What Is the Utility of Surgical Excision?

Pulmonary infarcts and tissue sequestration are common causes of failed antifungal therapy and fatal hemorrhage. The role of adjunctive surgery in the management of aspergillosis also has not been addressed in a conclusive way (Kontoyiannis and Bodey, 2002). In one study, early detection of *Aspergillus* lesions by chest CT combined with aggressive antifungal and surgical treatment appeared to confer a survival advantage (Caillot et al., 1997), and resection of infected pulmonary tissue appeared to be beneficial in another cohort of selected patients (Yeghen et al., 2000). Residual cavitary lesions, especially those containing fungus balls, after successful antifungal therapy may cause late exsanguinating hemorrhage or reactivation of infection during subsequent myelosuppressive chemotherapy. Removal of these lesions, if surgically feasible, should be considered and may provide a survival benefit (Offner et al., 1998). Surgical intervention may be life-saving for acute pulmonary hemorrhage, even when early in the disease process. However, potential benefits need to be weighed against risks, which include pulmonary function decline, surgical risks, creation of fungal empyemas by rupture within the pleural space, and, importantly, the risk of leukemia relapse in the postoperative period.

Current Controversy 7: What Is the Role of Immunomodulators in Management of IA?

As is the case with the other refractory opportunistic mycoses, in anecdotal clinical reports, the beneficial adjunctive role of immunomodulation using cytokines or the infusion of immune effector cells in various combinations (e.g., granulocyte-macrophage colony-stimulating factor, granulocyte colony-stimulating factor, gamma interferon, granulocyte-macrophage colony-stimulating factor-primed white cell transfusions) in refractory or recurrent IA cases is only suggestive. Many of these interventions appear at least safe, even for high-risk patients (Safdar et al., 2005, 2006). For some interventions, potential toxicities are not mild and may include alloimmunization, worsening of graft rejection or graft-versus-host disease, and pulmonary toxicities. Adequately powered safety and efficacy studies are necessary.

Current Controversy 8: How Does the Emerging Knowledge Regarding Immune Reconstitution and Immunopharmacology and Immunobiology of IA Influence Therapeutic Decisions?

Although the course of pathogen-specific immune reconstitution is ill-defined in neutropenic leukemic patients and in immunosuppressed, nonmyelosuppressed patients with IA, there is little doubt that in some cases there is a paradoxical worsening driven by an exuberant immune response to *Aspergillus* (Miceli et al., 2007). In addition, there is an increased appreciation that modern antifungal agents interact in vivo with host immune functions involved in defense against *Aspergillus* (Hohl and Feldmesser, 2007; Wheeler and Fink, 2006). The natures of such interactions are diverse and depend on the drug and the immunological status of the host. Given the prominent role of the host immune response in controlling IA, it might be that immune modulation by antifungal drugs may prove to be clinically significant. Elucidation of the immunopharmacology of these drugs may aid in designing therapeutic regimens for specific clinical scenarios associated with a defined immunological dysfunction (Ben-Ami et al., 2008).

Current Controversy 9: Are New Diagnostic Methods Best Suited as Adjunct Diagnostics in Early IA or as Surveillance Tools for Subclinical Disease?

Undisputably, delayed diagnosis remains a major impediment to successful treatment of IA. This is a fast-moving and increasingly complex area of investigation; with regard to the evolving epidemiology of opportunistic mold infections, developing diagnostic strategies to distinguish between IA and other mycoses is another important future research direction. In recent years, efforts have been directed towards identifying non-culture-based markers for early, reliable diagnosis of IA, using detection of *Aspergillus* components such as GM, β-1,3-D-glucan, and DNA (Mennink-Kersten and Verweij, 2006). Studies of profoundly immunosuppressed BMT recipients receiving fluconazole prophylaxis have shown high sensitivity (67 to 100%) and specificity (86 to 99%) rates for the GM assay. However, the performance of the GM assay in other settings appears suboptimal, including pediatric patients, patients receiving mold-active antifungal prophylaxis, patients with graft-versus-host disease, and recipients of solid organ transplants. Other factors, such as the pretest probability of infection, sequestration of *Aspergillus* lesions, patient's immune status, presence of anti-GM antibodies, antibacterial therapy, and diet, may also affect both the performance and interpretation of the GM assay (Marr et al., 2004a, 2005). A colorimetric assay for the detection of β-1,3-D-glucan, an integral cell wall component of most pathogenic fungi, has shown promising sensitivity (55 to 100%) and specificity (52 to 100%) in limited studies (Chamilos and Kontoyiannis, 2006). PCR analysis of *Aspergillus* DNA is also a promising method for early detection of IA and other opportunistic fungal infections (Chamilos and Kontoyiannis, 2006). The sensitivity of PCR has been variable but was excellent in some studies; however, low specificities may be problematic. Multiple unresolved issues accompany the use of PCR for diagnosis of IA, including the sample type, amplification strategy, protocol, and primer selection, and these differences account for the lack of a standardized, commercially available assay. A comparative prospective evaluation of non-culture-based assays would facilitate their incorporation into preemptive strategies for the optimal management of IA. Combinations of non-culture-based assays could enhance their performance and broaden the spectrum of detectable fungi. Whether antifungal responses are correlated with the rise of markers such as GM, as suggested by single-institution reports (Woods et al., 2007), is a controversial area where more studies are needed. Clearly, there is a need for innovation in the area of *Aspergillus* diagnostics.

Current Controversy 10: Other Unresolved Issues

Advancements in early diagnosis by CT and the introduction of effective agents have improved response rates of primary antifungal therapy. These changes allow the subsequent continuation of cytotoxic chemotherapy and/or performance of BMT in an increasing number of patients with hematological malignancies. These developments have increased the interest in secondary prophylaxis of IA, as the resumption of myelotoxic chemotherapy in these patients is associated with high rates of relapse in the absence of prophylaxis. However, the strategies for reducing risk are not well defined. The limited uncontrolled data regarding the role of secondary antifungal prophylaxis suggest that the new triazoles have a prominent role (Sipsas and Kontoyiannis, 2006). However, results of a large multicenter study emphasize that multiple host factors (e.g., disease risk) and transplant variables (e.g., type of conditioning therapy, type of donor, and stem cell source) are strong predictors of relapsed infection (Martino et al., 2006).

Finally, the role of local antifungal drug delivery in selected cases of cavitary pulmonary IA (Arthur et al., 2004), the role of the modern mycology laboratory in disease management (e.g., the role of *Aspergillus* susceptibility testing), and how to manage the toxicities of antifungal agents are areas where more experience is needed. For example, it is important to emphasize that the known toxicities and drug-drug interactions of agents employed in the treatment of IA typically have been derived from healthy volunteer studies. In view of the polypharmacy encountered in high-risk patients with IA who typically have multiple comorbidities and who receive the antifungal agents for several weeks or months, it is quite challenging to detect and manage side effects that are attributed to antifungals. In addition, the role of in vitro susceptibility testing in *Aspergillus* species remains investigational, as there are no susceptibility breakpoints that correlate with outcome. Such correlations are becoming increasingly difficult to make, if not impossible (Lionakis et al., 2005). On the other hand, the importance of accurate speciation for *Aspergillus* should be emphasized, as several non-fumigatus *Aspergillus* species, such as *A. terreus* (Steinbach et al., 2004), *A. ustus*, and *A. nidulans* (Kontoyiannis and

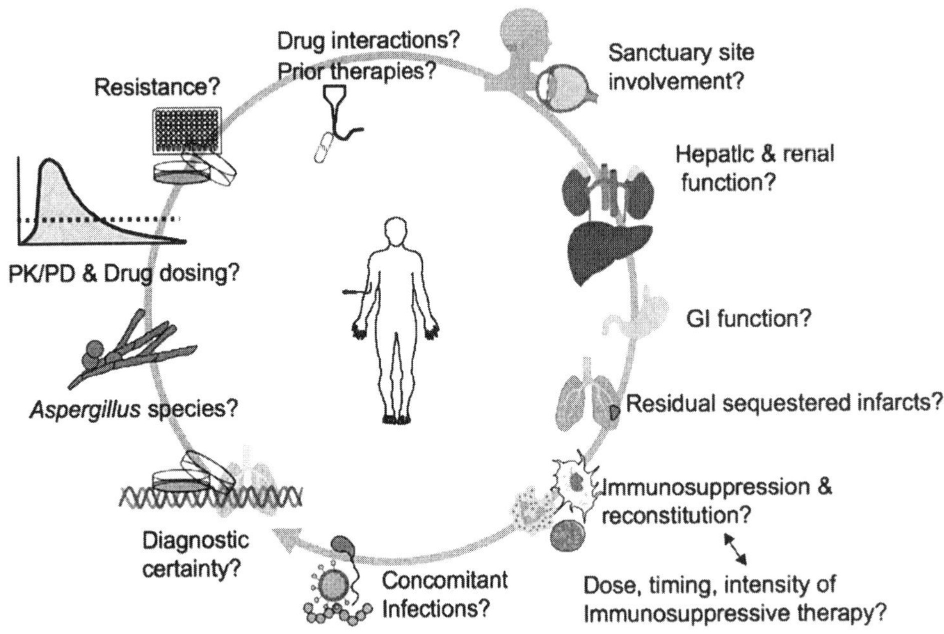

Figure 3. Factors influencing individualization of therapy in IA.

Bodey, 2002), demonstrate in vitro resistance to antifungals, especially AMB. "Cryptic" species that are misidentified as *A. fumigatus* but that have unique biological and resistance patterns, such as *A. lentulus*, have been identified as pathogens in multiple medical centers worldwide (Balajee et al., 2005).

CONCLUSIONS

We have come a long way in studying IA, generating some consensus regarding diagnostic approaches and antifungal therapies. Such evidence needs to be subjected to further validation in carefully conducted randomized studies. Figure 3 outlines the general principles in the management of acute IA and the importance of individualization of therapy. Currently, there does not exist a single drug or antifungal strategy of choice; decisions need to be made based on the host's state of immunosuppression and comorbidities, local epidemiology, previous antifungal exposures, acuity of illness, site of IA, and logistic considerations. There is still much to learn about treating this infection.

Acknowledgments. We thank R. E. Lewis for assistance with Fig. 3.

D.P.K. has received research support and honoraria from Merck & Co., Inc., Pfizer, Inc., Fujisawa Pharmaceutical Co., Ltd., and Enzon Pharmaceuticals and serves on the advisory board for Merck & Co., Inc., and Schering-Plough Research Institute. K.A.M. has received research support from Astellas, Merck, and Pfizer and serves as a consultant or on advisory boards for Astellas, Basilea, Enzon, F2G, Merck, Pfizer, and Schering-Plough Research Institute.

REFERENCES

Aliff, T. B., P. G. Maslak, J. G. Jurcic, M. L. Heaney, K. N. Cathcart, K. A. Sepkowitz, and M. A. Weiss. 2003. Refractory *Aspergillus* pneumonia in patients with acute leukemia. *Cancer* **97:**1025–1032.

Arthur, R. R., R. H. Drew, and J. R. Perfect. 2004. Novel modes of antifungal drug administration. *Expert Opin. Investig. Drugs* **13:** 903–932.

Balajee, S. A., J. L. Gribskov, E. Hanley, D. Nickle, and K. A. Marr. 2005. *Aspergillus lentulus* sp. nov., a new sibling species of *A. fumigatus*. *Eukaryot. Cell* **4:**625–632.

Becker, M. J., S. de Marie, M. Fens, W. C. J. Hop, H. A. Verbrugh, and I. Bakker-Woudenberg. 2002. Enhanced antifungal efficacy in experimental invasive pulmonary aspergillosis by combination of AmBisome with Fungizone as assessed by several parameters of antifungal response. *J. Antimicrob. Chemother.* **49:**813–820.

Ben-Ami, R., R. E. Lewis, and D. P. Kontoyiannis. 2008. Immunocompromised hosts: immunopharmacology of modern antifungals. *Clin. Infect. Dis.* **47:**226–235.

Boucher, H. W., A. H. Groll, C. C. Chiou, and T. J. Walsh. 2004. Newer systemic antifungal agents: pharmacokinetics, safety and efficacy. *Drugs* **64:**1997–2020.

Cagnoni, P. J., T. J. Walsh, M. M. Prendergast, D. Bodensteiner, S. Hiemenz, R. N. Greenberg, C. A. Arndt, M. Schuster, N. Seibel, V. Yeldandi, and K. B. Tong. 1997. Pharmacoeconomic analysis of liposomal amphotericin B versus conventional amphotericin B in the empirical treatment of persistently febrile neutropenic patients. *J. Clin. Oncol.* **18:**2476–2483.

Caillot, D. 2003. Intravenous itraconazole followed by oral itraconazole for the treatment of amphotericin B-refractory invasive pulmonary aspergillosis. *Acta Haematol.* **109:**111–118.

Caillot, D., O. Casasnovas, A. Bernard, J. F. Couaillier, C. Durand, B. Cuisenier, E. Solary, F. Piard, T. Petrella, A. Bonnin, G. Couillault, M. Dumas, and H. Guy. 1997. Improved management of invasive pulmonary aspergillosis in neutropenic patients using early thoracic computed tomographic scan and surgery. *J. Clin. Oncol.* **15:**139–147.

Caillot, D., A. Thiebaut, R. Herbrecht, S. de Botton, A. Pigneux, F. Bernard, J. Larche, F. Monchecourt, S. Alfandari, and L. Mahi. 2007. Liposomal amphotericin B in combination with caspofungin for invasive aspergillosis in patients with hematologic malignancies: a randomized pilot study (Combistrat trial). *Cancer* **110:**2740–2746.

Chamilos, G., and D. P. Kontoyiannis. 2006. Defining the diagnosis of invasive aspergillosis. *Med. Mycol.* **44:**S163–S172.

Chamilos, G., R. E. Lewis, and D. P. Kontoyiannis. 2008. Delaying amphotericin B-based front-line therapy significantly increases mortality in hematologic malignancy patients with zygomycosis. *Clin. Infect. Dis.* **47:**503–509.

Chamilos, G., M. Luna, R. E. Lewis, G. P. Bodey, R. Chemaly, J. J. Tarrand, A. Safdar, I. I. Raad, and D. P. Kontoyiannis. 2006. Invasive fungal infections in patients with hematologic malignancies in a tertiary care cancer center: an autopsy study over a 15-year period (1989–2003). *Haematologica* **91:**986–989.

Chamilos, G., E. M. Marom, R. E. Lewis, M. S. Lionakis, and D. P. Kontoyiannis. 2005. Predictors of pulmonary zygomycosis versus invasive pulmonary aspergillosis in patients with cancer. *Clin. Infect. Dis.* **41:**60–66.

Cornely, O. A., J. Maertens, M. Bresnik, R. Ebrahimi, A. J. Ullmann, E. Bouza, C. P. Heussel, O. Lortholary, C. Rieger, A. Boehme, M. Aoun, H. A. Horst, A. Thiebaut, M. Ruhnke, D. Reichert, N. Vianelli, S. W. Krause, E. Olavarria, and R. Herbrecht. 2007. Liposomal amphotericin B as initial therapy for invasive mold infection: a randomized trial comparing a high-loading dose regimen with standard dosing (AmBiLoad trial). *Clin. Infect. Dis.* **44:**1289–1297.

Denning, D. W., P. Ribaud, N. Milpied, D. Caillot, R. Herbrecht, E. Thiel, A. Haas, M. Ruhnke, and H. Lode. 2002. Efficacy and safety of voriconazole in the treatment of acute invasive aspergillosis. *Clin. Infect. Dis.* **34:**563–571.

Greene, R. E., H. T. Schlamm, J. W. Oestmann, P. Stark, C. Durand, O. Lortholary, J. R. Wingard, R. Herbrecht, P. Ribaud, T. F. Patterson, P. F. Troke, D. W. Denning, J. E. Bennett, B. E. de Pauw, and R. H. Rubin. 2007. Imaging findings in acute invasive pulmonary aspergillosis: clinical significance of the halo sign. *Clin. Infect. Dis.* **44:**373–379.

Groll, A. H., S. C. Piscitelli, and T. J. Walsh. 1998. Clinical pharmacology of systemic antifungal agents: a comprehensive review of agents in clinical use, current investigational compounds, and putative targets for antifungal drug development. *Adv. Pharmacol.* **44:**343–500.

Hachem, R. Y., D. P. Kontoyiannis, M. R. Boktour, C. Afif, C. Cooksley, G. P. Bodey, I. Chatzinikolaou, C. Perego, H. M. Kantarjian, and I. I. Raad. 2004. *Aspergillus terreus*: an emerging amphotericin B-resistant opportunistic mold in patients with hematologic malignancies. *Cancer* **101:**1594–1600.

Herbrecht, R., D. W. Denning, T. F. Patterson, J. E. Bennett, R. E. Greene, J. W. Oestmann, W. V. Kern, K. A. Marr, P. Ribaud, O. Lortholary, R. Sylvester, R. H. Rubin, J. R. Wingard, P. Stark, C. Durand, D. Caillot, E. Thiel, P. H. Chandrasekar, M. R. Hodges, H. T. Schlamm, P. F. Troke, and B. de Pauw. 2002. Voriconazole versus amphotericin B for primary therapy of invasive aspergillosis. *N. Engl. J. Med.* **347:**408–415.

Hohl, T. M., and M. Feldmesser. 2007. *Aspergillus fumigatus*: principles of pathogenesis and host defense. *Eukaryot. Cell* **6:**1953–1963.

Kontoyiannis, D. P., and G. P. Bodey. 2002. Invasive aspergillosis in 2002: an update. *Eur. J. Clin. Microbiol. Infect. Dis.* **21:**161–172.

Kontoyiannis, D. P., R. Hachem, R. E. Lewis, G. A. Rivero, H. A. Torres, J. Thornby, R. Champlin, H. Kantarjian, G. P. Bodey, and I. I. Raad. 2003a. Efficacy and toxicity of caspofungin in combination with liposomal amphotericin B as primary or salvage treatment of invasive aspergillosis in patients with hematologic malignancies. *Cancer* **98:**292–299.

Kontoyiannis, D. P., R. E. Lewis, M. S. Lionakis, N. D. Albert, G. S. May, and I. I. Raad. 2003b. Sequential exposure of *Aspergillus fumigatus* to itraconazole and caspofungin: evidence of enhanced in vitro activity. *Diagn. Microbiol. Infect. Dis.* **47:**415–419.

Kontoyiannis, D. P., and R. E. Lewis. 2006. Invasive zygomycosis: update on pathogenesis, clinical manifestations, and management. *Infect. Dis. Clin. North Am.* **20:**581–607.

Kontoyiannis, D. P., and R. E. Lewis. 2004. Toward more effective antifungal therapy: the prospects of combination therapy. *Br. J. Haematol.* **126:**165–175.

Kontoyiannis, D. P., M. S. Lionakis, R. E. Lewis, G. Chamilos, M. Healy, C. Perego, A. Safdar, H. Kantarjian, R. Champlin, T. J. Walsh, and I. I. Raad. 2005. Zygomycosis in a tertiary-care cancer center in the era of *Aspergillus*-active antifungal therapy: a case-control observational study of 27 recent cases. *J. Infect. Dis.* **191:**1350–1360.

Latgé, J. P. 1999. *Aspergillus fumigatus* and aspergillosis. *Clin. Microbiol. Rev.* **12:**310–350.

Lewis, R. E. 2008. What is the "therapeutic range" for voriconazole? *Clin. Infect. Dis.* **46:**212–214.

Lewis, R. E., G. Liao, J. Hou, G. Chamilos, R. A. Prince, and D. P. Kontoyiannis. 2007. A comparative analysis of amphotericin B lipid complex and liposomal amphotericin B kinetics of lung accumulation and fungal clearance in a murine model of acute invasive pulmonary aspergillosis. *Antimicrob. Agents Chemother.* **51:**1253–1258.

Lionakis, M. S., R. E. Lewis, G. Chamilos, and D. P. Kontoyiannis. 2005. *Aspergillus* susceptibility testing in patients with cancer and invasive aspergillosis: difficulties in establishing correlation between in vitro susceptibility data and the outcome of initial amphotericin B therapy. *Pharmacotherapy* **25:**1174–1180.

Maertens, J., I. Raad, G. Petrikkos, M. Boogaerts, D. Selleslag, F. B. Petersen, C. A. Sable, N. A. Kartsonis, A. Ngai, A. Taylor, T. F. Patterson, D. W. Denning, and T. J. Walsh. 2004. Efficacy and safety of caspofungin for treatment of invasive aspergillosis in patients refractory to or intolerant of conventional antifungal therapy. *Clin. Infect. Dis.* **39:**1563–1571.

Marr, K. A., S. A. Balajee, L. McLaughlin, M. Tabouret, C. Bentsen, and T. J. Walsh. 2004a. Detection of galactomannan antigenemia by enzyme immunoassay for the diagnosis of invasive aspergillosis: variables that affect performance. *J. Infect. Dis.* **190:**641–649.

Marr, K. A., M. Boeckh, R. A. Carter, H. W. Kim, and L. Corey. 2004b. Combination antifungal therapy for invasive aspergillosis. *Clin. Infect. Dis.* **39:**797–802.

Marr, K. A., M. Laverdiere, A. Gugel, and W. Leisenring. 2005. Antifungal therapy decreases sensitivity of the *Aspergillus* galactomannan enzyme immunoassay. *Clin. Infect. Dis.* **40:**1762–1769.

Martino, R., R. Parody, T. Fukuda, J. Maertens, K. Theunissen, A. Ho, G. J. Mufti, N. Kroger, A. R. Zander, D. Heim, M. Paluszewska, D. Selleslag, K. Steinerova, P. Ljungman, S. Cesaro, A. Nightnen, C. Cordonnier, L. Vazquez, M. Lopez-Duaerte, J. Lopez, R. Cabreara, M. Rovira, S. Neuburger, O. Cornely, A. E. Hunter, K. A. Marr, H. J. Dornbusch, and H. Einsele. 2006. Impact of the intensity of the prettransplantation conditioning regimen in patients with prior invasive aspergillosis undergoing allogeneic hematopoietic stem cell transplantation: a retrospective survey of the Infectious Diseases Working Party of the European Group for Blood and Marrow Transplantation. *Blood* **108:**2928–2936.

Mennink-Kersten, M. A., and P. E. Verweij. 2006. Non-culture-based diagnostics for opportunistic fungi. *Infect. Dis. Clin. North Am.* **20:**711–727.

Miceli, M. H., J. Maertens, K. Buve, M. Grazziutti, G. Woods, M. Rahman, B. Barlogie, and E. J. Anaissie. 2007. Immune reconstitution inflammatory syndrome in cancer patients with pulmonary aspergillosis recovering from neutropenia: proof of principle, description, and clinical and research implications. *Cancer* **110**:112–120.

Offner, F., C. Cordonnier, P. Ljungman, H. G. Prentice, D. Engelhard, D. De Bacquer, F. Meunier, and B. De Pauw. 1998. Impact of previous aspergillosis on the outcome of bone marrow transplantation. *Clin. Infect. Dis.* **26**:1098–1103.

Ostrosky-Zeichner, L., K. A. Marr, J. H. Rex, and S. H. Cohen. 2003. Amphotericin B: time for a new "gold standard". *Clin. Infect. Dis.* **37**:415–425.

Pascual, A., T. Calandra, S. Bolay, T. Buclin, J. Bille, and O. Marchetti. 2007. Voriconazole therapeutic drug monitoring in patients with invasive mycoses improves safety and efficacy outcomes. *Clin. Infect. Dis.* **46**:201–211.

Patterson, T. F., H. W. Boucher, R. Herbrecht, D. W. Denning, O. Lortholary, P. Ribaud, R. H. Rubin, J. R. Wingard, B. DePauw, H. T. Schlamm, P. Troke, and J. E. Bennett. 2005. Strategy of following voriconazole versus amphotericin B therapy with other licensed antifungal therapy for primary treatment of invasive aspergillosis: impact of other therapies on outcome. *Clin. Infect. Dis.* **41**: 1448–1452.

Patterson, T. F., W. R. Kirkpatrick, M. White, J. W. Hiemenz, J. R. Wingard, B. Dupont, M. G. Rinaldi, D. A. Stevens, and J. R. Graybill. 2000. Invasive aspergillosis: disease spectrum, treatment practices, and outcomes. *Medicine* **79**:250–260.

Raad, I. I., H. A. Hanna, M. Boktour, Y. Jiang, H. Torres, C. Afif, D. P. Kontoyiannis, and R. Y. Hachem. 2008. Novel antifungal agents as salvage therapy for invasive aspergillosis in patients with hematologic malignancies: posaconazole compared with high-dose lipid formulations of amphotericin B alone or in combination with caspofungin. *Leukemia* **22**:496–503.

Safdar, A., G. Rodriguez, N. Ohmagari, D. P. Kontoyiannis, K. V. Rolston, I. I. Raad, and R. E. Champlin. 2005. The safety of interferon-gamma-1b therapy for invasive fungal infections after hematopoietic stem cell transplantation. *Cancer* **103**:731–739.

Safdar, A., G. H. Rodriguez, B. Lichtiger, B. F. Dickey, D. P. Kontoyiannis, E. J. Freireich, E. J. Shpall, I. I. Raad, H. M. Kantarjian, and R. E. Champlin. 2006. Recombinant interferon γ1b immune enhancement in 20 patients with hematologic malignancies and systemic opportunistic infections treated with donor granulocyte transfusions. *Cancer* **106**:2664–2671.

Segal, B. H., R. Herbrecht, D. A. Stevens, L. Ostrosky-Zeichner, J. Sobel, C. Viscoli, T. J. Walsh, J. Maertens, T. F. Patterson, J. R. Perfect, B. Dupont, J. R. Wingard, T. Calandra, C. A. Kauffman, J. R. Graybill, L. R. Baden, P. G. Pappas, J. E. Bennett, D. P. Kontoyiannis, C. Cordonnier, M. A. Viviani, J. Bille, N. G. Almyroudis, L. J. Wheat, W. Graninger, E. J. Bow, S. M. Holland, B. J. Kullberg, W. E. Dismukes, and B. E. De Pauw. 2008. Defining responses to therapy and study outcomes in clinical trials of invasive fungal diseases: Mycoses Study Group and European Organization for Research and Treatment of Cancer consensus criteria. *Clin. Infect. Dis.* **47**:674–683.

Singh, N., A. P. Limaye, G. Forrest, N. Safdar, P. Munoz, K. Pursell, S. Houston, F. Rosso, J. G. Montoya, P. Patton, R. Del Busto, J. M. Aguado, R. A. Fisher, G. B. Klintmalm, R. Miller, M. M. Wagener, R. E. Lewis, D. P. Kontoyiannis, and S. Husain. 2006. Combination of voriconazole and caspofungin as primary therapy for invasive aspergillosis in solid organ transplant recipients: a prospective, multicenter, observational study. *Transplantation* **81**:320–326.

Sipsas, N. V., and D. P. Kontoyiannis. 2006. Clinical issues regarding relapsing aspergillosis and the efficacy of secondary antifungal prophylaxis in patients with hematological malignancies. *Clin. Infect. Dis.* **42**:1584–1591.

Steinbach, W. J., D. K. Benjamin, Jr., D. P. Kontoyiannis, J. R. Perfect, I. Lutsar, K. A. Marr, M. S. Lionakis, H. A. Torres, H. Jafri, and T. J. Walsh. 2004. Infections due to Aspergillus terreus: a multicenter retrospective analysis of 83 cases. *Clin. Infect. Dis.* **39**:192–198.

Stevens, D. A., and J. Y. Lee. 1997. Analysis of compassionate use itraconazole therapy for invasive aspergillosis by the NIAID Mycoses Study Group criteria. *Arch. Intern. Med.* **157**:1857–1862.

Trifilio, S., G. Pennick, J. Pi, J. Zook, M. Golf, K. Kaniecki, S. Singhal, S. Williams, J. Winter, M. Tallman, L. Gordon, O. Frankfurt, A. Evens, and J. Mehta. 2007a. Monitoring plasma voriconazole levels may be necessary to avoid subtherapeutic levels in hematopoietic stem cell transplant recipients. *Cancer* **109**:1532–1535.

Trifilio, S., S. Singhal, S. Williams, O. Frankfurt, L. Gordon, A. Evens, J. Winter, M. Tallman, J. Pi, and J. Mehta. 2007b. Breakthrough fungal infections after allogeneic hematopoietic stem cell transplantation in patients on prophylactic voriconazole. *Bone Marrow Transplant.* **40**:451–456.

Upton, A., K. Kirby, P. Carpenter, M. Boeckh, and K. A. Marr. 2007. Invasive aspergillosis following hematopoietic cell transplantation: outcomes and prognostic factors associated with mortality. *Clin. Infect. Dis.* **44**:531–540.

Verweij, P. E., E. Mellado, and W. J. Melchers. 2007a. Multiple-triazole-resistant aspergillosis. *N. Engl. J. Med.* **356**:1481–1483.

Verweij, P. E., L. van Die, and J. P. Donnelly. 2007b. Halo sign and improved outcome. *Clin. Infect. Dis.* **44**:1666–1667.

Walsh, T. J., E. J. Anaissie, D. W. Denning, R. Herbrecht, D. P. Kontoyiannis, K. A. Marr, V. A. Morrison, B. H. Segal, W. J. Steinbach, D. A. Stevens, J. van Burik, J. R. Wingard, and T. F. Patterson. 2008. Treatment of aspergillosis: clinical practice guidelines of the Infectious Diseases Society of America. *Clin. Infect. Dis.* **46**:327–360.

Walsh, T. J., J. W. Hiemenz, N. Seible, J. R. Perfect, G. Horwith, L. Lee, J. L. Silber, M. J. DiNubile, A. Reboli, E. Bow, J. Lister, and E. J. Anaissie. 1998. Amphotericin B lipid complex in immunocompromised patients with invasive fungal infections: analysis of safety and efficacy in 556 cases. *Clin. Infect. Dis.* **26**:1383–1396.

Walsh, T. J., I. Raad, T. F. Patterson, P. Chandrasekar, G. R. Donowitz, R. Graybill, R. E. Greene, R. Hachem, S. Hadley, R. Herbrecht, A. Langston, A. Louie, P. Ribaud, B. H. Segal, D. A. Stevens, J. A. van Burik, C. S. White, G. Corcoran, J. Gogate, G. Krishna, L. Pedicone, C. Hardalo, and J. R. Perfect. 2007. Treatment of invasive aspergillosis with posaconazole in patients who are refractory to or intolerant of conventional therapy: an externally controlled trial. *Clin. Infect. Dis.* **44**:2–12.

Wheeler, R. T., and G. R. Fink. 2006. A drug-sensitive genetic network masks fungi from the immune system. *PLoS Pathog.* **2**:e35.

Wingard, J. R. 2002. Lipid formulations of amphotericins: are you a lumper or a splitter? *Clin. Infect. Dis.* **35**:891–895.

Wingard, J. R., M. H. White, E. J. Anaissie, J. T. Rafalli, J. L. Goodman, and A. C. Arieta. 1999. A randomized double blind study of Ambisome and Abelcet in febrile neutropenic patients, abstr. 015. *Progr. Abstr. Focus Fungal Infect. IX*, San Diego, CA, 18 to 20 March 1999. Imedex, Alpharetta, GA.

Woods, G., M. H. Miceli, M. L. Grazziutti, W. Zhao, B. Barlogie, and E. Anaissie. 2007. Serum *Aspergillus* galactomannan antigen values strongly correlate with outcome of invasive aspergillosis: a study of 56 patients with hematologic cancer. *Cancer* **110**:830–834.

Yeghen, T., C. C. Kibbler, H. G. Prentice, L. A. Berger, R. K. Wallesby, P. H. M. McWhinney, F. C. Lampe, and S. Gillespie. 2000. Management of invasive pulmonary aspergillosis in hematology patients: a review of 87 consecutive cases at a single institution. *Clin. Infect. Dis.* **31**:859–868.

VIII. SPECIFIC PATIENT POPULATIONS

Aspergillus fumigatus and Aspergillosis
Edited by J.-P. Latgé and W. J. Steinbach
©2009 ASM Press, Washington, DC

Chapter 38

Invasive Aspergillosis in Solid Organ Transplant Recipients

HSIN-YUN SUN, PATRICIA MUÑOZ, EMILIO BOUZA, AND NINA SINGH

Invasive aspergillosis occurs in 1 to 15% of organ transplant recipients (Table 1), and the mortality rate in transplant recipients with invasive aspergillosis typically ranges from 65 to 92% (Gavalda et al., 2005; Morgan et al., 2005; Paterson and Singh, 1999; Singh et al., 1997, 2003). An estimated 9.3 to 16.9% of all deaths in transplant recipients in the first year are considered attributable to invasive aspergillosis (Paterson and Singh, 1999). Although the incidence of invasive aspergillosis has declined, the risk factors for invasive aspergillosis have been better characterized, and outcomes have improved in the current era, invasive aspergillosis remains a significant posttransplant complication in solid organ transplant (SOT) recipients. This chapter discusses the current status and evolving trends in the epidemiology, risk factors, diagnostic laboratory assays, and the approach to antifungal prophylaxis and treatment of invasive aspergillosis in SOT recipients.

EPIDEMIOLOGY AND RISK FACTORS

The net state of immunosuppression and intensity of the immunosuppressive regimen is a major determinant for the development of invasive aspergillosis in SOT recipients, regardless of the type of SOT. However, the incidence of invasive aspergillosis differs, and there are unique risk factors for *Aspergillus* spp. infections for the various types of organ transplant recipients (Tables 1 and 2).

Liver Transplant Recipients

Invasive aspergillosis occurs in 1 to 9.2% of liver transplant recipients (Briegel et al., 1995; Brown et al., 1996; Gavalda et al., 2005; Kusne et al., 1992; Morgan et al., 2005; Paterson and Singh, 1999). A number of well-characterized risk factors have been shown to portend a high risk of invasive aspergillosis after liver transplantation. Retransplantation and renal failure are among the most significant risk factors for invasive aspergillosis in these patients (Fortun et al., 2002; Gavalda et al., 2005; Singh et al., 2001; Singh and Paterson, 2005). Retransplantation confers a 30-fold-higher risk, and renal dysfunction, particularly the requirement of any form of renal replacement therapy, e.g., hemodialysis or continuous venovenous hemofiltration, is associated with a 15- to 25-fold-greater risk of invasive fungal infections in liver transplant recipients. With the introduction of the model for end-stage liver disease (MELD) score as the basis for prioritizing liver transplant allocation in the United States, the number of liver transplant recipients with renal dysfunction has increased significantly in recent years (Gonwa et al., 2006). Most invasive fungal infections in these high-risk patients occur within the first month posttransplant; the median time to onset of invasive aspergillosis after renal replacement therapy and retransplantation was 13 and 28 days, respectively, in one study (Singh et al., 2001, 2006c). Other factors associated with invasive aspergillosis in liver transplant recipients include transplantation for fulminate hepatic failure, cytomegalovirus (CMV) infection, and prolonged intensive care unit stay (Collins et al., 1994; George et al., 1997; Osawa et al., 2007) (Table 2).

Historically, invasive aspergillosis in liver transplant recipients has occurred in the early posttransplant period; the median time to onset after transplantation was 17 days in one study (Singh et al., 1997) and 16 days in another (Selby et al., 1997). More recently, however, *Aspergillus* infections have been shown to occur in the late posttransplant period. In a study that compared a

Hsin-Yun Sun • Dept. of Internal Medicine, National Taiwan University Hospital and National Taiwan University College of Medicine, Taipei, Taiwan. **Patricia Muñoz and Emilio Bouza** • Clinical Microbiology and Infectious Diseases Dept., Hospital General Universitario Gregorio Marañón, Spanish Study Group of Infection in Transplant Patients, and CIBER de Enfermedades Respiratorias, Madrid, Spain. **Nina Singh** • Dept. of Internal Medicine, Division of Infectious Diseases, University of Pittsburgh, Pittsburgh, PA 15240.

Table 1. Epidemiologic characteristics of invasive aspergillosis in transplant recipients[a]

Transplanted organ	Incidence of IA[b] (%)	% of IA cases with onset timing[c]:		% of IA cases with disseminated disease	% Mortality
		Early	Late		
Liver	0.3–9.2	15 77	23–55	0–67	33–100
Lung	3–23.3	21–83.3	16.7–79	8.3–31.6	20–83.3
Heart	0.3–8.6	68	32	25–36.2	35.7–78
Kidney	0.1–1.6	60	40	10–25	0–100
Pancreas	1.3–4.2	NA[d]	NA	NA	66.7–100
Small bowel	0–10	NA	NA	66	66

[a] Data were obtained from various published studies (Briegel et al., 1995; Brown et al., 1996; Fortun et al., 2002; Gavalda et al., 2005; Hellinger et al., 2005; Husain et al., 2004, 2006; Iversen et al., 2007; Kanj et al., 1996; Kwak et al., 2002; Linden et al., 2003; Mehrad et al., 2001; Montoya et al., 2003; Morgan et al., 2005; Munoz et al., 2003, 2004; Paterson and Singh, 1999; Singh et al., 1997, 2003, 2006c; Singh and Husain, 2003; Sole et al., 2005; Veroux et al., 2007; Westney et al., 1996).
[b] IA, invasive aspergillosis.
[c] Early onset, IA occurring <90 days after transplantation; late onset, IA occurring ≥90 days after transplantation.
[d] NA, not available.

cohort of patients with invasive aspergillosis from 1998 to 2002 with similar patients from 1990 to 1995, 55% of the infections in the later cohort, compared with 23% in the earlier cohort, occurred after 90 days of transplantation (Singh et al., 2003). Improved outcomes in the early postoperative period due to technical surgical advances and delayed onset of posttransplant risk factors, such as CMV infection and allograft dysfunction due to recurrent hepatitis C virus infection, were proposed to have led to the delayed occurrence of invasive aspergillosis in liver transplant recipients in the current era (Singh et al., 2003). CMV and hepatitis C virus infection are independent risk factors for late-onset invasive aspergillosis in liver transplant recipients (Fortun et al., 2002; Gavalda et al., 2005; Singh et al., 1997).

The mortality rate in liver transplant recipients with invasive aspergillosis has typically ranged from 83 to 88% (Denning, 1998; Fortun et al., 2002; Paterson and Singh, 1999). Requirement of dialysis and CMV infection are independent predictors of mortality in SOT recipients, including liver transplant recipients, with invasive aspergillosis (Singh et al., 2006a). More recent studies have reported better outcomes, with mortality ranging from 33.3 to 65% (Fortun et al., 2003a; Morgan et al., 2005; Singh et al., 2003). Mortality rates, however, remain high in patients who develop invasive aspergillosis after liver retransplantation (82.4%), particularly in those undergoing retransplantation after 30 days of primary transplantation (100%) (Singh et al., 2006c).

Lung and Heart-Lung Transplant Recipients

Invasive fungal infections occur in 15 to 35% of lung transplant recipients, with *Aspergillus* spp. accounting for nearly one-half of these (Kanj et al., 1996; Maurer et al., 1992; Mehrad et al., 2001; Paradowski, 1997). Up to 4% of the patients may develop tracheobronchitis, and 6 to 8% develop invasive pulmonary or disseminated aspergillosis. Ulcerative tracheobronchitis is a locally invasive form of disease involving the anastomotic site and the large airways (Kramer et al., 1991). The bronchial anastomotic site, because of transient devascularization, is uniquely susceptible to ischemic injury and necrosis. Nearly 15% of lung transplant patients will develop some degree of airway narrowing, bronchomalacia, or stenosis that often requires a bronchial stent. Airway stents may trap fungi and can lead to fungal tracheobronchitis, impaction, and pneumonia. Lesions of *Aspergillus* tracheobronchitis may be ulcerative, pseudomembranous, or nodular (de Pablo et al., 2000; Mehrad et al., 2001; Singh and Husain, 2003). Tracheobronchial or anastomotic lesions can result in bronchopleural fistulas. Lesions in the vicinity of large blood vessels can lead to bronchovascular fistulas and potentially fatal hemorrhage.

In a review summarizing the literature on invasive aspergillosis in lung transplant recipients, the median time to onset of infection was 120 days posttransplant; 49 and 68% of the infections occurred within 3 and 6 months of lung transplantation, respectively (Singh, 2000). In more recent studies, however, invasive aspergillosis has been shown to occur later in the posttransplant period, largely due to the routine employment of antifungal prophylaxis in the early posttransplant period (Husain et al., 2006).

Direct communication of the transplanted lung with the environment and impaired local host defense mechanisms, such as the cough reflex and mucociliary clearance, render airway colonization a common occurrence in these patients. Airway colonization has been shown to portend a higher risk for invasive aspergillosis in some but not all studies in lung transplant recipients. Patients with airway cultures positive for *Aspergillus* within 6 months of lung transplantation were 11-fold more likely to develop invasive aspergillosis (Paradowski

Table 2. Risk factors for invasive aspergillosis in organ transplant recipients[a]

Liver transplant recipients
Retransplantation
Kidney failure, particularly requiring renal replacement therapy
Longer intensive care unit stay
Corticosteroid use
Hepatitis C virus or CMV infection for late-onset invasive aspergillosis
Lung transplant recipients
Single-lung transplant
CMV infection
Rejection and augmented immunosuppression
Obliterative bronchitis
Heart transplant recipients
Isolation of *Aspergillus* species in respiratory tract cultures
Kidney transplant recipients
Graft failure requiring hemodialysis
High and prolonged duration of corticosteroids
SOT recipients
Early-onset invasive aspergillosis
• Posttransplant kidney failure or requirement for hemodialysis
• >1 episode of bacterial infection
• Prolonged intensive care unit stay
• CMV disease
• Use of vascular amines for >24 h after operation
Late-onset invasive aspergillosis
• Immunosuppression-related neoplasm
• >1 episode of bacterial infection
• Chronic graft rejection
• Kidney failure after SOT
• Use of tacrolimus and cyclosporine A for the same patient
• SOT at age >50 years
• Blood levels of tacrolimus >15 ng/ml or cyclosporine A >500 ng/ml at month 3
• Receipt of sirolimus in conjunction with tacrolimus for refractory rejection or cardiac allograft

[a] Data were obtained from various published sources (Fortun et al., 2002; Gavalda et al., 2005; Husni et al., 1998; Munoz et al., 2003; Osawa et al., 2007; Paradowski, 1997; Paterson and Singh, 1999; Singh et al., 1997, 2001, 2006b; Singh and Husain, 2003).

et al., 1997). In a recent study, the only independent risk factor for invasive aspergillosis in lung transplant recipients in the first posttransplant year was *Aspergillus* colonization of the recipient in the 6 months before transplantation (odds ratio, 11.2; 95% confidence interval [CI], 1.4 to 89) (Gavalda et al., 2005). Intraoperative risk factors for invasive aspergillosis in lung transplant recipients include anastomotic complications and airway or graft ischemia. In the posttransplant period, specific risk factors for *Aspergillus* infection include reperfusion injury, bronchial anastomotic leaks, and airway narrowing (Kramer et al., 1991). CMV infection, poor allograft function, and requirement of a higher degree of immunosuppressive therapy for rejection or bronchiolitis are also risk factors for fungal infections after lung transplantation (Dauber et al., 1990; Husni et al., 1998).

In single-lung transplant patients, invasive aspergillosis affects the native lung more frequently than the transplanted lung (Sandur et al., 1999; Westney et al., 1996). Single-lung transplant recipients with invasive aspergillosis are also more likely to have chronic pulmonary obstructive disease as an underlying illness, to develop invasive aspergillosis at a later time posttransplantation, and to have a higher incidence of invasive pulmonary aspergillosis than bilateral transplant recipients (Singh and Husain, 2003).

Patients undergoing lung transplantation for cystic fibrosis have a higher rate of colonization with *Aspergillus fumigatus* before (53%) and after (59%) transplantation than patients transplanted for other indications (28%) (Helmi et al., 2003). Preoperatively colonized cystic fibrosis recipients may develop tracheobronchial aspergillosis distal to the bronchial anastomoses or dehiscence of the anastomosis despite antifungal prophylaxis. However, the overall rate of invasive aspergillosis is higher in non-cystic fibrosis compared to cystic fibrosis recipients (10% versus 0%) (Helmi et al., 2003). Early surveillance bronchoscopy is recommended to detect tracheobronchial aspergillosis in patients with cystic fibrosis, particularly in the recipients with pretransplant colonization (Helmi et al., 2003).

Reported mortality rates in lung transplant recipients with invasive aspergillosis have ranged from 20 to 83.3% (Gavalda et al., 2005; Husain et al., 2004; Morgan et al., 2005; Paterson and Singh, 1999). Patients with bronchial anastomotic infections have lower mortality rates (~20%) than those with invasive pulmonary infections (22 to 70%) (McAdams et al., 2001; Palmer et al., 2001).

Heart Transplant Recipients

Most invasive fungal infections in heart transplant recipients are caused by *Aspergillus* spp. (Grossi et al., 2000). Independent risk factors for invasive aspergillosis after heart transplantation are reoperation (relative risk [RR], 5.8; 95% CI, 1.8 to 18; $P = 0.002$), CMV disease (RR, 5.2; 95% CI, 2 to 13.9; $P = 0.001$), posttransplant hemodialysis (RR, 4.9; 95% CI, 1.2 to 18; $P = 0.02$), and the existence of an episode of invasive aspergillosis in the institutional transplant program 2 months before or after the date of transplantation (RR, 4.6; 95% CI, 1.5 to 14.4; $P = 0.007$) (Munoz et al., 2004). Antifungal prophylaxis with itraconazole was independently protective against invasive aspergillosis (RR, 0.2; 95% CI, 0.07 to 0.9; $P = 0.03$) and was associated with significantly prolonged 1-year survival (RR, 0.5; 95% CI, 0.3 to 0.8; $P = 0.01$) (Munoz et al., 2004). In heart transplant recipients with invasive aspergillosis, the overall mortality ranges from 66 to 78% (Gavalda et al., 2005; Morgan et al., 2005; Paterson and Singh, 1999).

Renal Transplant Recipients

Invasive aspergillosis has been reported in ~0.7% and in up to 4% of kidney transplant recipients (Altiparmak et al., 2002; Brown et al., 1996; Cofan et al., 1996; Gallis et al., 1975; Gustafson et al., 1983; Munoz et al., 1996; Paterson and Singh, 1999; Peterson et al., 1982; Weiland et al., 1983). High doses and prolonged duration of corticosteroids, graft failure requiring hemodialysis, and potent immunosuppressive therapy have been shown to be risk factors for invasive aspergillosis after kidney transplantation (Gustafson et al., 1983; Panackal et al., 2003; Paterson and Singh, 1999). Despite a relatively lower overall incidence compared to other organ transplant recipients, invasive aspergillosis is a significant contributor to morbidity in kidney transplant recipients. The mortality rate in kidney transplant recipients with invasive aspergillosis ranges from 67 to 75% (Gavalda et al., 2005; Paterson and Singh, 1999).

PATHOPHYSIOLOGIC BASIS OF INFECTION

The host defense against *Aspergillus* is comprised primarily of phagocytic cells; however, emerging data also show that the $CD4^+$ T-helper (Th) cell response also affects the susceptibility to invasive aspergillosis (Cenci et al., 1999; Grazziutti et al., 1997; Roilides et al., 1998). Whereas a Th1 or proinflammatory response is protective (Cenci et al., 1999; Roilides et al., 1998), a Th2 response is associated with disease progression (Cenci et al., 1999). Immunosuppression in organ transplant recipients is associated with deficits in cell-mediated adaptive immunity and to a lesser extent in innate immune responses. Th1 cytokines are the primary mediators of allograft rejection, and downregulation of Th1 responses is a major mechanism by which immunosuppressive agents, such as calcineurin inhibitors, prevent allograft rejection (D'Elios et al., 1997; Ferraris et al., 2004; Gras et al., 2004; Sugiyama et al., 2004; Weimer et al., 2000). Although less potent than calcineurin inhibitors, corticosteroids also exert an inhibitory effect on Th1 cytokines (Hodge et al., 1999; Lionakis and Kontoyiannis, 2003). In addition, corticosteroids suppress phagocytic function and prevent monocyte recruitment to the site of infection (Balloy et al., 2005; Duong et al., 1998; Stevens, 2006). Thus, iatrogenic immunosuppression in transplant recipients is associated with a dominant Th2 response that may portend a risk for invasive aspergillosis. It should be noted that neutropenia is not a major risk factor for invasive aspergillosis in SOT recipients.

A higher risk for *Aspergillus* infections conferred by a number of well-recognized risk factors in transplant recipients may also be due to their effects on T-cell-mediated immune responses. OKT3, an anti-CD3 monoclonal antibody, has been shown to be an independent risk factor for invasive aspergillosis in liver transplant recipients (Kusne et al., 1992). A major risk factor for invasive aspergillosis after liver transplantation is renal dysfunction (Fortun et al., 2002; Singh, 2000; Singh et al., 2001). Renal failure and hemodialysis have been shown to impair T-cell proliferative responses and result in an increase in activation-induced T-cell death (Ankersmit et al., 2001). A heightened susceptibility of transplant recipients with CMV infection to opportunistic mycoses, including invasive aspergillosis, is believed to result primarily from cell-mediated immunosuppressive effects of CMV (George et al., 1997). CMV may also affect the respiratory burst of macrophages (Laursen et al., 2001).

CLINICAL PRESENTATION

The most common site of invasive aspergillosis in SOT recipients is the lung, with pulmonary involvement present in ~92% of the cases. Pulmonary symptoms include nonproductive cough, pleuritic pain, low-grade fever, hemoptysis, and occasionally dyspnea. Invasive pulmonary aspergillosis in SOT recipients frequently presents with single or multiple pulmonary nodules with or without cavitation (Grossi et al., 1992; Hummel et al., 1992; Loire et al., 1993; Munoz et al., 2000). In a multicenter study conducted by the Spanish Group for Infections in Transplant Patients, which included 4,338 SOT recipients, over half of the patients had bilateral lung involvement; cavitation was present in 26% (P. Munoz, personal communication). Halo or air crescent signs are distinctly unusual in SOT recipients with invasive aspergillosis.

Aspergillus may disseminate from the lungs to virtually any organ. Disseminated disease has been documented in 15 to 20% of lung recipients and 20 to 35% of heart recipients (Bonham et al., 1998; Paterson and Singh, 1999). Liver transplant recipients, however, are uniquely predisposed to extrapulmonary dissemination, which occurs in ~50 to 60% of the cases with invasive aspergillosis (Paterson and Singh, 1999; Singh et al., 1997; Torre-Cisneros et al., 1993). Disseminated invasive aspergillosis and central nervous system involvement are particularly more likely to occur in liver retransplant recipients in the late posttransplant period, i.e., after 30 days of primary transplantation (Singh et al., 2006c). The frequency of disseminated infections after liver transplantation, however, has declined in the current era (Singh et al., 2003). It has been proposed that the potent in vitro antifungal activities of calcineurin inhibitors and target of rapamycin-inhibitor agents against *Aspergillus* may have a protective effect against disseminated infections due to *Aspergillus* (Rasmussen et al., 1994; Singh and Heitman, 2004; Steinbach et al., 2006).

In lung and heart-lung transplant recipients, invasive aspergillosis may present as bronchial anastomosis dehiscence, vascular anastomosis erosion, bronchitis, tracheobronchitis, invasive lung disease, aspergilloma, empyema, disseminated disease, endobronchial stent obstruction, or mucoid bronchial impaction. Less common clinical presentations of invasive aspergillosis in SOT recipients include spondylodiscitis, mediastinitis, endopthalmitis, endocarditis, or small bowel infarction (Akerele and Lightman, 2007; Forestier et al., 2005; Levin et al., 2004; Safieddine et al., 2002; Sherman-Weber et al., 2004; Weclawiak et al., 2007). A potentially highly lethal form of invasive aspergillosis in kidney transplant recipients is pseudoaneurysm of the iliac artery, which is transmitted by graft contamination during the preservation phase (Garrido et al., 2003).

DIAGNOSIS

A substantial delay in establishing an early diagnosis remains a major impediment to the successful treatment of invasive aspergillosis. Cultures of respiratory tract secretions lack sensitivity, and a fungus may only be detected in clinical samples in late stages of the disease. On the other hand, a positive culture with *Aspergillus* from respiratory tract samples does not always indicate invasive disease. The significance of a positive culture from an airway sample also varies with the type of organ transplant. Isolation of *Aspergillus* spp. from the respiratory tract of liver transplant recipients is an infrequent event (~1.5%). However, it has a high positive predictive value, ranging from 41 to 72% for the subsequent development of invasive aspergillosis (Paterson and Singh, 1999). *Aspergillus* spp. can be detected in airway samples of ~25 to 30% of lung transplant recipients (Cahill et al., 1997; Mehrad et al., 2001; Singh and Husain, 2003). While positive airway cultures have a low positive predictive value for the diagnosis of invasive aspergillosis in lung transplant recipients, they portend a higher risk for subsequent invasive infection (Paterson and Singh, 1999). Recovery of *Aspergillus* spp. from an airway sample in lung transplant recipients mandates a bronchoscopic examination to exclude the presence of tracheobronchitis or invasive disease, since radiographic and imaging studies may be unremarkable at this stage.

In heart transplant recipients, the positive predictive value of culturing *Aspergillus* from respiratory track samples for the diagnosis of invasive aspergillosis was 60 to 70% (Munoz et al., 2003). When analyzed by species, the positive predictive value of recovering *A. fumigatus* was 78 to 91%, whereas it was 0% for other species. The positive predictive value increased from 88 to 100% when *Aspergillus* was recovered from a respiratory specimen other than sputum, and it decreased from 67 to 50 when it was recovered only from sputum (Munoz et al., 2003). The sensitivities of fungal and conventional media for the recovery of *Aspergillus* spp. were 95% to 100% and 33% to 38%, respectively (Munoz et al., 2003).

The utility of the galactomannan test for the early diagnosis of invasive aspergillosis has been assessed in a limited number of studies in SOT recipients (Table 3). In liver transplant recipients for whom archived sera were tested, the sensitivity of the test was 55.6% and the specificity was 93.9% (Fortun et al., 2001). A prospective study in 154 liver transplant recipients documented a specificity of 98.5% (Kwak et al., 2004). In lung transplant recipients, the galactomannan test had a specificity of 95%, but a relatively low sensitivity (30%), for the diagnosis of invasive aspergillosis (Husain et al., 2004). Although the test was able to detect the single case of systemic invasive aspergillosis and 29% of the cases of pulmonary invasive aspergillosis, it detected none of the cases of *Aspergillus* tracheobronchitis (Husain et al., 2004). A meta-analysis showed that the galactomannan assay may have greater utility in hematopoietic stem cell transplant recipients than in SOT recipients; the sensitivity and specificity of the test for hematopoietic stem cell transplant recipients were 22% and 84%, respectively (Pfeiffer et al., 2006).

The sensitivity of the galactomannan assay for the diagnosis of invasive aspergillosis in SOT recipients may be improved by testing bronchoalveolar lavage (BAL) fluid. In one study, BAL fluid had a sensitivity of 67% and specificity of 98% at the index cutoff value of ≥1 for the diagnosis of invasive aspergillosis in lung transplant recipients (Husain et al., 2007). In another study, BAL had a sensitivity of 100% and specificity of 91% at the same index cutoff value for the diagnosis of invasive aspergillosis in SOT recipients (Clancy et al., 2007).

False-positive galactomannan tests have been documented in up to 13% of liver and 20% of lung transplant recipients (Husain et al., 2004; Kwak et al., 2004). Liver transplant recipients undergoing transplantation for autoimmune liver disease and those requiring dialysis were significantly more likely to have false-positive galactomannan tests (Kwak et al., 2004). In a report of lung transplant recipients, false reactivity with an *Aspergillus* enzyme immunoassay was documented in 20% (14 of 70) of the patients (Husain et al., 2004). Most false-positive tests occurred in the early posttransplant period, i.e., within 3 days of lung transplantation in 43%, within 7 days in 64%, and within 14 days of transplantation in 79% of the patients (Husain et al., 2004). Patients undergoing lung transplantation for cystic fibrosis and chronic obstructive pulmonary disease were

Table 3. Performance characteristics of the *Aspergillus* galactomannan enzyme immunoassay in studies in SOT recipients[a]

Reference	Specimen	Cutoff value	Transplant organ	No. of patients	Sensitivity (%)	Specificity (%)	PPV (%)	NPV (%)	False-positive rate (%)	Days by which test preceded presentation
Fortun et al., 2001	Serum	1.0	Liver	33	55.6	93.9	71.4	88.6	NA	NA
Kwak et al., 2004	Serum	0.5	Liver	154	NA	98.5	NA	NA	13	NA
Husain et al., 2004	Serum	0.5	Lung	70	30	95	NA	NA	20	3 days (in 1 of 12 patients)
Clancy et al., 2007	BAL	1.0	Solid organ	81	100	90.8	41.7	100	9.2	2 days to 4 weeks
Husain et al., 2007	BAL	1.0	Lung	116	60	98	36.4	98.1	6.4	−16 to 30 days

[a] Abbreviations: PPV, positive predictive value; NPV, negative predictive value; NA, not available.

more likely to have positive tests in the early posttransplant period (Husain et al., 2004). False-positive galactomannan tests in 29% of liver transplant recipients in the first week posttransplantation were attributed to perioperative prophylaxis with β-lactam agents (Fortun et al., 2007).

The utility of β-1,3-D-glucan for the diagnosis of invasive aspergillosis has not been fully defined. The test, however, was useful for the diagnosis of invasive aspergillosis in living-donor liver allograft recipients in one study (Kawagishi et al., 2006). A pan-fungal PCR of the blood was shown to be helpful and preceded clinical signs of invasive fungal infections in kidney transplant recipients by 27 days (Badiee et al., 2007). However, PCR-based molecular diagnostic tests for *Aspergillus* are not commercially available and remain largely unstandardized, and their precise role in the diagnosis and management of invasive aspergillosis in SOT recipients remains to be determined.

MANAGEMENT OF INVASIVE ASPERGILLOSIS

Treatment

General principles for the treatment of invasive aspergillosis in SOT recipients remain the same as for other patient populations. Prompt initiation of antifungal therapy is critical for achieving optimal outcomes in SOT recipients with invasive aspergillosis (Table 4). Beginning in the early 1990s and for almost a decade, lipid formulations of amphotericin B have been the mainstay for the treatment for invasive aspergillosis in SOT recipients, largely because of a lower potential of nephrotoxicity. In a study consisting of 47 SOT patients with invasive aspergillosis who were treated with lipid formulations of amphotericin B (5 to 7.4 mg/kg of body weight/day), the overall 90-day mortality was 49% and the invasive aspergillosis-associated mortality was 43% (Singh et al., 2006a). In another study that compared the efficacy of amphotericin B lipid complex (median dose of 5.2 mg/kg/day) and amphotericin B deoxycholate (median dose of 1.1 mg/kg/day) for the treatment of invasive aspergillosis in SOT recipients (Linden et al., 2003), the overall and invasive aspergillosis-related mortality rates were 33 and 25% in the amphotericin B lipid complex group and 83 and 76% in the amphotericin B deoxycholate group (Linden et al., 2003). Outcomes associated with the use of lipid formulations of amphotericin B in SOT recipients with invasive aspergillosis are summarized in Table 5.

The availability of newer triazole agents and the echinocandins, with potent anti-*Aspergillus* activities and better tolerability profiles, has led to an expanded armamentarium of antifungal agents for the treatment of invasive aspergillosis. A large, randomized trial that compared voriconazole with amphotericin B deoxycholate as primary therapy of invasive aspergillosis included 11 SOT recipients (Herbrecht et al., 2002). At week 12, successful outcomes were documented in 52.8% of the patients in the voriconazole group and in 31.6% in the amphotericin B group. The survival rate at 12 weeks was 70.8% in the voriconazole group and 57.9% in the amphotericin B group (hazard ratio, 0.59; 95% CI, 0.40 to 0.88). Voriconazole-treated patients had significantly fewer severe drug-related adverse events, except for transient visual disturbances.

Since this study, a number of reports of employing voriconazole for the treatment of invasive aspergillosis in SOT recipients have appeared in the literature. Three studies comprising four to six SOT patients with invasive aspergillosis documented that the complete or partial response rate (or favorable response rate) with voriconazole was 100%, 100%, and 50% (Denning et al., 2002; Fortun et al., 2003b; Veroux et al., 2007). In another report that included 11 SOT recipients with central nervous system aspergillosis treated with voriconazole, the favorable response rate was 36% (Schwartz et al., 2005). Voriconazole was successfully

Table 4. Proposed treatment of invasive aspergillosis in organ transplant recipients[a]

Infection category	Primary therapy	Alternative therapies
Invasive disease	Voriconazole: 6 mg/kg i.v. every 12 h for 2 doses, followed by 4 mg/kg i.v. every 12 h; oral dosage is 200 mg every 12 h	Liposomal amphotericin B (3–5 mg/kg/day i.v.) Amphotericin B lipid complex (5 mg/kg/day i.v.) Caspofungin (70 mg i.v. day 1 and 50 mg/day i.v. thereafter) Micafungin (100–150 mg/day i.v.; dose not established) Posaconazole (200 mg QID initially, then 400 mg BID orally after stabilization of disease) Itraconazole (dosage depending upon formulation)
Tracheobronchial infection	Voriconazole: 6 mg/kg i.v. every 12 h for 2 doses, followed by 4 mg/kg i.v. every 12 h; oral dosage is 200 mg every 12 h	Liposomal amphotericin B (3–5 mg/kg/day i.v.) Amphotericin B lipid complex (5 mg/kg/day i.v.) Itraconazole (dosage depends upon formulation) w/ or w/o inhaled nebulized amphotericin B (6 mg/kg q8h) Caspofungin (70 mg i.v. day 1 and 50 mg/day i.v. thereafter) Micafungin (100–150 mg/day i.v.; dose not established) Posaconazole (200 mg QID initially, then 400 mg BID orally after stabilization of disease)

[a] Abbreviations: i.v., intravenous; BID, twice a day; QID, four times a day; q8h, every 8 h.

used in heart transplant recipients as first-line and salvage therapy for invasive aspergillosis (Weclawiak et al., 2007; Wieland et al., 2005). Intravitreal voriconazole has also been used in a lung transplant patient with *Aspergillus* endophthalmitis (Kramer et al., 2006). Voriconazole is now regarded as the drug of choice for primary treatment of invasive aspergillosis in all hosts, including SOT recipients, a recommendation endorsed by the recent Clinical Practice Guidelines of the Infectious Diseases Society of America (IDSA) for the treatment of invasive aspergillosis (level A-I recommendation) (Walsh et al., 2008).

Caspofungin is the only echinocandin currently approved by the U.S. Food and Drug Administration for the treatment of invasive aspergillosis. In a study that employed caspofungin as primary therapy for invasive aspergillosis in 12 SOT recipients, the response rate was 92% (Groetzner et al., 2008). Caspofungin has been used successfully as salvage therapy in invasive aspergillosis as a single agent (Carby et al., 2004) and in combination with other drugs (Forestier et al., 2005; Shlobin et al., 2005; Vagefi et al., 2008). To date limited experience exists with the use of posaconazole, micafungin, or anidulafungin for the treatment of invasive aspergillosis in SOT recipients (Denning et al., 2006; Lodge et al., 2004; Walsh et al., 2007).

In patients developing therapy-limiting toxicity or with contraindications to voriconazole, liposomal amphotericin B is considered an alternative primary therapy as per the IDSA guidelines. Based on the AmBiLoad study, in which 3 mg/kg/day showed similar efficacy to 10 mg/kg/day and less toxicity, higher doses are not recommended (Cornely et al., 2007). Amphotericin B lipid complex, itraconazole, caspofungin, posaconazole, or micafungin (level B-II recommendation) are rational choices for alternative therapy for invasive aspergillosis (Walsh et al., 2008).

The new guidelines of the IDSA also recommend voriconazole as initial therapy for the treatment of tracheobronchial aspergillosis (Walsh et al., 2008). Therapy should be initiated promptly in an attempt to prevent anastomotic disruption and loss of the lung allograft in lung transplant recipients. Bronchoscopic evaluation is an essential tool for initial diagnosis, and a computed tomography chest examination is recommended to exclude extension into the remainder of the pulmonary tree and the lungs. Aerosolized amphotericin B deoxycholate or lipid formulations of amphotericin B may have some benefits; however, this approach for the treatment of tracheobronchial infection has not been standardized and remains investigational (Walsh et al., 2008). There is little experience with caspofungin or other echinocandins in treating tracheobronchial infections.

The role of combination antifungal therapy for invasive aspergillosis has not been fully defined. Updated guidelines of the IDSA suggest reserving this option for salvage therapy (Walsh et al., 2008). A prospective, multicenter study in SOT recipients compared outcomes in 40 patients who received voriconazole plus caspofungin as primary therapy for invasive aspergillosis with those in 47 patients in an earlier cohort who received a lipid formulation of amphotericin B as primary therapy (Singh et al., 2006a). The two groups were well-matched, including the proportion with disseminated disease (10% versus 12.8%), proven invasive aspergillosis (55% versus 51.1%), or *A. fumigatus* (71.1% versus 80.9%). Overall survival at 90 days was 67.5%

Table 5. Outcomes with the newer triazoles, echinocandins, and lipid formulations of amphotericin B as treatment for invasive aspergillosis in organ transplant recipients[a]

Antifungal therapy and reference	No. of SOT patients/total no. of patients (%)	Type of therapy	Response rate	% Overall mortality (% IA related)
Voriconazole				
Herbrecht et al., 2002	9/144 (6.2)	Primary	NA	NA
Veroux et al., 2007	4/4 (100)	Primary	100	0 (0)
Denning et al., 2002	6/116 (5)	Primary	50	NA (NA)
Schwartz et al., 2005	11/81 (14)	Salvage	36	NA (NA)
Fortun et al., 2003	4/5 (80)	Salvage	100	0 (0)
Voriconazole + caspofungin				
Singh et al., 2006a	40/40 (100)	Primary	NA	32.5 (26)
Posaconazole				
Walsh et al., 2007	12/107(11.2)	Salvage	58	NA (NA)
AmB deoxycholate				
Herbrecht et al., 2002	5/133 (3.8)	Primary	NA	NA (NA)
White et al., 1997	34/261 (13)	Primary	23.6	NA (NA)
Linden et al., 2003	29/29 (100)	Primary	NA	83 (76)
AmB lipid complex				
Linden et al., 2003	12/12 (100)	Primary	NA	33 (25)
Chandrasekar and Ito, 2005	109/398 (27)	Both	52	NA (NA)
Linden et al., 2000	39/39 (100)	Salvage	47	NA (NA)
ABCD				
White et al., 1997	10/82 (12.2)	Salvage	40	NA (NA)
Any lipid AmB				
Singh et al., 2006a	47/47 (100)	Primary	NA	49 (43)
Caspofungin				
Groetzner et al., 2008	12/12 (100)	Primary	92	8 (8)
Petrovic et al., 2007	16/16 (100)	Salvage	50	44 (NA)
Kartsonis et al., 2005	4/48 (8)	Salvage	25	NA (NA)
Maertens et al., 2004	9/83 (11)	Salvage	25	NA (NA)

[a]Abbreviations: IA, invasive aspergillosis; NA, not available; AmB, amphotericin B; ABCD, amphotericin B colloidal dispersion.

among the cases and 51% in the control group. Mortality was attributable to invasive aspergillosis in 26% of the cases and in 43% of the controls ($P = 0.11$). Deaths tended to occur later in cases than in the control patients (mean, 49.5 versus 36.7 days; $P < 0.11$). In a multivariate Cox regression model, CMV infection and kidney failure were independently predictive of mortality at 90 days. Combination therapy was associated with a trend towards lower mortality (hazard ratio, 0.58; 95% CI, 0.30 to 1.14; $P < 0.117$) when controlled for CMV infection and kidney failure. When 90-day mortality was analyzed in subgroups of patients, combination therapy was independently associated with reduced mortality in patients with kidney failure (and in those with *A. fumigatus* infection), even when adjusted for other factors predictive of mortality in the study population (Singh et al., 2006a). No correlation was found between in vitro antifungal interactions and outcome. None of the patients required discontinuation of antifungal therapy for intolerance or adverse effects; however, patients in the combination therapy arm were more likely to develop an increase in calcineurin inhibitor agent level or gastrointestinal intolerance (Singh et al., 2006a).

A retrospective survey documented invasive pulmonary aspergillosis in 7.5% (19/251) of lung transplant recipients, of whom 47% (9/19) had disseminated aspergillosis (Sole et al., 2005). The mortality rate was 86% (12/14) in patients who received amphotericin B preparations (amphotericin B deoxycholate or a lipid formulation of amphotericin B) and 0% (0/3) in those who received voriconazole plus caspofungin. Two of 19 cases were diagnosed only at autopsy (Sole et al., 2005). The mortality rate in patients receiving voriconazole plus caspofungin was also lower than with lipid formulations of amphotericin B (0/3 versus 8/8; $P = 0.006$).

We believe that the potential benefits of combination therapy may be best realized when used as initial therapy, particularly in patients with more severe forms of the disease, such as disseminated invasive aspergillosis, or with poor prognostic factors, such as kidney failure. Although definitive clinical trials are pending, the combination of voriconazole and caspofungin for the treatment of invasive aspergillosis posed a lesser economic burden on institutional resources than 5 mg/kg/day of liposomal amphotericin B (Anonymous, 2002). A survey of antifungal therapeutic practices for invasive aspergillosis in liver transplant recipients documented that currently combination therapy is used as first-line treatment in 47% and as salvage therapy in 80% of the transplant centers in North America (Singh et al., 2008). In many transplant programs, including ours, combination therapy constitutes standard antifungal therapy for invasive aspergillosis in SOT recipients (Munoz et al., 2006; Singh et al., 2006a).

Surgical excision or debridement remains an integral part of the management of invasive aspergillosis for both diagnostic and therapeutic purposes (Hummel et al., 1993; Lodge et al., 2004; Loria, 1992; Metin et al., 2005; Rea et al., 2006; Saxena et al., 2007; Scherer et al., 2005; Shlobin et al., 2005). Specifically, surgery is indicated for persistent or life-threatening hemoptysis, for lesions in the proximity of great vessels or the pericardium, for sino-nasal infections, for single cavitary lung lesions which progress despite adequate treatment, and for lesions invading the pericardium, bone, or subcutaneous or thoracic tissue (Walsh et al., 2008). Pneumonectomy led to a successful outcome in a lung transplant recipient with progressive, refractory angioinvasive aspergillosis whose disease had worsened despite conventional antifungal therapy (Sandur et al., 1999). Surgical resection is also indicated for intracranial abscesses, depending upon the location, accessibility of the lesion, and neurologic sequelae.

The optimal duration of therapy for invasive aspergillosis depends upon the response to therapy and the patient's underlying disease(s) or immune status. Treatment is usually continued for 12 weeks; however, the precise duration of therapy should be guided by clinical response rather than an arbitrary total dose or duration. A reasonable course would be to continue therapy until all clinical and radiographic abnormalities have resolved and cultures, if they can be readily obtained, don't yield *Aspergillus*. Reversal of immunosuppression that includes withdrawal of corticosteroids and a decrease in calcineurin inhibitor agents is an important adjuvant measure to surgical and medical treatment of invasive aspergillosis. Close monitoring of cyclosporine A or tacrolimus levels and of allograft function is critical.

Drug Interactions of Antifungal Agents with Immunosuppressants

Drug interactions of a number of antifungal agents with immunosuppressants must be carefully considered when treating transplant recipients with invasive aspergillosis. The triazole agents are potent inhibitors of the CYP34A isoenzymes and have the potential to increase the levels of calcineurin inhibitor agents and sirolimus (Saad et al., 2006). Itraconazole has been shown to increase cyclosporine A or tacrolimus levels by 40 to 83% (Leather et al., 2006; Wimberley et al., 2001). A 50 to 60% reduction in the dose of calcineurin inhibitor agents may be necessary with the concurrent use of voriconazole (Saad et al., 2006). The use of sirolimus is contraindicated in patients receiving voriconazole. In some reports, however, the two agents have been safely coadministered, with sirolimus dose reduction by 75 to 90% (Marty et al., 2006; Mathis et al., 2004). Coadministration of posaconazole increased cyclosporine exposure and necessitated dosage reductions of 14 to 20% for cyclosporine (Sansone-Parsons et al., 2007). Posaconazole increased the maximum blood concentration and the area under the concentration-time curve for tacrolimus by 121% and 357%, respectively (Sansone-Parsons et al., 2007).

The pharmacokinetics of caspofungin is unaltered by coadministration of tacrolimus, but caspofungin may reduce tacrolimus concentrations by up to 20% and may increase cyclosporine A plasma concentrations by 35% (Sable et al., 2002). Elevated liver function tests in healthy volunteers receiving caspofungin and cyclosporine A led to the exclusion of cyclosporine recipients from the initial phase II/III clinical studies of caspofungin (Sable et al., 2002). In the clinical setting, however, coadministration of caspofungin with cyclosporine A has been well-tolerated (Anttila et al., 2003; Saner et al., 2006; Sanz-Rodriguez et al., 2004). Nevertheless, it is prudent to monitor hepatic enzymes in cyclosporine recipients treated with caspofungin. There is no interaction between caspofungin and mycophenolate mofetil.

Anidulafungin clearance is not affected by drugs that are substrates, inducers, or inhibitors of cytochrome P450 hepatic isoenzymes (Vazquez, 2005). Further, since the drug is negligibly excreted in the urine, drug-drug interactions due to competitive renal elimination are unlikely (Dowell et al., 2007a; Vazquez, 2005). Coadministration with tacrolimus resulted in no pharmacokinetic interaction between the two agents (Dowell et al., 2007b). When administered with cyclosporine A, a small (22%) increase in anidulafungin concentration was observed after 4 days of dosing with cyclosporine A and was not considered to be clinically relevant (Dowell et al., 2005). Micafungin is a weak substrate and a mild inhibitor of the CYP3A enzyme, but not of P-glycoproteins (Joseph et al., 2007). In healthy volunteers, micafungin was shown to be a mild inhibitor of cyclosporine levels (Hebert et al., 2005). In patients receiving sirolimus, serum concentrations of this agent were increased by 21% with concomitant use of micafungin (Chandrasekar and Sobel, 2006). No drug interactions have been noted between micafungin and mycophenolate mofetil or cyclosporine (Joseph et al., 2007).

Adjunctive Immunotherapeutic Agents

Enhancement of the host's immune status with immunomodulatory agents is a potentially attractive therapeutic adjunct in the management of invasive aspergillosis. Evidence from in vitro and animal studies has shown enhanced antifungal activity with cytokine or colony-stimulating factors and modulation of cellular immune responses (Gil-Lamaignere et al., 2005; Lieschke and Burgess, 1992; Vora et al., 1998). Granulocyte colony-stimulating factor (G-CSF) stimulates proliferation and maturation of committed myeloid precursor cells and also augments neutrophil functions, including chemotaxis, phagocytosis, and oxidative responses (Dale et al., 1995; Lieschke and Burgess, 1992). Granulocyte-macrophage colony-stimulating factor (GM-CSF) stimulates the proliferation and differentiation of multiple lineages of cells such as neutrophils, eosinophils, and monocyte progenitor cells (Root and Dale, 1999). G-CSF or GM-CSF has been shown to be effective for invasive aspergillosis as adjuvant therapy for invasive fungal infections in some studies in patients with hematologic malignancies (Rowe et al., 1995).

Although GM-CSF use in SOT recipients appears to be safe, there are no studies that have evaluated its efficacy as adjunctive antifungal therapy specifically in these patients. In vitro studies have also demonstrated a potential role of gamma interferon (IFN-γ) against *Aspergillus* (Gaviria et al., 1999; Nagai et al., 1995; Rex et al., 1991; Roilides et al., 1996), and case reports in patients other than SOT recipients have documented possible beneficial effects of the adjunctive use of IFN-γ in invasive fungal infections, including invasive aspergillosis (Mamishi et al., 2005; Pasic et al., 1996; Saulsbury et al., 2001; Summers et al., 2005). Guidelines of the IDSA suggest a role for IFN-γ as adjunctive antifungal therapy for invasive aspergillosis in an immunocompromised nonneutropenic host (Walsh et al., 2008). The use of this cytokine in organ transplant recipients is of concern, however, given the risk of potential graft rejection.

Antifungal Prophylaxis

At present, anti-*Aspergillus* prophylaxis is not routinely recommended in all SOT recipients. A more rational approach is to target antifungal prophylaxis towards high-risk patients, such as lung transplant recipients and high-risk liver transplant recipients. Clinical trials of antifungal prophylaxis in liver transplant recipients have comprised small sample sizes in single-center studies. An optimal approach to the prevention of invasive fungal infections in these patients, therefore, has not been defined.

A meta-analysis of antifungal prophylactic trials in liver transplant recipients documented a beneficial effect on morbidity and attributable mortality but an emergence of infections due to non-*albicans Candida* spp. in patients receiving antifungal prophylaxis (Cruciani et al., 2006). Since the risk factors and the period of susceptibility to invasive fungal infections are clearly definable, antifungal prophylaxis targeted towards these high-risk patients is also deemed a rational approach for the prevention of invasive aspergillosis after liver transplantation. Targeted antifungal prophylaxis using the lipid formulations of amphotericin B in doses ranging from 1 to 5 mg/kg/day has been shown to be effective in observational studies and may be considered in high-risk patients (Fortun et al., 2003a; Hellinger et al., 2005; Reed et al., 2007; Singhal et al., 2000). Currently, targeted prophylaxis in liver transplant recipients is employed most frequently during the initial hospital stay or for the first month posttransplant (Singh et al., 2008). The role of echinocandins or the newer triazoles as antifungal prophylaxis in high-risk liver transplant recipients has not yet been defined.

The incidence of invasive aspergillosis in lung transplant recipients has been shown to decrease with prophylaxis with aerosolized amphotericin B at 20 mg, three times per day or 0.6 mg/kg/day (Calvo et al., 1999; Guillemain et al., 1995; Monforte et al., 2001; Reichenspurner et al., 1997; Ruffini, 2001). Therefore, this prophylactic approach is employed by most lung transplant centers. In a survey of antifungal prophylactic strategies in lung transplant units in the United States, aerosolized amphotericin B was the most frequently employed antifungal prophylactic agent (Dummer et al., 2004). In order of preference, the prophylactic agents used in lung transplant centers in this study were inhaled amphotericin B (61%), itraconazole (46%), parenteral amphotericin formulations (25%), and fluconazole (21%); many centers used more than one agent (Dummer et al., 2004). Aerosolization of the lipid formulations of amphotericin B has been proposed to enhance drug delivery by causing less foaming during nebulization. Concentrations in the lung that were achievable with the lipid formulations were severalfold higher than those with amphotericin B deoxycholate (Singh and Paterson, 2005). It should be noted that unlike bilateral lung transplant recipients, in whom the lung deposition with nebulized formulations of amphotericin B is symmetrical and uniform, single-lung transplant recipients may have erratic and nonuniform drug deposition (Corcoran et al., 2006). Furthermore, drug concentrations in the proximal airways may be lower than in the distal airways (Monforte et al., 2001). In a study utilizing aerosolized amphotericin B lipid complex, pulmonary fungal infections developed in 2 of 51 of the lung transplant recipients and included anastomotic infections due to *Candida* in both cases (Palmer et al., 2001).

The duration of prophylaxis with nebulized lipid amphotericin B formulations varies and depends on the intensity of immunosuppression and the presence of specific risk factors. It is reasonable to employ such prophylaxis (6 mg/day) during the period of highest immunosuppression, i.e., the first 4 weeks posttransplant. Some authors have recommended the administration of nebulized amphotericin B for several weeks after transplantation or until the healing of bronchial anastomosis, while others recommend maintaining it for the patient's lifetime (Monforte et al., 2001).

Preemptive therapy, where antifungal prophylaxis is directed towards lung transplant recipients colonized with *Aspergillus* spp. immediately before or within 6 to 9 months after transplantation, is another approach to the prevention of invasive aspergillosis in lung transplant recipients (Paya, 2001; Singh, 2000). Approximately one-third of the lung transplant programs that employ antifungal prophylaxis prefer to administer it as preemptive therapy (Husain et al., 2006). In one study, preemptive therapy of colonized patients with oral itraconazole completely prevented the development of invasive disease (Hamacher et al., 1999). Itraconazole has also been employed in conjunction with nebulized liposomal amphotericin B in this setting (Husain et al., 2006). Antifungal preemptive therapy should be administered for at least 4 to 6 months.

Limited data exist on the use of newer triazoles as antifungal prophylaxis in lung transplant recipients. Voriconazole has substantial intrapulmonary penetration, with levels exceeding those in plasma in one study (Capitano et al., 2006). A study in lung transplant recipients documented a 1.5% rate of invasive aspergillosis with voriconazole prophylaxis at 1 year, compared to 23% in the control group that received targeted prophylaxis with itraconazole, with or without inhaled amphotericin B (Husain et al., 2006). However, twofold more patients (14%) in the voriconazole group discontinued prophylaxis because of adverse effects (Husain et al., 2006). The role of newer triazoles as antifungal prophylaxis in lung transplant recipients remains to be fully defined.

REFERENCES

Akerele, T., and S. Lightman. 2007. Ocular complications in heart, lung and heart-lung recipients. *Br. J. Ophthalmol.* **91:**310–312.

Altiparmak, M. R., S. Apaydin, S. Trablus, K. Serdengecti, R. Ataman, R. Ozturk, and E. Erek. 2002. Systemic fungal infections after renal transplantation. *Scand. J. Infect. Dis.* **34:**284–288.

Ankersmit, H. J., R. Deicher, B. Moser, I. Teufel, G. Roth, S. Gerlitz, S. Itescu, E. Wolner, G. Boltz-Nitulescu, and J. Kovarik. 2001. Impaired T cell proliferation, increased soluble death-inducing receptors and activation-induced T cell death in patients undergoing haemodialysis. *Clin. Exp. Immunol.* **125:**142–148.

Anonymous. 2002. Voriconazole. *Med. Lett. Drugs Ther.* **44:**63–65.

Anttila, V. J., A. Piilonen, and M. Valtonen. 2003. Co-administration of caspofungin and cyclosporine to a kidney transplant patient with pulmonary *Aspergillus* infection. *Scand. J. Infect. Dis.* **35:**893–894.

Badiee, P., P. Kordbacheh, A. Alborzi, S. Malekhoseini, F. Zeini, H. Mirhendi, and M. Mahmoodi. 2007. Prospective screening in liver transplant recipients by panfungal PCR-ELISA for early diagnosis of invasive fungal infections. *Liver Transplant.* **13:**1011–1016.

Balloy, V., M. Huerre, J. P. Latgé, and M. Chignard. 2005. Differences in patterns of infection and inflammation for corticosteroid treatment and chemotherapy in experimental invasive pulmonary aspergillosis. *Infect. Immun.* **73:**494–503.

Bonham, C. A., E. A. Dominguez, M. B. Fukui, D. L. Paterson, G. A. Pankey, M. M. Wagener, J. J. Fung, and N. Singh. 1998. Central nervous system lesions in liver transplant recipients: prospective assessment of indications for biopsy and implications for management. *Transplantation* **66:**1596–1604.

Briegel, J., H. Forst, B. Spill, A. Haas, B. Grabein, M. Haller, E. Kilger, K. W. Jauch, K. Maag, G. Ruckdeschel, et al. 1995. Risk factors for systemic fungal infections in liver transplant recipients. *Eur. J. Clin. Microbiol. Infect. Dis.* **14:**375–382.

Brown, R. S., Jr., J. R. Lake, B. A. Katzman, N. L. Ascher, K. A. Somberg, J. C. Emond, and J. P. Roberts. 1996. Incidence and significance of *Aspergillus* cultures following liver and kidney transplantation. *Transplantation* **61:**666–669.

Cahill, B. C., J. R. Hibbs, K. Savik, B. A. Juni, B. M. Dosland, C. Edin-Stibbe, and M. I. Hertz. 1997. *Aspergillus* airway colonization and invasive disease after lung transplantation. *Chest* **112:**1160–1164.

Calvo, V., J. M. Borro, P. Morales, A. Morcillo, R. Vicente, V. Tarrazona, F. Paris, et al. 1999. Antifungal prophylaxis during the early postoperative period of lung transplantation. *Chest* **115:**1301–1304.

Capitano, B., B. A. Potoski, S. Husain, S. Zhang, D. L. Paterson, S. M. Studer, K. R. McCurry, and R. Venkataramanan. 2006. Intrapulmonary penetration of voriconazole in patients receiving an oral prophylactic regimen. *Antimicrob. Agents Chemother.* **50:**1878–1880.

Carby, M. R., M. E. Hodson, and N. R. Banner. 2004. Refractory pulmonary aspergillosis treated with caspofungin after heart-lung transplantation. *Transplant. Int.* **17:**545–548.

Cenci, E., A. Mencacci, G. Del Sero, A. Bacci, C. Montagnoli, C. F. d'Ostiani, P. Mosci, M. Bachmann, F. Bistoni, M. Kopf, and L. Romani. 1999. Interleukin-4 causes susceptibility to invasive pulmonary aspergillosis through suppression of protective type I responses. *J. Infect. Dis.* **180:**1957–1968.

Chandrasekar, P. H., and J. I. Ito. 2005. Amphotericin B lipid complex in the management of invasive aspergillosis in immunocompromised patients. *Clin. Infect. Dis.* **40**(Suppl. 6):S392–S400.

Chandrasekar, P. H., and J. D. Sobel. 2006. Micafungin: a new echinocandin. *Clin. Infect. Dis.* **42:**1171–1178.

Clancy, C. J., R. A. Jaber, H. L. Leather, J. R. Wingard, B. Staley, L. J. Wheat, C. L. Cline, K. H. Rand, D. Schain, M. Baz, and M. H. Nguyen. 2007. Bronchoalveolar lavage galactomannan in diagnosis of invasive pulmonary aspergillosis among solid-organ transplant recipients. *J. Clin. Microbiol.* **45:**1759–1765.

Cofan, F., P. Inigo, M. J. Ricart, F. Oppenheimer, J. Vilardell, J. M. Campistol, and P. Carretero. 1996. Aspergilosis pulmonar invasiva en el trasplante renal y renopancreatico. *Nefrologia* **XVI:**253–260.

Collins, L. A., M. H. Samore, M. S. Roberts, R. Luzzati, R. L. Jenkins, W. D. Lewis, and A. W. Karchmer. 1994. Risk factors for invasive fungal infections complicating orthotopic liver transplantation. *J. Infect. Dis.* **170:**644–652.

Corcoran, T. E., R. Venkataramanan, K. M. Mihelc, A. L. Marcinkowski, J. Ou, B. M. McCook, L. Weber, M. E. Carey, D. L. Paterson, J. M. Pilewski, K. R. McCurry, and S. Husain. 2006. Aerosol deposition of lipid complex amphotericin-B (Abelcet) in lung transplant recipients. *Am. J. Transplant.* **6**:2765–2773.

Cornely, O. A., J. Maertens, M. Bresnik, R. Ebrahimi, A. J. Ullmann, E. Bouza, C. P. Heussel, O. Lortholary, C. Rieger, A. Boehme, M. Aoun, H. A. Horst, A. Thiebaut, M. Ruhnke, D. Reichert, N. Vianelli, S. W. Krause, E. Olavarria, and R. Herbrecht. 2007. Liposomal amphotericin B as initial therapy for invasive mold infection: a randomized trial comparing a high-loading dose regimen with standard dosing (AmBiLoad trial). *Clin. Infect. Dis.* **44**:1289–1297.

Cruciani, M., C. Mengoli, M. Malena, O. Bosco, G. Serpelloni, and P. Grossi. 2006. Antifungal prophylaxis in liver transplant patients: a systematic review and meta-analysis. *Liver Transplant.* **12**:850–858.

Dale, D. C., W. C. Liles, W. R. Summer, and S. Nelson. 1995. Review: granulocyte colony-stimulating factor—role and relationships in infectious diseases. *J. Infect. Dis.* **172**:1061–1075.

Dauber, J. H., I. L. Paradis, and J. S. Dummer. 1990. Infectious complications in pulmonary allograft recipients. *Clin. Chest Med.* **11**: 291–308.

D'Elios, M. M., R. Josien, M. Manghetti, A. Amedei, M. de Carli, M. C. Cuturi, G. Blancho, F. Buzelin, G. del Prete, and J. P. Soulillou. 1997. Predominant Th1 cell infiltration in acute rejection episodes of human kidney grafts. *Kidney Int.* **51**:1876–1884.

Denning, D. W. 1998. Invasive aspergillosis. *Clin. Infect. Dis.* **26**:781–803.

Denning, D. W., K. A. Marr, W. M. Lau, D. P. Facklam, V. Ratanatharathorn, C. Becker, A. J. Ullmann, N. L. Seibel, P. M. Flynn, J. A. van Burik, D. N. Buell, and T. F. Patterson. 2006. Micafungin (FK463), alone or in combination with other systemic antifungal agents, for the treatment of acute invasive aspergillosis. *J. Infect.* **53**: 337–349.

Denning, D. W., P. Ribaud, N. Milpied, D. Caillot, R. Herbrecht, E. Thiel, A. Haas, M. Ruhnke, and H. Lode. 2002. Efficacy and safety of voriconazole in the treatment of acute invasive aspergillosis. *Clin. Infect. Dis.* **34**:563–571.

de Pablo, A., P. Ussetti, M. Cruz Carreno, T. Lazaro, M. J. Ferreiro, A. Lopez, P. Mendaza, and J. Estada. 2000. Aspergillosis in pulmonary transplantation. *Enferm. Infecc. Microbiol. Clin.* **18**:209–214. (In French.)

Dowell, J. A., M. Stogniew, D. Krause, and B. Damle. 2007a. Anidulafungin does not require dosage adjustment in subjects with varying degrees of hepatic or renal impairment. *J. Clin. Pharmacol.* **47**: 461–470.

Dowell, J. A., M. Stogniew, D. Krause, T. Henkel, and B. Damle. 2007b. Lack of pharmacokinetic interaction between anidulafungin and tacrolimus. *J. Clin. Pharmacol.* **47**:305–314.

Dowell, J. A., M. Stogniew, D. Krause, T. Henkel, and I. E. Weston. 2005. Assessment of the safety and pharmacokinetics of anidulafungin when administered with cyclosporine. *J. Clin. Pharmacol.* **45**: 227–233.

Dummer, J. S., N. Lazariashvilli, J. Barnes, M. Ninan, and A. P. Milstone. 2004. A survey of anti-fungal management in lung transplantation. *J. Heart Lung Transplant.* **23**:1376–1381.

Duong, M., N. Ouellet, M. Simard, Y. Bergeron, M. Olivier, and M. G. Bergeron. 1998. Kinetic study of host defense and inflammatory response to *Aspergillus fumigatus* in steroid-induced immunosuppressed mice. *J. Infect. Dis.* **178**:1472–1482.

Ferraris, J. R., M. L. Tambutti, R. L. Cardoni, and N. Prigoshin. 2004. Conversion from cyclosporine A to tacrolimus in pediatric kidney transplant recipients with chronic rejection: changes in the immune responses. *Transplantation* **77**:532–537.

Forestier, E., V. Remy, O. Lesens, M. Martinot, Y. Hansman, B. Eisenmann, and D. Christmann. 2005. A case of *Aspergillus* mediastinitis after heart transplantation successfully treated with liposomal amphotericin B, caspofungin and voriconazole. *Eur. J. Clin. Microbiol. Infect. Dis.* **24**:347–349.

Fortun, J., P. Martin-Davila, M. E. Alvarez, A. Sanchez-Sousa, L. Gajate, R. Barcena, J. Nuno, and S. Moreno. 2007. *Aspergillus* galactomannan antigen in liver transplant recipients (LTR): a high false-positive rate in serum samples obtained during the first week post-transplantation, abstr. K-2167. *Abstr. 47th Intersci. Conf. Antimicrob. Agents Chemother*, Chicago, IL.

Fortun, J., P. Martin-Davila, M. E. Alvarez, A. Sanchez-Sousa, C. Quereda, E. Navas, R. Barcena, E. Vicente, A. Candelas, A. Honrubia, J. Nuno, V. Pintado, and S. Moreno. 2001. Aspergillus antigenemia sandwich-enzyme immunoassay test as a serodiagnostic method for invasive aspergillosis in liver transplant recipients. *Transplantation* **71**:145–149.

Fortun, J., P. Martin-Davila, S. Moreno, R. Barcena, E. de Vicente, A. Honrubia, M. Garcia, J. Nuno, A. Candela, M. Uriarte, and V. Pintado. 2003a. Prevention of invasive fungal infections in liver transplant recipients: the role of prophylaxis with lipid formulations of amphotericin B in high-risk patients. *J. Antimicrob. Chemother.* **52**:813–819.

Fortun, J., P. Martin-Davila, S. Moreno, E. De Vicente, J. Nuno, A. Candelas, R. Barcena, and M. Garcia. 2002. Risk factors for invasive aspergillosis in liver transplant recipients. *Liver Transplant.* **8**: 1065–1070.

Fortun, J., P. Martin-Davila, M. A. Sanchez, V. Pintado, M. E. Alvarez, A. Sanchez-Sousa, and S. Moreno. 2003b. Voriconazole in the treatment of invasive mold infections in transplant recipients. *Eur. J. Clin. Microbiol. Infect. Dis.* **22**:408–413.

Gallis, H. A., R. A. Berman, T. R. Cate, J. D. Hamilton, J. C. Gunnells, and D. L. Stickel. 1975. Fungal infection following renal transplantation. *Arch. Intern. Med.* **135**:1163–1172.

Garrido, J., J. L. Lerma, M. Heras, P. J. Labrador, P. Garcia, A. Bondia, L. Corbacho, and J. M. Tabernero. 2003. Pseudoaneurysm of the iliac artery secondary to *Aspergillus* infection in two recipients of kidney transplants from the same donor. *Am. J. Kidney Dis.* **41**: 488–492.

Gavalda, J., O. Len, R. San Juan, J. M. Aguado, J. Fortun, C. Lumbreras, A. Moreno, P. Munoz, M. Blanes, A. Ramos, G. Rufi, M. Gurgui, J. Torre-Cisneros, M. Montejo, M. Cuenca-Estrella, J. L. Rodriguez-Tudela, and A. Pahissa. 2005. Risk factors for invasive aspergillosis in solid-organ transplant recipients: a case-control study. *Clin. Infect. Dis.* **41**:52–59.

Gaviria, J. M., J. A. van Burik, D. C. Dale, R. K. Root, and W. C. Liles. 1999. Comparison of interferon-gamma, granulocyte colony-stimulating factor, and granulocyte-macrophage colony-stimulating factor for priming leukocyte-mediated hyphal damage of opportunistic fungal pathogens. *J. Infect. Dis.* **179**:1038–1041.

George, M. J., D. R. Snydman, B. G. Werner, J. Griffith, M. E. Falagas, N. N. Dougherty, R. H. Rubin, et al. 1997. The independent role of cytomegalovirus as a risk factor for invasive fungal disease in orthotopic liver transplant recipients. *Am. J. Med.* **103**:106–113.

Gil-Lamaignere, C., M. Simitsopoulou, E. Roilides, A. Maloukou, R. M. Winn, and T. J. Walsh. 2005. Interferon-gamma and granulocyte-macrophage colony-stimulating factor augment the activity of polymorphonuclear leukocytes against medically important zygomycetes. *J. Infect. Dis.* **191**:1180–1187.

Gonwa, T. A., M. A. McBride, K. Anderson, M. L. Mai, H. Wadei, and N. Ahsan. 2006. Continued influence of preoperative renal function on outcome of orthotopic liver transplant (OLTX) in the US: where will MELD lead us? *Am. J. Transplant.* **6**:2651–2659.

Gras, J., A. Cornet, D. Latinne, and R. Reding. 2004. Evidence that Th1/Th2 immune deviation impacts on early graft acceptance after pediatric liver transplantation: results of immunological monitoring in 40 children. *Am. J. Transplant.* **4**:444.

Grazziutti, M. L., J. H. Rex, R. E. Cowart, E. J. Anaissie, A. Ford, and C. A. Savary. 1997. *Aspergillus fumigatus* conidia induce a Th1-type cytokine response. *J. Infect. Dis.* 176:1579–1583.

Groetzner, J., I. Kaczmarek, T. Wittwer, J. Strauch, B. Meiser, T. Wahlers, S. Daebritz, and B. Reichart. 2008. Caspofungin as first-line therapy for the treatment of invasive aspergillosis after thoracic organ transplantation. *J. Heart Lung Transplant.* 27:1–6.

Grossi, P., R. De Maria, A. Caroli, M. S. Zaina, L. Minoli, et al. 1992. Infections in heart transplant recipients: the experience of the Italian heart transplantation program. *J. Heart Lung Transplant.* 11:847–866.

Grossi, P., C. Farina, R. Fiocchi, D. Dalla Gasperina, et al. 2000. Prevalence and outcome of invasive fungal infections in 1,963 thoracic organ transplant recipients: a multicenter retrospective study. *Transplantation* 70:112–116.

Guillemain, R., V. Lavarde, C. Amrein, P. Chevalier, A. Guinvarc'h, and D. Glotz. 1995. Invasive aspergillosis after transplantation. *Transplant. Proc.* 27:1307–1309.

Gustafson, T. L., W. Schaffner, G. B. Lavely, C. W. Stratton, H. K. Johnson, and R. H. Hutcheson, Jr. 1983. Invasive aspergillosis in renal transplant recipients: correlation with corticosteroid therapy. *J. Infect. Dis.* 148:230–238.

Hamacher, J., A. Spiliopoulos, A. M. Kurt, L. P. Nicod, et al. 1999. Pre-emptive therapy with azoles in lung transplant patients. *Eur. Respir. J.* 13:180–186.

Hebert, M. F., R. W. Townsend, S. Austin, G. Balan, D. K. Blough, D. Buell, J. Keirns, and I. Bekersky. 2005. Concomitant cyclosporine and micafungin pharmacokinetics in healthy volunteers. *J. Clin. Pharmacol.* 45:954–960.

Hellinger, W. C., H. Bonatti, J. D. Yao, S. Alvarez, L. M. Brumble, M. R. Keating, J. C. Mendez, D. J. Kramer, R. C. Dickson, D. M. Harnois, J. R. Spivey, C. B. Hughes, J. H. Nguyen, and J. L. Steers. 2005. Risk stratification and targeted antifungal prophylaxis for prevention of aspergillosis and other invasive mold infections after liver transplantation. *Liver Transplant.* 11:656–662.

Helmi, M., R. B. Love, D. Welter, R. D. Cornwell, and K. C. Meyer. 2003. *Aspergillus* infection in lung transplant recipients with cystic fibrosis: risk factors and outcomes comparison to other types of transplant recipients. *Chest* 123:800–808.

Herbrecht, R., D. W. Denning, T. F. Patterson, J. E. Bennett, R. E. Greene, J. W. Oestmann, W. V. Kern, K. A. Marr, P. Ribaud, O. Lortholary, R. Sylvester, R. H. Rubin, J. R. Wingard, P. Stark, C. Durand, D. Caillot, E. Thiel, P. H. Chandrasekar, M. R. Hodges, H. T. Schlamm, P. F. Troke, and B. de Pauw. 2002. Voriconazole versus amphotericin B for primary therapy of invasive aspergillosis. *N. Engl. J. Med.* 347:408–415.

Hodge, S., G. Hodge, R. Flower, and P. Han. 1999. Methylprednisolone up-regulates monocyte interleukin-10 production in stimulated whole blood. *Scand. J. Immunol.* 49:548–553.

Hummel, M., S. Schuler, U. Weber, G. Schwertlick, S. Hempel, D. Theiss, W. Rees, J. Mueller, and R. Hetzer. 1993. Aspergillosis with *Aspergillus* osteomyelitis and diskitis after heart transplantation: surgical and medical management. *J. Heart Lung Transplant.* 12:599–603.

Hummel, M., U. Thalmann, G. Jautzke, F. Staib, M. Seibold, and R. Hetzer. 1992. Fungal infections following heart transplantation. *Mycoses* 35:23–34.

Husain, S., E. J. Kwak, A. Obman, M. M. Wagener, S. Kusne, J. E. Stout, K. R. McCurry, and N. Singh. 2004. Prospective assessment of Platelia *Aspergillus* galactomannan antigen for the diagnosis of invasive aspergillosis in lung transplant recipients. *Am. J. Transplant.* 4:796–802.

Husain, S., D. L. Paterson, S. Studer, J. Pilewski, M. Crespo, D. Zaldonis, K. Shutt, D. L. Pakstis, A. Zeevi, B. Johnson, E. J. Kwak, and K. R. McCurry. 2006. Voriconazole prophylaxis in lung transplant recipients. *Am. J. Transplant.* 6:3008–3016.

Husain, S., D. L. Paterson, S. M. Studer, M. Crespo, J. Pilewski, M. Durkin, J. L. Wheat, B. Johnson, L. McLaughlin, C. Bentsen, K. R. McCurry, and N. Singh. 2007. *Aspergillus* galactomannan antigen in the bronchoalveolar lavage fluid for the diagnosis of invasive aspergillosis in lung transplant recipients. *Transplantation* 83:1330–1336.

Husni, R. N., S. M. Gordon, D. L. Longworth, A. Arroliga, P. C. Stillwell, R. K. Avery, J. R. Maurer, A. Mehta, and T. Kirby. 1998. Cytomegalovirus infection is a risk factor for invasive aspergillosis in lung transplant recipients. *Clin. Infect. Dis.* 26:753–755.

Iversen, M., C. M. Burton, S. Vand, L. Skovfoged, J. Carlsen, N. Milman, C. B. Andersen, M. Rasmussen, and M. Tvede. 2007. *Aspergillus* infection in lung transplant patients: incidence and prognosis. *Eur. J. Clin. Microbiol. Infect. Dis.* 26:879–886.

Joseph, J. M., R. Jain, and L. H. Danziger. 2007. Micafungin: a new echinocandin antifungal. *Pharmacotherapy* 27:53–67.

Kanj, S. S., K. Welty-Wolf, J. Madden, V. Tapson, M. A. Baz, R. D. Davis, and J. R. Perfect. 1996. Fungal infections in lung and heart-lung transplant recipients. Report of 9 cases and review of the literature. *Medicine* (Baltimore) 75:142–156.

Kartsonis, N. A., A. J. Saah, C. Joy Lipka, A. F. Taylor, and C. A. Sable. 2005. Salvage therapy with caspofungin for invasive aspergillosis: results from the caspofungin compassionate use study. *J. Infect.* 50:196–205.

Kawagishi, N., K. Satoh, Y. Enomoto, Y. Akamatsu, S. Sekiguchi, K. Fujimori, and S. Satomi. 2006. Risk factors and impact of beta-D glucan on invasive fungal infection for the living donor liver transplant recipients. *Tohoku J. Exp. Med.* 209:207–215.

Kramer, M., M. R. Kramer, H. Blau, J. Bishara, R. Axer-Siegel, and D. Weinberger. 2006. Intravitreal voriconazole for the treatment of endogenous *Aspergillus* endophthalmitis. *Ophthalmology* 113:1184–1186.

Kramer, M. R., D. W. Denning, S. E. Marshall, D. J. Ross, G. Berry, N. J. Lewiston, D. A. Stevens, and J. Theodore. 1991. Ulcerative tracheobronchitis after lung transplantation. A new form of invasive aspergillosis. *Am. Rev. Respir. Dis.* 144:552–556.

Kusne, S., J. Torre-Cisneros, R. Manez, W. Irish, M. Martin, J. Fung, R. L. Simmons, and T. E. Starzl. 1992. Factors associated with invasive lung aspergillosis and the significance of positive *Aspergillus* culture after liver transplantation. *J. Infect. Dis.* 166:1379–1383.

Kwak, E., K. Abu-Elmagd, J. Bond, M. F. Zak, M. McHenry, and S. Kusne. 2002. Invasive fungal infections in adult small bowel transplant recipients, abstr. K-1231. *42nd Intersci. Conf. Antimicrob. Agents Chemother.*, San Diego, CA.

Kwak, E. J., S. Husain, A. Obman, L. Meinke, J. Stout, S. Kusne, M. M. Wagener, and N. Singh. 2004. Efficacy of galactomannan antigen in the Platelia *Aspergillus* enzyme immunoassay for diagnosis of invasive aspergillosis in liver transplant recipients. *J. Clin. Microbiol.* 42:435–438.

Laursen, A. L., S. C. Mogensen, H. M. Andersen, P. L. Andersen, and S. Ellermann-Eriksen. 2001. The impact of CMV on the respiratory burst of macrophages in response to *Pneumocystis carinii*. *Clin. Exp. Immunol.* 123:239–246.

Leather, H., R. M. Boyette, L. Tian, and J. R. Wingard. 2006. Pharmacokinetic evaluation of the drug interaction between intravenous itraconazole and intravenous tacrolimus or intravenous cyclosporin A in allogeneic hematopoietic stem cell transplant recipients. *Biol. Blood Marrow Transplant.* 12:325–334.

Levin, T., B. Suh, D. Beltramo, and R. Samuel. 2004. *Aspergillus* mediastinitis following orthotopic heart transplantation: case report and review of the literature. *Transplant. Infect. Dis.* 6:129–131.

Lieschke, G. J., and A. W. Burgess. 1992. Granulocyte colony-stimulating factor and granulocyte-macrophage colony-stimulating factor (2). *N. Engl. J. Med.* 327:99–106.

Linden, P., P. Williams, and K. M. Chan. 2000. Efficacy and safety of amphotericin B lipid complex injection (ABLC) in solid-organ transplant recipients with invasive fungal infections. *Clin. Transplant.* 14:329–339.

Linden, P. K., K. Coley, P. Fontes, J. J. Fung, and S. Kusne. 2003. Invasive aspergillosis in liver transplant recipients: outcome comparison of therapy with amphotericin B lipid complex and a historical cohort treated with conventional amphotericin B. *Clin. Infect. Dis.* 37:17–25.

Lionakis, M. S., and D. P. Kontoyiannis. 2003. Glucocorticoids and invasive fungal infections. *Lancet* 362:1828–1838.

Lodge, B. A., E. D. Ashley, M. P. Steele, and J. R. Perfect. 2004. *Aspergillus fumigatus* empyema, arthritis, and calcaneal osteomyelitis in a lung transplant patient successfully treated with posaconazole. *J. Clin. Microbiol.* 42:1376–1378.

Loire, R., A. Tabib, and O. Bastien. 1993. Fatal aspergillosis after cardiac transplantation. About 26 cases. *Ann. Pathol.* 13:157–163. (In French.)

Loria, K. M., M. H. Salinger, T. G. Frohlich, M. D. Gendelman, F. V. Cook, and C. E. Arentzen. 1992. Primary cutaneous aspergillosis in a heart transplant recipient treated with surgical excision and oral itraconazole. *J. Heart Lung Transplant.* 11:156–159.

Maertens, J., I. Raad, G. Petrikkos, M. Boogaerts, D. Selleslag, F. B. Petersen, C. A. Sable, N. A. Kartsonis, A. Ngai, A. Taylor, T. F. Patterson, D. W. Denning, and T. J. Walsh. 2004. Efficacy and safety of caspofungin for treatment of invasive aspergillosis in patients refractory to or intolerant of conventional antifungal therapy. *Clin. Infect. Dis.* 39:1563–1571.

Mamishi, S., K. Zomorodian, F. Saadat, M. Gerami-Shoar, B. Tarazooie, and S. A. Siadati. 2005. A case of invasive aspergillosis in CGD patient successfully treated with amphotericin B and INF-gamma. *Ann. Clin. Microbiol. Antimicrob.* 4:4.

Marty, F. M., C. M. Lowry, C. S. Cutler, B. J. Campbell, K. Fiumara, L. R. Baden, and J. H. Antin. 2006. Voriconazole and sirolimus coadministration after allogeneic hematopoietic stem cell transplantation. *Biol. Blood Marrow Transplant.* 12:552–559.

Mathis, A. S., N. K. Shah, and G. S. Friedman. 2004. Combined use of sirolimus and voriconazole in renal transplantation: a report of two cases. *Transplant. Proc.* 36:2708–2709.

Maurer, J. R., D. E. Tullis, R. F. Grossman, H. Vellend, T. L. Winton, and G. A. Patterson. 1992. Infectious complications following isolated lung transplantation. *Chest* 101:1056–1059.

McAdams, H. P., J. J. Erasmus, and S. M. Palmer. 2001. Complications (excluding hyperinflation) involving the native lung after single-lung transplantation: incidence, radiologic features, and clinical importance. *Radiology* 218:233–241.

Mehrad, B., G. Paciocco, F. J. Martinez, T. C. Ojo, M. D. Iannettoni, and J. P. Lynch III. 2001. Spectrum of *Aspergillus* infection in lung transplant recipients: case series and review of the literature. *Chest* 119:169–175.

Metin, K. S., B. S. Ugurlu, B. Kabakci, N. O. Sariosmanoglu, E. Hazan, and O. Oto. 2005. Surgical resection for successful treatment of invasive pulmonary aspergillosis: report of 3 cases. *Scand. J. Infect. Dis.* 37:694–696.

Monforte, V., A. Roman, J. Gavalda, C. Bravo, L. Tenorio, A. Ferrer, J. Maestre, and F. Morell. 2001. Nebulized amphotericin B prophylaxis for *Aspergillus* infection in lung transplantation: study of risk factors. *J. Heart Lung Transplant.* 20:1274–1281.

Montoya, J. G., S. V. Chaparro, D. Celis, J. A. Cortes, A. N. Leung, R. C. Robbins, and D. A. Stevens. 2003. Invasive aspergillosis in the setting of cardiac transplantation. *Clin. Infect. Dis.* 37(Suppl. 3): S281–S292.

Morgan, J., K. A. Wannemuehler, K. A. Marr, S. Hadley, D. P. Kontoyiannis, T. J. Walsh, S. K. Fridkin, P. G. Pappas, and D. W. Warnock. 2005. Incidence of invasive aspergillosis following hematopoietic stem cell and solid organ transplantation: interim results of a prospective multicenter surveillance program. *Med. Mycol.* 43(Suppl. 1):S49–S58.

Munoz, P., L. Alcala, M. Sanchez Conde, J. Palomo, J. Yanez, T. Pelaez, and E. Bouza. 2003. The isolation of *Aspergillus fumigatus* from respiratory tract specimens in heart transplant recipients is highly predictive of invasive aspergillosis. *Transplantation* 75:326–329.

Munoz, P., J. Palomo, P. Guembe, M. Rodriguez Creixems, P. Gijon, and E. Bouza. 2000. Lung nodular lesions in heart transplant recipients. *J. Heart Lung Transplant.* 19:660–667.

Munoz, P., C. Rodriguez, E. Bouza, J. Palomo, J. F. Yanez, M. J. Dominguez, and M. Desco. 2004. Risk factors of invasive aspergillosis after heart transplantation: protective role of oral itraconazole prophylaxis. *Am. J. Transplant.* 4:636–643.

Munoz, P., N. Singh, and E. Bouza. 2006. Treatment of solid organ transplant patients with invasive fungal infections: should a combination of antifungal drugs be used? *Curr. Opin. Infect. Dis.* 19: 365–370.

Munoz, P., J. Torre, E. Bouza, A. Moreno, A. Echantz, J. Fortun, C. Lumbreras, J. M. Aguado, I. Losada, V. Cuervas, M. Gurgui, J. M. Cisneros, M. Montejo, and C. Farinas. 1996. Invasive aspergillosis in transplant recipients. A large multicenter study, p. 242. *Abstr. 36th Intersci. Conf. Antimicrob. Agents Chemother.*, New Orleans, LA.

Nagai, H., J. Guo, H. Choi, and V. Kurup. 1995. Interferon-gamma and tumor necrosis factor-alpha protect mice from invasive aspergillosis. *J. Infect. Dis.* 172:1554–1560.

Osawa, M., Y. Ito, T. Hirai, R. Isozumi, S. Takakura, Y. Fujimoto, Y. Iinuma, S. Ichiyama, K. Tanaka, and M. Mishima. 2007. Risk factors for invasive aspergillosis in living donor liver transplant recipients. *Liver Transplant.* 13:566–570.

Palmer, S. M., R. H. Drew, J. D. Whitehouse, V. F. Tapson, R. D. Davis, R. R. McConnell, S. S. Kanj, and J. R. Perfect. 2001. Safety of aerosolized amphotericin B lipid complex in lung transplant recipients. *Transplantation* 72:545–548.

Panackal, A. A., A. Dahlman, K. T. Keil, C. L. Peterson, L. Mascola, S. Mirza, M. Phelan, B. A. Lasker, M. E. Brandt, J. Carpenter, M. Bell, D. W. Warnock, R. A. Hajjeh, and J. Morgan. 2003. Outbreak of invasive aspergillosis among renal transplant recipients. *Transplantation* 75:1050–1053.

Paradowski, L. J. 1997. Saprophytic fungal infections and lung transplantation—revisited. *J. Heart Lung Transplant.* 16:524–531.

Pasic, S., M. Abinun, B. Pistignjat, B. Vlajic, J. Rakic, L. Sarjanovic, and N. Ostojic. 1996. *Aspergillus* osteomyelitis in chronic granulomatous disease: treatment with recombinant gamma-interferon and itraconazole. *Pediatr. Infect. Dis. J.* 15:833–834.

Paterson, D. L., and N. Singh. 1999. Invasive aspergillosis in transplant recipients. *Medicine* (Baltimore) 78:123–138.

Paya, C. V. 2001. Prevention of fungal and hepatitis virus infections in liver transplantation. *Clin. Infect. Dis.* 33(Suppl. 1):S47–S52.

Peterson, P. K., R. Ferguson, D. S. Fryd, H. H. Balfour, Jr., J. Rynasiewicz, and R. L. Simmons. 1982. Infectious diseases in hospitalized renal transplant recipients: a prospective study of a complex and evolving problem. *Medicine* (Baltimore) 61:360–372.

Petrovic, J., A. Ngai, S. Bradshaw, A. Williams-Diaz, A. Taylor, C. Sable, S. Vuocolo, and N. Kartsonis. 2007. Efficacy and safety of caspofungin in solid organ transplant recipients. *Transplant. Proc.* 39:3117–3120.

Pfeiffer, C. D., J. P. Fine, and N. Safdar. 2006. Diagnosis of invasive aspergillosis using a galactomannan assay: a meta-analysis. *Clin. Infect. Dis.* 42:1417–1427.

Rasmussen, C., C. Garen, S. Brining, R. L. Kincaid, R. L. Means, and A. R. Means. 1994. The calmodulin-dependent protein phosphatase catalytic subunit (calcineurin A) is an essential gene in *Aspergillus nidulans*. *EMBO J.* 13:2545–2552.

Rea, F., G. Marulli, M. Loy, L. Bortolotti, C. Giacometti, M. Schiavon, and F. Calabrese. 2006. Salvage right pneumonectomy in a patient with bronchial-pulmonary artery fistula after bilateral sequential lung transplantation. *J. Heart Lung Transplant.* **25**:1383–1386.

Reed, A., J. B. Herndon, N. Ersoz, T. Fujikawa, D. Schain, P. Lipori, A. Hemming, Q. Li, E. Shenkman, and B. Vogel. 2007. Effect of prophylaxis on fungal infection and costs for high-risk liver transplant recipients. *Liver Transplant.* **13**:1743–1750.

Reichenspurner, H., P. Gamberg, M. Nitschke, H. Valantine, S. Hunt, P. E. Oyer, and B. A. Reitz. 1997. Significant reduction in the number of fungal infections after lung-, heart-lung, and heart transplantation using aerosolized amphotericin B prophylaxis. *Transplant. Proc.* **29**:627–628.

Rex, J. H., J. E. Bennett, J. I. Gallin, H. L. Malech, E. S. DeCarlo, and D. A. Melnick. 1991. In vivo interferon-gamma therapy augments the in vitro ability of chronic granulomatous disease neutrophils to damage *Aspergillus* hyphae. *J. Infect. Dis.* **163**:849–852.

Roilides, E., C. Blake, A. Holmes, P. A. Pizzo, and T. J. Walsh. 1996. Granulocyte-macrophage colony-stimulating factor and interferon-gamma prevent dexamethasone-induced immunosuppression of antifungal monocyte activity against *Aspergillus fumigatus* hyphae. *J. Med. Vet. Mycol.* **34**:63–69.

Roilides, E., A. Dimitriadou-Georgiadou, T. Sein, I. Kadiltsoglou, and T. J. Walsh. 1998. Tumor necrosis factor alpha enhances antifungal activities of polymorphonuclear and mononuclear phagocytes against *Aspergillus fumigatus*. *Infect. Immun.* **66**:5999–6003.

Root, R. K., and D. C. Dale. 1999. Granulocyte colony-stimulating factor and granulocyte-macrophage colony-stimulating factor: comparisons and potential for use in the treatment of infections in nonneutropenic patients. *J. Infect. Dis.* **179**(Suppl. 2):S342–S352.

Rowe, J. M., J. W. Andersen, J. J. Mazza, J. M. Bennett, E. Paietta, F. A. Hayes, D. Oette, P. A. Cassileth, E. A. Stadtmauer, and P. H. Wiernik. 1995. A randomized placebo-controlled phase III study of granulocyte-macrophage colony-stimulating factor in adult patients (>55 to 70 years of age) with acute myelogenous leukemia: a study of the Eastern Cooperative Oncology Group (E1490). *Blood* **86**:457–462.

Ruffini, E., S. Baldi, M. Rapellino, A. Cavallo, A. Parola, F. Robbiano, N. Cappello, and M. Mancuso. 2001. Fungal infections in lung transplantation. Incidence, risk factors and prognostic significance. *Sarcoidosis Vasc. Diffuse Lung Dis.* **18**:181–190.

Saad, A. H., D. D. DePestel, and P. L. Carver. 2006. Factors influencing the magnitude and clinical significance of drug interactions between azole antifungals and select immunosuppressants. *Pharmacotherapy* **26**:1730–1744.

Sable, C. A., B. Y. Nguyen, J. A. Chodakewitz, and M. J. DiNubile. 2002. Safety and tolerability of caspofungin acetate in the treatment of fungal infections. *Transplant. Infect. Dis.* **4**:25–30.

Safieddine, N., B. M. Taylor, and C. M. Guiraurdon. 2002. Small bowel infarction secondary to aspergillosis in a post-cardiac transplant patient: a case report. *J. Heart Lung Transplant.* **21**:935–937.

Sandur, S., S. M. Gordon, A. C. Mehta, and J. R. Maurer. 1999. Native lung pneumonectomy for invasive pulmonary aspergillosis following lung transplantation: a case report. *J. Heart Lung Transplant.* **18**:810–813.

Saner, F., J. Gensicke, P. Rath, N. Fruhauf, Y. Gu, A. Paul, A. Radtke, M. Malago, and C. Broelsch. 2006. Safety profile of concomitant use of caspofungin and cyclosporine or tacrolimus in liver transplant patients. *Infection* **34**:328–332.

Sansone-Parsons, A., G. Krishna, M. Martinho, B. Kantesaria, S. Gelone, and T. G. Mant. 2007. Effect of oral posaconazole on the pharmacokinetics of cyclosporine and tacrolimus. *Pharmacotherapy* **27**:825–834.

Sanz-Rodriguez, C., M. Lopez-Duarte, M. Jurado, J. Lopez, R. Arranz, J. M. Cisneros, M. L. Martino, P. J. Garcia-Sanchez, P. Morales, T. Olive, M. Rovira, and C. Solano. 2004. Safety of the concomitant use of caspofungin and cyclosporin A in patients with invasive fungal infections. *Bone Marrow Transplant.* **34**:13–20.

Saulsbury, F. T. 2001. Successful treatment of *Aspergillus* brain abscess with itraconazole and interferon-gamma in a patient with chronic granulomatous disease. *Clin. Infect. Dis.* **32**:E137–E139.

Saxena, P., B. Clarke, and J. Dunning. 2007. *Aspergillus* endocarditis of the mitral valve in a lung-transplant patient. *Tex. Heart Inst. J.* **34**:95–97.

Scherer, M., H. G. Fieguth, T. Aybek, Z. Ujvari, A. Moritz, and G. Wimmer-Greinecker. 2005. Disseminated *Aspergillus fumigatus* infection with consecutive mitral valve endocarditis in a lung transplant recipient. *J. Heart Lung Transplant.* **24**:2297–2300.

Schwartz, S., M. Ruhnke, P. Ribaud, L. Corey, T. Driscoll, O. A. Cornely, U. Schuler, I. Lutsar, P. Troke, and E. Thiel. 2005. Improved outcome in central nervous system aspergillosis, using voriconazole treatment. *Blood* **106**:2641–2645.

Selby, R., C. B. Ramirez, R. Singh, I. Kleopoulos, S. Kusne, T. E. Starzl, and J. Fung. 1997. Brain abscess in solid organ transplant recipients receiving cyclosporine-based immunosuppression. *Arch. Surg.* **132**:304–310.

Sherman-Weber, S., P. Axelrod, B. Suh, S. Rubin, D. Beltramo, J. Manacchio, S. Furukawa, T. Weber, H. Eisen, and R. Samuel. 2004. Infective endocarditis following orthotopic heart transplantation: 10 cases and a review of the literature. *Transplant. Infect. Dis.* **6**:165–170.

Shlobin, O. A., L. K. Dropulic, J. B. Orens, J. F. McDyer, J. V. Conte, S. Y. Yang, and R. Girgis. 2005. Mediastinal mass due to *Aspergillus fumigatus* after lung transplantation: a case report. *J. Heart Lung Transplant.* **24**:1991–1994.

Singh, N. 2000. Antifungal prophylaxis for solid organ transplant recipients: seeking clarity amidst controversy. *Clin. Infect. Dis.* **31**:545–553.

Singh, N., P. M. Arnow, A. Bonham, E. Dominguez, D. L. Paterson, G. A. Pankey, M. M. Wagener, and V. L. Yu. 1997. Invasive aspergillosis in liver transplant recipients in the 1990s. *Transplantation* **64**:716–720.

Singh, N., R. K. Avery, P. Munoz, T. L. Pruett, B. Alexander, R. Jacobs, J. G. Tollemar, E. A. Dominguez, C. M. Yu, D. L. Paterson, S. Husain, S. Kusne, and P. Linden. 2003. Trends in risk profiles for and mortality associated with invasive aspergillosis among liver transplant recipients. *Clin. Infect. Dis.* **36**:46–52.

Singh, N., and J. Heitman. 2004. Antifungal attributes of immunosuppressive agents: new paradigms in management and elucidating the pathophysiologic basis of opportunistic mycoses in organ transplant recipients. *Transplantation* **77**:795–800.

Singh, N., and S. Husain. 2003. *Aspergillus* infections after lung transplantation: clinical differences in type of transplant and implications for management. *J. Heart Lung Transplant.* **22**:258–266.

Singh, N., A. P. Limaye, G. Forrest, N. Safdar, P. Munoz, K. Pursell, S. Houston, F. Rosso, J. G. Montoya, P. Patton, R. Del Busto, J. M. Aguado, R. A. Fisher, G. B. Klintmalm, R. Miller, M. M. Wagener, R. E. Lewis, D. P. Kontoyiannis, and S. Husain. 2006a. Combination of voriconazole and caspofungin as primary therapy for invasive aspergillosis in solid organ transplant recipients: a prospective, multicenter, observational study. *Transplantation* **81**:320–326.

Singh, N., A. P. Limaye, G. Forrest, N. Safdar, P. Munoz, K. Pursell, S. Houston, F. Rosso, J. G. Montoya, P. R. Patton, R. Del Busto, J. M. Aguado, M. M. Wagener, and S. Husain. 2006b. Late-onset invasive aspergillosis in organ transplant recipients in the current era. *Med. Mycol.* **44**:445–449.

Singh, N., and D. L. Paterson. 2005. *Aspergillus* infections in transplant recipients. *Clin. Microbiol. Rev.* **18**:44–69.

Singh, N., D. L. Paterson, T. Gayowski, M. M. Wagener, and I. R. Marino. 2001. Preemptive prophylaxis with a lipid preparation of amphotericin B for invasive fungal infections in liver transplant recipients requiring renal replacement therapy. *Transplantation* **71**:910–913.

Singh, N., T. L. Pruett, S. Houston, P. Munoz, T. V. Cacciarelli, M. M. Wagener, and S. Husain. 2006c. Invasive aspergillosis in the recipients of liver retransplantation. *Liver Transplant.* **12**:1205–1209.

Singh, N., M. M. Wagener, T. V. Cacciarelli, and J. Levitsky. 2008. Antifungal management practices in liver transplant recipients. *Am. J. Transplant.* **8**:426–431.

Singhal, S., R. W. Ellis, S. G. Jones, S. J. Miller, N. C. Fisher, J. G. Hastings, and D. J. Mutimer. 2000. Targeted prophylaxis with amphotericin B lipid complex in liver transplantation. *Liver Transplant.* **6**:588–595.

Sole, A., P. Morant, M. Salavert, J. Peman, and P. Morales. 2005. *Aspergillus* infections in lung transplant recipients: risk factors and outcome. *Clin. Microbiol. Infect.* **11**:359–365.

Steinbach, W. J., R. A. Cramer, Jr., B. Z. Perfect, Y. G. Asfaw, T. C. Sauer, L. K. Najvar, W. R. Kirkpatrick, T. F. Patterson, D. K. Benjamin, Jr., J. Heitman, and J. R. Perfect. 2006. Calcineurin controls growth, morphology, and pathogenicity in *Aspergillus fumigatus*. *Eukaryot. Cell* **5**:1091–1103.

Stevens, D. 2006. Th1/Th2 in aspergillosis. *Med. Mycol.* **44**:S229–S235.

Sugiyama, M., M. Funauchi, T. Yamagata, Y. Nozaki, B. S. Yoo, S. Ikoma, K. Kinoshita, and A. Kanamaru. 2004. Predominant inhibition of Th1 cytokines in New Zealand Black/White F_1 mice treated with FK506. *Scand. J. Rheumatol.* **33**:108–114.

Summers, S. A., A. Dorling, J. J. Boyle, and S. Shaunak. 2005. Cure of disseminated cryptococcal infection in a renal allograft recipient after addition of gamma-interferon to anti-fungal therapy. *Am. J. Transplant.* **5**:2067–2069.

Torre-Cisneros, J., O. L. Lopez, S. Kusne, A. J. Martinez, T. E. Starzl, R. L. Simmons, and M. Martin. 1993. CNS aspergillosis in organ transplantation: a clinicopathological study. *J. Neurol. Neurosurg. Psychiatry* **56**:188–193.

Vagefi, P. A., A. B. Cosimi, L. C. Ginns, and C. N. Kotton. 2008. Cutaneous *Aspergillus ustus* in a lung transplant recipient: emergence of a new opportunistic fungal pathogen. *J. Heart Lung Transplant.* **27**:131–134.

Vazquez, J. A. 2005. Anidulafungin: a new echinocandin with a novel profile. *Clin. Ther.* **27**:657–673.

Veroux, M., D. Corona, M. Gagliano, M. Sorbello, M. Macarone, M. Cutuli, G. Giuffrida, G. Morello, A. Paratore, and P. Veroux. 2007. Voriconazole in the treatment of invasive aspergillosis in kidney transplant recipients. *Transplant. Proc.* **39**:1838–1840.

Vora, S., S. Chauhan, E. Brummer, and D. A. Stevens. 1998. Activity of voriconazole combined with neutrophils or monocytes against *Aspergillus fumigatus:* effects of granulocyte colony-stimulating factor and granulocyte-macrophage colony-stimulating factor. *Antimicrob. Agents Chemother.* **42**:2299–2303.

Walsh, T. J., E. J. Anaissie, D. W. Denning, R. Herbrecht, D. P. Kontoyiannis, K. A. Marr, V. A. Morrison, B. H. Segal, W. J. Steinbach, D. A. Stevens, J. A. van Burik, J. R. Wingard, and T. F. Patterson. 2008. Treatment of aspergillosis: clinical practice guidelines of the Infectious Diseases Society of America. *Clin. Infect. Dis.* **46**:327–360.

Walsh, T. J., I. Raad, T. F. Patterson, P. Chandrasekar, G. R. Donowitz, R. Graybill, R. E. Greene, R. Hachem, S. Hadley, R. Herbrecht, A. Langston, A. Louie, P. Ribaud, B. H. Segal, D. A. Stevens, J. A. van Burik, C. S. White, G. Corcoran, J. Gogate, G. Krishna, L. Pedicone, C. Hardalo, and J. R. Perfect. 2007. Treatment of invasive aspergillosis with posaconazole in patients who are refractory to or intolerant of conventional therapy: an externally controlled trial. *Clin. Infect. Dis.* **44**:2–12.

Weclawiak, H., C. Garrouste, N. Kamar, M. D. Linas, P. Tall, C. Dambrin, D. Durand, and L. Rostaing. 2007. *Aspergillus fumigatus*-related spondylodiscitis in a heart transplant patient successfully treated with voriconazole. *Transplant. Proc.* **39**:2627–2628.

Weiland, D., R. M. Ferguson, P. K. Peterson, D. C. Snover, R. L. Simmons, and J. S. Najarian. 1983. Aspergillosis in 25 renal transplant patients. Epidemiology, clinical presentation, diagnosis, and management. *Ann. Surg.* **198**:622–629.

Weimer, R., A. Melk, V. Daniel, S. Friemann, W. Padberg, and G. Opelz. 2000. Switch from cyclosporine A to tacrolimus in renal transplant recipients: impact on Th1, Th2, and monokine responses. *Hum. Immunol.* **61**:884–897.

Westney, G. E., S. Kesten, A. De Hoyos, C. Chapparro, T. Winton, and J. R. Maurer. 1996. *Aspergillus* infection in single and double lung transplant recipients. *Transplantation* **61**:915–919.

White, M. H., E. J. Anaissie, S. Kusne, J. R. Wingard, J. W. Hiemenz, A. Cantor, M. Gurwith, C. Du Mond, R. D. Mamelok, and R. A. Bowden. 1997. Amphotericin B colloidal dispersion vs. amphotericin B as therapy for invasive aspergillosis. *Clin. Infect. Dis.* **24**:635–642.

Wieland, T., A. Liebold, M. Jagiello, G. Retzl, and D. E. Birnbaum. 2005. Superiority of voriconazole over amphotericin B in the treatment of invasive aspergillosis after heart transplantation. *J. Heart Lung Transplant.* **24**:102–104.

Wimberley, S. L., M. T. Haug III, K. M. Shermock, A. Qu, J. R. Maurer, A. C. Mehta, R. J. Schilz, and S. M. Gordon. 2001. Enhanced cyclosporine-itraconazole interaction with cola in lung transplant recipients. *Clin. Transplant.* **15**:116–122.

Aspergillus fumigatus and Aspergillosis
Edited by J.-P. Latgé and W. J. Steinbach

Chapter 39

Invasive Aspergillosis in Malignancy and Stem Cell Transplant Recipients

Elio Castagnola and Claudio Viscoli

Invasive mycoses represent a major cause of morbidity and mortality in patients with malignancy or undergoing hematopoietic stem cell transplantation (HSCT) (De Pauw and Verveij, 2005; van Burik and Weisdorf, 2005). However, the exact burden of invasive aspergillosis in patients with malignancy or in HSCT recipients may be difficult to assess, as it is frequently complicated to obtain an accurate microbiological diagnosis because of difficulties in performing invasive procedures due to severe thrombocytopenia or other life-threatening conditions. Moreover, fungi frequently do not grow in culture, and therefore often only a descriptive diagnosis of "infection due to filamentous fungi" can be made based on histological or morphological examination. To overcome these pitfalls and better understand invasive fungal infections and invasive aspergillosis, a set of definitions has been proposed with three levels of certainty of diagnosis: proven, probable, and possible invasive fungal infection (Ascioglu et al., 2002). Presently, this set of definitions is frequently used for epidemiologic purposes or clinical trial design. Proven fungal infection requires a positive culture from a sterile site and/or histopathological evidence. In the definition of probable infection, mycological evidence acquired by means of positive antigenemia (galactomannan or β-D-glucan) or direct examination or culture of specimens from sites that may be colonized (e.g., sputum, bronchoalveolar lavage fluid, or sinus aspirate) supports the diagnosis but does not prove it (Ascioglu et al., 2002). Finally, the definition of possible invasive mycoses is based only on clinical grounds (host factor, clinical picture) in the absence of any microbiological data (Ascioglu et al., 2002). It has been suggested that nearly 60% of these cases are actually true invasive aspergillosis when autopsy is performed (Subira et al., 2003). The absence of a complete speciation of an isolated *Aspergillus* strain (in either cases of true infection or colonization) further reduces the possibility of correctly estimating the role of *Aspergillus fumigatus* among the other isolated filamentous fungi. As a consequence, the real incidence of invasive mycoses due to *A. fumigatus* is probably underestimated and the presently available data probably represent only the tip of the iceberg for the epidemiological, clinical, and therapeutic information available for this specific infection in patients with malignancy or who have undergone HSCT.

SOURCE OF *ASPERGILLUS* AND RISK FACTORS FOR INVASIVE ASPERGILLOSIS

A. fumigatus is a ubiquitous mold that is found in every region of the world. It is believed that decomposing vegetable material represents the primary ecological niche and, therefore, rural areas are a major source for this organism. Possible domestic sources of *A. fumigatus* and other species include potted plants, flower arrangements, and carpets. The air spore count in the environment varies in a remarkable way, including seasonally. In the majority of healthy individuals *Aspergillus* spores are trapped in the upper respiratory tract and only a small proportion of them enter the lower airways, where spores can germinate and cause invasive disease. Colonization by *Aspergillus* is a necessary condition for the development of invasive disease in the presence of immunosuppression, and colonization may in fact precede the development of the malignancy or the HSCT procedure. Other sources of *Aspergillus* spores include building construction and renovation, which actually have been indicated as major causes of epidemic clusters in hospitalized patients (Vonberg and Gastmeier, 2006). In recent years hospital water supplies have also been identified as other possible sources of *A. fumigatus* (Anaissie et al., 2002). However, genotypic analyses of

Elio Castagnola • Infectious Diseases Unit, Dept. of Hematology and Oncology, "G. Gaslini" Children Hospital, 16147 Genoa, Italy. **Claudio Viscoli** • Division of Infectious Disease, University of Genoa, San Martino University Hospital, 16132 Genoa, Italy.

strains of *A. fumigatus* isolated from water and air at various locations inside and outside hospitals and from patients with proven or probable invasive aspergillosis have rarely shown that strains infecting patients were actually coming from the hospital environment (Warris et al., 2003).

There is growing evidence that in many patients *Aspergillus* colonization is present before hospitalization and the organisms are community acquired from different sources and imported into the hospital in the form of prior colonization or infection (Manuel and Kibbler, 1998). The rate of colonization by *Aspergillus* has been demonstrated to be age related (Silvestri et al., 1996), being lower in preschool children and increasing with age, with the highest rates in adolescents. This could explain the overall lower incidence of invasive aspergillosis in immunocompromised children compared to adults, in spite of the presence of similar risk factors (Castagnola et al., 2006), including colonization (Abbasi et al., 1999), and the identification of age of ≥10 years as a risk factor for the development of invasive aspergillosis during treatment for acute leukemia or following allogeneic HSCT in pediatric patients (Dvorak et al., 2005; Rosen et al., 2005). Similar to adult patients, both building renovation or construction (Castagnola et al., 2006; Cesaro et al., 2004) and water supplies (Warris et al., 2001) have also been documented as possible sources of *Aspergillus* in children. Gastrointestinal tract colonization due to food contamination has also been suggested as another possible portal of entry of *Aspergillus* after disruption of the intestinal mucosal barrier by chemotherapy-induced severe gastrointestinal mucositis (Manuel and Kibbler, 1998). All these data indicate that invasive aspergillosis in patients with malignancy or receiving HSCT is an endemic disease that is usually community acquired, with occasional epidemic outbreaks possibly associated with massive environmental exposures (in and out of the hospital) (Hajjeh and Warnock, 2001; Manuel and Kibbler, 1998).

The impairment of defense mechanisms is the second condition necessary for the development of invasive aspergillosis. Prolonged and profound granulocytopenia secondary to administration of antineoplastic chemotherapy and secondary neutropenia associated with failure of hematopoietic stem cell engraftment represent well-known risk factors for invasive aspergillosis (De Pauw and Verveij, 2005; Marchetti and Calandra, 2004; Mihu et al., 2008; Thursky, et al., 2004). Since granulocytes represent the main effector cells involved in phagocytosis and killing of *Aspergillus* spores, it is easily understood why prolonged and profound granulocytopenia secondary to administration of aggressive antineoplastic chemotherapy represents a well-known risk factor for invasive aspergillosis (De Pauw and Verveij, 2005; Marchetti and Calandra, 2004). In addition, recent epidemiological studies have shown that following allogeneic HSCT, invasive aspergillosis may develop as a late infection in nongranulocytopenic patients (Castagnola et al., 2006; Marr et al., 2002; Ochs et al., 1995). In these settings, lymphocytopenia is probably the most important risk factor, resulting in the absence or abnormal function of T lymphocytes, especially of the CD4 subtype (Romani, 2004). It has been demonstrated recently that the effective transfer of anti-*Aspergillus* T-cell clones in lymphocytopenic recipients of haploidentical HSCT may be effective in treating this invasive disease (Perruccio et al., 2005).

In allogeneic HSCT the degree of matching between donor and recipient and the consequent presence of severe acute or chronic extensive graft-versus-host disease (GVHD) have been associated with an increased risk of developing invasive aspergillosis per se or via the drugs administered for GVHD management (Marr et al., 2002; Mihu et al., 2008). Steroids represent a cornerstone of this therapy, and it has been demonstrated that the risk of invasive aspergillosis is a function of the dose and duration of steroid therapy (Cordonnier et al., 2006; Fukuda et al., 2003; Marr et al., 2002). Corticosteroids affect the host immune response to *Aspergillus* by preventing killing of phagocytosed *A. fumigatus* conidia by alveolar macrophages (Philippe et al., 2003) and by blunting alveolar macrophage production of proinflammatory cytokines (interleukin-1a and tumor necrosis factor alpha) and chemokines (macrophage inhibitory factor 1α) that are pivotal in recruiting neutrophils and monocytes (Brummer et al., 2003). Corticosteroids may also affect the type of T helper cell reaction to invasive aspergillosis. Peripheral blood mononuclear cells from healthy human subjects produce Th1-type cytokines in the presence of *A. fumigatus* in vitro (Grazziutti et al., 1997; Hebart et al., 2002). Corticosteroids are associated with induction of Th2 responses and poor outcomes (Balloy et al., 2005; Hebart et al., 2002; Roilides et al., 2001). Finally, the addition of hydrocortisone to *A. fumigatus* cultures is able to hasten significantly fungal growth in vitro (Ng et al., 1994). From a clinical point of view the administration of 0.5 mg/kg of body weight/day for a period of >30 days (Baddley et al., 2001) or >1 mg/kg for a period of >21 days (Cordonnier et al., 2006; Grow et al., 2002) has been associated with an increased risk of invasive aspergillosis, which starts within 2 weeks of steroid administration (>1 mg/kg/day of prednisone or equivalent) and has a dose-response effect extending to 6 weeks. Also, lower doses (0.25 to 1.0 mg/kg/day) for 2 to 10 weeks have been shown to impact the risk of invasive aspergillosis (Thursky et al., 2004). It is interesting that clinical data suggest that the effects of steroids may be additive to those of other drugs, such as ganciclovir, administered for management of other infectious complications in HSCT recipients (Thursky et al., 2004). In recent years, monoclonal antibodies directed either against various types of T or B cells or against tumor necrosis factor have become important tools for the management of GVHD and for the treatment of some lymphoproliferative disorders. Among them, alemtuzumab (anti-CD52), which is capable of inducing a severe combined immunodeficiency with impairment of B and T lymphocytes and maybe also other cells of the myeloid series, has been associated with an increase in the incidence of invasive aspergillosis (Martin et al., 2006; Mihu et al., 2008; Thursky et al., 2006). There is no reason to believe that these immunological risk factors do not have a role in children as well as adult patients (Castagnola et al., 2006; Dvorak et al., 2005; Rosen et al., 2005). Several risk factors may also be present simultaneously or consecutively in the same patient and may have a different impact on different patients.

Recently it has also been shown that the Toll-like receptor system could play an important role in the development of invasive aspergillosis in patients receiving chemotherapy (Lanciotti et al., 2008) or after HSCT (Kesh et al., 2005), and there is evidence suggesting that the involvement of Toll-like receptors during *A. fumigatus* infection is influenced by the immunological status of the host (Chignard et al., 2007). *A. fumigatus* may also favor invasive disease by the production of gliotoxin, which on one hand may suppress the adaptive immune defense system and on the other hand may increase polymorphonuclear leukocyte-mediated inflammation, which is likely to play an important role in tissue destruction in the setting of invasive aspergillosis (Orciuolo et al., 2007).

EPIDEMIOLOGY OF *A. FUMIGATUS* INFECTION IN PATIENTS WITH MALIGNANCY OR RECEIVING HSCT

Taking into account the conditions required for the development of invasive aspergillosis (simultaneous presence of colonization and risk factors), it is not difficult to understand why patients (and especially adults) receiving antineoplastic chemotherapy for acute leukemia or allogeneic HSCT represent the group with a higher incidence of invasive aspergillosis.

The epidemiology of invasive aspergillosis in these patient populations has been clearly described in extensive reviews (Anaissie and Nucci, 2003; De Pauw and Verveij, 2005; Denning, 1998; Donnelly and De Pauw, 2005; Junghanss and Marr, 2002; Lin et al., 2001; Manuel and Kibbler, 1998; Marchetti and Calandra, 2004; Marr and Bowden, 1999; Steinbach, 2005; Steinbach and Marr, 2003; van Burik and Weisdorf, 2005; Zaoutis et al., 2006). For this reason in the present chapter we will report and comment mainly on the most recent observations.

In patients with hematologic malignancies, a multicenter study performed in Italy from 1999 to 2003 on 11,802 adults showed that invasive aspergillosis was present in 2.6% of patients (Pagano et al., 2006). The incidence of invasive aspergillosis varied according to the underlying disease (and therefore the aggressiveness of antineoplastic chemotherapy) and was 7.9% in acute nonlymphoblastic leukemia, 4.3% in acute lymphoblastic leukemia, 2.3% in chronic myelogenous leukemia, and <1% in chronic lymphocytic leukemia, Hodgkin's disease, non-Hodgkin's lymphoma, or multiple myeloma. *A. fumigatus* was identified in 53% of the cases of proven aspergillosis.

On the other side of the Atlantic Ocean, a retrospective study on autopsies performed from 1989 to 2003 at M.D. Anderson Cancer Center in Texas (Chamilos et al., 2006) in patients receiving chemotherapy or HSCT showed that 17% of patients had invasive aspergillosis, but *A. fumigatus* was identified only in 9% of patients, even if its frequency increased from 0.6% in the period from 1989 to 1993, to 2.1% in the period 1994 to 1998, to 2.9% in the period from 1999 to 2003. This discrepancy could be due to regional epidemiological factors and/or to differences in the two study designs, i.e., a retrospective clinical survey versus an autopsy study. In recent years, invasive aspergillosis has been increasingly observed in patients with solid tumors as well, especially those with central nervous system neoplasias receiving high-dose steroids for the control of intracranial hypertension (Ohmagari et al., 2004). In these settings *A. fumigatus* was the most frequently isolated *Aspergillus* species. This aspect could be simply related to local factors, but it could also represent the beacon of a new patient population at risk.

After HSCT the epidemiology of invasive aspergillosis is more complex, since it varies according to the hematopoietic stem cell source, the type of conditioning regimen, and the type and intensity of immunosuppression after transplant. Autologous HSCT presents a very low incidence of invasive aspergillosis, estimated to be 0.3% (7 in 1,979) in the previously mentioned Italian study from 1999 to 2003. *A. fumigatus* was identified as the etiologic agent in only three of these cases. This observation is consistent with other reports describing a 0.5 to 2% incidence of invasive aspergillosis after autologous HSCT (Cornet et al., 2002; Fukuda et al., 2004; Jantunen et al., 2000; Morgan et al., 2005; Pagano et al., 2007; Zaoutis et al., 2006), mainly observed during neutropenia preceding engraftment (Boeckh and Marr, 2002; Pagano et al., 2007; Sepkowitz, 2003; van Burik and Weisdorf, 2005).

In allogeneic HSCT the overall incidence of invasive aspergillosis ranges from 2.9 to 15% of patients

(Alangaden et al., 2002; Baddley et al., 2001; Fukuda et al., 2003, 2004; Grow et al., 2002; Kojima et al., 2004; Martino et al., 2002; Morgan et al., 2005; Thursky et al., 2004; Zaoutis et al., 2006). These data have been confirmed by a recent multicenter retrospective study on 1,249 allogeneic HSCT patients, among which 79 cases of invasive aspergillosis were identified, for a rate of 6% of the transplant procedures. *A. fumigatus* was isolated in 16 (20%) of these cases, but in another 50 cases (63%) *Aspergillus* was not further speciated. Finally, the incidence of *A. fumigatus* invasive disease ranged from 2.2 to 31% in patients receiving allogeneic HSCT from alternative donors (matched unrelated or T-cell depleted) or after reduced-intensity conditioning regimens (Fukuda et al., 2003; Martino et al., 2006; Mihu et al., 2008; Morgan et al., 2005; Pagano et al., 2007). This wide range of incidence rates could be due to local factors or to the lack of further speciation of isolated strains.

The incidence of invasive aspergillosis after allogeneic HSCT is bimodal, with a first peak during the preengraftment phase and a second occurring later, frequently 3 months or more after the procedure (Boeckh and Marr, 2002; Cornet et al., 2002; Fukuda et al., 2004; Pagano et al., 2007; Sepkowitz, 2003; van Burik and Weisdorf, 2005; Wald et al., 1997). At present these late-onset infections represent the major challenge for clinicians. *A. fumigatus* has been identified as the main cause of late-onset proven or probable aspergillosis (Jantunen et al., 2000; Mihu et al., 2008; Saugier-Veber et al., 1993), with a cumulative risk of 15% at 12 months after the procedure (Baddley et al., 2001). *A. fumigatus* was also detected in 75% (9 of 12) of community-acquired pneumonia cases occurring a median of 181 days after allogeneic HSCT in patients treated for severe chronic GVHD (Alangaden et al., 2002). Interestingly, gram-negative bacteria were present as associated pathogens in nearly all cases.

An additional confounding factor is that cases of invasive aspergillosis occurring before HSCT can reactivate during the posttransplant immunosuppression (preengraftment neutropenia and GVHD management). Indeed, a history of invasive aspergillosis was documented in 45 (2%) of 2,319 patients receiving HSCT between 1992 and 2001 in a single center (Fukuda et al., 2004). Among this group, 13 (29%) developed invasive disease within 1 year after HSCT and in 9 cases the disease was considered to be a recurrence by anatomic site and timing. In 10 cases (77%), aspergillosis relapsed very early after HSCT (median, 26 days) and in 60% of cases this occurred before neutrophil recovery. The recurrence of invasive aspergillosis was associated with a shorter duration (<1 month) of antifungal therapy before transplantation, with persistence of radiological abnormalities and with the administration of a myeloablative conditioning regimen, especially if this included total body irradiation.

The role of the conditioning regimen for allogeneic HSCT patients as a risk factor for reactivation was also evaluated in 129 patients with a diagnosis of proven or probable invasive aspergillosis preceding the transplant (Martino et al., 2006). Progression of invasive aspergillosis was observed in 27 (17%) of these patients, with a cumulative incidence of 22% at 2 years after the transplant. The cumulative incidence was related to a longer duration of neutropenia after transplantation, advanced status of the underlying disease, less than 6 weeks from start of systemic anti-*Aspergillus* therapy, and with an allogeneic HSCT. Other important factors were conventional myeloablative conditioning (especially for early aspergillosis), cytomegalovirus disease (for late aspergillosis), donor source (bone marrow or cord blood), and presence of severe acute GVHD. In these settings *A. fumigatus* was identified as the causative agent in 31% of relapsing infections (Martino et al., 2006).

Invasive aspergillosis is a not frequent disease in children with malignancy or undergoing HSCT. In a prospective 5-year surveillance study in children with cancer, invasive aspergillosis occurred exclusively in the context of hematological malignancies, with an incidence of 6.8% (Groll et al., 1999). Similarly, a single-center 34-year retrospective study (Abbasi et al., 1999) showed that <1% (66 of ≈9,500) of immunocompromised children developed invasive aspergillosis, with 83% of cases observed during treatment for acute leukemia and only 0.5% after treatment for a solid tumor (including patients with lymphoma). *A. fumigatus* was identified as the etiologic agent in 15 cases. Similar results were reported in two different retrospective studies from a single center, showing an incidence of 0.9% for aspergillosis during aggressive treatment of acute leukemias (with a rate of 0.006 episodes/100 days at risk) (Castagnola et al., 2005) and 0.16% (with a rate of 0.0005/100 days at risk) in children aggressively treated for solid tumors (Haupt et al., 2001). Furthermore, a prospective multicenter survey in children with malignancy documented only 96 cases of invasive mycosis over 2 years, and proven or probable mold infections accounted for 33% of the episodes (32 of 96) (Castagnola et al., 2006). *A. fumigatus* was isolated in only 1 case, but in 10 cases the *Aspergillus* species was not further identified. Finally, invasive aspergillosis was observed in 7 of 703 infectious episodes that developed during 1,792 periods of granulocytopenia, accounting for a total of 28,001 days at risk, occurring in 366 children after chemotherapy or HSCT (1% of infectious episodes, 2% of the patients, 0.07 episodes per 30 days of neutropenia) (Castagnola et al., 2007). *A. fumigatus* was identified in only one of these episodes. In this study the very low frequency was observed in the absence of specific antifungal prophylaxis but under strict control of environmental conditions obtained by use of HEPA filters for all the wards (hematology, oncology, and the HSCT unit) and use of HEPA filter masks for patients when outside of the filtered areas (Benet et al., 2007; Humphreys, 2004; Nihtinen et al., 2007; Raad et al., 2002). Invasive mold infections have been reported after 1 to 3% of autologous HSCT procedures in children (Benjamin et al., 2002; Hovi et al., 2000), in all cases during the preengraftment period. After allogeneic HSCT the incidence of invasive aspergillosis ranges from 0 to 14% (Barker et al., 2005; Benjamin et al., 2002; Castagnola et al., 2008; Dvorak et al., 2005; Hovi et al., 2000; Steinbach et al., 2007), with the majority of cases observed during preengraftment granulocytopenia (Castagnola et al., 2006, 2008). However, *A. fumigatus* was identified as the causative agent in only a very few cases (Barker et al., 2005; Benjamin et al., 2002; Dvorak et al., 2005; Hovi et al., 2000, 2007; Steinbach, et al., 2007).

CLINICAL FEATURES

The upper and lower respiratory tracts represent the most frequent locations of *A. fumigatus* infection in all patient categories and age groups (Abbasi et al., 1999; Alangaden et al., 2002; Baddley et al., 2001; Castagnola et al., 2006; Denning, 1998; Groll et al., 1999; Hori et al., 2002; Mihu et al., 2008; Ohmagari et al., 2004). This is not surprising when considering the route of acquisition of the infection. The lungs are the most frequently involved organ, and a cough is present in nearly all cases, frequently associated with fever and other respiratory symptoms like dyspnea and chest pain (Abbasi et al., 1999). Recently, an immune reconstitution inflammatory syndrome has been described in patients with pulmonary aspergillosis who present with a worsening of respiratory clinical features and imaging changes in the absence of signs of dissemination to other organs. This clinical picture was observed after neutrophil recovery and coincided with microbiological and clinical response in 84% of the subjects (Miceli et al., 2007). However, it must be remembered that in patients with pulmonary aspergillosis a (too) rapid recovery of the granulocyte count may be associated with the development of severe complications, such as pneumothorax or fatal hemoptysis (Martino et al., 1990; Pagano et al., 1995; Todeschini et al., 1999).

Localization in the paranasal sinuses may frequently present with facial swelling or nasal discharge, but sometimes pain at the superior dental arch may represent the first reported symptom. From paranasal sinuses the infection may progress to the mouth by involving the hard palate or to the orbit and up to the central nervous system (rhinocerebral aspergillosis). *Aspergillus* species have angioinvasive properties and may disseminate from the primary lesions, usually within the lungs, to a variety of organs via hematogenous spread. The central nervous system represents the second most frequent location of invasive aspergillosis (Baddley et al., 2001; Hori et al., 2002; Jantunen et al., 2000; Saugier-Veber et al., 1993). The sudden occurrence of seizures, diminished consciousness, and/or focal neurological signs in a patient with pulmonary aspergillosis may represent the first signs of involvement of the central nervous system. However, cerebral aspergillosis may also occur in patients without pulmonary disease. Localization in the gastrointestinal tract is frequently associated with abdominal symptoms, as would be expected, such as abdominal pain, hematemesis, and melena (Hori et al., 2002). Skin lesions occur in disseminated disease and are represented by nodular, necrotic, crusted, or ecchymotic lesions with surrounding erythema or cellulitis (Abbasi et al., 1999; Denning, 1998; Hori et al., 2002).

Other localizations may be found in the liver, spleen, kidneys, adrenal glands, and bones. In some patients, *Aspergillus* can spread from the primary locations directly to contiguous structures, such as the pleura, heart, stomach, liver, and large vessels (Abbasi et al., 1999; Denning, 1998; Hori et al., 2002). Other localizations are rare and frequently not diagnosed in patients while living (Abbasi et al., 1999; Denning, 1998; Hori et al., 2002). In a systematic review of the literature (Lin et al., 2001), it was found that recipients of HSCT are most likely to develop disseminated aspergillosis (53%), while in patients with hematologic malignancies the most common clinical presentation is pulmonary disease (53%), although disseminated infection may account for a significant proportion (29%) of the cases. Relapsing disease usually affects the primary site, although dissemination can occur (Fukuda et al., 2004; Sipsas and Kontoyiannis, 2006).

Unfortunately, even if highly suggestive of invasive aspergillosis, neither one of these clinical features nor any radiological sign, such as halo or air crescent (Brodoefel et al., 2006; Caillot et al., 1997, 2001; Greene et al., 2007; Kojima et al., 2005; Levine et al., 2007; Sipsas and Kontoyiannis, 2006), is totally specific. However, when observed in a high-risk patient in association with a positive culture (even if from a nonsterile site) or with a positive galactomannan test, the level of certainty may be very high. The galactomannan test, which has a relatively low sensitivity but a high specificity, has been shown to be an important diagnostic tool. Its specificity approximates 92% (95% confidence interval, 90 to 93%) for documented aspergillosis in hematologic malignancies and 86% (95% confidence interval, 83 to 88%) in allogeneic HSCT patients (Pfeiffer et al., 2006; Steinbach et al., 2007).

MANAGEMENT AND PROGNOSIS

The prognosis of patients with malignancy or receiving HSCT who develop invasive aspergillosis is often discouraging. A large systematic review of the literature published in 2001 (Lin et al., 2001) showed a mortality rate of approximate 60% in patients with malignancy and 90% in allogeneic HSCT recipients. More recent studies in adults have substantially confirmed these findings, showing a mortality varying from 42% in cases occurring after chemotherapy (Pagano et al., 2006) to 17% after autologous HSCT (Pagano et al., 2007) and 77 to 90% after allogeneic HSCT (Pagano et al., 2007; Upton et al., 2007). After allogeneic HSCT the occurrence of late invasive aspergillosis has been associated with a high (>90%) mortality rate (Alangaden et al., 2002; Bjorklund et al., 2007; Hoyle and Goldman 1994; Mihu et al., 2008; Upton et al., 2007), probably as the result of severe immunosuppression in patients with severe acute or chronic extensive GVHD. Data for children seem to show a lower mortality rate (Castagnola et al., 2006) but confirm that the risk of death is higher among allogeneic HSCT recipients, with values ranging from 41 to 56% (Castagnola et al., 2006, 2008; Dvorak et al., 2005). The presence of a disseminated infection has been associated with the worst prognosis, both in children and adults (Abbasi et al., 1999; Cordonnier et al., 2006; Upton et al., 2007). Finally, the relapse of invasive aspergillosis during subsequent cycles of chemotherapy or after HSCT has also been associated with a higher mortality (Fukuda et al., 2004; Offner et al., 1998; Sipsas and Kontoyiannis, 2006).

The role of secondary prophylaxis in preventing relapses has never been studied systematically, even if a longer duration of antifungal therapy before HSCT has been associated with a better outcome (Fukuda et al., 2004; Martino et al., 2006). In general, early diagnosis appears to be associated with better clinical outcome (Caillot et al., 1997; Cornely et al., 2007a; Pagano et al., 2006; Sipsas and Kontoyiannis, 2006).

In recent years the availability of new antifungal drugs has at least partially improved the prognosis of invasive aspergillosis. Voriconazole was shown to be more effective than amphotericin B deoxycholate in a randomized trial in patients with proven or probable invasive aspergillosis (*A. fumigatus* was detected in 85 [77%] of the 110 infections in which the species was identified at baseline), with an overall survival of 60% at 3 months after the beginning of therapy (Herbrecht et al., 2002). Survival was substantially lower in proven than in probable infections (45 versus 60%). Similar results were found in a randomized double-blind trial that compared two different doses of liposomal amphotericin B (3 mg/kg/day in one group versus 10 mg/kg/day for 2 weeks followed by 3 mg/kg/day for the other group) (Cornely et al., 2007a). The success rates were 50 and 46%, respectively, for the two dosages, with better tolerability of the lower one. Unfortunately, in this study *Aspergillus* was fully identified as the causative agent in only eight cases and no further speciation was provided.

Comparison of different clinical trials is always problematic. In doing so, Denning (2007) noticed that the liposomal amphotericin B study enrolled types of patients who generally had a better prognosis and who probably received an earlier diagnosis and treatment. Moreover, between the two studies there were substantial differences in end points. Finally, in the second study many patients were included only on the basis of the halo sign, whose specificity for invasive aspergillosis is questionable. In the wake of all these considerations and in the absence of any direct comparative trials, it is reasonable to consider voriconazole as the first choice for patients with invasive aspergillosis, despite some hepatic toxicity and, especially, drug interactions. On the other hand, 3 mg/kg/day of liposomal amphotericin B has been demonstrated to be a good therapeutic option, especially when voriconazole cannot be administered. Data regarding other lipid formulations of amphotericin B are mainly available from retrospective studies (Chandrasekar and Ito, 2005), which showed 47% efficacy as primary treatment and 44% as salvage therapy in patients with invasive aspergillosis.

Oral posaconazole has been assessed as salvage therapy for various invasive fungal infections, including a cohort of 107 patients with invasive aspergillosis (Walsh et al., 2007). A 42% response rate was found in posaconazole-treated patients and compared favorably with respect to an external control group. For the first time in this trial, a correlation between serum levels of posaconazole and efficacy was demonstrated, suggesting that therapeutic drug monitoring might be indicated when using azole drugs. Among echinocandins, caspofungin has been approved for salvage therapy in patients with invasive aspergillosis, since it has been demonstrated to be well-tolerated and with a 39% response in patients with refractory infections (Maertens et al., 2004). However, in this study as in other salvage therapy trials (Maertens et al., 2006), the need for salvage therapy was established in the presence of progression of disease or failure to improve clinically despite receiving at least 7 days of "standard therapy" (mainly amphotericin B deoxycholate, liposomal amphotericin B, or itraconazole). It must be emphasized that the definition of refractory infection is quite arbitrary, since the clinical stabilization of a patient with invasive aspergillosis after "only" 7 days of treatment may not always be considered a "failure" requiring a new treatment.

Moreover, a retrospective analysis (Patterson et al., 2005) of data prospectively collected during a randomized clinical trial (Herbrecht et al., 2002) showed that an effective drug usually obtains better results when given as primary therapy compared with rescue treatment. In a recently reported phase II study (Viscoli et al., 2007), caspofungin given as first-line therapy achieved a 38% response rate among 52 patients with proven or probable aspergillosis (patients with halo sign only were not included) and who were severely neutropenic and with an advanced hematological malignancy.

The administration of combined antifungal therapy has been advocated to improve prognosis and survival (Aliff et al., 2003; Cesaro et al., 2007b; Kontoyiannis et al., 2003; Maertens et al., 2004, 2006; Marr et al., 2004; Upton et al., 2007). However, none of these studies included a comparative arm and all but one (Maertens et al., 2006) were retrospective. Interestingly, taking into account the pharmacokinetics of the different forms of amphotericin B, which have a long half-life, a salvage therapy trial might actually be considered as a combination therapy study (but also in this case the previous considerations on efficacy are rational). In any case, in the absence of randomized clinical trials, combined antifungal therapy cannot be recommended as a first-line therapy for invasive aspergillosis.

Finally, surgery in association with antifungals has been indicated as a possible therapeutic option to prevent relapsing invasive aspergillosis in patients needing additional antineoplastic therapy or undergoing HSCT (Caillot et al., 1997; Cesaro et al., 2007a; Habicht et al., 2001). Surgical resection is supposed to provide better control of the infection compared to antifungal therapy alone, especially in the presence of bulk lesions. In addition, surgical resection may allow an accurate diagnostic documentation and can prevent death due to massive hemoptysis, which may occur early in the course of pulmonary infection as the patient recovers from neutropenia (Pagano et al., 1995; Todeschini et al., 1999). A crucial point is obviously the possibility of postoperative complications that can delay the prompt resumption of chemotherapy. Indeed, postsurgery complications have been reported in 5 to 39% of patients in different series (Bernard et al., 1997; Cesaro et al., 2007a; Salerno et al., 1998; Temeck et al., 1994; Yeghen et al., 2000). In a pediatric series chemotherapy was resumed after a median interval of 19 days and autologous or allogeneic HSCT was performed after a median of 60 days from surgery (Cesaro et al., 2007a). In conclusion, although the role of surgery in preventing relapse of invasive aspergillosis remains undefined, it is likely that younger patients in hematological remission, in good general condition, and with single *Aspergillus* lesions might benefit from resection.

ROLE OF UNDERLYING DISEASE IN PROGNOSIS FOR INVASIVE ASPERGILLOSIS

The degree of immunosuppression and the status of the underlying disease could represent very important factors acting upon a patient's survival. For example, in the voriconazole study (Herbrecht et al., 2002), success was observed in 63% of neutropenic patients receiving chemotherapy but in only 32% of recipients of allogeneic HSCT. In the study by Cornely et al. in which two different dosages of liposomal amphotericin B were used, the success rate in the 3-mg/kg/day arm was 53% in patients in complete remission of the underlying disease, 67% in patients without neutropenia at baseline, and 47% after allogeneic HSCT (Cornely et al., 2007a). In the caspofungin first-line therapy trial the 12-week survival rates were 93% and 40% in patients with or without a "controlled" cancer, respectively (Viscoli et al., 2007). When surgical resection was associated with antifungal drug administration, the invasive mycosis relapsed only in cases in which the underlying disease was relapsing as well (Cesaro et al., 2007a). In the caspofungin salvage therapy study, the response was influenced by persistence of neutropenia (26% efficacy if the granulocyte count was $<500/mm^3$) and by the underlying immunocompromised status (only 14% efficacy in allogeneic HSCT patients) (Maertens et al., 2004). Persistence of neutropenia (Maertens et al., 2004; Upton et al., 2007) or lymphocytopenia (Perruccio et al., 2005) was associated with poor survival after allogeneic HSCT.

In a cohort of 391 allogeneic HSCT recipients with proven or probable invasive aspergillosis, administration of voriconazole was associated with better survival in a univariate analysis, but this effect was lost in a multivariate model where only transplant-related factors (i.e., a good degree of matching and a nonmyeloablative conditioning regimen) and other transplant-related complications (absence of renal or hepatic failure or other concomitant infections) were associated with better survival (Upton et al., 2007). Similarly, the prolongation of the patient observation period beyond the "classical" 3-month period showed that the 1-year survival was influenced by the status of the underlying disease more than by the administration of combined antifungal therapy (Marr et al., 2005). Lack of remission of the underlying hematological malignancy, allogeneic HSCT from unrelated donors or mismatched related donors, or the use of cord blood, systemic steroids, or high-dose cytosine-arabinoside, together with incomplete resolution of imaging findings, duration of antifungal treatment for <1 month, and persistent neutropenia (>4 weeks) are all factors associated with a poor prognosis for invasive aspergillosis, independent of the type of antifungal drugs administered (Sipsas and Kontoyiannis, 2006).

PREVENTION

Two recent randomized clinical trials showed that oral posaconazole is effective in reducing the incidence of invasive mycoses in adults with hematologic malignancy (Cornely et al., 2007b) or in HSCT recipients with severe GVHD (Ullmann et al., 2007), although caution should probably be used in patients with mucositis or intestinal GVHD in light of the somewhat erratic intestinal absorption of posaconazole (De Pauw and Donnelly, 2007; Ezzet et al., 2005; Gubbins et al., 2006). Paradoxically, thanks to these good results the management of aspergillosis among high-risk patients has only become more puzzling. Indeed, after posaconazole prophylaxis, is it logical to administer voriconazole (considered as the first choice) for therapy of invasive aspergillosis, perhaps while the patient is receiving empirical antifungal therapy with caspofungin or liposomal amphotericin B for persistent febrile neutropenia (Chapman, 2007a, 2007b)? Of course, no reasonable answer can be easily given, and the only recommendation is that the choice to administer posaconazole for prophylaxis of invasive aspergillosis should consider local epidemiological factors (i.e., the frequency of invasive aspergillosis in a given patient population) more than the results of a "well-conducted" clinical trial (in other words, the clinical trial says what you could, not what you must do). In this sense it should be noted that in both studies the number of patients that was needed to treat for preventing 1 event was 16. However, this number increases significantly in centers with lower incidence rates of invasive aspergillosis (De Pauw and Donnelly, 2007). In this sense it is interesting that both studies (Chapman, 2007a, 2007b) were powered for a 20% incidence of invasive mycoses in the arm not receiving posaconazole, but the actual proportion was near 10%. This "error" in the estimation of cases of invasive mycoses during treatment for acute leukemia or following allogeneic HSCT underlines the need for more accurate knowledge of the general and local epidemiology of invasive mycosis.

Prevention of colonization, when possible, represents probably the best policy in terms of efficacy, tolerability, and absence of drug interactions. The use of HEPA filters for all patient wards (hematology, oncology, and the HSCT unit) and use of HEPA filter masks for patients when outside the filtered areas have been demonstrated to be effective in reducing the incidence of invasive aspergillosis (Benet et al., 2007; Humphreys, 2004; Nihtinen et al., 2007; Raad et al., 2002), and this effectiveness could be more evident in children, who present an age-related rate of colonization (Dvorak et al., 2005; Rosen et al., 2005; Silvestri et al., 1996).

REFERENCES

Abbasi, S., J. L. Shenep, W. T. Hughes, and P. M. Flynn. 1999. Aspergillosis in children with cancer: a 34-year experience. *Clin. Infect. Dis.* **29:**1210–1219.

Alangaden, G. J., M. Wahiduzzaman, and P. H. Chandrasekar. 2002. Aspergillosis: the most common community-acquired pneumonia with gram-negative bacilli as copathogens in stem cell transplant recipients with graft-versus-host disease. *Clin. Infect. Dis.* **35:**659–664.

Aliff, T. B., P. G. Maslak, J. G. Jurcic, M. L. Heaney, K. N. Cathcart, K. A. Sepkowitz, and M. A. Weiss. 2003. Refractory *Aspergillus* pneumonia in patients with acute leukemia: successful therapy with combination caspofungin and liposomal amphotericin. *Cancer* **97:** 1025–1032.

Anaissie, E., and M. Nucci. 2003. Risk and epidemiology of infections after autologous hemopoietic stem cell transplantation, p. 39–50. *In* R. A. Bowden, P. Ljungman, and C. V. Paya (ed.), *Transplant Infections.* Lippincott Williams & Wilkins, Philadelphia, PA.

Anaissie, E. J., S. R. Penzak, and M. C. Dignani. 2002. The hospital water supply as a source of nosocomial infections: a plea for action. *Arch. Intern. Med.* **162:**1483–1492.

Ascioglu, S., J. H. Rex, B. de Pauw, J. E. Bennett, J. Bille, F. Crokaert, D. W. Denning, J. P. Donnelly, J. E. Edwards, Z. Erjavec, D. Fiere, O. Lortholary, J. Maertens, J. F. Meis, T. F. Patterson, J. Ritter, D. Selleslag, P. M. Shah, D. A. Stevens, and T. J. Walsh. 2002. Defining opportunistic invasive fungal infections in immunocompromised patients with cancer and hematopoietic stem cell transplants: an international consensus. *Clin. Infect. Dis.* **34:**7–14.

Baddley, J. W., T. P. Stroud, D. Salzman, and P. G. Pappas. 2001. Invasive mold infections in allogeneic bone marrow transplant recipients. *Clin. Infect. Dis.* **32:**1319–1324.

Balloy, V., M. Huerre, J. P. Latgé, and M. Chignard. 2005. Differences in patterns of infection and inflammation for corticosteroid treatment and chemotherapy in experimental invasive pulmonary aspergillosis. *Infect. Immun.* **73:**494–503.

Barker, J. N., R. E. Hough, J. A. van Burik, T. E. DeFor, M. L. MacMillan, M. R. O'Brien, and J. E. Wagner. 2005. Serious infections after unrelated donor transplantation in 136 children: impact of stem cell source. *Biol. Blood Marrow Transplant.* **11:**362–370.

Benet, T., M. C. Nicolle, A. Thiebaut, M. A. Piens, F. E. Nicolini, X. Thomas, S. Picot, M. Michallet, and P. Vanhems. 2007. Reduction of invasive aspergillosis incidence among immunocompromised patients after control of environmental exposure. *Clin. Infect. Dis.* **45:** 682–686.

Benjamin, D. K. J., W. C. Miller, S. Bayliff, L. Martel, K. A. Alexander, and P. L. Martin. 2002. Infections diagnosed in the first year after pediatric stem cell transplantation. *Pediatr. Infect. Dis. J.* **21:** 227–234.

Bernard, A., D. Caillot, J. F. Couaillier, O. Casasnovas, H. Guy, and J. P. Favre. 1997. Surgical management of invasive pulmonary aspergillosis in neutropenic patients. *Ann. Thorac. Surg.* **64:**1441–1447.

Bjorklund, A., J. Aschan, M. Labopin, M. Remberger, O. Ringden, J. Winiarski, and P. Ljungman. 2007. Risk factors for fatal infectious complications developing late after allogeneic stem cell transplantation. *Bone Marrow Transplant.* **40:**1055–1062.

Boeckh, M., and K. A. Marr. 2002. Infection in hematopoietic stem cell transplantation, p. 527–571. *In* R. H. Rubin and L. S. Young (ed.), *Clinical Approach to Infection in the Compromised Host.* Kluwer Academic/Plenum, New York, NY.

Brodoefel, H., M. Vogel, H. Hebart, H. Einsele, R. Vonthein, C. Claussen, and M. Horger. 2006. Long-term CT follow-up in 40 non-HIV immunocompromised patients with invasive pulmonary aspergillosis: kinetics of CT morphology and correlation with clinical findings and outcome. *Am. J. Roentgenol.* 187:404–413.

Brummer, E., M. Kamberi, and D. A. Stevens. 2003. Regulation by granulocyte-macrophage colony-stimulating factor and/or steroids given in vivo of proinflammatory cytokine and chemokine production by bronchoalveolar macrophages in response to *Aspergillus* conidia. *J. Infect. Dis.* 187:705–709.

Caillot, D., O. Casasnovas, A. Bernard, J. F. Couaillier, C. Durand, B. Cuisenier, E. Solary, F. Piard, T. Petrella, A. Bonnin, G. Couillault, M. Dumas, and H. Guy. 1997. Improved management of invasive pulmonary aspergillosis in neutropenic patients using early thoracic computed tomographic scan and surgery. *J. Clin. Oncol.* 15:139–147.

Caillot, D., J. F. Couaillier, A. Bernard, O. Casasnovas, D. W. Denning, L. Mannone, J. Lopez, G. Couillault, F. Piard, O. Vagner, and H. Guy. 2001. Increasing volume and changing characteristics of invasive pulmonary aspergillosis on sequential thoracic computed tomography scans in patients with neutropenia. *J. Clin. Oncol.* **19:** 253–259.

Castagnola, E., F. Bagnasco, M. Faraci, I. Caviglia, S. Caruso, B. Cappelli, C. Moroni, G. Morreale, A. Timitilli, G. Tripodi, E. Lanino, and R. Haupt. 2008. Incidence of bacteremias and invasive mycoses in children undergoing allogeneic hematopietic stem cell transplantation: a single center experience. *Bone Marrow Transplant.* **41:**339–347.

Castagnola, E., I. Caviglia, A. Pistorio, F. Fioredda, C. Micalizzi, C. Viscoli, and R. Haupt. 2005. Bloodstream infections and invasive mycoses in children undergoing acute leukaemia treatment: a 13-year experience at a single Italian institution. *Eur. J. Cancer* **41:** 1439–1445.

Castagnola, E., S. Cesaro, M. Giacchino, S. Livadiotti, F. Tucci, G. Zanazzo, D. Caselli, I. Caviglia, S. Parodi, R. Rondelli, P. E. Cornelli, R. Mura, N. Santoro, G. Russo, R. De Santis, S. Buffardi, C. Viscoli, R. Haupt, and M. R. Rossi. 2006. Fungal infections in children with cancer: a prospective, multicenter surveillance study. *Pediatr. Infect. Dis. J.* **25:**634–639.

Castagnola, E., V. Fontana, I. Caviglia, S. Caruso, M. Faraci, F. Fioredda, M. L. Garre, C. Moroni, M. Conte, G. Losurdo, F. Scuderi, R. Bandettini, P. Toma, C. Viscoli, and R. Haupt. 2007. A prospective study on the epidemiology of febrile episodes during chemotherapy-induced neutropenia in children with cancer or after hemopoietic stem cell transplantation. *Clin. Infect. Dis.* **45:**1296–1304.

Cesaro, S., G. Cecchetto, F. De Corti, P. Dodero, M. Giacchino, I. Caviglia, F. Fagioli, S. Livadiotti, F. Salin, D. Caselli, and E. Castagnola. 2007a. Results of a multicenter retrospective study of a combined medical and surgical approach to pulmonary aspergillosis in pediatric neutropenic patients. *Pediatr. Blood Cancer* **49:**909–913.

Cesaro, S., M. Giacchino, F. Locatelli, M. Spiller, B. Buldini, C. Castellini, D. Caselli, E. Giraldi, F. Tucci, G. Tridello, M. R. Rossi, and E. Castagnola. 2007b. Safety and efficacy of a caspofungin-based combination therapy for treatment of proven or probable aspergillosis in pediatric hematological patients. *BMC Infect. Dis.* **7:** 28.

Cesaro, S., T. Toffolutti, C. Messina, E. Calore, R. Alaggio, R. Cusinato, M. Pillon, and L. Zanesco. 2004. Safety and efficacy of caspofungin and liposomal amphotericin B, followed by voriconazole in young patients affected by refractory invasive mycosis. *Eur. J. Haematol.* 73:50–55.

Chamilos, G., M. Luna, R. Lewis, G. Bodey, R. Chemaly, J. Tarrand, A. Safdar, I. Raad, and D. Kontoyiannis. 2006. Invasive fungal infections in patients with hematologic malignancies in a tertiary care cancer center: an autopsy study over a 15-year period (1989–2003). *Haematologica* **91:**986–989.

Chandrasekar, P. H., and J. I. Ito. 2005. Amphotericin B lipid complex in the management of invasive aspergillosis in immunocompromised patients. *Clin. Infect. Dis.* **40**(Suppl. 6):S392–S400.

Chapman, S. W. 2007a. Azole prophylaxis to prevent invasive fungal infections in patients with profound neutropenia. *Curr. Infect. Dis. Rep.* **9:**446–447.

Chapman, S. W. 2007b. Azole prophylaxis to prevent invasive fungal infections in patients with severe graft-versus-host disease. *Curr. Infect. Dis. Rep.* **9:**445–446.

Chignard, M., V. Balloy, J. Sallenave, and M. Si-Tahar. 2007. Role of Toll-like receptors in lung innate defense against invasive aspergillosis. Distinct impact in immunocompetent and immunocompromised hosts. *Clin. Immunol.* **124:**238–243.

Cordonnier, C., P. Ribaud, R. Herbrecht, N. Milpied, D. Valteau-Couanet, C. Morgan, and A. Wade. 2006. Prognostic factors for death due to invasive aspergillosis after hematopoietic stem cell transplantation: a 1-year retrospective study of consecutive patients at French transplantation centers. *Clin. Infect. Dis.* **42:**955–963.

Cornely, O. A., J. Maertens, M. Bresnik, R. Ebrahimi, A. J. Ullmann, E. Bouza, C. P. Heussel, O. Lortholary, C. Rieger, A. Boehme, M. Aoun, H. A. Horst, A. Thiebaut, M. Ruhnke, D. Reichert, N. Vianelli, S. W. Krause, E. Olavarria, and R. Herbrecht. 2007a. Liposomal amphotericin B as initial therapy for invasive mold infection: a randomized trial comparing a high-loading dose regimen with standard dosing (AmBiLoad Trial). *Clin. Infect. Dis.* **44:**1289–1297.

Cornely, O. A., J. Maertens, D. J. Winston, J. Perfect, A. J. Ullmann, T. J. Walsh, D. Helfgott, J. Holowiecki, D. Stockelberg, Y. T. Goh, M. Petrini, C. Hardalo, R. Suresh, and D. Angulo-Gonzalez. 2007b. Posaconazole vs. fluconazole or itraconazole prophylaxis in patients with neutropenia. *N. Engl. J. Med.* 356:348–359.

Cornet, M., L. Fleury, C. Maslo, J. F. Bernard, and G. Brucker. 2002. Epidemiology of invasive aspergillosis in France: a six-year multicentric survey in the Greater Paris area. *J. Hosp. Infect.* **51:**288–296.

Denning, D. W. 1998. Invasive aspergillosis. *Clin. Infect. Dis.* **26:**781–803.

Denning, D. W. 2007. Comparison of 2 studies of treatment of invasive aspergillosis. *Clin. Infect. Dis.* **45:**1106–1108.

De Pauw, B., and P. E. Verveij. 2005. Infections in patients with hematologic malignancies, p. 3432–3441. *In* G. L. Mandell, J. E. Bennett, and R. Doolin (ed.), *Principles and Practice of Infectious Diseases.* Churchill Livingstone, Philadelphia, PA.

De Pauw, B. E., and J. P. Donnelly. 2007. Prophylaxis and aspergillosis: has the principle been proven? *N. Engl. J. Med.* **356:**409–411.

Donnelly, J. P., and B. E. De Pauw. 2005. Infections in the immunocompromised host: general principles, p. 3421–3432. *In* G. L. Mandell, J. E. Bennett, and R. Doolin (ed.), *Principles and Practice of Infectious Diseases.* Churchill Livingstone, Philadelphia, PA.

Dvorak, C. C., W. J. Steinbach, J. M. Brown, and R. Agarwal. 2005. Risks and outcomes of invasive fungal infections in pediatric patients undergoing allogeneic hematopoietic cell transplantation. *Bone Marrow Transplant.* **36:**621–629.

Ezzet, F., D. Wexler, R. Courtney, G. Krishna, J. Lim, and M. Laughlin. 2005. Oral bioavailability of posaconazole in fasted healthy subjects: comparison between three regimens and basis for clinical dosage recommendations. *Clin. Pharmacokinet.* **44:**211–220.

Fukuda, T., M. Boeckh, R. Carter, B. Sandmaier, M. Maris, D. Maloney, P. Martin, R. Storb, and K. Marr. 2003. Risks and outcomes of invasive fungal infections in recipients of allogeneic hematopoietic stem cell transplants after nonmyeloablative conditioning. *Blood* **102:**827–833.

Fukuda, T., M. Boeckh, K. A. Guthrie, D. K. Mattson, S. Owens, A. Wald, B. M. Sandmaier, L. Corey, R. F. Storb, and K. A. Marr. 2004. Invasive aspergillosis before allogeneic hematopoietic stem cell transplantation: 10-year experience at a single transplant center. *Biol. Blood Marrow Transplant.* 10:494–503.

Grazziutti, M. L., J. H. Rex, R. E. Cowart, E. J. Anaissie, A. Ford, and C. A. Savary. 1997. *Aspergillus fumigatus* conidia induce a Th1-type cytokine response. *J. Infect. Dis.* 176:1579–1583.

Greene, R. E., H. T. Schlamm, J. W. Oestmann, P. Stark, C. Durand, O. Lortholary, J. R. Wingard, R. Herbrecht, P. Ribaud, T. F. Patterson, P. F. Troke, D. W. Denning, J. E. Bennett, B. E. de Pauw, and R. H. Rubin. 2007. Imaging findings in acute invasive pulmonary aspergillosis: clinical significance of the halo sign. *Clin. Infect. Dis.* 44:373–379.

Groll, A. H., M. Kurz, W. Schneider, V. Witt, H. Schmidt, M. Schneider, and D. Schwabe. 1999. Five-year-survey of invasive aspergillosis in a paediatric cancer centre. Epidemiology, management and long-term survival. *Mycoses* 42:431–442.

Grow, W. B., J. S. Moreb, D. Roque, K. Manion, H. Leather, V. Reddy, S. A. Khan, K. J. Finiewicz, H. Nguyen, C. J. Clancy, P. S. Mehta, and J. R. Wingard. 2002. Late onset of invasive *Aspergillus* infection in bone marrow transplant patients at a university hospital. *Bone Marrow Transplant.* 29:15–19.

Gubbins, P. O., G. Krishna, A. Sansone-Parsons, S. R. Penzak, L. Dong, M. Martinho, and E. J. Anaissie. 2006. Pharmacokinetics and safety of oral posaconazole in neutropenic stem cell transplant recipients. *Antimicrob. Agents Chemother.* 50:1993–1999.

Habicht, J. M., P. Matt, J. R. Passweg, F. Reichenberger, A. Gratwohl, H. R. Zerkowski, and M. Tamm. 2001. Invasive pulmonary fungal infection in hematologic patients: is resection effective? *Hematol. J.* 2:250–256.

Hajjeh, R. A., and D. W. Warnock. 2001. Counterpoint: invasive aspergillosis and the environment—rethinking our approach to prevention. *Clin. Infect. Dis.* 33:1549–1552.

Haupt, R., M. Romanengo, T. Fears, C. Viscoli, and E. Castagnola. 2001. Incidence of septicaemias and invasive mycoses in children undergoing treatment for solid tumours: a 12-year experience at a single Italian institution. *Eur. J. Cancer* 37:2413–2419.

Hebart, H., C. Bollinger, P. Fisch, J. Sarfati, C. Meisner, M. Baur, J. Loeffler, M. Monod, J. P. Latgé, and H. Einsele. 2002. Analysis of T-cell responses to *Aspergillus fumigatus* antigens in healthy individuals and patients with hematologic malignancies. *Blood* 100: 4521–4528.

Herbrecht, R., D. W. Denning, T. F. Patterson, J. E. Bennett, R. E. Greene, J. W. Oestmann, W. V. Kern, K. A. Marr, P. Ribaud, O. Lortholary, R. Sylvester, R. H. Rubin, J. R. Wingard, P. Stark, C. Durand, D. Caillot, E. Thiel, P. H. Chandrasekar, M. R. Hodges, H. T. Schlamm, P. F. Troke, and B. de Pauw. 2002. Voriconazole versus amphotericin B for primary therapy of invasive aspergillosis. *N. Engl. J. Med.* 347:408–415.

Hori, A., M. Kami, Y. Kishi, U. Machida, T. Matsumura, and T. Kashima. 2002. Clinical significance of extra-pulmonary involvement of invasive aspergillosis: a retrospective autopsy-based study of 107 patients. *J. Hosp. Infect.* 50:175–182.

Hovi, L., U. M. Saarinen-Pihkala, K. Vettenranta, and H. Saxen. 2000. Invasive fungal infections in pediatric bone marrow transplant recipients: single center experience of 10 years. *Bone Marrow Transplant.* 26:999–1004.

Hovi, L., H. Saxen, U. M. Saarinen-Pihkala, K. Vettenranta, T. Meri, and M. Richardson. 2007. Prevention and monitoring of invasive fungal infections in pediatric patients with cancer and hematologic disorders. *Pediatr. Blood Cancer* 48:28–34.

Hoyle, C., and J. M. Goldman. 1994. Life-threatening infections occurring more than 3 months after BMT. 18 UK Bone Marrow Transplant Teams. *Bone Marrow Transplant.* 14:247–252.

Humphreys, H. 2004. Positive-pressure isolation and the prevention of invasive aspergillosis. What is the evidence? *J. Hosp. Infect.* 56: 93–100.

Jantunen, E., A. Piilonen, L. Volin, T. Parkkali, P. Koukila-Kahkola, T. Ruutu, and P. Ruutu. 2000. Diagnostic aspects of invasive *Aspergillus* infections in allogeneic BMT recipients. *Bone Marrow Transplant.* 25:867–871.

Junghanss, C., and K. A. Marr. 2002. Infectious risks and outcomes after stem cell transplantation: are nonmyeloablative transplants changing the picture? *Curr. Opin. Infect. Dis.* 15:347–353.

Kesh, S., N. Y. Mensah, P. Peterlongo, D. Jaffe, K. Hsu, V. D. B. M, R. O'Reilly, E. Pamer, J. Satagopan, and G. A. Papanicolaou. 2005. TLR1 and TLR6 polymorphisms are associated with susceptibility to invasive aspergillosis after allogeneic stem cell transplantation. *Ann. N. Y. Acad. Sci.* 1062:95–103.

Kojima, R., M. Kami, Y. Nannya, E. Kusumi, M. Sakai, Y. Tanaka, Y. Kanda, S. Mori, S. Chiba, S. Miyakoshi, K. Tajima, H. Hirai, S. Taniguchi, H. Sakamaki, and Y. Takaue. 2004. Incidence of invasive aspergillosis after allogeneic hematopoietic stem cell transplantation with a reduced-intensity regimen compared with transplantation with a conventional regimen. *Biol. Blood Marrow Transplant.* 10:645–652.

Kojima, R., U. Tateishi, M. Kami, N. Murashige, Y. Nannya, E. Kusumi, M. Sakai, Y. Tanaka, Y. Kanda, S. Mori, S. Chiba, M. Kusumoto, S. Miyakoshi, H. Hirai, S. Taniguchi, H. Sakamaki, and Y. Takaue. 2005. Chest computed tomography of late invasive aspergillosis after allogeneic hematopoietic stem cell transplantation. *Biol. Blood Marrow Transplant.* 11:506–511.

Kontoyiannis, D. P., R. Hachem, R. E. Lewis, G. A. Rivero, H. A. Torres, J. Thornby, R. Champlin, H. Kantarjian, G. P. Bodey, and I. I. Raad. 2003. Efficacy and toxicity of caspofungin in combination with liposomal amphotericin B as primary or salvage treatment of invasive aspergillosis in patients with hematologic malignancies. *Cancer* 98:292–299.

Lanciotti, M., S. Pigullo, T. Lanza, C. Dufour, I. Caviglia, and E. Castagnola. 2008. Possible role of toll-like receptor 9 polymorphism in chemotherapy-related invasive mold infections in children with hematological malignancies. *Pediatr. Blood Cancer* 50:944.

Levine, D., O. Navarro, G. Chaudry, J. Doyle, and S. Blaser. 2007. Imaging the complications of bone marrow transplantation in children. *Radiographics* 27:307–324.

Lin, S. J., J. Schranz, and S. M. Teutsch. 2001. Aspergillosis case-fatality rate: systematic review of the literature. *Clin. Infect. Dis.* 32:358–366.

Maertens, J., A. Glasmacher, R. Herbrecht, A. Thiebaut, C. Cordonnier, B. H. Segal, J. Killar, A. Taylor, N. Kartsonis, T. F. Patterson, M. Aoun, D. Caillot, and C. Sable. 2006. Multicenter, noncomparative study of caspofungin in combination with other antifungals as salvage therapy in adults with invasive aspergillosis. *Cancer* 107: 2888–2897.

Maertens, J., I. Raad, G. Petrikkos, M. Boogaerts, D. Selleslag, F. Petersen, C. Sable, N. Kartsonis, A. Ngai, A. Taylor, T. Patterson, D. Denning, and T. Walsh. 2004. Efficacy and safety of caspofungin for treatment of invasive aspergillosis in patients refractory to or intolerant of conventional antifungal therapy. *Clin. Infect. Dis.* 39: 1563–1571.

Manuel, R. J., and C. C. Kibbler. 1998. The epidemiology and prevention of invasive aspergillosis. *J. Hosp. Infect.* 39:95–109.

Marchetti, O., and T. Calandra. 2004. Infections in the neutropenic cancer patient, p. 1077–1092. *In* J. Cohen and E. G. Powderly (ed.), *Infectious Diseases.* Mosby, London, United Kingdom.

Marr, K. A., M. Boeckh, R. A. Carter, H. W. Kim, and L. Corey. 2004. Combination antifungal therapy for invasive aspergillosis. *Clin. Infect. Dis.* 39:797–802.

Marr, K. A., M. Boeckh, and H. Kim. 2005. Combination antifungal therapy for invasive aspergillosis. *Clin. Infect. Dis.* 40:1075–1076. (Author's reply.)

Marr, K. A., and R. A. Bowden. 1999. Fungal infections in patients undergoing blood and marrow transplantation. *Transplant. Infect. Dis.* 1:237–246.

Marr, K. A., R. A. Carter, M. Boeckh, P. Martin, and L. Corey. 2002. Invasive aspergillosis in allogeneic stem cell transplant recipients: changes in epidemiology and risk factors. *Blood* 100:4358–4366.

Martin, S. I., F. M. Marty, K. Fiumara, S. P. Treon, J. G. Gribben, and L. R. Baden. 2006. Infectious complications associated with alemtuzumab use for lymphoproliferative disorders. *Clin. Infect. Dis.* 43:16–24.

Martino, P., C. Girmenia, M. Venditti, A. Micozzi, G. Gentile, R. Raccah, E. Martinelli, E. Rendina, and F. Mandelli. 1990. Spontaneous pneumothorax complicating pulmonary mycetoma in patients with acute leukemia. *Rev. Infect. Dis.* 12:611–617.

Martino, R., R. Parody, T. Fukuda, J. Maertens, K. Theunissen, A. Ho, G. Mufti, N. Kroger, A. Zander, D. Heim, M. Paluszewska, D. Selleslag, K. Steinerova, P. Ljungman, S. Cesaro, A. Nihtinen, C. Cordonnier, L. Vazquez, M. López-Duarte, J. Lopez, R. Cabrera, M. Rovira, S. Neuburger, O. Cornely, A. Hunter, K. Marr, H. Dornbusch, and H. Einsele. 2006. Impact of the intensity of the pretransplantation conditioning regimen in patients with prior invasive aspergillosis undergoing allogeneic hematopoietic stem cell transplantation: a retrospective survey of the Infectious Diseases Working Party of the European Group for Blood and Marrow Transplantation. *Blood* 108:2928–2936.

Martino, R., M. Subira, M. Rovira, C. Solano, L. Vazquez, G. F. Sanz, A. Urbano-Ispizua, S. Brunet, and R. De la Camara. 2002. Invasive fungal infections after allogeneic peripheral blood stem cell transplantation: incidence and risk factors in 395 patients. *Br. J. Haematol.* 116:475–482.

Miceli, M. H., J. Maertens, K. Buve, M. Grazziutti, G. Woods, M. Rahman, B. Barlogie, and E. J. Anaissie. 2007. Immune reconstitution inflammatory syndrome in cancer patients with pulmonary aspergillosis recovering from neutropenia: proof of principle, description, and clinical and research implications. *Cancer* 110:112–120.

Mihu, C., E. King, O. Yossepovitch, Y. Taur, A. Jakubowski, E. Pamer, and G. Papanicolaou. 2008. Risk factors and attributable mortality of late aspergillosis after T-cell depleted hematopoietic stem cell transplantation. *Transplant. Infect. Dis.* 10:162–167.

Morgan, J., K. A. Wannemuehler, K. A. Marr, S. Hadley, D. P. Kontoyiannis, T. J. Walsh, S. K. Fridkin, P. G. Pappas, and D. W. Warnock. 2005. Incidence of invasive aspergillosis following hematopoietic stem cell and solid organ transplantation: interim results of a prospective multicenter surveillance program. *Med. Mycol.* 43(Suppl. 1):S49–S58.

Ng, T. T., G. D. Robson, and D. W. Denning. 1994. Hydrocortisone-enhanced growth of *Aspergillus* spp.: implications for pathogenesis. *Microbiology* 140:2475–2479.

Nihtinen, A., V. Anttila, M. Richardson, T. Meri, L. Volin, and T. Ruutu. 2007. The utility of intensified environmental surveillance for pathogenic moulds in a stem cell transplantation ward during construction work to monitor the efficacy of HEPA filtration. *Bone Marrow Transplant.* 40:457–460.

Ochs, L., X. O. Shu, J. Miller, H. Enright, J. Wagner, A. Filipovich, W. Miller, and D. Weisdorf. 1995. Late infections after allogeneic bone marrow transplantations: comparison of incidence in related and unrelated donor transplant recipients. *Blood* 86:3979–3986.

Offner, F., C. Cordonnier, P. Ljungman, H. G. Prentice, D. Engelhard, D. De Bacquer, F. Meunier, and B. De Pauw. 1998. Impact of previous aspergillosis on the outcome of bone marrow transplantation. *Clin. Infect. Dis.* 26:1098–1103.

Ohmagari, N., I. I. Raad, R. Hachem, and D. P. Kontoyiannis. 2004. Invasive aspergillosis in patients with solid tumors. *Cancer* 101: 2300–2302.

Orciuolo, E., M. Stanzani, M. Canestraro, S. Galimberti, G. Carulli, R. Lewis, M. Petrini, and K. Komanduri. 2007. Effects of *Aspergillus fumigatus* gliotoxin and methylprednisolone on human neutrophils: implications for the pathogenesis of invasive aspergillosis. *J. Leukoc. Biol.* 82:839–848.

Pagano, L., M. Caira, A. Candoni, M. Offidani, L. Fianchi, B. Martino, D. Pastore, M. Picardi, A. Bonini, A. Chierichini, R. Fanci, C. Caramatti, R. Invernizzi, D. Mattei, M. E. Mitra, L. Melillo, F. Aversa, M. T. Van Lint, P. Falcucci, C. G. Valentini, C. Girmenia, and A. Nosari. 2006. The epidemiology of fungal infections in patients with hematologic malignancies: the SEIFEM-2004 study. *Haematologica* 91:1068–1075.

Pagano, L., M. Caira, A. Nosari, M. T. Van Lint, A. Candoni, M. Offidani, T. Aloisi, G. Irrera, A. Bonini, M. Picardi, C. Caramatti, R. Invernizzi, D. Mattei, L. Melillo, C. de Waure, G. Reddiconto, L. Fianchi, C. G. Valentini, C. Girmenia, G. Leone, F. Aversa, et al. 2007. Fungal infections in recipients of hematopoietic stem cell transplants: results of the SEIFEM B-2004 study. *Clin. Infect. Dis.* 45:1161–1170.

Pagano, L., P. Ricci, A. Nosari, A. Tonso, M. Buelli, M. Montillo, L. Cudillo, A. Cenacchi, C. Savignana, L. Melillo, et al. 1995. Fatal haemoptysis in pulmonary filamentous mycosis: an underevaluated cause of death in patients with acute leukaemia in haematological complete remission. A retrospective study and review of the literature. *Br. J. Haematol.* 89:500–505.

Patterson, T. F., H. W. Boucher, R. Herbrecht, D. W. Denning, O. Lortholary, P. Ribaud, R. H. Rubin, J. R. Wingard, B. DePauw, H. T. Schlamm, P. Troke, and J. E. Bennett. 2005. Strategy of following voriconazole versus amphotericin B therapy with other licensed antifungal therapy for primary treatment of invasive aspergillosis: impact of other therapies on outcome. *Clin. Infect. Dis.* 41: 1448–1452.

Perruccio, K., A. Tosti, E. Burchielli, F. Topini, L. Ruggeri, A. Carotti, M. Capanni, E. Urbani, A. Mancusi, F. Aversa, M. F. Martelli, L. Romani, and A. Velardi. 2005. Transferring functional immune responses to pathogens after haploidentical hematopoietic transplantation. *Blood* 106:4397–4406.

Pfeiffer, C. D., J. P. Fine, and N. Safdar. 2006. Diagnosis of invasive aspergillosis using a galactomannan assay: a meta-analysis. *Clin. Infect. Dis.* 42:1417–1427.

Philippe, B., O. Ibrahim-Granet, M. C. Prevost, M. A. Gougerot-Pocidalo, M. Sanchez Perez, A. Van der Meeren, and J. P. Latgáe. 2003. Killing of *Aspergillus fumigatus* by alveolar macrophages is mediated by reactive oxidant intermediates. *Infect. Immun.* 71: 3034–3042.

Raad, I., H. Hanna, C. Osting, R. Hachem, J. Umphrey, J. Tarrand, H. Kantarjian, and G. Bodey. 2002. Masking of neutropenic patients on transport from hospital rooms is associated with a decrease in nosocomial aspergillosis during construction. *Infect. Control Hosp. Epidemiol.* 23:41–43.

Roilides, E., T. Sein, M. Roden, R. L. Schaufele, and T. J. Walsh. 2001. Elevated serum concentrations of interleukin-10 in nonneutropenic patients with invasive aspergillosis. *J. Infect. Dis.* 183:518–520.

Romani, L. 2004. Immunity to fungal infections. *Nat. Rev. Immunol.* 4:1–23.

Rosen, G. P., K. Nielsen, S. Glenn, J. Abelson, J. Deville, and T. B. Moore. 2005. Invasive fungal infections in pediatric oncology patients: 11-year experience at a single institution. *J. Pediatr. Hematol. Oncol.* 27:135–140.

Salerno, C. T., D. W. Ouyang, T. S. Pederson, D. M. Larson, J. P. Shake, E. M. Johnson, and M. A. Maddaus. 1998. Surgical therapy for pulmonary aspergillosis in immunocompromised patients. *Ann. Thorac. Surg.* 65:1415–1419.

Saugier-Veber, P., A. Devergie, A. Sulahian, P. Ribaud, F. Traore, H. Bourdeau-Esperou, E. Gluckman, and F. Derouin. 1993. Epidemiology and diagnosis of invasive pulmonary aspergillosis in bone marrow transplant patients: results of a 5 year retrospective study. *Bone Marrow Transplant.* 12:121–124.

Sepkowitz, K. A. 2003. Risk and epidemiology of infections after allogeneic hemopoietic stem cell transplantation, p. 31–38. *In* R. A. Bowden, P. Ljungman, and C. V. Paya (ed.), *Transplant Infections*. Lippincott Williams & Wilkins, Philadelphia, PA.

Silvestri, M., S. Oddera, G. A. Rossi, and P. Crimi. 1996. Sensitization to airborne allergens in children with respiratory symptoms. *Ann. Allergy Asthma Immunol.* **76:**239–244.

Sipsas, N. V., and D. P. Kontoyiannis. 2006. Clinical issues regarding relapsing aspergillosis and the efficacy of secondary antifungal prophylaxis in patients with hematological malignancies. *Clin. Infect. Dis.* **42:**1584–1591.

Steinbach, W. J. 2005. Pediatric aspergillosis: disease and treatment differences in children. *Pediatr. Infect. Dis. J.* **24:**358–364.

Steinbach, W. J., R. M. Addison, L. McLaughlin, Q. Gerrald, P. L. Martin, T. Driscoll, C. Bentsen, J. R. Perfect, and B. D. Alexander. 2007. Prospective *Aspergillus* galactomannan antigen testing in pediatric hematopoietic stem cell transplant recipients. *Pediatr. Infect. Dis. J.* **26:**558–564.

Steinbach, W. J., and K. A. Marr. 2003. Mold infections after hemopoietic stem cell transplantation, p. 466–482. *In* R. A. Bowden, P. Ljungman, and C. V. Paya (ed.), *Transplant Infections*. Lippincott Williams & Wilkins, Philadelphia, PA.

Subira, M., R. Martino, M. Rovira, L. Vazquez, D. Serrano, and R. De La Camara. 2003. Clinical applicability of the new EORTC/MSG classification for invasive pulmonary aspergillosis in patients with hematological malignancies and autopsy-confirmed invasive aspergillosis. *Ann. Hematol.* **82:**80–82.

Temeck, B. K., D. J. Venzon, C. A. Moskaluk, and H. I. Pass. 1994. Thoracotomy for pulmonary mycoses in non-HIV-immunosuppressed patients. *Ann. Thorac. Surg.* **58:**333–338.

Thursky, K., G. Byrnes, A. Grigg, J. Szer, and M. Slavin. 2004. Risk factors for post-engraftment invasive aspergillosis in allogeneic stem cell transplantation. *Bone Marrow Transplant.* **34:**115–121.

Thursky, K. A., L. J. Worth, J. F. Seymour, H. Miles Prince, and M. A. Slavin. 2006. Spectrum of infection, risk and recommendations for prophylaxis and screening among patients with lymphoproliferative disorders treated with alemtuzumab. *Br. J. Haematol.* **132:**3–12.

Todeschini, G., C. Murari, R. Bonesi, G. Pizzolo, G. Verlato, C. Tecchio, V. Meneghini, M. Franchini, C. Giuffrida, G. Perona, and P. Bellavite. 1999. Invasive aspergillosis in neutropenic patients: rapid neutrophil recovery is a risk factor for severe pulmonary complications. *Eur. J. Clin. Investig.* **29:**453–457.

Ullmann, A. J., J. H. Lipton, D. H. Vesole, P. Chandrasekar, A. Langston, S. R. Tarantolo, H. Greinix, W. Morais de Azevedo, V. Reddy, N. Boparai, L. Pedicone, H. Patino, and S. Durrant. 2007. Posaconazole or fluconazole for prophylaxis in severe graft-versus-host disease. *N. Engl. J. Med.* **356:**335–347.

Upton, A., K. A. Kirby, P. Carpenter, M. Boeckh, and K. A. Marr. 2007. Invasive aspergillosis following hematopoietic cell transplantation: outcomes and prognostic factors associated with mortality. *Clin. Infect. Dis.* **44:**531–540.

van Burik, J., and D. Weisdorf. 2005. Infections in recipients of hematopoietic stem cell transplantation, p. 3486–3501. *In* G. L. Mandell, J. E. Bennett, and R. Doolin (ed.), *Principles and Practice of Infectious Diseases*. Churchill Livingstone, Philadelphia, PA.

Viscoli, C., R. Herbrecht, H. Akan, L. Baila, C. Doyen, A. Gallamini, A. Giagounidis, O. Marchetti, R. Martino, L. Meerts, M. Paesmans, M. Shivaprakash, A. J. Ullman, and J. Maertens. 2007. Caspofungin (C) as first-line therapy of invasive aspergillosis (IA) in haematological patients (pts): a study of the EORTC Infectious Diseases Group. *J. Chemother.* **19**(Suppl. 3):36.

Vonberg, R. P., and P. Gastmeier. 2006. Nosocomial aspergillosis in outbreak settings. *J. Hosp. Infect.* **63:**246–254.

Wald, A., W. Leisenring, J. A. van Burik, and R. A. Bowden. 1997. Epidemiology of *Aspergillus* infections in a large cohort of patients undergoing bone marrow transplantation. *J. Infect. Dis.* **175:**1459–1466.

Walsh, T. J., I. Raad, T. F. Patterson, P. Chandrasekar, G. R. Donowitz, R. Graybill, R. E. Greene, R. Hachem, S. Hadley, R. Herbrecht, A. Langston, A. Louie, P. Ribaud, B. H. Segal, D. A. Stevens, J. A. van Burik, C. S. White, G. Corcoran, J. Gogate, G. Krishna, L. Pedicone, C. Hardalo, and J. R. Perfect. 2007. Treatment of invasive aspergillosis with posaconazole in patients who are refractory to or intolerant of conventional therapy: an externally controlled trial. *Clin. Infect. Dis.* **44:**2–12.

Warris, A., P. Gaustad, J. F. Meis, A. Voss, P. E. Verweij, and T. G. Abrahamsen. 2001. Recovery of filamentous fungi from water in a paediatric bone marrow transplantation unit. *J. Hosp. Infect.* **47:**143–148.

Warris, A., C. H. Klaassen, J. F. Meis, M. T. De Ruiter, H. A. De Valk, T. G. Abrahamsen, P. Gaustad, and P. E. Verweij. 2003. Molecular epidemiology of *Aspergillus fumigatus* isolates recovered from water, air, and patients shows two clusters of genetically distinct strains. *J. Clin. Microbiol.* **41:**4101–4106.

Yeghen, T., C. C. Kibbler, H. G. Prentice, L. A. Berger, R. K. Wallesby, P. H. McWhinney, F. C. Lampe, and S. Gillespie. 2000. Management of invasive pulmonary aspergillosis in hematology patients: a review of 87 consecutive cases at a single institution. *Clin. Infect. Dis.* **31:**859–868.

Zaoutis, T. E., K. Heydon, J. H. Chu, T. J. Walsh, and W. J. Steinbach. 2006. Epidemiology, outcomes, and costs of invasive aspergillosis in immunocompromised children in the United States, 2000. *Pediatrics* **117:**e711–e716.

Aspergillus fumigatus and Aspergillosis
Edited by J.-P. Latgé and W. J. Steinbach
©2009 ASM Press, Washington, DC

Chapter 40

Aspergillosis in Pediatric Patients

EMMANUEL ROILIDES AND PARASKEVI PANAGOPOULOU

The continuously rising number of immunocompromised pediatric patients in recent decades has led to a significant increase of invasive fungal infections. Invasive aspergillosis (IA) is one of the most challenging diseases and the cause of significant morbidity and mortality, with an associated cost of more than 25 million U.S. dollars during 2000 in the United States (Zaoutis et al., 2006). IA affects patients with primary or secondary immunodeficiencies, most commonly those with hematological malignancies (Abbasi et al., 1999; Steinbach et al., 2005), chronic granulomatous disease (CGD) (Winkelstein et al., 2000), and those receiving corticosteroids or other immunosuppressants.

If left untreated, IA has a very high mortality that can reach 100%, but even after treatment mortality remains disappointingly high. The most important step for the successful management of IA is early and accurate diagnosis that provides the time for initiation of appropriate therapy. When diagnosis is late the mortality rate reaches 80%, in contrast to early diagnosis, which is associated with a mortality of <30% (Herbrecht et al., 2002a). Early diagnosis of IA, however, remains difficult. Diagnosis should be based on epidemiological data, appropriate host factors, and clinical characteristics, as well as culture and nonculture mycological data.

Also of major importance is the choice of the antifungal agent(s) used for treatment of aspergillosis. Although agents such as amphotericin B lipid formulations, newer azoles, and echinocandins are increasingly used for the treatment of IA, the mortality is still high (Lin et al., 2001). The epidemiology and various presentations as well as specific considerations for early diagnosis and therapy of aspergillosis in children are reviewed here.

EPIDEMIOLOGY

The increase in the number of immunocompromised patients during the last decades has led to a parallel increase in the incidence of opportunistic infections, IA being one of the most feared. Although IA concerns mainly immunocompromised children, cases of IA in immuncompetent hosts have also been reported (Leroy et al., 2006). A study from one of the largest hematopoietic stem cell transplantation (HSCT) centers in the United States showed that the incidence of IA increased from 7.9% in 1992 to 16.9% in 1998 (Marr et al., 2002). IA additionally has an extremely high mortality rate that has increased by 357% since 1980 according to data from the Centers for Disease Control and Prevention (McNeil et al., 2001). However, these two studies did not focus on pediatric patients.

Although the majority of IA studies have also included young children and adolescents, they contained no specific analysis for these subgroups. Therefore, the exact incidence of IA in pediatric patients remains unclear. There are, however, a few studies that have focused on pediatric patients. Hovi et al. (2000) followed 148 pediatric HSCT patients (both allogeneic and autologous) for a 10-year period (1986 to 1996). During this period eight cases with proven IA were detected, with a calculated incidence of about 5%. Additionally, 48 cases with suspected invasive fungal infection were also detected. However, the results were not stratified between *Candida* and *Aspergillus*; thus, there was no specific analysis for IA in this study (Hovi et al., 2000).

In a report of 485 pediatric patients that underwent 510 HSCTs between 1990 and 1998, 26 cases of IA were documented during the first year after transplantation. The rate of IA in this patient cohort was 4.8% (Benjamin et al., 2002). The same study also revealed that severe graft-versus-host disease was a risk factor for IA (relative risk, 7.5; 95% confidence interval [CI], 3.0 to 18.4) and that the most critical time period for the development of IA was the first 30 days posttransplant (10 cases in the first 30 days, 13 cases from day 31 through day 100, and 3 cases from day 101 through day 365 posttransplant) (Benjamin et al., 2002).

Emmanuel Roilides • 3rd Dept. of Pediatrics, Aristotle University Medical School, Hippokration Hospital, Thessaloniki 54642, Greece. **Paraskevi Panagopoulou** • Dept. of Pediatric Oncology, Hippokration Hospital, Thessaloniki 54642, Greece.

A recent study from Taiwan reviewed all episodes of invasive fungal infections occurring in children with cancer in one hospital from 1987 to 2005. The study showed that 29 episodes of invasive fungal infections occurred in 26 cancer patients. The leading pathogens were *Candida* spp. (14 of the 29 episodes), followed by *Aspergillus* spp. (11 of the 29 episodes). The 12-month survival rate was 75% in patients with invasive candidiasis and 100% in those with IA (Yeh et al., 2007). A study from Japan with 412 patients including children showed that there were 14 cases of IA among children 0- to 9-year-old children and 24 cases among children 10- to 19-year children. However, it is impossible to calculate an incidence of pediatric IA, since the total number of children examined was not recorded (Kume et al., 2003).

A review of 1,941 patients from clinical trials, cohort studies, case-control studies, or case series with definite or probable IA between 1995 and 1999 also included children and has given some stratification of case fatality rates per decade of life. The case fatality rate in the youngest age cohort (<20 years old) was the highest, at 68.2%. The next highest case fatality rate was noted in the age group between 21 and 30 years old (59.3%). The investigators concluded that there was little variation in mortality by age, but the pediatric case fatality rate was considerably higher than in the other age groups (Lin et al., 2001).

Older reports of IA among pediatric patients indicated incidence and mortality rates that were influenced by the fact that the diagnostic and therapeutic tools of the time were significantly limited. A report from the Hospital for Sick Children in Toronto reviewed 39 cases of pediatric IA between 1979 and 1988. Twenty-four of these 39 patients had proven IA, and 15 had probable IA. The median age of the cases was 10 years (range, 22 days to 18 years), and 74% had a hematological malignancy or were HSCT recipients (Walmsley et al., 1993). Eighty-six percent of the patients had neutropenia (absolute neutrophil count [ANC], <500 cells/μl) with mean duration of neutropenia (ANC, <1,000 cells/μl) of 20 days. Most patients had an underlying condition that caused immunosuppression. In 41% of the patients, the infection was cutaneous aspergillosis, which was first suspected by a skin lesion that typically presented at sites of trauma related to arm boards or sites of intravenous access (69%). Resolution of skin lesions was noted in 56% of the patients and coincided with recovery from neutropenia. In another 41% of patients, IA was first suspected based on fever and an abnormal chest radiograph (15 cases) or pleuritic pain (1 case) despite broad-spectrum antibiotics. The overall survival rate was only 23.1% (Walmsley et al., 1993), and this was not significantly different from the 31.8% in patients <20 years old from the large case review by Lin et al (2001) but lower than that of many adult studies.

Another study from St. Jude's Children's Hospital revealed 66 cases of proven pediatric IA among 9,500 children with cancer who were treated between 1962 and 1996 (Abbasi et al., 1999). The median age was 11.2 years (range, 1.3 to 21.6 years), and the median interval between onset of underlying disease and IA was 16 months (range, 0 to 180 months). Sixty-six percent of patients had been hospitalized for a median of 36 days (1 to 52 days) before the onset of clinical disease, and clinical symptoms were present for a median of 11 days (0 to 69 days) before the diagnosis of IA. The incidence of IA in specific pediatric subpopulations was found to be 8% in patients with myelodysplastic syndrome, 7% in patients with CGD, 6% in children with choriocarcinoma, 4.6% in patients with aplastic anemia, 4% in patients with acute myelogenous leukemia (AML), 4% in children with chronic myelogenous leukemia, and 1% in patients with acute lymphocytic leukemia (ALL). The patient survival rate was 58% after 1 month, 25% after 2 months, and only 15% after 10 months. These survival rates are lower than those reported in most adult studies. Additionally, it was shown that pulmonary IA had a worse prognosis than nonpulmonary forms of the disease, with an overall median time of 29 days (3 to 312 days) between diagnosis and death (Abbasi et al., 1999).

The species responsible for IA in children have been shown to be different from those in adults. A study from the National Institute of Allergy and Infectious Diseases Mycoses Study Group reviewed 256 isolates of *Aspergillus* spp. from patients with IA from 24 medical centers and showed that *Aspergillus fumigatus* was identified in 67% of the isolates, while *Aspergillus flavus* was the second most common isolate (16%) (Perfect et al., 2001). These rates are similar to the species distribution documented in the large voriconazole randomized clinical trial, in which 77% of the isolates were *A. fumigatus* and 6% were *A. flavus* (Herbrecht et al., 2002a). In the pediatric voriconazole compassionate release study, the species distribution was predominantly *A. fumigatus* (62%), followed by *A. flavus* (14%) and *Aspergillus nidulans* (7%) (Walsh et al., 2002).

Older studies, however, showed different findings. Walmsley et al. showed that *A. flavus* was the predominant pathogen (65%) and *A. fumigatus* followed (15%) (Walmsley et al., 1993). The St. Jude's study showed that 72% of the isolates were *A. flavus*, followed by 38% A. *fumigatus* (Abbasi et al., 1999). A French pediatric study of amphotericin B lipid complex (ABLC) showed that the most common isolates were *A. fumigatus* (48%), *A. flavus* (26%), and *Aspergillus niger* (4%) (Herbrecht et al., 2001). Thus, the most common *Aspergillus* spp. in children are reported to be either *A. fumigatus* or *A.*

flavus. The difference from earlier studies in which *A. flavus* predominance was noted may be attributable to the different site of infection, given that in those studies a large percentage of cutaneous disease was documented whereas in the later studies mostly patients with pulmonary aspergillosis were included (Steinbach, 2005).

IA most commonly affects patients with hematological malignancies, those undergoing hematopoietic stem cell or solid organ transplantation (Abbasi et al., 1999; Steinbach, 2005), patients treated with immunosuppressive medications such as corticosteroids, and those with primary immunodeficiencies such as CGD (Winkelstein et al., 2000). Other situations that predispose to invasive fungal infections, including aspergillosis, less commonly have been found to be prematurity of neonates (Groll et al., 1998a), infection with human immunodeficiency virus (HIV) (Shetty et al., 1997), intravascular catheters, and serious burns (Groll and Walsh, 2001).

ASPERGILLOSIS IN SPECIFIC UNDERLYING PEDIATRIC SETTINGS

Hematological Malignancies

Acute leukemia is the most common underlying hematological malignancy in children with aspergillosis (Hasan and Abuhammour, 2006). Children with hematological malignancies are at increased risk for IA because they receive chemotherapeutic and immunosuppressive agents as well as broad-spectrum antibiotics. In patients with hematological malignancies, *Aspergillus* spp. cause a life-threatening disease with a mortality of 50 to 100%. Skin and bone infections are caused mainly by *A. flavus*, while pulmonary or sinus disease is mainly caused by *A. fumigatus* and *A. flavus* (Hasan and Abuhammour, 2006).

The major risk factor for IA in these patients is profound and prolonged neutropenia (ANC, $<500/mm^3$ for >14 days) (Gerson et al., 1984) as well as increased environmental fungal load, especially when construction work has taken place near the hospital. T-cell function also appears to play a role in host defense against this infection, since cases of IA in children with normal neutrophil counts have been reported (Walmsley et al., 1993). The mean time between the diagnosis of underlying disease and aspergillosis is 16 months.

The site most commonly involved is the lung, followed by the nasopharynx, sinuses, skin, and subcutaneous tissues. Overall pulmonary involvement in patients with hematological malignancies reaches 70% (Hasan and Abuhammour, 2006). However, invasive pulmonary aspergillosis can disseminate to other organs, such as the bones and central nervous system (CNS).

Pulmonary aspergillosis usually manifests with recurrent or prolonged fever in neutropenic patients. Nonproductive cough or pleuritic pain might be present, and chest radiograph may reveal pulmonary infiltrates. Diagnosis is largely based on radiological findings (plain films or computed tomography [CT] scans), since many diagnostic tools have limited sensitivity and specificity in pediatric patients.

IA of the sinuses occurs after invasion of the mucosa. Infection then spreads to adjacent structures, such as the palate, orbit, and the CNS (de Carpentier et al., 1994). Symptoms are nonspecific and include headache, fever, cough, and sore throat, and often CNS disease coexists with pulmonary disease (Gulen et al., 2006).

Another rare IA location is the gastrointestinal tract, e.g., the esophagus and small bowel. In a case of esophageal IA in a child with AML, the diagnosis was obtained through microscopic examination of an endoscopic biopsy specimen (Alioglu et al., 2007). Furthermore, IA of the small bowel in a child with autologous stem cell transplantation was diagnosed on the basis of clinical characteristics (abdominal pain) and laboratory evidence (galactomannan [GM] antigenemia and isolation of *A. fumigatus* from the stool) (Lehrnbecher et al., 2006).

Aspergilli have a tendency to invade blood vessels in immunologically compromised individuals. This invasion is then followed by infarction and necrosis of the local tissues or even systemic dissemination to other organs (Marr and Bowden, 1999). Especially in IPA it has been shown that tissue injury is greatly dependent on the underlying disease. Thus, patients with neutropenia and HSCT recipients present predominantly with angioinvasion and intraalveolar hemorrhage. In contrast, immunocompromised patients without neutropenia, mainly patients with CGD, present with neutrophilic and monocytic infiltrates as well as inflammatory necrosis (Stergiopoulou et al., 2007) but without true angioinvasion (Moskaluk et al., 1994). Differences in diagnosis of aspergillosis between hematological malignancies and primary immunodeficiencies are discussed in the Diagnosis section.

Primary Immunodeficiencies

CGD

CGD is a functional impairment of neutrophils that results in defective intracellular killing. This impairment is due to the absence or very low levels of superoxide-generating NADPH oxidase activity in phagocytes, leading to increased susceptibility to catalase-positive bacteria and fungi (e.g., *Staphylococcus aureus* and *A. fumigatus*) (Assari, 2006). What makes patients with CGD different from other immunodeficient patients is that the underlying disease cannot be resolved; therefore, IA tends to follow a more protracted course (Almyroudis et al., 2005). While the most common *Aspergillus* species recovered in CGD patients is *A. fumigatus*, there is an increased incidence of *A. nidulans* reported in CGD patients. Additionally, most *A. nidulans* infections have been reported in CGD patients (Segal et al., 1998). Moreover, *A. nidulans* infections tend to have a graver course and are resistant to antifungal therapy. The clinical presentation of IA in patients with CGD differs from other patient groups (e.g., neutropenic patients) and also has unique pathologic findings, such as a lack of true angioinvasion by *A. nidulans* (Moskaluk et al., 1994).

The reported incidence of *Aspergillus* infections among CGD patients has varied from 7% (Johnston and Baehner, 1971) to 40% (Mouy et al., 1989), and IA is the most common cause of death in these patients (Mouy et al., 1989; Liese et al., 2000; Winkelstein et al., 2000). A review from the United States reported an incidence of IA of 6.5% in CGD patients (Zaoutis et al., 2006). A 2000 report showed that among 368 patients with CGD in the U.S. registry, *Aspergillus* spp. were the most common organisms causing pneumonia (41% of 290 cases) and the second most common cause of osteomyelitis (22% of 90 cases) (Winkelstein et al., 2000).

The most common *Aspergillus* species isolated from CGD patients is *A. fumigatus*, followed by *A. nidulans*, but infection with *A. flavus* also has been reported (Antachopoulos et al., 2007). Segal et al. reviewed 23 cases of IA among 145 patients with CGD (an incidence of 15.9%) and found that *A. fumigatus* was the cause in 17 cases and *A. nidulans* was the cause in 6 cases (Segal et al., 1998). *A. nidulans* tends to be highly resistant to most antifungal agents; therefore, surgery may be the only therapeutic option. Invasive aspergillosis usually affects CGD patients during the first 2 decades of life and may even be their first manifestation.

The most common site of infection is the lungs, and up to one-third of the patients may be asymptomatic at diagnosis. About 20% of the patients may present with fever and other signs and symptoms that are nonspecific. Similarly, inflammatory markers are also nonspecific. Primary IPA usually disseminates to the CNS, orbit, internal organs, bones, and skin (Antachopoulos et al., 2007). When the causative organism is *A. fumigatus*, extrapulmonary primary sites of infection may occasionally be observed, such as bone, brain, liver, or lymph nodes. The bones constitute a common site of *Aspergillus* infection in CGD patients; the ribs or vertebrae are most commonly involved, and infection usually spreads from the lungs (Wilhelm et al., 2000). *A. nidulans* infections usually extend locally to the adjacent pleura, chest wall, and vertebrae (Antachopoulos et al., 2007). In a recent review of osteomyelitis cases caused by *Aspergillus* spp. in CGD patients, when small bones were involved and pulmonary infection was present, *A. nidulans* seemed to be the main causative organism (Dotis and Roilides, 2004). Other sites of osteomyelitis, such as the femur, have also been reported (Dotis et al., 2003). The site of infection in 12 of 14 cases of *A. nidulans* osteomyelitis was in the ribs or vertebrae and resulted from contiguous spread from the primary pulmonary lesion. In contrast, of 10 cases of *A. fumigatus* osteomyelitis, the infection was contiguously spread from the lungs to the ribs, sternum, or vertebrae in only 4 cases; in the remaining 6 cases there was no pulmonary lesion and the site of infection primarily involved remote bones, such as the skull, humerus, femur, and tibia (Dotis and Roilides, 2004).

A high index of suspicion is required for the timely diagnosis of IA in CGD patients. Even mild or nonspecific symptoms should prompt radiological evaluation of the lungs with high resolution. Differential diagnosis of pulmonary lesions should be made from *S. aureus*, *Nocardia* spp., and *Burkholderia cepacia*, as well as non-*Aspergillus* filamentous fungi, while mixed infections have also been noted (Winkelstein et al., 2003). The radiological findings include segmental and lobar consolidation, perihilar infiltrates, multiple small nodules, peripheral nodular masses, and pleural effusions. Due to differences in the inflammatory response from neutropenic patients and a lack of angioinvasion, "classic" radiological signs, such as halo, air crescent, and other signs of cavitation within areas of consolidation, are not typically seen in CGD patients (Thomas et al., 2003). Imaging studies for extrapulmonary infection sites may also include magnetic resonance imaging or radioisotope bone scans. Brain lesions often present as rim-enhancing areas in the magnetic resonance image, consistent with brain abscesses (Thomas et al., 2003).

Apart from imaging studies, the diagnosis should also be established through isolation and identification of the organism or through visualization of hyphae in stained tissue specimens. Bronchoalveolar lavage (BAL) fluid and lung biopsy specimens may also be used. Detection of *Aspergillus* GM antigen in serum is a promising noninvasive diagnostic tool, although some reports indicate high false-positive results in young infants (Antachopoulos et al., 2007). Importantly, in CGD patients the GM assay may have low sensitivity compared to results in other immunocompromised hosts (Verweij et al., 2000). For these reasons, GM testing may have limitations in pediatric CGD patients. Other diagnostic modalities, such as *Aspergillus* nucleic acid detection by PCR or the β-1,3-D-glucan assay, have either not been standardized or their sensitivity and specificity in pediatric CGD patients have not been studied yet (Roilides, 2006).

Newer triazoles and echinocandins are promising alternatives for the treatment of IA in patients with CGD. An open, randomized trial compared voriconazole and amphotericin B deoxycholate as primary therapy of IA and concluded that voriconazole is superior, so this is currently the antifungal treatment of choice (Herbrecht et al., 2002a). Case reports also show that it is more effective against A. *nidulans* in CGD patients (Antachopoulos et al., 2007). The lipid formulations of amphotericin B have comparable efficacy with amphotericin B deoxycholate, less nephrotoxicity, and fewer infusion-related reactions (Ostrosky-Zeichner et al., 2003). Finally, posaconazole and caspofungin were effective in cases refractory to or intolerant to other antifungal medications.

Additionally, in cases of osteomyelitis or aggressive *A. nidulans* pulmonary infection, other measures such as surgical debridement may also be necessary. Other interventions for refractory infections include granulocyte transfusion from healthy donors, which temporarily restores the patient's impaired phagocytic activity and potentially improves the outcome. Preliminary clinical data are encouraging (Antachopoulos et al., 2007).

Gamma interferon has also been used as adjunctive treatment in CGD patients with IA (Bernhisel-Broadbent et al., 1991; Mamishi et al., 2005; Saulsbury, 2001). Its clinical efficacy, however, has not been systematically evaluated with clinical trials, and the exact mechanism of action has not been fully elucidated.

Prophylactic administration of itraconazole reduced the frequency of severe invasive fungal infections in CGD patients (Gallin et al., 2003). In this double-blind, placebo-controlled trial CGD patients received prophylactic itraconazole. Patients between 5 and 13 years old with a body weight of <50 kg received a single daily dose of 100 mg, while all other patients received a dose of 200 mg. Adverse events that resolved after drug discontinuation included rash, moderate liver enzyme elevation, and headache. In another study eight patients who received prophylactic itraconazole for a mean period of 23 months at an average dose of 5.1 mg/kg of body weight had no adverse effects (Liese et al., 2000). Continuous administration of itraconazole, however, may provoke resistance. There is a report of three CGD patients undergoing long-term prophylaxis with itraconazole that was associated with infection from *A. fumigatus* resistant to itraconazole and other azoles (Verweij et al., 2007). Further research is needed in order to document whether continuous or intermittent prophylaxis is better and whether resistance is emerging.

The clinical course of IA in CGD patients tends to be graver and has a greater tendency for local extension or dissemination than in other patient groups, especially when the causative organism is *A. nidulans* in comparison to *A. fumigatus*. One-half of the CGD patients with *A. nidulans* osteomyelitis died, compared with none of those with *A. fumigatus* osteomyelitis (Dotis and Roilides, 2004). Although a high mortality rate is characteristic of IA in CGD patients, studies with newer agents (such as voriconazole and caspofungin) have given promising results (Sallmann et al., 2003; van 't Hek et al., 1998).

Hyper-IgE syndrome

Hyper-immunoglobulin E (IgE) syndrome, or Job's syndrome, is characterized by extremely elevated serum IgE concentrations and eosinophil counts resulting in recurrent infections of the skin, lungs, or bones, caused mainly by *Staphylococcus* spp. (Antachopoulos et al., 2007). Aspergilli tend to colonize preexisting pneumatoceles, leading to formation of aspergillomas, and patients present with cough and hemoptysis. Sometimes, invasion of the lung parenchyma occurs with dissemination to the CNS and formation of mycotic aneurysms. The GM assay may also be difficult to interpret in patients with hyper-IgE syndrome, yielding false positives, similar to patients with CGD. Management includes a combination of antifungal agents as well as surgical resection (Freeman et al., 2007).

HIV Infection

Another patient group susceptible to aspergillosis are HIV-infected children. According to several reports, the incidence of IA in pediatric HIV patients is 1.5 to 3%. Shetty et al. (1997) reviewed records for 473 HIV-infected children between 1987 and 1995 and found that 7 patients (1.5%) had developed IA. All patients had low CD4 counts reflecting severe immunosuppression (absolute CD4 count within 3 months of IA ranging from 0 to 338 cells/μl) and they all were class C3 according to the CDC classification for AIDS. Sustained neutropenia (>7 days) or corticosteroid therapy as a predisposing factor for IA was encountered in only two patients (28%). Neutropenia was attributable to antiretroviral therapy in one case and to bone marrow suppression by HIV in another. Five patients had IPA, all resulting in death, and two had cutaneous disease. The most common presenting symptoms in patients with IPA were fever, cough, and dyspnea. In another review, only 2 cases of IA (2.7%) were detected among 74 autopsies of pediatric HIV patients (Drut et al., 1997). In a third review, among 30 autopsies, 1 case of IA was found (3%) (Reik et al., 1995). The heterogeneous spectrum of abnormal radiological findings in HIV-associated IPA in adults, including normal findings during the initial stages of infection, appears to be also present in children (Muller et al., 1999; Khoo and Denning, 1994).

Premature Neonates

A special group of immunodeficient patients are neonates, especially those born prematurely. In these patients an accurate and timely diagnosis is more difficult due to their unique characteristics, such as an immature immune system and nonspecific presenting signs.

While the most common fungal pathogens in neonates are *Candida* spp., aspergillosis has been increasingly recognized. Premature infants are especially at risk for primary cutaneous aspergillosis, a life-threatening complication that may result in fulminant sepsis and subsequent multiorgan failure. Risk factors for IA in this age group include immaturity of the immune system (especially phagocytes), skin trauma from adhesive tape, and prolonged armboard use (Steinbach, 2005). In a review of 44 IA cases that occurred in the first 3 months of life, 25% of the patients had cutaneous aspergillosis, 22.7% had IPA, and 31.8% had disseminated disease (Groll et al., 1998a). The study confirmed prematurity as the main risk factor for this group (43.2%). At least 41% of the patients had received corticosteroid therapy before diagnosis. Interestingly, neutropenia is not a major risk factor for IA in neonates, since only 2.3% of the 44 neonatal patients were neutropenic. Other underlying diseases included CGD (14%) and a complex of diarrhea, dehydration, and malnutrition as well as invasive bacterial infections (23%). With regard to cutaneous disease, immaturity of the neonatal skin plays a major role in disease acquisition. Neonatal cutaneous aspergillosis is often the primary disease and is not secondary to hematogenous spread from a primary pulmonary site, as seen in adults. The most common isolate was *A. fumigatus* (41%), with *A. flavus* following (13.6%). Among patients who received medical and/or surgical treatment, the outcome was relatively favorable, with an overall survival rate of 73%.

The most frequently recorded treatment for neonatal aspergillosis has been amphotericin B. However, due to resistance to treatment in some patients, newer agents are increasingly used. Characteristic of this trend is the example of a neonate born after a 24-week gestation who developed aspergillosis while being treated with amphotericin B and fluconazole for candidiasis (Herron et al., 2003). In a recent report of two extremely low-birth-weight neonates with primary cutaneous aspergillosis, both refractory to amphotericin B, the use of systemic voriconazole supported by topical care was successful (Frankenbusch et al., 2006). In a premature neonate, IA presented as pulmonary, hepatic, and CNS aspergillosis during the first days of life. A hyperechogenic lesion adjacent to the lateral ventricle was diagnosed by ultrasound and initially considered to represent periventricular leukomalacia. Within several days the lesion increased in size, and it was then incorrectly considered to be an intraventricular hemorrhage. *A. fumigatus* was ultimately isolated in the tracheal aspirates, ascites, and in material recovered by open brain biopsy (Fuchs et al., 2006).

Immunocompetent Children and Various Diseases

IA has been reported in immuncompetent children under special circumstances. Leroy et al. (2006) reported a case of lethal IPA and CNS aspergillosis after a near-drowning. Kohli et al. (2007) reported the case of a 9-year-old immunocompetent boy recovering from dengue shock syndrome who developed IA associated with complete heart block and heart failure. The patient died, and myocardial biopsy revealed myocarditis and invading fungal elements with branching septate hyphae suggestive of aspergillosis. Renal biopsy also showed glomerular invasion with *Aspergillus* and patchy necrosis. Donoso et al. (2006) reported the case of an immunocompetent child with IA after open laparotomy 3 weeks after which the patient developed multiple organ failure syndrome and died despite treatment with voriconazole and amphotericin B as well as surgical debridement. In addition, IA has complicated the course of various diseases in childhood, such as Pearson syndrome and nephrotic syndrome (Warris et al., 1999; Roilides et al., 2003).

SPECIFIC *ASPERGILLUS*-RELATED DISEASES

Allergic Bronchopulmonary Aspergillosis

Allergic bronchopulmonary aspergillosis (ABPA) is a condition commonly complicating patients with asthma or cystic fibrosis (CF). The exact incidence of ABPA is unknown, ranging from 1 to 25% in patients with asthma or CF in various studies. European studies have shown an incidence of 7.8% among CF patients (2.1 to 13.6%) (Mastella et al., 2000). The most important pathogen is *A. fumigatus*.

Genetic factors, mucus quality and quantity, preactivation of epithelial cells, bronchial penetration of fungi, and host immune responses are some of the factors implicated in the pathogenesis of ABPA (Tillie-Leblond and Tonnel, 2005). Additionally, elevated total and specific IgE antibodies are found in these patients. ABPA leads to extensive bronchiectasis and fibrosis if left undiagnosed and untreated (Chatziagorou et al., in press).

The clinical picture consists of episodes of wheezing, cough, worsening asthma, expectoration of brown mucus plugs, fever, dyspnea, and malaise. However, patients can be asymptomatic. Laboratory findings include transient pulmonary infiltrates, positive *A. fumigatus* skin reaction, elevated total serum IgE level, elevated *Aspergillus*-specific IgE and IgG concentrations, eosinophilia, positive precipitins, and central bronchiectasis.

Diagnosis of ABPA is supported by the presence of precipitating antibodies to *A. fumigatus* in CF or asthma patients. It also has been shown that total IgE elevations predict disease exacerbations. Molecular biology techniques allow the identification of IgE antibodies against certain *Aspergillus* antigens (rAspf1 and rAspf2), which indicate sensitized patients (specificity of 100%, sensitivity of 88%) (de Almeida et al., 2006).

Early and appropriate treatment of ABPA is necessary for the optimal management and prevention of fibrosis. Components of management include environmental measures for elimination of exposure to *Aspergillus* conidia and systemic corticosteroids, such as prednisolone, for a period of at least 3 months. Antifungal agent results are conflicting. Itraconazole can be administered orally and has shown beneficial effects, and serum drug level monitoring is essential. Monitoring of serum total IgE levels is useful as a parameter of disease control. Intermittent checking of IgE is recommended, as recurrences may be detected early based on increasing IgE levels (Greenberger, 2002). Additionally, spirometry should be obtained at least annually, and an unanticipated decline of ≥15% in functional vital capacity might be an indication of ABPA exacerbation (Stevens et al., 2003).

Cerebral Aspergillosis

Dissemination of *Aspergillus* spp. is relatively common, with the CNS being one of the most frequent sites of IA after the lungs (Denning, 1998). Neurological symptoms eventually develop in one-half of patients with disseminated aspergillosis and are the presenting feature in one-third of them (Boes et al., 1994). The extent of gross neuropathological disease includes subtle abscesses, extensive hemorrhage, focal purulent meningitis, and bland infarctions to massive hemorrhagic necrosis causing herniation and death (Dotis et al., 2007). The mortality rate exceeds 80%. Despite perceived advances in early diagnosis and initiation of treatment with newer antifungal agents, CNS aspergillosis remains a devastating opportunistic infection, and the prognosis is discouragingly poor in adults (Schwartz et al., 2005).

Dotis et al. (2007) reviewed 90 pediatric and adolescent cases that were recorded up to 2005, with a mean age of 9 years (range, 18 days to 18 years). The patients most commonly presented with either single or multiple brain abscesses. Among infants, prematurity was the predominant underlying condition, while in children it was leukemia. *A. fumigatus* was isolated from 75.5% of the cases, and the overall mortality in published cases was 65.4%. Surgical debridement was independently associated with survival and had a major role in treatment, especially of brain abscesses. The authors concluded that CNS aspergillosis in infants and children predominantly presents as brain abscess(es) and the outcome is better than in adults. The introduction of novel *Aspergillus*-active azoles such as voriconazole, together with efforts for early and accurate diagnosis of IA in young patients, further improved the outcome (Dotis et al., 2007). More recently, Stiefel et al. (2007) reported an *Aspergillus* abscess in a boy with medulloblastoma that was successfully treated with voriconazole.

Cutaneous Aspergillosis

Cutaneous aspergillosis may be either primary or secondary, and this distinction is related to the underlying pathogenesis. Primary cutaneous aspergillosis refers to cases in which the initial infection begins in the skin (lesions are the result of direct inoculation of the fungus from the environment after trauma). Secondary cutaneous aspergillosis is the result of hematogenous spreading from contiguous structures. This distinction, however, is not always feasible, especially when there is no obvious compromise in skin integrity (Hope et al., 2005).

Primary cutaneous aspergillosis

The primary form of cutaneous aspergillosis is more common in young patients than in adults, since two-thirds (67%) of all reported cases concern neonates (Abbasi et al., 1999). The use of armboards for stabilization of intravenous infusion sites, occlusive dressings, and adhesive tape have been recognized as major risk factors, provided a significant local and systemic immunological impairment is present (Hope et al., 2005). Risk has also been related to intravenous silastic catheters in the context of prolonged neutropenia and HIV/AIDS. It is also noted at sites of significant burns, surgical wounds, or other traumatic injuries in otherwise-immunocompetent hosts (Hope et al., 2005).

The clinical picture starts from an erythematous or violaceous area that evolves to indurated edema and plaque formation, followed by necrosis and ulceration with the formation of a central eschar. Occasionally, hemorrhagic bullae, subcutaneous nodules, granulomas, pustular lesions, or vegetating plaques may develop (Walmsley et al., 1993; Abbasi et al., 1999; Hope et al., 2005). This infection can lead to progressive necrosis of the skin, which then may spread to the underlying tissues and even cause death unless extensive surgical debridement is undertaken (Walsh, 1998).

Diagnosis is usually possible with skin biopsy, histology, and culture, providing adequate and appropriate specimens obtained. The distinction, however, between primary and secondary aspergillosis is difficult and often based on clinical judgment (Hope et al., 2005). In children, a high cure rate that reaches 50% has been documented (Walmsley et al., 1993).

Secondary cutaneous aspergillosis

The secondary form of cutaneous aspergillosis is caused by hematogenous spread of the fungus and is a feature of disseminated aspergillosis that occurs only when immunosuppression is present. The skin lesions are usually found at multiple sites which are not related anatomically, with a clinical presentation similar to primary cutaneous aspergillosis. In order to document the secondary form of the disease, apart from the detection of the fungus at the sites of the infection it is necessary to demonstrate the existence of IA at a remote site as well as an appropriate temporal sequence between primary infection and cutaneous manifestation (Hope et al., 2005).

Acute and Chronic Invasive *Aspergillus* Sinusitis

Aspergillus sinusitis develops following the inhalation and deposition of conidia in the paranasal sinuses. The rapidly destructive sinusitis spreads to contiguous structures, such as the orbit, frontal bone, and carotids. Gulen et al. (2006) reported four cases of sinopulmonary aspergillosis in children with hematological malignancies. Clinical signs included fever, cough, respiratory distress, swallowing difficulty, headache, facial pain or edema, and hard palate necrosis. Radiological methods showed bilateral multiple nodular infiltrations, soft tissue densities filling all the paranasal sinuses, and bronchiectasis. Liposomal amphotericin B was administered to one patient, and the combination of liposomal amphotericin B (LAMB) with itraconozole was administered to three patients. The outcome was favorable in three cases, but one patient died due to respiratory failure (Gulen et al., 2006). Schuster et al. (2005) reported the case of a 12-year-old boy with ALL who developed infection of the lungs and paranasal sinuses with *A. flavus*. The patient was successfully treated with combination therapy of voriconazole and caspofungin during myeloablative bone marrow transplant. Despite 6 weeks of aplasia, a dramatic decrease of lesions highly suggestive of aspergillosis was observed after bone marrow transplantation (Schuster et al., 2005).

DIAGNOSTIC CONSIDERATIONS

Prompt initiation of antifungal therapy after early diagnosis has been shown to be associated with better outcome of IA. This has led to the frequent use of prophylactic or empirical antifungal therapeutic strategies, which can be both expensive and potentially toxic. It is therefore imperative to develop diagnostic assays with a high negative predictive value in order to avoid unnecessary prophylactic and empirical treatment (Roilides, 2006). Nonspecific laboratory findings and inflammatory markers are of little help in the diagnosis of IA, and diagnosis is mainly based on invasive assays or radiological findings. Although aspergilli grow and rapidly disseminate in vivo, they cannot be easily detected and/or cultured and they are absent from easily accessible sites such as blood. The clinical suspicion for IA remains especially important for early diagnosis. Therefore, research efforts have focused on the development of noninvasive surrogate markers.

It is important to stress that for the timely diagnosis of IA the following are required: (i) a very high index of suspicion in immunodeficient pediatric patients and (ii) careful evaluation of clinical and radiological findings suggestive of IA, as well as of mycological data, including serology and molecular biology.

Histology

Microscopic examination of a specimen is a rapid method of diagnosis of fungal infection, especially after addition of potassium hydroxide or staining with highlighting dyes, such as Gomori's methenamine silver stain, the fluorescent dyes calcofluor white or blankophor, and other stains. Obviously, the sensitivity of the method depends on the degree of infection (Roilides, 2006).

Culture

The cornerstone for diagnosis of fungal infections is mycological and histological investigation. Culture of appropriate specimens on mycological media such as Sabouraud dextrose agar or potato dextrose agar remains the gold standard of mycological diagnosis of IA in all patients. Appropriate samples, however, are difficult to obtain from small children, especially preterm neonates, due to their small size and immaturity of tissues, as well as their critical condition. This makes early and accurate diagnosis at this age even more difficult.

Imaging Studies

In adults, >50% of patients with IPA show cavitation and 40% show air crescent formation (Gefter et al., 1985). In pediatric patients, however, a 10-year review of 27 patients (mean age, 5 years) showed central cavitation of small nodules in one-fourth of children and no evidence of air crescent formation within any area of consolidation (Thomas et al., 2003). Other authors have reported varied rates of cavitation ranging between 22% (on chest radiography) (Allan et al., 1988) and 43% (on CT scan) (Taccone et al., 1993). Evidence shows that there is a wide spectrum of radiological disease presentation directly related to age. Perhaps cavitation and air crescent formation are more likely in the older child and adult than in the younger child (Steinbach, 2005). High-resolution CT of the lungs, serially performed, is a very sensitive method for the diagnosis of IPA (Thomas et al., 2003).

Serology

The GM assay is a noninvasive test that is approved for the detection of IA (Steinbach et al., 2007). Its use has been validated in adults with hematological malignancies who have undergone HSCT. Its role needs to be further defined in pediatric patients (Verdaguer et al., 2007). There are few studies for its use in children, and the existing studies have shown repeated differences in pediatric and adult values. Historically, false-positive results have been more common in pediatric patients than in adults (Steinbach et al., 2007).

Several GM studies have included at least some children in their larger cohorts, but results were not stratified for pediatric patients (Steinbach, 2005). In one of these prospective studies, from 1995 to 1998 among 450 adult allogeneic HSCT patients (3,883 samples) and 347 children with hematological malignancies (2,376 samples), the false-positive rate was 2.5% in adults and 10.1% in children. The sensitivity and specificity of the test based on an optical density of >1.5 in at least two sequential samples were 88.6% and 97.5%, respectively; the sensitivity increased to 100% in adult patients and the specificity dropped to 89.9% in children (Sulahian et al., 1996). In another study of patients with fever of unknown origin that included 48 pediatric patients, the false-positive rate was 0.9% in adults and 44.0% in children. Additionally, the specificity of the test was lower in children, 47.6%, compared with 98.2% in adult patients (Herbrecht et al., 2002b).

The complete answer to the question of increased false positivity in children remains unclear, despite numerous theories, such as the presence of *Bifidobacterium bifidum* in the neonatal intestinal microflora mimicking the epitope recognized by the EB-A2 antibody in the enzyme-linked immunosorbent assay kit or a GM-positive infant formula used in pediatric patients (Mennink-Kersten et al., 2004; Warris et al., 2001).

Another issue with GM testing is the increased rate of false-negative results in specific pediatric patients, such as those with CGD. For example, a nonneutropenic 4-year-old child with CGD and IA was diagnosed by lung biopsy despite having persistent false-negative serum GM test results (Verweij et al., 2000). In addition, 10 patients with CGD and 6 patients with Job's syndrome were diagnosed as having IA despite GM antigenemia being detected in only 26% of the cases, versus a detection rate of 80% in all other immunocompromised patients (Walsh et al., 2002). Nevertheless, serial *Aspergillus* GM monitoring in pediatric BMT patients is an effective and accurate means of early detection of IA (Baker, 2006).

A prospective study in 64 pediatric HSCT recipients with twice-weekly sampling for GM detection during the highest risk periods of neutropenia and GVHD showed that the assay had a specificity of 97.5% (95% CI, 96.2 to 98.4%). Administration of piperacillin-tazobactam influenced the result of the assay in certain patients, similar to results in adult patients. When samples from these patients were excluded, specificity increased to 98.4% (95% CI, 97.2 to 99.1%) by sample and to 91.5% (95% CI, 81.6 to 96.3%) by patient. Thus, GM appears to be a promising noninvasive assay for the timely diagnosis of IA in high-risk children, and earlier concerns of excessive false positivity were likely unfounded (Steinbach et al., 2007).

Another recent study of the efficacy of the GM assay for the diagnosis of IA in 121 HSCT patients showed that GM had fewer false-positive and false-negative results in pediatric patients than adults, with a sensitivity of 50% and a specificity of 94%. The study also revealed that a positive GM result preceded other diagnostic tools, e.g., culture (Foy et al., 2007). The case of a previously healthy immunocompetent child who developed IA after near-drowning should be noted. The child developed CNS aspergillosis, but diagnosis was delayed because conventional diagnostic tools were used. However, retrospective examination of serum GM was positive. Thus, serum GM may be helpful in diagnosing IPA especially after near-drowning and may contribute to an early appropriate treatment (Leroy et al., 2006).

β-1,3-D-Glucan is a component of the wall of most fungi except zygomycetes and *Cryptococcus neoformans*. The detection of β-D-glucan in the blood is a new test applied for the diagnosis of invasive fungal infections in adults but is not specific for IA (Roilides, 2006). There are no published reports on the clinical utility of the β-D-glucan assay specifically for infections due to *Aspergillus* spp. in pediatric patients.

PCR

Another useful tool for the diagnosis of IA is the detection of genetic material of *Aspergillus* by PCR. Several studies have been performed to evaluate the efficacy of PCR in comparison to other diagnostic tools such as GM, and they have included pediatric patients as well.

However, the assay has not been standardized in either adult or pediatric patients, and therefore no guidelines exist for its use (Roilides, 2006). While in older children PCR is probably as good as in adult patients, no studies have addressed this issue in neonates and young children. A total of 207 serum samples from 41 immunocompromised patients were examined with both PCR and GM (Challier et al., 2004). Patients with proven or probable IA had positive results with at least one of the two methods. The PCR method seems to be more specific, but a combination of the two methods improves the diagnostic ability for IA. The study included 21 adults and 20 children of ages <1 to 18 years, and PCR results presented no difference between adults and children; however, many children with possible IA had false-positive GM results. The combination of the two methods, measurement of GM twice weekly in combination with repeated performance of PCR, may allow timely diagnosis of IA, giving the opportunity for prompt initiation of specific therapy.

Another study prospectively evaluated the diagnostic potential of three noninvasive tests for IA that were used in a weekly screening strategy: GM, a real-time PCR assay for *Aspergillus* DNA, and a β-D-glucan assay (Kawazu et al., 2004). Among 149 consecutive treatment episodes in 96 patients with hematological disorders at high risk for IA, 9 proven IA cases, 2 probable IA cases, and 13 possible IA cases were diagnosed. GM testing was the most sensitive for predicting the diagnosis of IA in high-risk patients (Kawazu et al., 2004). Further studies are needed in order to test this strategy in children.

BAL

BAL fluid is widely used for evaluation of patients with suspected IPA. The use of BAL has been studied by several investigators and found to yield variable results, with sensitivities ranging from 25% to as much as 75% when analyzed in tissue-proven infection. Although isolation of *A. fumigatus* from BAL fluid is indicative of IPA in febrile neutropenic children with new pulmonary infiltrates, the absence of hyphal elements or negative culture does not exclude the diagnosis (Horvath and Dummer, 1996). Earlier diagnosis may be facilitated by assays that detect *Aspergillus* GM or DNA in BAL fluid. GM assay and quantitative real-time PCR were shown to have greater sensitivity than culture in detection of *A. fumigatus* in BAL fluid in experimental IPA (Francesconi et al., 2006). Use of these methods in conjunction with culture-based diagnostic methods applied to BAL could facilitate accurate diagnosis and more timely initiation of specific therapy.

THERAPY

The optimal therapy for pediatric IA is still unknown. Until recently, amphotericin B was the treatment of choice. However, current data report promising results with newer agents, such as voriconazole and caspofungin. In order to effectively treat aspergillosis in children it is important to develop a clear understanding of the pharmacokinetics and pharmacodynamics of antifungal drugs, which present essential differences from adults. Most data available on the treatment of pediatric IA come from adult studies that have included pediatric patients. Additional data come either from individual case reports or recent reviews (Blyth et al., 2007) that have highlighted the treatment of pediatric IA. The age of the patient, the specific underlying condition, the site of infection, and the species isolated, as well as the local trends of drug susceptibility, are factors that should determine the choice of the appropriate medication for prophylaxis as well as therapy of IA. Data also indicate the importance of a combination of antifungal agents for the management of severe invasive fungal infections as well as the potential benefit of adjunctive modalities, such as surgery (i.e., IPA and osteomyelitis) or immunomodulation (use of cytokines or growth factors).

Furthermore, special mention should be made of the preventive strategies that must be applied in order to decrease the incidence of aspergillosis in high-risk patients. These strategies include both environmental and personal measures for each patient, such as screening (e.g., serum GM) or prophylactic oral administration of antifungal agents (e.g., itraconazole or posaconazole).

Amphotericin B Formulations

For more than 30 years the only antifungal agents available for the management of IA in children were amphotericin B and flucytosine. Amphotericin B deoxycholate, the oldest established agent, has no indication for the treatment of IA in most pediatric patients. The lipid formulations of amphotericin B (LAMB, ABLC, and amphotericin B colloidal dispersion [ABCD]) share a reduced nephrotoxicity, which allows for the safe delivery of higher doses of amphotericin B (Groll et al., 1998b).

One pivotal open-label, emergency use, multicenter study for the use of ABLC (5 mg/kg/day) also included children. The decision to proceed to treatment with ABLC was based on previous failure of conventional antifungals or severe side effects (e.g., nephrotoxicity). The study included 111 treatment episodes in pediatric patients which were evaluated for safety, and efficacy analysis was performed on 54 children. The age range was 21 days to 16 years (mean age, 9.3 years), and the duration of treatment was 6 weeks. There was a significant elevation of total serum bilirubin but no change in the levels of creatinine, potassium, magnesium, liver transaminases, alkaline phosphatase, or hemoglobin. With regard to antifungal efficacy, 70% of the pediatric patients that received ABLC achieved a response (56% of patients with aspergillosis achieved a complete or partial response). More specifically, the percentages of complete or partial response were 50% for IPA, 29% for disseminated IA, 100% for sinusitis, and 67% for single-organ extrapulmonary IA. Thus, it appears that the use of ABLC is useful for the treatment of pediatric IA that is intolerant or refractory to conventional antifungal therapy (Walsh et al., 1999).

A study from France that was published 2 years later included 23 patients with IA who also received ABLC. The mean age was 9.7 years (range, 3 months to 18 years). Seventy-eight percent of the patients were either cured (52%) or improved (26%), and in 22%, therapy failed. Three patients who initially improved later relapsed, dropping the cure or improvement rate to 15 of 23 (65%) (Herbrecht et al., 2001). Evidence-based, but currently not licensed, indications for first-line therapy exist for LAMB and include the treatment of IA (Cornely et al., 2007). The recommended therapeutic doses are 3 to 5 mg/kg/day for LAMB and 5 mg/kg/day for ABLC and ABCD.

Azoles

Based on published pharmacokinetic and safety data, the initial dose for oral itraconazole in pediatric patients beyond the neonatal period is 5 mg/kg/day in two divided doses. Gulen et al. reported three cases of sinopulmonary aspergillosis in children with hematological malignancies who were successfully managed with a combination of LAMB and itraconazole (Gulen et al., 2006). Recently, an open-label, prospective study showed that the prophylactic administration of itraconazole in pediatric cancer patients at high risk for the development of IA was feasible, inexpensive, and effective (Simon et al., 2007).

Voriconazole is a synthetic, broad-spectrum triazole that inhibits the enzyme 14α-sterol demethylase, which is dependent on cytochrome P450 and disrupts the cell membrane and stops fungal growth (Scott and Simpson, 2007). A study investigated the use of voriconazole in 42 children younger than 16 years with refractory proven or probable IA (clinical or radiological progression of disease after more than 7 days of systemic antifungal therapy). Analysis revealed a complete or partial response in 43%, stable response in 7%, and intolerance to therapy in 10% of the patients. More specifically, the complete or partial response rate was 33% for IPA, 50% for CNS aspergillosis, 86% for disseminated IA, 29% for sinusitis, and 30% for single-organ (bone, liver, or skin) IA (Walsh et al., 2002).

Another study by Cesaro et al. (2003) reported the administration of voriconazole in seven pediatric patients with hematological diseases (age range, 2 to 13 years; mean, 5 years). First-line therapy in all patients was LAMB administered at a dose of 3 to 5 mg/kg/day. Subsequently, antifungal treatment was changed to voriconazole administered for a median of 8 weeks (range, 2 to 15 weeks). Two of the seven patients failed to respond to treatment, two showed a complete response, two showed a partial response, and one patient was stable. The voriconazole treatment was well-tolerated. Four patients died, including two with progressive aspergillosis, while three patients were alive 6, 5, and 4 months after the diagnosis of IA. Voriconazole proved to be an effective salvage treatment for IA in pediatric patients (Cesaro et al., 2003).

Special emphasis should be given to the appropriate dosing of voriconazole in children. A study compared two different administration regimens. One arm was an open study of 11 children ages 2 to 11 years from two centers in the United Kingdom. Six patients received single doses of 3 mg/kg and five patients received 4 mg/kg. The second arm was a multicenter study (eight centers) with 28 patients who received multiple doses (loading doses of 6 mg/kg every 12 h on day 1, followed by 3 mg/kg every 12 h on day 2 to day 4 and 4 mg/kg every 12 h on day 4 to day 8). Patients were divided into two age cohorts (2 to 6 and 6 to 12 years). In contrast to healthy adult volunteers, in whom voriconazole metabolism is nonlinear, elimination of voriconazole in children was linear after its administration at 3 and 4 mg/kg every 12 h. The observed variability in voriconazole pharmacokinetics was attributed more to body weight than to age (Walsh et al., 2004). Recent data suggest that, in order to achieve drug exposure comparable to that in adults treated with 4 mg/kg intravenously every 12 h, children should be dosed at 7 mg/kg every 12 h (Walsh et al., 2006).

More recently, Klein and Blackwood (2006) reported the case of a pediatric patient who developed cutaneous aspergillosis after bone marrow transplantation and was successfully treated with topical voriconazole in combination with systemic antifungal therapy. Another case report described the successful treatment of an *Aspergillus* brain abscess with voriconazole initially administered intravenously and then orally for 18 months (Stiefel et al., 2007).

Posaconazole is an orally bioavailable, extended-spectrum antifungal triazole used for the prophylaxis and treatment of refractory invasive fungal infections (Herbrecht, 2004). The dose of posaconazole in pediatric patients is not well-determined and requires further study of its safety and plasma pharmacokinetics. A study by Krishna et al. (2007) evaluated plasma posaconazole concentrations in 12 juvenile patients (age range, 8 to 17 years) with hematological malignancies who participated in a multicenter, phase III, open-label study that assessed the efficacy and safety of the drug. These concentrations were compared with the results from 194 adults who were also treated with posaconazole and who were intolerant or had invasive fungal infection refractory to conventional antifungal therapies. Eleven patients received an 800-mg/day oral posaconazole suspension in divided doses, and one patient received 400 mg/day on the day of sample collection. The study showed similar posaconazole plasma concentrations in adults and children, but full pharmacokinetic profiling in children remains unclear. Overall success rates and adverse event profiles were comparable (Krishna et al., 2007).

Echinocandins

Caspofungin, anidulafungin, and micafungin are a novel class of antifungal lipopeptides. Caspofungin, the first licensed agent of echinocandins, is a broad-spectrum antifungal agent effective against *Candida* and *Aspergillus* species with favorable pharmacokinetics and an excellent safety profile (Groll et al., 2006). Caspofungin inhibits the synthesis of a component of the fungal cell wall, β-1,3-D glucan, so it has a different target of action compared to the polyenes and triazoles. The combination of caspofungin with either amphotericin B or voriconazole may exert a synergistic effect (Cesaro et al., 2007).

Data on the effectiveness, safety, and dosage of caspofungin in children are limited. In adults the dose of caspofungin is 70 mg the first day followed by 50 mg/day thereafter. The initial pediatric pharmacokinetic study involved 39 patients (age range, 2 to 17 years). The dose was calculated with two methods: one based on weight (1 mg/kg/day) and one based on body surface area (70 or 50 mg/m^2/day) (Walsh et al., 2005). Pharmacokinetics appeared slightly different in children than in healthy men. The 50-mg/m^2/day regimen yielded similar plasma drug concentrations and an increased area under the time-concentration curve to adult patients (50 mg/day). Pharmacokinetic projections suggested that a loading dose of 70 mg/m^2/day followed by a dose of 50 mg/m^2/day is more appropriate in children than 1 mg/kg/day. Drug exposure of the pediatric population comparable to that of adults is achieved with doses of 50 mg/m^2/day (maximum, 70 mg/day) (Walsh et al., 2005). According to a recent report, therapeutic concentrations of caspofungin were maintained after a 48-h interruption of treatment in a child with IA (Castagnola et al., 2007).

A multicenter retrospective survey was conducted in order to obtain data on immunocompromised pediatric patients. The survey identified 64 patients (median age, 11.5 years; 25 females) with hematological malignancies (48 patients), marrow failure (9), solid tumors (3), other hematological disorders (2), or congenital immunodeficiency (2) who received caspofungin. Seventeen of them had proven, 14 had probable, and 17 had possible invasive fungal infection. In 16 patients caspofungin was administered empirically. Caspofungin was administered for a median of 37 days (range, 3 to 218 days) either as monotherapy (n = 20) or in combination (n = 44). The median daily maintenance dosage was 1.07 mg/kg (34.3 mg/m^2). Clinical adverse events were observed in 53.1% of the patients and were mild to moderate. There were no significant elevations of serum bilirubin or alkaline phosphatase, and creatinine values and liver enzymes were slightly higher at the end of treatment. Response rates were high, with overall survival at the end of treatment and 3 months later of 75 and 70%, respectively. Thus, it was shown that caspofungin was safe and well-tolerated and may be useful against invasive fungal infections in severely immunocompromised pediatric patients (Groll et al., 2006). In a recent retrospective study, Cesaro et al. (2007) analyzed the safety and efficacy of caspofungin-based combination therapy in 40 children and adolescents with malignant disease and IA. Twenty (50%) of the patients had documented IA and 20 had probable IA. Twenty-one patients (53%) showed a favorable response, and the 100-day survival was 70%, based on a dose of 70 mg/m^2 the first day followed by 50 mg/m^2/day. The combination therapy was well-tolerated. Thus, a caspofungin-based antifungal combination is an effective therapeutic option for children with IA (Cesaro et al., 2007).

Anidulafungin is an echinocandin with activity against *Aspergillus* spp. Adult dose is 100 mg/day for IA. The safety and pharmacokinetics of anidulafungin in children were studied in a multicenter study of neutropenic pediatric patients. Patients were divided into two age cohorts (2 to 11 and 12 to 17 years) and were enrolled into sequential groups to receive 0.75 or 1.5 mg/kg/day. Pharmacokinetic parameters were determined for 12 patients at each dose (0.75 or 1.5 mg/kg/day). Concentrations and drug exposures were similar for patients between age cohorts, and weight-adjusted clearance was consistent across age groups. No serious adverse events were noted. It was concluded that anidulafungin is well-tolerated in pediatric patients and can be dosed based on body weight. Pediatric patients receiving 0.75 or 1.5 mg/kg/day have anidulafungin concentration profiles similar to those of adults receiving 50 or 100 mg/day, respectively (Benjamin et al., 2006).

A multinational, noncomparative study investigated the activity of micafungin alone ($n = 2$) or in combination ($n = 56$) with other agents in 58 pediatric patients with proven or probable IA. The mean daily dose was 2.0 ± 1.2 mg/kg/day, and the mean duration of dosing was 67 ± 85 days. The overall response rate was 45% (Flynn et al., 2005).

Combination antifungal therapy may be advantageous compared to monotherapy, but there is a paucity of combination data in children. The combination of an echinocandin with a triazole or with a formulation of amphotericin B may be additive or synergistic in vitro and in vivo against experimental IA (Petraitis et al., 2003). However, the clinical experience with combination therapies against IA consists only of case reports. The combination of amphotericin B and voriconazole together with surgical debridement was successfully used for the treatment of an immunocompromised child with acute undifferentiated leukemia who developed presumptive *Aspergillus* infection disseminated to lung, liver, spleen, and bone (Shouldice et al., 2003). Sterba et al. (2005) described the case of an adolescent girl with ALL and liver failure who developed an *Aspergillus* brain abscess and was treated with a combination of amphotericin B local instillation and prolonged ABCD followed by surgical debridement and oral voriconazole. Schuster et al. (2005) reported the successful management of a 12-year-old boy in the third remission of ALL who developed a sinopulmonary IA and was successfully treated with a combination of voriconazole and caspofungin during myeloablative BMT. A case report of an 8-year-old allogeneic stem cell transplant recipient who developed a central venous catheter tunnel infection caused by *A. flavus* showed that combined treatment consisting of LAMB and caspofungin cured the deep fungal infection despite previous unsuccessful treatment with conventional and liposomal amphotericin B (Krivan et al., 2006). Leroy et al. (2006) proposed the combination of voriconazole and caspofungin as the treatment of choice for IA after near-drowning. Finally, Alioglu et al. (2007) reported a case of invasive eosophageal aspergillosis in a child with AML who was successfully treated with caspofungin in combination with LAMB. However, these combinations have not been studied in prospective randomized clinical trials for superiority compared to standard monotherapy. The results of clinical trials for antifungal combinations against IA in pediatric patients will help further guide the rational use of an expanding armamentarium of antifungal compounds.

REFERENCES

Abbasi, S., J. L. Shenep, W. T. Hughes, and P. M. Flynn. 1999. Aspergillosis in children with cancer: a 34-year experience. *Clin. Infect. Dis.* **29:**1210–1219.

Alioglu, B., Z. Avci, O. Canan, F. Ozcay, B. Demirhan, and N. Ozbek. 2007. Invasive esophageal aspergillosis associated with acute myelogenous leukemia: successful therapy with combination caspofungin and liposomal amphotericin B. *Pediatr. Hematol. Oncol.* **24:**63–68.

Allan, B. T., D. Patton, N. K. Ramsey, and D. L. Day. 1988. Pulmonary fungal infections after bone marrow transplantation. *Pediatr. Radiol.* **18:**118–122.

Almyroudis, N. G., S. M. Holland, and B. H. Segal. 2005. Invasive aspergillosis in primary immunodeficiencies. *Med. Mycol.* **43**(Suppl. 1):S247–S259.

Antachopoulos, C., T. J. Walsh, and E. Roilides. 2007. Fungal infections in primary immunodeficiencies. *Eur. J. Pediatr.* **166:**1099–1117.

Assari, T. 2006. Chronic granulomatous disease; fundamental stages in our understanding of CGD. *Med. Immunol.* **5:**4.

Baker, C. 2006. Serial *Aspergillus* antigen monitoring in pediatric bone marrow transplant patients. *J. Pediatr. Oncol. Nurs.* **23:**300–304.

Benjamin, D. K., Jr., W. C. Miller, S. Bayliff, L. Martel, K. A. Alexander, and P. L. Martin. 2002. Infections diagnosed in the first year after pediatric stem cell transplantation. *Pediatr. Infect. Dis. J.* **21:** 227–234.

Benjamin, D. K., Jr., T. Driscoll, N. L. Seibel, C. E. Gonzalez, M. M. Roden, R. Kilaru, K. Clark, J. A. Dowell, J. Schranz, and T. J. Walsh. 2006. Safety and pharmacokinetics of intravenous anidulafungin in children with neutropenia at high risk for invasive fungal infections. *Antimicrob. Agents Chemother.* **50:**632–638.

Bernhisel-Broadbent, J., E. E. Camargo, H. S. Jaffe, and H. M. Lederman. 1991. Recombinant human interferon-γ as adjunct therapy for *Aspergillus* infection in a patient with chronic granulomatous disease. *J. Infect. Dis.* **163:**908–911.

Blyth, C. C., P. Palasanthiran, and T. A. O'Brien. 2007. Antifungal therapy in children with invasive fungal infections: a systematic review. *Pediatrics* **119:**772–784.

Boes, B., R. Bashir, C. Boes, F. Hahn, J. R. McConnell, and R. McComb. 1994. Central nervous system aspergillosis. Analysis of 26 patients. *J. Neuroimag.* **4:**123–129.

Castagnola, E., B. Cappelli, M. Faraci, S. Fallani, M. I. Cassetta, and A. Novelli. 2007. Maintenance of therapeutic concentrations of caspofungin after temporary treatment interruption (48 hours) in a child with invasive aspergillosis. *Antimicrob. Agents Chemother.* **51:** 3775.

Cesaro, S., L. Strugo, R. Alaggio, G. Cecchetto, L. Rigobello, M. Pillon, R. Cusinato, and L. Zanesco. 2003. Voriconazole for invasive aspergillosis in oncohematological patients: a single-center pediatric experience. *Support. Care Cancer* **11:**722–727.

Cesaro, S., M. Giacchino, F. Locatelli, M. Spiller, B. Buldini, C. Castellini, D. Caselli, E. Giraldi, F. Tucci, G. Tridello, M. R. Rossi, and E. Castagnola. 2007. Safety and efficacy of a caspofungin-based combination therapy for treatment of proven or probable aspergillosis in pediatric hematological patients. *BMC Infect. Dis.* **7:**28.

Challier, S., S. Boyer, E. Abachin, and P. Berche. 2004. Development of a serum-based Taqman real-time PCR assay for diagnosis of invasive aspergillosis. *J. Clin. Microbiol.* **42:**844–846.

Chatziagorou, E., J. N. Tsanakas, T. J. Walsh, and E. Roilides. *Aspergillus* and the pediatric lung. *Pediatr. Respir. Rev.*, in press.

Cornely, O. A., J. Maertens, M. Bresnik, R. Ebrahimi, A. J. Ullmann, E. Bouza, C. P. Heussel, O. Lortholary, C. Rieger, A. Boehme, M. Aoun, H. A. Horst, A. Thiebaut, M. Ruhnke, D. Reichert, N. Vianelli, S. W. Krause, E. Olavarria, and R. Herbrecht. 2007. Liposomal amphotericin B as initial therapy for invasive mold infection: a randomized trial comparing a high-loading dose regimen with standard dosing (AmBiLoad Trial). *Clin. Infect. Dis.* **44:**1289–1297.

de Almeida, M. B., M. H. Bussamra, and J. C. Rodrigues. 2006. Allergic bronchopulmonary aspergillosis in paediatric cystic fibrosis patients. *Paediatr. Respir. Rev.* **7:**67–72.

de Carpentier, J. P., L. Ramamurthy, D. W. Denning, and P. H. Taylor. 1994. An algorithmic approach to *Aspergillus* sinusitis. *J. Laryngol. Otol.* **108:**314–318.

Denning, D. W. 1998. Invasive aspergillosis. *Clin. Infect. Dis.* **26:**781–803.

Donoso, F. A., A. J. Camacho, L. P. Alarcon, and R. P. Cruces. 2006. Invasive multisystemic aspergillosis in an immunocompetent child: case report. Rev. Chilena Infectol. **23:**69–72. (In Spanish.)

Dotis, J., P. Panagopoulou, J. Filioti, R. Winn, C. Toptsis, C. Panteliadis, and E. Roilides. 2003. Femoral osteomyelitis due to *Aspergillus nidulans* in a patient with chronic granulomatous disease. *Infection* **31:**121–124.

Dotis, J., and E. Roilides. 2004. Osteomyelitis due to *Aspergillus* spp. in patients with chronic granulomatous disease: comparison of *Aspergillus nidulans* and *Aspergillus fumigatus*. *Int. J. Infect. Dis.* **8:** 103–110.

Dotis, J., E. Iosifidis, and E. Roilides. 2007. Central nervous system aspergillosis in children: a systematic review of reported cases. *Int. J. Infect. Dis.* **11:**381–393.

Drut, R., V. Anderson, M. A. Greco, C. Gutierrez, B. de Leon-Bojorge, D. Menezes, A. Peruga, G. Quijano, C. Ridaura, M. Siminovich, P. V. Mayoral, M. Weissenbacher, et al. 1997. Opportunistic infections in pediatric HIV infection: a study of 74 autopsy cases from Latin America. *Pediatr. Pathol. Lab. Med.* **17:**569–576.

Foy, P. C., J. A. van Burik, and D. J. Weisdorf. 2007. Galactomannan antigen enzyme-linked immunosorbent assay for diagnosis of invasive aspergillosis after hematopoietic stem cell transplantation. *Biol. Blood Marrow Transplant.* **13:**440–443.

Francesconi, A., M. Kasai, R. Petraitiene, V. Petraitis, A. M. Kelaher, R. Schaufele, W. W. Hope, Y. R. Shea, J. Bacher, and T. J. Walsh. 2006. Characterization and comparison of galactomannan enzyme immunoassay and quantitative real-time PCR assay for detection of *Aspergillus fumigatus* in bronchoalveolar lavage fluid from experimental invasive pulmonary aspergillosis. *J. Clin. Microbiol.* **44:** 2475–2480.

Frankenbusch, K., F. Eifinger, A. Kribs, J. Rengelshauseu, and B. Roth. 2006. Severe primary cutaneous aspergillosis refractory to amphotericin B and the successful treatment with systemic voriconazole in two premature infants with extremely low birth weight. *J. Perinatol.* **26:**511–514.

Freeman, A. F., D. E. Kleiner, H. Nadiminti, J. Davis, M. Quezado, V. Anderson, J. M. Puck, and S. M. Holland. 2007. Causes of death in hyper-IgE syndrome. *J. Allergy Clin. Immunol.* **119:**1234–1240.

Fuchs, H., H. von Baum, M. Meth, N. Wellinghausen, W. Lindner, and H. Hummler. 2006. CNS manifestation of aspergillosis in an extremely low-birth-weight infant. *Eur. J. Pediatr.* **165:**476–480.

Gallin, J. I., D. W. Alling, H. L. Malech, R. Wesley, D. Koziol, B. Marciano, E. M. Eisenstein, M. L. Turner, E. S. DeCarlo, J. M. Starling, and S. M. Holland. 2003. Itraconazole to prevent fungal infections in chronic granulomatous disease. *N. Engl. J. Med.* **348:** 2416–2422.

Gefter, W. B., S. M. Albelda, G. H. Talbot, S. L. Gerson, P. A. Cassileth, and W. T. Miller. 1985. Invasive pulmonary aspergillosis and acute leukemia. Limitations in the diagnostic utility of the air crescent sign. *Radiology* **157:**605–610.

Gerson, S. L., G. H. Talbot, S. Hurwitz, B. L. Strom, E. J. Lusk, and P. A. Cassileth. 1984. Prolonged granulocytopenia: the major risk factor for invasive pulmonary aspergillosis in patients with acute leukemia. *Ann. Intern. Med.* **100:**345–351.

Greenberger, P. A. 2002. Allergic bronchopulmonary aspergillosis. *J. Allergy Clin. Immunol.* **110:**685–692.

Groll, A. H., G. Jaeger, A. Allendorf, G. Herrmann, R. Schloesser, and V. von Loewenich. 1998a. Invasive pulmonary aspergillosis in a critically ill neonate: case report and review of invasive aspergillosis during the first 3 months of life. *Clin. Infect. Dis.* **27:**437–452.

Groll, A. H., F. M. Muller, S. C. Piscitelli, and T. J. Walsh. 1998b. Lipid formulations of amphotericin B: clinical perspectives for the management of invasive fungal infections in children with cancer. *Klin. Paediatr.* **210:**264–273.

Groll, A. H., and T. J. Walsh. 2001. Uncommon opportunistic fungi: new nosocomial threats. *Clin. Microbiol. Infect.* **7**(Suppl. 2):8–24.

Groll, A. H., A. Attarbaschi, F. R. Schuster, N. Herzog, L. Grigull, M. N. Dworzak, K. Beutel, H. J. Laws, and T. Lehrnbecher. 2006. Treatment with caspofungin in immunocompromised paediatric patients: a multicentre survey. *J. Antimicrob. Chemother.* **57:**527–535.

Gulen, H., A. Erbay, F. Gulen, E. Kazanci, C. Vergin, E. Demir, and R. Tanac. 2006. Sinopulmonary aspergillosis in children with hematological malignancy. *Minerva Pediatr.* **58:**319–324.

Hasan, R. A., and W. Abuhammour. 2006. Invasive aspergillosis in children with hematologic malignancies. *Paediatr. Drugs* **8:**15–24.

Herbrecht, R., A. Auvrignon, E. Andres, R. Guillemain, A. Suc, D. Eyer, C. Pailler, V. Letscher-Bru, G. Leverger, and G. Schaison. 2001. Efficacy of amphotericin B lipid complex in the treatment of invasive fungal infections in immunosuppressed paediatric patients. *Eur. J. Clin. Microbiol. Infect. Dis.* **20:**77–82.

Herbrecht, R., D. W. Denning, T. F. Patterson, J. E. Bennett, R. E. Greene, J. W. Oestmann, W. V. Kern, K. A. Marr, P. Ribaud, O. Lortholary, R. Sylvester, R. H. Rubin, J. R. Wingard, P. Stark, C. Durand, D. Caillot, E. Thiel, P. H. Chandrasekar, M. R. Hodges, H. T. Schlamm, P. F. Troke, and B. de Pauw. 2002a. Voriconazole versus amphotericin B for primary therapy of invasive aspergillosis. *N. Engl. J. Med.* **347:**408–415.

Herbrecht, R., V. Letscher-Bru, C. Oprea, B. Lioure, J. Waller, F. Campos, O. Villard, K. L. Liu, S. Natarajan-Ame, P. Lutz, P. Dufour, J. P. Bergerat, and E. Candolfi. 2002b. *Aspergillus* galactomannan detection in the diagnosis of invasive aspergillosis in cancer patients. *J. Clin. Oncol.* **20:**1898–1906.

Herbrecht, R. 2004. Posaconazole: a potent, extended-spectrum triazole anti-fungal for the treatment of serious fungal infections. *Int. J. Clin. Pract.* **58:**612–624.

Herron, M. D., S. L. Vanderhooft, C. Byington, and J. D. King. 2003. Aspergillosis in a 24-week newborn: a case report. *J. Perinatol.* **23:** 256–259.

Hope, W. W., T. J. Walsh, and D. W. Denning. 2005. The invasive and saprophytic syndromes due to *Aspergillus* spp. *Med. Mycol.* **43**(Suppl. 1):S207–S238.

Horvath, J. A., and S. Dummer. 1996. The use of respiratory-tract cultures in the diagnosis of invasive pulmonary aspergillosis. *Am. J. Med.* **100:**171–178.

Hovi, L., U. M. Saarinen-Pihkala, K. Vettenranta, and H. Saxen. 2000. Invasive fungal infections in pediatric bone marrow transplant recipients: single center experience of 10 years. *Bone Marrow Transplant.* **26:**999–1004.

Johnston, R. B., Jr., and R. L. Baehner. 1971. Chronic granulomatous disease: correlation between pathogenesis and clinical findings. *Pediatrics* **48:**730–739.

Kawazu, M., Y. Kanda, Y. Nannya, K. Aoki, M. Kurokawa, S. Chiba, T. Motokura, H. Hirai, and S. Ogawa. 2004. Prospective comparison of the diagnostic potential of real-time PCR, double-sandwich enzyme-linked immunosorbent assay for galactomannan, and a (1→3)-β-D-glucan test in weekly screening for invasive aspergillosis in patients with hematological disorders. *J. Clin. Microbiol.* **42:** 2733–2741.

Khoo, S. H., and D. W. Denning. 1994. Invasive aspergillosis in patients with AIDS. *Clin. Infect. Dis.* **19**(Suppl. 1):S41–S48.

Kohli, U., J. Sahu, R. Lodha, N. Agarwal, and R. Ray. 2007. Invasive nosocomial aspergillosis associated with heart failure and complete heart block following recovery from dengue shock syndrome. *Pediatr. Crit. Care Med.* **8:**389–391.

Krishna, G., A. Sansone-Parsons, M. Martinho, B. Kantesaria, and L. Pedicone. 2007. Posaconazole plasma concentrations in juvenile patients with invasive fungal infection. *Antimicrob. Agents Chemother.* 51:812–818.

Krivan, G., J. Sinko, I. Z. Nagy, V. Goda, P. Remenyi, A. Batai, S. Lueff, B. Kapas, M. Reti, A. Tremmel, and T. Masszi. 2006. Successful combined antifungal salvage therapy with liposomal amphothericin B and caspofungin for invasive *Aspergillus flavus* infection in a child following allogeneic bone marrow transplantation. *Acta Biomed.* 77(Suppl. 2):17–21.

Kume, H., T. Yamazaki, M. Abe, H. Tanuma, M. Okudaira, and I. Okayasu. 2003. Increase in aspergillosis and severe mycotic infection in patients with leukemia and MDS: comparison of the data from the Annual of the Pathological Autopsy Cases in Japan in 1989, 1993 and 1997. *Pathol. Int.* 53:744–750.

Lehrnbecher, T., M. Becker, D. Schwabe, U. Kohl, S. Kriener, K. P. Hunfeld, H. Schmidt, P. Beyer, T. Klingebiel, P. Bader, and J. Sorensen. 2006. Primary intestinal aspergillosis after high-dose chemotherapy and autologous stem cell rescue. *Pediatr. Infect. Dis. J.* 25:465–466.

Leroy, P., A. Smismans, and T. Seute. 2006. Invasive pulmonary and central nervous system aspergillosis after near-drowning of a child: case report and review of the literature. *Pediatrics* 118:e509–e513.

Liese, J., S. Kloos, V. Jendrossek, T. Petropoulou, U. Wintergerst, G. Notheis, M. Gahr, and B. H. Belohradsky. 2000. Long-term follow-up and outcome of 39 patients with chronic granulomatous disease. *J. Pediatr.* 137:687–693.

Lin, S. J., J. Schranz, and S. M. Teutsch. 2001. Aspergillosis case-fatality rate: systematic review of the literature. *Clin. Infect. Dis.* 32:358–366.

Mamishi, S., K. Zomorodian, F. Saadat, M. Gerami-Shoar, B. Tarazooie, and S. A. Siadati. 2005. A case of invasive aspergillosis in CGD patient successfully treated with amphotericin B and IFN-gamma. *Ann. Clin. Microbiol. Antimicrob.* 4:4.

Marr, K. A., and R. A. Bowden. 1999. Fungal infections in patients undergoing blood and marrow transplantation. *Transplant. Infect. Dis.* 1:237–246.

Marr, K. A., R. A. Carter, F. Crippa, A. Wald, and L. Corey. 2002. Epidemiology and outcome of mould infections in hematopoietic stem cell transplant recipients. *Clin. Infect. Dis.* 34:909–917.

Mastella, G., M. Rainisio, H. K. Harms, M. E. Hodson, C. Koch, J. Navarro, B. Strandvik, S. G. McKenzie, et al. 2000. Allergic bronchopulmonary aspergillosis in cystic fibrosis. A European epidemiological study. *Eur. Respir. J.* 16:464–471.

McNeil, M. M., S. L. Nash, R. A. Hajjeh, M. A. Phelan, L. A. Conn, B. D. Plikaytis, and D. W. Warnock. 2001. Trends in mortality due to invasive mycotic diseases in the United States, 1980–1997. *Clin. Infect. Dis.* 33:641–647.

Mennink-Kersten, M. A. S. H., R. R. Klont, A. Warris, H. J. M. Op den Camp, and P. E. Verweij. 2004. *Bifidobacterium* lipoteichoic acid and false ELISA reactivity in *Aspergillus* antigen detection. *Lancet* 363:325–327.

Moskaluk, C. A., H. W. Pogrebniak, H. I. Pass, J. I. Gallin, and W. D. Travis. 1994. Surgical pathology of the lung in chronic granulomatous disease. *Am. J. Clin. Pathol.* 102:684–691.

Mouy, R., A. Fischer, E. Vilmer, R. Seger, and C. Griscelli. 1989. Incidence, severity, and prevention of infections in chronic granulomatous disease. *J. Pediatr.* 114:555–560.

Muller, F. M., A. H. Groll, and T. J. Walsh. 1999. Current approaches to diagnosis and treatment of fungal infections in children infected with human immunodeficiency virus. *Eur. J. Pediatr.* 158: 187–199.

Ostrosky-Zeichner, L., K. A. Marr, J. H. Rex, and S. H. Cohen. 2003. Amphotericin B: time for a new "gold standard". *Clin. Infect. Dis.* 37:415–425.

Perfect, J. R., G. M. Cox, J. Y. Lee, C. A. Kauffman, L. de Repentigny, S. W. Chapman, V. A. Morrison, P. Pappas, J. W. Hiemenz, and D. A. Stevens. 2001. The impact of culture isolation of *Aspergillus* species: a hospital-based survey of aspergillosis. *Clin. Infect. Dis.* 33:1824–1833.

Petraitis, V., R. Petraitiene, A. A. Sarafandi, A. M. Kelaher, C. A. Lyman, H. E. Casler, T. Sein, A. H. Groll, J. Bacher, N. A. Avila, and T. J. Walsh. 2003. Combination therapy in treatment of experimental pulmonary aspergillosis: synergistic interaction between an antifungal triazole and an echinocandin. *J. Infect. Dis.* 187:1834–1843.

Reik, R. A., M. M. Rodriguez, and G. T. Hensley. 1995. Infections in children with human immunodeficiency virus/acquired immunodeficiency syndrome: an autopsy study of 30 cases in south Florida, 1990–1993. *Pediatr. Pathol. Lab. Med.* 15:269–281.

Roilides, E., E. Pavlidou, F. Papadopoulos, C. Panteliadis, E. Farmaki, M. Tamiolaki, and J. Sotiriou. 2003. Cerebral aspergillosis in an infant with corticosteroid-resistant nephrotic syndrome. *Pediatr. Nephrol.* 18:450–453.

Roilides, E. 2006. Early diagnosis of invasive aspergillosis in infants and children. *Med. Mycol.* 44:S199–S205.

Sallmann, S., A. Heilmann, F. Heinke, M. L. Kerkmann, M. Schuppler, G. Hahn, M. Gahr, A. Rosen-Wolff, and J. Roesler. 2003. Capofungin therapy for *Aspergillus* lung infection in a boy with chronic granulomatous disease. *Pediatr. Infect. Dis. J.* 22:199–200.

Saulsbury, F. T. 2001. Successful treatment of *Aspergillus* brain abscess with itraconazole and interferon-gamma in a patient with chronic granulomatous disease. *Clin. Infect. Dis.* 32:E137–E139.

Schuster, F., C. Moelter, I. Schmid, U. B. Graubner, B. Kammer, B. H. Belohradsky, and M. Fuhrer. 2005. Successful antifungal combination therapy with voriconazole and caspofungin. *Pediatr. Blood Cancer* 44:682–685.

Schwartz, S., M. Ruhnke, P. Ribaud, L. Corey, T. Driscoll, O. A. Cornely, U. Schuler, I. Lutsar, P. Troke, and E. Thiel. 2005. Improved outcome in central nervous system aspergillosis, using voriconazole treatment. *Blood* 106:2641–2645.

Scott, L. J., and D. Simpson. 2007. Voriconazole: a review of its use in the management of invasive fungal infections. *Drugs* 67:269–298.

Segal, B. H., E. S. DeCarlo, K. J. Kwon-Chung, H. L. Malech, J. I. Gallin, and S. M. Holland. 1998. *Aspergillus nidulans* infection in chronic granulomatous disease. *Medicine* (Baltimore) 77:345–354.

Shetty, D., N. Giri, C. E. Gonzalez, P. A. Pizzo, and T. J. Walsh. 1997. Invasive aspergillosis in human immunodeficiency virus-infected children. *Pediatr. Infect. Dis. J.* 16:216–221.

Shouldice, E., C. Fernandez, B. McCully, M. Schmidt, R. Fraser, and C. Cook. 2003. Voriconazole treatment of presumptive disseminated *Aspergillus* infection in a child with acute leukemia. *J. Pediatr. Hematol. Oncol.* 25:732–734.

Simon, A., M. Besuden, S. Vezmar, C. Hasan, D. Lampe, S. Kreutzberg, A. Glasmacher, U. Bode, and G. Fleischhack. 2007. Itraconazole prophylaxis in pediatric cancer patients receiving conventional chemotherapy or autologous stem cell transplants. *Support. Care Cancer* 15:213–220.

Steinbach, W. J. 2005. Pediatric aspergillosis: disease and treatment differences in children. *Pediatr. Infect. Dis. J.* 24:358–364.

Steinbach, W. J., R. M. Addison, L. McLaughlin, Q. Gerrald, P. L. Martin, T. Driscoll, C. Bentsen, J. R. Perfect, and B. D. Alexander. 2007. Prospective *Aspergillus* galactomannan antigen testing in pediatric hematopoietic stem cell transplant recipients. *Pediatr. Infect. Dis. J.* 26:558–564.

Sterba, J., J. Prochazka, J. Ventruba, L. Kren, D. Valik, D. Burgetova, P. Mudry, J. Skotakova, and J. Blatny. 2005. Successful treatment of *Aspergillus* brain abscess in a child with acute lymphoblastic leukemia and liver failure. *Pediatr. Hematol. Oncol.* 22:649–655.

Stergiopoulou, T., J. Meletiadis, E. Roilides, D. E. Kleiner, R. Schaufele, M. Roden, S. Harrington, L. Dad, B. Segal, and T. J. Walsh. 2007. Host-dependent patterns of tissue injury in invasive pulmonary aspergillosis. *Am. J. Clin. Pathol.* 127:349–355.

Stevens, D. A., R. B. Moss, V. P. Kurup, A. P. Knutsen, P. Greenberger, M. A. Judson, D. W. Denning, R. Crameri, A. S. Brody, M. Light, M. Skov, W. Maish, and G. Mastella. 2003. Allergic bronchopulmonary aspergillosis in cystic fibrosis: state of the art. Cystic Fibrosis Foundation Consensus Conference. *Clin. Infect. Dis.* 37(Suppl. 3):S225–S264.

Stiefel, M., T. Reiss, M. S. Staege, J. Rengelshausen, J. Burhenne, A. Wawer, and J. L. Foell. 2007. Successful treatment with voriconazole of *Aspergillus* brain abscess in a boy with medulloblastoma. *Pediatr. Blood Cancer* **49**:203–207.

Sulahian, A., M. Tabouret, P. Ribaud, J. Sarfati, E. Gluckman, J. P. Latgé, and F. Derouin. 1996. Comparison of an enzyme immunoassay and latex agglutination test for detection of galactomannan in the diagnosis of invasive aspergillosis. *Eur. J. Clin. Microbiol. Infect. Dis.* **15**:139–145.

Taccone, A., M. Occhi, A. Garaventa, L. Manfredini, and C. Viscoli. 1993. CT of invasive pulmonary aspergillosis in children with cancer. *Pediatr. Radiol.* **23**:177–180.

Thomas, K. E., C. M. Owens, P. A. Veys, V. Novelli, and V. Costoli. 2003. The radiological spectrum of invasive aspergillosis in children: a 10-year review. *Pediatr. Radiol.* **33**:453–460.

Tillie-Leblond, I., and A. B. Tonnel. 2005. Allergic bronchopulmonary aspergillosis. *Allergy* **60**:1004–1013.

van 't Hek, L. G., P. E. Verweij, C. M. Weemaes, R. van Dalen, J. B. Yntema, and J. F. Meis. 1998. Successful treatment with voriconazole of invasive aspergillosis in chronic granulomatous disease. *Am. J. Respir. Crit. Care Med.* **157**:1694–1696.

Verdaguer, V., T. J. Walsh, W. Hope, and K. J. Cortez. 2007. Galactomannan antigen detection in the diagnosis of invasive aspergillosis. *Expert Rev. Mol. Diagn.* **7**:21–32.

Verweij, P. E., C. M. Weemaes, J. H. Curfs, S. Bretagne, and J. F. Meis. 2000. Failure to detect circulating *Aspergillus* markers in a patient with chronic granulomatous disease and invasive aspergillosis. *J. Clin. Microbiol.* **38**:3900–3901.

Verweij, P. E., E. Mellado, and W. J. Melchers. 2007. Multiple-triazole-resistant aspergillosis. *N. Engl. J. Med.* **356**:1481–1483.

Walmsley, S., S. Devi, S. King, R. Schneider, S. Richardson, and L. Ford-Jones. 1993. Invasive *Aspergillus* infections in a pediatric hospital: a ten-year review. *Pediatr. Infect. Dis. J.* **12**:673–682.

Walsh, T. J. 1998. Primary cutaneous aspergillosis: an emerging infection among immunocompromised patients. *Clin. Infect. Dis.* **27**: 453–457.

Walsh, T. J., N. L. Seibel, C. Arndt, R. E. Harris, M. J. Dinubile, and A. Reboli. 1999. Amphotericin B lipid complex in pediatric patients with invasive fungal infections. *Pediatr. Infect. Dis.* **18**:702–708.

Walsh, T. J., I. Lutsar, T. Driscoll, B. Dupont, M. Roden, P. Ghahramani, M. Hodges, A. H. Groll, and J. R. Perfect. 2002a. Voriconazole in the treatment of aspergillosis, scedosporiosis and other invasive fungal infections in children. *Pediatr. Infect. Dis. J.* **21**:240–248.

Walsh, T. J., M. O. Karlsson, T. Driscoll, A. G. Arguedas, P. Adamson, X. Saez-Llorens, A. J. Vora, A. C. Arrieta, J. Blumer, I. Lutsar, P. Milligan, and N. Wood. 2004. Pharmacokinetics and safety of intravenous voriconazole in children after single- or multiple-dose administration. *Antimicrob. Agents Chemother.* **48**: 2166–2172.

Walsh, T. J., P. C. Adamson, N. L. Seibel, P. M. Flynn, M. N. Neely, C. Schwartz, A. Shad, S. L. Kaplan, M. M. Roden, J. A. Stone, A. Miller, S. K. Bradshaw, S. X. Li, C. A. Sable, and N. A. Kartsonis. 2005. Pharmacokinetics, safety, and tolerability of caspofungin in children and adolescents. *Antimicrob. Agents Chemother.* **49**:4536–4545.

Warris, A., P. E. Verweij, R. Barton, D. C. Crabbe, E. G. Evans, and J. F. Meis. 1999. Invasive aspergillosis in two patients with Pearson syndrome. *Pediatr. Infect. Dis. J.* **18**:739–741.

Wilhelm, L., M. S. McLeary, and D. Janner. 2000. MR diagnosis of pulmonary and chest wall aspergillosis as an initial presentation of chronic granulomatous disease in a 7-month-old male. *Pediatr. Radiol.* **30**:719–720.

Winkelstein, J. A., M. C. Marino, R. B. Johnston, Jr., J. Boyle, J. Curnutte, J. I. Gallin, H. L. Malech, S. M. Holland, H. Ochs, P. Quie, R. H. Buckley, C. B. Foster, S. J. Chanock, and H. Dickler. 2000. Chronic granulomatous disease. Report on a national registry of 368 patients. *Medicine* (Baltimore) **79**:155–169.

Winkelstein, J. A., M. C. Marino, H. Ochs, R. Fuleihan, P. R. Scholl, R. Geha, E. R. Stiehm, and M. E. Conley. 2003. The X-linked hyper-IgM syndrome: clinical and immunologic features of 79 patients. *Medicine* (Baltimore) **82**:373–384.

Yeh, T. C., H. C. Liu, L. Y. Wang, S. H. Chen, and D. C. Liang. 2007. Invasive fungal infection in children undergoing chemotherapy for cancer. *Ann. Trop. Paediatr.* **27**:141–147.

Zaoutis, T. E., K. Heydon, J. H. Chu, T. J. Walsh, and W. J. Steinbach. 2006. Epidemiology, outcomes, and costs of invasive aspergillosis in immunocompromised children in the United States, 2000. *Pediatrics* **117**:e711–e716.

IX. FUTURE DIRECTIONS

Aspergillus fumigatus and Aspergillosis
Edited by J.-P. Latgé and W. J. Steinbach

Chapter 41

A Perspective on *Aspergillus fumigatus* Research for the Next Ten Years

JEAN-PAUL LATGÉ AND WILLIAM J. STEINBACH

The last 10 years of *Aspergillus* research have yielded tremendous gains fueled by collaborative efforts in research and clinical care, leading to several sequenced *Aspergillus* genomes and better management of patients with invasive aspergillosis. Of special interest are large collaborative studies in different fields, such as transcriptomics, diagnosis, animal models of aspergillosis, and large clinical trials for new antifungals. The future is indeed promising for important breakthroughs in *Aspergillus* research. The next 10 years will yield great advances, but we have to be careful of the many pitfalls that can plague our expectations. *Aspergillus fumigatus* remains a very resistant beast and one of the most difficult to manipulate at the molecular level, and there are only a limited number of laboratories studying this fungus. In patients, invasive aspergillosis remains one of the most difficult diagnoses, and treatment is often late and unsuccessful. This chapter presents some thoughts and unpublished data from our laboratories as well as questions that have been repeatedly discussed in meetings which should open new research avenues and may question some ongoing ones.

VIRULENCE FACTORS: A MOLECULAR DREAM OR WORDS TO FORGET?

Why is *A. fumigatus* pathogenic to human beings? This question automatically leads to the hypothesis of the presence of virulence factors specifically synthesized by *A. fumigatus* that are essential for invasion of its mammalian host. One goal in defining a virulence factor is the possibility of counteracting the pathway in order to protect the host against the fungus (Cegelski et al., 2008). However, this is not always possible. Melanin was recognized as a factor essential for fungal survival in the host (Jahn et al., 2000), but there is no possibility to therapeutically inhibit pigment biosynthesis, because all inhaled conidia in the atmosphere are pigmented.

The concept of microbial virulence factors comes from the bacteriology and plant pathology fields, where a microbial pathogen is strongly associated with its host and indeed needs its host to complete its biological cycle. Accordingly, the selective pressure is high for genes that will be essential for growth of the pathogen in vivo (van der Does and Rep, 2007; Speth et al., 2007; Schroeder and Hilbi, 2008). When a gene coding for such a virulence factor is disrupted, the fungus has the same growth rate as the wild-type strain in vitro but is unable to infect its host. Virulence factors defined as such do not exist in *A. fumigatus*. In contrast, the low virulence estimated in murine models of experimental aspergillosis has been most often correlated to a significant reduction of fungal growth in vitro. The extreme situation is total avirulence associated with an absence of germination in the lung due to the unavailability of molecules such as Zn or Fe, that are absolutely required for fungal growth (Moreno et al., 2007; Schrettl et al., 2007). Auxotrophy is, however, not a virulence determinant. Should we call *PABA, LYSF, ARGB*, or *PYRG* virulence genes because mutants with these genes deleted are not able to grow in vivo in the absence of *p*-aminobenzoic acid, lysine, arginine, uridine, and uracil? Certainly not.

Epidemiological studies also suggest that all strains of *A. fumigatus* isolated from the environment or patients are equally virulent (Chazalet et al., 1998). Again, in contrast to fungal plant pathogens, there is no host specificity noted in this species. Any strain of *A. fumigatus* is capable of growth in a compost heap or invasion in a bird or human being. Indeed, for *A. fumigatus*, ending up in a mammalian host is a deadly cul de sac for the fungus, since it will be internalized and killed by the resident phagocytes (Fig. 1). So, what makes this fungus one of the most threatening pathogens on Earth today?

Jean-Paul Latgé • Institut Pasteur, Aspergillus Unit, 25 rue du Docteur Roux, 75015 Paris, France. **William J. Steinbach** • Division of Pediatric Infectious Diseases, Duke University Medical Center, Durham, NC 27710.

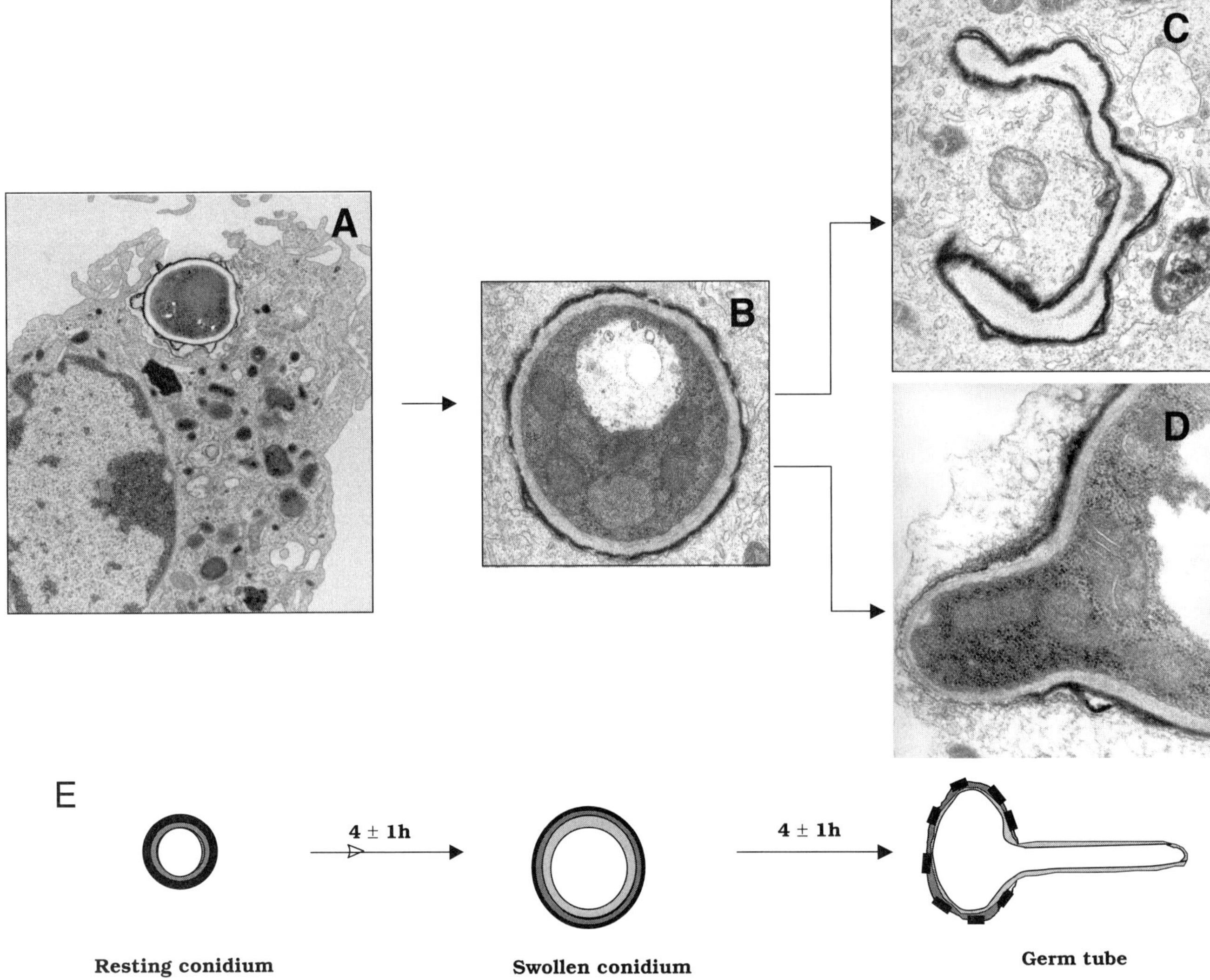

Figure 1. Germination of conidia of *A. fumigatus* in vitro and in vivo. (A) Binding to and internalization of a conidium by an alveolar macrophage. (B) The first stages of germination (swelling) are not affected in the phagolysosome of the alveolar macrophage. (C) Killing of the conidium in an alveolar macrophage of an immunocompetent mouse. (D) Germination of the conidium in an alveolar macrophage of a cortisone acetate-treated mouse. (E) Three characteristic stages of conidial germination: a resting conidium with a thick melanin outer layer; a swollen conidium characterized by plasticization of the existing cell wall layers and synthesis of a new inner layer; and germling formation with disruption of the outer melanin layer and establishment of a polarized cell surrounded by the neosynthesized cell wall emerging through the outer conidia glucan layer. Adapted from Philippe et al. (2003).

Here are several possible reasons, all of which require more future research to help discern the true pathogenicity factors.

To begin, *A. fumigatus* is ubiquitously present in our indoor and outdoor atmospheres, at an average conidial concentration of 10 to 100 conidia/m^3 (and up to billions in some specific environments, like compost centers). However, presence in the air is not the only explanation for *A. fumigatus* virulence. Previous studies in Europe and in North America have shown that *Aspergillus niger* is also found in high concentrations in the air but only extremely rarely causes invasive aspergillosis in humans. Phagocytosis assays have shown that *A. niger* conidia are more resistant to phagocyte killing than *A. fumigatus* (Fig. 2). The reason for the low virulence of *A. niger* in humans remains unexplored, but it may lead to a new concept for medical mycology that has been repeatedly demonstrated with fungal pathogens of plants. Virulence of *A. fumigatus* would be the normal situation, whereas the low pathogenicity of *A. niger* could depend on the presence of avirulence genes in this species.

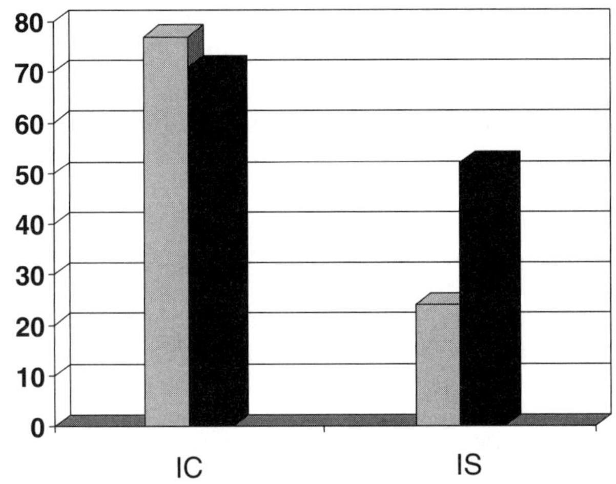

Figure 2. Percentage of killing of conidia of *A. fumigatus* (gray bars) and *A. niger* (black bars) by alveolar macrophages of immunocompetent (IC) and immunosuppressed (IS) mice recovered from bronchoalveolar lavage fluid 24 h after conidia inhalation.

A. fumigatus also grows very well at temperatures greater than 40°C and does not have any specific nutritional requirements for carbon and nitrogen sources. *A. fumigatus* can grow either on proteins such as collagen that will serve for both carbon and nitrogen sources or on simple minimal media containing a single hexose as a carbon source and a nitrate or ammonium salt as a nitrogen source. Once it has germinated, it can even grow in water! However, the richest media based on a protein hydrolysate and glucose produce the greatest mycelial growth. Most importantly, *A. fumigatus* can only establish itself in mammals with a severely compromised immune system. Indeed, any thermophilic fungus growing at a temperature above the human body temperature can become pathogenic in experimental infections (de Hoog et al., 2005). Intranasal inoculation of GRAS (generally recognized as safe) species, such as *Aspergillus oryzae*, can infect cortisone acetate-treated mice. In a comparative study with immunocompromised mice, the human pathogen *Aspergilllus flavus* was not more pathogenic than *A. oryzae* strains from the biotechnology industry (J. P. Latgé, unpublished data).

These last data emphasize a major weakness in the study of putative virulence factors of *A. fumigatus*: the animal model used to investigate virulence. Most studies are undertaken in heavily immunocompromised mice that receive a single inoculum of conidia. The conidial concentration received is also high compared to the multiple low conidial amounts continuously inhaled by humans that lead to a chronic infection, but a low-dose, chronic infection model has not yet been developed in laboratory animals. Recent studies analyzing the role of gliotoxin have shown that the virulence of the Δ*gliA* mutant is different in glucocorticoid-treated versus neutropenic mice (Spikes et al., 2008). Several studies have indeed shown that neutropenic mice will mainly die from fungal invasion of the lung, whereas cortisone-treated animals will die more from an exacerbation of the inflammation (Balloy et al., 2005). Such a dichotomy has an impact on treatment, since antifungals will be active mainly in neutropenic mice. The role of inflammation in the severity of invasive aspergillosis in humans has not really been investigated to date and remains a topic for future research. Due to the severity of the immunosuppression required in mice to obtain invasive aspergillosis, joint mixed infection is the method of choice to analyze the virulence of a mutant and its parental strains or the virulence of several wild-type strains. Using this method, we were able to show that *A. niger* and *A. fumigatus* have similar pathogenicities in cortisone acetate-treated mice. In a model where three strains of *A. fumigatus* with pigmented conidia and three wild-type strains with white conidia were inoculated together into mice, we were able, however, to show that all wild-type strains of *A. fumigatus* did not have the same virulence and that this was not only correlated with the presence or absence of the pigment on the conidia. In contrast, when tested alone, any of the six strains analyzed would kill the mice with no significant differences seen in the survival curve established for each strain (Sarfati et al., 2002) (Fig. 3). Only fitness analysis of the different strains in the lung would be able to illustrate this putative difference of virulence among strains. Although it is a well-known concept in yeast population studies, such as in *Candida albicans* (Anderson, 2005; Brown et al., 2007), strain fitness studies have not yet been undertaken in *A. fumigatus* in either in vitro or in vivo settings.

Difficulties in the analysis of *A. fumigatus* virulence also come from the fact that this fungus is not easily amenable to molecular biology manipulations. This species is both extremely robust and rustic and at the same time extremely flexible genetically. For example, in contrast to *Saccharomyces cerevisiae* or even *C. albicans*, there are extremely few mutants with a significant growth phenotype. Most of the time, to see a phenotype, the fungus has to be stressed, for example, by cultivating it in a very poor minimal medium at 50°C in the presence of inhibitors (Latgé, unpublished). Major compensatory genomic rearrangements occur if an essential gene is deleted: doubling of the *AGS1* gene or excision of the resistance cassette, as seen in the case of RNA interference assays with the essential *FKS1* gene or major deletions (>20 kb) following transformations of artificial diploids (Henry et al., 2007; A. Beauvais and S. Chabane, unpublished data).

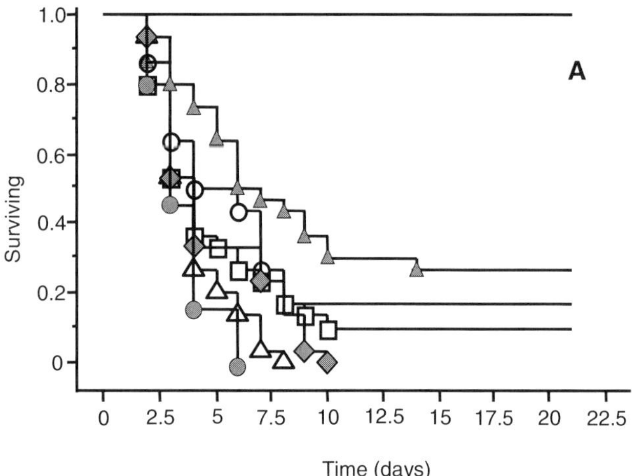

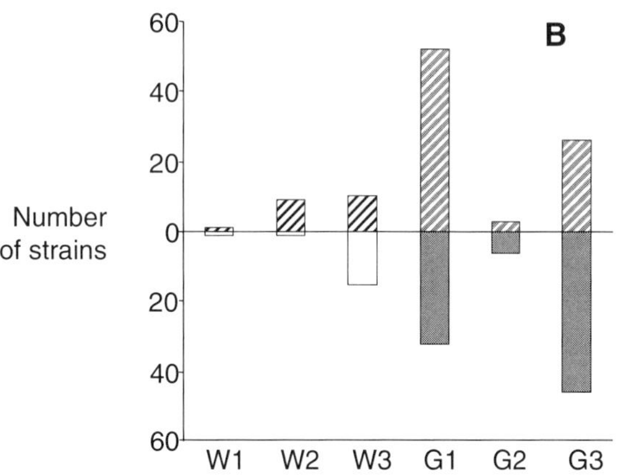

Figure 3. *A. fumigatus* strain virulence seen as differential fitness during joint infections. (A) Survival curve of cortisone acetate-treated mice infected intranasally with a unique strain, one of three strains with pigmentless conidia (W1 [○], W2 [□], W3 [△]) or one of three strains with green-pigmented conidia (G1 [♦], G2 [▲], G3 [●]; solid line, control). (B) Number of strains isolated from lungs of mice infected with the six strains together (hatched bars, intranasally; solid bars, intravenously). Note that strains G1 and G3 are the ones found in larger amounts (the most fit for growth in the lung) and that strain W3 is more pathogenic than the two other white strains and green strain G2. Adapted from Sarfati et al. (2002).

EARLY STAGES OF INFECTION AND DEFENSE REACTIONS ARE A MYSTERY

The natural history of invasive aspergillosis itself is poorly understood. For example, the roles of the respiratory epithelium and mucociliary clearance have been poorly analyzed. It remains unknown today if the site of infection is the epithelium of the bronchial tree or the alveoli. Are epithelial sites that have been damaged by radiotherapy and chemotherapy more prone to infection? Are there multiple sites of infection or is there a single infectious site? These essential questions have not been answered (or even asked) in the *Aspergillus* field. Likewise, the molecules released by phagocytes that can kill *A. fumigatus* in an immunocompetent host have not been fully identified. Some studies suggest that reactive oxidants are essential for killing *A. fumigatus* (Philippe et al., 2003). Indeed, alveolar macrophages from patients with chronic granulomatous disease did not kill a single conidium in vitro (B. Philippe and J.-P. Latgé, unpublished observations). Other data suggested that reactive oxidant intermediates are not so important, since mutants of *A. fumigatus* that are more susceptible to reactive oxidants are as virulent as the wild type in experimental mouse infections. For example, a *YAP1* mutant affected in a transcription factor that makes *A. fumigatus* very sensitive to reactive oxidant species is as pathogenic as the wild type in a mouse model, suggesting that host-reactive oxidant intermediates may not play a significant role in killing the fungus during infection in mice (Lessing et al., 2007). Cathepsin and elastase of phagocyte granules have been reported to be essential in microbial killing and more important than reactive oxidants in the killing of *A. fumigatus* (Segal, 2005). However, alveolar macrophages from elastase and cathepsin knockout mice are as efficient as those from wild-type mice in killing *A. fumigatus* (B. Philippe et al., unpublished data).

If the fate of the conidium in the immunocompetent host is poorly understood, knowledge of the killing mechanisms partially affected in the immunocompromised host is even poorer. Very few studies have analyzed the response of the phagocytes of an immunocompromised host to *A. fumigatus*. Surprisingly, in vitro alveolar macrophages of immunocompromised humans engulf *A. fumigatus* conidia at the same rate as immunocompetent ones and have almost the same capacity to kill *A. fumigatus* (Philippe et al., unpublished). No studies have analyzed the capacity and mechanisms of neutrophils from transplant recipients to kill conidial germ tubes. Analysis of the interactions between *A. fumigatus* and the innate immune system in an immunocompromised patient remains in its infancy. More studies have to be developed in this area, since they will lead to better knowledge of the early stages of infection and also help develop better diagnostic and treatment strategies for invasive aspergillosis.

The pathogenicity of *A. fumigatus* may be due to a better resistance or survival of the fungus against the defense reactions of the host, especially when the host is immunocompromised. Note that an immunocompetent mouse needs weeks to clear a conidial inoculum and that any outbred mouse analyzed always contains live fungus in its lungs (unpublished data). *A. fumigatus* is indeed a saprophytic fungus that has evolved for billions of years and has adapted to survive in hostile environments. The mechanisms developed by *A. fumigatus* to survive in nature can help it to also survive in the host, at least for some time. Melanin is one of these factors that is known to quench many external toxic molecules and is recognized as a virulence factor because it can quench the reactive oxidants produced by phagocytes. The genome also contains hundreds of genes coding for efflux pumps developed for avoiding toxic molecules (Tekaia and Latgé, 2005). Such ABC transporters, or major facilitators, could also protect the fungus from toxic peptides, such as defensins produced by the innate immune system.

ARE LARGE-SCALE BIOLOGICAL STUDIES GOING TO HELP US UNDERSTAND THE *A. FUMIGATUS* LIFE CYCLE?

The sequencing of the genome of *A. fumigatus* has opened the possibility of undertaking large-scale biological and genetic studies with this fungus. Having a complete library of mutants of *A. fumigatus* is, however, currently only a dream due to the limited number of laboratories in the world involved in the molecular analysis of *A. fumigatus* and the difficult task of obtaining mutants for this fungus. However, generation of a subset of mutants resulting from the disruption of the 500 to 600 genes identified in silico that are specific either for the genus *Aspergillus* or the species *A. fumigatus* seems a feasible venture. The disruption of the 350 transcription factors annotated in the genome is another possible alternative. Such studies may lead to the discovery of essential genes associated with a new *Aspergillus*-specific metabolic function that could become an interesting drug target.

Transcriptome arrays are now available, and their use will certainly pinpoint molecular pathways essential for the fungus to complete its life cycle in vitro and in vivo. For example, understanding the global changes associated with conidial swelling, which is the stage sensitive to killing by the host innate immune response, will be essential. Understanding the physiological mechanisms governing the establishment of polarized fungal growth in the lung environment will also help the management of *Aspergillus* infection. However, such in vivo studies will obviously be undertaken in immunocompromised mice and cannot be automatically translated to the human situation. The same issues can be raised for proteome studies. Moreover, to date less than 10% of the total *A. fumigatus* proteome is available (Kniemeyer et al., 2006).

Metabolomics studies seem to be the direction for future research to follow to understand the physiology and associated pathogenicity of *A. fumigatus*. The identification of *A. fumigatus*-specific metabolic pathways will help us decipher fungal growth in vivo and in vitro. For example, what are the nutrients essential for the growth of *A. fumigatus* that are not freely available in the lung environment? The biochemical mechanisms that allow the fungus to bypass this artificial auxotrophy may become novel antifungal targets. Zinc and iron are two examples of essential nutrients for which *A. fumigatus* has developed specific mechanisms to extract them from an environment where they are present freely but at too low a concentration to allow fungal growth.

Another set of physiological mechanisms on which future studies could focus is the way the fungus counteracts suicide pathways. If one interrupts these pathways by disrupting an essential gene of these pathways or with an inhibitor, the fungus will die. Only a few of the proteins that are able to block the accumulation of toxic metabolites during fungal growth in vivo have been identified. This was the case for the toxic propionyl coenzyme A, which is removed by the methylcitrate pathway and accumulated and killed *A. fumigatus* in a methylcitrate synthase mutant (Ibrahim-Granet et al., 2008). Fungal antioxidant molecules can represent another set of protectants since they seem more efficient in fighting the proper oxido-reductive internal reactions of the fungus rather than counteracting the reactive oxidant intermediates of the host (Temple et al., 2005; Herrero et al., 2007).

These "omes" approaches will certainly help our understanding of the fungal physiology and the infection itself. Exploring the specificity of *A. fumigatus* pathways remains an essential goal for all *A. fumigatus* laboratories. As pointed out in this book, many physiological and biochemical pathways are specific to *A. fumigatus*, such as cell wall composition, signal transduction cascades, growth polarization and lack of cytokinesis, transcription and translation machineries, intercommunication between cells of different ages in the same colony, and others. Within the genus *Aspergillus*, significant differences occur, such as the sexual cycle and the calcineurin or secondary metabolite pathways between *A. fumigatus* and other *Aspergillus* species. The differences are even greater between *A. fumigatus* and yeasts. Not taking into account these differences may have direct implications in fungal management. For example, standardization of antifungal inhibition in vitro uses a culture medium initially developed for *Candida*, the cell culture medium RPMI, which is probably the worst medium to grow *A. fumigatus* in vitro! These specificities are just beginning to be understood, and lessons acquired in fungi other than *A. fumigatus* will unfortunately not always be applicable to this species, as also highlighted by the differences in phenotypes of mutants with orthologous genes deleted.

What would be a logical approach to look for unique pathways in this fungus? Ecology can give us a clue of where we have to go. We have to remember that *A. fumigatus* does not live as a domesticated strain in a shaken flask containing a rich medium but grows as a colony in the soil, and in this very hostile environment it has to fight bacteria and amoebas. Deciphering the mechanisms it uses to survive engulfment by protozoa and slime molds in the soil may help us understand how it can survive phagocytosis (Mylonakis et al., 2007). A good knowledge of its resistance to aggressive and complex bacterial populations in the soil may also help our understanding of the growth of this fungal species in mixed populations of bacteria of the upper respiratory pathways of cystic fibrosis patients. Fungal virulence may just be seen as enhanced survival and a countermeasure to environmental predation. Moreover, soil may be the natural ecological niche to find the sexual stage of *A. fumigatus*, yet undiscovered in the laboratory despite repeated attempts. Working in its proper environment may also tell us why *A. fumigatus* is the most prominent thermophilic species in nature. If other members of the Fumigati clade have been discovered by molecular taxonomy, the true *A. fumigatus* remains indeed the majority species, with >95% of the species of the section Fumigati (S. Hong and T. Yaguchi, personal communications).

LABORATORY DIAGNOSIS: CAN WE IMPROVE IT?

The insufficient efficacy of the diagnostic methods available today is the Achilles' heel of patient care. Progress in this area is absolutely required to be able to treat *Aspergillus* infection while it is still confined to the early stages of fungal development in the lungs. Current molecular diagnosis is based upon methods of detecting circulating molecules indicative of the presence of *A. fumigatus* in the body. These molecules can be either polysaccharides, such as galactomannan or β-1,3-glucans, or nucleic acids such as DNA or RNA. In spite of technical progress made in improving the sensitivity of these tests over the last 10 years, it seems that the assay positivity in the clinical laboratory setting always occurs too late and at a time when the fungal burden is too high. This is understandable, since most of these molecules (in particular, nucleic acids) are not secreted molecules and are only released from a dead fungus. This means that the fungus has to grow extensively and the host has to destroy the fungus before some of the molecules can be detected. But how is this happening in a patient with a poorly active immune system? The kinetics and conditions of release of circulating molecules in the human body are biological questions that have not yet been answered. For example, how is β-1,3-glucan removed from the fungus, since humans do not have β-1,3-glucanases? Is it due to an autolytic process that can only start on a dead fungus? What is the value of the ratio of dead versus live fungal cells that will allow detection of free-floating DNA or RNA in the supernatant or a culture filtrate or, more importantly, in the biological fluids of infected patients? These essential underlying biological questions have to be answered before the technical PCR consensus groups that have been formed can begin standardizations to enter the common clinical venue.

Diagnosing *Aspergillus* infection is different from diagnosing a systemic viral or bacterial infection. The fungus remains in the lung and does not circulate in the body until the last stages of invasive aspergillosis. Is looking at fungal molecules the right option to diagnose *A. fumigatus* infections? It may be better to look at specific responses of the host against the fungus. For instance, Caillot et al. (1997) demonstrated a large amount of fibrinogen during *Aspergillus* infection. Although this parameter is not specific and sensitive enough for clinical use, it paves the way for a new diagnostic paradigm. Phagocytosed conidia of *A. fumigatus* are always killed by the innate immune response. However, in the immunocompromised patient, *A. fumigatus* conidia germinate. A specific immune response of the host towards these germinating conidia must be initiated even in immunocompromised humans. Identifying the panel of molecules of the immune system or a biological fluid parameter disturbance may be a new way to diagnose *A. fumigatus* infection earlier. In addition to improving diagnosis, such a study could have some prognostic value. Recent data have suggested that some chemokines can be used for such an integrative diagnostic approach (Mezger et al., 2008).

It is unlikely that there will be a single future diagnostic tool for the early and accurate diagnosis of invasive or chronic aspergillosis. The next steps in the diagnostic arena will focus on optimizing the many modalities available, including antibody serology, galactomannan, β-glucan, PCR, and radiology (including computed tomography, magnetic resonance imaging, and positron emission tomography scans) to augment clinical examination in the highest-risk patients. Early testing has already shown us that these many testing formats perform best at different time points in invasive aspergillosis, and in the future they need to be tested in combinations and sequences to optimize the timing of each to the overall contribution towards diagnosis.

ARE SOME PATIENTS AT HIGHER RISK THAN OTHERS?

The clinical classification of invasive aspergillosis will need to be refined in the future. It is quite apparent now that disease caused by the genus *Aspergillus* not only varies by infection with different species, but also the same organism can affect different hosts in a heterogeneous fashion. We now are just beginning to better ask the fundamental epidemiological question of "Who will develop invasive aspergillosis?" This returns to the pivotal diagnostic question of specific patient susceptibility and patient subpopulations, including patients who have undergone specific immunosuppressive treatments. It is clear from the epidemiologic work that patients who have received an allogeneic stem cell transplant are at greater risk of developing invasive aspergillosis than those who have received an autologous stem cell transplant (Marr et al., 2002). While in the grand scheme of immunosuppression each patient has received a new immune system, the details are critical. However, aspergillosis will need to be further characterized in the underlying patient groups. Just as acute leukemia is painstakingly characterized using specific molecular markers and known chromosomal rearrangements before antineoplastic therapy is begun in order to determine the best individualized therapy, treatment of invasive aspergillosis cannot continue to be delivered on a mass production basis. Therapy in the future will depend on optimal tailoring to the individual patient. We will need to better understand the specific risks for each patient by using host markers, such as possible single-nucleotide polymorphisms or other host-related surrogate markers (Carvalho et al., 2008).

Similar to an improved understanding of patient subpopulation epidemiology to best comprehend which specific transplant patient will develop invasive aspergillosis, the future should yield advances in pharmacogenomics in order to best predict an individual patient's specific response to an antifungal or immune modulator. It currently seems ridiculous to assume that every patient on the planet infected with invasive aspergillosis would metabolize and respond to a particular antifungal agent in the exact same fashion, yielding the same results and the exact same profile of toxicities. The future holds promise for deciphering a patient's potential host immune response as well as the response to individualized therapy. On the most rudimentary level, simple drug serum levels of an antifungal agent will depend on the hepatic and renal clearance of the antifungal which is individualized based not only on overall patient hepatic and renal function but also on the capabilities of each patient's overall metabolism. As examples, we see that voriconazole can be metabolized very slowly, often leading to toxic levels in approximately one-fourth of Asian patients (Johnson and Kaufmann, 2003). In the future we will need to better characterize vague statements such as this, since one-fourth of the population of Asia or patients with an Asian genetic background is currently approximately 1 to 2 billion potential patients. However, while serum drug level monitoring is an important step, that itself is too simplistic. Understanding a patient's relevant genetic host response to a therapeutic agent before delivering the agent is critical to future developments in this field.

NEW ANTI-*ASPERGILLUS* THERAPIES

There are currently 11 antifungals with anti-*Aspergillus* activity, and several others are under evaluation in early clinical and preclinical studies. Unfortunately, many pharmaceutical companies are gradually leaving the field of antifungal research and are no longer involved in the discovery and launch of new antifungal molecules directed at novel targets. The only new molecules in the market or in the pipelines of these companies remain sterol and β-1,3-glucan synthesis inhibitors. This situation will likely not change in the near future, especially since the risk of seeing the emergence of drug-resistant strains seems very limited. For unknown reasons, and fortunately for the benefit of patients, in spite of the tons of azole fungicides sprayed in nature against plant pathogens, very few azole-resistant *A. fumigatus* strains have appeared (about 20 strains in the entire world) and the resistance has not spread into clinics. This result leads to interesting issues. Since it is very easy to obtain in vitro strains of *A. fumigatus* which are resistant to azoles, the lack of occurrence of such strains in nature is not due to an intrinsic resistance of *A. fumigatus* to azoles. The lack of appearance of such resistant strains may simply indicate that they are not able to survive in nature or in a human host, and these survival studies should now be undertaken.

Looking for a new drug target is a very exciting scientific adventure, but today this seems an unrealistic commercial dream. Better management of patients with invasive aspergillosis will likely rely more on the better use of existing antifungal drugs rather than expecting new and more potent molecules. The first major issue is resolving the poor efficacy of the antifungal molecules in vivo, when the same molecules are very active in vitro. Pharmacokinetic studies using radiolabeled drugs must be undertaken. Could we improve the efficiency of these drugs by coupling them with *A. fumigatus* ligands such as anti-*A. fumigatus* antibodies? This is a completely virgin field of investigation.

One pressing issue for the future is the optimal way to use the many available agents to increase efficacy and yet avoid toxicities and high costs. One of the most debated topics is the issue of combination antifungal therapy against invasive aspergillosis. There are now many other medical disciplines which draw upon this approach in their treatment foundations, such as malignancy therapy, where disease modification is best achieved with well-studied therapeutic protocols involving multiple agents from different classes. The future will include the first-ever large-scale combination antifungal clinical trial and hopefully another trial involving immunotherapy.

IMMUNOTHERAPY: A STRATEGY TO REPLACE ANTIFUNGALS?

In animals, T cells do not play a major role against *A. fumigatus*, since nude mice are not more sensitive to infection than outbred Swiss mice. The resident alveolar macrophages are indeed sufficient to contain an *A. fumigatus* population even larger than the one inhaled daily by the animal (Philippe et al., 2003). When they interact with dendritic cells, however, T cells play a significant role in promoting a protective response in the animal (Bellochio et al., 2005). Assays in mice have been translated to humans and infusion of *Aspergillus*-primed T cells has been effective in controlling the infection (Perruccio et al., 2005; Montagnoli et al., 2008). Genetic polymorphisms associated with the T-cell pathway may also serve as prognostic signs for survival or accelerated infection in a patient population at risk for invasive aspergillosis.

Monoclonal antibodies also can be used to fight *A. fumigatus*. A similar approach has been used successfully in cancer therapy. These monoclonal antibodies could

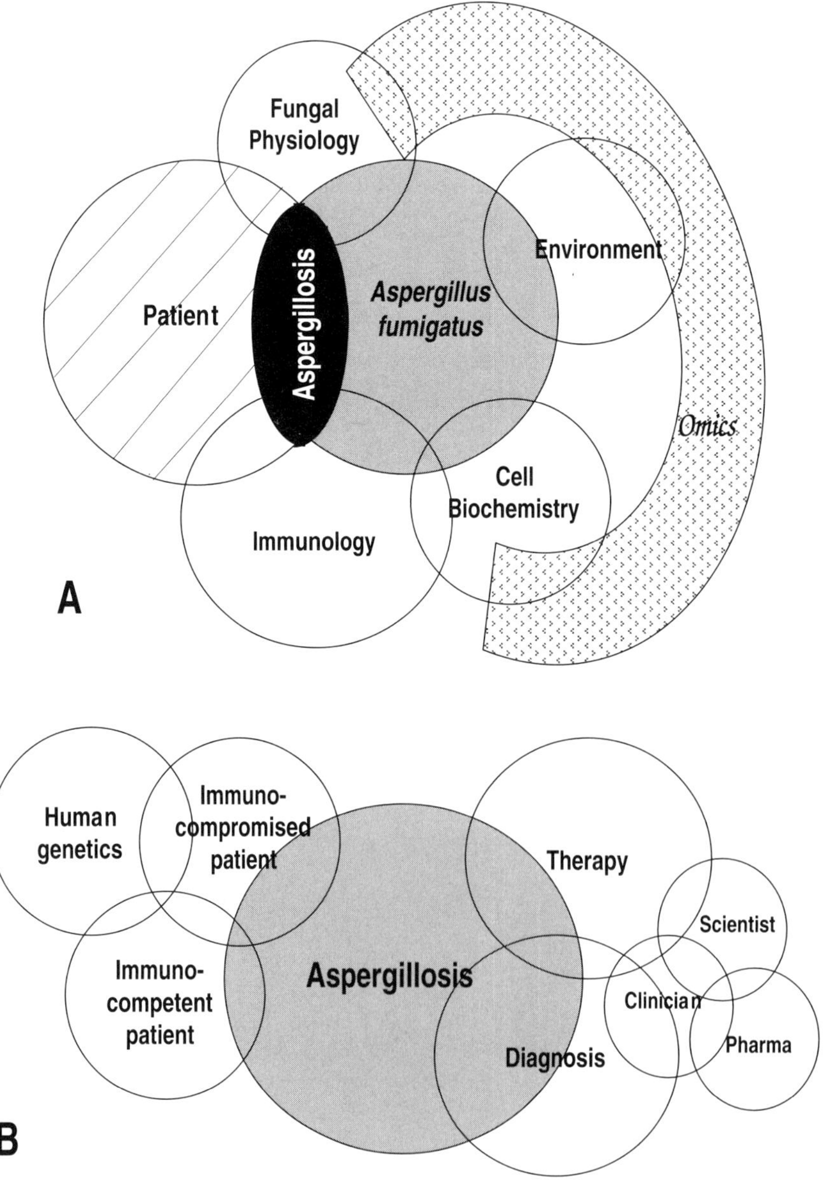

Figure 4. Two diagrams showing the interconnection between partners and disciplines in the study of aspergillosis in laboratory (A) and hospital (B) settings.

be used in two ways. First, monoclonal antibodies directed against *A. fumigatus* would be humanized and coupled to a radioactive element. When in contact with the fungus, this "atomic bomb" would kill the fungus (Dadachova et al., 2003; Dadachova and Casadevall, 2008). Second, monoclonal antibodies can by themselves inhibit fungal growth; however, this inhibition is not complete and monoclonal antibodies could not be used alone in practice (Torosantucci et al., 2005). These new immunological developments could be a viable alternative to antifungal drug therapy.

CONCLUSIONS

The number of severely immunocompromised patients will continue to increase over the next 10 years as the field of medicine develops the technology to treat an increasing number of underlying conditions, such as malignancy and failed organs. Similarly, other medical disciplines such as rheumatology and gastroenterology are increasingly starting to use immunosuppressive agents for their recalcitrant patients, further multiplying the number of patients at risk for invasive aspergillosis. The future holds countless opportunities for progress to be made in fundamental areas of *Aspergillus* research, such as basic epidemiology and pathogenesis, as well as clinical arenas, such as optimal antifungal therapy and immune modulation.

The future work involving *A. fumigatus* and aspergillosis has to include collaborations of partners from different fields to efficiently and appropriately study the disease (Fig. 4). We are starting to see growing evidence of this, as future clinical trials are originating from an international focus and enrolling patients on multiple continents. The biannual Advances Against Aspergillosis Conference has successfully brought together researchers and clinicians from around the world to address some of the field's most challenging questions. From it and collaborative efforts like this textbook we are now seeing partnerships between pharmaceutical companies who manufacture the antifungals and clinicians who treat patients with invasive aspergillosis working together in the design of effective clinical trials as well as novel investigations into such critical molecular topics as antifungal resistance.

REFERENCES

Anderson, J. B. 2005. Evolution of antifungal-drug resistance: mechanisms and pathogen fitness. *Nat. Rev. Microbiol.* **3**:547–556.

Balloy, V., M. Huerre, J. P. Latgé, and M. Chignard. 2005. Differences in patterns of infection and inflammation for corticosteroid treatment and chemotherapy in experimental invasive pulmonary aspergillosis. *Infect. Immun.* **73**:494–503.

Bellocchio, S., S. Bozza, C. Montagnoli, K. Perruccio, R. Gaziano, L. Pitzurra, and L. Romani. 2005. Immunity to *Aspergillus fumigatus*: the basis for immunotherapy and vaccination. *Med. Mycol. Suppl.* **1**:S181–S188.

Brown, A. J., F. C. Odds, and N. A. Gow. 2007. Infection-related gene expression in *Candida albicans*. *Curr. Opin. Microbiol.* **10**: 307–313.

Caillot, D., O. Casasnovas, A. Bernard, J. F. Couaillier, C. Durand, B. Cuisenier, E. Solary, F. Piard, T. Petrella, A. Bonnin, G. Couillault, M. Dumas, and H. Guy. 1997. Improved management of invasive pulmonary aspergillosis in neutropenic patients using early thoracic computed tomographic scan and surgery. *J. Clin. Oncol.* **15**:139–147.

Carvalho, A., A. C. Pasqualotto, L. Pitzurra, L. Romani, D. W. Denning, and F. Rodrigues. 2008. Polymorphisms in toll-like receptor genes and susceptibility to pulmonary aspergillosis. *J. Infect. Dis.* **197**:618–621.

Cegelski, L., G. R. Marshall, G. R. Eldridge, and S. J. Hultgren. 2008. The biology and future prospects of antivirulence therapies. *Nat. Rev. Microbiol.* **6**:17–27.

Chazalet, V., J. P. Debeaupuis, J. Sarfati, J. Lortholary, P. Ribaud, P. Shah, M. Cornet, H. Vu Thien, E. Gluckman, G. Brücker, and J. P. Latgé. 1998. Molecular typing of environmental and patient isolates of *Aspergillus fumigatus* from various hospital settings. *J. Clin. Microbiol.* **36**:1494–1500.

Dadachova, E., and A. Casadevall. 2008. Host and microbial cells as targets for armed antibodies in the treatment of infectious diseases. *Curr. Opin. Investig. Drugs* **9**:184–188.

Dadachova, E., A. Nakouzi, R. A. Bryan, and A. Casadevall. 2003. Ionizing radiation delivered by specific antibody is therapeutic against a fungal infection. *Proc. Natl. Acad. Sci. USA* **100**:10942–10947.

de Hoog, G. S., J. Guarro, J. Gené, and M. J. Figueras. 2005. *Atlas of Clinical Fungi*, p. 1126. Centraalbureau voor Schimmelcultures, Utrecht, The Netherlands.

Henry, C., I. Mouyna, and J. P. Latgé. 2007. Testing the efficacy of RNA interference constructs in *Aspergillus fumigatus*. *Curr. Genet.* **51**:277–284.

Herrero, E., J. Ros, G. Belli, and E. Cabiscol. 2008. Redox control and oxidative stress in yeast cells. *Biochim. Biophys. Acta.* **1780**: 1217–1235.

Ibrahim-Granet, O., M. Dubourdeau, J. P. Latgé, P. Ave, M. Huerre, A. A. Brakhage, and M. Brock. 2008. Methylcitrate synthase from *Aspergillus fumigatus* is essential for manifestation of invasive aspergillosis. *Cell. Microbiol.* **10**:134–148.

Jahn, B., F. Boukhallouk, J. Lotz, K. Langfelder, G. Wanner, and A. A. Brakhage. 2000. Interaction of human phagocytes with pigmentless *Aspergillus* conidia. *Infect. Immun.* **68**:3736–3739.

Johnson, L. B., and C. A. Kauffman. 2003. Voriconazole: a new triazole antifungal agent. *Clin. Infect. Dis.* **36**:630–637.

Kniemeyer, O., F. Lessing, O. Scheibner, C. Hertweck, and A. A. Brakhage. 2006. Optimisation of a 2-D gel electrophoresis protocol for the human-pathogenic fungus *Aspergillus fumigatus*. *Curr. Genet.* **49**:178–189.

Lessing, F., O. Kniemeyer, I. Wozniok, J. Loeffler, O. Kurzai, A. Haertl, and A. A. Brakhage. 2007. The *Aspergillus fumigatus* transcriptional regulator AfYap1 represents the major regulator for defence against reactive oxygen intermediates but is dispensable for pathogenicity in an intranasal mouse infection model. *Eukaryot. Cell* **6**:2290–2302.

Marr, K. A., R. A. Carter, F. Crippa, A. Wald, and L. Corey. 2002. Epidemiology and outcome of mould infections in hematopoietic stem cell transplant recipients. *Clin. Infect. Dis.* **34**:909–917.

Mezger, M., M. Steffens, M. Beyer, C. Manger, J. Eberle, M. R. Toliat, T. F. Wienker, P. Ljungman, H. Hebart, H. J. Dornbusch, H. Einsele, and J. Loeffler. 2008. Polymorphisms in the chemokine (C-

X-C motif) ligand 10 are associated with invasive aspergillosis after allogeneic stem-cell transplantation and influence CXCL10 expression in monocyte-derived dendritic cells. *Blood* **111**:534–536.

Montagnoli, C., K. Perruccio, S. Bozza, P. Bonifazi, T. Zelante, A. De Luca, S. Moretti, C. D'Angelo, F. Bistoni, M. Martelli, F. Aversa, A. Velardi, and L. Romani. 2008. Provision of antifungal immunity and concomitant alloantigen tolerization by conditioned dendritic cells in experimental hematopoietic transplantation. *Blood Cells Mol. Dis.* **40**:55–62.

Moreno, M. A., O. Ibrahim-Granet, R. Vicentefranqueira, J. Amich, P. Ave, F. Leal, J. P. Latgé, and J. A. Calera. 2007. The regulation of zinc homeostasis by the ZafA transcriptional activator is essential for *Aspergillus fumigatus* virulence. *Mol. Microbiol.* **64**:1182–1197.

Mylonakis, E., A. Casadevall, and F. M. Ausubel. 2007. Exploiting amoeboid and non-vertebrate animal model systems to study the virulence of human pathogenic fungi. *PLoS Pathog.* **3**:e101.

Perruccio, K., A. Tosti, E. Burchielli, F. Topini, L. Ruggeri, A. Carotti, M. Capanni, E. Urbani, A. Mancusi, F. Aversa, M. F. Martelli, L. Romani, and A. Velardi. 2005. Transferring functional immune responses to pathogens after haploidentical hematopoietic transplantation. *Blood* **106**:4397–4406.

Philippe, B., O. Ibrahim-Granet, M. C. Prévost, M. A. Gougerot-Pocidalo, M. Sanchez Perez, A. Van der Meeren, and J. P. Latgé. 2003. Killing of *Aspergillus fumigatus* by alveolar macrophages is mediated by reactive oxidant intermediates. *Infect. Immun.* **71**:3034–3042.

Sarfati, J., M. Diaquin, J. P. Debeaupuis, A. Schmidt, D. Lecaque, A. Beauvais, and J. P. Latgé. 2002. A new experimental murine aspergillosis model to identify strains of *Aspergillus fumigatus* with reduced virulence. *Nippon Ishinkin Gakkai Zasshi* **43**:203–213.

Schrettl, M., E. Bignell, C. Kragl, Y. Sabiha, O. Loss, M. Eisendle, A. Wallner, H. N. Arst, Jr., K. Haynes, and H. Haas. 2007. Distinct roles for intra- and extracellular siderophores during *Aspergillus fumigatus* infection. *PLoS Pathog.* **3**:1195–1207.

Schroeder, G. N., and H. Hilbi. 2008. Molecular pathogenesis of *Shigella* spp.: controlling host cell signalling, invasion, and death by type III secretion. *Clin. Microbiol. Rev.* **21**:134–156.

Segal, A. W. 2005. How neutrophils kill microbes. *Annu. Rev. Immunol.* **23**:197–223.

Speth, E. B., Y. N. Lee, and S. Y. He. 2007. Pathogen virulence factors as molecular probes of basic plant cellular functions. *Curr. Opin. Plant Biol.* **10**:580–586.

Spikes, S., R. Xu, C. K. Nguyen, G. Chamilos, D. P. Kontoyiannis, R. H. Jacobson, D. E. Ejzykowicz, L. Y. Chiang, S. G. Filler, and G. S. May. 2008. Gliotoxin production in *Aspergillus fumigatus* contributes to host-specific differences in virulence. *J. Infect. Dis.* **197**:479–486.

Tekaia, F., and J. P. Latgé. 2005. *Aspergillus fumigatus*: saprophyte or pathogen? *Curr. Opin. Microbiol.* **8**:385–392.

Temple, M. D., G. G. Perrone, and I. W. Dawes. 2005. Complex cellular responses to reactive oxygen species. *Trends Cell Biol.* **15**:319–326.

Torosantucci, A., C. Bromuro, P. Chiani, F. De Bernardis, F. Berti, C. Galli, F. Norelli, C. Bellucci, L. Polonelli, P. Costantino, R. Rappuoli, and A. Cassone. 2005. A novel glyco-conjugate vaccine against fungal pathogens. *J. Exp. Med.* **202**:597–606.

van der Does, H. C., and M. Rep. 2007. Virulence genes and the evolution of host specificity in plant-pathogenic fungi. *Mol. Plant Microbe Interact.* **20**:1175–1182.

INDEX